Electrodynamics
Classical and Quantum

for MSc (Phys), BSc (Hons), BTech, GATE, NET/SET, UGC-CSIR

Electrodynamics
Classical and Quantum

for **MSc (Phys), BSc (Hons), BTech, GATE, NET/SET, UGC-CSIR**

SL Kakani
MSc (Physics), PhD
Former Executive Director
Institute of Technology and Management
(Affiliated to Rajasthan Technical University, Kota)
Bhilwara 311001, Rajasthan, India
E-mail: slkakani28@gmail.com

Shubhra Kakani
MSc (Physics), PhD

CBS

CBS Publishers & Distributors Pvt Ltd

New Delhi • Bengaluru • Chennai • Kochi • Kolkata • Mumbai
Bhopal • Bhubaneswar • Hyderabad • Jharkhand • Nagpur • Patna • Pune • Uttarakhand • Dhaka (Bangladesh) • Kathmandu (Nepal)

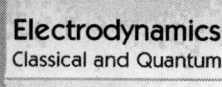

Electrodynamics
Classical and Quantum

ISBN: 978-93-89239-03-4

Copyright © Authors and Publisher

First Edition: 2020

Published by Satish Kumar Jain and produced by Varun Jain for

CBS Publishers & Distributors Pvt Ltd
4819/XI Prahlad Street, 24 Ansari Road, Daryaganj, New Delhi 110 002, India.
Ph: 23289259, 23266861, 23266867   Fax: 011-23243014   Website: www.cbspd.com
e-mail: delhi@cbspd.com; cbspubs@airtelmail.in

Corporate Office: 204 FIE, Industrial Area, Patparganj, Delhi 110 092
Ph: 4934 4934   Fax: 4934 4935   e-mail: publishing@cbspd.com; publicity@cbspd.com

Branches

- **Bengaluru:** Seema House, 2975, 17th Cross, K.R. Road,
  Banasankari 2nd Stage, Bengaluru 560 070, Karnataka
  Ph: +91-80-26771678/79   Fax: +91-80-26771680 .   e-mail: bangalore@cbspd.com
- **Chennai:** 7, Subbaraya Street, Shenoy Nagar, Chennai 600 030, Tamil Nadu
  Ph: +91-44-26680620, 26681266   Fax: +91-44-42032115   e-mail: chennai@cbspd.com
- **Kochi:** 42/1325, 1326, Power House Road, Opposite KSEB Power House,
  Ernakulam 682 018, Kochi, Kerala
  Ph: +91-484-4059061-65   Fax: +91-484-4059065   e-mail: kochi@cbspd.com
- **Kolkata:** 6/B, Ground Floor, Rameswar Shaw Road, Kolkata-700 014, West Bengal
  Ph: +91-33-22891126, 22891127, 22891128   e-mail: kolkata@cbspd.com
- **Mumbai:** 83-C, Dr E Moses Road, Worli, Mumbai-400018, Maharashtra
  Ph: +91-22-24902340/41   Fax: +91-22-24902342   e-mail: mumbai@cbspd.com

Representatives

- **Bhopal**   0-8319310552   • **Bhubaneswar** 0-9911037372   • **Hyderabad** 0-9885175004   • **Jharkhand**   0-9811541605
- **Nagpur**   0-9421945513   • **Patna**   0-9334159340   • **Pune**   0-9623451994   • **Uttarakhand** 0-9716462459
- **Dhaka (Bangladesh)** 01912-003485   • **Kathmandu (Nepal)** 977-9818742655

Printed at: Mudrak, Noida, UP, India

Preface

Electrodynamics is a complex subject and uses higher mathematics. Therefore, efforts have been made to write a student-friendly book which will help the students grasp the subject easily. This book is intended to serve as a textbook for BSc (Hons/Pass), MSc (Physics) and BTech students of various Indian as well as foreign universities.

This book comprises 18 chapters and 9 appendices. The mathematical background of this subject involves vector algebra, vector calculus, scalar and vector fields elaborated in Chapters 1 and 2. Chapter 3 explains the essentials of electrostatics and Chapter 4 deals with electrostatics in dielectrics. Chapter 5 covers the solution of boundary value problems in electrostatic fields. Since the electrostatic and magnetostatic fields are inter-related magnetostatics is covered in Chapter 6. Chapter 7 is devoted to the study of electromagnetic induction and Maxwell's equations. Electromagnetic waves are discussed in Chapter 8. Chapter 9 deals with scattering and dispersion of electromagnetic waves. Chapters 10–13 are devoted to the study of transmission lines, waveguides, electromagnetic radiation and radiating systems (antenna), and plasma physics respectively. Chapter 14 focuses on relativistic electrodynamics. Chapter 15 deals with interaction of charged particles with matter. Chapter 16 is devoted to the study of multipole radiations. Chapter 17 is devoted to the study of Hamiltonian formulation of field equations. Chapter 18 covers basics of quantum electrodynamics.

The subject matter has been tailored systematically from the fundamental concepts, keeping in mind the actual difficulties of the students. Efforts have been made to make the presentation lucid and comprehensive. The obscure and difficult points have been explained in a simple language and with necessary self-explanatory diagrams in a manner that the students may feel no difficulty in understanding the subject.

Each chapter contains a good number of worked out examples and exercises comprising short answer and multiple choice questions that appeared in GATE, NET/SET and UGC-CSIR and various other examinations. Glimpses in the form of important points given in each chapter provide an insight into the entire chapter.

This book is not just another addition to the books already available in the market, it offers a rich structural framework of principles, key equations, well labelled figures and typical worked out problems with a good number of multiple choice questions with answers.

We are thankful to CBS Publishers and Distributors, New Delhi. We would like to put on record the sincere efforts of Mr YN Arjuna, Senior Vice President Publishing, Editorial and Publicity, and his team comprising Ms Ritu Chawla, Ms Sanjubala Tripathy, Mr Manish Raj and Mr Pramod Kumar, for bringing out the book in the present form.

We welcome comments, suggestions and corrections from the readers.

SL Kakani
Shubhra Kakani

Contents

12 ELECTROMAGNETIC RADIATION AND RADIATING SYSTEMS   545–591

13 PROPAGATION OF ELECTROMAGNETIC WAVES THROUGH IONIZED MEDIA—PLASMA PHYSICS 603–636

14 RELATIVISTIC ELECTRODYNAMICS 637–717

Vector Analysis

1.1 INTRODUCTION

The use of vector algebra and vector calculus makes the language of electromagnetics most precise and concise. Physical quantities are divided into two main classes: (i) scalars (ii) vectors.

i. Scalar Quantities

Scalar quantities are those which can be completely characterized by their magnitude and algebraic sign. These quantities do not involve direction. Mass, length, time, pressure, work, temperature, charge, potential, etc. are a few examples of scalar quantities.

ii. Vector Quantities

A vector is a quantity that is completely characterized by its magnitude and direction. Displacement, velocity, momentum, force, current density, electric and magnetic field intensities, magnetic induction, etc. are a few examples of vector quantities.

A vector is represented on a diagram by an arrow line whose length is proportional to the magnitude and the arrow indicating the direction of the vector (Fig. 1.1). Vectors are further classified as *localized and free* vectors. For the complete specification of a localized vector, one has to specify the point at which the vector acts, whereas there is no such restriction for free vectors.

We must note that a directed quantity is a vector provided that it obeys the law (commutative) of vector addition, e.g. a large angular displacement about a given axis is not a vector because its addition to another angular displacement about a different axis is not commutative.

To distinguish vectors from scalars, *bold-face* type letters, e.g. A, B, a, b may be used for vectors. The magnitude of a vector A is denoted by $|A|$ or A. The other suitable or usually used notations for vectors are $\vec{A}, \vec{B}, \vec{a}, \vec{b}$ or A, B, a, b, etc. In electromagnetics, one usually come across scalar and vector fields. These are discussed in Chapter 2.

Fig. 1.1: Representation of a vector

1.2 VECTOR ALGEBRA

Addition of Vectors

Let us consider two vectors A and B representing, for example, two successive displacements of a point. One can obtain their sum $C = A + B$ by placing the tail of B at the head of

A, then the sum $C = A + B$ is the vector from the tail of A to the head of B (Fig. 1.2). (This rule generalizes the obvious procedure for combining two displacements). Vector addition is *commutative*, i.e.

$$A + B = B + A \qquad (1.1)$$

Vector sum is independent of the order in which the vectors are taken.

Fig. 1.2: Addition of two vectors

The vector sum is *associative*, i.e. the sum of vectors A, B and C is same when we add A to the sum of B and C or we add B to the sum of A and C or we add C to the sum of A and B, i.e.

$$D = A + (B + C)$$
$$= B + (A + C) = C + (A + B) \qquad (1.2)$$

The same is evident from Fig. 1.3.

Fig. 1.3: Illustration showing vector sum is associative

Subtraction of Vectors

Let there be two vectors A and B and we have to subtract B from A. Subtraction is just the same thing as adding $-B$ to A, i.e.

$$C = A - B = A + (-B) \qquad (1.3)$$

The negative of a vector means a vector of equal magnitude but opposite direction. Obviously, to subtract B from A, we draw $-B$ from the end point of A and draw A from the initial point of A to the end point of $-B$ (Fig. 1.4).

Fig. 1.4: Subtraction of vectors (considering different points)

If $A = B$, then $A - B = 0$. Thus **0** is called *null vector*.

One can also subtract B from A as shown in Fig. 1.5. In this process, one has to draw both the given vectors A and B from the same point, and then draw C from the end point of B to the end point of A. Thus,

$$C = A + (-B)$$

Fig. 1.5: Subtraction of vectors (same point)

Multiplication of a Vector by a Scalar

Multiplication of a vector A by a positive scalar a multiplies the *magnitude* but leaves the direction unchanged (Fig. 1.6). If a is negative, the direction is reversed, we may write

$$C = aA \qquad (1.4)$$

Scalar multiplication is *distributive*, i.e.

$$a(A + B) = aA + aB \qquad (1.5)$$

$aA = C$
(Here $a = 2, C = 2A$)

Fig. 1.6: Multiplication of a vector **A** by scalar

To divide a vector by a scalar, the vector is multiplied by the reciprocal of the scalar.

When a null vector is multiplied by a finite scalar number, then one obtains the null vector, i.e.

$$n0 = 0 \qquad (1.6)$$

When a vector is multiplied by 0, one obtains null vector

$$0A = 0 \qquad (1.7)$$

1.3 UNIT VECTOR

A vector A may be expressed as $A\hat{a}$, where A is the magnitude of vector A and $\hat{a}$ is a vector of unit magnitude in direction that of A. The vector $\hat{a}$ is known as an *unit vector*. Obviously,

$$\hat{a} = \frac{A}{A} \qquad (1.8)$$

Thus, any vector in the direction of unit vector may be represented by the product of the unit vector and the scalar magnitude of the vector. In cartesian coordinates x, y, z system, usually $\hat{i}, \hat{j}$ and $\hat{k}$ are taken as unit vectors parallel to x, y and z axes respectively. These are called the *orthogonal unit vectors*. In some texts, unit vectors parallel to x, y and z axes in cartesian coordinate system are expressed by $\hat{a}_x, \hat{a}_y$ and $\hat{a}_z$ respectively. Unit vectors $\hat{i}, \hat{j}$ and $\hat{k}$ are also termed *basis vectors*.

1.4 RESOLUTION OF A VECTOR

If $C = A + B$, then A and B are said to be the components of C. In other words, C may be resolved into its component vectors A and B. Obviously, a given vector can be resolved into components in a variety of ways. The components of a vector A are any number of vectors whose algebric sum is A. The components most frequently used, are those parallel to x-, y- and z-axes of the right-handed cartesian coordinate system and they are called *rectangular components of a vector*.

Let the origin O be one extremity of vector A. Now draw a rectangular parallelepiped with the three edges which meet at O lying along the three axes x, y and z and such that A is the diagonal through O (Fig. 1.7). Let $\hat{i}, \hat{j}$ and $\hat{k}$ be unit vectors parallel to x-, y- and z-axes respectively. The vector A can be expanded in terms of these basis vectors as

$$A = A_x\hat{i} + A_y\hat{j} + A_z\hat{k} \qquad (1.9)$$

The number A_x, A_y and A_z are called components of A. Geometrically, they are the projections of A along the three coordinate axes.

By geometry of the figure, we have

$$A_x = A_x\hat{i}, A_y = A_y\hat{j} \text{ and } A_z = A_z\hat{k}$$

Thus

$$A^2 = A_x^2 + A_y^2 + A_z^2$$

or

$$A = \sqrt{A_x^2 + A_y^2 + A_z^2} \qquad (1.10)$$

If α, β, γ be the angles between the vector A and the positive directions of x-, y- and z-axes respectively, then cosine of these angles are called the *direction cosines*. From Fig. 1.7, one can write

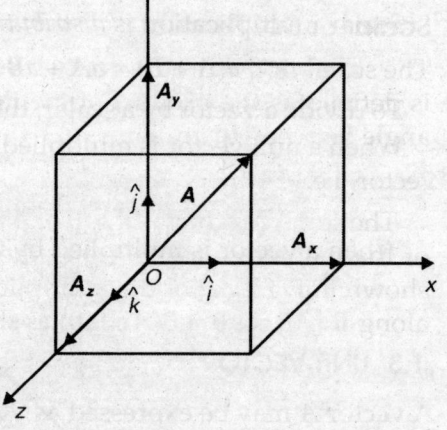

$$\cos\alpha = \frac{A_x}{|A|} = \frac{A_x}{A}$$

$$\cos\beta = \frac{A_y}{|A|} = \frac{A_y}{A} \qquad \left.\right\} \qquad (1.11)$$

and

$$\cos\gamma = \frac{A_z}{|A|} = \frac{A_y}{A}$$

Fig. 1.7: Resolution of a vector

Obviously, $\quad A_x = A\cos\alpha, A_y = A\cos\beta$ and $A_z = A\cos\gamma \qquad (1.11a)$

and $\quad A = A[\hat{i}\cos\alpha + \hat{j}\cos\beta + \hat{k}\cos\gamma] \qquad (1.12)$

The unit vector $\hat{a}$ corresponding to vector A is given by

$$\hat{a} = \frac{A}{A} = \hat{i}\cos\alpha + \hat{j}\cos\beta + \hat{k}\cos\gamma \qquad (1.13)$$

Using Eq. (1.11a), one obtains from Eq. (1.10)

$$\cos^2\alpha + \cos^2\beta + \cos^2\gamma = 1 \qquad (1.14)$$

Obviously, *the sum of the squares of three direction cosines is equal to unity.* Further any three numbers proportional to the direction cosines of a line having the same proportional constant are called *direction ratios.* If a, b and c be the direction ratios of a line and $\cos\alpha$, $\cos\beta$ and $\cos\gamma$ be its *direction cosines*, then

$$\frac{\cos\alpha}{a} = \frac{\cos\beta}{b} = \frac{\cos\gamma}{c}$$

$$= \frac{\sqrt{\cos^2\alpha + \cos^2\beta + \cos^2\gamma}}{\sqrt{a^2 + b^2 + c^2}} = \frac{1}{\sqrt{a^2 + b^2 + c^2}} \qquad (1.15)$$

$$\cos \alpha = \frac{a}{\sqrt{a^2 + b^2 + c^2}}$$

$$\cos \beta = \frac{b}{\sqrt{a^2 + b^2 + c^2}}$$

and $\quad \cos \gamma = \dfrac{c}{\sqrt{a^2 + b^2 + c^2}}$

(1.16)

1.5 PRODUCT OF VECTORS

The product of two vectors is defined in two different ways:

Scalar or Dot Product

The scalar or dot product of two vectors A and B (written as $A \cdot B$ and read as A dot B) is defined as the product of the magnitude of A and B and the cosine of the smaller angle between them. Thus,

$$A \cdot B = AB \cos \theta \qquad (1.17)$$

The scalar product $AB \cos \theta$ of A and B may be considered as either the multiplication of the magnitude of $|A|$ and the projection of B along A [$AB \cos \theta = A(B \cos \theta)$ as shown in Fig. 1.8a] or the multiplication of the magnitude of B and the projection of A along B [$AB \cos \theta = B(A \cos \theta)$ as shown in Fig. 1.8b].

Fig. 1.8: Scalar product of two vectors

It is clear that the scalar multiplication of two vectors gives a scalar. Obviously, the commutative law of multiplication holds, i.e.

$$A \cdot B = B \cdot A = AB \cos \theta \qquad (1.18)$$

If A and B are parallel to each other and pointing in the same direction, then $\theta = 0$. Thus

$$A \cdot A = A^2 \qquad (1.19)$$

If this property is applied to unit vectors $\hat{i}, \hat{j}, \hat{k}$ which are mutually perpendicular to each other, then

$$\hat{i} \cdot \hat{i} = \hat{j} \cdot \hat{j} = \hat{k} \cdot \hat{k} = 1 \qquad (1.20)$$

In case, if two vectors are perpendicular to each other, then $\cos \dfrac{\pi}{2} = 0$. This means

$$A \cdot B = AB \cos \frac{\pi}{2} = 0 \qquad (1.21)$$

This is the condition for two orthogonal vectors. If we apply this property to unit vectors, $\hat{i}, \hat{j}, \hat{k}$, then

$$\hat{i} \cdot \hat{j} = \hat{j} \cdot \hat{k} = \hat{k} \cdot \hat{i} = 0 \tag{1.22}$$

Obviously,

$$\begin{aligned} A \cdot B &= (A_x \hat{i} + A_y \hat{j} + A_z \hat{k}) \cdot (B_x \hat{i} + B_y \hat{j} + B_z \hat{k}) \\ &= A_x B_x + A_y B_y + A_z B_z \end{aligned} \tag{1.23}$$

Since the scalar product of two vectors is a scalar, the physical scalar quantity like work, gravitational potential energy, electric potential, electric power, etc. can be described as scalar products of two vectors. Equation (1.23) reveals that the scalar product of two vectors is equal to the sum of the products of their corresponding x, y and z components.

The scalar magnitude of A and B are $\sqrt{A_x^2 + A_y^2 + A_z^2}$ and $\sqrt{B_x^2 + B_y^2 + B_z^2}$. Therefore, the angle between A and B is given by

$$\begin{aligned} \cos(A, B) &= \frac{A \cdot B}{AB} \\ &= \frac{A_x B_x + A_y B_y + A_z B_z}{\sqrt{A_x^2 + A_y^2 + A_z^2}\sqrt{B_x^2 + B_y^2 + B_z^2}} \end{aligned} \tag{1.24}$$

Vector or Cross Product

The vector product of two vectors (A and B) written with a cross between them and read "A cross B", is a vector having magnitude equal to the product of the magnitudes of the two vectors and the sine of the angle between them, and a direction perpendicular to the plane containing the two vectors. Thus, the vector product of two vectors A and B is defined as

$$C = A \times B = AB \sin(A, B)\hat{n} \tag{1.25}$$

where A and B are the magnitudes of A and B, (A, B) is the angle between them, and $\hat{n}$ is a unit vector perpendicular to the plane of A and B.

Since the sine of an angle between 0 and π is not negative, the vector product cannot be negative. We must remember that the scalar product can be negative.

The direction of $C(= A \times B)$ is perpendicular to the plane containing A and B and in the sense of advance of a right hand screw rotated from A (first vector) to B (second vector) through the smaller angle between them (Fig. 1.9). Obviously, if a right-handed screw whose axis is perpendicular to the plane formed by A and B is rotated from A to B through the *smaller* angle between them, then the direction of advance of the screw gives the direction of C.

The vector product of two vectors is a vector. Torque, angular momentum, flow of electromagnetic energy, force on a moving charge in a magnetic field, etc. are few examples of vector product of two vectors.

Fig. 1.9: Vector product of two vectors

Properties of Vector Product

(a) The vector product is not commutative, i.e.

$$A \times B \neq B \times A \qquad (1.26)$$

We have

$$A \times B = AB \sin(A, B)\,\hat{n}$$

If the order of the vectors is reversed, then the vector product will be

$$B \times A = BA \sin(B, A)\,\hat{n}$$

Now $\sin(B, A) = \sin\{-(A, B)\} = -\sin(A, B)$, because $\sin(-\theta) = -\sin\theta$. This means

$$A \times B = -B \times A, \text{ i.e. } A \times B \neq B \times A \qquad (1.26a)$$

Obviously, the vector product of two vectors is not commutative and the order of the vectors must be strictly maintained.

(b) The vector product is distributive, i.e.

$$A \times (B + C) = A \times B + A \times C \qquad (1.27)$$

However, the order of the vectors must remain unchanged.

Let us express A, B, and C in terms of their x, y and z components.

$$A \times (B+C) = (A_x\hat{i} + A_y\hat{j} + A_z\hat{k}) \times [(B_x\hat{i} + B_y\hat{j} + B_z\hat{k}) + (C_x\hat{i} + C_y\hat{j} + C_z\hat{k})]$$

$$= (A_x\hat{i} + A_y\hat{j} + A_z\hat{k}) \times [(B_x + C_x)\hat{i} + (B_y + C_y)\hat{j} + (B_z + C_z)\hat{k}]$$

$$= \begin{vmatrix} \hat{i} & \hat{j} & \hat{k} \\ A_x & A_y & A_z \\ (B_x + C_x) & (B_y + C_y) & (B_z + C_z) \end{vmatrix}$$

$$= \begin{vmatrix} \hat{i} & \hat{j} & \hat{k} \\ A_x & A_y & A_z \\ B_x & B_y & B_z \end{vmatrix} + \begin{vmatrix} \hat{i} & \hat{j} & \hat{k} \\ A_x & A_y & A_z \\ C_x & C_y & C_z \end{vmatrix}$$

$$= A \times B + A \times C$$

One can easily see that above is true for any number of vectors.

(c) If either vector is multiplied by a scalar (say p), the vector product is also multiplied by a scalar, i.e.

$$(pA) \times B = (A \times pB) = pAB \sin(A, B)\,\hat{n}$$

(d) The magnitude of the vector product of two vectors mutually at right angles is equal to the product of the magnitude of the vectors. Thus,

$$A \times B = AB \sin(A, B)\hat{n} = AB\,(\sin 90°)\,\hat{n} = AB\,\hat{n}.$$

Obviously, the two vectors A and B and their vector product $A \times B$ are mutually at right angles.

(e) The vector product of two parallel vectors is a null vector (or zero). Thus, when two vectors A and B are parallel to each other, then the angle between them is $0°$ or $180°$. We have

$$A \times B = AB \sin(A, B)\,\hat{n} = AB\,(\sin 0°)\,\hat{n} = 0 \text{ (null vector)}$$

Thus, if $A \times B = 0$ and A and B are not null vectors, then A and B are parallel.

(f) The vector product of a vector by itself is a null vector, i.e.

$$A \times A = 0 \tag{1.28}$$

We have $A \times A = AA \sin(A, A)\,\hat{n} = AA\,(\sin 0°)\,\hat{n} = 0$

(g) The vector products of unit orthogonal vectors $\hat{i}, \hat{j}, \hat{k}$ have the following relations in the right-handed coordinate system:

$$\hat{i} \times \hat{i} = \hat{j} \times \hat{j} = \hat{k} \times \hat{k} = 0 \tag{1.29}$$

and

$$\left.\begin{array}{l} \hat{i} \times \hat{j} = -\hat{j} \times \hat{i} = \hat{k} \\[4pt] \hat{j} \times \hat{k} = -\hat{k} \times \hat{j} = \hat{i} \\[4pt] \hat{k} \times \hat{i} = -\hat{i} \times \hat{k} = \hat{j} \end{array}\right\} \tag{1.30}$$

The magnitude of each of the unit vectors $\hat{i}, \hat{j}$ and $\hat{k}$ is 1, and they are along the positive directions of the right-handed cartesian axes x, y and z respectively (Fig. 1.10). Obviously, the angle between any two of them is $90°$. Therefore, one obtains

$$\hat{i} \times \hat{j} = (1)(1) \sin 90°\,\hat{n} = \hat{n}$$

where, $\hat{n}$ is a unit vector perpendicular to the plane of $\hat{i}$ and $\hat{j}$, i.e. it is just the third unit vector $\hat{k}$. Thus,

$$\hat{i} \times \hat{j} = \hat{k}$$

Now
$$\hat{i} \times \hat{j} = -\hat{j} \times \hat{i}$$

$\therefore$
$$\hat{i} \times \hat{j} = -\hat{j} \times \hat{i} = \hat{k}$$

Fig. 1.10: Unit vectors in right-handed coordinate system

Similarly, one finds $\hat{j} \times \hat{k} = -\hat{k} \times \hat{j} = \hat{i}$ and $\hat{k} \times \hat{i} = -\hat{i} \times \hat{k} = \hat{j}$. Further, $\hat{i} \times \hat{i} = (1)(1) \sin 0°\,\hat{n} = 0 = 0$. Similarly, $\hat{j} \times \hat{j} = 0$ and $\hat{k} \times \hat{k} = 0$.

(h) The vector product of two vectors in terms of their x, y and z components can be expressed as a determinant. We have

$$A = A_x\hat{i} + A_y\hat{j} + A_z\hat{k}$$

and

$$B = B_x\hat{i} + B_y\hat{j} + B_z\hat{k}$$

$$\therefore \qquad A \times B = (A_x\hat{i} + A_y\hat{j} + A_z\hat{k}) \times (B_x\hat{i} + B_y\hat{j} + B_z\hat{k})$$

$$= (A_yB_z - A_zB_y)\hat{i} + (A_zB_x - A_xB_z)\hat{j} + (A_xB_y - A_yB_x)\hat{k}$$

$$(1.31)$$

This cumbersome expression can be written more neatly as a determinant, i.e.

$$A \times B = \begin{vmatrix} \hat{i} & \hat{j} & \hat{k} \\ A_x & A_y & A_z \\ B_x & B_y & B_z \end{vmatrix} \qquad (1.32)$$

$$= \hat{i} \begin{vmatrix} A_y & A_z \\ B_y & B_z \end{vmatrix} - \hat{j} \begin{vmatrix} A_x & A_z \\ B_x & B_z \end{vmatrix} + \hat{k} \begin{vmatrix} A_x & A_y \\ B_x & B_y \end{vmatrix} \qquad (1.33)$$

1.6 TRIPLE PRODUCTS

Since the cross product of two vectors is itself a vector, it can be dotted or crossed with a third vector to form a triple product.

Scalar Triple Product

Geometrically, $|A \cdot (B \times C)|$ is the volume of the parallelepiped generated by A, B and C. Since $|B \times C|$ is the area of the base, and $|A \cos\theta|$ is the altitude (Fig. 1.11), evidently

Fig. 1.11: Vector $A \cdot (B \times C)$

$$A \cdot (B \times C) = B \cdot (C \times A)$$

$$= C \cdot (A \times B) \qquad (1.34)$$

Note that "alphabetical" order is preserved in view of Eq. (1.26a), the "nonalphabetical" triple products

$$A \cdot (C \times B) = B \cdot (A \times C) = C \cdot (B \times A)$$

have the opposite sign. In component form,

$$A \cdot (B \times C) = \begin{vmatrix} A_x & A_y & A_z \\ B_x & B_y & B_z \\ C_x & C_y & C_z \end{vmatrix} \qquad (1.35)$$

We must note that the dot and cross products can be interchanged

$$A \cdot (B \times C) = (A \times B) \cdot C \qquad (1.36)$$

Like a vector A, the product $B \times C$ is also a vector. Obviously, their scalar product $A \cdot (B \times C)$ gives a scalar. It is equal in magnitude to the volume of the parallelepiped formed by the three vectors A, B and C. We must note that the placement of the parenthesis is critical, i.e. $(A \cdot B) \times C$ is a nonsensical expression. One cannot make a cross product from a scalar and a vector.

Vector Triple Product

By the property of the vector product, $A \times (B \times C)$ is a vector perpendicular to A and also to $(B \times C)$. Since $(B \times C)$ is perpendicular to the plane of B and C and hence $A \times (B \times C)$ must lie in the plane of B and C and perpendicular to A. We must note that

$$A \times (B \times C) \neq (A \times B) \times C$$

i.e. the *associative law* does not apply to vector triple products. Obviously, the parenthesis in vector triple product is very important and one cannot avoid them in vector triple product as we can do in scalar triple product. One can easily show that

$$A \times (B \times C) = B(A \cdot C) - C(A \cdot B) \tag{1.37}$$

$$A \times B \times C = (A_x\hat{i} + A_y\hat{j} + A_z\hat{k}) \times \begin{vmatrix} \hat{i} & \hat{j} & \hat{k} \\ B_x & B_y & B_z \\ C_x & C_y & C_z \end{vmatrix}$$

$$= (A_x\hat{i} + A_y\hat{j} + A_z\hat{k}) \times \ \hat{i}(B_yC_z - B_zC_y) - \hat{j}(B_xC_z - B_zC_x) + \hat{k}(B_xC_y - B_yC_x)$$

$$= \begin{vmatrix} \hat{i} & \hat{j} & \hat{k} \\ A_x & A_y & A_z \\ B_yC_z - B_zC_y & B_zC_x - B_xC_z & B_xC_y - B_yC_x \end{vmatrix}$$

$$= \hat{i}(A_yB_xC_y - A_yB_yC_x - A_zB_zC_x + A_zB_xC_z)$$
$$\quad - \hat{j}(A_xB_xC_y - A_xB_yC_x - A_zB_yC_z + A_zB_zC_y)$$
$$\quad + \hat{k}(A_xB_zC_x - A_xB_xC_z - A_yB_yC_z + A_yB_zC_y)$$

$$= \hat{i}B_x(A_yC_y + A_zC_z + A_xC_x) - \hat{i}C_x(A_yB_y + A_zB_z + A_xB_x)$$
$$\quad + \hat{j}B_y(A_xC_x + A_zC_z + A_xC_x) - \hat{j}C_y(A_xB_x + A_zB_z + A_yB_y)$$
$$\quad + \hat{k}B_z(A_xC_x + A_yC_y + A_zC_z) - \hat{k}C_z(A_xB_x + A_yB_y + A_zB_z)$$

$$= (\hat{i}B_x + \hat{j}B_y + \hat{k}B_z)(A_xC_x + A_yC_y + A_zC_z)$$
$$\quad - (\hat{i}C_x + \hat{j}C_y + \hat{k}C_z)(A_xB_x + A_yB_y + A_zC_z)$$

$$= B(A \cdot C) - C(A \cdot B)$$

One can put the vector triple product in the determinant form as follows:

$$A \times (B \times C) = \begin{vmatrix} B & C \\ A \cdot B & A \cdot C \end{vmatrix} \tag{1.38}$$

One can easily verify that the following cyclic relation in vector triple product holds good:

$$A \times (B \times C) + B \times (C \times A) + C \times (A \times B) = 0 \tag{1.39}$$

1.7 VECTOR DERIVATIVES

i. Differentiation of a Vector to a Scalar

Let us consider a particle moving along a curve (Fig. 1.12) so that at any instant it is at the point P, whose position vector is r relative to origin O. Obviously, r is a function of scalar variable t. As t changes to $t + \delta t$, r changes to $r + \delta r$. Clearly, $\dfrac{\delta r}{\delta t}$ is the average rate of change of r with t, which is a vector in the direction of δr. One can define the differential coefficient, or the derivative of the position vector r w.r.t. to time t as

Fig. 1.12: Particle moving along a curve

$$\frac{dr}{dt} = \lim_{\delta t \to 0} \frac{\delta r}{\delta t} \qquad (1.40)$$

Obviously, the derivative of a vector with respect to a scalar variable is itself a vector.

Here $\dfrac{dr}{dt}$ is the velocity V of the particle and V is a vector. The derivative of V with respect to time, i.e. $\dfrac{dV}{dt}$ is the acceleration a which is also a vector.

If we express r in rectangular coordinates, then

$$r = x\hat{i} + y\hat{j} + z\hat{k}$$

$$\therefore \qquad \frac{dr}{dt} = V = \frac{dx}{dt}\hat{i} + \frac{dy}{dt}\hat{j} + \frac{dz}{dt}\hat{k}$$

$$\text{and} \qquad \frac{d^2r}{dt^2} = a = \frac{d^2x}{dt^2}\hat{i} + \frac{d^2y}{dt^2}\hat{j} + \frac{d^2z}{dt^2}\hat{k}$$

ii. Differentiation of Sum of Vectors

We can see that the general rules of ordinary differential calculus hold good for differentiation of sum and product of vector quantities also. If A and B are functions of scalar variable t, then we have:

(a) $$\frac{d}{dt}(A + B) = \frac{dA}{dt} + \frac{dB}{dt} \qquad (1.41)$$

One can easily prove it.

Let $$V = A + B$$

When t changes to $t + \delta t$, then A changes to $A + \delta A$ and B to $B + \delta B$. We have

$$V + \delta V = (A + \delta A) + (B + \delta B)$$

$$\therefore \qquad \delta V = \delta A + \delta B$$

Dividing by δt, one obtains

$$\frac{\delta V}{\delta t} = \frac{\delta A}{\delta t} + \frac{\delta B}{\delta t}$$

Taking the limit $\delta t \to 0$, one obtains

$$\frac{dV}{dt} = \frac{dA}{dt} + \frac{dB}{dt} = \frac{d}{dt}(A + B)$$

(b) If m is a scalar function of t, then

$$\frac{d}{dt}(mA) = m\frac{dA}{dt} + \frac{dm}{dt}A \qquad (1.42)$$

Let $\qquad\qquad V = mA$

When t changes to $t + \delta t$, then m changes to $m + \delta m$ and A to $A + \delta A$. Obviously,

$$V + \delta V = (m + \delta m)(A + \delta A)$$
$$= mA + m\delta A + \delta mA + \delta m\delta A$$

Here we ignore the smallar term $\delta m\delta A$. Thus, we have

$$\delta V = m\delta A + \delta mA$$

Dividing by δt, one obtains

$$\frac{\delta V}{\delta t} = m\frac{\delta A}{\delta t} + \frac{\delta m}{\delta t}A$$

Taking the limit $\delta t \to 0$, one obtains

$$\frac{dV}{dt} = \frac{d}{dt}(mA) = m\frac{dA}{dt} + \frac{dm}{dt}A$$

$$(1.43)$$

(c) $\qquad \dfrac{d}{dt}(A \cdot B) = A \cdot \dfrac{dB}{dt} + \dfrac{dA}{dt} \cdot B$

Let $C = A \cdot B$. When t changes to $t + \delta t$, then A changes to $A + \delta A$ and B changes to $B + \delta B$. Obviously,

$$C + \delta C = (A + \delta A) \cdot (B + \delta B)$$
$$= A \cdot B + A \cdot \delta B + \delta A \cdot B + \delta A \cdot \delta B.$$

Here we ignore the smaller term $\delta A \cdot \delta B$. Thus,

$$\delta C = A \cdot \delta B + \delta A \cdot B$$

or $\qquad\qquad \dfrac{\delta C}{\delta t} = A \cdot \dfrac{\delta B}{\delta t} + \dfrac{\delta A}{\delta t} \cdot B$

Taking the limit $\delta t \to 0$, one obtains

$$\frac{dC}{dt} = \frac{d}{dt}(A \cdot B) = A \cdot \frac{dB}{dt} + \frac{dA}{dt} \cdot B$$

We may note that the order of factors in any of the above terms is immaterial.

(d) $\qquad \dfrac{d}{dt}(A \times B) = A \times \dfrac{dB}{dt} + \dfrac{dA}{dt} \times B \qquad (1.44)$

Let $C = A \times B$. When t changes to $t + \delta t$, then A changes to $A + \delta A$ and B to $B + \delta B$. We have

$$C + \delta C = (A + \delta A) \times (B + \delta B)$$
$$= A \times B + A \times \delta B + \delta A \times B + \delta A \times \delta B$$

$$= C + A \times \delta B + \delta A \times B$$

We have neglected the smaller term $\delta A \times \delta B$. Thus,

$$\delta C = A \times \delta B + \delta A \times B$$

or

$$\frac{\delta C}{\delta t} = A \times \frac{\delta B}{\delta t} + \frac{\delta A}{\delta t} \times B$$

Taking the limit $\delta t \to 0$, one obtains

$$\frac{dC}{dt} = \frac{d}{dt}(A \times B) = A \times \frac{dB}{dt} + \frac{dA}{dt} \times B$$

We must note that the order of factors in each term should remain unchanged.

1.8 PARTIAL DIFFERENTIATION OF VECTORS

Consider that A is a vector which depends on more than one scalar variable such as cartesian coordinates x, y, z of a point in space. Thus, $\frac{\partial A}{\partial x}, \frac{\partial A}{\partial y}$ and $\frac{\partial A}{\partial z}$ are the partial derivatives of vector A with respect to x (when y and z remain constant), y (when x and z remain constant) and z (when x and y remain constant) respectively. Let us consider that if x, y and z change simultaneously by small increments dx, dy and dz, then the total change or total derivative of A will be

$$dA = \frac{\partial A}{\partial x}dx + \frac{\partial A}{\partial y}dy + \frac{\partial A}{\partial z}dz$$

$$= \left[\frac{\partial}{\partial x}dx + \frac{\partial}{\partial y}dy + \frac{\partial}{\partial z}dz\right] A$$

$$= \left[\frac{\partial}{\partial x}\hat{i} + \frac{\partial}{\partial y}\hat{j} + \frac{\partial}{\partial z}\hat{k}\right] \cdot (dx\hat{i} + dy\hat{j} + dz\hat{k})A$$

$$= (\nabla \cdot dr)A$$

Here $\nabla = \frac{\partial}{\partial x}\hat{i} + \frac{\partial}{\partial y}\hat{j} + \frac{\partial}{\partial z}\hat{k}$ is called *del operator* and $r = x\hat{i} + y\hat{j} + z\hat{k}$ is the position vector of the point (x, y, z) with respect to origin. We must note that ∇ is a vector quantity possessing all the properties of a ordinary vector. However, $\nabla \cdot \nabla = \nabla^2 = \frac{\partial^2}{\partial x^2} + \frac{\partial^2}{\partial y^2} + \frac{\partial^2}{\partial z^2}$ is a scalar quantity and known as *Laplacian operator*.

ILLUSTRATIVE EXAMPLES

Example 1.1: Two vectors A and B are such that $A + B = |A - B|$. Find the angle between them.

Solution: Let θ be the angle between A and B. Then angle between A and $-B$ is $\pi - \theta$. Thus,

$$|A + B| = \sqrt{A^2 + B^2 + 2AB\cos\theta}$$

$$|A - B| = |A +(-B)| = \sqrt{A^2 + B^2 + 2AB\cos(\pi - \theta)}$$

$$= \sqrt{A^2 + B^2 - 2AB\cos\theta}$$

Since $|A + B| = |A - B|$, we have $A^2 + B^2 + 2AB\cos\theta = A^2 + B^2 - 2AB\cos\theta$ or $4AB\cos\theta = 0$

But $\qquad\qquad\qquad\qquad\qquad A \neq 0$ and $B \neq 0$

$\therefore \qquad\qquad\qquad\qquad\qquad \cos\theta = 0 \quad$ or $\theta = 90°$

Example 1.2: Two vectors A and B are such that $A - B = C$ and $A - B = C$. Find the angle between them.

Solution: We have

$$C = A + (-B)$$

This yields $\qquad\qquad\qquad\qquad C^2 = A^2 + B^2 - 2AB\cos\theta \qquad\qquad\qquad (1)$

$\because \qquad\qquad\qquad\qquad\qquad\qquad C = A - B$

$\therefore \qquad\qquad\qquad\qquad\qquad C^2 = A^2 + B^2 - 2AB \qquad\qquad\qquad\qquad (2)$

Comparing Eqs. (1) and (2), one obtains

$$\cos\theta = 1 \quad\therefore\ \theta = 0°$$

Example 1.3: If $A = 4\hat{i} - \hat{j}$, $B = -3\hat{i} + 2\hat{j}$ and $C = -3\hat{j}$, find the vector sum, its scalar magnitude and direction. Also find a unit vector parallel to $A + B + C$.

Solution: Let $\quad r = A + B + C$

$$= (4\hat{i} - \hat{j}) + (-3\hat{i} + 2\hat{j}) + (-3\hat{j})$$

$$= \hat{i} - 2\hat{j}$$

Now $\qquad |r| = |A + B + C|$

$$= |\hat{i} - 2\hat{j}| = \sqrt{(1)^2 + (-2)^2} = \sqrt{5}$$

The angle which the vector r makes with the positive x-axis, measured counterclockwise is

$$\theta = \tan^{-1}\frac{r_y}{r_x} = \tan^{-1}\left(-\frac{2}{1}\right)$$

$$= \tan^{-1}(-2) = -63°$$

As θ is negative, i.e. this is the angle which r makes with the x-axis measured clockwise. This means the angle which it makes with the x-axis measured counterclockwise is $360° - \theta = 297°$. A unit vector parallel to r,

$$\hat{r} = \frac{r}{|r|} = \frac{\hat{i} - 2\hat{j}}{\sqrt{5}} = \frac{\hat{i}}{\sqrt{5}} - \frac{2}{\sqrt{5}}\hat{j}.$$

Example 1.4: If $A = 3\hat{i} + 2\hat{j} + 4\hat{k}$ and $B = 4\hat{i} + 2\hat{j} + \hat{k}$, find the magnitudes and direction cosines of $A + B$ and $A - B$.

Solution: We have

$$A + B = 7\hat{i} + 4\hat{j} + 5\hat{k} \qquad\qquad\qquad\qquad\qquad (1)$$

and
$$A - B = -\hat{i} + 3\hat{k} \qquad (2)$$
$$|A + B| = \sqrt{(7)^2 + (4)^2 + (5)^2} = \sqrt{(90)}$$

and
$$|A - B| = \sqrt{(-1)^2 + (3)^2} = \sqrt{10}$$

$\therefore$
$$\cos\alpha = \frac{|A+B|_x}{|A+B|} = \frac{7}{\sqrt{90}}$$

$$\cos\beta = \frac{|A+B|_y}{|A+B|} = \frac{4}{\sqrt{90}}$$

$$\cos\gamma = \frac{|A+B|_z}{|A+B|} = \frac{5}{\sqrt{90}}$$

Similarly, one obtains the direction cosines of $A - B$ as $-\dfrac{1}{\sqrt{10}}, 0, \dfrac{3}{\sqrt{10}}$.

Example 1.5: If l_1, m_1, n_1 and l_2, m_2, n_2 are the direction cosines of two vectors, show that the angle θ between them is given by

$$\cos\theta = l_1 l_2 + m_1 m_2 + n_1 n_2.$$

Solution: Let A and B be two vectors and the angle between them be θ. We have
$$A \cdot B = AB \cos\theta$$

$\therefore$
$$\cos\theta = \frac{A \cdot B}{AB} = \frac{A_x B_x + A_y B_y + A_z B_z}{AB}$$

$$= \frac{A_x B_x}{AB} + \frac{A_y B_y}{AB} + \frac{A_z B_z}{AB}$$

Now, $\dfrac{A_x}{A}, \dfrac{A_y}{A}$ and $\dfrac{A_z}{A}$ are direction cosines of A, whereas $\dfrac{B_x}{B}, \dfrac{B_y}{B}$ and $\dfrac{B_z}{B}$ are direction cosines of B. Thus,

$$\cos\theta = l_1 l_2 + m_1 m_2 + n_1 n_2$$

Example 1.6: If the nonparallel vectors A and B are equal in magnitude, show that vector $(A + B)$ is perpendicular to vector $(A - B)$.

Solution:
$$(A + B) \cdot (A - B) = A \cdot A - A \cdot B + B \cdot A - B \cdot B$$
$$= A \cdot A - B \cdot B \quad (\because A \cdot B = B \cdot A)$$
$$= A^2 - B^2$$
$$= 0 \quad (\because A = B)$$

Since A is not parallel to B, the vector $A - B$ cannot be zero. This means $(A + B)$ is perpendicular to $(A - B)$.

Example 1.7: Can $A \times B = A \cdot B$?

Solution: This is possible only when $A = B = 0$. If either A or B is zero and the other is nonzero, then $A \times B$ will be a null vector, while $A \cdot B$ will be 0, i.e. a null scalar.

Example 1.8: Given two vectors $A = 4\hat{i} + 3\hat{j}$ and $B = -2\hat{i} + 6\hat{j}$, determine $A \times B$ and a unit vector in the direction of B.

Solution: $A \times B = \begin{vmatrix} \hat{i} & \hat{j} & \hat{k} \\ 4 & 3 & 0 \\ -2 & 6 & 0 \end{vmatrix} = 30\hat{k}$

A unit vector in the direction of B is given by

$$\hat{b} = \frac{B}{B} = \frac{-2\hat{i} + 6\hat{j}}{\sqrt{(-2)^2 + (6)^2}} = \frac{1}{\sqrt{40}}(-2\hat{i} + 6\hat{j})$$

Example 1.9: Find the constant p such that the vector $2\hat{i} - \hat{j} + \hat{k}, \hat{i} + 2\hat{j} - 3\hat{k}$ and $3\hat{i} + p\hat{j} + 5\hat{k}$ are coplanar.

Solution: The scalar triple product of vectors is

$$= \begin{vmatrix} 2 & -1 & 1 \\ 1 & 2 & -3 \\ 3 & p & 5 \end{vmatrix} = 2(10 + 3p) + 1(5 + 9) + 1(9 - 6)$$

$$= 28 + 7p$$

Since the vectors are coplanar, their scalar triple product is zero, i.e.

$$28 + 7p = 0$$

$$\therefore \qquad p = -4$$

Example 1.10: If $\hat{r}$ is a unit vector, show that $\dfrac{d\hat{r}}{dt}$ is perpendicular to $\hat{r}$.

Solution: Since $\hat{r}$ is a unit vector, its scalar product by itself is 1, i.e.

$$\hat{r} \cdot \hat{r} = 1$$

$$\therefore \qquad \frac{d}{dt}(\hat{r} \cdot \hat{r}) = 0$$

or $\qquad\qquad \hat{r} \cdot \dfrac{d\hat{r}}{dt} + \dfrac{d\hat{r}}{dt} \cdot \hat{r} = 0$

or $\qquad\qquad\qquad\qquad 2\hat{r} \cdot \dfrac{d\hat{r}}{dt} = 0$

or $\qquad\qquad\qquad\qquad \hat{r} \cdot \dfrac{d\hat{r}}{dt} = 0$

Therefore, $\hat{r}$ and $\dfrac{d\hat{r}}{dt}$ are perpendicular to each other.

GLIMPSES

- *Scalar* quantity is one which has only magnitude but no direction.
- *Vector* quantity is one which has both magnitude and direction.
- Dot product of two vectors $A \cdot B = AB \cos\theta$, where θ is the angle between A and B.

- *Cross product* of two vectors $A \times B$ is defined as $A \times B = |A||B|\sin\theta\,\hat{n}$, where $\hat{n}$ is the unit vector normal to the plane containing A and B.

- $\hat{i}, \hat{j}, \hat{k}$ denote three vectors of unit length drawn respectively in the directions of x-, y- and z-axes of a right-handed rectangular system. In terms of these vectors, any vector r may be expressed as

$$r = x\hat{i} + y\hat{j} + z\hat{k}$$

Unit vectors are always mutually perpendicular to each other.

- *Scalar triple product* of three vectors $(A \times B) \cdot C$ is a number, i.e. a scalar. The value of the said triple product depends on the cyclic order of the factors and is independent of the position of the dot and cross product between them.

- *Vector triple product* of three vectors A, B and C is the cross product of A and $(B \times C)$ and written as $A \times (B \times C)$ which is itself a vector. The vector $A \times (B \times C)$ is perpendicular to $B \times C$ and therefore coplanar with B and C

$$A \times (B \times C) = (A \cdot C)B - (A \cdot B)C$$

- The *vector product or cross product of* $(A \times B)$ and $(C \times D)$ is called a vector product of four vectors, which is denoted by $(A \times B) \times (C \times D)$. This is a vector which is perpendicular to $(A \times B)$ and therefore coplanar with A and B. Similarly, this is perpendicular to $(C \times D)$ and therefore coplanar with C and D.

$$(A \times B) \times (C \times D) = [A \cdot (C \times D)]B - [B \cdot (C \times D)]A$$

REVIEW QUESTIONS

1. What is a unit vector? Show that a unit vector in three dimensions can be expressed as

$$\hat{a} = \hat{i}\cos(a, x) + \hat{j}\cos(a, y) + \hat{k}\cos(a, z)$$

2. If A makes angles α, β and γ respectively with the x-, y- and z-axes of a cartesian coordinate system, prove that

$$\cos^2\alpha + \cos^2\beta + \cos^2\gamma = 1$$

3. Define scalar and vector products of two vectors. Give physical examples.

4. Prove that $A \times B = \begin{vmatrix} \hat{i} & \hat{j} & \hat{k} \\ A_x & A_y & A_z \\ B_x & B_y & B_z \end{vmatrix}$

5. If $\hat{i}, \hat{j}$ and $\hat{k}$ are unit vectors along x-, y- and z-axes respectively, show that $\hat{i} \cdot \hat{i} = 1$, $\hat{i} \cdot \hat{j} = 0$, $\hat{i} \times \hat{i} = 0$ and $\hat{i} \times \hat{j} = \hat{k}$

6. Show that the component of A along B may be expressed as $(A \cdot B)B / B^2$.

7. Suppose that A, B and C are any three non-coplanar vectors. They are not necessarily at right angles. Show that

$$A \cdot (B \times C) = B \cdot (C \times A)$$

8. Show that

$$A \times (B \times C) = B(A \cdot C) - C(A \cdot B)$$

PROBLEMS

1. If $A = 2\hat{i} - 4\hat{j} - 3\hat{k}$ and $B = \hat{i} + 2\hat{j} + \hat{k}$, find $|A + B|$ and $|A - B|$ and a unit vector parallel to $|A + B|$.

$$\text{Ans. } \sqrt{17}; \sqrt{53}; \frac{1}{\sqrt{17}}(3\hat{i} - 2\hat{j} + \hat{k})$$

2. If $A = \hat{i} + \hat{j} - 2\hat{k}, B = 2\hat{i} + \hat{j} + \hat{k}$ and $C = \hat{i} - 2\hat{j} + 2\hat{k}$, find magnitude and direction of vectors $(A + B + C)$ and $(A - B + C)$.

$$\text{Ans. } \sqrt{17}; \sqrt{5}; \ \frac{4}{\sqrt{17}}, 0, \frac{1}{\sqrt{17}}, \ 0, -\frac{2}{\sqrt{5}}, -\frac{1}{\sqrt{5}}$$

3. If α, β are the direction cosines of a vector with respect to x- and y-axes and if C is its z-component, find the magnitude of the vector.

$$\text{Ans. } C / \sqrt{(1 - \alpha^2 - \beta^2)}$$

4. Show that the vectors $5\hat{i} + 7\hat{j} - 3\hat{k}$ and $2\hat{i} - b\hat{j} + c\hat{k}$ will be parallel to each other, if $b = -\dfrac{14}{5}$ and $c = -\dfrac{6}{5}$.

5. For what value of n the vectors $4\hat{i} + 3\hat{j} - 7\hat{k}$ and $5\hat{i} - 2\hat{j} - n\hat{k}$ are orthogonal.

[*Ans.* $n = -2$]

6. Determine unit vectors which are normal to x-axis and lie in the plane $3x - 2y + 3z = 1.5$.

[*Ans.* $\pm(3\hat{j} + 2\hat{k})/\sqrt{13}$]

7. Show that the cross product of each two of the following vectors is parallel to the third

$$(\hat{i} + \hat{j} + \hat{k}), (\hat{i} - 2\hat{j} + \hat{k}), (\hat{i} - \hat{k}).$$

8. Prove the following:

   (a) $A \times (B \times C) + B \times (C \times A) + C \times (A \times B) = 0$

   (b) $(A - B) \times (A + B) = 2(A \times B)$

9. Show that if A, B and C are coplanar vectors,

   $(A \times B) \times (C \times D) = 0$

10. Write down the expression for the time derivative or $r\hat{r}$.

$$\text{Ans. } r\frac{d\hat{r}}{dt} + \hat{r}\frac{dr}{dt}$$

SHORT ANSWER QUESTIONS

1. Is a quantity having a magnitude and direction necessarily a vector?
   Ans. No. A directed quantity is a vector provided it obeys the commutative law of vector addition.

2. If a vector has zero magnitude, is it meaningful to call it a vector?

Ans. Yes, it is a null vector.

3. A zero vector is multiplied by a finite number. A finite vector is multiplied by zero. Is the result in two cases same?

Ans. Yes, both are equal to null or zero vector.

4. Give one example each in physics where 'dot' and 'cross' product of two vectors occur as physical quantities.

Ans. Dot product: Work, Cross product: Torque

5. The two vectors *A* and *B* are at right angles to each other. What will be their scalar product?

Ans. Zero

6. When the two vectors *A* and *B* are parallel to each other, what will be their scalar product?

Ans. $A \cdot B = AB \cos 0° = AB$

7. Find $A \times A$.

Ans. Null vector

8. What is the necessary and sufficient condition for *A*, *B* and *C* to be coplanar?

Ans. Their scalar triple product $A \cdot B \times C$ must be zero

9. Can the dot and cross product in a scalar triple product $A \cdot B \times C$ be interchanged?

Ans. Yes

MULTIPLE CHOICE QUESTIONS

1. If three vectors *A*, *B* and *C* are 12, 5 and 13 respectively in magnitude such that $C = A + B$, then the angle between *A* and *B* is
   (a) 30°   (b) 60°   (c) 90°   (d) 120°   **[c]**

2. Two nonzero vectors *A* and *B* are such that $|A + B| = |A - B|$. The angle between them is
   (a) 180°   (b) 90°   (c) 60°   (d) 0°   **[b]**

3. Two vectors *A* and *B* are such that $A + B = C$ and $A^2 + B^2 = C^2$. The angle between them is
   (a) 180°   (b) 120°   (c) 90°   (d) 0°   **[c]**

4. If $A = 4\hat{i} + 3\hat{j} - 2\hat{k}$ and $B = 8\hat{i} + 6\hat{j} - 4\hat{k}$, the angle between *A* and *B* is
   (a) 90°   (b) 60°   (c) 45°   (d) 0°   **[d]**

5. Two vectors *A* and *B* lie in a plane. Another vector *C* lies outside this plane. Then $A + B + C$
   (a) can be zero
   (b) cannot be zero
   (c) lies in the plane containing $A + B$
   (d) lies in the plane containing $A - B$ **[b]**

6. The scalar product of vectors $A = 2\hat{i} + 5\hat{k}$ and $B = 3\hat{j} + 4\hat{k}$ is

   (a) 20   (b) $5\sqrt{33}$ (c) 26   (d) 23   **[a]**

7. A vector A is along the positive x-axis and if B is another vector such that $A \times B$ is zero, then B could be

   (a) $4\hat{j}$ (b) $-4\hat{i}$ (c) $-(\hat{i} + \hat{j})$ (d) $(\hat{j} + \hat{k})$ **[b]**

8. If $A = 5\hat{i} + 7\hat{j} - 3\hat{k}$ and $B = 2\hat{i} + 2\hat{j} - a\hat{k}$ are perpendicular vectors, the value of a is

   (a) 8   (b) –8   (c) –7   (d) –2   **[b]**

9. Three vectors A, B and C satisfy the relation $A \cdot B = 0$ and $A \cdot C = 0$. The vector A is parallel to

   (a) $B \cdot C$ (b) $B \times C$ (c) B (d) C **[b]**

10. A particle moves from position $r_1 = 3\hat{i} + 2\hat{j} - 6\hat{k}$ to position $r_2 = 14\hat{i} + 13\hat{j} + 9\hat{k}$ under the action of a force $F = 4\hat{i} + \hat{j} + 3\hat{k}$. The work done by the force is

   (a) 200 units   (b) 100 units   (c) 75 units   (d) 50 units   **[b]**

11. Identify which of the following is not a vector.

   (a) work   (b) momentum   (c) force   (d) acceleration   **[a]**

12. The unit vector parallel to the resultant of vectors $r_1 = 2\hat{i} - 3\hat{j} + 4\hat{k}$ and $r_2 = 3\hat{i} - 4\hat{j} + 2\hat{k}$

   (a) $\dfrac{5\hat{i} - 7\hat{j}}{\sqrt{110}}$ (b) $\dfrac{5\hat{i} - 7\hat{j} + 6\hat{k}}{\sqrt{110}}$ (c) $\dfrac{7\hat{j} - 6\hat{k}}{\sqrt{110}}$ (d) none of the above **[b]**

13. The dot product of vectors $A = 2\hat{i} - \hat{j} + \hat{k}$ and $b = \hat{i} - 3\hat{j} - 5\hat{k}$ is

   (a) 10   (b) 2   (c) 7   (d) 0   **[d]**

14. If $A = 3\hat{i} + 2\hat{j} - \hat{k}$ and $B = \hat{i} + 4\hat{j} + \hat{k}$ then $A \times B$ is equal to

   (a) $2\hat{i} + 4\hat{j} - \hat{k}$ (b) $6\hat{i} + 4\hat{j} - 10\hat{k}$ (c) $6\hat{i} + 4\hat{j}$ (d) $4\hat{j} - 10\hat{k}$ **[b]**

15. If vectors $2\hat{i} + \hat{j} + \hat{k}$ and $\hat{i} + 2\hat{j} - 3\hat{k}$ and $3\hat{i} - p\hat{j} + 5\hat{k}$ are coplanar, then p is equal to

   (a) 6   (b) 2   (c) 4   (d) 8   **[c]**

16. A point in cartesian coordinates is represented by $P(2, 6, 3)$, the radial component r in cylindrical coordinate will be _____radial component of A in cylindrical coordinates

   (a) equal to   (b) greater than   (c) less than   (d) unrelated to   **[b]**

17. Let a point in spherical and cylindrical coordinates be (r, θ, ϕ) and (ρ, ϕ, z). The component r in cylindrical coordinates is related to spherical coordinates as

   (a) $z \tan^{-1} \phi$ (b) ρ (c) $(\rho^2 + z^2)^{1/2}$ (d) $z \tan \phi$ **[c]**

18. The vector transformation between cylindrical and spherical coordinates is represented as

(a) $\begin{bmatrix} A_r \\ A_\theta \\ A_\phi \end{bmatrix} = \begin{bmatrix} \sin\phi & 0 & \cos\phi \\ \cos\phi & 0 & -\sin\phi \\ 0 & 1 & 0 \end{bmatrix} \begin{bmatrix} A_\rho \\ A_\phi \\ A_z \end{bmatrix}$

(b) $\begin{bmatrix} A_r \\ A_\theta \\ A_\phi \end{bmatrix} = \begin{bmatrix} \sin\phi & \cos\phi & 0 \\ 0 & 0 & 1 \\ \cos\phi & -\sin\phi & 0 \end{bmatrix} \begin{bmatrix} A_\rho \\ A_\phi \\ A_z \end{bmatrix}$

(c) $\begin{bmatrix} A_r \\ A_\theta \\ A_\phi \end{bmatrix} = \begin{bmatrix} \sin\theta & 0 & \cos\theta \\ \cos\theta & 0 & -\sin\phi \\ 0 & 1 & 0 \end{bmatrix} \begin{bmatrix} A_\rho \\ A_\phi \\ A_z \end{bmatrix}$

(d) $\begin{bmatrix} A_r \\ A_\theta \\ A_\phi \end{bmatrix} = \begin{bmatrix} \sin\theta & \cos\theta & 0 \\ 0 & 0 & 1 \\ \cos\theta & -\sin\theta & 0 \end{bmatrix} \begin{bmatrix} A_\rho \\ A_\phi \\ A_z \end{bmatrix}$ **[c]**

19. Let a point in spherical and cylindrical coordinates be (r, θ, ϕ) and (ρ, ϕ, z). The radial component ρ in cylindrical coordinates is related to spherical components as
    (a) $r \sin\theta$ (b) $r \cos\theta$
    (c) $r \cos\phi$ (d) $r \sin\phi$ **[a]**

20. $\hat{i} \times (\hat{j} \times \hat{k})$ is equal to
    (a) 0   (b) 1
    (c) 2   (d) none of the above   **[a]**

Scalar and Vector Fields

2.1 FIELDS

A physical quantity can be expressed as a continuous function of the position of a region of space. If the cartesian coordinates of a point P are (x, y, z) and the physical quantity at this is A, represented as

$$A = A(x, y, z) \tag{2.1}$$

then this function is called a point function and the region in which it specifies the physical quantity is known as a *field*.

Fields are of two kinds: *scalar field* and *vector field*, depending upon the nature of the physical quantity concerned.

2.2 SCALAR FIELDS

When a 'scalar' (physical quantity) is expressed from point to point in a region of space by a continuous point function,

$$\phi = \phi(x, y, z)$$

which gives the value of the quantity at each point, then the region is a *scalar field* and the function ϕ is a *scalar point function*. Distribution of density, temperature, electric potential, gravitational potential in space, etc. are a few familiar examples of scalar fields.

One can represent scalar quantities graphically by drawing such surfaces at which the value of the quantity remains constant. We must note that one point in space corresponds to only one value of the quantity, i.e. scalar fields are single-valued functions at each point, i.e. two level surfaces cannot intersect (otherwise ϕ should have two values at the point of intersection which contradicts our definition) (Fig. 2.1). These surfaces are called *equiscalar surfaces*. If we put a point charge at any place, then the electric potential around the charge will depend on the position of the point. Electric potential is a scalar quantity and hence the field around the charge is known as *scalar potential field*. If we join all such points

Fig. 2.1: Graphical representation of scalar field

at which the potential is constant, then such a surface is called an *equipotential surface*. Figure 2.1 shows S_1 and S_2 as equipotential surfaces, with constant potentials ϕ_1 and ϕ_2 respectively.

2.3 VECTOR FIELDS

When a vector physical quantity is expressed from point to point in a region of space by a continuous vector function $A(x, y, z)$, then the region is said to be a *vector field* and the function A is called a *vector point function*. At any given point in the field, the function A is represented by a vector of definite magnitude and direction which changes continuously from point to point. Distribution of velocity in a fluid and distribution of gravitational or magnetic field intensity in space are few examples of vector fields.

Vector fields are graphically represented by lines and these lines are called *field lines* or *flux lines*. These lines are drawn in the field in such a way that the tangent at any point on the line represents the direction of the vector field at that point. The magnitude of A at any point on a flux line is given by the number of flux lines crossing unit area perpendicular to their direction drawn at that point. Two flux lines cannot intersect (otherwise the direction of the vector would become indefinite at the point of intersection).

Fig. 2.2: Flux going out of an area element

Let us consider that the number of field lines, or flux going out of an area element *ds* perpendicular to field lines at any point be $d\phi$ (Fig. 2.2).

Obviously, the magnitude of the intensity of the vector field will be

$$|A| = \frac{d\phi}{ds} \tag{2.2}$$

$$\therefore \qquad A = \left(\frac{d\phi}{ds}\right)\hat{n} \tag{2.3}$$

where, $\hat{n}$ is the unit vector perpendicular to the area *ds*. Thus,

$$\hat{n}\,ds = ds$$

Now, $$A \cdot \hat{n} = \frac{d\phi}{ds}\,\hat{n} \cdot \hat{n} = \frac{d\phi}{ds}$$

or $$d\phi = (A \cdot \hat{n})ds = A \cdot ds \tag{2.4}$$

We must remember that point function $\phi(x, y, z)$ or $A(x, y, z)$ at every point of the field is single-valued.

2.4 PARTIAL DERIVATIVE

Let us consider that a physical quantity f be a positional function of the coordinates x, y and z of the point of observation, where x, y and z are independent variables, then

$$f = f(x, y, z) \tag{2.5}$$

Keeping y and z constant, let us change the value of x by an infinitesimal amount Δx and the value of f changes by an amount Δf, then for constant y and z the rate of change of f with respect to x, i.e. $\left(\dfrac{\Delta f}{\Delta x}\right)$ is known as *partial derivative of f with respect to x*.

Usually it is expressed as $\left(\dfrac{\partial f}{\partial x}\right)_{y,z}$ or simply $\dfrac{\partial f}{\partial x}$. Obviously, Δx is considered infinitely small, i.e. $\Delta x \to 0$ and hence

$$\frac{\partial f}{\partial x} = \lim_{\Delta x \to 0} \frac{f(x+\Delta x, y, z) - f(x, y, z)}{\Delta x} \tag{2.6}$$

Similarly, if one keeps x and z constant and the value of y is changed by an infinitesimal amount Δy, then

$$\frac{\partial f}{\partial y} = \lim_{\Delta y \to 0} \frac{f(x, y+\Delta y, z) - f(x, y, z)}{\Delta y} \tag{2.7}$$

and if one keeps x and y constant and the value of z is changed by an infinitesimal amount Δz, then

$$\frac{\partial f}{\partial z} = \lim_{\Delta z \to 0} \frac{f(x, y, z+\Delta z) - f(x, y, z)}{\Delta z} \tag{2.8}$$

Now, consider that all the three independent variables x, y and z are simultaneously changed by infinitesimal amounts, say δx, δy and δz respectively and the value of f is changed by an amount δf. When $\delta x \to 0$, $\delta y \to 0$ and $\delta z \to 0$, i.e. all the three vanish simultaneously, $\delta f \to 0$, i.e. δf will also vanish. Obviously, in this state, the total change in the value of f, i.e. δf is known as total derivative. Thus,

$$\delta f = \left(\frac{\partial f}{\partial x}\right)\delta x + \left(\frac{\partial f}{\partial y}\right)\delta y + \left(\frac{\partial f}{\partial z}\right)\delta z$$

One can write the total derivative for the limiting change as

$$df = \left(\frac{\partial f}{\partial x}\right)dx + \left(\frac{\partial f}{\partial y}\right)dy + \left(\frac{\partial f}{\partial z}\right)dz \tag{2.9}$$

The partial derivatives $\dfrac{\partial f}{\partial x}, \dfrac{\partial f}{\partial y}$ and $\dfrac{\partial f}{\partial z}$ are called *first partial derivatives* or *partial derivatives of the first order*. By differentiating these derivatives once more, one can obtain the *second partial derivatives* (or partial derivatives of the second order), e.g. $\dfrac{\partial^2 f}{\partial x^2}, \dfrac{\partial^2 f}{\partial y \partial x}, \dfrac{\partial^2 f}{\partial x \partial y}, \dfrac{\partial^2 f}{\partial y^2}, \dfrac{\partial^2 f}{\partial z^2}$, etc. Obviously, $\dfrac{\partial^2 f}{\partial x^2}$ is the partial derivative of $\dfrac{\partial f}{\partial x}$ with respect to x, $\dfrac{\partial^2 f}{\partial x \partial y}$ is the partial derivative of $\dfrac{\partial f}{\partial x}$ with respect y, etc. If all the derivatives concerned are continuous, then $\dfrac{\partial^2 f}{\partial x \partial y} = \dfrac{\partial^2 f}{\partial y \partial x}$, i.e then the order of differentiation is immaterial. By differentiating the second partial derivatives again with respect to x, y and z respectively, one obtains the third partial derivatives and so on.

If A is a vector depending on more than one scalar variables such as cartesion coordinates x, y and z of a point in space, then $\dfrac{\partial A}{\partial x}, \dfrac{\partial A}{\partial y}$ and $\dfrac{\partial A}{\partial z}$ are partial derivatives of A with respect to x (when y and z remain constant), y (when x and z remain constant) and z (when x and y remain constant respectively. If now x, y and z change simultaneously by small increments dx, dy and dz, then the total change or total derivative of A will be

$$dA = \frac{\partial A}{\partial x} dx + \frac{\partial A}{\partial y} dy + \frac{\partial A}{\partial z} dz$$

$$= \left(\frac{\partial}{\partial x} dx + \frac{\partial}{\partial y} dy + \frac{\partial}{\partial z} dz \right) A$$

$$= \left(\frac{\partial}{\partial x}\hat{i} + \frac{\partial}{\partial y}\hat{j} + \frac{\partial}{\partial z}\hat{k} \right) \cdot (dx\hat{i} + dy\hat{j} + dz\hat{k})A \tag{2.10}$$

where $\hat{i}, \hat{j}$ and $\hat{k}$ are unit vectors in the directions of x-, y- and z-axes respectively.

$\frac{\partial}{\partial x}\hat{i} + \frac{\partial}{\partial y}\hat{j} + \frac{\partial}{\partial z}\hat{k}$ is defined as the *vector differential operator*, written as ∇ and read as *del*.

Thus,

$$\nabla = \frac{\partial}{\partial x}\hat{i} + \frac{\partial}{\partial y}\hat{j} + \frac{\partial}{\partial z}\hat{k} \tag{2.11}$$

Further if r is the position vector of a point (x, y, z) with respect to origin, then

$$r = \hat{i}x + \hat{j}y + \hat{k}z \tag{2.12}$$

so that

$$dr = \hat{i}\,dx + \hat{j}\,dy + \hat{k}\,dz \tag{2.13}$$

Equation (2.10) can be expressed as

$$dA = (\nabla \cdot dr)A \tag{2.14}$$

Here ∇ is a vector quantity possessing all the properties of an ordinary vector. Since it is an *operator*, its magnitude has no physical significance.

The differential operator $\nabla \cdot \nabla = \nabla^2$ is defined as

$$\nabla^2 = \frac{\partial^2}{\partial x^2} + \frac{\partial^2}{\partial y^2} + \frac{\partial^2}{\partial z^2} \tag{2.15}$$

∇^2 is a scalar and known as *Laplacian operator*.

2.5 GRADIENT OF A SCALAR FUNCTION

A scalar field can be mapped out by a series of level surfaces. Let us consider that in a scalar field two surfaces S_1 and S_2 represent surfaces with constant scalar quantities ϕ and $\phi + \delta\phi$ respectively (Fig. 2.3). Let r be the position vector of any point P on S, relative to the origin O.

The position vector of another point Q on S_2 is $r + \delta r$. Obviously, the displacement of Q with respect to P is $PQ = \delta r$. The rate of change of scalar function in the direction of PQ is $\frac{\delta\phi}{\delta r}$. Now, draw a perpendicular PR from P on surface S_2.

The minimum distance between the surfaces S_1 and S_2, i.e. $PR = \delta n$, will be in the direction of their normal. Therefore, the rate of change of scalar function ϕ in the

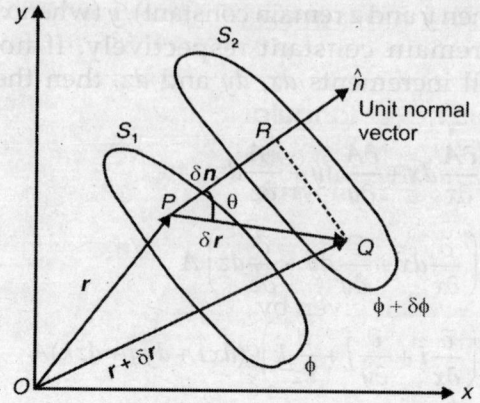

Fig. 2.3: Mapping of a scalar field

direction of $PR = \dfrac{\delta\phi}{\delta n}$. If the distance δr is infinitesimal, then one can consider ΔPQR, a right-angled triangle. If θ is the angle between δr and δn, then

$$PR = PQ \cos\theta \quad \text{or} \quad \delta n = \delta r \cos\theta$$

$\therefore$

$$PQ = \frac{\delta\phi}{\delta n}\cos\theta \tag{2.16}$$

Obviously, Eq. (2.16) represents the rate of change of ϕ in the direction of PQ. Since the maximum value of $\cos\theta$ is 1 and hence the maximum rate of change of

$$\phi = \left(\frac{\delta\phi}{\delta r}\right)_{max} = \left(\frac{\delta\phi}{\delta n}\right)$$

$\therefore$

$$\lim_{\delta r \to 0}\left(\frac{\delta\phi}{\delta r}\right)_{max} = \frac{d\phi}{dn}$$

Obviously, the rate of change of a scalar function is maximum in a direction perpendicular to the equiscalar surface.

Thus, *in a scalar field, the maximum rate of change of the scalar function is known as the gradient of the scalar field.*

Gradient is a vector quantity and its direction is perpendicular to the equiscalar surface, i.e. the direction in which rate of change is maximum. If $\hat{n}$ is the unit vector normal to the equiscalar surface, then

$$\text{grad } \phi = \left(\frac{d\phi}{dn}\right)\hat{n} = \left(\frac{d\phi}{dr}\right)_{max}\hat{n} \tag{2.17}$$

Obviously, the gradient of a scalar field ϕ depends only on the way the values of ϕ are distributed in space. It is entirely independent of the position and direction of the coordinate axes. Thus, grad ϕ is an invariant vector field with respect to transformation of coordinates.

To make the above statement more clear, let us consider an example. Suppose ϕ is the potential in an electric field. The intensity of the field at any point is in the direction of the greatest rate of fall of potential, i.e. normal to the equipotential surface and is equal to the rate, i.e.

$$E = -\frac{\partial\phi}{\partial n}\hat{n} = -\text{grad } \phi \tag{2.18}$$

The negative sign is used because the direction of the field is opposite to the direction of the increase in potential.

Gradient of a Scalar Field in Rectangular or Cartesian Coordinate System

Suppose the scalar field ϕ is a function of cartesian coordinates x, y and z, then with the use of partial differentiation, one can write

$$d\phi = \frac{\partial \phi}{\partial x}dx + \frac{\partial \phi}{\partial y}dy + \frac{\partial \phi}{\partial z}dz \qquad (2.19)$$

Now, the gradient of ϕ is a vector given by

$$\nabla \phi = \frac{\partial \phi}{\partial n}\hat{n}$$

where, $\hat{n}$ is a unit vector. Let us take its scalar product with an element dr of radius vector r, one obtains

$$(\text{grad } \phi) \cdot dr = \frac{\partial \phi}{\partial n}\hat{n} \cdot dr$$

$$= \frac{\partial \phi}{\partial n}dr\cos\theta \qquad (\because \hat{n} \cdot dr = dr\cos\theta)$$

$$= \frac{\partial \phi}{\partial n}dn \qquad \left(\because \frac{\partial \phi}{\partial n} \text{ is the rate of change of } \phi \text{ in the direction of } \hat{n}\right)$$

$$= d\phi$$

$$= \left(\frac{\partial \phi}{\partial x}\right)dx + \left(\frac{\partial \phi}{\partial y}\right)dy + \left(\frac{\partial \phi}{\partial z}\right)dz$$

$$= \left(\frac{\partial \phi}{\partial x}\hat{i} + \frac{\partial \phi}{\partial y}\hat{j} + \frac{\partial \phi}{\partial z}\hat{k}\right) \cdot (dx\hat{i} + dy\hat{j} + dz\hat{k})$$

$$= \left(\frac{\partial \phi}{\partial x}\hat{i} + \frac{\partial \phi}{\partial y}\hat{j} + \frac{\partial \phi}{\partial z}\hat{k}\right) \cdot dr$$

$$\therefore \qquad \nabla \phi = \frac{\partial \phi}{\partial x}\hat{i} + \frac{\partial \phi}{\partial y}\hat{j} + \frac{\partial \phi}{\partial z}\hat{k} = \text{grad } \phi \qquad (2.20)$$

Obviously, $\nabla \phi$ is *such a vector which represents the maximum rate of change of scalar function* ϕ.

It is worthwhile to note that the gradient of a scalar field is always a vector field, however, its converse is always not true. All vector fields cannot be expressed as gradient of scalar fields.

A vector field which can be expressed as gradient of a scalar field is called a *lamellar or non-curl field*.

A pure electrostatic field is a lamellar vector field because it can be expressed as (negative) gradient of a scalar potential field. If V is the electric potential, then one can write the electric field as

$$E = -\nabla V \qquad (2.21)$$

A scalar field $\phi(x, y, z)$ evaluated at a particular point is independent of the coordinates of that point. For example, the temperature at a point is not dependent on whether coordinates (x, y, z) or (x', y', z') are used. This means a scalar field is invariant with respect to the transformation of coordinates. Its gradient which is a vector field, must therefore be an invariant with respect to the transformation also.

2.6 LINE INTEGRAL OF A VECTOR FIELD

Let us consider two points P and Q in a vector field. One can divide the path between P and Q into displacement elements of lengths dr_1, dr_2, dr_3,..., etc. Since these are very small, one can consider them as straight lines. dr, one of the such elements, is shown in Fig. 2.4. Let A be the vector at R in a direction making an angle θ with dr. If A varies in magnitude and direction from point to point along the curve PQ, then the integral

$$\int_P^Q A \cdot dr = \int_P^Q (A\cos\theta)dr = \int_P^Q (A_x dx + A_y dy + A_z dz) \tag{2.22}$$

is defined as the line integral of vector A along the curve PQ. A few familiar examples of line integral are as follows:

i. If A denotes a force acting on a particle which moves along the curve PQ, then the integral $\int_P^Q A \cdot dr$ represents the work done by the force in moving from P to Q.

ii. If A represents the electric field intensity at any point, then the line integral represents the potential difference between P and Q.

iii. If A represents the velocity at any point in a fluid, then the line integral of A round a closed path is known as circulation of the fluid.

Fig. 2.4: Line integral of a vector

If the value of the line integral depends only upon the coordinates of the two points in the vector field and not upon the actual path taken between them, then the vector field $A(x, y, z)$ is called a *conservative* or *non-curl* or *lamellar field*. Line integral, round a closed path in a lamellar vector field is zero.

2.7 SURFACE INTEGRAL

Let S be a surface drawn in a vector field, and ds be an infinite small element of the surface (Fig. 2.5). Let $\hat{n}$ be a unit positive vector normal to ds. The surface element of area ds is represented by a vector ds whose magnitude is $|ds|$ and direction is that of $\hat{n}$. Thus,

$$ds = \hat{n}ds$$

ds is called the **area vector**.

Now, if A is a vector at the middle of the element ds in a direction making an angle θ with $\hat{n}$, then the integral

$$\iint A \cdot ds \quad \text{or} \quad \int_S A \cdot \hat{n}ds = \int_S (A\cos\theta)ds \tag{2.23}$$

is defined as the *surface integral* or *flux* of A across the surface S.

If A denotes the velocity of a moving fluid in which a fixed surface S is drawn, then the surface integral $\int_S A \cdot ds$ gives the amount of fluid flowing per unit time normally through the surface S. In case of a closed surface, a positive surface integral means that the fluid is flowing outwards, if negative then inwards.

Fig. 2.5: Normal vector flux

Other examples of surface integrals are $\int_S A \times ds$ and $\int_S \phi \, ds$, where ϕ is a scalar function.

2.8 VOLUME INTEGRAL

Let $dV = dx \, dy \, dz$ denote an element of volume V. Volume element dV is a scalar. If A is a vector function inside it, then the integral $\int_V A dV$ is called the volume integral, $\iiint_V A dV$ (we will write it as $\int_V A dV$) of A over the volume V, i.e.

$$\int_V A dV = \int_V (A_x \hat{i} + A_y \hat{j} + A_z \hat{k}) dx \, dy \, dz \qquad (2.24)$$

In case of a scalar function ϕ, the integral $\int_V \phi \, dV$ represents the volume integral over V (Fig. 2.6).

Fig. 2.6: Volume integral

2.9 DIVERGENCE OF A VECTOR FIELD

The divergence of a vector field at any point is defined as the amount of flux per unit volume diverging from that point. Since the divergence is the amount of flux, it is essentially a scalar. One can express the divergence of a vector field $A\,(x, y, z)$ as

$$\nabla \cdot A = \frac{\partial A_x}{\partial x} + \frac{\partial A_y}{\partial y} + \frac{\partial A_z}{\partial z} = \text{div} A \qquad (2.25)$$

Equation (2.25) is the expression for divergence in cartesian coordinates. Divergence of a vector field has important applications in hydrodynamics. Now, we derive the expression for divergence in cartesian coordinates.

Let us consider a small rectangular parallelepiped whose centre is $C(x, y, z)$ and whose sides have lengths dx, dy and dz parallel to the coordinate axes x, y and z respectively as shown in Fig. 2.7.

Fig. 2.7: Parallelepiped

Suppose the vector A has components A_x, A_y and A_z along x-, y- and z-axes respectively.

The value of the x-component of A at the centre of the face $EFGH$ is

$$A_x - \frac{1}{2}\frac{\partial A_x}{\partial x} dx \qquad (2.26a)$$

and that at the centre of the face $PQRS$ is

$$A_x + \frac{1}{2}\frac{\partial A_x}{\partial x} dx \qquad (2.26b)$$

Now, the volume of the fluid flowing per unit time through a face is given by the product of the area of the face and the normal component of the velocity upon it. We call it the 'flux' through the face. The area of each of the faces $EFGH$ and $PQRS$ is $dy\,dz$, one finds,

$$\text{flux entering the face } EFGH = \left(A_x - \frac{1}{2}\frac{\partial A_x}{\partial x} dx \right) dy\,dz \qquad (2.27)$$

and flux leaving the face $PQRS = \left(A_x + \frac{1}{2}\frac{\partial A_x}{\partial x} dx \right) dy\,dz \qquad (2.28)$

Obviously, the excess of the flux leaving the parallelepiped over that entering it in the x-direction is

$$\left(A_x + \frac{1}{2}\frac{\partial A_x}{\partial x} dx \right) dy\,dz - \left(A_x - \frac{1}{2}\frac{\partial A_x}{\partial x} dx \right) dy\,dz = \frac{\partial A_x}{\partial x} dx\,dy\,dz \qquad (2.29)$$

Similarly, one obtains that the net flux leaving the parallelepiped in the y- and z-directions are $\dfrac{\partial A_y}{\partial y} dx\,dy\,dz$ and $\dfrac{\partial A_z}{\partial z} dx\,dy\,dz$ respectively.

Therefore, the total net flux diverging from (i.e. leaving) the parallelepiped is

$$\left(\frac{\partial A_x}{\partial x} + \frac{\partial A_y}{\partial y} + \frac{\partial A_z}{\partial z} \right) dx\,dy\,dz$$

Since $dx\,dy\,dz$ is the volume of the elementary parallelepiped, and hence the amount of flux diverging per unit volume, which is defined as divergence of A, is

$$\nabla \cdot A = \operatorname{div} A = \frac{\partial A_x}{\partial x} + \frac{\partial A_y}{\partial y} + \frac{\partial A_z}{\partial z} \qquad (2.30)$$

2.10 PHYSICAL MEANING OF DIVERGENCE OF A VECTOR

$\nabla \cdot A$ at a point gives the amount of flux diverging per unit volume from that point. If A denotes the velocity of the moving fluid, then $\nabla \cdot A$ gives the rate at which the fluid is diverging from the point per unit volume.

If the divergence of a vector is positive at a point in a fluid, then either the fluid is expanding and its density at that point is falling with time or the point is acting as a source of fluid. In case, if the divergence of a vector is negative, then either the fluid is contracting and its density is rising or the point is a negative source, i.e. *sink*.

If the flux entering any element of field space is exactly balanced by that leaving it, then the quantity $\nabla \cdot A = 0$. Obviously, there is no source or sink, nor its density is changing, i.e. the fluid is incompressible. If the fluxes entering and leaving an element are equal, the lines of flow of vector A should form either closed curves (e.g. in case of magnetic field due to a current) or extend to infinity. A vector which satisfies this condition is called *solenoidal*.

In the case of non-material fluxes (e.g. magnetic or electric flux), the existence of divergence means the presence of a source or sink of flux at that point.

2.11 GAUSS'S DIVERGENCE THEOREM

This theorem states that *the volume integral of the divergence of a vector field A taken over any volume V is equal to the surface integral of A taken over the closed surface S surrounding the volume,* i.e.

$$\iiint_V (\nabla \cdot A)\,dV = \iint_S A \cdot dS \qquad (2.31)$$

Proof: In cartesian coordinates,

$$\nabla \cdot A = \left(\hat{i}\frac{\partial}{\partial x} + \hat{j}\frac{\partial}{\partial y} + \hat{k}\frac{\partial}{\partial z} \right) \cdot (A_x\hat{i} + A_y\hat{j} + A_z\hat{k})$$

$$= \frac{\partial A_x}{\partial x} + \frac{\partial A_y}{\partial y} + \frac{\partial A_z}{\partial z}$$

and $dV = dx\,dy\,dz$

Also
$$A \cdot dS = (A_x \hat{i} + A_y \hat{j} + A_z \hat{k}) \cdot (dS_x \hat{i} + dS_y \hat{j} + dS_z \hat{k})$$
$$= A_x dS_x + A_y dS_y + A_z dS_z$$
$$= A_x dy dz + A_y dx dz + A_z dx dy$$

Using the above results, one can express Gauss's theorem in Cartesian coordinates as

$$\iiint_V \left(\frac{\partial A_x}{\partial x} + \frac{\partial A_y}{\partial y} + \frac{\partial A_z}{\partial z} \right) dx\, dy\, dz = \iint_S (A_x dS_x + A_y dS_y + A_z dS_z) \tag{2.32}$$

Let us consider the first part of the LHS integral of Eq. (2.32). As shown in Fig. 2.8, integrate it w.r.t. x along a stripe of cross-section $dy dz$ from $P_1 (x_1, y, z)$ to $P_2(x_2, y, z)$, i.e.

$$\iiint_V \frac{\partial A_x}{\partial x} dx\, dy\, dz = \iint_S |A_x|_{P_1}^{P_2} dy\, dz$$
$$= \iint_S [A_x(P_2) - A_x(P_1)] dy\, dz$$
$$= \iint_S [A_x(x_2, y, z) - A_x(x_1, y, z)] dy\, dz$$

Fig. 2.8: A stripe of cross-section *dydz*

Evidently, at P_1, $A_x\, dy\, dz = -A_x\, dS_x$
and at P_2, $A_x\, dy\, dz = A_x\, dS_x$
($\because$ at P_1 and P_2, the x-components of area have opposite directions)

$$\therefore \qquad \iiint_V \frac{\partial A_x}{\partial x} dx\, dy\, dz = \iint_S A_x dS_x \tag{2.32a}$$

We should not confuse that we have written for

$$\iiint_V A_x(P_2) - A_x(P_1) dy\, dz = \iint_S (A_x dS_x + A_x dS_x)$$

only $\iint_S A_x dS_x$, because if we take all the stripes then each integral $\iint_S A_x dS_x$ will account for the half integral. Hence $\iint_S A_x dS_x$ accounts for the total surface integral for x-component. Similarly, one obtains

$$\iiint_V \frac{\partial A_y}{\partial y} dx\, dy\, dz = \iint_S A_y dS_y \tag{2.32b}$$

and
$$\iiint_V \frac{\partial A_z}{\partial z} dx\, dy\, dz = \iint_S A_z dS_z \tag{2.32c}$$

Adding Eqs. (2.32a) to (2.32c), one obtains

$$\iiint_V \left(\frac{\partial A_x}{\partial x} + \frac{\partial A_y}{\partial y} + \frac{\partial A_z}{\partial z} \right) dx\, dy\, dz = \iint_S \left(A_x dS_x + A_y dS_y + A_z dS_z \right)$$

or
$$\iiint_V \nabla \cdot A\, dV = \iint_S A \cdot ds$$

or
$$\iiint_V (\text{div}\, A) dV = \iint_S A \cdot ds \tag{2.33}$$

This proves the *Gauss's divergence theorem*.

Now we apply this theorem to an electric field E. Let us consider that there exist throughout the volume V enclosed by the surface S, a charge distribution of varying density ρ, then

$$\frac{1}{\varepsilon_0} \int_V \rho dV = \int_S E \cdot dS = \int_V (\nabla \cdot E) dV \tag{2.34}$$

The two volume integrals in Eq. (2.34) are equal, whatever be the volume over which integration takes place. It, therefore, follows that integrands themselves must be equal, i.e.

$$\nabla \cdot E = \frac{\rho}{\varepsilon_0} \tag{2.35}$$

Equation (2.35) is sometimes referred to as the *differential form of Gauss's law in electrostatics*. According to this, the divergence of the electric field at a point equals ρ/ε_0 in the immediate neighbourhood of that point. If the point is situated in a material of dielectric constant K, then RHS of Eq. (2.35) must be divided by K.

The differential or point form of Eq. (2.35) is valid only where the derivatives exist. The integral form, i.e. Eq. (2.34) is valid everywhere.

2.12 LAPLACIAN OPERATOR

Let ϕ be a scalar potential function. If ϕ represents an electric field E,

then
$$E = -\nabla \phi = -\text{grad}\, \phi \tag{2.36}$$

According to Gauss's law in electrostatics, the divergence of an electric field emerging from a closed surface is equal to $\dfrac{1}{\varepsilon_0}$ times the charge density enclosed in the surface, i.e.

$$\nabla \cdot E = \frac{\rho}{\varepsilon_0} \tag{2.37}$$

Here ε_0 is the permittivity of vacuum. Substituting Eq. (2.36) in Eq. (2.37), one obtains

$$-\nabla \cdot \nabla \phi = \frac{\rho}{\varepsilon_0} \tag{2.38}$$

But $\qquad \nabla \cdot \nabla \phi = \text{div} \, (\text{grad} \, \phi)$

$$= \left(\hat{i} \frac{\partial}{\partial x} + \hat{j} \frac{\partial}{\partial y} + \hat{k} \frac{\partial}{\partial z} \right) \cdot \left(\hat{i} \frac{\partial}{\partial x} + \hat{j} \frac{\partial}{\partial y} + \hat{k} \frac{\partial}{\partial z} \right) \phi$$

$$= \left(\frac{\partial^2}{\partial x^2} + \frac{\partial^2}{\partial y^2} + \frac{\partial^2}{\partial z^2} \right) \phi = \nabla^2 \phi \qquad (2.39)$$

Thus $\nabla^2 = \dfrac{\partial^2}{\partial x^2} + \dfrac{\partial^2}{\partial y^2} + \dfrac{\partial^2}{\partial z^2}$ is known as *Laplacian operator* or simply the *Laplacian*. ∇^2
$= \nabla \cdot \nabla$ is usually pronounced as del squared. In some books ∇^2 is also denoted by ∇.

2.13 POISSON'S AND LAPLACE'S EQUATIONS

Substituting Eq. (2.39) in Eq. (2.38), one obtains

$$\nabla^2 \phi = -\frac{\rho}{\varepsilon_0} \qquad (2.40)$$

Equation (2.40) is known as *Poisson's equation*. One can calculate the electric potential at a point due to a charge distribution with the help of this equation. In a medium of dielectric constant K, Eq. (2.40) takes the form

$$\nabla^2 \phi = -\frac{\rho}{K\varepsilon_0} \qquad (2.41)$$

In the charge free region, volume charge density $\rho = 0$, so Eq. (2.40) or Eq. (2.41) becomes

$$\nabla^2 \phi = 0 \qquad (2.42)$$

Equation (2.42) is known as *Laplace's equation*. By solving this equation, one can calculate the potential in an electric field, which is devoid of the source charge.

In *spherical polar coordinates*, Poisson's equation becomes

$$\frac{1}{r^2} \frac{\partial}{\partial r} \left(r^2 \frac{\partial \phi}{\partial r} \right) + \frac{1}{r^2} \frac{1}{\sin \theta} \frac{\partial}{\partial \theta} \left(\sin \theta \frac{\partial \phi}{\partial \theta} \right) + \frac{1}{r^2 \sin^2 \theta} \frac{\partial^2 \phi}{\partial \theta^2} = \frac{\rho}{K\varepsilon_0} \qquad (2.43)$$

In *cylindrical polar coordinates*, Poisson's equation becomes

$$\frac{1}{r} \frac{\partial}{\partial r} \left(r \frac{\partial \phi}{\partial r} \right) + \frac{1}{r^2} \frac{\partial^2 \phi}{\partial \theta^2} + \frac{\partial^2 \phi}{\partial z^2} = \frac{\rho}{K\varepsilon_0} \qquad (2.44)$$

2.14 CURL OF A VECTOR FUNCTION

We have seen that divergence of a vector field is a scalar. Now, we define a vector called curl A, also written as $\nabla \times A$, associated with a vector field A.

If a vector field is derived as the gradient of a scalar field, then the line integral of the vector around any closed path in the field is zero. Such a field is called *lamellar* or *non-curl field*. There are, however, such vector fields for which the line integral round any closed path is non-zero and have a finite value. Such vector fields cannot be derived as gradient of any scalar field, and they exhibit the property of curl.

Suppose a non-lamellar vector field is represented by several lines of flow as shown in Fig. 2.9. Let us consider a plane rectangular area in this field. When the plane *abcd* is perpendicular to the field (position 1), none of the lines of the flow lies along its boundary so that the line integral round it is zero. When the plane *abcd* is parallel to the field (position 2), lines of flow lie along the upper and lower edges of the plane. Since the value of the vector at the upper edge is different from that at the lower edge, the line integral round the boundary has a finite value. This value will be different for different orientations of the plane and shall be maximum for a particular orientation. This maximum line integral computed per unit area along the boundary is called the curl of the vector field. Thus, one can define the curl of a vector field as follows:

The curl of a (non-lamellar) vector field at any point is defined as the maximum line integral of the vector component per unit area along the boundary of an infinitesimal area at that point. Obviously, curl of a vector field is a vector.

Fig. 2.9: Rectangular plane in a vector field

Curl in Cartesian Coordinates

We now calculate curl *A*, where *A* is a function of cartesian coordinates *x, y* and *z*. Let us consider a rectangular shaped infinitely small element of area *dxdy* in the *xy*-plane as shown in Fig. 2.10.

Let the magnitude of the vector *A* at a point *O* is $|A|$ and A_x, A_y and A_z be its components along *x*-, *y*- and *z*-axes respectively. Let the sides *dx*, *dy* of the rectangle be parallel to the *x*- and *y*-axes respectively. So the normal to the area is along the *z*-axis. The arrow heads on the sides of the rectangle show the directions in which the components of *A* act.

Since the rectangle is infinitely small, the average value of *A* along any side of it may be taken as the value at the middle point of that side. Obviously, the average values along the four sides will be as follows:

Fig. 2.10: Rectangular plane in a vector field

$$\text{along } ab: \quad A_x - \frac{1}{2}\frac{\partial A_x}{\partial y}dy$$

$$\text{along } bc: \quad A_y + \frac{1}{2}\frac{\partial A_y}{\partial x}dx$$

$$\text{along } dc: \quad A_x + \frac{1}{2}\frac{\partial A_x}{\partial y}dy$$

and $$\text{along } ad: \quad A_y - \frac{1}{2}\frac{\partial A_y}{\partial x}dx$$

Here we have considered only the first order of small quantities *dx*, *dy* in the Taylor's expansion.

Therefore, the closed path line integral around the boundary *abcd* of the rectangle is given by

$$= \left[\left(A_x - \frac{1}{2}\frac{\partial A_x}{\partial y}dy \right) - \left(A_x + \frac{1}{2}\frac{\partial A_x}{\partial y}dy \right) \right]dx$$

$$+ \left[\left(A_y + \frac{1}{2}\frac{\partial A_y}{\partial x}dx \right) - \left(A_y - \frac{1}{2}\frac{\partial A_y}{\partial x}dx \right) \right]dy = \left(\frac{\partial A_y}{\partial x} - \frac{\partial A_x}{\partial y} \right)dx\,dy$$

On dividing the value of the circulation by the area *dxdy* of the rectangle, one obtains the magnitude of the component of curl *A* along z-axis as

$$(\text{curl } A)_z = \left(\frac{\partial A_y}{\partial x} - \frac{\partial A_x}{\partial y} \right)\hat{k}$$

Similarly, one obtains the components of curl *A* along *x*- and *y*-axes as

$$(\text{curl } A)_x = \left(\frac{\partial A_z}{\partial y} - \frac{\partial A_y}{\partial z} \right)\hat{i}$$

and

$$(\text{curl } A)_y = \left(\frac{\partial A_x}{\partial z} - \frac{\partial A_z}{\partial x} \right)\hat{j}$$

Adding the three components, one obtains

$$\text{curl } A = \left(\frac{\partial A_z}{\partial y} - \frac{\partial A_y}{\partial z} \right)\hat{i} + \left(\frac{\partial A_x}{\partial z} - \frac{\partial A_z}{\partial x} \right)\hat{j} + \left(\frac{\partial A_y}{\partial x} - \frac{\partial A_x}{\partial y} \right)\hat{k} \qquad (2.45)$$

One can express $\nabla \times A$ in the form of a determinant as

$$\text{curl } A = \begin{vmatrix} \hat{i} & \hat{j} & \hat{k} \\ \dfrac{\partial}{\partial x} & \dfrac{\partial}{\partial y} & \dfrac{\partial}{\partial z} \\ A_x & A_y & A_z \end{vmatrix} \qquad (2.46)$$

We must note that Eq. (2.46) is independent of axes. The name curl indicates that a vector field at a point in space has circulation. The determinant in Eq. (2.46) is equal to vector product of del operator (∇) and vector *A*, i.e.

$$\text{curl } A = \nabla \times A \qquad (2.47)$$

2.15 STOKES' CURL THEOREM

This theorem states that *the flux of the curl of a vector field **A** over any surface of any shape is equal to the line integral of the vector field **A** over the boundary of that surface.* We must note that the sense of line integral is related with the positive normal to the surface.

Mathematically, one can express it as

$$\iint_S \text{curl } A \cdot ds = \oint A \cdot dl \qquad (2.48)$$

Here the line integration is being carried out for the boundary of the surface.

The path *C* bounds the surface *S* as shown in Fig. 2.11. Let the surface *S* be divided into large number of small areas $ds_1, ds_2, ds_3, \ldots$, etc. Since all the areas are to be traversed

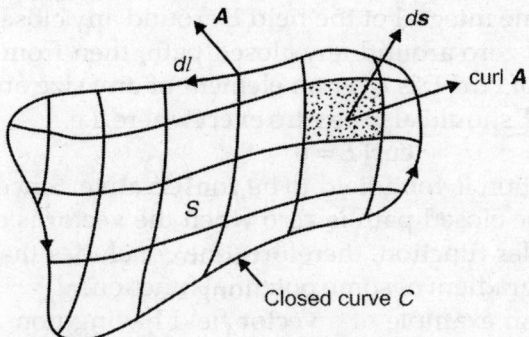

Fig. 2.11: Surface S with closed boundary C

by the curves bounding them in the same sense, the sum of the line integrals around these areas is equal to the line integral around the closed path C, i.e.

$$\oint A \cdot dl = \sum_i \oint_C A \cdot dl$$

From the definition of curl A, at a point inside inside ds_i, we have

$$\text{curl } A \cdot ds_i = \oint_{C_i} A \cdot dl$$

Since the number of these areas is very large, and hence in the limiting case

$$\oint A \cdot dl = \left[\lim_{\partial s \to 0} \sum_i \frac{\partial}{\partial s_i} \oint A \cdot dl \right] \partial s_i \qquad (2.49)$$

Since the curl is the line integral per unit area in the limiting case when area tends to zero, we have

$$\oint A \cdot dl = \iint_S \text{curl } A \cdot ds \qquad (2.50)$$

Equation (2.50) represents the *Stoke's curl theorem*. Obviously, *the line integral of a vector field along the limiting boundary of the area enclosed by a closed path is equal to the surface integral of the curl of the vector field*. This theorem is used to convert line integral into surface integral and *vice versa*.

2.16 PHYSICAL MEANING OF CURL OF A VECTOR FIELD

The existence of the curl of a vector field is always due to its rotation or circulation. A vector field having non-zero curl has circulation and has curly or rotational field lines about the point of non-zero curl. In German literature, curl is written as *rot* which is derived from rotation. The following examples illustrate clearly the physical meaning of curl.

i. In an electrostatic field E, the curl E is zero everywhere. The line integral of the field E between two points A and B is independent of the path. Therefore, line integral along A to B is negative of the line integral along B to A or in other words,

$$\int_A^B E \cdot dl = -\int_A^B E \cdot dl$$

Hence $\qquad \int_A^B E \cdot dl + \int_B^A E \cdot dl = 0$

Obviously, the line integral of the field E around any closed path is zero. Also if the circulation is zero around any closed path, then from Stoke's theorem, the surface integral of curl E is over an element of any size and shape. This clearly reveals that curl E should also be zero everywhere, i.e.

$$\text{curl } E = 0 \qquad (2.51)$$

The above condition is for a field to be conservative. Since the line integral of a vector around the closed path is zero when the vector is derived by taking the gradient of a scalar function, therefore, one concludes that electric field can be described as the gradient of some potential function.

ii. Let us consider an example of a vector field having non-zero curl. In the subsequent chapters, we shall see that the curl of the magnetic induction vector is proportional to the current density vector at the given point, i.e.

$$\nabla \times B = \mu_0 J \qquad (2.52)$$

where μ_0 is the permeability of free space ($\mu_0 = 4\pi \times 10^{-7} \, \text{NA}^{-2}$). However, we must note that Eq. (2.52) holds good only for the magnetic field in a vacuum in the absence of time varying electric fields.

A comparison of Eqs. (2.51) and (2.52) shows that the electrostatic and magnetic fields are of an appreciably different nature. The curl of an electrostatic field is zero, i.e. an electrostatic field is conservative and can be characterized by the scalar potential ϕ. The curl of a magnetic field at points where current differs from zero is not zero. Accordingly the circulation of B is proportional to the current enclosed by a loop. This is why we cannot ascribe to a magnetic field a scalar potential that would be related to B by an equation similar to Eq. (2.36). This potential would not be unique upon each circumvention of the loop and return to the initial point would receive an increment equal to $\mu_0 I$. A field whose curl differs from zero is called a *vortex* or a *solenoidal* field.

Since $\nabla \cdot B$ is zero everywhere, and hence B can be represented as curl of another vector A. Thus,

$$B = \nabla \times A \qquad (2.53)$$

The divergence of curl is always zero. A is called the *vector potential*.

2.17 GREEN'S THEOREM

Green's theorem is a corollary of the divergence theorem. The theorem states that if ϕ and ψ are two scalar point functions such that these functions and their first derivatives are continuously differentiable, then

$$\iiint_V (\phi \nabla^2 \psi + \nabla \phi \cdot \nabla \psi) dV = \iint_S (\phi \nabla \psi) \cdot dS \qquad (2.54)$$

and

$$\iiint_V (\phi \nabla^2 \psi - \psi \nabla^2 \phi) dV = \iint_S (\phi \nabla \psi - \psi \nabla \phi) \cdot dS \qquad (2.55)$$

Equations (2.54) and (2.55) are respectively known as first and second form of Green's theorem. One can derive it as follows:

We start with Gauss's divergence theorem which states

$$\iiint_V \nabla \cdot A \, dV = \iint_S A \cdot dS \qquad (2.56)$$

Let us take $A = \phi \nabla \psi$

or
$$\hat{i}A_x + \hat{j}A_y + \hat{k}A_z = \phi\left(\hat{i}\frac{\partial \psi}{\partial x} + \hat{j}\frac{\partial \psi}{\partial y} + \hat{k}\frac{\partial \psi}{\partial z}\right)$$

Obviously
$$A_x = \phi\frac{\partial \psi}{\partial x}, \quad A_y = \phi\frac{\partial \psi}{\partial y} \quad \text{and} \quad A_z = \phi\frac{\partial \psi}{\partial z}$$

Now
$$\nabla \cdot A = \frac{\partial A_x}{\partial x} + \frac{\partial A_y}{\partial y} + \frac{\partial A_z}{\partial z}$$

$$= \frac{\partial}{\partial x}\left(\phi\frac{\partial \psi}{\partial x}\right) + \frac{\partial}{\partial y}\left(\phi\frac{\partial \psi}{\partial y}\right) + \frac{\partial}{\partial z}\left(\phi\frac{\partial \psi}{\partial z}\right)$$

$$= \left(\phi\frac{\partial^2 \psi}{\partial x^2} + \frac{\partial \phi}{\partial x}\frac{\partial \psi}{\partial x}\right) + \left(\phi\frac{\partial^2 \psi}{\partial y^2} + \frac{\partial \phi}{\partial y}\frac{\partial \psi}{\partial y}\right) + \left(\phi\frac{\partial^2 \psi}{\partial z^2} + \frac{\partial \phi}{\partial z}\frac{\partial \psi}{\partial z}\right)$$

$$= \phi\left(\frac{\partial^2 \psi}{\partial x^2} + \frac{\partial^2 \psi}{\partial y^2} + \frac{\partial^2 \psi}{\partial z^2}\right) + \frac{\partial \phi}{\partial x}\frac{\partial \psi}{\partial x} + \frac{\partial \phi}{\partial y}\frac{\partial \psi}{\partial y} + \frac{\partial \phi}{\partial y}\frac{\partial \psi}{\partial z}$$

$$= \phi\nabla^2\psi + \nabla\phi \cdot \nabla\psi$$

Making use of these results in Eq. (2.56), one obtains

$$\iiint_V (\phi\nabla^2\psi + \nabla\phi \cdot \nabla\psi)dV = \iint_S (\phi\nabla\psi) \cdot dS \tag{2.57}$$

This is Green's first equation.

Interchanging ϕ and ψ in Eq. (2.57), one obtains

$$\iiint_V (\psi\nabla^2\phi + \nabla\psi \cdot \nabla\phi)dV = \iint_S (\psi\nabla\phi) \cdot dS \tag{2.58}$$

Subtracting Eq. (2.58) from Eq. (2.57), one obtains

$$\iiint_V (\phi\nabla^2\psi - \psi\nabla^2\phi)dV = \iint_S (\phi\nabla\psi - \psi\nabla\phi) \cdot dS \tag{2.59}$$

This is Green's second equation. This is also referred to as *symmetrical theorem*.
On the RHS of Eq. (2.59), we have

$$\nabla\psi \cdot ds = \frac{\partial \psi}{\partial n}ds$$

and
$$\nabla\phi \cdot ds = \frac{\partial \phi}{\partial n}ds$$

where $\dfrac{\partial \psi}{\partial n}$ and $\dfrac{\partial \phi}{\partial n}$ are the directional derivatives of ψ and ϕ respectively along the
outward normal to dS. In the light of these results, Eq. (2.59) can also be expressed as

$$\iiint_V (\phi\nabla^2\psi - \psi\nabla^2\phi)dV = \iint_S \left(\phi\frac{\partial \psi}{\partial n} - \psi\frac{\partial \phi}{\partial n}\right)dS \tag{2.60}$$

Equation (2.60) is an alternative form of *Green's symmetrical theorem*.

One can easily derive Gauss's divergence theorem from Green's theorem. We have
Green's symmetrical theorem given by Eq. (2.59). If ϕ is constant, then we have

$$\phi\iiint_V \nabla^2\psi \, dV = \phi\iint_S \nabla\psi \cdot dS$$

or
$$\iiint_V (\nabla \cdot \nabla \psi) dV = \iint_S \nabla \psi \cdot dS$$

Putting $\nabla \psi = A$, one obtains
$$\iiint_V (\nabla \cdot A) dV = \iint_S A \cdot dS$$

which is *Gauss's divergence theorem*.

ILLUSTRATIVE EXAMPLES

Example 2.1: A field is expressed by the potential function $V = 3x^2 z - xy^3 + z$. Calculate the potential at the point $(1, -2, +2)$.

Solution: We have
$$V = 3x^2 z - xy^3 + z$$
Potential at the point $(1, -2, +2)$ is obtained as
$$V_{(1, -2, +2)} = 3(1)^2 \times (2) - (1)(-2)^3 + 2 = 16 \text{ V}$$

Example 2.2: A vector field A is represented by the function $A = \hat{i}(2x^2 y - x^4) + \hat{j}(yz^2) - \hat{k}(xy^2)$, then obtain the values of $\dfrac{\partial A}{\partial x}, \dfrac{\partial^2 A}{\partial x^2}$ and $\dfrac{\partial^2 A}{\partial y \partial z}$.

Solution: We have

$$A = \hat{i}(2x^2 y - x^4) + \hat{j} yz^2 - \hat{k} xy^2$$

$$\therefore \quad \frac{\partial A}{\partial x} = \frac{\partial}{\partial x}[\hat{i}(2x^2 y - x^4) + \hat{j} yz^2 - \hat{k} xy^2]$$

$$= \hat{i}(4xy - 4x^3) + \hat{j} 0 - \hat{k} y^2$$

Now
$$\frac{\partial^2 A}{\partial x^2} = \frac{\partial}{\partial x}[\hat{i}(4xy - 4x^3) - \hat{k} y^2]$$

$$= \hat{i}(4y - 12x^2) - \hat{k} 0$$

and
$$\frac{\partial^2 A}{\partial y \partial z} = \frac{\partial}{\partial z}\left(\frac{\partial A}{\partial y}\right)$$

$$= \frac{\partial}{\partial z}\left[\frac{\partial}{\partial y}\{\hat{i}(2x^2 y - x^4) + \hat{j} yz^2 - \hat{k} xy^2\}\right]$$

$$= \frac{\partial}{\partial z}[\hat{i}(2x^2) + \hat{j} z^2 - \hat{k} 2xy] = 2z\hat{j}.$$

Example 2.3: Find the value of div $(r^n r) = \nabla \cdot (r^n r)$.

Solution: $\nabla \cdot (r^n r) = \left(\hat{i}\dfrac{\partial}{\partial x} + \hat{j}\dfrac{\partial}{\partial y} + \hat{k}\dfrac{\partial}{\partial}\right) \cdot [r^n (\hat{i} x + \hat{j} y + \hat{k} z)]$

$$= \frac{\partial}{\partial x}(xr^n) + \frac{\partial}{\partial y}(yr^n) + \frac{\partial}{\partial z}(zr^n).$$

$$= r^n + xnr^{n-1}\frac{\partial r}{\partial x} + r^n + ynr^{n-1}\frac{\partial r}{\partial y} + r^n + znr^{n-1}\frac{\partial r}{\partial z}$$

$$= 3r^n + nr^{n-1}\left(x\frac{\partial r}{\partial x} + y\frac{\partial r}{\partial y} + z\frac{\partial r}{\partial z}\right)$$

$$= 3r^n + nr^{n-1}\left(x\cdot\frac{x}{r} + y\cdot\frac{y}{r} + z\cdot\frac{z}{r}\right)$$

$$= 3r^n + nr^{n-1}\left(\frac{x^2 + y^2 + z^2}{r}\right)$$

$$= (3 + n)\, r^n.$$

Example 2.4: Find (a) $\nabla \cdot r$, where r is the position vector and

$$\text{(b) } \nabla \cdot [\hat{i}(x+y) + \hat{j}(x-y) + \hat{k}(4z)]$$

Solution: (a) $\nabla \cdot r = \left(\hat{i}\dfrac{\partial}{\partial x} + \hat{j}\dfrac{\partial}{\partial y} + \hat{k}\dfrac{\partial}{\partial z}\right)\cdot(\hat{i}x + \hat{j}y + \hat{k}z)$

$$= \frac{\partial x}{\partial x} + \frac{\partial y}{\partial y} + \frac{\partial z}{\partial z} = 3$$

(b) $\nabla \cdot [\hat{i}(x+y) + \hat{j}(x-y) + \hat{k}(4z)]$

$$= \left(\hat{i}\frac{\partial}{\partial x} + \hat{j}\frac{\partial}{\partial y} + \hat{k}\frac{\partial}{\partial z}\right)\cdot[\hat{i}(x+y) + \hat{j}(x-y) + \hat{k}(4z)]$$

$$= \frac{\partial}{\partial x}(x+y) + \frac{\partial}{\partial y}(x-y) + \frac{\partial}{\partial z}(4z)$$

$$= 1 - 1 + 4 = 4$$

Example 2.5: If r is the position vector of a point, calculate the gradient of $\dfrac{1}{r}$.

Solution: $\qquad r = \hat{i}x + \hat{j}y + \hat{k}z$

$\therefore \qquad\quad |r| = \sqrt{x^2 + y^2 + z^2}$

Now $\qquad \nabla\left(\dfrac{1}{r}\right) = \nabla(x^2 + y^2 + z^2)^{-1/2}$

or $\qquad \nabla\left(\dfrac{1}{r}\right) = \left(\hat{i}\dfrac{\partial}{\partial x} + \hat{j}\dfrac{\partial}{\partial y} + \hat{k}\dfrac{\partial}{\partial z}\right)(x^2 + y^2 + z^2)^{-1/2}$

$$= \hat{i}\frac{\partial}{\partial x}(x^2 + y^2 + z^2)^{-1/2} + \hat{j}\frac{\partial}{\partial y}(x^2 + y^2 + z^2)^{-1/2}$$

$$+ \hat{k}\frac{\partial}{\partial z}(x^2 + y^2 + z^2)^{-1/2}$$

$$\because \quad \frac{\partial}{\partial x}(x^2 + y^2 + z^2)^{-1/2} = -\frac{1}{2}(x^2 + y^2 + z^2)^{-3/2}(2x) = \frac{-x}{r^3}$$

Similarly,
$$\frac{\partial}{\partial y}(x^2 + y^2 + z^2)^{-1/2} = \frac{-y}{r^3}$$

and
$$\frac{\partial}{\partial z}(x^2 + y^2 + z^2) = \frac{-z}{r^3}$$

$$\therefore \quad \nabla\left(\frac{1}{r}\right) = -\frac{\hat{i}x + \hat{j}y + \hat{k}z}{r^3} = \frac{-r}{r^3}$$

Example 2.6: The electric field due to a point charge is expressed as $E = \dfrac{Q}{r^2}\hat{r}$. Show that the divergence of electric field due to that point charge is zero.

Solution: $\nabla \cdot E = \nabla \cdot \left(\dfrac{Q}{r^2}\hat{r}\right) = \nabla \cdot \left(\dfrac{Q}{r^3}\hat{r}r\right) = \nabla \cdot \left(\dfrac{Q}{r^3}r\right)$

$$= \left(\hat{i}\frac{\partial}{\partial x} + \hat{j}\frac{\partial}{\partial y} + \hat{k}\frac{\partial}{\partial z}\right) \cdot Q\left(\frac{\hat{i}x + \hat{j}y + \hat{k}z}{(x^2 + y^2 + z^2)^{3/2}}\right)$$

$$(\because \ r = \hat{i}x + \hat{j}y + \hat{k}z \quad \therefore \ |r| = \sqrt{x^2 + y^2 + z^2})$$

$$= Q\left[\frac{\partial}{\partial x}\left\{\frac{x}{(x^2 + y^2 + z^2)^{3/2}}\right\}\right] + Q\left[\frac{\partial}{\partial y}\left\{\frac{y}{(x^2 + y^2 + z^2)^{3/2}}\right\}\right]$$

$$+ Q\left[\frac{\partial}{\partial z}\left\{\frac{z}{(x^2 + y^2 + z^2)^{3/2}}\right\}\right]$$

$$= Q\left[\frac{1}{(x^2 + y^2 + z^2)^{3/2}} - \frac{3x^2}{(x^2 + y^2 + z^2)^{5/2}} + \frac{1}{(x^2 + y^2 + z^2)^{3/2}}\right.$$

$$\left. - \frac{3y^2}{(x^2 + y^2 + z^2)^{5/2}} + \frac{1}{(x^2 + y^2 + z^2)^{3/2}} - \frac{3z^2}{(x^2 + y^2 + z^2)^{5/2}}\right]$$

$$= 0$$

Example 2.7: Show that the potential function $V = q(x^2 + y^2 + z^2)^{1/2}$ does not satisfy the Laplace's equation.

Solution: The Laplace's equation is $\nabla^2 V = 0$, i.e.

$$\frac{\partial^2 V}{\partial x^2} + \frac{\partial^2 V}{\partial y^2} + \frac{\partial^2 V}{\partial z^2} = 0 \tag{1}$$

Using $V = q(x^2 + y^2 + z^2)^{1/2}$, LHS of (1) becomes

$$\text{LHS} = q\frac{\partial}{\partial x}\left[\frac{\partial}{\partial x}(x^2 + y^2 + z^2)^{1/2}\right] + q\frac{\partial}{\partial y}\left[\frac{\partial}{\partial y}(x^2 + y^2 + z^2)^{1/2}\right]$$

$$+q\frac{\partial}{\partial z}\left[\frac{\partial}{\partial z}(x^2+y^2+z^2)^{1/2}\right]$$

$$= q\frac{\partial}{\partial x}\left[x(x^2+y^2+z^2)^{-1/2}\right]+q\frac{\partial}{\partial y}\left[y(x^2+y^2+z^2)^{-1/2}\right]$$

$$+q\frac{\partial}{\partial z}\left[z(x^2+y^2+z^2)^{-1/2}\right]$$

$$= q\left\{-\frac{1}{2}(x^2+y^2+z^2)^{-3/2}2x+(x^2+y^2+z^2)^{-1/2}\right.$$

$$-\frac{1}{2}(x^2+y^2+z^2)^{-3/2}2y+(x^2+y^2+z^2)^{-1/2}$$

$$\left.-\frac{1}{2}(x^2+y^2+z^2)^{-3/2}2z+(x^2+y^2+z^2)^{-1/2}\right\}$$

$$= q\left\{(x^2+y^2+z^2)^{-3/2}(x^2+y^2+z^2)+3(x^2+y^2+z^2)^{-1/2}\right\}$$

$$= 4q(x^2+y^2+z^2)^{-1/2}\neq 0$$

Obviously, the given potential function does not satisfy the Laplace's equation.

Example 2.8: Show that curl $r = 0$, where r is a position vector.

Solution:

$$\nabla\times r = \left(\hat{i}\frac{\partial}{\partial x}+\hat{j}\frac{\partial}{\partial y}+\hat{k}\frac{\partial}{\partial z}\right)\times(\hat{i}x+\hat{j}y+\hat{k}z)$$

$$= \begin{vmatrix} \hat{i} & \hat{j} & \hat{k} \\ \dfrac{\partial}{\partial x} & \dfrac{\partial}{\partial y} & \dfrac{\partial}{\partial z} \\ x & y & z \end{vmatrix} = 0$$

Example 2.9: A potential function in an electric field is given by $\phi = (2x^2 - y^2 + 3z^2)$ V. Show that the intensity of electric field at the point $(1, -1, 0)$ is $-(4\hat{i} - 2\hat{j})$ V/m.

Solution: $\quad E = -\nabla\phi = -\left(\hat{i}\dfrac{\partial\phi}{\partial x}+\hat{j}\dfrac{\partial\phi}{dy}+\hat{k}\dfrac{\partial\phi}{\partial z}\right)$

Now, $\quad \dfrac{\partial\phi}{\partial x} = 4x, \dfrac{\partial\phi}{\partial y} = -2y \quad$ and $\quad \dfrac{\partial\phi}{\partial z} = 6z$

$\therefore \qquad E = -(\hat{i}4x - \hat{j}2y + \hat{k}6z)$

$\therefore \quad E$ at the point $(1, -1, 0) = -(\hat{i}4 + 2\hat{j}) = (-4\hat{i} - 2\hat{j})$ V/m

Example 2.10: Show that curl grad $\phi = 0$, where ϕ is any scalar function.

Solution: $\nabla\times\nabla\phi = \left(\hat{i}\dfrac{\partial}{\partial x}+\hat{j}\dfrac{\partial}{\partial y}+\hat{k}\dfrac{\partial}{\partial z}\right)\times\left(\hat{i}\dfrac{\partial\phi}{\partial x}+\hat{j}\dfrac{\partial\phi}{\partial y}+\hat{k}\dfrac{\partial\phi}{\partial z}\right)$

$$= \begin{vmatrix} \hat{i} & \hat{j} & \hat{k} \\ \dfrac{\partial}{\partial x} & \dfrac{\partial}{\partial y} & \dfrac{\partial}{\partial z} \\ \dfrac{\partial \phi}{\partial x} & \dfrac{\partial \phi}{\partial y} & \dfrac{\partial \phi}{\partial z} \end{vmatrix}$$

$$= \hat{i}\left(\frac{\partial^2 \phi}{\partial y \partial z} - \frac{\partial^2 \phi}{\partial z \partial y}\right) + \hat{j}\left(\frac{\partial^2 \phi}{\partial z \partial x} - \frac{\partial^2 \phi}{\partial x \partial z}\right) + \hat{k}\left(\frac{\partial^2 \phi}{\partial x \partial y} - \frac{\partial^2 \phi}{\partial y \partial x}\right)$$

$$= 0 \qquad\qquad \left(\because \ \frac{\partial^2 \phi}{\partial y \partial z} = \frac{\partial^2 \phi}{\partial z \partial y} \ \text{ and so on} \right)$$

Example 2.11: Show that div grad $(R^n) = n(n+1)R^{n-1}$, where R is a position vector.

Solution: div grad $(R^n) = \nabla \cdot \nabla (R^n)$

Now
$$\nabla(R^n) = \left(\hat{i} \frac{\partial}{\partial x} + \hat{j} \frac{\partial}{\partial y} + \hat{k} \frac{\partial}{\partial z}\right) R^n$$

$$= \hat{i} nR^{n-1} \frac{\partial R}{\partial x} + \hat{j} nR^{n-1}\frac{\partial R}{\partial y} + \hat{k}nR^{n-1}\frac{\partial R}{\partial z}$$

$$= nR^{n-1}\left(\hat{i}\frac{\partial R}{\partial x} + \hat{j}\frac{\partial R}{\partial y} + \hat{k}\frac{\partial R}{\partial z}\right)$$

$$\therefore \qquad\qquad R = \hat{i}x + \hat{j}y + \hat{k}z$$
$$\therefore \qquad\qquad R^2 = R \cdot R = x^2 + y^2 + z^2$$

$$\therefore \qquad 2R\frac{\partial R}{\partial x} = 2x$$

and
$$\frac{\partial R}{\partial x} = \frac{x}{R}$$

Similarly
$$\frac{\partial R}{\partial y} = \frac{y}{R} \ \text{ and } \ \frac{\partial R}{\partial z} = \frac{z}{R}$$

$$\therefore \qquad \nabla R^n = nR^{n-1}\left(\hat{i}\frac{x}{R} + \hat{j}\frac{y}{R} + \hat{k}\frac{z}{R}\right)$$

$$\therefore \qquad \nabla \cdot \nabla R^n = \left(\hat{i}\frac{\partial}{\partial x} + \hat{j}\frac{\partial}{\partial y} + \hat{k}\frac{\partial}{\partial z}\right) \cdot nR^{n-2}(\hat{i}x + \hat{j}y + \hat{k}z)$$

$$= \frac{\partial}{\partial x}(nR^{n-2}x) + \frac{\partial}{\partial y}(nR^{n-2}y) + \frac{\partial}{\partial z}(nR^{n-2}z)$$

$$= n\left\{ R^{n-2} + x(n-2)R^{n-3}\frac{\partial R}{\partial y} + R^{n-2}\right.$$

$$\left. + y(n-2)R^{n-3}\frac{\partial R}{\partial y} + R^{n-2} + z(n-2)R^{n-3}\frac{\partial R}{\partial z}\right\}$$

$$= 3nR^{n-2} + n(n-2)R^{n-3}\left\{x\frac{\partial R}{\partial x} + y\frac{\partial R}{\partial y} + z\frac{\partial R}{\partial z}\right\}$$

$$= 3nR^{n-2} + n(n-2)R^{n-3}\left\{x\frac{x}{R} + y\frac{y}{R} + z\frac{z}{R}\right\}$$

$$= 3nR^{n-2} + n(n-2)R^{n-3}\left\{\frac{x^2 + y^2 + z^2}{R}\right\}$$

$$= 3nR^{n-2} + n(n-2)R^{n-3}R$$

$$= nR^{n-2}(3 + n - 2) = n(n+1)R^{n-2}$$

Example 2.12: The electric field E in a certain space is given by

$$E_x = (ax + by + c), \quad E_y = 0, \quad E_z = 0$$

Use Gauss's theorem to evaluate the charge enclosed in a cube with one of the sides of length L along the x-axis, the two faces of the cube perpendicular to the x-axis lying in the planes $x = l$ to $x = l + L$.

Solution: Applying Gauss's theorem

$$\iint_S E \cdot dS = \iiint_V (\nabla \cdot E)dV$$

$$= \iiint_V \left(\frac{\partial E_x}{\partial x} + \frac{\partial E_y}{\partial y} + \frac{\partial E_z}{\partial z}\right)dV$$

Here $\dfrac{\partial E_x}{\partial x} = a$, while E_y and E_z is each zero.

Thus,
$$\iint_S E \cdot dS = \iiint_V a\,dV = aV = aL^3 \quad (\because V = L^3) \tag{1}$$

According to Gauss's law in electrostatics, we have

$$\iint_S E \cdot dS = \frac{q}{\varepsilon_0} \tag{2}$$

where, q is the charge within the surface S which encloses the volume V.

Equating Eqs (1) and (2), one obtains

$$\frac{q}{\varepsilon_0} = aL^3$$

or
$$q = \varepsilon_0\, aL^3.$$

Example 2.13: The electric field at any point in a given space is directed radially outward along the line joining the point to a fixed point ($r = 0$) and is given by $(ar + br^2)$. Evaluate the charge enclosed in a sphere of radius R with the fixed point ($r = 0$) as centre.

Solution: The magnitude of radial electric field, $E = ar + br^2$.

The value at the surface of the sphere of radius R is

$$E = aR + bR^2$$

The surface integral of the normal component of E over the closed surface S of the sphere is

$$\iint_S E \cdot dS = \iint_S E\,dS \qquad (\because E \text{ and } dS \text{ are in the same direction})$$

$$= \iint_S (aR + bR^2)\, dS$$

$$= (aR + bR^2) \iint_{dS} dS$$

$$= (aR + bR^2)(4\pi R^2) \qquad (1)$$

But by Gauss's law in electrostatics

$$\iint_S E \cdot dS = \frac{q}{\varepsilon_0} \qquad (2)$$

where, q is the charge enclosed by the surface S. Equating Eqs (1) and (2), one obtains

$$\frac{q}{\varepsilon_0} = 4\pi R^3 (a + bR)$$

$$\therefore \qquad q = 4\pi\varepsilon_0 R^3 (a + bR)$$

Example 2.14: If $f = \psi\nabla\phi$, show that $f \cdot \nabla \times f = 0$

Solution: We have

$$\nabla \times f = \nabla \times (\nabla\phi)$$

$$= (\nabla \psi) \times (\nabla \psi) + \psi \times (\nabla \phi)$$

$$= (\nabla \psi) \times (\nabla \phi)$$

$$\therefore \qquad f \cdot (\nabla \times f) = \psi(\nabla\phi) \cdot \{(\nabla\psi) \times (\nabla\phi)\} = 0$$

i.e. a scalar triple product in which two vectors are equal is zero.

GLIMPSES

- To every point in space, finite or infinite, there corresponds a definite value of some physical property, the region is called a *field*. If this property is a scalar, the field is called a *scalar field* and if this property is a vector, the field is known as *vector field*.
- Rate of change of a scalar in a given field is called the gradient.
- The *gradient of a scalar field* ϕ is a vector field whose magnitude at any point is equal to the maximum rate of increase of ϕ at that point, and whose direction is along the normal to the level surface at that point, i.e. in the direction of this maximum rate of increase.

$$\nabla\phi = \frac{\partial\phi}{\partial x}\hat{i} + \frac{\partial\phi}{\partial y}\hat{j} + \frac{\partial\phi}{\partial z}\hat{k}$$

- A *vector field* which can be expressed as gradient of a scalar field is called a *lamellar* or *non-curl field*.
- The net outflow of flux per unit volume is called the *divergence*. Divergence of vector A is represented as $\nabla \cdot A$ or div A.

$$\nabla \cdot A = \frac{\partial A_x}{\partial x} + \frac{\partial A_y}{\partial y} + \frac{\partial A_z}{\partial z}$$

Since the divergence is the amount of flux, it is essentially a scalar. If $\nabla \cdot A$ is positive at a point, then either the fluid is expanding and its density at that point is falling with time, or point is a *source* of the fluid. If $\nabla \cdot A$ is negative, then either the fluid is contracting and its density is rising at that point, or the point is a negative vector, i.e. sink. From a charged body there is continuous outflow of flux, e.g. sun rays coming out continuously from sun. When a vacuum tube is broken there is inflow of air.

- *Vector fields* for which the line integral round any closed path is non-zero and have a finite value, cannot be derived as gradient of any scalar field and they show the property of curl. *Curl* can be defined as the net circulation per unit area. Curl of vector A is defined as $\nabla \times A$.

$$\nabla \times A = \left(\frac{\partial}{\partial x}\hat{i} + \frac{\partial}{\partial y}\hat{j} + \frac{\partial}{\partial z}\hat{k} \right) \times (A_x\hat{i} + A_y\hat{j} + A_z\hat{k})$$

$$= \begin{vmatrix} \hat{i} & \hat{j} & \hat{k} \\ \frac{\partial}{\partial x} & \frac{\partial}{\partial y} & \frac{\partial}{\partial z} \\ A_x & A_y & A_z \end{vmatrix}$$

In hydrodynamics, curl is sensed as rotation of a fluid and hence it is sometimes written as *rotation* also. Consider wind whirlpool which is upwards. When a piece of paper is released here, it rotates and finally moves upward. This is an example of curl.

- The operator ∇^2 is called the *Laplacian operator* and defined by the equation

$$(\nabla \cdot \nabla)f = \nabla \cdot (\nabla f)$$

$$\nabla^2 f = \frac{\partial^2 f}{\partial x^2} + \frac{\partial^2 f}{\partial y^2} + \frac{\partial^2 f}{\partial z^2}$$

This is scalar type Laplacian, which is defined as divergence of gradient of a scalar function.

- The curl of a (non-lamellar) vector field at any point is defined as the maximum line integral of the vector computed per unit area along the boundary of an infinitesimal area at that point. It is essentially a vector.

- A vector A is said to be *solenoidal* if $\nabla \cdot A = 0$.

- *Gauss's divergence theorem* states that the surface integral of the normal the surface integral of the normal component of a vector A taken over a closed surface S is equal to the volume integral of the divergence of A taken over the volume V enclosed by the surface S, i.e.

$$\iint_S A \cdot dS = \iiint_V \nabla \cdot A \, dV$$

This theorem gives us a method of reducing triple integrals to double integrals.

- If ϕ and ψ are two scalar point functions such that these functions and their first derivatives are continuously differentiable, then we have

$$\iiint_V (\phi \nabla^2 \psi + \nabla \phi \cdot \nabla \psi) dV = \iint_S (\phi \nabla \psi) \cdot dS$$

and
$$\iiint_V (\phi \nabla^2 \psi - \psi \nabla^2 \phi) dV = \iint_S (\phi \nabla \psi - \psi \nabla \phi) \cdot dS$$

These equations are respectively known as *first and second form of Green's theorem*.

- The *Stokes' theorem* states that the line integral of a vector A taken around a closed curve C which bounds a surface S is equal to the surface integral of the $\nabla \times A$ taken over S, i.e.

$$\oint_C A \cdot dr = \iint_S \nabla \times A \cdot dS = \iint_S (\nabla \times A) \cdot dS$$

When $\nabla \cdot A = 0$, the line integral of A over a closed path $\left(\oint A \cdot dr \right)$ is also zero, and the field is then *lamellar*.

REVIEW QUESTIONS

1. Describe scalar and vector fields. How can they be represented graphically?
2. What do you mean by the field of any physical quantity? What are scalar and vector fields? Give examples.
3. Define gradient of a scalar field. Show that the gradient of a scalar field is a vector. Give an example.
4. Prove that the gradient of a scalar field is normal to the surface, $\phi = $ constant.
5. Express the gradient of a scalar field $\phi(x, y, z)$ in terms of cartesian coordinates.
6. Show that when a vector field can be derived as the gradient of a scalar field, the line integral of the vector taken round any closed path in the vector field is zero.
7. Explain the meaning of line integral of a vector field using at least one example.
8. Explain the physical meaning of the divergence of a vector field A. Express divergence A in orthogonal coordinates and show that div $A = \nabla \cdot A$.
9. Define curl of a vector field and give its physical significance. Derive an expression for a vector field and show that curl $A = \nabla \times A$.
10. What is a Laplacian operator? Express divgradϕ in cartesion coordinates, where ϕ is a scalar quantity.
11. Derive Laplace's and Poisson's equations starting from the differ-ential form of Gauss's law.
12. State and prove Green's theorem.

PROBLEMS

1. For a position vector $R = \hat{i}x + \hat{j}y + \hat{k}z$, find the values of

   (a) grad $\dfrac{1}{R}$ (b) div $\left(\dfrac{R}{R^3} \right)$ and   (c) curl $\left(\dfrac{R}{R^3} \right)$

$$\left[\textit{Ans. } (a) - \frac{R}{R^3} \ \ (b) \ 0 \ (c) \ 0 \right]$$

2. A potential field is represented by the equation, $\phi = 4yz^3 + 3xyz - z^2 + 2$. Calculate the potential at the point $(1, -1, -2)$. 　　　　　　　　　　　　　　　[*Ans.* −36 V]

3. A vector field is represented by the equation, $A = \hat{i}x^2yz - \hat{j}2xz^3 + \hat{k}xz^2$. Calculate the values of the partial derivatives:

(a) $\dfrac{\partial A}{\partial x}$

(b) $\dfrac{\partial A}{\partial z}$

[*Ans.* (a) $\hat{i}2xyz - \hat{j}2z^3 + \hat{k}z^2$ (b) $\hat{i}x^2z - \hat{j}6xz^2 + \hat{k}2xz$]

4. The temperature of a body at any point changes according to the relation, $T = 4x^2 + 3y^2 - 2z^2$. Show that the grad T at the point (1, 2, 3) is equal to $-8\hat{i} + 12\hat{j} - 12\hat{k}$.

5. Show that the function $\phi = x^2 - y^2$ satisfies the Laplace's equation.

6. Show that the following potential functions satisfy the Laplace's equation.

   (a) $V = x^2 - y^2 + z$
   (c) $V = x^2 + y^2 - 2z$

   (b) $V = ax + by + cz$
   (d) $V = ax^2 - ay^2 + 4z$

7. Calculate curl of the following fields:

   (a) $E = a[3x^2yz\hat{i} + \{x^3z + yz^2(yz)^{1/2}\}\hat{j} + \{x^3y + y^2z(yz)^{1/2}\}\hat{k}]$

   (b) $B = y\hat{i} - x\hat{j}$ and

   (c) $B = a\left\{\dfrac{y\hat{i}}{(x^2 + y^2)^{3/2}} - \dfrac{x\hat{k}}{(x^2 + y^2)^{3/2}}\right\}$

$$\left[\textit{Ans.}\ (a)\,0\ \ (b)\,-2\hat{k}\ \ (c)\ \dfrac{a\hat{k}}{(x^2 + y^2)^{3/2}}\right]$$

8. Calculate the divergence at a point (x, y, z) for the fields in problem 7 above.

$$\left[\textit{Ans.}\ (a)\ 6axyz + \dfrac{3}{4}a(yz)^{1/2}\ \ (b)\,0\ (c)\,0\right]$$

9. If $A = xz^3\hat{i} - 2x^2yz\hat{j} + 2yz^4\hat{k}$, show that curl A at the point (1, –1, 1) is $(3\hat{i} + 4\hat{k})$.

10. Evaluate $\iint_S F \cdot ds$, where $F = \hat{i}4xz - \hat{j}y^2 + \hat{k}yz$, and S is the surface of the cube bounded by $x = 0, x = 1, y = 0, y = 1, z = 0$ and $z = 1$.   [*Ans.* 3]

11. If r is position vector of any point on surface S enclosing a volume V, find $\iint_S r \cdot dS$.

[*Hint:* By Gauss's divergence theorem, changing the surface integral into volume integral, we obtain

$$\iint_S r \cdot dS = \iiint_V (\nabla \cdot r)dV$$

$$\therefore\quad \iint_S r \cdot dS = \iiint_V \left[\left(\hat{i}\dfrac{\partial}{\partial x} + \hat{j}\dfrac{\partial}{\partial y} + \hat{k}\dfrac{\partial}{\partial}\right) \cdot (\hat{i}x + \hat{j}y + \hat{k}z)\right]$$

$$= \iiint_V \left(\dfrac{\partial x}{\partial x} + \dfrac{\partial y}{\partial y} + \dfrac{\partial z}{\partial z}\right)dV$$

$$= 3\iiint_V dV = 3V]$$

12. Show that $\iint_S \hat{n}ds = 0$ for any closed surface S.

13. Show that $\iint_S \text{curl } F \cdot dS = 0$ for any closed surface S.

14. A fluid of density $\rho(x, y, z, t)$ moves with velocity $v(x, y, z, t)$. If there are no sources or sinks, prove that

$$\nabla \cdot J + \frac{\partial \rho}{\partial t} = 0, \quad \text{where } J = ev.$$

15. Show that the necessary and sufficient condition for $\oint_C A \cdot dr = 0$ to be true for every curve C is that curl $A = 0$.

16. Prove Stokes' theorem for the vector $A(x + y, 2x - z, y + z)$ taken over the triangle cut from the plane $3x + 2y + z = 6$ by the coordinate planes.

SHORT ANSWER QUESTIONS

1. How many kinds of fields are there?
   Ans. Two: (i) scalar, and (ii) vector

2. Give two examples of vector field.
   Ans. (i) distribution of velocity in a fluid (ii) distribution of gravitational, magnetic or electric field in space.

3. If $\phi(x, y, z)$ is a scalar field, is $\hat{i}\frac{\partial \phi}{\partial x} + 2\hat{j}\frac{\partial \phi}{\partial y} + 3\hat{k}\frac{\partial \phi}{\partial z}$ a vector field?

   Ans. We know that a scalar field $\phi(x, y, z)$ evaluated at a particular point is independent of the coordinates of that point. Obviously, scalar field is invariant with respect to the transformation of coordinates. Its gradient, which is a vector field, must also be invariant with respect to the transformation. Now, on applying coordinate transformation, the vector $\hat{i}\frac{\partial \phi}{\partial x} + 2\hat{j}\frac{\partial \phi}{\partial y} + 3\hat{k}\frac{\partial \phi}{\partial z}$ does not retain its form.

   Thus, it cannot be the vector field corresponding to the given scalar field $\phi(x, y, z)$.

4. What are lamellar fields?
   Ans. Certain vector fields can be expressed as gradient of a scalar field ϕ and such fields are known as lamellar fields.

5. What is name of the theorem which provides a method of reducing volume integrals to surface integrals?
   Ans. Gauss's theorem of divergence.

MULTIPLE CHOICE QUESTIONS

1. In two dimension, the divergence transforms as
   (a) a scalar   (b) a vector
   (c) a scalar and a vector   (d) none of the above   [a]

2. The curl of a vector field A is
   (a) scalar   (b) vector
   (c) vector as well as scalar   (d) none of the above   [b]

3. The divergence of a curl is always
   (a) infinite   (b) zero
   (c) finite   (d) none of the above   [b]

4. The curl of a gradient is always
   (a) zero
   (b) infinite
   (c) finite
   (d) none of the above [a]

5. If E is the intensity of electric field and V is the electric potential, then the relation between these two is:
   (a) $E = \nabla \times V$
   (b) $E = \nabla \cdot V$
   (c) $E = \nabla V$
   (d) $E = \nabla^2 V$ [c]

6 The theorem which relates the line integral of a vector field along the limiting boundary of the area enclosed by a closed path to the surface integral of the curl of the vector field is known as
   (a) Green's theorem
   (b) Gauss's divergence theorem
   (c) Stokes' curl theorem
   (d) none of the above [b]

7. A vector field which can be expressed as a gradient of a scalar field is called a
   (a) gradient
   (b) curl
   (c) lamellar field
   (d) divergence [c]

8. If $\phi = r^n = (x^2 + y^2 + z^2)^{n/2}$, then $\nabla\phi$ is equal to
   (a) $n r^{n-2} r$
   (b) $n r^{n-1} r$
   (c) $n r^{n-3} r$
   (d) $n r^{n-4} r$ [a]

9. If $\nabla \cdot A = 0$, the vector A is called
   (a) lamellar vector
   (b) solenoidal vector
   (c) linear vector
   (d) none of the above [b]

10. If $\nabla \cdot f = 0$, the vector f is called
    (a) lamellar vector
    (b) solenoidal vector
    (c) rotational vector
    (d) irrotational vector [d]

11. If the vector $f = 3x\hat{i} + (x+y)\hat{j} - az\hat{k}$ is solenoidal, the value of a is
    (a) 2
    (b) 4
    (c) 8
    (d) 6 [b]

$$\left[\textbf{Hint. } \nabla \cdot f = \frac{\partial}{\partial x}(3x) + \frac{\partial}{\partial y}(x+y) + \frac{\partial}{\partial z}(-az) = 3 + 1 - a = 0 \quad \therefore \quad a = 4 \right]$$

12. For scalar field ϕ and vector field A, $\nabla \cdot (\phi A)$ will be
    (a) zero
    (b) $\phi \nabla \cdot A$
    (c) $\phi \nabla \cdot A + A(\nabla \cdot \phi)$
    (d) $\phi \nabla \cdot A + A \cdot \nabla \phi$ [d]

13. The gradient of scalar field is always
    (a) a scalar
    (b) a vector
    (c) a numeric
    (d) sometimes a scalar and sometimes a vector [b]

14. Grad $\left(\dfrac{1}{r}\right)$ is equal to

   (a) $\dfrac{r}{r^2}$ (b) $-\dfrac{r}{r^2}$

   (c) $-\dfrac{r}{r^3}$ (d) $\dfrac{xyz}{r^2}$ [c]

15. If r is a position vector and A is a constant vector, then curl $(A \times r)$ is

   (a) 0 (b) A

   (c) $2A$ (d) $3\,(A \times r)$ [c]

3

Electrostatics in Vacuum

3.1 INTRODUCTION

Physics is based on the concept that all matter is made up of atoms and these atoms in turn comprises electrically charged particles whose behaviour is governed by electrodynamics.

There are two kinds of electric charges: *positive* and *negative*. Like charges repel each other while unlike charges attract each other. The algebraic sum of electric charges remains constant in a closed system. This is known as the *law of conservation of electric charges*.

The electric charge on any body consists of a whole number of elementary charges each of which equals to 1.6×10^{-19} C. The smallest stable subatomic particle having a negative elementary charge is called the *electron*. The mass of an electron is 9.1×10^{-31} kg. The smallest stable subatomic particle having a positive charge is the *proton*. The mass of a proton is 1.67×10^{-27} kg. Electrons and protons are found in the atoms of any substance. A neutral body contains charges of opposite sign which are equal in absolute value. Electric charges are said to be point charges if the linear dimensions of the bodies on which the charges are concentrated are very much smaller than any other lengths pertinent to the problem under consideration. A body exhibiting positive electrification has a positive electric charge, and one with negative electrification has a negative electric charge. The net charge on a body is the algebraic sum of its positive and negative charges. A particle having a nonzero net charge is often called an *ion*. Since matter in bulk does not exhibit gross electrical forces, one may assume that it is composed of equal amounts of positive and negative charges. *The net or total charge does not change for any process occuring within an isolated system.* No exception has been found to this rule, i.e. *principle of conservation of charge*.

Charge Quantization

It is found experimentally that the electric charge of a body, large or small, is always an integral multiple of certain minimum amount of charge. This basic charge is the charge on an electron, which is denoted by e and has magnitude 1.6×10^{-19}C. Thus, the charge on an electron is $-e$, on a proton is $+e$ and that on α (alpha) particle is $+2e$.

The experimental fact that electric *charges occur in discrete amounts instead of continuous amount is called quantization of electric charge*. The quantization of electric charge means that the total charge (q) of a body is always an integral multiple of a basic quantum of charge (e),

i.e. $\qquad\qquad\qquad\qquad q = ne \qquad\qquad$ [where $n = 0, \pm 1, \pm 2, \pm 3, ...$]

We may note that while dealing with *macroscopic* charges ($q = ne$), we can ignore the quantization of electric charge. This is because the basic charge e is very small and n is very large in most practical situations, so q behaves as it were continuous, i.e. as if a large amount of charge were flowing. However, quantization of charge becomes important at the *microscopic* level, where the charges involved are of the order of a few tens or hundreds of e.

If the net charge within a certain region of space, decreases with time, charge conservations implies that an equal amount of charge must appear somewhere else. However, charge must pass along some continuous path from one point to the other. This is termed *local conservation of charge*. The transport of charge constitutes a *current* and the conservation of charge is governed by *continuity equation* (see Section 7.11).

System of Units

i. *MKS System:* This system is widely used and principal convenience of MKS system is that it introduces the common technical units: volt (V), ampere (A), coulomb (C), etc. for voltage, current and charge respectively. This system is particularly suitable for treating applications that involve lumped circuit parameters and fields. Moreover, these units are indeed of a convenient magnitude for laboratory-scale effects in case of engineering applications.

ii. *Gaussian System or Symmetrical System:* This system of units uses the electric quantities from the e.s.u. system and magnetic quantities from the e.m.u. system and we can see that as a consequence, the velocity of light appears explicitly in some of the equations that interrelate the electrical and magnetic quantities.

In Gaussian system, the electric field E and the magnetic field B have the same units and this is why, this system is more suitable to the requirements of *relativistic invariance*. Furthermore, this system of units is more adequate to the treatment of microscopic problems involving the electromagnetic interaction of the fundamental constitutents of matter, e.g. electrons, atoms, molecules, etc.

3.2 COULOMB'S LAW

The results of the electric interaction between two charged particles at rest in the observer's inertial frame of reference or, atmost, moving with a very small velocity constitute what is called *electrostatics*. The electrostatic interaction between two charged particles is given by *Coulomb's law*.

Charles A de Coulomb (1736–1806), from experimental observations concluded that the *electrostatic interaction between two charged particles is proportional to the square of the distance between them, and its direction is along the line joining the two charges.*

Let '1' and '2' be two particles carrying charges q_1 and q_2 respectively and separated by a distance r_{12} (Fig. 3.1). According to Coulomb's law, the electric force exerted by particle '1' on particle '2' is

Fig. 3.1: Coulomb's law

$$F_{12} \propto \frac{q_1 q_2}{r_{12}^2} \quad \text{i.e.} \quad F_{12} = K \frac{q_1 q_2}{r_{12}^2} = -F_{21} \qquad (3.1)$$

where K is a constant of proportionality and F_{21} is the force exerted by the particle '2' on the particle '1'. Expressing Eq. (3.1) in vector notation, one obtains

$$F_{12} = K \frac{q_1 q_2}{|r_{12}|^2} \hat{r}_{12} = K \frac{q_1 q_2}{|r_{12}|^3} r_{12} \qquad (3.2)$$

where $\hat{r}_{12}$ is the unit vector along r_{12}.

Coulomb's law is very similar to the law of gravitational interaction. The sign of charges in Eq. (3.2) decides whether the force is *attractive* or *repulsive*. When both the charges are similar (i.e. positive or negative), F_{12} is positive and represents the force of repulsion; while if one is positive and other is negative, F_{12} is negative and represents the force of attraction.

The numerical value of the constant K in the MKSC system is equal to $10^{-7} c^2 = 8.9874 \times 10^9$, where c is the velocity of light. For practical purposes, we may say that $K = 9 \times 10^9$. Then when the distance r_{12} is measured in metre, and the force in newton, Eq. (3.1) becomes

$$F_{12} = 9 \times 10^9 \frac{q_1 q_2}{r_{12}^2} = -F_{21} \qquad (3.3)$$

Once we have decided on the value of K, the unit charge is fixed. This unit is called a coulomb and is designated by C. *The coulomb is that charge which when placed one metre from an equal charge in vacuum, repels it with a force of* 8.981×10^9 *newton.* Formula given by Eq. (3.3), holds only for two charged particles in vacuum, i.e. for two charged particles in the absence of any other charge or matter. We must note that according to Eq. (3.2), the unit of K is $N m^2 C^{-2}$ or $m^3 kg\, s^{-2} C^{-2}$. The constant K is then equal to

$$K = \frac{1}{4\pi\varepsilon_0}$$

where, ε_0 is the vacuum *permittivity* or *permittivity of free space*. It is also sometimes referred to as capacitivity in vacuo, or the dielectric constant of vacuum. Thus,

$$\varepsilon_0 = \frac{10^7}{4\pi c^2} = 8.854 \times 10^{-12} N^{-1} m^{-2} C^2 \quad (\text{or } m^{-3} kg^{-1} s^2)$$

Now, the Coulomb's law is written as

$$F_{12} = \frac{1}{4\pi\varepsilon_0} \frac{q_1 q_2}{|r_{12}|^3} r_{12} \qquad (3.4)$$

The forces of electrostatic interaction are central and long range. One can obtain an idea about the relative magnitudes of electrostatic and gravitational forces by comparing the force of attraction between two electrons due to gravitation and the force of repulsion due to electrostatic repulsion.

$$\frac{\text{Electrostatic repulsion between two electrons}}{\text{Gravitational attraction between two electrons}} = \frac{1}{4\pi\varepsilon_0} \frac{e^2}{r^2} \frac{r^2}{G m^2}$$

$$= \frac{e^2}{4\pi\varepsilon_0 G m^2} \qquad (3.5)$$

$$= 9 \times 10^9 \times \frac{(1.6 \times 10^{-19})^2}{(6.67 \times 10^{-11})(9.1 \times 10^{-31})^2}$$

$$= 4.17 \times 10^{42}$$

Obviously, the electrical interaction is of the order of magnitude required to produce the binding between atoms to form molecules, or the binding between the electrons and protons to form atoms.

This reveals that the chemical processes (and in general the behaviour of matter in bulk) are due to electrical interactions between atoms and molecules.

Coulomb's law is based primarily on experiments. At this stage, one can raise a question whether the law is exactly of the inverse square form, i.e. is the force proportional to $\frac{1}{r^n}$, is n exactly equal to 2? Cavendish found that $n = 2 \pm 0.02$. Plimpton and Lawton in 1936, reported that n differs from 2 by not more than one part in 10^9. Lamb and Rutherford in 1947 found from their measurements of energy levels of the hydrogen atom that the exponent in Coulomb's law is correct to one part in 10^9 at distances of the order of 10^{-10} m. Evidence from nuclear experimental observation has shown that the electrostatic forces vary approximately according to the inverse square law even at distances of the order of 10^{-15} m.

3.3 ELECTRIC FIELD

Any region where an electric charge experiences a force is called an *electric field*. The force is due to the presence of other charges in that region. For example, a charge q placed in a region where there are other charges $q_1, q_2, q_3, \ldots$, etc. (Fig. 3.2) experiences a force $F = F_1 + F_2 + F_3 + \cdots$, and we say that it is in electric field produced by the charges $q_1, q_2, q_3, \ldots$. We must note that the charge q, of course, also exerts forces on q_1, $q_2, q_3, \ldots$, but we are not concerned with them now. Since the force that each charge $q_1, q_2, q_3, \ldots$ produces on the charge q is proportional to q, the resultant force F is also proportional to q. Obviously, the force on a particle placed in an electric field is proportional to the charge on the particle.

Fig. 3.2: Resultant electric field at P due to the presence of several charges in the region

The intensity of the electric field at a point is equal to the force per unit charge placed at that point. Thus,

$$E = \frac{F}{q} \quad \text{or} \quad F = qE \qquad (3.6)$$

The unit of electric field intensity is newton/coulomb or NC^{-1}.

If q is positive, the force F acting on the charge has the same direction as the field E, but if q is negative, the force F has the direction opposite to E (Fig. 3.3).

Electric field $\xrightarrow{\quad E \quad}$

Positive charge $\quad \oplus \xrightarrow{F = qE}$

Negative charge $\quad \xleftarrow{F = qE} \ominus$

Fig. 3.3: Direction of the force produced by an electric field on the positive and negative charge

Obviously, if we apply an electric field to a region where positive and negative particles or ions are present, the field will tend to move the positively and the negatively charged bodies in opposite directions, resulting in a charge separation, an effect sometimes referred to as *polarization*.

Let us write Eq. (3.3) in the form $F = q_2\left(\dfrac{q_1}{4\pi\varepsilon_0 r_{12}^2}\right)$. This gives the force exerted by the charge q_1 on the charge q_2 placed at a distance r_{12} from q_1. Using Eq. (3.6), we may say that the electric field E at the point where q_2 is placed is such that $F = q_2 E$. Obviously, on comparing both expressions of F, we conclude that the electric field at a distance r_{12} from the charge q_1 is $E = \dfrac{q_1}{4\pi\varepsilon_0 r_{12}^2}$, or in the vector form,

$$E = \frac{q_1}{4\pi\varepsilon_0 r_{12}^2}\hat{a}_r \tag{3.7}$$

where, $\hat{a}_r$ is the unit vector in the radial direction away from the charge q, since F is along this direction. Equation (3.7) is valid for both positive and negative charges with the direction of E relative to $\hat{a}_r$ given by the sign of q_1. Obviously, E is directed away from a positive charge and towards a negative charge (Fig. 3.4).

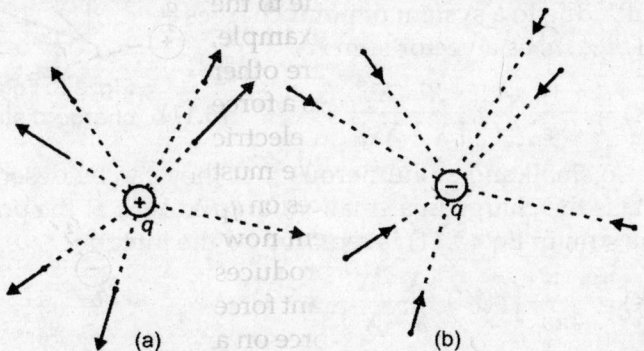

Fig. 3.4: Electric field produced by (a) a positive charge and (b) a negative charge

Let $q_1 = q_0$ be sufficiently small so that it does not distort the field whose intensity is to be measured. The resultant force F exerted on a test charge q by a field set up by a system of fixed charges $q_1, q_2, q_3, \ldots, q_n$ is equal to the vector sum of the forces F_i exerted on the test charge by each of the fields of charges q_i, i.e.

$$F = \sum_{i=1}^{n} F_i \tag{3.8}$$

This leads to the *principle of superposition* of electric fields given as

$$E = \sum_{i=1}^{n} E_i \tag{3.8a}$$

3.4 PRINCIPLE OF SUPERPOSITION

If there are more that two particles present with charges, say $q_1, q_2, q_3, \ldots$, the total force on any one particle is the vector sum of forces it experiences due to all other particles present there separately. This is called the *principle of superposition*. There are three charges q_1, q_2 and q_3 as shown in Fig. 3.5. The force on q_3 is

$$F = F_{13} + F_{23}$$

$$= \frac{1}{4\pi\varepsilon_0} \frac{q_1 q_3}{|r_{13}|^3} r_{13} + \frac{1}{4\pi\varepsilon_0} \frac{q_2 q_3}{|r_{23}|^3} r_{23}$$

Generalising, one finds the force acting on a charge q_j due to a number of other charges present in the region is

Fig. 3.5: Force due to continuous charge distribution

$$F_j = \frac{1}{4\pi\varepsilon_0} \sum_{i \neq j} \frac{q_i q_j}{|r_{ij}|^3} r_{ij} \tag{3.9}$$

Let r_i and r_j be the vectors representing the location of charges q_i and q_j respectively (Fig. 3.6). One can write the Eq. (3.9) in terms of these position vectors as

$$F_j = \frac{1}{4\pi\varepsilon_0} \sum_{i \neq j} \frac{q_i q_j}{|r_j - r_i|^3} (r_j - r_i) \tag{3.10}$$

The experimentally observed linear superposition of forces due to many charges means that one can write the electric field at X due to a system of point charges q_i located at X_i, $i = 1, 2, \ldots, n$, as a vector sum

Fig. 3.6: Force due to continuous charge distribution

$$E(X) = \frac{1}{4\pi\varepsilon_0} \sum_{i=1}^{n} q_i \frac{X - X_i}{|X - X_i|^3} \tag{3.11}$$

If the charges are so small and so numerous that they can be described by a charge density $\rho(X')$ [if Δq is the charge in a small volume $\Delta x \Delta y \Delta z$ at the point X', then $\Delta q = \rho(X') \Delta x \Delta y \Delta z$], the sum in Eq. (3.11) is replaced by the integral

$$E(X) = \frac{1}{4\pi\varepsilon_0} \int \rho(X') \frac{X - X'}{|X - X'|^3} d^3 x' \tag{3.12}$$

where, $d^3 x' = dx' dy' dz'$ is a three-dimensional volume element at X'.

We have seen that the principle of superposition has facilitated considerably the mathematical handling of electromagnetic theory. However, this principle fails in nuclear interactions and this is one of the reasons why the nuclear theory is somewhat more complicated and troublesome than the theory of atomic interactions.

3.5 DIRAC DELTA FUNCTION

It is sometimes convenient to regard a point charge as a 'fictitious' continuous charge distribution. One can achieve this with the help of Dirac delta function, which is written as $\delta(x-a)$ and is a mathematically improper function having the following properties:

   i. $\delta(x - a) = 0$ for $x \neq a$

  ii. $\int \delta(x - a) dx = 1$, if the region of integration includes $x = a$

                $= 0$, otherwise

 iii. $\int f(x) \delta(x - a) dx = f(a)$

iv. $\int f(x)\delta'(x-a)dx = -f'(a)$, where prime denotes differentiation with respect to the argument.

v. $\delta(f(x)) = \sum_i \dfrac{1}{\left|\dfrac{df}{dx}(x_i)\right|}\delta(x-x_i)$, where $f(x)$ is assumed to have only simple zeros, located at $x = x_i$. In more than one dimension, one merely takes the product of delta functions in each dimension. In three dimensions, for example with cartesian coordinates, $\delta^3(r) = \delta(x)\,\delta(y)\,\delta(z)$, where $r = \hat{i}x + \hat{j}y + \hat{k}z$.

vi. $\delta(x - X) = \delta(x_1 - X_1)\,\delta(x_2 - X_2)\,\delta(x_3 - X_3)$ is a delta function that vanishes everywhere except at $x = X$.

vii. $\int_{\Delta V}\delta(x - X)d^3x = \begin{cases} 1 \text{ if } \Delta V \text{ contains } x = X \\ 0 \text{ if } \Delta V \text{ does not contain } x = X \end{cases}$

We must remember that a delta function has the dimensions of an inverse volume, whatever number of dimensions the space has.

A discrete set of point charges can be described with a charge density by means of delta functions. For example,

$$\rho(X) = \sum_{i=1}^{n} q_i(X - X_i) \tag{3.13}$$

represents a distribution of n point charges q_i, located at the points X_i. Using Eq. (3.13) in Eq. (3.12), integrating, and using the properties of the delta function, yields the discrete sum given by Eq. (3.11). For details see Appendix B.

3.6 LINES AND TUBES OF FORCE

Just as in the case of a gravitational field, an electric field may be represented by lines of force, which are lines that, at each point, are tangent to the direction of the electric field at that point, i.e. *lines of force* or *field lines* are parallel to the direction of the field everywhere. Since there is a single direction for the electric field at every point of the field, and hence there is usually just one field line through any given point, i.e. two field lines can never intersect. The lines of force in Fig. 3.7a depict the electric field of a positive charge, and those in Fig. 3.7b depict the electric field of a negative charge. We can see that they are straight lines passing through the charges.

Figures 3.8 and 3.9 show the lines of force near a pair of charges of equal magnitude, for one positive and one negative and two equal positive charges respectively.

In both the above figures, the lines of force of the resultant electric field produced by the two charges have also been represented.

Let us consider an element of area *ds* in the field (small enough). Suppose that the electric intensity is the same in magnitude and direction at every point of *ds*, so that the field of lines or lines of force at all points on the boundary will be approximately all parallel.

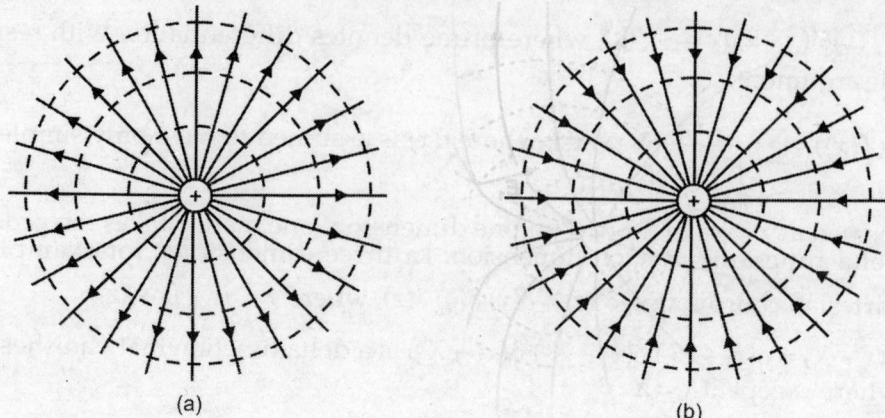

Fig. 3.7: Lines of force and equipotential surfaces of the electric field of (a) a positive charge and (b) a negative charge

Fig. 3.8: Lines of force and equipotential surfaces of the electric field of two equal but opposite charges

One obtains the "tubular" surface as shown in Fig. 3.10 and it is called a *tube of flux*. We must note that the normal cross-section of the tube of force forms a part of an *equipotential surface*.

3.7 ELECTRIC FLUX

The concept of electric flux is of great usefulness in many physical problems. To understand the concept of electric flux, let us imagine that a fluid is flowing with a

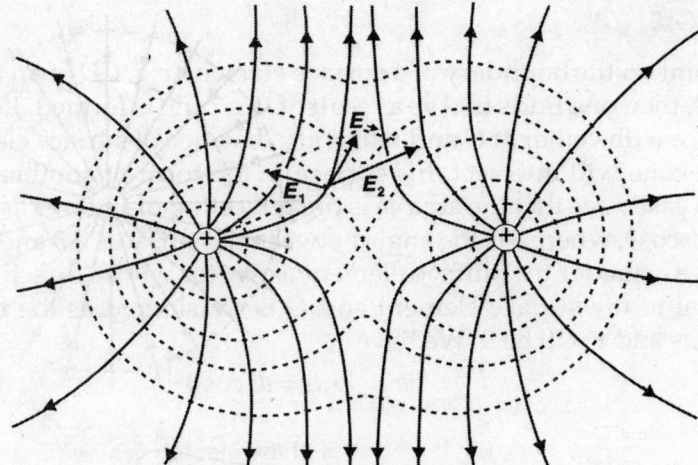

Fig. 3.9: Lines of force and equipotential surfaces of the two identical positive charges

Fig. 3.10: Tube of flux   (a)   (b)

                Fig. 3.11

speed v through a small flat surface ds in a direction normal to the surface as shown in Fig. 3.11a. The rate of flow of the fluid, i.e. the volume of the fluid crossing the area per unit time is termed the flux of fluid, and it is equal to vds. Now, if the normal to the surface is not parallel to the direction of flow of the fluid and makes an angle with the surface as shown in Fig. 3.11b, then one finds the projected area in a plane perpendicular to v is $ds \cos\theta$. Thus, the flux F is

$$F = vds\cos\theta \qquad (3.14)$$

Let us represent the surface by vector $ds\,\hat{a}_n$ having magnitude ds and direction along the normal to the surface, $\hat{a}_n$ being the unit vector in this direction. One can express the flux as

$$F = v \cdot \hat{a}_n \, ds \qquad (3.15)$$

In an electrostatic field, there is nothing actually flowing. One can mathematically define a quantity analogous to the flux of a fluid. This is called as *electric flux* and defined as

$$\text{Electric flux, } \phi = E \cdot \hat{a}_n \, ds \qquad (3.16)$$

The concepts of electric flux and tubes are sometimes useful in drawing the electric field lines about charges.

3.8 SOLID ANGLE

When every point on the boundary of a surface element, e.g. *CD* of area *ds* is joined to a given point *O*, then one finds that as a result of it, a cone is formed. Let us imagine a spherical surface with center at *O* and radius *r* (= distance of surface element from *O*), then the above cone will intersect this surface in the form of another surface *AB* of area *da*. We can easily see that the area *da* is the projection of the area *ds* perpendicular to *OP*, i.e. *da* = *ds*cosθ, where θ is the angle between the surfaces *AB* and *CD* (Fig. 3.12). Obviously, if we consider the surface element as vector whose direction is along the outward normal to the surface element and *O* is considered as the origin then the angle between *ds* and *r* will be θ. We have

$$da = \hat{r} \cdot ds = ds \cos\theta$$

Fig. 3.12: Concept of a solid angle

The angle formed by this surface element *da* at the point *O* in three dimensions is termed a *solid angle*. The solid angle formed by the area *ds* at the point *O* is obtained as the ratio of the normal area to the square of the distance, i.e.

$$d\Omega = \frac{da}{r^2} = \frac{ds\cos\theta}{r^2} \tag{3.17}$$

Solid angle has no dimensions like an angle, but it is represented by the unit, steradian. If *r* = 1, then *d*Ω = *da*. Thus, the solid angle formed by any surface at point *O* is equal in magnitude to the area intersected by the cone or pyramid formed by joining every point on that surface to the point *O*, on a spherical surface of unit radius and having centre at *O*.

Let us now determine the solid angle formed by an element of area *PQRS* situated on the spherical surface of radius *r*, at the centre *O* (Fig. 3.13). Let us draw planes *TOM* and *TON* making angles φ and φ + *d*φ respectively with the plane *ZOX*, i.e. the angle formed by the side *SP* of area element *PQRS* with the z-axis is *d*φ. From Fig. 3.13, it is evident that lines

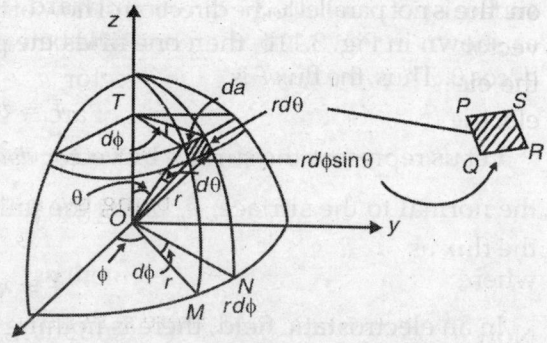

Fig. 3.13: Concept of a solid angle

OP and *OQ* make angles θ and θ + *d*θ with the z-axis, and hence the angle formed at the centre *O* by the side *PQ* of the area element is *d*θ.

The radius of the spherical surface $PQRS$ is r, and hence
$$PQ = rd\theta$$

From $\triangle OPT$, we have
$$\frac{TP}{OP} = \sin\theta$$

$\therefore$
$$TP = OP\sin\theta = r\sin\theta \qquad (3.17a)$$

or
$$SP = (r\sin\theta)d\phi$$

$\therefore$ Area of element $PQRS = PQ \times PS = (rd\theta)r\sin\theta\,d\phi$

or
$$da = r^2\sin\theta d\theta d\phi \qquad (3.17b)$$

$\therefore$
$$\text{Solid angle } d\Omega = \frac{da}{r^2} = \frac{r^2\sin\theta d\theta d\phi}{r^2}$$
$$= \sin\theta d\theta d\phi \qquad (3.17c)$$

Now, the solid angle formed at the centre by the entire spherical surface is equal to the sum of solid angles formed at the centre by all the area elements situated on the surface. Since the values of θ and ϕ vary from 0 to π and 0 to 2π respectively for the entire spherical surface, and hence solid angle subtended by spherical surface at its centre is

$$\Omega = \int_0^{2\pi}\int_0^{\pi}\sin\theta d\theta d\phi = 2\pi\int_0^{\pi}\sin\theta d\theta$$
$$= 2\pi(-\cos\theta)_0^{\pi} = 4\pi$$

3.9 GAUSS'S LAW FOR AN ELECTRIC FIELD (INTEGRAL FORM)

One can use the relation given by Eq. (3.16) to obtain an important law in electrostatics, i.e. Gauss's law. This law relates the electric flux ϕ through any closed surface to the net amount of charge within the surface.

Let us now imagine a closed surface S surrounding a charge q_i as shown in Fig. 3.14. Let ds be an element of area around the point P on the surface, and $\hat{a}_n$ be an outward unit vector normal to it. Let θ be the angle between the electric field at P and unit vector $\hat{a}_n$. The electric flux through the element of area ds is

$$d\phi = E_i \cdot \hat{a}_n ds = E_i\cos\theta ds$$
$$= \frac{1}{4\pi\varepsilon_0}\frac{q_i}{r^2}\cos\theta ds$$

Fig. 3.14: Point charge q_i surrounded by a closed surface S

where r is the distance of the element ds from the charge q_i.

Now $d\Omega = \dfrac{ds\cos\theta}{r^2}$ is the solid angle subtended by the area element ds at q_i. Thus,

$$d\phi = E_i \cdot \hat{a}_n ds = \frac{1}{4\pi\varepsilon_0}q_i d\Omega \qquad (3.18)$$

Thus, the total flux through the entire surface S is

$$\phi = \int_S E_i \cdot \hat{a}_n ds = \frac{1}{4\pi\varepsilon_0}q_i\int d\Omega = \frac{q_i}{\varepsilon_0} \qquad (3.18a)$$

Obviously, the electric flux through the surface S is proportional to the charge and independent of the distance r (in case of a spherical surface, r is radius). Therefore, if we draw several concentric spherical surfaces $S_1, S_2, S_3,...$ (Fig. 3.15) around the charge q_i, the electric flux through all of them is same and equal to $\dfrac{q}{\varepsilon_0}$. The result is due to $\dfrac{1}{r^2}$ dependence of the field. The same is also true in the case of gravitational field.

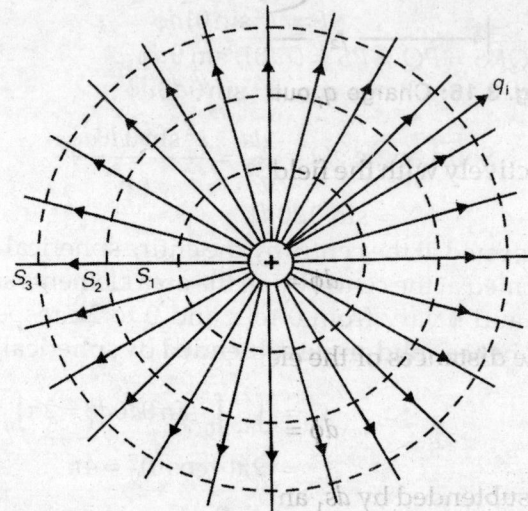

Fig. 3.15: The electric flux through concentric spheres surrounding the same charge

If there is any arbitrary distribution of charges within the surface S, then by the principle of superposition,

$$E = \sum E_i \qquad (3.19)$$

Equation (3.19) leads to the relation

$$\int_S E_i \cdot \hat{a}_n ds = \frac{\sum q_i}{\varepsilon_0} = \frac{Q}{\varepsilon_0} \qquad (3.20)$$

where, Q is the total charge ($= \sum q_i$) within the closed surface S. This result is valid for any closed surface, irrespective of the position of the charge within the surface.

In case of a continuous distribution of charge within the surface S, Eq. (3.20) changes to

$$\int_S E_i \cdot \hat{a}_n ds = \frac{1}{\varepsilon_0} \int \rho dV \qquad (3.21)$$

If no charges are present inside the closed surface S, or if the net charge is zero, the total electric flux through it is zero. The charges outside the closed surface do not contribute to the total flux. One can easily verify it. Let us suppose that the charge q_i is outside the closed surface S (Fig. 3.16). The electric field vector E passes through two elements of area ds_1 and ds_2 cut out by a cone with its apex at charge q_i as shown in Fig. 3.16. $\hat{a}_{n_1}$ and $\hat{a}_{n_2}$ are the outward drawn unit vectors normal to ds_1 and ds_2 making

Fig. 3.16: Charge q_i outside the closed surface S

angles θ_1 and θ_2 respectively with the field. The net outward flux through area ds_1 and ds_2 is

$$d\phi = \frac{1}{4\pi\varepsilon_0}\left[\frac{q_i}{r_1^2}\cos(180° + \theta_1)ds_1 + \frac{q_i}{r_2^2}\cos\theta_2 ds_2\right]$$

where, r_1 and r_2 are the distances of the elements ds_1 and ds_2 from charge q_i. Thus,

$$d\phi = \frac{q_i}{4\pi\varepsilon_0}\left[-\frac{\cos\theta_1}{r_1^2}ds_1 + \frac{\cos\theta_2}{r_2^2}ds_2\right]$$

Now, the solid angle subtended by ds_1 and ds_2 at q_i are the same, i.e.

$$\frac{\cos\theta_1}{r_1^2}ds_1 = \frac{\cos\theta_2}{r_2^2}ds_2 = d\Omega$$

Hence $d\phi = 0$.

One can see that this holds good for all the cones drawn from q_i through the surface, and hence the total flux $\phi = 0$.

These results constitute what is known as *Gauss's law for the electric field*. One can state it as follows:

$$\left.\begin{array}{l}\oint_S E \cdot \hat{a}_n ds = \dfrac{Q}{\varepsilon_0} \text{ for charge } Q \text{ inside } S \\[2mm] \qquad\qquad = 0 \text{ for charge } Q \text{ outside } S\end{array}\right\}\qquad(3.22)$$

Relation (3.22) is the *integral form of Gauss's law for an electric field*.

Gauss's law is particularly useful when we wish to compute the electric field produced by charge distribution having certain geometrical symmetries.

3.10 GAUSS'S LAW FOR AN ELECTRIC FIELD (DIFFERENTIAL FORM)

One can also express the Gauss's law in a differential form. We have the relation,

$$\mathbf{div}\, A = \lim_{dV \to 0} \frac{\oint_S A \cdot \hat{a}_n ds}{dV} \qquad (3.23)$$

where, S is the surface enclosing the volume element. Integrating over the finite volume, one obtains

$$\int_V \text{div} A \, dV = \oint A \cdot \hat{a}_n \, ds \tag{3.24}$$

We are familiar with Eq.(3.24), which is the well-known *divergence theorem* usually used for transforming a volume integral into a surface integral and *vice versa*. Using Eq. (3.24), the Gauss's law can be expressed as

$$\oint_S E \cdot \hat{a}_n \, ds = \oint_V \text{div} E \, dV = \frac{1}{\varepsilon_0} \int_V \rho \, dV$$

i.e.

$$\int_V \left(\text{div} E - \frac{\rho}{\varepsilon_0} \right) dV = 0$$

The above is true for any arbitrary volume V. Thus,

$$\text{div} E - \frac{\rho}{\varepsilon_0} = 0$$

or

$$\nabla \cdot E = \frac{\rho}{\varepsilon_0} \tag{3.25}$$

Relation (3.25) is the *differential form of Gauss's law for an electric field*. In this form, the law reveals that the sources of electric displacement are free electric charges (since $\rho \neq 0$), i.e. the electric flux lines either leave or enter a closed surface, therefore, there must be either a source or sink of flux lines. We know from convention that flux lines start from positive charge as sources and terminate in negative charges as sinks. This shows that positive and negative charges exist independently. Obviously, Gauss's law in differential form expresses a local relation between E and ρ. Thus, we may say that electric charges are the sources of the electric field, and that their distribution and magnitude determine the electric field at each point of space.

We have adopted Coulomb's law in electrostatics as the fundamental law. We have also seen that the Gauss's law in electrostatics is a consequence of the fact that the electric force between charged particles is inversely proportional to the square of the distance between them, i.e. Coulomb's inverse square law. One can easily see that any other law, say $\frac{1}{r^n}$ with $n \neq 2$ would not give Gauss's law. Obviously, we can also take Gauss's law as the fundamental law of electrostatics.

3.11 APPLICATIONS OF GAUSS'S LAW

This law provides a powerful method for evaluating the electric field intensity E and potential variation due to simple charge distribution whose *symmetry* is such that the field is constant and normal over the entire Gaussian surface. We shall discuss some useful applications of Gauss's law in this section. We must remember that *symmetry* is crucial to the application of Gauss's law. There are only three kinds of symmetry which are sufficient:

1. *Spherical symmetry:* One can make Gaussian surface a concentric sphere.
2. *Cylindrical symmetry:* One can make Gaussian surface a coaxial cylinder (Fig. 3.17).
3. *Plane symmetry:* One can make use of Gaussian 'pillbox', which straddles the surface (Fig. 3.18).

Fig. 3.17: Infinite cylinder of charge

Fig. 3.18: Infinite plane sheet of charge

Field Due to an Infinite Layer of Positive Charge with Uniform Surface Density (σ)

Let us consider that $ABCD$ be a plane as shown in Fig. 3.19. Symmetry leads that the field lines are perpendicular to the plane. Now, we consider a right circular cylinder PQR with cross-section ds. One obtains from Gauss's law,

$$(E_1 - E_2)ds = \frac{\sigma ds}{\varepsilon_0}$$

$$\therefore \qquad E_1 - E_2 = \frac{\sigma}{\varepsilon_0} \qquad (3.26)$$

Obviously, the field changes across a charge

layer, and the charge is equal to $\dfrac{\sigma}{\varepsilon_0}$.

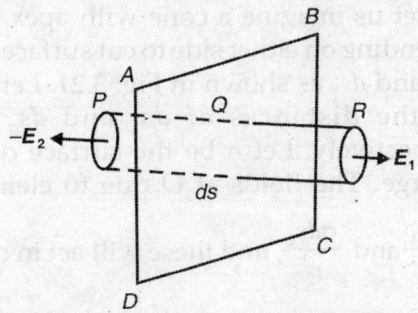

Fig. 3.19: Field due to infinite layer of charge

Field Outside an Isolated Charged Sphere

Let us suppose that a sphere A be filled with a uniform charge distribution (Fig. 3.20). We are interested in finding the electric field at a point P outside the sphere.

Construct a spherical surface B concentric with the charged sphere A and passing through the point P. Let r be the radius of the imaginary sphere B, then its surface area is $4\pi r^2$. By symmetry, we can say that the electric field intensity E is same at every point of the surface. Obviously, the flux outwards through the surface is

Fig. 3.20: Point charge surrounded by spherical Gaussian surface

$$\oint_S E \cdot \hat{a}_n ds = 4\pi r^2 E = \frac{Q}{\varepsilon_0} \text{ (Guass's law)}$$

where, Q is the total charge within the sphere. Thus,

$$E = \frac{1}{4\pi\varepsilon_0}\frac{Q}{r^2} \qquad (3.27)$$

Equation (3.27) is same as the field produced by the point charge Q at the centre of the sphere. If there is continuous charge distribution within the sphere, we have

$$Q = \frac{4}{3}\pi a^3 \rho$$

where, ρ is the charge density and a the radius of the sphere. Thus,

$$E = \frac{1}{4\pi\varepsilon_0}\frac{4\pi a^3 \rho}{3r^2} = \frac{a^3 \rho}{3\varepsilon_0 r^2} \qquad (3.28)$$

Electric Field due to a Spherical Shell of Charge

Following the arguments applied in the case of a field outside an isolated charged sphere, one can see that the field outside a thin spherical shell of charge is same as if the total charge on the shell is concentrated at the centre. Now, we are interested in finding out the field at a point O inside the shell.

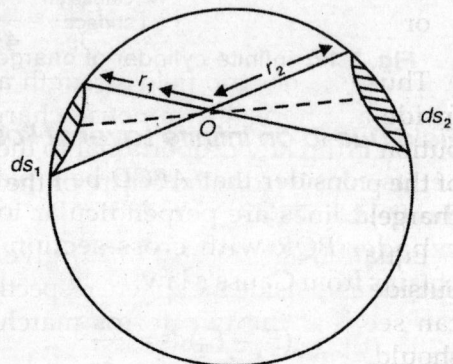

Let us imagine a cone with apex at O and extending on either side to cut surface elements ds_1 and ds_2 as shown in Fig. 3.21. Let r_1 and r_2 be the distances of ds_1 and ds_2 from O respectively. Let σ be the surface density of charge. The fields at O due to elements are

Fig. 3.21: Electric field due to spherical shell of charge

$\dfrac{\sigma ds_1}{r_1^2}$ and $\dfrac{\sigma ds_2}{r_2^2}$, and these will act in opposite directions. However,

$$\frac{ds_1}{r_1^2} = \frac{ds_2}{r_2^2} = d\Omega$$

($d\Omega$ is the solid angle subtended at O by ds_1 and ds_2).

Thus, the contribution to the electric field due to the two elements being equal and opposite cancels completely. Extending this to cover the entire shell by balancing off field contributions of opposite differential areas, one can see that each pair of differential areas gives a zero contribution. Thus, net field at O is zero.

Electric Field Strength at an Internal Point

Let P be an internal point at a distance r from the centre of the charge distribution O as shown in Fig. 3.22. Imagine a sphere of radius $r(= OP)$ concentric with the spherical charge. Let ρ be the volume charge density (i.e. charge per unit volume),

$$\rho = \frac{\text{Charge}}{\text{Volume}} = \frac{Q}{\frac{4}{3}R^3}$$

We must note that the charge in the shell of thickness $(R - r)$ does not contribute to the field

Fig. 3.22: Electric field due to a uniformly charged sphere at P since the point P lies inside the shell. If E is the field at P, the flux through the imaginary or Gaussian surface is

$$\frac{ds_1}{r_1^2} = \frac{ds_2}{r_2^2} = d\Omega$$

$$4\pi r^2 E = \frac{1}{\varepsilon_0} \frac{4}{3} \pi r^3 \rho \quad \text{(Gauss's law)}$$

or
$$E = \frac{r\rho}{3\varepsilon_0} = \frac{Qr}{4\pi\varepsilon_0 R^3} \qquad (3.29)$$

Thus, the electric field strength at point P inside a spherical symmetric charge distribution is directly proportional to the distance of the point P from the centre of the spherical charge, i.e. $E \propto r$.

Equations (3.28) and (3.29) give the field outside and inside the sphere respectively. One can see that the two forms match, as they should, when $r = a$.

The variation of magnitude of electric field strength with the distance (r) from the centre of the spherical charge distribution is represented by the curve shown in Fig. 3.23.

Fig. 3.23: Variation of E with r due to a uniformly charged sphere

Electric Field Strength due to Uniformly Charged Infinite Cylinder

Let us consider a charged cylinder of infinite length of radius r as shown in Fig. 3.24. We are interested in finding the field at a point distant d from the axis. For this, let us imagine a closed cylindrical surface coaxial with the cylinder and passing through P, its two end faces being perpendicular to the axis of the cylinder. One can easily see by symmetry that the field will be normal to the axis directed away from the axis and is the same at equal distances from the axis. If l is the height of the cylindrical Gaussian surface, the flux through this surface is $4\pi lE$.

Fig. 3.24: Uniformly charged infinite cylinder

must remember that the end faces do not contribute to the field since E is tangential to these faces. Applying Gauss's law, we have

$$2\pi l d E = \frac{\lambda l}{\varepsilon_0}$$

where, λ is the charge per unit length. Thus,

$$E = \frac{\lambda}{2\pi\varepsilon_0 d} \qquad (3.30)$$

Obviously, the electric field strength at a distance d from an infinite cylinder uniformly charged with linear charge density λ has magnitude $\left(\dfrac{\lambda}{2\pi\varepsilon_0}\right)$ which is inversely proportional to d. We can further see that the result is independent of the radius of charged cylinder and holds also for the rectilinear distribution of charges.

One can obtain the above result by direct integration. Let us consider an element of finite length of constant line density of charge λ C/m as shown in Fig. 3.25. Now, the field at a point P due to a line element dz is

$$dE_d = \frac{\lambda\,dz\cos\theta}{4\pi\varepsilon_0\,r^2}$$

Here we have assumed that the element coincides with the z-axis. The distance of the point P from the element dz is r and the perpendicular distance from the z-axis is d. One obtains the total field as

Fig. 3.25: Element of finite length with constant line density of charge

$$E_d = \frac{\lambda}{4\pi\varepsilon_0}\int_{-\infty}^{\infty}\frac{\cos\theta}{r^2}dz$$

We have $z = d\tan\theta$ and $r = d\sec\theta$

$$\therefore \qquad E_d = \frac{\lambda}{4\pi\varepsilon_0}\int_{-\pi/2}^{\pi/2}\frac{\cos\theta}{d}d\theta = \frac{\lambda}{2\pi\varepsilon_0 d}$$

Electrostatic Field between Two Concentric Spheres with Equal and Opposite Charges

Figure 3.26 exhibits two concentric spheres 1 and 2 which have equal and opposite charges. Imagine a Gaussian surface drawn through P. Obviously, the field at P due to the outer sphere is zero and that due to the inner

sphere is $E = \dfrac{1}{4\pi\varepsilon_0}\dfrac{q}{r^2}$.

Let us now consider the case of two non-concentric charged spheres as shown in Fig. 3.27. The figure shows the redistribution of charge and field lines. Draw a cone with P as the apex. We can see that the two charge densities on ds_1 and ds_2 are not equal, i.e. $dE_2 > dE_1$, and hence the two different areas yield a net dE in the negative direction. One can find a net E_- due to the outer sphere in the negative x-direction by covering the entire sphere. The field due to the inner sphere at P is E_+. Thus, the effective field at P is $E = E_+ + E_-$.

Fig. 3.26: Electrostatic field between two concentric spheres

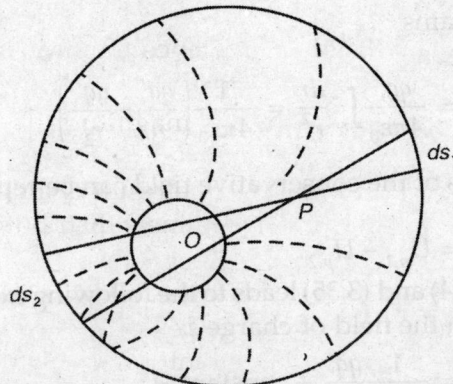

Fig. 3.27: Charge distribution and field lines for two non-concentric charged spheres

3.12 ELECTROSTATIC POTENTIAL

A charged particle placed in an electric field has potential energy because of its interaction with the field. One can define the electric potential at a point as the potential energy per unit charge placed at that point. The electric potential is usually designated by V and the potential energy of a charge q by U_p, we have

$$V = \frac{U_p}{q} \quad \text{or} \quad U_p = qV \tag{3.31}$$

The electric potential is measured in joule/coulomb or $J\,C^{-1}$, a unit called volt, abbreviated as V. In terms of international units $V = m^2\,kg\,s^{-2}\,C^{-1}$.

Let us consider the field produced by a stationary point charge q. At any point in this field, the point charge q' experiences the force,

$$F = \frac{1}{4\pi\varepsilon_0}\frac{qq'}{r^2}\hat{a}_r = F(r)\hat{a}_r \tag{3.32}$$

Here $F(r)$ is the magnitude of the force F and $\hat{a}_r$ is the unit vector in the direction of the position vector r determining the position of the charge q' relative to the charge q.

The force given by Eq. (3.32) is a central one. A central field of forces is conservative. Consequently, the work done by the forces of the field on the test charge q' when it is moved from one point to another does not depend on the path. This work is

$$W_{12} = \int_1^2 F(r)\hat{a}_r \cdot dl \tag{3.33}$$

where, dl is the elementary displacement of the charge q'. Figure 3.28 shows that the scalar product $\hat{a}_r \cdot dl$ equals the increment of the magnitude of the position vector r, i.e. dr. Equation (3.33) can therefore be written in the form

$$W_{12} = \int_1^2 F(R)dr$$

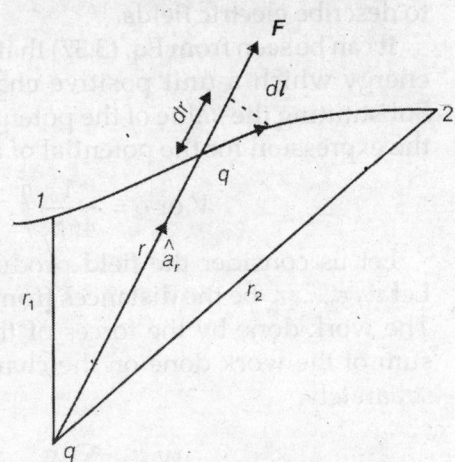

Fig. 3.28: Work done on a test charge

Using Eq. (3.32), one obtains

$$W_{12} = \frac{qq'}{4\pi\varepsilon_0} \int_{r_1}^{r_2} \frac{dr}{r^2} = \frac{1}{4\pi\varepsilon_0} \left(\frac{qq'}{r_1} - \frac{qq'}{r_2} \right) \tag{3.34}$$

The work of the forces of the conservative field can be represented as a decrement of the potential energy

$$W_{12} = U_{p,1} - U_{p,2} \tag{3.35}$$

A comparison of Eqs. (3.34) and (3.35) leads to the following expression for the potential energy of the charge q' in the field of charge q:

$$U_p = \frac{1}{4\pi\varepsilon_0} \frac{qq'}{r} + \text{constant} \tag{3.35a}$$

The value of the constant in the expression for the potential energy is usually chosen so that when the charge moves away to infinity (i.e. when $r = \infty$), the potential energy vanishes. When this condition is observed, one obtains

$$U_p = \frac{1}{4\pi\varepsilon_0} \frac{qq'}{r} \tag{3.36}$$

Let us use the charge q' as a test charge for studying the field. From Eq. (3.36), the potential energy which the test charge has depends not only on its magnitude q', but also on the quantities q and r determining the field. Obviously, one can use this energy to describe the field just like we used the force acting on the test charge for this purpose.

Different test charges q'_t, q''_t, etc. will have different energies U'_p, U''_p, etc. at the same point in the field. But the ratio $\dfrac{U_p}{q_t}$ will be same for all the charges [Eq. (3.36)]. The quantity

$$V \text{ or } \phi = \frac{U_p}{q_t} \tag{3.37}$$

is called the *field potential* at a given point and used together with the field strength E to describe electric fields.

It can be seen from Eq. (3.37) that the field potential numerically equals the potential energy which a unit positive charge would have at the given point in the field. Substituting the value of the potential energy in Eq. (3.37) from Eq. (3.36), one obtains the expression for the potential of a point charge

$$V \text{ or } \phi = \frac{1}{4\pi\varepsilon_0} \frac{q}{r} \tag{3.38}$$

Let us consider the field produced by a system of N point charges $q_1, q_2, ..., q_N$. Let $r_1, r_2, ..., r_N$ be the distances from each of the charges to the given point in the field. The work done by the forces of this field on the charge q' will equal the algebraic sum of the work done on the charge q' by the forces set up by each of the charges separately.

$$W_{12} = \sum_{i=1}^{N} W_i$$

From Eq. (3.34), each work W_i is given by

$$W_i = \frac{1}{4\pi\varepsilon_0}\left(\frac{q_iq'}{r_{i,1}} - \frac{q_iq'}{r_{i,2}}\right)$$

where, $r_{i,1}$ is the distance from the charge q_i to the initial position of the charge q' and $r_{i,2}$ is the distance from q_i to the final position of the charge q'. Thus,

$$W_{12} = \frac{1}{4\pi\varepsilon_0}\sum_{i=1}^{N}\frac{q_iq'}{r_{i,1}} - \frac{1}{4\pi\varepsilon_0}\sum_{i=1}^{N}\frac{q_iq'}{r_{i,2}}$$

Comparing this equation with Eq.(3.35), one obtains the following expression for the potential energy of the charge q' in the field of a system of charges:

$$U_p = \frac{1}{4\pi\varepsilon_0}\sum_{i=1}^{N}\frac{q_iq'}{r_i}$$

from which it can be seen that

$$V \text{ or } \phi = \frac{1}{4\pi\varepsilon_0}\sum_{i=1}^{N}\frac{q_i}{r_i} \tag{3.39}$$

In general, for a set of point charges q_j the potential at a point r_i is given by the algebraic sum of the individual potentials, i.e.

$$V \text{ or } \phi = \frac{1}{4\pi\varepsilon_0}\sum_{j}\frac{q_j}{r_{ij}} \tag{3.39a}$$

In case of a continuous distribution of charges,

$$V(r_i) = \frac{1}{4\pi\varepsilon_0}\int\frac{\rho d\tau_j}{r_{ij}} \tag{3.39b}$$

Comparing Eq. (3.39) with Eq. (3.38), we arrive at the conclusion that the *potential of the field produced by a system of charges equals the algebraic sum of the potentials produced by each of the charges separately.* Whereas the field strengths are added vectorially in the superposition of fields, the potentials are added algebraically. This is why it is usually much simpler to calculate the potentials than the electric field strengths.

Examination of Eq. (3.37) shows that the charge q at a point of the field with the potential V or ϕ has the potential energy

$$U_p = qV = q\phi \tag{3.40}$$

Hence, the work of the field forces on the charge q can be expressed through the potential difference as

$$W_{12} = V_{p,1} - V_{p,2} = q(V_1 - V_2) = q(\phi_1 - \phi_2) \tag{3.41}$$

Obviously, the work done on a charge by the forces of a field equals the product of the magnitude of the charge and the difference between potentials at the initial and final points (i.e. the potential decrement).

If the charge q is removed from a point having the potential V or ϕ to infinity (where by convention the potential vanishes), then the work of the field forces will be

$$W_\infty = q\phi = qV \tag{3.42}$$

Obviously, *the potential numerically equals the work done by the forces of the field on a unit positive charge when the latter is moved from the given point to infinity*. Work of the same magnitude must be done against the electric field forces to move a unit positive charge from infinity to the given point of a field.

3.13 RELATION BETWEEN ELECTRIC FIELD STRENGTH AND POTENTIAL

An electric field can be described either with the aid of the vector quantity E, or with the aid of the scalar quantity V or ϕ. There must evidently be a definite relation between these quantities. If we bear in mind that E is proportional to the force acting on a charge and ϕ to the potential energy of the charge, it is easy to see that this relation must be similar to that between the potential energy and the force.

The force F is related to the potential energy by the expression

$$F = -\nabla U_p \tag{3.43}$$

For a charged particle in an electrostatic field, we have

$$F = qE$$

and

$$U_p = q\phi = qV$$

Introducing these values in Eq. (3.43), one obtains

$$qE = -\nabla(q\phi) = -q\nabla(\phi)$$

or

$$E = -\nabla\phi = -\operatorname{grad}\phi \tag{3.44}$$

This is the relation between the field strength and potential. Since E has a finite value at any point of the field, the potential ϕ is a continuous function of the coordinates of the point of the field. The rectangular components of the electric field E are given by

$$E_x = -\frac{\partial\phi}{\partial x}, \; E_y = -\frac{\partial\phi}{\partial y} \; \text{ and } \; E_z = -\frac{\partial\phi}{\partial z} \tag{3.45}$$

Thus

$$E = -\left[\hat{i}\frac{\partial}{\partial x} + \hat{j}\frac{\partial}{\partial y} + \hat{k}\frac{\partial}{\partial z}\right]\phi$$

In general, the component along the direction corresponding to displacement ds is

$$E_s = -\frac{\partial\phi}{\partial s} \tag{3.46}$$

Equations (3.45) and (3.46) are used to find the electric potential ϕ or V when the field E is known, and conversely.

Let us consider a simple case of a uniform electric field (Fig. 3.29).

The first of Eq. (3.45) gives, for x-axis which is parallel to the field, $E = -\dfrac{d\phi}{dx}$. Since E is constant, and we assume $\phi = 0$ at $x = 0$, by integration, one obtains

$$\int_0^\phi d\phi = -\int_0^x E dx = -E\int_0^x dx \text{ or } \phi = -Ex \tag{3.47}$$

Equation (3.47) is very useful relation represented graphically in Fig. 3.30.

Fig. 3.29: Uniform electric field

Fig. 3.30: Variation of E and ϕ for a uniform electric field

We may note that because of the negative sign in Eq.(3.46) or Eq. (3.47), the electric field points in the direction in which the electric potential decreases. When we consider two points x_1 and x_2, Eq. (3.47) gives $\phi_1 = -Ex_1$ and $\phi_2 = -Ex_2$. On subtracting, we have $\phi_2 - \phi_1 = -E(x_2 - x_1)$, or calling $d = x_2 - x_1$, we obtain

$$E = -\frac{\phi_2 - \phi_1}{d} = \frac{\phi_1 - \phi_2}{d} \qquad (3.48)$$

Although Eq.(3.48) is valid only for uniform electric fields, it can be used to estimate the electric field between two points separated by a distance d, if the potential difference $\phi_1 - \phi_2$ between them is known. If the potential difference $\phi_1 - \phi_2$ is positive, the field points in the direction from x_1 to x_2, and if it is negative it points in the opposite direction. Equation (3.48) [Eq. (3.45) or Eq. (3.46) also] indicates that the electric field can also be expressed in volt/metre, a unit which is equivalent to newton/coulomb given before. One can easily verify this in the following way:

$$\frac{\text{volt}}{\text{metre}} = \frac{\text{joule}}{\text{coulomb} \cdot \text{m}} = \frac{\text{newton} \cdot \text{m}}{\text{coulomb} \cdot \text{m}} = \frac{\text{newton}}{\text{coulomb}} = \text{NC}^{-1}$$

For common usage, the term volt/metre, abbreviated as Vm^{-1} is preferred to NC^{-1}.

We have seen that Eq. (3.44) allows us to find the field strength at every point from the known values of ϕ. One can also solve the reverse problem, i.e. find the potential difference between two arbitrary points of a field according to the given values of E. For this purpose, one can take advantage of the circumstance that the work done by the forces of a field on the charge q when it is moved from point 1 to point 2 can be calculated as (Fig. 3.28)

$$W_{12} = \int_1^2 qE\,dl$$

At the same time in accordance with Eq. (3.41), this work can be written as

$$W_{12} = q(\phi_1 - \phi_2)$$

Equating these two expressions, one obtains

$$\phi_1 - \phi_2 = \int_1^2 E \cdot dl \tag{3.49}$$

The integral can be taken along any line joining points 1 and 2 because the work of the field forces is independent of the path for circumven-tion along closed contour $\phi_1 = \phi_2$, and Eq. (3.49) becomes

$$\oint E \cdot dl = 0 \tag{3.50}$$

The circle on the integral sign indicates that integration is performed over a closed contour. In vector calculus, the line integral of a vector round a closed curve is called the *circulation* or the *curl of the vector field*. Equation (3.50) indicates that the circulation of the electric field is zero. Its sources are given by Eq. (3.25). Other examples of such fields are the gravitational field, the magnetic field in region free from electric current and the velocity field of the flow of incompressible fluids free from viscosity. Such fields are termed *irrotational fields*. We must note that Eq. (3.50) is not applicable to the field of moving charges (i.e. a field changing with time), i.e. this field is not a potential one.

Now, we take curl of vectors in Eq. (3.44) and obtain

$$\nabla \times E = -\nabla \times \nabla\phi = 0 \tag{3.51}$$

Using Stokes' theorem, one can transform line integral in Eq. (3.50) to a surface integral as

$$\oint E \cdot dl = \int_S \nabla \times E \cdot \hat{a}_n dS$$

$$\therefore \qquad \int_S \nabla \times E \cdot \hat{a}_n dS = 0$$

Since S is arbitrary, $\nabla \times E = 0$. Thus, one obtains the following two important laws of electrostatics:

i. $\nabla \cdot E = \dfrac{\rho}{\varepsilon_0}$ (3.52a)

ii. $\nabla \times E = 0$ (3.52b)

Equation (3.52a) follows from Coulomb's inverse square law. Equation (3.52b) does not depend upon this law. All predictions of electrostatics follow from these laws and hence Eqs. (3.52a) and (3.52b) can be considered as fundamental laws of electrostatics.

3.14 EQUIPOTENTIAL SURFACES

A locus of the points in an electrostatic field having the same potential is called an equipotential surface. Its equation has the form

$$\phi(r) = c$$

where, c is a constant which represents an equipotential surface. Different values of c generate a family of such surfaces.

The potential does not change in movement along an equipotential surface over the distance $dr (d\phi = 0)$. Hence, according to Eq. (3.46), the tangential component of E to the surface equals zero. Let us consider a displacement dr on an equipotential surface, then

$$\phi(r) - \phi(r + dr) = 0$$

i.e. $$\phi(r) - \left\{ \phi(r) + \frac{\partial\phi}{\partial r} \cdot dr \right\} = 0$$

$$\therefore \qquad \frac{\partial\phi}{\partial r} \cdot dr = 0$$

i.e. $$\nabla\phi \cdot dr = 0$$

or $$E \cdot dr = 0 \qquad\qquad (3.53)$$

This shows that if dr is on the surface, E must be normal to the surface, i.e. the field lines are normal or orthogonal to the equipotential surface.

An equipotential surface can be drawn through any point of a field, consequently one can construct an infinitely great number of such surfaces. They are conventionally drawn so that the potential difference for two adjacent surfaces is the same everywhere. Thus, the density of the equipotential surfaces allows us to assess the magnitude of the field strength. Indeed, the denser are the equipotential surfaces, the more rapidly does the potential changes when moving along a normal to the surface. Hence, $\nabla\phi$ is at the given place, and therefore E is greater too.

Figure 3.7 shows equipotential surfaces (more exactly, their intersections with the plane of the drawing) for the field of a point charge. Figures 3.8 and 3.9 show the equipotential surfaces along with the lines of force for two equal but opposite charges and two identical charges respectively. In accordance with the nature of the dependence of E on r, equipotential surfaces become denser, the nearer we approach a charge.

Equipotential surfaces for a homogeneous field are a collection of equispaced planes at right angles to the direction of field.

3.15 ELECTROSTATIC ENERGY

The energy of electrostatic systems is solely the potential energy arising from the interaction between the charges.

Let us consider a charge q_1 be situated at a certain point in space and charge q_2 which was initially at infinity, be brought up to a distance r_{12} from q_1. We have the potential of the charge q_1 at r_{12} as $\dfrac{q_1}{4\pi\varepsilon_0 r_{12}}$. Obviously, this is the work done in bringing a unit charge from infinity to the point where the charge q_2 is situated and specified by r_{12}, and given as $\dfrac{q_1 q_2}{4\pi\varepsilon_0 r_{12}}$. Now a third charge q_3 is added to the system. We will have to perform work against the field of q_1 and q_2. If r_{13}, r_{23} are the distances of q_3

from q_1 and q_2 respectively, then the additional contribution to the potential energy is as follows:

$$\frac{q_1 q_3}{4\pi\varepsilon_0 r_{13}} + \frac{q_2 q_3}{4\pi\varepsilon_0 r_{23}}$$

Proceeding to build up the assembly in this way, one finds the total energy of the assembly is

$$W = \frac{q_2}{4\pi\varepsilon_0}\left(\frac{q_1}{r_{12}}\right) + \frac{q_3}{4\pi\varepsilon_0}\left(\frac{q_1}{r_{13}} + \frac{q_2}{r_{23}}\right) + \frac{q_4}{4\pi\varepsilon_0}\left(\frac{q_1}{r_{14}} + \frac{q_2}{r_{24}} + \frac{q_3}{r_{34}}\right) + \cdots$$

$$= \frac{1}{4\pi\varepsilon_0}\sum_i q_i \sum_{j<i}\frac{q_i}{r_{ij}} \tag{3.54}$$

The restriction $j < i$ ensures that the interaction between every pair is counted only once. One can write Eq. (3.54) in a different form as follows:

$$W = \frac{1}{4\pi\varepsilon_0}\frac{1}{2}q_1\left(\frac{q_2}{r_{12}} + \frac{q_3}{r_{13}} + \cdots\right) + \frac{1}{4\pi\varepsilon_0}\frac{1}{2}q_2\left(\frac{q_1}{r_{21}} + \frac{q_3}{r_{23}}\cdots\right)$$

$$+ \frac{1}{4\pi\varepsilon_0}\frac{1}{2}q_3\left(\frac{q_1}{r_{31}} + \frac{q_2}{r_{32}} + \frac{q_4}{r_{34}} + \cdots\right) + \cdots$$

$$= \frac{1}{4\pi\varepsilon_0}\frac{1}{2}\sum_{i=1} q_i \sum_{j=1}\frac{q_j}{r_{ij}}, \ i \neq j \tag{3.55}$$

The factor $\dfrac{1}{2}$ appears in Eq. (3.55) because in this expression each pair is counted twice.

Since $\dfrac{1}{4\pi\varepsilon_0}\displaystyle\sum_{j\neq i}\frac{q_j}{r_{ij}}$ is the potential ϕ_i produced by all the charges of the system except i^{th} one at the point at which q_i is situated, we have

$$W = \frac{1}{2}\sum_i q_i \phi_i \tag{3.56}$$

When there is a continuous distribution of charges and charges are not localized, we have

$$W = \frac{1}{2}\int \phi_\rho d\tau \tag{3.57}$$

3.16 ELECTRIC DIPOLE

An electric dipole is defined as a system of two point charges $+q$ and $-q$ identical in value and opposite in sign, the distance between which is much smaller than that to the points at which the field of the system is being determined. The straight line passing through both the charges is called the *dipole axis* (Fig. 3.31). There are several familiar examples of such systems in physics. For example:

  i. When we place an atom or a molecule in an electric field, the positive and negative charges feel opposite forces and are displaced slightly forming a dipole (Fig. 3.32).

Fig. 3.31: An electric dipole   **Fig. 3.32:** An atom in electric field

ii. The structure of some molecules, e.g. H_2O is such that even though there is no external field, the positive charge is slightly separated from the negative charge (Fig. 3.33). One can easily see that though this structure is not equivalent to the one we defined as dipole, however, one can treat them as such when one is considering a field they produce at a very long distance.

iii. During contraction the *heart muscle* becomes essentially a *dipole*, and physicians performing electrocardiography are measuring among other things, the strength and orientation of that dipole. *Technological devices*, including radios and TV antennas also make use of the dipole configuration.

Fig. 3.33: Structure of water molecule

Let us first calculate the potential and then the field strength of a dipole. Suppose the two point charges $+Q$ and $-Q$ are placed on the z-axis, one at a distance $\dfrac{d}{2}$ and other at $-\dfrac{d}{2}$ from the origin O.

When the point P at which the magnitude of the potential and field due to dipole are to be determined is close to either of the charges, the potential is almost the same as if the charges were independent of one another and equipotential surfaces are nearly spherical. If the point $P(r, \theta, \phi)$ is at a distance r from the origin and $r \gg d$, then the potential of the positive charge at P is

$$\phi_1(r_1) = \frac{1}{4\pi\varepsilon_0}\frac{Q}{r_1}$$

The potential of the negative charge at P is

$$\phi_2(r_2) = \frac{1}{4\pi\varepsilon_0}\left(\frac{-Q}{r_2}\right)$$

Then, the total potential ϕ at P is

$$\phi(r) = \phi_1(r_1) + \phi(r_2) = \frac{Q}{4\pi\varepsilon_0}\left(\frac{1}{r_1} - \frac{1}{r_2}\right) \tag{3.58}$$

where, r_1 and r_2 are the distances of P from $+Q$ and $-Q$ respectively.

From Fig. 3.31, we have

$$r_1^2 = r^2 + \left(\frac{d}{2}\right)^2 - 2\left(\frac{d}{2}\right)r\cos\theta$$

where, θ is the angle between the z-axis and the position vector of P.

$$\therefore \qquad r_1^2 = r^2\left[1 + \frac{d^2}{4r^2} - \frac{d\cos\theta}{r}\right]$$

$$\therefore \qquad \frac{1}{r_1} = \frac{1}{r}\left[1 + \frac{d^2}{4r^2} - \frac{d\cos\theta}{r}\right]^{-1/2} = \frac{1}{r}\left[1 + \frac{d\cos\theta}{2r}\cdots\right] \approx \frac{1}{r} + \frac{d\cos\theta}{2r^2}$$

We have neglected terms in d^2 and higher order terms. The said approximation is justified if we let $\left(\frac{d}{r}\right) \to 0$ keeping the product qd finite. Similarly, one can show

$$\frac{1}{r_2} \approx \frac{1}{r} - \frac{d\cos\theta}{2r^2}$$

Substituting the values of $\frac{1}{r_1}$ and $\frac{1}{r_2}$ in Eq. (3.58), one obtains

$$\phi(r) = \frac{Q}{4\pi\varepsilon_0}\frac{d\cos\theta}{r^2} \qquad (3.59)$$

It is interesting to note that:

i. the potential due to a dipole falls off as the square of the distance r, whereas the potential from a single point charge varies only as $\frac{1}{r}$. This comes from the fact that the charges in a dipole appear close together for an observer at some distance away and that their fields cancel more and more as the distance r is increased.

ii. the potential of the dipole is also a function of the angle θ and this suggests that at a fixed radius r, the potential is maximum on the axis of the dipole ($\theta = 0$) and is zero on normal to the axis ($\theta = 90°$). One can easily see that when $\theta = 90°$, the point P is exactly equidistant from the two charges so that their effects cancel.

The vector along the axis of the dipole pointing in the direction $-Q$ to $+Q$ and having magnitude Qd is termed the *dipole moment* and represented by p. Since

$$p \cdot r = pr\cos\theta = Qd(r\cos\theta)$$

$$\therefore \qquad \phi(r) = \frac{p\cos\theta}{4\pi\varepsilon_0 r^2} = \frac{p \cdot r}{4\pi\varepsilon_0 |r|^3} \qquad (3.60)$$

One can also express dipole potential as a gradient. Since

$$\nabla\left(\frac{1}{r}\right) = -\frac{r}{|r|^3}$$

Therefore $\qquad \phi(r) = -\frac{p}{4\pi\varepsilon_0} \cdot \nabla\left(\frac{1}{r}\right) = -p \cdot \nabla(\phi_0) \qquad (3.61)$

where, $\phi_0 = \frac{1}{4\pi\varepsilon_0 r}$ is the potential of a unit charge.

Knowing potential, one can obtain electrostatic field by taking the gradient of ϕ. Thus, for an electric dipole oriented along the z-axis (Fig. 3.31), one obtains the three components of the field strength of a dipole as

$$E_z = -\frac{\partial \phi}{\partial z} = \frac{p}{4\pi\varepsilon_0}\frac{\partial}{\partial z}\left(\frac{r\cos\theta}{r^3}\right) = -\frac{p}{4\pi\varepsilon_0}\frac{\partial}{\partial z}\left(\frac{z}{r^3}\right)$$

$$= -\frac{p}{4\pi\varepsilon_0}\left(\frac{1}{r^3} - \frac{3z^2}{r^5}\right) = \frac{p}{4\pi\varepsilon_0 r^3}\left(\frac{3z^2}{r^2} - 1\right)$$

$$= \frac{p}{4\pi\varepsilon_0 r^3}(3\cos^2\theta - 1)$$

$$E_x = \frac{p}{4\pi\varepsilon_0}\frac{3zx}{r^5} \quad \text{and} \quad E_y = \frac{p}{4\pi\varepsilon_0}\frac{3xy}{r^5} \tag{3.62}$$

It is sometimes convenient to express the field of a dipole in spherical polar coordinates. We have

$$E = -\left[\hat{a}_r\frac{\partial \phi}{\partial r} + \hat{a}_\theta\frac{\partial \phi}{\partial \theta} + \hat{a}_\phi\frac{1}{r\sin\theta}\frac{\partial \phi}{\partial \phi}\right]$$

Using Eq. (3.60), one obtains

$$E_r = -\frac{\partial \phi}{\partial r} = \frac{2p\cos\theta}{4\pi\varepsilon_0 r^3} \tag{3.63}$$

$$E_\theta = -\frac{1}{r}\frac{\partial \phi}{\partial \theta} = \frac{p}{4\pi\varepsilon_0}\frac{\sin\theta}{r^3} \tag{3.64}$$

and

$$E_\phi = -\frac{1}{r\sin\theta}\frac{\partial \phi}{\partial \phi} = 0 \tag{3.65}$$

$$\therefore E = \hat{a}_r\frac{2Qd\cos\theta}{4\pi\varepsilon_0 r^3} + \hat{a}_\theta\frac{Qd\sin\theta}{r^3} \tag{3.66}$$

The two components E_r and E_θ of the field of an electric dipole are shown in Fig. 3.34. The lines of force are indicated in Fig. 3.35. Although in an electric dipole the two charges are equal and opposite, giving a zero net charge, the fact that they are slightly displaced is enough to produce a non-vanishing electric field.

One can also express the field of an electric dipole in the following way:

$$E = -\nabla\phi = -\frac{1}{4\pi\varepsilon_0}\nabla\left(\frac{p\cdot r}{|r|^3}\right)$$

$$= -\frac{1}{4\pi\varepsilon_0}\left[\frac{1}{|r|^3}\nabla(p\cdot r) + p\cdot r\nabla\left(\frac{1}{|r|^3}\right)\right]$$

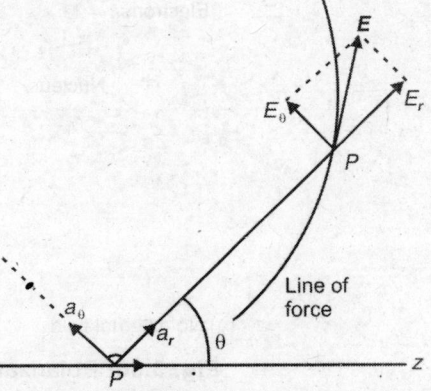

Fig. 3.34: Field of a dipole

Fig. 3.35: Lines of force of the field of an electric dipole

Since $\nabla(p \cdot r) = p$ and $\nabla\left(\dfrac{1}{|r|^3}\right) = -\dfrac{3r}{|r|^5}$

$$\therefore \quad E = \frac{1}{4\pi\varepsilon_0}\left[\frac{3(p \cdot r)r}{|r|^5} - \frac{p}{|r|^3}\right] = \frac{1}{4\pi\varepsilon_0 r^3}\left[\frac{3(p \cdot r)r}{r^2} - p\right] \quad (3.67)$$

In atoms, the center of mass of the electrons coincides with the nu-cleus, and therefore the average electric dipole moment of the atom is zero (Fig. 3.36a). But if an external field is applied, the electronic motion is distorted, and the center of mass of the electrons is displaced through a distance x relative to the nucleus (Fig. 3.36b). The atom is thus polarized and becomes an electric dipole of moment p. This moment is proportional to the external field E.

(a) No external field   (b) External field

Fig. 3.36: Polarization of an atom under an external field

Molecules, on the other hand, may have a permanent electric dipole moment. Such molecules are called *polar molecules*. For example, in HCl molecule (Fig 3.37), the electron of the H atom spends more time moving around the Cl atom than around the H atom. Therefore, the center of the negative charges does not coincide with that of the positive charges, and the molecule has a dipole moment directed from the Cl atom to the H atom, that is, one may write $H^+ Cl^-$. The electric dipole moment of the HCl molecule is $p = 3.43 \times 10^{-30}$ Cm. In the CO molecule, the charge distribution is only slightly asymmetric and the electric dipole moment is relatively small, about 0.4×10^{-30} Cm, with the carbon atom corresponding to the positive and the oxygen atom to the negative end of the molecule.

Fig. 3.37: Polar diatomic molecules

In the H_2O molecule, where the two H–O bonds are at an angle slightly over 90° (Fig. 3.38), the electrons try to crowd around the oxygen atom, which thereupon

Fig. 3.38: Electric dipole of H_2O molecule

Fig. 3.39: The CO_2 molecule has no electric dipole

becomes slightly negative relative to the H atoms. Each H–O bond thus contributes to the electric dipole moment, whose resultant, because of symmetry, lies along the axis of the molecule and has a value equal to 6.2×10^{-30} C m. But in CO_2 molecule, the atoms are in a straight line (Fig. 3.39), and the resultant electric dipole moment is zero because of the symmetry. Obviously, the electric dipole moments can provide useful information about the structure of molecules. The values of p for several polar molecules are given in Table 3.1. We must note that dipole moments of CO_2, H_2, CH_4, C_2H_6 and CCl_4 molecules are zero.

Table 3.1: Electric dipole moments for selected polar molecules

Molecule	P (C m)
HCl	3.43×10^{-30}
HBr	2.60×10^{-30}
HI	1.26×10^{-30}
CO	0.40×10^{-30}
H_2O	6.2×10^{-30}
H_2S	5.3×10^{-30}
SO_2	5.3×10^{-30}
NH_3	5.0×10^{-30}
C_2H_5OH	3.66×10^{-30}

3.17 ELECTRIC DIPOLE IN A UNIFORM ELECTRIC FIELD

When an electric dipole consisting of two charges $-q$ and $+q$ distance d apart is placed in a uniform electric field (Fig. 3.40), a force is produced on each charge of the dipole. The resultant force is

$$F = qE - qE' = q(E - E')$$

Consider that electric field is along the x-axis and the dipole is oriented parallel to the field. Now, considering

magnitudes only, $E - E' = \dfrac{dE}{dx}d$, and therefore $F = P\dfrac{dE}{dx}$.

This shows that *an electric dipole oriented parallel to the electric field tends to move in the direction in which the field increases*. One obtains the opposite result if the dipole is oriented antiparalled to the field. We must note that if the electric field is uniform, the resultant force on the electric dipole is zero.

Fig. 3.40: Electric dipole in an external uniform electric field

The potential energy of the dipole is

$$U_p = q\phi - q\phi' = q(\phi - \phi') = -qd\left(-\frac{\phi - \phi'}{d}\right)$$

If θ is the angle between the dipole and the electric field, then the component $E\cos\theta$ of the field E is parallel to d. Therefore,

$$U_p = -pE\cos\theta = -p\cdot E \tag{3.68}$$

The potential energy is minimum when $\theta = 0°$, indicating that the dipole is in equilibrium when it is oriented parallel to the field. If we neglect the slight difference between E and E', the force qE and $-qE'$ on the charges comprising the dipole form essentially a couple whose torque is

$$\tau = d \times (qE) = (qd) \times E = p \times E \tag{3.69}$$

The torque of the electric field tends to align the dipole parallel to the field. The magnitude of the torque is $\tau = pE\sin\theta$, and its direction is indicated in Fig. 3.40. One can

easily see that $\tau_z = -\dfrac{\partial U_p}{\partial\theta}$. Using Eq. (3.68), one obtains $\tau_z = -pE\sin\theta$. The difference in

sign of τ is due to the fact that τ gives the magnitude of the torque while τ_z gives the component of the torque along the direction z, perpendicular to the plane in which the angle θ is measured, and oriented in the sense of advance of right-handed screw rotated in the sense in which θ increases. The negative sign of τ_z confirms that the torque tends to decrease the angle θ.

These properties of a dipole placed in an electric field have very important applications. For example, the electric field of an ion in solution polarizes the molecules of the solvent (electrolysis) which surrounds the ions and they become oriented in the form indicated in Fig. 3.41.

These oriented molecules more or less attached to the ion, increase their effective mass and decrease effective charge, which is partially screened by the molecules. The

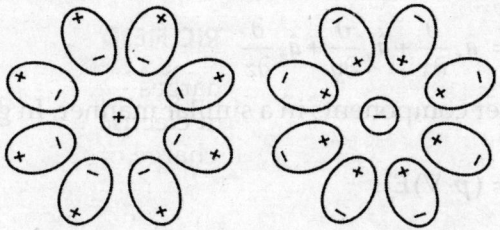

Fig. 3.41: Polarization effects of an ion in solution (electrolyte)

net effect is that the mobility of the ion in an external field is decreased. Also when a gas or a liquid whose molecules are permanent dipoles is placed where an electric field exists, the molecules, as a result of the torques due to the electric field, tend to align with their dipoles parallel. We say that the substance has been polarized.

3.18 ELECTRIC DIPOLE IN A NON-UNIFORM ELECTRIC FIELD

Let us consider a dipole in a non-uniform electric field (Fig. 3.42). A translational force is exerted on the dipole. We are interested in finding the expression for force.

Suppose the origin of the coordinate system coincides with the charge $-q$. The force acting on the dipole in the positive x-direction is

$$F_x^{(+)} = q(E_x + dE_x)$$

The force acting on the dipole in the negative x-direction is

$$F_x^{(-)} = qE_x$$

Fig. 3.42: Electric dipole in a non-uniform electric field

The net force component in the positive x-direction is

$$F_x = qdE_x$$

As the field is non-uniform and hence

$$dE_x = \left(\frac{\partial E_x}{\partial x}\right)dx + \left(\frac{\partial E_x}{\partial y}\right)dy + \left(\frac{\partial E_x}{\partial z}\right)dz$$

$$\therefore \quad F_x = q\left(dx\frac{\partial E_x}{\partial x} + dy\frac{\partial E_x}{\partial y} + dz\frac{\partial E_x}{\partial z}\right)$$

$$= q\left(dx\frac{\partial}{\partial x} + dy\frac{\partial}{\partial y} + dz\frac{\partial}{\partial z}\right)E_x$$

$$= q(dl \cdot \nabla)E_x = (P \cdot \nabla)E_x \qquad (3.70)$$

Because $\qquad dl = \hat{a}_x dx + \hat{a}_y dy + \hat{a}_z dz$

and
$$\nabla = \hat{a}_x \frac{\partial}{\partial x} + \hat{a}_y \frac{\partial}{\partial y} + \hat{a}_z \frac{\partial}{\partial z}$$

One can obtain the other components in a similar manner. In general, the force on the electric dipole is

$$F = (p \cdot \nabla)E \qquad (3.71)$$

3.19 ELECTRIC DOUBLE LAYERS

When one studies biological and celloid problems, then it is observed that on a given surface there is a double layer of surface charges consisting of two neighbouring layers of charges of equal magnitude but opposite sign separated by an infinitesimal distance dx as shown in Fig. 3.43a. If the surface charge density is σ, then the strength of the layer is

$$D = \sigma dx$$

Fig. 3.43: Electric double layers

Figure 3.43b shows two surfaces with two small elemental areas on it. The two charges $\pm \sigma ds$ separated by dx are equivalent to dipole moment $P = \sigma ds dx$. Obviously, one may think it as a layer of dipoles rather than two layers with positive and negative charges. The direction of the dipole moment of the layer is normal to the surface. One obtains the potential of the dipole layer at a point distant r as

$$\phi(r) = \frac{1}{4\pi\varepsilon_0} \int \hat{a}_n \frac{Dds \cdot r}{|r|^3} = \frac{D}{4\pi\varepsilon_0} \int \frac{\hat{a}_n ds \cdot r}{|r|^3} = \frac{D}{4\pi\varepsilon_0} = d\Omega \qquad (3.72)$$

where, $d\Omega$ is the element of the solid angle subtended at the point by ds.

3.20 HIGHER ELECTRIC MULTIPOLES

It is possible to define higher order or multipole electric moments. For example, a charge distribution such as that shown in Fig. 3.44 constitutes an *electric quadrupole*. Its total charge Σq_i and electric dipole moment p are zero and hence gives a zero value of potential. Actually, however, the field of a quadrupole, although it is much weaker than that of a dipole (with the same values of q and d), differs from zero.

Let us consider the case of a *linear quadrupole* as shown in Fig. 3.45. One can regard the distribution of charge of a linear quadrupole (Fig. 3.45) as derived from two equal and opposite dipole moments p placed in a line and separated by a distance d. The potential due to the quadrupole at a point P is given by

$$\phi_P = \phi_1 - \left(\phi_1 + d\frac{\partial\phi_1}{\partial x} \right) = -d\frac{\partial\phi_1}{\partial x}$$

Fig. 3.44: Electric quadrupole   **Fig. 3.45:** Linear quadrupole

where, ϕ_1 is the potential due to the first dipole. Therefore,

$$\phi_P = -d\frac{\partial}{\partial x}\left(\frac{p\cos\theta}{4\pi\varepsilon_0 r^2}\right) = -\frac{pd}{4\pi\varepsilon_0}\frac{\partial}{\partial x}\left(\frac{\cos\theta}{r^2}\right)$$

$$= -\frac{pd}{4\pi\varepsilon_0}\left[-\frac{1}{r^2}\sin\theta\frac{\partial\theta}{\partial x} - \frac{2\cos\theta}{r^3}\frac{\partial r}{\partial x}\right]$$

$$= -\frac{pd}{4\pi\varepsilon_0}\left[\frac{\sin^2\theta}{r^3} - \frac{2\cos^2\theta}{r^3}\right] \quad \left(\because \frac{\partial\theta}{\partial x} = -\frac{\sin\theta}{r}, \frac{\partial r}{\partial x} = \cos\theta\right)$$

$$= \frac{pd}{4\pi\varepsilon_0 r^3}(3\cos^2\theta - 1) \tag{3.73}$$

The quadrupole moment, defined analogously to the dipole moment, is given by

$$Q = pd = qd^2$$

$$\therefore \quad \phi_P = \frac{Q}{4\pi\varepsilon_0 r^3}(3\cos^2\theta - 1) \tag{3.74}$$

Obviously, the potential of the field set up by a quadrupole is proportional to $\frac{1}{r^3}$, whereas the field decreases as $\frac{1}{r}$.

The electric quadrupole moment is zero for a spherical distribution of charge, positive for an elongated or prolate charge distribution, and negative for an oblate or flattened charge distribution (Fig. 3.46).

Therefore, the electric quadrupole moment gives an indication of the degree of departure from the spherical form of a charge distribution. We generally consider atomic

Fig. 3.46: Electric quadrupole of ellipsoidal charge distribution

nuclei as spherical. However, careful measurements indicate that certain nuclei have relatively large electric quadrupole moments. This has been interpreted as indicating that such nuclei are greatly deformed, and thus the electric field they produce departs from a point charge. This, in turn, affects the energy of the electronic motion.

We must note that the potential of a point charge decreases as r^{-1} and the field as r^{-2}. Similarly, for an electric dipole the potential decreases as r^{-2} and the field as r^{-3}. For an electric quadrupole, the potential varies as r^{-3} and the field as r^{-4}. Similar results are obtained for higher order multipoles. We conclude then that the higher the multipole order, the smaller the range within which its electric field has any noticeable effect.

3.21 ELECTRIC POTENTIAL DUE TO AN ARBITRARY DISTRIBUTION OF CHARGE

Let us consider an assembly of point charges distributed within a region in a complicated manner as shown in Fig. 3.47. We are interested in determining the potential due to this distribution at a point P distant r from the origin O.

Let us consider that a charge q_i be at a distance d_i from the origin. The potential at P due to this charge is

$$\phi_i(r_i) = \frac{q_i}{4\pi\varepsilon_0 r_i} = \frac{q_i}{4\pi\varepsilon_0 |r - d_i|} \quad (3.75)$$

Fig. 3.47: Assembly of point charges

where, r_i is the distance of charge q_i from P. If $r \gg d_i$, one may write to the first approximation

$$\phi_i(r_i) = \frac{q_i}{4\pi\varepsilon_0 r}$$

for each charge q_i.

Thus, the total potential due to the entire distribution of charges is

$$\phi(r) = \frac{\Sigma q_i}{4\pi\varepsilon_0 r} = \frac{Q}{4\pi\varepsilon_0 r} \quad (3.76)$$

Obviously, the potential is same as that obtained by a point charge Q at the origin, where

$$Q = \sum_i q_i$$

Let us consider the case when there are equal number of positive and negative charges in the region. Clearly, in this case, $Q = 0$, i.e. the object under consideration is neutral. At once, one may conclude that there will not be anything like potential around the object. However, there may be objects which are neutral, and yet the charges are not all at one point, and hence one should expect some effect of these separate charges at points not far away from the object we are interested in finding the potential due to such a charge distribution. For this purpose, one can use Eq. (3.75) without making the approximation that d_i being much smaller than r. However, we can introduce better approximation. Let us assume that r_i differs from r by projection of d_i on r, i.e by OM as shown in Fig. 3.47. Thus,

$$r_i = r - (d_i \cdot \hat{a}_r) \tag{3.77}$$

where, $\hat{a}_r$ is the unit vector in the direction of r. Thus,

$$\frac{1}{r_i} = \frac{1}{r\left(1 - \dfrac{d_i \cdot \hat{a}_r}{r}\right)} = \frac{1}{r}\left(1 + \frac{d_i \cdot \hat{a}_r}{r}\right)$$

Here we have neglected the higher terms.

$$\therefore \qquad \phi(r) = \sum_i \frac{q_i}{4\pi\varepsilon_0 r}\left(1 + \frac{d_i \cdot \hat{a}_r}{r}\right)$$

$$= \frac{Q}{4\pi\varepsilon_0 r} + \sum_i q_i \frac{d_i \cdot \hat{a}_r}{4\pi\varepsilon_0 r^2} \tag{3.78}$$

We must note that the first term is same as in Eq. (3.76) and equal to zero, since $Q = 0$. Therefore,

$$\phi(r) = \sum_i q_i \frac{d_i \cdot \hat{a}_r}{4\pi\varepsilon_0 r^2} \tag{3.79}$$

Defining $p = \Sigma q_i d_i$ as the dipole moment of charge distribution, one obtains

$$\phi(r) = \frac{p \cdot \hat{a}_r}{4\pi\varepsilon_0 r^2} \tag{3.80}$$

Obviously, for any distribution of charges which is as a whole neutral, the potential is a dipole potential. Examples of such charge distributions are H_2O and CO_2 molecules.

ILLUSTRATIVE EXAMPLES

Example 3.1: A charge q coulomb is distributed uniformly throughout a non-conducting spherical volume of radius R metre. Show that the potential at a distance r form the centre where $r \leq R$ is given by

$$\phi = \frac{1}{4\pi\varepsilon_0}\frac{q(3R^2 - r^2)}{2R^3}$$

Solution: Let us consider that charge q coulomb be distributed uniformly throughout a non-conducting sphere of radius R metre. The field strength at point distance $r > R$ and $r < R$ from the centre of the spherical charge are, respectively, given by

$$E_0 = \frac{1}{4\pi\varepsilon_0}\frac{q}{r^3}r \quad \text{and} \quad E_i = \frac{1}{4\pi\varepsilon_0}\frac{q}{R^3}r$$

The potential at a distance $r < R$ from the centre is given by

$$\phi = -\int_\infty^r E\cdot dr = -\left[\int_\infty^R E_0\cdot dr + \int_R^r E_i\cdot dr\right]$$

$$= -\left[\int_\infty^R \frac{1}{4\pi\varepsilon_0}\frac{qr}{r^3}\cdot dr + \int_R^r \frac{1}{4\pi\varepsilon_0}\frac{q}{R^3}r\cdot dr\right]$$

$$= -\left[\frac{1}{4\pi\varepsilon_0}q\int_\infty^R r^{-2}dr + \frac{1}{4\pi\varepsilon_0}\int_R^r \frac{q}{R^3}r\cdot dr\right]$$

$$= -\frac{1}{4\pi\varepsilon_0}q\left[\left\{-\frac{1}{r}\right\}_\infty^R + \left\{\frac{r^2}{2R^3}\right\}_R^r\right]$$

$$= -\frac{1}{4\pi\varepsilon_0}q\left[-\frac{1}{R} + \frac{r^2}{2R^2} - \frac{1}{2R}\right]$$

$$= -\frac{1}{4\pi\varepsilon_0}q\left[\frac{r^2}{2R^3} - \frac{3}{2R}\right] = \frac{q}{4\pi\varepsilon_0}\left[\frac{3R^2 - r^2}{2R^3}\right]$$

Example 3.2: Two particles each of mass m and having a charge Q are suspended by string of length l from a common point. Show that the angle θ, which each string makes with the vertical, is given by the relation

$$\frac{\tan^3\theta}{1+\tan^2\theta} = \frac{Q^2}{16\pi\varepsilon_0 mgl^2}$$

Solution: The two particles carry like charges, and hence the force F between the charges will be repulsive (Fig. 3.48). The magnitude of the force F is given by

$$F = \frac{Q\times Q}{4\pi\varepsilon_0(AB)^2} = \frac{Q^2}{4\pi\varepsilon_0(AB)^2} \tag{1}$$

Now $\qquad AB = 2AM = 2l\sin\theta \tag{2}$

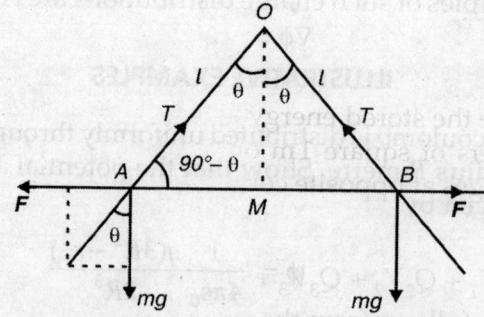

Fig. 3.48: Two charged particles suspended by string

Using Eq. (2), Eq. (1) becomes

$$F = \frac{Q^2}{4\pi\varepsilon_0 (2l\sin\theta)^2} = \frac{Q^2}{16\pi\varepsilon_0 l^2 \sin^2\theta} \tag{3}$$

Applying Lame's theorem at point A, one obtains

$$\frac{F}{\sin(180° - \theta)} = \frac{mg}{\sin(90° + \theta)}$$

or

$$\frac{F}{\sin\theta} = \frac{mg}{\cos\theta} \tag{4}$$

Substituting the value of F from Eq. (4) in Eq. (3), one obtains

$$\frac{Q^2}{16\pi\varepsilon_0 l^2 \sin^2\theta} = \frac{mg}{\cos\theta}\sin\theta$$

or

$$\frac{Q^2}{16\pi\varepsilon_0 mgl^2} = \frac{\sin^3\theta}{\cos\theta} = \frac{\tan^3\theta}{\sec^2\theta} = \frac{\tan^3\theta}{1 + \tan^2\theta}$$

Example 3.3: Show that the potential difference between two points distant r_1 and r_2 from an infinitely long charge of linear charge density (i.e. charge per unit length) λ is given by

$$\nabla\phi = \frac{1}{4\pi\varepsilon_0}2\lambda\log_e\left(\frac{r_2}{r_1}\right)$$

Solution: The field strength at a distance r from an infinite linear charge of density λ is given by

$$E = \frac{1}{4\pi\varepsilon_0}\frac{2\lambda}{r} \tag{1}$$

The potential at a distance r from the axis is

$$\phi_r = -\int_{r_{ref}}^{r} E \cdot dr$$

Here r_{ref} denotes the reference distance for zero potential. We must note that the reference distance cannot be taken as infinity since the wire itself extends to infinity.

As per the problem, we have to find the potential difference between two points having distance r_1 and r_2 from the wire, and hence

$$\nabla\phi = -\int_{r_2}^{r_1}\frac{1}{4\pi\varepsilon_0}\frac{2\lambda}{r}dr = \frac{1}{4\pi\varepsilon_0}2\lambda\log_e\frac{r_2}{r_1}$$

Example 3.4: Calculate the stored energy in a system of four identical point charges, $Q = 4\,\text{nC}$, at the corners of square $1\,\text{m}$ on a side. What is the stored energy in the system when two charges at opposite corners are in place?

Solution: We have

$$2W = Q_1V_1 + Q_2V_2 + Q_3V_3 + Q_4V_4 = 4Q_1V_1$$

where the last equality follows from the symmetry of the system.

Now,

$$V_1 = \frac{Q_2}{4\pi\varepsilon_0 r_{12}} + \frac{Q_3}{4\pi\varepsilon_0 r_{13}} + \frac{Q_4}{4\pi\varepsilon_0 r_{14}}$$

$$= \frac{4\times 10^{-9}}{4\pi\varepsilon_0}\left(\frac{1}{1}+\frac{1}{1}+\frac{1}{\sqrt{2}}\right) = 97.5\,\text{V}$$

$$\therefore \quad W = 2Q_1 V_1 = 2\times(4\times 10^{-9})\times(97.5) = 780\,\text{nJ}$$

For only two charges in place,

$$2W = Q_1 V_1 = 4\times 10^{-9}\times\frac{4\times 10^{-9}}{4\pi\varepsilon_0\sqrt{2}} = 102\,\text{nJ}$$

Example 3.5: Show that the potential at a point distant r_1 and r_2 respectively from centres of a long parallel pair of wires of negligible cross-section and having equal and opposite linear charge density λ Cm is given by

$$\phi = \frac{\lambda}{4\pi\varepsilon_0}\log_e\left(\frac{r_2}{r_1}\right)$$

Solution: The electric field strength at distance r from the line charge is

$$E = \frac{1}{4\pi\varepsilon_0}\frac{2\lambda}{r}$$

By definition, we have electric potential

$$\phi = -\int_{r_{\text{ref}}}^{r} E\cdot dr$$

where, r_{ref} is the reference distance for zero potential.

In this problem, we cannot choose reference point at infinity as charge itself extends to infinity. Therefore, we choose the reference point at r_0 where potential is ϕ_0. Thus, potential difference between r_0 and r is

$$\phi - \phi_0 = -\int_{r_0}^{r} E\cdot dr = -\int_{r_0}^{r}\frac{1}{4\pi\varepsilon_0}\frac{2\lambda}{r}dr$$

$$\phi - \phi_0 = \frac{1}{4\pi\varepsilon_0}2\lambda\log_e\frac{r_0}{r} = \frac{\lambda}{2\pi\varepsilon_0}\log_e\frac{r_0}{r}$$

Thus, the potential at distance r is

$$\phi = \frac{\lambda}{2\pi\varepsilon_0}\log_e\frac{r_0}{r}+\phi_0$$

Let us consider the case of two lines of charge of linear charge densities λ and $(-\lambda)$. If the point under consideration (say P) lies at distance r_1 and r_2 from wires, then potential ϕ_1 at P due to one line charge of charge density $(+\lambda)$ will be

$$\phi_1 = \frac{\lambda}{2\pi\varepsilon_0}\log_e\left(\frac{r_0}{r_1}\right)+\phi_0$$

The potential due to other line charge of density $(-\lambda)$ will be

$$\phi_2 = \frac{-\lambda}{2\pi\varepsilon_0}\log_e\frac{r_0}{r_2} + \phi_0$$

∴ The net potential due to both the wires will be

$$\phi = \phi_1 + \phi_2 = \frac{\lambda}{2\pi\varepsilon_0}\log_e\frac{r_2}{r_1} + 2\phi_0$$

Let the potential at reference point be zero. This is possible in the case if the reference point is at equal distance from both the wires, i.e. we have to choose $\phi_0 = 0$. Thus,

$$\phi = \frac{\lambda}{2\pi\varepsilon_0}\log_e\frac{r_2}{r_1}$$

Example 3.6: Show that in free space the normal component of field on any surface is discontinuous to the extent of $\dfrac{\rho}{\varepsilon_0}$, where ρ is the surface charge density.

Solution: Construct a volume element in the shape of a pill (pillbox) across the surface (Fig. 3.49). Now apply Gauss's law to the volume of the pillbox as the flat faces approach the surface. The flux through the lateral surface approaches zero and E_{n_1}, E_{n_2} approach the normal components of the field in the two sides of the surface, we have

$$(E_{n_1} - E_{n_2})\Delta S = \frac{\rho}{\varepsilon_0}\Delta S$$

or
$$E_{n_1} - E_{n_2} = \frac{\rho}{\varepsilon_0}$$

Fig. 3.49: Pili across the surface

Example 3.7: Show that the energy of a sphere of charge with radius R in which the charge is uniformly distributed is

$$W = \frac{4\pi\rho^2}{15\varepsilon_0}R^5$$

where, ρ is the density of charge.

Solution: We know that energy is equal to the work done in bringing a unit charge from infinity to the point (here it is sphere) under consideration. Let us imagine that sphere is formed by assembling various thin shells of charge. Now we consider a small sphere of charge with radius r as shown in Fig. 3.50. ρ is the density of charge. The total charge on the surface of the sphere is $\dfrac{4}{3}\pi r^3\rho$. Imagine that

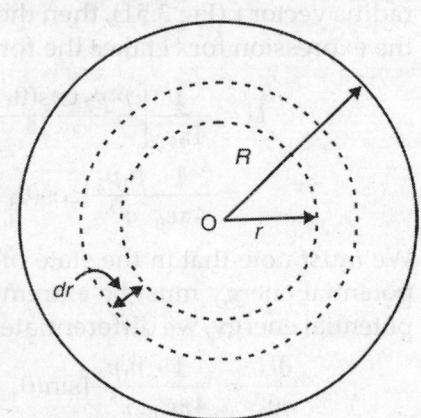

Fig. 3.50: Charge distributed uniformly in a sphere

a small layer of charge dq in the form of thin shell of thickness dr is deposited on the sphere. Therefore,

$$dq = \rho 4\pi r^2 dr$$

Now, the work done in bringing this charge from infinity is

$$dW = \text{potential at } r \times dq$$

$$= \frac{1}{4\pi\varepsilon_0} \frac{\left(\frac{4}{3}\right)\pi r^3 \rho}{r} \rho 4\pi r^2 dr = \frac{4\pi\rho^2}{3\varepsilon_0} r^4 dr$$

One can obtain the total energy required to assemble charges so as to build up a sphere of radius R as follows:

$$W = \frac{4\pi\rho^2}{3\varepsilon_0} \int_0^R r^4 dr = \frac{4\pi\rho^2}{15\varepsilon_0} R^5$$

Example 3.8: Write down the interaction potential energy of two short electric dipoles separated by a certain distance. If one of the dipoles is inclined at an angle θ_1 to the radius vector joining them, show that in the state of equilibrium, the other dipole will make an angle θ_2 with it given by

$$\tan\theta_2 = -\frac{1}{2}\tan\theta_1$$

Solution: The interaction potential energy of two short dipoles is

Fig. 3.51: Interaction energy of two short dipoles separated by a distance r

$$U = \frac{1}{4\pi\varepsilon_0}\left[\frac{\boldsymbol{p_1}\cdot\boldsymbol{p_2}}{r^3} - \frac{3}{r^5}(\boldsymbol{p_1}\cdot\boldsymbol{r})(\boldsymbol{p_2}\cdot\boldsymbol{r})\right]$$

Let the dipole moment $\boldsymbol{p_1}$ of one of the dipole be inclined at an angle θ_1 to the radius vector $\boldsymbol{r}$ and dipole moment $\boldsymbol{p_2}$ of second dipole be inclined at angle θ_2 with the same radius vector $\boldsymbol{r}$ (Fig 3.51), then the angle between the two poles would be $\theta_1 - \theta_2$. Now the expression for U takes the form

$$U = \frac{1}{4\pi\varepsilon_0}\left[\frac{p_1 p_2 \cos(\theta_1 - \theta_2)}{r^3} - \frac{3}{r^5}(p_1 r \cos\theta_1)(p_2 r \cos\theta_2)\right]$$

$$= \frac{1}{4\pi\varepsilon_0}\frac{p_1 p_2}{r^3}[\cos(\theta_1 - \theta_2) - 3\cos\theta_1\cos\theta_2] \tag{1}$$

We must note that in the state of equilibrium of dipoles, the force must be zero and potential energy must be extremum. Now, for applying the condition of extremum potential energy, we differentiate Eq. (1) w.r.t θ_2 and then equate to zero. One obtains,

$$\frac{\partial U}{\partial\theta_2} = \frac{1}{4\pi\varepsilon_0}\frac{p_1 p_2}{r^3}[\sin(\theta_1 - \theta_2) + 3\cos\theta_1\sin\theta_2] = 0$$

Since $\frac{1}{r^3}$ is not equal to zero, we have

$$\sin(\theta_1 - \theta_2) + 3\cos\theta_1\sin\theta_2 = 0$$

or $\quad \sin\theta_1\cos\theta_2 - \cos\theta_1\sin\theta_2 + 3\cos\theta_1\sin\theta_2 = 0$

or $\quad\quad\quad\quad\quad \sin\theta_1\cos\theta_2 + 2\cos\theta_1\sin\theta_2 = 0$

or $\quad\quad\quad\quad\quad\quad\quad \dfrac{\sin\theta_1}{2\cos\theta_1} = \dfrac{-\sin\theta_2}{\cos\theta_2}$

$\therefore \quad\quad\quad\quad\quad\quad\quad\quad \tan\theta_2 = -\dfrac{1}{2}\tan\theta_1$

Example 3.9: Calculate the resultant dipole moment for the system of charges shown in Fig. 3.52, and hence determine the electric potential and field at a point quite far away from the charges.

Solution: We have the x-component of dipole moment due to charge $-q$ at point $A(a, 0)$ as

$$\int_V \rho x\, dV = -q \cdot a = -qa$$

Similarly, the x-component of dipole moment due to charge $+2q$

at point $B\left(\dfrac{a}{2}, \dfrac{a\sqrt{3}}{2}\right)$ is $2q\dfrac{a}{2} = +qa$

Fig. 3.52

The x-component of dipole moment due to charge $-2q$

at point $C\left(-\dfrac{a}{2}, \dfrac{a\sqrt{3}}{2}\right)$ is $-2q\left(\dfrac{-a}{2}\right) = +qa$

The x-component of dipole moment due to charge $+q$

at pont $D(-a, 0)$ is $+q(-a) = -qa$

The x-component of dipole moment due to charge $-2q$

at point $E\left(-\dfrac{a}{2}, \dfrac{-a\sqrt{3}}{2}\right)$ is $-2q\left(-\dfrac{a}{2}\right) = +qa$

Lastly, the x-component of dipole moment due to charge $+2q$

at point $F\left(+\dfrac{a}{2}, \dfrac{-a\sqrt{3}}{2}\right)$ is $2q\dfrac{a}{2} = +qa$

Thus, the x-component of the dipole moment for the given system of charges is $p_x = +2qa$.

Similarly, one obtains the contribution of all the charges to y-component of dipole moment as

$$p_y = \int_V \rho\, dV = 0 + 2q\left(\dfrac{a\sqrt{3}}{2}\right) - 2q\left(\dfrac{a\sqrt{3}}{2}\right) + 0 - 2q\left(-\dfrac{a\sqrt{3}}{2}\right) + 2q\left(-\dfrac{a\sqrt{3}}{2}\right) = 0$$

∴ Total dipole moment,

$$p = \hat{i}p_x + \hat{j}p_y = 2qa\hat{i}$$

∴ The potential at a distant point P due to dipole moment,

$$\phi_P = \frac{1}{4\pi\varepsilon_0}\frac{p \cdot r}{r^3} = \frac{1}{4\pi\varepsilon_0}\frac{2qa\cos\theta}{r^2}$$

The field components are

$$E_r = -\frac{\partial\phi}{\partial r} = -\frac{1}{4\pi\varepsilon_0}\frac{4qa\cos\theta}{r^3}$$

$$E_\theta = -\frac{1}{r}\frac{\partial\phi}{\partial\theta} = \frac{1}{4\pi\varepsilon_0}\frac{2qa\sin\theta}{r^3}$$

∴

$$E = \hat{r}E_r + \hat{\theta}E_\theta$$

$$= \frac{1}{4\pi\varepsilon_0}\frac{2qa}{r^3}[2\cos\theta\hat{r} + \sin\theta\hat{\theta}].$$

Example 3.10: An infinite circular cylindrical volume (Fig. 3.53) carries a uniform charge density except for the cylindrical cavity which is also infinitely long. The cavity carries no charge. Determine the field inside the cavity and show that it is constant.

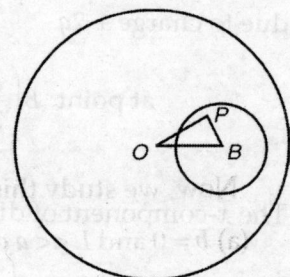

Solution: Applying Gauss's law, one can write

$$E(P) = \frac{\rho(OP)}{2\varepsilon_0} - \frac{\rho}{2\varepsilon_0}(BP)$$

$$= \frac{\rho}{2\varepsilon_0}(OB) = \text{constant}.$$

Fig. 3.53: Field inside a cylindrical cavity

Example 3.11: Calculate the potential due to a thin charged rod of length L at the point along and perpendicular to its length.

Solution: Let us suppose that the thin rod has a uniform charge of charge density λ per unit length. In order to find potential at point P (Fig. 3.54), let us divide the rod

Fig. 3.54: Potential due to a thin charged rod

into a series of small elements, each of which may be considered as a point charge. Now, the potential at point P due to one element of length Δx is given by

$$\Delta V = \frac{1}{4\pi\varepsilon_0}\frac{\lambda\Delta x}{r}$$

Total potential due to entire rod, $V = \Sigma\Delta V = \Sigma\lambda\,\Delta x/(4\pi\varepsilon_0 r)$

Now, assuming $\Delta x \to 0$, one can replace the summation by an integration

$$\therefore \quad V = \int_1^{a+L}\frac{\lambda\,dx}{4\pi\varepsilon_0(b^2+x^2)^{1/2}}$$

$$= \frac{\lambda}{4\pi\varepsilon_0}\int_1^{a+L}\frac{dx}{(b^2+x^2)^{1/2}}$$

or

$$V = \frac{\lambda}{4\pi\varepsilon_0}\left[\log_e\left\{x+(b^2+x^2)^{1/2}\right\}\right]_a^{a+L}$$

$$= \frac{\lambda}{4\pi\varepsilon_0}\log_e\left[\frac{a+L\{b^2+(a+L)^2\}^{1/2}}{a+(b^2+a^2)^{1/2}}\right]$$

Now, we study this result to following limiting cases:

(a) $b = 0$ and $L \ll a$ or at a far distance on its axis

$$V = \frac{\lambda}{4\pi\varepsilon_0}\log_e\frac{2(a+L)}{2a} = \frac{\lambda}{4\pi\varepsilon_0}\log_e\left(1+\frac{L}{a}\right)$$

$$\therefore \quad \log_e(1+z) = z - \frac{z^2}{2} + \frac{z^3}{3}\cdots = z, \quad \text{if } z < 1$$

Hence $V = \dfrac{\lambda}{4\pi\varepsilon_0}\dfrac{L}{a} = \dfrac{1}{4\pi\varepsilon_0}\dfrac{q}{a}$ ($\because \lambda L = q$, the total charge on the rod)

We see that the relation is same as due to point charge. Clearly, at a large distance, charged rod behaves as a point charge.

(b) $a = 0$, $L \ll b$ or at the points on the line perpendicular to axis. We have

$$V = \frac{\lambda}{4\pi\varepsilon_0}\log_e\left[\frac{L}{b}+\left(1+\frac{L^2}{b^2}\right)^{1/2}\right]$$

$$= \frac{\lambda}{4\pi\varepsilon_0}\log_e\left(1+\frac{L}{b}\right)$$

$$= \frac{\lambda}{4\pi\varepsilon_0}\frac{L}{b} = \frac{q}{4\pi\varepsilon_0 b}$$

which is again as due to point charge.

Example 3.12: For a region comprising electric field intensity $E = -\dfrac{16}{r^2}\hat{a}_r$, find the potential difference V_{AB} in spherical polar coordinate system where A is at $(2, \pi, \pi/2)$ and B is at $(4, 0, \pi)$.

$$W = -\frac{1}{4\pi\varepsilon_0 |r|^3}\left[p_1 \cdot p_2 - \frac{3(p_1 \cdot r)(p_2 \cdot r)}{|r|^2}\right]$$

Solution: $\quad dl = dr\,\hat{a}_r$

$$\therefore \qquad V = -\int E \cdot dl = -\int_4^2\left(-\frac{16}{r^2}\right)dr$$

$$= 16\int_4^2 \frac{dr}{r^2} = 16\left[\frac{r^{-2+1}}{-2+1}\right]_4^2$$

$$= -16\left[\frac{1}{r}\right]_4^2 = -16\left[\frac{1}{2} - \frac{1}{4}\right] = -\frac{16}{4} = -4\,\text{V}$$

Example 3.13: Two water molecules each having a dipole moment 6.2×10^{-30} C m, point in the same direction along the line joining their centres. Show that the potential energy due to their dipole–dipole interactions when their centres are 3.1×10^{-10} m apart is -0.0145 eV. **[MDSA]**

Solution: Let p_1 and p_2 be the dipole moments of two dipoles. Potential energy due to their dipole interaction is

$$W = -\frac{1}{4\pi\varepsilon_0 |r|^3}\left[p_1 \cdot p_2 - \frac{3(p_1 \cdot r)(p_2 \cdot r)}{|r|^2}\right]$$

Here $p_1 = p_2 = 6.2 \times 10^{-30}$ C m and $r = 3.1 \times 10^{-10}$ m.

Since dipoles are oriented in the same direction as r, we have

$$W = \frac{1}{4\pi\varepsilon_0(3.1\times10^{-10})^3}\left[(6.2\times10^{-30})^2 - 3(6.2\times10^{-30})^2\right]$$

$$= -0.023 \times 10^{-18}\,\text{J}$$

$$= 232.26 \times 10^{-21}\,\text{J}$$

$$= -0.0145\,\text{eV}$$

Example 3.14: Forty nanocoulomb of charge is uniformly distributed around a circular ring of radius 2 m. Find the potential at a point on the axis 5 m from the plane of the ring. Compare with the result where all the charges are at the origin in the form of a point charge. **[BTech]**

Solution: We have

$$V = \int \frac{\rho_l dl}{4\pi\varepsilon_0 R}$$

Here $\qquad \rho_l = \dfrac{40 \times 10^{-9}}{2\pi \times 2} = \dfrac{10^{-8}}{\pi} \, \text{Cm}^{-1}$

From Fig. 3.55, $R = \sqrt{29}\,\text{m}, \; dl = (2\,\text{m})$

$\therefore \qquad V = \displaystyle\int_0^{2\pi} \dfrac{(10^{-8}/\pi)(2)}{4\pi(10^{-9}/36\pi)\sqrt{29}}$

$\qquad\qquad = 66.9\,\text{V}$

Fig. 3.55: Circular ring (uniformly charged)

When the charge is concentrated at the origin,

$$V = \dfrac{40 \times 10^{-9}}{4\pi\varepsilon_0 \times 5} = 72.0 \; \text{V}$$

Example 3.15: Calculate the work done in moving a 5 µC charge from the origin $(0, 0, 0)$ to the point P $(4, -2, 6)$ through the field $E = (5xyz\hat{i} + 3x^2z\hat{j} + 2xy^2\hat{k})\,\text{Vm}^{-1}$ via straight lines $x = -5y, z = 3x$.

Solution: We have

$$E = 5xyz\hat{i} + 3x^2z\hat{j} + 2xy^2\hat{k}$$

$$dl = dx\hat{i} + dy\hat{j} + dz\hat{k}$$

$$E \cdot dl = 5xyz\,dx + 3x^2z\,dy + 2xy^2\,dz$$

Further, $\qquad x = -5y; \; z = 3x$

$\therefore \qquad dx = -5dy; \quad dz = 3$

$\therefore \quad$ Total work done $= -Q\displaystyle\int E \cdot dl$

$$= -Q\int_{(0,0,0)}^{(4,-2,6)} (5xyz\,dx + 3x^2z\,dy + 2xy^2\,dz)$$

$$= -Q\int_0^4 \left(-3x^3\,dx - \frac{9}{5}x^3\,dx + \frac{6}{25}x^3\,dx \right)$$

$$= -Q\int_0^4 (x^3\,dx)\left[-\frac{114}{25} \right] = 1459.2 \; \mu\text{J}.$$

Example 3.16: Calculate the potential at $r_A = 5\,\text{m}$ and $r_B = 15\,\text{m}$ due to a point charge Q = 500 pC at the origin and zero reference at infinity.

Solution: Due to point charge

$$V_{AB} = \dfrac{Q}{4\pi\varepsilon_0}\left(\dfrac{1}{r_A} - \dfrac{1}{r_B} \right)$$

In order to find the potential difference, the zero reference is not needed.

Thus, $\qquad V_{AB} = \dfrac{500 \times 10^{-12}}{4\pi(10^{-9}/36\pi)}\left(\dfrac{1}{5} - \dfrac{1}{15} \right) = 0.60 \; \text{V}$

Now, we want to calculate V_1 (at $r_A = 5$ m) and V_2 (at $r_B = 15$ m) and for this purpose zero reference at infinity may be used.

$$V_1 = \frac{Q}{4\pi\varepsilon_0}\left(\frac{1}{5}\right) = 0.90 \text{ V}$$

$$V_2 = \frac{Q}{4\pi\varepsilon_0}\left(\frac{1}{15}\right) = 0.30 \text{ V}$$

$$\therefore \quad V_{AB} = V_1 - V_2 = 0.60 \text{ V}.$$

Example 3.17: Infinite uniform line charge of 5 nC/m lying along positive and negative x- and y-axes in free space is shown in Fig. 3.56. Find the electric field intensity at $(0, 0, 4)$ and $(0, 3, 4)$.

Fig. 3.56: Infinite uniform line charge of 5 nC/m lying along positive and negative x and y directions

Solution: For $(0, 0, 4)$,

$$E = E_x + E_y = \frac{5\times10^{-9}\hat{a}_z}{2\pi\varepsilon_0\times4} + \frac{5\times10^{-9}\hat{a}_z}{2\pi\varepsilon_0\times4} = \frac{2\times5\times10^{-9}}{2\pi\varepsilon_0\times4}\hat{a}_z = 45\hat{a}_z$$

For $(0, 3, 4)$ $E_y = \frac{\lambda}{2\pi\varepsilon_0\times4}\hat{a}_z = \frac{9\times10^9}{2}\times5\times10^{-9}\hat{a}_z = 22.5\hat{a}_z$

$$E_x = \frac{\lambda}{2\pi\varepsilon_0\times5}\left(\frac{3\hat{a}_y + 4\hat{a}_z}{5}\right)$$

$$= \frac{5\times10^{-9}\times18\times10^{-9}}{5\times5}(3\hat{a}_y+4\hat{a}_z) = \left(\frac{18\times3}{5}\hat{a}_y + \frac{18\times4}{5}\hat{a}_z\right)$$

$$\therefore \quad E = E_x + E_y = 10.8\hat{a}_y + 36\hat{a}_z.$$

Example 3.18: Determine the electric field intensity at the perpendicular bisector of a L section of length 60 cm and 40 cm respectively, where the line charge is having a density 10 C/m along y-axis and -5 C/m along x-axis as shown in Fig. 3.57.

Fig. 3.57: The line charge in form of L section having a density 10 C/m along y-axis and –5 C/m along x-axis

Solution: $E = E_{PA} + E_{PB}$

$$= \frac{10 \times 0.6}{4\pi\varepsilon_0 \times 0.2} \frac{\hat{a}_x}{\sqrt{(0.2)^2 + (0.3)^2}} + \frac{5 \times 0.4(-\hat{a}_y)}{4\pi\varepsilon_0 \times 0.3\sqrt{(0.2)^2 + (0.3)^2}}$$

$$= \left[\frac{6}{0.2}\hat{a}_x - \frac{2}{0.3}\hat{a}_y\right] \frac{9 \times 10^9}{\sqrt{(0.2)^2 + (0.3)^2}}$$

$$= 4.99 \times 10^{10}(15\hat{a}_x - 0.33\hat{a}_y) = 14.97 \times 10^{10}(5\hat{a}_x - 1.11\hat{a}_y)$$

Example 3.19: Determine the electric field intensity due to an infinite line charge of density 5 C/m placed at y = 3 and z = –3 and a surface charge distribution of density 3 C/m² placed at z = 4 at the point (2, –1, 0) as shown in Fig. 3.58.

Fig. 3.58: An infinite line charge of density 5 C/m placed at y = 3 and z = –3 and a surface charge distribution of density 3 C/m² placed at z = 4

Solution: Electric field intensity at point P due to infinite surface charge sheet is given by

$$E_{P(s)} = \frac{\sigma}{2\varepsilon_0}(-\hat{a}_z) = \frac{-3}{2\varepsilon_0}\hat{a}_z$$

where, $E_{P(s)}$ is electric field at point P due to surface charge sheet.

Now

$$\rho = \sqrt{0^2 + (-1-3)^2 + [0-(-3)]^2} = \sqrt{16+9} = 5$$

$\therefore$

$$E_{P(l)} = \frac{\lambda(-4\hat{a}_y + 3\hat{a}_z)}{2\pi\varepsilon_0\rho^2} = \frac{5(-4\hat{a}_y + 3\hat{a}_z)}{2\pi\varepsilon_0 5 \times 5} = \frac{18 \times 10^9}{5}(-4\hat{a}_y + 3\hat{a}_z)$$

where, $E_{P(l)}$ is the electric field at point P due to line charge sheet.

Then

$$E = E_{P(s)} + E_{P(l)} = -1.44 \times 10^{10}\hat{a}_y + 1.08 \times 10^{10}\hat{a}_z - \frac{3}{2\varepsilon_0}\hat{a}_z$$

$$= -1.44 \times 10^{10}\hat{a}_y - 1.58 \times 10^{11}\hat{a}_z.$$

Example 3.20: Determine electric field intensity in the region P, Q and R as shown in Fig. 3.59.

Fig. 3.59: Two parallel sheets facing each other having opposite surface charge distribution

Solution: At the point P, $\quad E_P = \frac{\sigma}{2\varepsilon_0}(-\hat{a}_y) + \frac{\sigma}{2\varepsilon_0}\hat{a}_y = 0$

At the point Q, $\quad E_Q = \frac{2\sigma}{2\varepsilon_0}\hat{a}_y = \frac{\sigma}{\varepsilon_0}\hat{a}_y$

At the point R, $\quad E_R = \frac{\sigma}{2\varepsilon_0}(-\hat{a}_y) + \frac{\sigma}{2\varepsilon_0}(\hat{a}_y) = 0$

Example 3.21: Determine the electric flux density at $r = 6$ cm for three coaxial cylindrical charged sheets with following charge densities as shown in Fig. 3.60.

$$\sigma_1 = 5 \text{ nC/m}^2 \quad \text{at} \quad r = 2 \text{ cm}$$
$$\sigma_2 = -2 \text{ nC/m}^2 \quad \text{at} \quad r = 4 \text{ cm}$$
$$\sigma_3 = 3 \text{ nC/m}^2 \quad \text{at} \quad r = 5 \text{ cm}$$

Fig. 3.60: Three coaxial cylindrical sheets

Solution: By applying Gauss' law to Gaussian surface G_4 and using principle of discrete distribution, the flux density at point P at a distance 0.06 m from the coaxial charge distribution is given by

$$\int D_P \cdot da = Q_{enclosed}$$

$$= (\sigma_1 a_1 + \sigma_2 a_2 + \sigma_3 a_3)$$

or $\quad D_P a = 5 \times 10^{-9} \times 2\pi l \times 0.02 - 2 \times 10^{-9} \times 2\pi l \times 0.04 - 3 \times 10^{-9} \times 2\pi l \times 0.05$

$$D_P \times 0.06 \times 2\pi l = 2\pi l \times 10^{-9} [0.1 - 0.08 - 0.15]$$

or $\quad D_P = \dfrac{10^{-9}[0.1 - 0.23]}{0.06} = -\left(\dfrac{0.13 \times 10^{-9}}{0.06}\right) = -2.167 \times 10^{-9}$

or $\quad D_P = -2.167 \times 10^{-9} \hat{a}_p$.

Example 3.22: An infinitely thin, long straight wire carries the homogeneous line charge κ (charge per unit lenght).

    i. What is the spatial charge density $e(r)$

    ii. Calculate using Gauss's theorem, the electric field strength and its potential.

Solution:

    i. The wire defines the z-axis, $\rho(r)$ is then surely independent of φ and z. We therefore choose as ansatz (cylindrical coordinates ρ', φ, z):

$$\rho(r) = \alpha(\rho') \, \delta(\rho')$$

(Z_l: cylinder of the height l, radius R, wire = axis)

$$\Rightarrow \kappa l = \int_{Z_l} d^3 r \, \rho(r) = \int_0^l dz \int_0^{2\pi} d\varphi \int_0^R \rho' \, d\rho' \, \rho(r) = 2\pi l \int_0^R \rho' \, d\rho' \, \alpha(\rho') \, \delta(\rho')$$

Only $\alpha(\rho') = a/\rho'$ does not lead to a contradiction:

$$\kappa l = 2\pi\, la \int_0^R d\rho'\, \delta(\rho') = \pi la$$

$$\Rightarrow a = \frac{\kappa}{\pi}$$

$$\Rightarrow \rho(r) = \frac{\kappa}{\pi}\frac{\delta(\rho')}{\rho'}$$

ii. Electric field:

$$E(r) = \frac{1}{4\pi\,\epsilon_0}\int d^3r'\,\frac{\rho(r')}{|r-r'|^3}(r-r')$$

$$= \frac{\kappa}{4\pi\,\epsilon_0}\frac{1}{\pi}\int_0^\infty \rho'\,d\rho'\,\frac{1}{\rho}\delta(\rho')\int_0^{2\pi}d\varphi'\int_{-\infty}^{+\infty}dz'\,\frac{r-r'}{|r-r'|^3}$$

$$= \frac{\kappa}{4\pi\,\epsilon_0}\frac{1}{\pi}\frac{1}{2}\int_0^{2\pi}d\varphi'\int_{-\infty}^{+\infty}dz'\,\frac{\rho\,e_p + (z-z')e_z}{\left[\rho^2 + (z-z')^2\right]^{3/2}}$$

$$= \frac{\kappa}{4\pi\,\epsilon_0}\int_{-\infty}^{+\infty}dy\left[\frac{\rho\,e_p}{\left(\rho^2+y^2\right)^{3/2}} - \underbrace{\frac{y\,e_z}{\left(\rho^2+y^2\right)^{3/2}}}_{e_z\frac{d}{dy}\frac{1}{\sqrt{\rho^2+y^2}}}\right]$$

$$= \frac{\kappa\rho}{4\pi\,\epsilon_0}\,e_\rho\underbrace{\int_{-\infty}^{+\infty}dy\,\frac{1}{\left(\rho^2+y^2\right)^{3/2}}}_{\frac{1}{\rho^2}\frac{y}{\sqrt{y^2+\rho^2}}\Big|_{-\infty}^{+\infty} = \frac{2}{\rho^2}}$$

Therefore it follows the electric field strength

$$E(r) = \frac{\kappa}{2\pi\,\epsilon_0\,\rho}\,e_\rho$$

and for the potential:

$$\varphi(r) = \frac{-\kappa}{2\pi\,\epsilon_0}\ln\rho + \text{const.}$$

Example 3.23: An infinitely long circular cylinder is homogeneously charged. Calculate the electric field strength and the corresponding potential.

Solution: We choose cylindrical coordinates (Fig. 3.61).

$$\rho, \varphi, z.$$

and exploit the cylindrical symmetry of the problems:

$$E(r) = E(\rho)e_\rho$$

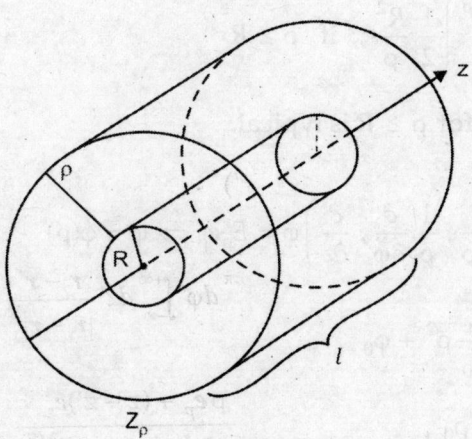

Fig. 3.61

For the charge density we have:

$$\overline{\rho}(r) = \begin{cases} \rho_0 & \text{for } \rho \leq R, \\ 0 & \text{otherwise} \end{cases}$$

Let Z_ρ be a cylinder with the lenght l, cylinder axis: z-axis, ρ: radius. Applying the Gauss theorem, leads to:

$$\int_{S(Z_\rho)} E \cdot df = \frac{1}{\varepsilon_0} \int_{Z_\rho} \overline{\rho}(r) d^3 r$$

We calculate the various contributions separately:

Front faces:

$$E \perp df \Rightarrow \text{no contribution to the flux,}$$

Now,

$$df = \rho \, d\varphi \, dz \, e_\rho$$

$$\Rightarrow \quad E \cdot df = \rho E_\rho(\rho) d\varphi \, dz$$

$$\Rightarrow \quad \int_{S(Z_\rho)} E \cdot df = 2\pi l \rho E_\rho(\rho)$$

$\rho \geq R$:

$$\int_{Z_p} \overline{\rho}(r) d^3 r = \rho_0 2\pi \int_0^R \rho' \, d\rho' \int_0^l dz' = \rho_0 \pi R^2 l$$

$\rho \leq R$:

$$\int_{Z_p} \bar{\rho}(r) d^3 r = \rho_0 2\pi \int_0^\rho \rho' \, d\rho' \int_0^l dz' = \rho_0 \pi \, \rho^2 l$$

Finally this yields:

$$E(r) = \frac{\rho_0}{\epsilon_0} e_\rho \begin{cases} \dfrac{1}{2} \rho, & \text{if } \rho \leq R, \\[2ex] \dfrac{1}{2} \dfrac{R^2}{\rho}, & \text{if } \rho \geq R \end{cases}$$

The $(1/\rho)$-dependence for $\rho \geq R$ is typical.

Potential:

$$E = -\left(\frac{\partial}{\partial \rho}, \frac{1}{\rho} \frac{\partial}{\partial \varphi}, \frac{\partial}{\partial z} \right) \varphi = E_\rho e_\rho \Rightarrow \varphi = \varphi(\rho)$$

inside:

$$\varphi(\rho) = -\frac{\rho_0}{4\varepsilon_0} \rho^2 + \varphi_0$$

outside:

$$\varphi(\rho) = -\frac{R^2 \rho_0}{2\varepsilon_0} \ln \rho + \varphi_1$$

The choice of the reference point is still free, e.g. $\varphi(\rho = R) \stackrel{!}{=} 0$

Then we have:

$$\varphi_0 = \frac{\rho_0}{4\varepsilon_0} R^2 ; \quad \varphi_1 = \frac{\rho_0 R^2}{2\varepsilon_0} \ln R$$

Thus it remains:

$$\varphi(r) = \varphi(\rho) = \frac{\rho_0 R^2}{2\varepsilon_0} \begin{cases} \dfrac{1}{2} \left(1 - \dfrac{\rho^2}{R^2} \right) & \text{for } \rho \leq R \\[2ex] \ln \dfrac{R}{\rho} & \text{for } R \leq \rho \end{cases}$$

Example 3.24: A dipole with moment p is located at r and point charge q at the origin of the coordinates (Fig. 3.62).

   i. Calculate the potential energy of the dipole

   ii. Calculate the force which acts on the dipole

   iii. Prove whether or not the third Newtonian axiom is fulfilled?

Solution:

   i. The potential energy of a dipole in the electric field amounts to:

$$V_D(r) = -p \cdot E(r)$$

Fig. 3.62: Dipole p and point charge q shifted relative to each other by vector r

The point charge creates field

$$E(r) = \frac{q}{4\pi\,\varepsilon_0}\,\frac{r}{r^3}$$

This gives:

$$V_D(r) = -\frac{q}{4\pi\,\varepsilon_0}\,\frac{p\cdot r}{r^3} = \frac{q}{4\pi\,\varepsilon_0}\,p\cdot\nabla\frac{1}{r}$$

ii. The force on the dipole can be derived from the potential energy:

$$F_D(r) = -\nabla V_D(r) = -\frac{q}{4\pi\,\varepsilon_0}\,\nabla\!\left(p\cdot\nabla\frac{1}{r}\right)$$

We use the formula:

$$\nabla(a\cdot b) = (b\cdot\nabla)a + (a\cdot\nabla)b + b\times\text{curl}\,a + a\times\text{curl}\,b$$

and obtain with $p = $ const:

$$\nabla\!\left(p\cdot\nabla\frac{1}{r}\right) = (p\cdot\nabla)\nabla\frac{1}{r} + \underbrace{p\times\text{curl}\nabla\frac{1}{4}}_{=0} = -\sum_i p_i\,\frac{\partial}{\partial x_i}\,\frac{r}{r^3}$$

$$= -\sum_i p_i\left(\frac{e_i}{r^3} - 3\frac{r}{r^4}\frac{x_i}{r}\right) = 3\frac{r(r\cdot p)}{r^5} - \frac{p}{r^3}$$

Thus,

$$F_D(r) = -\frac{q}{4\pi\,\varepsilon_0}\,\frac{3r(r\cdot p) - pr^2}{r^5}$$

iii. The field of the dipole at position **0** of the point charge is

$$E_D(r) = -\frac{1}{4\pi\,\varepsilon_0}\,\frac{3(-r)\,[(-r)\cdot p] - pr^2}{r^5}$$

From this a force results which the dipole exerts on the point charge:

$$F_p(r) = qE_D(r) = -F_D(r)$$

Obviously, the third Newtonian axiom is thus fulfilled.

Example 3.25: A hollow sphere with radius R carries the charge density

$$\rho(r) = \sigma_0\cos^2\theta\;\delta(r - R)$$

Calculate:
- i. the total charge q
- ii. the dipole moment p
- iii. the component's Q_{ij} of the quadrupole tensor
- iv. the electrostatic potential $\phi(r)$ and the electric field $E(r)$ upto quadrupole term.

Solution: We have charge density

$$e(r) = \sigma_0 \cos^2 \theta \, \delta(r - R)$$

a. Monopole:

$$q = \int d^3 r \rho(r) = \sigma_0 R^2 2\pi \int_{-1}^{+1} d\cos\theta \cos^2\theta$$

$$= \frac{4\pi}{3} R^2 \sigma_0$$

b. Dipole:

$$p = \int d^3 r \, r \, \rho(r)$$

$$= \sigma_0 R^3 \int_{-1}^{+1} d\cos\theta \int_0^{2\pi} d\varphi \cos^2\theta(\sin\theta\cos\varphi, \sin\theta\sin\varphi, \cos\theta)$$

$$= 2\pi\sigma_0 R^3 \int_{-1}^{+1} d\cos\theta(0, 0, \cos^3\theta)$$

$$\therefore \quad p = 0$$

c. Quadrupole:

$$Q_{xx} = \int d^3 r \rho(r)(3x^2 - r^2)$$

$$= \sigma_0 R^4 \int_0^{2\pi} d\varphi \int_{-1}^{+1} d\cos\theta \cos^2\theta \, (3\sin^2\theta\cos^2\varphi - 1)$$

We use

$$\int_0^{2\pi} d\varphi \cos^2\varphi = \int_0^{2\pi} d\varphi \cos^2\varphi = \pi$$

Applying integration by parts, we have

$$Q_{xx} = \pi\sigma_0 R^4 \int_{-1}^{+1} d\cos\theta \, (3(\cos^2\theta - \cos^4\theta) - 2\cos^2\theta)$$

$$= \pi\sigma_0 R^4 \left(\frac{2}{3} - \frac{6}{5} \right)$$

$$= -\frac{8}{15} \pi\sigma_0 R^4$$

$$Q_{yy} = \int d^3 r \rho(r) \, (3y^2 - r^2)$$

$$= \sigma_0 R^4 \int_0^{2\pi} d\varphi \int_{-1}^{+1} d\cos\theta \cos^2\theta \, (3\sin^2\theta\sin^2\varphi - 1)$$

$$= Q_{xx}$$

$$Q_{zz} = \int d^3 r \rho(\mathbf{r})\,(3z^2 - r^2)$$

$$= \sigma_0 R^4 \int_0^{2\pi} d\varphi \int_{-1}^{+1} d\cos\theta\, \cos^2\theta\, (3\cos^2\theta - 1)$$

$$= 2\pi\sigma_0 R^4 \left(\frac{6}{5} - \frac{2}{3} \right)$$

$$= \frac{16}{15}\pi\sigma_0 R^4$$

$$Q_{xy} = \int d^3 r \rho(\mathbf{r})(3xy)$$

$$= 3\sigma_0 R^4 \int_0^{2\pi} d\varphi\, d\cos\theta\, \sin^2\theta\, \cos\varphi\,\sin\varphi$$

$$= 0 = Q_{yx}$$

The φ-integration makes the term to vanish. The same holds also for the following elements.

$$Q_{xz} = \int d^3 r \rho(\mathbf{r})(3xz)$$

$$= 3\sigma_0 R^4 \int_0^{2\pi} d\varphi \int_{-1}^{+1} d\cos\theta\, \sin\theta\, \cos\varphi\,\cos\theta$$

$$= 0 = Q_{zx}$$

$$Q_{yz} = \int d^3 r \rho(\mathbf{r})(3yz)$$

$$= 3\sigma_0 R^4 \int_0^{2\pi} d\varphi \int_{+1}^{+1} d\cos\theta\, \sin\theta\, \sin\varphi\,\cos\theta$$

$$= 0 = Q_{yz}$$

Quadrupole tensor:

$$Q = \frac{8}{15}\pi\sigma_0 R^4 \begin{pmatrix} -1 & 0 & 0 \\ 0 & -1 & 0 \\ 0 & 0 & 2 \end{pmatrix}$$

iv. Scalar potential and electric field:
We take for the scalar potential

$$4\pi\varepsilon_0\, \varphi(\mathbf{r}) = \frac{4\pi}{3} R^2 \sigma_0 \frac{1}{2} + \frac{1}{2r^5} \frac{8}{15}\pi\sigma_0 R^4 (-x^2 - y^2 + 2z^2) + \cdots$$

$$= \frac{4\pi}{3} R^2 \sigma_0 \left(\frac{1}{r} + \frac{1}{5} R^2 \frac{1}{r^3} (2\cos^2\theta - \sin^2\theta) + \cdots \right)$$

This means

$$\varphi(\mathbf{r}) = \frac{\sigma_0 R^2}{3\varepsilon_0} \left(\frac{1}{r} + \frac{1}{r^3} \frac{R^2}{5} (3\cos^2\theta - 1) + \cdots \right)$$

From the scalar potential we drive with

$$\nabla \equiv \left(\frac{\partial}{\partial r}, \frac{1}{r}\frac{\partial}{\partial \theta}, \frac{1}{r\sin\theta}\frac{\partial}{\partial \varphi} \right)$$

electric field

$$E_r = \frac{\sigma_0 R^2}{3\varepsilon_0}\left(\frac{1}{r^2} + \frac{3}{r^4}\frac{R^2}{5}(3\cos^2\theta - 1) + \cdots \right)$$

$$E_\theta = \frac{\sigma_0 R^2}{3\varepsilon_0}\left(\frac{R^2}{5r^4}(-6\cos\theta\sin\theta) + \cdots \right)$$

$$E_\varphi = 0$$

We finally obtain:

$$E_{(r)} = \frac{\sigma_0 R^2}{3\varepsilon_0}\left(\left(\frac{1}{r^2} + \frac{3}{r^4}\frac{R^2}{5}(3\cos^2\theta - 1) + \cdots \right)e_r - \left(\frac{3R^2}{5r^4}\sin 2\theta + \cdots \right)e_\theta \right)$$

Example 3.26: Four charges q are arranged in a cartesian system of coordinate at the points.

$$(0, d, 0), \quad (0, -d, 0) \quad (0, 0, d), \quad (0, 0, -d)$$

and four charges $-q$ at the points

$$(-d, 0, 0), \quad \left(-\frac{d}{2}, 0, 0 \right), \quad (d, 0, 0), \quad (2d, 0, 0)$$

Calculate the dipole moment p and quadrupole tensor Q of this charge arrangement.

Solution: Charge density

$$\rho(r) = q\{\delta(x)\,\delta(z)\,[\delta(y-d) + \delta(y+d)] + \delta(x)\,\delta(y)\,[\delta(z-d) + \delta(z+d)]$$

$$-\delta(y)\,\delta(z)\,[\delta(x+d) + \delta(x+d/2) + \delta(x-d) + \delta(x-2d)]\}$$

Dipole moment:

$$p = \int d^3 r\rho(r)\cdot r = q\begin{pmatrix} +d + d/2 - d - 2d \\ d - d \\ d - d \end{pmatrix}$$

$$\Rightarrow \quad p = -qd\begin{pmatrix} 3/2 \\ 0 \\ 0 \end{pmatrix}$$

Quadrupole tensor:

$$Q_{ij} = \int d^3r\rho(r)\,(3x_ix_j - r^2\delta_{ij})$$

$$Q_{ij} = 0 \qquad\qquad [\forall i \neq j]$$

$$Q_{xx} = \int d^3r\rho(r)\left(2x^2 - y^2 - z^2\right)$$

$$= q\left(-d^2 - d^2 - d^2 - d^2 - 2d^2 - 2d^2/4 - 2d^2 - 8d^2\right)$$

$$= -qd^2\left(16 + \frac{1}{2}\right) = \frac{33}{2}qd^2$$

$$Q_{yy} = \int d^3r\rho(r)\left(2y^2 - x^2 - z^2\right)$$

$$= q\left(2d^2 + 2d^2 - d^2 - d^2 + d^2 + d^2/4 + d^2 + 4d^2\right)$$

$$= qd^2\left(8 + \frac{1}{4}\right) = \frac{33}{2}qd^2 = -\frac{1}{2}$$

$$Q_{zz} = \int d^3r\rho(r)\left(2z^2 - x^2 - y^2\right)$$

$$= q\left(-d^2 - d^2 + 2d^2 + 2d^2 + d^2 + d^2/4 + d^2 + 4d^2\right)$$

$$= qd^2\left(8 + \frac{1}{4}\right) = Q_{yy}$$

It follows:

$$Q_{zz} = Q_{yy} = -\frac{1}{2}Q_{xx} = \frac{33}{4}qd^2$$

$\therefore$ The trace is zero.

Example 3.27: Find the electric field at a distance Z above the centre of a square loop (side s) (Fig. 3.63) carrying a uniform line charge λ. **[MDS]**

Solution: Given:

$$L = \frac{s}{2} \text{ and } z = \sqrt{z^2 + (s/2)^2}$$

(distance from centre of edge to P)
therefore, the field of one edge

$$E_1 = \frac{1}{4\pi\varepsilon_0}\frac{\lambda s}{\sqrt{z^2 + \dfrac{s^2}{4}}\sqrt{z^2 + \dfrac{s^2}{4} + \dfrac{s^2}{4}}}$$

Fig. 3.63

We see that there are 4 sides, and we want vertical component, so we have to multiply by:

$$4 \cos \theta = 4 \frac{z}{\sqrt{z^2 + \dfrac{s^2}{4}}} \qquad \therefore E_1 = \frac{1}{4\pi\varepsilon_0} \frac{4\lambda s z}{\left(z^2 + \dfrac{s^2}{4}\right)\sqrt{z^2 + \dfrac{s^2}{2}}} \hat{z}$$

Example 3.28: Find the electric field at a distance Z above one end of a straight line segment of length L, which carries a uniform line charge λ. [RU]

Solution: From Fig. 3.64, we get

$$E_z = \frac{1}{4\pi\varepsilon_0} \int_0^L \frac{\lambda dx}{r^2} \cos \theta$$

$$r = \sqrt{z^2 + x^2}; \quad \cos \theta = \frac{z}{r}$$

Fig. 3.64

Now, we have

$$\therefore \quad E_z = \frac{1}{4\pi\varepsilon_0} \lambda \int_0^L \frac{dx}{(z^2 + x^2)^{3/2}} \hat{z} = \frac{1}{4\pi\varepsilon_0} \lambda z \left[\frac{1}{z^2} \frac{x}{\sqrt{z^2 + x^2}} \right]_0^L = \frac{1}{4\pi\varepsilon_0} \frac{\lambda}{z} \frac{L}{\sqrt{z^2 + x^2}}$$

$$E_x = -\frac{1}{4\pi\varepsilon_0} \int_0^L \frac{\lambda dx}{r^2} \sin \theta$$

$$\therefore \quad E_x = -\frac{1}{4\pi\varepsilon_0} \lambda \int_0^L \frac{x dx}{(z^2 + x^2)^{3/2}} = -\frac{1}{4\pi\varepsilon_0} \lambda \left[-\frac{1}{\sqrt{z^2 + x^2}} \right]_0^L$$

$$= -\frac{1}{4\pi\varepsilon_0} \lambda \left[\frac{1}{z} - \frac{1}{\sqrt{z^2 + L^2}} \right]$$

Therefore, total electric field

$$E = E_z + E_x = \frac{1}{4\pi\varepsilon_0}\frac{\lambda}{z}\left[\left(-1 + \frac{z}{\sqrt{z^2 + L^2}}\right)\hat{i} + \left(\frac{L}{\sqrt{z^2 + L^2}}\right)\hat{k}\right]$$

Example 3.29: The space between two concentric spheres with radius R_1 and R_2 $(R_1 < R_2)$ is charged with density

$$\rho(r) = \begin{cases} \dfrac{\alpha}{r^2} & \text{for } R_1 < r < R_2 (\alpha > 0), \\ 0 & \text{otherwise} \end{cases}$$

Calculate the total charge.

Solution: Let the sphere may carry the total charge Q:

$$Q = \int d^3r\, \rho(r) = \rho_0 \int_0^R r^2 dr \int_0^\pi \sin\theta\, d\theta \int_0^{2\pi} d\phi$$

$$= \rho_0 \frac{4\pi}{3} R^3,$$

$$\rho(r) = \begin{cases} \dfrac{Q}{\left(4\pi/3\right)R^3}, & \text{if } 0 \leq r \leq R, \\ 0 & \text{otherwise} \end{cases}$$

Example 3.30: An infinitely extended plane carries the homogeneous surface charge density σ (charge per unit area). Calculate the electric field and its potential.

Solution: Charge density:

$$\rho(r) = \sigma\delta(z)$$

Field strength:

$$E(r) = \iint_{-\infty}^{+\infty} dx'\, dy'\, \frac{(x - x', y - y', z)}{\left[(x - x')^2 + (y - y')^2 + z^2\right]^{3/2}}$$

$$= \frac{\sigma}{4\pi\varepsilon_0} z\hat{k} \int_{-\infty}^{+\infty} dx \int_{-\infty}^{+\infty} dy \frac{1}{\left(x^2 + y^2 + z^2\right)^{3/2}}$$

$$= \frac{\sigma}{4\pi\varepsilon_0} z\hat{k} \int_{+\infty}^{+\infty} dx \left(\frac{1}{x^2 + z^2} \frac{1}{\sqrt{x^2 + y^2 + z^2}}\right)\Bigg|_{-\infty}^{+\infty}$$

$$= \frac{\sigma}{4\pi\varepsilon_0} z\hat{k} \int_{-\infty}^{+\infty} dx \left(\frac{1}{x^2 + z^2}\right)$$

For $z \neq 0$ it holds for the integral:

$$\frac{1}{z} \arctan \frac{x}{z} \Big|_{-\infty}^{+\infty} = \frac{1}{z} \frac{z}{|z|} \pi$$

$$\Rightarrow E(r) = \frac{\sigma}{2\varepsilon_0} \frac{z}{|z|} \hat{k}$$

and $V(r) = -\frac{\sigma}{2\varepsilon_0} |z| + \text{constant.}$

Example 3.31: For the hydrogen atom in its ground state it holds approximately. The nuclear charge is centered point-like at the origin and the average electron charge density is given by

$$\rho_c(r) = \frac{e}{\pi a^3} \exp\left(-\frac{2r}{a}\right)$$

where a is Bohr radius. Calculate the electric field strength $E(r)$ and its potential V. Discuss the limiting cases $r \ll a, r \gg a$. [MDSA]

Solution: We have from Gauss's law

$$\int_S E \cdot ds = \frac{1}{\varepsilon_0} \int_V \rho(r')d^3r'$$

Charge density:

$$\rho(r') = \rho(r') = \frac{e}{4\pi r'^2} \delta(r') \qquad \text{[for point like nuclear charge } (z = 1)]$$

$$- \rho(r) = \frac{e}{4\pi a^3} e^{-(2r'/a)} \qquad \text{(for electron in its ground state)}$$

We see that this is spherically-symmetric charge distribution. Therefore, we have

$$E(r) = E_r(r)\hat{r}$$

Let us choose:

volume V_r sphere with radius r, origin at the center of the sphere,

$$ds = r^2 \sin\theta \, d\theta \, d\phi \hat{r};$$

area element on $S(V)$

Now, we have

$$4\pi r^2 E_r(r) = \frac{e}{\varepsilon_0} - \frac{e}{\varepsilon_0 \pi a^3} 4\pi \int_0^r dr' \, r'^2 e^{(-2r'/a)}$$

$$\int_0^r dx \; x^2 \; e^{-\beta x} = -\frac{d^2}{d\beta^2} \int_0^r dx \; e^{-\beta x}$$

$$= \frac{d^2}{d\beta^2} \left[-\frac{1}{\beta} \left(e^{-\beta r} - 1 \right) \right]$$

$$= \frac{d}{d\beta} \left[\frac{1}{\beta^2} \left(e^{-\beta r} - 1 \right) + \frac{r}{\beta} e^{-\beta r} \right]$$

$$= -\frac{2}{\beta^3} \left(e^{-\beta r} - 1 \right) - \frac{2r}{\beta^2} e^{-\beta r} - \frac{r^2}{\beta} e^{-\beta r}$$

$$= \frac{2}{\beta^3} - e^{-\beta r} \left(\frac{2}{\beta^3} + \frac{2r}{\beta^2} + \frac{r^2}{\beta} \right)$$

$$\int_0^r dr' \; e^{-(2r'/a)} = \frac{a^3}{4} - \frac{a}{2} e^{-(2r/a)} \left(\frac{a^2}{2} + ar + r^2 \right)$$

$$\Rightarrow 4\pi r^2 \; E_r(r) = \frac{2e}{\varepsilon_0 a^2} e^{-(2r/a)} \left(\frac{a^2}{r} + ar + r^2 \right)$$

$$= \frac{e}{\varepsilon_0} e^{-(2r/a)} \left(1 + \frac{2r}{a} + \frac{2r^2}{a^2} \right)$$

Thus, the electric field

$$E(r) = \hat{r} \; \frac{e}{4\pi\varepsilon_0} e^{-(2r/a)} \left(\frac{1}{r^2} + \frac{2}{ra} + \frac{2}{a^2} \right)$$

and the potential can be obtained by using the relation

$$E_r(r) = -\frac{dV}{dr}$$

We have

$$V(r) = \int_\infty^r E_r , \; dr'$$

$$\int_\infty^r dr' \left(e^{(-2r'/a)} \right) \frac{2}{a^2} = -\frac{1}{a} e^{-2r'/a}$$

$$= -\frac{1}{a} e^{-(2r/a)}$$

$$\int_\infty^r dr' \left(\frac{2}{r'a} + \frac{1}{r'^2} \right) e^{-(2r'/a)} = -\int_\infty^r dr' \frac{d}{dr'} \left[\frac{1}{r'} e^{-(2r'/a)} \right]$$

$$= -\frac{1}{r} e^{-(2r/a)}$$

Thus, we obtain a screened Coulomb potential:

$$V(r) = \frac{e}{4\pi\varepsilon_0} e^{-2r/a} \left(\frac{1}{r} + \frac{1}{a} \right)$$

i. $r \ll a$:

$$V(r) \approx \frac{e}{4\pi\varepsilon_0} r$$

(pure Coulomb potential of the nucleus)

ii. $r \gg a$:

$$V(r) \approx \frac{e}{4\pi\varepsilon_0 a} e^{-(2r/a)}$$

Obviously, the total potential of the H-atom vanishes exponentially for long distances.

Example 3.32: An electric dipole p_1 is at the origin of the coordinates and point in the z-direction. A second electric dipole p_2 is at the position $(x_0, 0, z_0)$. Which direction does p_2 take in the field of p_1? [MDSA]

Solution: The dipole $p_1 = p_1 \hat{k}$ produces the potential

$$V_1(r) = \frac{1}{4\pi\varepsilon_0} \frac{r \cdot p_1}{r^3} = \frac{1}{4\pi\varepsilon_0} \frac{p_1 z}{r^3}$$

and therewith the electric field:

$$E_x^{(1)} = -\frac{\partial V_1}{\partial x} = \frac{3p_1}{4\pi\varepsilon_0} \frac{xz}{r^5}$$

$$E_y^{(1)} = -\frac{\partial V_1}{\partial y} = \frac{3p_1}{4\pi\varepsilon_0} \frac{yz}{r^5}$$

$$E_z^{(1)} = -\frac{\partial V_1}{\partial z} = \frac{p_1}{4\pi\varepsilon_0} \left(\frac{3z^2}{r^5} - \frac{1}{r^3} \right)$$

The potential energy of the dipole p_2 in the field of dipole p_1 can be calculated from

$$PE_D^{(2)} = \bigcup_D^{(2)} = -p_2 \cdot E^{(1)}$$

The potential energy becomes minimal for the direction we are looking for, i.e. p_2 orients itself parallel to $E^{(1)}$ as shown in Fig. 3.65.

Fig. 3.65

$$E_x^{(1)}(x_0, 0, z_0) = \frac{3p_1}{4\pi\varepsilon_0} \frac{x_0 z_0}{r_0^5}$$

$$E_y^{(1)}(x_0, 0, z_0) = 0$$

$$E_z^{(1)}(x_0, 0, z_0) = \frac{p_1}{4\pi\varepsilon_0} \frac{1}{r_0^5}\left(2z_0^2 - x_0^2\right)$$

$$\tan\alpha = \frac{E_z^{(1)}(x_0, 0, z_0)}{E_x^{(1)}(x_0, 0, z_0)} = \frac{2z_0^2 - x_0^2}{3x_0 z_0}$$

Example 3.33: A given charge distribution $\rho(r)$ possesses axial symmetry about the z-axis.

i. Show that the quadrupole tensor is diagonal

ii. Verify $Q_{xx} = Q_{yy} = -\frac{1}{2}Q_{zz}$

iii. Calculate the potential and the electric field strength of the quadrupole as a function of Q_{zz}. [PU]

Solution:

i. Spherical coordinates: r, θ, ϕ

Axial symmetry: $\rho(r) = \rho(r, \theta); \dfrac{\partial\rho}{\partial\phi} = 0$

$$x = r\sin\theta\cos\phi,$$
$$y = r\sin\theta\sin\phi,$$
$$z = r\cos\theta$$

$$Q_{xy} = \int d^3r\, \rho(r)(3xy) = 3\int_0^\infty dr\, r^4 \int_{-1}^{+1} d\cos\theta \sin^2\theta\rho(r,\theta) \int_0^{2\pi} d\phi\cos\phi\sin\phi$$

$$= 0 = Q_{yx} \qquad \left(\because \int_0^{2\pi} d\phi\cos\phi\sin\phi = \frac{1}{2}\sin^2\phi \Big|_0^{2\pi} = 0\right)$$

$$Q_{xz} = 3 \int_0^\infty dr\, r^4 \int_{-1}^{+1} d\cos\theta \sin\theta \cos\theta\, \rho(r,\theta) \int_0^{2\pi} d\phi \cos\phi$$

$$= 0 = Q_{zx} \qquad\qquad \left(\because \int_0^{2\pi} d\phi \cos\phi = 0 \right)$$

$$Q_{yz} = 3 \int_0^\infty dr\, r^4 \int_{-1}^{+1} d\cos\theta \sin\theta \cos\theta\, \rho(r,\theta) \int_0^{2\pi} d\phi \sin\phi$$

$$= 0 = Q_{zy} \qquad\qquad \left(\because \int_0^{2\pi} d\phi \sin\phi = 0 \right)$$

ii. $\qquad Q_{xx} = \int d^3r\, \rho(r)(3x^2 - r^2)$

$$= \int d^3r\, \rho(r)\left(2x^2 - y^2 - z^2 \right)$$

$$Q_{yy} = \int d^3r\, \rho(r)\left(2y^2 - x^2 - z^2 \right)$$

Combining, we obtain

$$Q_{xx} - Q_{yy} = 3 \int d^3r\, \rho(r)(x^2 - y^2)$$

$$= 3 \int_0^\infty dr\, r^4 \int_{-1}^{+1} d\cos\theta \sin^2\theta\, \rho(r,\theta) \int_0^{2\pi} d\phi \underbrace{\left(\cos^2\phi - \sin^2\phi \right)}_{\cos 2\phi} \frac{1}{2} \sin 2\phi \Big|_0^{2\pi} = 0$$

$$Q_{xx} = Q_{yy}$$

We further follows from the tracelessness:

$$Q_{zz} = Q_0 = -(Q_{xx} + Q_{yy})$$

$$Q_{xx} = Q_{yy} = -\frac{1}{2} Q_0$$

iii. $\qquad 4\pi\varepsilon_0 V_Q(r) = \dfrac{1}{2r^5} \sum_{i,j} Q_{ij}\, x_i x_j = \dfrac{Q_0}{2r^5}\left(z^2 - \frac{1}{2}x^2 - \frac{1}{2}y^2 \right)$

$$= \frac{Q_0}{r^3}\left(\cos^2\theta - \frac{1}{2}\sin^2\theta \right) = -\frac{Q_0}{4r^3}\left(1 - 3\cos^2\theta \right)$$

$$\Rightarrow V_Q(r) = -\frac{Q_0}{16\pi\varepsilon_0} \frac{1 - 3\cos^2\theta}{r^3}$$

Using $\nabla = \left(\dfrac{\partial}{\partial r},\, \dfrac{1}{r}\dfrac{\partial}{\partial\theta},\, \dfrac{1}{r\sin\theta}\dfrac{\partial}{\partial\phi} \right)$, one obtains

$$\frac{\partial}{\partial r}\left(\frac{1-3\cos^2\theta}{r}\right) = -3\frac{1-3\cos^2\theta}{r^4}$$

$$\frac{1}{r}\frac{\partial}{\partial\theta}\left(\frac{1-3\cos^2\theta}{r^3}\right) = \frac{3}{r^4}2\cos\theta\sin\theta = \frac{3\sin 2\theta}{r^4}$$

$$\therefore\quad E_Q(r) = -\nabla V_Q(r) = -\frac{3Q_0}{16\pi\varepsilon_0}\frac{1}{r^4}\left[(1-3\cos^2\theta)\hat{r} - \sin 2\theta\,\hat{\theta}\right]$$

GLIMPSES

1. *Electric charge* is an intrinsic property of the electrons and protons that along with uncharged neutrons, make up ordinary matter.
2. *Charge* comes in *two varieties*, which Benjamin Franklin designated as *positive* and *negative*. The object's *net charge* is the algebraic sum of its constituent charges.
3. Like charges *repel* and opposites *attract*—a fact that constitutes a qualitative description of the *electric force*.
4. The magnitude of an electron or a proton charge is *elementary charge 'e'*.
5. Charge is *quantized*, i.e. it comes only in discrete amounts. Elementary particle

   theories show that the basic unit of charge is actually $\frac{1}{3}e$. Such *fractional charges*

   reside on *quarks*, the building blocks of protons, neutrons, and many other particles. Quarks always join to produce particles with integral multiples of the full elementary charge, and it seems impossible to isolate individual quarks.
6. The *SI unit* of charge is coulomb (C). $1\,C$ is about 6.25×10^{18} elementary charges making the elementary charge approximately $1.60 \times 10^{-19}\,C$.
7. Electric charge is a *conserved* quantity, meaning that the net charge in a closed region remains constant. Charged particles may be created or annihilated, but always in pairs of equal and opposite charge. The net charge always remains same.
8. *Coulomb's law* states that the force between two charges act along the line joining them, with magnitude directly proportional to the product of the charges and inversely proportional to the square of the distance between them.

$$F_{12} = K\frac{q_1 q_2}{r^2}\hat{r}$$

   where, F_{12} is the force that charge q_1 exerts on charge q_2 and r is the distance between then. The proportionality constant K has SI value $9.0 \times 10^9\,N\,m^2\,C^{-2}$. The Coulomb's law accounts for the fact that like charges repel and opposite charges attract.
9. *Coulomb's law for electric force* is similar to *Newton's law for gravitational force*. Both show the same inverse square decrease with distance, and both are proportional

to the product of the interacting *charges* or *masses*. But there is an important difference, *viz.* there is only one kind of mass, and the gravitational force is always attractive.

10. The electric force between individual particles is vastly stronger than the gravitational force between the same particles, and it is only because of nearly complete cancellation of positive and negative charge in bulk matter that *gravity becomes important on the macroscopic scale.*

11. Coulomb's law is strictly true only for *point charges,* i.e. charged objects of negligible size.

12. When a number of charges are interacting, the total force on a given charge is the vector sum of the forces exerted on it due to all other charges. The force between two charges is not affected by the presence of other charges. This is called the *principle of superposition of electrostatic forces.*

13. The *electric field (E)* at any point is the force per unit charge that would be experienced by a charge at that point.

$$(E) = \frac{F}{q}$$

14. The *electric field intensity* at a point in an electric field is defined as the force acting per unit positive test charge placed at that point.

$$E = \lim_{q \to 0} \frac{F}{q}$$

SI unit of electric field intensity is NC^{-1}.

15. The field of a point charge is

$$E = \frac{F}{q_{test}} = K \frac{q}{r^2} \hat{r}$$

16. The *field of charge distribution or a set of discrete point charges* is the vector sum of the fields of the individual point charges making up the distribution.

$$\sum_i E_i = \sum_i K \frac{q_i}{r_i^2} \hat{r}_i$$

Here E_i is the field of the point charge q_i located at distance r_i from the point where we are evaluating the field called, approximately, the field point.

17. An *electric dipole* consists of two charges equal in magnitude, opposite in sign, and separated by a certain distance. The study of electric dipole is important for understanding the electrical behaviour of certain compounds. When a dipole is placed in an electric field, there is a torque on the dipole, tending to rotate it such that its axis becomes aligned with the electric field. The magnitude of the torque depends on the strength of the external field, the magnitude of the charges, the separation of the charges, and the angle between the field direction and the axis of the dipole. For a given field strength, magnitude and separation of charges, maximum torque occurs when the axis of the dipole is at right angles to the field direction.

18. *Electric potential* at a point having position vector r, due to a dipole of moment p at the origin is given by

$$V = \frac{1}{4\pi\varepsilon_0} \frac{p \cdot r}{r^3} = \frac{1}{4\pi\varepsilon_0} \frac{p\cos\theta}{r^2}$$

At points on the axial line of the dipole ($\theta = 0°$ or $180°$)

$$V_{axial} = \pm\frac{1}{4\pi\varepsilon_0} \frac{p}{r^2}$$

At points on the equatorial line of the dipole ($\theta = 90°$), $V_{equa} = 0$.

19. In general, a dipole's electric properties are characterized completely by its *electric dipole moment $p = q \times 2a$*, defined as the product of the magnitude q of the charges and separation $2a$ between the two charges making up the dipole.

20. A dipole in an electric field experiences a torque that tends to align it with the field, $\tau = p \times E$.

21. Potential energy (P.E.) of a dipole in a *uniform electric field* is equal to the amount of work done in turning the dipole from orientation θ_1 to θ_2 in the field E,

$$U = -pE (\cos\theta_1 - \cos\theta_2)$$

If initially the dipole is perpendicular to the field E, $\theta_1 = 90°$ and $\theta_2 = \theta$ (say), then $U = pE \cos\theta = -p \cdot E$. When $\theta = 0°$, i.e. the potential energy of the dipole is minimum. The dipole is in *stable equilibrium*,

when $\theta = 90°$, $U = 0$
when $\theta = 180°$, $U = pE$

i.e. the P.E. of the dipole is maximum, and the dipole is in *unstable equilibrium*.

22. Potential at an external point due to a continuous charge distribution may be expressed as the sum of the potentials due to monopole, dipole, quadrupole, octupole, etc. The quadrupole moment is $Q = qd^2$ and potential is

$$V_p = \frac{Q}{4\pi\varepsilon_0 r^3}(3\cos^2\theta - 1)$$

The field decreases as $1/r^4$.

REVIEW QUESTIONS

1. Define the flux of a vector field and intensity of electric field.
2. State the principle of superposition as applied to electric potential.
3. State Gauss's law in electrostatics. Apply it to calculate the electric field intensity due to a uniformly charged sphere (non-conducting) at points: (a) outside the sphere, (b) on the surface of the sphere and (c) inside the sphere.
4. Apply Gauss's law to find the electric field intensity at a point near a uniformly charged wire of infinite length. Hence, calculate the force per unit length on a similarly charged wire parallel to it and placed at a distance d from the first one.
5. (a) Plot the magnitude of electric field against the distance for a conducting sphere of radius a carrying charge Q.
   (b) In what region does this field differ from its value for a uniformly charged sphere of the same radius and charge?

6. Define dipole moment of an electric dipole Derive an expression for the potential at a point (r, θ) due to a dipole moment p.

7. Define potential at a point in electrostatic field. Show that $E = -\nabla V$.

8. Derive an expression for the electric field due to an electric dipole and show that the absolute value of the electric field of an electric dipole decreases monotonically as the angle θ between the dipole moment and radius to the observer increases from 0 to $\dfrac{\pi}{2}$ radians when the distance of the observer is constant.

9. Define electric dipole, quadrupole and octupole. Show that the potential of charge distribution over a finite volume can always be expressed as a series of multipole potentials, referred to a suitable origin of coordinate system.

10. What do you understand by electric dipole and quadrupole moment? Obtain expressions for electric potential and field at a point in space due to a (a) dipole and (b) quadrupole.

11. An electric quadrupole is formed by a charge $-q$ at the origin and charge $+q$ at the points $(\pm a, 0, 0)$. Show that the potential ϕ at a distance r $(r \gg a)$ is approximately given by

$$\phi = \frac{1}{4\pi\varepsilon_0} \frac{qa^2(3\cos^2\theta - 1)}{r^3}$$

where, θ is the angle between r and the line through the charge and that the value of electric field E on the axis of quadrupole for a point distant r from its centre (assume $r \gg a$) is given by

$$E = \frac{1}{4\pi\varepsilon_0} \frac{6qa^2}{r^4}$$

12. Show that the interaction energy of two dipoles of moments p_1 and p_2 is given by

$$U = \frac{1}{4\pi\varepsilon_0} \left[\frac{p_1 \cdot p_2}{r^2} - \frac{3}{r^5}(p_1 \cdot r)(p_2 \cdot r) \right]$$

where, r is the radius vector joining the centres of the two dipoles.

PROBLEMS

1. The potential at a certain distance from a point charge is 600 V and the electric field is 200 N/C. (a) What is the distance of the point charge? (b) What is the magnitude of the point charge?

[**Ans.** (a) 3 m (b) 2×10^{-7} C]

2. Two particles each of mass m and charge q are suspended by strings each of length l from the same point. Show that the inclination θ of each string to the vertical is given by

$$\frac{\cos\theta}{\sin^3\theta} = 16\pi\varepsilon_0 mgl^2 / q^2$$

3. The flux entering a closed surface is $2 \times 10^3 \, \text{N m}^2/\text{C}$ and leaving the closed surface is $8 \times 10^3 \, \text{N m}^2/\text{C}$. Calculate the charge enclosed by the surface.

[*Ans.* $5.31 \times 10^{-8} \text{C}$]

4. An infinite line charge has a charge density $2 \, \mu\text{C}/\text{m}$. Calculate the electric field at a point which is at a normal distance of 10 cm from the line charge.

[*Ans.* $36 \times 10^4 \, \text{N/C}$]

5. Show that the rectangular components of the electric field produced by a charge q

at the distance r are $E_x = \dfrac{qx}{4\pi\varepsilon_0 r^3}$, etc.

6. Four equal small spheres, each of mass m and carrying a charge q are suspended by light strings each of length l, from the same point. Show that the side of the square that the spheres will form due to the force of repulsion is given by

$$\frac{a}{\sqrt{2l}} = \frac{q^2}{4\pi\varepsilon_0 mga^2}(\sqrt{2}+0.5)\frac{\sqrt{l^2-a^2/2}}{l}$$

If the charge on each sphere and length of each string is increased by a factor k, what should be the mass of each sphere so as to make the side of the square ka?

[*Ans.* $m = m'$]

7. Consider a conducting spherical shell of radius a and charge ρ per unit area. Find by integration the potential at a point distant r from its centre. Then, show that

$$E = \frac{\rho}{\varepsilon_0}\frac{a^2}{r^2} \quad \text{for } r > a$$

$$= 0 \quad \text{for } r < a$$

8. Find the electric potential and field along the points on the axis of a disc of radius R carrying a charge σ per unit area.

[*Hint:* Divide the disc into rings and add the contribution of all rings.]

$$\left[Ans. \ \left(\frac{\sigma}{2\varepsilon_0}\right)(\sqrt{R^2+r^2}-r); \left(\frac{a}{4\varepsilon_0}\right)\log\left(\frac{R^2}{r^2}+1\right) \right]$$

9. A spherical region of radius a is filled with positive charge of volume density ρ which varies with the distance r from the centre of the sphere but is constant for a given r. Find the total charge in the sphere if

(a) $\dfrac{\rho}{\rho_0} = \dfrac{a}{r}$ and (b) $\dfrac{\rho}{\rho_0} = \left(1-\dfrac{r^2}{a^2}\right)$ [*Ans.* (a) $2\pi\rho_0 a^3$ (b) $\dfrac{8\pi}{15}\rho_0 a^3$]

10. A wire of length L carries a charge λ per unit length (Fig. 3.66). (a) Show that the electric field at a point distant R from the wire is given by

$$E_\perp = \left(\frac{\lambda}{4\pi\varepsilon_0 R}\right)(\sin\theta_2 - \sin\theta_1)$$

$$E_{\parallel} = \left(\frac{\lambda}{4\pi\varepsilon_0 R}\right)(\cos\theta_2 - \cos\theta_1)$$

where, $E_\perp$ and $E_\parallel$ are the components of E perpendicular and parallel to the wire and θ_1 and θ_2 are the angles that the lines from the point to the ends of the wire make with the perpendicular to the wire. (b) Find the field when the point is equidistant from both ends. The signs of angles θ_1 and θ_2 are as indicated in the figure.

$$\left[\textbf{\textit{Ans.}} \text{ (b) } E_\perp = \left(\frac{2}{4\pi\varepsilon_0 R}\right)\sin\theta; \ E_\parallel = 0\right]$$

Fig. 3.66

11. Two identical spheres of radius r are separated by a distance $d \gg r$. (a) What is the potential energy of the system if we put charge $\dfrac{Q}{2}$ on each sphere? (b) What is the potential energy if we put all the charge Q on one sphere? (c) What are the electrical potentials on each sphere in cases (a) and (b)?

[**Ans.** (a) $\dfrac{KQ^2}{4d}$ (b) $\dfrac{KQ^2}{2r}$ (c) $\dfrac{1}{2}\dfrac{KQ}{r}$ for case (a) and $\dfrac{KQ}{r}$ for case (b) where $K = \dfrac{1}{4\pi\varepsilon_0}$]

12. A square sheet is uniformly charged with charge density σ. Show that the potential at the centre of the square is $\phi_0 = \dfrac{\sigma a}{\pi\varepsilon_0}\log(1+\sqrt{2})$, where a is the length of the side of the square.

13. Two electric dipoles A and B are placed in such a way that the direction of A passes through B while the direction of B is per-pendicular to that of A. Show that the actual force exerted by A on B is not in the same direction as the actual force exerted by B on A. Does this not contradict well known Newton's third law of motion?

14. Obtain the interaction energy between two electric dipoles. Use the result to estimate the interaction energy between two water molecules. Also discuss relative orientation effects.

15. For a region comprising electric field intensity $E = -\dfrac{16}{r^2}\hat{a}_r$, find the potential difference V_{AB} in spherical coordinate system, where A is at $(2, \pi, \pi/2)$ and B is at $(4, 0, \pi)$.

[*Hint:* $dl = dr\,\hat{a}_r$

Then $V = -\int E \cdot dl = -\int_4^2 \left(-\dfrac{16}{r^2}\right)dr = 16\int_4^2 \dfrac{dr}{r^2} = 16\left[\dfrac{r^{-2+1}}{-2+1}\right]_4^2$

$= -16\left[\dfrac{1}{r}\right]_4^2 = -16\left(\dfrac{1}{2} - \dfrac{1}{4}\right) = -\dfrac{16}{4} = -4\text{V}]$

16. A thin line charge of length $2a$ is coincident with y-axis with its centre at the origin, uniformly distributed along the line with a line charge density λ. Determine potential at any point $P(x, y)$ as shown in Fig. 3.67.

Fig. 3.67: A thin line charge of length $2a$ coincident with the y-axis

[*Hint:* $dl = dy$, $dq = \lambda dl = \lambda dy$. Since $Q = \int \lambda dl$ and $dl = dy$
so the potential at point P is given by

$V = \dfrac{Q}{4\pi\varepsilon_0 (OP)} = \int_{-a}^{a} \dfrac{\lambda dy}{4\pi\varepsilon_0 \sqrt{(x_0 - 0)^2 + (y_0 - y)^2}}$

$= \dfrac{\lambda}{4\pi\varepsilon_0} \int_{-a}^{a} \dfrac{dy}{x_0^2 + (y_0 - y)^2} = \dfrac{\lambda}{4\pi\varepsilon_0}\left[\left(-\dfrac{1}{x_0}\right)\sinh^{-1}\left(\dfrac{y_0 - y}{x_0}\right)\right]_{-a}^{a}$

$= \dfrac{\lambda}{4\pi\varepsilon_0} \dfrac{1}{x_0}\left[\sinh^{-1}\left(\dfrac{y_0 + a}{x_0}\right) - \sinh^{-1}\left(\dfrac{y_0 - a}{x_0}\right)\right]$

$V = \dfrac{\lambda}{4\pi\varepsilon_0 x}\left[\sinh^{-1}\left(\dfrac{y + a}{x}\right) - \sinh^{-1}\left(\dfrac{y - a}{x}\right)\right].$

17. Determine the change in potential in moving along x-axis from $x = b > a$ to $x = -b$ in a field due to two long parallel oppositely charged uniform line charges of densities λ and $-\lambda$ located parallel to z-axis at $x = \pm a$ (Fig. 3.68).

Fig. 3.68: Two infinite line charges separated by distance $2a$ and points of observation P and Q equidistant from the origin O

[*Hint:* We know that the potential at point P due to an infinite line charge is given by

$$V = \frac{\lambda}{2\pi\varepsilon_0}\ln\frac{k}{\rho}$$

where, k is the distance between reference point and line charge and ρ is the distance between point of consideration and line charge. Using this concept and taking origin as the point of reference, the potential due to two given oppositely line charges at point P is expressed as

$$V_P = \frac{\lambda}{2\pi\varepsilon_0}\ln\frac{a}{b-a} - \frac{\lambda}{2\pi\varepsilon_0}\ln\frac{a}{b+a}$$

$$= \frac{\lambda}{2\pi\varepsilon_0}\ln\left(\frac{b+a}{b-a}\right)$$

Similarly, the potential at point Q is given by

$$V_Q = \frac{\lambda}{2\pi\varepsilon_0}\ln\frac{a}{b-a} + \frac{\lambda}{2\pi\varepsilon_0}\ln\frac{a}{b+a}$$

$$= \frac{\lambda}{2\pi\varepsilon_0}\ln\frac{b-a}{b+a}$$

Thus, the change in potential in moving from P to Q, i.e. V_{PQ} is given by

$$V_P - V_Q = \frac{\lambda}{2\pi\varepsilon_0}\ln\left(\frac{b+a}{b-a}\right) - \frac{\lambda}{2\pi\varepsilon_0}\ln\left(\frac{-b-a}{b+a}\right)$$

$$= \frac{\lambda}{2\pi\varepsilon_0}\ln\left(\frac{\frac{b+a}{b-a}}{\frac{b-a}{b+a}}\right) = \frac{\lambda}{2\pi\varepsilon_0}\ln\left(\frac{b+a}{b-a}\right)^2$$

$$V_{PQ} = 2\frac{\lambda}{2\pi\varepsilon_0}\ln\left(\frac{b+a}{b-a}\right) = \frac{\lambda}{\pi\varepsilon_0}\ln\left(\frac{b+a}{b-a}\right).]$$

SHORT ANSWER QUESTIONS

1. Can the net or total charge for any process occuring within an isolated system change?

Ans. No

2. Can an electric field be represented by lines of force?

   Ans. Yes

3. When several charges are present, what will be the resultant electric field?

   Ans. The vector sum of the electric fields produced by each charge.

4. How many electrons are there in one coulomb charge?

   Ans. $\left[\dfrac{1}{1.602 \times 10^{-19}} = 6.25 \times 10^{18} \right]$

5. Does Coulomb's law of electric force obey Newton's third law of motion?

   Ans. Yes

6. Write the dimensional formula for ε_0.

   Ans. $M^{-1} L^{-3} T^4 A^2$

7. Is the electric force between two charges greater than the gravitational force between them? If so, by what factor?

   Ans. $\dfrac{F_e}{F_g} \approx 10^{39}$, i.e. electric force ($F_e$) is about 10^{39} times greater than the gravitational interaction.

8. Which is bigger, -1 coulomb or charge on the electron?

   Ans. 1 coulomb

9. Why do not two electric lines of force intersect each other?

   Ans. Let us suppose that two lines of force interest each other. Then two tangents can be drawn from the point of intersection making for two directions of the resultant electric force, which is not possible.

10. What is an electric dipole?

    Ans. It is a system of two equal and opposite electric charges (total charge, i.e. $q_1 + q_2 = 0$) separated by a distance. Obviously, an electric dipole or just a dipole, consists of a pair of equal and opposite point charges q and $-q$ (where $q > 0$) separated by a very small distance a. The magnitude of the dipole moment $|p| = qa$ and p is a vector quantity.

11. Does the work done to move a charge from one point to another depends on the path it follows between them?

    Ans. No. The work done to move a charge between two points depends on the position of the points only, otherwise the principle of conservation of energy would be violated.

12. Two identical metallic spheres of exactly equal masses are taken. One of them is given a positive charge $+Q$ coulomb and the other an equal negative charge. What will be their masses after charging?

    Ans. Different. When positive charge is given it involves removal of electrons, whereas a negative charge involves addition of electrons.

13. Can a point in space have a non-zero potential even when electric field in the space is zero?

Ans. Yes. Electric field intensity at any point is given by $E = -\dfrac{d\phi}{dx}$. If potential ϕ is constant in space, the potential gradient in any direction is zero, hence the intensity is zero. Obviously, potential can exist where there is no electric field.

MULTIPLE CHOICE QUESTIONS

1. The number of lines of force that radiate outward from one coulomb of positive charge is

   (a) infinite

   (b) $\varepsilon_0 (= 8.85 \times 10^{-12})$

   (c) $\dfrac{1}{\varepsilon_0} (= 1.13 \times 10^{11})$

   (d) zero   **[c]**

2. A positively charged particle of mass m kg and charge Q coulomb travels from rest through a potential difference of V volt. Its kinetic energy in joule is

   (a) $m\dfrac{Q}{V}$ (b) $\dfrac{m}{QV}$ (c) mQV (d) QV **[d]**

3. Two charges of $1\,C$ and $5\,C$ are placed at some distance in air. The ratio of the forces acting on each of them is

   (a) $1:1$ (b) $5:1$ (c) $1:5$ (d) $1:25$ **[a]**

4. The work done in moving a positive charge on an equipotential surface is

   (a) finite and positive   (b) infinite

   (c) finite and negative   (d) zero   **[d]**

5. A and B are two points in an electric field. If $8\,J$ of work is done in taking $2\,C$ of electric charge from A to B, then the potential difference between A and B is

   (a) $16\,V$ (b) $4\,V$ (c) $2\,V$ (d) zero   **[b]**

6. Two point charges $+2\,C$ and $+6\,C$ repel each other with a force of $12\,N$. If a charge of $-4\,C$ is given to each of these charges, the force will be

   (a) $4\,N$ (attractive)   (b) $4\,N$ (repulsive)

   (c) $8\,N$ (repulsive)   (d) $8\,N$ (attractive)   **[a]**

7. Two thin infinite parallel plates have uniform charge densities $+\sigma$ and $-\sigma$. The electric field in the space between them is

   (a) $\dfrac{\sigma}{\varepsilon_0}$ (b) $\dfrac{\sigma}{2\varepsilon_0}$ (c) $\dfrac{2\sigma}{\varepsilon_0}$ (d) zero   **[a]**

8. A hollow metal sphere is charged so that the potential on its surface is $10\,V$. The potential at the centre of the sphere is

   (a) $10\,V$

   (b) $0\,V$

   (c) same as at the point $5\,cm$ away from the surface

   (d) same as at the point $25\,cm$ away from the surface   **[a]**

9. When placed in a uniform electric field, a dipole experiences
   (a) only a torque
   (b) only a net force
   (c) both a net force and a torque
   (d) neither a net force nor a torque [a]

10. An electric dipole placed in a uniform electric field will have minimum potential energy when the dipole moment is inclined to the field at an angle
    (a) $\dfrac{\pi}{2}$ (b) $\dfrac{3\pi}{2}$ (c) π (d) 0   [d]

11. Three charges $2q, -q, -q$ are located at the vertices of an equilateral triangle. At the circumcentre of the triangle,
    (a) field is non-zero but potential is zero
    (b) both field and potential are zero
    (c) both field and potential are non-zero
    (d) field is zero but potential is non-zero [b]

12. The work done in carrying a charge q once round a circle of radius r with a charge Q at the centre is
    (a) zero   (b) $\dfrac{qQ}{4\pi\varepsilon_0}\left(\dfrac{1}{2\pi r}\right)$ (c) $\dfrac{qQ}{4\pi\varepsilon_0}$ (d) $\dfrac{qQ}{4\pi\varepsilon_0}\dfrac{1}{\pi r}$ [a]

13. A charge q is placed at the centre of the line joining two equal charges Q. The system of the three charges will be in equilibrium if q is equal to
    (a) $+\dfrac{Q}{2}$ (b) $-\dfrac{Q}{2}$ (c) $-\dfrac{Q}{4}$ (d) $+\dfrac{Q}{4}$ [c]

14. The electric potential in a region along the x-axis varies with x according to the relation $V(x) = 5 + 4x^2$.
    (a) The force experienced by the above charge is along the +ve x-axis
    (b) A uniform electric field exists in this region along the x-axis
    (c) The force experienced by a charge of 1 C placed at $x = -1$ m is 8 N
    (d) Potential difference between the points $x = 1$ and $x = -3$ is 32 V [a, c, d]

15. Bond length of HCl molecule is 1.28 Å. Its electric dipole moment will be
    (a) 16×10^{-19} C m   (b) 1.28×10^{-10} C m
    (c) 2.05×10^{-29} C m   (d) 2.05×10^{-10} C m   [c]

16. The radius of two spheres are R_1 and R_2 and their charges are q_1 and q_2 respectively. They are at the same potential. The ratio $\dfrac{q_1}{q_2}$ will be equal to
    (a) $\dfrac{R_1}{R_2}$ (b) $\dfrac{R_2}{R_1}$ (c) $\dfrac{R_1^2}{R_2^2}$ (d) $\dfrac{R_2^2}{R_1^2}$ [a]

17. The SI unit of electric dipole moment is
    (a) C m   (b) C/m
    (c) C/m^2 (d) N/C   [a]

18. State which of the following is correct:
    (a) joule = volt + amp   (b) joule = volt × amp
    (c) joule = coulomb × volt   (d) joule = coulomb + volt   [c]

19. When a potential difference is applied across human heart, its behaviour can be modeled as that of electric dipole. Abnormal heart can be detected by mapping
    (a) electric field   (b) electric flux lines
    (c) equipotential surfaces   (d) all of the above   [d]

20. Two concentric hollow spheres of radii r_1 and r_2 ($r_1 > r_2$) have respective charges Q_1 and Q_2 distributed uniformly over their surfaces. The electric flux density D at a Gaussian surface of radius r such that ($r_1 > r > r_2$) will be

    (a) $Q_1/(4\pi r^2)$ (b) $Q_2/(4\pi r_2^2)$

    (c) $Q_2/(4\pi r^2)$ (d) $Q_1/(4\pi r_1^2)$ [c]

21. The electric field strength E of a dipole varies
    (a) inversely as the distance
    (b) inversely as the cube of the distance
    (c) inversely as the square of the distance
    (d) directly as the square of the distance   [b]

22. The electric field due to very short electric dipole at a point P on its axial line at a distance r from it is E_P and at a point Q on the equatorial line at a distance $2r$ from it is E_Q. Then
    (a) $E_P = -E_Q$ (b) $E_P = -8E_Q$
    (c) $E_P = 8E_Q$ (d) $E_P = -16E_Q$ [d]

$$\left[Hint: \quad E_P = \frac{2p}{4\pi\varepsilon_0 r^3} \quad \text{and} \quad E_Q = -\frac{p}{4\pi\varepsilon_0 (2r)^3} \right]$$

$$\therefore \quad E_P = -16 E_Q$$

23. An electric dipole placed with its axis in the direction of a uniform electric field experiences
    (a) a force but no torque
    (b) a force as well as a torque
    (c) a torque but no force
    (d) neither a force nor a torque   [d]
    [*Hint:* Electric field E exerts a force qE on the charge $+q$ and a force $-qE$ on the charge $-q$ of the dipole. Since these forces are equal and opposite, they add to zero.]

24. An electric dipole placed with its axis inclined at an angle to the direction of uniform electric field experiences
    (a) a torque but no force
    (b) a force as well as a torque

(c) a force but no torque

(d) neither a force nor a torque [a]

[*Hint:* A torque acts on the dipole which tends to align it along the field.]

25. An electric dipole placed in a non-uniform electric field experiences
    (a) a force as well as a torque
    (b) a force but no torque
    (c) a torque but no force
    (d) neither a force nor a torque [a]

    [*Hint:* In a non-uniform electric field, a dipole experiences a force which gives it a translational motion and a torque which gives it a rotational motion.]

26. Electric dipole of charges of magnitude e are placed inside a cube. The total electric flux coming out of the cube will be
    (a) zero   (b) e/ε_0
    (c) $16e/\varepsilon_0$ (d) $8e/\varepsilon_0$ [a]

    [**Hint:** A dipole consists of two equal and opposite charges separated by a certain distance. Hence, the total charge enclosed in the cube is zero. Therefore, the electric flux is zero.]

27. A point Q lies on the perpendicular bisector of an electric dipole of moment p. If the distance of Q from the dipole is r (much larger than the size of the dipole), then the electric field at Q is proportional to
    (a) p and r^{-3} (b) p and r^{-2}
    (c) p^{-1} and r^{-1} (d) p^2 and r^{-3} [a]

    [*Hint:* The electric field at a point far away on the perpendicular bisector of a dipole of dipole moment p is given by (for $r \gg a$),

    $$E = \frac{p}{4\pi\varepsilon_0 r^3} \Rightarrow E \propto pr^{-3}]$$

28. n small metal drops of same size are charged to V volt each. If they coalesce to form a single charge drop, then its potential will be

    (a) $\dfrac{V}{n}$ (b) Vn

    (c) $Vn^{2/3}$ (d) $Vn^{1/3}$ [c]

    [**Hint:** $\dfrac{4}{3}\pi R^3 = n \times \dfrac{4}{3}\pi r^3$, or $R = n^{1/3}r$

    where, R is the radius of single charge drop and r the radius of small drop.
    Potential of small drop,

    $$V = \frac{1}{4\pi\varepsilon_0} \frac{q}{r}$$

    Potential of big drop,

    $$V' = \frac{1}{4\pi\varepsilon_0} \frac{nq}{r}$$

    $\therefore$ $V' = n^{2/3}V$]

29. A Gaussian surface within a metallic spherical shell of inner and outer radii R_1 and R_2 contains charge Q placed at the centre. The normal component of electric flux density D at the Gaussian surface will be

(a) $Q/[4\pi(R_1 - R_2)^2]$ (b) zero

(c) $Q/4\pi R_1^2$ (d) $Q/4\pi R_2^2$ **[b]**

30. A uniform electric field pointing in postive x-direction exists in a region. Let A be the origin, B be the point on the x-axis at $x = +1$ cm and C be the point on the y-axis at $y = +1$ cm. Then the potential at points A, B and C satisfy

(a) $V_A < V_B$ (b) $V_A > V_B$

(c) $V_A < V_C$ (d) $V_A > V_C$ **[b]**

[*Hint:* Electric potential increases in the direction of electric field, so $V_A > V_C$.]

Electrostatics in Dielectrics—Capacitance and Dielectric Materials

4.1 INTRODUCTION

So far, we have discussed electrostatic fields in free space and the field produced was exclusively by the free charges. Now, we shall examine the phenomena which results from the introduction of a dielectric materiel into an electrostatic field. However, if matter is present, the magnitude of electric field decreases. The medium plays an important part in determining the absolute value of the field. The electromagnetic fields that result in matter are, in general, very complex displaying considerable variation from particle to particle over distances of the order of separation of atoms. In dealing with this phenomena that range over large distances, we shall not be concerned with the detailed fluctuations in the field on the atomic scale. We shall try to present a consistent macroscopic model of matter that does predict in every instance the correct fields. We must note that the macroscopic fields within matter are not necessarily the fields actually exerted on any of the particles in matter.

On the basis of electric behaviour, media can be classified into two distinct groups:

i. *Conductors:* We are familiar with conductors which contain some electrons which are free to move in the presence of a field. These free electrons can be imagined as charge carriers. The conductivity of these substances is of the order of 10^8 mho/m. Examples of conductors are metals, etc. There are some substances, both elements and compounds, the resistance of which vanishes at a certain temperature, known as *critical temperature* (T_c). These substances are known as *superconductors*. For example, (T_c) for mercury is 4.2 K. The highest known T_c so far is 134 K for mercury cuprate.

ii. *Insulators or dielectrics:* These are the substances in which the electrons are strongly bound to the atoms or molecules composing the material and cannot be detached by the application of an electric field to these materials. The forbidden energy gap (~7 eV) is very large compared to the thermal energy and hence the excitation of electrons from valence band to conduction band is normally not possible and free charge carriers are not available. The conductivity of these substances is very small of the order of 10^{-16} mho/m. In the absence of an external field, the dipole moments of the molecules of a dielectric are usually either equal to zero (non polar molecules) or distributed in space by directions chaotically (polar molecules). In both cases, the total dipole moment of a dielectric equals zero. However, there are substances that can have a dipole moment in the absence of an external field. These substances are called *ferroelectrics*. These substances exhibit spontaneous polarization in the absence of an external field, e.g. Rochelle salt. Obviously, under the action of an external electric field, no steady motion of electrons is possible and at the most they may be slightly displaced from their positions in response to the field applied. Such insulating media are also known as dielectric media. Sulphur, mica, porcelain, etc. are few examples of dielectrics. We must note that if the applied field is too large, catastrophic breakdown may occur in these substances. A good insulator breaks down in fields of about 10^9 volt/m.

In addition to conductors and insulators, we have another class of materials, the so called *semiconductors* which have properties intermediate between conductors and insulators. In these substances, electrons in the conduction band and holes in the valence band act as charge carriers. There is a forbidden gap of the order of 1 eV between the conduction and valence bands. The conductivity of these substances is less than that of conductors and is of the order of 10^{-1}–10^{-2} mho/m. The present day technology depends on such materials.

4.2 CONDUCTORS IN ELECTROSTATIC FIELD

Let us consider that a conductor is placed in an electric field (Fig. 4.1). Figure 4.1 shows that a slab of conducting material is placed in an electric field. One can see that under the influence of the field, the electrons can move freely in the material but cannot leave the surface because strong forces act to prevent them from leaving the surface. This means that they leave a net charge on the surface. The charges thus induced on the surface of the slab generate a electric field inside it (conductor) which is in a direction opposite to the external electric field. We must note that the electrons continue to move so long as there is any field left in the conductor and until the induced charge ensures that the resultant field inside the conductor is zero. Obviously, no electrostatic field can exist within the body of a conductor, i.e. $E = 0$ inside the conductor. Since $E = -\nabla\phi = 0$, it shows that the conductor

Fig. 4.1: A conductor in an electric field

is an *equipotential region*. Further, because $E = 0$, $\nabla = \dfrac{\rho}{\varepsilon} = 0$,

i.e. $\rho = 0$. This means, there cannot be a volume distribution of charges within the conductor, i.e. the charges must reside entirely on the surface of a conductor. As stated earlier, the surface of a conductor upon which charges are at rest, is an equipotential surface. Since the surface being an equipotential surface, the field is normal to the surface of the conductor. If this was not the case, there would be an electrostatic field along the surface resulting into a current.

4.3 ELECTRIC FIELD AT THE SURFACE OF A CHARGED CONDUCTOR

Let us imagine a small disc enclosing an area ds on the surface of a conductor as shown in Fig. 4.2. From the figure, it is evident that the volume enclosed by the outer surface of the disc is half in the conductor and half in the air. Further, the flat surfaces of the disc are parallel to the surface of the conductor. If E is the electric field, then the flux through the surface outside is Eds. The flux through the other surface is zero since the field inside the conductor is zero. Applying Gauss's law, we have

$$Eds = \frac{\sigma}{\varepsilon_0} ds$$

or $$E = \frac{\sigma}{\varepsilon_0} \qquad (4.1)$$

Fig. 4.2: A small disc on the surface of a conductor

where, σ is surface charge density.

4.4 DIELECTRIC AND PERMITTIVITY

An important characteristic of a dielectric is its permittivity or dielectric constant ε. If two point charges are placed in an electrically homogeneous and isotropic medium, then according to Coulomb's law the force between the two point charges is given by

$$F = \frac{1}{4\pi\varepsilon} \frac{q_1 q_2}{r^2} \hat{r} \qquad (4.2)$$

Here ε is called the permittivity of the medium. We must note that ε (permittivity) is not always a constant. The permittivity of a dielectric is always greater than the permittivity of vacuum, hence it is often convenient to use the relative permittivity ε_r of the dielectric expressed as the ratio of its permittivity to that of vacuum. Thus,

$$\varepsilon_r = \frac{\varepsilon}{\varepsilon_0} = \frac{\text{Permittivity of the medium}}{\text{Permittivity of the vacuum}}$$

The value of relative permittivity (ε_r) or dielectric constant or specific inductivity capacity is different for different media. One can find the explanation of this effect in terms of the electric nature of dielectric media.

Dielectric materials become polarized in an electric field, with the result that the electric flux density D is greater than it would be under free space conditions with the same field intensity. In the macroscopic view, however, polarization P can account for the increase in flux density, the equation being

$$D = \varepsilon_0 E + P$$

This equation permits E and P to have different directions as they do in certain crystalline dielectrics. In an isotropic linear material, E and P are parallel at each point, which is expressed as

$$P = \chi_e \varepsilon_0 E \text{ (isotropic material)}$$

where, χ_e is the electric susceptibility, a dimensionless constant. Then

$$D = \varepsilon_0 (1 + \chi_e) E = \varepsilon_0 \varepsilon_r E \text{ (isotropic material)}$$

where, $\varepsilon_r = 1 + \chi_e$ is also a pure number.

Since $D = \varepsilon E$, we have

$$\varepsilon_r = \frac{\varepsilon}{\varepsilon_0} \qquad (4.3)$$

where, ε_r is relative permittivity. A detailed study of dielectrics is presented in subsequent sections.

4.5 CAPACITOR

It is a device used for storing energy. It is a system of two oppositely charged conductors having charges of equal magnitude and having such shapes and relative position that the field set up by the system is concentrated (localized) in a limited region of space. In case of parallel plate capacitors, the conductors are called plates.

Let us suppose that the two conductors in a capacitor carry charges $\pm Q$. These charges are distributed over the surfaces with the surface densities σ_+ and σ_- respectively. We must note that these may not be constant over the surfaces of the conductors, but the potentials ϕ_+ and ϕ_- are, since the surfaces of the conductors are equipotential surfaces. Let V be the potential difference between the two conductors, then

$$V = \phi_+ - \phi_- \tag{4.4}$$

In an isolated charged conductor, not subject to the action of external fields, potential difference is always proportional to the charge Q, i.e.

$$Q = CV \tag{4.5}$$

or
$$C = \varepsilon_0 A / d \tag{4.5a}$$

where, C is a constant and called the capacitance (or capacity) of the capacitor. The capacitance of a capacitor is the mutual capacitance of its plates (or conductors in the general case) and is related to the ability of a capacitor to hold charges. The capacitance depends on the geometry, i.e. the shape and size of the conductors. Capacitors are made in the form of two conductors placed close to each other. The conductors forming a capacitor are called its plates. To prevent external bodies from influencing the capacitance of a capacitor, the plates are shaped and arranged relative to each other so that the field set up by the charges accumulating on them is concentrated inside the capacitor. This condition is satisfied by two plates arranged close to each other, two coaxial (plane), cylindrical, and spherical capacitors are encountered. Since the field is confined inside a capacitor, the electric displacement lines begin on one plate and terminate on the other. Consequently, the extraneous charges produced on the plates have the same magnitude and are opposite in sign. Now we will calculate the capacitance of a few capacitors.

Capacitors are essential in modern technology. They range from the billions of 25 fF capacitors that store individual bits of information in computer's memory, to millifarad range capacitors that smooth the 50 Hz AC power to provide steady current to stereo amplifier, to ultracapacitors measuring hundreds of farads that store electric energy for short bursts of power in gas–electric hybrid and fuel–cell vehicles.

Practical Capacitors

Equation (4.5a) shows that the way to achieve large capacitance is with large plate area and small spacing. This is true in general whether or not a capacitor has the parallel plate geometry. Inexpensive capacitors are often made from two long strips of aluminium foil reoperated by thin plastic insulation. This foil "sandwich" is rolled into a compact cylinder, wires are attached, and the whole thing is dipped in protective coating. Very large capacitances are achieved with electrolyte capacitors in which a thin insulating layer develops chemically under the influence of the applied voltage. Capacitors are among the hardest components to fabricate on integrated circuit (IC) chips, but small capacitor units can be made by alternating conductive material with an insulating layer.

Parallel Plate Capacitor

It is the most commonly used type of capacitor. The conductors forming a parallel plate capacitor are called its plates. Let us consider two plates each of area A, separated by a small distance d as shown in Fig. 4.3. Let equal and opposite charges $\pm Q$ be put on the plates. One can ignore the edge effect, if the lateral dimensions of the plates are much larger than this separation and the field between the plates be assumed to be uniform. We know that the potential difference between the plates is the amount of work required to take a unit charge from one plate to the other. We have

$$V = Ed = \frac{\sigma}{\varepsilon_0}d = \frac{Qd}{\varepsilon_0 A} \qquad \left(\because \sigma = \frac{Q}{A} \right)$$

$$\therefore \qquad C = \frac{Q}{V} = \frac{\varepsilon_0 A}{d} \qquad (4.6)$$

Fig. 4.3: A parallel plate capacitor

It must be noted that the accuracy of determining the capacitance of a real parallel plate capacitor by Eq. (4.6) is greater, the smaller is the separation distance d in comparison with the linear dimensions of the plates.

The formula for the capacitance of a multiplate capacitor differs from that for a parallel plate capacitor in a way that the term A is replaced by $A(n-1)$, where n is the number of plates.

Capacitance of an Isolated Sphere

Let us consider that a capacitor is formed from two spheres. Let the radius of sphere be a and carry a charge Q. Let us imagine that another sphere is of infinite radius carrying a charge $-Q$. The potential (V) of the capacitor is given by

$$V = -\int_\infty^a E\,dr = -\int_\infty^a \frac{Q}{4\pi\varepsilon_0 r^2}\,dr = \frac{Q}{4\pi\varepsilon_0 a} \qquad (4.7)$$

$$\therefore \qquad C = \frac{Q}{V} = 4\pi\varepsilon_0 a$$

Coaxial Cable or Cylindrical Capacitor

Consider a cylindrical capacitor, i.e. two coaxial cylinders of infinite length. Let their radius be r_1 and r_2 as shown in Fig. 4.4a.

Fig. 4.4a: Cylindrical capacitor

If the charge per unit length on the inner cylinder is $+\lambda$, then E is given by the relation

$$E = \frac{\lambda}{2\pi\varepsilon_0 r}$$

The potential difference between the two coaxial cylinders is given by

$$V = \phi_{r_1} - \phi_{r_2} = -\int_{r_2}^{r_1} E\,dr = -\frac{\lambda}{2\pi\varepsilon_0}\int_{r_2}^{r_1}\frac{dr}{r}$$

$$= -\frac{\lambda}{2\pi\varepsilon_0}(\log r_1 - \log r_2) = \frac{\lambda}{2\pi\varepsilon_0}\log\frac{r_2}{r_1} \tag{4.8}$$

If we consider that l be the length of cylinders, then the capacitance of a cylindrical capacitor is given by

$$C = \frac{\lambda l}{\dfrac{\lambda}{2\pi\varepsilon_0}\log\dfrac{r_2}{r_1}} = \frac{2\pi\varepsilon_0 l}{\log\dfrac{r_2}{r_1}} \tag{4.9}$$

The accuracy of determining the capacitance of a real capacitor by Eq. (4.9) is greater, the smaller is the separation distance $d = r_2 - r_1$ in comparison with l and r_1.

Apart from the capacitance, every capacitor is characterized by the maximum voltage U_{max} that may be applied across its plates without the danger of a breakdown. When this voltage is exceeded, a spark jumps across the space between the plates. The result is the failure of the capacitor. This voltage is called the *breakdown voltage*. The breakdown voltage depends upon the thickness of the dielectric, its properties and the shape of the conductors in the capacitor.

The capacitance can be increased by connecting the conductors with like charges of two or more capacitors in *parallel*. The equivalent capacitance for such a connection is

$$C = \sum_{i=1}^{n} C_i \tag{4.10}$$

where, C_i is the capacitance of the ith capacitor.

When capacitors are connected in *series*, their oppositely charged conductors are connected together. In this case, the equivalent capacitance is

$$\frac{1}{C} = \sum_{i=1}^{n} \frac{1}{C_i} \tag{4.11}$$

The equivalent capacitance is always less than the minimum capacitance of any capacitor in the network.

4.6 MULTIPLE DIELECTRIC CAPACITORS

When two dielectrics are present in a capacitor with the interface parallel to E and D as shown in Fig. 4.4b, the equivalent capacitance can be obtained by treating the arrangement as two capacitors in parallel (Fig. 4.4c).

$$C_1 = \frac{\varepsilon_0 \varepsilon_{r_1} A_1}{d}, \quad C_2 = \frac{\varepsilon_0 \varepsilon_{r_2} A_2}{d}$$

$$C_{eq.} = C_1 + C_2 = \frac{\varepsilon_0}{d}\left(\varepsilon_{r_1} A_1 + \varepsilon_{r_2} A_2\right)$$

Fig. 4.4b and c: Multiple dielectric capacitors (parallel)

When two dielectrics are present such that the interface is normal to D and E as shown in Fig. 4.4d, the equivalent capacitance can be obtained by treating the arrangement as two capacitors in series (Fig. 4.4e).

$$C_1 = \frac{\varepsilon_0 \varepsilon_{r_1} A}{d_1}, \quad C_2 = \frac{\varepsilon_0 \varepsilon_{r_2} A}{d_2},$$

$$\frac{1}{C_{eq.}} = \frac{1}{C_1} + \frac{1}{C_2} = \frac{\varepsilon_{r_2} d_1 + \varepsilon_{r_1} d_2}{\varepsilon_0 \varepsilon_{r_1} \varepsilon_{r_2} A}$$

Fig. 4.4d and e: Multiple dielectric capacitors (series)

One can extend the result to any number of dielectrics such that the interfaces are all normal to D and E. The reciprocal of the equivalent capacitance is the sum of the reciprocals of the individual capacitance.

4.7 ENERGY STORED IN A CAPACITOR

We have seen that the charge q on a conductor can be considered as a system of point charges Δq. Moreover, the charge in the case of conductors resides only on the surfaces, the potential energy of a conductor can be found from

$$W = \frac{1}{2}\int \sigma \phi \, ds \qquad (4.12)$$

(we have replaced summation by integration in Chapter 3).

Assume that the potential of a capacitor plate carrying the charge $+q$ is ϕ_+ and that of a plate carrying the charge $-q$ is ϕ_-. Consequently, each of the elementary charges Δq into which the charge $+q$ can be divided is at a point with the potential ϕ_+, and each of the charges into which the charge $-q$ can be divided is at a point with the potential ϕ_-. By Eq. (4.12), the potential energy of such a system of charges is

$$W = \frac{1}{2}\int \sigma_+ \phi_+ \, ds + \frac{1}{2}\int \sigma_- \phi_- \, ds \qquad (4.13)$$

where σ_+ and σ_- are the surface densities of charges on the two conductors. Therefore,

$$W = \frac{1}{2}Q(\phi_+ - \phi_-) = \frac{1}{2}QV \qquad (4.14)$$

One can also express Eq. (4.14) in terms of capacity as

$$W = \frac{1}{2}CV^2 = \frac{1}{2}\frac{Q^2}{C} \qquad (4.15)$$

One can also express the energy [Eq. (4.15)] in a different way by considering it to be distributed over the space between the charges occupied by their electric field. Let us consider two equipotential surfaces close enough between two conductors (Fig. 4.5).

Let us consider that two surfaces differ in their potential by a small amount dV. If now two conducting surfaces with potentials exactly equal to that of equipotential surfaces are inserted so as to coincide with the surfaces, one finds that the condition of the problem remain unaltered. Thus, the two conducting surfaces will

Fig. 4.5: Equipotential surfaces between two conductors

form a capacitor with capacity $\varepsilon_0 \dfrac{ds}{dz}$, where ds is the area of each plate and dz their separation. Thus, the energy of the capacitor

$$dW = \frac{1}{2}CV^2 = \frac{1}{2}\varepsilon_0\frac{ds}{dz}(dV)^2 = \frac{1}{2}\varepsilon_0\frac{ds}{dz}(Edz)^2$$

$$= \frac{1}{2}\varepsilon_0 E^2 dsdz = \frac{1}{2}\varepsilon_0 E^2 d\tau$$

where, $d\tau$ is the volume occupied by the capacitor. Thus,

$$W = \frac{1}{2}\varepsilon_0\int E^2 d\tau$$

This shows that one can regard the energy as distributed throughout the field and hence the energy density at a point being $\frac{1}{2}\varepsilon_0 E^2$.

One can derive the above result in a more rigorous way. Let us consider an assembly of conductors enclosed by a surface so sufficiently far away that the field on it is negligible (Fig. 4.6). Let us suppose that there exists within the surface a volume distribution of charge of density ρ and surface distribution of density σ over the conductors. One obtains the electrostatic energy as

$$W = \frac{1}{2}\int \rho\phi d\tau + \frac{1}{2}\int \sigma\phi ds_1 \qquad (4.16)$$

where, ϕ is the potential of the surface.

The surface integral is taken over the surfaces of all the conductors and the volume integral over the entire region occupied by the field, i.e. the volume bounded by the surface as shown in Fig. 4.6. Applying Gauss's theorem in the differential form

$\nabla \cdot E = \dfrac{\rho}{\varepsilon_0}$, one obtains using divergence theorem,

$$\frac{1}{2}\int \rho\phi d\tau = \frac{1}{2}\int \varepsilon_0(\nabla \cdot E)\phi d\tau = \frac{\varepsilon_0}{2}\int \{\nabla \cdot (\phi E) - E \cdot \nabla\phi\}d\tau$$

$$= \frac{\varepsilon_0}{2}\int \nabla \cdot (\phi E)d\tau + \frac{\varepsilon_0}{2}\int E^2 d\tau$$

$$= \frac{\varepsilon_0}{2}\int \phi E \cdot \hat{a}_n ds + \frac{\varepsilon_0}{2}\int E^2 d\tau$$

Fig. 4.6: Volume bounded by the surface

The first integral is to be performed over the closed surface S_0 enclosing the volume as well as over the surfaces of the conductors. Thus,

$$\frac{1}{2}\int \rho\phi d\tau = \frac{\varepsilon_0}{2}\int \phi E \cdot \hat{a}_n\, dS_0 + \frac{\varepsilon_0}{2}\int \phi E \cdot \hat{a}_{n_1}\, dS_1 + \frac{\varepsilon_0}{2}\int E^2 dt$$

over the surface
of the conductor

Here, $\hat{a}_n$ and $\hat{a}_n$ are unit vectors normal to the respective surfaces.

We know that ϕ varies as r^{-1} and E as r^{-2} while dS increases as r^2. The first integral is proportional to $1/r$ and hence vanishes when r is very large, i.e. on the surface S_0.

The integration in the second integral is over the surfaces of the medium. $\hat{a}_{n_1}dS_1$ is the vector drawn outwards from the medium and hence into the conducting surface (Fig. 4.6). We have $E = \dfrac{-\sigma}{\varepsilon_0}$ in this direction,

$$\frac{1}{2}\int \rho\phi d\tau = -\frac{1}{2}\int \sigma\phi dS_1 + \frac{\varepsilon_0}{2}\int E^2 d\tau$$

Thus, one obtains the total energy as

$$W = -\frac{1}{2}\int \sigma\phi dS_1 + \frac{\varepsilon_0}{2}\int E^2 d\tau + \frac{1}{2}\int \sigma\phi dS_1 = \frac{\varepsilon_0}{2}\int E^2 d\tau \qquad (4.17)$$

Since $E = 0$ within a conductor, one can regard this energy as distributed throughout the surrounding medium with density $\varepsilon_0 \dfrac{E^2}{2}$.

Equation (4.17) has certain limitations. For example, let us compute the electrostatic energy of a point charge q with the help of Eq. (4.17). The energy will be

$$W = \frac{\varepsilon_0}{2}\int \frac{q^2}{(4\pi\varepsilon_0)^2 r^4} 4\pi r^2 dr = \frac{q^2}{8\pi\varepsilon_0}\int_0^\infty \frac{dr}{r^2} = \frac{q^2}{8\pi\varepsilon_0}\left[-\frac{1}{r}\right]_0^\infty = \infty$$

This shows that the self energy of a point charge is infinite, which is a physically absurd result. This means that the concept of energy density is not compatible with the concept of point charge.

4.8 NON-CONDUCTING MEDIUM IN AN ELECTRIC FIELD

How does a non-conducting medium, i.e. a dielectric behaves in an electric field? If a dielectric is introduced in an electric field, then the field and the dielectric itself undergo appreciable changes.

Cavandish and later on Faraday studied the effects of introducing a dielectric material between the plates of a capacitor. They observed that the capacity of a conductor depends not only on shape and size of the conducting plates but also on the nature of the dielectric material by which they are separated. It is found that the capacitance of a capacitor increases on inserting a slab of dielectric between its plates. We have $C = \dfrac{Q}{V}$.

Since the capacitance C increases on inserting a dielectric, where Q remaining the same, the potential difference V falls, i.e. the field is weakened. If the capacitance is increased by a factor ε_r, the field is weakened by a factor $\dfrac{1}{\varepsilon_r}$. Thus, the electric field in a dielectric medium is given by

$$E = \frac{q}{4\pi\varepsilon_0\varepsilon_r} \frac{r}{|r|^3} \tag{4.18}$$

The parameter ε_r, the *relative permittivity* of the dielectric is called the dielectric constant of the medium. ε_r depends on the nature of the dielectric medium. The value of ε_r for vacuum is 1, for water is 81, for air is 1.00057 and for most substances it varies between 1 and 10.

4.9 POLARIZATION

Let us study the effect of an electric field on a piece of matter. We know that atoms do not have permanent electric dipole moments because of their spherical symmetry, but when they are placed in an electric field, they become polarized, acquiring induced electric dipole moments in the direction of the field. This results from the perturbation of the motion of the electrons produced by the applied electric field.

On the other hand, many molecules do have permanent electric dipole moments. When a molecule has a permanent electric dipole moment, it tends to be oriented parallel to the applied electric field because of the torque it experiences. As a consequence of either of these two effects, a piece of matter placed in an electric field becomes electrically *polarized*, that is, its molecules (or atoms) become electric dipoles, oriented in the direction of the local electric field (Fig. 4.7), either because of the distortion of the electronic motion or the orientation of their permanent dipoles. A medium that can be polarized by an electric field is called a *dielectric medium*. The polarization gives rise to a net positive charge on negative side of the piece of matter and a net negative charge on the positive side. The piece of matter then becomes a large electric dipole that tends to move in the direction in which the field increases. The polarization of a dielectric material due to some externally applied electric field may occur as a consequence of three effects:

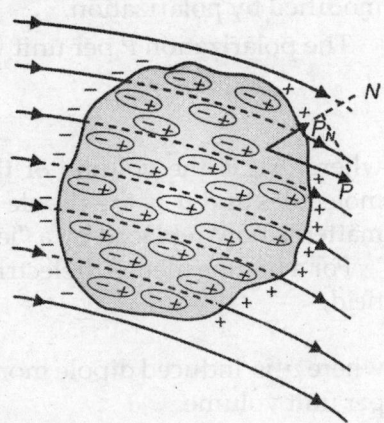

Fig. 4.7: Polarization of matter by an electric field

i. *Electron polarization:* This occurs when the externally applied field causes a shift in the atom's positive and negative charges.

ii. *Ionic polarization:* This occurs in molecules composed of positively and negatively charged ions.

iii. *Orientational polarization:* This occurs in materials that possess permanent, microscopic separation of charge centres and such materials possess persistent polarization. In the absence of an applied field, these permanent dipoles are randomly oriented. However, in the presence of an applied electric field these permanent dipoles tend to rotate to align with the direction of the applied field. Materials (such as water) possessing these permanent dipoles in the absence of an external applied field are termed *polar molecules*. Obviously, there are two types of materials: (i) *polar* and (ii) *non-polar*. In a non-polar dielectric, the centres of positive and negative charges normally coincide, whereas in a polar dielectric the centres of charges are always displaced from each other.

Now, we are interested in finding how the dielectric constant ε_r is related to the intrinsic properties of the materials.

When an electric field is applied on an atom, a dipole moment is induced in it. In a non-polar atom, the induced dipole moment is in the direction of the applied field and directly proportional to it, i.e.

$$p \propto E$$

or

$$p = \alpha E \tag{4.19}$$

Here α is a constant known as atomic polarizability. We have

$$\alpha = \frac{p}{E} = \frac{Cm}{V/m} = Fm^2$$

The above statement is not true for all materials especially crystals for which α is a matrix. For dielectric materials used as insulators, α is a constant. Such materials when placed in an external field get polarized. The polarized matter gives rise to its own field which modifies the external field which in turn modifies polarization. Under equilibrium conditions we have the polarized matter and the original external field modified by polarization.

The polarization P per unit volume of the dielectric is defined as

$$P = \frac{1}{\Delta V} \sum_{1}^{N} p_i \tag{4.20}$$

where, ΔV is the volume of the dielectric containing a large number of polarized molecules and p_i is the dipole moment of the ith molecule. Obviously, the polarized matter is now replaced by a "jelly" of dipoles.

For a homogeneous dielectric with non-polar molecules in a homogeneous electric field,

$$P_e = n_0 p_e \tag{4.21}$$

where p_e is induced dipole moment of one molecule and n_0 is the number of molecules per unit volume.

In an isotropic dielectric, dipole moment p is induced in the direction of the electric field E. In such a medium, the polarization P is directly proportional to the intensity of electric field E, i.e.

$$P \propto E$$

or

$$P = \varepsilon_0 \chi E \tag{4.22}$$

The quantity χ is a dimensionless scalar constant, i.e. independent of E and known as *electrical susceptibility*. For most substances, χ is a positive quantity.

In an anisotropic or nonisotropic dielectric, the directions of P and E, generally, do not coincide. In this case, the relation between P and E is described by the following equations:

$$P_x = \varepsilon_0(\chi_{xx}E_x + \chi_{xy}E_y + \chi_{xz}E_z)$$

$$P_y = \varepsilon_0(\chi_{yx}E_x + \chi_{yy}E_y + \chi_{yz}E_z) \qquad (4.23)$$
$$P_z = \varepsilon_0(\chi_{zx}E_x + \chi_{zy}E_y + \chi_{zz}E_z)$$

The combination of the nine quantities forms a symmetrical tensor of rank two, called the *tensor of dielectric susceptibility*. This tensor characterizes the electrical properties of an anisotropic dielectric.

Let us consider a slab of material of thickness l and surface S placed perpendicular to a uniform field E (Fig. 4.8). The polarization P, being parallel to E, is also perpendicular to S. The volume of the slab is lS, and therefore its total electric dipole moment is $P(lS) = (PS)l$. But l, just the separation between the positive and negative charges that appear on the two surfaces. Since by definition the electric dipole moment is equal to product of charge and distance, we conclude that the total electric charge that appears on each of the surfaces is PS, and therefore the charge per unit area σ_p on the faces of polarized slab is P, or $\sigma_p = P$. Although this result has been obtained for a particular geometrical arrangement, it is of general validity, and *the charge per unit area on the surface of a polarized piece of matter is equal to the component of the polarization P in the direction of the normal to the surface of the body*. So, in Fig. 4.7, the charge per unit area on the surface at A is $P_N = P \cos \theta$.

Fig. 4.8: A slab of material placed perpendicular to a uniform electric field

4.10 ELECTRIC DISPLACEMENT

We have seen that a polarized dielectric has certain charges on its surface (and also, unless the polarization is uniform throughout its volume). These polarization charges however, are "frozen" in the sense that they are bound to specific atoms or molecules and are not free to move through the dielectric. In other materials such as metal or an ionized gas, there may be electric charges capable of moving through the material, and therefore we call them *free charges*.

To understand a clear distinction between free charges and polarization charges, let us consider a slab of a dielectric material placed between two conducting parallel plates (Fig. 4.9), carrying equal and opposite free charges. The surface charge density on the left plate is $+\sigma_{free}$ and on the right plate is $-\sigma_{free}$. These charges produce an electric field that polarizes the slab so that polarization charges appear on each surface of the slab. These polarization charges have a sign opposite to that

Fig. 4.9: Dielectric placed between oppositely charged plates. Charges on the plates are free charges and charges on the dielectric surface are polarization charges

on the plates. Therefore, the polarization charges on the faces of the dielectric slab partially balance the free charges on the conducting plates. If P is the magnitude of the polarization in the slab, the surface charge density on the left face of the slab is $-P$, while on the right face is $+P$. The effective or net surface charge density on the left face is $\sigma = \sigma_{free} - P$, with an equal and opposite result on the right face. These net surface charges give rise to a uniform electric field given by $E = \dfrac{\sigma}{\varepsilon_0}$. Thus, using the effective value of σ, we have

$$E = \frac{1}{\varepsilon_0}(\sigma_{free} - P) \quad \text{or} \quad \sigma_{free} = \varepsilon_0 E + P \tag{4.24}$$

which is an expression that gives the free charges on the surface of a conductor surrounded by a dielectric in terms of the electric field in the dielectric and the polarization of the dielectric. When E and P are vectors in the same direction, the above result suggests the introduction of a new vector field which is called the *electric displacement*, and defined as

$$D = \varepsilon_0 E + P \tag{4.25}$$

Obviously, D is expressed in $C\,m^{-2}$, since these are the units of the two terms that appear on the RHS of Eq. (4.25). Obviously, $(\varepsilon_0 E + P)$ or D is a physical quantity which depends on free charge density only.

For isotropic dielectrics, the polarization vector P is proportional to the field intensity E and coincides with E in direction. Hence,

$$D = K\varepsilon_0 E \tag{4.26}$$

where K is the relative permittivity or dielectric constant (ε_r) is a scalar quantity, i.e.

$$K \text{ or } \varepsilon_r = \frac{\text{Permittivity of medium } (\varepsilon)}{\text{Permittivity of vacuum } (\varepsilon_0)}$$

$$= \frac{\text{Electric field in vacuum} (E_0)}{\text{Electric field in the medium } (E)}$$

Using Eq. (4.22), Eq. (4.24) can be expressed as

$$\sigma_{free} = \varepsilon_0 E + \varepsilon_0 \chi E$$

or

$$\sigma_{free} = \varepsilon_0 E(1 + \chi)$$

or

$$E = \frac{\sigma_{free}}{\varepsilon_0 (1 + \chi)} \tag{4.27}$$

Equation (4.27) shows that the field is reduced by a factor $1/(1 + \chi)$.

Using Eq. (4.22), Eq. (4.25) can be expressed as

$$D = \varepsilon_0 E + \chi \varepsilon_0 E = \varepsilon_0 E(1 + \chi) = \varepsilon E$$

where the coefficient $\varepsilon = \dfrac{D}{E} = (1 + \chi)\varepsilon_0$ is called the permittivity of the medium. The relative permittivity or the dielectric constant $(K$ or $\varepsilon_r)$ is defined as

$$K \text{ or } \varepsilon_r = \frac{\varepsilon}{\varepsilon_0} = 1 + \chi$$

This equation represents the relation between the dielectric constant ε_r and *electrical susceptibility* χ of the medium, or

$$\varepsilon_r = \varepsilon_0(1 + \chi)$$

For most substances K is larger than one and it is a pure number (Table 4.1). We must note that the quantities ε_r, χ, ε and P are all a measure of the extent of polarization produced in a dielectric medium under the action of an external field.

Table 4.1: Values of ε_r or K at room temperature

Material	ε_r or K
Carbon tetrachloride	2.24
Transformer oil	2.4
Paraffin wax	2.25
Polyethylene	2.30
Nylon	2.50
Porcelain	6.00
Mica	7.00
Water (0°C)	88.00
Water (100°C)	48.00

Let us consider a parallel plate capacitor (Fig. 4.10) with negative charge on the top and positive charge on the bottom plate. Let A be the area of each plate, σ the surface charge density and d their separation. We have,

$$E = \frac{\sigma}{\varepsilon_0}, \quad V = \frac{\sigma}{\varepsilon_0}d, \quad C = \frac{Q}{V} = \frac{\varepsilon_0 A}{d}$$

Now, suppose we introduce a dielectric slab between the plates of the capacitor. Figure 4.10 exhibits a gap between the plates and the walls of the dielectric. We must note that such a gap will always be present though its width may be only of molecular dimensions. As shown in the figure, a positive charge will be induced on the top and negative charge on the bottom side of the slab. This will produce a macroscopic field inside the material which is opposed to the field between the plates, thus effectively reducing it. Thus, we have the effective field as

Fig. 4.10: Dielectric slab between the plates of a capacitor

$$E = \frac{\sigma - P}{\varepsilon_0} \quad (\because P = \varepsilon_0 \chi E)$$

$$\therefore \quad V = Ed = \frac{\sigma d}{\varepsilon_0(1 + \chi)} \tag{4.28}$$

and

$$C = \frac{Q}{V} = \frac{\sigma A}{\sigma d / \varepsilon_0(1 + \chi)}$$

$$= \frac{\varepsilon_0 A}{d}(1 + \chi) = \frac{\varepsilon_0 \varepsilon_r A}{d} \tag{4.29}$$

Thus, the capacitance is increased by a factor ε_r.

4.11 LAWS OF ELECTROSTATICS IN THE PRESENCE OF DIELECTRICS

We have seen in the preceding sections that the field in a dielectric medium is not the same as would be expected from Coulomb's law. Obviously, the law, therefore, is not of universal validity. Experi-mentally, it has been proved for air and is found not to be true for dielectrics. This means that the laws of electrostatics so far established need modification in dielectrics.

So far, we have considered uniform polarization. However, there may be situations where the polarization is not uniform. This may be due to either dielectric being non-uniform or the variation of the field even inside a uniform dielectric. Now, we consider a situation in which the polarization P is not the same everywhere.

Let us consider that an unpolarized slab of a material is placed in an electric field. Due to polarization process, a certain amount of net charge will pass through a given amount of area ds due to polarization process. The amount of the charge is $P \cdot \hat{a}_n \, ds$ (product of the component of P normal to ds and the magnitude of ds). If P is not uniform, then the displacement of charges can result in a volume charge density. The total charge displaced out of any volume V by the polarization is obtained as

$$\int_S P \cdot \hat{a}_n \, ds$$

where, S is the surface that binds the volume. Let q' be the equal excess charge of the opposite sign left behind, then

$$q' = \int_S P \cdot \hat{a}_n \, ds \qquad (4.30)$$

If ρ' be the density of the polarization charges within the volume, then

$$q' = \int_V \rho' \, d\tau$$

$$\therefore \qquad \int_V \rho' \, d\tau = -\int_V P \cdot \hat{a}_n \, ds \qquad (4.31)$$

Making use of the divergence theorem, one obtains

$$\int_V \rho' \, d\tau = -\int_V \nabla \cdot P \, d\tau \qquad (4.32)$$

Thus
$$\rho' = -\nabla \cdot P \qquad (4.33)$$

Obviously, the polarization charge density ρ' $(C\,m^3)$ is given by the negative divergence of polarization density P.

We have Gauss's law in free space as

$$\nabla \cdot E = \frac{\rho}{\varepsilon_0}$$

One can write an equivalent law for dielectric medium by remembring that besides free charges there are also polarization charges present. One obtains,

$$\nabla \cdot E = \frac{\rho + \rho'}{\varepsilon_0} = \frac{\rho}{\varepsilon_0} - \frac{\nabla \cdot P}{\varepsilon_0}$$

$$\therefore \qquad \nabla \cdot \left(E + \frac{P}{\varepsilon_0} \right) = \frac{\rho}{\varepsilon_0}$$

or
$$\nabla \cdot (\varepsilon_0 E + P) = \rho \qquad (4.34)$$

or
$$\nabla \cdot D = \rho \tag{4.35}$$

where, $D = \varepsilon_0 E + P$.

Equation (4.35) is the modified form of Gauss's law and also called as Gauss's theorem in differential form which also includes the effects of polarization charges. Obviously, the flux of D out of a closed surface S equals the total free charge enclosed by S. Obviously, D has the same dimensions as P. Equation (4.35) is also the Maxwell's first equation which suggests that the evolution of electric displacement vector is due to free charges.

Since
$$P = \varepsilon_0 \chi E$$

$\therefore$
$$D = \varepsilon_0 E + \varepsilon_0 \chi E = (1 + \chi)\varepsilon_0 E = \varepsilon_0 \varepsilon_r E = \varepsilon E \tag{4.36}$$

where, $\varepsilon = \varepsilon_0 \varepsilon_r$ is the *permittivity* of the medium. We must remember that $\nabla \times E = 0$ remains unchanged in dielectrics.

4.12 ENERGY OF THE ELECTROSTATIC FIELD IN THE PRESENCE OF A DIELECTRIC

We have already obtained the expression for the potential energy, when the charges are distributed with the volume density ρ and surface density σ as

$$W = \frac{1}{2}\int_V \rho\phi dV + \frac{1}{2}\int_{S_1} \sigma\phi ds_1$$

The above result is not affected by the presence of a dielectric. We have from Gauss's theorem in the presence of a dielectric, $\nabla \cdot D = \rho$. Thus,

$$\frac{1}{2}\int_V \rho\phi d\tau = \frac{1}{2}\int_V \phi(\nabla \cdot D)d\tau = \frac{1}{2}\int_V \{\nabla \cdot (\phi D) - D \cdot \nabla\phi\}d\tau$$

$$= \frac{1}{2}\int_S \phi D \cdot \hat{a}_n ds + \frac{1}{2}\int_V D \cdot E d\tau \tag{4.37}$$

In obtaining the above result, we have used the following vector identity

$$\nabla \cdot (\phi D) = \phi\nabla \cdot D + D \cdot \nabla\phi \tag{4.38}$$

and the divergence theorem. We can see that the first integral in Eq. (4.37) must be taken over a closed surface bounding the whole volume (which is zero) and also over the surfaces of the conductors. Since integration is over the surface of the medium, the vector $\hat{a}_n ds$ is outwards from the medium, i.e. into the conducting surface (Fig. 4.6).

One can also show that the normal component of D is

$$D \cdot \hat{a}_n ds_1 = -\sigma ds_1$$

$\therefore$
$$\frac{1}{2}\int_V \rho\phi d\tau = -\frac{1}{2}\int_{S_1} \sigma\phi ds_1 + \frac{1}{2}\int_V D \cdot E d\tau$$

$\therefore$
$$W = \frac{1}{2}\int_V \rho\phi d\tau + \frac{1}{2}\int_{S_1} \sigma\phi ds_1 = \frac{1}{2}\int_V D \cdot E dt \tag{4.39}$$

Equation (4.39) shows that one may regard the energy as being distributed throughout all space. The energy density is $\frac{1}{2}D \cdot E$.

4.13 BOUNDARY CONDITIONS

In a single medium the electric field is continuous. In case the field is not constant, there occurs an infinitesimal amount of change in the field in an infinitesimal distance. However, at the boundary between two different media the electric field may change abruptly both in magnitude and direction. Thus, it becomes necessary to establish the field relations at such boundaries. Moreover, the solution of Poisson's or Laplace's equation should always satisfy certain conditions at the boundaries between different media in the field. These conditions pertaining to potential, tangential component of the electric field intensity, normal component of the displacement density vector and refraction lines of forces are called *boundary conditions*.

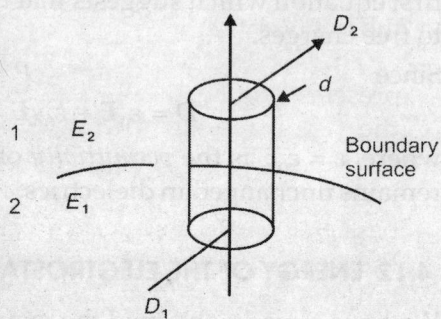

Fig. 4.11: A disc enclosing a part of boundary surface between two media

Let us imagine a disc enclosing a part of the boundary surface between two media and whose axis is normal to the boundary as shown in Fig. 4.11.

Let ε_1 and ε_2 be the relative permittivities of two media marked as 1 and 2 respectively. Assuming that the thickness of the disc is very small, the only contribution to the outward flux from the disc comes from its flat surfaces. We have from Gauss's theorem,

$$\nabla \cdot D = \rho$$

$$\therefore \qquad \int_V \nabla \cdot D d\tau = \int_V \rho d\tau \qquad (4.40)$$

where, ρ is the density of free charges. Since there is no free charge on the surface, we have

$$\int_V \nabla \cdot D d\tau = 0$$

·By divergence theorem, one obtains

$$\int_V \nabla \cdot D d\tau = \int_S D \cdot \hat{a}_n ds \qquad (4.41)$$

$$\therefore \qquad \int_S D \cdot \hat{a}_n ds = D_1 \cdot \hat{a}_{n_1} ds + D_2 \cdot \hat{a}_{n_2} ds = 0$$

where, D_1 and D_2 are the values of D in the two media and $\hat{a}_{n_1}$ and $\hat{a}_{n_2}$ are the unit vectors. Also,

$$\hat{a}_{n_1} = -\hat{a}_{n_2} = \hat{a}_n$$

Thus, $\qquad (D_1 \cdot \hat{a}_n - D_2 \cdot \hat{a}_n) ds = 0 \quad$ or $\quad D_1 \cdot \hat{a}_n = D_2 \cdot \hat{a}_n \qquad (4.42)$

Obviously, the normal component of D in two media are equal.

Now, we construct at the boundary a rectangular path $ABCD$ (Fig. 4.12) with sides $AB = CD = dl$ parallel to the surface and $BC = DA = dt$ perpendicular to it.

Let E_1 and E_2 be the electric fields in the two media. We know that electric field is conservative and hence no net work is done by taking a unit charge round the rectangular path $ABCDA$. Obviously,

$$\oint E \cdot dl = E_{\parallel(1)} dl - E_{\parallel(2)} dl + \text{contribution from } BC \text{ and } DA = 0 \qquad (4.43)$$

where, $E_{\|(1)}$ and $E_{\|(2)}$ are the components of electric field E in the two media parallel to the surface. Let us assume that dl, i.e. AB, CD is vanishingly small, one finds

$$E_{\|(1)}dl = E_{\|(2)}dl = 0$$

$$\therefore \qquad E_{\|(1)} = E_{\|(2)} \qquad\qquad (4.44)$$

Obviously, the tangential component of the electric field E, i.e. $E_\|$ is continuous across the boundary.

Thus, we have obtained two conditions represented by Eqs. (4.42) and (4.44) to be satisfied by the displacement vector D and electric field E respectively.

The results we have obtained signify that when passing through the interface between two dielectrics, the normal component of the vector D and the tangential component of vector E change continuously. The tangential component of the vector D and the normal component of the vector E, however, are disrupted when passing through the interface.

Fig. 4.12: A rectangular path $ABCD$ at the boundary

One can also express the above conditions in terms of potential. Let us take two media in contact with potentials ϕ_1 and ϕ_2 respectively. Then at all connecting points of the two media 1 and 2, we have

$$\phi_1 = \phi_2$$

and

$$\varepsilon_1 \frac{\partial \phi_1}{\partial n} = \varepsilon_2 \frac{\partial \phi_2}{\partial n} \qquad\qquad (4.45)$$

We must note that Eq. (4.42) for the displacement vector was arrived at based on the assumption that there is no free charge on the boundary. However, if there is any free charge present on the boundary, say with surface density σ, one finds that Eq. (4.42) changes to

$$D_1 \cdot \hat{a}_n - D_2 \cdot \hat{a}_2 = \sigma \qquad\qquad (4.46)$$

Obviously, the normal component of D abruptly changes by σ.

4.14 POTENTIAL AND FIELD DUE TO A POLARIZED SPHERE

Consider a sphere of dielectric material which has been polarized along the positive z-axis (Fig. 4.13). Due to polarization, the positive charges are concentrated at the top and the negative charges at the bottom. Thus, a polarized sphere is equivalent to a combination of two spheres, one having $+Q$ charge and other having $-Q$ charge such that the centre of gravity of the positive charge has been displaced slightly upwards while the centre of the negative charge has been displaced slightly downwards. If d is the distance between these centres (Fig. 4.14), then the moment of the dipole is $p_0 = Qd$ (considering the polarized sphere to be equivalent to dipole). If r is the radius of the sphere, then volume of the sphere $= \frac{4}{3}\pi r^3$ and if P is the density of polarization, then

dipole moment of the sphere is $p_0 = \frac{4}{3}\pi r^3 P.$

Fig. 4.13: A sphere of dielectric material

Fig. 4.14: A polarized sphere

Obviously, the dipole moment of a polarized sphere is equivalent to the vector sum of the dipole moments of atoms or molecules situated inside the sphere.

The potential V at a point A on the surface of the sphere is

$$V = \frac{1}{4\pi\varepsilon_0}\frac{p_0\cos\theta}{r^2} = \frac{1}{4\pi\varepsilon_0}\frac{\frac{4}{3}\pi r^3 P\cos\theta}{r^2} = \frac{1}{3\varepsilon_0}Pr\cos\theta \tag{4.47}$$

Now $r\cos\theta = z_A$, the z-coordinate of the point A.

$$\therefore \qquad V = \frac{1}{4\pi\varepsilon_0}\frac{4}{3}\pi p z_A = \frac{1}{3\varepsilon_0}p z_A = K z_A \tag{4.48}$$

where, $K = \dfrac{p}{3\varepsilon_0}$ is a constant. Obviously, the nature

of the field is same as due to a dipole (Fig. 4.15).

The potential inside the sphere must satisfy the Laplace's equation, $\nabla^2 V = 0$, i.e.

$$\frac{\partial^2 V}{\partial x^2} + \frac{\partial^2 V}{\partial y^2} + \frac{\partial^2 V}{\partial z^2} = 0$$

Putting the value of $V = K z_A$, one finds that the equation is satisfied. Then according to the *uniqueness theorem*, Laplace's equation can have only one solution, i.e. $V = K z_A$. This relation gives the value of potential everywhere inside the polarized sphere. The electric field inside the sphere, i.e. along z-axis is

Fig. 4.15: Field due to a polarized sphere, which is same as due to a dipole

$$E_z = -\frac{\partial V}{\partial z} = -\frac{\partial}{\partial z}(K z_A) = -K = -\frac{p}{3\varepsilon_0} \tag{4.49}$$

Equation (4.49) shows that the field inside the sphere is uniform and constant and directed opposite to the direction of polarization as shown in Fig. 4.16.

Thus, $E = \dfrac{p}{3\varepsilon_0}$. Obviously, the electric field inside the polarized sphere is constant.

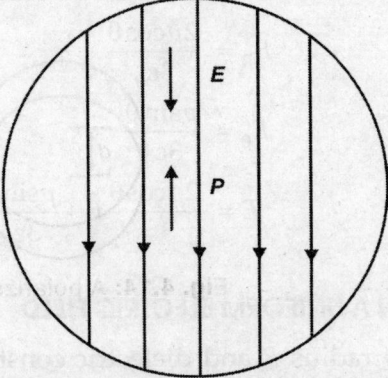

Fig. 4.16: Field inside the sphere

Potential and field due to a polarized sphere when the point is outside the sphere (R > r):

We have seen that a polarized sphere behaves like a dipole with dipole moment p. Thus, the potential at a point having coordinates (R, θ) situated outside the polarized sphere, will be

$$V = \frac{p \cos\theta}{4\pi\varepsilon_0 R^2} = \frac{\frac{4}{3}\pi r^3 p \cos\theta}{4\pi\varepsilon_0 R^2}$$

$$= \frac{pr^3 \cos\theta}{3\varepsilon_0 R^2} \tag{4.50}$$

The components of the electric field intensity at (R, θ) will be

$$E_R = -\frac{\partial V}{\partial R} = -\frac{\partial}{\partial R}\left(\frac{pr^3 \cos\theta}{3\varepsilon_0 R^2}\right) = \frac{2pr^3 \cos\theta}{3\varepsilon_0 R^3} \tag{4.51}$$

or

$$E_\theta = -\frac{1}{R}\frac{\partial V}{\partial \theta} = -\frac{1}{R}\frac{\partial}{\partial \theta}\left(\frac{pr^3 \cos\theta}{3\varepsilon_0 R^2}\right) = \frac{pr^3 \sin\theta}{3\varepsilon_0 R^3} \tag{4.52}$$

$\therefore$ Total intensity $\quad E = \left[\dfrac{2pr^3 \cos\theta}{3\varepsilon_0 R^3}\hat{R} + \dfrac{pr^3 \sin\theta}{3\varepsilon_0 R^3}\hat{\theta}\right]$

$$= \frac{pr^3}{3\varepsilon_0 R^3}\left[(2\cos\theta)\hat{R} + (\sin\theta)\hat{\theta}\right] \tag{4.53}$$

where, $\hat{R}$ and $\hat{\theta}$ are unit vectors.

Point on the surface of the polarized sphere (r = R):

Putting $r = R$ in Eqs. (4.50), (4.51) and (4.52), one obtains

$$V = \frac{pR \cos\theta}{3\varepsilon_0} = \frac{pz}{3\varepsilon_0} \qquad (\because R\cos\theta = z) \tag{4.54}$$

Obviously, the potential of a polarized sphere depends on the z-coordinate only. The components of the intensity of electric field on the surface of polarized sphere are

$$E_R = \frac{2p\cos\theta}{3\varepsilon_0} \tag{4.55}$$

or

$$E_\theta = \frac{p\sin\theta}{3\varepsilon_0} \tag{4.56}$$

∴ Total intensity

$$E = \frac{2p\cos\theta}{3\varepsilon_0}\hat{R} + \frac{p\sin\theta}{3\varepsilon_0}\hat{\theta} \tag{4.57}$$

4.15 DIELECTRIC SPHERE IN A UNIFORM ELECTRIC FIELD

Let us consider a sphere of radius R and dielectric constant K or ε_r, be placed in a uniform field E_0. The centre of the sphere is at the origin and the electric field is along z-axis as shown in Fig. 4.17.

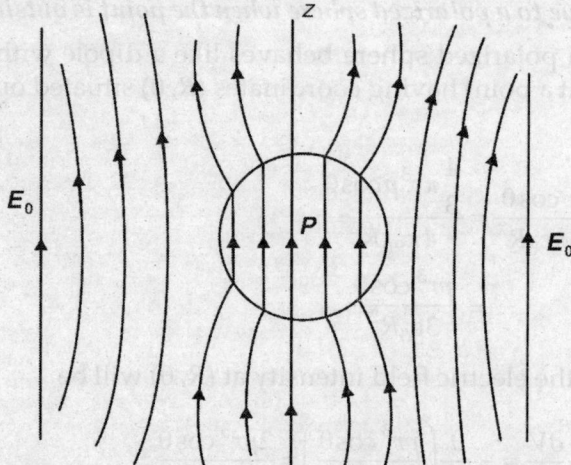

Fig. 4.17: Dielectric sphere in a uniform field

Due to this field, polarization is induced in the sphere along z-axis which alters the field intensity inside and outside the sphere. However, the field at a large distance from the sphere remains same, i.e. equal to E_0.

Electric Field Inside the Dielectric Sphere

We can easily see that the resultant electric field E inside the dielectric sphere is equal to the vector sum of the superposed electric fields, i.e. external field E_0 and the field produced due to the polarization P, i.e. E_P. Thus,

$$E_{in} = E_0 + E_P \tag{4.58}$$

The electric field inside the dielectric sphere due to induced polarization P will be

$$E_P = -\frac{P}{3\varepsilon_0} \tag{4.59}$$

∴

$$E_{in} = E_0 - \frac{P}{3\varepsilon_0} \tag{4.60}$$

Since the induced polarization P in a dielectric is proportional to E_{in}, therefore, one obtains

$$P = \varepsilon_0 \chi E_{in}$$

where, for a dielectric $\chi = \varepsilon_r - 1$, is the electrical susceptibility.

$$\therefore \qquad P = \varepsilon_0 (\varepsilon_r - 1) E_{in} \qquad (4.61)$$

Using Eq. (4.61), Eq. (4.60) becomes

$$E_{in} = E_0 - \frac{\varepsilon_0 (\varepsilon_r - 1) E_{in}}{3\varepsilon_0} = \frac{3}{2 + \varepsilon_r} E_0 \qquad (4.62)$$

Obviously, the direction of electric field inside the dielectric will be the same as that of the external field E_0. Moreover, for a dielectric, ε_r is always greater than 1, the magnitude of E_{in} is always less than E_0. Using Eq. (4.62), Eq. (4.61) becomes

$$P = \frac{3\varepsilon_0 (\varepsilon_r - 1)}{(\varepsilon_r + 2)} E_0 = \frac{3\varepsilon_0 (K - 1)}{(K + 2)} E_0 \qquad (4.63)$$

Equation (4.63) suggests that polarization P in the dielectric is induced along the direction of the applied field.

Electric Field Outside the Dielectric Sphere

The resultant field outside the dielectric sphere E_{out} would be equal to the vector sum of electric field E_0 and the electric field E produced due to polarized sphere, i.e.

$$E_{out} = E_0 + E \qquad (4.64)$$

One can find the intensity of electric field at a point (R, θ) outside the polarized sphere with polarization P [using Eq. (4.53)] as

$$E = \frac{2PR^3 \cos\theta}{3\varepsilon_0 r^3} \hat{r} + \frac{PR^3 \sin\theta}{3\varepsilon_0 r^3} \hat{\theta} \qquad (4.65)$$

where, $\hat{r}$ and $\hat{\theta}$ are unit vectors along directions r and θ respectively.

4.16 GASEOUS NON-POLAR DIELECTRICS

A dielectric in electromagnetic theory is considered to be a continuous medium and the macroscopic quantities such as charge density and electric polarization are defined as continuous functions. A dielectric is really an assembly of positive and negative charges. Therefore, it is essential to relate the electrical properties of the constituent discrete particles and the corresponding properties of the dielectric (continuous medium). We have mentioned earlier that if the field is not too large, the strength of the dipole induced in an atom, ion or molecule is proportional to the actual field acting on the particle. If we call this field E_{local}, then

$$p = \alpha \varepsilon_0 E_{local} \qquad (4.66)$$

where, p is the induced dipole moment and $\alpha \varepsilon_0$ is the constant of proportionality.

We must note that the local field E_{local} acting on a particular particle is not the same as the macroscopic field, which is the field averaged over a region large enough to contain many particles, and calculated by taking the average of the vector sum of the electric fields set up by all the dipoles in the medium including that of the particle for which one require vector E_{local}. The local field E_{local} is equal to the maroscopic field E due to the dipoles more than a few atomic diameters away, which we can treat as a continuous medium with polarization P, plus the contribution from nearby dipoles excluding the field of the dipole at which one requires the value of E_{local}, since the field due to this dipole does not act on itself. We must note that this is particularly true in the case of liquids and solids. In the case of gases, the molecules, for most of the time, are far apart, and hence the short range fields are negligible. Thus, E_{local} is the same as the macroscopic field E. Obviously, in the case of gases

$$p = \alpha \varepsilon_0 E$$

and

$$P = Np = N\alpha \varepsilon_0 E = \varepsilon_0 \chi E \qquad (4.67)$$

$\therefore$

$$\alpha = \frac{\chi}{N} \qquad (4.68)$$

Here α is called the *polarizability* of the atom, ion or molecule. It measures the resistance of the particle to the displacement of its electron cloud. The scalar χ is the electrical susceptibility of the medium.

4.17 GASEOUS POLAR DIELECTRICS

We now study the dielectric behaviour of molecules which have permanent dipole moments. We have mentioned in the previous section that the local field acting on a molecule in a gas is the same as the external field E. This field induces an extra dipole moment in the molecule giving it the same kind of polarizability as in non-polar molecules. Besides, the field also tends to line up the individual dipoles. However, this does not happen effectively due to the thermal motion of molecules. The potential energy of a dipole when placed in an electric field is given by

$$W = -p \cdot E = -pE \cos\theta \qquad (4.69)$$

The potential energy varies with the orientation of the dipole. The energy is minimum when $\theta = 0$, i.e. when the dipole is aligned along the field. One can get an idea about the order of magnitude of this energy by calculating the energy required to reverse the dipole, so as to make it point in the direction opposite to that of the field. Obviously, this energy is equal to $2pE$. We have atomic spacing $\sim 10^{-10}$ m and let $E = 10^6$ V/m, the maximum that can be achieved in gaseous dielectric. Therefore,

$$p = e \times \text{atomic spacing} = 1.6 \times 10^{-19} \times 10^{-10} = 1.6 \times 10^{-29} \text{ C m}$$

and

$$2pE = 2 \times 1.6 \times 10^{-29} \times 10^6 \text{ J} = 2 \times 10^{-4} \text{ eV} \qquad (4.70)$$

We know that the average kinetic energy of a gaseous molecule at room temperature is of the order of 0.04 eV. Obviously, the energy given by Eq. (4.70) is much less than the average kinetic energy. This reveals that the random thermal motion will not be affected much by the presence of the field E and there will still be molecules with their dipoles pointing in all directions. However, the dipoles, in general, tend to line up in a direction in which their potential energy is minimum, i.e. in the direction of the field. Obviously, there will be a slight excess of molecules pointing in this direction. Let us calculate the net polarization caused by this polarization. One can do it by using the statistical methods.

Let us consider that N be the number of molecules per unit volume. We know, from statistical mechanics that in the state of thermal equilibrium, the relative number of molecules with the potential energy W is proportional to

$$N = N_0 \exp\left(-\frac{W}{kT}\right)$$

where, k is Boltzmann constant and T is the absolute temperature.

We know that in the absence of an electric field, all orientations of molecules are equally probable. Thus, the number of molecules per unit volume having their dipoles within a solid angle $d\Omega$ will be equal to $N\dfrac{d\Omega}{4\pi}$. Now, the solid angle in the range of orientation between θ and $\theta + d\theta$ as shown in Fig. 4.18, is given by the following equation:

$$d\Omega = \frac{2\pi \sin\theta\, r d\theta}{r^2} = \frac{2\pi \sin\theta\, d\theta}{r} \qquad (4.71)$$

$$\therefore \qquad N\frac{d\Omega}{4\pi} = \frac{1}{2} N \sin\theta\, d\theta \qquad (4.72)$$

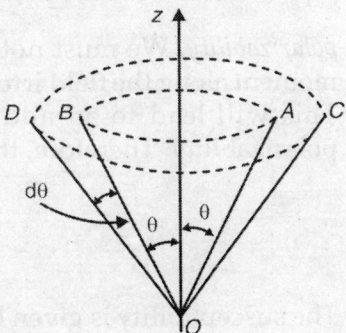

Fig. 4.18: Solid angle in the range θ and $\theta + d\theta$

Now, consider that a field E is applied along the z-axis. Obviously, the dipoles making angle θ with the field acquire an additional potential energy $W = -pE\cos\theta$. The probability distribution will depend upon the Boltzmann factor $\exp\left(-\dfrac{W}{kT}\right)$. Thus, the number of molecules per unit volume with their dipoles in the range between θ and $\theta + d\theta$ is given by

$$N(\theta)\, d\theta = C \exp\left(-\frac{W}{kT}\right) d\Omega \qquad (4.73)$$

Since $\dfrac{W}{kT}$ is very small and hence one may write this as

$$N(\theta)\, d\theta = C\left(1 - \frac{W}{kT}\right) d\Omega = C2\pi \sin\theta\left(1 + \frac{pE\cos\theta}{kT}\right) \qquad (4.74)$$

Therefore, the total number of molecules per unit volume is obtained as

$$N = \int_0^\pi N(\theta)\, d\theta = 2\pi C \int_0^\pi \sin\theta\left(1 + \frac{pE\cos\theta}{kT}\right) d\theta = 4\pi C$$

$$\therefore \qquad C = \frac{N}{4\pi} \quad \text{and} \quad N(\theta) = \frac{N}{2}\sin\theta\left(1 + \frac{pE\cos\theta}{kT}\right) \qquad (4.75)$$

Equation (4.75) clearly reveals that there will be more molecules in the field direction ($\cos\theta = 1$) than in the opposite direction. This is in agreement with our assumption that there will be net alignment when the field E is applied.

We find that the component of dipole moment pointing in the direction θ, along the field is $p \cos \theta$. Thus, the net dipole moment per unit volume along the field is

$$P = \int_0^\pi N(\theta) p \cos \theta \, d\theta = \frac{N}{2} \int_0^\pi \sin \theta \left(1 + \frac{pE \cos \theta}{kT}\right) = \frac{Np^2 E}{3kT} \tag{4.76}$$

This equation reveals the temperature dependence of the polarization taking place in dielectrics. As the temperature T increases, the polarization of the dielectrics is seen to decrease. This is physically visualized in terms of increase of orientational time with temperature which inhibits the increase of orientation of the electric dipoles along the applied field. Moreover, Eq. (4.76) also represents the contribution of the permanent dipole moments to the polarization. The quantity $\dfrac{p^2}{3kT}$ is usually called as *orientational polarizability*. We must note that the molecules will also acquire an induced dipole moment along the field irrespective of the direction of the permanent dipole moment. This will lead to an additional polarization $N\alpha_0 \varepsilon_0 E$, where α_0 is deformation polarizability. Therefore, the net polarization is obtained as

$$P = \left(N\alpha_0 \varepsilon_0 \frac{Np^2}{3kT}\right) E \tag{4.77}$$

The susceptibility is given by

$$\chi = N\left(\alpha_0 + \frac{p^2}{3kT\varepsilon_0}\right) \tag{4.78}$$

One can easily see from Eq. (4.78) that the measurement of susceptibilities at different temperatures will provide information about permanent and induced dipole moments. Further, we note that the plot of χ versus $1/T$ will be a straight line. The slope of this plot will give the permanent dipole moment and intercept at $1/T = 0$ will give χ due to the induced polarization. We can also get some information about the shape of molecules from the knowledge of susceptibilities and the dipole moments, e.g. CS_2 and H_2O both molecules contain two identical atoms bound to a common partner. One can see that there is electron transfer across C–S bond as well as O–H bond and dipole moments associated with each bond. However, the dipole moment of H_2O is 6.2×10^{-30} C m whereas that of CS_2 is zero. One can conclude that since dipole moments of CS_2 cancel each other and hence it must be a linear molecule, i.e. S–C–S whereas H_2O is not a linear molecule.

In the case of polar molecules, it is found that the frequency response depends on the moment of inertia of the molecules. The moment of inertia of molecules determine the amount of time they take to turn toward the direction of the applied electric field. Obviously, if one applies electromagnetic fields in the microwave range, the polar contribution to the dielectric constant begins to fall because the heavy molecules do not respond to these high frequencies. In contrast to this, one finds that the electronic polarizability remains unaffected up to optical frequencies and the contribution to dielectric constant does not change with increasing frequency.

It is to be noted that if we consider the effect of collisions on the orientation time of the electric dipoles, the total polarization in the dielectric dipoles in the dielectric medium can be expressed as

$$P_T = \frac{Np^2E}{3kT(1 - j\omega\tau_r)} \qquad (4.79)$$

where, τ_r is the relaxation time of electric dipoles in dielectric medium. Thus, we have

or

$$\frac{P_T}{E} = \frac{Np^2}{3kT(1 - j\omega\tau_r)} = N\varepsilon_0\alpha(\omega) \qquad (4.80)$$

where, $\alpha(\omega)$ is the atomic polarizability of the dielectrics

$$\alpha(\omega) = \frac{p^2}{3\varepsilon_0 kT(1 - j\omega\tau_r)} \qquad (4.81)$$

This gives the complex form of *orientational polarizability*. The variation of real and imaginary parts of polarizability with frequency is shown in Fig. 4.19. The real part of α, say $\alpha'(\omega)$, is constant at low frequencies and decreases steadily with increasing frequency. The slope becomes maximum at $\omega = 1/\tau_r$.

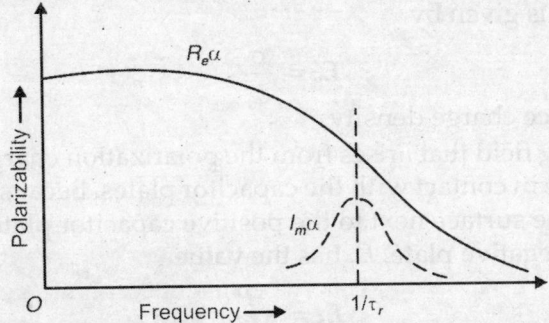

Fig. 4.19: Variation of polarizability with frequency

4.18 MOLECULAR FIELD IN A DIELECTRIC: THE CLAUSIUS–MOSSOTTI RELATION

The electric field which is responsible for polarizing a molecule of the dielectric is called the *molecular field* E_m. This is the electric field at a molecular position in the dielectric; it is produced by all external sources and by all polarized molecules in the dielectric with an exception of the one molecule at the point under consideration. It is clear that E_m need not be the same as the macroscopic electric feld because this is related to the force on a test charge which is large in comparison with molecular dimensions.

In order to compute the molecular field, we cut out a small piece of the dielectric, leaving a spherical cavity surrounding the point at which the molecular field is to be computed. The dielectric which is on left will be treated as a continuum from the macroscopic point of view. Now, we put the dielectric back into the cavity, molecule by molecule, except for the molecule at the centre of the cavity where the molecular field is to be computed. The molecules which have just been replaced are to be treated, not as a continuum, but as individual dipoles. We can summarize the above statement that for the calculation of molecular field, we regard the dielectric as consisting of two parts: (i) a sphere of radius r (centred at O, where the molecule at which the molecular field is to be computed, is located) within which the continuous nature of the dielectric must be taken into account, and (ii) the rest of the dielectric slab which is sufficiently distanced from O, may be regarded as a continuous medium.

Fig. 4.20: A spherical cavity

Let us now suppose that the dielectric sample has been polarized by placing it in a uniform electric field between the plates of a parallel plate condenser (Fig. 4.20). Suppose the density of polarization is uniform and along the field which produces it.

The molecular field at O consists of four components:

$$E_m = E_0 + E_1 + E_2 + E_3. \tag{4.82}$$

E_0 is the field due to the charged condenser plates that would exist at O if the dielectric were absent. If the dimensions of the condenser plates are large compared with their separations, then E_0 is given by

$$E_0 = \frac{\sigma}{\varepsilon_0} \tag{4.83}$$

where, σ is the surface charge density.

E_1 is the depolarizing field that arises from the polarization charges on the surfaces of the dielectric that are in contact with the capacitor plates. Because the density of these bound charges on the surface next to the positive capacitor plate is $-P$ and $+P$ on the surface next to the negative plate, E_1 has the value

$$E_1 = -\frac{P}{\varepsilon_0} \tag{4.84}$$

E_2 is the field at O due to the polarization bound charges on the surface S of the cavity which is formed when the spherical section of the dielectric is removed. This value of E_2 will be the same as the field inside a uniformly polarized dielectric sphere having density of polarization as $-P$. Thus,

$$E_2 = \frac{P}{3\varepsilon_0} \tag{4.85}$$

Alternatively, E_2 may be computed independently by summing the contributions of the field at O of all elements of polarization charge on the surface S of the cavity. The bound charge density on the spherical surface S is given by the component of the polarization normal to S, that is, $-P\cos\theta$.

Thus, the charge on the surface element ds that lies between θ and $\theta + d\theta$ is $-P\cos\theta$ $2\pi r \sin\theta \cdot r \, d\theta$.

Because of the symmetry of the arrangement, all components of the field that are not normal to the capacitor plates cancel one another. Therefore, the polarization charges on ds give rise to a field at O, in the direction of P, given by

$$dE_2 = \frac{(P\cos\theta)(\cos\theta)2\pi r^2 \sin\theta \, d\theta}{4\pi\varepsilon_0 r^2}.$$

The value of E_2 is then given by

$$E_2 = \int_0^\pi dE_2$$

$$= \frac{2\pi P}{4\pi\varepsilon_0} \int_0^\pi \cos^2\theta \sin\theta\, d\theta$$

$$= \frac{P}{3\varepsilon_0}$$

or, vectorically,

$$E_2 = \frac{P}{3\varepsilon_0} \qquad\qquad (4.86)$$

The last term in Eq. (4.82), i.e. E_3 is the field at O due to the electric dipoles inside the spherical section. There are a number of important cases for which this term vanishes. If there are a great many dipoles in the cavity, and if these are oriented parallel but randomly distributed in position, and also if there is no correlation between the positions of the dipoles, then $E_3 = 0$. This is the situation which might prevail in a gas or a liquid. Similarly, if the dipoles in the cavity are located at the regular atomic positions of a cubic crystal, then again $E_3 = 0$. In the general case, E_3 is not zero, and if the material contains several species of molecules, E_3 may differ as the various molecular positions. However, we restrict our discussion to the rather large class of materials in which $E_3 = 0$.

Combining Eqs. (4.83), (4.84) and (4.85), one obtains the molecular field,

$$E_m = \frac{\sigma}{\varepsilon_0} - \frac{P}{\varepsilon_0} + \frac{P}{3\varepsilon_0} \qquad\qquad (4.87)$$

But $D = \sigma$, the free charge density on the condenser plates.

$$\therefore \qquad E_m = \frac{D-P}{\varepsilon_0} + \frac{P}{3\varepsilon_0} \qquad\qquad (4.87a)$$

or

$$E_m = E + \frac{P}{3\varepsilon_0} \qquad\qquad (4.88)$$

where, $E = (D-P)/\varepsilon_0$ is the macroscopic electric field in the dielectric, since $D = \varepsilon_0 E + P$.

Equation (4.88) gives the electric field E_m that acts upon a single molecule of the dielectric. Under the action of this field, the molecule displays an electric moment p in the direction of the field which is dependent upon the field strength, i.e.

$$p = \alpha E_m \qquad\qquad (4.88a)$$

where, α is the molecular polarizability. It there are n molecules per unit volume of the dielectric, then the polarization (electric moment/volume) is obtained as

$$P = np = n\alpha E_m$$

$$= n\alpha\left(E + \frac{P}{3\varepsilon_0}\right) \qquad\qquad (4.88b)$$

This equation may be rewritten in terms of the dielectric constant K or ε_r, since $P = (K - 1)\,\varepsilon_0 E = (\varepsilon_r - 1)\,\varepsilon_0 E$. In this way, Eq. (4.88b) becomes,

$$\frac{n\alpha}{3\varepsilon_0} = \frac{(K-1)}{(K+2)} = \frac{\varepsilon_r - 1}{\varepsilon_r - 2} \tag{4.89}$$

If the molecular weight of the substance is M and its density ρ, and N is the Avogadro number, then $n = \dfrac{N\rho}{M}$ and Eq. (4.89) becomes

$$\frac{N\alpha}{3\varepsilon_0} = \frac{M}{\rho} \cdot \frac{(K-1)}{(K+2)} = \frac{M}{\rho} \frac{(\varepsilon_r - 1)}{(\varepsilon_r + 1)} \tag{4.90}$$

The quantity $N\alpha/3\varepsilon_0$ is sometimes called the *molar polarizability*. This equation indicates that molecular polarizability of a dielectric of known density can be determined by measuring its dielectric constant. Again since α is approximately proportional to the cube of the radius of the molecule, so if K is found and n is known for a gas at a given temperature and pressure, the approximate diameter of the atom may be found using Eq. (4.89) for monoatomic gases. For the monoatomic gases, the results are in agreement with the kinetic theory methods of determining atomic dimensions.

Equation (4.89) or (4.90) is known as the *Clausius–Mossotti relation*. Remember that, this relation is applicable to those dielectrics for which $E_3 = 0$ such as gases and non-polar liquids.

Relation (4.89) shows that for any given substance, $\left(\dfrac{\varepsilon_r - 1}{\varepsilon_r + 2}\right)$ or $\left(\dfrac{K-1}{K+2}\right)$ should be proportional to the density of the substance. (At optical frequencies, $\varepsilon/\varepsilon_0 = \mu^2$, where μ is the index of refraction. With μ^2 replacing ε_r in Eq. (4.89), the equation is sometimes called the *Lorentz–Lorentz equation*). Equation (4.89) is approximately valid for liquids and solids, especially if the dielectric constant is large. Equation (4.89) reveals that local field is larger than the macroscopic field. One can also find the increase in size. We must note that Eq. (4.89) is not very accurate. Table 4.2 provides the relative permittivity of some substances at NTP.

Table 4.2: Relative permittivity of some substances (at NTP)

Substance	Gas ε_r	Gas ρ(density)	Liquid ε_r	Liquid ρ(density)
O_2	1.0005	0.0014	1.51	1.190
CS_2	1.0029	0.0034	2.64	1.293
CCl_4	1.0030	0.0030	2.24	1.590
CO_2	1.0010	0.0019	1.61	0.975

4.19 POLARIZATION IN CHANGING ELECTRIC FIELDS

So far we have considered the behaviour of dielectric materials in static electric fields. Now, we consider the effects of electric fields that are varying in time, like the field in a capacitor used in an alternating current circuit. When the frequency of the applied field is low, i.e. the field changes slowly in time, the polarization P follows E at every instant and the ratio of P to E is almost constant. We must note that slow variation in electric field is a relative term and depends on the physical process involved.

In dielectrics, the response to an electric field involves two processes: (i) induced polarization and (ii) orientation of permanent dipoles. We have seen that the induced polarization is due to the distortion of electric charge distribution. The motion of electrons in atoms and molecules are characterized by very high frequencies ($\sim 10^{16}$ Hz) or very low periods $\sim 10^{-16}$ s due to their very small mass and hence the period of the applied field (from DC to microwave) is comparatively quite large and these fields may be considered slowly varying fields. Thus, in polar substances in which polarization plays a predominant role, the polarization keeps in step with the applied field. Thus, the susceptibility χ and hence the dielectric constant $\varepsilon_r = (1 + \chi)$ are independent of frequency.

The orientation of a polar molecule is a process quite different from the mere distortion of the electron cloud. The whole molecular framework has to rotate. The frictional drag tends to make the rotation lag behind the torque and to reduce the amplitude of the resulting polarization, where on the time scale this effect sets in, varies enormously from one polar substance to another. In water, the "response time" for dipole reorientation is about 10^{-11} s. The dielectric constant remains around 80 up to frequencies of the order of 10^{10} c/s. Above 10^{11} c/s, dielectric constant falls to a modest value typical of a non-polar liquid. The dipoles simply cannot follow so rapid an alternation f in the field. In other substances, especially solids, the characteristic time can be much longer. In ice just below the freezing

Fig. 4.21: Dielectric constant versus frequency curves for water and ice

point, the response time for electric polarization is around 10^{-5} s. Figure 4.21 shows experimental curves for dielectric constant versus frequency for water and ice.

4.20 THE BOUND CHARGE (POLARIZATION) CURRENT

The charges in the molecules of a dielectric are called *bound*. The action of a field can only cause bound charges to be displaced slightly from their equilibrium positions. They cannot leave the molecule containing them.

The charges that are within the boundaries of a dielectric, but are not inside its molecules and also the charges outside a dielectric are called *extraneous charges*. Normally such charges are called *free charges*.

Bound charges differ from free charges only in that they cannot leave the confines of the molecules within which they are lying. Otherwise, they have the same properties as all other charges. In particular, they are sources of an electric field. When a varying electric field is applied on a dielectric, the centres of electron clouds start oscillating with the same frequency. Let the centre of electron cloud in an atom be displaced by a distance ds in time dt, then the change in dipole moment due to the displacement ds is given by

$$dp = qds \qquad (4.91)$$

where, q is the charge on electron cloud.

If N is the number of atoms per unit volume in the dielectric, then the change in polarization P in time dt is given by

$$dP = Ndp = Nqds \qquad (4.92)$$

Therefore, the rate of polarization in dielectric due to changing electric field is

$$\frac{dP}{dt} = Nq\frac{ds}{dt} = Nqv \qquad (4.93)$$

where, $v = \dfrac{ds}{dt}$ is the velocity.

A charge cloud of density Nq moving with velocity v produces a conduction current of density $J\ C/s\,m^2$ given by

$$J = Nqv \qquad (4.94)$$

A comparison of Eqs (4.93) and (4.94) shows that current is produced due to change in polarization with respect to time, i.e. current is produced due to change in the induced bound charges on the surface of the dielectric. Such a current is source of magnetic field. This current is called *bound charge current*. Therefore, the bound charge current density,

$$J = \frac{dP}{dt} \qquad (4.95)$$

The only difference between an "ordinary" conduction current density and current

density $\dfrac{dP}{dt}$ is that former involves free charges in motion while the latter involves

bound charges in motion. For the development of magnetic field in dielectric medium, three different current densities which contribute are as follows:

i. Free current density J_{free}

ii. Displacement current density due to change in electric field, $J_d = \varepsilon_0 \dfrac{\partial E}{\partial t}$

iii. Bound charge current density produced due to change in polarization as a result

of change in electric field, $J_p = \dfrac{dP}{dt}$.

4.21 TRANSIENT BEHAVIOUR OF RC CIRCUIT

The term transient is usually used to designate the result produced by making a sudden change in an electrical circuit, such as switching on or off an electromotive force, the release of the charge on a capacitor, the sudden change in the magnetic flux linking a circuit, etc. A study of the performance of a network produced by these typical changes may be effected by solving the network equations. Before discussing transient behaviour of a RC circuit, a brief idea of Kirchhoff's laws and their applications to any mesh (completely closed electric circuit) is essential.

Kirchhoff's laws: The usual statements of Kirchhoff's laws are as follows:

1. The algebraic sum of the currents which meet at a junction point in an electric circuit is zero.

2. The algebraic sum of the electromotive forces and potential drops around any closed path or mesh of an electric circuit is zero.

With the application of Kirchhoff's second law, the electromotive forces in any mesh may be added to form the differential equation for the mesh. Alternatively, by the application of Kirchhoff's first law, the currents entering any branch point or junction in the network may be added to zero to form the differential equation for the branch point.

Now, we consider behaviour of a RC circuit. Join a resistance R, capacitor C and a morse key K in series with a battery of emf E as shown in Fig. 4.22. The capacitor can be charged and discharged with time by pressing or releasing the key K.

Fig. 4.22: Charging and discharging of a capacitor in RC series circuit

Charging of a Capacitor

When morse key K is pressed, then current flows through the resistance R and capacitor C gets charged. Let current I flow through the resistance R for time t and accumulate a charge Q on the capacitor C. Due to storage of this charge, a potential difference, $V = \dfrac{Q}{C}$ is developed across the capacitor C and a potential difference IR across the resistance R, then at any instant for this circuit, we have

$$E = \frac{Q}{C} + IR \qquad (4.96)$$

When the voltage across the capacitor becomes equal to the battery voltage E, a maximum charge Q_0 is stored in the capacitor and the current in the circuit becomes zero. Thus, at $Q = Q_0$, we have

$$E = \frac{Q_0}{C} \quad \text{and} \quad I = 0$$

From Eq. (4.96), we have

$$\frac{Q_0}{C} = \frac{Q}{C} + IR$$

or

$$\frac{dQ}{dt} = \frac{Q_0 - Q}{RC} \qquad \left(\because I = \frac{dQ}{dt} \right)$$

or

$$\frac{dQ}{Q_0 - Q} = \frac{dt}{RC} \qquad (4.96a)$$

Integrating the above equation, one obtains

$$\int \frac{dQ}{Q_0 - Q} = \int \frac{dt}{RC}$$

or
$$-\log_e(Q_0 - Q) = \frac{t}{RC} + A \qquad (4.97)$$

where, A is the constant of integration. Initially at $t = 0$, the charge on the capacitor is zero, i.e. $Q = 0$. Using this, Eq. (4.97) gives

$$A = -\log_e Q_0$$

$$\therefore \qquad -\log_e(Q_0 - Q) = \frac{t}{RC} - \log_e Q_0$$

or
$$\log_e\left(\frac{Q_0 - Q}{Q_0}\right) = -\frac{t}{RC}$$

or
$$Q = Q_0(1 - e^{-\frac{t}{RC}}) \qquad (4.98)$$

Equation (4.98) represents the *charging of the capacitor*. The charge on the capacitor in this charging process increases exponentially with time. The product RC which controls the charging rate has the dimensions of time and is called *time constant* of the circuit.

Taking $CR = t$ gives

$$\frac{t}{CR} = 1$$

$$\therefore \qquad Q = Q_0(1 - e^{-1}) = 0.632 Q_0$$

Obviously, the time constant (RC) is the time taken by the capacitor to get charged to 0.632 times its maximum value. Here C is measured in farads, R in ohms and t in seconds. Putting $RC = t$ in Eq. (4.96a), one obtains

$$\frac{dQ}{dt} = \frac{Q_0 - Q}{t}$$

or
$$\frac{dQ}{dt} \propto \frac{1}{t} \qquad (4.99)$$

Equation (4.99) suggests that the rate of increase of charge is inversely proportional to t, the time constant. The circuit with large time constant will take more time in charging the capacitor to the maximum value and *vice versa*.

Decay of Current
At the time of charging the capacitor, the current is given by

$$I = \frac{dQ}{dt}$$

$$= \frac{Q_0}{RC}(e^{-\frac{t}{RC}}) \qquad \left[\because Q = Q_0(1 - e^{-\frac{t}{RC}})\right]$$

$$\therefore \qquad I = I_0 e^{-\frac{t}{RC}} \qquad \left(\begin{array}{l} I_0 = \dfrac{Q_0}{RC} \text{ is the maximum} \\ \text{current which flows at } t = 0 \end{array}\right) \qquad (4.100)$$

Obviously, the current decreases exponentially during charging and in time $t = RC$, the current reduces to $\dfrac{1}{e}$ times, i.e. 37% of the maximum current I_0 (Fig. 4.23).

Fig. 4.23: The variation of charge Q with time t during charging and discharging of a capacitor in a series RC circuit

Discharging of a Capacitor

When a maximum charge Q_0 is accumulated in the capacitor, then if key K is released, the battery gets disconnected and the capacitor starts discharging through the resistance R (Fig. 4.24). Obviously, during discharge there is no source of emf in the circuit, so we have

$$\frac{Q}{C} + RI = 0 \qquad (4.101)$$

or

$$\frac{Q}{C} + R\frac{dQ}{dt} = 0 \qquad \left(\because I = \frac{dQ}{dt} \right)$$

$\therefore$

$$\frac{dQ}{Q} = -\frac{dt}{RC}$$

Fig. 4.24: Discharging of a capacitor

On integrating, one obtains

$$\log_e Q = -\frac{t}{RC} + B \qquad (4.102)$$

where, B is the constant of integration. Initially, i.e. at time $t = 0$, $Q = Q_0$, we have

$$\log_e Q_0 = B$$

$\therefore$

$$\log_e Q = -\frac{t}{RC} + \log_e Q_0$$

or

$$\log_e \frac{Q}{Q_0} = -\frac{t}{RC}$$

$\therefore$

$$Q = Q_0 e^{-\frac{t}{RC}} \qquad (4.103)$$

Equation (4.103) shows that during discharging the capacitor, the charge on the capacitor decreases exponentially with time (Fig. 4.25).

Time Constant

If $t = RC$, then from Eq. (4.103)

$$Q = \frac{Q_0}{e} = 0.37 Q_0$$

Obviously, *the time in which 63% of the charge decays or 37% remains on the capacitor is called the time constant of the circuit. In other words, the time in which the charge on the capacitor reduces to $\frac{1}{e}$ of its initial value (Q_0) is called the time constant.*

Fig. 4.25: Transient curves for RC series circuit (discharging)

Decay Constant

While discharging the capacitor, the current in the circuit is given by

$$I = \frac{dQ}{dt} = -\frac{Q_0}{RC} e^{-\frac{t}{RC}}$$

$$\therefore \quad I = -I_0 e^{-\frac{t}{RC}} \tag{4.104}$$

At $t = 0$, the current is maximum but in a direction opposite to the current at the start of charging the capacitor (Fig. 4.25). Further, again the current decreases in magnitude, exponentially with time.

Energy Stored in the Capacitor during Charging

We have seen that when an emf E is introduced suddenly into a RC circuit, the capacitor starts charging. At any time t, if the charge on the capacitor is Q, then potential difference between the plates at that time will be $\frac{Q}{C}$. This potential difference will be opposed in polarity to the supplied emf. Hence, the current flows against the potential difference between the plates of the capacitor and does external work. This work done to charge the capacitor is stored between the plates of the capacitor in the form of electrical potential energy U.

To obtain the qualitative expression for the storage of energy, consider an extra increment of charge dQ, the small amount of additional work needed will be

$$dU = \frac{Q}{C} dQ$$

Obviously, this work done will increase the potential energy of the system by $UdQ\left(= \frac{Q}{C} dQ\right)$. If the process is continued until a total charge Q_0 has been transferred, the total work done will be

$$U = \int dU = \int_0^{Q_0} \frac{Q}{C} dQ = \frac{1}{2} \frac{Q_0^2}{C} = \frac{1}{2} CE^2 \tag{4.105}$$

Obviously, Eq. (4.105) gives the total energy of the charged capacitor.

The Dissipation of Energy during Charging of a Capacitor through a Resistance

Let us examine the energy dissipated through the resistance R during the charging of a capacitor. We have

$$Q = Q_0(1 - e^{-\frac{t}{RC}}) \quad \text{and} \quad I = \frac{dQ}{dt} = \frac{Q_0}{RC}\exp\left(-\frac{t}{RC}\right)$$

If δW is the energy dissipated in time δt, then

$$\delta W = I^2 R \delta t = \frac{Q_0^2 R}{C^2 R^2}\exp\left(-\frac{t}{RC}\right)$$

On integrating the above, one obtains the total energy dissipated during charging when the capacitor has charged to its maximum value.

$$W = \int dW = \frac{Q_0^2}{RC}\int_0^\infty \exp\left(-\frac{2t}{RC}\right)dt = \frac{1}{2}\frac{Q_0}{C} = \frac{1}{2}Q_0 E$$

which is independent of R. The energy supplied by the DC source is $Q_0 E$ and that stored

in the capacitor is $\frac{1}{2}Q_0 E$. Obviously, the energy supplied by the source is accounted for.

4.22 SOLID DIELECTRICS (ELECTRETS)

One can prepare such materials in solid state which have a permanent polarization and exists even when a field is not applied, e.g. if some wax is melted and a strong electric field is applied to it when it is still in the liquid form, the dipoles get partly lined up and remain that way when the liquid freezes. The solid material thus formed has, like a magent, a permanent dipole moment. These materials are called *solid dielectrics* or *electrets*. Barium titanate is the most familiar example of electret. An electret, however, soon gets discharged, as it attracts free charges from the air.

It is also reported that a permanent internal polarization also exist in the same complicated crystals. However, this is not noticed because the external fields are discharged, as in electrets.

One can work out a theory of the dielectric constant for crystals that do not have permanent dipole moment, on the same lines as for non-polar liquids.

ILLUSTRATIVE EXAMPLES

Example 4.1: Calculate the induced dipole moment per unit volume (density of polarization) of helium gas if placed in a field of 6×10^5 V/m. The atomic polarizability of helium is $0.21 \times 10^{-23} \, m^3$ and density of helium is 2.6×10^{25} atoms/m^3. Also calculate the separation between centres of positive and negative charges in each atom.

Solution: We have

$$p = \varepsilon_0 \alpha E$$
$$\alpha = 0.21 \times 10^{-23} \, m^3, \quad \varepsilon_0 = 8.85 \times 10^{-12} \, mF^{-1}$$
$$E = 6 \times 10^5 \, Vm^{-1}$$
$$p = 8.85 \times 10^{-12} \times 0.21 \times 10^{-23} \times 6 \times 10^5$$
$$= 1.1 \times 10^{-29} \, Cm$$

∴ The dipole moment per unit volume is

$$P = Np = 2.6 \times 10^{25} \times 1.1 \times 10^{-29} = 2.86 \times 10^{-4}$$

The atoms of helium contains 2 electrons and 2 protons, hence dipole moment of atom is $2ed$.

∴
$$2ed = p = 1.1 \times 10^{-29}$$

or
$$d = \frac{1.1 \times 10^{-29}}{2 \times 1.6 \times 10^{-19}} = 0.34 \times 10^{-10} \, \text{m}.$$

Example 4.2: An uncharged capacitor is connected to a battery. Show that half the energy supplied by the battery is lost as heat during charging the capacitor.

Solution: Let the capacitance of the capacitor be C and the emf of the battery be V. Now, the charge on the capacitor is $Q = CV$. Thus, the work done by the battery is

$$W = QV$$

Let charge at any instant on the capacitor be q and its potential V. Work done in further charging through dq is

$$dW = Vdq$$

Total work done in charging the capacitor is

$$W = \int dW = \int Vdq$$

∴
$$W = \int_0^Q \frac{q}{C} dq = \frac{1}{2} \frac{Q^2}{C} = \frac{1}{2} QV \qquad \left(\because \, V = \frac{q}{c} \right)$$

Energy lost in the form of heat $= QV - \dfrac{1}{2} QV = \dfrac{1}{2} QV$

Hence the result.

Example 4.3: An atom of oxygen on being polarized produces a dipole moment of 5×10^{-35} C m. If the distance of the centre of negative charge cloud from the nucleus is 4×10^{-17} m, calculate the polarizability of oxygen atom.

Solution: The polarizing field E tries to push the negative cloud of eight electrons away from the nucleus but the positive charge on the nucleus attracts it. In the equilibrium state the two forces balance each other. If d be the equilibrium distance between positive and negative charges, then

$$8eE = \frac{1}{4\pi\varepsilon_0} \frac{8e \times 8e}{d^2} \quad \text{or} \quad E = \frac{1}{4\pi\varepsilon_0} \frac{8e}{d^2}$$

∴
$$p = \alpha E$$

∴
$$\alpha = \frac{p}{E} = \frac{p \times 4\pi\varepsilon_0 d^2}{8e} = \frac{5 \times 10^{-35} \times (4 \times 10^{-17})^2}{8 \times 1.6 \times 10^{-19} \times 9 \times 10^9} \, \text{m}^3$$

$$= 0.69 \times 10^{-59} \, \text{m}^3$$

Example 4.4: The dielectric constant of a medium is 1.0005. Calculate its atomic and molecular polarizability. Given, Avogadro number $N_A = 6 \times 10^{23}$ per gm-mole and molar volume = 22.4 litre/mole.

Solution: From Clausius–Mossotti's relation,

$$\frac{N_A \alpha}{3\varepsilon_0} = \frac{\varepsilon_r - 1}{\varepsilon_r + 2} V_A \qquad \left| \begin{array}{l} \varepsilon_r = 1.0005 \\ N_A = 6 \times 10^{23} \text{ atoms/mole} \\ V_a = 22.4 \text{ litre/mole} \end{array} \right.$$

Molar polarizability,

$$\alpha_m = N_A \alpha = \frac{5 \times 10^{-4} \times 22.4 \times 10^{-3} \times 3 \times 8.85 \times 10^{-12}}{3.0005}$$

$$= 9.9 \times 10^{-17} \text{ F m}^2/\text{mole}$$

Atomic polarizability,

$$\alpha = \frac{\alpha_m}{N_A} = \frac{9.9 \times 10^{-18}}{6 \times 10^{23}} = 1.65 \times 10^{-40} \text{ F m}^2.$$

Example 4.5: Calculate the dipole moment of a water molecule assuming that all the ten electrons in the molecule circulate symmetrically about the oxygen atom and the OH distance is 0.96 Å and the angle between OH bonds is 104°.

Solution: The effective centre of protons of hydrogen atoms lies at the middle of the line joining the H-atoms (Fig. 4.26). Obviously, a water molecule can be assumed as a dipole of charge 2e and a separation distance a.

Fig. 4.26

Given $e = 1.6 \times 10^{-19}$ C

OH distance $= 0.96 \times 10^{-10}$ m; $\theta = 52°$

$$\cos 52° = 0.616$$

We have,

$$p = (2e) \, a = (2e)(\text{OH})\cos\theta$$
$$= 2 \times 1.6 \times 10^{-19} \times 0.96 \times 10^{-10} \times \cos 52°$$
$$= 2 \times 1.6 \times 10^{-19} \times 0.96 \times 10^{-19} \times 0.616$$
$$= 1.9 \times 10^{-29} \text{C m}.$$

Example 4.6: A dielectric sphere is placed in a uniform electric field E_0. Polarization is uniform in the sphere. The field inside the sphere is 1/3rd of the applied field. Determine dielectric constant and electrical susceptibility of the material.

Solution: In a dielectric sphere, we have

$$E_{in} = \frac{3}{2 + \varepsilon_r} E_0$$

where,

$$E_{in} = \frac{E_0}{3}$$

$\therefore$

$$\frac{E_0}{3} = \frac{3}{2 + \varepsilon_r} E_0$$

or

$$2 + \varepsilon_r = 9$$

$\therefore$

$$\varepsilon_r = 7$$

and

$$\chi = \varepsilon_r - 1 = 7 - 1 = 6.$$

Example 4.7: An artificial dielectric of a large number of metal spheres of macroscopic size is arranged in a three dimensional (lattice) structure as shown in Fig. 4.27. Show that the permittivity of the dielectric $= 1 + 3VN$, where V is the volume of the metal sphere and N number of spheres per unit volume.

Fig. 4.27: Three dimensional (lattice) structure of an artificial dielectric

Solution: Let E be a uniform field applied to the dielectric. This field will induce charges on the individual metallic spheres (Fig. 4.28). Obviously, each sphere becomes analogous to a polarized atom and may be represented by a dipole moment $p = ql$ (Fig. 4.29). If N be the number of spheres per unit volume, then the polarization (P) is obtained as

$$P = Nql$$

We have,

$$D = \varepsilon E = \varepsilon_0 E + P$$

$\therefore$

$$\varepsilon = \varepsilon_0 + \frac{P}{E} = \varepsilon_0 + \frac{Nql}{E} \tag{1}$$

Fig. 4.28: Charges induced on a metallic sphere

Fig. 4.29

The potential at a point due to the uniform field is

$$V = V_0 + V_d = -Er\cos\theta + \frac{ql\cos\theta}{4\pi\varepsilon_0 r^2} \tag{2}$$

Here r denotes the distance of the point from the origin which, for convenience, is taken at the centre of the dipole and θ is the angle between the axis of the dipole and the radial line (Fig. 4.29). Now, the potential due to the dipole,

$$V_d = \frac{p\cos\theta}{4\pi\varepsilon_0 r^2} = \frac{ql\cos\theta}{4\pi\varepsilon_0 r^2} \tag{3}$$

The total potential at the point is

$$V = V_0 + V_d = -Er\cos\theta + \frac{ql\cos\theta}{4\pi\varepsilon_0 r^2} \tag{4}$$

We know that the metal sphere has equal amount of positive and negative charges on it and hence its potential is zero. If a is the radius of the sphere, then

$$0 = -Ea\cos\theta + \frac{ql\cos\theta}{4\pi\varepsilon_0 a^2} \tag{5}$$

$$\therefore \qquad \frac{ql}{E} = 4\pi\varepsilon_0 a^3 \tag{6}$$

Using Eq. (6), Eq. (1) becomes

$$\varepsilon = \varepsilon_0 + 4\pi\varepsilon_0 N a^3$$

and

$$\varepsilon_r = \frac{\varepsilon}{\varepsilon_0} = 1 + 4\pi N a^3 = 1 + 3VN \tag{7}$$

where, $V = \frac{4}{3}\pi a^3$ is the volume of the sphere. Equation (7) shows that the permittivity (ε_r) of the dielectric under consideration depends on both the number of spheres and their sizes. Such dielectrics are called artificial dielectrics consisting of discrete metal particles of macroscopic size.

Example 4.8: A dielectric slab of flat surface with relative permittivity 3 is exposed to a uniform field with its surface normal to the field. The slab has a charge density of $1.5\ C/m^2$ and occupies a volume of $0.8\ m^3$ and is uniformly polarized. Determine the polarization in the slab and the total dipole moment of the slab.

Solution: We have

$$D = \varepsilon E = \varepsilon_0 \varepsilon_r E$$

$$\therefore \qquad |E| = \frac{|D|}{\varepsilon_0 \varepsilon_r} = \frac{1.5}{3\varepsilon_0} \quad \text{or} \quad \varepsilon_0 E = 0.5$$

Now using

$$D = \varepsilon_0 E + P$$

or
$$P = D - \varepsilon_0 E$$
$$= \varepsilon_0 \varepsilon_r E - \varepsilon_0 E = \varepsilon_0 (\varepsilon_r - 1) E$$
$$\therefore \quad P = 0.5(3-1) = 0.5 \times 2 = 1\,\text{C/m}^2$$

Total dipole moment = polarization × volume
$$= 1 \times 0.8 = 0.8\,\text{Cm}.$$

Example 4.9: Show that by a change of dielectric medium, field lines are refracted according to Snell's

law $\dfrac{\varepsilon_1}{\varepsilon_2} = \dfrac{\tan \alpha_1}{\tan \alpha_2}$, where α_1 and α_2 are the angles

between the directions of the field and common normal to the boundary (Fig. 4.30) and ε_1 and ε_2 are the relative permittivities of the two media.

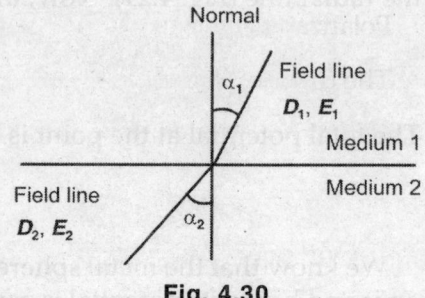

Fig. 4.30

Solution: By applying the first boundary condition, we have

$$D_1 \cdot \hat{a}_n = D_2 \cdot \hat{a}_n$$
or
$$D_1 \cos \alpha_1 = D_2 \cos \alpha_2$$
or
$$\varepsilon_1 E_1 \cos \alpha_1 = \varepsilon_2 E_2 \cos \alpha_2 \qquad (1)$$

The second boundary condition gives
$$E_1 \sin \alpha_1 = E_2 \sin \alpha_2 \qquad (2)$$

Dividing Eq. (2) by Eq. (1), one obtains
$$\frac{\varepsilon_1}{\varepsilon_2} = \frac{\tan \alpha_1}{\tan \alpha_2}$$

Example 4.10: Find the angle of incidence of an electric field on the boundary between a conductor and a dielectric.

Solution: Medium 2 is a conductor and hence $D_2 = \varepsilon_2 E_2 = 0$ under static conditions. According to the boundary conditions, we have
$$D_{n_1} = \rho_s$$
or
$$E_1 E_{n_1} = \rho_s$$
and
$$E_{t_1} = E_{t_2} = 0$$
$$\therefore \quad \alpha_1 = \tan^{-1} \left(\frac{E_{t_1}}{E_{t_2}} \right) = \tan^{-1}(0) = 0$$

Obviously, a static electric field line or flux tube at a dielectric–conductor boundary is always perpendicular to the conductor surface (where no currents are present). This result is of fundamental importance in *field mapping*.

Example 4.11: A water drop of volume 30 mm³ is polarized uniformly. Calculate the maximum field strength due to the polarized drop at a distance of 10 cm from its centre and polarization density. The dipole moment of water molecule is 6.1×10^{-30} Cm.

Solution: A uniformly polarized sphere is equivalent to a dipole of dipole moment p times the volume of the sphere for external points. The molecular weight of water is 18 and Avogadro number is 6×10^{23}. Therefore, the number of molecules in $1\,m^3$ of water

is $\dfrac{6 \times 10^{23} \times 10^6}{18} = \dfrac{10^{29}}{3}$.

Polarization density $P = \rho N = 6.1 \times 10^{-30} \times \dfrac{10^{29}}{3} = 2.03 \times 10^{-1} = 0.203\ C/m^2$

The dipole moment of the drop,

$$p = P \times volume$$
$$= 0.203 \times 30 \times 10^{-9}\ Cm$$
$$= 6.09 \times 10^{-9}\ Cm$$

One can obtain the field components at any point (r, θ) due to a short dipole as

$$E_r = \frac{1}{4\pi\varepsilon_0}\left(\frac{2p\cos\theta}{r^3}\right) \quad \text{and} \quad E_\theta = \frac{1}{4\pi\varepsilon_0}\left(\frac{p\sin\theta}{r^3}\right)$$

The maximum field corresponds to $\theta = 0$ and is given by

$$E_{max} = \frac{1}{4\pi\varepsilon_0}\frac{2p}{r^3} = \frac{9 \times 10^9 \times 2 \times 6.09 \times 10^{-9}}{(0.1)^3} \quad (r = 0.1\,m)$$
$$= 1.09 \times 10^5\ N/C.$$

Example 4.12: A slab of dielectric material with dielectric constant ε_r, thickness b and area a is to be kept in between two parallel plates of area A. The distance between the plates is d. Initially a potential difference of V_0 is applied between two ends of the plates without dielectric. Determine: (i) capacity C_0 of the capacitor without dielectric (ii) free charge on the plates Q_0 (iii) electric field intensity in free space (E_0) (iv) electric field in dielectric and (v) potential difference between the plates (V).

Solution:

(i) We have the electric field between the plates of the capacitor as

$$E_0 = \frac{V_0}{d} = \frac{\sigma}{\varepsilon_0} = \frac{Q_0}{A\varepsilon_0}$$

$$\therefore \qquad C_0 = \frac{Q_0}{V_0} = \frac{\varepsilon_0 A}{d}$$

(ii) Free charge on the plates, Q_0 is

$$Q_0 = \frac{A\varepsilon_0 V_0}{d}$$

(iii) Electric field intensity in the free space is

$$E_0 = \frac{V_0}{d}$$

(iv) Electric field in the dielectric is

$$E = \frac{E_0}{\varepsilon_r} = \frac{V_0}{\varepsilon_r d}$$

(v) Potential difference between the plates of the capacitor with dielectric is

$$V = E_0(d-b) + Eb$$
$$= \frac{V_0}{d}(d-b) + \frac{V_0 b}{\varepsilon_r d}$$
$$= \frac{V_0}{d}\left[(d-b) + \frac{b}{\varepsilon_r}\right]$$

Example 4.13: A coaxial cable consists of two materials of different relative permittivities ε_1 and ε_2 (Fig. 4.31). The radii of inner and outer conductors are r_1 and r_2 respectively and the radius of the interface between the dielectric is r. Considering that both dielectric materials breakdown in an electric field greater than E_{r_2}, find the value of r which will allow the cable to carry the maximum possible voltage. Find this voltage.

Solution: We know that when the inner dielectric is on the point of breakdown, the field on the surface of the inner conductor must be E_{r_2}. One can easily verify that the field $E_1(d)$ at a point in the region $r_1 < a < r_2$ is

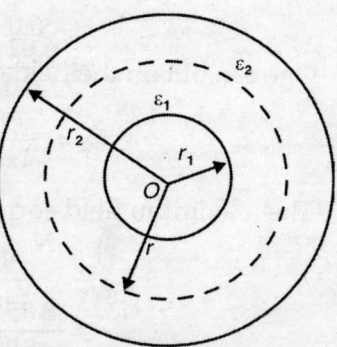

Fig. 4.31: A coaxial cable

$$E_1(a) = E_{r_2}\frac{r_1}{a}$$

According to boundary relations, we must see that the perpendicular component of the electric displacement be continuous at r_2. The field in the outer region, therefore, has to satisfy the condition

$$\varepsilon_2 E_2(r_2) = \varepsilon_1 E_1(r_2) = \varepsilon_1 E_r \frac{r_1}{r_2}$$

We know that the highest voltage occurs when both dielectrics are off the point of breakdown. Thus,

$$E_2(r_2) = E_r \frac{\varepsilon_1 r_1}{\varepsilon_2 r_2} = E_r$$

$\therefore$
$$r_2 = \frac{\varepsilon_1}{\varepsilon_2}r_1$$

The field at a point a in the outer region is

$$E_2(a) = E_r \frac{r_2}{a}$$

$\therefore$
$$V = -\int_{r_1}^a E \cdot da = -\int_r^{r_2} E_r \frac{r_2}{r}da - \int_{r_2}^{r_1} E_r \frac{r_1}{a}da$$

$$= E_r \left[r_2 \log \frac{r}{r_2} + r_1 \log \frac{r_2}{r_1} \right]$$

Example 4.14: Using the following data, calculate the individual dipole moment p of a carbon tetrachloride (CCl_4) molecule. Also calculate the average electron displacement. Relative permittivity (ε_r) = 2.24, molecular weight = 156, density = 1.6 gm/cm^3 and field strength = 10^7 V/m.

Solution:

$$\text{Molecular density } (N) = \frac{\text{Avogadro's number}}{\text{Molecular weight}} \times \text{density}$$

$$= \frac{6.02 \times 10^{23}}{156} \times 1.6 = 6.17 \times 10^{21} \text{ molecules/cm}^3$$

$$= 6.17 \times 10^{27} \text{ molecules/m}^3$$

$$p = \frac{P}{N} = \frac{\varepsilon_0 \chi E}{N} = \frac{\varepsilon_0 (\varepsilon_r - 1) E}{N} \qquad (\because \ \varepsilon_r = 1 + \chi)$$

$$= \frac{8.85 \times 10^{-12} \times 1.24 \times 10^7}{6.17 \times 10^{27}} = 1.77 \times 10^{-32} \text{ Cm}$$

There are 74 electrons in CCl_4 and if d is average electron displacement, we have

$$p = 74 \, de$$

$$\therefore \qquad d = \frac{p}{74e} = \frac{1.77 \times 10^{-32}}{74 \times 1.6 \times 10^{-19}} = 1.5 \times 10^{-15} \text{ m}$$

Example 4.15: Derive an expression for the induced dipole moment of an atom by assuming a simple classical model of an atom. Show that polarizability is independent of the field.

Solution: According to classical model of an atom, a point positive charge (Ze) called nucleus, is surrounded by a spherically symmetric cloud of negative charge ($-Ze$) in which the density is uniform up to the radius (r_0) and zero at larger radii.

When an atom is placed in an electric field E, the nucleus is displaced in the direction of E by a certain distance, say d. Obviously, the force acting on the nucleus in the direction of E is $Ze\,E$. We know that an electrostatic force between the nucleus and the charge cloud will tend to restore the initial configuration. The negative charge attracting the nucleus is, by Gauss's law, a part of the cloud of negative charge within the sphere of radius d and this charge is

$$\frac{4}{3} \pi d^3 \rho = \frac{4}{3} \pi d^3 \frac{Ze}{\frac{4}{3} \pi r_0^3} = \frac{Zed^3}{r_0^3}$$

where, ρ is the density of charge.

$$\therefore \qquad \frac{(Ze)\left(\dfrac{Zed^3}{r_0^3}\right)}{4\pi\varepsilon_0 d^2} = ZeE$$

or $\qquad\qquad\qquad Zed = 4\pi\varepsilon_0 r_0^3 E$

$\therefore \qquad p \text{ (atomic dipole moment)} = Zed = 4\pi\varepsilon_0 r_0^3 E$

or $\qquad\qquad\qquad p = \alpha\varepsilon_0 E \quad$ (where $\alpha = 4\pi r_0^3$)

Obviously, the polarizability is independent of the field. The dielectrics which satisfy this condition are said to be *linear dielectrics*.

Example 4.16: A capacitor of capacity $1\,\mu F$ is charged to $200\,V$. If the charge is leaked through a resistance of $10^6\,\Omega$, calculate the time in which the potential difference will become $100\,V$.

Solution: The amount of charge on the capacitor after time t during the process of discharging is

$Q = Q_0 e^{-\frac{t}{RC}}$	$V = 100\,V$
$\therefore \quad \dfrac{Q}{Q_0} = \dfrac{CV}{CV_0} = e^{-\frac{t}{RC}}$	$V_0 = 200\,V$
or $\quad \dfrac{V}{V_0} = e^{-\frac{t}{RC}}$	$C = 10^{-6}\,F$
or $\quad \dfrac{100}{200} = e^{-t}$	$R = 10^6\,\Omega$
$\therefore \quad t = \log_e 2 = 0.693\,s$	$\therefore \quad RC = 10^{+6} \times 10^{-6} = 1\,s$

Example 4.17: A capacitor of capacity $0.5\,\mu F$ is discharged through a resistance of $10\,M\Omega$. Find the time taken for half the charge on the capacitor to escape. Given $\log_e 2 = 0.6931$.

Solution: We have

$Q = Q_0 e^{-\frac{t}{RC}}$	$\dfrac{Q}{Q_0} = \dfrac{1}{2}$
or $\quad \dfrac{Q}{Q_0} = e^{-\frac{t}{RC}}$	$C = 0.5\,\mu F = 0.5 \times 10^{-6}\,F$
or $\quad \dfrac{1}{2} = e^{-\frac{t}{RC}}$	$R = 10^7\,\Omega$
$\therefore \quad e^{\frac{t}{RC}} = 2$	
or $\quad \dfrac{t}{RC} = \log_e 2 = 0.6931$	

$$\therefore \qquad t = 0.6931 \times 10^7 \times 0.5 \times 10^{-6} = 3.4655\,s$$

Example 4.18: After what value of time constant, will the energy stored in the capacitor in a RC circuit reach one-half its equilibrium value?

Solution: The energy stored in the capacitor at any time t when the charge on it is Q is given by $\cup = \dfrac{1}{2}\dfrac{Q^2}{C}$. We have,

$$Q = Q_0(1 - e^{-\frac{t}{RC}})$$

Here $\dfrac{Q_0^2}{2C}$ is the equilibrium energy $\cup_\infty$

$$\therefore \qquad \cup = \cup_\infty \left[1 - \exp\left(-\frac{t}{RC}\right)\right]^2$$

From the given data $\cup = \dfrac{1}{2}\cup_\infty$, one obtains

$$\frac{1}{\sqrt{2}} = 1 - \exp\left(-\frac{t}{RC}\right)$$

or

$$e = 0.3$$

or

$$-\frac{t}{RC} = -1.2$$

or

$$t = 1.2\,RC = 1.2 \text{ times the time constant.}$$

Example 4.19: A capacitor is charged to a certain potential through a resistance of 2 MΩ. If it reaches 3/4th of its final potential in 0.5 s, calculate its capacity. Given $\log_e 2 = 0.6931$.

Solution: We have

$$Q = Q_0(1 - e^{-\frac{t}{RC}})$$

or

$$CV = CV_0(1 - e^{-\frac{t}{RC}})$$

or

$$\frac{V}{V_0} = 1 - e^{-\frac{t}{RC}}$$

$$\therefore \qquad \frac{3}{4} = 1 - \exp\left(-\frac{0.5}{2\times10^6}\right)$$

or

$$\frac{1}{4\times10^6} = 2\log_e 2 = 2\times0.6931$$

$$\therefore \qquad C = 0.18 \text{ μF}$$

Example 4.20: After how many time constants, the charge on a capacitor will be 99% of the value at the equilibrium state?

Solution: We have

$$Q = Q_0(1 - e^{-\frac{t}{RC}}) \qquad \qquad \frac{Q}{Q_0} = \frac{99}{100}$$

$$\therefore \qquad \frac{99}{100} = 1 - e^{-\frac{t}{RC}}$$

or

$$t = RC\log_e 100$$
$$= 4.6052\,RC$$

Thus, after 4.6052 time constants, the charge will be 99% of the equilibrium state value.

Example 4.21: A parallel plate capacitor with area $0.30\,m^2$ and separation $5.5\,mm$ contains three dielectrics with interfaces normal to E and D as follows: $\varepsilon_{r_1} = 3.0$, $d_1 = 1.0\,mm$; $\varepsilon_{r_2} = 4.0$, $d_2 = 2.0\,mm$; $\varepsilon_{r_3} = 6.0$, $d_3 = 2.5\,mm$. Find the capacitance.

Solution: Each dielectric is treated as making up one capacitor in a set of three capacitors in series.

$$C_1 = \frac{\varepsilon_0 \varepsilon_{r_1} A}{d_1} = \frac{\varepsilon_0 (3.0)(0.30)}{10^{-3}} = 7.96\,nF$$

Similarly, $C_2 = 5.31\,nF$ and $C_3 = 6.37\,nF$

$$\therefore \quad \frac{1}{C_{eq}} = \frac{1}{7.96 \times 10^{-9}} + \frac{1}{5.31 \times 10^{-9}} + \frac{1}{6.37 \times 10^{-9}} \text{ or } C_{eq} = 2.12\,nF$$

Example 4.22: In the free space region $x < 0$, the electric field intensity is $E_1 = 3\hat{a}_x + 5\hat{a}_y - 3\hat{a}_z \,V\,m^{-1}$. The region $x > 0$ is a dielectric for which $\varepsilon_{r_2} = 3.6$. Show that the angle θ_2 which the field in the dielectric makes with the plane $x = 0$ is $8.13°$.

Solution: We can find the angle made by E from

$$E \cdot \hat{a}_x = |E_1| \cos(90° - \theta_1)$$

or

$$3 = \sqrt{43} \sin \theta_1$$

$\therefore$

$$\theta_1 = 27.2°$$

Now,

$$\tan \theta_2 = \frac{1}{\varepsilon_{r_2}} \tan \theta_1 = 0.1428$$

or

$$\theta_2 = 8.13°.$$

Example 4.23: Find the capacitance of an isolated spherical shell of radius a.

Solution: The potential of such a conductor with a zero reference of infinity is

$$V = \frac{Q}{2\pi\varepsilon_0 a}$$

or

$$C = \frac{Q}{V} = 2\pi\varepsilon_0 a.$$

Example 4.24: Determine the electric flux intensity for a medium separated by two conductors in such a way that half of this is filled with oil of relative permittivity $\varepsilon_r = 10$ and the other half is filled with air. The distance of separation between the two plates is $10\,cm$ and is subjected to a potential difference of $100\,V$.

Solution: The boundary condition for electric flux density for charge free interface separating two linear isotropic homogeneous (LIH) media is given by

$$D_2 = D_1$$

or $$\varepsilon_2 E_2 = \varepsilon_1 E_1$$

or $$\varepsilon_0 \varepsilon_{r_2} E_2 = \varepsilon_0 \varepsilon_{r_1} E_1$$

or $$\varepsilon_{r_2} E_2 = \varepsilon_{r_1} E_1$$

or $$1 E_2 = 10 E_1$$

∴ $$E_2 = 10 E_1$$

When capacitors are in series, voltage gets divided and current remains same. Thus,

$$V = V_1 + V_2 = E_1 d_1 + E_2 d_2$$

or $$100 = E_1 (0.05) + E_2 (0.05)$$

or $$100 = E_1 (0.05) + 10 E_1 (0.05)$$

∴ $$E_1 = \frac{100}{0.05 + 0.50} = 181.81 \, \text{Vm}^{-1}$$

Now, $$E_2 = 10 E_1 = 10 \times 181.81 = 1818.1 \, \text{Vm}^{-1}$$

$$D_1 = \varepsilon_0 \varepsilon_{r_1} E_1 = 8.854 \times 10^{-12} \times 181.81 = 0.0016 \, \mu\text{Cm}^{-2}$$

and $$D_2 = \varepsilon_0 \varepsilon_{r_2} E_2 = 8.854 \times 10^{-12} \times 10 \times 1818.1 = 0.16 \, \mu\text{Cm}^{-2}$$

GLIMPSES

1. In technological applications, we often store energy using a pair of electrical conductors that carry equal but opposite charges. Such a system constitutes a *capacitor*.

   Capacitors are excellent devices for short term storage and supply of electrical energy because they can deliver their stored energy much more quickly than a battery that might contain a lot more total energy.

2. The *capacitance* of a capacitor is the charge which it can store per unit volume, i.e. $C = Q/V$, where Q is the charge stored, and V is the potential across the capacitor.

3. The unit of C is coulomb/volt, known as farad.

4. Capacitance depends on the physical arrangement of the conductors that make up a capacitor, and it is a constant for given capacitor.

5. A parallel plate capacitor consists of two metal plates separated by a gap, normally filled with dielectric material, e.g. paper, mica, and some resins. The capacitance of parallel plate capacitor is given by

$$C = \frac{K \varepsilon_0 A}{d}$$

where, A is area of each plate, d is the distance between them and K is dielectric constant of the material between them.

6. $K = \dfrac{C_m}{C_0}$, where C_0 is the capacitance of the capacitor when the medium is air (or vacuum) and C_m its capacitance when the medium is a dielectric other than air.

7. A *spherical capacitor* consists of a charged solid of radius a surrounded by concentric hollow sphere of radius b. Its capacitance is given by $C = 4\pi\varepsilon_0 \left(\dfrac{ab}{b-a} \right)$.

8. A *cylindrical capacitor* consists of two coaxial cylinders and its capacitance is given by $C = \dfrac{2\pi\varepsilon_0 Kl}{\log_e(b/a)}$, where l is the length of each cylinder and a and b are the radii of the inner and outer cylinders respectively.

9. If the space between the plates of a parallel plate capacitor is filled with two media of thickness d_1 and d_2 having dielectric constants K_1 and K_2, then the capacitance of a capacitor is given by

$$C = \frac{\varepsilon_0 A}{\dfrac{d_1}{K_1} + \dfrac{d_2}{K_2}}$$

10. In the case of a charged capacitor, the *energy stored* is given by $U = \dfrac{Q^2}{2C} = \dfrac{1}{2} QV$, $= \dfrac{1}{2} CV^2$ where Q is the charge on each plate of the capacitor, V is the potential difference between plates and C is the capacitance of the capacitor. This potential energy resides in the electric field in the medium between the plates.

11. If two charged bodies carrying charges Q_1 and Q_2 and having capacitances C_1 and C_2 are connected with each other, then their potential after the sharing of charges is given by

$$V = \frac{Q_1 + Q_2}{C_1 + C_2} = \frac{C_1 V_1 + C_2 V_2}{C_1 + C_2}$$

where, V_1 and V_2 are the initial potentials of the charged bodies. The loss of energy is given by

$$dE = \frac{1}{2} \frac{C_1 C_2}{(C_1 + C_2)} (V_1 - V_2)^2$$

12. The plates of a capacitor carry equal and opposite charges and hence they exert an *attractive force* on each other which is given by $F = \dfrac{Q^2}{2K\varepsilon_0 A}$. The force per unit area of the plates is

$$f = \frac{F}{A} = \frac{Q^2}{2K\varepsilon_0 A^2} = \frac{\sigma^2}{2K\varepsilon_0}.$$

13. *Dielectric materials* become polarized in an electric field, with the result that the electric flux density D is greater than it would be under free-space conditions with the same field intensity.

14. In the *macroscopic* view, polarization P can account for the increase in the electric flux density, the equation being $D = \varepsilon_0 E + P$.

This equation permits D and P to have different directions as they do in certain crystalline dielectrics. In an isotropic linear material, E and P are parallel at each point, which is expressed by

$$P = \chi_e \varepsilon_0 E \text{ (isotropic material)}$$

where, χ_e (electrical susceptibility) is a dimensionless constant. Then

$$D = \varepsilon_0(1 + \chi_e)E = \varepsilon_0 \varepsilon_r E \text{ (isotropic material)}$$

where, $\varepsilon_r = 1 + \chi_e$ is also a pure number. Since $D = \varepsilon E$, we have $\varepsilon_r = \varepsilon/\varepsilon_0$, where ε_r is called relative permittivity.

15. Boundary conditions at the interface of two dielectrics are:

   i. The tangential component of E is continuous across a dielectric interface

$$E_{r_1} = E_{r_2} \quad \text{and} \quad \frac{D_{r_1}}{\varepsilon_{r_1}} = \frac{D_{r_2}}{\varepsilon_{r_2}}$$

   ii. The normal component of D has a *discontinuity of magnitude* $|\rho_s|$ across a dielectric interface.

$$D_{s_1} - D_{s_2} = -\rho_s \quad \text{and} \quad \varepsilon_{r_1} E_{n_1} - \varepsilon_{r_2} E_{n_2} = -\frac{\rho_s}{\varepsilon_0}$$

Generally, the interface will have no free charge, i.e. $\rho_s = 0$. Thus,

$$D_{n_1} = D_{n_2} \quad \text{and} \quad \varepsilon_{r_1} E_{n_1} = \varepsilon_{r_2} E_{n_2}.$$

REVIEW QUESTIONS

1. Define dielectric polarization. What relation does it has with electric field?
2. Find relations between P, E and D. Explain the physical significance of these vectors. What is meant by electric susceptibility?
3. What do you mean by the capacitance of a capacitor? On what factors does it depend? Show that the energy density of electric field in the region between the plates of a parallel plate capacitor is given by $U = \frac{1}{2}\varepsilon E^2$.
4. Derive Clausius–Mossotti relation. Discuss its limitations.
5. What do you mean by polar molecules? How does polarization of such molecules depend upon temperature?
6. What are polar and non-polar dielectrics? What happens when a gaseous dielectric is subjected to an electrostatic field?
7. Derive an expression for the interaction energy of two dipoles.
8. Derive the general form of Gauss's law in the presence of a dielectric.
9. Explain clearly what do you understand by polarization in changing fields.
10. Explain what do you understand by transient behaviour of a RC circuit. Discuss clearly charging and discharging of a capacitor through resistance.
11. What is time constant of a RC circuit? Show graphically the variation of charge and discharge with time.
12. State Gauss's theorem and prove that the divergence of electrical displacement in a dielectric is equal to the free charge density.

13. A dielectric sphere is placed in a uniform electric field of intensity E_0. Prove that the polarizability (P) produced is given by

$$P = 3E_0 \left(\frac{\varepsilon_r - 1}{\varepsilon_r + 2} \right) \varepsilon_0$$

where, ε_r is relative permittivity and ε_0 is the permittivity of free space.

14. Define electrical susceptibility(χ) and atomic polarizability(α). Prove that

$$\chi = \frac{n\alpha}{1 - \dfrac{n\alpha}{3}}, \text{ where } n \text{ is the number of atoms per unit volume of the medium.}$$

15. Define atomic polarizability. Show that if an empty parallel plate capacitor of capacity C_0 is filled with a dielectric, its capacitance changes to $C = \varepsilon_r C_0$, where ε_r is the dielectric constant of the material filled.

16. If E is the intensity of electric field applied on a homogeneous and isotropic dielectric, prove that the intensity of electric field acting on an atom of the dielectric is $E_{eff} = E + \dfrac{P}{3\varepsilon_0}$, where P is polarizability of the dielectric.

17. Consider a dielectric sphere in a uniform electric field E_0. Prove that the electric field inside the sphere is given by $E = \dfrac{3}{2 + \varepsilon_r} E_0$, where ε_r is dielectric constant of the sphere.

18. Derive the relations for current during charging and discharging of a capacitor in a RC circuit.

19. Assume that a capacitor is fully charged in a RC circuit, and the source of emf is disconnected and capacitor is allowed to discharge through the resistance. Derive a relationship for the charge Q on the capacitor and time t and also plot a graph for Q versus t.

20. Write short notes:
   i. Dielectrics
   ii. Gauss's law in dielectrics
   iii. Bound charge current
   iv. Polar and non-polar molecules.

PROBLEMS

1. The electrical susceptibility of hydrogen at S.T.R. is 5×10^{-4}. Calculate dipole moment of hydrogen when an electric field of 10^5 V/m is applied on it.

   [*Ans.* 1.65×10^{-35} C m]

2. Using the given data, calculate the induced dipole moment and atomic polarizability when a uniform field of 6×10^5 V/m is applied on a sample of helium. Given $\chi = 6 \times 10^{-5}$, $\varepsilon_0 = 8.85 \times 10^{-12}$ F/m and Avogadro number $N = 6 \times 10^{23}$ molecules.

   [*Ans.* 1.19×10^{-35} C m; 1.98×10^{-41} F m²]

3. Calculate the energy of interaction between the electron cloud of a hydrogen atom and the proton (which is the nucleus of the hydrogen atom). The charge density inside the cloud is

$$\rho = \frac{e}{\pi a^3} \exp\left(-\frac{2r}{a}\right), \text{ where } a \text{ is the Bohr radius.} \qquad \left[\textbf{Ans. } U = -\frac{e^2}{4\pi\varepsilon_0 a}\right]$$

4. Two large parallel plates each of area A, are separated by a distance d, their potentials being O and V respectively. A third plate carrying a charge q is introduced midway between them. Show that the potential of this plate is $\left(V + \frac{qd}{\varepsilon_0 A}\right)/2$.

5. A parallel plate capacitor has a plate separation t. The capacitance with air only between the plates is C. When a slab of thickness t_1, and relative permittivity ε_r is placed on one of the plates, the capacitance is C_1. Show that

$$C_1 = C \frac{\varepsilon_r t}{[t + \varepsilon_r (t - t_1)]}.$$

6. Show that the energy expended in polarizing a dielectric is $\dfrac{(\varepsilon_r - \varepsilon_0)E^2}{2}$.

7. The electric field strength of a mass of porcelain ($\varepsilon_{r_1} = 6$) in air is 1000 V/cm. At the inner surface of the porcelain, the field makes an angle of 45° to the normal and emerges into the air. Show that the angle of emergence is 9.5° and the strength of the external field is 4300 V/cm.

8. A spherical capacitor consists of two concentric spheres of radii a and $r(a < r)$. A spherical shell of permittivity ε, bounded by spheres of radii b and c lies between them, with its centre coinciding with that of the spheres. Find the capacity (C) of the capacitor.

$$\left[\textbf{Ans. } \frac{1}{C} = \frac{1}{4\varepsilon_0}\left[\frac{1}{a} - \frac{1}{r} + \frac{\varepsilon_r - \varepsilon_0}{\varepsilon}\left(\frac{1}{b} - \frac{1}{c}\right)\right]\right]$$

9. Find the energy expended in polarizing a dielectric. $\qquad \left[\textbf{Ans. } \left(\frac{\varepsilon_r - \varepsilon_0}{2}E^2\right)\right]$

10. A conducting sphere of radius r and having charge q is placed on the flat boundary between two dielectrics of permittivities ε_1 and ε_2. Show that the potential of this system and charge distribution on the surfaces of the sphere are

$$V = \frac{q}{[2\pi(\varepsilon_1 + \varepsilon_2)r]}, \quad \sigma_1 = \frac{q\varepsilon_1}{[2\pi(\varepsilon_1 + \varepsilon_2)r]},$$

and $\qquad \sigma_2 = \dfrac{q\varepsilon_2}{[2\pi(\varepsilon_1 + \varepsilon_2)r^2]}$ respectively.

11. A three megaohm resistor and a one microfarad capacitor are connected in series with a battery of 4 volt emf. After one second, what are the rates at which (i) the charge of the capacitor is increasing, (ii) energy is being stored in the capacitor, (iii) heat is appearing in the resistor, and (iv) the energy is delivered by the battery.

[*Ans.* (i) 9.5×10^{-7} C/s (ii) 1.1×10^{-6} W
(iii) 2.7×10^{-6} W (iv) 3.8×10^{-6} W]

12. A capacitor is being charged from a DC source through a resistance of 2 MΩ. If it takes 0.5 s for the charge to reach 3/4th of its initial value, what is the capacity of the capacitor? [*Ans.* 0.18 F]

13. A dipole moment of 5×10^{-35} C m is developed when an atom of oxygen is polarized. If the centre of negative charge cloud is situated at a distance of 4×10^{-17} m from the nucleus, then calculate atomic polarizability of oxygen.

[*Ans.* 6.25×10^{-50} Fm^2]

14. A sphere of mass m and radius r floats on a liquid of density ρ and permittivity ε. One quarter of it is immersed in the liquid when it is not charged. Show that the charge on the sphere when it sinks to a depth exactly half of its diameter is given by

$$Q = \left[\frac{8\pi\varepsilon_0 r^2 (1+\varepsilon_r)^2}{\varepsilon_r - 1} \right]^{1/2}$$

SHORT ANSWER QUESTIONS

1. What form of energy is stored in the capacitor?
Ans. Electrostatic energy

2. What is the formula for energy stored in a capacitor?

Ans. $\left[E = \frac{1}{2}CV^2 = \frac{1}{2}QV = \frac{1}{2}\frac{Q^2}{C} \right]$

3. What is energy density? Write its SI unit.
Ans. Energy density is defined as the energy stored per unit volume of the medium between the plates of the capacitor. Energy density $= \frac{1}{2}\varepsilon_0 E^2$, where E is electric intensity and ε_0 is electric permittivity of free space. The SI unit of energy density is J m^{-3}.

4. Define the dielectric constant of a dielectric.
Ans. The dielectric constant of a dielectric is defined as the ratio of external field applied across the dielectric to the net field available in the presence of the dielectric, i.e. $K = \frac{E_0}{E - E'}$ where E_0 is the external field applied and E' is the induced (internal) field of the dielectric.

5. What are the factors on which the capacitance of a parallel plate capacitor depends?

Ans. (i) Area of the plates, A (ii) separation between the plates, and (iii) the electric permittivity of the medium between the plates, ε_0 Thus, $C = \dfrac{A\varepsilon}{d}$. For vacuum $\varepsilon = \varepsilon_0$ and for a medium of dielectric constant K, $\varepsilon = K\varepsilon_0$.

6. What is the effect of temperature on the dielectric constant of a dielectric?

   Ans. When the temperature rises, K decreases because the induced field decreases.

7. How will the formula for the capacitance of a parallel plate capacitor modify, when a dielectric slab is inserted between the plates of the capacitor?

   Ans. The capacitance increases, $C' = \dfrac{A\varepsilon_0}{(d-t) + \dfrac{t}{K}}$, where t is the thick ness of the dielectric slab.

8. How will the formula for the capacitance of a parallel plate capacitor modify when a conducting slab is inserted between the plates of a capacitor?

   Ans. Capacitance of the capacitor increases, $C' = \dfrac{A\varepsilon_0}{\left(1 - \dfrac{t}{d}\right)}$, where t is the thickness of the conducting slab; when $t = d$, C' is infinite.

9. What is electric polarization?

   Ans. It is defined as the induced surface charge per unit area.

10. When a conductor is placed in an electric field, what will be its dielectric constant?

    Ans. Infinite, because the intensity inside the conductor is zero.

11. What will happen when a dielectric is introduced between the plates of a capacitor at a constant potential difference?

    Ans. The capacitance of capacitor increases. Since the plates are maintained at a constant potential difference, the charge on the plates increases ($Q = CV$). The extra charge flows to the plates from the source to which the capacitor is connected.

12. What is the SI unit of atomic polarizability?

    Ans. m^3

13. What is the SI unit of electrical susceptibility?

    Ans. Electrical susceptibility has no units.

14. What is macroscopic polarization density?

    Ans. It is defined as the electric dipole moment per unit volume and it is found by taking the vector sum of the microscopic dipole moments in unit volume.

15. Give two examples of non-polar molecules.

    Ans. H_2O_2 and CO_2

16. Give two examples of polar molecules.

    Ans. NaCl and H_2O

17. What are fictitious or bound charges?

    Ans. Bound charges are those charges which are included in the composition of the atoms and molecules of the dielectric as well as the charges of ions in crystalline dielectrics having an ionic lattice. All the other charges are free.

18. What do you understand by *breakdown voltage* of a capacitor?

Ans. Each type of a capacitor has its breakdown voltage which is the potential difference between its plates (or conductors) at which an electric discharge passes through the layer of dielectric. The breakdown voltage depends upon the thickness of the dielectric, its properties and the shape of the conductors in the capacitor.

19. What are dielectrics?

    Ans. These are substances which do not conduct electric current. They have no free electric charges.

20. On what factors does the polarizability of a molecule depend?

    Ans. Its volume.

21. Why does the molecules of polar dielectrics like CH_3O, NH_3, HCl, CH_3Cl, etc. possess a permanent dipole moment?

    Ans. This is due to the asymmetry in the arrangement of electron clouds and nucleus of these molecules. The centres of the positive and negative charges in such molecules do not coincide.

22. What do you understand by orientational polarization of a dielectric with polar molecules?

    Ans. This is also called dipole polarization and consists of aligning of the axes of permanent atomic dipoles along the direction of the intensity vector of the field. It increases with the intensity of the field and decreases with the rise in temperature.

MULTIPLE CHOICE QUESTIONS

1. A point charge in space is attracted towards a dielectric material because of the
   (a) phenomenon of dielectric polarization
   (b) dielectric hysteresis
   (c) fact that dielectric is an insulator
   (d) none of the above

   [a]

2. The electric flux and field intensity inside a conducting sphere is
   (a) maximum
   (b) zero
   (c) uniform
   (d) none of the above

   [b]

3. Two parallel plate capacitors of capacitance C and $2C$ are connected in parallel and charged to a potential difference V. The battery is then disconnected and the region between the plates of capacitor C is completely filled with a material of dielectric constant K. The potential difference across the capacitors now becomes

   (a) $\dfrac{3V}{K+2}$ (b) $\dfrac{3V}{K}$ (c) $\dfrac{V}{K+2}$ (d) $\dfrac{V}{3K}$ **[a]**

   [*Hint:* The original capacitance of the system $= 2C + C = 3C$. Total charge is $3CV$. Capacitance after inserting dielectric $= KC + 2C = C'$. We know that charge remains the same and potential now will be $\dfrac{3CV}{C'} = \dfrac{3CV}{KC + 2C} = \dfrac{3V}{K+2}$.]

4. An air filled parallel plate capacitor has a capacitance of 10^{-12} F. The separation of the plates is doubled and wax is inserted between them, which increases the capacitance to 2×10^{-12} F. The dielectric constant of wax is

   (a) 8   (b) 4   (c) 3   (d) 2   **[b]**

   [*Hint:* $C = \varepsilon_0 \dfrac{A}{d} = 10^{-12}$ F, when dielectric is introduced and d is doubled, the

   capacitance becomes $C' = \varepsilon_0 \dfrac{KA}{2d} = 2 \times 10^{-12}$ F

   $\therefore \ K = \dfrac{d}{\varepsilon_0 A} \times 4 \times 10^{-12} = \dfrac{4 \times 10^{-12}}{1 \times 10^{-12}} = 4$]

5. When a dielectric of dielectric constant K is introduced between the plates of a charged parallel plate capacitor, the electric field at a point between the plates

   (a) increases   (b) decreases

   (c) becomes KE (d) becomes E/K **[b, d]**

   [*Hint:* The induced charges produce an electric field in a direction opposite to the field produced by the charged plates.]

6. If a dielectric is placed in an electric field, the field strength

   (a) decreases   (b) increases

   (c) becomes zero   (d) remain unchanged   **[a]**

7. A parallel plate capacitor of plate area A and plate separation d is charged to a potential difference V and then the battery is disconnected. A slab of dielectric constant K is then inserted between the plates so as to fill the space between the plates. If Q, E and W denote respectively, the magnitude of charge on each plate, the electric field between the plates (after the slab is inserted) and the work done on the system in the process of inserting the slab, then

   (a) $Q = \varepsilon_0 \dfrac{AV}{d}$ (b) $Q = \varepsilon_0 KA \dfrac{V}{d}$

   (c) $C = \dfrac{V}{Kd}$ (d) $W = \varepsilon_0 \dfrac{AV^2}{2d}\left(1 - \dfrac{1}{K}\right)$ **[a, c, d]**

   [*Hint:* Charge remains the same, $Q = CV = \varepsilon_0 \dfrac{AV}{d}$, potential and electric field are

   reduced by a factor K, $\therefore \ E = \dfrac{V}{Kd}$. Energy of the system before the dielectric is inserted is given as

   $U_1 = \dfrac{1}{2}\dfrac{Q^2}{C} = \dfrac{\varepsilon_0^2 A^2 V^2}{2d^2}\dfrac{d}{\varepsilon_0 A} = \dfrac{\varepsilon_0 A V^2}{2d}$. After the dielectric is inserted, the energy becomes

   $U_2 = \dfrac{1}{2}\dfrac{Q^2}{C'}$, where $C' = \varepsilon_0 \dfrac{KA}{d} = \dfrac{1}{2}\dfrac{\varepsilon_0^2 A^2 V^2}{2d^2}\dfrac{d}{\varepsilon_0 KA} = \dfrac{\varepsilon_0 A V^2}{2Kd}$.

   $\therefore \ U_1 - U_2 = \dfrac{\varepsilon_0 A V^2}{2d}\left(1 - \dfrac{1}{K}\right)$]

8. The dielectric of a charged capacitor experiences
   (a) tensile force
   (b) compressive force
   (c) lateral force
   (d) no force   **[d]**

9. The relation between the permittivity of the artificial dielectric (ε_r), number of spheres per unit volume (N) and the size of the sphere (v) is
   (a) $\varepsilon_r = 1 + 3vN$
   (b) $\varepsilon_r = vN$
   (c) $\varepsilon_r = 1 + \dfrac{N}{v}$
   (d) $\varepsilon_r = \dfrac{N}{1+v}$ **[a]**

10. If α is atomic polarizability, N is the number of atoms per unit volume and χ is the electric susceptibility, then the Clausius–Mossotti relation can be expressed as
    (a) $\alpha = \dfrac{3\varepsilon_0}{N}\dfrac{\varepsilon_r - 1}{\varepsilon_r + 2}$
    (b) $\dfrac{Nd}{3\varepsilon_0} = \dfrac{\chi}{\chi + 3}$
    (c) $\alpha = \dfrac{\chi}{\chi + 3}$
    (d) $d = \varepsilon_0 \dfrac{\chi}{\chi + 3}$ **[b]**

11. The Gauss's law in a dielectric medium is
    (a) $\nabla \cdot D = 0$
    (b) $\nabla \cdot D = \rho_{\text{free}}$
    (c) $\nabla \cdot D = \chi$
    (d) $\nabla \cdot D = \varepsilon_r$ **[a]**

12. The bound charge current density is given by
    (a) $J_p = \dfrac{dP}{dt}$
    (b) $J_p = \dfrac{dE}{dt}$
    (c) $J_p = \dfrac{\varepsilon_r}{E}$
    (d) $J_p = \dfrac{E}{\varepsilon_0}$ **[a]**

13. A charged capacitor with capacitance 0.1 μF is discharged through a resistance 10^7 Ω. The charge will reduce to 36.8% of the original value in
    (a) 4 s   (b) 3 s   (c) 2 s   (d) 1 s   **[d]**
    [*Hint:* $\tau = RC = 10^7 \times 0.1 \times 10^{-6} = 1$ s]

14. A dielectric sphere is placed in a uniform electric field of intensity E_0. The polarizability produced is given by
    (a) $P = 3E_0 \left(\dfrac{\varepsilon_r - 1}{\varepsilon_r + 2} \right) \varepsilon_0$
    (b) $P = 3\varepsilon_0 E_0$
    (c) $P = \dfrac{3E_0}{\varepsilon_r + 2}$
    (d) none of the above   **[a]**

15. The capacitance per unit length of a cable with an inside conductor of radius 0.75 cm and cylindrical shield of radius 2.25 cm if the dielectric has $\varepsilon_r = 2.70$ is
    (a) 1.37 pF/m
    (b) 13.7 pF/m
    (c) 1.25 pF/m
    (d) 0.137 pF/m   **[c]**

16. The polarization P in a dielectric material with $\varepsilon_r = 2.8$ and
    $D = 3.0 \times 10^{-7} \hat{a}$ C m^{-2} is
    (a) $0.6 \hat{a}$ C m^{-2}
    (b) $0.9 \hat{a}$ C m^2
    (c) $1.25 \hat{a}$ C m^2
    (d) $1.93 \times 10^{-7} \hat{a}$ C m^2 **[d]**

[**Hint:** Assuming the material to be homogeneous and isotropic $P = \chi_e \varepsilon_0 E$. Since $D = \varepsilon_0 \varepsilon_r E$ and $\chi_e = \varepsilon_r - 1$

$$\therefore P = \left(\frac{\varepsilon_r - 1}{\varepsilon_r} \right) D = 1.93 \times 10^{-7} \hat{a} \, \text{Cm}^2$$

17. The capacitance of a coaxial capacitor of length L, where the inner conductor has the radius a and the outer has radius b is

(a) $C = \dfrac{2\pi\varepsilon_0\varepsilon_r L}{\ln(b/a)}$

(b) $C = \dfrac{8\pi\varepsilon_0\varepsilon_r L^2}{\ln(b/a)}$

(c) $C = \dfrac{2\pi\varepsilon_0\varepsilon_r L}{\ln\sqrt{b/a}}$

(d) $C = \dfrac{2\pi L}{\ln\sqrt{b/a}}$　　　　　**[a]**

18. An ideal capacitor is charged to a voltage V_0 and connected at $t = 0$ across an ideal inductor. The circuit now consists of a capacitor and inductor alone. If $\omega_0 = \dfrac{1}{\sqrt{LC}}$, the voltage across the capacitor at time $t > 0$ is given by　　　　　**[GATE]**

(a) $\dfrac{V_0}{8} \cos\omega_0 t$

(b) $V_0 \cos\omega_0 t$

(c) $\dfrac{V_0}{2} \sin\omega_0 t$

(d) $\dfrac{V_0}{4} e^{-\omega_0 t} \cos\omega_0 t$　　　　　**[c]**

19. Under high electric fields in a semiconductor with increasing electric field
    (a) the mobility of charge carriers decreases
    (b) the mobility of charge carriers increases
    (c) the velocity of charge carriers saturates
    (d) the velocity of charge carriers increases　　　　　**[GATE] [a, c]**

20. In a multistage RC coupled amplifier, the coupling capacitor
    (a) limits the low frequency response
    (b) limits the high frequency response
    (c) does not affect the frequency response
    (d) blocks the DC component without affecting the frequency response
    　　　　　[GATE] [a]

21. Given the potential function in free space to be
    $$V(x) = (50x^2 + 50y^2 + 50z^2) \, V,$$
    the magnitude (in Vm^{-1}) and the direction of the electric field at a point $(1, -1, 1)$, where the dimensions are in metre, are

(a) $100; \, (\hat{i} + \hat{j} + \hat{k})$

(b) $\dfrac{100}{\sqrt{3}}; \, (\hat{i} - \hat{j} + k)$

(c) $\dfrac{100}{\sqrt{3}}; \, \left(\dfrac{-\hat{i} + \hat{j} - k}{\sqrt{3}} \right)$

(d) $\dfrac{100}{\sqrt{3}}; \, \left(\dfrac{-\hat{i} - \hat{j} - k}{\sqrt{3}} \right)$　　　　　**[GATE] [c]**

Solution to Boundary Value Problems in Electrostatic Fields

5.1 INTRODUCTION

The methods discussed in the previous chapters for the calculation of electrostatic fields are useful only in special cases. A more general method, however, will involve the solution of Poisson's equation, $\nabla^2\phi = -\rho/\varepsilon_0$ or if the space charge density $\rho = 0$, the solution of Laplace's equation $\nabla^2\phi = 0$. Since the solutions of these equations are carried out with the help of boundary conditions, these problems are usually termed *boundary value problems*.

5.2 POISSON'S AND LAPLACE'S EQUATIONS

Remembering that the components of the electric field E are expressed in terms of the electric potential ϕ by $E_x = -\dfrac{\partial\phi}{\partial x}$ and similar expressions for E_y and E_z, one may write

$$\frac{\partial E_x}{\partial x} = \frac{\partial}{\partial x}\left(-\frac{\partial\phi}{\partial x}\right) = -\frac{\partial^2\phi}{\partial x^2}$$

with similar results for E_y and E_z. Making the substitution in Gauss's theorem in differential form, i.e.

$$\nabla \cdot E = \frac{\rho}{\varepsilon_0} \quad \text{(in free space)} \tag{5.1}$$

or $\quad \dfrac{\partial E_x}{\partial x} + \dfrac{\partial E_y}{\partial y} + \dfrac{\partial E_z}{\partial z} = \dfrac{\rho}{\varepsilon_0} \quad$ (in free space)

one obtains

$$\frac{\partial^2\phi}{\partial x^2} + \frac{\partial^2\phi}{\partial y^2} + \frac{\partial^2\phi}{\partial z^2} = -\frac{\rho}{\varepsilon_0}$$

or $\qquad\qquad\qquad \nabla \cdot \nabla\phi = -\dfrac{\rho}{\varepsilon_0}$

or $\qquad\qquad\qquad \nabla^2\phi = -\dfrac{\rho}{\varepsilon_0} \tag{5.2}$

The operator $\nabla \cdot \nabla = \nabla^2$ is known as the Laplacian and Eq. (5.2) is known as *Poisson's equation*. We can use Eq. (5.2) to obtain the electric potential when we know the charge distribution and conversely, so long as the charge distribution is time independent.

In free space where there are no charges, $\rho = 0$, Eq. (5.1) becomes $\nabla \cdot E = 0$ and Eq. (5.2) gives us

$$\frac{\partial^2 \phi}{\partial x^2} + \frac{\partial^2 \phi}{\partial y^2} + \frac{\partial^2 \phi}{\partial z^2} = 0$$

or

$$\nabla^2 \phi = 0 \tag{5.3}$$

This equation is called *Laplace's equation*. It is one of the most important equations in mathematical physics and appears in many problems outside the theory of the electromagnetic field, such as fluid motion and elasticity.

∇^2 (the Laplacian) is a scalar operator. If has the following forms in different coordinate systems:

i. Cartesian system:

$$\nabla^2 = \frac{\partial^2}{\partial x^2} + \frac{\partial^2}{\partial y^2} + \frac{\partial^2}{\partial z^2} \tag{5.4}$$

ii. Spherical polar coordinate system:

$$\nabla^2 = \frac{1}{r^2} \frac{\partial}{\partial r}\left(r^2 \frac{\partial}{\partial r}\right) + \frac{1}{r^2 \sin\theta} \frac{\partial}{\partial \theta}\left(\sin\theta \frac{\partial}{\partial \theta}\right) + \frac{1}{r^2} \frac{1}{\sin^2\phi} \frac{\partial^2}{\partial \phi^2} \tag{5.5}$$

iii. Cylindrical polar coordinates:

$$\nabla^2 = \frac{1}{r} \frac{\partial}{\partial r}\left(r \frac{\partial}{\partial r}\right) + \frac{1}{r^2} \frac{\partial^2}{\partial \phi^2} + \frac{\partial^2}{\partial z^2} \tag{5.6}$$

The solution of a boundary value problem is usually simplified by proper selection of a coordinate system, for instance, a problem involving a rectangular object may be most readily handled with rectangular coordinates, a cylindrical object by cylindrical coordinates, an elliptical object by elliptical–hyperbolic coordinates, etc. In other words, the problem can be solved directly by solving Poisson's or Laplace's equation, subject to the condition that potential (ϕ) thus obtained satisfies certain boundary conditions appropriate to the configuration of electrodes. Such problems are termed boundary value problems. In cases where the boundaries cannot be simply expressed in any coordinate system, efforts must be made to other methods such as graphical, point to point, numerical techniques with the help of computer or experimental methods. There are a few problems which, however, because of the high degree of symmetry involved, cannot be solved exactly. These boundary value problems may be handled by the use of series of known functions. Present chapter is devoted to the study of different boundary value problems.

5.3 EARNSHOW'S THEOREM

According to this theorem, *a charged body cannot rest in stable equilibrium under the influence of electrostatic forces alone.* This theorem says that electrostatic containment is also out of question. However, a hot plasma ($T \approx 10^7$ K) can be confined magnetically. Such type of confinement of plasma is widely used in controlled thermonuclear reactions. We know that for any extreme value of potential ϕ it must satisfy the condition $\nabla^2(\phi) = 0$. For ϕ to be maximum, $\nabla^2 \phi$ must be negative whereas for ϕ to be minimum,

$\nabla^2\phi$ must be positive. Obviously, in the region where there are no charges, $\nabla^2\phi = 0$ and hence, the potential cannot have a maximum or minimum value. This shows that for a positive charge to be stable, it must be at a point of minimum potential, i.e. $\nabla^2\phi$ must be positive. However, Poisson's equation $\nabla^2\phi = -\rho/\varepsilon_0$ tells that $\nabla^2\phi$ is negative indicating that ϕ is maximum, and hence the charge is not stable. In case of a negative charge, the condition for stability requires that $\nabla^2\phi$ is negative, while Poisson's equation gives $\nabla^2\phi = \rho/\varepsilon_0$. This means that there can be no points of stable equilibrium in an electrostatic field. This does not mean that it is not possible to balance a charge by electrostatic forces.

5.4 UNIQUENESS THEOREM

Laplace's equation does not itself determine ϕ; in addition, a suitable set of boundary conditions must be known. This means that the solution of Laplace's equation or Poisson's equation must satisfy certain conditions at the boundaries, i.e. at some preselected places in the space, the potential must have unique values at the boundaries, since these potentials only are responsible for charge, electric intensity, etc. which are unique. In other words, if a function ϕ satisfies Laplace's equation, $\nabla^2\phi = 0$, and also the boundary condition, then it is a unique (the only) solution. No other function will satisfy these requirements. Thus, if ϕ is real with its first and second derivatives continuous in a region and

  (i) the value of ϕ on the closed surface is specified (*Dirichlet condition*) or

 (ii) the normal derivative of potential ϕ, $\dfrac{\partial\phi}{\partial n}$, i.e. the field vector is specified

everywhere on the closed surface (*Neumann condition*) then the solution of Laplace's equation or Poisson's equation is unique. The proof that a proposed set of boundary conditions will suffice is usually presented in the form of a *uniqueness theorem*. This is a fundamental theorem of potential theory. There are many such theorems in electrostatics, all sharing the same basis format.

To prove the uniqueness property, imagine that the solution is not unique and there are, on the contrary, two such possible solutions ϕ_1 and ϕ_2 of Laplace's equation within the boundary region but satisfying the boundary condition $\phi_1 = \phi_2$ and

$$\frac{\partial\phi_1}{\partial n} = \frac{\partial\phi_2}{\partial n}$$

At the boundary, let $\phi = \phi_1 - \phi_2$.

Thus, $\qquad\qquad\qquad\qquad\qquad \phi = 0 \quad$ and $\quad \dfrac{\partial\phi}{\partial n} = 0$ at the boundary $\qquad\qquad$ (5.7)

Since ϕ_1 and ϕ_2 both are solutions of Laplace's equation,

$$\nabla^2\phi_1 = 0$$

and $\qquad\qquad\qquad\qquad\qquad\qquad \nabla^2\phi_2 = 0$

or $\qquad\qquad\qquad\qquad\qquad \nabla^2(\phi_1 - \phi_2) = \nabla^2\phi = 0 \qquad\qquad\qquad\qquad\qquad$ (5.8)

throughout the entire region. Obviously, ϕ is also a solution of Laplace's equation. Making use of Green's theorem, viz.

$$\int_S \psi \frac{\partial \phi}{\partial n} ds = \int_V [\psi \nabla^2 \phi + \nabla \phi \cdot \nabla \psi] dV \qquad (5.9)$$

where, ψ and ϕ are arbitrary scalar functions. $\frac{\partial \phi}{\partial n}$ is the normal derivative of ϕ on the surface and ds and dV are the surface and volume elements respectively. Now, putting $\psi = \phi$, one obtains

$$\int_S \phi \frac{\partial \phi}{\partial n} ds = \int_V [\phi \nabla^2 \phi + |\nabla \phi|^2] dV \qquad (5.10)$$

We see that LHS of Eq. (5.10) vanishes for any of the boundary conditions given by Eq. (5.7). Moreover $\nabla^2 \phi = 0$ by Eq. (5.8). Thus, we have

$$\int_V |\nabla \phi|^2 \, dV = 0 \qquad (5.11)$$

Further, we have

$$\nabla \phi = 0 \qquad (5.12)$$

Consequently, $\phi = \phi_1 - \phi_2 =$ constant (say c) throughout the volume. This constant must apply even to the boundary where we know Dirichlet condition is satisfied. This means $c = 0$ on the surface [by Eq. (5.6)]. This shows that it is zero throughout the region, i.e. ϕ_1 and ϕ_2 are identical potential distributions, i.e. $\phi_1 = \phi_2 = 0$, i.e. the **solution is unique.** If on the other hand, Neumann condition is satisfied on the surface,

$$\frac{\partial \phi}{\partial n} = 0, \text{ i.e } \phi = \phi_1 = \text{a constant.}$$

Again the constant is arbitrary and hence one can take it to be zero. Obviously, the solution is again unique and hence

$$\phi_1 = \phi_2$$

We must note that if Dirichlet condition is satisfied over a part of the surface S and Neumann condition over the remaining part, the LHS of Eq. (5.10) still vanishes. The uniqueness theorem remains valid for linear dielectrics. However, if the system involves non-linear dielectrics, then the region may be divided into sub-regions each having uniform polarization density and for each sub-region, uniqueness theorem holds separately.

5.5 GREEN'S RECIPROCITY THEOREM

With the help of this theorem, one can transform the solution of a known problem into the solution of desired unknown problem.

Let us consider a set of n point charges q_i's placed at points where the potentials due to the other charges are given by a set of numbers ϕ_j's. One finds that the potential at j^{th} point due to charges q_i's at other points are given by

$$\phi_j = \frac{1}{4\pi\varepsilon_0} \sum_{i=1}^{n} {}' \frac{q_i}{r_{ij}} \qquad (5.13)$$

The prime on the summation is used to indicate that $i \neq j$.

Let us now consider a different set of charges q_i' placed at the same points giving rise to the corresponding potential ϕ_j', then

$$\phi_j' = \frac{1}{4\pi\varepsilon_0} \sum_{i=1}^{n}{}' \frac{q_i'}{r_{ij}} \tag{5.14}$$

Multiplying Eq.(5.13) by q_j, and Eq. (5.14) by q_j' and summing over the index j, one obtains

$$\sum_{j=1}^{n} \phi_j q_j' = \sum_{j=1}^{n}\sum_{i=1}^{n}{}' \frac{q_i q_j'}{r_{ij}} \tag{5.15}$$

$$\sum_{j=1}^{n} \phi_j' q_j = \sum_{j=1}^{n}\sum_{i=1}^{n}{}' \frac{q_i' q_j}{r_{ij}} \tag{5.16}$$

We know that i and j are summation indices and hence one may interchange them in one product of q's, one obtains

$$\sum_{j=1}^{n} \phi_j q_j' = \sum_{j=1}^{n} \phi_j q_j \tag{5.17}$$

This is *Green's reciprocity theorem*. One may generalize it from a set of n point charges to a set of n conductors at potential ϕ_j's and carrying charges q_j's.

5.6 SOLUTION OF POTENTIAL PROBLEMS BY GREEN'S FUNCTION

One can solve large number of potential problems with the help of Green's function. The Green's function is simply the solution of potential problem for given geometrical arrangement of grounded conducting boundaries when the only charge present is a unit positive charge located at the point r'. According to this theorem, the Green's function for a particular geometrical arrangement is a symmetrical function of coordinates of a unit charge located at the point r' and the coordinates of the point of observation r, we have

$$G(r, r') = G(r', r) \tag{5.18}$$

This symmetry property merely represents the physical interchangeability of the positions of the point charge and observation point.

One can solve the two types of problems by the use of Green's function method: (i) when the potential distribution ϕ_s over a certain boundary is known, and (ii) when the charge distribution is known in a region within a conducting boundary.

The solution of above both types of problems can be obtained by means of usual Green's theorem

$$\int (\phi \nabla^2 \psi - \psi \nabla \phi)\,dv = \int (\phi \nabla \psi - \psi \nabla \phi)\,ds \tag{5.19}$$

where, ψ and ϕ are arbitrary functions of position which are required to be non-singular throughout the volume V.

Let ϕ be the desired solution of the problems and $\psi = G$ be the Green's function for the geometry of problem, i.e. the solution of the problem of a unit positive charge with grounded surface S. The form of the Green's function $G(r, r')$ is

$$G(r, r') = \frac{1}{4\pi\varepsilon_0} \frac{1}{|r - r'|} + F(r, r') \tag{5.20}$$

The function $F(r, r')$ represents the potential due to the induced charges on S and satisfies Laplace's equation inside the volume V, i.e.

$$\nabla^2 F(r, r') = 0 \qquad (5.21)$$

From Eq. (5.20), one gets

$$\nabla^2 G(r, r') = \frac{1}{4\pi\varepsilon_0}\nabla^2\left[\frac{1}{|r-r'|}\right] + \nabla^2 F(r, r')$$

or $\qquad \nabla^2 G(r, r') = \dfrac{1}{4\pi\varepsilon_0}\nabla^2\left[\dfrac{1}{|r-r'|}\right] \quad [\because \nabla^2 F(r, r') = 0] \qquad (5.22)$

We must remember that

$$\nabla^2\left[\frac{1}{|r-r'|}\right] = 4\pi\delta(r - r') \qquad (5.23)$$

From Eqs. (5.22) and (5.23), one obtains

$$\nabla^2 G(r, r') = -\frac{\delta(r - r')}{\varepsilon_0} \qquad (5.24)$$

Substituting $\psi = G$ in Eq. (5.19), one obtains

$$\int(\phi\nabla^2 G - G\nabla^2\phi)dV = \int(\phi\nabla G - G\nabla\phi)ds$$

or $\qquad \int[\phi(r')\nabla^2 G(r, r') - G(r, r')\nabla^2\phi(r')]dV$

$$= \int\phi(r')\nabla G(r, r') - G(r, r')\nabla\phi(r')]ds$$

Using Eq. (5.24), one obtains

$$\int\left[-\phi(r')\frac{\delta(r-r')}{\varepsilon_0}G(r, r')\nabla^2\phi(r')\right]dV$$

$$= \int[\phi(r')\nabla G(r, r') - G(r, r')\nabla\phi(r')]ds$$

or $\qquad \int\left[\phi(r')\dfrac{\delta(r-r')}{\varepsilon_0} + G(r, r')\nabla^2\phi(r')\right]dV$

$$= \int[G(r, r')\nabla\phi(r') - \phi(r') - \phi(r')\nabla G(r, r')]ds \qquad (5.25)$$

Since $\int\delta(r - r')dV = 1$ and $G(r, r') = 0$ for r' on surface S, we have

$$\frac{\phi(r)}{\varepsilon_0} + \int G(r, r')\nabla^2\phi(r')dV = \int-\phi(r')\nabla G(r, r')ds$$

or $\qquad \phi(r) = -\varepsilon_0\int G(r, r')\nabla^2\phi(r')dV - \varepsilon_0\int\phi(r')\nabla G(r, r')ds \qquad (5.26)$

Now we shall consider the two cases mentioned earlier.

i. *The surface surrounding the point r' is grounded,* making $\phi(r') = 0$ and $\nabla^2\phi(r') = -\rho(r)/\varepsilon_0$ due to the charge distribution $\rho(r')$ throughout V. Now, Eq. (5.26) reduces to

$$\phi(r) = -\varepsilon_0 \int G(r, r')\nabla^2\phi \, dV = \int G(r, r')\rho(r')dV \qquad (5.27)$$

This equation merely represents the principle of superposition applied to density of point source within the volume V.

ii. *When there are no sources of ϕ throughout the volume V* so that $\nabla^2\phi(r') = 0$, we have

$$\phi(r) = -\varepsilon_0 \int_S \phi(r')\nabla G(r, r')ds \qquad (5.28)$$

Equation (5.28) gives the potential within a given region enclosed by a boundary such that its different parts are raised to a given set of potentials.

5.7 THE METHOD OF ELECTROSTATIC IMAGES

Some typical problems in electrostatics can be solved without specifically solving a differential equation. This method is known as the *method of images*. This method, originated by Lord Kelvin in 1448, however, does not require a knowledge of the actual charge distribution, but involves the conversion of an electrostatic field into another equivalent field which is simpler to calculate. The general principle underlying the method of images is as follows:

i. For a system of n point charges $q_1, q_2,..., q_n$, the potential at any point is given by

$$\phi = \frac{1}{4\pi\varepsilon_0}\sum_{i=1}^{n}\frac{q_i}{r_i} \qquad (5.29)$$

ii. The surface of zero potential, i.e. grounded conductor is the locus of points for which the potential is given by

$$\phi = \frac{1}{4\pi\varepsilon_0}\sum_{i=1}^{n}\frac{q_i}{r_i} = 0 \qquad (5.30)$$

iii. In a problem involving the system of charges $q_1, q_2,..., q_j$ and the grounded conductors, the latter may be replaced by the system of charges $q_{j+1}, q_{j+2},..., q_j$ constituting the image if the image charge and the system of real charges combine to produce zero potential over the surface occupied by the grounded conductor. Equation (5.30) can be rewritten as

$$\phi = \frac{1}{4\pi\varepsilon_0}\sum_{i=1}^{j}\frac{q_i}{r_i} + \frac{1}{4\pi\varepsilon_0}\sum_{i=j+1}^{n}\frac{q_i}{r_i} = 0 \qquad (5.31)$$

Equation (5.31) provides the condition of potential function over the surface occupied by the grounded conductor replaced by electrical image constituted by the system of charges $q_1, q_2,..., q_j$.

iv. If the conductor under consideration is at a constant potential $\phi = 0$, the system of charges and its image must combine to produce the actual potential ϕ. One can write Eq. (5.31) as

$$\phi = \frac{1}{4\pi\varepsilon_0}\sum_{i=1}^{n}\frac{q_i}{r_i} + \frac{1}{4\pi\varepsilon_0}\sum_{i=j+1}^{n}\frac{q_i}{r_i} \qquad (5.32)$$

Equation (5.32) represents the condition of potential function over the surface occupied by the conductor replaced by electrical image constituted by the system of charges $q_{j+1}, q_{j+2},..., q_n$.

Now we shall illustrate the method of images by some typical examples.

i. A point charge and an infinite conducting plane

Let a point charge +q be placed at a perpendicular distance d from a conducting plate of infinite extent (Fig. 5.1a). Obviously, the charge q will induce charges in the plate. The plate is conducting and hence potential at all points on it must be constant. Let us assume this potential to be zero, i.e. the induced charges produce at the plate a potential equal and opposite to that of the inducing charge. We are interested in finding the field at a point in the region on the right hand side of the plate and also in the induced charge distribution on conducing plate.

Fig. 5.1: (a) Conducting plate (b) Charge −q placed behind the conducting plate

We now consider another situation. Let us consider that a charge −q is placed exactly at a distance d behind the plane of the conducting plate as shown in Fig. 5.1b. The potential at any point C on the plane is $\dfrac{q}{4\pi\varepsilon_0 r} - \dfrac{q}{4\pi\varepsilon_0 r} = 0$. This means, if the conducting plate were removed and the charge −q is placed at B, every point on the plane previously occupied by the conducting plate would still be at zero potential. Obviously, the introduction of the charge −q at B and the removal of the conducting plate does not affect the flux on the right hand side of the plane. This clearly reveals that the conditions of original problem, then remain unaltered by making this change. We must remember that the introduction of charge −q does not describe a situation compatible with the actual situation, i.e. the field on the left hand side of the plane is no longer zero. However, one should not worry about it because we are interested only in the region on the right hand side of the plane. Now, knowing that the problem remains unaltered by placing a charge −q at B, we can see by the *uniqueness theorem*, that the field produced by the charge +q at A and −q at B is identical with that produced by the charge +q at A and the induced charge on the plate. Obviously, one can solve the problem by ignoring the conducting plate and assuming a charge −q at B. The charge −q at B is called the *image* of the charge +q at A in analogy with optical images. We must note that the charge −q has no real existence and it is only a *virtual image*. The electric potential at P (Fig. 5.1b) due to charge +q at A and the induced charges on the conductor is

$$\phi_P = \frac{q}{4\pi\varepsilon_0 R} - \frac{q}{4\pi\varepsilon_0 \sqrt{R^2 + 4d^2 + 4dR\cos\theta}} \tag{5.33}$$

where, θ is the angle between BA and AP and $AP = R$. If $R = \infty$, we get $\phi = 0$. Obviously, the boundary condition $\phi = 0$ at $R = \infty$ is also satisfied if the conducting plate (with zero potential) is replaced by a charge $-q$ at B.

The electric field components E_R and E_θ at point P can be obtained by differentiating Eq. (5.33) partially with respect to R and θ respectively, i.e.

$$E_R = -\frac{\partial \phi_P}{\partial R} = \frac{q}{4\pi\varepsilon_0 R^2} - \frac{q(R + 2d\cos\theta)}{4\pi\varepsilon_0 (R^2 + 4d^2 + 4Rd\cos\theta)^{3/2}} \tag{5.34}$$

$$E_\theta = -\frac{1}{R}\frac{\partial \phi_P}{\partial \theta} = \frac{2qd\sin\theta}{4\pi\varepsilon_0 (R^2 + 4d^2 + 4Rd\cos\theta)^{3/2}} \tag{5.35}$$

Now, we calculate the field E normal to the plate at C. We have

$$E \text{ at } P = E_R \cos\theta - E_\theta \sin\theta$$

$$= \frac{q}{4\pi\varepsilon_0}\left[\frac{\cos\theta}{R^2} - \frac{R\cos\theta + 2d}{(R^2 + 4d^2 + 4Rd\cos\theta)^{3/2}}\right]$$

$$\therefore \qquad E \text{ at } C = -\frac{q}{4\pi\varepsilon_0}\frac{2d}{R^3}$$

Surface charge density

If σ is the surface density of the induced charge on the conducting plane at C, we have from Gauss's law

$$E = -\frac{\sigma}{\varepsilon_0}$$

$$\therefore \qquad \sigma = \frac{qd}{2\pi R^3} \tag{5.36}$$

Obviously, the surface density of induced charge varies inversely as the cube of the distance of the position on the plate from the charge $+q$.

The force F exerted on the charge $+q$ by the induced charge on the conducting plate is the same as that exerted by the image on $+q$, i.e.

$$F = \frac{1}{4\pi\varepsilon_0}\frac{(q)(-q)}{(2d)^2} = -\frac{q^2}{16\pi\varepsilon_0 d^2} \tag{5.37}$$

The negative sign indicates that the charge is attracted towards the plate.

We must note that with the simple geometry as above, it was possible to determine the potential function by introducing one image charge. Thus, the problem of charge distribution on the surface and the potential and field produced by the system can be easily solved by introducing an image or a set of images suitably placed in a region external to the region under consideration and ignoring the actual surface of electrification, i.e. replacing boundary conditions by images. However, we must remember that the success of the method depends upon one's ability to guess a simple equivalent image distribution, i.e. the distribution that keeps the surface still an equipotential surface with requisite potential.

ii. A point charge in the vicinity of a grounded conducting sphere

The *sphere-gap voltmeter* and the breakdown tests of liquid and solid insulation make use of spherical electrodes, wherein a knowledge of the electric field intensity may sometimes be required. The image method employed to evaluate this field, a group of image charges together with the originally specified charges have to be considered to produce equipotential surfaces at the conductors. The problem is straightforward only in cases where the geometry is simple. Such is the case, however, for a point charge Q in the vicinity of a conducting sphere.

Let us consider a conducting sphere of radius a, maintained at zero potential, i.e. grounded and a charge $+q$ at the point A at a distance d from the centre of the sphere, such that $d > a$ (Fig. 5.2a). Since the sphere is grounded, hence the boundary conditions are:

    i. The electric potential ϕ at every point on the surface of the sphere is zero, i.e. $\phi = 0$.

    ii. The electric potential at infinitely far point is zero, i.e. $\phi = 0$.

Let us consider that the centre of the sphere coincides with the origin. Now, the charge $+q$ is outside the sphere and one is interested in the region outside the sphere and hence the image must be within the sphere. By symmetry, we observe that the image charge must be placed on the line joining A to the centre of the sphere. Figure 5.2b represents this problem with the boundary conditions

$$\phi(R, \theta)_{|R| = a} = 0 \tag{5.38}$$

(a)   (b)

Fig. 5.2: (a) A conducting sphere (b) Image charge

As shown in the Fig. 5.2b, B represents the position of the image charge. We have $OB = d'$. The potential at point P is obtained as

$$\phi_P = \frac{q}{4\pi\varepsilon_0 |R - d|} + \frac{q'}{4\pi\varepsilon_0 |R - d'|}$$

$$= \frac{q}{4\pi\varepsilon_0 |R\hat{a}_R - d\hat{a}_d|} + \frac{q'}{4\pi\varepsilon_0 |R\hat{a}_R - d'\hat{a}_d|} \tag{5.39}$$

where, $\hat{a}_R$ and $\hat{a}_d$ are the unit vectors in the directions OP and OA respectively.

$$\therefore \qquad \phi_P = \frac{q}{4\pi\varepsilon_0 R\left|\hat{a}_R - \frac{d}{R}\hat{a}_d\right|} - \frac{q'}{4\pi\varepsilon_0 d'\left|\hat{a}_d - \frac{R}{d'}\hat{a}_d\right|} \tag{5.40}$$

We have on the surface of the sphere $R = a$.

$$\therefore \quad V_P\big|_{|R|=a} = \frac{q}{4\pi\varepsilon_0 a\left|\hat{a}_R - \frac{d}{a}\hat{a}_d\right|} - \frac{q'}{4\pi\varepsilon_0 d'\left|\hat{a}_d - \frac{a}{d'}\hat{a}_R\right|} = 0$$

The above holds true if

$$\frac{q}{a} = -\frac{q'}{d'} \quad \text{and} \quad \frac{d}{a} = \frac{a}{d'}, \quad \therefore \quad d' = \frac{a^2}{d}$$

and
$$q' = -\frac{qd'}{a} = -\frac{qa}{d} \tag{5.41}$$

We must remember that A and B points are inverse points with respect to the sphere. Alternatively, at $|R| = a$ one obtains

$$|R - d|^{-1}_{|R|=a} = (R^2 + d^2 - 2Rd\cos\theta)^{-1/2}_{|R|=a}$$

$$= (a^2 + a^2 - 2ad\cos\theta)^{-1/2}$$

$$= d^{-1}(1 - 2h\cos\theta + h^2)^{-1/2} \tag{5.42}$$

where
$$h = \frac{d}{a}$$

Similarly, one obtains

$$|R - d'|^{-1}_{|R|=a} = a^{-1}(1 - 2h'\cos\theta + h'^2)^{-1/2} \tag{5.43}$$

where
$$h' = \frac{d'}{a}.$$

We have

$$(1 - 2hx + h^2)^{-1/2} = \sum_{n=0}^{\infty} h^n P_n(x)$$

where $|h| < 1$ and $|x| \leq 1$. Thus, we have

$$|R - d|^{-1}_{|R|=a} = \sum_{n=0}^{\infty} a^n d^{-(n+1)} P_n(\cos\theta) \tag{5.44}$$

and
$$|R - d|^{-1}_{|R|=a} = \sum_{n=0}^{\infty} d'^n a^{-(n+1)} P_n(\cos\theta) \tag{5.45}$$

$$\therefore \quad \phi_{(a,\,\theta)} = \frac{1}{4\pi\varepsilon_0} \sum_{n=0}^{\infty} qa^n d^{-(n+1)} + q'd'^n a^{-(n+1)} P_n(\cos\theta) = 0 \tag{5.46}$$

We know that Legendre polynomials are linearly independent and hence the coefficient of each Legendre polynomial in the expansion vanishes. Now, for $n = 0$, we obtain

$$qd^{-1} + q'a^{-1} = 0 \quad \text{or} \quad q' = -\left(\frac{a}{d}\right)q \tag{5.47}$$

and for $n = 1$, one obtains

$$qad^{-2} + q'd'a^{-2} = 0$$

or
$$qad^{-2} - \left(\frac{a}{d}\right) qd'a^{-2} = 0 \quad \text{[using Eq. (5.47)]}$$

$$\therefore \qquad d' = \frac{a^2}{d} \tag{5.48}$$

We can easily see that for $n = 1$, the conditions given by Eqs. (5.47) and (5.48) reduce all succeeding coefficients to zero.

Now, substituting the values of q' and d' from Eqs. (5.47) and (5.48) respectively in Eq. (5.40), one obtains the desired expression for the potential, i.e.

$$\phi_P = \frac{q}{4\pi\varepsilon_0 R \left| \hat{a}_R - \frac{d}{R}\hat{a}_d \right|} + \frac{\left(-\dfrac{a}{d}\right)q}{4\pi\varepsilon_0 \dfrac{a^2}{d} \left| \hat{a}_d - \dfrac{dR}{a^2}\hat{a}_R \right|}$$

$$= \frac{q}{4\pi\varepsilon_0 R \left| \hat{a}_R - \frac{d}{R}\hat{a}_d \right|} + \frac{q}{4\pi\varepsilon_0 a \left| \hat{a}_d - \dfrac{dR}{a^2}\hat{a}_R \right|} \tag{5.49}$$

We must note that as the charge q is brought nearer and nearer to the surface, q' becomes larger and larger and moves away from the centre.

One can obtain the charge density from the relation

$$E_x = -\frac{\partial \phi}{\partial x} = -\frac{\sigma}{\varepsilon_0}$$

Integrating this density over the whole surface of the sphere, one obtains the total induced charge

$$q' = \int_0^\pi \sigma(2\pi a^2)\sin\theta d\theta$$

$$= -\frac{a}{d}q \tag{5.50}$$

This shows that the total induced charge on the real conducting sphere is thus the same as the magnitude of the image charge which replaced the sphere. This must be true for Gauss's law.

iii. A charge q placed in the vicinity of a charged and insulated sphere

Let us consider that the conducting sphere is insulated and given a charge Q. Let the charge q be at a distance d from the centre as in the case (ii) above. One can find the potential at a point due to this system by the method of linear superposition. Let us assume that the sphere is grounded, i.e. its potential is zero and the induced charge q' equal to the magnitude of the image charge of q is distributed over it.

Now insulate the sphere by disconnecting it from the ground, and give it an additional amount of charge $Q - q'$. Obviously, the total charge on the sphere now is Q. Moreover, the forces due to the point charge q' are balanced by the charge Q and hence the added charge $Q - q'$ will be uniformly distributed over the surface of the sphere. The potential at a point outside the sphere is now given by the sum of potentials due to q, q' and added charge $Q - q'$. The contribution of the potential due to $Q - q'$ will be the same as that of a point charge of magnitude $Q - q'$ at the centre. Finally, one obtains

$$\phi_P = \frac{q}{4\pi\varepsilon_0 |R - d|} - \frac{qa}{4\pi\varepsilon_0 d |R - (a^2/d)\,\hat{a}_d|} + \frac{Q + qa/d}{4\pi\varepsilon_0 |R|} \tag{5.51}$$

iv. When the sphere is maintained at a fixed potential

Let us consider that the sphere is maintained at a potential ϕ. Obviously, the charge on the sphere = potential × capacity = ϕa, where a is the capacity of the sphere. The potential at a point P due to this sphere

$$\phi_P = \frac{q}{4\pi\varepsilon_0 |R - d|} - \frac{qa}{4\pi\varepsilon_0 d |R - (a^2/d)\hat{a}_d|} + \frac{\phi a}{4\pi\varepsilon_0 |R|} \tag{5.52}$$

v. Conducting sphere placed in a uniform field

Let us consider that a conducting sphere of radius a is placed in a uniform electric field E_0 with centre at the origin O (Fig. 5.3). Due to induced charges on the surface of the sphere, the field immediately around the sphere will become distorted as shown in Fig. 5.3.

Fig. 5.3: Conducting sphere in a uniform field

One can regard a uniform field as the field produced by the appropriate positive and negative charges at infinity. For example, let us consider a charge $-Q$ at $z = +R$ and a charge $+Q$ at $Z = -R$. Obviously, the field produced at a point near the origin O is approximately $\frac{2Q}{4\pi\varepsilon_0 R^2}$ and this will be parallel to z-axis. In the limiting case when $Q \to \infty$, keeping $\frac{Q}{R^2}$ constant, the field is uniform and given by $E_0 = \frac{2Q}{4\pi\varepsilon R^2}$.

One can easily see that the potential at a point P distant r from O due to the system consisting of charges $\pm Q$ at $z = \mp R$ and the induced charges on the sphere is the same as that produced by the charge $+Q$ at $-R$, $-Q$ at $+R$, the image charge $-\frac{Qa}{R}$ at $-\frac{a^2}{R}$ and the image charge $\frac{Qa}{R}$ at $+\frac{a^2}{R}$. We have

$$\phi_P = \frac{Q}{4\pi\varepsilon_0 (r^2 + R^2 + 2rR\cos\theta)^{1/2}} - \frac{Q}{4\pi\varepsilon_0 (r^2 + R^2 - 2rR\cos\theta)^{1/2}}$$

+ potentials due to image charges.

Since $r \ll R$ and hence one can easily see that the first two terms give the contribution equal to (expanding and neglecting higher terms) $-\dfrac{2Q}{4\pi\varepsilon_0 R^2} r\cos\theta$.

Let us assume the image charges $-\dfrac{Qa}{R}$ and $+\dfrac{Qa}{R}$ be separated by a distance $\dfrac{2a^2}{R}$ and

forming a dipole. One finds that their contribution to the potential at P is $\dfrac{p\cos\theta}{4\pi\varepsilon_0 r^2}$,

where $p = \dfrac{2Qa^3}{R^2}$ is the dipole moment.

$$\therefore \qquad \phi_P = -\frac{2Q}{4\pi\varepsilon_0 R^2} r\cos\theta + \frac{(2Qa^3/R^2)\cos\theta}{4\pi\varepsilon_0 r^2}$$

$$= -E_0 r\cos\theta + \frac{E_0 a^3 \cos\theta}{r^2}$$

$$= -E_0\left(r - \frac{a^3}{r^2}\right)\cos\theta \qquad\qquad (5.53)$$

The induced charge density is obtained as

$$\sigma = \varepsilon_0 E = -\varepsilon_0 \left.\frac{\partial\phi_P}{\partial r}\right|_{r=a} = 3\varepsilon_0 E_0 \cos\theta \qquad\qquad (5.54)$$

From Eq. (5.53) it is evident that the resultant potential at any point is due to the uniform field intensity E_0 (first term) plus the potential of a dipole moment (second term). So if we replace the sphere by a dipole of this moment (potential due to a dipole

$= \dfrac{p\cos\theta}{4\pi\varepsilon_0 r^2}$, comparing it with

the second term in Eq. (5.53), one obtains $p = 4\pi\varepsilon_0 E_0 a^3$) located at the centre, the field outside the surface previously occupied by the sphere will remain unchanged. We further note that the surface integral of the charge density in Eq. (5.54) vanishes so that there is no difference between a grounded and an insulated sphere.

5.8   IMAGES IN DIELECTRICS

The method of images can also be used to determine potentials and fields in the presence of dielectrics. We shall discuss few typical examples.

i. A point charge q placed at a perpendicular distance d from a semi infinite dielectric material bounded by a plane surface

Let us consider that the surface coincides with the plane $z = 0$ (Fig. 5.4). Let O be the origin and A be the point on the z-axis where the charge q is placed. The field of charge q polarizes the dielectric which in turn affects the field in vacuum on the right hand side. Let us assume that the field produced by the polarization charges at a point on the right hand side is the same as that produced by charge q' at A' (the image of q). Now we are interested in finding the value of q' which must satisfy our assumption.

Fig. 5.4: A point charge at a $\perp^r$ distance d from a semi-infinite dielectric material

One can write the potential at a point P in vacuum as

$$\phi_P = \frac{q}{4\pi\varepsilon_0 r_1} + \frac{q'}{4\pi\varepsilon_0 r_2}$$

Writing the above in cartesian coordinates, we have

$$\phi_P = \frac{1}{4\pi\varepsilon_0}\left[\frac{q}{[x^2+y^2+(z-d)^2]^{1/2}} + \frac{q'}{[x^2+y^2+(z+d)^2]^{1/2}}\right] \qquad (5.55)$$

Now, the question arises, what would be the potential at a point P' in the dielectric medium due point charge q? Obviously, it would not be the same as q would produce at P' in vacuum. Let us assume that this potential be same as that produced by a point charge q'' at A if the space was vacuum, i.e. there is no dielectric, obviously, one finds this potential at $P'(x',y',z')$ as

$$\phi'_P = \frac{q''}{4\pi\varepsilon_0[x'^2+y'^2+(z'-d)^2]^{1/2}} \qquad (5.56)$$

The potential in vacuum at P when this point is in the plane $z = 0$, is

$$\phi_P|_{z=0} = \frac{1}{4\pi\varepsilon_0}\left[\frac{q+q'}{(x^2+y^2+d^2)^{1/2}}\right] \qquad (5.57)$$

While that P' when this is at the same point but in the dielectric medium

$$\phi_{P'}|_{z=0} = \frac{q''}{4\pi\varepsilon_0(x'^2+y'^2+d^2)^{1/2}} \qquad (5.58)$$

On the boundary, we have $\phi_P = \phi'_P$. This satisfies the first boundary condition that the tangential component of field E must be same on either side of boundary. This yields

$$q + q' = q'' \qquad (5.59)$$

The second boundary condition requires that the normal component of **D** (displacement vector) must be continuous at the boundary. In accordance with this condition, we have

$$\frac{\partial\phi_P}{\partial z} = \varepsilon\frac{\partial\phi_{P'}}{\partial z} \quad \text{at } z = 0 \qquad (5.60)$$

Now, we have

$$\frac{\partial\phi_P}{\partial z} = \frac{1}{4\pi\varepsilon_0}\left[-\frac{q(z-d)}{[x^2+y^2+(z-d)^2]^{3/2}} - \frac{q'(z+d)}{[x^2+y^2+(z+d)^2]^{3/2}}\right]$$

$$= \frac{1}{4\pi\varepsilon_0} \frac{d(q-q')}{(x^2+y^2+d^2)^{3/2}} \quad \text{at } z = 0 \tag{5.61}$$

Similarly, one obtains

$$\frac{\partial\phi_{P'}}{\partial z} = \frac{1}{4\pi\varepsilon_0} \frac{q''d}{(x^2+y^2+d^2)^{3/2}} \quad \text{at } z' = 0 \tag{5.62}$$

Making use of Eqs. (5.61) and (5.62), Eq. (5.60) yields

$$\frac{d(q-q')}{(x^2+y^2+d^2)^{3/2}} = \frac{\varepsilon q''d}{(x^2+y^2+d^2)^{3/2}}$$

or

$$q - q' = \varepsilon q'' \tag{5.63}$$

Solving Eqs. (5.59) and (5.63), one obtains

$$q' = -\frac{q(\varepsilon-1)}{\varepsilon+1} \text{ and } q'' = \frac{2q}{\varepsilon+1} \tag{5.64}$$

These combination of charges produce fields which satisfy the boundary conditions. The force of attraction between the dielectric and the point charge is obtained as

$$F = \frac{qq'}{4\pi\varepsilon_0(2d)^2} = \frac{q^2(\varepsilon-1)}{16\pi\varepsilon_0 d^2(\varepsilon+1)} \tag{5.65}$$

ii. A dielectric sphere in a uniform field

Let us consider a dielectric sphere of radius a and dielectric constant ε_1 placed in a uniform dielectric field E_0 existing in a medium with dielectric constant ε_2 as shown in Fig. 5.5. There are no free charges inside and outside the surface and hence the potential ϕ_1 (outside) and ϕ_2 (inside) the sphere will satisfy Laplace's equation, i.e.

$$\nabla^2\phi_1 = 0 \quad \text{and } \nabla^2\phi_2 = 0 \tag{5.66}$$

Also, the potentials must satisfy the boundary conditions, i.e.

   i. $\phi(r, \theta)$ must be continuous at $r = a$ for all values of θ, i.e. $\phi_1 = \phi_2$ for $r = a$.

   ii. The normal component of D must be continuous on $r = a$ for all values of θ. We can express this as

$$\varepsilon_1\frac{\partial\phi_1}{\partial r} = \varepsilon_2\frac{\partial\phi_2}{\partial r} \quad \text{on } r = a$$

Fig. 5.5: A dielectric sphere in a uniform field

   iii. At $r = 0$, $\phi_1(r, \theta)$ must be finite

   iv. $\phi_2(r, \theta) = \phi_0 = -E_0 z = -E_0 r\cos\theta$ as $r \to \infty$

One can easily see that the potential which is the solution of Laplace's equation can be expressed as

$$\phi = \sum_{i=1}^{\infty}[A_l\phi^l + B_l r^{-(l+1)}]P_l(\cos\theta) \tag{5.67}$$

where, l is a positive integer and $P_l(\cos\theta)$ are the Legendre's polynomials or *zonal harmonics* (refer to Section 5.11).

In accordance with the condition (iii) above, the potential at the origin be finite. This means that no term of the form $Br^{-(l+1)}v$ should be included in the expression for ϕ_1, as it gives infinite contribution to the boundary at $r = 0$ position. This means

$$\phi_1 = \sum_{l=0}^{\infty} A_l r^l P_l(\cos\theta) \tag{5.68}$$

$$\phi_2 = \sum_{l=0}^{\infty} [C_l r^l + D_l r^{-(l+1)}] P_l \cos\theta \tag{5.69}$$

The constants appearing in the above equations can be determined with the help of boundary conditions. The potential ϕ_2 at infinity is

$$\phi_2 = C_0 P_0(\cos\theta) + C_1 r P_1(\cos\theta) + C_2 r^2 P_2(\cos\theta) + \cdots \tag{5.70}$$

Condition (iv) implies that the above must be equal to $-E_0 r P_1(\cos\theta)$. Obviously, all terms in Eq. (5.70) vanish except the second term and hence

$$l = 1 \quad \text{and} \quad C_1 = -E_0 \tag{5.71}$$

One obtains from boundary condition (i) $\phi_1 = \phi_2$ on $r = a$ for all values of θ

$$\therefore \qquad A_1 a = -E_0 a + \frac{D_1}{a^2} \tag{5.72}$$

Condition (ii) yields

$$\varepsilon_1 \frac{\partial\phi_1}{\partial r} = \varepsilon_2 \frac{\partial\phi_2}{\partial r} \quad \text{or } r = a \text{ for all } \theta$$

$$\therefore \qquad \varepsilon_1 A_1 = -\varepsilon_2\left(E_0 + \frac{2D_1}{a^2}\right) \tag{5.73}$$

Solving Eqs. (5.72) and (5.73), one obtains

$$D_1 = \frac{\varepsilon_1 - \varepsilon_2}{\varepsilon_1 + 2\varepsilon_2} E_0 a^2 \quad \text{and} \quad A_1 = \frac{-3\varepsilon_0}{\varepsilon_1 + 2\varepsilon_2} E_0 \tag{5.74}$$

$$\therefore \qquad \phi_1 = -\frac{3\varepsilon_2}{\varepsilon_1 + 2\varepsilon_2} E_0 r \cos\theta \quad \text{and} \quad \frac{3\varepsilon_0}{\varepsilon_1 + 2\varepsilon_2} E_{0z} \tag{5.75}$$

and

$$\phi_2 = -E_0 r \cos\theta + \frac{\varepsilon_1 - \varepsilon_2}{\varepsilon_1 + 2\varepsilon_2} \frac{E_0 a^3 \cos\theta}{r^2} \tag{5.76}$$

One obtains the field inside the sphere as

$$E_{in} = -\frac{\partial\phi_1}{\partial z} = \frac{3\varepsilon_2}{\varepsilon_1 + 2\varepsilon_2} E_0 \tag{5.77}$$

This field, i.e. E_{in} is constant, uniform and parallel to the field applied. If

$$\varepsilon_1 > \varepsilon_2, \quad \text{then } E_{in} < E_0 \tag{5.78}$$

This is due to the induced surface polarization charges which give rise to an opposing field.

One can obtain the external field from Eq. (5.76). E_{ext} is equal to E_0 plus a dipole field of moment given by

$$p = \frac{\varepsilon_1 - \varepsilon_2}{\varepsilon_1 + 2\varepsilon_2} a^2 E_0 \qquad (5.79)$$

This shows that a dielectric sphere in a uniform field acts as a simple dipole.

iii. A spherical cavity in a dielectric medium

Figure 5.6 exhibits a spherical cavity in a dielectric medium $ABCD$ with dielectric constant ε. One can solve the problem in the same manner as in the case of the dielectric sphere, i.e. case (ii). Now putting $\varepsilon_1 = 1$ and $\varepsilon_2 = \varepsilon$ in Eq. (5.77), one obtains

$$E_{in} = \frac{3\varepsilon}{1 + 2\varepsilon} E_0 \qquad (5.80)$$

If $\varepsilon > 1$ then $E_{in} > E_0$.

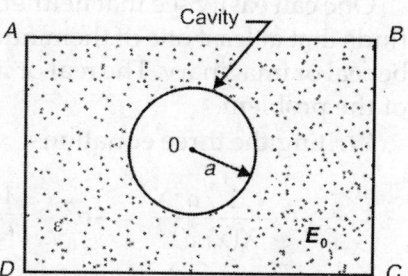

Fig. 5.6: A spherical cavity in a dielectric medium

The outside field is equal to the applied field E_0 plus the field due to the dipole at the origin of dipole moment given by

$$p = \frac{1-\varepsilon}{1+2\varepsilon} a^2 E \quad \text{[from Eq. (5.79), using } \varepsilon_1 = 1 \text{, and } \varepsilon_2 = 2\text{]}$$

The outside field, i.e. $E_0 + p$ is oriented oppositely to the applied field.

5.9 SOLUTION OF LAPLACE'S EQUATION IN CARTESIAN COORDINATES

We have the Laplace's equation

$$\nabla^2 \phi = 0 \qquad (5.81)$$

In rectangular coordinates, we can express Eq. (5.81) as

$$\nabla^2 \phi = \frac{\partial^2 \phi}{\partial x^2} + \frac{\partial^2 \phi}{\partial y^2} + \frac{\partial^2 \phi}{\partial z^2} = 0 \qquad (5.82)$$

One can solve this equation by the method of separation of variables. Let us assume that solution of Eq. (5.82), $\phi(x, y, z)$ can be represented as the product of the three functions, viz. $X(x)$, $Y(y)$ and $Z(z)$, each one of which depends on one coordinate only. Thus, we have

$$\phi(x, y, z) = X(x)Y(y)Z(z) \qquad (5.83)$$

Substituting Eq. (5.83) in Eq. (5.82) and dividing throughout by $X(x)\,Y(y)\,Z(z)$, one obtains

$$\frac{1}{X(x)}\frac{d^2X}{dx^2} + \frac{1}{Y(y)}\frac{d^2Y}{dy^2} + \frac{1}{Z(z)}\frac{d^2Z}{dz^2} = 0 \qquad (5.84)$$

We must note that we have replaced partial derivatives by total derivaties. This can be done in the present case as each term involves a function of one coordinate only.

For Eq. (5.84) to be valid for arbitrary values of independent coordinates, one finds that each term on the left must be separately equal to a constant. One gets

$$\frac{1}{X(x)}\frac{d^2X}{dx^2} = \alpha^2; \quad \frac{1}{Yy}\frac{d^2Y}{dy^2} = \beta^2 \text{ and } \frac{1}{Z(z)}\frac{d^2Z}{dz^2} = \gamma^2 \tag{5.85}$$

subject to the condition that

$$\alpha^2 + \beta^2 + \gamma^2 = 0 \tag{5.86}$$

One can easily see that neither all these constants can be real nor all imaginary. One finds that at least one of them must be real and one imaginary. The third constant may be real or imaginary. The real or imaginary nature of constants is decided by the nature of the problem.

Writing the three equations

$$\frac{1}{X(x)}\frac{d^2X}{dx^2} = -\alpha^2, \quad \frac{1}{Y(y)}\frac{d^2Y}{dy^2} = -\beta^2 \text{ and } \frac{1}{Z(z)}\frac{d^2Z}{dz^2} = \gamma^2 \tag{5.87}$$

(Note that we have taken constants α^2 and β^2 as $-\alpha^2$ and $-\beta^2$).

One obtains the solutions of three differential equations [Eq. (5.87)] as

$$\left.\begin{array}{l} X(x) = A_1 e^{i\alpha x} + A_2 e^{-i\alpha x} \\ Y(y) = B_1 e^{i\beta y} + B_2 e^{-i\beta y} \\ Z(z) = C_1 e^{rz} + C_2 e^{-rz} \end{array}\right\} \tag{5.88}$$

It is evident that the first two solutions are oscillatory, while the last solution is exponential. The constants involved in Eq. (5.88) can be found from boundary conditions to be satisfied by the potential. The potential ϕ is the product of three functions as shown by Eq. (5.83).

We must note that Eq. (5.88) is one particular solution of Eq. (5.84). One can also find other values of α, β and γ which could give valid solutions of Eq. (5.84). One can write the general solution as

$$\phi(x, y, z) = \sum_{m,n,p}^{\infty} (A_1^m e^{i\alpha m^x} + A_2^m e^{-i\alpha_m x})$$

$$\times (B_1^n e^{i\beta_n y} + B_2^n e^{-i\beta_n y})(C_1^p e^r p^x + C_2^p e^{-r} p^x) \tag{5.89}$$

5.10 LAPLACE'S EQUATION IN SPHERICAL POLAR COORDINATES

When one is dealing with a problem having axial symmetry, it is generally convenient to use spherical polar coordinates r, θ and ϕ and may choose the axis of symmetry as the polar axis $\theta = 0$.

Laplace's equation for the potential function V is

$$\nabla^2 V = 0 \tag{5.90}$$

In the spherical polar coordinates (r, θ, ϕ), Laplace's equation is written in the form

$$\frac{1}{r^2}\frac{\partial}{\partial r}\left(r^2\frac{\partial V}{\partial r}\right) + \frac{1}{r^2\sin\theta}\frac{\partial}{\partial\theta}\left(\sin\theta\frac{\partial V}{\partial\theta}\right) + \frac{1}{r^2\sin^2\theta}\frac{\partial^2 V}{\partial\phi^2} = 0 \tag{5.91}$$

The potential has been denoted by V to avoid confusion with the coordinate ϕ.

It is obvious from Eq. (5.91) that the potential function V is a function of (r, θ, ϕ); hence using method of separation of variables, we may write

$$V = RS \tag{5.92}$$

where, R is a function of r only and S is a function of θ and ϕ, i.e. both the angular coordinates. Using Eq. (5.92), Eq. (5.91) becomes

$$\frac{S}{r^2}\frac{\partial}{\partial r}\left(r^2\frac{\partial R}{\partial r}\right) + \frac{R}{r^2\sin\theta}\frac{\partial}{\partial\theta}\left(\sin\theta\frac{\partial S}{\partial\theta}\right) + \frac{R}{r^2\sin^2\theta}\frac{\partial^2 S}{\partial\phi^2} = 0 \tag{5.93}$$

Dividing Eq. (5.93) throughout by $\dfrac{RS}{r^2}$, one obtains

$$\frac{1}{R}\frac{\partial}{\partial r}\left(r^2\frac{\partial R}{\partial r}\right) + \frac{1}{S\sin\theta}\frac{\partial}{\partial\theta}\left(\sin\theta\frac{\partial S}{\partial\theta}\right) + \frac{1}{S\sin^2\theta}\frac{\partial^2 S}{\partial\phi^2} = 0 \tag{5.94}$$

One can see that the first term is a function of r only while the remaining two terms are independent of r. The above equation is satisfied if we take

$$\frac{1}{R}\frac{\partial}{\partial r}\left(r^2\frac{\partial R}{\partial r}\right) = m \tag{5.95}$$

and $$\frac{1}{S\sin\theta}\frac{\partial}{\partial\theta}\left(\sin\theta\frac{\partial S}{\partial\theta}\right) + \frac{1}{S\sin^2\theta}\frac{\partial^2 S}{\partial\phi^2} = -m \tag{5.96}$$

where m is a constant. The solutions of Eqs. (5.95) and (5.96) take simpler forms, if one takes the constant m as $l(l+1)$ where the constant l is still arbitrary. We get

$$\frac{1}{R}\frac{\partial}{\partial r}\left(r^2\frac{\partial R}{\partial r}\right) = l(l+1) \tag{5.97}$$

$$\frac{1}{\sin\theta}\frac{\partial}{\partial\theta}\left(\sin\theta\frac{\partial S}{\partial\theta}\right) + \frac{1}{\sin^2\theta}\frac{\partial^2 S}{\partial\phi^2} = -l(l+1)S \tag{5.98}$$

First, we may find the solution of radial equation [Eq. (5.97)]. This may be expressed as

$$\frac{\partial}{\partial r}\left(r^2\frac{\partial R}{\partial r}\right) - l(l+1)R = 0 \tag{5.99}$$

Let us substitute

$$R(r) = \frac{F(r)}{r} \tag{5.100}$$

Then Eq. (5.99) becomes

$$\frac{\partial^2 F(r)}{\partial r^2} - \frac{l(l+1)}{r^2}F(r) = 0 \tag{5.101}$$

From the form of Eq. (5.101) it is apparent that a single power of r (rather than of power series) will satisfy it. One finds the solution to be

$$F(r) = Ar^{l+1} + Br^{-l} \qquad (5.102)$$

where, l is yet undetermined and A and B are arbitrary constants. Using Eq. (5.102), we note that

$$R(r) = Ar^l + \frac{\beta}{r^{l+1}} \qquad (5.103)$$

Now, we may try for the solution of Eq.(5.98). Any solution of Eq. (5.98), S_l is a function of θ and ϕ and is called a surface harmonic of degree l. Adopting the same technique, let the solution of Eq. (5.98) be

$$S = P(\theta)Q(\phi) \qquad (5.104)$$

Obviously, $P(\theta)$ is a function of θ only and $Q(\phi)$ is a function of ϕ only. Substituting Eq. (5.104) in Eq. (5.98), one obtains

$$\frac{Q}{\sin\theta}\frac{d}{d\theta}\left(\sin\theta\frac{dP}{d\theta}\right) + \frac{P}{\sin^2\theta}\frac{d^2Q}{d^2\phi} + l(l+1)PQ = 0 \qquad (5.105)$$

Dividing Eq. (5.105) throughout by $\dfrac{PQ}{\sin^2\theta}$, one obtains

$$\frac{\sin\theta}{P}\frac{d}{d\theta}\left(\sin\theta\frac{dP}{d\theta}\right) + l(l+1)\sin^2\theta + \frac{1}{Q}\frac{d^2Q}{d\phi^2} = 0 \qquad (5.106)$$

One finds that the variables are again separable. The first two terms in Eq. (5.106) are functions of θ only and the last term is a function of ϕ only. Let us take

$$\frac{1}{Q}\frac{d^2Q}{d\phi^2} = -m^2 \qquad (5.107)$$

where, m^2 is a constant. The solution of Eq. (5.107) is

$$Q_m = Ce^{\pm im\phi} \qquad (5.108)$$

where, C is a constant. In order that potential V be single-valued, it is essential that

$$e^{\pm im\phi} = e^{\pm im(\phi+2\pi)} \qquad (5.109)$$

This is possible only if m is an integer. One can normalize the function Q_m by choosing the constant in such way that $\int_0^{2\pi} Q_m^* Q_m d\phi = 1$. For this to be satisfied, constant C must be equal to $\dfrac{1}{\sqrt{2\pi}}$. We must note that the function Q_m is also orthogonal, i.e.

$$\int_0^{2\pi} Q_m^* Q_A d\phi = 0 \qquad \text{if } m \neq n$$

$$\therefore \qquad \int_0^{2\pi} Q_m^* Q_n d\phi = \delta_{mn} \qquad (5.110)$$

The solution of Laplace's equation (5.91) is

$$V = \left(Arl + \frac{B}{r^{l+1}}\right)S_l. \qquad (5.111)$$

5.11 LEGENDRE'S EQUATION

Using Eq. (5.107), one obtains the θ part of the Eq. (5.106) as

$$\frac{\sin\theta}{P}\frac{d}{d\theta}\left(\sin\theta\frac{dP}{d\theta}\right)+l(l+1)\sin^2\theta = m^2$$

Dividing throughout by $\dfrac{\sin^2\theta}{P}$, one obtains

$$\frac{1}{\sin\theta}\frac{d}{d\theta}\left(\sin\theta\frac{dP}{d\theta}\right)+l(l+1)P-\frac{m^2}{\sin^2\theta}P = 0 \tag{5.112}$$

Put $x = \cos\theta$, one obtains

$$\frac{d}{dx}\left[(1-x^2)\frac{dP}{dx}\right]+\left[l(l+1)-\frac{m^2}{1-x^2}\right]P = 0 \tag{5.113}$$

$$\left(\because \frac{d}{d\theta}=\frac{dx}{d\theta}\cdot\frac{d}{dx}=-\sin\theta\frac{d}{dx}\right)$$

Equation (5.113) is known as the *generalized Legendre's equation*. Putting $m = 0$, one obtains the ordinary Legendre's equation, i.e.

$$\frac{d}{dx}\left[(1-x^2)\frac{dP}{dx}\right]+l(l+1)P = 0 \tag{5.114}$$

or

$$(1-x^2)P''-2xP'+l(l+1)P = 0 \tag{5.115}$$

Let us try for the solution of Eq. (5.115). One can try its solution by series integration. Let us assume that the solution of Eq. (5.115) is

$$P(x) = a_0x^\lambda + a_1x^{1+\lambda} + a_2x^{2+\lambda} + \cdots = \sum_{k=0}^{\infty} a_k x^{k+\lambda} \tag{5.116}$$

The constants a_k and λ are to be determined substituting Eq. (5.116) in Eq. (5.115), one obtains

$$\sum_{k=0}^{\infty} a_k(k+\lambda)(k+\lambda-1)x^{k+\lambda-2}$$

$$-\sum_{k=0}^{\infty} a_k\left[(k+\lambda)(k+\lambda-1)+2(k+\lambda)-l(l+1)\right]x^{k+\lambda} = 0 \tag{5.117}$$

Equation (5.117) must hold good for every value of x. Obviously, the coefficient of every power of x must vanish. One can find the lowest power of x by putting $k = 0$ in Eq. (5.117). One obtains this to be $\lambda-2$. This means the coefficient of $x^{\lambda-2}$ must be zero, i.e.

$$a_0\lambda(\lambda-1) = 0 \tag{5.118}$$

Equation (5.118) is known as *indicial equation*. Since $a_0 \neq 0$, because it is assumed to be the coefficient of lowest power and hence, either

$$\lambda = 0 \quad \text{or} \quad \lambda = 1$$

One can find the coefficient of $x^{j+\lambda}$ from Eq. (5.117) where j is any value k, by putting $k = j + 2$ in the first term and $k = j$ in the second. One obtains

$$a_{j+2}(j+\lambda+2)(j+\lambda+1) - a_j[(j+\lambda)(j+\lambda+1) - l(\lambda+1)]$$

and this must be zero. Thus,

$$a_{j+2} = \frac{(j+\lambda)(j+\lambda+1) - l(l+1)}{(j+\lambda+2)(j+\lambda+1)} a_j \tag{5.119}$$

Obviously, if a_j is known, one can find a_{j+2}. Starting from an arbitrary value of a_0, one can easily compute $a_2, a_4,...$, etc. By assigning any arbitrary value to a_1, one can also compute $a_3, a_5, a_7,...$, etc.

For this computation, let us take $\lambda = 0$, since this provides all the necessary independent solutions. One finds

$$a_{j+2} = \frac{j(j+1) - l(l+1)}{(j+2)(j+1)} a_j \tag{5.120}$$

From this, one obtains

$$a_2 = -\frac{l(l+1)}{\underline{|2}} a_0$$

$$a_4 = \frac{6 - l(l-1)}{r-3} a_2 = -\frac{6 - l(l+1)l(l+1)}{4 \cdot 3 \underline{|2}} a_0 = \frac{(l-2)l(l+1)(l+3)}{\underline{|4}} a_0$$

$$a_6 = -\frac{(l-4)(l-2)l(l+1)(l+3)(l+5)}{\underline{|6}}$$

$$\cdots\cdots \qquad \cdots\cdots$$

$$a_3 = \frac{2 - l(l+1)}{3 \cdot 2} a_1 = \frac{(l-1)(l+2)}{\underline{|3}} a_1$$

$$a_5 = \frac{(l-3)(l-1)(l+2)(l+4)}{\underline{|5}} a_1$$

$$\cdots\cdots \qquad \cdots\cdots$$

$$\left.\vphantom{\begin{array}{c}1\\1\\1\\1\\1\\1\\1\\1\end{array}}\right\} \tag{5.121}$$

Thus, one obtains

$$P_1 = a_0\left\{1 - \frac{l(l+1)}{\underline{|2}} x^2 + \frac{(l-2)l(l+1)(l+3)}{\underline{|4}} x^4 \right. \tag{5.122}$$

$$P_2 = a_1\left\{x - \frac{(l-1)(l+2)}{\underline{|3}} x^3 + \frac{(l-3)l(l-1)(l+2)(l+4)}{\underline{|5}} x^5 \right. \tag{5.123}$$

Equations (5.122) and (5.123) both are solutions of Legendre's equation [Eq. (5.115). One can write the general solution as

$$U = AP_1 + BP_2 \qquad (5.124)$$

Of course, we have found the solutions of the Legendre's equation, but whether the solutions are of any interest depend on their converging properties. We must note that a series converges if the ratio of two successive terms, i.e. $\dfrac{a_{j+2}}{a_j} x^2$ is smaller than unity for large j.

One can see from Eq. (5.119) that

$$\frac{a_{j+2}}{a_j} = 1 \quad \text{as } j \to \infty$$

Obviously, the series will converge if $x^2 < 1$, i.e. if the values of x lie in the range $-1 < x < 1$. We must remember that x in our equation stands for $\cos \theta$, its values range from $+1$ to -1. It is worthwhile to mention that for $x = \pm 1$, the series diverges and the solution becomes unacceptable unless the series terminates and becomes a polynomial. One can see from Eq. (5.122) and Eq. (5.123) that P_1 terminates for even values of l and P_2 for odd values of l. The polynomials in each case have x^l as their highest power of x, the next highest being x^{l-2}, and so on, down to $x^0(x)$ for l even(odd). By convention, these polynomials are normalized to have the value as unity at $x = \pm 1$ and are called the *Legendre polynomials* of order l, $P_l(x)$. One can bring normalization by taking

$$a_0 = \frac{1}{1 - \dfrac{l(l+1)}{\lfloor 2} + \cdots \text{up to the coefficients of the highest power of } x} \qquad (5.125)$$

$$a_1 = \frac{1}{1 - \dfrac{(l-1)(l+2)}{3} + \cdots \text{up to the coefficients of the highest power of } x} \qquad (5.126)$$

The first few Legendre polynomials are [Eqs. (5.122) and (5.123)]

for
$$l = 0, \quad a_0 = 1 \quad \text{and} \quad P_0(x) = 1$$
$$l = 1, \quad a_1 = 1 \quad \text{and} \quad P_1(x) = x$$
$$l = 2, \quad a_0 = -\frac{1}{2} \quad \text{and} \quad P_2(x) = \frac{1}{2}(3x^2 - 1)$$
$$l = 3, \quad a_1 = -\frac{3}{2} \quad \text{and} \quad P_3(x) = \frac{1}{2}(5x^3 - 3x)$$
$$l = 4, \quad a_0 = -\frac{1}{9} \quad \text{and} \quad P_4(x) = \frac{1}{8}(35x^4 - 30x^2 + 3)$$

and so on.

The general formula for $P_l(x)$ is

$$P_l(x) = \sum_{r=0}^{N} \frac{(-1)^r \lfloor (2l - 2r)}{2^l \lfloor r \rfloor (l-r) \rfloor (l-2r)} x^{l-2r} \qquad (5.127)$$

with $N = \dfrac{l}{2}$ if l is even and $N = \dfrac{l-1}{2}$ if l is odd. By manipulation of power series solutions given by Eqs. (5.116) and (5.120), one can obtain a compact representation of the Legendre's polynomials known as *Rodrigues formula*:

$$P_l(x) = \frac{1}{2^l \lfloor l} \frac{d^l}{dx^l}(x^2 - 1)^l \tag{5.128}$$

One can easily see that Eq. (5.128) yields the same expression for $P_l(x)$ as the one obtained from Eq. (5.127) for a given value of l.

One can express the Legendre's polynomials in a different way also. If one expands $(1 - 2xs + s^2)^{-1/2}$ by Maclaurin's theorem, one finds that the coefficients of various powers of s are Legendre's polynomials, i.e.

$$(1 - 2xs + s^2)^{-1/2} = \sum_{l=0}^{\infty} P_l(x)s^l \tag{5.129}$$

The function $(1 - 2xs + s^2)^{-1/2}$ is called the *generating function* of Legendre's polynomials.

The Legendre's polynomials form a complete orthogonal set of functions on the interval $-1 \le x \le 1$. Let us prove it. To prove the orthogonality, we write the Legendre's equation as

$$\frac{d}{dx}[(1 - x^2)P_l'(x)] + l(l+1)P_l(x) = 0 \tag{5.130}$$

Multiplying by $P_q(x)$ and integrating over the interval $-1 \le x \le 1$, one obtains

$$\int_{-1}^{1} P_q(x)\frac{d}{dx}[(1 - x^2)P_l'(x)]dx + l(l+1)\int_{-1}^{1} P_q(x)P_l(x)ds = 0$$

$$\therefore \quad \left[P_q(x)\{(1 - x^2)P_l'(x)\} \right]_{-1}^{1} - \int_{-1}^{1}(1 - x^2)P_l'(x)P_q'(x)dx$$

$$+ l(l+1)\int_{-1}^{1} P_q(x)P_l(x)dx = 0$$

The first term of the above equation vanishes at both limits. One obtains

$$-\int_{-1}^{1}(1 - x^2)P_l'(x)P_q'(x)dx + l(l+1)\int_{-1}^{1} P_q(x)P_l(x)\,dx = 0 \tag{5.131}$$

Interchanging l and q, one obtains

$$-\int_{-1}^{1}(1 - x^2)P_q'(x)P_l'(x)dx + q(q+1)\int_{-1}^{1} P_l(x)P_q(x)dx = 0 \tag{5.132}$$

Subtracting Eq. (5.132) from Eq. (5.131), one obtains

$$[l(l+1) - q(q+1)]\int_{-1}^{1} P_l(x)P_q(x)dx = 0 \tag{5.133}$$

Case (*i*): When $l \ne q$, one obtains

$$\int_{-1}^{1} P_l(x)P_q(x) = 0 \tag{5.134}$$

This shows that the Legendre's polynomials of different order are orthogonal.

Case (ii): When $l = q$, the integral is finite. The value of integral $\int_{-1}^{1} [P_l(x)]^2 dx$ can be obtained easily by using the generating function $(1 - 2xs + s^2)^{-1/2}$. We have

$$(1 - 2xs + s^2)^{-1} = \left[\sum_{l=0}^{\infty} P_l(x) s^l \right]^2 \tag{5.135}$$

$$\therefore \int_{-1}^{1} (1 - 2xs + s^2)^{-1} dx = \int_{-1}^{1} \left[\sum_{l=0}^{\infty} P_l(x) s^l \right]^2 dx$$

Integrating LHS, one obtains

$$\frac{1}{s} \log \frac{1+s}{1-s} = \int_{-1}^{1} \left[\sum_{l=0}^{\infty} s^l P_l(x) \right]^2 dx$$

The product terms in the summation on the RHS side vanish due to orthogonality condition, one obtains

$$\frac{1}{s} \log \frac{1+s}{1-s} = \sum_{l=0}^{\infty} s^{2l} \int_{-1}^{1} [P_l(x)]^2 dx \tag{5.136}$$

Now, expanding LHS and equating the coefficients of the power s^{2l}, one obtains

$$\int_{-1}^{1} [P_l(x)]^2 dx = \frac{2}{2l+1}, \quad l = 0, 1, 2, ... \tag{5.137}$$

One can express the orthogonality condition as

$$\int_{-1}^{1} P_l(x) P_q(x) dx = \frac{2}{2l+1} \delta_{ql} \tag{5.138}$$

and the orthonormal functions are

$$U_l(x) = \sqrt{\frac{2l+1}{2}} P_l(x) \tag{5.139}$$

Obviously, when $m = 0$, i.e. in the problems in which there is azimuthal symmetry, one can write the solution of Laplace's equation as

$$V = R(r)S(\theta) = \sum_{l=0}^{\infty} (A_l r^l + B_l r^{-(l+1)}) \sqrt{\frac{2l+1}{2}} P_l(\cos\theta) \tag{5.140}$$

Since the Legendre's polynomials form a complete set of orthogonal functions, any function $f(x)$ on the interval $-1 \le x \le 1$ can be expanded in terms of them, i.e.

$$f(x) = C_0 + C_1 P_1(x) + C_2 P_2(x) + \cdots = \sum_{l=0}^{\infty} C_l P_l(x) \tag{5.141}$$

In order to determine the coefficients C_l, one should multiply both sides by $P_m(x)$ and then integrate. Thus,

$$\int_{-1}^{1} f(x)P_m(x)dx = \sum_{l=0}^{\infty} C_l \int_{-1}^{1} P_l(x)P_m(x)dx = \frac{2C_m}{2m+1}$$

$$\therefore \qquad C_m = \frac{2m+1}{2}\int_{-1}^{1} f(x)P_m(x)dx \qquad (5.142)$$

5.12 ASSOCIATED LEGENDRE'S FUNCTIONS

So far we have dealt with problems possessing azimuthal symmetry. Now, we discuss the solution of Laplace's equation when the potential problem has azimuthal variation, i.e. $m \neq 0$. We have seen that $y = P_l(x)$ is a solution of the equation

$$(1-x^2)\frac{d^2y}{dx^2} - 2x\frac{dy}{dx} + l(l+1)y = 0 \qquad (5.143)$$

Differentiating Eq. (5.143) m times, one obtains

$$(1-x^2)\frac{d^{m+2}y}{dx^{m+2}} - 2x(m+1)\frac{d^{m+1}y}{dx^{m+1}} + (l-m)(l+m+1)\frac{d^m y}{dx^m} = 0 \qquad (5.144)$$

Let us put

$$v = \frac{d^m y}{dx^m} = \frac{d^m P_l(x)}{dx^m} \qquad (5.145)$$

One obtains Eq. (5.144) as

$$(1-x^2)\frac{d^2v}{dx^2} - 2x(m+1)\frac{dv}{dx} + (l-m)(l+m+1)v = 0 \qquad (5.146)$$

Equation (5.146) is satisfied by Eq. (5.145). Now we make a further substitution

$$y = v(1-x^2)^{m/2} \qquad (5.147)$$

One obtains

$$(1-x^2)\frac{d^2y}{dx^2} - 2x\frac{dy}{dx} + \left[l(l+1) - \frac{m^2}{1-x^2}\right]y = 0 \qquad (5.148)$$

One can see that Eq.(5.148) is the same as generalized Legendre's equation (5.113) and its solution is

$$Y = v(1-x^2)^{m/2} = (1-x^2)^{m/2}\frac{d^m P_l(x)}{dx^m} = P_l^m(x) \qquad (5.149)$$

The functions $P_l^m(x)$ are known as *associated Legendre's polynomials*.

The orthogonality condition for associated Legendre's polynomials can be obtained as

$$\int_{-1}^{1} P_{l'}^m(x)P_l^m(x)dx = \frac{2}{2l+1}\frac{\lfloor l+m}{\lfloor l-m}\delta_{l'l} \qquad (5.150)$$

From the normalized condition given by Eq. (5.150), it is clear that the suitably normalized functions, denoted by $Y_{lm}(\theta, \phi)$ are

$$Y_{lm}(\theta, \phi) = \sqrt{\frac{(2l+1)\lfloor l-m}{4\pi\lfloor l+m}}P_l^m(\cos\theta)e^{im\phi} \qquad (5.151)$$

One can write the general solution of Laplace's equation

$$V(r, \theta, \phi) = \sum_{l=0}^{\infty} \sum_{m=-l}^{l} [A_{lm} r^l + B_{lm} r^{-(l+1)}] Y_{lm}(\theta, \phi) \tag{5.152}$$

For a few small values of l and $m \geq 0$, the list below shows the explicit form of the $Y_{lm}(\theta, \phi)$.

$$l = 0 \quad \left\{ Y_{00} = \frac{1}{\sqrt{4\pi}} \right.$$

$$l = 1 \quad \left\{ \begin{array}{l} Y_{11} = \sqrt{\dfrac{3}{8\pi}} \sin\theta \, e^{i\phi} \\[3mm] Y_{10} = \sqrt{\dfrac{3}{4\pi}} \cos\theta \end{array} \right.$$

$$l = 2 \quad \left\{ \begin{array}{l} Y_{22} = \dfrac{1}{4} \dfrac{15}{2\pi} \sin^2\theta \, e^{2i\phi} \\[3mm] Y_{21} = -\sqrt{\dfrac{15}{8\pi}} \sin\theta \cos\theta \, e^{i\phi} \\[3mm] Y_{20} = \sqrt{\dfrac{5}{4\pi}} \left(\dfrac{3}{2} \cos^2\theta - \dfrac{1}{2} \right) \end{array} \right.$$

$$l = 3 \quad \left\{ \begin{array}{l} Y_{33} = -\dfrac{1}{4} \sqrt{\dfrac{35}{4\pi}} \sin^3\theta \, e^{3i\phi} \\[3mm] Y_{32} = \dfrac{1}{4} \sqrt{\dfrac{105}{2\pi}} \sin^2\theta \cos\theta \, e^{2i\phi} \\[3mm] Y_{31} = -\dfrac{1}{4} \sqrt{\dfrac{21}{4\pi}} \sin\theta (5\cos^2\theta - 1) e^{i\phi} \\[3mm] Y_{30} = \sqrt{\dfrac{7}{4\pi}} \left(\dfrac{5}{2} \cos^3\theta - \dfrac{3}{2} \cos\theta \right) \end{array} \right.$$

For $m = 0$, we have

$$Y_{l0}(\theta, \phi) = \sqrt{\frac{2l+1}{4\pi}} P_l(\cos\theta) \tag{5.153}$$

5.13 LAPLACE'S EQUATION IN CYLINDRICAL COORDINATES

If a solution of the form $V = R(r)\, Q(\phi)\, Z(z)$ is assumed, then the Laplace's equation

$$\frac{1}{r} \frac{\partial}{\partial r} \left(r \frac{\partial V}{\partial r} \right) + \frac{1}{r^2} \frac{\partial^2 V}{\partial \phi^2} + \frac{\partial^2 V}{\partial z^2} = 0$$

becomes

$$\frac{QZ}{r} \frac{\partial}{\partial r} \left(r \frac{\partial R}{\partial r} \right) + RZ \frac{\partial^2 Q}{\partial \phi^2} + RQ \frac{\partial^2 Z}{\partial z^2} = 0 \tag{5.154}$$

Here R is a function of r only, Q is a function of ϕ only and Z is a function of z only.

Dividing Eq. (5.154) by $\dfrac{RQZ}{r^2}$, one obtains

$$\frac{r}{R}\frac{\partial}{\partial r}\left(r\frac{\partial R}{\partial r}\right)+\frac{1}{Q}\frac{\partial^2 Q}{\partial \phi^2}+\frac{r^2}{Z}\frac{\partial^2 Z}{\partial z^2}=0$$

or

$$\frac{r}{R}\frac{\partial}{\partial r}\left(r\frac{\partial R}{\partial r}\right)+\frac{r^2}{Z}\frac{\partial^2 Z}{\partial z^2}=-\frac{1}{Q}\frac{\partial^2 Q}{\partial \phi^2} \tag{5.155}$$

We see that RHS of Eq. (5.155) is a function of ϕ only, while LHS is independent of ϕ. The said equation is satisfied if each side is equal to a constant. Let

$$\frac{1}{Q}\frac{\partial^2 Q}{\partial \phi^2}=-v^2 \tag{5.156}$$

where, v is a constant. The solution of Eq. (5.156) is

$$Q(\phi)=E^{\pm iv\phi} \tag{5.157}$$

To ensure that $Q(\phi)$ is single-valued, v should take only integral values

$$\frac{r}{R}\frac{\partial}{\partial r}\left(r\frac{\partial R}{\partial r}\right)+\frac{r^2}{Z}\frac{\partial^2 Z}{\partial z^2}=v^2 \tag{5.158}$$

Dividing throughout by v^2 and transposing, one obtains

$$\frac{1}{rR}\frac{\partial}{\partial r}\left(r\frac{\partial R}{\partial r}\right)-\frac{v^2}{r^2}=\frac{1}{Z}\frac{\partial^2 Z}{\partial z^2} \tag{5.159}$$

Now, putting

$$\frac{1}{Z}\frac{\partial^2 Z}{\partial z^2}=k^2 \tag{5.160}$$

where k is a constant.

We may note that while in Eq. (5.156) the constant is prefixed with a negative sign, in Eq. (5.160) it is prefixed with a positive sign. This is not a cause of worry. This is simply for convenience. The solution of Eq. (5.160) is

$$Z(z)=e^{\pm kz} \tag{5.161}$$

The r part of the equation now turns as

$$\frac{1}{rR}\frac{\partial}{\partial r}\left(r\frac{\partial R}{\partial r}\right)-\frac{v^2}{r^2}=-k^2 \tag{5.162}$$

or

$$\frac{r}{R}\frac{\partial}{\partial r}\left(r\frac{\partial R}{\partial r}\right)+k^2 r^2-v^2=0 \tag{5.163}$$

Putting $x = kr$, equation takes the form

$$\frac{\partial^2 R}{\partial x^2} + \frac{1}{x}\frac{\partial R}{\partial x} + \left(1 - \frac{v^2}{x^2}\right)R = 0 \tag{5.164}$$

Equation (5.164) is a form of Bessel's differential equation. Its solutions are in the form of power series called Bessel's functions. As in the case of Legendre's equation, we now try to find the solution of Eq. (5.164) in the form

$$R(x) = \sum_{\lambda=0}^{\infty} a_\lambda x^{k+\lambda} \tag{5.165}$$

Substituting this in Eq. (5.164), one obtains

$$\left[\sum_{\lambda=0}^{\infty} a_\lambda(k+\lambda)(k+\lambda-1) + \sum_{\lambda=0}^{\infty} a_\lambda(k+\lambda) - v^2\sum_{\lambda=0}^{\infty} a_\lambda\right]x^{k+\lambda-2}$$

$$+ \sum_{\lambda=0}^{\infty} a_\lambda x^{k+\lambda} = 0$$

or

$$\sum_{\lambda=0}^{\infty} a_\lambda\{(k+\lambda)^2 - v^2\}x^{k+\lambda-2} + \sum_{\lambda=0}^{\infty} a_\lambda x^{k+\lambda} = 0 \tag{5.166}$$

The coefficient of lowest power $\lambda = 0$ gives the indicial equation as

$$a_0(k^2 - v^2) = 0$$

∴
$$k = \pm v \tag{5.167}$$

Now, we consider next highest power $\lambda = 1$. Its coefficient is $a_1\{(k+1)^2 - v^2\}$ and it must be zero. Thus,

$$a_1\{(k+1)^2 - v^2\} = 0$$

Since $k = \pm v$, we have $a_1(2v \pm 1) = 0$

Since v is an integer and hence $2v \pm 1 \neq 0$. Thus,

$$a_1 = 0 \tag{5.168}$$

The coefficient of x^{k+j} is

$$a_{j+2}\{(k+j+2)^2 - v^2\} + a_j = 0$$

or

$$a_{j+2} = \frac{a_j}{(k+j+2)^2 - v^2} \tag{5.169}$$

Equation (5.169) gives the relation between the coefficients of alternate terms. Since $a_1 = 0$, all odd terms should vanish.

If $k = v$, the even numbered coefficients are

$$a_0, \frac{a_0}{2^2(v+1)}, \frac{1}{2^4(v+1)(v+2)}\frac{a_0}{\lfloor 2}, \cdots, \frac{(-1)^s}{2^{2s}(v+1)(v+2)\cdots(v+s)}\frac{a_0}{\lfloor s}, \cdots$$

$$R(r) = a_0 \left[x^v - \frac{1}{2^2(v+1)} x^{v+2} + \frac{1}{2^4(v+1)(v+2)\underline{|2}} x^{v+4} + \cdots \right.$$

$$\left. + \frac{(-1)^s}{2^{2s}(v+1)\cdots(v+s)\underline{|s}} x^{v+2s} \cdots \right]$$

$$= \sum_{s=0}^{\infty} a_0 \frac{(-1)^s x^{v+2s}}{2^{2s}\underline{|x}(v+1)\cdots(v+s)} \tag{5.170}$$

One can define a_0 as

$$a_0 = \frac{1}{2^v \Gamma(v+1)} \tag{5.171}$$

Hence,
$$R(r) = J_v(x) = \sum_{s=0}^{\infty} \frac{(-1)^s x^{v+2s}}{2^v \Gamma(v+1)2^{2s}\underline{|s}(v+1)\cdots(v+s)}$$

$$= \sum_{s=0}^{\infty} \frac{(-1)^s}{(s+1)\Gamma(v+s+1)}\left(\frac{x}{2}\right)^{v+2s} \tag{5.172}$$

The series $J_v(x)$ is known as a Bessel function of the first kind of order v. The ratio test reveals that the function converges for any value of x. If $k = -v$, we have

$$R(r) = J_{-v}(x)\sum_{s=0}^{\infty} \frac{(-1)^s}{(s+1)\Gamma(s-v+1)}\left(\frac{x}{2}\right)^{2s-v} \tag{5.173}$$

One can now write the general solution of the Bessel's equation as
$$R = A J_{v(x)} + B J_{-v(x)} \tag{5.174}$$

where A and B are constants. This is indeed true when v is not an integer. In that case $J_{v(x)}$ and $J_{-v(x)}$ as defined by Eqs. (5.173) and (5.174) are linearly independent solutions. However, when v is an integer (as it is in the present case), one can show that the two solutions are linearly dependent. If v is an integer, the first v terms in the denominator of Eq. (5.173) for which $s = 0, 1, 2,\ldots, (v-1)$ vanish as $\dfrac{1}{\Gamma(s-v+1)} = 0$. Thus, Eq. (5.173) can be written as

$$J_{-v(x)} = \frac{(-1)^s}{\Gamma(s+1)\Gamma(s-v+1)}\left(\frac{x}{2}\right)^{2s-v} \tag{5.175}$$

Putting $p = s - v$, one obtains

$$J_{-v(x)} = \sum_{p=0}^{\infty} \frac{(-1)^{p+v}}{\Gamma(p+1)\Gamma(p+v+1)}\left(\frac{x}{2}\right)^{2p+v}$$

$$= (-1)^v \sum_{p=0}^{\infty} \frac{(-1)^p}{\Gamma(p+1)\Gamma(p+v+1)}\left(\frac{x}{2}\right)^{2p+v}$$

$$= (-1)^v J_v(x) \tag{5.176}$$

Thus, when v is an integer, Eq. (5.174) is not a general solution of the second order Bessel's equation, and it is necessary to find another linearly independent solution. As per convention, this solution is taken to be $N_v(x)$, the *Neumann function* or also known as *Bessel's function of second kind*, defined by

$$N_v(x) = \frac{J_v(x)\cos v\pi - J_{-v}(x)}{\sin v\pi} \tag{5.177}$$

One can easily verify that this satisfies Bessel's equation. One can also show that $N_v(x)$ is independent of $J_v(x)$. Now, the complete solution of Bessel's equation in cylindrical coordinates is

$$V(r, \phi, z) = \sum_{m,v}^{\infty} \left[A_{mv} J_v(k_m r) + B_{mv} N_v(k_m r) \right] e^{\pm iv\phi} e^{\pm kz} \tag{5.178}$$

where we have taken account of the fact that various values of k may give acceptable solutions.

If $v = n$, an integer, $N_n(x)$ is defined as the limit of quotient in Eq. (5.177) as $v \to n$.

The function $N_v(x)$ behaves like $\ln r$ near $r = 0$ (Fig. 5.7). Therefore, it is not involved in the solution wherever the potential is known to be finite at $r = 0$.

Fig. 5.7: Behaviour of (a) Bessel function $J_n(x)$ (b) Function $N_v(x)$

Some useful properties of Bessel's functions are:

i. Bessel's functions satisfy the following orthogonality condition:
   If $x = k_m \rho$ is the mth root of $J_v(x)$, i.e. $J_v(k_m \rho) = 0$, then in the interval $0 \le r \le \rho$

$$\int_0^\rho J_v(k_m r) J_v(k_m r) r\, dr = \frac{\rho^2}{2} J_{v+1}^2(k_m \rho) \delta_{mm'} \tag{5.179}$$

ii. Since $J_v(k_m r)$ form a complete orthogonal set, any function $f(r)$ can be expanded in the interval $0 \le r \le \rho$ in terms of them, i.e.

$$f(r) = \sum_{m=1}^{\infty} D_{mv} J_v(k_m r) \tag{5.180}$$

where
$$D_{mv} = \frac{2}{\rho^2 J_{v+1}^2(k_{m\rho})} \int_0^\rho f(r) J_v(k_m r) r\, dr \tag{5.181}$$

iii. $J_0'(k_m r) = \frac{dJ_0(k_m r)}{dr} = -k_m J_1(k_m r) r\, dr \tag{5.182}$

iv. $J_0'(k_m r) = \frac{v}{r} J_v(k_m r) - k_m J_{v+1}(k_m r) \tag{5.183}$

v. $\int J_1(k_m r)\,dr = -\dfrac{1}{k_m} J_0(k_m r)$ (5.184)

vi. $\int (k_m r) J_0(k_m r)\,dr = r J_1(k_m r)$ (5.185)

vii. For integral order n and large argument $x\,(=k_m r)$, the Bessel functions behave like damped sine wave (Fig. 5.7a):

$$J_n(k_m r) \approx \sqrt{\frac{2}{\pi k_m r}}\,\cos\!\left(k_m r - \frac{\pi}{4} - \frac{n\pi}{2}\right)$$

and
$$N_v(k_m r) \approx \sqrt{\frac{2}{\pi k_m r}}\,\sin\!\left(k_m r - \frac{\pi}{4} - \frac{n\pi}{2}\right)$$

(5.186)

5.14 SPHERICAL PRODUCT SOLUTION

Of particular interest in spherical coordinates are those problems in which V may vary with r and θ but not with ϕ. For product solution $V = R(r)\,\Theta\,(\theta) = R\,\Theta$, Laplace's equation becomes

$$\left(\frac{r^2}{R}\frac{d^2 R}{dr^2} + \frac{2r}{R}\frac{dR}{dr}\right) + \left(\frac{1}{\Theta}\frac{d^2\Theta}{d\theta^2} + \frac{1}{\Theta\tan\theta}\frac{d\Theta}{d\theta}\right) = 0$$

(5.187)

We choose the separation constant as $n(n+1)$, where n is an integer, for reasons which will become apparent. One can write the two separated equations as

$$r^2\frac{d^2 R}{dr^2} + 2r\frac{dR}{dr} - n(n+1)R = 0$$

(5.188)

and
$$\frac{d^2\Theta}{d\theta^2} + \frac{1}{\tan\theta}\frac{d\Theta}{d\theta} + n(n+1)\Theta = 0$$

(5.189)

This equation in r has the solution

$$R = C_1 r^n + C_2 r^{-(n+1)}$$

(5.190)

The equation in θ possesses (unlike Bessel's equation) a polynomial solution of degree in the variable $\xi = \cos\theta$, given by

$$P_n \xi = \frac{1}{2^n\,\lfloor n}\frac{d^n}{d\xi^n}(\xi^2 - 1)^n, \quad n = 0,1,2,\dots$$

(5.191)

The polynomial $P_n(\xi)$ is the Legendre's polynomial of order n. There is a second, independent solution $Q_n(\xi)$ which is logarithmically infinite at $\xi = \pm 1$ (i.e. $\theta = 0, \pi$).

ILLUSTRATIVE EXAMPLES

Example 5.1: Two uniformly and oppositely charged wires run through a metal cylinder of radius R. With the help of method of images, show that there will be no force on the wires if they have a separation of $l = 2R\sqrt{\sqrt{5}-2}$.

Solution: One can obtain the field internal to the cylinder by having images charges at a distance $d = \dfrac{R^2}{l}$ from the axis of the cylinder as shown in Fig. 5.8. Here l is the separation between the wires. The field of the charge $+Q$ is obtained as

$$E = \frac{Q}{2\pi\varepsilon_0}\left(\frac{1}{\dfrac{R^2}{l}+l} - \frac{1}{2l} + \frac{1}{\dfrac{R^2}{l}-l}\right)$$

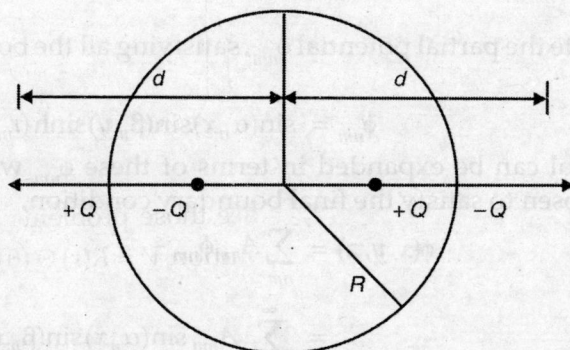

Fig. 5.8: Metal cylinder through which two oppositely charged wires run

When there is no force on the wire, we have $E = 0$. This yields

$$l^4 + 4l^2R^2 - R^4 = 0, \quad l > 0$$

or

$$l = 2R\sqrt{\sqrt{5}-2}.$$

Example 5.2: Consider a rectangular box as shown in Fig. 5.9 with dimensions (a, b, c) in the x, y and z directions respectively. If all the surfaces of the box are at zero potential except the surface $z = c$ which is at potential $\phi_0(x, y)$, then determine the potential everywhere inside the box.

Solution: We have the Laplace's equation

$$\nabla^2\phi = 0$$

The general solution of Laplace's equation is

$$\phi = X(x)\, Y(y)\, Z(z)$$

Fig. 5.9: A rectangular box

Making use of boundary condition that $\phi = 0$ for $x = 0$, $y = 0$, and $z = 0$, one finds

$$X(x) = \sin \alpha x$$
$$Y(y) = \sin \beta y$$

and

$$Z(z) = \sinh\sqrt{\{(\alpha^2 + \beta^2)z\}}$$

In order that $\phi = 0$ at $x = a$ and $y = b$, it is essential that

$$\alpha_n a = n\pi \quad \text{and} \quad \beta_m b = m\pi$$

$$\text{or} \qquad \alpha_n = \frac{n\pi}{a} \quad \text{and} \quad \beta_m = \frac{m\pi}{b}$$

$$\text{and} \qquad r_{nm} = \sqrt{\alpha^2 + \beta^2} = \pi\sqrt{\frac{n^2}{a^2} + \frac{m^2}{b^2}}$$

Now, one can write the partial potential ϕ_{nm}, satisfying all the boundary conditions, except one

$$\phi_{nm} = \sin(\alpha_n x)\sin(\beta_m y)\sinh(r_{nm}z)$$

Now, the potential can be expanded in terms of these ϕ_{nm} with initial arbitrary coefficients to be chosen to satisfy the final boundary condition,

$$\phi(x, y, z) = \sum_{nm} A_{nm}\phi_{nm}$$

$$= \sum_{n,m-1}^{\infty} A_{nm} \sin(\alpha_n x)\sin(\beta_m y)\sinh(\gamma_{nm}c)$$

We have the final boundary condition,

$$\phi = \phi_0(x, y) \quad \text{at } z = c$$

$$\therefore \qquad \phi_0(x, y) = \sum_{n,m-1}^{\infty} A_{nm} \sin(\alpha_n x)\sin(\beta_m y)\sinh(\gamma_{nm}c)$$

Obviously, this is just a double Fourier series for the function $\phi_0(x, y)$. One obtains the coefficients A_{nm} as

$$A_{nm} = \frac{4}{ab\sinh(\gamma_{nm}c)}\int_0^a dx\int_0^b dy\phi_0(x,y)(\alpha_m x)\sin(\beta_m y)$$

Example 5.3: Two infinitely long grounded metal plates at $y = 0$ and $y = b$ are connected at $x = \pm 1$ by metal strips maintained at potential ϕ_0 (Fig. 5.10). A thin silver of insulation at each corner is used to prevent them from shorting out. Show that the potential inside the resulting rectangular pipe is

$$\phi(x, y) = \frac{4\phi_0}{\pi} \sum_{k=1,3,5}^{\infty} \frac{1}{k}\frac{\cosh kx}{\cosh k}\sin ky$$

Solution: We have the Laplace's equation

$$\frac{\partial^2\phi}{\partial x^2} + \frac{\partial^2\phi}{\partial y^2} = 0$$

The boundary conditions are $\phi = 0$ when $y = 0$ and $\phi = \phi_0$ when $x = \pm 1$. Thus, we have

$$X = B_1 e^{kx} + B_2 e^{-kx}$$

Fig. 5.10: Two infinitely grounded metal plates

and $\qquad\qquad \dot{Y} = A_1 \sin ky + A_2 \sin ky$

Applying the boundary conditions, one obtains $A_2 = 0$ and $k = \dfrac{n\pi}{b}$, where $n = 1, 2, 3, ...$

Obviously, here we cannot set $B_1 = 0$, as the region in question does not extend to $x = \infty$ and hence e^{kx} is a perfectly acceptable contribution to the potential. On the other hand, the situation is obviously even with respect to x, i.e. $\phi(-x, y) = \phi(x, y)$. It gives $B_1 = B_2$.

$$\therefore \qquad\qquad B_1 e^{kx} + B_1 e^{-kx} = 2B_1 \cosh kx$$

Absorbing $2B_1$ into A_1 and A_2, one obtains

$$\phi(x, y) = \cosh kx\,(A_1 \sin ky + A_2 \cos ky)$$

Applying the boundary condition, one obtains $A_2 = 0$.

$$\therefore \qquad\qquad \phi(x, y) = C \cosh kx \sin ky$$

The general solution is

$$\phi(x, y) = \sum_{k=1}^{\infty} C_k \cosh kx \sin ky$$

Again, one finds

$$\phi(1, y) = \sum_{k=1}^{\infty} C_k \cosh k \sin ky = \phi_0$$

Making use of Fourier analysis, one obtains $C_n \cosh n = \dfrac{4\phi_0}{n\pi}$ when n is odd, and zero if n is even.

$$\therefore \qquad\qquad \phi(x, y) = \frac{4\phi_0}{\pi} \sum_{k=1,3,5}^{\infty} \frac{1}{k} \frac{\cosh kx}{\cosh k} \sin ky$$

Example 5.4: A set of wires parallel to the y-axis in the plane $z = 0$ is arranged as shown in Fig. 5.11. The wires are equally spaced at $x = \pm n\left(\dfrac{a}{2}\right)$, where $n = 0, 1, 2, 3, 4, ...$. The wires at odd positions (i.e. for n odd) are at potential ϕ_0 while those at even positions (i.e. for n even) are at potential $-\phi_0$. Show that the potential at all points in space is given by

Fig. 5.11: A set of wires parallel to y-axis in plane $z = 0$

$$\phi(x, z) = \sum_k A_k e^{-\frac{2\pi k}{a}} \cos \frac{2\pi kx}{a} \quad \text{for } z > 0$$

Solution: We can see that the potential ϕ is independent of y and varies periodically with x. Potential is an even function of x, one can represent it by a cosine series

$\cos\left(\dfrac{2\pi kx}{a}\right)$. Although the functional dependence of ϕ on z is not known; but from the above one can expect it to be exponential. Obviously, one expect the general form of ϕ as

$$\phi = \sum_{k=-\infty}^{\infty} f_k(z)\cos\frac{2\pi kx}{a} \tag{1}$$

We have to determine the form of $f(z)$. If ϕ is a valid potential, then it will have to satisfy Laplace's equation $\dfrac{\partial^2\phi}{\partial x^2} + \dfrac{\partial^2\phi}{\partial y^2} = 0$ in the region above the wires where there are no charges. Obviously,

$$\sum_{k=-\infty}^{\infty}\left[-\frac{4\pi^2 k^2}{a^2}f_k(z)\cos\frac{2\pi kx}{a} + \frac{\partial^2 f_k(z)}{\partial z^2}\cos\frac{2\pi kx}{a}\right] = 0$$

The above equation must be true for every value of k. Thus, one obtains

$$-\frac{4\pi^2 k^2}{a^2}f_k(z) + \frac{\partial^2 f_k(z)}{\partial z^2} = 0$$

Solving the above, one obtains

$$f_k(z) = A_k e^{\pm\frac{2\pi k}{a}z}$$

where, A_k is a constant.

Nature of the problem demands that for $z > 0$, one should retain only the negative sign of the exponent. Thus, one obtains

$$f_k(z) = A_k e^{-\frac{2\pi k}{a}z}$$

Making use of the above, one obtains from Eq. (1)

$$\phi(x, z) = \sum_k A_k e^{-\frac{2\pi k}{a}z}\cos\frac{2\pi kx}{a} \quad \text{for } z > 0 \tag{2}$$

At $z = 0$, ϕ [given by Eq. (2)] must coincide with the prescribed potential. One can determine A_k from this requirement.

Example 5.5: Consider the situation shown at the left of Fig. 5.12. The potential at the plane $z = 0$ is constant, so this is equivalent to the situation shown at the right of Fig. 5.12. Keeping in mind that one can determine the electric field for all $z > 0$ and uniqueness allows only one admissible solution, show that the surface charge density is

$$\rho = \frac{2\varepsilon_0 ql}{(s^2 - l^2)^{3/2}} \quad \text{for } l = z.$$

Fig. 5.12: Image charges used to simulate a conducting plane

Solution: The induced surface charge density on the conductor is found from

$E = -\left(\dfrac{\partial \phi}{\partial z}\right)_{z=0}$. One can easily show that

$$\rho = \varepsilon_0 E = \frac{2\varepsilon_0 ql}{(s^2 - l^2)^{3/2}} \quad \text{for } l = z.$$

Example 5.6: A right circular cone of semivertical angle α and height h has a uniform polarization P parallel to its axis of symmetry. Find the polarization charges everywhere.

Solution: Let us assume that the polarization vector P is directed from the vertex V to the base of the cone as shown in Fig. 5.13. We have

Fig. 5.13: Circular cone

$$\sigma_P = \mathbf{P} \cdot \hat{n} = P$$

∴ Charge on the base, $q_B = \sigma_P \pi r^2 = P\pi h^2 \tan^2 \alpha$

On the curved surface, we have

$$\sigma_P = \mathbf{P} \cdot \hat{n} = P\cos(\pi/2 + \alpha) = -P\sin\alpha$$

Charge on the curved surface, $q_c = -P\sin\alpha \times \pi rl$

$$= -P\sin\alpha\, \pi\, (h\tan\alpha)\,(h\sec\alpha)$$
$$= -P\pi h^2 \tan^2 \alpha$$

∴ Total charge on the cone $= P\pi h^2 \tan^2 \alpha - P\pi h^2 \tan^2 \alpha = 0.$

Example 5.7: Using Laplace's equation in cylindrical coordinates, obtain potential between coaxial conducting cylinders.

Solution: Let us consider a cylinder of radius a, surrounded by another coaxial cylinder of radius b. Let the outer cylinder be grounded and the inner cylinder be charged to a potential V_a. Obviously, in this problem infinity is not one of the boundaries. Instead the space under consideration is bounded by two given coaxial cylinders. One can easily conclude that the arrangement is similar to that of a cylindrical capacitor. We have the boundary conditions as

$$\left.\begin{array}{l} V = 0 \text{ when } r = b \\ \text{and } V = V_a \text{ when } r = a \end{array}\right\} \tag{1}$$

where, r is the distance from the axis. Laplace's equation in cylindrical coordinates is

$$\nabla^2 V = \frac{1}{r}\frac{\partial}{\partial r}\left(r\frac{\partial V}{\partial r}\right) + \frac{1}{r^2}\frac{\partial^2 V}{\partial \theta^2} + \frac{\partial^2 V}{\partial z^2} = 0 \tag{2}$$

Assuming that the cylinders are long, one can neglect the end effects. This means that potential is independent of both θ and z. Thus, Eq. (2) takes the form

$$\frac{1}{r}\frac{\partial}{\partial r}\left(r\frac{\partial V}{\partial r}\right) = 0 \tag{3}$$

Integration of Eq. (3) yields

$$r\frac{\partial V}{\partial r} = \text{constant}, C_1 (\text{say})$$

i.e.

$$\frac{\partial V}{\partial r} = \frac{C_1}{r}$$

Again integrating, one obtains

$$V = C_1 \log_e r + C_2 \qquad (4)$$

where, C_2 is another constant of integration. One can see from Eq. (4) that the potential V varies logarithmically. One can evaluate the constants C_1 and C_2 by applying boundary conditions given by Eq. (1).

Making use of first boundary condition, i.e. $V = 0$ at $r = b$ in Eq. (4), one obtains

$$0 = C_1 \log_e b + C_2 \qquad (5)$$

Now, using second boundary condition, i.e. $V = V_a$ at $r = a$ in Eq. (4), one obtains

$$V_a = C_1 \log_e a + C_2 \qquad (6)$$

Solving Eqs. (5) and (6), one obtains

$$\left. \begin{array}{c} C_1 = \dfrac{V_a}{\log_e(a/b)} \\[4mm] \text{and} \qquad C_2 = -\dfrac{V_a \log_e b}{\log_e(a/b)} \end{array} \right\} \qquad (7)$$

Using Eq. (7), Eq. (4) yields

$$V = \frac{V_a}{\log_e(a/b)}\log_e r - \frac{V_a \log_e b}{\log_e(a/b)}$$

$$= \frac{V_a}{\log_e(a/b)}\log_e(r/b)$$

Example 5.8: Find the potential at any point inside the cylinder as shown in Fig. 5.14.

Solution: The cylinder has radius a and height h, the top and bottom surfaces being at $z = h$ and $z = 0$. The boundary conditions are (i) $V = 0$ on the side and the bottom of the cylinder (ii) $V = V_0 (r, \theta)$ at the top.

The Laplace's equation in cylindrical coordinates is

$$\nabla^2 V = \frac{1}{r}\frac{\partial}{\partial r}\left(r\frac{\partial V}{\partial r}\right) + \frac{1}{r^2}\frac{\partial^2 V}{\partial \theta^2} + \frac{\partial^2 V}{\partial Z^2} = 0 \qquad (1)$$

The solution of Eq. (1) by the method of separation of variables is

$$V(r, \theta, z) = R(r)\Theta(\theta)Z(z) \qquad (2)$$

where $\qquad R(r) = AJ_m(kr) + BN_m(kr) \qquad (3)$

Fig. 5.14: Potential inside the cylinder

$$\Theta(\theta) = Ce^{\pm im\theta} \tag{4}$$

and
$$Z(z) = De^{\pm kz} \tag{5}$$

In order that the potential V be single-valued and vanish at $z = 0$, one can express Eq. (4) as

$$\Theta(\theta) = C\sin m\theta + D\cos m\theta \tag{6}$$

$$Z(z) = \sinh kz \tag{7}$$

If the potential V is finite at $r = 0$, then $B = 0$ in Eq. (3). Obviously, the radial solution takes the form

$$R(r) = AJ_m(kr) \tag{8}$$

Now, the requirement of the potential at $r = a$ means that k can take only special values, i.e.

$$k_{mn} = \frac{x_{mn}}{a}, \; n = 1, 2, 3 \tag{9}$$

where, x_{mn} are the roots of $J_{mn}(x_{mn}) = 0$

Combining Eqs. (2), (6), (7), (8) and (9), one obtains that the general form of the solution is

$$V(r, \theta, z) = \sum_{m=0}^{\infty} \sum_{n=1}^{\infty} J_n(k_{mn}r)\sinh(k_{mn}z)[C_{mn}\sin m\theta + D_{mn}\cos m\theta]$$

At $z = h$, $V(r, \theta, z) = V_0(r, \theta)$, we have

$$V_0(r, \theta) = \sum_{mn} \sinh(k_{mn}h)J_m(k_{mn}r)[C_{mn}\sin m\theta + D_{mn}\cos m\theta]$$

One can easily see that this is a Fourier series in θ and a Bessel–Fourier series in r. One obtains the coefficients C_{mn} and D_{mn} as

$$C_{mn} = \frac{2\,\text{cosech}(k_{mn}h)}{\pi a^2 J_{m+1}^2(k_{mn}a)} \int_0^{2\pi} d\theta \int_a^a V_0(r, \theta)J_m(k_{mn}r)\sin m\theta \cdot r dr$$

and
$$D_{mn} = \frac{2\,\text{cosech}(k_{mn}h)}{\pi a^2 J_{m+1}^2(k_{mn}a)} \int_0^{2\pi} d\theta \int_0^a V_0(r, \theta)J_{mn}(k_{mn}r)\cos m\theta \, r dr$$

Example 5.9: Consider a conducting sphere of radius a composed of two hemispheres separated by a small insulated ring. The hemispheres are kept at potentials $\pm V_0$. Calculate the electrostatic potential both inside and outside the sphere using spherical harmonics.

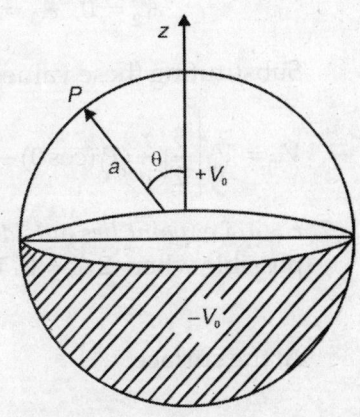

Solution: Figure 5.15 shows a conducting sphere of radius a made up of two hemispheres separated by a small insulated ring. The two hemispheres are at potential $+V_0$ and $-V_0$ (Fig. 5.15).

There is azimuthal symmetry and hence we have

$$V = \sum \left[A_n r^n + B_n r^{-(n+1)} \right] P_n(\cos\theta) \tag{1}$$

Fig. 5.15: A conducting sphere

Let us consider the following two situations:

1. $r < a$, i.e. *point lies inside the sphere:* Since there are no charges at the origin and hence V is finite at the origin. This is possible only when $\Sigma B_n = 0$. One can express the inside potential V_{in} as

$$V_{in} = \sum A_n r^n P_n(\cos\theta) \qquad (2)$$

At the surface where $r = a$, the potential $(V_{in})_s$ is obtained as

$$(V_{in})_s = \sum A_n a^n P_n(\cos\theta) \qquad (3)$$

To determine the values of constants A_n, let us multiply Eq. (3) on both sides by $P_m(\cos\theta)$ and integrate between the limits -1 to $+1$, and obtain

$$\int_{-1}^{+1} (V_{in})_s P_m(\cos\theta) d(\cos\theta) = \int_{-1}^{+1} \sum A_n a^n P_n(\cos\theta) P_m(\cos\theta) d(\cos\theta)$$

or $\qquad -\int_\pi^0 (V_{in})_s P_n(\cos\theta) \sin\theta\, d\theta = A_n a^n \dfrac{2}{2n+1}$ (for $n = m$)

$$\because \quad \int_{-1}^1 P_n(\cos\theta) P_m(\cos\theta) d(\cos\theta) = \frac{1}{2n+1}$$

$$\therefore \quad A_n = \frac{(2n+1)}{2a^n} \int_0^\pi (V_0)_s P_n(\cos\theta) \sin\theta\, d\theta$$

$$= \frac{(2n+1)}{2a^n} \left[\int_0^{\pi/2} (V_0) P_n(\cos\theta) \sin\theta\, d\theta + \int_{\pi/2}^\pi (-V_0) P_n(\cos\theta) \sin\theta\, d\theta \right]$$

Now $\qquad A_0 = \dfrac{V_0}{2} \left[(-\cos\theta)_0^{\pi/2} - (-\cos\theta)_{\pi/2}^\pi \right] = 0$

and $\qquad A_1 = \dfrac{3V_0}{2a} \left[\left(-\dfrac{\cos^2\theta}{2} \right)_0^{\pi/2} + \left(\dfrac{\cos^2\theta}{2} \right)_{\pi/2}^\pi \right] = \dfrac{3V_0}{2a}$

Similarly, one obtains

$$A_2 = 0, \ A_3 = -\frac{7}{8}\frac{V_0}{a^3}, \ A_4 = 0, \ A_5 = \frac{11}{16}\frac{V_0}{a^5}, \text{ etc.}$$

Substituting these values in Eq. (2), one obtains

$$V_{in} = V_0 \left[\frac{3}{2}\left(\frac{r}{a}\right) P_1(\cos\theta) - \frac{7}{8}\left(\frac{r}{a}\right)^3 P_3(\cos\theta) + \frac{11}{16}\left(\frac{r}{a}\right)^5 P_5(\cos\theta) + \cdots \right]$$

2. $r > a$, i.e. *point lies outside the sphere:* In this situation, $r \to \infty$, $V = 0$. This is only possible when $\Sigma A_n = 0$. Thus,

$$V_{out} = \sum B_n r^{n-1} P_n(\cos\theta)$$

At the surface

$$(V_{out})_s = \sum B_n a^{-(n+1)} P_n(\cos\theta)$$

Multiplying on both sides by $P_m(\cos\theta)$ and integrating between the limits -1 to $+1$, one obtains

$$\int_{-1}^{1}(V_{out})_s P_m(\cos\theta)d(\cos\theta) = \int_{-1}^{1}\sum B_n a^{-(n+1)} P_n(\cos\theta)P_m(\cos\theta)d(\cos\theta)$$

or $\quad -\int_{\pi}^{0}(V_{out})_s P_m(\cos\theta)\sin\theta\, d\theta = B_n a^{-(n+1)}\left[\dfrac{2}{2n+1}\right]$

$$\therefore\quad B_n = \frac{2n+1}{2}a^{(n+1)}\left[\int_{0}^{\pi/2}(+V_0)P_m(\cos\theta)\sin\theta\, d\theta\right.$$

$$\left.+\int_{\pi/2}^{\pi}(-V_0)P_m(\cos\theta)\sin\theta\, d\theta\right]$$

Putting $n = 0$, one obtains

$$B_0 = \frac{V_0}{2}a\left[(-\cos\theta)_0^{\pi/2}+(\cos\theta)_{\pi/2}^{\pi}\right]=0$$

Similarly, one obtains

$$B_1 = \frac{3}{2}V_0 a^2,\quad B_2 = 0,$$

$$B_3 = -\frac{7}{8}V_0 a^4,\quad B_4 = 0,\quad B_5 = \frac{11}{16}V_0 a^6,\text{ etc.}$$

Thus, we have

$$V_{out} = V_0\left[\frac{3}{2}\left(\frac{a}{r}\right)^2 P_1(\cos\theta)-\frac{7}{8}\left(\frac{a}{r}\right)^4 P_3(\cos\theta)+\frac{11}{16}\left(\frac{a}{r}\right)^6 P_5(\cos\theta)+\cdots\right]$$

Example 5.10: Show that the potential of a spherical cap of angle α charged to a uniform surface density σ_0 as shown in Fig. 5.16 is given by

$$V_P = \frac{a\sigma_0}{2\varepsilon_0}\left[(1-\cos\alpha)\left(\frac{a}{r}\right)+\sum_{n=1}^{\infty}\frac{P_{n-1}(\cos\alpha)-P_{n+1}(\cos\alpha)}{2n+1}\times\left(\frac{a}{r}\right)^{n+1}P_n(\cos\theta)\right]$$

Solution: One can consider that the spherical cap is the surface cut from a sphere by a right circular cone of semi-vertical angle α.

To obtain the solution of the problem, one can regard the entire sphere as charged to a surface density σ, where

$$\sigma = \sigma_0 \quad\text{from } \theta = 0 \text{ to } \theta = \alpha$$

$$= 0 \quad\text{from } \theta = \alpha \text{ to } \theta = \pi$$

The problem has azimuthal symmetry and therefore, one can express σ in terms of harmonics

$$\sigma = a_0 + a_1 P_1(\cos\theta) + a_2 P_2(\cos\theta)+\cdots$$

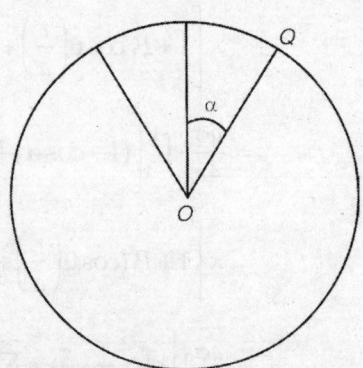

Fig. 5.16: Potential of a spherical cap

where,

$$a_l = \frac{2l+1}{2} \int_{\theta=\pi}^{\theta=0} \sigma P_l(\cos\theta) d(\cos\theta)$$

$$= \frac{2l+1}{2} \int_{\theta=\pi}^{\theta=\alpha} \sigma P_l(\cos\theta) d(\cos\theta) + \frac{2l+1}{2} \int_{\theta=\alpha}^{\theta=0} \sigma P_l(\cos\theta) d(\cos\theta)$$

$$= \frac{2l+1}{2} \sigma_0 \int_{\theta=\alpha}^{\theta=0} P_l(\cos\theta)(d\cos\theta) \quad (\because \ \sigma = 0 \text{ from } \theta = \pi \text{ to } \theta = \alpha)$$

$$\therefore \quad a_0 = \frac{1}{2}\sigma_0 \int_{\theta=\alpha}^{\theta=0} d(\cos\theta) = \frac{1}{2}\sigma_0(1-\cos\alpha)$$

One can easily show that

$$\int_{\theta=\alpha}^{\theta=0} P_l(\cos\theta) d(\cos\theta) = \left[\frac{P_{l+1}(\cos\theta) - P_{l-1}(\cos\theta)}{2l+1} \right]_{\theta=\alpha}^{\theta=0}$$

$$\therefore \quad \sigma = \frac{1}{2}\left[(1-\cos\alpha) + \sum_{l=1}^{\infty}\{P_{l-1}(\cos\alpha) - P_{l+1}(\cos\alpha)\} P_l(\cos\theta) \right]$$

The potential at a point P on the polar axis, distant r from the centre is obtained as

$$V_P = \frac{1}{4\pi\varepsilon_0} \iint \frac{\sigma ds}{PQ} = \frac{1}{4\pi\varepsilon_0} \iint \frac{\sigma ds}{(a^2 + r^2 - 2ar\cos\theta)^{1/2}}$$

where, ds is an element of area at Q on the surface of the sphere and a is the radius of the sphere. If $r < a$, we have

$$V_P = \frac{1}{4\pi\varepsilon_0} \iint \frac{\sigma}{a}\left(1 + \frac{r^2}{a^2} - \frac{2r}{a}\cos\theta\right)^{-1/2} ds$$

$$= \frac{1}{4\pi\varepsilon_0} \iint \frac{\sigma}{a} \sum_{n=0}^{\infty} P_n(\cos\theta)\left(\frac{r}{a}\right)^n ds$$

$$= \frac{1}{4\pi\varepsilon_0} \int \frac{\sigma_0}{2a}\left[(1-\cos\alpha) + \sum_{l=1}^{\infty}\{P_{l-1}(\cos\alpha) - P_{l+1}(\cos\alpha)\} P_l(\cos\theta) \right]$$

$$\times \left[1 + P_1\cos\theta\left(\frac{r}{a}\right) + P_2(\cos\theta)\left(\frac{r}{a}\right)^2 + \cdots \right] 2\pi a^2 \sin\theta\, d\theta$$

$$= \frac{a\sigma_0}{4\varepsilon_0} \int_{-1}^{1}\left[(1-\cos\alpha) + \sum_{l=1}^{\infty}\{P_{l-1}(\cos\alpha) - P_{l+1}(\cos\alpha)\} P_l(\cos\theta) \right]$$

$$\times \left[1 + P_1\left(\cos\theta\frac{r}{a}\right) + P_2(\cos\theta)\left(\frac{r}{a}\right)^2 + \cdots \right] d(\cos\theta)$$

$$= \frac{a\sigma_0}{2\varepsilon_0}\left[(1-\cos\alpha) + \sum_{n=1}^{\infty} \frac{P_{n-1}(\cos\alpha) - P_{n+1}(\cos\alpha)}{2n+1}\left(\frac{r}{a}\right)^n \right]$$

Thus, the potential at a point $(r, 0)$ is obtained as

$$V(r, \theta) = \frac{a\sigma_0}{2\varepsilon_0}\left[(1-\cos\alpha) + \sum_{n=1}^{\infty} \frac{P_{n-1}(\cos\alpha) - P_{n+1}(\cos\alpha)}{2n+1}\left(\frac{r}{a}\right)^n P_n(\cos\theta)\right]$$

For $r > a$, we have

$$V_P = \frac{a\sigma_0}{2\varepsilon_0}\left[(1-\cos\alpha)\left(\frac{a}{r}\right) + \sum_{n=1}^{\infty} \frac{P_{n-1}(\cos\alpha) - P_{n+1}(\cos\alpha)}{2n+1}\left(\frac{a}{r}\right)^{n+1} P_n(\cos\theta)\right]$$

Example 5.11: Determine the field intensity and capacitance per metre length of a single core cable.

Solution: We have

$$E = -\nabla\phi = -\hat{a}_r \frac{\partial\phi}{\partial r}$$

$$= -\hat{a}_r \frac{\partial}{\partial r}\left[\frac{\phi_1}{\log_e(a/b)}\left\{\log_e\left(\frac{r}{b}\right)\right\}\right]$$

$$= \hat{a}_r \frac{\phi_1}{\log_e(a/b)}\left[\frac{b}{r} \cdot \frac{1}{b}\right] = a_r \frac{\phi_1}{\log_e(b/a)}\left(\frac{1}{r}\right)$$

$$\frac{\text{Capacitance}}{\text{m}} = \frac{\sigma A}{\phi_1} = \frac{\varepsilon E A}{\phi_1}$$

$$= \frac{\varepsilon}{\phi_1}\left[\frac{\phi_1}{r\log_e(b/a)}\right]2\pi r \times 1$$

$$= \frac{2\pi\varepsilon}{\log_e(b/a)} \frac{\text{farad}}{\text{m}}$$

Example 5.12: An infinitely long conducting cylinder is placed in a uniform field of intensity E_0 in such a way that its z-axis is perpendicular to the field and the centre coincides with the origin (Fig. 5.17). Determine the potential.

Solution: We have

Fig. 5.17: An infinitely long conducting cylinder

$$V = A_0 + B_0 \log r + \sum_{n=1}^{\infty}(A_n r^n + B_n r^{-n})\cos n\theta + \sum_{n=1}^{\infty}(C_n r^n + D_n r^{-n})\sin n\theta \qquad (1)$$

The boundary conditions are $V = 0$ at $r = a$ and $V = -E_0 r\cos\theta$ at $r \gg a$ or ∞.

As desired by the problem, the potential is finite at $r = \infty$ and zero at $r \to 0$. This means $B_0 = 0$. Now, when $r \gg a$, the potential is only a function of $\cos\theta$, i.e. C_n and $D_n = 0$. Now substituting these values in Eq. (1), one obtains

$$V = A_0 + \sum_{n=1}^{\infty}(A_n r^n + B_n r^{-n})\cos n\theta \qquad (2)$$

For $r \gg a$, Eq. (2) can be expressed as

$$-E_0 r \cos\theta = A_0 + \sum_{n=1}^{\infty} (A_n r^n) \cos n\theta \tag{3}$$

Equation (3) is valid for any value of r, hence comparing the coefficients of different powers of r on both sides, one obtains .

$$A_0 = A_2 = A_3 = 0 \quad \text{and} \quad A_1 = -E_0$$

For $n = 1$, Eq. (2) reduces to

$$V = \left[-E_0 r + \frac{B_1}{r} \right] \cos\theta \tag{4}$$

Applying the boundary condition $V = 0$ at $r = a$ in Eq. (4), one obtains

$$0 = \left[-E_0 a + \frac{B_1}{a} \right] \cos\theta \quad \text{or} \quad B_1 = E_0 a^2$$

$$\therefore \qquad V = \left[E_0 r + \frac{E_0 a^2}{r} \right] \cos\theta$$

Example 5.13: Show that the potential at all points in space exterior to a conducting sphere of radius a placed in a uniform field E_0 is given by

$$V(r, \theta) = -E_0 r \cos\theta \left(1 - \frac{a^3}{r^3} \right)$$

Solution: Let us choose the axis in such a way that the polar axis coincides with the direction of the field. The problem has azimuthal symmetry, the symmetry axis being, say z-axis, $\theta = 0$. We have the potential on the symmetry axis given by

$$V(r, \theta) = \sum_{l=0}^{\infty} \left[A_l r^l + \frac{B_l}{r^{l+1}} \right] \quad [\because \ P_l(\cos\theta) = 1] \tag{1}$$

with $r = z$. If one evaluates this potential by some means at an arbitrary point and the potential function is expanded in power series, then one can obtain the potential at any point in space by multiplying each power of r^l and $\dfrac{1}{r^{l+1}}$ by $P_l(\cos\theta)$. We have

$$E_0 = -\nabla(\phi_0) \tag{2}$$

$$\therefore \qquad \phi_0 = -E_0 z = -E_0 r \cos\theta = -E_0 r P_1(\cos\theta) \tag{3}$$

However, this does not satisfy the condition that $\phi = 0$ as $r \to \infty$. However, we assume that for a uniform field of infinite extent, the source of which must lie at infinity and, hence, ϕ will not vanish as $r \to \infty$.

One can easily see that the field immediately around the sphere will be distorted owing to induced charges on the surface; but at large distances it will be equal to E_0 (Fig. 5.18). The total potential at a point $V(r, \theta)$ is

$$V(r, \theta) = \phi_0 + \text{the potential due to the charges}$$
$$\text{induced in the conducting sphere} \tag{3a}$$

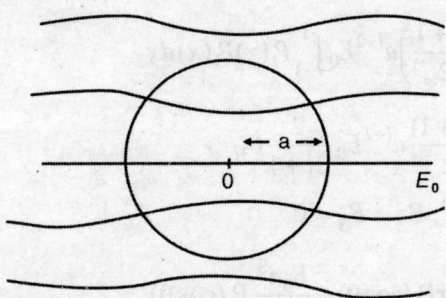

Fig. 5.18: A conducting sphere placed in a uniform field

We know that at very large values of r, however, the potential due to the induced charges may be insignificant and

$$V(r, \theta) = \phi_0$$

Now
$$V(r, \theta) = \sum_{l=0}^{\infty} \left(A_l r^l + \frac{B_l}{r^{l+1}} \right) P_l(\cos\theta) \qquad (4)$$

One can neglect the term involving B_l in Eq. (4) at large value of r, i.e.

$$V(r, \theta) = \sum_{l=0}^{\infty} A_l r^l P_l(\cos\theta)$$

$$= A_0 P_0(\cos\theta) + A_1 r P_1(\cos\theta) + A_2 r^2 P_2(\cos\theta) + \cdots \qquad (5)$$

This must be identical with ϕ_0 as given by Eq. (3)

$$A_1 r P_1(\cos\theta) = -E_0 r P_1(\cos\theta)$$

Hence, $A_1 = E_0$ and all other A_l vanish. Obviously, from Eq. (4)

$$V(r, \theta) = A_1 r P_1(\cos\theta) + \sum_{l=0}^{\infty} \frac{B_l}{r^{l+1}} P_l(\cos\theta)$$

$$= -E_0 r P_1(\cos\theta) + \sum_{l=0}^{\infty} \frac{B_l}{r^{l+1}} P_l(\cos\theta) \qquad (6)$$

The sphere under question placed in the uniform field being a conductor, the induced charges generate an electric field within it. This cancels the external field giving zero field within the space and, hence, the potential on the surface ($r = a$) is zero.

$$\therefore \qquad V(a, \theta) = -E_0 a P_1(\cos\theta) + \sum_{l=0}^{\infty} B_l a^{-(l+1)} P_l(\cos\theta) = 0$$

i.e.
$$E_0 a P_1(\cos\theta) = \sum_{l=0}^{\infty} B_l a^{-(l+1)} P_l(\cos\theta) \qquad (7)$$

One can obtain the coefficients B_l as

$$B_l a^{-(l+1)} = \frac{2l+1}{2} \int_{-1}^{1} E_0 a P_1(x) P_l(x) dx$$

$$\therefore \qquad B_l = \left(\frac{2l+1}{2}\right) a^{l+2} E_0 \int_{-1}^{1} P_l(x) P_1(x)\, dx$$

$$= \frac{(2l+1)}{2} a^{l+2} E_0 \frac{2}{2l+1} \delta_{1l}$$

$$\therefore \qquad B_1 = E_0 a^3,\ B_2 = B_3 = 0 \qquad\qquad (8)$$

Hence
$$V(r,\theta) = -E_0 r P_1(\cos\theta) + \frac{E_0 a^3}{r^2} P_1(\cos\theta)$$

$$= -E_0 r \cos\theta \left(1 - \frac{a^3}{r^3}\right) \qquad\qquad (9)$$

One can also obtain the field components using $E_r = -\dfrac{\partial V}{\partial r}$, and $E_\theta = -\dfrac{1}{r}\dfrac{\partial V}{\partial r}$.

Example 5.14: Two parallel plates have equal and opposite charges and are separated by a dielectric 5 mm thick of dielectric constant 3. If dielectric intensity in the dielectric is 10^6 Vm^{-1}, calculate: (a) the displacement D in the dielectric (b) the polarization P in the dielectric (c) the bound charge per unit area on the surfaces of the dielectric (d) the free charge per unit area on the conducting plate and (e) the energy in the dielectric.

Solution:

(a) Displacement $D = K\varepsilon_0 E$

or $\qquad D = 3 \times 8.854 \times 10^{-12} \times 10^6 = 26.55\ \mu C\,m^{-2}$

(b) Polarization vector $P = \varepsilon_0 (K-1)E$

or $\qquad P = 8.854 \times 10^{-12} (3-1) \times 10^6 = 17.71\ \mu C\,m^{-2}$

(c) Since polarization vector P is along E and thus $\perp^r$ to the surface of the plate. Thus, the bound charge per unit area on the surface of the dielectric

$$|\sigma'| = P_n = P = 17.71\ \mu C\,m^{-2}$$

(d) Free charge per unit area

$$|\sigma| = D = 26.56\ \mu C\,m^{-2}$$

(e) Energy density in dielectric

$$= \frac{1}{2} K\varepsilon_0 E^2 = \frac{1}{2} \times 3 \times 8.854 \times 10^{-12} \times 10^{12} = 13.28\ Jm^{-3}.$$

Example 5.15: Consider the two dimenstional boundary-value problem sketched in Fig. 5.19. The region V is free of charges. On three of the edges of V it is $\varphi = 0$, while on the fourth side of the rectangle it holds

$$\varphi_0(y) = \sin\left(\frac{\pi}{y_0} y\right)$$

Determine the scalar potential everywhere in V.

Solution: Laplace equation in two dimensions:

$$\nabla\varphi = \left(\frac{\partial^2}{\partial x^2} + \frac{\partial^2}{\partial y^2}\right)\varphi = 0$$

Separation ansatz:

$$\varphi(x,y) = f(x)\, g(y)$$

Insertion into the Laplace equation:

$$\underbrace{\frac{1}{f}\frac{d^2 f}{dx^2}}_{\text{only dependent on } x} + \underbrace{\frac{1}{g}\frac{d^2 g}{dy^2}}_{\text{only dependent on } y} = 0$$

Fig. 5.19: Two dimensional boundary-value problem for the scalar potential in an otherwise charge-free space

$$\Rightarrow \frac{1}{f}\frac{d^2 f}{dx^2} = \alpha^2 = -\frac{1}{g}\frac{d^2 g}{dy^2}$$

Structure of the solution:

$$f(x) = a e^{\alpha x} + b e^{-\alpha x}$$

$$g(y) = \bar{a}\cos(\alpha y) + \bar{b}\sin(\alpha y)$$

Boundary conditions:

$$\varphi(x=0, y) = 0 \Rightarrow b = -a$$
$$\varphi(x, y=0) = 0 \Rightarrow \bar{a} = 0$$

$$\varphi(x, y=y_0) = 0 \Rightarrow \alpha \to \alpha_n = \frac{n\pi}{y_0}; \quad n \in \mathbb{N}$$

Intermediate result:

$$\varphi(x,y) = \sum_n c_n \sin\left(\frac{n\pi}{y_0}y\right)\sinh\left(\frac{n\pi}{y_0}x\right)$$

Further boundary condition:

$$\varphi_0 = \varphi(x=x_0, y) = \sum_n c_n \sin\left(\frac{n\pi}{y_0}y\right)\sinh\left(\frac{n\pi}{y_0}x_0\right) \overset{!}{=} \sin\left(\frac{\pi}{y_0}y\right)$$

Orthogonality relation:

$$\frac{2}{y_0}\int_0^{y_0}\sin\left(\frac{m\pi}{y_0}y\right)\sin\left(\frac{\pi}{y_0}y\right)dy = \delta_{m1}$$

$$= \sum_n c_n \sinh\left(\frac{n\pi}{y_0}x_0\right)\frac{2}{y_0}\int_0^{y_0}\sin\left(\frac{n\pi}{y_0}y\right)\sin\left(\frac{m\pi}{y_0}y\right)dy$$

$$= \sum_n c_n \sinh\left(\frac{n\pi}{y_0} x_0\right) \delta_{nm}$$

$$\Rightarrow c_m = \frac{\delta_{m1}}{\sinh\left(\dfrac{m\pi}{y_0} x_0\right)}$$

Thus, we have the following relation:

$$\varphi(x, y) = \frac{\sinh\left(\dfrac{\pi}{y_0} x\right)}{\sinh\left(\dfrac{\pi}{y_0} x_0\right)} \sin\left(\frac{\pi}{y_0} y\right)$$

GLIMPSES

1. Poisson's equation: $\nabla^2 \phi = -\rho/\varepsilon_0$. If there are no charges in the region, $\rho = 0$ and $\nabla^2 \phi = 0$. This is valid only in the region of free charges.

2. ∇^2 (Laplacian) is a scalar operator and has the following form in three coordinate systems generally used:

   i. Cartesian system: $\dfrac{\partial^2}{\partial x^2} + \dfrac{\partial^2}{\partial y^2} + \dfrac{\partial^2}{\partial z^2}$

   ii. Spherical polar coordinate system:

   $$\frac{1}{r^2} \frac{\partial}{\partial r}\left(r^2 \frac{\partial}{\partial r}\right) + \frac{1}{r^2 \sin\theta} \frac{\partial}{\partial \theta}\left(\sin\theta \frac{\partial}{\partial \theta}\right) + \frac{1}{r^2 \sin^2\theta} \frac{\partial^2}{\partial \phi}$$

   iii. Cylindrical polar coordinates:

   $$\frac{1}{r} \frac{\partial}{\partial r}\left(r \frac{\partial}{\partial r}\right) + \frac{1}{r^2} \frac{\partial^2}{\partial \phi^2} + \frac{\partial^2}{\partial z^2}$$

3. Earnshaw's theorem states that a charged body cannot rest in stable equilibrium under the influence of electrostatic forces alone.

4. Any solution to Laplace's equation or Poisson's equation which also satisfies the boundary conditions must be the only solution that exists. This is uniqueness theorem.

5. Two important properties of the potential in a charge-free region can be obtained from Laplace's equation:

   i. At the centre of an induced circle or sphere, the potential V or ϕ is equal to the average (mean) of the values it assumes on the circle or sphere. This is mean value theorem.

   ii. The potential V or ϕ cannot have a maximum (or a minimum) within the region. It follows from (ii) that any maximum of V must occur on the boundary of the region. Since V must obey Laplace's equation, $\dfrac{\partial^2 V}{\partial x^2} + \dfrac{\partial^2 V}{\partial y^2} + \dfrac{\partial^2 V}{\partial z^2} = 0$ so do

$\dfrac{\partial V}{\partial x}, \dfrac{\partial V}{\partial y}$ and $\dfrac{\partial V}{\partial z}$. Thus, the cartesian components of the electric field intensity

take their maximum values on the boundary. This is maximum value theorem.

6. Generalized Legendre's equation:

$$\frac{d}{dx}\left[(1-x^2)\frac{dP}{dx}\right] + \left[l(l+1) - \frac{m^2}{1-x^2}\right]P = 0$$

where P is a function of θ only. If we put $m = 0$ in the above equation, we get ordinary Legendre's equation.

REVIEW QUESTIONS

1. State and prove uniqueness theorem of electrostatics regarding electric potential. With the help of this theorem, prove that the electric field inside the hollow conductor is zero.

2. State and prove Green's theorem. How this theorem is applied for the solution of potential problem?

3. Define Green's function with reference to electrostatic potential problems. Prove that this function satisfies the symmetry

$$G(x, x') = G(x', x).$$

4. State Poisson's equation. Show that the potential function

$$\phi(r) = \frac{1}{4\pi\varepsilon_0}\int \frac{\rho(r')}{|r - r'|}dV'$$

represents the solution of Poisson's equation.

5. Explain the method of images for the solution of electrostatic problems. A point charge is situated near an infinite plane earthed conductor. Applying the method of images, calculate: (i) surface charge density induced on the plane and (ii) the force between the plane and the charge.

6. A point charge is placed outside a conducting sphere. Find by the method of electrical images the intensity of electrical field outside the sphere and also the surface density of induced charge on the sphere when the sphere is at zero potential.

7. Explain the method of images for the solution of electrostatic problems. Obtain the image system for a point charge situated at a distance c from the centre of a conducting sphere of radius $b(< c)$ at zero potential.

8. A point Q is placed at a distance d from the centre of an earthed conducting sphere of radius a. Find the position of the image and show that the induced charge on the sphere is given by

$$Q' = -\frac{Qa}{d}.$$

9. Write down Laplace's equation in spherical polar coordinates and obtain its solution. What will be the form of solution for spherically symmetric problems.

10. Deduce the solution of Laplace's equation for a problem having azimuthal symmetry. Use it to obtain potential inside and outside a dielectric sphere in a uniform field.

PROBLEMS

1. A charge q is placed at a distance d from the centre of an insulated, uncharged conducting sphere of radius a. Find the rise in potential due to the presence of the charge and show that there is an attractive force between the sphere and the charge equal to

$$\frac{q^2}{4\pi\varepsilon_0} \frac{a^3(2d^2 - a^2)}{(d^2 - a^2)^2}.$$

2. A field is produced by a point charge q in the presence of an uninsulated spherical conductor of radius a whose centre is at a distance d from the charge. Find the force with which the charge is attracted to the sphere and show that the force is changed due to repulsion by insulating the sphere and connecting it to a large distance conductor of potential V_0 provided that

$$V_0 > \frac{1}{4\pi\varepsilon_0} \frac{qd^3}{(d^2 - a^2)^2}$$

3. Show that the following relation holds good for Legendre's polynomials

$$\int P_l(\cos\theta) d(\cos\theta) = \frac{P_{l+1}(\cos\theta) - P_{l-1}(\cos\theta)}{2l+1}$$

where the symbols have usual meanings.

4. A rectangular parallelepiped is bounded by the planes $x = 0$ and $x = a$; $y = 0$ and $y = b$ and $z = 0$ and $z = c$. The side at $z = c$ is kept at a constant potential ϕ_0 while the other sides are kept at zero potential. Solve Laplace's equation for the problem and show that the potential inside the enclosure is given by

$$\sum_{m,n=1,3,5,...etc.} \frac{16\phi_0}{mn\pi^2} \left(\frac{m\pi x}{a}\right) \sin\left(\frac{n\pi y}{b}\right) \times \frac{\sinh(\gamma_{mn}z)}{\sinh(\gamma_{mn}C)}$$

where $\qquad \gamma_{mn}^2 = \left[\frac{m^2\pi^2}{a^2} + \frac{n^2\pi^2}{b^2}\right].$

5. A dielectric sphere is kept in a uniform field. Solve the appropriate Laplace's equation when the potential and electric fields are continuous at the surface of the dielectric sphere. Show that the potential at a radial distance r outside the sphere is

$$V = -\left[1 - \left(\frac{\varepsilon_r - 1}{\varepsilon_r + 2}\right)\frac{a^3}{r^3}\right]E_0 r \cos\theta$$

where, θ is the angle which the line joining the outside point with the centre of the sphere makes with the electric field.

6. A hemisphere of radius a is charged uniformly with a charge Q. Assuming that the centre of the base of the hemisphere is at the origin, determine the potential at a point $r \gg a$.

$$\left[Ans.\ V \sim \frac{Q}{4\pi\varepsilon_0}\left(1 + \frac{3a}{2r}\cos\theta\right)\right]$$

7. An electron is at distance 10 Å from a plane conductor. Calculate the force experienced by the electron and the work done in moving it to infinite distance from the conductor.

$$[Ans.\ F = -5.75 \times 10^{-11}\,N,\ W = 5.75 \times 10^{-20}\,J]$$

8. Find the capacitance (C) between two long coaxial cones placed tip to tip with infinite gap.

$$\left[Ans.\ Capacitance/m = \frac{\pi\varepsilon_0}{\log\cot(\theta/2)}F/m\right]$$

9. A thunder cloud is stationary above ground level. Considering the earth as a perfect conductor and the thunder cloud as an electric dipole with its axis vertical, show that the electric field at a point on the ground is proportional to $3\sin 5\alpha - \sin^3\alpha$, where α is the elevation of the cloud from the point.

10. Determine the potential at any point in space due to a charge q uniformly distributed around a circular ring of radius a, its axis being the z-axis and its centre at $z = b$.

$$[Ans.\ V(r,\theta) = q\sum_{l=0}^{\infty}\frac{r^l}{c^{l+1}}P_l(\cos\alpha)P_l(\cos\theta)\ \text{and}$$

$$r < c,\ V(r,\theta) = q\sum_{l=0}^{\infty}\frac{c^l}{r^{l+1}}P_l(\cos\alpha)P_l(\cos\theta)]$$

11. Find the potential and field intensity between anode and cathode (earthed) of a working vacuum diode, assuming uniform space charge.

$$\left[Ans.\ \phi_x = \left(\frac{V}{l} + \frac{\rho l}{2\varepsilon_0}\right)x - \frac{\rho x^2}{2\varepsilon_0}\ \text{and}\ E_x = -i\left(\frac{V}{l} + \frac{\rho l}{2\varepsilon_0}\right) + i\frac{\rho x}{\varepsilon_0}\right]$$

12. An infinite long cylinder of radius a with its axis parallel to z-axis is cut in an infinite dielectric medium of permittivity ε. If a uniform electric field E_0 is applied in the $+x$ direction, find a solution for the resultant electric field both within and outside the cylindrical cavity.

$$[Ans.\ V_i(\rho,\phi) = -\frac{2}{\varepsilon_r + 1}\rho\cos\phi\ \text{and}$$

$$V_0(\rho,\phi) = -\left[1 + \left(\frac{\varepsilon_r - 1}{\varepsilon_r + 1}\right)\frac{a^2}{\rho^3}\right]\frac{E_0}{\varepsilon_r}\rho\cos\phi$$

and the components of electric field E just outside the cavity are

$$E_\rho = -\frac{\partial V_0}{\partial \rho} = \left[1 - \left(\frac{\varepsilon_r - 1}{\varepsilon_r + 1}\right)\frac{a^2}{\rho^2}\right]\frac{E_0}{\varepsilon_0}\cos\phi$$

$$E_\phi = \frac{1}{\rho}\frac{\partial V_0}{\partial \phi} = \left[1 - \left(\frac{\varepsilon_r - 1}{\varepsilon_r + 1}\right)\frac{a^2}{\rho^2}\right]\frac{E_0}{\varepsilon_r}\cos\phi]$$

13. A grounded conducting sphere of radius a is placed in a potential field

$V(r,\theta,\phi) = \dfrac{r^2 \sin 2\theta}{4\pi\varepsilon_0}e^{i\phi}$ with its centre at the origin. Show that the potential at a

point outside the sphere is given by

$$V = \frac{1}{2\pi\varepsilon_0}\left(r^2 - \frac{a^5}{r^3}\right)\sin\theta\cos\theta\, e^{i\phi}.$$

14. A spherical cavity of radius a is cut out of a conducting block of metal maintained at zero potential. A charge q is placed at a distance d from the centre of gravity. Calculate the force on it.

$$\left[\textbf{Ans. } F = \frac{q^2 ad}{[4\pi\varepsilon_0(a^2 - d^2)^2]}\right]$$

15. A conducting sphere of radius a is kept at zero potential. An electric dipole of moment p pointing away from the sphere is at a distance d from the centre of the sphere. Calculate the dipole moment and charge on the image at the inverse point.

$$[\textbf{Ans. } \text{Dipole moment} = \frac{pa^3}{d^3} \text{ and charge} = \frac{pa}{d^3}]$$

16. A conducting circular cylinder of radius a and infinite length is placed in a uniform field with its axis at right angles to the field. If the cylinder is maintained at zero potential, then show that the potential at a point outside the cylinder is given by

$$V = -E_0\left(r - \frac{a^2}{r}\right)\cos\theta$$

17. An electron is placed at a height h above an infinite dielectric slab of relative permittivity ε. Show that the force on the electron is given by

$$F = \frac{e^2}{4\pi\varepsilon_0}\left(\frac{\varepsilon - 1}{\varepsilon + 1}\right)\frac{1}{(2h)^2}.$$

SHORT ANSWER QUESTIONS

1. Give a simple example of method of images.
   Ans. A point charge located in front of an infinite plane conductor at zero potential.
2. Write Laplace's equation in spherical polar coordinates.

Ans. $\left[\dfrac{1}{r^2}\dfrac{\partial}{\partial r}\left(r^2\dfrac{\partial V}{\partial r}\right)+\dfrac{1}{r^2\sin\theta}\dfrac{\partial}{\partial\theta}\left(\sin\theta\dfrac{\partial V}{\partial\theta}\right)+\dfrac{1}{r^2\sin^2\theta}\dfrac{\partial^2 V}{\partial\phi^2}=0\right]$

3. A function (say Q_m) is normalized and also orthogonal. How will you express it mathematically?

   Ans. $\left[\displaystyle\int_0^{2\pi}Q_m^*Q_n d\phi=\delta_{mn}\right]$

4. Write Laplace's equation in cylindrical coordinates.

   Ans. $\left[\dfrac{1}{r}\dfrac{\partial}{\partial r}\left(r\dfrac{\partial V}{\partial r}\right)+\dfrac{1}{r^2}\dfrac{\partial^2 V}{\partial\phi^2}+\dfrac{\partial^2 V}{\partial z^2}=0\right]$

5. Does the Green's function with reference to the electrostatic potential problems satisfy symmetry problem? If yes, how will you express it mathematically?
   Ans. Yes, $G(x, x') = G(x', x)$.

6. What do you understand by method of images and image charges?
   Ans. The method of images concerns itself with the problem of one or more point charges in the presence of boundary surfaces, e.g. conductors either grounded or held at fixed potentials. Under favourable conditions it is possible to infer from the geometry of the situation that a small number of suitably placed charges of appropriate magnitudes, external to the region of interest, can stimulate the required boundary conditions. These charges are called image charges, and the replacement of the actual problem with boundaries by an enlarged region with images charges but not boundaries is called the method of images.

7. Prove that within a charge free region the potential cannot attain a maximum value.
   Ans. Let us suppose that a maximum were attained at an interior point P. Then a very small sphere could be centered on P, such that the potential V_c at P exceeds the potential at each point on the sphere. Then V_c should also exceed the average value of the potential over the sphere. But this is in contradiction to the mean value theorem.

8. Find the potential function for the region between the parallel discs shown in Fig. 5.19 (neglect fringing).
   Ans. Here, V is not a function of r and ϕ. Laplace's equation reduces to $\dfrac{d^2 V}{dz^2}=0$. The solution is $V = AZ + B$. The parallel circular discs have a potential function identical to that for any pair of parallel planes. For another choice of axes, the linear potential function might be $Ay + B$ or $Ax + B$.

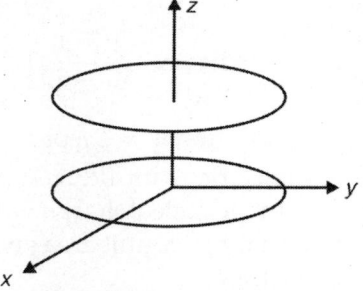

Fig. 5.19: Two parallel discs

9. The region between two concentric right circular cylinders contains a uniform charge density ρ. Find V using Poisson's equation.

Ans. Neglecting fringing, Poisson's equation can be written as

$$\frac{1}{r}\frac{d}{dr}\left(r\frac{dV}{dr}\right) = -\frac{\rho}{\varepsilon}$$

or

$$\frac{d}{dr}\left(r\frac{dV}{dr}\right) = -\frac{\rho r}{\varepsilon}$$

Integrating, one obtains

$$r\frac{dV}{dr} = -\frac{\rho r^2}{2\varepsilon} + A$$

or

$$\frac{dV}{dr} = -\frac{\rho r}{2\varepsilon} + \frac{A}{r}$$

or

$$V = -\frac{\rho r^2}{4\varepsilon} + A\log r + B$$

10. A potential in cylindrical coordinates is a function of r and ϕ but not z. Obtain the separated differential equations for R and ϕ, where $V = R(r)\,\Phi(\phi)$ and solve them. The region is charge free.

Ans. Laplace's equation becomes

$$\Phi\frac{d^2R}{dr^2} + \frac{\Phi}{\rho}\frac{dR}{dr} + \frac{R}{r^2}\frac{d^2\Phi}{d\phi^2} = 0$$

or

$$\frac{r^2}{R}\frac{d^2R}{dr^2} + \frac{r}{R}\frac{dR}{dr} = -\frac{1}{\Phi}\frac{d^2\Phi}{d\phi^2}$$

LHS is a function of r only, while RHS is a function of Φ only; therefore, both sides are equal to a constant, a^2

or

$$\frac{r^2}{R}\frac{d^2R}{dr^2} + \frac{r}{R}\frac{dR}{dr} = a^2$$

or

$$\frac{d^2R}{dr^2} + \frac{1}{r}\frac{dR}{dr} - \frac{a^2R}{r^2} = 0$$

with solution $R = C_1 r^a + C_2 r^{-a}$.

Also, we have $-\dfrac{1}{\Phi}\dfrac{d^2\Phi}{d\phi^2} = a^2$

with solution $\Phi = C_3\cos a\phi + C_4\sin a\phi$

11. A dielectric is introduced between the plates of a parallel plate capacitor. Show that the induced charge varies with the dielectric as $K = (1 - \sigma'/\sigma)^{-1}$, where σ is free charge per unit area and σ' is the bound charge per unit area on the surface of dielectric.

Ans. We have $q' = q\left(1 - \dfrac{1}{K}\right)$, where $+q$ and $-q$ are the charges on the plates of the capacitor in both the cases, with and without the dielectric, q' is the induced charge.

$$\therefore \qquad \sigma' = \sigma\left(1 - \frac{1}{K}\right)$$

or

$$\sigma = \frac{1}{\left(1 - \dfrac{\sigma'}{\sigma}\right)}$$

In the metal dielectric, induced charge is same as that of the free charge but of opposite sign, i.e. $\sigma' = \sigma$. Hence

$$K = \frac{1}{0} = \infty.$$

MULTIPLE CHOICE QUESTIONS

1. A point charge is kept somewhat midway between two infinitely extended earthed metallic planes, then
   (a) there will be an infinite number of images of alternate polarity and the charge will tend to occupy a position exactly midway between the planes
   (b) there will be no image at all
   (c) there will be an infinite number of images of alternate polarity and the charge will be in equilibrium
   (d) there will be an infinite number of images of the same polarity and the charge will be in equilibrium   **[a]**

2. A point charge is placed near an earthed sphere. The induced image charge depends on position and magnitude of point charge in the following manner:
   (a) increases with magnitude and decreases with distance
   (b) increases with both distance and magnitude
   (c) decreases with both distance and magnitude
   (d) increases with distance and decreases with magnitude   **[a]**

3. A long cylindrical dielectric is introduced in a uniform electric field in a direction perpendicular to the longitudinal axis. The potential inside the cylinder and that outside (at a long distance) the cylinder exhibit that
   (a) the inside potential is greater
   (b) the inside potential is smaller
   (c) both are equal
   (d) none of the above is correct   **[b]**

4. A grounded conducting sphere of radius a is placed in a uniform electric field E_0 with its centre at the origin. The induced charge density is
   (a) $3\varepsilon_0 E_0$ \qquad (b) $3\varepsilon_0 E_0 \cos\theta$ \qquad (c) zero \qquad (d) $3\varepsilon_0 E_0 \sin\theta$ \qquad **[b]**

5. An infinite circular cylinder of relative permittivity ε_r is placed in a uniform field E_0 with its axis perpendicular to the field. Field inside the cylinder is
   (a) $\dfrac{2}{\varepsilon_0} E_0$ \hspace{3cm} (b) $\dfrac{2}{\varepsilon_0 + 1} E_0$
   (c) $\dfrac{E_0}{\varepsilon_0}$ \hspace{3.5cm} (d) zero \hspace{4cm} **[b]**

6. In a charge free region, Poisson's equation is

(a) $\nabla^2 V = -\dfrac{\rho}{\varepsilon}$

(b) $\nabla^2 V = 0$

(c) $\nabla^2 V = E$

(d) $\nabla^2 V = \dfrac{\rho}{\varepsilon} P$ **[b]**

7. In cartesian coordinates a potential is a function of x only. At $x = -2.0$ cm, $V = 25.0$ V and $E = 1.5 \times 10^3 (-\hat{a}_x)$ V/m throughout the region. V at $x = 3.0$ cm is

(a) 10 V

(b) 100 V

(c) 1000 V

(d) 1 V **[b]**

8. The electric field strength at a point P due to a point charge $+q$ coulomb located at the origin, is 100 μV/m. If the point charge is now enclosed by a perfectly conducting metal sheet sphere whose centre is at origin, then the electric field strength at point P outside the sphere, becomes

(a) zero

(b) 100 μV/m

(c) -100 μV/m

(d) 50 μV/m **[GATE] [b]**

9. An ideal capacitor is charged to a voltage V_0 and connected at $t = 0$ across an ideal inductor L. (The circuit now consists of a capacitor and inductor alone). If $\omega_0 = 1/\sqrt{LC}$, the voltage across the capacitor at $t > 0$ is given by

(a) V_0

(b) $V_0 \cos \omega_0 t$

(c) $V_0 \sin \omega_0 t$

(d) $V_0 e^{-\omega_0 t} \cos \omega_0 t$ **[GATE] [c]**

10. The electric field E (in V/m) at the point $(1, 1, 0)$ due to a point charge of $+1$ μC located at $(-1, 1, 1)$ (coordinates in metres) is

(a) $\dfrac{10^{-6}}{20\sqrt{5}\,\pi\varepsilon_0}(2\hat{i} - \hat{k})$

(b) $\dfrac{10^{-6}}{2\pi\varepsilon_0}(2\hat{i} - \hat{k})$

(c) $\dfrac{-10^{-6}}{20\sqrt{5}\,\pi\varepsilon_0}(2\hat{i} - \hat{k})$

(d) $\dfrac{-10^{-6}}{2\pi\varepsilon_0}(2\hat{i} - \hat{k})$ **[GATE] [d]**

11. In the finite plane, $y = 6$, there exists a uniform surface charge density of 1600 μC/m². The associated electric field strength is

(a) $30\hat{i}$ V/m

(b) $30\hat{j}$ V/m

(c) $30\hat{k}$ V/m

(d) $60\hat{i}$ V/m **[GATE] [b]**

12. The plane $y = 3$ m contains a uniform charge distribution of density $\sigma = \dfrac{10^{-8}}{6\pi}$ C/m². The electric field at all points is

(a) $30\hat{j}$ V/m

(b) $30\hat{i}$ V/m

(c) $10\hat{k}$ V/m

(d) $50\hat{i}$ V/m **[a]**

13. A dielectric slab of flat surface with $\varepsilon_r = 4$ is disposed with surface normal to flux density 1.5 C/m². Slab occupied a volume of 0.08 m³ and is uniformly polarized. The polarization in the slab and the total dipole moment are

(a) 3 C/m², 0.1 C m

(b) 1.125 C/m², 0.09 C m

(c) 1.5 C/m², 0.5 C m

(d) 0.5 C/m², 0.2 C m **[b]**

14. Given field $E = -\dfrac{16}{r^2}\hat{a}_r$ in spherical coordinates. The potential of the point $A\,(2m, \pi,$ $\pi/2)$ with respect to $B(4m, 0, \pi)$ is

   (a) $-1\,V$ (b) $-2\,V$

   (c) $-3\,V$ (d) $-4\,V$ **[d]**

15. A total charge of $\dfrac{40}{3}$ nC is uniformly distributed in the form of a circular disc of radius $2\,m$. The potential due to this charge at a point on the axis $2\,m$ from the disc is

   (a) $15\,V$ (b) $30\,V$

   (c) $49.7\,V$ (d) $33.7\,V$ **[c]**

6

Magnetostatics

6.1 INTRODUCTION

So far we have dealt with the electric charges that are stationary. Now, we shall consider the motion of charges and the forces associated with them. The concept of electric field has been developed starting from the experimental observation that a charge brought into the vicinity of other charges experiences a force. One can also experimentally determine that a current element (e.g. a small loop carrying current) will be acted on by force if it is brought in the vicinity of another current or system of currents. The region in which such forces exist is often called as a *region of magnetic field*. This concept may appear to exclude the well-known magnetic effects arising from permanent magnets, but these effects may be included conceptually if one thinks of them as arising from groups of atomic currents in the ferromagnetic material. This aspect will be discussed later in this chapter. Obviously, similar to a static electric charge which has an electric field, an electric current, on the other hand, will possess a magnetic field.

6.2 ELECTRIC CURRENT

An electric current consists of a stream of charged particles or ions. This applies to the ions in an accelerator of any kind, to those in an electrostatic solution, to those in an ionized gas, or plasma, or to the electrons in metallic conductors. We have already studied that when a conductor is placed in an electric field, the electrons acquire an average velocity in the direction opposite to the field. This results into a net drift of charges and one can say that there is an electric current flowing in the conductor. Obviously, for an electric current to be produced, an electric field must be applied to move the charged particles in a well-defined direction. We must remember that the conductor carrying the current is electrically neutral since the number of inducting electrons is balanced by the number of positively charged atoms.

Let us consider that there are N charges per unit volume at a point P and each carries a charge q and moves with mean velocity u. The amount of charge crossing unit area perpendicular to mean drift velocity per unit time

$$J = Nqu \qquad (6.1)$$

where, J is called the current density and it is defined as the electric current (I) held normal to the local velocity of the current carriers. It is a vector quantity and the directions of J and u are identical, i.e. the direction of J is the direction in which the positive charges move. If ds is an element of area in the conductor (or say material), the amount of charge flowing across that area in unit time is $J \cdot \hat{a}_n ds$, where $\hat{a}_n$ is the unit vector normal to ds. Obviously, this is the flux of J across ds.

As stated above, the electric charge in motion makes an electric current. The net charge that passes through the surface per unit time is called the electric current and is usually denoted by I. Thus, the intensity of electric current is

$$I = \frac{Nq}{t} = \frac{Q}{t} \tag{6.2}$$

Actually, Eq. (6.2) gives the average current in time t; the instantaneous current is

$$I = \frac{dQ}{dt} = \int J \cdot \hat{a}_n ds \tag{6.3}$$

The electric current is expressed in coulomb per second (C/s), a unit called the ampere (abbreviated as A). An ampere is the intensity of electric current corresponding to a charge of one coulomb passing through a section of material.

Let us consider a closed surface S enclosing a volume V. If ρ is the volume density of charge, then the total charge within the volume is $\int \rho dV$. Since the charge is conserved, i.e. charge can neither be created nor destroyed, it follows that the net flow of the charge, if any, out of this volume must be equal to the rate of decrease of the total charge inside the volume. This can be expressed as

$$\int J \cdot \hat{a}_n ds = -\frac{d}{dt} \int \rho dV \tag{6.4}$$

Since the surface S is fixed in space and hence the rate of change of the charge within the volume is solely due to the time variation of ρ. Obviously, we have

$$\int_S J \cdot \hat{a}_n ds = \int \frac{\partial \rho}{\partial t} dV \tag{6.5}$$

Making use of divergence theorem, we transform the surface integral into a volume integral,

$$\int_S J \cdot \hat{a}_n ds = \int_V \nabla \cdot J dV = -\int \frac{\partial \rho}{\partial t} dV$$

$$\therefore \quad \int_V \left(\nabla \cdot J + \frac{\partial \rho}{\partial t} \right) dV = 0 \tag{6.6}$$

Since the above equation is true for any volume V, we can equate the integrands on both sides and obtain the so called *equation of continuity*,

$$\nabla \cdot J + \frac{\partial \rho}{\partial t} = 0 \tag{6.7}$$

We must remember that the above equation is based on the law of conservation of charge. In most practical situations of interest, the charge density is constant in time at each point. This means $\frac{\partial \rho}{\partial t}$ is zero at each point. This is called a *steady state*. Thus, for steady state conditions, the equation of continuity gives

$$\nabla \cdot J = 0 \tag{6.8}$$

Equation (6.8) is valid in the region which does not contain a source or a sink of current.

6.3 SURFACE CURRENT DENSITY

In electrostatics, we have seen that a very thin slab of thickness t containing a volume charge density ρ may considered to be a surface charge density of value

$\sigma = \rho t \dfrac{C}{m^2}$. Similarly, in current electricity one can define a

Fig. 6.1: Surface electric current in *xy* plane

surface current density K as a vector field on the surface. The direction of K is the direction of velocity of the current carriers (assumed positively charged) at that point and the magnitude of K is the amount of charge crossing a unit length held normal to the local direction of K. The total rate at which the charge is crossing the curve of length L from A to B (Fig. 6.1) can be expressed as

$$\frac{dQ}{dt} = \int_A^B K\, dl \sin\theta = \int_A^B |K \times dl| \tag{6.9}$$

One can see from Fig. 6.1 that $dl\sin\theta$ is the projection of dl normal to K. Obviously, the integral of $K\, dl \sin\theta$ over appropriate limits, gives the current crossing the given curve.

6.4 ELECTRICAL CONDUCTIVITY: OHM'S LAW

We have read that when an electric field is applied to a dielectric, a polarization of the dielectric results. But if the field is applied in a region when free charges exist, the charges are set in motion and an electric current instead of polarization of the medium results. The charges are accelerated by the field and therefore gain energy.

When free charges are present within a body, such as electrons in a metal, their motion is hindered by the interaction with the positive ions that form the crystal lattice of the metal. Let us consider, for example, a metal with the positive ions regularly arranged in three dimensions, as shown in Fig. 6.2. The free electrons move in an electric field that exhibits the same periodicity as the lattice, and during their motion they are very frequently scattered by the field. One can describe this type of electronic motion with the help of methods of quantum

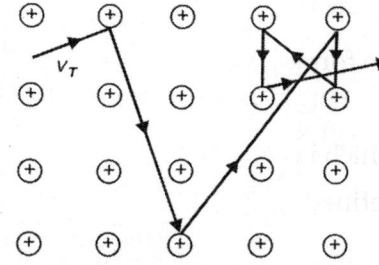

Fig. 6.2: Electron motion through the crystal lattice of a metal. v_T represents the thermal velocity of the electrons

mechanics. Because the electrons are moving in all directions, no net charge transport or electric current results. However, if an external electric field is applied, a drift motion is superposed on the natural random motion of the electrons and an electric current results. Experimentally, it is found that for many materials, the current density in the steady state is linearly proportional to the applied field E, i.e.

$$J \propto E \text{ or } J = \sigma E \tag{6.10}$$

where, σ is a constant called the *electrical conductivity* of the material. It is expressed in $\Omega^{-1}m^{-1}$ or $m^{-3}kg^{-1}sC^2$. σ is a constant for a material at a given temperature. The reciprocal of the conductivity of a material is called its *resistivity* (ρ). It is also called the *specific resistance* of the material. The resistivity ρ and therefore σ is a characteristic property of a given material at a given temperature. In terms of ρ, Eq. (6.10) can be written as

$$E = \rho J \qquad (6.11)$$

Equation (6.10) states that the proportionality constant σ between J and E is the same for different values of E and and that the two vectors are parallel in an isotropic, homogeneous material. This is found to be true for metallic conductors over a wide range of experimental conditions. It forms the basis of Ohm's law.

One obtains the steady current I

$$I = \int_S J \cdot \hat{a}_n dS = \sigma \int_S E \cdot \hat{a}_n ds \qquad (6.12)$$

If the conductor has uniform cross-sectional area A, we have

$$I = \sigma A E \qquad (6.13)$$

Let V be the potential difference maintained across the ends of a conductor of length l, then the electric field within the conductor will be uniform and given by

$$E = \frac{V}{l} \qquad (6.14)$$

$\therefore$

$$I = \frac{\sigma A V}{l}$$

This leads to

$$\frac{V}{I} = \frac{l}{\sigma A} = \frac{\rho l}{A} \qquad (6.15)$$

The ratio $\dfrac{V}{I}$ is called the *resistance* of the conductor and Eq. (6.15) is often written in the form

$$V = IR \qquad (6.16)$$

which is known as *Ohm's law*. The unit of resistance, $R = \dfrac{\rho l}{A}$ is ohm (symbol Ω) and it is defined as

$$1 \text{ ohm} = \frac{1 \text{ volt}}{1 \text{ ampere}} = \frac{1 \text{ V}}{1 \text{ A}} \qquad (6.17)$$

Experimentally, it is found that the resistivity ρ of a material generally increases with temperature. The fractional change in ρ for 1°C rise in temperature compared to the value of ρ at 0°C is called the *temperature coefficient of resistivity*. It is generally denoted by α. Thus, we have

$$\rho_t = \rho_0 (1 + \alpha t) \qquad (6.18)$$

where, t is the temperature in degrees centigrade. One can easily see that that the resistance $R = \dfrac{\rho l}{A}$ also has the same temperature coefficient α and one can write

$$R_t = R_0 (1 + \alpha t) \qquad (6.19)$$

α is about 0.004°C^{-1} for many metals. The materials obeying Ohm's law [Eq. (6.16)] are said to be *ohmic*. Table 6.1 gives the value of σ, ρ and α for some common materials.

Table 6.1: Conductivity (σ), resistivity (ρ) and temperature coefficient of resistivity (α) for some substances

Material	σ (at 20°C) in mho/m	ρ (at 20°C) in Ωm	α °C^{-1}
Aluminium	3.54×10^7	0.282×10^{-7}	0.004
Copper	5.8×10^7	0.172×10^{-7}	0.004
Silver	6.14×10^7	0.162×10^{-7}	0.0038
Mercury	0.01×10^7	100×10^{-7}	0.0009
Tungsten	1.81×10^7	0.552×10^{-7}	0.0045
Brass (66% Cu + 33% Zn)	1.39×10^7	0.719×10^{-7}	0.002
Germanium (300 K)	0.022	45	Semiconductor
Silicon (300 K)	4.35×10^{-6}	2.3×10^5	Semiconductor
Glass	1.1×10^{-12}	9×10^{14}	
Paraffin oil	1×10^{-14}	1×10^{14}	

Ohm's law essentially states that the resistance R is independent of the current density in the material. However, there are many instances when Ohm's law fails. One may cite the example of a space charge limited diode. I–V characteristics for a vacuum diode has the form shown in Fig. 6.3. When the anode is positive with respect to the cathode, the diode is said to be forward biased. Obviously, there cannot be a current in the reversed biased condition because the electrons are repelled by the cathode. We know that the diode is not a metallic conductor of the type for which Eq. (6.10) is normally used. Still one can represent its behaviour by its I–V characteristics, which indicates that in the reversed biased condition, this device offers an *infinite resistance* to the passage of current. We all know that this property of the diode is utilized in rectifiers to convert alternating current into direct current. One can easily see that the potential at a point in the space charge limited region between the electrodes is given by

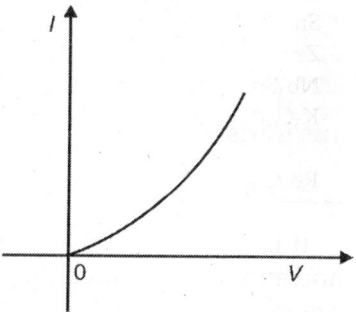

Fig. 6.3: I–V characteristic of a vacuum diode

$$V = \left(\frac{3xm^{1/4}}{2^{5/4} e^{1/4} \varepsilon_0^{1/2}} \right)^{4/3} J^{2/3} \qquad (6.20)$$

The current density J is given by

$$J = \frac{I}{A}$$

$\therefore$
$$V = \left(\frac{3xm^{1/4}}{2^{5/4} e^{1/4} \varepsilon_0^{1/2}} \right) \frac{I^{2/3}}{A^{2/3}}$$

or
$$I = \frac{4}{9} \varepsilon_0 A \left(\frac{2e}{m} \right)^{1/2} \frac{V^{3/2}}{x^2} \qquad (6.21)$$

i.e.
$$I \propto V^{3/2} \qquad (6.22)$$

Obviously, Eq. (6.22) is not in conformity with Ohm's law. One can also find such non-ohmic behaviour at semiconductor junctions.

There are also few elements and alloys which exhibit zero resistivity, i.e. infinite conductivity at very low temperatures. The temperature at which these materials exhibit the property of zero resistivity is called critical temperature (T_c) and the phenomenon is known as *superconductivity*. This phenomenon was first observed in the laboratory in 1911 by Kamerlingh Onnes. Today the highest known critical temperature (T_c) is 134K of $HgBa_2Ca_2Cu_3O_8$. Table 6.2 lists some super-conducting metals and alloys along with their transition temperatures.

Table 6.2: Some superconducting materials and their transition temperatures

Material	$T_c(K)$	Material	$T_c(K)$
Nb	9.3	$La_{2-x}M_xCuO_{4-y}$ (M = Ba, Sr, Ca) $x \simeq 0.15, y \rightarrow$ small	38
Pb	7.2		
Ta	4.5	$HgBa_2Ca_2Cu_3O_{1+x}$	133
Sn	3.7	Y_9Co_7	3
Zr	0.8	UBe_{13}	0.85
Nb_3Sn	18.0	UPd_2Al_3	2.0
K_3C_{60}	19.3	$BaPb_xBi_{1-x}O_{3-y}$ ($0.05 \leq x \leq 0.3$)	13
Rb_3C_{60}	29.6		

It is now believed that many more high temperature superconductors must exist and await discovery. Superconductors find several applications. Some important fields are magnets, electronics and small devices, computers and information processing, energy related fields, e.g. magnetic energy storage, production by magnetic fusion and MHD, electric power transmission and transportation, e.g. high speed trains (MAGLEV), ship drive system, etc.

6.5 MAGNETIC INTERACTION

Another type of interaction observed in nature is the one called as magnetic interaction. Magnetic phenomena were known since ancient times. Before the means to produce steady electric currents became available, the study of magnetism was confined to naturally occurring magnetic objects such as lodestones. Lodestones have the property of attracting small pieces of iron. The property is exhibited in the natural state by iron, cobalt and manganese, and by many compounds of these metals. The regions of a body where magnetism appears to be concentrated are called *magnetic poles*. A magnetized body is called a *magnet*.

The earth itself is a huge magnet. When we suspend a magnetized rod at any point on the earth's surface and allow it to rotate freely about the vertical, the rod orients itself so that the same end always points toward the north geographic pole. This result shows that the earth is exerting an additional force on magnetized rod which it does not exert on unmagnetized rods. This experiment also suggests that there are two kinds of magnetic poles, viz. north (*N*) and south (*S*). The interaction between like magnetic poles is repulsive and that between unlike magnetic poles is attractive.

When we attempt to measure the strength of a magnetic pole by defining a *magnetic mass* or charge, and investigate the dependence of the magnetic interaction on the distance between the poles, a fundamental difficulty arises. Although we have been able to isolate positive and negative electric charges and associate a definite amount of electric charge with the fundamental particles constituting all atoms, we have not been able to isolate a magnetic pole or identify a fundamental particle having only one kind of magnetism, either *N* or *S*. Magnetized bodies always exhibit poles in pairs, equal and opposite. On the other hand, the notions of magnetic pole and magnetic mass have been found unnecessary for the description of magnetism. Electric and magnetic interactions are very closely related, and in fact are only two different aspects of one property of matter, its electric charge; *magnetism is a manifestation of electric charges in motion*. Electric and magnetic interactions are considered together under the general name of electromagnetic interaction. A magnetic field is one of the forms of an electromagnetic field. Its distinguishing feature is that it acts only on moving particles and bodies having an electric charge, as well as on magnetized bodies regardless of their state of motion. A magnetic field is produced by current carrying conductors, by moving electrically charged particles and bodies, by magnetized bodies, or by a variable electric field (by displacement currents). Experiments by Biot and Savart, Ampere and others helped to establish the laws of magnetism as we know them today.

6.6 MAGNETIC FORCE ON A MOVING CHARGE

Since we observe interactions between magnetized bodies, we may say in analogy with the gravitational and electrical cases, that a magnetized body produces a *magnetic field* around it. When we place an electric charge at rest in a magnetic field, no special force or interaction is observed on the charge. But when an electric charge moves in a region where there is a magnetic field, a new force is observed on the charge in addition to those due to its gravitational and electrical interactions.

By measuring, at the same point in a magnetic field, the force experienced by different charges moving in different ways, one may obtain a relation between the force, the charge and its velocity. In this way we may obtain that *the force exerted by a magnetic field on a moving charge is proportional to the electric charge and its velocity, and the direction of the force is perpendicular to the velocity of the charge.*

The force characteristic of a magnetic field is the vector of magnetic induction *B*. It is defined through its action on a moving charge *q* moving with velocity *v* in a region where *B* exists. Such a charged particle experiences a force given by

$$F = qv \times B \qquad (6.23)$$

Obviously, *B* describes a property that is characteristic of the magnetic field, and one may call it the magnetic field strength. Another name, imposed by usage, is magnetic induction.

When *F* is measured in newton, *q* in coulomb and *v* in m/s, the dimensions of *B* are Ns/Cm. This unit of *B* in MKSA system of units is called 1 weber/m^2 = 1 tesla = 1 Ns/Cm. The more familiar unit in CGS system, viz. 1 gauss, equals 10^{-4} tesla. The earth produces a magnetic induction of about 0.36×10^{-4} tesla on its surface.

When the particle moves in a region where there are electric and magnetic fields, the total force is the sum of the electric force qE and the magnetic force $qv \times B$, that is,

$$F = q(E + v \times B) \qquad (6.24)$$

This expression is called the *Lorentz force*.

Because of the property of vector product, Eq. (6.23) gives a force perpendicular to the velocity v, but also perpendicular to the magnetic field B. Equation (6.23) also implies that when v is parallel to B, the force F is zero. In fact it is observed that at each point in every magnetic field there is a certain direction of motion in which no force is exerted on the moving charge. This direction is defined as the direction of the magnetic field at the point. In Fig. 6.4, the relation between the three vectors v, B and F is illustrated for both a positive and a negative charge. The direction of the force can be determined by the right hand rule as shown in Fig. 6.4.

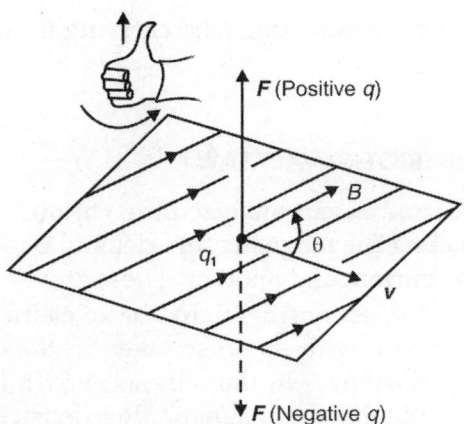

Fig. 6.4: Vector relation between magnetic force, magnetic field, and charge velocity. The force F is perpendicular to the plane containing B and v

If θ is the angle between v and B, the magnitude of F is

$$F = . qvB \sin\theta \qquad (6.25)$$

The maximum force occurs when $\theta = \dfrac{\pi}{2}$ or v is perpendicular to B, resulting in

$$F = qvB \qquad (6.26)$$

The minimum force F equal to zero, occurs when $\theta = 0$ or when v is parallel to B.

Because the magnetic force $F = qv \times B$ is perpendicular to the velocity, its work is zero, and therefore, it does not produce any change in the kinetic energy of the particle. Although the magnetic force is not conservative but when a particle moves in combined electric and magnetic fields, its total energy (i.e. kinetic + potential due to its different interactions) remains constant.

The Lorentz force law [Eq. (6.24)] is found to be true even for particles moving at speeds close to the speed of light. This means that the charge on a body is independent of its speed. Obviously, the charge, like rest mass, is relativistically invariant.

6.7 FORCE ON A CURRENT

Let us now consider the situation, when we have a number of moving charges, for example, conduction electrons in metals. Let the number of electrons per unit volume be N, then the force per unit volume will be

$$F = Nqv \times B = J \times B \qquad (6.27)$$

(if electric field E is assumed to be zero). The total force

$$F_{tot} = \int_v J \times B d\tau = \int_s \int_l (J \times B)(dl \cdot \hat{a}_n ds) \qquad (6.28)$$

where, ds is the cross-section of the conductor, and dl an element of its length, with its direction the same as that of the current. We have that J and dl are parallel, and hence

$$F_{tot} = \int_s \int_l (dl \times B)(J \cdot \hat{a}_n ds) = \int I dl \times B \qquad (6.29)$$

If the conducting wire carrying the current is in the form of a closed loop, we have

$$F = \oint I dl \times B = I \oint dl \times B \qquad (6.30)$$

6.8 BIOT–SAVART LAW

Biot and Savart analyzed the various experiments and formulated appropriate laws relating the magnetic flux density B to the current and also the law of force between one current and another. These are:

i. Let us consider that dl be an element of length of a wire with its sense taken in the direction of the current I_1 in the wire as shown in Fig. 6.5. One obtains the magnetic flux density dB due to this element at a point P specified by the location vector r as

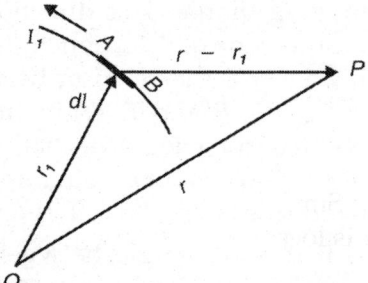

$$dB = \frac{\mu_0}{4\pi} \frac{I_1 dl_1 \times (r - r_1)}{|r - r_1|^3} \qquad (6.31)$$

where, r_1 is the position or location vector of the element dl. Obviously, the flux density is directly proportional to the current flowing in the circuit and to the length of the element of the wire and inversely proportional to the square of the distance of the point from the element of the wire. Thus, we have

Fig. 6.5: Magnetic field due a current element

$$B = \frac{\mu_0 I_1}{4\pi} \int \frac{dl_1 \times (r - r_1)}{|r - r_1|^3} \qquad (6.32)$$

The integration extends for the whole length of the current element. The constant μ_0 is termed the *permeability of free space* and its value is exactly $4\pi \times 10^{-7}$ NA^{-2} or $4\pi \times 10^{-7}$ H/m ($= 1.3566 \times 10^{-6}$ m kg C^{-2}). Equations (6.31) and (6.32) are statements of the Biot–Savart law.

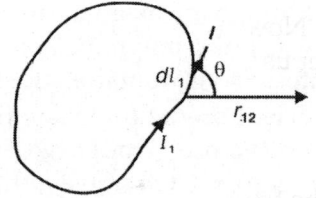

For a closed loop of current as shown in Fig. 6.6, the resultant magnetic induction at P is

Fig. 6.6: Closed loop of current

$$B = \frac{\mu_0}{4\pi} \oint \frac{I_1 dl_1 \times r_{12}}{r_{12}^3} \qquad (6.33)$$

where, the integration is taken over the entire closed *loop circuit*.

ii. If there are two closed circuits as shown in Fig. 6.7, the force on circuit 2 due to circuit 1 is

$$F_2 = \oint I_2 dl_2 \times B_2 \qquad (6.34)$$

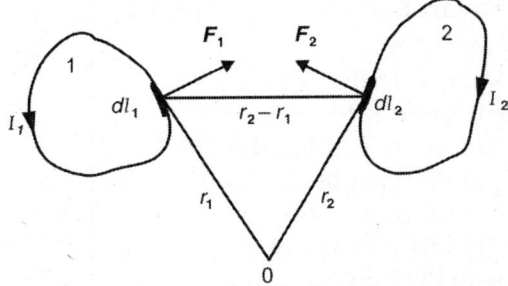

Fig. 6.7: Two closed loop circuits

where, B_2 is the magnetic flux density due to circuit at the position where circuit 2 is located as shown in (Fig. 6.7). Let us assume that B_2 be constant over the region occupied by the circuit. Thus, we have

$$F_2 = \oint_2 I_2 dl_2 \times \frac{\mu_0}{4\pi} \oint_1 \frac{I_1 dl_1 \times (r_2 - r_1)}{|r_2 - r_1|^3}$$

$$= \frac{\mu_0}{4\pi} I_1 I_2 \oint_2 \oint_1 \frac{dl_2 \times [dl_1 \times (r_2 - r_1)]}{|r_2 - r_1|^3} \tag{6.35}$$

Similarly, one obtains the force exerted by the circuit 2 at the position where circuit 1 is located,

$$F_1 = \frac{\mu_0}{4\pi} I_1 I_2 \oint_1 \oint_2 \frac{dl_1 \times [dl_2 \times (r_1 - r_2)]}{|r_1 - r_2|^3} \tag{6.36}$$

However, Eqs. (6.35) and (6.36) seem to give rise to an unacceptable solution. We must expect that the forces F_2 and F_1 be equal and opposite. However, we have

$$\frac{dl_2 \times [dl_1 \times (r_2 - r_1)]}{|r_2 - r_1|^3} \neq \frac{dl_1 \times [dl_2 \times (r_1 - r_2)]}{|r_1 - r_2|^3}$$

i.e. $$F_2 \neq F_1$$

Now the question arises, where this violates Newton's law. It seems that it does. Let us now expand the integrand in Eq. (6.35).

$$F_2 = \frac{\mu_0}{4\pi} I_1 I_2 \oint_2 \oint_1 \left[\frac{\{dl_2 \cdot (r_2 - r_1)\} dl_1}{|r_2 - r_1|^3} - \frac{(dl_1 \cdot dl_2)(r_2 - r_1)}{|r_2 - r_1|^3} \right]$$

The first integral vanishes,

$$\oint_2 \frac{dl_2 (r_2 - r_1)}{|r_2 - r_1|^3} = -\oint \nabla_2 \left(\frac{1}{|r_2 - r_1|} \right) \cdot dl_2 = 0$$

(We must remember that ∇_2 operates on the terms involving r_2). One obtains

$$F_2 = -\frac{\mu_0}{4\pi} I_1 I_2 \oint_1 \oint_2 \frac{(dl_1 \cdot dl_2)(r_2 - r_1)}{|r_2 - r_1|^3} = -F_1 \tag{6.37}$$

When we compare with the expression obtained from Eq. (6.36) by similar transformations, we find that it is in conformity with Newton's law.

Now we apply Biot–Savart law to calculate *the force between parallel wires carrying currents.*

Figure 6.8 shows two parallel wires distant a apart and carrying current I_1 and I_2 respectively. Let us assume that the first wire is along the z-axis and second wire is parallel to first wire and passing through point $(a, 0, 0)$ as shown in Fig. 6.8. Let us take a point $P(a, 0, z)$ and calculate the field at this point due to the current I_1 in the first wire. Let r be the distance of the point P from the origin.

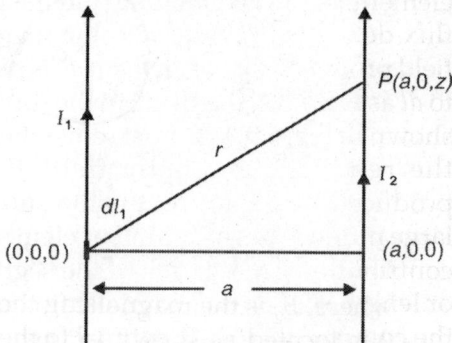

Fig. 6.8: Parallel wires carrying currents

Let us consider an element dl_1 of the first wire at the origin (Fig. 6.8). One easily see that its components are $(0, 0, dz)$ and those of r are $(a, 0, z)$. Obviously, the components of $dl_1 \times r$ are $(0, adz, 0)$. This clearly reveals that the magnetic field dB at P due to the element dl is parallel to y-axis. Obviously, B also will be parallel to y-axis components $B_x = 0$, $B_z = 0$ and

$$B_y = \frac{\mu_0}{4\pi} I_1 \int_{-\infty}^{\infty} \frac{|dl \times r|}{|r|^3} = \frac{\mu_0}{4\pi} I_1 \int_{-\infty}^{\infty} \frac{adz}{(a^2 + z^2)^{3/2}}$$

Now, putting $z = a \tan \theta$, one obtains

$$B_y = \frac{\mu_0}{4\pi} I_1 \int_{-\pi/2}^{\pi/2} \frac{a \times a \sec^2 \theta}{a \sec^3 \theta} d\theta = \frac{2\mu I_1}{4\pi a} = \frac{\mu_0 I_1}{2\pi a} \qquad (6.38)$$

From Eq. (6.38), it is evident that B is directly proportional to I_1 and inversely proportional to a. Moreover, B will always be normal to the plane containing the wire and the radius vector r and hence the lines of constant B form closed circles with their centres on the wire in planes perpendicular to the current.

Let us now find out the force dF acting on an element dl_2 of the second wire due to the current in the first wire. Force is given by Eq. (6.29). Now, we see that B is in positive y-direction and dl_2 in the z-direction, the force acting on dl_2 will be in the negative x-direction, its component in this direction being

$$dF_x = -\frac{\mu_0 I_1 I_2 dl_2}{2\pi a} \qquad (6.39)$$

This result indicates that if the currents are in the same direction, the force is one of attraction. If the currents are in the opposite direction, the force is one of repulsion. This result confirms the Ampere's observations.

The above result can be extended to currents of any configuration. These interactions among currents are of great practical importance for electric motors and other engineering applications.

6.9 MAGNETIC FIELD INDUCTION ON THE AXIS OF A CIRCULAR LOOP

Let us consider a circular loop of radius a, carrying current I ampere with its direction perpendicular to the plane of the paper (Fig. 6.9). Let P be a point on the axis of this circular loop at a distance r from an element dl. Let us calculate the magnetic flux density at P. Since dB, the magnetic field produced by the element dl, is normal to dl as well as r, it will act in the direction shown in Fig. 6.9. We must remember that the resultant magnetic field **B** at P produced by the current is the sum of a large number of very small or elementary contributions dB by each of the segments

Fig. 6.9: Magnetic field due to a current carrying circular loop

or length elements dl composing the circuit. On integrating round the coil, the sum of the components of dB, normal to the axis, is zero. One obtains the component parallel to the axis as

$$dB = \frac{\mu_0}{4\pi} \frac{Idl}{r^2} \cos\phi$$

where, ϕ is the angle between the magnetic field dB and the axis. Now integrating the above, we obtain

$$B = \frac{\mu_0 I}{4\pi r^2} \oint \cos\phi \, dl = \frac{\mu_0 I \cos\phi}{4\pi r^2} 2\pi a \quad \left(\because \oint dl = 2\pi a \right)$$

or
$$B = \frac{\mu_0 Ia}{2r^2} \frac{a}{r} \quad \left(\because \cos\phi = \frac{a}{r} \right)$$

$$= \frac{\mu_0 Ia^2}{2r^3} \tag{6.40}$$

Noting that $r = (a^2 + z^2)^{1/2}$, we can write the magnetic field for points on the axis of a circular current as

$$B = \frac{\mu_0 Ia^2}{2(a^2 + z^2)^{3/2}} \tag{6.41}$$

The magnetic field of a circular current has been represented in Fig. 6.10. If the coil contains n turns, then

$$B = \frac{\mu_0 n Ia^2}{2(a^2 + z^2)^{3/2}} \tag{6.41a}$$

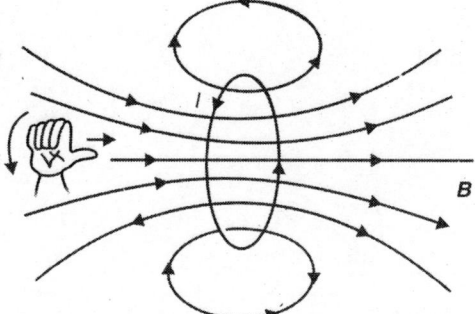

Fig. 6.10: Magnetic lines of force due to circular current

Different Cases

i. *The magnetic field due to induction at the centre of the circular coil:* At the centre of the circular coil, $Z = 0$, hence

$$B = \frac{\mu_0 I}{2a} \frac{Wb}{m^2} \tag{6.42}$$

If the coil contains n turns, then

$$B = \frac{\mu_0 n I}{2a} \tag{6.42a}$$

ii. *The magnetic field due to a small circular coil:* An interesting case occurs when the circuit is very small, so that the radius a can be neglected in comparison with the distance Z. Then Eq. (6.41) reduces to

$$B = \frac{\mu_0 I a^2}{2Z^3} \tag{6.43}$$

The magnetic dipole moment of the circuit is $M = I(\pi a^2)$. Then

$$B = \frac{\mu_0 M}{2\pi Z^3} = \frac{\mu_0 (2M)}{4\pi Z^3} \tag{6.44}$$

When we compare Eq. (6.44) with the dipole expression, with $\theta = 0$, i.e. $E_r = \frac{1}{4\pi\varepsilon_0}\left(\frac{2p}{r^3}\right)$, we see that the magnetic field along the axis of the small current is identical to the electric field of an electric dipole along its axis if we make $\left(\frac{\mu_0}{4\pi}\right)M$ correspond to $\frac{p}{4\pi\varepsilon_0}$. For that reason the circuit is called a *magnetic dipole*. Therefore, we can apply electric dipole results for a magnetic dipole so that the magnetic field off-axis may be computed (Fig. 6.11). One obtains

$$\left. \begin{array}{l} B_r = \dfrac{\mu_0}{4\pi}\dfrac{2M\cos\theta}{r^3} \\[3mm] B_\theta = \dfrac{\mu_0}{4\pi}\dfrac{M\sin\theta}{r^3} \end{array} \right\} \tag{6.45}$$

and

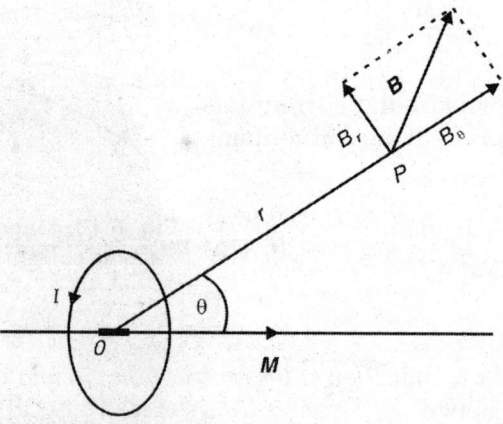

Fig. 6.11: Magnetic field at point P due to magnetic dipole current

We have seen that the lines of force of an electric field go from the negative to positive charges or perhaps, in some cases, from or to infinity. However, in the case of magnetism (Fig. 6.10), we may see that the lines of force of a magnetic field are closed lines, linked with the current. The reason for this is that the magnetic field is not originated by magnetic poles. A field of this kind, which does not have point sources, is called *solenoidal*.

6.10 MAGNETIC FIELD DUE TO SOLENOID

Solenoid is a cylindrically shaped coil consisting of a long wire wound usually on a non-magnetic frame in a closed-packed helix and carrying an electric current. A solenoidal current is a current composed of several coaxial circular loops of the same radius, all carrying the same current (Fig. 6.12). Such an arrangement may be used in electromag-netic relay and in some of the electrical instruments.

Figure 6.13 shows a solenoid of length L having N turns uniformly wound round a cylinder of radius a and carrying a current I. One can obtain its magnetic field by adding the magnetic fields of each of the component circular currents. The field is indicated by lines of magnetic force in Fig. 6.12, where some fluctuations in the space between loops have been smoothed out. Let us calculate the field at a point P on the axis of the solenoid. The point P is at a distance z_0 from O. Since the

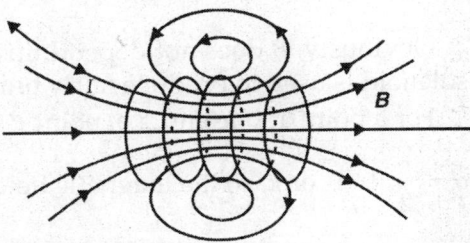

Fig. 6.12. Magnetic lines of force due to a solenoidal current

turns in the solenoid are closely wound, one may regard the current flowing uniformly round the cylinder. Let us consider that the length of the solenoid is divided into elements dz, such as the one shown at a distance z from O. The current in this elemental section dz of the coil is $\dfrac{N}{L}dzI$ and the field at axial point P due to this element is

$$dB = \frac{1}{2}\frac{\mu_0 NI}{L}\frac{a^2 dz}{[(z_0-z)^2+a^2]^{3/2}} \quad \text{[see Eq. (6.41)]} \tag{6.46}$$

$$\therefore \quad B = \frac{1}{2}\frac{\mu_0 NI}{L}\int_0^L \frac{a^2 dz}{[(z_0-z)^2+a^2]^{3/2}}$$

$$= \frac{1}{2}\frac{\mu_0 NI}{L}\left[\frac{z_0 - z}{\sqrt{(z_0-z)^2+a^2}}\right]_0^L$$

$$= \frac{\mu_0 NI}{2L}\left[\frac{z_0}{\sqrt{z_0^2+a^2}}+\frac{L-z_0}{\sqrt{(L-z_0)^2+a^2}}\right]$$

$$= \frac{\mu_0 NI}{2L}[\cos\alpha+\cos\beta] \tag{6.47}$$

where α and β are the angles subtended by the ends E' and F' of the solenoid at point P with axis (Fig. 6.13).

Fig. 6.13: Computation of the magnetic field at a point P located along the axis of a solenoidal current

If the solenoid is *very long*, i.e. for an *infinite solenoid* $\alpha = \beta = 0$

$$B = \frac{\mu_0 NI}{L} \tag{6.48}$$

Obviously, B does not depend upon the position of the point P and, hence a long solenoid is used to produce fairly uniform fields in limited regions around its centre.

For a point at one end, i.e. point P is at one end of the long solenoid, (say F'), then

$\alpha = \dfrac{\pi}{2}, \beta = 0$ or π, so the magnetic field at an end point of long solenoid is obtained as

$$B = \frac{\mu_0 NI}{2L} \tag{6.49}$$

or one-half the value at the centre. A long solenoid is used to produce fairly uniform magnetic fields in limited regions around its centre. The variation in magnitude of B from point to point over the complete length of a long solenoid is shown in Fig. 6.14.

Fig. 6.14: Variation of B over the complete length of a long solenoid

6.11 LAWS OF MAGNETOSTATICS

i. $\nabla \cdot B = 0$

One can express the expression for the magnetic flux density [Eq. (6.32)] in terms of current density. Thus, we have

$$B = \frac{\mu_0 I}{4\pi} \int \frac{dl_1 \times (r - r_1)}{|r - r_1|^3} = \frac{\mu_0}{4\pi} \int \frac{J(r_1) \times (r - r_1)}{|r - r_1|^3} \, dV$$

Here we have used

$$I dl_1 = J(r_1) \cdot \hat{a}_n ds = J(r_1) dv$$

$\therefore \qquad \boldsymbol{B} = -\dfrac{\mu_0}{4\pi}\int_V\left[\boldsymbol{J}(r_1)\times\nabla\left(\dfrac{1}{|\,r-r_1\,|}\right)\right]dV \qquad\qquad (6.50)$

Taking divergence, one obtains

$$\nabla\cdot\boldsymbol{B} = -\dfrac{\mu_0}{4\pi}\int_V\nabla\cdot\left[\boldsymbol{J}(r_1)\times\nabla\left(\dfrac{1}{|\,r-r_1\,|}\right)\right]dV$$

Using the vector identity $\nabla\cdot(\boldsymbol{A}\times\boldsymbol{C}) = \boldsymbol{C}\cdot(\nabla\times\boldsymbol{A}) - \boldsymbol{A}\cdot(\nabla\times\boldsymbol{C})$, one obtains

$$\nabla\cdot\boldsymbol{B} = -\dfrac{\mu_0}{4\pi}\int_V\nabla\left(\dfrac{1}{|\,r-r_1\,|}\right)\cdot\{\nabla\times\boldsymbol{J}(r_1)\}\,dV$$

$$+\dfrac{\mu_0}{4\pi}\int\boldsymbol{J}(r_1)\cdot\left\{\nabla\times\nabla\left(\dfrac{1}{|\,r-r_1\,|}\right)\right\}dV \qquad (6.51)$$

We can see that ∇ operates on r and hence the first integral in Eq. (6.51) is zero. The second term in Eq. (6.51) contains a factor of curl grad $\left(\dfrac{1}{|\,r-r_1\,|}\right)$, which is identically zero. Obviously,

$$\nabla\cdot\boldsymbol{B} = 0 \qquad\qquad (6.51a)$$

The divergenceless character of $\boldsymbol{B}$ is an important property. Equation (6.51) is one of the four equations of Maxwell that forms the basis of electromagnetic theory; it reflects the absence of magnetic monopoles. One can understand this by noting that the divergence of a vector field is a measure of its source density. We have already seen this result as Poisson's equation in electrostatics, i.e. $\nabla\cdot\boldsymbol{E} = \dfrac{\rho}{\varepsilon_0}$. Similarly, $\nabla\cdot\boldsymbol{B}$ would give us the local density of magnetic charges or monopoles. Since $\boldsymbol{B}$ is always divergenceless, it implies that no volume element contains magnetic charges of one sign only. We must note that the mathematical description of $\boldsymbol{B}$ is arranged in such a manner that $\nabla\cdot\boldsymbol{B}$ is zero. This is necessary to confirm with the fact that magnetic monopoles have not been observed so far. Equation (6.51) is the *first law of magnetostatics* corresponding to the electrostatic relation $\nabla\times\boldsymbol{E} = 0$. This relation shows that the magnetic field is **solenoidal** in contrast to the electric field which is **irrotational**.

ii. Ampere's law for the magnetic field (Ampere's circuital law)

Now, we determine the value of $\nabla\times\boldsymbol{B}$. This will help us to complete the analogy with electrostatics. We have

$$\boldsymbol{B} = \dfrac{\mu_0}{4\pi}\int_V\dfrac{\boldsymbol{J}(r_1)\times(r-r_1)}{|\,r-r_1\,|^3}\,dV$$

$$= -\dfrac{\mu_0}{4\pi}\int_V\left[\boldsymbol{J}(r_1)\times\nabla\left(\dfrac{1}{|\,r-r_1\,|}\right)\right]dV$$

$$= \frac{\mu_0}{4\pi} \nabla \times \int \frac{J(r_1)}{|r - r_1|} dV \qquad (6.52)$$

Taking curl of the above equation, one obtains

$$\nabla \times B = \frac{\mu_0}{4\pi} \nabla \times \left(\nabla \times \int \frac{J(r_1)}{|r - r_1|} dV \right)$$

Using the vector identity $\nabla \times (\nabla \times A) = \nabla (\nabla \cdot A) - \nabla^2 A$, one obtains

$$\nabla \times B = \frac{\mu_0}{4\pi} \nabla \left(\nabla \cdot \int \frac{J(r_1)}{|r - r_1|} dV \right) - \frac{\mu_0}{4\pi} \nabla^2 \int \frac{J(r_1)}{|r - r_1|} dV$$

$$= \frac{\mu_0}{4\pi} \nabla \int J(r_1) \cdot \nabla \left(\frac{1}{|r - r_1|} \right) dV - \frac{\mu_0}{4\pi} \nabla \int J(r_1) \nabla^2 \left(\frac{1}{|r - r_1|} \right) dV \qquad (6.53)$$

We have

$$\nabla^2 \left(\frac{1}{|r - r_1|} \right) = -4\pi\delta(r - r_1) \qquad (6.54)$$

and

$$\nabla \left(\frac{1}{|r - r_1|} \right) = -\nabla_1 \left(\frac{1}{|r - r_1|} \right) \qquad (6.55)$$

where, ∇_1 operates only on r_1.

Using Eqs. (6.54) and (6.55), Eq. (6.53) reduces to

$$\nabla \times B = -\frac{\mu_0}{4\pi} \nabla \int J(r_1) \cdot \nabla_1 \left(\frac{1}{|r - r_1|} \right) dV + \frac{\mu_0}{4\pi} \int J(r_1) 4\pi\delta(r - r_1) dV$$

$$= -\frac{\mu_0}{4\pi} \nabla \int J(r_1) \cdot \nabla_1 \left(\frac{1}{|r - r_1|} \right) dV + \mu_0 J(r)$$

$$= -\frac{\mu_0}{4\pi} \nabla \int \left[\nabla_1 \cdot \frac{J(r_1)}{|r - r_1|} - \frac{\nabla_1 \cdot J(r_1)}{|r - r_1|} \right] dV + \mu_0 J(r) \qquad (6.56)$$

We must remember that $\nabla(\phi A) = A \times \nabla\phi + \phi\nabla A$.

One can easily show with the help of divergence theorem by taking a large enough surface outside the current carrying region that the first term in Eq. (6.56) vanishes, and second is identically zero since $\nabla \cdot J = 0$ for steady state. Thus, one obtains

$$\nabla \times B = \mu_0 J \qquad (6.57)$$

This is *second law of magnetostatics*. This corresponds to $\nabla \cdot E = \dfrac{\rho}{\varepsilon_0}$ in electrostatics. One can easily find the integral from the above. We have

$$\int_S \nabla \times B \cdot \hat{a}_n ds = \mu_0 \int J(r) \cdot \hat{a}_n \, ds \qquad (6.58)$$

From Stokes' theorem,

$$\int_S \nabla \times \boldsymbol{B} \cdot \hat{a}_n ds = \oint \boldsymbol{B} \cdot d\boldsymbol{I} \tag{6.59}$$

Using Eq. (6.59), Eq. (6.58) reduces to

$$\oint \boldsymbol{B} \cdot d\boldsymbol{I} = \mu_0 \int \boldsymbol{J}(r) \cdot \hat{a}_n ds = \mu_0 I \tag{6.60}$$

where, S is any surface bounded by l and I stands for the total current linked by the path l.

This is Ampere's law for the magnetic field, i.e. *Ampere's circuital law*. The sign of integral in Eq. (6.60) depends upon the direction in which the current is encircled. The sign is positive, if the path followed for line integral is parallel to $\boldsymbol{B}$ and negative, if the path followed is antiparallel. According to Ampere's circuital law, the line integral of flux density round any closed path is equal to μ_0 times the current flowing through the area enclosed by the path. The fact that the circulation of the magnetic field $\boldsymbol{B}$ generally is not zero indicates that the magnetic field does not have a magnetic potential in the same sense that the electric field has an electric potential. The law further indicates that the magnetic field of a current is *non-conservative*. Ampere's law is particularly useful when we wish to compute the magnetic field produced by current distributions having certain geometrical symmetries.

We must note that in the integral statement of Ampere's law and also in the differential statement, the magnetic field $\boldsymbol{B}$ need not be that field only which is produced by the encircled current I (or the local current density $\boldsymbol{J}$). It may include the field produced by any other neighbouring currents. The closed line integral for the total $\boldsymbol{B}$ gives μ_0 times the current linked with the path of integration. One may compare this with Gauss's law in electrostatics. We have read that the surface integral $\oint_S$ of E over a closed surface gives $\dfrac{1}{\varepsilon_0}$ times q, the charge enclosed within that surface. The E that figures in the calculation of flux ϕ_E can be due to all charges, q included. The charges outside the closed surface make zero contribution to the integral that gives ϕ_E. One can summarize the laws of electrostatics and magnetostatics as follows:

$$\left. \begin{array}{ll} \text{Electrostatics} & \text{Magnetostatics} \\ \nabla \times E = 0 & \nabla \cdot B = 0 \\[2mm] \nabla \cdot E = \dfrac{\rho}{\varepsilon_0} & \nabla \times B = \mu_0 J \end{array} \right\} \tag{6.61}$$

6.12 MAGNETIC POTENTIALS

Magnetic Scalar Potential

Ampere's law states that the line integral of $\boldsymbol{B}$ for a closed path is equal to μ_0 times the current I encircled by the path. Obviously, when the current linkage is nonzero, the magnetic field along such a path must be considered as non-conservative. However, there can be situations where the current encircled is zero, e.g. when one is confined to a region outside of the current sources. In such a region, there is no possibility of any closed path being linked with a current. Obviously, this makes $\oint \mathbf{B} \cdot d\mathbf{l} = 0$ for all closed paths. But this is the situation of a conservative field. This means that the magnetic field in such a region can be described as a conservative field. Obviously, one can therefore express the magnetic field for such situations as the gradient of a scalar potential field.

In electrostatics, we have seen that once the electrostatic potential V is found, one can easily find $\mathbf{E}$ using the relation $\mathbf{E} = -\nabla V$. (Since $\nabla \times \mathbf{E} = 0$ and hence one can express $\mathbf{E}$ as the negative gradient of scalar quantity V, i.e. electrostatic potential). In case of magnetism, if $\mathbf{J} = 0$, one finds an analogous situation, i.e. $\nabla \times \mathbf{B} = 0$. We may, therefore, express $\mathbf{B}$ as a gradient of scalar quantity, ϕ_m (say). Thus, we have

$$\mathbf{B} = -\nabla \phi_m \tag{6.62}$$

Let us consider ϕ_m as *magnetic scalar potential*. Using the relation $\nabla \cdot \mathbf{B} = 0$, Eq. (6.62) yields,

$$\nabla \cdot \mathbf{B} = -\nabla \cdot \nabla \phi_m = -\nabla^2 \phi_m = 0$$

This shows that ϕ_m satisfies Laplace's equation. Let us consider a situation shown in Fig. 6.15. There is a circuit carrying a current I. One obtains the magnetic flux density at the point P due to the current I flowing in the circuit as

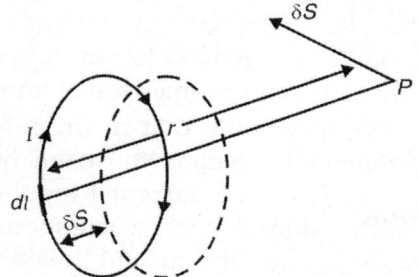

$$\mathbf{B} = \frac{\mu_0}{4\pi} I \int \frac{d\mathbf{l} \times \mathbf{r}}{|\mathbf{r}|^3}$$

Fig. 6.15: Magnetic flux density at P due to a circuit carrying current I

Let us consider that the point P is displaced through a distance δS, then the change in the potential will be

$$\delta \phi_m = -\mathbf{B} \cdot \delta S = -\delta S \cdot \frac{\mu_0 I}{4\pi} \int \frac{d\mathbf{l} \times \mathbf{r}}{|\mathbf{r}|^3}$$

$$= -\frac{\mu_0 I}{4\pi} \int \frac{\delta S \cdot (d\mathbf{l} \times \mathbf{r})}{|\mathbf{r}|^3} \tag{6.63}$$

We have taken δS inside the sign of integration as it is constant during integration.

Let us now consider that the point P is fixed and the circuit is displaced through a small amount $-\delta S$. Obviously, the change in the potential should be the same as above. In such a small displacement, one can easily see that an element $d\mathbf{l}$ of the circuit sweeps an area $\delta l \delta S \sin \theta$, where θ is the angle between $d\mathbf{l}$ and δS. One can easily express the area swept out as $-\delta S \times d\mathbf{l}$. One finds the angle subtended at P by this area as

$$d\Omega = \frac{\mathbf{r} \cdot (\delta S \times d\mathbf{l})}{|\mathbf{r}|^3} = -\frac{\delta S \cdot (d\mathbf{l} \times \mathbf{r})}{|\mathbf{r}|^3} \tag{6.64}$$

Using Eq. (6.64), one can express Eq. (6.63) as

$$\delta\phi_m = \frac{\mu_0 I}{4\pi} d\Omega$$

where, $d\Omega$ is the solid angle subtended by the area element at P. Obviously, the potential at P is obtained as

$$\phi_m = \frac{\mu_0 I}{4\pi} \Omega \tag{6.65}$$

Here Ω is the solid angle subtended at P by the boundary of the given current loop. Let us compare Eq. (6.64) with the potential of an electrostatic double layer, i.e.

$\phi(r) = \dfrac{D}{4\pi\varepsilon_0} d\Omega$. This reveals that the magnetic scalar potential has mathematically the

same properties as the electrostatic double layer potential.

We must note that the concept of magnetic scalar potential is of practical utility only to derive magnetic field in the absence of continuous current distributions. The scalar potential ϕ_m can be used if line integrals encircling any currents are considered or if the fields within current carrying media are desired. Now,

$$\int B \cdot dl = -\int \nabla\phi_m \cdot dl = -\phi_m \tag{6.66}$$

The quantity $-\phi_m$ is sometimes called as *magnetomotive force.*

Vector Potential

We have seen that the electrostatic field can be expressed as the negative gradient of a scalar potential ϕ, i.e. $E = -\nabla\phi$. Since the curl of a gradient is always zero, the above formulation immediately leads to $\nabla \times E = 0$, which is an expression of the conservative nature of the electrostatic field.

The question arises: can one define the magnetic induction B in terms of some potential function? One can easily see that the answer is yes, one can. The guiding equation for this purpose is

$$\nabla \cdot B = 0.$$

We know that

$$\text{div}(\text{curl } A) = \nabla \cdot \nabla \times A = 0$$

Obviously, one can express B as curl of A, i.e.

$$B = \nabla \times A \tag{6.67}$$

The vector A which satisfies Eq. (6.67) is known as the *vector potential.*

We find that A is not uniquely defined by Eq. (6.67). Let us add to A any function whose curl is zero, say gradient of a scalar ψ and still B is same.

$$\nabla \times (A + \nabla\psi) = \nabla \times A + \nabla \times (\nabla\psi) = \nabla \times A = B$$

We must remember that in electrostatics, the scalar potential V was not completely specified by the definition $E = -\nabla V$. If V is the potential for some problem, one would need a different potential $V' = V + b$, where b is a constant, gives the same field. Thus,

$$-\nabla V' = -\nabla(V + b) = -\nabla V - \nabla b = -\nabla V = E$$

One can have A more specific by imposing an additional restriction to it, without affecting B. One can easily take,

$$\nabla \cdot A = 0 \tag{6.68}$$

This is possible because it makes calculations easier than with any other choice in magnetostatics. In electrodynamics, one have to choose it differently.

Now, we have $B = \nabla \times A$ and $\nabla \cdot A = 0$. These two together define the vector potential A which satisfies the fundamental equation $\nabla \cdot B = 0$.

It is more appropriate to drop the word *magnetic* from *magnetic vector potential*, because in the case of time dependent phenomena, the potential A appears in the description of the electric field E also.

Now, we try to find a relation in magnetostatics corresponding to Poisson's equation $\nabla^2 V = \dfrac{\rho}{\varepsilon_0}$ in electrostatics. We have

$$\nabla \times B = \mu_0 J \text{ (Ampere's law)}$$

Also
$$\nabla \times B = \nabla \times \nabla \times A = \nabla(\nabla \cdot A) - \nabla^2 A = -\nabla^2 A$$

$$\nabla^2 A = -\mu_0 J \tag{6.69}$$

We can see that Eq. (6.69) is similar to Poisson's equation except that A is a vector. However, this should not cause any difficulty since each component of A must satisfy the differential equation. Thus,

$$\nabla^2 A_x = \mu_0 J_x, \nabla^2 A_y = -\mu_0 J_y \quad \text{and} \quad \nabla^2 A_z = -\mu_0 J_z \tag{6.70}$$

One finds the solution as

$$A_x = \frac{\mu_0}{4\pi} \int \frac{J_x}{|r|} dV, \text{ etc.} \tag{6.71}$$

One can write the general solution as

$$A = \frac{\mu_0}{4\pi} \int \frac{J}{|r|} dV \tag{6.72}$$

Obviously, one can compute the field produced by a current by first determining A and then substituting this value in $B = \nabla \times A$.

6.13 MAGNETIC DIPOLE

While discussing the field of an electric dipole having dipole moment p, we have considered it as a combination of two point charges, $+q$ and $-q$, separated by a distance a such that $p = qa$. However, this approach is ruled out for a magnetic dipole m unless one is willing to assume the existence of magnetic charges. This means m cannot be broken down into simpler objects.

One can easily show the equivalence of a current loop and a magnetic dipole by showing that the induction produced by a loop has the same form as the electric field produced by an electric dipole. In the case of the electric dipole, we have seen that the dumb-bell model, i.e. $+q$ and $-q$ separated by a distance a, produces a dipole field only for distance $r \gg a$. In the case of magnetic dipole, the corresponding condition is that

r, the distance from the loop should be large compared with the dimensions of the loop. The actual shape of the loop, assumed to be in one plane, does not come in the picture.

Let us consider a simple loop of wire in rectangular form carrying current I and suspended in vertical plane in an horizontal magnetic field B in such a way that the normal to the plane to the coil makes an angle θ with the direction of the field as shown in Fig. 6.16.

Let the cross-section of the wire be a. If N be the number of free electrons per unit volume, e the charge on the electron and u be the velocity of electrons, then the current flowing in the wire is

$$I = Neua \qquad (6.73)$$

Fig. 6.16: A rectangular wire loop suspended in vertical plane in a horizontal field **B**

The force acting on a small element dl of the side PS of the loop (Fig. 6.16) is obtained as

$$dF_1 = Neau\, dl_1 \times B = Idl_1 \times B \qquad (6.74)$$

We can easily see that dl_1 is in the direction of the current. Thus, the total force on the side PS is

$$F_1 = Il_1 \times B = Il_1 B\hat{a}_y \quad (\because \ PS \text{ is perpendicular to } B) \qquad (6.75)$$

Let us assume that current I is parallel to z-axis and field B is parallel to x-axis. The force F acts in the positive y-direction (Fig. 6.17).

One can see that Fig. 6.17 presents a view looking down on the coil. From Fig. 6.17, it is clear that force F_2, equal and opposite to F_1, acts on the side QR of the loop. We can easily say that in the same way there will be forces acting on the other two sides of the loop, i.e. PQ and SR which are equal and opposite. Obviously, the net force on the loop is zero and hence there will be no translational motion of the loop. However, there will be a torque τ acting on the coil. This torque will tend to turn the coil about the vertical axis. This torque is obtained as

$$|\tau| = Il_1 Bl_2 \sin\theta = IdsB\sin\theta \qquad (6.76)$$

where, $l_1\, l_2 = ds$ is the area enclosed by the loop. Expressing the torque in the vector notation, we have

Fig. 6.17: Force due to field **B** parallel to x-axis and current I parallel to z-axis

$$\tau = I\hat{a}_n ds \times B \qquad (6.77)$$

We can see the that torque tends to decrease the angle θ. Let $\cup$ be the potential energy of the loop, then the torque tending to increase θ is $-\dfrac{\partial \cup}{\partial \theta}$. This means

$$-\frac{\partial \cup}{\partial \theta} = -IdsB\sin\theta$$

or
$$\cup = -IdsB\cos\theta + c$$

where, c is constant of integration. When $\theta = \dfrac{\pi}{2}$, we have $\cup = 0$. Thus,

$$\cup = -IdsB\cos\theta = I\hat{a}_n ds \cdot B \qquad (6.78)$$

where, $\phi = B \cdot \hat{a}_n ds$ is the magnetic flux through the circuit. Thus, one obtains

$$\cup = -I\phi \qquad (6.79)$$

We must remember that ϕ and ϕ_m are different. Comparing Eq. (6.78) with potential energy of an electric dipole of dipole moment p in an electric field E, i.e. $\cup = -p \cdot E$, one finds that the magnetic field produced by a current loop is similar in form to the electrostatic field of an electric dipole. Thus, magnetic dipole moment of the loop is given by

$$m = I\hat{a}_n ds \qquad (6.80)$$
$$\therefore \qquad \cup = -m \cdot B \qquad (6.81)$$

and the torque is given by

$$\tau = m \times B \qquad (6.82)$$

We must note that the above result is true for a small loop of any shape.

Dipole in a Non-uniform Magnetic Field

When a current loop which is equivalent to a magnetic dipole is placed in a non-uniform magnetic field, then in addition to the torque given by Eq. (6.82), the dipole also experiences a translatory force.

We have already obtained the expression for the magnetic potential energy of a magnetic dipole m in a magnetic field B as

$$\cup = -m \cdot B$$

Now, the force F is related to the potential energy $\cup$ by the relation

$$F = -\nabla\cup$$
$$\therefore \qquad F = \nabla(m \cdot B)$$
$$= (B \cdot \nabla)m + (m \cdot \nabla)B + B \times (\nabla \times m) + m(\nabla \times B) \qquad (6.83)$$

We know that m is not a function of the space coordinates, and hence one gets

$$(B \cdot \nabla)m = 0, \nabla \times m = 0$$
$$\therefore \qquad F = m \times (\nabla \times B) + (m \cdot \nabla)B$$

But $\nabla \times B = 0$ and hence

$$F = (m \cdot \nabla)B \qquad (6.84)$$

Equation (6.84) is the general formula for force F on a magnetic dipole m in a non-uniform field B.

6.14 APPLICATIONS OF VECTOR POTENTIAL

Vector Potential due to a Small Current Loop

Let us consider a small current loop of radius a and carrying a current I (Fig. 6.18).

Let us assume that the origin of the frame of reference coincides with the centre of the coil O and its z-axis is perpendicular to the coil. Let us consider an element dl of the circuit. Let r be the distance of the point P from the element dl. ϕ is the azimuthal angle of the element *dl*. One can easily see that the components of the element $dl = ad\phi$ are $-a\sin\phi d\phi$, $a\cos\phi d\phi$ and O. We have

$$A = \frac{\mu_0}{4\pi}\int\frac{J}{|r|}dV = \frac{\mu_0 I}{4\pi}\int\frac{dl}{|r|}$$

Now, the *x* component is obtained as

$$A_x = \frac{\mu_0 I}{4\pi}\int\frac{-a\sin\phi d\phi}{|r|}$$

We have $r^2 = (x-a\cos\phi)^2 + (y-a\sin\phi)^2 + z^2$. Since $a \ll r_0$, we have

$$\frac{1}{r} = \frac{1}{r_0} + \frac{ax\cos\phi + ay\sin\phi}{r_0^2} + \cdots$$

where $r_0 = OP$. Therefore,

Fig. 6.18: Magnetic vector potential due to small current loop

$$A_x = -\frac{\mu_0 Ia}{4\pi}\int_0^{2\pi}\left(\frac{1}{r_0} + \frac{ax\cos\phi + ay\sin\phi}{r_0^3}\right)\sin\phi d\phi$$

$$= -\frac{\mu_0 Ia^2\pi y}{4\pi r_0^3} \tag{6.85}$$

Similarly, one obtains

$$A_y = -\frac{\mu_0 Ia^2\pi x}{4\pi r_0^3}\quad\text{and}\quad A_z = 0 \tag{6.86}$$

We have the magnitude of the dipole moment equivalent to the circuit (small current loop) as

$$m = \text{area}\times\text{current} = ma^2 I \tag{6.87}$$

The direction of dipole moment *m* is normal to the plane of the coil, i.e. along the z-axis. Obviously, $m_x = 0$, $m_y = 0$ and $m_z = \pi a^2 I$. One can easily see that the components of vector potential **A** given by Eqs. (6.85) and (6.86) are proportional to those of the vector $m \times r$. Dropping the subscripts, one can write,

$$A = \frac{\mu_0}{4\pi r^3}(m\times r) = -\frac{\mu_0}{4\pi}\left[m\times\nabla\left(\frac{1}{r}\right)\right] \tag{6.88}$$

Magnetic induction **B** can be determined by taking the curl of Eq. (6.88). We have

$$B(r) = \nabla\times A = \nabla\times\left[\frac{\mu_0}{4\pi r^3}(m\times r)\right]$$

$$= \frac{\mu_0}{4\pi}\left[-(m\cdot\nabla)\frac{r}{r^3} + m\nabla\cdot\left(\frac{r}{r^3}\right)\right]$$

(for constant *m*)

But $\qquad \nabla\cdot\left[\dfrac{r}{r^3}\right] = 0 \tag{6.89}$

and
$$(m \cdot \nabla)\frac{r}{r^3} = \frac{m}{r^3} - 3\frac{(m \cdot r)r}{r^5}$$

$$\therefore \qquad B(r) = \frac{\mu_0}{4\pi}\left[\frac{3(m \cdot r)r}{r^5} - \frac{m}{r^3}\right] \qquad (6.90)$$

One can also determine the scalar magnetic potential ϕ_m by the relation
$$B = -\nabla\phi_m \qquad (6.91)$$

Equation (6.90) can be written as
$$B = -\nabla\left(\frac{\mu_0}{4\pi}\frac{m \cdot r}{r^3}\right) \qquad (6.92)$$

Comparing Eqs. (6.91) and (6.92), one obtains
$$\phi_m = \frac{\mu_0}{4\pi}\left(\frac{m \cdot r}{r^3}\right) \qquad (6.93)$$

Alternative Method

Let us consider a filament circuit C carrying current I as shown in Fig. 6.19. We have the vector potential A at a point $P(x, y, z)$ due to the current loop as

$$A = \frac{\mu_0 I}{4\pi}\oint\frac{dl'}{r'} \qquad (6.94)$$

Fig. 6.19: Filament circuit C carrying current I

where, dl' (x', y', z') is an element of the circuit C. Making use of Stokes' theorem,

$$\oint\frac{dl'}{r'} = \int_S \hat{a}_n \times \nabla'\left(\frac{1}{r'}\right)dS$$

one obtains

$$A = \frac{\mu_0 I}{4\pi}\int_S \hat{a}_n \times \nabla'\left(\frac{1}{r'}\right)ds = \frac{\mu_0 I}{4\pi}\int_S\left(\hat{a}_n \times \frac{\hat{a}_{r'}}{r'^2}\right)dS \qquad (6.95)$$

where, $\hat{a}_n$ and $\hat{a}_{r'}$ are the unit vectors perpendicular to the surface enclosed by the circuit C and along r' respectively. Since the dimensions of the current loop are very small compared with the distance to the point, one can consider the factor $\dfrac{\hat{a}_{r'}}{r'^2}$ very nearly constant during integration. Now taking $\dfrac{\hat{a}_{r'}}{r'^2}$ as $\dfrac{\hat{a}_r}{r^2}$, one obtains

$$A = -\frac{\mu_0 I}{4\pi}\frac{\hat{a}_r}{r^2}\times\int_S \hat{a}_s dS = -\frac{\mu_0 I}{4\pi r^2}\hat{a}_r\times\hat{a}_n S$$

$$= \frac{\mu_0}{4\pi}\left(m\times\frac{\hat{a}_r}{r^2}\right) = \frac{\mu_0}{4\pi}\left(\frac{m\times r}{|r|^3}\right) \quad (\because m = \hat{a}_n I s) \qquad (6.96)$$

Now, one can obtain the magnetic flux density B. We have

$$B = \nabla \times A = \frac{\mu_0}{4\pi} \nabla \times \left(\frac{m \times r}{|r|^3} \right)$$

$$= \frac{\mu_0}{4\pi} \left[m \nabla \cdot \left(\frac{r}{|r|^3} \right) - (m \cdot \nabla) \left(\frac{r}{|r|^3} \right) \right]$$

$$\therefore \qquad \nabla \cdot \left(\frac{r}{|r|^3} \right) = 0 \text{ and hence}$$

$$B = -\frac{\mu_0}{4\pi} (m \cdot \nabla) \left(\frac{r}{|r|^3} \right)$$

Also $\qquad \nabla \left(m \cdot \frac{r}{|r|^3} \right) = (m \cdot \nabla) \frac{r}{|r|^3} + m \times \left(\nabla \times \left| \frac{r}{|r|^3} \right| \right)$

The last term vanishes, and hence one obtains

$$(m \cdot \nabla) \frac{r}{|r|^3} = \nabla \left(m \cdot \frac{r}{|r|^3} \right) = \left[\frac{m}{|r|^3} - \frac{3(m \cdot r)r}{|r|^3} \right]$$

and magnetic induction B as

$$B = \frac{\mu_0}{4\pi r^3} \left[\frac{3(m \cdot r)r}{r^2} - m \right] \qquad (6.97)$$

One can easily verify that Eq. (6.97) is similar to the relation for a field of an electric dipole.

Magnetic Induction due to an Infinite Current Carrying Wire

Let us consider a long wire whose axis coincides with z-axis and carries a current. Obviously, the only component of current density and hence magnetic vector potential is z-component (Fig. 6.20).

(i) *Inside the wire:* In cylindrical coordinates (r, θ, z), the Poisson's equation is

$$\nabla^2 A_z = -\mu J_z \qquad (6.98)$$

where, μ is the permeability inside the wire. Let R be the radius of the wire, then

$$J_z = \frac{I}{\pi R^2}$$

We have

$$\nabla^2 \phi = \frac{1}{r} \frac{\partial}{\partial r} \left(r \frac{\partial \phi}{\partial r} \right) + \frac{1}{r^2} \frac{\partial}{\partial \theta} \left(\sin \theta \frac{\partial \phi}{\partial \theta} \right) + \frac{\partial^2 \phi}{\partial z^2}$$

Since A_z is independent of θ and z, Eq. (6.98) can be expressed as

$$\frac{1}{r} \frac{\partial}{\partial r} \left(r \frac{\partial A_z}{\partial r} \right) = -\frac{\mu I}{\pi r^2} \qquad (6.99)$$

Fig. 6.20: Magnetic induction due to an infinite current carrying wire

Integrating, one obtains

$$r\frac{\partial A_z}{\partial r} = -\frac{\mu I r^2}{2\pi R^2} + b$$

where, b is a constant of integration. We have $r = 0$, $r\frac{\partial A_z}{\partial r} = 0$ and therefore $b = 0$.
Thus,

$$r\frac{\partial A_z}{\partial r} = -\frac{\mu I r^2}{2\pi R^2}$$

or
$$\frac{\partial A_z}{\partial r} = -\frac{\mu_0 I r}{2\pi r^2} \tag{6.100}$$

Integrating again, one obtains

$$A_z = -\frac{\mu_0 I r^2}{4\pi R^2} + b_1 \tag{6.101}$$

If one assumes that $A_z = 0$ at $r = R$, then

$$b_1 = \frac{\mu_0 I R^2}{4\pi R^2} = \frac{\mu I}{4\pi}$$

Therefore, Eq. (6.101) becomes

$$A_z = \frac{\mu_0 I}{4\pi}\left[1 - \frac{r^2}{R^2}\right] \tag{6.102}$$

We have $\boldsymbol{B} = \nabla \times \boldsymbol{A}$, one can express the cylindrical components of $\nabla \times \boldsymbol{A}$ as

$$\left.\begin{aligned}
(\nabla \times A)_r &= \frac{1}{r}\frac{\partial A_z}{\partial \theta} - \frac{\partial A_z}{\partial z} \\
(\nabla \times A)_\theta &= \frac{\partial A_r}{\partial z} - \frac{\partial A_z}{\partial r} \\
(\nabla \times A)_z &= \frac{\partial A_\theta}{\partial r} + \frac{A_\theta}{r} - \frac{1}{r}\frac{\partial A_r}{\partial \theta}
\end{aligned}\right\} \tag{6.103}$$

One obtains,

$$B_\theta = \frac{\mu I r}{2\pi R^2} \tag{6.104}$$

(ii) Outside the wire: Poisson's equation takes the form as

$$\frac{1}{r}\frac{\partial}{\partial r}\left(r\frac{\partial A_z}{\partial r}\right) = 0 \tag{6.105}$$

Integrating, one obtains

$$r\frac{\partial A_z}{\partial r} = C \,(\text{constant})$$

or
$$\frac{\partial A_z}{\partial r} = \frac{C}{r} \qquad (6.106)$$

Integrating again, one obtains
$$A_z = C \log_e r + C' \text{ (constant)}$$

We have $r = R, A_z = 0$

$$\therefore \qquad C' = -C \log_e R$$

$$\therefore \qquad A_z = C \log_e r - C \log_e R = C \log_e \left(\frac{r}{R}\right) \qquad (6.107)$$

Magnetic field induction $B = \nabla \times A$

$$\therefore \qquad B_\theta = (\nabla \times A)_\theta = \frac{\partial A_r}{\partial z} - \frac{\partial A_z}{\partial r} = -\frac{\partial A_z}{\partial r}$$

As B_θ must be continuous at $r = R$, we have

$$\frac{\partial A_z}{\partial r} = \frac{\mu I}{4\pi}\left(0 - \frac{2r}{R}\right)_{r=R} = -\frac{\mu I}{2\pi R} \qquad \text{[from Eq. (6.102)]}$$

Also from Eq. (6.106), $\left(\dfrac{\partial A_z}{\partial r}\right)_{r=R} = \dfrac{C}{R}$

$$\frac{C}{R} = -\frac{\mu I}{2\pi R} \quad \text{or} \quad C = -\frac{\mu I}{2\pi}$$

Substituting this value in Eq. (6.107), one obtains

$$A_z = -\frac{\mu I}{2\pi}\log_e\left(\frac{r}{R}\right) \qquad (6.108)$$

$$\therefore \qquad B_\theta = -\frac{\partial A_z}{\partial r} = \frac{\mu I}{2\pi R} \qquad (6.109)$$

6.15 MAGNETIC FIELD IN A SUBSTANCE

We assumed in the preceding chapter that conductors carrying current are in vacuum. If the conductors carrying current are in a medium, the magnetic field changes. The explanation is that any substance is a magnet, i.e. is capable of acquiring a magnetic moment under the action of a magnetic field (of becoming magnetized). The magnetized substance sets up the magnetic field B' which is superposed onto the field B produced by the current. Both fields produce the resultant field

$$B = B_0 + B' \qquad (6.110)$$

The true (microscopic) field in a magnetic medium varies greatly within the limits of intermolecular distances. Here B denotes the averaged (macroscopic) field. This clearly reveals that a material substance acquires a *magnetic polarization* when placed in a magnetic field, just as a dielectric medium acquires an electric polarization.

The response of the materials to an applied magnetic field depends on the properties of the individual atoms and molecules, and on their interactions. The orbital motion of electrons in atoms and molecules provide currents which give rise to the magnetic dipoles. There are several materials, e.g. water, Cu, Pb, NaCl, S, graphite, etc. in which the small electric currents associated with the orbital motion and spin of the electrons average to zero. When such substances are placed in a magnetic field, minute electron currents are generated by induction in the clouds of electrons, the direction of which is such that the magnetic field associated with them opposes the inducing field **B**. Substances which exhibit such a behaviour are called *diamagnetic substances*. *Diamagnetism* is found only in substances whose atoms have no magnetic moment (the vector sum of the orbital and spin magnetic moments of the atom electrons is zero). We must remember that diamagnetic properties are present in all atoms, however, they are observed only when other effects are not present to mask the comparatively weak diamagnetic effects.

There are materials which may possess an intrinsic magnetic dipole moment on account of the fact that the currents from the orbital motions and spins of the electrons do not average to zero. Although electron spins tend to pair and cancel one another, there are atoms in which pairing is incomplete. When such materials are placed in an external magnetic field, the magnetic moments of the atoms and induced magnetism enhance the external field. Such substances are called *paramagnetic materials*. Examples of paramagnetic substances are Al, Na, $CuCl_2$, $NiSO_4$, etc.

Paramagnetic effect is more pronounced at low temperatures. This effect arise due to the alignment of permanent dipoles due to an external magnetic field. When the external field is removed, thermal agitations disturb the atomic dipoles and produce random orientations. This means that even in the presence of an external field, lowering of the temperature of the substance should enhance the paramagnetic effects that are observed. This is observed to be true, because the average thermal kinetic energy of an atom is proportional to the absolute temperature of the substance.

There are a third and important kind of substances capable of having magnetization in the absence of an external field. According to the name of their most widespread representative, i.e. ferrum (iron), they have been called *ferromagnetic*. These are counterparts of electrets and do not shed their magnetic properties even at high temperatures. In addition to iron, they include nickel, cobalt, gadolinium, their alloys and compounds, and also certain alloys and compounds of manganese and chromium with non-ferromagnetic elements. All these substances display *ferromagnetism* only in the crystalline state.

Ferromagnetics are strong magnetic substances. Their magnetization exceeds that of dia- and paramagnetics, which belong to the category of weakly magnetized substances, an enormous number of times (up to 10^{10}). For these substances, the atomic dipoles get aligned by the application of an external magnetic field and they remain locked in the aligned position, even after the external field is removed. Obviously, these substances retain a memory of their history.

It follows from experiments involving the study of gyromagnetic phenomena that the intrinsic (spin) magnetic moments of electrons are responsible for the magnetic properties of ferromagnetics. In definite conditions, exchange forces may appear in crystals that make the magnetic moments of the electrons become lined up parallel to one another. The result is the setting up of regions of *spontaneous magnetization*, also called *domains*. Within the confines of each domain, a ferromagnetic is spontaneously magnetized to saturation and has a definite magnetic moment. The directions of these moments are different for different domains (Fig. 6.21), so that in the absence of an external field the total moment of an entire body is zero. Domains have dimensions of the order of 1 to 10 µm.

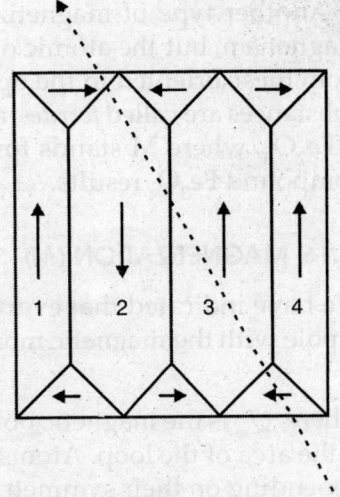

Fig. 6.21: Domain structure

The action of a field on domains at different stages of the magnetization process is different. The resulting net macroscopic field in the vicinity of ferromagnetic material can be expected to depend on the relative sizes of these domains and on the orientation of the molecular dipoles within each domain. First, with weak fields the displacement of the domain boundaries is observed. There is a net alignment of the domains in the field direction resulting in a very large paramagnetic effect. With an increase in the field strength, these domains which are already aligned become enlarged at the expense of the others. This process goes on further and further until the domains which have a smaller energy in magnetic field completely absorb the domains that are less advantageous from the energy point of view. In the next stage, the magnetic moments of the domains turn in the direction of the field. The moments of the electrons within the confines of a domain turn simultaneously without violating their strict parallelism to one another. These processes (excluding slight displacements of the boundaries between the domains in very weak fields) are irreversible, and this is exactly what causes *hysteresis*.

There is a definite temperature for every ferromagnetic at which the regions of spontaneous magnetization (domains) break up and the substance loses it ferromagnetic properties. This temperature is called the *Curie point*. It is 768°C for iron and 365°C for nickel. At a temperature above the Curie point, a ferromagnetic becomes an ordinary paramagnetic whose magnetic susceptibility obeys the *Curie–Weiss law*.

$$\chi = \frac{c}{T - T_c} \qquad (6.111)$$

When a ferromagnetic is cooled below its Curie point, domains once appear in it.

Exchange forces sometimes result in the appearance of so-called *antiferromagnetics* (chromium, manganese, etc.). In these substances, the intrinsic magnetic moments of the electrons are spontaneously oriented antiparallel to one another. Such an orientation involves adjacent atoms in pairs. The result is that antiferromagnetics have an extremely low magnetic susceptibility and behave like very weak paramagnetics. There is also a temperature T_N for antiferromagnetics at which the antiparallel orientation of the spins vanishes. This temperature is known as the *Neel point* or *Neel temperature*. Some antiferromagnetics, for example, erbium, dysprosium, alloys of manganese and copper have two such points (an upper and a lower Neel point), the antiferromagnetic properties being observed only at the intermediate temperatures. Above the upper point, the substance behaves like a paramagnetic, and at temperatures below the lower Neel point it becomes a ferromagnetic.

Another type of magnetization is *called ferrimagnetism*. It is similar to antiferro-magnetism, but the atomic or ionic magnetic moments in one direction are different from these oriented in the opposite direction resulting in a net magnetization. These substances are called *ferrites*, and can be generally represented by the chemical formula MFe_2O_3, where M stands for Co, Ni, Cu, Mg, Zn, Cd, etc. Note that if M is Fe, the compound Fe_3O_4 results.

6.16 MAGNETIZATION (*M*)

We have indicated that every atom or molecule may be regarded as a tiny magnetic dipole with the magnetic moment

$$Q_m l = -\hat{a}_n I ds \qquad (6.112)$$

where, Q_m is the magnetic pole strength, l is the pole separation, I is the current and ds is the area of the loop. Atoms may or may not exhibit a net magnetic dipole moment, depending on their symmetry or on the relative orientation of their electronic orbits. Since most molecules are not spherically symmetric, they may exhibit a permanent magnetic dipole moment because of special orientation of the electronic orbits. For example, diatomic molecules have axial symmetry, and may possess a magnetic dipole moment parallel to the molecule axis. Even so, matter in bulk, with the exception of ferromagnetic materials, does not exhibit a net magnetic moment, because of random orientation of the molecules, a situation similar to that found in the electric polarization of matter. However, the presence of an external magnetic field distorts the electronic motion, giving rise to a net magnetic polarization or *magnetization* of the material.

The *magnetization vector M* of a material is defined as the magnetic dipole moment of the medium per unit volume. If *m* is the magnetic dipole moment contributed by each atom or molecule and *n* is the number of atoms or molecules per unit volume, the magnetization is $M = nm$.

The magnetic moment of an elementary current is expressed in Am^2, and therefore the magnetization M is expressed in $Am^2/m^3 = Am^{-1}$ or $m^{-1}s^{-1}C$, and is equivalent to current per unit length. We must note that the magnetization M is not necessarily uniform throughout the substance. Thus,

$$M = \frac{dm}{dV} \qquad (6.113)$$

Let us consider an elementary volume with sides δx, δy and δz at a point (x, y, z) within a magnetized substance (Fig. 6.22). One obtains the component of the magnetic moment associated with this volume in the z-direction as

$$M_z \delta x \delta y \delta z$$

Using Eq.(6.112), one obtains

$$M_z \delta x \delta y \delta z = I \delta x \delta y \qquad (6.114)$$

where, I is the current flowing in the loop in a plane parallel to xy plane as shown in Fig. 6.22.

$$\therefore \qquad I = M_z \delta z \qquad (6.115)$$

One can obtain the current in the adjacent loop at $(x + \delta x, y, z)$ as

$$I' = I + \frac{\partial I}{\partial x} \delta x = M_z \delta z + \frac{\partial M}{\partial x} \delta x \delta z$$

Fig. 6.22: An elementary volume within a magnetized substance

Obviously, the current in the interface AB of the two elements has a component in the y-direction and it is given by

$$I - I' = -\frac{\partial M_z}{\partial x}\delta x\delta z$$

Similarly, by considering a current loop in the yz plane, one finds another component of current in the y-direction. Its magnitude is $\dfrac{\partial M}{\partial z}\delta x\delta z$. Thus, one obtains the total current in the y-direction as

$$\left(\frac{\partial M_x}{\partial z} - \frac{\partial M_z}{\partial x}\right)\delta x\delta z \qquad (6.116)$$

One can express the above in terms of the y-component of the magnetization current (J_M) as

$$J_{M_y}\delta x\delta z = \left(\frac{\partial M_x}{\partial z} - \frac{\partial M_z}{\partial x}\right)\delta x\delta z$$

$$\therefore \qquad J_{M_y} = \left(\frac{\partial M_x}{\partial z} - \frac{\partial M_z}{\partial x}\right) \qquad (6.116a)$$

Similarly, one obtains the other components as

$$J_{M_x} = \frac{\partial M_z}{\partial y} - \frac{\partial M_y}{\partial z} \qquad (6.116b)$$

and $$J_{M_z} = \frac{\partial M_y}{\partial x} - \frac{\partial M_x}{\partial y} \qquad (6.116c)$$

One can easily see that the expressions appearing on the right hand side of Eqs. (6.116a) to (6.116c) are the components of $\nabla \times M$. Therefore,

$$J_M = \nabla \times M = \begin{vmatrix} \hat{a}_x & \hat{a}_y & \hat{a}_z \\ \dfrac{\partial}{\partial x} & \dfrac{\partial}{\partial y} & \dfrac{\partial}{\partial z} \\ M_x & M_y & M_z \end{vmatrix} \qquad (6.117)$$

Obviously, the magnetization current density is the curl of the magne-tization. Moreover, a current density equal to that given by Eq. (6.117) can produce the same magnetic field as the field produced by the magnetization M. We must note that magnetization current flow exists only if M is varying. If M is uniform, we have $J_M = 0$. The magnetization current I_M is a bound current in the sense that one cannot tap the I_M of a bar magnet to set up a perpetual source of current at no cost. One can compare this with the polarization charge (Q_p) on an electret. One cannot wipe out the Q_p from the surface of an electret to make an inexhaustible supply of electric charge.

Alternative Method

One can obtain the above result using vector potential. We have

$$A = \frac{\mu_0}{4\pi} \int \left\{ M(R) \times \nabla \left(\frac{1}{|r|} \right) \right\} dV \tag{6.118}$$

Making use of the vector identity

$$\nabla \times (\phi B) = \phi \nabla \times B - B \times \nabla \phi$$

One can express Eq. (6.118) in the form

$$A = \frac{\mu_0}{4\pi} \int \frac{\nabla \times M}{|r|} dV - \frac{\mu_0}{4\pi} \int \left(\nabla \times \frac{M}{|r|} \right) dV \tag{6.119}$$

We can also see that the following vector relation holds good

$$\int_S B \times \hat{a}_n dS = -\int_V \nabla \cdot B dV \tag{6.120}$$

$$\int (\nabla \times B) dV = -\int B \times \hat{a}_n dS$$

If B is a constant vector, then

$$C \cdot \int_S B \times \hat{a}_n dS = \int_S (C \times B) \cdot \hat{a}_n dS$$

$$= \int_V \nabla \cdot (C \times B) dV \quad \text{(divergence theorem)}$$

$$= -C \cdot \int_V \nabla \times B dV$$

where, C is an arbitrary vector.

Making use of Eq. (6.120), Eq. (6.119) takes the form

$$A = \frac{\mu_0}{4\pi} \int \frac{\nabla \times M}{|r|} dV + \frac{\mu_0}{4\pi} \int_s \frac{M \times \hat{a}_n ds}{|r|}$$

We know that M is localized. This means that the surface integral taken over a surface outside the region in which the current flows, vanishes. Thus,

$$A = \frac{\mu_0}{4\pi} \int \frac{\nabla \times M}{|r|} dV \tag{6.121}$$

We have already obtained that

$$A = \frac{\mu_0}{4\pi} \int \frac{J(r)}{|r|} dV \qquad (6.122)$$

Comparing Eqs. (6.121) and (6.122), one obtains

$$J_m = \nabla \times M \qquad (6.123)$$

6.17 MAGNETIC FIELD INTENSITY (*H*)

When a medium is electrically conducting and also magnetizable, then there will be present both a real current density (J) and the magnetization current density (J_M). Obviously, one will have to take into account both of these. Thus,

$$\nabla \times B = \mu_0 (J + J_M) = \mu_0 (J + \nabla \times M) \qquad (6.124)$$

$$\therefore \qquad \nabla \times (B - \mu_0 M) = \mu_0 J \qquad (6.125)$$

The vector within the parenthesis is such that its curl depends only on the conduction current density J at the point, and not on the equivalent current. Thus, analogous to the displacement vector D which in an electrostatic system is defined in such a way as to make D depend only on the free charges, it is convenient to define a vector H, usually called the *magnetic field intensity*, which depends only on total external currents. Whence it follows that

$$H = \frac{B}{\mu_0} - M \qquad (6.126)$$

or

$$B = \mu_0 (M + H) \qquad (6.127)$$

In accordance with Eq. (6.126), Eq. (6.125) can be expressed as

$$\nabla \times H = J \qquad (6.128)$$

Obviously, the curl of vector H equals the vector of the density of the macroscopic currents. H is expressed in Am^{-1} or $m^{-1}s^{-1}C$ which are the units of the two terms that appear on the right hand side of Eq. (6.126), and Eq. (6.128) is a new form of Ampere's law which holds both in vacuum and a medium and, hence, is more general than obtained earlier. In vacuum, we have

$$B = \mu_0 H$$

$$\nabla \times B = \mu_0 J$$

which is a particular case of Eq. (6.128).

The expression in integral form corresponding to Eq. (6.128) is

$$\int_S \nabla \times H \cdot \hat{a}_n dS = \oint H \cdot dl = \int J \cdot \hat{a}_n dS$$

or

$$\oint H \cdot dl = I \qquad (6.129)$$

The line integral $\oint H \cdot dl$ is called the *magnetomotance* and Eq. (6.129) is completely a general expression, except that I must be interpreted as total current if both conduction and displacement current exist. Equation (6.129) expresses the theorem on the circulation of the vector H, i.e. the circulation of the magnetic field strength vector H around a loop equals the algebraic sum of the macroscopic currents enclosed by this loop.

6.18 MAGNETIC SUSCEPTIBILITY AND PERMEABILITY

Equation (6.126) may be written in a somewhat different form for certain classes of materials. This follows from the fact that experimentally for most non-ferromagnetic materials, magnetization is proportional to and in the same direction as the magnetic field. Thus,

$$M = \chi_m H \tag{6.130}$$

where, the quantity χ_m is called the *magnetic susceptibility* of the material, and is a pure number (dimensionless) independent of the units chosen for M and H. χ_m is a function of temperature. Paramagnetic substances have small positive values of the susceptibility, while diamagnetic substances have negative values. Table 6.3 gives the susceptibility values for some common substances.

Table 6.3: χ_m of some materials

Paramagnetic		Diamagnetic	
Material	χ_m	Material	χ_m
Aluminium	2.3×10^{-5}	Antimony	-7×10^{-5}
Magnesium	1.2×10^{-5}	Gold	-3.6×10^{-5}
Neodymium	3.0×10^{-3}	Copper	-9.8×10^{-6}
Tungsten	6.8×10^{-5}	Germanium	-1.5×10^{-6}
Oxygen (NTP)	19.2×10^{-7}	Water	-9.1×10^{-6}

Substituting Eq. (6.130) in Eq. (6.127), one obtains

$$B = \mu_0(H + \chi_m H) = \mu_0(1 + \chi_m) H = \mu H \tag{6.131}$$

where

$$\mu = \frac{B}{H} = \mu_0(1 + \chi_m) \tag{6.132}$$

is called the *permeability* of the medium, and is expressed in the same unit as μ_0, that is, m kg^{-2} C. The ratio

$$\frac{\mu}{\mu_0} = \mu_r \tag{6.133}$$

is called the *relative permeability* of the medium and is a pure number independent of the system of units. μ differs from unity only by few parts for non-ferromagnetic substances. For paramagnetic substances, $\mu > 1$ and for diamagnetic $\mu < 1$. For ferromagnetic substances μ has a large value around 1000.

We must note that the relation (6.132), i.e. $B = \mu H$ is not valid for certain materials, such as crystals, solid under strain, iron or steel which has been subjected to rolling or hammering or materials which have a definite structure. Moreover, an anisotropic magnetic material must be described in terms of a tensor permeability as

$$\left. \begin{array}{l} B_x = \mu_{xx}H_x + \mu_{xy}H_y + \mu_{xz}H_z \\ B_y = \mu_{yx}H_x + \mu_{yy}H_y + \mu_{yz}H_z \\ B_z = \mu_{zx}H_x + \mu_{zy}H_y + \mu_{zz}H_z \end{array} \right\} \tag{6.134}$$

Obviously, μ is a tensor for anisotropic materials in the relationship $B = \mu H$, but $B = \mu_0(H + M)$ remains valid, although B, H and M are no longer parallel in general. The most common anisotropic magnetic material is a single *ferromagnetic* crystal. Thin magnetic films also exhibit anisotropy. Most applications of ferromagnetic materials, however, involve polycrystalline arrays that are much easier to make.

One can easily show that the magnetization is equal to

$$M = \frac{Nm^2}{3kT} \cdot B = \frac{Nm^2\mu_0 H}{3kT} \tag{6.135}$$

where, m is the permanent magnetic dipole moment associated with each molecule and N is the number of molecules per unit volume. One obtains the magnetic susceptibility as

$$\chi_m = \frac{Nm^2\mu_0}{3kT} \tag{6.136}$$

6.19 MAGNETIC BOUNDARY CONDITIONS

In a single medium the magnetic field is continuous, that is, the field, if not constant, changes only by an infinitesimal amount in an infinitesimal distance. However, at the boundary between two different media, the magnetic field may change abruptly both in magnitude and direction. Obviously, it is important to know the relations for magnetic fields at the interface between two magnetic media.

We have divergence theorem

$$\int_V \nabla \cdot B \, dV = \int_S B \cdot \hat{a}_n dS \tag{6.137}$$

We must note that the condition $\nabla \cdot B = 0$ remains unaltered by the presence of magnetic materials. Therefore,

$$\oint B \cdot \hat{a}_n dS = 0 \tag{6.138}$$

This is *Gauss's theorem*. This shows that the flux of field B out of any closed surface is zero. Consider a small disc of height h that sits astride the boundary between two media with permeabilities μ_1 and μ_2 as shown in Fig. 6.23. If h is very small, i.e. $h \to 0$, the integral $\int B \cdot \hat{a}_n dS = 0$ has contribution from the top and bottom ends only, i.e.

$$\int_1 B \cdot \hat{a}_n dS - \int_2 B \cdot \hat{a}_n dS = 0 \tag{6.139}$$

We have $B \cdot \hat{a}_n ds = B_\perp ds$, where $B_\perp$ is the component of B normal to ds. Therefore,

$$\int_1 B \cdot \hat{a}_n dS = \int B_{(1)\perp} dS$$

and

$$\int_2 B \cdot \hat{a}_n dS = \int B_{(2)\perp} dS$$

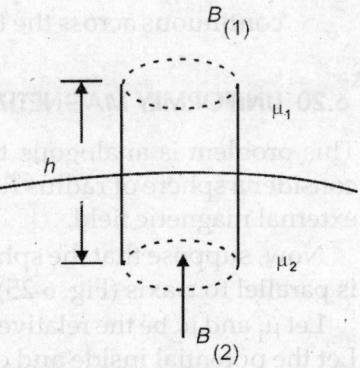

Fig. 6.23: A small disc sits astride the boundary between two media

This means $\qquad\qquad \int B_{(1)\perp} dS = \int B_{(2)\perp} dS \qquad\qquad$ (6.140)

We must note that Eq. (6.140) is true whatever be the size of ds. Therefore,

$$B_{(1)\perp} = B_{(2)\perp} \qquad\qquad (6.141)$$

According to Eq. (6.141), *the normal component of the flux density B is continuous across the boundary between two media.* This relation applies at the boundary of any two media for both static and time changing fields.

One can obtain the condition for H by means of Ampere's law. Let us consider a small circuit $ABCDA$ as shown in Fig. 6.24. $AB = CD = dl$ and the sides BC and DA are extremely small. Now,

Fig. 6.24: A small circuit $ABCD$

$$\oint H \cdot dl = I \qquad\qquad (6.142)$$

where I is the current. Since BC and DA are extremely small and hence they tend to zero. We have

$$H_{(1)\parallel} dl = H_{(1)\parallel} dl = I \qquad\qquad (6.143)$$

where, $H_{(1)\parallel}$ and $H_{(2)\parallel}$ are the tangential components of H in the two media. Obviously, there is a discontinuity in the component of the magnetic field H equal to the surface current. When there is no surface current, one obtains

$$H_{(1)\parallel} = H_{(2)\parallel} \qquad\qquad (6.144)$$

According to Eq. (6.144), the tangential components of H are continuous across the boundary between two media provided the boundary has no current sheet of infinitesimal thickness. Let us compare the boundary conditions in magnetostatics and electrostatics. One finds:

i. The normal component of D is continuous only if there is no charge change, whereas the normal component of B is strictly continuous.

ii. The tangential component of E is strictly continuous, while that of H is continuous across the boundary only if there is no current.

6.20 UNIFORMLY MAGNETIZED SPHERE IN AN EXTERNAL FIELD

This problem is analogous to that of a dielectric sphere in an external field. Let us consider a sphere of radius R with a uniform magnetization M along z-axis without an external magnetic field.

Now, suppose that the sphere is placed in a uniform field H_0, the direction of which is parallel to z-axis (Fig. 6.25).

Let μ_1 and μ_2 be the relative permeabilities inside and outside the sphere respectively. Let the potential inside and outside the sphere be

$$\phi_1 = -H_1 r \cos\theta \ (r < a) \qquad\qquad (6.145)$$

and

$$\phi_2 = -H_0 r \cos\theta + Ar^{-2}\cos\theta \ (r > a) \qquad\qquad (6.146)$$

where, θ is the angle which the radius vector r makes with the field direction, i.e. z-axis and H_1 is the field inside the sphere. We can see that the term containing r^{-2} is not included in ϕ_1, as the inclusion of such a term in ϕ_1 would make it infinite at $r = 0$.

Fig. 6.25: Uniformly magnetized sphere in an external magnetic field

We have from boundary conditions that the tangential component of H be continuous at the boundary. This makes $\phi_1 = \phi_2$ at $r = a$. Hence,

$$-H_a a \cos \theta = -H_a a \cos \theta + A a^{-2} \cos \theta$$

or
$$H_1 = H_0 - A a^{-3} \qquad (6.147)$$

Let us assume that the magnetization M_1 inside the sphere is parallel to the field H_0 and made up of two components: a permanent component M_0 and a component induced by the field M_1. This is given by

$$M' = \chi_m H_1 = (\mu_1 - 1) H_1$$
$$\therefore \qquad M_1 = (\mu_1 - 1) H_1 + M_0$$

But we have

$$B_1 = \mu_0(H_1 + M_1) = \mu_0[H_1 + (\mu_1 - 1) H_1 + M_0]$$
$$= \mu_0 \mu_1 H_1 + \mu_0 M_0 \qquad (6.148)$$

and
$$B_2 = \mu_0 \mu_2 H_0 \qquad (6.149)$$

One obtains the radial components of B as

$$-\mu_0 \mu_1 \left(\frac{\partial \phi_1}{\partial r} \right) + \mu_0 M_0 \cos \theta = \mu_0(\mu_1 H_1 + M_0) \cos \theta \qquad (6.150)$$

(inside the sphere)

and
$$\mu_0 \mu_2 \left(\frac{\partial \phi_2}{\partial r} \right) = \mu_0 \mu_2 (H_0 + 2A r^{-3}) \cos \theta \qquad (6.151)$$

(outside the sphere)

In accordance with the boundary conditions, these two components must be equal to one another at $r = a$. Thus, we obtain

$$\mu_0(\mu_1 H_1 + M_0) \cos \theta = \mu_0 \mu_2 (H_0 + 2A a^{-3}) \cos \theta$$
$$\therefore \qquad \mu_1 H_1 + M_0 = \mu_2 H_0 + 2\mu_2 A a^{-3}$$

But
$$A = \frac{H_0 - H_1}{a^{-3}}$$

$$\therefore \qquad \mu_1 H_1 + M_0 = \mu_2 H_0 + 2\mu_2 (H_0 - H_1)$$

or
$$(\mu_1 + 2\mu_2) H_1 = 3\mu_2 H_0 - M_0$$

$$\therefore \quad H_1 = \frac{3\mu_2}{\mu_1 + 2\mu_2}\mu_0 - \frac{1}{\mu_1 + 2\mu_2}M_0 \qquad (6.152)$$

The difference between the fields inside and outside the sphere, i.e.

$$H_d = H_1 - H_0 \qquad (6.153)$$

is called the *demagnetizing field*, i.e. magnetization produces a reverse field inside the sphere. For a paramagnetic substance, H_1 is slightly less than H_0 and the induction is greater than the free space value $\mu_0 H_0$. This means that the lines of induction are crowded together in a paramagnetic sphere. In a diamagnetic substance, the reverse is the case.

Table 6.4 provides a comparison of static electric and magnetic field equations.

Table 6.4: Comparison of static electric and magnetic field equations

Property	Electric field	Magnetic field
i. Force	$F = qE$	$dF = idl \times B$
ii. Basic equation for the field	$\nabla \times E = 0$	$\nabla \cdot B = 0$
iii. Field and potential relation	$E = -\nabla\phi$ $\phi = \dfrac{1}{4\pi\varepsilon_0}\displaystyle\int \frac{\rho dV}{r}$	$B = \nabla \times A$ $A = \dfrac{\mu_0}{4\pi}\displaystyle\int \frac{J}{\|r\|}dV$
iv. Constitutive relations for field	$D = \varepsilon E$	$B = \mu H$
v. Sources of the fields	$\nabla \cdot D = \rho$	$\nabla \times H = J$
vi. Derivation from scalar potentials	$E = -\nabla\phi$	$H = -\nabla U$ (in current free region)

ILLUSTRATIVE EXAMPLES

Example 6.1: Show that the magnetic field at the focus of a wire carrying current I in the form of the parabola is given by

$$B = \frac{\mu_0 I}{4d}$$

where, d is the distance from focus to the apex.

Solution: We have magnetic field for a wire at A as

$$B = \frac{\mu_0 I}{4\pi}\int \frac{r \times dr}{\|r\|^3}$$

The wire is in the form of a parabola as shown in Fig. 6.26. We have

$$\frac{r \times dr}{\|r\|^3} = \frac{rd\theta}{r^2} = \frac{d\theta}{r}$$

The field at the focus P of the parabola is

$$B = \frac{\mu_0 I}{4\pi}\int_{-\pi}^{\pi} \frac{d\theta}{r}$$

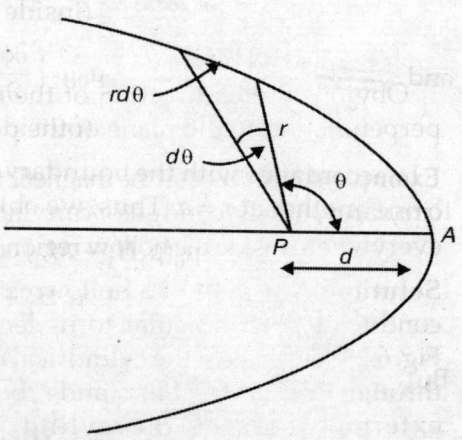

Fig. 6.26: Magnetic field at the focus of parabolic wire

In the case of parabola, we have $r(1 + \cos\theta) = 2d$

$$\therefore \quad \boldsymbol{B} \text{ at } P = \frac{\mu_0 I}{4\pi} \int_{-\pi}^{\pi} \frac{1+\cos\theta}{2d} d\theta = \frac{\mu_0 I}{4d}$$

Example 6.2: Use Ampere's theorem to obtain an expression for the magnetic field at the centre of a coil in the form of a square of side $2d$ carrying current I.

Solution: $ABCD$ is a square of side $2d$ metres carrying current I. Let P be the centre of this square. The magnetic field at the centre P due to side CD carrying current I is given by

$$B = \frac{\mu_0 I}{4\pi d}(\sin 45° + \sin 45°) = \frac{\sqrt{2}\mu I}{4Hd} \quad (\because \; \alpha = \beta = 45° \text{ and } R = d)$$

Fig. 6.27: Magnetic field at the centre of square coil

By symmetry the magnetic field at P due to each side will be $\dfrac{\sqrt{2}\,\mu I}{4\pi d}$ tesla.

$\therefore$ The magnetic field at the centre of the square, i.e. at P is given by

$$B = 4\sqrt{2}\frac{\mu I}{4\pi d} = \frac{\sqrt{2}\mu I}{\pi d} \text{ tesla}$$

Obviously, the direction of the magnetic field is perpendicular to the plane of the paper.

Example 6.3: Show that the magnetic field due to hollow circular cylindrical conductor carrying current vanishes everywhere inside the hollow region.

Solution: A section of a hollow circular cylindrical conductor perpendicular to its length is shown in Fig. 6.28. The axis of the cylindrical conductor passes through the point O. Let r_1 and r_2 be the internal and external radii of the cylindrical conductor respectively. Let us consider a circular path of radius r around O inside the hollow conductor. By symmetry the magnetic field produced due to the

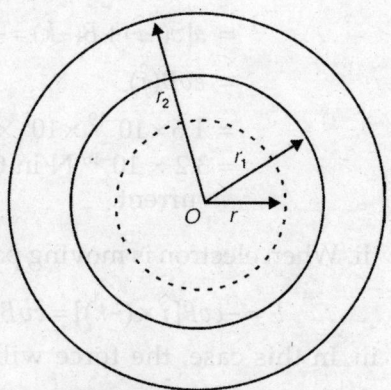

Fig. 6.28: Hollow circular cylindrical conductor

current flowing in the conductor at any point on this path will be tangential and equal in magnitude everywhere. Therefore, the line integral of B along this path is

$$\oint B \cdot dl = B2\pi r$$

But $\oint B \cdot dl = m \times$ (current enclosed by the circular path of radius r), i.e.

$$B = \mu \times 0 = 0$$

Obviously, the magnetic field of a hollow circular cylindrical conductor carrying current vanishes everywhere inside the hollow region.

Example 6.4: A current of 20 A is flowing in a long straight wire. An electron is moving with a velocity of 10^7 m/s. How much force will act on the electron if wire is 2 cm away from it and the velocity of electron is: (i) towards the wire, (ii) parallel to the wire, and (iii) perpendicular to the directions given by (i) and (ii).

Solution: Magnetic field at a distance r from a current carrying wire is

$$B = \frac{\mu_0 I}{2\pi r} \quad \begin{vmatrix} \mu_0 = 4\pi \times 10^{-7}\,\text{Wb m}^{-2} \\ I = 20\,\text{A} \\ r = 2 \times 10^{-2}\,\text{m} \end{vmatrix}$$

$$= \frac{4\pi \times 10^{-7} \times 20}{2\pi \times 2 \times 10^{-2}} = 2 \times 10^{-4}\,\text{Wb m}^{-2}$$

The magnetic field B is perpendicular to the plane containing wire and r. If wire is in y-direction and r in x-direction, B will be in z-direction (Fig. 6.29).

$\therefore$ Force on electron is

$$F = q(V \times B)$$

i. When electron is moving towards the wire, then

$$v = v(-\hat{i})$$

$\therefore$

$$= a[v(-\hat{i}) + B(-\hat{k}) = -eVB(-\hat{j})$$

$$= evB(\hat{j})$$

$$= 1.6 \times 10^{-19} \times 10^7 \times 2 \times 10^{-4}\hat{j}$$

$$= 3.2 \times 10^{-16}\,\text{N in the direction of current.}$$

Fig. 6.29: Current carrying long straight wire

ii. When electron is moving parallel to the wire, then $v = v(\hat{j})$

$\therefore \quad F = -evB[\hat{j} \times (-\hat{k})] = evB\hat{i} = 3.2 \times 10^{-16}\,\text{N} \perp^r$ to the wire.

iii. In this case, the force will be zero as the motion of the electron, i.e. v is in $\hat{k}$-direction.

Example 6.5: The vector potential corresponding to a magnetic field is given by $A = -\hat{k}\dfrac{I}{C}\ln(x^2 + y^2)$. Determine the field.

Solution: We have

$$B = \begin{vmatrix} \hat{i} & \hat{j} & \hat{k} \\ \dfrac{\partial}{\partial x} & \dfrac{\partial}{\partial y} & \dfrac{\partial}{\partial z} \\ 0 & 0 & -\dfrac{I}{C}\ln(x^2+y^2) \end{vmatrix}$$

$$= \hat{i}\dfrac{\partial}{\partial y}\left[-\dfrac{I}{C}\ln(x^2+y^2)\right] - \hat{j}\dfrac{\partial}{\partial y}\left[-\dfrac{I}{C}\ln(x^2+y^2)\right]$$

$$= -\dfrac{I}{C}\dfrac{2(y\hat{i} - x\hat{j})}{(x^2+y^2)}.$$

Example 6.6: A toroid is wound uniformly. It has N turns, and its inner and outer radii are a and b respectively. Find the magnetic induction inside the toroid when the current is I and determine the ratio b/a so that B in the ring does not vary by more than 25%.

Solution: Let us consider a closed circular path of radius r concentric with the toroid, $a < r < b$. If B is the magnetic induction along this path, then by Ampere's law

$$2\pi r B = \mu_0 NI$$

or

$$B = \mu_0 \dfrac{NI}{2\pi r}$$

$$\therefore \quad B_{\max} = \dfrac{\mu_0 NI}{2\pi a} \quad \text{and} \quad B_{\min} = \dfrac{\mu_0 NI}{2\pi b}$$

Hence

$$B_{\max} - B_{\min} = \dfrac{\mu_0 NI}{2\pi}\left(\dfrac{1}{a} - \dfrac{1}{b}\right)$$

$$= \dfrac{\mu_0 NI}{2\pi}\left(\dfrac{b-a}{ab}\right) = \dfrac{1}{4}\left(\dfrac{\mu_0 NI}{2\pi a}\right)$$

or

$$\dfrac{b-a}{ab} = \dfrac{1}{4a} \quad \therefore \quad \dfrac{b}{a} = \dfrac{4}{3}.$$

Example 6.7: An infinitely long cylindrical conductor of radius a carries a steady current of I A. Show that the fields inside and outside the conductors are

$$H_{\phi(\text{in})} = \dfrac{Ir}{2\pi a^2} \quad \text{(for } r < a)$$

and

$$H_{\phi(\text{out})} = \dfrac{I}{2\pi r} \quad \text{(for } r > a)$$

Solution: The problem can be solved with the help of Ampere's circuital law. From consideration of symmetry, it is evident that the magnetic intensity has only a ϕ component. Considering field point P at r (< 0) inside the cylinder, we obtain

$$H_\phi \times 2\pi r = \text{current enclosed} = \frac{Ir^2}{a^2}$$

Since the current is uniformly distributed over the cross-section, therefore, one obtains

$$(H_\phi)_{\text{in}} = \frac{Ir}{2\pi a^2} \quad (r < a)$$

When the field point is outside the conductor, i.e. $r > a$ the current enclosed is I. Thus,

$$(H_\phi)_{\text{out}} = \frac{I}{2\pi r} \quad (r > a)$$

Example 6.8: Show that the current required to balance the earth's gravitational force of a horizontal wire of mass m and length l, and oriented from east to west in the earth's magnetic field is given by $I = \dfrac{mg}{lB}$, where B is the magnitude of earth's magnetic field directed north–south.

Solution: The earth's magnetic field is directed north–south and has a magnitude B. The force due to interaction of current and flux = IlB. This force is equal to the weight mg. Therefore,

$$IlB = mg$$

or

$$I = \frac{mg}{Bl}.$$

Example 6.9: An infinitely long straight wire carrying current I_1 and a rectangular loop carrying current I_2 are situated in the same plane with two long sides of the rectangle parallel to the infinitely long wire. Show that the force between them is given by $F = \dfrac{\mu_0 a I_1 I_2 b}{2\pi d(d+b)}$.

Fig. 6.30: An infinitely long current carrying wire and a current carrying rectangular wire in the same plane

Solution: Figure 6.30 is drawn in accordance with the problem. We see that the conductor AB experiences a force of attraction towards the infinitely long conductor, and conductor CD experiences a force of repulsion, whereas conductors BC and DA do not experience any force. The flux density acting on AB due to the infinitely long conductor

$= \dfrac{\mu_0 I}{2\pi d}$ in a direction penetrating the paper. Therefore, one obtains the force on conductor

AD as $F_1 = \dfrac{I_2 \mu_0 I_1 a}{2\pi d}$ directed towards the infinitely long conductor.

Similarly, one can obtain the force on conductor CD as

$$F_2 = \frac{I_2 \mu_0 I_1 a}{2\pi(d+b)} \text{ away from the conductor.}$$

$$\therefore \quad F = \frac{\mu_0 I_1 I_2 a}{2\pi}\left[\frac{1}{d} - \frac{1}{(d+b)}\right] = \frac{\mu_0 I_1 I_2 ab}{2\pi d(d+b)}$$

Thus, the net force $F = F_1 - F_2 = \dfrac{\mu_0 I_1 I_2 ab}{2\pi d(d+b)}$

directed towards the current carrying current.

Example 6.10: Show that the magnetizing force at the centre of a concentrated circular coil is 1.047 times as great as magnetization force at the centre of the concentrated regular hexagonal coil as shown in Fig. 6.31 of same ampere turns in which the distance between the opposite sides is equal to the diameter of the circular coil.

Solution: We know that the magnetic flux density due to a finite straight conductor at its

perpendicular bisector $= \dfrac{\mu_0 I}{2\pi\rho}\cos\alpha\,\hat{a}_\phi$.

Now, for six sides

$$\boldsymbol{B} = \frac{\mu_0 I}{2\pi\rho}6N\cos\alpha\,\hat{a}_\phi$$

(for hexagonal coil having N number of turns)

$$\therefore \quad \boldsymbol{H} = \frac{\boldsymbol{B}}{\mu_0} = \frac{6NI}{2\pi\rho}\cos\alpha\,\hat{a}_\phi$$

Fig. 6.31: A concentrated circular coil and concentrated regular hexagonal coil having centre C

Similarly, we can write the expression for magnetic flux density at the centre of the circular coil carrying N number of turns

$$\boldsymbol{B}_{cir} = \frac{\mu_0 NI}{2\rho}\hat{a}_\phi$$

or $$\boldsymbol{H}_{cir} = \frac{\boldsymbol{B}_{cir}}{\mu_0} = \frac{\mu_0 NI}{2\rho}\hat{a}_\phi$$

or $$\left|\frac{\boldsymbol{H}_{cir}}{\boldsymbol{H}}\right| = \frac{\pi}{6}\frac{1}{\cos\alpha} = \frac{\pi}{6}\frac{1}{\cos 60°} = \frac{\pi}{6}\times 2 = \frac{\pi}{3} = 1.0466 \simeq 1.047.$$

Example 6.11: Calculate $\nabla \times \boldsymbol{B}$ in cartesian coordinates due to a current filament along the z-axis with current I in the $\hat{a}_z$ direction.

Solution: $$B = \frac{\mu_0 I}{2\pi r}\hat{a}_\phi = \frac{\mu_0 I}{2\pi}\left(\frac{-y\hat{a}_y + x\hat{a}_y}{x^2 + y^2}\right)$$

$$\nabla \times \boldsymbol{B} = \mu_0 \begin{vmatrix} \hat{a}_x & \hat{a}_y & \hat{a}_z \\[6pt] \dfrac{\partial}{\partial x} & \dfrac{\partial}{\partial y} & \dfrac{\partial}{\partial z} \\[10pt] \dfrac{-y}{x^2+y^2} & \dfrac{x}{x^2+y^2} & 0 \end{vmatrix}$$

$$= \left[\frac{\partial}{\partial x}\left(\frac{x}{(x^2+y^2)}\right) - \frac{\partial}{\partial y}\left(-\frac{y}{x^2+y^2}\right) \right]\hat{a}_z = 0$$

except at $x = y = 0$. This is consistent with $\nabla \times \boldsymbol{B} = \mu_0 \boldsymbol{J}$.

Example 6.12: Determine the torque acting on a hexagonal loop of side a lying on yz plane carrying current I in the anticlockwise direction and subjected to a uniform magnetic field $\boldsymbol{B} = B\hat{a}_z$.

Solution: Let us consider a hexagonal loop placed in yz plane as shown in Fig. 6.32.

Fig. 6.32: A hexagonal loop of side a lying on yz plane

$$m = IA\hat{a}_\phi = I \times 6\left[\frac{\sqrt{3}}{4}a^2\right]\hat{a}_\phi = \frac{3\sqrt{3}}{2}Ia^2\hat{a}_\phi$$

where $\qquad A = a^2$

Now $\qquad \tau = (m \times B) = \dfrac{3\sqrt{3}}{2}Ia^2\hat{a}_\phi \times B\hat{a}_z$

or $\qquad \tau = \dfrac{3\sqrt{3}}{2}Ia^2 B\hat{a}_\rho$.

Example 6.13: A current of 6 A flows both inwards (along the positive x-axis and outwards (along the positive y-axis) as shown in Fig. 6.33. Find the magnetizing force H at point $P\,(0, 0, 1)$.

Fig. 6.33: The current flows inwards along the positive x-axis and outwards along the positive y-axis

Solution: The magnetic flux density due to a semi-infinite straight current carrying conductor at a distance ρ from it is given by

$$B_P = \frac{\mu_0 I}{4\pi\rho}\hat{a}_n \tag{1}$$

where,
$$\hat{a}_n = \hat{I} \times \hat{a}_\rho$$

Therefore, the flux density due to two semi-infinite straight current carrying conductors at point P at par with Eq. (1) is given by

$$B_P = \frac{\mu_0 I}{4\pi\rho}\hat{a}_y + \frac{\mu_0 I}{4\pi\rho}\hat{a}_x$$

$$\frac{B_P}{\mu_0} = \frac{6}{4\pi \times 1}\hat{a}_y + \frac{6}{4\pi \times 1}\hat{a}_x$$

or
$$H = \frac{6}{4\pi}(\hat{a}_x + \hat{a}_y)$$

or
$$H = \frac{6}{4\pi}\sqrt{1^2 + 1^2} = \frac{6\sqrt{2}}{4\pi}$$

or
$$H = \frac{3}{\sqrt{2}\pi} = 0.675 \text{ A T/m.}$$

Example 6.14: Show that the magnetic flux density at a point between the regions of two infinite parallel sheets carrying current in the opposite direction (Fig. 6.34) with a surface current density K carried by each sheet is constant in magnitude.

Solution: The magnetic flux density at a point P due to an infinite surface sheet carrying a current normal to it (Fig. 6.35) is given by

Fig. 6.34: Two infinite parallel sheets carrying current in the opposite direction

$$B = \frac{\mu_0 K}{2}\hat{a}_n \tag{1}$$

where $\hat{a}_n = \hat{K} \times \widehat{OP}$

Let us now determine flux density B at par with Eq. (1) for the given problem as follows:

$\hat{K}$ = unit vector along K

$$B = \left[\frac{\mu_0 K}{2} + \frac{\mu_0 K}{2}\right]\hat{a}_z = \mu_0 K \hat{a}_z$$

or
$$\frac{B}{\mu_0} = K\hat{a}_z$$

or
$$H = K\hat{a}_z$$

Fig. 6.35: An infinite surface sheet carrying a current having a point P normal to it

Example 6.15: A current-carrying plane loop of conductor generate a magnetic induction $B(r)$. The current-element at p interacts with the B-field which is created by other current elements. Calculate the total force which the conductor loop exerts on itself. Consider the conductor as a 'thread of current'.

Solution:

$$F = \oint_C (j(r) \times B(r))d^3r = I\oint_C dr \times B(r)$$

$B(r)$ is effected by j:

$$B(r) = \frac{\mu_0}{4\pi} \oint_C d^3r'j(r') \times \frac{r - r'}{|r - r'|^3} = \frac{\mu_0}{4\pi} I\oint_C dr' \times \frac{r - r'}{|r - r'|^3}$$

Self-force:

$$F = \frac{\mu_0}{4\pi} I^2 \oint_C \oint_C dr \times \left(dr' \times \frac{r - r'}{|r - r'|^3} \right)$$

$$= \frac{\mu_0}{4\pi} I^2 \oint_C \oint_C \left(dr' \left(dr \cdot \frac{r - r'}{|r - r'|^3} \right) - \frac{r - r'}{|r - r'|^3} (dr' \cdot dr) \right)$$

For the first summand we have:

$$\oint_C dr \cdot \frac{r - r'}{|r - r'|^3} = -\oint dr \cdot \nabla_r \frac{1}{|r - r'|}$$

Using Stoke's theorem, we obtain:

Self-force:

$$= -\int_{F_C} df \cdot \underbrace{\text{curl grad} \frac{1}{|r - r'|}}_{0} = 0$$

It remains:

$$F = \frac{\mu_0}{4\pi} I^2 \oint_C \oint_C dr \cdot dr' \frac{r - r'}{|r - r'|^3}$$

When interchanging the variable r and r' the force F turns to $-F$. That is possible only if
$$F = 0$$

Conclusion:

The current-loop does not exert any force on itself.

Example 6.16: A coaxial cable consists of an infinitely long straight wire with a circular cross-section of the radius ρ_1, which is coaxially surrounded by a metal cylinder and inner radius ρ_2 and outer radius ρ_3 (Fig. 6.36). Inner and outer conductors carry the currents I_1 and I_2 respectively, which are homogeneously distributed over the respective conductor cross-section.

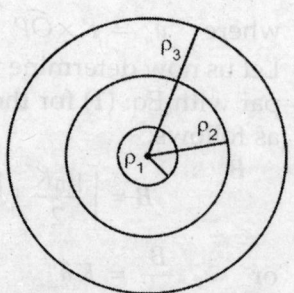

Fig. 6.36: Current carrying coaxial cable

i. Calculate the B-field in the whole space. Sketch the field behaviour for $I_1 = 0$ and $I_2 < 0$ as well as for $I_1 > 0$ and $I_2 = 0$.

ii. What does happen for the special case $I_1 = -I_2 = I$? Sketch for this special case, too, the field behaviour.

Solution: The symmetry considerations lead to the ansatz (Fig. 6.37)

$$\mathbf{B} = B_\varphi(\rho)\mathbf{e}_\varphi$$

F_ρ: Front surface (radius ρ) of a coaxial cylinder.

Maxwell equation:

$$\oint_{\partial F_\rho} d\mathbf{r} \cdot \mathbf{B} = 2\pi\rho B_\varphi(\rho) = \mu_0 \int_{F_\rho} d\mathbf{f} \cdot \mathbf{j}$$

Fig. 6.37

Applying Ampere's law, we have

$$\int_{F_\rho} d\mathbf{f} \cdot \mathbf{j} = \begin{cases} I_1 \dfrac{\rho^2}{\rho_1^2} ; & 0 \leq \rho \leq \rho_1, \\[2mm] I_1 ; & \rho_1 \leq \rho \leq \rho_2, \\[2mm] I_1 + I_2 \dfrac{\rho^2 - \rho_2^2}{\rho_3^2 - \rho_2^2} ; & \rho_2 \leq \rho \leq \rho_3, \\[2mm] I_1 + I_2 ; & \rho_3 \leq \rho. \end{cases}$$

The **B**-field is therewith already determined:

$$B_\varphi(\rho) = \frac{\mu_0}{2\pi} \begin{cases} \dfrac{I_1}{\rho_1^2}\rho ; & 0 \leq \rho \leq \rho_1 \\[2mm] I_1 \dfrac{1}{\rho} ; & \rho_1 \leq \rho \leq \rho_2 \\[2mm] \left(I_1 + I_2 \dfrac{\rho^2 - \rho_2^2}{\rho_3^2 - \rho_2^2} \right)\dfrac{1}{\rho} ; & \rho_2 \leq \rho \leq \rho_3 \\[2mm] (I_1 + I_2)\dfrac{1}{\rho} ; & \rho_3 \leq \rho \end{cases}$$

Special case: $I_1 = -I_2 = I$ (Fig. 6.38)

$$B_\varphi(\rho) = \frac{\mu_0 I}{2\pi} \begin{cases} \dfrac{1}{\rho_1^2}\rho ; & 0 \leq \rho \leq \rho_1, \\[2mm] \dfrac{1}{\rho} ; & \rho_1 \leq \rho \leq \rho_2, \\[2mm] \dfrac{\rho_3^2 - \rho^2}{\rho_3^2 - \rho_2^2}\dfrac{1}{\rho} ; & \rho_2 \leq \rho \leq \rho_3, \\[2mm] 0 ; & \rho_3 \leq \rho. \end{cases}$$

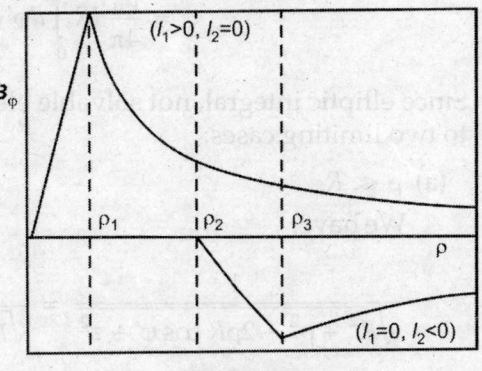

Fig. 6.38

Example 6.17: Calculate the vector potential $A(r)$ and the magnetic induction $B(r)$ of a circular conducting loop (thread of current). The current density reads in cylindrical coordinates (ρ, ϕ, z):

$$j(r) = I\delta(\rho - R)\delta(z)\hat{e}_\phi$$

The calculation of $A(r)$ leads to an elliptic integral which cannot be solved elementarily. Estimate this integral for the limits $\rho \ll R$ and $\rho \gg R$ by the use of suitable Taylor expansions. Show that for $\rho \gg R$ a dipole field emerges.

Find the corresponding magnetic dipole moment.

Solution: Because of the special form of the current density

$$j(r) = I\delta(\rho - R)\delta(z)e_\varphi$$

from symmetry reasons the vector potential will have the structure

$$A(r) = A(\rho, z)e_\varphi$$

with the unit-vector:

$$e_\varphi = (-\sin\varphi, \cos\varphi, 0)$$

y-component:

$$A_y = A(\rho, z)\cos(\varphi)$$

$$= \frac{\mu_0}{4\pi} \int d^3r' \frac{I\delta(\rho' - R)\delta(z')\cos\varphi'}{|r - r'|}$$

The evaluation for $\varphi = 0$ gives:

$$r = (x, y, z) = (\rho, 0, z)$$

$$r' = (x', y', z') = (R\cos\varphi', R\sin\varphi', 0)$$

$$A_y \rightarrow A(\rho, z) = \frac{\mu_0}{4\pi} \int_0^\infty \rho'd\rho' \int_0^{2\pi} d\varphi' \int_{-\infty}^{+\infty} dz' \frac{I\delta(\rho' - R)\delta(z')\cos\varphi'}{\sqrt{(R\cos\varphi' - \rho)^2 + R^2\sin^2\varphi' + z^2}}$$

$$= \frac{\mu_0}{4\pi} IR \int_0^{2\pi} d\varphi' \cos\varphi' \frac{1}{\sqrt{R^2 + \rho^2 - 2\rho R\cos\varphi' + z^2}}$$

Since elliptic integral, not solvable elementarily, therefore, we restrict out discussion to two limiting cases:

(a) $\rho \ll R$

We have,

$$\frac{1}{\sqrt{R^2 + \rho^2 - 2\rho R\cos\varphi' + z^2}} = \frac{1}{\sqrt{R^2 + z^2}}\left(1 + \frac{\rho^2 - 2\rho R\cos\varphi'}{R^2 + z^2}\right)^{-1/2}$$

$$\approx \frac{1}{\sqrt{R^2 + z^2}}\left(1 - \frac{\rho^2 - 2\rho R\cos\varphi'}{2(R^2 + z^2)}\right)$$

The first summand does not contribute.

For the second summand the relation $\int\limits_{0}^{2\pi} d\varphi' \cos^2 \varphi' = \pi$ leads to

$$A(\rho, z) \simeq \frac{1}{4} \mu_0 I R^2 \frac{\rho}{(R^2 + z^2)^{3/2}}$$

From that we derive in the net step the *components of the magnetic induction*. We need the curl in cylindrical coordinates:

$$\mathrm{curl} A = \frac{1}{\rho} \begin{vmatrix} e_\rho & \rho e_\varphi & e_z \\ \partial/\partial\rho & \partial/\partial\varphi & \partial/\partial z \\ 0 & \rho A(\rho, z) & 0 \end{vmatrix} = \frac{1}{\rho} \left(-\frac{\partial}{\partial z} \rho A(\rho, z), 0, \frac{\partial}{\partial \rho} \rho A(\rho, z) \right)$$

It follows therewith:

$$B_\rho = \frac{1}{4} \mu_0 I R^2 \frac{3z\rho}{\left(R^2 + z^2\right)^{\frac{5}{2}}}$$

$$B_\varphi = 0,$$

$$B_z = \frac{1}{4} \mu_0 I R^2 \frac{2}{\left(R^2 + z^2\right)^{\frac{3}{2}}}$$

(b) $\rho \gg R$

$$\frac{1}{\sqrt{R^2 + \rho^2 - 2\rho R \cos \varphi' + z^2}} = \frac{1}{\sqrt{\rho^2 + z^2}} \left(1 + \frac{R^2 - 2\rho R \cos \varphi'}{\rho^2 + z^2} \right)^{-\frac{1}{2}}$$

$$\simeq \frac{1}{\sqrt{\rho^2 + z^2}} \left(1 - \frac{R^2}{2\left(\rho^2 + z^2\right)} \right) + \frac{\rho R \cos \varphi'}{\left(\rho^2 + z^2\right)^{\frac{3}{2}}}$$

The first summand does not contribute to $A(\rho, z)$ in contrast to the second:

$$A(\rho, z) \approx \frac{1}{4} \mu_0 I R^2 \frac{\rho}{\left(\rho^2 + z^2\right)^{\frac{3}{2}}}$$

Cylindrical coordinates

$$r = \rho e_\rho + z e_z$$

$$\Rightarrow \left(\rho^2 + z^2\right)^{\frac{3}{2}} = r^3$$

$$e_z \times r = \rho e_z \times e_\rho = \rho e_\varphi$$

We get therewith the vector potential,

$$A(r) = \frac{\mu_0}{4\pi} \frac{m \times r}{r^3}$$

of a magnetic dipole of a circular conductor loop

$$m = \pi R^2 I e_z$$

Thus, the magnetic induction in the above relation is well-known for the *dipole field*:

Example 6.18: A small ferromagnet with the dipole moment m is fixed $x_0 = x_0 \hat{e}_x$ in such a way that it is freely rotatable in the xy plane. On it acts a homogeneous field $B_0 = B_0 \hat{e}_x$. In addition there is the field B_1 of a linear thread of current of the current density I. Determine the angle α between the dipole moment and the x-axis.

Solution: Potential energy of a magnetic dipole m in the external field B.

$$V = -m \cdot B_0$$

Equilibrium if V minimal $\Rightarrow$

$$m \uparrow\uparrow B_0 \text{ (in } x\text{-direction)}$$

Current-carrying wire $\Rightarrow$

$$B_1(r) = \mu_0 \frac{I e_\varphi}{2\pi\rho}$$

$$e_\varphi = (-\sin\varphi, \cos\varphi, 0)$$

Total field

$$B(r) = B_0 \begin{pmatrix} 0 \\ 0 \\ 0 \end{pmatrix} + \mu_0 \frac{I}{2\pi(x^2 + y^2)} \begin{pmatrix} -y \\ x \\ 0 \end{pmatrix}$$

at point $x_0 = (x_0, 0, 0)$:

$$B(x_0) = B_0 e_x + \mu_0 \frac{I}{2\pi x_0} e_y$$

The dipole orients itself parallel to B, thus enclosing with the x-axis the angle α:

$$\tan\alpha = \mu_0 \frac{I}{2\pi x_0 B_0}$$

Small angle:

$$\tan\alpha \simeq \alpha = \mu_0 \frac{I}{2\pi x_0 B_0}$$

Example 6.19: Two long parallel wires A and B placed at a separation of 4 cm, carry currents in opposite direction. It the current through each wire is 10 A, find the magnetic field at a point midway between the wire. **[MDSA]**

Solution: The condition is depicted in Figure 6.39. Let P be the point midway between the wires. The magnitude of magnetic field due to long straight wire is given by

$$B = \frac{\mu_0}{2\pi} \frac{I}{r}$$

Here $I = 10A$,

$$r = 4/2 = 2.4 \text{ cm} = 2.4 \times 10^{-2} \text{ m}$$
$$\mu_0 = 4\pi \times 10^{-7}$$

$$\therefore \quad B = \frac{4\pi \times 10^{-7} \times 10}{2\pi \times 2.410}$$

$$B = 8 \times 10^{-4} \text{T}$$

Fig. 6.39

This gives the magnetic field due two each wire at the point P. According to right hand thumb rule the magnetic field due to each wire at P is pointing vertically downward so the total field at P due to both wires is

$$B_t = 2B = 16 \times 10^{-4} \text{ T}$$

Example 6.20: Five wires which are perpendicular to the plane of paper carry current as shown in Fig. 6.40 Evaluate the line integral of magnetic field B for the close path.

[PU, KU]

Solution:

According to Ampere's law, we have

$$\oint B \cdot dr = \mu_0 I_{net}$$

The current directed towards us is shown as ⊙ is positive and directed away from us shown as ⊗ is negative for the path (Fig. 6.40).

Thus the net current is

$$I_{net} = 2I - 3I - I = -2I$$

The current $2I$ and I outside the path will not contribute to the line integral, therefore

$$\oint B \cdot dr = -2\mu_0 I$$

Fig. 6.40

GLIMPSES

1. A static magnetic field can originate from either a constant current or a permanent magnet.

2. Biot–Savart law is the relationship defining the differential magnetic field strength dB produced by a differential current element idl.

3. Biot–Savart law states that the magnetic field dB produced at a point P by a differential current element idl is proportional to the product idl and sine of the angle θ between the element idl and the line joining P to the element and is inversely proportional to the square of the distance r between P and the element, i.e.

$$dB = \frac{\mu_0}{4\pi} \frac{idl \sin\theta}{r^2}$$

μ_0 is the permeability of free space. In vector form, the relation is expressed as

$$dB = \frac{\mu_0}{4\pi} \frac{idl \times r}{r^3} (\text{Am}^{-1})$$

The direction of dB will be perpendicular to the plane containing dl and r, current elements have no separate existence. All elements making up the complete current filament contribute to B and must be included. The summation leads to the integral form of the Biot–Savart law as

$$B = \int_L \frac{\mu_0}{4\pi} \frac{idl \times r}{r^3} \text{ (line current)}$$

$$= \int_S \frac{\mu_0}{4\pi} \frac{k \times r}{r^3} dS \text{ (surface current)}$$

$$= \int_V \frac{\mu_0}{4\pi} \frac{j \times r}{r^3} dV \text{ (volume current)}$$

where, $idl = KdS = JdV$

4. *Maxwell's cork-screw rule:* If the direction of the current through the conductor is represented by the linear motion of the cork-screw motion, then the direction of the magnetic field can be represented by the direction of rotation of the cork-screw.

5. *Right hand clasp rule:* If a conductor carrying current is clasped in the right hand so that the thumb indicates the current direction, then the direction of the magnetic field is represented by the finger tips round the wire.

6. *Ampere's law:* The line integral of the tangential component of the magnetic field around a closed path is equal to the current enclosed by the path, i.e.

$$\oint B \cdot dl = \mu_0 I$$

This is quite similar to the use of Gauss's law to find D, given the charge distribution.

In order to utilize Ampere's law to determine B, there must be a considerable degree of symmetry in the problem. Two conditions must be met:

 i. At each point of the closed path, B is either tangential or normal to the path.

 ii. B has the same value at all points of the path where B is tangential.

Biot–Savart law can be used in selecting a path which meets the above conditions. In most cases a proper path will be evident.

The differential form of Ampere's law is $\nabla \times B = \mu_0 J$

where, J is the current density, $I = \int J \cdot dS$.

The differential form of Ampere's law establishes a local relation between current density J and magnetic induction B at the same point.

$\nabla \times B = \mu_0 J$ is one of the Maxwell's equations for static fields. If B is known throughout a region, then $\nabla \times B$ will be J for the region.

7. The magnetic field strength H depends only on (moving) charges and is independent of the medium. The force field associated with H is the magnetic flux density B which is given by $B = \mu_0 H$, where $\mu = \mu_0 \mu_r$ is the permeability of the medium. The unit of B is tesla (T)

$$1T = 1N/Am$$

The free space permeability $(\mu_0) = 4\pi \times 10^{-7} H\,m^{-1}$. μ_r is relative permeability and it is a pure number very near to unity.

8. Magnetic flux (ϕ) through a surface is defined as

$$\phi = \oint_S B \cdot dS$$

The sign of ϕ may be positive or negative depending upon the choice of the surface normal in dS. The unit of ϕ is weber (Wb)

$$1T = 1Wb\,m^{-2}, \quad 1H = Wb\,A^{-1}$$

9. A vector magnetic potential, A is defined such that

$$\nabla \times A = B$$

The above definition is consistent with the requirement $\nabla \cdot B = 0$. The units of A are $Wb\,m^{-1}$ or $T\,m$.

If the additional condition, $\nabla \cdot A = 0$ is imposed, then A can be determined from the known currents in the region of interest. For the three standard current configurations, the expressions are as follows:

Current filament: $A = \oint \dfrac{\mu I\,dl}{4\pi r}$

Sheet current: $A = \int_S \dfrac{\mu K\,dS}{4\pi r}$

Volume current: $A = \int_V \dfrac{\mu J\,dV}{4\pi r}$

Here r is the distance from the current element to the point at which the vector potential A is being calculated.

The above expressions cannot be applied if the current distribution itself extends to infinity.

10. Stokes' theorem states that the integral of the tangential component of a vector field F around C is equal to the integral of the normal component of $\nabla \times F$ over S:

$$\oint F \cdot dl = \int_S (\nabla \times F) \cdot dS$$

If F is chosen to be the vector potential A, Stokes' theorem gives

$$\oint A \cdot dl = \int_S B \cdot dS = \phi$$

11. A charge particle in motion in a magnetic field experiences a force at right angles to its velocity (v), with a magnitude proportional to the charge (q), the velocity (v), and the magnetic flux density.

$$F = q\,(v \times B)$$

Obviously, the direction of a particle in motion can be changed by a magnetic field. The magnitude of the velocity, v, and consequently the kinetic energy, will remain the same. This is in contrast to an electric field (E), where the force $F = qE$ does work on the particle and therefore changes its kinetic energy.

If the magnetic field B is uniform throughout a region and the particle has initial velocity normal to the field, the path of the particle is a circle of certain radius r.

The force of the field is of magnitude, $F = |q| vB$ and is directed towards the centre of the circle. The centripetal acceleration is of magnitude $\omega^2 r = \dfrac{v^2}{r}$. Then, by Newton's second law,

$$|q| vB = \frac{mv^2}{r} \quad \text{or } r = \frac{mv}{|q| B}$$

12. A current carrying conductor in an external field experiences a force

$$F = ILB \sin \theta$$

The magnetic force is actually exerted on the electrons that make up the current I. However, since the electrons are confined to the conductor, the force is effectively transferred to the heavy lattice. This transferred force can do work on the conductor as a whole.

13. The magnetic moment m of a planar current loop and the torque (τ) on a planar coil are related as $\tau = m \times B$.

The concept of magnetic moment is essential for an understanding of the behaviour of orbiting a charged particle, e.g. a positive charge q moving in a circular orbit at a velocity v, or an angular velocity ω, is equivalent to a current $I = \left(\dfrac{\omega}{2\pi}\right) q$, and so gives rise to a magnetic moment $m = \dfrac{\omega}{2\pi} q \hat{a}_n$, where $\hat{a}_n$ is unit normal.

In the presence of a magnetic field B there will be a torque $\tau = m \times B$, which tends to turn the current loop until m and B are in the same direction, in which orientation the torque will be zero.

REVIEW QUESTIONS

1. State and explain Ampere's circuital law.
2. State Biot–Savart law and deduce Ampere's circuital law from it for the steady current.
3. Show that the divergence of magnetic induction is always zero.
4. Define the magnetic vector potential for steady current and show that it satisfies the Poisson's equation $\nabla^2 A = -\mu_0 J$.
5. State Ampere's circuital law. Discuss why and how it was modified to include the displacement current.
6. How would you define the magnetic field at a point? Describe some general properties of the magnetic field.
7. State Ampere's law for magnetic field of a rectilinear current. Using it, discuss the magnetic field produced by a current along a circular cylinder of infinite length.
8. Prove that the line integral of magnetic field in a currentless closed path is always zero.
9. What do you understand by scalar and vector potentials? What is their significance?
10. Derive expressions for the magnetic vector potential and magnetic induction due to a small element of current carrying circuit at a large distance.

11. What do you understand by magnetic permeability and susceptibility of a substance?

12. Distinguish between para, dia and ferromagnetism.

13. What do you understand by the intensity of magnetization m. Establish the relation $J = \nabla \times m$, where J is current density vector in a non-uniformly magnetized material at a point where intensity of magnetization vector is m.

14. Obtain the magnetic boundary conditions and compare with elec-trostatic boundary conditions.

15. Show that the lines of magnetic field strength H are refracted at a change of media. Also, show that $\mu_1 \cot \theta_1 = \mu_2 \cot \theta_2$, where μ_1, μ_2 are permeabilities of the media and θ_1, θ_2 are the angles made with the normal.

PROBLEMS

1. A current carrying loop consists of three straight edges, each of length a and a semicircular portion of radius $a/2$, all in one plane as shown in Fig. 6.41. The current is I. (i) Calculate the magnetic field at the point C which is the mid point of PQ and (ii) the semi-circular portion is bent about the line PQ so that its plane is at $90°$ to the plane of the remaining loop. What is the magnetic field at point C for the same current.

   [*Ans.* (i) $\dfrac{\mu_0 I}{a}\left(\dfrac{1}{4} + \dfrac{\sqrt{5}}{2\pi}\right)$ (ii) $0.435\,\mu_0 I/a$ at an angle of $35°$ to the normal to the square area].

2. The current flowing in a conductor varies in time in accordance to the relation $I = I_0 \exp(-\alpha t)$, where I_0 and α are constants. Show that the charge Q which accumulated on the conductor after a time t_0 is given by $Q = I_0 \alpha^{-1}[1 - \exp(-\alpha t_0)]$.

3. Show that the magnetic vector potential for two long, straight parallel wires carrying the same current I, in opposite directions is given by $A =$

   $\dfrac{\mu_0 I}{2\pi}\log_e\left(\dfrac{r_2}{r_1}\right)\hat{a}_n$, where r_1 and r_2 are the distances from the field point to the wires and $\hat{a}_n$ is a unit vector parallel to the wires.

Fig. 6.41: A current carrying loop

4. Show that in a medium in which a uniform electric current of constant density exists, the magnetic field is $B = \dfrac{1}{2}\mu_0 J \times r$.

5. A closed surface in the form of a hemisphere has radius a. Its base is in the xy-plane and it is kept in a uniform field $B = B\hat{k}$. Determine the magnetic flux over the surface by direct integration.

   [*Ans.* Zero]

6. The radius and number of turns of a circular coil are 6 cm and 20 respectively. If a current of 1.5 A is passed through the coil, then calculate the magnetic field at a point on the axis of the coil at a distance of 8 cm from its centre. [**Ans.** 6×10^{-5} tesla]

7. A current of 10 A is flowing in a long straight wire. Find the magnetic field at a distance of 2 cm from the wire. [**Ans.** 10^{-4} tesla]

8. A long, cylindrical conductor of radius a carries a current along its length. The current density at a distance r from the axis of the conductor is given by $J(r) = J_0 r/a$, where J_0 being a constant. Show that the magnetic field for $r \le a$ and $r \ge a$ are given by:
   (i) $B(r \le a) = \mu_0 J_0 r^2/3a$ and (ii) $B(r \ge a) = \mu_0 a^2 J_0/3r$ respectively.

9. Using Biot–Savart law, show that the magnetic flux density due to a circular coil of radius a and carrying current I at a point distant z from its centre and lying on its axis is given by

$$B = \frac{\mu_0 I a^2}{2(a^2 + z^2)^{3/2}}.$$

10. Two infinite thin wires [one at $(0,0,0)$ and other at $(R,0,0)$] parallel to z-axis carry current I in opposite directions, where R is very large. Calculate the magnetic potential at (r, ϕ, z).

$$\left[\textbf{Ans. } \phi_m = \frac{I}{2\pi}(\pi - \phi)\right]$$

11. A circular conductor of radius $r_0 = 1$ cm has an internal field

$$H = \frac{10^4}{4}\left(\frac{1}{a^2}\sin ar - \frac{r}{a}\cos ar\right)\hat{a}_\phi \text{ A/m}.$$

If $a = \dfrac{\pi}{4r_0}$, show that the total current in the conductor is

$$\frac{8}{\sqrt{2}}\left(\frac{4}{\pi} - 1\right)\text{A}.$$

12. A vector potential is given by

$$A = x^2 y \hat{a}_x + y^2 x \hat{a}_y - 4xyz\hat{a}_z$$

Determine the field B. [**Ans.** $4zx\hat{a}_x - 4yz\hat{a}_y + (y^2 - x^2)\hat{a}_z$]

13. A current is flowing in a long circular conductor of radius a. The current is distributed in the wire in such a way that the current density at a distance r from the axis is given by $J = J_0\left(1 + \dfrac{r^2}{a^2}\right)$. Show that the total current in the wire is

$$I = \frac{3}{2}\pi J_0 a^2.$$

14. In problem 13, calculate the magnetic flux density both inside and outside the wire.

$$\left[\textbf{Ans. } B_{\text{inside}} = \frac{\mu_0 J_0 r}{2}\left(1 + \frac{r^2}{2a^2}\right) \text{ and } B_{\text{outside}} = \frac{3\mu_0 J_0 a^2}{4r}\right]$$

15. A circular coil of radius a carries a current I. Show that the magnetic scalar potential ϕ_m at a point P on its axis at a distance x from the coil is given by

$$\phi_m = \frac{I}{2}\left[1 - \frac{x}{\sqrt{x^2 - a^2}}\right].$$

16. A uniformly charged sphere whose radius is a and charge density ρ, rotates with a constant angular velocity ω. Calculate the magnetic flux density B at the centre of the sphere.

$$[\textbf{\textit{Ans.}} \; B = \frac{\mu_0 \rho \omega a^2}{3}]$$

17. The molecular weight of a compound is 400 and its density is $2 \times 10^3 \, kg/m^3$. Its magnetic susceptibility is 2.56×10^{-4} at 293 K. Show that the permanent dipole associated with each molecule is $m = 2.87 \times 10^{-23} \, A\,m^2$. Given Avogadro's number is 6.02×10^{23}.

SHORT ANSWER QUESTIONS

1. What will be the force on a particle carrying a charge q and moving with a velocity v in a region of magnetic induction B?
   Ans. $F = q(v \times B)$. We must note that this equation enables one to measure B without reference to its sources which are moving charges or currents.

2. What is the unit of B?
   Ans. $Wb/m^2 = N/A\,m = $ tesla.

3. What is the magnetic flux over a closed surface?
   Ans. Zero

4. How will you define an ampere?
   Ans. An ampere is that current which when flowing through two thin, parallel conductors of infinite length placed in vacuum, produces a mutual force of $2 \times 10^{-7} \, N/m$ between the conductors. $1\,A$ is the same as a rate of flow of $1\,C/s$.

5. What is an ideal solenoid?
   Ans. An ideal solenoid is essentially a circular current loop, where the current is not filamentary but in the form of a broad sheet flowing on the curved surface of a cylinder. In practice, one can consider a closely wound, long solenoid (length $\gg$ radius) with n turns per unit length as an ideal solenoid. Basically, ideal solenoid is one in which current always flows perpendicular to its length.

6. What will be the magnetic induction at interior points of an ideal solenoid defined in Q5?
   Ans. $B = \mu_0 nI$

7. How will you define magnetic field B due to any type of current source?
   Ans. $B = \nabla \times A$, where A is a vector potential given by $\frac{\mu_0 I}{4\pi} \int \frac{dl}{r}$.

8. When the right hand side of Ampere's law equation $\int B \cdot dl = \mu_0 I$ is non-zero, what does it signifies?
   Ans. B is a non-conservative field, i.e. $B = \nabla \times A$, where A is magnetic vector potential.

9. Is the dimensions of magnetic scalar potential ϕ_m same as that of an ampere?

   Ans. Yes

10. What will be the scalar potential for a magnetic dipole?

   Ans. $\phi_m(r, \theta) = \dfrac{1}{4\pi} \dfrac{m\cos\theta}{r^2}$

11. The atoms of certain elemental substances and the molecules of some compounds possess an intrinsic magnetic dipole moment. What are these substances called?

   Ans. Paramagnetic substances

12. For diamagnetic and paramagnetic substances, what is the relation between magnetic moment density *M* and magnetic intensity *H*?

   Ans. $M = \chi H$

13. What is relative permeability?

   Ans. $\mu_r = \dfrac{\mu}{\mu_0} = 1 + \chi$

14. Which one of the following three groups of magnetic materials exhibit the phenomenon of hysteresis?

   (i) diamagnetic (ii) paramagnetic (iii) ferromagnetic

   Ans. Ferromagnetic

15. In the region $0 < r < 0.5\,\text{m}$, in cylindrical coordinates, the current density is

   $J = 4.5e^{-2r}\hat{k}\,(\text{A}\,\text{m}^{-2})$ and $J = 0$ elsewhere. Use Ampere's law to find *B*.

   Ans. The current density is symmetrical about the origin, a circular path may be

   used in Ampere's law, with the enclosed current given by $\oint J \cdot dS$. Thus, for

   $r < 0.5$ m,

   $$B_\phi = (2\pi r) = \mu_0 \int_0^{2\pi} \int_0^r 4.5 e^{-2r} r\,dr\,d\phi$$

   $$B = \frac{1.125}{r}\mu_0 (1 - e^{2r} - 2re^{-2r})\hat{k}\,(\text{A}\,\text{m}^{-1})$$

   For any $r \geq 0.5$ m, the enclosed current is the same, 0.594π A, then

   $$B_\phi (2\pi r) = 0.594\,\mu_0\pi$$

   or $$B = \frac{0.297\mu_0}{r}\hat{k}\,(\text{A}\,\text{m}^{-1})$$

16. Calculate $\nabla \times B$ in cartesian coordinates due to a current filament along the z-axis with current *I* in the $\hat{k}$ direction.

   Ans. We have $B = \dfrac{\mu_0 I}{2\pi r}\hat{a}_\phi = \dfrac{\mu_0 I}{2\pi}\left(\dfrac{-y\hat{i} + x\hat{j}}{x^2 + y^2}\right)$

$$\nabla \times \boldsymbol{B} = \mu_0 \begin{vmatrix} \hat{i} & \hat{j} & \hat{k} \\ \dfrac{\partial}{\partial x} & \dfrac{\partial}{\partial y} & \dfrac{\partial}{\partial z} \\ -\dfrac{y}{x^2 + y^2} & \dfrac{x}{x^2 + y^2} & 0 \end{vmatrix}$$

$$= \mu_0 \left| \dfrac{\partial}{\partial x}\left(\dfrac{x}{x^2+y^2}\right) - \dfrac{\partial}{\partial y}\left(\dfrac{-y}{x^2+y^2}\right)\right| \hat{k} = 0$$

except at $x = y = 0$. This is consistent with $\nabla \times \boldsymbol{B} = \mu_0 \boldsymbol{J}$.

MULTIPLE CHOICE QUESTIONS

1. Biot–Savart law states the relation between magnetic intensity and
   (a) electric intensity   (b) electric current
   (c) work done   (d) vector potential   **[b]**

2. Which one of the following statement is incorrect in connection with Ampere's circuital law of magnetostatics?
   (a) There is no scalar magnetic field in regions including a current carrying conductor
   (b) The vector field at a point depends on current strength and inverse distance
   (c) The field due to a filament conductor is directed along it
   (d) The work done by a unit north pole in enclosing a current is independent of the shape and size of the path   **[c]**

3. A conducting circular loop of radius r carries a constant current I. It is placed in a uniform magnetic field B such that B is perpendicular to the plane of the loop. The magnetic force acting on the loop is given by
   (a) IrB (b) zero
   (c) πIrB (d) $2\pi IrB$ **[b]**

   [*Hint:* Magnetic field produced by the current I and B are along the same direction].

4. A small circular flexible loop of wire of radius r carries a current I. It is placed in a uniform magnetic field B. The tension in the loop will be doubled if
   (a) I is doubled   (b) both B and I are doubled
   (c) r is doubled   (d) both B and I are halved
        [(a) and (c)]

   [*Hint:* Tension in loop is BIr. Hence, T is doubled if either B is doubled or r is doubled.]

5. Two thin long parallel wires separated by a distance b are carrying a current I ampere each. The magnitude of the force per unit length exerted by one wire on the other is
   (a) $\mu_0 I^2/b^2$ (b) $\mu_0 I^2/2\pi b$
   (c) $\mu_0 I/2\pi b$ (d) $\mu_0 I/2\pi b^2$ **[b]**

6. A square loop carrying current is introduced in a uniform magnetic field inclined at angle α with the perpendicular to the plane. The coil experiences a torque proportional to
   (a) $\sin \alpha$
   (b) $\cos \alpha$
   (c) $\cot \alpha$
   (d) $\sin \alpha \cos \alpha$ [a]
   [*Hint:* $\tau = M \times B = MB \sin \alpha \perp$ to the plane in the direction of M]
7. The magnetic vector potential due to a single conductor carrying current is
   (a) infinite
   (b) zero
   (c) finite and definite
   (d) none of the above [a]
8. The chief characteristic of ferromagnetic substances is
   (a) permanent magnetization
   (b) no hysteresis under the action of a magnetic field
   (c) atomic dipoles are not aligned by the application of external field
   (d) none of the above [a]
9. The unit of magnetic permeability is
   (a) henry metre
   (b) weber
   (c) weber per square metre
   (d) henry/metre [d]
10. The magnetic field inside a current carrying toroidal solenoid is B If its radius is doubled and the current through it is also doubled, the magnetic field inside the solenoid will be
    (a) B
    (b) $B/2$
    (c) $2B$
    (d) $4B$ [a]
    [*Hint:* $B = \dfrac{NI}{2\pi r} \mu_0$ if I and r are doubled, B remains unchanged. N is total number of turns.]
11. A long solenoid has 20 turns/cm. The current necessary to produce a magnetic field of 20 mT inside the solenoid is approximately
    (a) 8 A
    (b) 4 A
    (c) 2 A
    (d) 1 A [a]
12. Force between two long straight parallel current carrying wires is F. If the current in one of them is doubled, the force between them will be
    (a) $2F$
    (b) $\sqrt{2}F$
    (c) $4F$
    (d) $2\sqrt{2}F$ [a]
13. A circular loop of radius r carrying current i_0 in counterclockwise direction is placed in a magnetic field B pointing towards the plane of the conductor. The force on point P due to small length dl of the coil will be
    (a) parallel to B
    (b) radially outwards
    (c) tangential at P
    (d) radially inwards [d]
14. A charged particle moving with velocity v in a magnetic field B experiences force F. Which of the following is a false statement?
    (a) F can perform work
    (b) F is a deflecting force
    (c) B and F are normal to each other
    (d) F depends on v [a]

15. Two parallel wires carry current along opposite directions. The resultant force experienced by the two wires is
    (a) respusive   (b) attractive
    (c) zero   (d) none of the above   [a]

16. The electrons in a cathode ray tube (CRO) of TV is deflected by a force F equal to
    (a) $F = qE$ (b) $F = 0$
    (c) $F = q(v \times B)$ (d) $F = q(E + v \times B)$ [d]
    [*Hint:* In CRO both electric and magnetic fields act.]

17. An electron beam is subjected to electric and magnetic fields in x-direction. The path of the electron will be
    (a) circular motion in plane perpendicular to x-direction
    (b) elliptical motion along x-direction
    (c) helical motion along x-direction
    (d) straight line motion along x-direction   [c]

18. An electric field on a plane is described by its potential
    $$V = 20 (r^{-1} + r^{-2})$$
    where, r is the distance from the source. The field is due to
    (a) a monopole   (b) a dipole
    (c) both a monopole and a dipole   (d) a quadrupole   [GATE] [c]

19. A material has conductivity of 10^{-2} mho/m and relative permittivity of 4. The frequency at which conduction current in the medium is equal to the displacement current is
    (a) 45 MHz   (b) 90 MHz
    (c) 450 MHz   (d) 900 MHz   [GATE] [a]

20. In Bohr model of the hydrogen atom, the electron circulates around the nucleus in a path of radius 5.1×10^{-11} m at a frequency $v = 6.8 \times 10^{15}$ rev/s. The value of B set up at the centre is
    (a) 5.6 Wb/m^2 (b) 7.8 Wb/m^2
    (c) 13.4 Wb/m^2 (d) 21.5 Wb/m^2 [c]
    [*Hint:* Current $i = ev = 1.6 \times 10^{-19} \times 10^{15} = 1.09 \times 10^{-3}$ A. The electron circulating around the nucleus is equivalent to a current carrying circular loop, where the loop radius is equal to the radius of the Bohr orbit. Thus, B at the centre of the orbit is

    $$B = \frac{\mu_0 i a^2}{2(x^2 + a^2)^{3/2}} = \frac{\mu_0 i}{2a} \quad (\because x = 0)$$

    $$= \frac{4\pi \times 10^{-7} \times 1.09 \times 10^{-3}}{2 \times 5.1 \times 10^{-11}} = 13.4 \text{ Wb/m}^2]$$

21. In MCQ 20, the equivalent magnetic dipole moment is
    (a) 2.3×10^{-24} A m^2 (b) 1.1×10^{-24} A m^2
    (c) 8.9×10^{-21} A m^2 (d) 8.9×10^{-24} A m^2 [d]
    [*Hint:* $p = Nia = 1 \times 1.09 \times 10^{-3} \times 3.14 \times (5.1 \times 10^{-11})^2$
    $= 8.9 \times 10^{-24}$ A m$^2]$

22. An infinitely long conductor of radius a is placed such that its axis is along the z-axis. The vector magnetic potential due to direct current I_0 flowing along $\hat{a}_z$ in the conductor is given by

(a) $A = -\dfrac{I_0}{4\pi a^2}\mu_0\,\hat{a}_z\,\text{Wb/m}$ (b) $A = -\dfrac{I_0}{4\pi a^2}\mu_0(x^2+y^2)\hat{a}_z\,\text{Wb/m}$

(c) $A = -\dfrac{I_0}{4\pi a^2}\mu_0\,x^2\hat{a}_z\,\text{Wb/m}$ (d) none of the above   **[b]**

23. Mutual inductance between the closed wound and coaxial solenoids of length l_1 and l_2 $(l_1 \gg l_2)$, turns N_2 and N_1 and radii r_1 and r_2 with $r_1 \approx r_2$ is

(a) $M_{12} = \dfrac{\mu N_1 N_2}{l_1}\pi r_1^2$ (b) $M_{12} = \dfrac{\mu N_1 N_2}{l_1 l_2}r_1$

(c) $M_{12} = \dfrac{\mu N_1 N_2}{l_1 l_2}\pi r_1^2$ (d) $M_{12} = \dfrac{\mu(N_1 N_2)}{l_1}\pi r_1^2$ **[a]**

24. The force per unit length of wires between two straight, infinite, parallel wires carrying currents i_1 and i_2, separated by a distance d and placed in air is

(a) $F = \dfrac{\mu_0 i_1 i_2}{2\pi d}$ (b) $F = \dfrac{\mu_0}{2\pi}\sqrt{\dfrac{i_1 i_2}{d}}$

(c) $F = \dfrac{\mu_0}{2\pi}\dfrac{\sqrt{i_1 i_2}}{d}$ (d) $F = \dfrac{\mu_0}{2\pi}\dfrac{i_1 i_2}{d^2}$ **[a]**

25. A square loop of wire of edge a carries a current i. If μ_0 is the permeability of free space, the value of the magnetic induction (B) at the centre of the loop is

(a) $B = \dfrac{1}{\sqrt{2}}\dfrac{\mu_0 i}{\pi a}$ (b) $B = \sqrt{2}\,\dfrac{\mu_0 i}{\pi a}$

(c) $B = 2\sqrt{2}\,\dfrac{\mu_0 i}{\pi a}$ (d) $B = 3\sqrt{2}\,\dfrac{\mu_0 i}{\pi a}$ **[b]**

26. The magnetization of a superconductor in a field of H is
(a) extremely small   (b) $-H$
(c) -1 (d) zero   **[d]**

27. The dimensions of the flux density is
(a) $MT^{-1}Q^{-1}$ (b) $MT^{-1}Q^{-2}$
(c) $MT^{1}Q^{1}$ (d) $MT^{-1}Q^{-2}$ **[a]**

Electromagnetic Induction and Maxwell's Equations

7.1 INTRODUCTION

One of the many electromagnetic phenomena familiar to us is *electromagnetic induction*, which was discovered, almost simultaneously, around 1830 by British physicist and chemist Michael Faraday and American scientist Joseph Henry independently. They discovered that an electric current is produced in a closed conducting loop when the flux of magnetic induction through the surface enclosed by this loop changes. This phenomenon is called *electromagnetic induction*, and the current produced an *induced current*. In this chapter, we shall investigate the behaviour of time-dependent electric and magnetic fields. In terms of fields, one may say that a time-varying magnetic field produces an electromotive force which may establish a current in a suitable closed circuit. Obviously, there exists a closer connection between the *E* and *B* fields, which is exhibited in phenomena of electromagnetic induction.

7.2 ELECTROMOTIVE FORCE

We are familiar that when a steady current flows in a single conductor, a potential difference exists between its two ends. This potential difference is maintained by a source of *electromotive force* (emf). There are many devices which provide emf such as cell, battery, dynamo, etc. We all know that these sources of emf convert some other kind of energy into electrical energy. We know that a battery uses stored chemical energy to drive a current in the circuit. If the battery maintains a constant potential difference of V volts between its terminals, its emf is said to be V volts. When the terminals of the cell or battery are connected to a conductor, an electric field is established in the conductor. This electric field is static. The line integral of the electric field over a path between two points (say P and Q) is equal to the potential difference between the two points. If these two points are the terminals of the battery, one obtains the emf of the battery as

$$V = \int_P^Q E \cdot dl \qquad (7.1)$$

We know that when the current flows in the conductor, the battery has to do work in order to keep the potential difference between its terminals as constant. If a charge q moves from one terminal of the battery to the other terminal through the conductor, the amount of work done by the battery is

$$Vq = \int_P^Q qE \cdot dl \qquad (7.2)$$

Therefore, the emf of the battery is

$$V = \frac{1}{q}\int_P^Q q\boldsymbol{E}\cdot d\boldsymbol{l} \tag{7.3}$$

V is expressed in volt (V) or $m^2\,kg\,s^{-2}\,C^{-1}$. Obviously, V has the dimensions of energy per charge. However, we must remember that the term emf is not a force.

Let us consider a metal rod moving with constant velocity u in a direction perpendicular to a uniform field B as shown in Fig. 7.1.

As discussed in Chapter 6, there will be a magnetic force on each electron in the rod given by

$$\boldsymbol{F} = -e(\boldsymbol{u}\times\boldsymbol{B})$$

The free electrons will move towards the end of the bar and a field E will be set as

$$\boldsymbol{E} = -\boldsymbol{u}\times\boldsymbol{B} \tag{7.4}$$

The potential difference between the two ends MN of the bar is

$$V_{MN} = \int_N^M \boldsymbol{E}\cdot d\boldsymbol{l} = uBL \tag{7.5}$$

Fig. 7.1: Metal rod in a uniform magnetic field

However, this potential difference will not produce any current flow. If, however, the metal rod forms a part of a circuit as shown in Fig. 7.2, a current I will flow.

One obtains the line integral of the force on a charge q round the circuit as

$$F = \oint(q\boldsymbol{E}\cdot d\boldsymbol{l}) = quBL \tag{7.6}$$

Thus, the induced emf in the closed circuit, due to the motion of the conductor is

$$V = \frac{1}{q}\oint(q\boldsymbol{E}\cdot d\boldsymbol{l}) = uBL \tag{7.6a}$$

One may call it a motional emf because it depends on the velocity of the conductor and not on its position. Obviously, we have two kinds of voltage: (i) electrostatic potential difference due to stationary charges and (ii) electromotive force due to moving charges. We can see that uBL is the magnetic flux through the area swept by the rod in unit time. If ϕ is total magnetic flux through the circuit, uBL is the rate of change of flux through the circuit. One can express

Fig. 7.2: Metal rod forming a part of the circuit

$$|\text{emf}| = uBL = \frac{d\phi}{dt} \tag{7.7}$$

7.3 FARADAY–HENRY LAW

One can summarize the observations of M Faraday and J Henry as:

  i. When a magnet is moved about in the neighbourhood of a conductor in the form of a closed circuit but with no source of emf, i.e. battery, then current is produced in the conductor so long as the movement lasts. However, the current disappears on cessation of movement of the magnet.

ii. One observes the same effect if the magnet is kept still and the circuit containing the conductor is moved. Obviously, for current to be produced in the conductor, there must be a relative motion. We must note that current can be produced without any mechanical movement.

iii. A transient current is induced in a loop of conducting wire if the stationary current in an adjacent circuit is switched on and off. This means that the transient current flowed when the flux through the circuit is altered. Obviously, the changing flux induced an electric field and, hence an emf in the circuit which causes the current to flow.

Faraday and Henry called this phenomenon *electromagnetic induction*. One can sum up Faraday's results in the form of a *flux rule*. We may state this in words as follows:

When the magnetic flux through a closed circuit is changing, an emf is induced in the circuit, the magnitude of which is proportional to the time rate of change of the magnetic flux through the circuit, one may write

$$|e| \propto \frac{d\phi}{dt} \qquad (7.8)$$

where e is the induced emf and ϕ, the magnetic flux.

The direction of the induced emf is determined by *Lenz's law*. According to this law, *the direction of the induced emf is such that the magnetic flux associated with the current generated by it, counteracts the cause generating it*. In other words, the induced current creates a magnetic flux which prevents the variation of the magnetic flux generating the induced emf. Therefore, one may write

$$e = -\frac{d\phi}{dt} \qquad (7.9)$$

ϕ is expressed in Wb or $m^2 \, kg \, s^{-1}C^{-1}$ and hence $\frac{d\phi}{dt}$ must be expressed in Wb/s or m^2

$kg \, s^{-2}/C^{-1}$. We must note that the induced emf is often called the *back emf*. Lenz's law expresses an important physical fact, viz. the tendency of a system to counteract the change in its state (electromagnetic inertia). In general, Eq. (7.9) can also be expressed in the following way,

$$e = \oint_L \boldsymbol{E} \cdot d\boldsymbol{l} = -\frac{d\phi}{dt} = -\frac{d}{dt}\int_S \boldsymbol{B} \cdot \hat{a}_n ds \qquad (7.10)$$

Referring to Fig. 7.3, if we divide the area limited by L into infinitesimal area elements, the magnetic flux $\phi = \int_S \boldsymbol{B} \cdot \hat{a}_n ds$. Also the emf V implies the existence of an electric field E such that $V = -\oint \boldsymbol{E} \cdot d\boldsymbol{l}$. Thus, we may write Eq. (7.10).

Fig. 7.3: Division of area into infinitesimal area elements

The electric field in the above equation is produced by the changing magnetic flux through the surface. The line integral of E is taken around the closed curve that forms the boundary of the surface S. This line integral is also called the emf around the circuit (Fig. 7.3a).

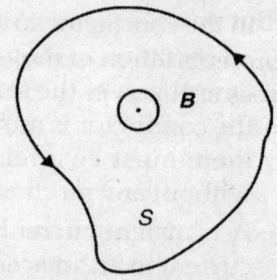

For convenience, we have taken the loop to be in one plane only. Faraday's law does not require this. It is applicable to an open surface of an arbitrary shape. We have seen in Chapter 6 that the surface integral of B is the same for all open surfaces bounded by a given closed curve. This is a consequence of divergence character of the field B. In Fig. 7.3a, we have therefore taken the simplest surface, viz. a plane surface bounded by the loop. We must note that there is no loss of generality in this. Then Eq. (7.10) is equivalent to saying:

Fig. 7.3a: Surface S is an open surface. Magnetic field is changing over this surface

A time dependent magnetic field implies the existence of an electric field such that the circulation of the electric field along an arbitrary closed path is equal to negative of the time rate of change of the magnetic flux through a surface bounded by the path.

This is another way of stating the Faraday–Henry law of electromagnetic induction. It gives us a deeper insight into the physical content of the phenomenon of electromagnetic induction, i.e. the fact that an electric field must exist wherever a magnetic field is changing with time, the two fields are related by Eq. (7.10). The electric field can be determined by measuring the force on a charge at rest in the region where the magnetic field is varying.

7.4 INDUCTION LAW FOR MOVING CIRCUITS

Let us consider a circuit of any arbitrary shape moving with in-stantaneous velocity v in uniform magnetic field B. An element $dl(PQ)$ will move in time dt a distance vdt to a position $P'Q'$. The displacement of element in time $dt = vdt$.

Let the velocity of the conducting electrons of element, relative to wire along the length of wire is u, then relative velocity is $v + u$.

Fig. 7.4: Moving circuit

The magnetic force on each electron of charge $-e$,

$$F_m = -e(v + u) \times B \qquad (7.11)$$

The component of this force along the element δl is F_{11} $[-e(v+u) \times B] \cdot \hat{a}_l$, where $\hat{a}_l$ is unit vector along PQ. As u and $\hat{a}_l$ are along the same direction, we have

$$u \times B \cdot \hat{a}_l = B \cdot \hat{a}_l \times u = 0$$

$$\therefore \qquad F_{11} = -ev \times B \cdot \hat{a}_l \qquad (7.12)$$

Equation (7.12) shows that there is an electric field $E = v \times B$ induced in the wire. Its component along the wire is $v \times B \cdot \hat{a}_l$.

The induced emf is the line integral of this field round the circuit, i.e. the induced emf is given by

$$\varepsilon = \oint (v \times B) \cdot \hat{a}_l dl \qquad (7.13)$$

But $vdt \times \hat{a}_l dl$ = area $PP'Q'Q$ swept by element dl in time dt, hence $(vdt \times \hat{a}_l dl)$. $\boldsymbol{B}$ represents the flux change across with the element in time dt. Obviously, the flux change across entire area

$$d\phi = \oint (vdt \times \hat{a}_l dl) \cdot \boldsymbol{B}$$

$$\therefore \quad \frac{d\phi}{dt} = \oint (v \times \hat{a}_l dl) \cdot \boldsymbol{B} = \oint \boldsymbol{B} \cdot (v \times \hat{a}_l) dl$$

$$= -\oint (v \times \boldsymbol{B}) \cdot \hat{a}_l dl \tag{7.14}$$

Comparing Eqs. (7.13) and (7.14), one obtains

$$\varepsilon = -\frac{d\phi}{dt} \tag{7.15}$$

This is *Faraday–Henry law*. Obviously, the law also holds for circuits moving in uniform magnetic field $\boldsymbol{B}$.

7.5 INTEGRAL AND DIFFERENTIAL FORMS OF FARADAY–HENRY LAW

We have read that the induced emf is equal to the line integral of the induced electric field E around the coil. Thus, we have

$$\varepsilon = \oint E \cdot \hat{a}_l dl$$

The magnetic flux through the coil is

$$\phi = \oint_S B \cdot \hat{a}_n dS$$

where the integral is taken over any area bounded by the circuit. One can express Eq. (7.15) as

$$\oint E \cdot \hat{a}_l dl = -\frac{d}{dt} \int B \cdot \hat{a}_n dS$$

Since the surface S neither changes its shape nor position with time, one can express the above equation as

$$\oint E \cdot \hat{a}_l dl = \int_S \frac{dB}{dt} \hat{a}_n dS \tag{7.16}$$

Equation (7.16) is the *integral form* of Faraday–Henry law.

Making use of Stokes' theorem, one obtains

$$\oint E \cdot \hat{a}_l dl = \int_S \nabla \times E \cdot \hat{a}_n dS = -\int \frac{\partial B}{\partial t} \cdot \hat{a}_n dS$$

$$\therefore \quad \int \left(\nabla \times E + \frac{\partial B}{\partial t} \right) \cdot \hat{a}_n dS = 0 \tag{7.17}$$

We have replaced the total derivative by partial derivative as one is only concerned with the changes in field B with time at the fixed position of the elemental area dS.

Equation (7.17) must hold for any arbitrary surface, we have

$$\nabla \times E = -\frac{\partial B}{\partial t} \qquad (7.18)$$

This is called the *differential form* of Faraday–Henry law of electro-magnetic induction. A knowledge of the time rate of change of B at any point thus enables us to find the curl of the electric field at that point One can compare this situation with the differential form of Gauss's law which relates the charge density at any point with the divergence of electric field at that point. Equation (7.18) together with $\nabla \cdot E = \frac{\rho}{\varepsilon_0}$, shows that the electric field has a non-conservative part due to changing magnetic flux density as well as a conservative part due to electric charge density.

So far we have seen that there are two kinds of sources of electro-magnetic field. The first kind of source is associated with a system such as electrostatics in which energy is conserved during a cyclic process within the system. The sources of such a *irrotational* or conservative system are well described by the relation $\nabla \cdot E = \frac{\rho}{\varepsilon_0}$, where $E = -\nabla \phi$. We have seen that such a field has no curl because $\nabla \times (\nabla \phi) = 0$.

The second kind of electromagnetic field source is associated with a system in which there is energy transferred in a cyclic process (e.g. magnetic field of a solenoid). One can specify such a field by the curl sources and has no divergence. In general, one finds that electromagnetic fields have both types of sources, and hence, the complete specification of such a vector field should include both types of sources. We must note that such a specification is not only necessary but sufficient. It is important to note that *any vector field is uniquely determined if its divergence and curl sources are given*. This is *Helmholtz theorem*.

Taking divergence of Eq. (7.18), one obtains

$$\nabla \cdot \nabla \times E = -\frac{\partial}{\partial t}(\nabla \cdot B) = 0 \qquad (7.18a)$$

Obviously, $\nabla \cdot B$ is necessarily independent of time at every point in space and this condition can reasonably be satisfied if one assumes

$$\nabla \cdot B = 0 \qquad (7.19)$$

This means, B is always *solenoidal*.

One can draw two following important conclusions from Faraday's law [Eq. (7.18)].

i. The electric field E is no longer a conservative field when the magnetic field varies with time. This means that energy can flow between electric and magnetic forms through time-varying fields.

ii. No free magnetic monopoles can exist. As stated earlier, all magnetic poles occur in pairs, positive and negative.

We have seen that how Faraday–Henry law of induction shows that the electric and magnetic fields are inter-related. When we consider the time-dependence of these fields, then their independent nature disappears. Obviously, one can look upon these fields as a single field, i.e. *electromagnetic field*.

7.6 SELF INDUCTANCE

Let us consider a circuit carrying current I (Fig. 7.5). According to Ampere's law, the current produces a magnetic field which, at each point is proportional to I. One may compute the magnetic flux through the circuit due to its own magnetic field, and call it the *self flux*. The flux, designated by ϕ_I, is then proportional to the current I, and we may write

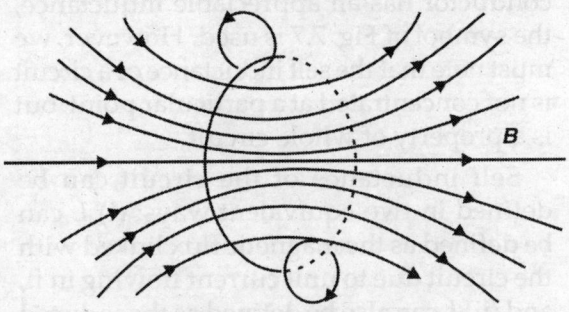

Fig. 7.5: Self inductance

$$\phi_I = LI \tag{7.20}$$

where, L is constant and depends on the geometrical shape of the conductor and is called the *self inductance* of the circuit. It is expressed in WbA^{-1}, a unit called *henry*, in honour of Joseph Henry, and abbreviated as H, i.e. $H = \dfrac{Wb}{A} = m^2\,kg\,C^{-2}$.

If the current I changes with time, the magnetic flux ϕ_I through the circuit also changes, and according to the law of electromagnetic induction, an emf is induced in the circuit. This special case of electromagnetic induction is called *self induction*.

Let us consider a situation in which the circuit is stationary but the magnetic field, and hence the flux changes with time. Suppose that the current in the circuit is time dependent. Obviously, the flux in this situation is a function of time, i.e. $\phi = \phi(t)$. Therefore,

$$\frac{d\phi}{dt} = \frac{d\phi}{dI}\cdot\frac{dI}{dI} = L\frac{dI}{dt} \qquad \text{[using Eq. (7.20)]}$$

$$\therefore \quad L = \frac{d\phi}{dt} \tag{7.21}$$

Using Eq. (7.21), Eq. (7.15) can now be written as

$$\varepsilon_L = -L\frac{dI}{dt} \tag{7.22}$$

The minus sign indicates that ε_L is opposed to change in the current. So if the current increases, $\dfrac{dI}{dt}$ is positive and ε_L acts in the same direction as the current (Fig. 7.6).

Fig. 7.6: Direction of self induced emf in a circuit

Therefore, ε_L always acts in a direction that opposes the change in the current. When we wrote Eq. (7.22), we assumed the circuit to be rigid and therefore we considered L as constant while computing the time derivative. If the shape of the circuit is variable, L is not constant, and instead of Eq. (7.22) we must write

$$\varepsilon_L = -\frac{d}{dt}(LI) \tag{7.23}$$

In diagrams drawn to indicate that a conductor has an appreciable inductance, the symbol of Fig. 7.7 is used. However, we must note that the self inductance of a circuit is not concentrated at a particular point, but is a property of whole circuit.

Fig. 7.7: Representation of self inductance in a circuit

Self inductance of the circuit can be defined in two equivalent ways: (i) L can be defined as the magnetic flux linked with the circuit due to unit current flowing in it, and (ii) L can also be defined as the induced emf in the circuit per unit rate of change of current in it.

When the current in the circuit is first switched on, the back emf opposes the growth of current, so that the current flows against the back emf and does work against it. Figure 7.8a shows an electric circuit containing a resistance and a self inductance. When an emf V is applied to a circuit by closing a switch (Fig. 7.8), the current does not instantaneously attain the value V/R corresponding to Ohm's law. This process is due to the self induced emf ε_L which opposes the change in the current and is present while the current increases from zero up to its final constant value. The total e.m.f applied to the circuit is then

Fig. 7.8: (a) Electric circuit contains a resistance and a self inductance. (b) Establishment of current in a circuit

$V - \varepsilon_L = V - L\left(\dfrac{dI}{dt}\right)$. Ohm's law is now

$$IR = V - \varepsilon_L \quad \text{or} \quad IR = V - L\left(\frac{dI}{dt}\right) \tag{7.24}$$

This may be written as $R(I - V/R) = -L\left(\dfrac{dI}{dt}\right)$ or, separating the variables I and t

$$\frac{dI}{I - V/R} = -\frac{R}{L}dt$$

Integrating and taking care that at $t = 0$, the current is also zero ($I = 0$),

we have

$$\int_0^I \frac{dI}{I - V/R} = -\frac{R}{L}\int_0^t dt$$

or

$$I = \frac{V}{R}\left(1 - e^{-\frac{Rt}{L}}\right) = I_0\left(1 - e^{-\frac{Rt}{L}}\right) \tag{7.25}$$

The second term in the parenthesis decreases with time and the current asymptotically approaches the value V/R which is given by Ohm's law (Fig. 7.8b). If R/L is large, the current reaches this value very fast but if R/L is small it may take a long time before the current stabilizes. When switch is made off, then the only emf that remains is $\varepsilon_L = -L\dfrac{dI}{dt}$ and Ohm's law becomes

$$RI = -L\frac{dI}{dt} \quad \text{or} \quad \frac{dI}{I} = -\frac{R}{L}dt$$

If we count time ($t = 0$) from the instance when V is removed from the circuit, the initial current is V/R. Integrating, one obtains

$$\int_{V/A}^{I} \frac{dI}{I} = -\frac{R}{L}\int_{0}^{t} dt$$

or

$$I = \left(\frac{V}{R}\right)e^{-\frac{Rt}{L}} = I_0 e^{-\frac{Rt}{L}} \tag{7.26}$$

The current decreases exponentially, as shown in Fig. 7.9. The larger the resistance R, or the smaller the inductance L, the faster is the drop in the current. The time required for the current to drop to $\dfrac{1}{e}$, or approximately 37% of the initial value, is $\tau = \dfrac{L}{R}$. This time is called *relaxation time*.

Fig. 7.9: Decay of the current in a circuit after emf has been removed

Let us now calculate the work done against back emf in a short time dt when the current is growing in a circuit (Fig. 7.8). We have

$$dW = -\varepsilon I\,dt = L\frac{dI}{dt} \cdot I\,dt$$

Hence, the total work done in bringing the current from zero to a steady maximum value I_0 is

$$W = L\int_{0}^{I_0} I\,dI = \frac{1}{2}LI_0^2 \tag{7.27}$$

This work is stored as the energy of the magnetic field. Equations (7.20), (7.23) and (7.27) enable one to define the self inductance (L) of a circuit in the following ways:

(a) We have $\phi = LI$ $\therefore$ $L = \dfrac{\phi}{I}$; if $I = 1\,\text{A}$, $L = \phi\,\text{H}$. Thus, the self inductance of a circuit

(in henry) is defined as the magnetic flux (in weber) linked with the circuit when 1 ampere current flows through it.

(b) We have $\varepsilon = -L\dfrac{dI}{dt}$ $\therefore$ $L = \varepsilon / -\dfrac{dI}{dt}$; when $\dfrac{dI}{dt} = 1$ A/s, we have $L = \varepsilon$. Thus, the self inductance of a circuit is the emf induced in the circuit when the rate of decay of current in circuit is unity. We must note that emf is in volt, current in ampere and L in henry.

(c) We have $W = \dfrac{1}{2}LI_0^2$ $\therefore$ $L = \dfrac{2W}{I_0^2}$. Obviously, the self inductance of a circuit is

defined as twice the work done against the induced emf in establishing unit current in the coil. We must note that work is in joule, current in ampere and L in henry.

We must note that the above three definitions of self inductance are identical only when the permeability of the medium round which the coil is wound remains constant. The role of self inductance in an electric circuit is same as that of inertia in mechanical motion of bodies. Obviously, the self inductance of a coil is a measure of its ability to oppose change in current through it.

Self Inductance of a Solenoid

Consider a long air cored solenoid having length l, cross-sectional area A and the number of turns per unit length n. If I ampere is the current flowing through the solenoid, then the magnetic field inside the solenoid is

$$A = \mu_0 nI \ \frac{\text{W}}{\text{m}^2}$$

The magnetic flux linked with the solenoid is $\phi = \mu_0 nI \times$ effective area of solenoid $= \mu_0 nI(nlA)$ weber ($\therefore$ effective area of the solenoid = area of each turn $\times$ total number of turns in solenoid). Thus,

$$\phi = \mu_0 n^2 AlI \qquad (7.28)$$

But
$$\phi = LI$$

$\therefore$
$$LI = \mu_0 n^2 AlI \quad \text{or} \quad L = \mu_0 n^2 Al \text{ (henry)}$$

7.7 MUTUAL INDUCTANCE (COUPLED CIRCUITS)

Let us consider a coil such as coil (1) in Fig. 7.10a carrying a current I. If a second coil (2) (Fig. 7.10a) is brought near this coil there will be a magnetic flux ϕ_2 through coil (2) due to the current in coil (1). Since ϕ_2 is linearly proportional to I,

$$\phi_2 = L_{21} I_1 \qquad (7.29)$$

where, L_{21} is a constant and represents the magnetic flux through circuit (2) per unit current in circuit (1). Similarly, if a current I_2 circulates in circuit (2), a magnetic field is produced, and it, in turn, produces a magnetic flux ϕ_1 through circuit (1) which is proportional to I_2 (Fig. 7.10b). Hence, we may write

$$\phi_1 = L_{12} I_2 \qquad (7.30)$$

Fig. 7.10: Mutual induction

where, L_{12} is again a constant. Now, question arises what is the relationship between these two constants? The potential energy of the system is given by

$$\cup = -\phi_2 I_2 = -L_{21}I_1I_2 = -\phi I_1 = -L_{12}I_2I_1$$

$$\therefore \qquad L_{21} = L_{12} = M \tag{7.31}$$

Calculations show and experiments confirm Eq. (7.31) in the absence of ferromagnetics, the magnetic flux through circuit (1) due to the unit current in circuit (2) is the same as before. This common coefficient (M) is called the *mutual inductance* of the two circuits and it is same in both the cases. This remarkable property of mutual inductance is usually called the *reciprocity theorem*. Obviously, mutual inductance is symmetrical. The coefficient M depends on the shapes of the circuits and their relative orientation. It is also measured in henry, since it corresponds to $\dfrac{\text{Wb}}{\text{A}}$.

Now

$$\cup = -\phi_1 I_1 = -I_1 \int B_1 \cdot \hat{a}_n ds$$

$$= -I_1 \int \nabla \times A_{12} \cdot \hat{a}_n ds = -I_1 \int A_{12} \cdot \hat{a}_n dl_1$$

(by Stokes' theorem)

$$= -I_1 \int \left(\frac{\mu}{4\pi} \int \frac{I_2 \hat{a}_{12} dl_2}{|r|} \right) \cdot \hat{a}_{l_1} dl_1$$

$$= -\frac{\mu}{4\pi} I_1 I_2 \iint \frac{\hat{a}_{l_1} dl_1 \cdot \hat{a}_{l_2} dl_2}{r} = -L_{12}I_1I_2$$

$$\therefore \qquad L_{12} = L_{21} = M = \frac{\mu}{4\pi} \iint \frac{\hat{a}_{l_1} dl_1 \cdot a_{l_2} dl_2}{r} \tag{7.32}$$

Relation (7.32) is known as *Neumann's formula*.

If the current I_1 is variable, the flux ϕ_2 through circuit (2) changes and an emf V_{M_1} is induced in this circuit. This emf is given by

$$V_{M_1} = -M \frac{dI_1}{dt}$$

In writing this equation, we assume that the circuits are rigid and fixed so that M is constant. Similarly, if the current I_2 is variable, an emf V_{M_1} is induced in circuit (1), given by

$$V_{M_1} = -M\frac{dI_2}{dt} \tag{7.33}$$

This is why M is called *mutual inductance*, since it describes the mutual effect or influence between the two circuits. In addition, if the circuits are moved relative to each other, resulting in a change in M, emfs are also induced in them.

We see that there will be an exchange of energy between the circuits. Common and practical applications of this process are the *transformer* and the *induction generator*. Another application of mutual induction, in a broader sense, is the transmission of a signal from one place to another producing a variable current in one circuit, called the *transmitter*. This current, in turn, acts on another circuit coupled to it, called the *receiver*. This is the case for telegraph, radio, television, radar, etc.

The most important and fundamental aspect of mutual induction is that *energy can be exchanged between two circuits via the electromagnetic field*, one can say that the electromagnetic field produced by the currents in the circuits acts as a carrier of energy, transporting the energy through space from one circuit to the other.

We note that mutual induction between two circuits is a macroscopic phenomenon, resulting from elementary interactions between the moving charges which constitute their respective currents. Obviously, we may conclude from the phenomenon of mutual induction that the electromagnetic interaction between two charged particles can be described as an exchange of energy via their mutual electromagnetic field.

Cores with Air Gaps

Magnetic circuits with small air gaps are very common. The gaps are generally kept as small as possible since the NI drop of the air gap is often much greater than the drop in the core. The flux finger points outward at the gap, so that the area at the gap exceeds the area of the adjacent core. Provided that the gap length ℓ_a is less than $\frac{1}{10}$th the smaller dimension of the core, an apparent area, S_a of the air gap can be calculated. For a rectangular core of dimensions a and b,

$$S_a = \frac{1}{10}(a+\ell_a)(b+\ell_a)$$

If the total flux in the air gap is known, H_a and $H_a\ell_a$ can be computed directly.

$$H_a = \frac{1}{\mu_0}\left(\frac{\phi}{S_a}\right), \quad H_a\ell_a = \frac{\ell_a\phi}{\mu_0 S_a}$$

For a uniform iron core of length ℓ_a with a single air gap, Ampere's law reads as

$$NI = H_i\ell_i + H_a\ell_a = H_i\ell_i + \frac{\ell_a\phi}{\mu_0 S_a}$$

If the flux ϕ is known, it is not difficult to compute the *NI* drop across the air gap, obtain B_i, take H_i from the appropriate *B–H* curve and compute the *NI* drop in the core, $H_i\ell_i$. The sum is the *NI* required to establish the flux ϕ. However, with *NI* given, it is a matter of trial and error to obtain B_i and ϕ, as will be seen in the problems. Graphical methods of solution are also available.

Multiple Coils

Two or more coils on a core could be wound such that their mmfs either aid one another or oppose. Consequently, a method of indicating polarity is given in Fig. 7.11. An assumed direction for the resulting flux ϕ could be incorrect, just as an assumed current in a dc circuit with two or more voltage sources may be incorrect. A negative result simply means that the flux is in the opposite direction.

(a)   (b)

Fig. 7.11: Multiple coils

7.8 ENERGY OF THE MAGNETIC FIELD

We have seen that to maintain current in a circuit, energy must be supplied. Consider a simple circuit as shown in Fig. 7.12 in which R is the resistance, an inductive coil with self-inductance L and ε is the emf of the battery. Let us consider that the current flowing in the circuit at any time

Fig. 7.12: Magnetic energy stored in an inductor

t be $I(t)$. We have the voltage drop across the coil as $L\dfrac{dI}{dt}$

and hence the forward emf is $-\varepsilon - L\dfrac{dI}{dt}$. Using Ohm's law, we have

$$\varepsilon - L\frac{dI}{dt} = IR \tag{7.34}$$

Now, we consider the work done by the emf ε in moving a small amount of charge Q through the circuit

$$dW = \varepsilon dQ = \varepsilon I dt$$

$$\therefore \quad \frac{dW}{dt} = I\varepsilon = IL\frac{dI}{dt} + I^2R \tag{7.35}$$

Therefore, total work done by the battery in time interval T in which the current changes from 0 to I_T is obtained as

$$W = \int_0^T I\varepsilon\, dt = L\int_0^T I\frac{dI}{dt}dt + R\int_0^T I^2 dt = \frac{1}{2}LI_T^2 + R\int_0^T I^2 dt \tag{7.36}$$

The first term on the RHS of Eq. (7.36) is the energy stored in the inductance in time T and the second term is the energy dissipated as heat in the resistance. Obviously, the magnetic energy required to increase current from zero to the value I is thus $\frac{1}{2}LI^2$.

Magnetic Energy Stored in a Series of Inductances

Let us consider a number of coils connected in series. Assume that the initial current in each coil is zero and that they attain their equilibrium value in all the circuits simultaneously at $t = T$. Obviously, at any instant t, in the interval $0 \le t \le T$, the current in each circuit $I_k(t)$ and the flux through it $\phi_k(t)$ will be some fraction α of their ultimate values, i.e.

$$I_k(t) = \alpha I_k \quad \text{and} \quad \phi_k(t) = \alpha\phi_k$$

The induced emf in the kth circuit is

$$\varepsilon_k = \frac{d\phi_k(t)}{dt}$$

$\therefore$ **Total work** done by applied source for k^{th} circuit,

$$W_k = \int_0^t \varepsilon_k I_k(t)dt = I_k\phi_k\int_0^T \alpha\frac{d\alpha}{dt}dt$$

$$= I_k\phi_k\int_0^1 \alpha d\alpha = \frac{1}{2}I_k\phi_k \tag{7.37}$$

On summing for all circuits, we have

$$W = \frac{1}{2}\sum_k I_k\phi_k \tag{7.37a}$$

We know that the magnetic flux depends on the self as well as mutual inductances, and hence we have

$$\phi_k = L_k I_k + \sum_{j=k} M_{kj}I_j \tag{7.38}$$

$$\therefore \quad W = \frac{1}{2}\sum_k L_k I_k^2 + \frac{1}{2}\sum_k \sum_j M_{kj}I_k I_j \tag{7.39}$$

Thus, we have for a pair of coils

$$W = \frac{1}{2}L_1 I_1^2 + \frac{1}{2}L_2 I_2^2 + M I_1 I_2 \tag{7.40}$$

where $M = M_{12} = M_{21}$

One can express Eq. (7.39) in a more general and standard form. For any circuit

$$\phi_k = \int_k B \cdot \hat{a}_n ds = \oint \nabla \times A \cdot \hat{a}_n ds = \oint_k A \cdot \hat{a}_l dl \qquad (7.41)$$

Now, Eq. (7.37a) can be expressed as

$$W = \frac{1}{2} \sum_k L_1 I_1^2 + \frac{1}{2} L_2 I_2^2 + M I_1 I_2$$

Let us now make a transition from discrete to the continuous case by using the relation $I \hat{a}_l dl = j d\tau$ and taking the integral over whole space, since contributions arise only from regions where j is finite, i.e. by using the relation

$$\sum_k \oint_k \rightarrow \int_V$$

$$\therefore \qquad W = \frac{1}{2} \int_V (A \cdot J) d\tau \qquad (7.42)$$

But $J = \nabla \times H$

$$\therefore \qquad W = \frac{1}{2} \int_V \{A \cdot (\nabla \times H)\} d\tau \qquad (7.43)$$

Using vector identity,

$$\nabla \cdot (A \times H) = H \times (\nabla \times A) - A \times (\nabla \times H)$$

Rewrite Eq. (7.43) as

$$W = \frac{1}{2} \int_V \{H \cdot (\nabla \times A)\} d\tau - \frac{1}{2} \int_V \{\nabla \cdot (A \cdot H)\} d\tau \qquad (7.44)$$

Using Gauss's divergence theorem to change volume integral in second term on RHS of Eq. (7.44), one obtains

$$W = \frac{1}{2} \int_V \{H \cdot B\} d\tau - \frac{1}{2} \int_V (A \times H) \cdot \hat{a}_n ds \qquad (7.45)$$

As the volume integral is to be taken for whole space, the surface integral must be taken over the sphere at infinity. Because H and A fall off rapidly at large distance $(H \sim r^{-3}, A \sim r^{-2})$, the surface integral vanishes as $r \rightarrow \infty$. Hence, Eq. (7.45) reduces to

$$W = \frac{1}{2} \int_V (H \cdot B) d\tau \qquad (7.46)$$

Equation (7.46) shows that the magnetic energy may be regarded as distributed throughout the region occupied by the field with density $\frac{1}{2}(H \cdot B)$. As $B = \mu H$ for linear medium, we have the density $= \frac{1}{2} \mu H^2 = \frac{1}{2} \frac{B^2}{\mu}$. We must note that this relation has been obtained by assuming that medium is linear. If the medium is non-linear, then the analysis will have to be modified suitably.

7.9 HYSTERESIS

We shall now consider the behaviour of ferromagnetic substances such as iron, cobalt, gadolinium and their alloys and compounds under the action of a magnetic field. As pointed out earlier, whether magnetized or not, these substances are made up of *magnetic domains,* each of which has an average volume of the order of 10^{-15} m^3 and are completely magnetized. Ordinarily the magnetic axes of these domains are oriented in various directions. In an unmagnetized iron crystal, the domains are parallel to the direction of easy magnetization which in an iron crystal is six. The application of the external field to ferromagnetics causes the walls of the domains to shift in such a way as to increase the sizes of the domains with magnetic axes in the direction of field and decrease the sizes of the domains which have unfavourable direction of the axes. Obviously, the pattern of the domains changes and the specimen exhibits magnetization. Increasing the intensity of the applied field increases the magnetization of the specimen. The magnetization of weakly magnetized substance (dia- and paramagnetic) varies linearly with the field strength. The magnetization of ferromagnetics depends on H in an intricate way, i.e. the relation between H and the flux density B of the field produced by a ferromagnetic specimen is not a simple direct proportion. In order to understand this relationship, let us consider a magnetization experiment and examine its results.

Consider a toroidal coil wound around an iron core in the form of ring as shown in Fig. 7.13. Let us suppose that a second winding is placed on the ring with its terminals connected to a fluxmeter or ballistic galvanometer (B.G.) [*see* Appendix 7.1]. With the iron core specimen unmagnetized at the begining of the experiment, H may be increased in small increments, say ΔH. One can determine the corresponding increment ΔB in B from the deflection of the B.G. A typical magnetization curve for a ferromagnetic material is shown in Fig. 7.14. From the figure, it is evident that as the magnetization current in the winding is steadily increased from zero until the magnetic intensity corresponds to the value corresponding to the point X_1 on the x-axis, the flux density is given by OY_1. We see that the curve is not a straight line and hence it is not possible to speak of a single value of the permeability (μ) of the substance. If now the magnetic intensity is increased further to OX_2 (at which point the magnetization curve levels off, i.e. ferromagnetic specimen (say iron) saturates) and then decreased to OX_1, one should expect the flux density to fall to OY_1. However, the magnetic state of the sample

Fig. 7.13: Rowland ring method of obtaining magnetization curve

Fig. 7.14: Magnetization curve for a ferromagnetic material

Fig. 7.15: Hysteresis loop

follows the path $OAPC$ (Fig. 7.15) and the flux density when H is reduced to OX_1 is given by OY_2 rather than OY_1. Obviously, the flux density is different, although the magnetic intensity is the same. This shows that the magnetization curve is not reversible, i.e. an increment $-\Delta H$ will not produce an increment $-\Delta B$ of the same magnitude as was associated with $+\Delta H$ at the same point. Obviously, the flux density does not depend on the magnetic intensity alone but on the *magnetic history* of the sample as well. This particular behaviour, i.e. lag of the magnetization of a ferromagnetic substance behind the strength of the applied field is called hysteresis. This shows that once the atomic magnetic dipoles have been aligned in a particular direction, they get locked in that state and it takes a reverse field H to go back to the original value.

If after the core is saturated, H is reduced to zero, some magnetization still remains in the ferromagnetic specimen and B is reduced to a value B_r. The magnitude of B_r is called the *remanence* or *retentivity* of the sample. Obviously, the specimen has now become a permanent magnet. One can see from the curve that a further change in H to H_c will be required to reduce B to zero. H_c is called the *coercive force* or the *coercivity*. A still further change of H to $-H_b$ will cause the iron core to saturate again with B in the opposite direction. We can see that reversing H again will reverse the process and when iron is again saturated in the original direction, a closed loop is formed. This loop is called the *hysteresis loop*. If we alternate the current between large positive and negative values, the B–H plot traces very nearly the same curve. It can be shown that the area enclosed by the hysteresis loop gives the work done in taking a unit volume of the material through the cycle represented by the loop. Though the main features of the hysteresis loop are common to all ferromagnetic materials, the actual shape of the loop can show considerable variation. Figure 7.16a shows the B–H curve for magnetically hard iron. One can easily see that this curve is characterized by large values of remanence and coer- cive force. Figure 7.16b shows the hysteresis loop for a specially made alloy, which can be taken almost to saturation level magnetization by a small magnetic intensity of the order of a few ampere-turns. We note that the coercive force for this alloy is small. One can use the pieces of such an alloy for recording a binary code of memory. For example, the two possible directions of magnetization of a pin made of such an alloy can indicate 1 and 0 respectively. A series of such pins can then be magnetized in the required directions to read any number. For example, 1011 could be represented by a sequence ↑↓↑↑↑ which indicates the number 23 in the binary code. Erasing a recorded number and writing another would then involve switching on magnetic fields in the appropriate directions to magnetize the pins in the required directions. We must note that the magnetization would remain even when the fields are switched off. Instead of pins, we can use small circular rings of magnetic material, which can be magnetized either clockwise or anticlockwise direction to denote 1 and 0 respectively. With such an arrangement one could express the number 23 as a sequence ∩∩∩∩∩. The possible way of demagnetizing a sample of ferromagnetic material is shown in Fig. 7.16c. We can see from the B–H curve that switching off the magnetic intensity does not lead to zero magnetization. One can demagnetize a sample by taking it through a series of hysteresis loops of diminishing amplitudes. Starting from a state indicated by Q in Fig. 7.16c, the sample goes through cycles of smaller and smaller size and the magnetization is sufficiently small to be considered as zero. One can also demagnetize a sample by heating it to a sufficiently high temperature, where the ferromagnetic properties of the material vanish and in the absence of an external magnetic intensity, the sample magnetization becomes zero.

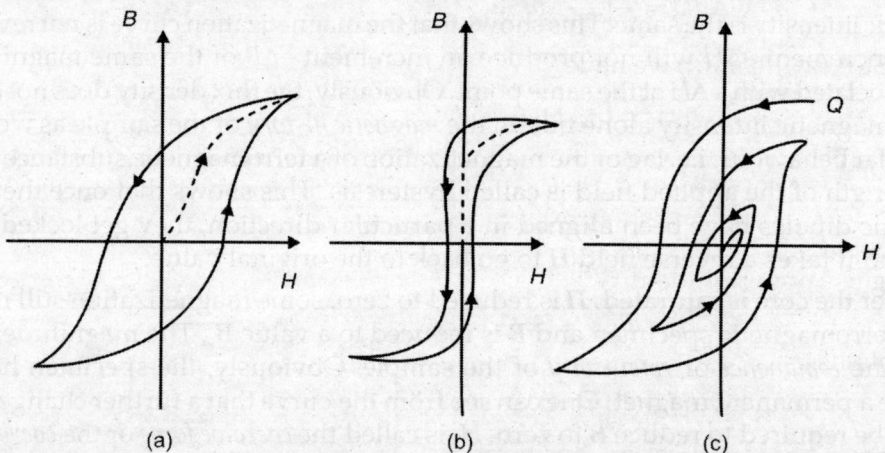

Fig. 7.16: (a) B–H loop for magnetically hard iron, (b) B–H loop for a specially made alloy having a small coercive force, (c) Method of demagnetization of a sample of iron

Steinmetz obtained an empirical relationship between the area of a hysteresis loop and the maximum value of B attained in a loop (B_m). The area of a hysteresis loop represents the work done in taking unit volume of the sample through one hysteresis cycle and as such one can express the *Steinmetz relationship* as

$$\text{work done per unit volume} = \eta(B_m)^n \qquad (7.47)$$

where, n has the value 1.6 for a large number of materials and η is the Steinmetz coefficient which is characteristic of a given material. One can estimate the order of magnitude knowing n and η.

When a specimen of a ferromagnetic material is taken through a cycle of magnetization, it results into an expenditure of irrecoverable energy. As the elementary magnets within the iron core try to line up with H, first one way and then the other, the tendency to turn around causes to develop mechanical stresses in the iron specimen; those, in turn produce heating. One can easily show that the energy dissipated per unit volume $\dfrac{dU}{d\tau}$ in each magnetization cycle, is proportional to the area enclosed by the hysteresis loop.

Let us consider the case when the current in the coil is I and the magnetic flux through the circuit is ϕ. The instantaneous induced emf in the circuit is equal to $-\dfrac{d\phi}{dt}$. Now, the rate at which work is being done against the induced emf by the source supplying the current is obtained as

$$\frac{dU}{dt} = I\frac{d\phi}{dt} \tag{7.48}$$

or

$$dU = I\frac{d\phi}{dt}dt$$

Using $\phi = \int_S \boldsymbol{B} \cdot d\boldsymbol{S}$, one obtains

$$dU = -\int \frac{\partial \boldsymbol{B}}{\partial t} \cdot \boldsymbol{I}dSdt = -\int \nabla \times \boldsymbol{E} \cdot \boldsymbol{I}dSdt \tag{7.49}$$

Using Stokes' theorem, we have

$$dU = -\int \boldsymbol{E} \cdot \boldsymbol{I}dldt$$

Since $Idl = Jd\tau$, we have

$$dU = \int \boldsymbol{E} \cdot \boldsymbol{J}d\tau dt = -\int \boldsymbol{E} \cdot \nabla \times \boldsymbol{H}d\tau dt \tag{7.50}$$

Making use of vector identity,

$$\nabla \cdot (\boldsymbol{A} \times \boldsymbol{B}) = \boldsymbol{B} \cdot \nabla \times \boldsymbol{A} - \boldsymbol{A} \cdot \nabla \times \boldsymbol{B}$$

One obtains

$$dU = -\int \boldsymbol{H} \cdot \nabla \times \boldsymbol{E}d\tau dt + \int \nabla \cdot (\boldsymbol{E} \times \boldsymbol{H})d\tau dt \tag{7.51}$$

Using divergence theorem, one transforms the second integral in Eq. (7.51). Finally one obtains,

$$dU = -\int \boldsymbol{H} \cdot \nabla \times \boldsymbol{E}d\tau dt + \int (\boldsymbol{E} \times \boldsymbol{H}) \cdot d\boldsymbol{S}dt \tag{7.52}$$

The second integral in Eq. (7.52) vanishes, as E and H vanish over a surface very far away from the circuit. Thus, we have

$$dU = -\int \boldsymbol{H} \cdot \nabla \times \boldsymbol{E}d\tau dt = \int \boldsymbol{H} \cdot \frac{\partial \boldsymbol{B}}{\partial t}d\tau \delta\tau$$

$$= \boldsymbol{H} \cdot d\boldsymbol{B}d\tau \tag{7.53}$$

or

$$W = \frac{dU}{d\tau} = \int \boldsymbol{H} \cdot d\boldsymbol{B} \tag{7.54}$$

Obviously, the work done per unit volume of the material per cycle is equal μ_0 times the area of M–H loop or the area of the B–H loop. The work is measured in J/m^3 per cycle and is dissipated in the form of heat. The above result further reveals that if a ferromagnetic material is to be subjected to a field which is continuously reversing its direction, e.g. transformer, it is desirable that the hysteresis loop of the material shall be narrow to minimize losses.

7.10 AMPERE'S LAW FOR MAGNETIC CIRCUITS

A coil consisting of N turns and current I around a ferromagnetic core produces a magnetomotive force (mmf or F) given by NI. The units of mmf are ampere or ampere turns. Ampere's law, applied around the path in the centre of the core (Fig. 7.17a), gives

$$F \text{ or mmf} = \oint H \cdot dl$$
$$= \int_1 H \cdot dl + \int_2 H \cdot dl + \int_3 H \cdot dl$$
$$= H_1 l_1 + H_2 l_2 + H_3 l_3$$

Now comparing with Kirchhoff's law around a single closed loop with three resistors and an emf E,

$$E = E_1 + E_2 + E_3$$

suggests that F can be viewed as NI rise and the Hl terms considered NI drops, in analogy to the voltage rise V and voltage drops V_1, V_2 and V_3. The analogy is developed in Figs 7.17b and c. Flux (ϕ) in Fig. 7.17b is analogous to current I, and reluctance ($\mathfrak{R}$) is analogous to resistance R. One can obtain an expression for reluctance as follows:

$$NI \text{ drop} = Hl = BA\left(\frac{l}{\mu A}\right) = \phi\mathfrak{R} \quad \text{or} \quad \mathfrak{R} = \frac{l}{\mu A}(H^{-1})$$

If the reluctances are known, then

$$F = NI = \phi(\mathfrak{R}_1 + \mathfrak{R}_2 + \mathfrak{R}_3)$$

for the magnetic circuit shown in Fig. 7.17b. Obviously, μ_r must be known for each material before its reluctance can be calculated. And after B or H is known, will the value of μ_r be known. This is in contrast to the relation

$$R = l/\sigma A$$

where, σ is the conductivity and is independent of the current.

Fig. 7.17: Ampere's law for magnetic circuits

7.11 MAXWELL'S EQUATIONS

Maxwell's theory and field equations are systematic generalizations of fundamental laws describing electric and electromagnetic phenomena. Until Maxwell's work, the known basic laws of electricity and magnetism were:

i. Gauss's law for electrostatics

$$\nabla \cdot D = \rho \tag{7.55}$$

ii. Gauss's law for magnetic field

$$\nabla \cdot B = 0 \tag{7.56}$$

iii. Faraday's law for electromagnetic induction

$$\nabla \times E = -\frac{\partial B}{\partial t} \tag{7.57}$$

iv. Ampere's law of magnetomotive force

$$\nabla \times H = J \tag{7.58}$$

Equations (7.55) to (7.57) enable us to calculate the divergence and curl of the electric field in any situation. If the charge density is zero, and there are no time varying magnetic fields, one obtains

$$\nabla \cdot E = 0 \quad \text{and} \quad \nabla \times E = 0$$

as special cases of Eqs. (7.55) and (7.57) respectively.

Mathematically, it can be shown that if the divergence and the curl of a vector field are known everywhere, that field can be calculated at any point. Equations (7.55) and (7.57) enable us to do this for the electric field. Obviously, Eqs. (7.55) to (7.57) are general equations and valid for static as well as dynamic field. Equation (7.58), i.e. Ampere's law was derived from steady state observations and one will have to examine its validity for time varying fields.

Taking the divergence of both sides of Eq. (7.58), one obtains

$$\nabla \cdot (\nabla \times H) = \nabla \cdot J = 0 \tag{7.59}$$

However, Eq. (7.59) is true for steady-state phenomena only. However, when the currents are changing with time, the result is incompatible with the principle of conservation of charge as evident in the *equation of continuity*

$$\nabla \cdot J + \frac{\partial \rho}{\partial t} = 0 \tag{7.60}$$

Maxwell appreciated this and suggested that Ampere's law must be modified when dealing with time-dependent fields. Maxwell realised that the difficulty arose from an incomplete definition of total current density in Eq. (7.58) by putting forward an argument that the current density consists of two parts: (i) J_f, the free current density developed due to the motion of free charge carriers and (ii) J_d developed due to the time varying electric field. Thus, the total current density is

$$J = J_f + J_d \tag{7.61}$$

Using Eq. (7.61) and taking divergence of Eq. (7.58), one obtains

$$\nabla \cdot \nabla \times H = \nabla \cdot (J_f + J_d) = 0$$

or

$$\nabla \cdot J_d = -\nabla \cdot J_f \tag{7.62}$$

We have from Eq. (7.58),

$$\nabla \cdot J_f = \frac{\partial \rho}{\partial t} = 0$$

$$\therefore \qquad \nabla \cdot J_d = \frac{\partial \rho}{\partial t} \qquad\qquad (7.63)$$

Using Eq. (7.55) in Eq. (7.63), one obtains

$$\nabla \cdot J_d = \frac{\partial}{\partial t}(\nabla \cdot D)$$

or $\qquad\qquad \nabla \cdot J_d = \nabla \cdot \dfrac{\partial D}{\partial t}$ ($\because$ space and time coordinates

$$\therefore \qquad J_d = \frac{\partial D}{\partial t} \qquad\qquad (7.64)$$

Obviously, J_d is developed due to change in D or electric field E with respect to time. It is called *displacement current*. Maxwell replaced J (we have also denoted it by J_f) by $J + \dfrac{\partial D}{\partial t}$ in Ampere's law [Eq. (7.59)] and obtained

$$\nabla \times H = J + \frac{\partial D}{\partial t} \qquad\qquad (7.65)$$

This is Ampere–Maxwell law. This expresses a relation between the electric current at a point in space and the electric and magnetic fields at the same point in space and magnetic field at the same point. In free space, where there are no currents, $J = 0$ and Eq. (7.65) becomes

$$\nabla \times H = \frac{\partial D}{\partial t} \qquad\qquad (7.66)$$

or $\qquad\qquad \nabla \times H = \mu_0 \varepsilon_0 \dfrac{\partial E}{\partial t}$ $\qquad\qquad (7.66a)$

This equation shows the relationship between the magnetic field and the time rate of change of the electric field at the same point.

We may observe in Eq. (7.65) that the effect of a time dependent electric field is to add to the current, a density term $\dfrac{\partial D}{\partial t}$. Maxwell interpreted this as an additional current and called it the *displacement current*. Maxwell's reasoning was the following. In a circuit displacement current in a capacitor C (Fig. 7.18), the current I is interrupted by the capacitor C. In order to "close" the circuit, there must be current from one plate to the other, and this current is just $\left(\dfrac{\partial D}{\partial t}\right)S$, where $E(D = \varepsilon_0 E)$ is the electric field within the capacitor and S its surface area. However, the term *displacement current* is

Fig. 7.18: Maxwell's electric displacement current in a capacitor

misleading and Maxwell's picture unnecessary, since there is no such current between the plates of the capacitor, and Eq. (7.65) simply expresses a correlation between D, H and J at the same point in space.

To have an idea of the relative size of the two types of currents in conductors, let us consider a copper wire in which there is an electric field $E = E_0 \exp(-i\omega t)$, we have

$$J = \sigma E = E_0 \exp(-i\omega t)$$

$$|J| = \sigma |E_0|$$

and

$$D = \varepsilon_0 E = \varepsilon_0 E_0 \exp(-i\omega t)$$

$$\therefore \quad \frac{\partial D}{\partial t} = -i\omega \varepsilon_0 E_0 e^{-i\omega t}$$

Hence

$$\left|\frac{\partial D}{\partial t}\right|^2 = \omega^2 \varepsilon_0^2 E_0^2$$

$$\therefore \quad \frac{J}{\left|\frac{\partial D}{\partial t}\right|} = \frac{\sigma}{\omega \varepsilon_0}$$

For copper, we have $\sigma = 5.9 \times 10^7 (\Omega\,\mathrm{m})^{-1}$. Therefore $\dfrac{\sigma}{\omega \varepsilon_0} \sim \dfrac{10^{19}}{\omega}$. The ratio is very large for all frequencies. This shows that displacement currents are not as significant as the currents due to the motion of free charges in the study of the continuous circuits. However, displacement currents have far reaching consequences in other respects. It is mainly due to the idea of the displacement current that led to uncover a rich variety of new electromagnetic phenomena.

We are now in a position to write down all the four equations of Maxwell in *differential* as well as in *integral form*. These are:

i. *Maxwell's first equation (Gauss's law in electrostatics):*

$$\nabla \cdot E = \frac{\rho}{\varepsilon_0} \quad \text{or} \quad \nabla \cdot D = \rho \tag{7.67}$$

The integral form of the above law is

$$\oint E \cdot dS = \frac{1}{\varepsilon_0} \text{ (net charge enclosed)} \tag{7.67a}$$

In terms of displacement vector D, we can write

$$\oint D \cdot dS = \text{charge enclosed} \tag{7.68}$$

ii. *Maxwell's second equation: Solenoidal nature of the magnetic induction B*

$$\nabla \cdot B = 0 \quad \text{or} \quad \oint B \cdot dS = 0 \tag{7.69}$$

This indicates that B is a solenoidal field. Obviously, the source density of the field B is always zero. This does not, of course, mean that the sources of B does not exist. In other words, lines of B are closed loops, i.e. magnetic monopole cannot be accomodated in this formalism.

iii. *Maxwell's third equation (Faraday's law of electromagnetic induction):*

$$\nabla \times E = -\frac{\partial B}{\partial t} \quad \text{or} \quad \oint E \cdot dl = -\frac{\partial}{\partial t} \oint B \cdot dS \tag{7.70}$$

This equation essentially states that a time varying magnetic field produces an electric field.

iv. *Maxwell's fourth equation (Ampere's law):*

$$\nabla \times H = J + \frac{\partial D}{\partial t} \tag{7.71}$$

or

$$\nabla \times B = \mu_0 J + \mu_0 \frac{\partial D}{\partial t} \tag{7.71a}$$

and in integral form

$$\oint B \cdot dl = \mu_0 \int J \cdot dS + \mu_0 \varepsilon_0 \frac{\partial}{\partial t} \int E \cdot dS \tag{7.71b}$$

The importance of the *displacement current density*, J_d becomes apparent when we consider the above equation for the special case when $J = 0$, i.e. when there are no true currents present. Equation (7.71b) then takes the form

$$\nabla \times B = \mu_0 \frac{\partial D}{\partial t} = \mu_0 \varepsilon_0 \frac{\partial E}{\partial t} = \frac{1}{c^2} \frac{\partial E}{\partial t} \left(\because c^2 = \frac{1}{\mu_0 \varepsilon_0} \right) \tag{7.71c}$$

Together with Eq. (7.70), we have now the pair of equations

$$\nabla \times E = -\frac{\partial B}{\partial t}$$

and

$$\nabla \times B = \frac{1}{c^2} \frac{\partial E}{\partial t}$$

We can easily see that except for the negative sign and the factor c^2, the two equations are almost symmetrical in E and B. These equations show that a time-varying magnetic (electric) field produces an electric (magnetic) field. Now, one can understand that the concept of the dis-placement current was essential to arrive at the above two equations. In the next chapter it will be shown that the idea of displacement current makes it possible to understand the formation of electromagnetic waves in free space, which is one of the most exciting and beautiful phenomena in nature.

Maxwell's equations are used in integral or differential form, depending on the problem to be solved. These equations are compatible with the principle of relativity in the sense that they remain invariant under a Lorentz transformation.

The synthesis of electromagnetic interactions expressed by Maxwell's equation is one of the greatest achievements in physics, and that is what places these interactions in a unique position.

We must recognize, however, that Maxwell's equations, as have been presented, have their limitations. They work very well when dealing with electromagnetic interactions between large aggregates of charges, such as radiating antennas, electric circuits, and even beams of ionized atoms or molecules. But it has been found that the electromagnetic interactions between fundamental particles (especially at high energies) must be treated in a somewhat different way and according to the laws of quantum mechanics, constituting a technique called *quantum electrodynamics*. Even granted these limitations, the results derived from Maxwell's equations are an excellent approximation for describing electromagnetic interactions between elementary particles. This method is called classical electrodynamics.

7.12 DECAY OF FREE CHARGE

We can use Maxwell's equations for one of the important deduction regarding the decay of free charges.

Writing Eq. (7.71a) as

$$\nabla \times H = J + \frac{\partial D}{\partial t} = \sigma E + \varepsilon \frac{\partial E}{\partial t} \tag{7.72}$$

Taking the divergence on both the sides of Eq. (7.72) and assuming that σ and ε are constants, one obtains

$$\nabla \cdot \nabla \times H = \sigma \nabla \cdot E + \varepsilon \frac{\partial}{\partial t}(\nabla \cdot E) = \frac{\sigma \rho}{\varepsilon} + \frac{\partial \rho}{\partial t} = 0$$

or

$$\frac{d\rho}{\rho} = -\frac{\sigma}{\varepsilon} dt \quad (\because \nabla \cdot D = \rho)$$

Integrating the above equation, one obtains

$$\rho = \rho_0 \exp\left(-\frac{t}{\tau}\right) \tag{7.73}$$

where

$$\tau = \frac{\varepsilon}{\sigma} \tag{7.74}$$

τ is known as relaxation time. Relation (7.73) shows that any distribu-tion of charge decays exponentially at a rate which is independent of any other electromagnetic disturbances that may be taking place.

Obviously, any excess of charge density in the interior of a copper conductor $\left(\sigma = \frac{58 \times 10^6}{\Omega m}\right)$ will disappear with time constant

$$\tau = \frac{\varepsilon_0}{\sigma} = \frac{8.85 \times 10^{-12}}{58 \times 10^6} = 1.5 \times 10^{-19} \, \text{s}.$$

7.13 POTENTIALS OF ELECTROMAGNETIC FIELD

One can obtain the complete description of an electromagnetic field by solving Maxwell's four equations. If one writes these equations in a suitable form, then the process becomes simpler. For this purpose, we introduce new quantities called *electromagnetic potentials*. With the help of these quantities one can easily reduce the number of equations. We have seen that the electrostatic fields can be expressed in terms of scalar potential (ϕ) as $E = -\nabla\phi$ and the magnetic fields in terms of a vector potential (A) as $B = \nabla \times A$. Let us consider potentials in electromagnetic fields when electric and magnetic fields are time-varying.

One finds that in the time-dependent field, the Maxwell equation

$$\nabla \cdot B = 0 \tag{7.75}$$

still holds and hence one can express B in terms of a vector potential A, i.e.

$$B = \nabla \times A \tag{7.76}$$

Let us now consider another Maxwell's equation which does not involve any currents or charges, i.e.

$$\nabla \times E = -\frac{\partial B}{\partial t} = -\frac{\partial}{\partial t}(\nabla \times A) = -\nabla \times \frac{\partial A}{\partial t}$$

$$\therefore \qquad \nabla \times \left(E + \frac{\partial A}{\partial t} \right) = 0 \qquad (7.76a)$$

We know that curl of the gradient of a scalar function vanishes and the quantity within the brackets can be expressed as a gradient of a scalar function ϕ, i.e.

$$E + \frac{\partial A}{\partial t} = -\nabla \phi \qquad \text{or} \qquad E = -\nabla \phi - \frac{\partial A}{\partial t} \qquad (7.77)$$

Obviously, we have solved the two Maxwell's homogeneous equations in terms of A and ϕ. Once A and ϕ are determined, one can obtain B and E.

We must note that Eq. (7.76) does not completely define A. Let us add the gradient of any arbitrary scalar function to the vector potential, i.e. we change A to

$$A' = A + \nabla \psi \qquad (7.78)$$

We must note that the magnetic field remains unchanged. In order that the addition of $\nabla \psi$ should not affect the electric field, the scalar potential ϕ must also be simultaneously transformed to ϕ, where

$$\phi' = \phi - \frac{\partial \psi}{\partial t} \qquad (7.79)$$

One can easily verify this by substituting A' and ϕ' in Eq. (7.77). Thus, any physical law that can be expressed in terms of the electromagnetic potentials A and ϕ remains unaffected by the transformations of the type given by Eqs. (7.78) and (7.79). These transformations are called gauge transformations. Obviously, equations involving these potentials must be gauge invariant.

The condition $\nabla \cdot A = 0$ in electrostatics together with $B = \nabla \times A$ specifies A. However, one will have to make a different choice in elec-tromagnetism. In order to specify A, one has to impose an additional condition on A which does not change the physics, i.e. it must remain consistent with the transformations in Eqs. (7.78) and (7.79). Obviously, E and B must remain unaffected. For this purpose, let us consider the remaining two non-homogeneous equations of Maxwell, viz. Eqs. (7.67) and (7.70). Now, substituting Eq. (7.77) in Eq. (7.67), one obtains

$$\nabla \cdot D = \nabla \cdot \varepsilon_0 E = \varepsilon_0 \nabla \cdot \left(-\nabla \phi - \frac{\partial A}{\partial t} \right) = \rho$$

or $\qquad -\nabla^2 \phi - \frac{\partial}{\partial t}(\nabla \cdot A) = \dfrac{\rho}{\varepsilon_0} \qquad (7.80)$

One can express Eq. (7.71a) as

$$\nabla \times H - \frac{\partial D}{\partial t} = J \qquad \text{or} \qquad \nabla \times \frac{B}{\mu} - \varepsilon_0 \frac{\partial E}{\partial t} = J$$

or $\qquad \dfrac{1}{\mu}(\nabla \times \nabla \times A) - \varepsilon_0 \dfrac{\partial}{\partial t}\left(-\nabla \phi - \dfrac{\partial A}{\partial t} \right) = J$

or $\qquad \nabla \times \nabla \times A - \mu \varepsilon_0 \dfrac{\partial}{\partial t}\left(-\nabla \phi - \dfrac{\partial A}{\partial t} \right) = \mu J$

or $\qquad -\nabla^2 A + \nabla(\nabla \cdot A) + \mu_0 \varepsilon_0 \dfrac{\partial}{\partial t}\nabla \phi + \mu_0 \varepsilon_0 \dfrac{\partial^2 A}{\partial t^2} = \mu J \qquad (7.81)$

In obtaining Eq. (7.81), we have used the vector identity

$$\nabla \times \nabla \times A = \nabla (\nabla \times A) - \nabla^2 A$$

Now, we determine the condition for A. Let us choose A and ϕ such that

$$\nabla \cdot A = -\mu_0 \varepsilon_0 \frac{\partial \phi}{\partial t} = -\frac{1}{c^2} \frac{\partial \phi}{\partial t} \tag{7.82}$$

One can see that with the substitution of Eq. (7.82), the two middle terms of Eq. (7.81) cancel and one obtains

$$\nabla^2 A - \mu_0 \varepsilon_0 \frac{\partial^2 A}{\partial t^2} = -\mu_0 J \tag{7.83}$$

Using Eq. (7.82), Eq. (7.80) becomes

$$-\nabla^2 \phi - \frac{\partial}{\partial t}(\nabla \cdot A) = -\nabla^2 \phi + \mu_0 \varepsilon_0 \frac{\partial^2 \phi}{\partial t^2} = \rho / \varepsilon_0$$

or

$$-\nabla^2 \phi + \mu_0 \varepsilon_0 \frac{\partial^2 \phi}{\partial t^2} = -\frac{\rho}{\varepsilon_0} \tag{7.84}$$

We have obtained two independent equations one for A [Eq. (7.80)] and other for ϕ [Eq. (7.84)]. We note that A is connected with J and ϕ with scalar ρ. Moreover, both the equations have the same form, i.e. both potentials satisfy the same equation. This shows that the condition introduces complete symmetry between the scalar and vector potentials. For the steady-state, the time derivatives vanish and one obtains

$$\nabla^2 A = -\mu_0 J \quad \text{and} \quad \nabla^2 \phi = -\frac{\rho}{\varepsilon_0} \tag{7.85}$$

The condition given by Eq. (7.85) is called *Lorentz gauge condition*. The gauge used in magnetostatics, viz. $\nabla \cdot A = 0$ is known as *Coulomb gauge*.

Transformation equations (7.78) and (7.79) show that the electric field E and the magnetic field B are invariant. The potentials, thus, transformed will have to satisfy the Lorentz condition. This means, the gauge function ψ which so far remained arbitrary must satisfy a certain condition.

The original and the transformed potentials have to satisfy Lorentz condition. We have

$$\nabla \cdot A + \mu_0 \varepsilon_0 \frac{\partial \phi}{\partial t} = 0 \tag{7.86}$$

and

$$\nabla \cdot A' + \mu_0 \varepsilon_0 \frac{\partial \phi'}{\partial t} = 0 \tag{7.87}$$

or

$$\nabla \cdot (A + \nabla \psi) + \mu_0 \varepsilon_0 \frac{\partial}{\partial t}\left(\phi - \frac{\partial \psi}{\partial t}\right) = 0$$

or

$$\nabla \cdot A + \nabla^2 \psi + \mu_0 \varepsilon_0 \frac{\partial \phi}{\partial t} - \frac{\partial^2 \psi}{\partial t^2} = 0$$

Hence, we get

$$\nabla^2 \psi - \mu_0 \varepsilon_0 \frac{\partial^2 \psi}{\partial t^2} = 0 \qquad (7.88)$$

Thus, one obtains the restricted gauge transformations as

$$A' \rightarrow A + \nabla \phi \qquad (7.89)$$

$$\phi' \rightarrow \phi - \frac{\partial \psi}{\partial t} \qquad (7.90)$$

where ψ satisfies the condition

$$\nabla^2 \psi - \mu_0 \varepsilon_0 \frac{\partial^2 \psi}{\partial t^2} = 0 \qquad (7.91)$$

Obviously, this preserves the Lorentz condition.

We can easily see that the Lorentz gauge condition is not quite arbitrary as it appears at first sight. One can relate it to the basic principles of electromagnetism, viz. Coulomb's law, Biot–Savart's law and principle of conservation of energy.

We have seen while deriving Ampere's law from Biot–Savart's law for stationary current distribution, the Coulomb gauge condition $\nabla \cdot A = 0$ appears a natural way. One can easily extend this derivation for quasistationary conditions.

One obtains the following expression using Biot–Savart's law

$$A(r_2) = \frac{\mu_0}{4\pi} \int_{V_1} \frac{J(r_1)}{|r_2 - r_1|} d\tau \qquad (7.92)$$

where, the subscripts 1 and 2 refer to the source and field coordinates. Thus, we have

$$\nabla_2 \cdot A(r_2) = \frac{\mu_0}{4\pi} \int_{V_1} J(r_1) \cdot \nabla_2 \left[\frac{1}{|r_2 - r_1|} \right] d\tau \qquad (7.93)$$

We must note that ∇_2 operates on r_2 only.

We have

$$\nabla_2 \left[\frac{1}{|r_2 - r_1|} \right] = -\nabla_1 \left[\frac{1}{|r_2 - r_1|} \right]$$

or $\quad \nabla_2 \cdot A(r_2) = -\dfrac{\mu_0}{4\pi} \displaystyle\int_{V_1} J(r_1) \cdot \nabla_1 \left[\frac{1}{|r_2 - r_1|} \right] d\tau$

$$= \frac{\mu_0}{4\pi} \left[\int_{V_1} \left\{ \frac{1}{|r_2 - r_1|} \right\} \nabla_1 \cdot J(r_1) d\tau_1 \right] - \int_{V_1} \nabla_{v_1} \left\{ \frac{J(r_1)}{|r_2 - r_1|} \right\} \qquad (7.94)$$

With the help of divergence theorem, one can write the second integral for bounded source distribution as

$$\int_{V_1} \nabla_1 \cdot \left\{ \frac{J(r_1)}{|r_2 - r_1|} \right\} d\tau_1 = \int \frac{J(r_1)}{|r_2 - r_1|} \cdot \hat{a}_n dS = 0$$

$$\therefore \quad \nabla_2 \cdot A(r_2) = \frac{\mu_0}{4\pi} \int_{V_1} \left\{ \frac{1}{|r_2 - r_1|} \right\} \nabla_1 \cdot J(r_1) d\tau_1 \tag{7.95}$$

One obtains from Coulomb's law

$$\phi(r_2) = \frac{1}{4\pi\varepsilon_0} \int_{V_1} \frac{\partial(r_1)}{|r_2 - r_1|} d\tau_1$$

$$\therefore \quad \nabla_2 \cdot A(r_2) = -\mu_0\varepsilon_0 \frac{\partial}{\partial t} \left\{ \frac{1}{4\pi\varepsilon_0} \int_{V_1} \frac{\partial(r_1)}{|r_2 - r_1|} d\tau_1 \right\}$$

$$= -\mu_0\varepsilon_0 \frac{\partial\phi(r_2)}{\partial t} \tag{7.96}$$

Generalizing, one obtains

$$\nabla \cdot A = -\mu_0\varepsilon_0 \frac{\partial\phi}{\partial t} \tag{7.97}$$

Obviously, Eq. (7.97) is the *Lorentz gauge condition*. Now, we have

$$\nabla_2 \times B(r_2) = \nabla_2 \times \{\nabla_2 \times A(r_2)\} = \nabla_2\{\nabla_2 \cdot A(r_2)\} - \nabla_2^2 A(r_2)$$

One finds that the first term of the above is

$$-\nabla_2 \left[\mu_0\varepsilon_0 \frac{\partial\phi(r_2)}{\partial t} \right] = \mu_0 \frac{\partial}{\partial t}[-\varepsilon_0\nabla_2\phi(r_2)]$$

$$= \mu_0 \frac{\partial}{\partial t}(\varepsilon_0 E) = \mu_0 \frac{\partial D}{\partial t} \tag{7.98}$$

The second term $= -\nabla_2^2 A(r_2) = -\frac{\mu_0}{4\pi} \int_{V_1} J(r_1)\nabla_2^2 \left\{ \frac{1}{r_2 - r_1} \right\} d\tau$

$$= \mu_0 J(r_2) \tag{7.99}$$

Thus, we have $\qquad \nabla \times B = \mu_0 \left(J + \frac{\partial D}{\partial t} \right) \tag{7.99a}$

Obviously, this treatment also leads to the displacement.

7.14 ELECTROMAGNETIC FIELD ENERGY AND FIELD MOMENTUM

The general expressions obtained earlier for the electrostatic and magnetostatic field energies are

$$W_E = \frac{1}{2} \int_V (E \cdot D) d\tau$$

$$W_M = \frac{1}{2} \int_V (B \cdot H) d\tau \tag{7.100}$$

and

Now, we can find the expression for the electromagnetic energy in time dependent situations.

The Lorentz force on a moving charge q is given by

$$F = q(E + v \times B) \tag{7.101}$$

The rate of doing work on this charge is

$$F \cdot v = q(E + v \times B) \cdot v = q (E \times v) \qquad (7.102)$$

The magnetic field does no work as magnetic force and velocity of charge are perpendicular.

If there exists a continuous distribution of charge, the total rate at which the work is done in a given volume is

$$\int \rho (E \cdot V) d\tau = \int (E \cdot J) d\tau \qquad (7.103)$$

Substituting for J from Eq. (7.70), one obtains

$$\int (E \cdot J) d\tau = \int \left[E \cdot (\nabla \times H - \frac{\partial D}{\partial t} \right] d\tau \qquad (7.104)$$

But

$$E \cdot (\nabla \times H) = \nabla \cdot (H \times E) + H \cdot (\nabla \times E)$$

$$= \nabla \cdot (H \times E) - H \cdot \frac{\partial B}{\partial t}$$

Therefore, one obtains

$$\int (E \cdot J) d\tau = \int \left[\nabla \cdot (H \times E) - E \cdot \frac{\partial D}{\partial t} - H \cdot \frac{\partial B}{\partial t} \right] d\tau$$

$$= \int \left[[\nabla \cdot (H \times E)] d\tau - \int \left(E \cdot \frac{\partial D}{\partial t} - H \cdot \frac{\partial B}{\partial t} \right) \right] d\tau$$

One can transform the first integral by using divergence theorem. Thus, one obtains

$$\int_V (E \cdot J) d\tau = \int_S (H \times J) \cdot \hat{a}_n dS - \int_V \left(E \cdot \frac{\partial D}{\partial t} + H \cdot \frac{\partial B}{\partial t} \right) d\tau$$

We must note that S is the surface boundary of the volume V. We get

$$-\int_V \left(E \cdot \frac{\partial D}{\partial t} + H \cdot \frac{\partial B}{\partial t} \right) d\tau = -\int_V (E \cdot J) d\tau + \int_S (E \times H) \cdot \hat{a}_n dS \qquad (7.105)$$

If ε and μ are assumed to be constant, one obtains for the linear media

$$\int_V \left(E \cdot \frac{\partial D}{\partial t} + H \cdot \frac{\partial B}{\partial t} \right) d\tau = \frac{\partial}{\partial t} \int_V \frac{1}{2} (E \cdot D + B \cdot H) d\tau$$

$$\therefore \qquad -\frac{\partial}{\partial t} \int_V \frac{1}{2} [E \cdot D + B \cdot H] d\tau = \int (E \cdot J) d\tau + \int (E \cdot J) \cdot \hat{a}_n dS \qquad (7.106)$$

We know that an electromagnetic field consists of electric and magnetic fields and hence it is reasonable to assume that the sum of the energies given in Eq. (7.100) represents the total electromagnetic energy.

Obviously, one may write $\varepsilon_M = \frac{1}{2} [E \cdot D + B \cdot H]$ as the electromagnetic energy density. This means that one can interpret LHS of Eq. (7.106) as the rate at which the energy stored in the electromagnetic field diminishes. The first term on the RHS gives the work done by the field forces on charges contained in the volume.

The dimensions of the vector $[E \times H]$ are of $\dfrac{\text{energy}}{\text{area} \times \text{time}}$ and hence we can interpret that the last term represents the flow of energy out of the boundary per unit time. Obviously, the vector $[E \times H]$ which gives the rate at which the energy flows across unit area of boundary, is called the *Poynting vector*. Poynting vector is usually represented by the symbol N, i.e.

$$E \times H = N \qquad (7.107)$$

Obviously, N represents the power being transmitted through unit area held, normal to N. Writing Eq. (7.106) in differential form, we have

$$\frac{d\varepsilon_M}{dt} + \nabla \cdot N = -E \cdot J \qquad (7.108)$$

If conductivity of the medium is zero, we have $J = \sigma E = 0$. Thus,

$$\frac{d\varepsilon_M}{dt} + \nabla \cdot N = 0 \qquad (7.109)$$

We can see that Eq. (7.109) has exactly the same form as the equation of continuity given by Eq. (7.60). We note that here ε_M behaves as volume density of conserved quantity, whereas N is current density. One can easily see that this analogy leads us to same conclusion as the identification of the flux of electromagnetic energy with the Poynting vector $N = E \times H$.

Equation (7.106) represents the law of conservation of energy, i.e. the decrease of electromagnetic energy per unit time in a certain volume V is equal to the work done by the field forces per unit time plus the flux flow outward per unit time. This theorem is called as *Poynting theorem*.

We know that energy and momentum are closely related. One can easily show that an electromagnetic field possesses momentum. We have the force on a region containing both charges and currents as

$$F = \int_V (\rho E + J \times B)\, d\tau \qquad (7.110)$$

Let the sum of the momenta of all the particles be P_{mech}, then

$$\frac{dP_{\text{mech}}}{dt} = \int_V (\rho E + J \times B)\, d\tau \qquad (7.111)$$

We have from Maxwell's equations

$$\rho = \nabla \cdot D \quad \text{and} \quad J = \nabla \times H - \frac{\partial D}{\partial t}$$

Using the above equations, Eq. (7.111) becomes

$$\frac{dP_{\text{mech}}}{dt} = \int_V \left\{ (\nabla \cdot D)E + \left(\nabla \times H - \frac{\partial D}{\partial t} \right) \times B \right\} d\tau$$

$$= \int_V \left\{ (\nabla \cdot D)E + B \times \frac{\partial D}{\partial t} - B \times (\nabla \times H) \right\} d\tau$$

Since

$$\frac{\partial}{\partial t}(D \times B) = D \times \frac{\partial B}{\partial t} + \frac{\partial D}{\partial t} \times B, \quad \text{we have}$$

$$\frac{dP_{mech}}{dt} = \int_V \left[(\nabla \cdot D)E + \left(D \times \frac{\partial B}{\partial t} \right) - \frac{\partial}{\partial t}(D \times B) - B \times (\nabla \times H) \right] d\tau$$

we have $\nabla \cdot B = 0$. This means addition of $(\nabla \cdot B)H$ to the square bracket does not alter the result.

$$\frac{dP_{mech}}{dt} + \frac{d}{dt} \int_V (D \times B) d\tau = \int_V [(\nabla \cdot B)E + (\nabla \cdot B)E + (\nabla \cdot B)H$$

$$- \{D \times (\nabla \times E)\} - \{B \times (\nabla \times H)\}] d\tau \qquad (7.112)$$

Here we have used $\nabla \times H = -\frac{\partial B}{\partial t}$.

We can see that the second term integral on LHS of Eq. (7.112) represents momentum. Further, we note that it is not associated with the mass of particles and consists only of fields, one can identify it as the electromagnetic moment (P_{field}). Writing $[D \times B] = g$, we find that g is *electromagnetic momentum density*. One can convert RHS of Eq. (7.112) into a surface integral and we find that it represents the momentum flow. One can easily see from Eq. (7.112) that the total momentum of the closed system consisting of a field and particles is conserved.

The relation between g and N is

$$g = [D \times B] = [\varepsilon E + \mu H] = \mu \varepsilon N \qquad (7.113)$$

ILLUSTRATIVE EXAMPLES

Example 7.1: A plane circuit composed of N turns, each of area S, is placed perpendicular to an alternating uniform magnetic field that varies with time. The equation of the field is $B = B_0 \sin \omega t$. Show that the induced emf in the circuit is oscillatory and given by

$$V = -NSB_0 \, \omega \cos \omega t.$$

Solution: The flux through one turn of the circuit,

$$\phi = SB = SB_0 \sin \omega t$$

∴ Total flux through N turns,

$$\phi' = \phi N = NSB_0 \sin \omega t$$

The induced emf

$$V = -\frac{d\phi'}{dt} = -NSB_0 \omega \cos \omega t$$

Obviously, the induced emf is oscillatory or alternating with the same frequency as the magnetic field.

Example 7.2: A circuit is composed of two coaxial cylindrical metallic sheets of radii a and b, each carrying a current I, but in the opposite direction (Fig. 7.19). The space between the cylin-ders is filled with a substance whose permeability is μ. Show that the self inductance per unit length of the circuit is

$$L = \frac{\mu}{2\pi} \log \left(\frac{b}{a} \right)$$

Solution: One can show that magnetic field for

this current arrangement is given by $B = \dfrac{\mu I}{2\pi r}$ in

the region within two cylinders, and zero elsewhere. Here μ is the permeability of the medium filling the space between the two cylinders. To obtain the self inductance, one will have to compute the magnetic flux through any section of the conductor, such as $PQRS$, having a length l. Let us divide this section into strips of width dr, the area of each strip $= l\,dr$. The magnetic field B is perpendicular to $PQRS$. Therefore,

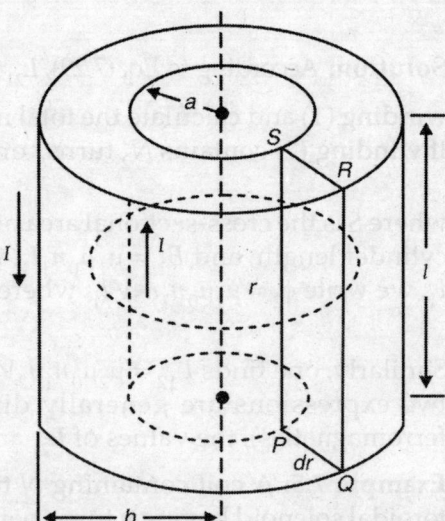

$$\phi = \int_{PQRS} B\,dS = \int_a^b \left(\frac{\mu I}{2\pi r}\right)(l\,dr)$$

$$= \frac{\mu I l}{2\pi}\int_a^b \frac{dr}{r} = \frac{\mu I l}{2\pi}\log\left(\frac{b}{a}\right)$$

Fig. 7.19: Two coaxial metallic sheets

$\therefore$ Self inductance of a portion of length l is

$$L = \frac{\phi}{I} = \frac{\mu l}{2\pi}\log\left(\frac{b}{a}\right)$$

$\therefore$ Self inductance per unit length $\dfrac{L}{l} = L' = \dfrac{\mu}{2\pi}\log\left(\dfrac{b}{a}\right)$.

Example 7.3: A superconducting ring of radius a and inductance L is in uniform magnetic field B. In the initial position, the plane of the ring is parallel to the vector B, and the current in the ring is equal to zero. The ring is turned to the position perpendicular to vector B. Find the current in the ring in the final position and the magnetic induction at its centre.

Solution: Initially the ring plane is parallel to B. Hence, the flux linked is zero. When ring plane is perpendicular to the magnetic field, flux linked is $\phi = \pi a^2 B$. This induces the current I in ring, such that

$$LI = \Delta\phi = \pi a^2 B$$

$\therefore$ $\qquad I = \pi a^2 B/L$

This current produces magnetic field at the centre of ring

$$B_1 = \frac{\mu_0 I}{2a} = \frac{\mu_0 \pi a^2 B}{2La} = \frac{\mu_0 \pi a B}{2L}$$ which is opposite in direction to B.

The resultant magnetic induction at this point is given by

$$B_{\text{res}} = B - B_1 = B\left(1 - \frac{\pi\mu_0 a}{2L}\right)$$

Example 7.4: A long ferromagnetic cylinder of volume V has two windings (one over the other). One winding contains n_1 turns per unit length, while the other contains n_2 turns. Ignoring the edge effects, find their mutual inductance.

Solution: According to Eq. (7.29), $L_{21} = \dfrac{\phi_2}{I_1}$. This means that we must create current I_1, in winding (1) and calculate the total magnetic flux through all the turns of winding (2). If winding (2) contains N_2 turns, one obtains

$$\phi = N_2 B_1 S$$

where S is the cross-sectional area of the cylinder. Consider that $N_2 = n_2 l$, where l is the cylinder length, and $B_1 = \mu_1 \mu_0 n_1 I_1$, where μ_1 is the magnetic permeability for current I_1, we write $\phi_2 = \mu_1 \mu_0 n_1 n_2 V l_1$, where $V = lS$. Hence

$$L_{21} = \mu_1 \mu_0 n_1 n_2 V$$

Similarly, one finds $L_{12} = \mu_2 \mu_0 n_1 n_2 V$. We must note that values of μ_1 and μ_2 in the last two expressions are generally different (they depend on currents I_1 and I_2 in ferromagnetics), the values of L_{21} and L_{12} do not coincide.

Example 7.5: A coil containing N turns is wrapped around the central portion of a toroidal solenoid having n turns per unit length and a cross-section of area S. Calculate the mutual inductance of the system (Fig. 7.20).

Fig. 7.20: A ring wrapped around the central portion of a toroidal solenoid

Solution: One can solve the problem either by finding the magnetic flux through the solenoid when a current flows along the coil, or alter-natively, by finding the magnetic flux through the coil when a current flows along the solenoid. We will follow the second procedure, which is easier. We have seen that in the case of a toroidal solenoid, the mag-netic field is confined to its interior and has a value given by $B = \mu_0 nI$. The magnetic flux through any cross-section of the solenoid is

$$\phi = BS = \mu_0 nIS$$

where S is the cross-sectional area of the solenoid. This is same as the flux through any turn of the coil, even if its cross-section is larger. Therefore, the magnetic flux through the coil is

$$\phi' = N\phi = \mu_0 nNSI$$

∴ Mutual inductance $M = \mu_0 nNS$

We must note that this arrangement is widely used in the laboratory when a standard mutual inductance is required.

Example 7.6: An all metal aeroplane drives down vertically at 30 km/s at a place where the horizontal component of the earth field is 0.4 oersted. If the wing span is 30 m, show that the resulting potential difference between the tips is 360 V.

Solution: The vertical distance covered in one second

$$= 30 \, \text{km} = 30 \times 10^3 \, \text{m}$$

The distance between the wing tips = 30m. The area swept by the wing span in one second

$$\frac{dA}{dt} = 30 \times 10^3 \times 30 = 9 \times 10^5 \, \text{m}^2/\text{s}$$

$\therefore$ The potential difference between the wing tips,

$$V = -\frac{d\phi}{dt} = -\frac{d}{dt}(B \cdot A) = \frac{d}{dt}(BA) = B\frac{dA}{dt} \quad \text{(numerically)}$$

$$= (0.4 \times 10^{-4} \, \text{W/m}^2) \times (9 \times 10^5 \, \text{m}^2/\text{s}) = 36 \, \text{V}$$

Example 7.7: A car ignition coil consists of two insulated coils, one of 16000 turns and the other of 400 turns, wound over each other. The length of each coil is 10 cm and the turns have the radius of 3cm. A current of 3 A is allowed to pass through the primary coil and switched off in 10^{-4} s. Show that the voltage induced in the secondary coil is 681.4 V.

Solution: We have the magnetic flux density inside the solenoid

$$B = \mu_0 N_1 I_1$$

where I_1 is the current flowing through the coil and N_1 is the number of turns per unit length since the magnetic flux through each turn of the loop is $\mu_0 N_1 I_1 \pi r^2$ and hence the total flux is

$$\phi_2 = \frac{\mu_0 N_1 N_2}{l} \pi r^2 I_1$$

where l is the length of the coil and N_2 is the number of turns per unit length of the second coil. The induced emf is given by

$$e = -\frac{d\phi_2}{dt} = \frac{-\mu_0 N_1 N_2}{l} \pi r^2 \frac{dI_1}{dt} = -M\frac{dI_1}{dt}$$

We have $\dfrac{dI_1}{dt} = 3 \times 10^4 \, \text{A/s}$

$\therefore$

$$e = 4\pi \times 10^{-7} \times 16 \times 10^4 \times 4 \times 10^3 \times 10^{-1} \pi \times (0.03)^2 \times 3 \times 10^4$$

$$= 681.4 \, \text{V}.$$

Example 7.8: Show that the self inductance of two long parallel wires is given by

$$L = \frac{\mu}{\pi} \log_e \frac{d-r}{r}$$

where r is the radius of each wire and d is the distance between their axes through a medium of absolute permeability μ.

Solution: Figure 7.21 shows two infinitely long wires A and B each of radius r. d is the distance between their axes and I ampere is the current flowing in each wire in opposite directions. Let us consider a thin strip of thickness dx between the wires at a distance x from the wire A.

Fig. 7.21: Self inductance of two long parallel wires

The magnetic field at any point on the strip is given by

$$B = \frac{\mu I}{2\pi x} + \frac{\mu I}{2\pi(d-x)}$$

∴ Magnetic flux linked with the unit length of the strip is

$$d\phi = B dx \qquad (\therefore \text{ For length } l \text{ we have } d\phi = Bldx)$$

$$= \left[\frac{\mu I}{2\pi x} + \frac{\mu I}{2\pi(d-x)} \right] dx$$

∴ Magnetic flux linked with the unit length of the system is

$$\phi = \frac{\mu I}{2\pi} \int_r^{d-r} \left(\frac{1}{x} + \frac{1}{x-d} \right) dx \doteq \frac{\mu I}{2\pi} \left[\log_e x - \log_e (d-x) \right]_r^{d-r}$$

$$= \frac{\mu I}{2\pi} \left[\log_e \frac{d-r}{r} - \log_e \frac{r}{d-r} \right]$$

$$= \frac{\mu I}{\pi} \log_e \frac{d-r}{r} \tag{1}$$

We have $\quad \phi = LI \tag{2}$

Comparing Eqs. (1) and (2),

$$L = \frac{\mu}{\pi} \log_e \frac{d-r}{r}$$

Example 7.9: Show that the mutual inductance between two coplanar concentric rings of radii R_1 and R_2 respectively is given by

$$M = \mu_0 n_1 n_2 \pi R_2^2 / 2R_1$$

where n_1 and n_2 are turns of wire in outer and inner rings respectively.

Solution: Figure 7.22 shows the arrangement of two rings. If the current flowing in the outer ring is I_1 ampere, then the magnetic field at the centre of the ring due to this current is

$$B_1 = \frac{\mu_0 I_1}{2R_1}$$

The magnetic flux through the inner ring of radius R_2 is

$$\phi = \frac{\mu_0 I_1}{2R_1} \pi R_2^2 \tag{1}$$

We have assumed that $R_2 \ll R_1$ so that the magnetic field B_1 remains nearly uniform over this area.

We have the flux through the inner ring as

$$\phi = MI_1 \qquad (2)$$

where M is the mutual inductance between the two rings. Comparing Eqs. (1) and (2), one obtains

$$M = \frac{\mu_0 \pi R_2^2}{2R_1} \text{ H}$$

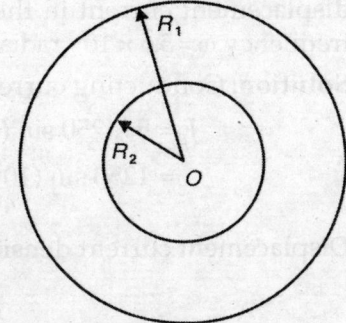

The outer ring has n_1 number of turns of wire and the inner ring has n_2 turns. The magnetic field at the centre of the ring will be $n_1 B_1$ and the flux through the inner ring $n_1 B_1 \pi R_2^2 n_2$. Thus, the mutual inductance between two rings is $n_1 B_1 \pi R_2^2 n / 2R_1$.

Fig. 7.22: Two coplanar concentric rings

Example 7.10: The self inductances of the primary and secondary coils of a transformer are 20 and 10 henry respectively. If the magnetic flux leakage linked with transformer coils is negligible, then find the rate of change of current in primary to produce an emf of 200 volts in the secondary.

Solution: When the magnetic flux leakage is negligible, the mutual inductance is given by

$$M = \sqrt{L_p L_s}$$

where L_p and L_s are self inductances of primary and secondary of the transformer. The emf induced in the secondary is

$$e_s = -M \frac{dI_p}{dt}, \text{ where } I_p \text{ is the current in the primary}$$

$$\therefore \qquad M = \sqrt{20 \times 10} = \sqrt{200} \text{ henry} \quad \left| \begin{array}{l} e_s = 200 \text{ V} \\ L_p = 20 \text{ henry} \end{array} \right.$$

$$\therefore \qquad \frac{dI_p}{dt} = \frac{e_s}{M} = \frac{200}{\sqrt{200}} = 14.14 \text{ A/s} \quad \left| \begin{array}{l} L_s = 10 \text{ henry} \end{array} \right.$$

Example 7.11: An antenna fixed vertically on a car is of length 1 m. If the car is travelling from east to west with a velocity of 60 km/hr, then calculate the induced emf between the two ends of the antenna (horizontal component of earth's magnetic field is 0.4×10^{-4} Wb/m^2).

Solution: Induced emf

$$\varepsilon = -\frac{\partial \phi}{\partial t}$$

and

$$\phi = B \cdot A$$

$$\therefore \qquad e = B \cdot \frac{\partial A}{\partial t} = Blv$$

If the magnetic field is uniform and constant,

$$\therefore \qquad e = Blv \qquad \qquad \left| \begin{array}{l} B = 0.4 \times 10^{-4} \text{ tesla} \end{array} \right.$$

$$= 0.4 \times 10^{-4} \times 1 \times \frac{50}{3} \quad \left| \begin{array}{l} l = 1 \text{ m} \\ v = 60 \text{ km/hr} \end{array} \right.$$

$$= 6.67 \times 10^{-4} \text{ volt.} \qquad \left| \begin{array}{l} = 50/3 \text{ m/s} \end{array} \right.$$

Example 7.12: For a medium, conductivity σ is 5 mho/m and dielectric constant is 1. If an electric field $E = 250 \sin(10^{10} t)$ is applied, then find conducting current and displacement current in the medium. Show that both the currents will be equal at frequency $\omega = 5.6 \times 10^{11}$ radian/s.

Solution: Conducting current density $J_f = \sigma E$

$$J_f = 5 \times 250 \sin(10^{10} t) \qquad\qquad \sigma = 5 \text{ mho/m}$$
$$= 1250 \sin(10^{10} t) \text{ A/m}^2 \qquad E = 250 \sin(10^{10} t)$$

Displacement current density $J_d = \varepsilon \dfrac{\partial E}{\partial t}$

$$J_d = \varepsilon_0 \varepsilon_r \frac{\partial E}{\partial t}$$

$$= 8.85 \times 10^{-12} \times \frac{\partial}{\partial t}(250 \sin 10^{10} t)$$

$$= 8.85 \times 10^{-12} \times 250 \times 10^{10} \cos 10^{10} t$$

$$= 8.85 \times 10^{-12} \times 1 \times 250 \times 10^{10} \cos(10^{10} t)$$

$$= 22.1 \cos(10^{10} t) \text{ A/m}^2$$

When the magnitudes of both the currents are same, then

$$J_f = I_d$$

or $$\sigma E = \varepsilon_0 \varepsilon_r \omega E$$

$\therefore$ $$\omega = \frac{\sigma}{\varepsilon_0 \varepsilon_r} = \frac{5}{8.85 \times 10^{-12} \times 1} = 5.6 \times 10^{11} \text{ rad/s}$$

Example 7.13: If the radius of the sun is 7×10^8 m and power radiated by it is 3.8×10^{26} W, then calculate the magnitude of Poynting vector at the surface of the sun.

Solution: We know that Poynting vector S, represents the power radiated per unit area. Surface area of the sun is $4\pi R^2$, where R is the radius of the sun. If P is the total power radiated by sun, then

$$P = S \times 4\pi R^2$$

or $$S = \frac{P}{4\pi R^2} = \frac{3.8 \times 10^{26}}{4 \times 3.14 \times (7 \times 10^8)^2} = 6.1 \times 10^7 \text{ W/m}^2$$

Example 7.14: A parallel plate capacitor is formed by two discs, the space between which is filled with a homogeneous, poorly conducting medium. The capacitor was charged and then disconnected from the power source. Ignoring edge effects, show that the magnetic field inside the capacitor is absent.

Solution: Magnetic field will be absent since the total current (conduc- tion current plus displacement current) is equal to zero. We now prove this. Let us consider the current density. Suppose that at a certain instance, density of conduction current is J. Obviously, $J \propto D$, and $D = \sigma_n$, where σ is surface charge density on the positive plate and n is the normal (Fig. 7.23).

Fig. 7.23: A parallel plate capacitor

The presence of conduction current leads to a decrease in the surface charge density σ, and hence in **D** as well. This means that conduction current will be accompanied by the displacement current whose density is

$$J_D = \frac{\partial \mathbf{D}}{\partial t} = \left(\frac{\partial \sigma}{\partial t} \right) n = -Jn = -J$$

Hence, it follows that

$$J_t = J + J_D = 0$$

Example 7.15: The space between the plates of a parallel plate capacitor in the form of circular discs is filled with a homogeneous poorly conducting medium with a conductivity σ and permittivity ε. Ignoring edge effects, find the magnitude of vector *H* between the plates at a distance *r* from their axes, if the electric field strength between the plates varies with time in accordance with $E = E_m \cos \omega t$.

Solution: From Maxwell's equation for circulation of vector *H*, we have

$$2\pi r H = \left(J_n + \varepsilon_r \varepsilon_0 \frac{\partial E}{\partial t} \right) \pi r^2$$

Taking into account Ohm's law, $J_n = \sigma E_n(t)$, one obtains

$$H = \frac{r}{2} \left(E_n + \frac{\varepsilon_r \varepsilon_0}{\sigma} \frac{\partial E_n}{\partial t} \right)$$

$$= \frac{r E_m}{2} (\sigma \cos \omega t - \varepsilon_r \varepsilon_0 \sin \omega t)$$

Let us transform the expression in the parentheses to cosine. For this purpose, we multiply and divide the expression by $f = \sqrt{\sigma^2 + (\varepsilon_r \varepsilon_0 \omega)^2}$ and then introduce angle δ through the relation $\frac{\sigma}{f} = \cos \delta$, $\varepsilon_r \varepsilon_0 \frac{\omega}{f} = \sin \delta$. This gives $H = \frac{1}{2} r E_m \sqrt{\sigma^2 + (\varepsilon_r \varepsilon_0 \omega)^2}$ cos $(\omega t + \delta)$

Example 7.16: Protons having the same velocity *v* form a beam of a circular cross-section with current *I*. Find the direction and magnitude of Poynting vector *S* outside the beam at a distance *r* from its axis.

Solution: Figure 7.24 shows that *S* is parallel to *v*. Let us find the magnitude of *S*. We know $S = EH$, where *E* and *H* depend on *r*. According to the Gauss's theorem, we have

$$2\pi r E = \frac{\lambda}{\varepsilon_0}$$

where λ is the charge per unit length of the beam. Besides, it follows from the theorem on circulation of vector *H* that

$$2\pi r H = I$$

Fig. 7.24

Now $I = \lambda v$, we obtain

$$S = EH = \frac{I^2}{4\pi^2 \varepsilon_0 v r^2}$$

Example 7.17: A parallel plate air capacitor whose plates are made in the form of discs of radius a are connected to a source of varying harmonic voltage of frequency ω. Find the ratio of the maximum values of magnetic and electric field energy inside the capacitor.

Solution: Let the voltage across the capacitor vary in accordance with the law $V = V_m \cos \omega t$ and the distance between the capacitor plates be h. Then the electric energy of the capacitor is equal to

$$W_e = \frac{\varepsilon_0 E^2}{2} \pi a^2 h = \frac{\varepsilon_0 \pi a^2}{2h} V_m^2 \cos^2 \omega t \tag{1}$$

The magnetic energy can be determined by the formula

$$W_m = \int_V \frac{B^2}{2\mu_0} dV \tag{2}$$

The quantity B required for evaluating this integral can be found from the theorem on the circulation of vector H: $2\pi r H = \pi r^2 \frac{\partial D}{\partial t}$. Since $H = \frac{B}{\mu_0}$ and $\frac{\partial D}{\partial t} = -\varepsilon_0 \left(\frac{V_m}{h} \right) \omega \sin \omega t$, one obtains

$$B = \frac{1}{2} \varepsilon_0 \mu_0 \frac{r \omega V_m}{h} | \sin \omega t | \tag{3}$$

Substituting Eq. (3) in Eq. (2), where for dV we must take an elementary volume in the form of a ring for which $dV = 2\pi r h dr$. On performing integration, one obtains

$$W_m = \frac{\pi}{16} \frac{\mu_0 \varepsilon_0^2 \omega^2 a V_m^2}{h} \sin^2 \omega t \tag{4}$$

The ratio of the maximum values of magnetic energy [Eq. (4)] and electric energy [Eq. (1)] is obtained as

$$\frac{(W_m)_{max}}{(W_e)_{max}} = \frac{1}{8} \mu_0 \varepsilon_0 a^2 \omega^2$$

For example, for $a = 6\,\text{m}$, $\omega = 10000/\text{s}$, this ratio is equal to 5×10^{-15}.

Example 7.18: Two infinite parallel wires separated by a distance a carry equal current i in opposite directions. A square loop of wire of length a on a side lies in the plane of wires at a distance a from one of the parallel wires. If the current i is increasing at the rate di/dt: (a) calculate the emf induced in the square loop, and (b) is the induced current clockwise or counter clockwise?

Solution: The magnetic field produced by an infinite straight current carrying wire A at a distance r from the wire is given by

Fig. 7.25

$$B = \frac{\mu_0 i}{2\pi r} \tag{1}$$

However, it is perpendicular to the wire into the page on RHS. Due to this field, the magnetic field passing through the loop is

$$\phi_1 = \int_{2a}^{3a} \frac{\mu_0 i}{2\pi r}(a \cdot dr) = \frac{\mu_0 ia}{2\pi} \log_e \frac{3a}{2a} = \frac{\mu_0 ia}{2\pi} \log_e (3/2) \tag{2}$$

The magnetic flux due to the current in wire B which is nearer to the loop is given by

$$\phi_2 = \int_a^{2a} \frac{\mu_0 i}{2\pi r}(a \cdot dr) = \frac{\mu_0 ia}{2\pi} \log_e \left(\frac{ra}{a}\right) = \frac{\mu_0 ia}{2\pi} \log_e 2 \tag{3}$$

It is perpendicular, but pointing out of the page. Hence, the net flux is obtained by subtracting Eq. (2) from Eq. (3), one obtains

$$\phi < \phi_2 - \phi_1 = \frac{\mu_0 ia}{2\pi}[\log_e 2 - \log_e 3/2] = \frac{\mu_0 ia}{2\pi} \log_e \left(\frac{4}{3}\right)$$

$$\therefore \text{ Induced emf in the square, } E = -\frac{d\phi}{dt} = -\frac{\mu_0 a}{2\pi} \log_e \left(\frac{4}{3}\right)\frac{di}{dt}$$

We may note that this induced emf of the induced current will oppose the change in magnetic flux. Clearly, the magnetic field produced by the induced current will direct the page, and the induced current is clockwise (according to the right hand rule).

Example 7.19: There are two inertial systms Σ and Σ'. Electromagnetic field be E, B in Σ and E', B' in Σ'. The field E has a constant direction in the whole space. Σ' is moving relative to Σ with the constant velocity v_0 paralled to $E - (v_0 = \alpha E)$. Show that the component of E' in the direction of E is equal to E.

Solution:

General: Σ Lorentz force on the charge q:
$$F = q\,(E + v \times B)$$

$\Sigma':r' = r - R; R = v_0 t$. Thus,

$$v' = v - v_0 : \text{particle velocity in } \Sigma'$$

Lorentz force:

$$F' = q(E' + v' \times B') = q[E' + (v - v_0) \times B']$$

Σ, Σ': inertial systems $\Leftrightarrow F = F'$. Thus, we find:

$$E + v \times B = E' + (v - v_0) \times B'$$

Now, Σ: particle at rest, i.e. $v = 0$

$$\Rightarrow \quad E' = E + v_0 \times B'$$

According to the presumption: $v_0 \uparrow\uparrow E \Leftrightarrow v_0 = \alpha E$

$$\Rightarrow \quad E' = E + \alpha E \times B'$$

Thus it follows for the component of E' in the direction of E:

$$\frac{E' \cdot E}{E} = \frac{E^2}{E} = E$$

Example 7.20: Show, starting from the Maxwell equations, that the fields E and B fulfill in the vacuum the inhomogeneous wave equations of the form

$$\left(\nabla^2 - \frac{1}{c^2}\frac{\partial^2}{\partial t^2}\right) E = \Box E = \lambda_1(r, t)$$

$$\left(\nabla^2 - \frac{1}{c^2}\frac{\partial^2}{\partial t^2}\right)B = \Box B = \lambda_2(r, t)$$

Determine λ_1 and λ_2.

Solution: Wave equations of the electromagnetic field in the vacuum.

$$\text{curl}H = j + \dot{D}$$

$$\Rightarrow \quad \text{curl}B = \mu_0 j + \mu_0\varepsilon_0\dot{E} = \mu_0 j + \frac{1}{c^2}\dot{E}$$

Then:

$$\text{curlcurl}B = \text{graddiv}B - \nabla^2 B = -\nabla^2 B$$

$$= \mu_0\text{curl}j + \frac{1}{c^2}\text{curl}\dot{E}$$

$$\frac{\partial}{\partial t}\text{curl}E = -\frac{\partial^2}{\partial t^2}B$$

$$\Rightarrow \quad \left(\nabla^2 - \frac{1}{c^2}\frac{\partial^2}{\partial t^2}\right)B = \Box B = -\mu_0\text{curl}j = \lambda_2(r, t)$$

In the same manner one proceeds:

$$\text{curl}E = -\dot{B} = -\mu_0\dot{H}$$

$$\Rightarrow \quad \text{curlcurl}E = \text{graddiv}\dot{E} - \nabla^2 E$$

$$= -\mu_0\text{curl}\dot{H}$$

$$\text{curl}\dot{H} = \frac{\partial}{\partial t}\text{curl}H = j + \varepsilon_0\dot{E}$$

$$\text{div}E = \frac{1}{\varepsilon_0}\text{div}D = \frac{1}{\varepsilon_0}\rho$$

$$\Rightarrow \quad \frac{1}{\varepsilon_0}\nabla^2\rho + \mu_0\frac{\partial}{\partial t}j = \nabla^2 E - \mu_0\varepsilon_0\frac{\partial^2}{\partial t^2}E$$

$$\Rightarrow \quad \left(\nabla^2 - \frac{1}{c^2}\frac{\partial^2}{\partial t^2}\right)E = \Box E = \frac{1}{\varepsilon_0}\nabla^2\rho + \mu_0\frac{\partial}{\partial t}j = \lambda_1(r, t)$$

Example 7.21: A homogeneously charged sphere with the total charge q and the radius R is rotating with constant angular velocity ω around an axis through the center of the sphere. Calculate

a. The electric field E in the whole space,
b. The current density j caused by the sphere rotation,
c. The vector potential A in the whole space,
d. The magnetic induction B in the while space,

e. The field momentum density $\bar{p}_{field} = D \times B$,

f. The angular momentum of the electromagnetic field

$$L_{\text{Field}} = \int d^3 r (r \times \bar{p}_{Field})$$

Solution:

a. The rotation does not change the homogeneous charge density of the sphere:

$$\rho(r, t) \equiv \rho(r) = \begin{cases} q / \frac{4\pi}{3} R^3, & \text{if } r \leq R \\ 0, & \text{if } r > R \end{cases}$$

The *electric field*, too, is therewith time-independent and identical to that of the homogeneously charged sphere '*at rest*'. This field is as follows:

$$E(r) = \frac{q}{4\pi\varepsilon_0} e_r \begin{cases} r/R^3, & \text{if } r \leq R \\ 1/r^2, & \text{if } r > R \end{cases}$$

b. Let the axis of rotation coincide with the z-axis and the center of the sphere with the origin of coordinates. Then the point of the sphere at $r(r < R)$ has the velocity

$$v(r) = \omega \, \hat{e}_z \times r$$

The time-independent charge density leads to a time-independent **current density**:

$$j(r) = \rho(r) \, v(r) = \frac{3q}{4\pi R^3} \omega \, (\hat{e}_z \times r) \, \Theta \, (R - r)$$

With

$$\hat{e}_z \times \hat{e}_r = (0, 0, 1) \times (\sin\theta\cos\varphi, \sin\theta\sin\varphi, \cos\theta)$$

$$= (-\sin\theta\sin\varphi, \sin\theta\cos\varphi, 0)$$

$$= \sin\theta \, \hat{e}_\varphi$$

it is also valid:

$$j(r) = \frac{3q\omega}{4\pi R^3} r \sin\theta \, \Theta \, (R - r) \, \hat{e}_r$$

Cartesian components of the current density:

$$j_x(r) = -\frac{3q\omega}{4\pi R^3} r \, \Theta \, (R - r) \sin\theta \sin\varphi$$

$$j_y(r) = \frac{3q\omega}{4\pi R^3} r \, \Theta \, (R - r) \sin\theta \cos\varphi$$

$$j_z(r) = 0$$

For the next part of this exercise we proceed with further transformation. With the spherical harmonic, we have

$$Y_{11}(\theta, \varphi) = -\sqrt{\frac{3}{8\pi}} \sin\theta \, e^{i\varphi}$$

it can obviously be written:

$$j_x + ij_y = -i\sqrt{\frac{8\pi}{3}}\frac{3q\omega}{4\pi R^3}\, r\, \Theta\,(R - r)\, Y_{11}(\theta, \varphi)$$

c. j, the **vector potential** will be time-independent:

$$A(r, t) = \frac{\mu_0}{4\pi}\int d^3r'\,\frac{j(r')}{|r - r'|} = A(r)$$

Because of $j_z = 0$, $A_z = 0$. For the calculation of the combination, we have

$$A_x + iA_y = \frac{\mu_0}{4\pi}\int d^3r'\,\frac{j_x(r') + ij_y(r')}{|r - r'|}$$

$$= -i\frac{\mu_0}{4\pi}\sqrt{\frac{8\pi}{3}}\frac{3q\omega}{4\pi R^3}\int d^3r'\,\frac{r'\,\Theta\,(R - r')}{|r - r'|}\,Y_{11}(\theta', \varphi')$$

Applying in the integrand the formula $(r_< = \min(r, r');\, r_> = \max(r, r'))$, we obtain

$$\frac{1}{|r - r'|} = \frac{4\pi}{r_>}\sum_{l,m}\frac{1}{2l + 1}\left(\frac{r_<}{r_>}\right)^l Y_{lm}^*(\theta', \varphi')\, Y_{lm}(\theta', \varphi')$$

In order to exploit the orthonormality of the spherical harmonics:

$$A_x + iA_y = -i\frac{\mu_0}{4\pi}\sqrt{\frac{8\pi}{3}}\frac{3q\omega}{4\pi R^3}\,4\pi\sum_{l,m}\frac{1}{2l + 1}\,Y_{lm}(\theta, \varphi)$$

$$\cdot\int d^3r'\,r'\,\Theta\,(R - r')\,\frac{r_<^l}{r_>^{l+1}}\,Y_{11}(\theta', \varphi')\,Y_{lm}^*(\theta', \varphi')$$

$$= -i\mu_0\sqrt{\frac{8\pi}{3}}\frac{3q\omega}{4\pi R^3}\sum_{l,m}\frac{1}{2l + 1}\,Y_{lm}(\theta, \varphi)$$

$$\cdot\int_0^R dr'\,r'^3\,\frac{r_<^l}{r_>^{l+1}}\underbrace{\int_0^{2\pi}d\varphi'\int_{-1}^{+1}d\cos\theta'\,Y_{11}(\theta', \varphi')\,Y_{lm}^*(\theta', \varphi')}_{\delta_{1l}\delta_{1m}}$$

$$= -i\mu_0\sqrt{\frac{8\pi}{3}}\frac{3q\omega}{4\pi R^3}\frac{1}{3}\,Y_{11}(\theta, \varphi)\int_0^R dr'\,r'^3\,\frac{r_<}{r_>^2}$$

For the remaining integral

$$f(r, R) = \int_0^R dr'\,r'^3\,\frac{r_<}{r_>^2}$$

We have to distinguish whether the point of observation r lies inside or outside the sphere:

$$f(r, R) = \begin{cases} \int_r^R dr' \dfrac{r'^4}{r^2} = \dfrac{R^5}{5r^2}, & \text{if } r \leq R \\[3mm] \dfrac{1}{r^2} \int_0^r dr' \, r'^4 + r \int_r^R dr' \, r' = \dfrac{r}{2}\left(R^2 - \dfrac{3}{5}r^2\right), & \text{if } r \leq R \end{cases}$$

If follows therewith:

$$A_x + iA_y = i\mu_0 \frac{q\omega}{4\pi R^3} \sin\vartheta \, e^{i\varphi} f(r, R)$$

In a nutshell, the following is thus valid for the Cartesian components of the vector potential:

$$A_x(r, \theta, \varphi) = -\mu_0 \frac{q\omega}{4\pi R^3} \sin\theta \sin\varphi \, f(r, R)$$

$$A_y(r, \theta, \varphi) = \mu_0 \frac{q\omega}{4\pi R^3} \sin\theta \sin\varphi \, f(r, R)$$

$$A_z(r, \theta, \varphi) = 0$$

$$= -\sin\varphi \, \boldsymbol{e}_x + \cos\varphi \, \boldsymbol{e}_y = \boldsymbol{e}_\varphi$$

The vector potential can eventually be compressed to:

$$\boldsymbol{A}(\boldsymbol{r}) = A_\varphi(r, \theta) \, \boldsymbol{e}_\varphi$$

$$A_\varphi(r, \theta) = \mu_0 \frac{q\omega}{4\pi R^3} \sin\theta \, f(r, R)$$

d. The curl reads in arbitrary curvilinear coordinates y_1, y_2, y_3

$$\text{curl}\boldsymbol{A} = \frac{1}{b_{y_1} b_{y_2} b_{y_3}} \begin{vmatrix} b_{y_1} \boldsymbol{e}_{y_1} & b_{y_2} \boldsymbol{e}_{y_2} & b_{y_3} \boldsymbol{e}_{y_3} \\[2mm] \dfrac{\partial}{\partial y_1} & \dfrac{\partial}{\partial y_2} & \dfrac{\partial}{\partial y_3} \\[2mm] b_{y_1} A_{y_1} & b_{y_2} A_{y_2} & b_{y_3} A_{y_3} \end{vmatrix}$$

Thereby it is

$$b_{y_i,} = \left| \frac{\partial \boldsymbol{r}}{\partial y_i} \right|$$

In the case of spherical coordinates ($b_r = 1$; $b_\theta = r$; $b_\varphi = r \sin \theta$) it remains to be evaluated for the *magnetic induction*:

$$\mathbf{B} = \text{curl} \mathbf{A} = \frac{1}{r^2 \sin \theta} \begin{vmatrix} \mathbf{e}_r & r\mathbf{e}_\theta & r \sin \theta \, \mathbf{e}_\varphi \\ \dfrac{\partial}{\partial r} & \dfrac{\partial}{\partial \theta} & \dfrac{\partial}{\partial \varphi} \\ 0 & 0 & r \sin \theta \, A_\varphi \end{vmatrix}$$

$$= B_r \mathbf{e}_r + B_\theta \mathbf{e}_\theta$$

Radial component:

$$B_r = \frac{1}{r^2 \sin \theta} \left(\frac{\partial}{\partial \theta} \left(r \sin \theta \, A_\varphi(r, \theta) \right) \right)$$

$$= \frac{1}{r^2 \sin \theta} \left(\frac{\partial}{\partial \theta} \left(r \sin \theta \, \mu_0 \frac{q\omega}{4\pi R^3} \sin \theta \, f(r, R) \right) \right)$$

$$= \mu_0 \frac{q\omega}{4\pi R^3} \frac{f(r, R)}{r \sin \theta} (2 \sin \theta \cos \theta)$$

With $f(r, R)$ for part 3, one gets:

$$B_r(r, \theta) = \mu_0 \frac{q\omega \cos \theta}{4\pi R^3} \begin{cases} \dfrac{2R^5}{5r^3}, & \text{if } r \geq R \\ R^2 - \dfrac{3}{5}r^2, & \text{if } r \leq R \end{cases}$$

Angular component:

$$B_\theta = -\frac{1}{r \sin \theta} \frac{\partial}{\partial r} \left(r \sin \theta \, A_\varphi(r, \theta) \right)$$

$$= -\frac{1}{r \sin \theta} \frac{\partial}{\partial r} \left(r \sin \theta \, \mu_0 \frac{q\omega}{4\pi R^3} \sin \theta \, f(r, R) \right)$$

$$= -\mu_0 \frac{q\omega}{4\pi R^3} \frac{\sin \theta}{r} \frac{\partial}{\partial r} (r f(r, R))$$

$$= -\mu_0 \frac{q\omega}{4\pi R^3} \frac{\sin \theta}{r} \frac{\partial}{\partial r} \begin{cases} \dfrac{R^5}{5r}, & \text{if } r \geq R \\ \dfrac{r}{2}\left(R^2 - \dfrac{3}{5}r^2 \right), & \text{if } r \leq R \end{cases}$$

$$= -\mu_0 \frac{q\omega}{4\pi R^3} \frac{\sin \theta}{r} \begin{cases} -\dfrac{R^5}{5r^2}, & \text{if } r \geq R \\ rR^2 - \dfrac{12}{10}r^3, & \text{if } r \leq R \end{cases}$$

We have therewith found for the θ-component of the magnetic induction:

$$B_\theta(r, \theta) = \mu_0 \frac{q\omega \sin \theta}{4\pi R^3} \begin{cases} \dfrac{R^5}{5r^3}, & \text{if } r \geq R \\[2mm] -R^2 + \dfrac{6}{5} r^2, & \text{if } r \leq R \end{cases}$$

e. Field-momentum density:

$$\boldsymbol{p}_{field} = \boldsymbol{D} \times \boldsymbol{B} = \varepsilon_0 \, E_r \hat{\boldsymbol{e}}_r \times (B_r \hat{\boldsymbol{e}}_r + B_\theta \hat{\boldsymbol{e}}_\theta)$$

$$= \varepsilon_0 E(r) B_\theta(r, \theta) \, \boldsymbol{e}_\varphi = \bar{p}_{field} \hat{\boldsymbol{e}}_\varphi$$

Thereby we can use the results from the parts 1 and 4.:

$$p_{field}(r, \theta) = \mu_0 \left(\frac{q}{4\pi} \right)^2 \frac{\omega \sin \theta}{R^3} \begin{cases} \dfrac{R^5}{5r^5}, & \text{if } r \geq R \\[2mm] \dfrac{6}{5} \dfrac{r^3}{R^3} - \dfrac{r}{R}, & \text{if } r \leq R \end{cases}$$

f. Field-angular momentum:

$$\boldsymbol{L}_{field} = \int d^3r (\boldsymbol{r} \times (\boldsymbol{p}_{field})) = - \int d^3r \, r \, p_{field}(r, \theta) \, \boldsymbol{e}_\theta$$

Because of

$$\boldsymbol{e}_\theta = (\cos \theta \cos \varphi, \, \cos \theta \sin \varphi, -\sin \theta)$$

the φ-integration,

$$\int_0^{2\pi} d\varphi \cos \varphi (\sin \varphi) = 0$$

Let us consider that x- and y-components of the angular momentum vanish, i.e.

$$\boldsymbol{L}_{field} = L \boldsymbol{e}_z$$

Now, we have

$$L = 2\pi \int_0^\infty r^2 dr \int_{-1}^{+1} d\cos \theta \, r \sin \theta \, p_{field}(r, \theta)$$

$$= \mu_0 \frac{q^2 \omega}{8\pi R^3} \int_{-1}^{+1} d\cos \theta \, (1 - \cos^2 \theta) \left(\int_0^R r^3 dr \left(\frac{6}{5} \frac{r^3}{R^3} - \frac{r}{R} \right) + \int_R^\infty r^3 dr \frac{R^5}{5r^5} \right)$$

$$= \mu_0 \frac{q^2 \omega}{8\pi R^3} \left(2 - \frac{2}{3} \right) \left(\frac{6}{5} \frac{R^4}{7} - \frac{R^4}{5} + \frac{R^4}{5} \right)$$

Therewith a relatively simple result arises eventually for the angular momentum of the electromagnetic field:

$$L_{\text{field}} = \mu_0 \frac{q^2 R}{35\pi} \omega \, e_z$$

It thus has the same direction as the angular velocity of the rotation homogeneously charged sphere!

GLIMPSES

1. The inductance L of a conductor may be defined as the ratio of the linking magnetic flux (ϕ) to the current producing the flux. For static (or, atmost, low frequency) current I and a coil containing N turns,

$$L = \frac{N\phi}{I}$$

The unit of L is henry, $1\,H = 1\,Wb/A$. Inductance (L) is also given by $L = \lambda/I$, where λ the flux linkage is $N\phi$ for coils with N turns or simply ϕ for other conductor arrangements.

2. There is no equivalent in a magnetic field to the point charge, and consequently no parallel development for its stored energy. However, a more sophisticated approach yields the completely analogous expression

$$W_H = \frac{1}{2} \int_{\text{vol}} \boldsymbol{B} \cdot \boldsymbol{H} \, dV$$

Comparing with $W_H = \frac{1}{2} L I^2$ from circuit analysis, one obtains

$$L = \int_{\text{vol}} \frac{\boldsymbol{B} \cdot \boldsymbol{H}}{I^2} dV$$

3. If the magnetic flux ϕ linking surface S bounded by a closed contour C varies with time (t), then the induced voltage V around C exists by *Faraday's law*,

$$V = -\frac{d\phi}{dt}$$

The *induced voltage V* is a multivalued function of position and is associated with a *nonconstructive field* (electromotive force).

Faraday's law holds, in particular, when the flux through a circuit element is changing because the current in that same element is changing:

$$E \text{ or } V = -\frac{d\phi}{di}\frac{di}{dt} = -L\frac{di}{dt}$$

In circuit theory, L is called the *self inductance* of the element and V is called the *voltage of self inductance* or *back-voltage* in the inductor.

4. Magnetic flux occurs within a conductor cross-section as well as external to the conductor. This internal flux gives rise to an *internal inductance*, which is often small compared to the external inductance and is usually ignored.

5. The property of an electric circuit or component that causes an emf to be generated in it as a result of a change in the current flowing through the circuit (*self inductance*) or of a change in the current flowing through a neighbouring circuit with which it is magnetically linked (*mutual inductance*). In both cases the changing current is associated with a changing magnetic field, the linkage with which in turn induces the emf. In the

Fig. 7.26: Mutual induction

case of self inductance L, the induced emf, E or V generated is given by E or $V = -L\, dI/dt$, where I is the instantaneous current and minus sign indicates that the emf induced is in opposition to the change in current. In case of mutual inductance,

M, the emf, E_1 induced in one circuit is given by $E_1 = -M\dfrac{dI_2}{dt}$, where I_2 is the

instantaneous current in the other circuit.

A part of ϕ_{12} of the magnetic flux produced by the currents I_1 through coil 1 links the N_2 turns of coil 2. The emf of mutual induction in coil 2 is given by (Fig. 7.26)

$$E_2 = N_2 \frac{d\phi_{12}}{dt} \quad \text{(–ve sign omitted)}$$

In terms of *mutual inductance* $M_{12} = N_2 \phi_{12}/I_1$

$$E_2 = N_2 \frac{d\phi_{12}}{dI_1} \frac{dI_1}{dt} = M_{12} \frac{dI_1}{dt}$$

This mutual inductance will be a product of the permeability μ of the region between the coils and a geometrical length, just like inductance L. If we reverse the roles of coils 1 and 2, then

$$E_1 = M_{21} \frac{dI_2}{dt}$$

Thus, $\hspace{4cm} M_{12} = M_{21}$

6. *Electromotive force (emf)* is greatest potential difference that can be generated by a particular source of electric current. In practice this may be observable only when the source is not supplying current, i.e. the source is in open circuit, because of its internal resistance.

7. *Magnetic circuit* is a completely closed path described by a given set of lines of magnetic flux set up in an electrical machine or apparatus by a magnetizing force which is a current coil or a permanent magnet. It consists chiefly of some ferromagnetic material.

   The magnetic circuits may be classified as: (i) undivided and (ii) divided. The type of the circuit to be prepared depends mainly on the purpose and design of the electrical apparatus. Magnetic circuits are also known as *iron core magnetics*.

8. *Magnetization curves:* The magnetic properties of ferromagnetic substances are usually studied by drawing curves relating the magnetization of the material to the strength and variations of the magnetization field. These curves are also known as *B–H* curves. The relative permeability can be computed from the *B–H* curve by use of $\mu_r = B/\mu_0 H$.

9. *Hysteresis* is a phenomenon in which two physical quantities are related in a manner that depends on whether one is increasing or decreasing in relation to other.

   The repeated measurement of "stress" against "strain" with the stress first increasing and then decreasing will produce for some specimen a graph that has the shape of a closed loop. This is known as *hysteresis cycle*. The most familiar hysteresis cycle, however, is produced by plotting the magnetic flux density (*B*) within a ferromagnetic material against the applied magnetic field strength (*H*). The area of hysteresis loop is proportional to energy loss (hysteresis loss) occurring during the cycle. The existence of *permanent magnets* is due to hysteresis. The phenomenon necessiates a dissipation of energy when the substance is subjected to a cycle of magnetic changes. This is known as *magnetic hysteresis loss*.

   The dissipation of energy that occurs due to *dielectric hysteresis*, when the dielectric is subjected to varying (in particular, an alternating) electric field is called *dielectric hysteresis loss*. The dissipation of energy through *elastic hysteresis* is called *elastic hysteresis loss*.

10. *Displacement current* (i_D) is the rate of change of electric flux through a dielectric when the applied electric field is varying. Displacement current does not involve the motion of the current carriers (as in the conductor) but rather the formation of electric dipoles (*i.e. dielectric polarization*), thus setting up the electric stress. Maxwell's recognition that displacement current in a dielectric gives rise to magnetic effects equivalent to those produced by an ordinary conduction current is the basis of his electromagnetic theory of light.

    Displacement current i_D through a specified surface is obtained by integration of the normal component of current density J_D over the surface.

    $$i_D = \int_S J_D \cdot dS = \int_S \frac{\partial D}{\partial t} \cdot dS$$

    $$= \frac{d}{dt} \int_S D \cdot dS$$

11. *Boundary relations for magnetic fields*
    (i) The normal component of **B** is continuous across an interface, i.e.
    $$B_1 \cos \theta_1 = B_2 \cos \theta_2$$
    or
    $$(B_1)_n = (B_2)_n$$

    (ii) The tangential component of magnetic field **H** is continuous across a current-free interface. Thus,
    $$H_1 \sin \theta_1 = H_2 \sin \theta_2$$
    or
    $$(H_1)_t = (H_2)_t$$

    between the angles made by H_1 and H_2 with a current-free interface. These relations were obtained assuming static conditions.

12. *Maxwell's equations*

General set

Point form	General form
$\nabla \times H = J_c + \dfrac{\partial D}{\partial t}$	$\oint H \cdot dl = \int_S \left(J_c + \dfrac{\partial D}{\partial t} \right) \cdot dS$ (Ampere's law)
$\nabla \times E = -\dfrac{\partial B}{\partial t}$	$\oint E \cdot dl = \int_S \left(-\dfrac{\partial B}{\partial t} \right) \cdot dS$ (Faraday's law; S fixed)
$\nabla \cdot D = \rho$	$\oint_S D \cdot dS = \int_V \rho dV$ (Gauss's law)
$\nabla \cdot B = 0$	$\oint_S B \cdot dS = 0$ (non coexistence of magnetic monopole)

Free space set

Point form	Integral form
$\nabla \times H = \dfrac{\partial D}{\partial t}$	$\oint H \cdot dl = \int_S \left(\dfrac{\partial D}{\partial t} \right) \cdot dS$
$\nabla \times E = -\dfrac{\partial B}{\partial t}$	$\oint E \cdot dl = \int_S \left(-\dfrac{\partial B}{\partial t} \right) \cdot dS$
$\nabla \cdot D = 0$	$\int_S D \cdot dS = 0$
$\nabla \cdot B = 0$	$\oint_S B \cdot dS = 0$

The first and second point form Maxwell's equations in the free space set can be used to show that time variable E and H fields cannot exist independently, e.g. if E is a function of time, then $D = \varepsilon_0 E$ will also be a function of time, so that $\dfrac{\partial D}{\partial t}$ will be non-zero. Consequently, $\nabla \times H$ is non-zero, and so a non-zero H must exist. In a similar way, the second equation can be used to show that if H is a function of time, then there must be an E field present.

REVIEW QUESTIONS

1. State and explain Faraday's laws of electromagnetic induction.
2. State and explain the laws of Faraday and Lenz relating to electromagnetic induction. Deduce a relationship between the quantity of electricity flowing through a circuit and a magnetic flux change producing it.
3. Deduce an expression for emf induced in varying magnetic field. What is the difference between electric field so produced and that produced by charges?
4. A conducting rod is moving with a velocity v through a uniform magnetic field of induction B. Show that there appears inside the rod an electric field given by $E = -(v \times B)$.

5. Deduce Faraday's law of electromagnetism, $\nabla \times E = -\dfrac{\partial B}{\partial t}$.

6. Explain the meaning of self and mutual induction. Derive an expression for self induction of a solenoid.

7. Find the self inductance of a long solenoid and calculate the energy stored per unit volume in a magnetic field.

8. Show that the energy stored in a magnetic field per unit volume in free space is

$$\frac{B^2}{2\mu_0}.$$

9. State Ampere's circuital law. Discuss why and how it was modified to include displacement current.

10. Write down Maxwell's field equations and prove Poynting's theorem relating to the flow of energy at a point in space in an electromagnetic field.

11. Write down Maxwell's field equations in differential and integral forms and explain their physical meaning.

12. Show that the total current flowing out of some volume must be equal to the rate of decrease of charge within the volume, assum-ing that no sources and sinks are present within the volume.

13. Obtain the Poynting theorem for the conservation of energy in an electromagnetic field and discuss the physical meaning of each term in the resulting equation.

14. Show that the law of conservation of energy for electromagnetic fields in non-conducting medium is

$$\frac{d\varepsilon_m}{dt} + \nabla \cdot N = 0$$

where ε_m is the energy density and N is Poynting vector.

15. State Maxwell's equations for a system of charges and currents. Show that the energy density and Poynting vector of electromag-netic field are given by

(i) $\varepsilon_m = \dfrac{1}{2}[E \cdot D + B \cdot H]$ and (ii) $N = [E \times H]$

PROBLEMS

1. A car having a vertical aerial of 1 m is moving with a uniform speed of 80 km/h in the east–west direction. If the value of $H = 0.36$ oersted, show that the induced emf in the aerial is 8×10^{-4} V.

2. A rectangular loop of wire with dimensions a and b and lying in the xy plane, is moving with a uniform velocity v in the direction of the positive y-axis in a static magnetic vector field pointing in the z-direction and varying sinusoidally in the y-direction as $B_z = B_0 \sin ky$. Show that the emf in the moving loop is

$$-2B_0 av \sin \frac{kb}{2} \cos kvt.$$

3. A railway line 1.2m wide runs along the magnetic meridian. The vertical component of earth's magnetic field is 0.5 oersted. Show that the emf that will exist between rails when a train runs on the line at a speed of 60 km/h is 1mV.

4. If a potential difference of 90V is developed in a coil, when the current in it is changing at a rate of 3 A/s, calculate the self inductance of the coil. [*Ans.* 30 henry]

5. A long solenoid of cross-sectional area 4×10^{-4} m^2 is wound with 1000 turns/m and the current in its windings is increased at the rate of 100 amp/s. Show that the emf induced in it is $16\pi \times 10^{-6}$ V.

6. A coil of 20 turns is wound on a solenoid of length 1m, area of cross-section 10 sq. cms and having 1000 turns. Show that the mutual inductance of the system is 25.1 mH.

7. A horizontal radial magnetic field $B = 1.1 \dfrac{\text{Wb}}{\text{m}^2}$ exists in the annular air gap between a vertical hollow cylinder and a coaxial cylindrical core. A thin circular ring of aluminium wire falls in the gap with the plane of the ring horizontal and its centre on the axis of the coaxial cylindrical system, under the influence of gravity and electromagnetic forces. If the resistivity of the aluminium is $\sigma^{-1} = 3.1 \times 10^{-8} \Omega$m and its density is $\rho = 2.7 \times 10$ kg/m^3, then show that the terminal velocity of the ring is 6.8×10^{-4} m/s.

8. The magnetic flux through a circular coil, perpendicular to its plane is varying according to the following relation $\phi = (6t^2 + 7t + 1)$, where ϕ is in milli webers and t is in seconds. Show that the emf induced in the coil when $t = 2$ s is 0.03 V.

9. A railway track has two parallel rails at separation 1m. A train passes at a speed of 90 km/h. Calculate the emf induced between the rails. Horizontal component of earth's field is 0.5 gauss and angle of dip $\phi = 45°$. [*Ans.* 1.25 mV]

10. A small magnet of magnetic moment m is rotating about its centre with an angular velocity ω. Show that $\omega \times m = \dfrac{dm}{dt}$ and further show that the motion of the magnet gives rise to an electric field $E = \dfrac{\mu_0}{4\pi r^3} [(r \cdot m)\, \omega - (r \cdot \omega)m]$.

11. Starting from the equation of continuity show for a conducting medium obeying Ohm's law, $J = \sigma E$ and using Gauss's law that $\dfrac{\partial \rho}{\partial t} + \dfrac{\sigma}{\varepsilon} = 0$. From this show that if initially there is charge density ρ_0 at any point inside the conducting medium, the charge density at a later time is given by $\rho = \rho_0 \exp\left[\left(-\dfrac{\sigma}{\varepsilon} t\right)\right]$, where σ is conductivity and $\dfrac{\varepsilon}{\sigma}$ is called the relaxation time.

12. Obtain equation of continuity $\nabla \cdot J + \dfrac{\partial \rho}{\partial t} = 0$ from Maxwell's equations.

13. A conducting rod MN of length L rotates with angular velocity ω about an axis perpendicular to its length and passing through a point C, where $MC = L/3$. A uniform magnetic induction B is parallel to ω. Show that the potential difference $V_M - V_N = B\omega L^2/6$.

SHORT ANSWER QUESTIONS

1. A rectangular closed loop moves horizontally in a uniform magnetic field. Will there be any induced current in the loop if the loop is completely in the magnetic field?

   Ans. There will be no induced current in the loop.

2. What is an ideal inductor?

   Ans. An ideal inductor is one that has zero resistance.

3. Name the quantity that plays an identical role in an electrical circuit as is played by inertia in mechanics.

   Ans. Inductance.

4. A quantitative measure of coupling of two coils is given by the coefficient of coupling $K = \sqrt{\dfrac{M}{L_1 L_2}}$, where L_1 and L_2 are the self inductances of the coils. Can K exceed unity?

   Ans. No, K cannot be more than unity.

5. Is the electric field defined through

   (i) $\nabla \times E = -\dfrac{\partial B}{\partial t}$ and (ii) $\oint_c E \cdot dI = -\dfrac{\partial}{\partial t} \int_s B \cdot dS$ conservative?

   Ans. No, the electric fields defined through (i) and (ii) are non conservative as can be seen from the fact that their line integral for a closed path are not zero.

6. A straight wire is carrying a uniform current density over circular cross-section. If we consider only the magnetic flux within the wire, then what will be the self inductance per unit length?

   Ans. $\dfrac{\mu_0}{8\pi}$ [*Hint:* See Q.No. 15]

7. What does the Poynting vector $N = [E \times H]$ represent?

   Ans. The power transmitted through unit area held normal to N.

8. Does Lenz's law violate the law of conservation of energy?

   Ans. No. Lenz's law is in confirmity with the law of conservation of energy.

9. Does Lenz's law hold for an open circuit?

   Ans. Yes, it can still hold.

10. How will you define displacement current at any point?

    Ans. $J = \dfrac{\partial D}{\partial t}$

11. What is a magnetic dipole?

    Ans. A bar magnet or a small filamentary current loop is usually referred to as magnetic dipole. We now determine the magnetic flux density B at the point $P(r, \theta, \phi)$ due to a circular loop carrying current I, A at P

    $= A_P = \dfrac{\mu_0 m}{4\pi r^2} \hat{a}_r$ and $B = \nabla \times A = \dfrac{\mu_0 m}{4\pi r^3} (2\cos\theta\, \hat{a}_r + \sin\theta\, \hat{a}_r)$.

Fig. 7.27

12. The current density in a given region is $J = J_0\hat{i}\exp(-x^2)$. What will be the change in the charge density at the point $(1, 0, 0)$?

    Ans. $2J_0 \exp(-1)$

13. Find the inductance of an ideal solenoid with 300 turns, $l = 0.50$ m, and a circular cross-section of radius 0.02 m.

    Ans. Axial field,

    $$B = \mu_0 ni = \frac{\mu_0 Ni}{l} \text{ and flux } \phi = B\pi r^2 N$$

    $$\therefore \quad \phi = \frac{\mu_0 Ni}{l} \times \pi r^2 N \text{ or } L = \frac{\phi}{I} = \frac{\mu_0 \pi N^2 r^2}{l}$$

    $$\text{or } L = 4\pi \times 10^{-7} \times \pi (300)^2 \times (2\times 10^{-2})^2 \times \frac{1}{0.5} = 284 \ \mu H.$$

14. A solenoid with $N_1 = 1000$, $r_1 = 1.0$ cm, and $l_1 = 50$ cm is concentric within a second coil of $N_2 = 2000$, $r_2 = 2.0$ cm and $l_2 = 50$ cm. Assuming free space conditions, calculate the mutual inductance.

    Ans. For long coils of small cross-section, H may be assumed constant inside the coil and zero for points outside the coils with the first coil carrying a current I_1.

    We have $H = \left(\dfrac{1000}{0.50}\right)I_1$ Am^{-1} (in the axial direction)

    $B = 2000\,\mu_0\, I_1$ (Wb m^{-2}), $\phi = BA = (2000\,\mu_0\, I_1) \times (\pi \times 10^{-4}) \times 2000$ Wb.

    Since H and B are zero outside the coils, this is the only flux linking the second coil.

    $$\therefore \quad M_{12} = \left(\frac{\phi}{I_1}\right) = 2000 \times 4\pi \times 10^{-7} \times 2000 \times \pi \times 10^{-4} = 1.58 \text{ mH}.$$

15. Use the energy integral to find the internal inductance per unit length of a cylindrical conductor of radius a.

    Ans. At a distance $r \le a$ from the conductor axis,

    $$B = \frac{\mu_0 Ir}{2\pi a^2} \text{ and energy density } U = \frac{1}{2}\frac{B^2}{\mu_0}$$

    $$\therefore \quad U = \frac{1}{2}\frac{\mu_0^2 I^2 r^2}{4\pi^2 a^4 \mu_0} = \frac{\mu_0 I^2 r^2}{8\pi^2 a^4}$$

    $$\therefore \text{ Total energy} = \int U\, dU = \int \frac{\mu_0 I^2 r^2}{8\pi^2 a^4} \cdot 2\pi r\, dr$$

    We have taken unit length

    $$\therefore \quad E = \frac{\mu_0 I^2}{4\pi a^4}\int_0^a r^3 dr = \frac{\mu_0 I^2}{4\pi a^4}\cdot\frac{a^4}{4} = \frac{\mu_0 I^2}{16\pi}$$

    But $E = \dfrac{1}{2}LI^2$

    $$\therefore \quad L = \frac{2E}{I^2} = \frac{\mu_0}{8\pi}.$$

Fig. 7.28

16. Determine the reluctance (the ratio of magnetomotive force to the total flux) of an air gap in a dc machine where the apparent area is $S_0 = 4.26 \times 10^{-2}\,\text{m}^2$ and the gap length $l_a = 5.6$ mm.

    Ans. $R = \dfrac{l_a}{\mu_0 S_a} = \dfrac{5.6 \times 10^{-3}}{(4\pi \times 10^{-7})(4.26 \times 10^{-2})} = 1.05 \times 10^5\,\text{H}^{-1}.$

17. A circular cross-section conductor of radius 1.5 mm carries a current $i_c = 5.5 \sin (4 \times 10^{10}\,t)$ (μA). Find the amplitude of the displacement current density. Given $\sigma = 35$ MS/m and $\varepsilon_r = 1$.

    Ans. $\dfrac{J_C}{J_D} = \dfrac{\sigma}{\omega \varepsilon} = \dfrac{3.50 \times 10^7}{(4 \times 10^{10})(10^{-9}/36\pi)} = 9.90 \times 10^7$

    $\therefore$ $J_D = \dfrac{5.5 \times 10^{-6}}{\pi \times (1.5 \times 10^{-3})^2}/(9.90 \times 10^7) = 7.86 \times 10^{-3}\,\mu\text{A/m}^2.$

18. Moist soil has a conductivity of 10^{-3} S/m and $\varepsilon_r = 2.5$. Find J_C and J_D where $E = 6.0 \times 10^{-6} \sin 9.0 \times 10^9\,t$ (V/m).

    Ans. $J_C = \sigma E = 10^{-3} \times 6.0 \times 10^{-6} \sin 9.0 \times 10^9\,t$ (A m^{-2})

    Now, $D = \varepsilon_0\,\varepsilon_r\,E$

    $\therefore$ $J_D = \dfrac{\partial D}{\partial t} = \varepsilon_0\,\varepsilon_r\,\dfrac{\partial E}{\partial t} = 1.20 \times 10^{-6} \cos 9.0 \times 10^9\,t$ $(\text{A/m}^2).$

19. What do you understand by hysteresis loop of a magnetic material?

    Ans. Phenomenon due to which energy is spent in overcoming the cycle of magnetization due to its reversal is called hysteresis. It is so called because **B** always falls behind **H** throughout the hysteresis loop.

20. What is the physical significance of magnetic vector potential?

    Ans. Magnetic vector potential (**A**) is basically defined by

    $$B = \nabla \times A$$

    or $\nabla \times B = \nabla \times \nabla \times A$

    We have $\mu_0 J = \nabla (\nabla \cdot A) - \nabla^2 A$

    Now, $\nabla^2 A = \mu_0 J$ (making $\nabla \times A = 0$)

    $\therefore$ $A = \dfrac{\mu_0}{4\pi} \displaystyle\int_V \dfrac{J\,dV}{r}$ (for volume current)

    and $A = \dfrac{\mu_0}{4\pi} \displaystyle\int_l \dfrac{I\,dl}{r}$ (for line current)

    A follows the direction of current. It is largely used to determine **E**. Using

    $E = -\dfrac{\partial A}{\partial t}$ and knowing E, H is determined. The concept of vector potential is used

    to determine the power radiated by Hertzian dipole antenna. It is also used to determine magnetic dipole moment and torque developed in magnetic field.

21. Write exact or approximate inductances of some common non coaxial arrangement of standard conductor configurations (Fig. 7.29).

$$L = \frac{\mu_0 N^2 a}{2\pi} \ln \frac{r_2}{r_1} \; (H)$$

N turns

(a): Toroid, square cross-section

$$L = \frac{\mu_0 N^2 S}{2\pi r} \; (H)$$

(Assuming average flux density at average radius *r*)

(b): Toroid, general cross-section

Radius *a*

$$\frac{L}{l} = \frac{\mu_0}{\pi} \cosh^{-1} \frac{d}{2a} \; (H/m)$$

For $d \gg a$,

$$\frac{L}{l} \approx \frac{\mu_0}{\pi} \ln \frac{d}{a} \; (H/m)$$

(c): Parallel conductors of radius *a*

Radius *a*

$$\frac{L}{l} = \frac{\mu_0}{2\pi} \cosh^{-1} \frac{d}{2a} \; (H/m)$$

$$\approx \frac{\mu_0}{2\pi} \ln \frac{d}{a} \; (H/m)$$

(d): Cylindrical conductor parallel to a ground plane

$$L = \frac{\mu_0 N^2 S}{l} \; (H)$$

(e): Long solonoid of small cross-sectional area *S*

Fig. 7.29

MULTIPLE CHOICE QUESTIONS

1. The dimensional formula for induced emf is
   (a) $ML^2T^{-2}A^{-1}$ (b) $ML^2T^{-3}A^{-1}$ (c) $ML^2T^{-3}A^{-2}$ (d) $ML^2T^{-2}A^{-2}$ **[b]**

2. The dimensional formula for the coefficient of self inductance is
   (a) ML^2T^{-2} (b) $ML^2T^{-2}A^{-1}$ (c) $ML^2T^{-2}A^{-2}$ (d) $ML^2T^{-3}A^{-2}$ **[c]**

3. The dimensional formula for magnetic flux is
   (a) ML^2T^{-2} (b) $ML^2T^{-2}A^{-1}$ (c) $ML^2T^{-2}A^{-2}$ (d) $MLT^{-2}A^{-2}$ **[b]**

4. Whenever the flux linked with a circuit changes, there is an induced emf in the circuit. This emf would last in the circuit
   (a) for a very short time
   (b) forever
   (c) for a very long time
   (d) as long as the flux in the circuit changes   **[d]**

5. The unit of magnetic permeability is
   (a) weber   (b) weber/m^2 (c) henry-metre   (d) henry/m   **[d]**

6. The condition for electric field E to be conservative is
   (a) $\oint E \cdot dI = 0$ (b) $\nabla \times E = 0$ (c) $\oint_S E \cdot dS = \dfrac{q}{\varepsilon_0}$ (d) $\nabla \times E = \delta$ **[a]**

7. Magnetic flux density B and vector potential A are related by
   (a) $A = \nabla \times B$ (b) $A = \nabla \cdot B$ (c) $B = \nabla \cdot A$ (c) $B = \nabla \times A$ **[d]**

8. Maxwell's curl equation for static magnetic field is
   (a) $\nabla \times B = 0$ (b) $\nabla \times B = \mu_0 J$

   (c) $\nabla \cdot B = \mu_0 J$ (c) $\nabla \cdot B = \dfrac{\mu_0}{J}$ **[b]**

9. Displacement vector D and electric field strength E are related by
   (a) $D = \dfrac{E}{\varepsilon}$ (b) $D = \varepsilon E$
   (c) $D = \varepsilon E^2$ (d) none of the above   **[b]**

10. A rectangular loop of wire is placed in a uniform magnetic field B acting normally to the plane of the loop. We attempt to put it out of the field with velocity v. The power required for this purpose is
    (a) $\dfrac{Blv}{R}$ (b) $\dfrac{B^2l^2v^2}{R}$ (c) $\dfrac{Blv}{R^2}$ (d) $\dfrac{B^2l^2v^2}{R^2}$ **[b]**

[*Hint:* $\phi = B \cdot A = Blx$, where l is the width of the loop and x is the length within the field region. Due to motion of the loop, induced emf,

$$E = -\frac{d\phi}{dt} = Bl\frac{dx}{dt} = Blv$$

This induced emf sets up a current in the loop.

Now, $i = \dfrac{E}{R}$, where R is the resistance of the loop.

or $\quad i = Blv/R$ (clockwise in accordance with Lenz's law)

This current produces a force on each arm of the loop. Forces F_3 and F_2 are equal and opposite having same line of action. Net force $F_1 = il \times B$ or $F_1 = ilB = B^2l^2v/R$

$\therefore$ Work done for displacement dx against this force

$$= F_1 dx = \left(\dfrac{B^2l^2v}{R}\right)dx = W$$

$\therefore$ Rate of doing work $= \dfrac{dW}{dt} = \dfrac{B^2l^2v}{R}\dfrac{dx}{dt} = \dfrac{B^2l^2v^2}{R}$]

11. The electromagnetic momentum density, g is

 (a) $g = [D \times E]$ (b) $g = [D \times B]$ (c) $g = [B \times A]$ (d) $g = \left[\dfrac{B}{H}\right]$ **[b]**

12. The dimension of the vector $[E \times H]$ is

 (a) $\dfrac{\text{energy}}{\text{area} \times \text{time}}$ (b) $\dfrac{\text{energy}}{\text{time}}$

 (c) $\dfrac{\text{energy}}{\text{area}}$ (d) $\dfrac{\text{energy}}{\text{length} \times \text{area}}$ **[a]**

13. Gauge transformation equations are

 (a) $A' = A + \nabla\psi$ and $\phi' = \phi - \dfrac{\partial\psi}{\partial t}$

 (b) $A' = A + \psi$ and $\phi = \psi$
 (c) $A = A'$ and $\phi' = \phi$
 (d) none of the above   **[a]**
 (Symbols have their usual meanings.)

14. In a medium the ratio $J_C/J_D = 10^{-10}$. The medium can be consi-dered as
 (a) lossy dielectric   (b) lossy conductor
 (c) perfect conductor   (d) perfect dielectric   **[a]**

15. Maxwell inserted the expression for J_D in Ampere's law to satisfy
 (a) Faraday's law
 (b) Gauss's law
 (c) Ampere's law for time varying case
 (d) equation of continuity   **[d]**

16. A parallel plate capacitor has area A, separated by a distance d, and contains dielectric of permittivity ε. When a voltage $V_0 \sin \omega t$ is applied to this plate, the magnitude of J_D and J_C are
 (a) $|J_D| = 0$ (b) $|J_D| = |J_C|$ (c) $|J_D| < |J_C|$ (d) $|J_D| > |J_C|$ **[b]**

17. The displacement current J_D is
    (a) dominant at low frequencies
    (b) dominant at high frequencies
    (c) dominant in time independent case
    (d) none of the above   **[a]**

18. For a circular loop of constant surface area, the Faraday's law gives $\nabla \times E = -dB/dt$. The electric field in this case is
    (a) equal to ∇V
    (b) non-conservative
    (c) conservative
    (d) zero   **[b]**

19. Faraday's law is valid for both open and closed loops. The Lenz's law is valid for
    (a) only closed loop
    (b) only open loop
    (c) both closed as well as open loops
    (d) none of the above   **[a]**

20. Which one of the fundamental equation was modified by Maxwell to form the basis of electromagnetic theory?
    (a) Ampere's law
    (b) Faraday's law
    (c) Gauss's law of magnetostatic
    (d) Gauss's law of electrostatic   **[a]**

21. Which one is not Maxwell's equation?
    (a) $\nabla \times H = J + dD/dt$ (b) $\nabla \cdot J = -d\rho/dt$
    (c) $\nabla \times E = -dB/dt$ (d) $\nabla \cdot D = \rho$ **[b]**

22. For a dielectric conductor interface consisting of surface charge density ρ_s, the boundary condition is
    (a) $E_{n_1}/\varepsilon_s = \rho_s$ (b) $E_{r_1} = E_{r_2}$
    (c) $\varepsilon_1 E_{r_1} = \varepsilon_2 E_{r_2}$ (d) $E_{n_1}/\varepsilon_1 - E_{n_2}/\varepsilon_2 = \rho_s$ **[a]**

23. For a dielectric conductor interface, the boundary condition which is not satisfied is
    (a) $D_{n_1} = 0$ (b) $H_{r_1} = H_{r_2}$
    (c) $E_{r_1} = E_{r_2}$ (d) $B_{n_1} = B_{n_2}$ **[b]**

24. A parallel plate capacitor is being discharged. What is the direction of the energy flow in terms of the Poynting vector in the space between the plates?   **[GATE]**
    (a) along the wire in the positive z-axis direction
    (b) radially inward $(-\hat{r})$
    (c) radially outward $(+\hat{r})$
    (d) circumferential (ϕ) **[c]**

Fig. 7.30

25. A plane electromagnetic wave is given by   **[GATE]**

    $$E = E_0(\hat{x} + e^{i\delta}\hat{y}) \exp\{i(kz - \omega t)\}$$

    At a given location the number of times E vanishes in one second is

    (a) an integer near $\dfrac{\omega}{\pi}$ when $\delta = n\pi$ and zero when $\delta \neq n\pi$, n is an integer

    (b) an integer near $\dfrac{\omega}{\pi}$ and is independent of δ

    (c) an integer near $\dfrac{\omega}{2\pi}$ when $\delta = n\pi$ and zero when $\delta \neq n\pi$, n is an integer

    (d) an integer near $\dfrac{\omega}{2\pi}$ and is independent of δ **[a]**

26. A rod of length L with uniform charge density per unit length is in xy-plane and rotating about z-axis passing through one of its edges with an angular velocity ω as shown in Fig. 7.31 where $(\hat{r}, \hat{\phi}, \hat{z})$ refer to the unit vectors. A_n is a vector potential at a distance d from the origin along the z-axis for $d \ll L$ and J the current density due to the motion of the rod. Which one of the following statements is correct?   **[GATE]**

    (a) J along r, A along z, $|A| \propto \dfrac{1}{d}$

    (b) J along $\hat{\phi}$, A along $\hat{\phi}$, $|A| \propto \dfrac{1}{d^2}$

    (c) J along r, A along z, $|A| \propto \dfrac{1}{d^2}$

    (d) J along $\hat{\phi}$, A along $\hat{\phi}$, $|A| \propto \dfrac{1}{d}$

Fig. 7.31

                               [d]

27. A conducting loop L of surface area S is moving with a velocity v in a magnetic field $B(r, t) = B_0 t^2$. B_0 is a positive constant of suitable dimensions. The emf induced ε in the loop is given by **[GATE]**

(a) $-\int_S \dfrac{\partial B}{\partial t} \cdot dS$

(b) $\oint (v \times B) \cdot dL$

(c) $-\int \dfrac{\partial B}{\partial t} \cdot dS - \oint_L (v \times B) \cdot dL$

(d) $-\oint \dfrac{\partial B}{\partial t} \cdot dS + \oint_L (v \times B) \cdot dL$

[a]

28. The magnetic field (in A m^{-1}) inside long solid cylindrical conductor of radius $a = 0.1$ m is

$$H = \frac{10^4}{r}\left[\frac{1}{\alpha^2}\sin(\alpha r) - \frac{r}{\alpha}\cos(\alpha r)\right]\hat{\phi} \quad \text{when } \alpha = \frac{\pi}{2a}$$

What is the total current (in A) in the conductor? **[GATE]**

(a) $\dfrac{\pi}{2a}$

(b) $\dfrac{800}{\pi}$

(c) $\dfrac{400}{\pi}$

(d) $\dfrac{300}{\pi}$

[b]

29. Which one of the following expressions for current density J can generate vector potential $A = (y^2\hat{i} + x^2\hat{j})$? **[GATE]**

(a) $\dfrac{2}{\mu_0}(x\hat{i} + y\hat{j})$

(b) $-\dfrac{2}{\mu_0}(\hat{i} + \hat{j})$

(c) $-\dfrac{2}{\mu_0}(\hat{i} - \hat{j})$

(d) $\dfrac{2}{\mu_0}(x\hat{i} - y\hat{j})$

[b]

30. In non conducting medium characterized by $\varepsilon = \varepsilon_0$ and $\mu = \mu_0$ and conductivity $\sigma = 0$, the electric field (in volt/m) is given by **[GATE]**

$$E = 20 \sin[10^8 t - kz]\,\hat{j}$$

The magnetic field H (in A m^{-1}) is given by

(a) $20k\cos(10^8 - kz)\hat{i}$

(b) $\dfrac{20k}{10^8 \mu_0}\sin(10^8 t - kz)\hat{j}$

(c) $\dfrac{20k}{10^8 \mu_0}\sin(10^8 t - kz)\hat{i}$

(d) $-20k\cos(10^8 t - kz)\hat{j}$

[c]

31. Consider a conducting loop of radius a and total loop resistance R placed in a region with magnetic field B thereby enclosing a flux ϕ_0. The loop is connected to an electronic circuit as shown in Fig. 7.32. The capacitor being initially uncharged, if the loop is pulled out of the region of magnetic field at constant speed u, the final output voltage V_{out} is independent of

Fig. 7.32

[GATE]

(a) ϕ_0

(b) u

(c) R

(d) c

[a]

Electromagnetic Waves

8.1 INTRODUCTION

We established in Chapter 7 that a varying electric field sets up a magnetic field which, generally speaking, is also varying. This varying magnetic field sets up an electric field, and so on. Thus, if we use oscillating charges to produce a varying (alternating) electromagnetic field, then in the space surrounding the charges a sequence of mutual transformations of an electric field and a magnetic field propagating from point to point will appear. This process will be periodic in both time and space and consequently, will be a wave. This may propagate in vacuum with velocity

$$c = \frac{1}{\sqrt{\varepsilon_0 \mu_0}} \approx 3 \times 10^8 \, \text{m/s}$$

which corresponds to the velocity of light in vacuum. When we were dealing with the phenomenon of electromagnetic induction in Chapter 7, we indicated the possibility of transmitting a signal from one place to another using a time dependent electromagnetic field. Near the end of 19th century, the German physicist proved beyond any doubt that the electromagnetic field does propagate in vacuum with a velocity equal to c. The property of these electromagnetic waves discovered by Hertz have been examined experimentally with great care. The large information that we have accumulated about the properties of electromagnetic waves, such as facts about their production, propagation, and absorption, has opened the door to marvellous world of communication that we know today. Before Hertz performed his experiment, the existence of electromagnetic waves had been predicted by Maxwell as a result of a careful analysis of the equations of the electromagnetic field. Maxwell's equations lead to the important conclusion about the existence of a new (in principle) physical phenomenon. Electromagnetic field may exist independently without electric charges and currents. A change in its state in this case necessarily has a wave nature. Such fields are called *electromagnetic waves*. In vacuum, these waves always propagate at a velocity equal to the velocity of light c. Now we shall show that the existence of electromagnetic waves follows from Maxwell's equations.

8.2 PLANE ELECTROMAGNETIC WAVES IN NON-CONDUCTING MEDIA

Let us assume that ε, μ and σ are constants. The curl of Maxwell's equation $\nabla \times E + \dfrac{\partial B}{\partial t}$ = 0 reads as

$$\nabla \times \nabla \times E = -\nabla \times \frac{\partial B}{\partial t} = -\mu \frac{\partial}{\partial t}(\nabla \times H) \qquad (8.1)$$

or
$$\nabla(\nabla \cdot E) - \nabla^2 E = -\mu \frac{\partial J}{\partial t} - \mu \varepsilon \frac{\partial^2 E}{\partial t^2} \qquad (8.2)$$

In obtaining the above equation we have used the well known vector identity to transform the LHS and Maxwell's equation $\nabla \times H = J + \dfrac{\partial D}{\partial t}$ to transform the RHS of Eq. (8.1). Rewriting Eq. (8.2), we have

$$\nabla^2 E - \mu \varepsilon \frac{\partial^2 E}{\partial t^2} - \mu \sigma \frac{\partial E}{\partial t} = \nabla \left(\frac{\rho}{\varepsilon_0} \right) \qquad (8.3)$$

In a region in which there is no free charge, i.e. $\rho = 0$, one obtains

$$\nabla^2 E - \mu \varepsilon \frac{\partial^2 E}{\partial t^2} - \mu \sigma \frac{\partial E}{\partial t} = 0 \qquad (8.4)$$

Assuming $E(r, t) = E(r) \exp(-i\omega t)$, one can compare the relative magnitudes of the second and the third term in Eq. (8.4). Obviously,

$$\frac{\left| \mu \sigma \dfrac{\partial E}{\partial t} \right|}{\left| \mu \varepsilon \dfrac{\partial^2 E}{\partial t^2} \right|} = \frac{\sigma}{\varepsilon \omega} = \frac{1}{\omega \tau} = \frac{T}{2\pi\tau} \qquad (8.5)$$

We must note that in obtaining Eq. (8.5) we have replaced $\dfrac{\varepsilon}{\sigma}$ by τ (relaxation time) and $\omega = \dfrac{2\pi}{T}$, T being the period of oscillation. If $\tau \ll T$, which is the case for a conducting medium, the third term in Eq. (8.4) is dominant and one can write Eq. (8.4) as

$$\nabla^2 E - \mu \sigma \frac{\partial E}{\partial t} = 0 \qquad (8.6)$$

which is a *diffusion equation*.

If $\tau \gg T$, one can neglect the term involving σ in Eq. (8.4). One obtains for the non-conducting medium

$$\nabla^2 E - \mu \varepsilon \frac{\partial^2 E}{\partial t^2} = 0 \qquad (8.7)$$

One can obtain similar equations for H, viz.

$$\nabla^2 H - \mu \varepsilon \frac{\partial^2 H}{\partial t^2} - \mu \sigma \frac{\partial H}{\partial t} = 0 \qquad (8.8)$$

and
$$\nabla^2 H - \mu \varepsilon \frac{\partial^2 H}{\partial t^2} = 0 \qquad (8.9)$$

One can see that these equations are of the type of wave equation. We have the general wave equation as

$$\nabla^2 \psi - \frac{1}{v^2} \frac{\partial^2 \psi}{dt^2} = 0 \tag{8.10}$$

where, v is the velocity of the propagation of the wave. Comparing Eqs. (8.7), (8.9) and (8.10), we note that the velocity in our case is

$$v = \frac{1}{\sqrt{\varepsilon \mu}} \tag{8.11}$$

This reveals that the fields generated by moving charges can leave the source and travel through space in the form of waves. For vacuum or free space, $\varepsilon = \varepsilon_0 = 8.8542 \times 10^{-12}$ C^2/N m^2 and $\mu = \mu_0 = 4\pi \times 10^{-7}$ H/m $(= Ns^2/C^2)$. Thus,

$$v = \frac{1}{\sqrt{\varepsilon_0 \mu_0}} = 2.99794 \times 10^8 \text{ m/s} \approx 3 \times 10^8 \text{ m/s} \tag{8.12}$$

Thus, v represents the speed of propagation of electromagnetic waves in free space. This agrees very closely with the experimental value of the velocity of light in free space ($c = 2.99794 \times 10^8$ m/s). This result is of immense importance in physics. This unites the subjects of optics and electromagnetism.

We should note that Eq. (8.10) is a scalar equation, while Eqs. (8.6) and (8.9) are vector equations. This reveals that latter equations are valid for each component of E and H.

The simplest type of wave that is a solution of Eq. (8.6) or Eq. (8.9) is a plane wave. A plane wave is one in which the wave amplitude, the field vector component, is constant over all points of a plane normal to the direction of propagation. This plane constitutes a wavefront which advances with a velocity v in a direction normal to itself. The field vector components that lie in given plane are functions of perpendicular distance of the plane from the origin and also of time. If one chooses the coordinate system in such a way that the direction of wave propagation coincides with, say x-axis, we can see that the wave function in this case does not depend on y and z. We can now write the one dimensional wave equation as

$$\frac{\partial^2 \psi}{\partial x^2} - \frac{1}{v^2} \frac{\partial^2 \psi}{\partial t^2} = 0 \tag{8.13}$$

Equation (8.13) has a general solution

$$\psi(x, t) = A \exp[i(kx - \omega t)] + B \exp[-j(kx + \omega t)] \tag{8.14}$$
$$= f(x - vt) + g(x + vt)$$

A and B appearing in Eq. (8.14) are constants (usually complex) and $k = \dfrac{\omega}{v}$, where v is the *phase velocity* of the wave. Equation (8.14) represents waves travelling to the right and left with velocity v. Let us assume that the plane wave fields are of the form

$$E(x, t) = E_0 \exp[i(kx - \omega t)] \tag{8.15}$$

$$H(x, t) = H_0 \exp[i(kx - \omega t)] \tag{8.16}$$

We have assumed that E and H are in phase. One can find the justifica tion for the said assumption in the following argument. Let us consider that α be the phase difference between E and H, i.e.

$$E = E_0 \exp[i(kx - \omega t)]$$

and
$$H = H_0 \exp[i(kx - \omega t + \alpha)]$$

Now, the fields E and H have to satisfy the Maxwell's equation, $\nabla \times E = -\dfrac{\partial B}{\partial t}$.

Therefore, we have

$$\frac{\partial E_z}{\partial x} = \frac{\partial B_y}{\partial t}$$

and
$$\frac{\partial E_y}{\partial x} = -\frac{\partial B_z}{\partial t}$$

$\left(\dfrac{\partial}{\partial y} \text{ and } \dfrac{\partial}{\partial z} \text{ components of } E \text{ are zero since } E \text{ travels in } x\text{-direction} \right)$.

From the second equation, we have

$$ik E_{oy} \exp[i(kx - \omega t)] = i\omega\mu H_{oz} \exp[i(kx - \omega t + \alpha)]$$

Considering the real parts, one obtains

$$k E_{oy} \cos(kx - \omega t) = i\omega\mu H_{oz} \exp[i(kx - \omega t + \alpha)]$$

We may easily see that the above is true for all x and t. We can further see that the only value of α which satisfies this condition is zero. This means E and H are in phase (Fig. 8.1). One finds the ratio of amplitudes of E and H as

$$\frac{E}{H} = \frac{\mu\omega}{k} = \mu v = Z_0 \tag{8.17}$$

Fig. 8.1: Electromagnetic wave

Z_0 is called the *intrinsic impedance* of the medium, i.e. it has the dimensions of impedance. For free space,

$$Z_0 = \mu_0 v_0 = 376\,\Omega \quad \left(\because \frac{E_0}{H_0} = \mu_0 v_0 \right).$$

If E and H are in time phase, Z_0 is a pure resistance. This is the case for free space and all lossless dielectric media. To emphasize the fact that impedance is a pure resistance, one can speak of the intrinsic resistance (instead of impedance) of free space and of lossless dielectric media. If E and H are not in time phase as in conducting media, the ratio of E to H is complex, so that the more general term intrinsic impedance must be used in connection with such media.

As the wave propagates, the two vectors E and H may both change their direction in yz plane without violating any of the conditions established so far.

For the propagation of wave in any arbitrary direction, one obtains

$$E(r, t) = E_0 \exp [i(k \cdot r - \omega t)] \tag{8.18}$$

and

$$H(r, t) = H_0 \exp [i(k \cdot r - \omega t)] \tag{8.19}$$

where, E_0 and H_0 are vector constants in time and $k = \hat{a}_k |k|$ is the propagation vector, and $\hat{a}_k$ is the unit vector in the direction of propagation. We note that E and H are real and hence we are interested only in the real part of Eqs. (8.18) and (8.19). Moreover, E and H obtained above must also satisfy Maxwell's equations. However, we must note that this is not automatic. Maxwell's equations do not completely determine the electromagnetic field. Using Eq.(8.18), Maxwell's equation $\nabla \cdot D = \rho$ takes the form

$$\nabla \cdot E_0 \exp [i(k \cdot r - \omega t)] = 0 \quad (\because \ \rho = 0)$$

This means

$$k \cdot E = 0 \tag{8.20}$$

Similarly, substituting Eq. (8.19) in Maxwell's equation $\nabla \cdot B = 0$, one obtains

$$k \cdot H = 0 \tag{8.21}$$

Equations (8.20) and (8.21) reveal that both E and H are perpendicular to the propagation vector k. This means that such a wave is *transverse wave*. Obviously, electromagnetic plane waves are wholly *transverse* in character. Now, we substitute Eq. (8.18) in

Maxwell's equation $\nabla \times E + \dfrac{\partial B}{\partial t} = 0$, one obtains

$$\begin{aligned}
\nabla \times E &= \nabla \times E_0 \exp (i(k \cdot r - \omega t)] \\
&= \nabla \exp [i(k \cdot r - \omega t)] \times E_0 \\
&= ik \times E = -\frac{\partial B}{\partial t} = -\mu \frac{\partial H}{\partial t} = i\omega\mu H
\end{aligned} \tag{8.22}$$

This means $k \times E = \omega\mu H$.

Equation (8.22) reveals that H is perpendicular both to k and E. This means, E and H which relate to the propagation of the wave, in addition to being perpendicular to the direction of propagation are also perpendicular to one another. The vector $E \times H$ points along the direction of propagation. We can see that the vectors, E, H and k constitute a right-handed orthogonal set (Fig. 8.2). This right-handed relation is an intrinsic property of an electromagnetic wave independent of the choice of coordinate system.

We have seen that the velocity of the electromagnetic waves in a non-conducting natural non-ferromagnetic medium is the same as the velocity of light in free space.

Fig. 8.2: Orientation of the electric and magnetic vector fields relative to the direction of propagation of plane electromagnetic wave

The vectors E and B always oscillate in phase, the instantaneous values of E and B at any point being connected through the relation $E = vB$ or $\sqrt{\varepsilon\varepsilon_0}\ E = \sqrt{\mu\mu_0}\ H$. This means that E and H (or B) simultaneously attain their maximum values, vanish.

Maxwell's brilliant success in the development of the electromagnetic theory of light was due to his understanding of the possibility of existence of electromagnetic waves, following from differential equations discussed in Chapter 7.

Electromagnetic waves cover a wide range of frequencies or wavelengths, and may be classified according to their main source. The classification does not have very sharp boundaries, since different sources may produce waves in overlapping ranges of frequencies. Figure 8.3. relates the various sections of the electromagnetic spectrum in terms of energy, frequency and wavelength. Light is simply a form of electromagnetic radiation. X-rays, ultraviolet rays, infrared rays, radiowaves and microwaves are all electromagnetic radiations. They differ from each other only in the order of magnitude of their wavelength, i.e. frequency. Obviously, all these travel with the same speed, i.e. speed of light in free space.

So far we have considered only the propagation of electromagnetic waves in vacuum. Experiments reveal that the velocity of propagation of an electromagnetic wave through

Fig. 8.3: The electromagnetic spectrum

matter is different from its value of propagation in vacuum. The phase velocity of electromagnetic waves in a material medium is given by

$$v = \frac{1}{\sqrt{\varepsilon\mu}} = \frac{1}{\sqrt{\varepsilon_0 \varepsilon_r \mu_0 \mu_r}} = \frac{v_0}{\sqrt{\varepsilon_r \mu_r}} \qquad (8.23)$$

which is less than v_0, i.e. $v < v_0$.

The ratio of the velocity of electromagnetic waves in vacuum, c and in matter, v is called the *absolute index of refraction*, designated by n. It is a useful concept for describing the properties of materials in relation to electromagnetic waves. Thus,

$$n = \frac{c}{v} = \frac{\sqrt{\varepsilon\mu}}{\sqrt{\varepsilon_0 \mu_0}} \qquad (8.24)$$

But $\dfrac{\varepsilon}{\varepsilon_0} = \varepsilon_r$, and $\dfrac{\mu}{\mu_0} = \mu_r$, where ε_r and μ_r are the relative permittivity and permeability of the medium. Then,

$$n = \frac{c}{v} = \sqrt{\varepsilon_r \mu_r} \qquad (8.25)$$

We must note that Eq. (8.25) applies only if n, μ_r and ε_r are determined at the same frequency. In general, μ_r differs very little from 1 for the majority of substances that transmit electromagnetic waves, and one can write as a satisfactory approximation,

$$n = \sqrt{\varepsilon_r} \qquad (8.26)$$

Equation (8.26) is termed Maxwell's relation and is verified experimentally in case of air, hydrogen, benzene, carbon, etc. having non-polar molecules. But for substances containing polar molecules, e.g. water, glass, etc. ε_r is dependent on frequency and decreases with increasing frequency. Obviously, refractive index exhibits variation with frequency—a phenomenon termed *dispersion*.

8.3 POLARIZATION

The direction E in Eq. (8.18) is constant in time. A wave represented by Eq. (8.18) is said to be *linearly polarized* wave. The plane of electric wave vector E is taken as the plane of polarization. We can see that in the case of Eq. (8.15), it is yz plane. One can express a linearly polarized monochromatic plane wave as a superposition of two linearly independent solutions of the wave equation, e.g.

$$E = (\hat{a}_y E_{oy} + \hat{a}_z E_{oz}) \exp[i(kx - \omega t)] \qquad (8.27)$$

where, $\hat{a}_y$ and $\hat{a}_z$ are the unit vectors along y and z directions respectively. These unit vectors are called *polarization vectors*. We must note that the amplitudes E_{oy} and E_{oz} are complex amplitudes. One can always express any complex quantity as the product of a real quantity and a complex phase factor. Thus, we have

$$\left. \begin{array}{l} E_{oy} = E_0^y \exp(i\alpha) \\[2mm] E_{oz} = E_0^z \exp(i\beta) \end{array} \right\} \qquad (8.28)$$

and

E_0^y and E_0^z in Eq. (8.28) are the real amplitudes. One finds the two independent solutions as

$$E_y = \hat{a}_y E_0^y \exp[i(kx - \omega t + \alpha)]$$

$$E_z = \hat{a}_z E_0^z \exp[i(kx - \omega t + \beta)] \tag{8.29}$$

$$\therefore \qquad E = E_y + E_z$$

$$= (\hat{a}_y E_0^y \exp(i\alpha) + \hat{a}_z E_0^z \exp(i\beta)) \exp[i(kx - \omega t)] \tag{8.30}$$

or $$E = (\hat{a}_y E_0^y + \hat{a}_z E_0^z \exp[i(\beta - \alpha)]) \exp[i(kx - \omega t + \alpha)] \tag{8.31}$$

Particular cases of interest are:

i. $\alpha = \beta$, i.e. E_y and E_z have the same phase or their phase differ by an integral multiple of π, i.e. $\beta = \alpha \pm m\pi$, where $m = 0, 1, 2, \ldots$, then

$$E = (\hat{a}_y E_0^y + \hat{a}_z E_0^z) \exp[i(kx - \omega t + \alpha) \tag{8.32}$$

Equation (8.32) represents a linearly polarized wave. One finds that the resultant polarization vector oscillates along a line making an angle $\theta = \tan^{-1} \dfrac{E_0^z}{E_0^y}$ (Fig. 8.4).

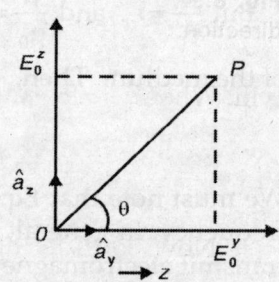

Fig. 8.4: Polarization vector

ii. We can see that amplitudes of two vectors are equal,

i.e. $E_0^y = E_0^z = E_0$ but their phases differ by $\dfrac{\pi}{2}$. Therefore,

$$E = E_0(\hat{a}_y \pm i\hat{a}_z) \exp[i(kx - \omega t + \alpha) \tag{8.33}$$

One can write the components of E (only the real part) as

$$E_y = E_0 \cos(kx - \omega t + \alpha)$$

$$E_z = \pm E_0 \sin(kx - \omega t + \alpha) \tag{8.34}$$

Obviously, Eq. (8.33) is a wave progressing along the x-axis in which the vector E has a constant length E_0 and is a function of time only changing its direction continuously. One obtains from Eq. (8.34),

$$\frac{E_y^2}{E_0^2} + \frac{E_z^2}{E_0^2} = 1 \tag{8.35}$$

This means that the vector traces out a circle at frequency ω. We know that such a wave is said to be circularly polarized (Fig. 8.5). One can determine the direction of rotation by the sign of E_z in Eq. (8.34). If we assume that the observer is facing the incoming wave and the sign is plus, then the rotation is counter clockwise. Such a wave has positive helicity and is said to have left circular polarization. When the rotation is clockwise, i.e. the sign is minus, the wave has the negative helicity, i.e. the wave is right circularly polarized.

Fig. 8.5: Circularly polarized electromagnetic wave. The **E** and **H** fields rotate around the direction of propagation

iii. When $E_0^y \neq E_0^z$ and $\beta = \alpha \pm \dfrac{\pi}{2}$, one obtains the components of E as

$$E_y = E_0^y \cos[i(kx - \omega t + \alpha)]$$

$$E_z = \pm E_0^z[i\sin(kx - \omega t + \alpha \pm \pi/2)]$$

Now, considering the real parts, one obtains

$$E_y = E_0^y \cos(kx - \omega t + \alpha)$$

$$E_z = \pm E_0^z \sin(kx - \omega t + \alpha)$$

$$\therefore \quad \frac{E_y^2}{(E_0^y)^2} + \frac{E_z^2}{(E_0^z)^2} = 1 \tag{8.36}$$

We can see that resultant vector in this case traces an ellipse and one finds that the wave is elliptically polarized.

An arbitrary plane wave can be represented as the combination of two plane waves linearly polarized in mutually perpendicular planes.

The amplitude a of a linearly polarized monochromatic electromagnetic wave is the maximum value of $|E|$. Obviously, a is $|E|_{\max}$.

The intensity I of an electromagnetic wave is a quantity numerically equal to the energy transported by the wave per unit time through a surface of unit area perpendicular to the direction of propagation. The intensity I is related to the Poynting vector N as

$$I = |N| = \frac{1}{T}\left|\int_0^T N dt\right| \tag{8.36a}$$

where, T is the period of the wave.

For a linearly polarized plane monochromatic wave, $I \propto a^2$. where a is the amplitude of the wave. For an arbitrary plane wave in a homogeneous non-absorbing medium, I = constant.

An electromagnetic wave is said to be *spherical* if its intensity depends only on the distance r from a certain point called the *wave centre*. It follows from the law of conservation of energy that for a spherical wave in a homogeneous non-absorbing medium $I = \text{constant}/r^2$.

Real electromagnetic waves are not monochromatic, if for no other reason than that they always are of limited extent in space and of limited duration in time. Such waves can be represented as a totality of monochromatic waves which is called a *wave train* or *packet*. As it is propagated in a dispersive medium, the shape of a wave train is distorted because of the different phase velocities of monochromatic components of the train. The conception of the phase velocity is therefore insufficient to characterize the propagation of a wave train and its rate of energy transport, i.e. the rate of propagation of a signal.

8.4 ENERGY FLUX IN A PLANE ELECTROMAGNETIC WAVE

We have studied about the Poynting vector $N = E \times H$ in Chapter 7. The Poynting vector represents the rate at which the energy is radiated across unit area. Now we will calculate the Poynting vector for a plane wave in which the field vector E and H are expressed in terms of complex amplitudes.

The Poynting vector $N = E \times H$ gives the instantaneous rate of energy flow. We know that E and H vary harmonically with time and therefore one can obtain the average energy flow by taking the averag of $N = E \times H$ over a complete period, i.e.

$$<N> = <\text{real part } E \times \text{real part } H> \tag{8.37}$$

Since E and H are assumed to be complex quantities and therefore one can express them as

and
$$\left. \begin{array}{l} E = (E_1 + iE_2)\exp(-i\omega t) \\ H = (H_1 + iH_2)\exp(-i\omega t) \end{array} \right\} \tag{8.38}$$

where E_1, E_2, H_1 and H_2 are all real. Now

$$\text{real } E = E_1 \cos\omega t + E_2 \sin\omega t$$
$$\text{real } H = H_1 \cos\omega t + H_2 \sin\omega t$$
$$\therefore \quad \text{real } E \times \text{real } H = (E_1 \times H_1)\cos^2\omega t + (E_2 \times H_2)\sin^2\omega t$$
$$+ \{(E_1 \times H_1) + (E_2 \times H_1)\}\sin\omega t \cos\omega t$$

Over a complete period of oscillation, we have

$$< \cos^2 \omega t > = < \sin^2 \omega t > = \frac{1}{2} \text{ and } < \sin\omega t \cos\omega t > = 0$$

$$\therefore \qquad < \text{real } E \times \text{real } H > = \frac{1}{2}\{(E_1 \times H_1) + (E_2 \times H_2)\} \tag{8.39}$$

Now we compute real $(E \times H^*)$. We have

$$E = (E_1 + iE_2)\exp(-i\omega t) = (E_1 + iE_2)(\cos\omega t - i\sin\omega t)$$
and
$$H^* = (H_1 - iH_2)\exp(i\omega t) = (H_1 - iH_2)(\cos\omega t + i\sin\omega t)$$
$$\therefore \quad \text{real } (E \times H^*) = (E_1 \times H_1)\cos^2\omega t + (E_1 \times H_1)\cos\omega t \sin\omega t$$
$$+ (E_2 \times H_2)\cos^2\omega t - (E_2 \times H_1)\cos\omega t \sin\omega t$$
$$- (E_1 \times H_2)\cos\omega t \sin\omega t + (E_1 \times H_1)\sin^2\omega t$$
$$+ (E_2 \times H_1)\cos\omega t \sin\omega t + (E_2 \times H_2)\sin^2\omega t$$
$$= (E_1 \times H_1) + (E_2 \times H_2) \tag{8.40}$$

Obviously $< \text{real } E \times \text{real } H > = \frac{1}{2} \text{real} (E \times H^*)$

$\therefore$ $\qquad\qquad <N> = \frac{1}{2} \text{real} (E \times H^*)$ $\qquad\qquad$ (8.41)

We note that the time factors are automatically cancelled in the product. Thus, one can obtain the time averaged energy from the field vectors.

We have $k \times E = \omega\mu H$ [from Eq. (8.22)]

$\therefore$ $\qquad\qquad H^* = \dfrac{k \times E^*}{\omega\mu}$ $\qquad\qquad$ (8.42)

$\therefore$ $\qquad E \times H^* = \dfrac{1}{\omega\mu} [E \times (k \times E^*)]$

$\qquad\qquad\qquad = \dfrac{1}{\omega\mu} [(E \cdot E^*) k - (E \cdot k) E^*]$

$\qquad\qquad\qquad = \dfrac{1}{\omega\mu} | E_0^2 | \, k$

$\therefore$ Average of N, i.e. $<N> = \dfrac{| E_0^2 |}{2\omega\mu} k$

$\qquad\qquad\qquad = \dfrac{1}{2} | E_0 |^2 \left(\dfrac{\varepsilon}{\mu} \right)^{1/2} \hat{a}_k$ $\qquad\qquad$ (8.43)

The above result has been obtained by using $k = k\hat{a}_k$, where $\hat{a}_k$ is the unit vector in the direction of propagation and $k = \dfrac{\omega}{v} = \omega\sqrt{\varepsilon\mu}$

Let us consider that the wave is moving in the x-direction, i.e. energy flow is in the x-direction and is the same at all points for which z-coordinate is the same. A rectangular box with length c along the x-axis and a and b being its other two sides is shown in Fig. 8.6. We have the energy density in an electromagnetic field $= \dfrac{1}{2} (E \cdot D + B \cdot H)$. One obtains the total energy as

Fig. 8.6: A rectangular box along x-axis

$$\cup = \frac{1}{2} \int_v (E \cdot D + B \cdot H) d\tau \qquad\qquad (8.44)$$

Let us assume that the fields E and H act along y-and z-axis respectively. We obtain

$$E = \hat{a}_y E_0 \exp[i(kx - \omega t)] = \hat{a}_y E_0 \cos(kx - \omega t)$$

and $\quad H = \hat{a}_z H_0 \cos(kx - \omega t)$

$$\therefore \quad \cup = \frac{1}{2} \int_0^c \int_0^a \int_0^b [\varepsilon_0 E_0^2 \cos^2(kx - \omega t) + \mu_0 H_0^2 \cos^2(kx - \omega t)] dx\, dy\, dz$$

$$= \frac{ab}{2} \int_0^c [\varepsilon_0 E_0^2 \cos^2(kx - \omega t) + \mu_0 H_0^2 \cos^2(kx - \omega t)] dx$$

Let T be the period of oscillation. One obtains the average energy in the box as

$$< \cup > = \frac{ab}{2T} \int_0^T \int_0^c [\varepsilon_0 E_0^2 \cos^2(kx - \omega t) + \mu H_0^2 \cos^2(kx - \omega t)] dx\, dt$$

$$= \frac{ab}{2} \int_0^c \frac{1}{2} [\varepsilon_0 E_0^2 + \mu_0 H_0^2] dx = \frac{abc}{4}(\varepsilon_0 E^2 + \mu_0 H_0^2) \tag{8.45}$$

Now, we obtain the time averaged energy density associated with a wave as

$$< \cup' > = \frac{1}{4}(E \cdot D^* + B \cdot H^*) = \frac{1}{4}(\varepsilon E^2 + \mu H^2)$$

$$= \frac{1}{4} \left[\varepsilon |E_0|^2 + \mu \frac{k \times E}{\omega \mu} \cdot \frac{k \times E^*}{\omega \mu} \right]$$

$$= \frac{1}{4}[\varepsilon |E_0|^2 + \varepsilon |E_0|^2] = \frac{1}{2}\varepsilon |E_0|^2 \tag{8.46}$$

$$\therefore \quad < N > = \frac{1}{2}\sqrt{\frac{\varepsilon}{\mu}}\hat{a}_k |E_0|^2 = < \cup' > \frac{\hat{a}_k}{\sqrt{\varepsilon\mu}} = v < \cup' > \hat{a}_k \tag{8.47}$$

Equation (8.47) reveals that the time averaged energy flow is in the direction of propagation of the wave and is equal to the phase velocity of the wave multiplied by the average energy density. Obviously, the energy flows with the same velocity as the wave itself. One can now summarize important results regarding electromagnetic waves in free space as follows:

i. In free space, electromagnetic waves travel with the speed of light.

ii. The electromagnetic field vectors E and H are mutually perpendicular and they are also perpendicular to the direction of propagation of electromagnetic waves. This shows that electromagnetic waves are *transverse* in nature.

iii. E and H field vectors are in the same phase.

iv. The direction of the flow of electromagnetic energy is along the direction of wave propagation and the energy flow per unit area per second is given by Eq. (8.47).

v. The electrostatic energy density is equal to the magnetic energy density and the energy density associated with the electromagnetic wave in free space propagates with the speed of light.

$$\text{Energy density} = \frac{1}{2}\varepsilon E^2 + \frac{1}{2}\mu H^2 = \varepsilon E^2 = \mu H^2 \text{ J/m}^3.$$

8.5 RADIATION PRESSURE AND MOMENTUM OF AN ELECTROMAGNETIC WAVE

Let us consider a plane polarized wave travelling in the z-direction. We must note that the only non-zero quantities involved are E_x, H_y and J_x and only variation in z-coordinate is permitted. In this case, Maxwell's equations

$$\nabla \times E = -\frac{\partial B}{\partial t} \quad \text{and} \quad \nabla \times H = J + \frac{\partial D}{\partial t}$$

reduces to

$$\frac{\partial E_x}{\partial z} = -\frac{\partial B_y}{\partial t} \quad \text{and} \quad -\frac{\partial H_y}{\partial z} = J_x + \frac{\partial D_x}{\partial t} \tag{8.48}$$

We note that any force on a body is due to the action of the wave on currents and charges induced within the body. One obtains the force per unit area as

$$F = \rho E + J \times B \tag{8.49}$$

We have $\nabla \cdot E = 0$, $\rho = 0$. Thus, the only component of the force is

$$F_z = J_x B_y = -B_y \left(\frac{\partial H_y}{\partial z} + \frac{\partial D}{\partial t} \right)$$

$$= -B_y \frac{\partial H_y}{\partial z} + D_x \frac{\partial B_y}{\partial t} - \frac{\partial}{\partial t}(D_x B_y)$$

$$= -B_y \frac{\partial H_y}{\partial z} - D_x \frac{\partial E_x}{\partial z} - \frac{\partial}{\partial t}(D_x B_y)$$

$$= -\frac{1}{2}\frac{\partial}{\partial z}(\varepsilon_0 E_x^2 + \mu_0 H_y^2) - \mu_0 \varepsilon_0 \frac{\partial}{\partial t}(E_x H_y)$$

$$= -\frac{1}{2}\frac{\partial}{\partial z}(\varepsilon_0 E_x^2 + \mu_0 H_y^2) - \mu_0 \varepsilon_0 \frac{\partial N}{\partial t}$$

Let us consider that the currents are confined to a region $z_1 < z < z_2$, then the total force per unit area, i.e. mean pressure on the region is obtained as

$$F_z = \int_{z_1}^{z_2} F_z dz = -\frac{1}{2}\int_{z_1}^{z_2} \frac{\partial}{\partial z}(\varepsilon_0 E_x^2 + \mu_0 H_y^2)dz - \mu_0 \varepsilon_0 \int_{z_1}^{z_2} \frac{\partial N}{\partial t} dz$$

$$= -\frac{1}{2}\left[\varepsilon_0 E_x^2 + \mu_0 H_y^2 \right]_{z_1}^{z_2} - \mu_0 \varepsilon_0 \int_{z_1}^{z_2} \frac{\partial N}{\partial t} dz \tag{8.50}$$

We can see that the fields E_x, H_y in the first two terms of Eq. (8.50) refer to the values just outside the region considered. Obviously, we can consider these two terms as an action from the outside of the region and clearly these represents an external force F_e.

We also note the term $\frac{1}{2}(\varepsilon_0 E_x^2 + \mu_0 H_y^2)$ represents the pressure due to radiation. One can easily show that

$$\frac{1}{2}(\varepsilon_0 E_x^2 + \mu_0 H_y^2) = \frac{N}{C}$$

Now, we can express Eq. (8.50) as

$$F_z = F_e - \frac{dp}{dt}$$

where,

$$p = \mu_0 \varepsilon_0 \int_{z_1}^{z_2} N \, dz \qquad (8.51)$$

In accordance with the Newton's second law of motion, one can interpret p as momentum. Thus, one can attribute to the electromagnetic wave a momentum equal to

$$\mu_0 \varepsilon_0 N = \frac{N}{c^2} \qquad (8.52)$$

We can see that $\frac{N}{c^2}$ has the dimensions of $m^{-2} \, kg \, s^{-1}$, which corresponds to momentum per unit volume. Since electromagnetic radiation propagates with velocity c, we may use the relation between energy (E), momentum (p); $p = \frac{vE}{c^2}$ (with $v = c$). Since momentum must have
the direction of propagation, we may write

$$p = \frac{E}{c} \hat{a}_n = \frac{N}{c^2} \qquad (8.53)$$

where, $\hat{a}_n$ is the unit vector in the direction of propagation.

If an electromagnetic wave has momentum, it also has angular momentum. The angular momentum per unit volume is

$$L = r \times p = r \times \frac{N}{c^2} \qquad (8.54)$$

This is called the *orbital* angular momentum of radiation. In addition, electromagnetic radiation possesses an intrinsic angular momentum spin, similar to the spin of fundamental particles. For circularly polarized plane waves, it can be shown that the spin has a component along the direction of propagation equal to $\pm E/\omega$ ($E \rightarrow$ energy), depending on whether the polarization is clockwise or counterclockwise. For a linearly polarized wave, the average value of the component of the spin along the direction of propagation is zero.

Therefore, when a charged particle absorbs or emits electromagnetic radiation, not only does its energy and momentum change but also its angular momentum changes accordingly, the result has been verified experimentally both directly and indirectly. In summary, we conclude that *an electromagnetic wave carries momentum and angular momentum as well as energy.*

The result is not surprising. An electromagnetic interaction between two electric charges means an exchange of energy and momentum between the charges. This is accomplished by means of the electro-magnetic field, which is the carrier of the energy and momentum exchanges.

8.6 PLANE ELECTROMAGNETIC WAVES IN A CONDUCTING MEDIUM

We have Maxwell's equations as

$$\left. \begin{array}{c} \nabla \cdot D = \rho \\[4pt] \nabla \cdot B = 0 \\[4pt] \nabla \times E = -\dfrac{\partial B}{\partial t} \\[6pt] \text{and} \quad \nabla \times H = J + \dfrac{\partial D}{\partial t} \end{array} \right\} \qquad (8.55)$$

Let us assume that medium is homogeneous and isotropic and characterized by permittivity ε, permeability μ and conductivity σ, but not any charge or any current other than that determined by Ohm's law. Thus, we have

$$D = \varepsilon E, \quad B = \mu M, \quad J = \sigma E \quad \text{and } \rho = 0$$

Using the above, Maxwell's equations given by Eq. (8.55) takes the form

$$
\left.
\begin{aligned}
\nabla \cdot E &= 0 \qquad &\text{(i)} \\
\nabla \cdot H &= 0 \qquad &\text{(ii)} \\
\nabla \times E &= -\mu \frac{\partial H}{\partial t} \qquad &\text{(iii)} \\
\text{and} \quad \nabla \times H &= \sigma E + \varepsilon \frac{\partial E}{\partial t} \qquad &\text{(iv)}
\end{aligned}
\right\}
\qquad (8.56)
$$

Taking curl of Eq. (8.56 iii), one obtains

$$\nabla \times \nabla \times E = -\mu \frac{\partial}{\partial t}(\nabla \times H)$$

Substituting $\nabla \times H$ from Eq. (8.56 iv), one obtains

$$\nabla \times \nabla \times E = -\mu \frac{\partial}{\partial t}\left(\sigma E + \varepsilon \frac{\partial E}{\partial t} \right)$$

or

$$\nabla \times \nabla \times E = -\sigma\mu \frac{\partial E}{\partial t} - \varepsilon\mu \frac{\partial^2 E}{\partial t^2} \qquad (8.57)$$

Similarly, if we take the curl of Eq. (8.56iv) and substitute $\nabla \times E$ from Eq. (8.56iii), one obtains

$$\nabla \times \nabla \times H = -\sigma\mu \frac{\partial H}{\partial t} - \varepsilon\mu \frac{\partial^2 H}{\partial t^2} \qquad (8.58)$$

Now using vector identity

$$\nabla \times \nabla \times A = \nabla (\nabla \cdot A) - \nabla^2 A$$

and keeping in mind Eqs. (8.56i) and (8.56ii) (i.e. $\nabla \cdot E = 0$ and $\nabla \cdot H = 0$), Eqs. (8.57) and (8.58) take the form

$$\nabla^2 E - \sigma\mu \frac{\partial E}{\partial t} - \varepsilon\mu \frac{\partial^2 E}{\partial t^2} = 0 \qquad (8.59)$$

$$\nabla^2 H - \sigma\mu \frac{\partial H}{\partial t} - \varepsilon\mu \frac{\partial^2 H}{\partial t^2} = 0 \qquad (8.60)$$

Equations (8.59) and (8.60) represent wave equations governing electromagnetic fields E and H in a homogeneous isotropic conducting medium of conductivity σ. We can see that these equations are vector equations of identical form. This means that each of the six components of E and H separately satisfies the same scalar wave equation of the form

$$\nabla^2 \psi - \sigma\mu \frac{\partial \psi}{\partial t^2} - \varepsilon\mu \frac{\partial^2 \psi}{\partial t^2} = 0 \qquad (8.61)$$

where, ψ is a scalar, and holds good for any one of components of E and H. Equation (8.61) contains damping terms proportional to the conductivity of the medium. Writing Eq. (8.61) for E and H, we have

$$\nabla^2 E - \varepsilon\mu \frac{\partial^2 E}{\partial t^2} - \sigma\mu \frac{\partial E}{\partial t} = 0 \tag{8.62}$$

and

$$\nabla^2 H - \varepsilon\mu \frac{\partial^2 H}{\partial t^2} - \sigma\mu \frac{\partial H}{\partial t} = 0 \tag{8.63}$$

Let us find the plane wave solution of Maxwell's equations for a conducting medium. Let us assume that field vectors E and H vary harmonically with time, i.e.

$$E(r, t) = E_0 \exp[i(k \cdot r - \omega t)] \tag{8.64}$$

and

$$H(r, t) = H_0 \exp[i(k \cdot r - \omega t)] \tag{8.65}$$

Substituting Eq. (8.64) in Eq. (8.62), one obtains

$$-k^2 E(r,t) + \varepsilon\mu\omega^2 E(r,t) + i\sigma\mu\omega E(r,t) = 0$$

or

$$[k^2 - \varepsilon\mu\omega^2 - i\sigma\mu\omega] E(r,t) = 0 \tag{8.66}$$

$$\therefore \qquad k^2 = \varepsilon\mu\omega^2 \left[1 + \frac{i\sigma}{\varepsilon\omega}\right] = [\varepsilon\mu\omega^2 - i\sigma\mu\omega] \tag{8.67}$$

We can see that the first term corresponds to the displacement current and the second term to the conduction current.

We know that for any wave there is a fundamental relationship between the wave number k and wave frequency ω. This relationship is known as *dispersion relation*. The dispersion relation for electromagnetic wave in vacuum is

$$k = \frac{\omega}{v} \tag{8.68}$$

For a conducting medium, k is given by Eq. (8.67). We can see that k is complex. For a free space, $\sigma = 0$ and we have $k^2 = \varepsilon\mu\omega^2 = \dfrac{\omega^2}{v^2}$. This relation will provide us with information regarding the nature of the propagation of electromagnetic waves inside a medium.

As stated above, k is complex in a conducting medium. We may write assuming that σ is real

$$k = \alpha + i\beta \tag{8.69}$$

so that

$$k^2 = \alpha^2 - \beta^2 + 2i\alpha\beta \tag{8.70}$$

Comparing Eq. (8.70) with Eq. (8.67), one obtains

$$\left.\begin{array}{c} \alpha^2 - \beta^2 = \varepsilon\mu\omega^2 \\ 2\alpha\beta = \sigma\mu\omega \end{array}\right\} \tag{8.71}$$

Solving these equations, one obtains

$$\alpha = \omega \sqrt{\frac{\varepsilon\mu}{2}} \left[1 + \left\{1 + \left(\frac{\sigma}{\varepsilon\omega}\right)^2\right\}^{1/2}\right]^{1/2} \tag{8.72}$$

$$\beta = \omega \sqrt{\frac{\varepsilon\mu}{2}} \left[-1 + \left\{ 1 + \left(\frac{\sigma}{\varepsilon\omega} \right)^2 \right\}^{1/2} \right]^{1/2} \qquad (8.73)$$

In order to make the solutions yield the proper form for k in free space, we have to consider positive square root only.

Since $k = \alpha + i\beta$, one can write Eqs. (8.64) and (8.65) as

$$E = E_0 \exp(-\beta r) \exp[i(\alpha r - \omega t)] \qquad (8.74)$$

and

$$H = H_0 \exp(-\beta r) \exp[i(\alpha r - \omega t)] \qquad (8.75)$$

From Eqs. (8.74) and (8.75), it is obvious that a plane wave cannot propagate in a conducting medium without attenuation. The field amplitudes are spatially attenuated due to the presence of term $(-\beta r)$. The quantity β is a measure of attenuation and is known as *absorption coefficient*. Also in last exponential terms in Eqs. (8.74) and (8.75), we have replaced k by α, therefore we conclude that the field vectors are propagated in the conducting medium with speed $\left(v = \dfrac{\omega}{k} \right)$ given by

$$v = \frac{\omega}{\alpha} = \frac{1}{\sqrt{\mu\varepsilon}} \left[\frac{\sqrt{\left\{ 1 + \left(\dfrac{\sigma}{\varepsilon\omega} \right)^2 \right\}} + 1}{2} \right]^{-1/2} \qquad (8.76)$$

Now, we can consider the form of propagation vector $k = \alpha + i\beta$ in the two particular cases of interest:

i. $\dfrac{\sigma}{\omega} \ll 1$ (for a poor conductor): We obtain $\alpha = \sqrt{\mu\varepsilon\omega}$ and

$$\beta = \frac{\sigma}{2} \sqrt{\frac{\mu}{\varepsilon}}$$

$$\therefore \qquad k + \alpha + i\beta = \sqrt{\mu\varepsilon\omega} + i\frac{\sigma}{\varepsilon}\sqrt{\frac{M}{\varepsilon}} \qquad (8.77)$$

This is correct to first order in $\dfrac{\sigma}{\omega\varepsilon}$. In this limit $\alpha \gg \beta$ and the attenuation of wave determined by β is independent of frequency, aside from the possible variation of conductivity.

ii. $\dfrac{\sigma}{\omega\varepsilon} \gg 1$ (for a good conductor): Obviously α and β are approximately equal, i.e.

$$\alpha \approx \beta = \omega\sqrt{\mu\varepsilon}\sqrt{\frac{\sigma/\omega\varepsilon}{2}} = \sqrt{\frac{\mu\sigma\omega}{2}} \qquad (8.78)$$

$$\therefore \qquad k \approx \beta + i\beta = (1+i)\sqrt{\frac{\mu\sigma\omega}{2}} \qquad (8.79)$$

We have only lowest order term in $\dfrac{\omega\varepsilon}{\sigma}$.

One can immediately obtain a qualitative picture of the behaviour of the medium at high and low frequencies from Eq. (8.73). As $\omega \to \infty$, $\beta \to 0$, i.e. at very high frequencies, the electromagnetic waves will be able to travel through the medium. At lower frequencies β will be finite and obviously there will be appreciable damping of the electromagnetic wave.

8.7 SKIN DEPTH OR PENETRATION DEPTH

We can see that the term $i\sigma\mu\omega$ in Eq. (8.67) arises from the term involving $\dfrac{\partial E}{\partial t}$ in Eq. (8.62), i.e. from the conduction current; while the term $\varepsilon\mu\omega^2$ arises from the term involving $\dfrac{\partial^2 E}{\partial t^2}$ in the same equation, i.e. from the displacement current. In almost all conducting media, the conduction current dominates the displacement current and obviously, one can neglect the middle term in Eq. (8.62). Thus, for a good conducting medium

$$\nabla^2 E = \sigma\mu \frac{\partial E}{\partial t} \tag{8.79a}$$

One finds the attenuated solution of Eq. (8.79a) as

$$E = E_0 \exp(-\beta r) \exp[i(\alpha r - \omega t)] \tag{8.80}$$

For a good conductor, if the frequency is not too high, we have $\dfrac{\sigma}{\varepsilon\omega} \gg 1$. From Eqs. (8.72) and (8.73), we have

$$\alpha = \beta = \sqrt{\frac{\omega\sigma\mu}{2}} = \frac{1}{\delta}$$

where,

$$\delta = \sqrt{\frac{2}{\mu\sigma\omega}} \tag{8.81}$$

The term $\delta = \dfrac{1}{\beta} = \sqrt{\dfrac{2}{\mu\sigma\omega}}$ measures the depth at which electromag- netic wave entering a conductor is damped to $\dfrac{1}{e} = 0.369$ of its initial amplitude at the surface. This depth is known as the *skin depth* or the *penetration depth*. Using Eq. (8.81), one can write Eq. (8.80) as

$$E = E_0 \exp\left(-\frac{r}{\delta}\right) \exp\left[i\left(\frac{r}{\delta} - \omega t\right)\right] \tag{8.82}$$

It is obvious from Eq. (8.81) that the skin depth decreases with increasing the frequency and becomes zero as the conductivity approaches infinity. δ is small for good conductors at high frequency currents. For copper at 60 cycles, δ is 8.5×10^{-3} m, but at 1 MHz, it has dropped to 6.6×10^{-5} m. At 30 GHz frequency, δ is found to be 3.8×10^{-7} m. This shows that why in high frequency circuits, current flows only on the surface of the conductors. The importance of the skin depth is that it measures the depth to which an electromagnetic wave can penetrate a conducting medium. Obviously, the conducting sheets which are used as electromagnetic shields must be thicker than the skin depth. The magnitudes of skin depth in different types of materials are listed below:

Material	Frequency	δ (penetration depth)
Sea water	30 kHz	0.1 m
Aluminium	50 Hz	1.25×10^{-2} m
Silver	100 MHz	10^{-7} m

We must note that the penetration depth is inversely proportional to the square root of the frequency, therefore a thin sheet of conducting material can act as a *low-pass filter* for electromagnetic waves. One can even make a good conductor from a poor conductor by thin coating of silver or copper. The relatively higher value of δ in the case of sea water explains why radio communication with submerged submarine becomes difficult at depth of several metres.

8.8 ELECTROMAGNETIC WAVES IN BOUNDED MEDIA

When plane waves are incident on a boundary between two media, some of its incident energy crosses the boundary and some is reflected. We are all familiar with the reflection and refraction of light waves at the surface separating two media of different refractive indices. Now, we will show how electromagnetic theory provides a simple explanation for it. This will also explain how optics is contained within the framework of Maxwell's theory of electrodynamics. Here in this section, we restrict to the study of plane boundaries. When boundaries are curved, they tend to scatter incident plane waves into several different directions simultaneously.

Reflection and Refraction of Plane Electromagnetic Waves at a Plane Surface

Let us consider two non-conducting ($\sigma = 0$) dielectric media marked as "1" and "2". These are characterized by μ_1, ε_1 and μ_2, ε_2 and separated by a plane boundary, plane x as shown in Fig. 8.7.

Let us consider that a plane electromagnetic wave is incident obliquely on the plane boundary. One can see that there will be, in general, both reflected wave and transmitted wave. Let us determine fraction of the energy of the incident wave reflected and transmitted.

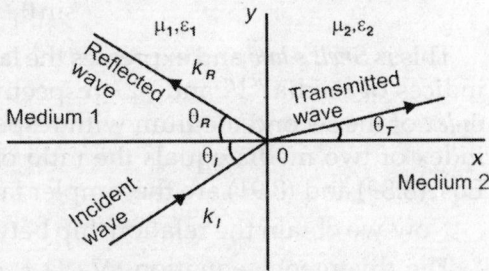

Fig. 8.7: Reflection and refraction of plane electromagnetic wave at a plane interface

One can express the fields for incident, reflected and transmitted waves as

$$E_I = E_{oI} \exp[i(k_I \cdot r - \omega t)], H_I = \frac{k_I \times E_I}{\omega_I \mu_1} \tag{8.83}$$

$$E_R = E_{oR} \exp[i(k_R \cdot r - \omega t)], H_R = \frac{k_R \times E_R}{\omega_R \mu_1} \tag{8.84}$$

$$E_T = E_{oT} \exp[i(k_T \cdot r - \omega t)], H_T = \frac{k_T \times E_T}{\omega_T \mu_2} \tag{8.85}$$

where, the subscripts I, R, and T in Eqs. (8.83) to (8.85) represent incident, reflected and transmitted waves. We must note that E_{oI}, E_{oR} and E_{oT} are time independent scalar amplitudes and these may be complex. One can find the relationship between these quantities by making the total fields obey the boundary conditions in the plane $x = 0$. We can easily see that the tangential components of E and H can be continuous across the boundary at all points and at all times only if the exponentials are the same at the boundary for all three fields. Obviously,

$$\omega_I = \omega_R = \omega_T \tag{8.86}$$

This means that the frequency remains unchanged in the reflected and transmitted waves, and

$$k_I \cdot r = k_R \cdot r = k_T \cdot r \tag{8.87}$$

Equation (8.87) reveals that all the propagation vectors are coplanar. Let us choose r to lie in the boundary plane, i.e. $\hat{a}_n \cdot r = 0$, where $\hat{a}_n$ is a unit vector normal to the plane, and in the plane of the propagation vector, one finds

$$k_I \sin\theta_I = k_R \sin\theta_R = k_T \sin\theta_T \tag{8.88}$$

We can see that k_I and k_R vectors are in the same medium. This means $|k_I| = |k_R|$. Thus, we have

$$\theta_I = \theta_R \tag{8.89}$$

This expresses the law of reflection.
Now we have

$$k_I \sin\theta_I = k_T \sin\theta_T$$

or $$\frac{\sin\theta_I}{\sin\theta_T} = \frac{k_I}{k_T} = \sqrt{\frac{\varepsilon_2\mu_2}{\varepsilon_1\mu_1}} \qquad (\because \ k = \omega\sqrt{\varepsilon\mu}) \tag{8.90}$$

One can assume $\mu_1 = \mu_2$ for non-magnetic materials. We have

$$\frac{\sin\theta_I}{\sin\theta_T} = \sqrt{\frac{\varepsilon_2}{\varepsilon_1}} = \frac{n_2}{n_1} = n_{12} \tag{8.91}$$

This is *Snell's law* and expresses the law of refraction. Here n_1 and n_2 are the refractive indices of media "1" and "2" respectively and n_{12} is known as the *relative refractive index* of the second medium with respect to the first one. Thus, the relative refractive index of two media equals the ratio of their absolute refractive indices. Obviously, Eqs. (8.89) and (8.91) are the simpler laws of geometrical optics.

Now we obtain the relationship between the various field vectors.

The divergence equations $\nabla \cdot D = \rho$ and $\nabla \cdot B = 0$ can be obtained by applying divergence operator to the remaining Maxwell's equations involving E and H. One can see that the boundary conditions on D_n and B_n are automatically satisfied provided the conditions on E_I and H_T are met. One can express the conditions as

$$(E_I + E_R) \times \hat{a}_n = E_T \times \hat{a}_n \tag{8.92}$$

and $$(H_I + H_R) \times \hat{a}_n = H_T \times \hat{a}_n \tag{8.93}$$

Equation (8.93) can be expressed as

$$(k_I \times E_I + k_I \times E_R) \times \hat{a}_n = (k_T \times E_T) \times \hat{a}_n \qquad (\because \ \mu_1 = \mu_2) \tag{8.94}$$

We now consider two separate situations: (i) E is polarized perpendicular to the plane of incidence, i.e. the plane defined by k and $\hat{a}_n$ and (ii) E is polarized parallel to the plane of incidence.

One can obtain the general case of arbitrary polarization by appropriate linear combination of the above two situations.

(i) E polarized perpendicular to the plane of incidence

Figure 8.8 exhibits the field vectors corresponding to reflection and refraction with E polarized perpendicular to the plane of incidence. We can see that the electric field vectors are directed away from the viewer. The conditions given by Eqs. (8.92) and (8.94) yields

Fig. 8.8: Reflection and refraction when E polarized $\perp^r$ to the plane of incidence

$$E_{oI} + E_{oR} = E_{oT} \tag{8.95}$$

and $\qquad k_1 E_{oI} \cos\theta_I - k_1 E_{oR} \cos\theta_R = k_T E_{oT} \cos\theta_T$

i.e. $\qquad (E_{oI} - E_{oR}) \cos\theta_I = \dfrac{k_T}{k_I} E_{oT} \cos\theta_T \tag{8.96}$

where, E_{oI}, E_{oR} and E_{oT} represent the scalar amplitudes of the incident, reflected and transmitted electromagnetic waves respectively. Solving Eqs. (8.95) and (8.96), one obtains

$$\frac{E_{oR}}{E_{oI}} = \frac{\cos\theta_I - \dfrac{\sin\theta_I}{\sin\theta_T}\cos\theta_T}{\cos\theta_I + \dfrac{\sin\theta_I}{\sin\theta_T}\cos\theta_T} = \frac{\sin(\theta_T - \theta_I)}{\sin(\theta_T - \theta_I)} \tag{8.97}$$

and $$\frac{E_{oT}}{E_{oI}} = \frac{2\cos\theta_I}{\cos\theta_I + \dfrac{k_T}{k_I}\cos\theta_T} = \frac{2\cos\theta_I \sin\theta_T}{\sin(\theta_I + \theta_T)} \tag{8.98}$$

We obtain the ratio of the amplitudes of the reflected and incident waves from Eq. (8.98). Equations (8.97) and (8.98) represent *Fresnel's equations* for non-conducting media when electric vector E is perpendicular to the plane of incidence. We have the following situations:

(a) $n_2 > n_1$, i.e. $\dfrac{n_1}{n_2} < 1$: The ratio in Eq. (8.98) is negative. This indicates that the reflection of the electromagnetic wave results in a phase change of π, i.e. the electric vector of the reflected wave oscillates 180° out of phase with that in the incident wave. The ratio $\dfrac{E_{oT}}{E_{oI}}$ is always positive.

One can define the reflection coefficient R as the energy flux reflected from the interface divided by the flux incident on it. Therefore,

$$R_\perp = \frac{\hat{a}_n \cdot <N_R>}{\hat{a}_n \cdot <N_I>} = \frac{|E_R \times H_R|}{|E_I \times H_I|} = \frac{E_{oR}^2}{E_{oI}^2} \tag{8.99}$$

where, N_R and N_I are the Poynting vectors and subscript $\perp$ indicates that E is polarized perpendicular to the plane of incidence. Thus, we have

$$R_\perp = \frac{\sin^2(\theta_T - \theta_I)}{\sin^2(\theta_T + \theta_I)} \tag{8.100}$$

Similarly, one can obtain the transmission coefficient as

$$
\begin{aligned}
T_\perp &= \frac{\hat{a}_n \cdot <N_T>}{\hat{a}_n \cdot <N_I>} = \frac{|E_{oT}|^2}{|E_{oI}|^2} \frac{n_2 \cos\theta_T}{n_1 \cos\theta_I} \\[2mm]
&= \frac{4\cos^2\theta_I \sin^2\theta_T}{\sin^2(\theta_I + \theta_T)} \frac{\sin\theta_I \cos\theta_T}{\sin\theta_T \cos\theta_I} \\[2mm]
&= \frac{4\cos\theta_I \cos\theta_T \sin\theta_I \sin\theta_T}{\sin^2(\theta_I + \theta_T)} \\[2mm]
&= \frac{\sin 2\theta_I \sin 2\theta_T}{\sin^2(\theta_I + \theta_T)}
\end{aligned}
\tag{8.101}
$$

From Eqs. (8.100) and (8.101), we have

$$R_\perp = T_\perp = 1 \tag{8.102}$$

From Eqs. (8.97) and (8.98), we have for normal incidence

$$R_\perp = \left(\frac{n_1 - n_2}{n_1 + n_2}\right)^2 \quad \text{and} \quad T_\perp = \frac{n_2}{n_1}\left(\frac{2n_1}{n_1 + n_2}\right)^2 \tag{8.103}$$

(b) $\frac{n_1}{n_2} > 1$: When an electromagnetic wave is incident on the interface of two dielectrics from a denser medium, then it can be shown that refracted ray is deviated away from the normal, and reflected and incident wave are in the same phase.

(ii) E polarized parallel to the plane of incidence

In this case, electric field vectors of all the three waves are parallel to the plane of incidence (Fig. 8.9).

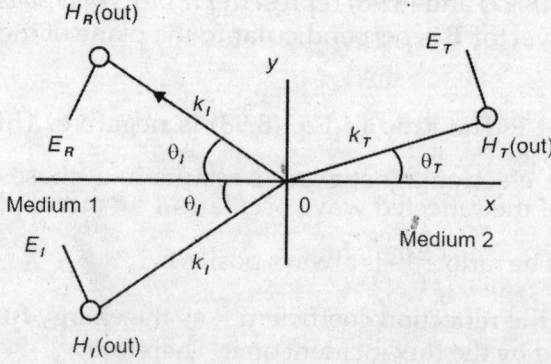

Fig. 8.9: Reflection and refraction of electromagnetic wave with polarization parallel to the plane of incidence

The boundary conditions give

$$E_{oI} \cos\theta_I - E_{oR} \cos\theta_R = E_{oT} \cos\theta_T$$

or
$$(E_{oI} - E_{oR}) \cos\theta_I = E_{oT} \cos\theta_T \qquad (\because \quad \theta_I = \theta_R) \tag{8.104}$$

and
$$k_I E_{oI} + k_I E_{oR} = k_T E_{oT} \tag{8.105}$$

or
$$E_{oI} + E_{oR} = \frac{n_2}{n_1} E_{oT} \tag{8.106}$$

Solving Eqs. (8.104) and (8.105), one obtains

$$\frac{E_{oR}}{E_{oI}} = \frac{\cos\theta_I - \left(\dfrac{n_1}{n_2}\right)\cos\theta_T}{\cos\theta_I + \cos\theta_T} = \frac{\tan(\theta_I - \theta_T)}{\tan(\theta_I + \theta_T)} \tag{8.107}$$

Relations (8.106) and (8.107) represent *Fresnel's equations* for the case when **E** vector is polarized parallel to the plane of incidence.

Comparison of Eqs. (8.97) and (8.107) indicates that there is an important distinction between the two states of polarization of **E**. Considering only simple case, i.e. when $n_2 = n_1$, i.e. $\theta_T = \theta_I$, we note from Eq. (8.97) that $\dfrac{E_{oR}}{E_{oI}} \neq 0$ for any angle of incidence θ_I

lying between 0 and $\dfrac{\pi}{2}$. However, when E is polarized in plane of incidence, Eq. (8.107) gives

$$\frac{E_{oR}}{E_{oI}} = 0 \tag{8.108}$$

Obviously, $R_{||} = 0$ when $\theta_I + \theta_T = \dfrac{\pi}{2}$.

Thus, when the wave is incident at an angle $\theta_I = \dfrac{\pi}{2} - \theta_T$, it crosses the interface without suffering reflection. This angle is called as *Brewster's angle* (θ_B) or polarizing angle. One can find the magnitude of θ_B by applying Snell's law, i.e.

$$\frac{\sin\theta_B}{\sin\theta_T} = \frac{\sin\theta_B}{\sin\left(\dfrac{\pi}{2} - \theta_B\right)} = \tan\theta_B = \frac{n_2}{n_1}$$

$$\therefore \qquad \theta_B = \tan^{-1}\left(\frac{n_2}{n_1}\right) \tag{8.109}$$

Thus, if an unpolarized electromagnetic wave consisting of a mixture of both types of radiation falls on the interface at Brewster's angle, only the radiation with its electric vector at right angles to the plane of incidence will be reflected and the reflected radiation will be polarized.

An important application of Brewster's angle can be found in gas lasers. The windows in a gas laser are arranged at Brewster's angle. In a gas laser, mirror outside the glass windows are as shown in Fig. 8.10. It is reported that at normal incidence, about 92% of the incident intensity is transmitted through glass windows and about 8% is lost in each transport. We know that in a laser there are a large number of traverses and obviously, very little of light will be left after a few traverses. To overcome this difficulty, the windows in gas lasers are arranged at Brewster's angle.

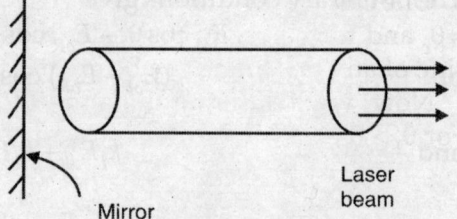

Fig. 8.10: Mirror position in a gas laser

This helps in perfect transmission of the electric field component polarized parallel to the plane of incidence, i.e. it suffers negligible loss even after many traverses. The emerging laser beam is 100% purely linearly polarized as the component polarized perpendicular to the plane of incidence is partly reflected and partly transmitted each time it strikes the surface and after a number of traverses it is completely eliminated.

8.9 TOTAL INTERNAL REFLECTION

Let us consider that a plane monochromatic electromagnetic wave is incident on the interface from a denser medium (say glass) to the rarer medium, i.e. $n_1 > n_2$. We have from Snell's law,

$$\sin\theta_T = \frac{n_1}{n_2}\sin\theta_I \qquad (8.110)$$

$$= \sqrt{\frac{\varepsilon_1}{\varepsilon_2}}\sin\theta_I$$

Obviously, $\theta_T > \theta_I$. We have assumed the media to be non-magnetic, i.e. $n_1 = \sqrt{\dfrac{\varepsilon_1}{\varepsilon_0}}$ and $n_2 = \sqrt{\dfrac{\varepsilon_2}{\varepsilon_0}}$. Let θ_I be gradually increased starting from zero. This means θ_T will also increase until its attains the value $\dfrac{\pi}{2}$. Let the value of θ_I corresponding to this situation be θ_c. From Eq. (8.110), we have

$$\sin\theta_c = \frac{n_2}{n_1} \qquad (8.111)$$

When $\theta_I = \theta_c$, there will only be reflected wave and no transmitted wave. The angle θ_c is called the *critical angle*. Obviously, when the wave incident at $\theta_I = \theta_c$, the refracted wave is propagated parallel to the surface.

Now let us see what happens if θ_c is increased further, i.e. when $\theta_I > \theta_c$. Let us express θ_T in terms of θ_I and θ_c, we have

$$\cos\theta_T = \sqrt{1-\sin^2\theta_T}$$

But $\qquad\qquad \sin\theta_T = \dfrac{n_1}{n_2}\sin\theta_1 = \dfrac{\sin\theta_I}{\sin\theta_c}$ [using Eq. (8.111)]

$$\therefore \qquad\qquad \cos\theta_T = \sqrt{1-\frac{\sin^2\theta_I}{\sin^2\theta_c}} \qquad (8.112)$$

One can see that the value of $\cos\theta_T$ decreases as θ_I is increased. It becomes zero at $\theta_I = \theta_c$ and for values of θ_I greater than θ_c, $\cos\theta_T$ becomes an imaginary number, since sine of any real angle can be greater than unity.

Now, we determine the amplitude of the reflected electric vector when $\theta_I > \theta_c$. For $\theta_I > \theta_c$, one can write

$$\cos\theta_T = \sqrt{1 - \frac{\sin^2\theta_I}{\sin^2\theta_c}} = i\sqrt{\frac{\sin^2\theta_I}{\sin^2\theta_c} - 1} = iQ \qquad (8.113)$$

where,

$$Q = \sqrt{\frac{\sin^2\theta_I}{\sin^2\theta_c} - 1}$$

When E is polarized perpendicular to the plane of incidence, we have

$$\frac{E_{oR}}{E_{oI}} = \frac{\cos\theta_I - \dfrac{n_2}{n_1}\cos\theta_T}{\cos\theta_I + \dfrac{n_2}{n_1}\cos\theta_T} = \frac{\cos\theta_I - \dfrac{n_2}{n_1}iQ}{\cos\theta_I + \dfrac{n_2}{n_1}iQ}$$

$$\therefore \qquad \left|\frac{E_{oR}}{E_{oI}}\right|^2 = 1, \quad \text{i.e. } |E_{oR}| = |E_{oI}|$$

Similarly, one finds when E is polarized parallel to the plane of incidence,

$$\left|\frac{E_{oR}}{E_{oI}}\right|^2 = 1, \quad \text{i.e. } |E_{oR}| = |E_{oI}|$$

Obviously, the wave is totally reflected. This phenomenon is called *total internal reflection*.

We must note that there is a change of phase on reflection, i.e. if the incident wave is polarized in a plane intermediate between the plane of incidence and the plane normal to it, the two components will not be in phase after reflection and the electromagnetic wave will be elliptically polarized.

One can conclude that when $\theta_I > \theta_c$, there is no refracted wave and all the energy is reflected. To verify it, we compute the average rate of energy flow across the boundary.

We have the rate of energy flow $= <N> \cdot \hat{a}_n$

$$= \frac{1}{2}R_e(E_T \times H_T^*)\cdot\hat{a}_n = \frac{1}{2}R_e\left(E_T \times \frac{k_T \times E_T^*}{\mu\omega}\right)\cdot\hat{a}_n$$

$$= \frac{1}{2\mu\omega}R_e[(E_T \cdot E_T^*)k_T - (E_T \cdot k_T)E_T^*]\cdot\hat{a}_n$$

$$= \frac{1}{2\mu\omega}R_e(E_T \cdot E_T^*)k_T \cdot \hat{a}_n \quad \text{(since } E_T \text{ is perpendicular to } k_T\text{)}$$

$$= \frac{1}{2\mu\omega}R_e|E_{oT}|^2 k_T \cos\theta_T = \frac{1}{2\mu\omega}R_e iQ k_T |E_{oT}|^2 \qquad (8.114)$$

One can see that Eq. (8.114) is purely imaginary. This means $<N> \times \hat{a}_n = 0$. Obviously, this justifies our assumption.

Now, the question arises, if $n_1 > n_2$ and the wave is incident at an angle greater than θ_c, will there be field on the other side of the boundary? Although there is no energy flow across the surface, but the field does exist on the other side of the surface. We can easily verify it. Let us consider the equation for the refracted wave. We have

$$E_T = E_{oT} \exp [i(k_T \cdot r - \omega t) \tag{8.115}$$

Equation (8.115) takes the following form for the coordinate axes shown in Fig. 8.7,

$$E_T = E_{oT} \exp [i(k_T x \cos \theta_T + k_T y \sin \theta_T - \omega t)]$$
$$= E_{oT} \exp [-k_T x Q \exp \{i(k_T y \sin \omega_T - \omega t)\} \tag{8.116}$$

Equation (8.116) clearly reveals that the field does exist on the far side of the surface in the medium 2, but it is rapidly attenuated as shown in Fig. 8. 11.

Fig. 8.11: Attenuation of field in medium 2

The skin depth or penetration depth, i.e. how far the wave penetrates the medium is given by

$$x = \frac{1}{k_T Q} = \frac{1}{k_T} \left(\frac{\sin^2 \theta_I}{\sin^2 \theta_c} - 1 \right)^{-1/2} \tag{8.117}$$

Let us consider that the wave is going from glass to air. We have μ for glass = 1.5. Thus, the critical angle for the glass is $\theta_c = \sin^{-1} 2/3 \approx 42°$. Obviously, if light is incident internally from the glass at an angle greater than 42°, say 45°, then there will be total internal reflection and we have

$$x = \frac{1}{k_T} \left(\frac{1}{2} \times \frac{9}{4} - 1 \right)^{-1/2} = \frac{2\sqrt{2}}{k_T} = \frac{\lambda}{2\pi} 2\sqrt{2} = 0.45\lambda$$

This means that the field becomes negligible beyond distances of the order of a few wavelengths. Attenuation results in an increase in the molecular internal energy, mainly rotational molecular motion and the translational kinetic energy. This means that the molecular charges in the medium, which oscillate due to interaction with the incident wave give rise to a radiation field. The forward wave in this field interferes destructively with the original wave and gives rise to a small transmission.

The question arises, how can one account for the decaying field on the far side of the boundary, when there is no transport of energy across the boundary. We must note that the energy does flow in the second medium since the component of the field in the medium is finite, but during the later part of the cycle the flow is in the opposite direction and the energy is returned to the first medium.

8.10 REFLECTION FROM A CONDUCTING PLANE: METALLIC REFLECTION

Let us consider the reflection and refraction at the interface separating two conducting media, medium '1' be characterized by permittivity ε_1, permeability μ_1, conductivity σ_1 and refractive index n_1, while the medium '2' by permeability μ_2, permittivity ε_2, refractive index n_2 and conductivity σ_2. Let a plane monochromatic electromagnetic wave be incident on the interface from medium '1' to medium '2'. Since the case of oblique incidence is more involved, we shall confine to the simple case of normal incidence only.

One can write for incident, reflected and transmitted waves as

$$\left. \begin{array}{l} E_I = E_{oI} \exp\{k_I \cdot r - \omega t\} \\[2mm] H_I = \dfrac{k_I \times E_I}{\omega \mu_1} \end{array} \right\} \tag{8.118}$$

$$E_R = E_{oR} \exp[i(k_I \cdot r - \omega t)]; \quad H_R = \dfrac{k_I \times E_R}{\omega \mu_1} \tag{8.118a}$$

and $\qquad E_T = E_{oT} \exp[i(k_T \cdot r - \omega t)]; \; H_T = \dfrac{k_T \times E_T}{\omega \mu_2} \tag{8.118b}$

In conducting medium '2', the propagation vector k_T is given by

$$k_T^2 = \varepsilon_2 \mu_2 \omega^2 \left[1 + \frac{i\sigma}{\varepsilon_2 \omega}\right] \tag{8.119}$$

The boundary conditions give

$$E_{oI} = E_{oR} = E_{oT} \tag{8.120}$$

$$k_I(E_{oI} + E_{oR}) = k_T E_{oT} \tag{8.121}$$

We know that k_T is complex, and hence E_{oR} and E_{oT} cannot both be real. This means, one should expect phase shifts other than 0 or π in the reflected and transmitted waves. Solving Eqs. (8.120) and (8.121), one obtains

$$E_{oR} = \frac{k_T - k_I}{k_T + k_I} E_{oI} \tag{8.122}$$

and $\qquad E_{oT} = \dfrac{2k_I}{k_T + k_I} E_{oI} \tag{8.123}$

Substituting $k_I = \omega\sqrt{\varepsilon_1 \mu_1}$ and k_T from Eq. (8.119), one obtains

$$E_{oR} = \frac{\omega\sqrt{\varepsilon_2 \mu_2}\left(1 + \dfrac{i\sigma}{\omega\varepsilon_2}\right)^{1/2} - \omega\sqrt{\varepsilon_1 \mu_1}}{\omega\sqrt{\varepsilon_2 \mu_2}\left(1 + \dfrac{i\sigma}{\varepsilon_2\omega}\right)^{1/2} + \omega\sqrt{\varepsilon_1 \mu_1}} E_{oI} \tag{8.124}$$

and $\qquad E_{oT} = \dfrac{2\omega\sqrt{\varepsilon_1 \mu_1}}{\omega\sqrt{\varepsilon_2 \mu_2}\left(1 + \dfrac{i\sigma}{\omega\varepsilon_2}\right)^{1/2} + \omega\sqrt{\varepsilon_1 \mu_1}} \tag{8.125}$

Special Cases

i. *Perfect conductor:* In the case of a perfect conductor, we have $\sigma = \infty$. This means

$$E_{oI} = E_{oR} \quad \text{and} \quad E_{oT} = 0 \tag{8.126}$$

Obviously, the *reflection is complete.*

ii. *Very good conductor:* When the conductor is not a perfect conductor but a very

good conductor, we have $\dfrac{\sigma}{\varepsilon_2 \omega} \gg 1$. The approximation adopted earlier yields,

$$k_T = \alpha + i\beta = (1+i)\sqrt{\frac{\omega\sigma\mu_2}{2}} = \frac{1+i}{\delta} \tag{8.127}$$

where, $\delta = \sqrt{\dfrac{2}{\omega\sigma\mu_2}}$ is the skin depth. Using Eq. (8.127), Eq. (8.122) gives

$$E_{oR} = \frac{\dfrac{1+i}{\delta} - \omega\sqrt{\varepsilon_1\mu_1}}{\dfrac{1+i}{\delta} + \omega\sqrt{\varepsilon_1\mu_1}} E_{oI} = \frac{\left[\dfrac{1}{\delta} - \omega\sqrt{\varepsilon_1\mu_1}\right] + \dfrac{i}{\delta}}{\left[\dfrac{1}{\delta} + \omega\sqrt{\varepsilon_2\mu_2}\right] + \dfrac{i}{\delta}} E_{oI} \tag{8.128}$$

One obtains the reflection coefficient R as

$$R = \frac{|E_{oR}|^2}{|E_{oI}|^2} = \frac{[1 - \omega\sqrt{\varepsilon_1\mu_1}\,\delta]^2 + 1}{[1 + \omega\sqrt{\varepsilon_1\mu_1}\,\delta]^2 + 1}$$

Since $\dfrac{\sigma}{\varepsilon_2\omega} \gg 1$, $\omega\sqrt{\varepsilon_1\mu_1}\,\delta \ll 1$ and therefore, we have

$$R \approx 1 - 2\omega\sqrt{\varepsilon_1\mu_1}\,\delta \tag{8.129}$$

$$= 1 - 2\sqrt{\frac{2\omega\varepsilon_1\mu_1}{\sigma\mu_2}} \tag{8.130}$$

If $\mu_1 = \mu_2$, i.e. magnetic permeabilities are equal, then one obtains

$$R = 1 - 2\sqrt{\frac{2\omega\varepsilon_1}{\sigma}} \tag{8.131}$$

$$= 1 - \frac{2}{n}, \quad \text{where} \quad n = \sqrt{\frac{\sigma}{2\omega\varepsilon_1}} \tag{8.132}$$

This equation shows that at low frequencies for substances of high conductivity, the reflection coefficient (R) is close to unity and obviously, almost all the energy will be reflected. This shows that why metals are opaque to light. The little energy that flows into the metal is rapidly dissipated by the heat loss associated with the induced currents. One can obtain the measure of the energy transmitted into the conducting medium by calculating the transmission coefficient (T). T gives the ratio of the energy transmitted to the energy incident. We have

$$T = 1 - R = \sqrt{\frac{2\omega\varepsilon_1}{\sigma}} = \frac{2}{n} \tag{8.133}$$

For example, taking $\qquad\qquad \sigma = 6 \times 10^7 \, (\Omega\,\text{m})^{-1}, \, \nu = 10^{10}/\text{s}$

we obtain $\quad T = 2\sqrt{\dfrac{2 \times 2\pi \times 10^{10} \times 8.85 \times 10^{-12}}{6 \times 10^7}} = 2.7 \times 10^{-4}\,\text{s}$

This is extremely small for direct measurements to be made. We must note that the deviation of reflection coefficient R from unity becomes significant for extremely good conductors like copper, silver and aluminium, only for radiations of very short wavelength. These materials absorb waves strongly, i.e. the skin depth is very small. Concluding we can say that all waves which are strongly absorbed are also strongly reflected by metals, i.e. all *good conductors are good absorbers and good reflectors.*

The total energy in the medium is

$$E = E_I + E_R$$

where, $E_I = \hat{a}_n E_{oI} \exp[i(k_I x - \omega t)]$ and $E_R = -\hat{a}_n E_{oI} \exp[i(k_I x - \omega t)]$. We must note that the amplitudes in both the waves are taken equal because the reflection coefficient for metallic surfaces is almost unity. Therefore, we have

$$E = \hat{a}_n E_{oI} e^{-i\omega t} \{e^{ik_I x} - e^{-ik_I x}\} \qquad (8.134)$$

Considering the real part only, we have

$$E = 2\hat{a}_n E_{oI} \sin \omega t \, \sin k_I x \qquad (8.135)$$

Obviously, the field in the medium is represented by a standing wave.

ILLUSTRATIVE EXAMPLES

Example 8.1: The earth receives 2 cal/min/sq. cm solar energy. What are the amplitudes of electric and magnetic fields of radiation?

Solution: We have

$$|N| = |E \times H| = EH \sin 90° = EH \qquad (1)$$

The energy flux per unit area per second at the surface of the earth is

$$2\,\text{cal/min/sq. cm} = \frac{2 \times 4.2 \times 10^4}{60}\,\text{J/m}^2/\text{s}^2 \qquad (2)$$

Comparing Eqs. (1) and (2) yields

$$EH = \frac{2 \times 4.2 \times 10^4}{60} = 1400 \qquad (3)$$

But $\qquad \dfrac{E}{H} = \sqrt{\dfrac{\mu_0}{\varepsilon_0}} = \sqrt{\dfrac{4\pi \times 10^7}{8.85 \times 10^{-12}}} = 376.6 \qquad (4)$

Multiplying Eqs. (3) and (4), one obtains

$$E^2 = \frac{1400 \times 376.6}{\sqrt{1400 \times 376.6}} \quad \text{or} \quad E = 726.1\,\text{V/m} \qquad (5)$$

Using Eq. (5) in Eq. (3) yields

$$H = \frac{1400}{E} = \frac{1400}{726.1} = 1.928\,\text{amp-turn/m.}$$

$\therefore \qquad E_0 = E\sqrt{2} = 726.1\sqrt{2} = 1027\,\text{V/m}$

and $\qquad H_0 = H\sqrt{2} = 1.928 \times \sqrt{2} = 2.73\,\text{amp-turn/m}$

Example 8.2: In a material for which $\sigma = 5.0\,\omega^{-1}\mathrm{m}^{-1}$ and $\sigma_r = 1$, the electric field intensity is $E = 250\sin 10^{10}t$ Vm^{-1}. Find the conduction and displacement current densities, and the frequency at which they have equal magnitudes.

Solution: We have seen that there are some materials which are neither good conductors nor perfect dielectrics, so both conduction current and displacement current exist. Now, assuming time dependence $e^{j\omega t}$ for E, the total current

$$J = J_C + J_D = \sigma E + \frac{\partial}{\partial t}(\varepsilon E) = \sigma E + J\omega\varepsilon E \tag{1}$$

Now, for the given material

$$J_C = \sigma E = 5 \times 250 \sin 10^{10}t = 1250 \sin 10^{10}t \, \mathrm{A\,m^{-2}}$$

Assuming that field direction does not vary with time, we have

$$J_D = \frac{\partial D}{\partial t} = \varepsilon_0\varepsilon_r\frac{\partial E}{\partial t}$$

$$= 8.85 \times 10^{-12} \times 1 \times \frac{\partial}{\partial t}(250 \sin 10^{10}t)$$

$$= 8.85 \times 10^{-12} \times 250 \times 10^{10} \cos 10^{10}t$$

$$= 22.1 \cos 10^{10}t \, \mathrm{A\,m^{-2}}$$

From Eq. (1), $\quad \dfrac{J_C}{J_D} = \dfrac{\sigma}{\omega\varepsilon}$ \hfill (2)

For $J_C = J_D$, the condition required is

$$\sigma = \omega\varepsilon$$

or $\qquad \omega = \dfrac{\sigma}{\varepsilon} = \dfrac{5.0}{8.85 \times 10^{-12}} = 5.65 \times 10^{11}\,\mathrm{rad/s}$

Thus $\qquad \nu = \dfrac{\omega}{2\pi} = \dfrac{5.65 \times 10^{11}}{2 \times 3.14} = 89.9\,\mathrm{GHz}.$

Example 8.3: A plane electromagnetic wave travelling in positive z-direction in an unbounded lossless dielectric medium with relative permeability $\mu_r = 1$ and relative permittivity $\varepsilon_r = 3$ has a peak electric field intensity $E_0 = 6$ V/m. Determine: (i) the speed of the wave, (ii) the intrinsic impedance of the medium, (iii) the peak magnetic intensity (H_0) and (iv) the peak Poynting vector $N(z, t)$.

Solution: We have $E_0 = \sqrt{(E_{ox}^2 + E_{oy}^2)} = 6$ V/m, $\varepsilon_r = 3$ and $\mu_r = 1$.

i. The speed of electromagnetic wave

$$v = \frac{1}{\sqrt{\mu\varepsilon}} = \frac{1}{\sqrt{\mu_r\mu_0\varepsilon_r\varepsilon_0}} = \frac{1}{\sqrt{\mu_r\varepsilon_r}} \times \frac{1}{\sqrt{\mu_0\varepsilon_0}} = \frac{c}{\sqrt{\mu_r\varepsilon_r}}$$

$$= \frac{3 \times 10^8}{\sqrt{1 \times 3}} = 1.73 \times 10^8\,\mathrm{m\,s^{-1}}$$

ii. Impedance of the medium

$$Z = \sqrt{\frac{\mu}{\varepsilon}} = \sqrt{\frac{\mu_r \mu_0}{\varepsilon_r \varepsilon_0}} = \sqrt{\left(\frac{\mu_0}{\varepsilon_0}\right)}\sqrt{\frac{\mu_r}{\varepsilon_r}}$$

$$= \sqrt{\left(\frac{4\pi \times 10^{-7}}{8.86 \times 10^{-12}}\right)\sqrt{\frac{1}{3}}} = \frac{376.6}{\sqrt{3}} = 217.6\,\Omega$$

iii. Peak value of the magnetic field

$$H_0 = \frac{E_0}{Z} = \frac{6}{217.6} = 2.76 \times 10^{-2}\,\text{A/m}$$

iv. Poynting vector $N = [E \times H]$

$$\therefore \quad \text{Peak Poynting vector} = E_0 H_0 = \frac{E_0^2}{Z} = \frac{(6)^2}{217.6} = 0.165\,\text{W/m}^2.$$

Example 8.4: A surface wave whose magnetic field strength is perpendicular to its line of propagation, travels along the interface of two dielectrics with permittivities ε_1 and $-|\varepsilon_2|$. Show that the dispersion relation is

$$k = \left[\frac{\omega^2 \varepsilon_1 |\varepsilon_2|}{c^2(|\varepsilon_2| - \varepsilon_1)}\right]^{1/2}$$

Solution: Let us consider that the interface of two dielectrics lies in the xy plane at $z = 0$. Let the wave be propagating along the x-axis and the magnetic field acts along the y-axis. Let the medium with permittivity ε_1, lies in the region $z > 0$ and that with permittivity $-|\varepsilon_2|$ in the region $z < 0$. One can describe the wave by the solution of the equation

$$\nabla^2 H_y + \frac{\omega^2}{c^2} H_y = 0$$

This has the form

$$H_{1y} = H_{01}e^{ik_1 x - f_1 z}, \quad \text{where } f_1 = \sqrt{k^2 - \frac{\omega^2 \varepsilon_1}{c^2}} \quad \text{for } z > 0 \tag{1}$$

and

$$H_{2y} = H_{02}e^{ik_1 x + f_2 z}$$

where,

$$f_2 = \sqrt{k^2 + \frac{\omega^2 |\varepsilon_2|}{c^2}} \quad \text{for } z < 0 \tag{2}$$

From boundary conditions at $z = 0$, one obtains $H_{1y} = H_{2y}$ and $E_{1x} = E_{2x}$. Applying these boundary conditions, we obtain

$$H_{01} = H_{02} \quad \text{and} \quad \frac{f_1^2}{\varepsilon_1} = \frac{f_2^2}{|\varepsilon_2|} \tag{3}$$

Using Eqs. (1) and (2), Eq. (3) yields

$$\frac{k^2}{\varepsilon_1^2} - \frac{\omega^2}{c^2 \varepsilon_1} = \frac{k^2}{|\varepsilon_2|^2} + \frac{\omega^2}{c^2 |\varepsilon_2|}$$

$$\therefore \qquad k = \sqrt{\frac{\omega^2 \varepsilon_1 |\varepsilon_2|}{c^2 (|\varepsilon_2| - \varepsilon_1)}}$$

Example 8.5: The parameters for aluminium are $\mu_r = 1$, $\varepsilon_r = 1$ and $\sigma = 3.54 \times 10^7$ mho/m. Show that the skin depth in aluminium is 0.01 mm for the frequency 71.5 MHz.

Solution: $\delta = \sqrt{\dfrac{2}{\mu\sigma\omega}} = \sqrt{\dfrac{2}{\mu\sigma 2\pi v}} = \sqrt{\dfrac{1}{\mu\sigma\pi v}} = \sqrt{\dfrac{1}{\mu\sigma\omega}}$

or $\qquad \delta^2 = \dfrac{1}{\mu\sigma\omega} = \dfrac{1}{4\pi \times 10^{-7} \times 3.54 \times 10^7 \times 2\pi \times 71.5 \times 10^6}$

or $\qquad \delta^2 = \dfrac{1}{8\pi^2 \times 3.54 \times 71.5 \times 10^6}$

$\therefore \qquad \delta = 0.01$ mm

Example 8.6: Calculate the energy flux dW passing during time dt through a unit area perpendicular to the direction of propagation of the electromagnetic wave.

Solution: If we know the values of E and B in the region of location of the area, we have

$$dW = wcdt$$

where, w is the energy density $= \varepsilon_0 \dfrac{E^2}{2} + \mu_0 \dfrac{H^2}{2}$. For an electromagnetic wave, we have

$$\varepsilon_0 E^2 = \mu_0 H^2$$

This means that the electric energy density in the electromagnetic wave at any instant is equal to the magnetic energy density at the same point, so one can write for the energy density

$$\omega = \varepsilon_0 E^2$$

Then $\qquad dW = \varepsilon_0 E^2 cdt = \sqrt{\dfrac{\varepsilon_0}{\mu_0}} E^2 dt$

One can also obtain dW with the help of Poynting vector. We have

$$dW = Ndt = EH\,dt = \sqrt{\dfrac{\varepsilon_0}{\mu_0}} E^2 dt$$

Thus, both the expressions lead to the same result.

Example 8.7: Given $\mu = \mu_0$, $\varepsilon_r = 2.3$ and $\sigma = 2.56 \times 10^{-4}$ mho/m, find the phase velocity and magnitude of the attenuation constant of plane waves in polyethylene at 10 GHz frequency.

Solution: Phase velocity,

$$v = \frac{1}{\sqrt{\mu\varepsilon}} = \frac{\sqrt{\varepsilon_0/\varepsilon}}{\sqrt{\mu_0\varepsilon_0}} = c\sqrt{\frac{\varepsilon_0}{\varepsilon}}$$

$$= 3\times10^8 \times \sqrt{\frac{1}{2.3}} = 1.97\times10^8 \text{ m/s}$$

We have $\dfrac{\sigma}{\omega\varepsilon} = \dfrac{2.56\times10^{-4}}{2\pi\times10^{10}\times2.3\times8.85\times10^{-12}} = 2\times10^{-4}$

Obviously, this is much smaller than 1. We have

$$\beta = \omega\sqrt{\frac{\varepsilon\mu}{2}}\left[-1+\left\{1+\left(\frac{\sigma}{\varepsilon\omega}\right)^2\right\}^{1/2}\right]^{1/2}$$

$$\approx \omega\sqrt{\frac{\varepsilon\mu}{2}\left(\frac{\sigma^2}{2\varepsilon^2\omega^2}\right)^{1/2}} = \frac{1}{2}\sigma\sqrt{\frac{\mu}{\varepsilon}} = \frac{1}{2}\frac{\sigma}{\omega\varepsilon}\omega\sqrt{\mu\varepsilon} = \frac{1}{2}\frac{\sigma}{\omega\varepsilon}\frac{\omega}{v}$$

$$= \frac{1}{2}\times2\times10^{-4}\times2\pi\times10^{10}\times\frac{1}{1.97\times10^8} = 3.1\times10^{-2}.$$

Example 8.8: Calculate the degree of polarization for ordinary light (index 1.5) at an angle of 45°.

Solution: We have from Snell's law,

$$\frac{\sin\theta_I}{\sin\theta_R} = \frac{n_2}{n_1} = n \text{ (say)}$$

where, θ_I and θ_R are the angles of incidence and refraction respectively.

$$\therefore \quad \sin\theta_R = \frac{\sin\theta_I}{n} = \frac{\sin 45°}{1.5} = \frac{1}{\sqrt{2}\times1.5} \quad \begin{vmatrix} \theta_I = 45° \\ n = 1.5 \end{vmatrix}$$

$$\therefore \quad \theta_R = \sin^{-1}\left(\frac{2}{3\sqrt{2}}\right) = \sin^{-1}\left(\frac{2\sqrt{2}}{6}\right) = \sin^{-1}(0.4713) = 28.1°$$

$$\therefore \quad R_\perp = \frac{\sin^2(\theta_I - \theta_R)}{\sin^2(\theta_I + \theta_R)} = \frac{\sin^2(45° - 28.1°)}{\sin^2(45° + 28.1°)} = \frac{\sin^2 16.9°}{\sin^2 73.1°}$$

$$= \frac{(0.2294)^2}{(0.9568)^2} = 0.09337$$

$$R_\parallel = \frac{\tan^2(\theta_I - \theta_R)}{\tan^2(\theta_I + \theta_R)} = \frac{\tan^2 16.9°}{\tan^2 73.1°} = \frac{(0.3038)^2}{(3.2914)^2} = 0.00852$$

$$\therefore \quad \text{Degree of polarization} = \frac{R_\perp - R_\parallel}{R_\perp + R_\parallel} = \frac{0.09337 - 0.00852}{0.09337 + 0.00852}$$

$$= \frac{0.08485}{0.10189} = 0.833 = 83.3\%.$$

Example 8.9: A lossless dielectric medium has $\sigma = 0$, $\mu_r = 1$ and $\varepsilon_r = 4$. An electromagnetic wave has magnetic field component expressed as

$$H = -0.1\cos(\omega t - z)\hat{i} + 0.5\sin(\omega t - z)\hat{j} \ \text{A m}^{-1}$$

Find: (a) the phase constant β (b) the angular velocity (c) the wave impedance, and (d) the component of the electric field intensity of the wave.

Solution: We have $\sigma = 0$, $\mu_r = 1$, $\varepsilon_r = 4$, $\alpha = 0$ for lossless dielectric.

(a) $H = H_m \cos(\omega t - \beta z)$

Comparing with the given equation, we obtain $\beta = 1 \ \text{rad/m}$

$$\beta = \omega\sqrt{\mu\varepsilon} = \omega\sqrt{\mu_0\varepsilon_0}\ \sqrt{\mu_r\varepsilon_r} = \frac{\omega}{c}\sqrt{4} = \frac{2\omega}{c}$$

(b) $\omega = \dfrac{\beta c}{2} = \dfrac{1 \times 3 \times 10^8}{2} = 1.5 \times 10^8 \ \text{rad/s}$

(c) Wave impedance $\eta = \sqrt{\dfrac{\mu}{\varepsilon}} = \sqrt{\dfrac{\mu_0}{\varepsilon_0}}\sqrt{\dfrac{\mu_r}{\varepsilon_r}}$

$$= \frac{120\pi}{2} = 60\pi \ \Omega$$

(d) $H = -0.1\cos(\omega t - z)\hat{i} + 0.5\sin(\omega t - z)\hat{j} \ \text{A m}^{-1}$ (1)

Obviously, the wave is travelling in z-direction and has components of H in $\hat{i}$ and $\hat{j}$ directions, H_x and H_y respectively varying w.r.t. z. Now, we want to find E. We use Maxwell's equation for lossless medium

$$\nabla \times H = \varepsilon\frac{\partial E}{\partial t} \qquad (\because \ \sigma = 0)$$

We have $\nabla \times H = \begin{vmatrix} \hat{i} & \hat{j} & \hat{k} \\ \dfrac{\partial}{\partial x} & \dfrac{\partial}{\partial y} & \dfrac{\partial}{\partial z} \\ H_x & H_y & H_z \end{vmatrix} = -\dfrac{\partial H_y}{\partial z}\hat{i} + \dfrac{\partial H_x}{\partial z}\hat{j}$

$$\nabla \times H = 0.5\cos(\omega t - z)\hat{i} - 0.1\sin(\omega t - z)\hat{j} \quad [\text{using Eq. (1)}]$$

Now $\qquad E = \dfrac{1}{\varepsilon}\int(\nabla \times H)dt = \dfrac{0.5}{\varepsilon\omega}\sin(\omega t - z)\hat{i} + \dfrac{0.1}{\varepsilon\omega}\cos(\omega t - z)\hat{j}$

or $\qquad E = 94.12\sin(\omega t - z)\hat{i} + 18.83\cos(\omega t - z)\hat{j} \ \text{V/m}.$

Example 8.10: A 1 MHz (300 m wavelength) plane electromagnetic wave travelling normally in a dispersive, lossless medium has a phase velocity, at this frequency, of $3 \times 10^8 \ \text{m/s}$. The phase velocity as a function of wavelength is given by $v = k\sqrt{\lambda}$, where k is a constant. Show that the group velocity is $1.5 \times 10^8 \ \text{m/s}$.

Solution: The group velocity v_g is given by

$$v_g = v - \lambda \frac{dv}{d\lambda} = v - \frac{k}{2}\sqrt{\lambda} = k\sqrt{\lambda} - \frac{k}{2}\sqrt{\lambda}$$

$$= v\left(1 - \frac{1}{2}\right) = \frac{v}{2} = \frac{3 \times 10^8}{2} = 1.5 \times 10^8 \text{ m/s.}$$

Example 8.11: For sea water with $\sigma = 5$ mho s^{-1} and $\varepsilon_r = 80$, find the distance up to which a radio signal can be transmitted at 25 kHz and 25 MHz if the range is taken to be the distance at which 90% of the wave amplitude is attenuated.

Solution: Let us consider that a electromagnetic wave is propagating in the x-direction with a propagation constant λ. Then for a plane wave linearly polarized in y-direction, we have

$$E_y = \varepsilon_0 \exp(-\gamma x), \text{ where } \gamma = \alpha + j\beta$$
$$E_y = E_0 e^{-\alpha x} e^{-j\beta x}$$

∴

Attenuation, therefore, is

$$e^{-\alpha x} = 0.1$$

This yields $x = \dfrac{2.30}{\alpha}$.

At $f = 25$ kHz

$$\alpha = \omega\sqrt{\frac{\varepsilon\mu}{2}}\left[1 + \left\{1 + \left(\frac{\sigma}{\varepsilon\omega}\right)^2\right\}^{1/2}\right]^{1/2} = 0.715$$

∴

$$x = \frac{2.33}{0.715} = 3.22 \text{ m}$$

At $f = 25$ MHz,
one obtains $\alpha = 22.6$

∴

$$x = \frac{2.33}{22.6} = 10.4 \text{ cm.}$$

Example 8.12: For a lossy dielectric material having $\mu_r = 1$, $\varepsilon_r = 48$, and $\sigma = 20$ Sm^{-1}, calculate the attenuation constant, phase constant and intrinsic impedance at a frequency of 16 GHz.

Solution: Attenuation constant:

$$\gamma = i\omega\sqrt{\mu\varepsilon}\sqrt{1 - i\frac{\sigma}{\omega\varepsilon}} \qquad (1)$$

$$\begin{cases} \mu_r = 1 \\ \varepsilon_r = 48 \\ \sigma = 20 \text{ Sm}^{-1} \\ f = 16 \times 10^9 \text{ Hz} \end{cases}$$

Let us first calculate

$$\frac{\sigma}{\omega\varepsilon} = \frac{20 \times 10^{12}}{2\pi \times 16 \times 10^9 \times 48 \times 8.856} = 0.47$$

Now, substituting the values in Eq. (1), we obtain

$$\gamma = i(2\pi) \times 16 \times 10^9 \sqrt{4\pi \times 10^{-7} \times 48 \times 8.856 \times 10^{-12} \times [1 - i(0.47)]}$$

$$= i\, 2323.25 \sqrt{1.0966} \angle -24.23°$$

$$= 2432.88 \angle 77.89° = 510.4 + i\, 2378.7 \text{ m}^{-1}$$

$$\therefore \quad \alpha = 510.4 \text{ Np/m, and } \beta = 2378.7 \text{ rad/m}$$

Intrinsic impedance:

$$\eta = \sqrt{\frac{i\omega\mu}{\sigma + i\omega\varepsilon}} = \sqrt{\frac{\mu}{\varepsilon\left(1 + \dfrac{\sigma}{i\omega\varepsilon}\right)}}$$

$$= \sqrt{\frac{4\pi \times 10^{-7}}{48 \times 8.854 \times 10^{-12}}} \times \sqrt{\frac{1}{1 - j(0.47)}}$$

$$= \frac{54.377}{\sqrt{1.0966} \angle 24.23°}$$

$$= 51.93 - 12.12° \, \Omega$$

This means that the electric field (E_y) leads the magnetic field (H_z) by 12.12° at every point.

Example 8.13: Prove that average Poynting vector, P_{av}, is given as $P_{av} = E_{rms} \times H_{rms}$. Further, prove that $I = P_{av} = \dfrac{1}{2} c\varepsilon_0 E_0^2 = \dfrac{1}{2} c\mu_0 H_0^2$, where I is the intensity of an electromagnetic wave.

Solution: Time average of Poynting vector,

$$P_{av} = \frac{1}{2}\sqrt{\frac{\mu_0}{\varepsilon_0}} H_0^2 = \frac{1}{2}\frac{\mu_0}{\sqrt{\mu_0 \varepsilon_0}} H_0^2 = \frac{1}{2} c\mu_0 H_0^2$$

Also,

$$H_0 = \sqrt{\frac{\varepsilon_0}{\mu_0}} E_0$$

$$\therefore \quad P_{av} = \frac{1}{2}\sqrt{\frac{\varepsilon_0}{\mu_0}} E_0^2 = \frac{1}{2}\frac{\varepsilon_0}{\sqrt{\mu_0 \varepsilon_0}} E_0^2 = \frac{1}{2} c\varepsilon_0 E_0^2 = \frac{1}{2} c\mu H_0^2$$

Example 8.14: Determine the propagation constant γ for a material having $\mu_r = 1$, $\varepsilon_r = 8$, and $\sigma = 0.25$ pS/m, if the wave frequency is 1.6 MHz.

Solution: In this case, [MHTU]

$$\frac{\sigma}{\omega\varepsilon} = \frac{0.25 \times 10^{-12}}{2\pi(1.6 \times 10^6)(8)(10^{-9}/36\pi)} = 10^{-9} \approx 0$$

so that

$$\alpha = 0 \text{ and } \beta = \omega\sqrt{\mu\varepsilon} = 2\pi f \frac{\sqrt{\mu_r \varepsilon_r}}{c} = 9.48 \times 10^{-2} \text{ rad/m}$$

and $\gamma = \alpha + j\beta = j9.48 \times 10^{-2}$ m^{-1}. The material behaves like a perfect dielectric at the given frequency. Conductivity of the order of 1 pS/m indicates that the material is more like an insulator than a conductor.

Example 8.15: A plane wave travelling in the +z direction in free space ($z < 0$) is normally incident at $z = 0$ on a conductor ($z > 0$) for which $\sigma = 61.7$ MS/m, $\mu_r = 1$. The free space E wave has a frequency $f = 1.5$ MHz and an amplitude of 1.0 V/m. Prove that at the interface it is given by

$$E(0, t) = 1.0 \sin 2\pi f t a_y \, (\text{V/m}) \qquad\qquad \text{[RTU]}$$

Find $H(z, t)$ for $z > 0$.

Solution: For $z > 0$, and in complex form,

$$E(z, t) = 1.0 e^{-\alpha z} e^{j(2\pi f t - \beta z)} a_y \, (\text{V/m})$$

where, the imaginary part will ultimately be taken. In the conductor,

$$\alpha = \beta = \sqrt{\pi f \mu \sigma} = \sqrt{\pi(1.5 \times 10^6)(4\pi \times 10^{-7})(61.7 \times 10^6)} = 1.91 \times 10^4$$

$$\eta = \sqrt{\frac{\omega \mu}{\sigma}} \angle 45° = 4.38 \times 10^{-4} e^{j\pi/4}$$

Then, since $E_y/(-H_x) = \eta$,

$$H(z, t) = -2.28 \times 10^3 e^{-\alpha z} e^{j}(2\pi f t - \beta z - \pi/4) a_x \, (\text{A/m})$$

or, taking the imaginary part,

$$H(z, t) = -2.28 \times 10^3 e^{-\alpha z} \sin(2\pi f t - \beta z - \pi/4) a_x \, (\text{A/m})$$

where f, α and β are as given above.

Example 8.16: Calculate the intrinsic impedance η, the propagation constant γ, and the wave velocity u for a conducting medium in which $\sigma = 58$ MS/m, $\mu_r = 1$ at a frequency $f = 100$ MHz. [UPTU]

Solution:

$$\gamma = \sqrt{\omega \mu \sigma} \angle 45° = 2.14 \times 10^5 \angle 45° \, \text{m}^{-1}$$

$$\eta = \sqrt{\frac{\omega \mu}{\sigma}} \angle 45° = 3.69 \times 10^{-3} \angle 45° \, \Omega$$

$$\alpha = \beta = 1.51 \times 10^5, \quad \delta = \frac{1}{\alpha} 6.61 \text{mm}$$

$$u = \omega \delta = 4.15 \times 10^3 \, \text{m/s}.$$

Example 8.17: Determine the amplitudes of the reflected and transmitted E and H at the interface shown in Fig. 8.12 if $E_0^i = 1.5 \times 10^{-3}$ V/m in region 1, in which $\varepsilon_{r_1} = 8.5$, $\mu_{r_1} = 1$ and $\sigma_1 = 0$. Region 2 is free space. Assume normal incidence.

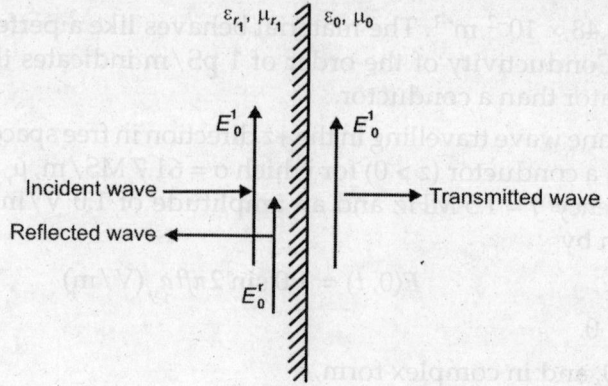

Fig. 8.12

Solution:

$$\eta_1 = \sqrt{\frac{\mu_0 \mu_{r_1}}{\varepsilon_0 \varepsilon_{r_1}}} = 129\,\Omega, \quad \eta_2 = 120\pi\,\Omega = 377\,\Omega$$

$$E_0^r = \frac{\eta_2 - \eta_1}{\eta_2 + \eta_1} E_0^i = 7.35 \times 10^{-4}\,\text{V/m}$$

$$E_0^t = \frac{2\eta_2}{\eta_2 + \eta_1} E_0^i = 2.24 \times 10^{-3}\,\text{V/m}$$

$$H_0^i = \frac{E_0^i}{\eta_1} = 1.16 \times 10^{-5}\,\text{A/m}$$

$$H_0^r = \frac{\eta_1 - \eta_2}{\eta_1 + \eta_2} H_0^i = -5.69 \times 10^{-6}\,\text{A/m}$$

$$H_0^t = \frac{2\eta_1}{\eta_1 + \eta_2} H_0^i = 5.91 \times 10^{-6}\,\text{A/m}$$

Example 8.18: A normally incident E field has amplitude $E_0^i = 1.0\,\text{V/m}$ in free space just outside of seawater in which $\varepsilon_r = 80$, $\mu_r = 1$ and $\sigma = 2.5\,\text{S/m}$. For a frequency of 30 MHz, at what depth will the amplitude of E be $1.0\,\text{mV/m}$?

Solution: Let free space be region 1 and the seawater be region 2.

$$\eta_1 = 377\,\Omega \qquad \eta_2 = 9.73\,\angle 43.5^\circ\,\Omega$$

Then the amplitude of E just inside the seawater is E_0^t.

$$\frac{E_0^t}{E_0^i} = \frac{2\eta_2}{\eta_1 + \eta_2}$$

or

$$E_0^t = 5.07 \times 10^{-2}\,\text{V/m}$$

From

$$\gamma = \sqrt{j\omega\mu(\sigma + j\omega\varepsilon)} = 24.36\,\angle 46.53^\circ\,\text{m}^{-1},$$

$$\alpha = 24.36 \cos 46.53^\circ = 16.76\,\text{Np/m}$$

we have
$$1.0 \times 10^{-3} = (5.07 \times 10^{-2})e^{-16.6z}$$
$$\Rightarrow z = 0.234 \, \text{m}.$$

Example 8.19: A monochromatic plane wave in free space with electric field amplitude of 1 V/m is normally incident on a fully reflecting mirror. The pressure exerted on the mirror is _____ $\times 10^{-12}$ Pa (up to two decimal places) ($\varepsilon_0 = 8.854 \times 10^{-12}$ F/m).

[GATE 17]

Solution: Pressure exerted on perfect reflector

$$p = \frac{1}{2} \varepsilon_0 E^2 \times 2 = \frac{1}{2} \times 8.854 \times 10^{-12} \times 1 \times 2$$
$$= 8.85 \times 10^{-12} \, \text{Pa}$$

Example 8.20: Consider a metal with free electron density of 6×10^{22} cm^{-3}. The lowest frequency electromagnetic radiation to which this metal is transparent is 1.38×10^{16} Hz. If this metal had a free electron density of 1.8×10^{23} cm^{-3} instead, the lowest frequency electromagnetic radiation to which it would be transparent is _____ $\times 10^{16}$ Hz (up to two decimal places).

[GATE 17]

Solution: Cut-off frequency

$$\omega_\rho \propto n^{1/2}$$

$$\omega_\rho = 1.38 \times 10^{16} \left[\frac{1.8 \times 10^{23}}{6 \times 10^{22}} \right]^{1/2}$$

$$= 2.39 \times 10^{16} \, \text{Hz}$$

Example 8.21: Find the momentum p of the particle and verify that the direction of $p_\perp = (p_x, p_y, 0)$ coincides at each point of time with the direction of B.

Solution:

$$p = \left(-\frac{qE}{\omega} \sin(kz - \omega t), \frac{qE}{\omega} \cos(kz - \omega t), mv_0 \right)$$

On comparison, we obtain

$$p_\perp = (p_x, p_y, 0) = \frac{q}{k} B.$$

Example 8.22: Let the particle be at $t = 0$ at the origin of coordinates. Solve the equation of motion.

Solution: Equation of motion of the particle

$$x(t) = -\frac{qE}{m\omega} \left(-\frac{1}{\omega \left(\frac{v_0}{u} - 1 \right)} \cos \left(\omega \left(\frac{v_0}{u} - 1 \right) t \right) \right) + x_0$$

$$x(t = 0) = 0 = \frac{qE}{m\omega^2} \frac{1}{\frac{v_0}{u} - 1} + x_0$$

x-component:

$$x(t) = \frac{qE}{m\omega^2} \frac{u}{v_0 - u} \left(\cos\left(\frac{v_0 - u}{u} \omega t \right) - 1 \right)$$

y-component:

$$y(t) = \frac{qE}{m\omega} \frac{1}{w\left(\frac{v_0}{u} - 1 \right)} \sin\left(\omega\left(\frac{v_0}{u} - 1 \right) t \right) + y_0$$

$$y(t = 0) = 0 = y_0$$

It remains:

$$y(t) = \frac{qE}{m\omega^2} \frac{u}{v_0 - u} \sin\left(\frac{v_0 - u}{u} \omega t \right)$$

z-component

$$z(t) = v_0 t$$

Example 8.23: A transverse electromagnetic wave in an insulating uncharged medium ($\rho_f = 0$), $i_f = 0$, $\sigma = 0$) is:

(a) Linearly polarized

$$E = E_0 \sin(kz - \omega t)$$

(b) Circularly polarized

$$E = E_0 \left[\cos\left(kz - \omega t \right)\hat{e}_x + \sin(kz - \omega t)\hat{e}_y \right]$$

and propagates in the z-direction. Calculate
(i) The magnetic induction $B(r, t)$
(ii) The Poynting vector $S(r, t)$
(iii) The radiation pressure on a plane which is inclined by the angle θ relative to the propagation ($k = k\hat{e}_z$).

Solution:
(i) **Magnetic induction:**

$$\text{curl} E = -\dot{B}$$

(a)

$$\text{curl} E = \left(\frac{\partial E_z}{\partial y} - \frac{\partial E_y}{\partial z}, \frac{\partial E_x}{\partial z} - \frac{\partial E_z}{\partial x}, \frac{\partial E_y}{\partial x} - \frac{\partial E_x}{\partial y} \right)$$

$$= \left(-\frac{\partial}{\partial z} E_y, \frac{\partial}{\partial z} E_x, 0 \right) = k(-E_{0y}, E_{0x}, 0) \cos(kz - \omega t)$$

$$\Rightarrow \dot{B} = (E_{0y}, -E_{0x})k \cos(kz - \omega t)$$

$$\Rightarrow B = \frac{k}{\omega}(-E_{0y}, E_{0x}, 0) \sin(kz - \omega t) = \frac{1}{\omega}(k \times E)$$

(b)

$$\text{curl}E = \left(-\frac{\partial}{\partial z}E_y, \frac{\partial}{\partial z}E_x, 0\right) = -E_0 k\left[\cos(kz - \omega t)\hat{e}_x + \sin(kz - \omega t)\hat{e}_y\right]$$

$$\Rightarrow \quad \dot{B} = E_0 k\left[\cos(kz - \omega t)\hat{e}_x + \sin(kz - \omega)\hat{e}_y\right]$$

$$\Rightarrow \quad B = E_0\frac{k}{\omega}\left[-\sin(kz - \omega t)\hat{e}_x + \cos(kz - \omega t)\hat{e}_y\right]$$

$$= \frac{k}{\omega}(-E_y, E_x, 0) = \frac{1}{\omega}(k \times E)$$

(ii) Poynting vector: ($S(r, t)$)

$$S(r, t) = E \times H = \frac{1}{\mu_r \mu_0}E \times B \quad \textit{(energy-current density)}$$

It holds for both the cases (a) and (b):

$$S(r, t) = \frac{1}{\mu_r \mu_0}\frac{1}{\omega}E \times (k \times E)$$

$$= \frac{1}{\omega \mu_r \mu_0}(kE^2 - \underbrace{E(E \cdot k)}_{=0}) = \frac{1}{u \mu_r \mu_0}E^2 e_z$$

Thus, we have

(a)

$$S(r, t) = \sqrt{\frac{\varepsilon_r \varepsilon_0}{\mu_r \mu_0}}E_0^2 \sin^2(kz - \omega t)\hat{e}_z$$

(b)

$$S(r, t) = \sqrt{\frac{\varepsilon_r \varepsilon_0}{\mu_r \mu_0}}E_0^2\,\hat{e}_z$$

(iii) Radiation pressure:

Radiation pressure ~ momentum transfer to the area ~ normal component ($n \cdot F$) of the force F exerted on the unit area.

Density of the field momentum:

$$\hat{p}_{field} = D \times B = \varepsilon_r \mu_r \varepsilon_0 \mu_0 S = \frac{1}{u^2}S$$

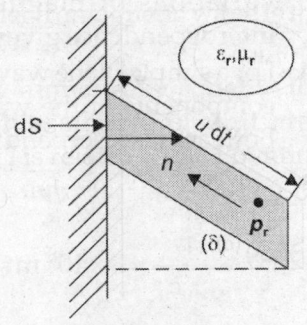

u is the phase velocity of the electromagnetic wave. All wave fronts in the inclined cylinder in Fig. 8.13, whose volume is

$$\Delta V = u\,dt\,\cos\theta\,dS$$

Fig. 8.13

reach within the time dt the area-element dS. The plane shall be totally absorbing.

The waves create a field momentum $\hat{p}_{field} \Delta V$ on dS in time dt.

Force = momentum per time:

$$F = \hat{p}_{field} \ u \cos \theta \ dS$$

Radiation pressure:

$$p_s = \frac{n \cdot F}{dS} = u \cos \theta \ n \cdot \hat{p}_{Field} = \frac{\cos \theta}{u} n \cdot S$$

Solution

(a)

$$p_s = \frac{1}{u} |S| \cos^2 \theta = \varepsilon_r \varepsilon_0 E_0^2 \ \sin^2 \omega t \cos^2 \theta \qquad \text{(wall at } z = 0)$$

(b)

$$p_s = \frac{1}{u} |S| \cos^2 \theta = \varepsilon_r \varepsilon_0 E_0^2 \ \cos^2 \theta$$

GLIMPSES

1. *Electromagnetic radiation* is the energy resulting from the acceleration of electric charge and associated electric fields and magnetic fields. The energy can be regarded as waves propagated through space (requiring no supporting medium) involving oscillating electric and magnetic fields at right angles to each other and to the direction of propagation. In vacuum, the waves travel with a constant speed (the speed of light) of 2.9979×10^8 m/s; if material is present, they are slower. These waves are electromagnetic waves.

2. Maxwell developed a mathematical theory that showed a close relationship between all electric and magnetic phenomena. Additionally, *Maxwell's theory* predicted that electric and magnetic fields can move through space as waves. According to Maxwell's theory, the changing electric and magnetic fields produced by oscillating charges result in electromagnetic disturbance that travel through space as waves. The waves sent out by oscillating charges are viewed as fluctuating electric and magnetic fields, hence, they are called electromagnetic waves.

3. The space variations of electric and magnetic field components are related to time variations of magnetic and electric field components respectively. This interdependence gives rise to the phenomenon of *electromagnetic wave propagation*.

4. For a simple plane wave travelling in a dielectric medium, the electric and magnetic components of the wave are identical in form and perpendicular to each other. Both are also perpendicular to the direction of travel of the wave, and travel with velocity $v = 1/\sqrt{\mu\varepsilon}$. When the wave is in free space, the velocity becomes

$$c = \frac{1}{\sqrt{\mu_0 \varepsilon_0}} = 3 \times 10^8 \text{ m s}^{-1}, \text{ with } B = \mu H, \ E_y = v\mu H_z = \frac{\mu}{\sqrt{\mu\varepsilon}} H_z = \sqrt{\frac{\mu}{\varepsilon}} H_z$$

or $H_z = \sqrt{\dfrac{\varepsilon}{\mu}}$.

5. The ratio of E and H in a wave is denoted by η, and is called intrinsic impedance: $\eta = \dfrac{E}{H} = \sqrt{\dfrac{\mu}{\varepsilon}}$. For free space, $\eta = \sqrt{\dfrac{\mu_0}{\varepsilon_0}} \cong 377$ or $120\,\pi$. The amplitude of E vector is dominating as compared to that of the magnetic field vector.

6. A uniform plane wave is a *transverse electromagnetic wave*, or TEM *wave*. Uniform plane waves do not exist in practice because they cannot be produced by finite sized antennas. At large distances from physical antennnas and ground, however the waves can be approximated as uniform plane waves. Furthermore, the principle of guiding of EM waves along transmission lines and wave guides, and the principles of many other wave phenomena can be studied basically in terms of uniform plane waves.

7. If the material is lossy dielectric medium, it will have finite conductivity (σ). Considering the electric and magnetic fields very sinusoidally, we obtain from Maxwell's equations

$$\nabla^2 E = \gamma^2 E \qquad \text{where} \qquad \gamma^2 = i\omega\mu\,(\sigma + i\omega\varepsilon)$$

γ is called *propagation constant.*

$$\gamma = \alpha + i\beta$$

α is called *attenuation constant*, and has units of Neper per meter (Np/m) and β is phase constant in rad/m

$$\alpha = \omega \sqrt{\dfrac{\mu\varepsilon}{2}\left[\sqrt{1 + \left(\dfrac{\sigma}{\omega\varepsilon}\right)^2} - 1\right]}$$

$$\beta = \omega \sqrt{\dfrac{\mu\varepsilon}{2}\left[\sqrt{1 + \left(\dfrac{\sigma}{\omega\varepsilon}\right)^2} + 1\right]}$$

8. Wave equation in a perfect dielectric ($\sigma = 0$) medium

$$\nabla^2 E - \mu\varepsilon \dfrac{\partial^2 E}{\partial t^2} = 0 \ \text{ and } \ \nabla^2 H - \mu\varepsilon \dfrac{\partial^2 H}{\partial t^2} = 0$$

The solutions of these plane polarized equations are given as

$$E_y = E_0\, e^{ik(vt-x)} = E_0\, e^{i(\omega t - kx)}$$

$$H_z = H_0\, e^{ik(vt-x)} = H_0\, e^{i(\omega t - kx)}$$

Here E_0 and H_0 are maximum values of the electric and magnetic field vectors respectively, $k = 2\pi/\lambda$ and $\omega = kv$, where v is the wave velocity.

9. Electromagnetic waves are *transverse* in nature, i.e. there is no longitudinal component of field vectors in EM waves, we have

$$\dfrac{\partial E_z}{\partial x} = \mu \dfrac{\partial H_y}{\partial t} \quad \text{and} \quad \dfrac{\partial E_y}{\partial x} = -\mu \dfrac{\partial H_z}{\partial t}$$

Similarly, $\dfrac{\partial H_z}{\partial x} = -\varepsilon \dfrac{\partial E_y}{\partial t}$ and $\dfrac{\partial H_y}{\partial x} = \varepsilon \dfrac{\partial E_z}{\partial t}$

From these equations it is clear that E_y, E_z, H_y and H_z are related to each other and their spatial and time variations are non-zero.

10. Wave equation for a medium with finite conductivity, i.e. $\sigma \neq 0$

$$\nabla^2 E - \mu\varepsilon\ddot{E} - \mu\sigma\dot{E} = 0$$

$$\nabla^2 H - \mu\varepsilon\ddot{H} - \mu\sigma\dot{H} = 0$$

11. In electromagnetism, materials are divided roughly into two classes: (i) conductors, and (ii) dielectrics or insulators.

The dividing line between the said two classes is not sharp and some media are considered as conductors in one part of radio frequency range, as dielectrics (with loss) in another part of the range. For good conductors, $\sigma \gg \omega\varepsilon$ in the range of practical frequencies.

12. Velocity of EM waves in a good dielectric medium is given as

$$v \cong v_0 \left(1 + \frac{\sigma^2}{8\omega^2\varepsilon^2}\right)^{-1} \cong v_0 \left(1 - \frac{\sigma^2}{8\omega^2\varepsilon^2}\right)$$

where $v_0 = 1/\sqrt{\mu\varepsilon}$ is the velocity of the wave in a perfect dielectric ($\sigma = 0$, lossless) medium. However, for a good dielectric ($\sigma \neq 0$ but σ has small value) some loss occurs which manifest itself with the reduced velocity of EM wave through such a dielectric medium.

13. In a good conducting medium, the velocity of EM waves is given by

$$v = \frac{\omega}{\beta} = \sqrt{\frac{2\omega}{\mu\sigma}}$$

i.e. v is inversely proportional to the conductivity of the medium and hence the wave gets attenuated strongly while propagating through such a medium.

14. The penetration depth or skin depth (δ) for a conductor is the depth at which the wave gets attenuated to $1/e$ times of its original value. Thus,

$$\delta = \frac{1}{\alpha} \cong \sqrt{\frac{2}{\omega\mu\sigma}}$$

15. The characteristic impedance (η) for a good dielectric medium is given as

$$\eta = \sqrt{\frac{\mu}{\varepsilon}} \left(1 + i\frac{\sigma}{2\omega\varepsilon}\right)$$

16. The characteristic impedance (η) for a good conducting medium is given as

$$\eta = \sqrt{\frac{\omega\mu}{\sigma}}\, e^{i\pi/4}$$

17. Poynting theorem

$$\int E \cdot J \; dV = -\frac{\partial}{\partial t} \int_V \left(\frac{\mu}{2} H^2 + \frac{\varepsilon}{2} E^2 \right) dV - \oint (E \times H) \, da$$

18. $P = E \times H$ is the vector at any point, which is a measure of energy flow per unit area at that point. The direction of flow is perpendicular to both E and H in the direction of vector $P = E \times H$.

REVIEW QUESTIONS

1. Derive the electromagnetic wave equation from Maxwell's field equations. Consider plane wave solutions of this equation and prove that the energy density associated with such a wave in a stationary homogeneous non-conducting medium propagates with the same speed with which the fields do.

2. Explain the propagation of electromagnetic waves in a conducting medium and hence define skin depth and explain why radio communication with submerged submarine becomes increasingly difficult at several skin depths.

3. Discuss propagation of electromagnetic waves in an isotropic dielectric medium. Show that the waves are of transverse nature. Also show that the wave energy is equally shared between electric and magnetic fields.

4. Discuss the propagation of plane electromagnetic waves in an isotropic dielectric medium. Show that the electric and magnetic field vectors (E and H) are mutually perpendicular.

5. Explain the propagation of electromagnetic waves in a conducting medium. Explain why in high frequency circuits, current flows only on the surface of conductors.

6. Explain: (i) why the speed of electromagnetic waves in isotropic dielectric is more than that in free space and (ii) why electromagnetic wave can easily penetrate in a dielectric medium but cannot do so in a conducting medium.

7. Using electromagnetic equations, obtain an expression for field energy and momentum densities.

8. A plane wave travelling in a homogeneous isotropic dielectric is incident obliquely at the plane surface of another dielectric. Determine the reflection and transmission coefficients if the incident wave is polarised with its electric vector (*i*) perpendicular to the plane of incidence and (*ii*) in the plane of incidence.

9. Obtain the boundary conditions satisfied by the electromagnetic field vectors E, D, B and H on the plane interface between two media.

10. Derive Fresnel's equations for reflection and refraction of electromagnetic waves at a plane boundary separating two dielectric media. Discuss important consequences of these relations.

11. Discuss metallic reflection and refraction. Obtain an expression for the reflection power of a metallic surface.

12. Derive Fresnel's equations for reflection and refraction of electromagnetic wave at a plane boundary separating two media when the incident wave is polarised with E vector parallel to the plane of incidence. Hence show that there is an angle of incidence for which there is no reflected wave. What is the name of this angle?

13. Show on the basis of electromagnetic theory that all good conductors are good absorbers and good reflectors.

14. Derive Fresnel's equations for reflection and refraction of electromagnetic waves at a plane boundary separating two dielectric media. Discuss the phenomenon of total internal reflection.

15. What are the pecularities of metallic reflection? Give its theory in brief and describe method for its verification.

PROBLEMS

1. A plane electromagnetic wave travelling in z-direction in an unbounded lossless dielectric medium $k_m = 1$ and $k_e = 1$ has a peak electric field strength $E = 6$ V/m where k_m is relative permeability of the medium and k_e is relative permittivity (or dielectric constant) of the medium. Determine: (i) the velocity of wave, (ii) intrinsic impedance of the medium, (iii) peak value of magnetic field strength and (iv) peak value of Poynting vector.

[*Ans.* (i) $v = 1.7 \times 10^8$ m/s (ii) 217.65 Ω (iii) 2.76×10^2 A/m
(iv) 0.65 W/m^2]

2. Given the equation of a wave is $E = E_0 \cos(500x - 6000t)$. Show that the wavelength is $\dfrac{\pi}{250}$ while the speed of propagation is 12 (with proper units).

3. A laser beam of monochromatic light has a uniform intensity over its cross-sectional area of 0.2 mm^2. If the total power of the beam is 500 mV, calculate the amplitude of the magnetic field vector of the plane wave that can represent the beam.

[*Ans.* $B_0 = 1.447 \times 10^{-4}$ T]

4. Assuming that the radiation due to sunlight is falling normally onto a non-reflecting slab and that it is of the order of 1 kW/m^2, evaluate the mean force exerted on the slab.   [*Ans.* $\dfrac{1}{3} \times 10^{-5}$ Nm^{-2}]

[*Hint.* $\dfrac{1}{2}\left(\dfrac{\varepsilon_0}{\mu_0}\right)^{1/2} E_0 = 10^3$ or $\dfrac{1}{2}\varepsilon_0 E_0^2 = 10^3 \sqrt{\mu_0 \varepsilon_0} = \dfrac{1}{3} \times 10^{-5}$

Hence, the mean force per unit area $= \dfrac{1}{3} \times 10^{-5}$ Nm^{-2}]

5. Show that the average value of the Poynting vector of a plane harmonic wave is $\dfrac{1}{2} c \varepsilon_0 E_0^2$ or $\dfrac{\varepsilon_0 B_0}{2\mu_0}$.

6. Radio waves received by a radio set have an electric field of maximum amplitude equal to 0.1 V/m. Assuming that the wave can be considered as plane, calculate: (i) the amplitude of the magnetic field, (ii) the average intensity of the wave,

(iii) the average energy density, and (iv) assuming that the radio set is 1 km from the broadcasting station and that the station radiates energy isotropically, determine the power of the station.

$$[Ans. \text{ (i) } \frac{1}{3} \times 10^{-9} T \text{ (ii) } 1.33 \times 10^{-6} \text{ W/m}^2 \text{ (iii) } 4.42 \times 10^{-15} \text{ J/m}^3 \text{ (iv) } 167 \text{ W}]$$

7. A wave is moving in a conducting medium. Show that the average energy in the magnetic field is greater than that in the electric field.

8. A monochromatic linearly polarized plane wave is moving in an isotropic non-conducting medium. Show that the time average of its energy density is distributed equally between the magnetic and the electric fields.

9. The wavelength of low frequency radio waves in seawater is 3×10^3 m, the electric conductivity of which is ~ $4/\Omega$ m. Find the skin depth.

$$[Ans. 0.796 \text{ m}]$$

10. A good conductor is roughly defined as one for which $\dfrac{\sigma}{\omega\varepsilon}$ is equal to or greater than 10. If σ and ε for seawater are 4 mho/m and $\left(\dfrac{9 \times 10^{-9}}{4\pi}\right)$ F/m, determine the frequency range for which the sea water is good conductor.

$$[Ans. \nu \leq 88 \text{ mHz}]$$

11. A good dielectric is roughly defined as one for which $\dfrac{\sigma}{\omega\varepsilon}$ is equal to or less than 0.1. If conductivity σ and permeability ε of fresh water are respectively 10^{-3} mho/m and $\dfrac{9 \times 10^{-9}}{4\pi}$ F/m, determine the frequency range for which the fresh water is a good dielectric.

$$[Ans. \nu \geq 2.2 \text{ mHz}]$$

12. A laser beam has a diameter of 10^{-3} m and its power is 1 watt. Show that the amplitude of the magnetic field in the beam in free space is 10^{-4}.

13. Electromagnetic waves are incident normally on a metal surface of refractive index n and extinction coefficient defined by the relation $\dfrac{n_2}{n_1} = {}_1 n_2 = n(1 + i_k)$. Show that the ratio of the reflected intensity to incident intensity is $\dfrac{n^2 + k^2 - 2n + 1}{n^2 + k^2 + 2n + 1}$.

14. Compute the coefficients of reflection and transmission for electromagnetic waves in the visible region for crown glass at an angle of incidence equal to 30°. Given $n_{glass} = 1.52$ and $n_{air} \approx 1$.

15. Linearly polarized light wave is incident on a glass plate ($n = 1.5$) with an angle of incidence equal to 45°. Determine the coefficients of reflection and refraction if the internal field of the incident wave is (a) in the plane of the incidence, and (b) normal to the plane of incidence.

$$[Ans. \text{ (a)} -0.0791 \text{ (b) } 0.7194]$$

16. A plane wave travelling in free space has an average Poynting vector equal to $1 \text{ W}/\text{m}^2$. Show that the average energy density is $3.33 \times 10^{-9} \text{ J}/\text{m}^3$.

17. In a perfect dielectric medium, the electric field progressing in the z-direction is given by $E_x = E_{xo} \cos(\omega t - \beta z)$ and the associated magnetic field by $H_y = \dfrac{E_{xo}}{\eta} \cos(\omega t - \beta z)$, where E_{xo} is the peak value of E_x at $t = 0$ and $z = 0$ and η is the intrinsic impedance of the dielectric. Show that the average power through any area S normal to the z-axis is given by $P_{z,\omega} = \dfrac{1}{2}\dfrac{E_{xo}^2}{\eta} S$.

18. A plane electromagnetic wave falls perpendicular on a plane surface separating a medium of index n_1 from a medium of index n_2. Show that the coefficients of reflection and refraction are $R = \dfrac{(n_1 - n_2)}{(n_1 + n_2)}$ and $T = \dfrac{2n_1}{(n_1 + n_2)}$ respectively.

SHORT ANSWER QUESTIONS

1. What is absolute index of refraction?

   Ans. $n = \dfrac{c}{v} = \dfrac{1}{\sqrt{\varepsilon_0 \mu_0}} \sqrt{\varepsilon \mu} = \sqrt{\dfrac{\varepsilon \mu}{\varepsilon_0 \mu_0}} = \sqrt{\varepsilon_r \mu_r}$.

2. Are gamma rays electromagnetic radiations?

   Ans. Yes, these waves are of nuclear origin.

3. Are there electromagnetic waves in cosmic radiations?

   Ans. Yes, there are electromagnetic waves of shorter wavelengths or longer frequencies.

4. Why does not one see the portion other than visible one of the electromagnetic spectrum?

   Ans. The retina of the eye is sensitive only to colours in the visible region, i.e. wavelength lying between 3900 Å and 7800 Å. This region corresponds to visible part of the spectrum.

5. On what factors the velocity of electromagnetic wave depends?

   Ans. Permittivity and permeability of medium, $c = \dfrac{1}{\sqrt{\varepsilon \mu}}$.

6. Do electromagnetic waves have same velocity in all transparent media?

   Ans. No, as the refractive index is different for different media $n = \dfrac{c}{v}$, v is different for different media.

7. What is a Poynting vector?

   Ans. The cross or vector product of electric field intensity E and magnetic field intensity H is known as Poynting vector and denoted by N. Thus, $N = E \times H$.

   It is interpreted as instantaneous power density that is measured in watt/m^2.

8. What is the average value of the Poynting vector $N = [E \times H]$ for the wave represented by $E = \hat{i}E_o \sin(kz - \omega t)$?

Ans. $<N> = \hat{k}\dfrac{c}{2}\varepsilon_0 E^2$

9. What is skin depth?

Ans. The depth of penetration δ is defined as the depth to which the wave has been attenuated to $1/e$ or approximately 37% of the original value

$$\delta = \frac{1}{\alpha} = \frac{1}{\omega\sqrt{\dfrac{\mu\varepsilon}{2}\sqrt{\left(1 + \dfrac{\sigma}{\omega\varepsilon}\right)^2 - 1}}} \approx \sqrt{\frac{2}{\pi\nu\mu\sigma}}$$

$$\left(\because \text{For a good conductor, } \frac{\sigma}{\omega\varepsilon} \gg 1\right)$$

10. What do you understand by the dispersion of an electromagnetic wave when propagating through matter?

Ans. When an electromagnetic wave propagates through matter, it suffers dispersion, i.e a pulse containing several frequencies will be distorted because each component will travel with a different velocity.

11. Calculate the coefficients of reflection and transmission of energy of the EM waves incident normallly on the water surface having $\varepsilon_r = 81$.

Ans. $\sqrt{\varepsilon_r \mu_r} = \sqrt{\varepsilon_r} = \sqrt{81} = 9$; water being a non-magnetic material.

$$\therefore \qquad R = \left(\frac{1-n}{1+n}\right)^2 = \left(\frac{1-9}{1+9}\right)^2 = 0.64$$

and

$$T = \frac{4n}{(1+n)^2} = \frac{4 \times 9}{(1+9)^2} = \frac{36}{100} = 0.36.$$

12. Calculate the Poynting vector from a 100 W lamp at a distance of 1 m from it.

Ans. $P = \dfrac{\text{Power}}{\text{Area}}$. The energy is radiated as a spherical wavefront. Thus, the area through which it is crossing is equal to the area of sphere of radius 1m. Thus,

$$P = \frac{100}{4\pi \times (1)^2} = 7.958 \text{ Wm}^{-2}.$$

13. A travelling wave is described by $y = 10\sin(\beta z - \omega t)$. Sketch the wave at $t = 0$ and $t = t_1$ when it has advanced $\lambda/8$. The velocity is 3×10^8 m/s and the angular frequency ω is 10^6 rad/s. Repeat for $\omega = 2 \times 10^6$ rad/s and the same t_1. [BTech]

Ans. The wave advances λ in one period, $T = 2\pi/\omega$. Hence,

$$t_1 = \frac{T}{8} = \frac{\pi}{4\omega}$$

Now, $\qquad \dfrac{\lambda}{8} = ct_1 = (3 \times 10^8)\dfrac{\pi}{4(10^6)} = 236\,\text{m}$

The wave is shown at $t = 0$ and $t = t_1$ in Fig. 8.14a. At twice the frequency, the wavelength λ is one-half, and the phase shift constant β is twice the former value (Fig. 8.14b). At t_1, the wave has also advanced 236 m, but this distance is now $\lambda/4$.

Fig. 8.14

14. In free space $E(z, t) = 10^3 \sin(\omega t - \beta z)\,\hat{a}_y$ (V/m), obtain $H(z, t)$. [BTech]

    Ans. Examination of the phase, $\omega t - bz$, shows that the direction of propagation is $+z$. Since $E \times H$ must also be in the $+z$ direction, H must have the direction $-\hat{a}_x$. Consequently,

    $$\frac{E_y}{-H_x} = \eta_0 = 120\pi\,\Omega \quad \text{or} \quad H_x = -\frac{10^3}{120\pi}\sin(\omega t - \beta z)\,(\text{A/m})$$

    and $\qquad H(z, t) = -\dfrac{10^3}{120\pi}\sin(\omega t - \beta z)a_x\,(\text{A/m})$.

15. For the wave of problem 14, determine the propagation constant γ, given that the frequency $f = 95.5$ MHz. [BTech]

    Ans. In general, $\gamma = \sqrt{j\omega\mu(\sigma + j\omega\varepsilon)}$. In free space, $\sigma = 0$, so that

    $$\gamma = j\omega\sqrt{\mu_0\varepsilon_0} = j\left(\frac{2\pi f}{c}\right) = j\frac{2\pi(95.5 \times 10^6)}{3 \times 10^8} = j(2.0)\,\text{m}^{-1}$$

    Note that this result shows the attenuation factor is $\alpha = 0$ and the phase shift constant is $\beta = 2.0$ rad/m.

16. Examine the field $E(z, t) = 10\sin(\omega t + \beta z)\hat{a}_x + 10\cos(\omega t + \beta z)\hat{a}_y$ in the $z = 0$ plane, for

    $\omega t = 0,\ \pi/2,\ \dfrac{3\pi}{4}$ and π. [BTech]

    Ans. The computations are presented in Table 1.

Table 1

ωt	$E_x = 10 \sin \omega t$	$E_y = \cos \omega t$	$E = E_x \hat{a}_x + E_y \hat{a}_y$
0	0	10	$10\hat{a}_y$
$\dfrac{\pi}{4}$	$\dfrac{10}{\sqrt{2}}$	$\dfrac{10}{\sqrt{2}}$	$10\left(\dfrac{\hat{a}_x + \hat{a}_y}{\sqrt{2}}\right)$
$\dfrac{\pi}{2}$	$\dfrac{10}{\sqrt{2}}$	$\dfrac{-10}{\sqrt{2}}$	$10\left(\dfrac{\hat{a}_x + \hat{a}_y}{\sqrt{2}}\right)$
$\dfrac{3\pi}{4}$	$\dfrac{10}{\sqrt{2}}$	$\dfrac{-10}{\sqrt{2}}$	$10\left(\dfrac{\hat{a}_x + \hat{a}_y}{\sqrt{2}}\right)$
π	0	-10	$10(-\hat{a}_y)$

As shown in Fig. 8.15, $E(x, t)$ is circularly polarized. In addition, the wave travels in the $-\hat{a}_x$ direction.

Fig. 8.15

17. Show that in the EM waves the electrostatic energy density is equal to the magnetic energy density.

    Ans. At a point the electrostatic energy is given as

    $$u_e = \frac{1}{2}\mathbf{B} \cdot \mathbf{H} = \frac{1}{2}\mu H^2 \tag{1}$$

    Now, $$\frac{E}{H} = \sqrt{\mu/\varepsilon} \quad \text{or} \quad E = \sqrt{\mu/\varepsilon}\, H \tag{2}$$

    At a point the magnetic energy density is given as

    $$u_m = \frac{1}{2}\mathbf{B} \cdot \mathbf{H} = \frac{1}{2}\mu H^2 \tag{3}$$

    Now, from Eqs. (1) and (2)

    $$u_e = \frac{1}{2}\varepsilon\left(\frac{\mu}{\varepsilon}H\right)^2 = \frac{1}{2}\mu H^2 = u_m.$$

18. A plane wave with $E = 10e^{j(\omega t - \beta z)}\hat{a}_y$ is incident normally on a thick plane conductor lying in the xy-plane. Its conductivity is $6 \times 10^6\,\mathrm{S\,m^{-1}}$ and surface impedance is $5 \times 10^{-4}\,\angle 45°\,\Omega$. Determine the propagation constant and skin depth in the conductor.

   Ans. 1500 m, 0.66 mm   **[GATE]**

19. The three regions shown in Fig. 8.16 are all lossless and non magnetic. Find:

    (a) wave impedance in media 2 and 3
    (b) d such that medium 2 acts as a quarter wave ($\lambda/4$) transformer
    (c) reflection coefficient (R) and voltage standing wave ratio (VSWR) at the interface of media 1 and 2, when $d = \lambda/4$.

                                                                          [GATE]

    Ans. (a) 188.41Ω, 125.61Ω (b) 7.5 cm   (c) –0.143, 1.33

Fig. 8.16

20. The electric field vector of a wave is given as

$$E = E_o \exp\left[i(\omega t - 3x - 4y)8\hat{a}_x + \frac{(6\hat{a}_y + 5\hat{a}_z)}{\sqrt{125}} \right] \mathrm{V\,m^{-1}}.\ \text{Its frequency is 10 GHz.}$$

    (a) Investigate whether this wave is plane wave.
    (b) Determine its propagation constant.
    (c) Calculate its phase velocity in y-direction.

    Ans. (a) plane wave (b) 5 rad/s (c) $1.25 \times 10^{10}\,\mathrm{m\,s^{-1}}$.

MULTIPLE CHOICE QUESTIONS

1. The ratio $\sqrt{\dfrac{\mu}{\varepsilon}}$ for a medium is called

   (a) characteristic impedance
   (c) refractive index
   (b) reactance
   (d) frequency

                                                                          [a]

2. For an electromagnetic wave propagated in a good dielectric having $\left(\dfrac{\sigma}{\omega\varepsilon}\right) \gg 1$, the attenuation factor α and phase shift factor β are given by

   (a) $\alpha = \dfrac{\sigma}{2}\sqrt{\dfrac{\mu}{\varepsilon}}, \quad \beta = \sqrt{\omega\mu\varepsilon}$

   (b) $\alpha = \sqrt{\dfrac{\mu}{\varepsilon}}, \quad \beta = \dfrac{1}{2}\sqrt{\omega\mu\varepsilon}$

   (c) $\alpha = \sqrt{\dfrac{\mu}{\varepsilon}}, \quad \beta = \sqrt{\dfrac{\mu\omega}{\varepsilon}}$

   (d) $\alpha = \dfrac{\sigma}{2}, \quad \beta = \sqrt{\mu\varepsilon}$ **[a]**

3. The value of $\sqrt{\dfrac{\mu}{\varepsilon}}$ for free space is about

   (a) $380\,\Omega$ (b) $38\,\Omega$ (c) $3.8\,\Omega$ (d) $3800\,\Omega$ **[a]**

4. The intrinsic impedance of free space is

   (a) $z_0 = \sqrt{\dfrac{\mu_0}{\varepsilon_0}}$

   (b) $z_0 = \dfrac{\mu_0}{\varepsilon_0}$

   (c) $z_0 = \mu_0\varepsilon_0$

   (d) $z = \sqrt{\mu_0\varepsilon_0}$ **[a]**

5. In a uniform plane, wave E and H are related by

   (a) $\dfrac{E}{H} = \sqrt{\dfrac{\mu}{\varepsilon}}$

   (b) $\dfrac{E}{H} = \sqrt{\dfrac{\varepsilon}{\mu}}$

   (c) $\dfrac{E}{H} = \dfrac{\mu}{\varepsilon}$

   (d) $\dfrac{E}{H} = \dfrac{\varepsilon}{\mu}$ **[a]**

6. In a perfect dielectric, wave propagation occurs
   (a) with small attenuation
   (b) with large attenuation
   (c) with zero attenuation
   (d) none of the above   **[c]**

7. An electromagnetic wave incident on a perfect conductor is
   (a) fully transmitted
   (b entirely reflected
   (c) partially reflected and partially transmitted
   (d) none of the above   **[b]**

8. For a good conductor, the depth of the penetration of electromagnetic waves is given by

   (a) $\delta = \sqrt{\dfrac{2}{\omega\sigma\mu}}$

   (b) $\delta = \sqrt{\dfrac{1}{\omega\sigma\mu}}$

   (c) $\delta = \sqrt{\dfrac{2}{\omega\sigma^2\mu}}$

   (d) $\delta = \dfrac{1}{\omega\sigma}$ **[a]**

9. For a plane good conductor, skin depth varies
   (a) directly as the conductivity
   (b) inversely as the conductivity
   (c) inversely as the square root of conductivity
   (d) directly as the square root of conductivity [c]

10. For a good conductor, the skin depth varies
    (a) inversely as frequency ω (b) inversely as $\sqrt{\omega}$
    (c) directly as ω (d) directly as $\sqrt{\omega}$ [b]

11. The dependence of the phase velocity of an electromagnetic wave in a medium on the frequency of the wave is called
    (a) reflection   (b) refraction
    (c) polarization   (d) dispersion [d]

12. The relation between the amplitude of a linearly polarized monochromatic electromagnetic wave and the vector E is
    (a) $q = |E|_{max}$ (b) $a = |E^2|$
    (c) $a = \left|\dfrac{E}{2}\right|_{max}$ (d) $a = |E|_{max}^{1/2}$ [a]

13. The skin depth for a good conductor is
    (a) $\sqrt{\dfrac{\omega\varepsilon\sigma}{2}}$ (b) $\sqrt{2\omega\mu\sigma}$
    (c) $2/\sqrt{\omega\mu\sigma}$ (d) $\sqrt{2/\omega\mu\sigma}$ [d]

14. The depth of penetration is the depth in which EM wave has been attenuated to
    (a) 100% of the original value
    (b) 40% of the original value
    (c) e times the original value
    (d) $1/e$ times the original value [d]

15. The intrinsic impedance of a good conductor is given by
    (a) $\sqrt{\dfrac{i\omega\mu}{\sigma}}$ (b) $\sqrt{\dfrac{i\omega\mu}{\sigma + i\omega\varepsilon}}$
    (c) $\sqrt{i\omega\mu\sigma\left(1 + i\dfrac{\omega\varepsilon}{\sigma}\right)}$ (d) $\sqrt{\dfrac{\mu}{\varepsilon}}\left(1 + \dfrac{i\sigma}{2\omega\varepsilon}\right)$ [a]

16. In a good conductor, for electromagnetic wave operating at 1 GHz, the E and H will be out of phase at distance $Z/\lambda = 1.0$ by
    (a) 90°   (b) 45°
    (c) 30°   (d) 15° [b]

17. At $f = 2$ GHz, a material has $\sigma = 255\,\mathrm{S\,m^{-1}}$, $\varepsilon_r = 80$. The material at this f acts as a
    (a) conductor   (b) insulator
    (c) perfect dielectric   (d) none of the above [a]

18. Two magnetic media have refractive indices $n_1(=\sqrt{\varepsilon_1})$ and $n_2(=\sqrt{\varepsilon_2})$. The ratio of reflected and incident fields E_r/E_i is

    (a) n_2/n_1

    (b) n_1/n_2

    (c) $\sqrt{\dfrac{n_1}{n_2}}$

    (d) $\dfrac{(n_1/n_2-1)}{(n_1/n_2+1)}$ **[d]**

19. An electromagnetic wave is to pass through an interface separating two media having dielectric constants ε_1 and ε_2 respectively. If $\varepsilon_1 = 4\varepsilon_2$, the wave will be totally reflected if angle of incidence is

    (a) 60°

    (b) 30°

    (c) 45°

    (d) 15°   **[b]**

20. The peak value of electric field in a non-magnetic medium is $12\,\text{V m}^{-1}$. If $\varepsilon_r = 100/\pi^2$, the time averaged power per m^2 will be

    (a) $9/\pi^2\,\text{W}$

    (b) $6/\pi^2\,\text{W}$

    (d) $3\pi^2\,\text{W}$

    (d) $6/\sqrt{\pi^2}\,\text{W}$ **[b]**

21. For a ground in a rural area, $\varepsilon_r = 14$ and conductivity $\sigma = 10^{-2}\,\text{Sm}^{-1}$. At a microwave frequency of 30 GHz, the ground will act like a

    (a) dielectric

    (b) conductor

    (c) partial dielectric

    (d) none of the above   **[a]**

22. In a copper conductor, the electromagnetic wave at 100 MHz penetrates at a depth of $7\,\mu\text{m}$. The wavelength of the EM wave is

    (a) $21\,\mu\text{m}$

    (b) $44\,\mu\text{m}$

    (c) $66\,\mu\text{m}$

    (d) $14\,\mu\text{m}$ **[b]**

23. An electromagnetic wave has electric field component along y-direction and magnetic field component along x-direction. The EM wave is propagating along

    (a) z-direction

    (b) y-direction

    (c) $-z$-direction

    (d) $-y$-direction   **[c]**

24. Which one of the following is incorrect statement about EM waves:

    (a) in free space, the EM waves travel with the speed of light

    (b) the electrostatic energy density is equal to the magnetic energy density

    (c) the direction of flow of electromagnetic energy is along the direction of propagation of wave

    (d) the energy density associated with the EM waves in free space propagates with a speed less than the speed of light.   **[d]**

25. The intrinsic impedance of a lossy dielectric medium is given by

    (a) $\dfrac{j\omega\mu}{\sigma}$

    (b) $\dfrac{j\omega\mu}{\mu}$

    (c) $\sqrt{\dfrac{j\omega\mu}{\sigma+j\omega\mu}}$

    (d) $\sqrt{\dfrac{\mu}{\varepsilon}}$ **[GATE] [c]**

26. The skin depth for a material at 1 GHz with conductivity of 10^6 mho/m and permeability of $4\pi \times 10^{-7}$ H/m is
    (a) 15.9 μm   (b) 20.9 μm
    (c) 25.9 μm   (d) 30.9 μm   **[GATE] [a]**

27. The time average Poynting vector in watt/m² for a wave with $E = 24e^{j(\omega t + \beta z)} \hat{a}_y$ V/m in free space is

    (a) $-\dfrac{2.4}{\pi}\hat{a}_z$ (b) $\dfrac{2.4}{\pi}\hat{a}_z$

    (c) $\dfrac{4.8}{\pi}\hat{a}_z$ (d) $-\dfrac{4.8}{\pi}\hat{a}_z$ **[a]**

28. The depth of penetration of a wave in a lossy dielectric increases with increasing
    (a) conductivity   (b) permeability
    (c) wavelength   (d) permittivity   **[GATE] [a]**

29. If a plane electromagnetic wave satisfies the equation $\dfrac{\partial^2 E_x}{\partial z^2} = c^2 \dfrac{\partial^2 E_x}{\partial t^2}$, the wave propagates in the
    (a) x-direction
    (b) y-direction
    (c) z-direction
    (d) xy-plane at an angle 45° between the x-and z-directions   **[GATE] [c]**

30. A plane wave is characterized by $E = \left(0.5\hat{a}_x + \dfrac{e^{j\pi}}{2}\hat{a}_y \right)e^{j(\omega t - kz)}$. This wave is

    (a) linearly polarized   (b) circularly polarized
    (c) elliptically polarized   (d) unpolarized   **[GATE] [a]**

31. A plane electromagnetic wave propagating in free space is incident normally on a large slab of lossless non-magnetic dielectric material with $\varepsilon > \varepsilon_0$. Maxima and minima are observed in front of the slab. The maximum electric field is to be 5 times the minimum field. The intrinsic impedance of the medium should be
    (a) 120π Ω   (b) 60π Ω
    (c) 600π Ω   (d) 24π Ω   **[GATE] [d]**

32. When a plane wave travelling in a free space is incident normally on a medium having $\varepsilon_r = 4.0$, then fraction of power transmitted into the medium is given by

    (a) $\dfrac{8}{9}$ (b) $\dfrac{1}{2}$

    (c) $\dfrac{1}{3}$ (d) $\dfrac{5}{6}$ **[GATE] [a]**

33. A plane wave of wavelength λ is travelling in a direction making an angle 30° with positive x-axis and 90° with positive y-axis. The E field of the plane wave can be represented as (E_0 is a constant)

   (a) $E = \hat{y} E_0 \exp\left[j\left(\omega t - \frac{\sqrt{3}\pi}{\lambda} - \frac{\pi}{\lambda}z \right) \right]$

   (b) $E = \hat{y} E_0 \exp\left[j\left(\omega t - \frac{\pi}{\lambda}x - \frac{\sqrt{3}\pi}{\lambda}z \right) \right]$

   (c) $E = \hat{y} E_0 \exp\left[j\left(\omega t + \frac{\sqrt{3}\pi}{\lambda}x + \frac{\pi}{\lambda}z \right) \right]$

   (d) $E = \hat{y} E_0 \exp\left[j\left(\omega t - \frac{\pi}{\lambda}x + \frac{\sqrt{3}\pi}{\lambda}z \right) \right]$ [GATE] [a]

34. The incoming solar radiation at a place on the surface of the earth is 1.2 kW/m². The amplitude of the electric field approximately corresponding to this incident power is

   (a) 350 V/m   (b) 650 V/m
   (c) 950 V/m   (d) 1250 V/m   [c]

35. The skin depth at 10 MHz for a conductor is 1 cm. The phase velocity of an electromagnetic wave in the conductor at 1000 MHz is

   (a) 3×10^6 m/s   (b) 6×10^6 m/s
   (c) 9×10^6 m/s   (d) 1.5×10^6 m/s   [GATE] [b]

36. The wavelength of a wave with propagation constant $(0.1\pi + j0.2\pi)\,\text{m}^{-1}$ is

   (a) $\dfrac{2}{\sqrt{0.05}}$ m   (b) 10 m

   (c) 20 m   (d) 30 m   [GATE] [b]

37. The electric field component of a plane electromagnetic wave traveling in vacuum is given by $E(z, t) = E_0 \cos(kz - \omega t)\hat{i}$. The Poynting vector for the wave is

   (a) $(c\varepsilon_0/2)\, E_0^2 \cos^2(kz - \omega t)\hat{j}$

   (b) $(c\varepsilon_0/2)\, E_0^2 \cos^2(kz - \omega t)\hat{k}$

   (c) $c\varepsilon_0 E_0^2 \cos^2(kz - \omega t)\hat{j}$

   (d) $c\varepsilon_0 E_0^2 \cos^2(kz - \omega t)\hat{k}$ [GATE 17] [d]

   [**Hint:** $E = E_0 \cos(kz - \omega t)\hat{i}$

   $B = B_0 \cos(kz - \omega t)\hat{j}$

   $= \dfrac{E_0}{c} \cos(kz - \omega t)\hat{j}$

Poynting vector is $\qquad S = \dfrac{1}{\mu_0} \cdot (E \times B)$

$$= \dfrac{1}{\mu_0} \dfrac{E_0^2}{c} \cdot \cos^2 (kz - \omega t)\hat{k}$$

But $\qquad\qquad\qquad \dfrac{1}{\mu_0} = \varepsilon_0 \cdot c^2$

$\therefore \qquad\qquad\qquad S = c\varepsilon_0 E_0^2 \cos^2 (kz - \omega t)\hat{k}$ $\qquad\qquad$ **[d]**

38. A plane wave $(\hat{x} + i\hat{y})E_0 \, \exp[i(kz - \omega t)]$ after passing through an optical element emerges as $(\hat{x} + i\hat{y})E_0 \, \exp[i(kz - \omega t)]$, where k and ω are the wavevector and the angular frequency, respectively. The optical element is a
    (a) quarter wave plate
    (b) half wave plate
    (c) polarizer
    (d) Faraday rotator $\qquad\qquad\qquad\qquad\qquad$ **[GATE 17] [b]**
    [*Hint:* (b) Wave equation is

    $$E = (\hat{x} + i\hat{y})E_0 \; e^{i(kz - \omega t)}$$

    When the wave is passing through a half-wave plate, it will be introduced a phase shift term $e^{i\pi} = -1$, between x and y components, i.e. $E' = (\hat{x} - i\hat{y})E_0 \; e^{i(kz - \omega t)}$.

9

Scattering and Dispersion of Electromagnetic Waves

9.1 INTRODUCTION

We have read that the electromagnetic waves undergo geometrical reflection, refraction and propagate through waveguides. In addition to specular reflections, the electromagnetic waves incident on targets are scattered in various directions. The scattering of electromagnetic radiation by a free electron has certain pecularities in comparison to the scattering by bound electrons or molecules. Scattering is a double process by which an electron absorbs energy from an electromagnetic wave and reradiates it as scattering radiation. We must keep in mind that an electromagnetic wave carries energy and momentum, and if some energy E is removed from the wave, a corresponding amount of momentum $p = \dfrac{E}{c}$ must also be removed from the wave. We must note that a free electron cannot absorb an amount of energy E and at the same time acquire a momentum $p = \dfrac{E}{c}$ because the relation between kinetic energy and momentum for an electron is $E_K = c\sqrt{m_e c^2 + p_e^2 c^2} - m_e c^2$ at high energies and $E_K = \dfrac{p^2.}{2m_e}$ at lower energies. Either of these is incompatible with the relation $p = \dfrac{E}{c}$, if $E = E_K$, as required by conservation of energy. However, in the case of bound electron, the energy and momentum absorbed are shared by both the electron and the ion forming the remaining part of the atom, and it is possible to split both energy and momentum in correct proportions. However, the ion, having a much larger mass, carries (along with some momentum) only a small fraction of the energy available, and usually one does not consider it at all. In the case of a free electron, there is no other particle with which the electron shares the energy and the momentum and no absorption or scattering should be possible.

Experiments, however, tells a different story. When we analyse the electromagnetic radiation that has passed through a region where free electrons are present, we observe that, in addition to the incident radiation there is another sort of radiation present of *different* frequency. This new radiation is interpreted as the radiation scattered by the free electrons. The frequency of the scattered radiation is *smaller* than the incident frequency and accordingly the wavelength of the scattered radiation is *longer* than the incident wavelength. The wavelength of the scattered radiation is also different for each direction of scattering. This interesting phenomenon is called the *Compton effect* after AH Compton, who first observed and analyzed it in the early 1920s. Compton effect is an example of *inelastic scattering*. Other examples of inelastic scattering are *Raman effect* and *Tyndall effect*. Rayleigh scattering is an example of *elastic scattering*. In this process, the photons of the radiation are reflected, i.e. they bounce off the atoms and molecules with unchanged energy and momentum. Theoretical analysis shows that one can "explain" the scattering of electromagnetic radiation by a free electron by identifying the process with collision of a free electron and a particle of zero rest mass, which is a photon of energy $h\nu$. However, this is quantum explanation of scattering of electromagnetic waves by a free charge.

So far we have considered only the propagation of electromagnetic waves in vacuum. Experiments reveal that the velocity of propagation of an electromagnetic wave through matter is different from its velocity of propagation in vacuum. When an electromagnetic wave propagates through matter, even if there are no free charges and current, it induces certain charges and current in the substance as a result of polarization and magnetization of matter as discussed in previous chapters. If the substance is homogeneous and isotropic, we have seen that the net effect of the polarization and magnetization of the medium by the electromagnetic wave is to replace the constants ε_0 and μ_0 in the Maxwell's equations by the electric permittivity ε and the magnetic permeability μ, characteristics of the material. Obviously, the velocity of wave now

becomes $v = \dfrac{1}{\sqrt{\varepsilon\mu}}$. The ratio $\dfrac{c}{v}$ is called the absolute index of refraction of the substance, designated by n. We have

$$n = \frac{c}{v} = \frac{1}{\sqrt{\varepsilon_0\mu_0}}\sqrt{\varepsilon\mu} = \sqrt{\frac{\varepsilon\mu}{\varepsilon_0\mu_0}} \tag{9.1}$$

But $\dfrac{\varepsilon}{\varepsilon_0} = \varepsilon_r$ (relative permittivity) and $\dfrac{\mu}{\mu_0} = \mu_r$ (relative permeability).

Then

$$n = \frac{c}{v} = \sqrt{\varepsilon_r\mu_r} \tag{9.2}$$

In general, μ_r differs very little from 1 for the majority of substances that transmit electromagnetic waves and we can write as a satisfactory approximation

$$n = \sqrt{\varepsilon_r} \tag{9.3}$$

This relation affords a simple experimental method for determining the relative permittivity of the substance if the index of refraction is obtained independently. One obtains the following relation between n, ε_r, number of electrons per unit volume (N) and wave frequency (ω):

$$n^2 = \varepsilon_r = 1 + \frac{Ne^2}{m\varepsilon_0}\left(\sum_\alpha \frac{f_\alpha}{\omega_\alpha^2 - \omega^2}\right) \tag{9.4}$$

The quantities designated by f_i are called the *oscillator strengths* of the substance and the summation extends over all the frequencies. They are all positive and smaller than one, and represent the relative proportion in which each of the frequencies of the spectrum contribute to the polarizability of the atom. They satisfy the relation, $\sum_\alpha f_\alpha = 1$.

Obviously, the index of refraction depends on the wave frequency, and hence also on the wavelength (Fig. 9.1). ω_1, ω_2,... are the characteristic frequencies of the emission spectrum of the substance.

Fig. 9.1: Variation of the index of refraction with frequency and wavelength

Consequently, the phase velocity $v = \dfrac{c}{n}$ of the electromagnetic wave in matter also depends on the frequency of the radiation. Therefore, electromagnetic wave suffers dispersion when propagating in a medium, that is, a pulse containing several frequencies will be distorted because each component will travel with a different velocity. The group velocity v_g is given by

$$v_g = v + k\frac{dv}{dk}$$

We have

$$\frac{dv}{dk} = \frac{dv}{d\omega}\cdot\frac{d\omega}{dk} = v_g\frac{dv}{d\omega}$$

Since $v_g = \dfrac{d\omega}{dk}$ and $v = \dfrac{c}{n}$, we have

$$\frac{dv}{d\omega} = -\frac{c}{n^2}\frac{dn}{d\omega}$$

$$\therefore \qquad v_g = v - \frac{v_g ck}{n^2}\cdot\frac{dn}{d\omega}$$

Solving for v_g, one obtains

$$v_g = \frac{v}{1+\left(\dfrac{ck}{n^2}\right)\left(\dfrac{dn}{d\omega}\right)} = \frac{v}{n+\omega\left(\dfrac{dn}{d\omega}\right)} \qquad \left(\text{where } k = \frac{\omega}{v} = \frac{\omega n}{c}\right) \qquad (9.5)$$

When $\dfrac{dn}{d\omega}$ is positive, the group velocity is smaller than the phase velocity. Such a situation is called *normal dispersion*. But if $\dfrac{dn}{d\omega}$ is negative, then the group velocity is larger than the phase velocity and *anomalous dispersion* results. The possibility exists in this case that the group velocity is larger than c, and that an electromagnetic pulse can be transmitted at a velocity larger than c. This is apparently in contradiction to the results obtained from Lorentz transformation and the principle of relativity.

A careful analysis of the transmission of an electromagnetic signal made by Brillouin, Sommerfeld and others, revealed that it is impossible to transmit a signal with a velocity greater than c. Figure 9.2 shows the variation of the phase velocity v, the group velocity v_g

Fig. 9.2: Phase, group and signal velocities of an electromagnetic pulse in a dispersion medium

and the signal velocity v_σ near the characteristic frequency ω. The signal velocity practically coincides with the group velocity except near the characteristic frequency and is never larger than c even in the region of anomalous dispersion.

We have said that electromagnetic waves appear to propagate in matter with a phase velocity different from their propagation velocity in vacuum. But that difference seems to stem from the fact that the permittivity and permeability of matter are different from those of vacuum. This difference (in the permittivity and permeability) is, in turn, a consequence of the electric and magnetic permeability of matter under the action of the incoming electromagnetic wave. Thus, when an electromagnetic wave falls on a piece of matter, it induces oscillations in the charged particles of the atoms or molecules which then emit *secondary* or *scattered waves*. These scattered waves are superposed on the original wave, giving a resultant wave. The phases of the secondary waves are, in general, different from that of the original wave, since a force oscillator is not always in phase with the driving force. A detailed analysis indicates that this phase difference affects the resultant wave in such a way that this wave appears to have a phase velocity different from that which a wave in vacuum has. This result is particularly satisfying since, from the atomic point of view, all charges, both free and sound, are equivalent, and the electromagnetic waves they emit must all propagate with the velocity c. It is the wave resulting from the superposition of their individual waves, which have different phases, that as a consequence of phase differences, appears to have a different velocity of propagation.

9.2 SCATTERING CROSS-SECTION AND SCATTERING AMPLITUDE

The scattering process is described in terms of a quantity known as scattering cross-section. Let us consider that a beam of electromagetic radiation moving along the z-direction with a flux of $I_0/\text{cm}^2/\text{sec}$. When the radiation is incident on a target, it is scattered in a solid angle $d\Omega$. The conservation of intensity of electromagnetic wave flux in the scattering process demands that

$$I d\Omega = I_0 d\sigma \tag{9.6}$$

where $d\sigma$ is the effective cross-sectional area which scattering target presents, perpendicular to the incident flux of electromagnetic radiation. The effective area (perpendicular to the incident beam of radiation) governs the solid angle in which incident radiations are scattered. The ratio $\dfrac{d\sigma}{d\Omega}$ is known as differential scattering and presents the basic feature of the scattering process. The total *scattering cross-section* is defined as

$$\sigma = \int_0^{2\pi} \frac{d\sigma}{d\Omega} d\Omega \tag{9.7}$$

The scattering process holds good equally for waves and particles. It can best be explained and defined from quantum–mechanical considerations.

9.3 SCATTERING FROM A FREE CHARGE PARTICLE (THOMSON SCATTERING)

The incidence and scattering of electromagnetic radiation on a free charge particle of charge q and mass m constitutes a fundamental process of scattering phenomena. Let us suppose that plane monochromatic polarized electromagnetic wave is incident on a free charged particle of mass m and carrying a charge q. The wave is of the form

$$E = E_0 \exp \left[i (k \cdot r - \omega t) \right] \tag{9.8}$$

The force experienced by the charged particle is given by

$$F = qE \tag{9.9}$$

or

$$m \frac{dv}{dt} = qE_0 \exp \left[i (k \cdot r - \omega t) \right] \tag{9.10}$$

Obviously, Eq. (9.10) is the equation of motion of the charged particle. The displacement of the charged particle under the influence of the incident field is small compared to the wavelength. This means that the velocity acquired by the particle as a result of this interaction is much smaller than c ($v \ll c$). Therefore, we have restricted to the non-relativistic limit and ignored the second term in the expression for the Lorentz force given by

$$F = q(E + v \times B) \tag{9.11}$$

The oscillating charge in the presence of electromagnetic wave constitutes an electric dipole whose dipole moment is given by

$$p(t) = qr(t)$$

Therefore, we have

$$\ddot{p} = q\ddot{r}(t) = \frac{q^2}{m} E(t) \tag{9.12}$$

The time average of the power

$$\left\langle \frac{dW}{d\Omega} \right\rangle = \frac{q^2 (\dot{u})^2}{16\pi^2 \varepsilon_0 c^3} \sin^2 \theta \tag{9.13}$$

$$= \frac{\left\langle \ddot{p}^2 \right\rangle \sin^2 \theta}{16\pi^2 \varepsilon_0 c^3} = \frac{q^4 E_0^4 \sin^2 \theta}{16\pi^2 m^4 \varepsilon_0^3} \tag{9.14}$$

We note that the rate of energy radiation is independent of frequency of the incident electromagnetic field. One can obtain the Poynting vector of the incident electromagnetic radiation as

$$N_i = \frac{1}{2}E_0 H_0 \hat{k} = \frac{E_0^2}{2Z_0}\hat{k} = \frac{1}{2}\varepsilon_0 c E_0^2 \hat{k} \quad \left(\text{Here } z_0 = \sqrt{\frac{\mu_0}{\varepsilon_0}}\right) \tag{9.15}$$

A target or an obstacle appearing in the direction of incident electromagnetic energy flow intercepts a part of the electromagnetic energy which is redirected in different directions. This phenomenon is termed *scattering of electromagnetic waves*. This is characterized by certain cross-section. We have differential scattering cross-section

$$\frac{d\sigma}{d\Omega} = \frac{\text{Scattered energy per unit solid angle per unit time}}{\text{Incident energy per unit area per unit time}}$$

$$= \frac{\left\langle\frac{dW}{d\Omega}\right\rangle}{\text{Incident flux}} = \frac{\left\langle\frac{dW}{d\Omega}\right\rangle}{\sqrt{\frac{\varepsilon_0}{\mu_0}}E_0^2} = \left(\frac{q^2}{4\pi\varepsilon_0 mc^2}\right)^2 \sin^2\theta \tag{9.16}$$

where θ is the angle between the induced dipole moment p and the direction of the outgoing propagation of wave, i.e. θ is the angle between E and the unit vector $\hat{a}_r$ as shown in Fig. 9.3.

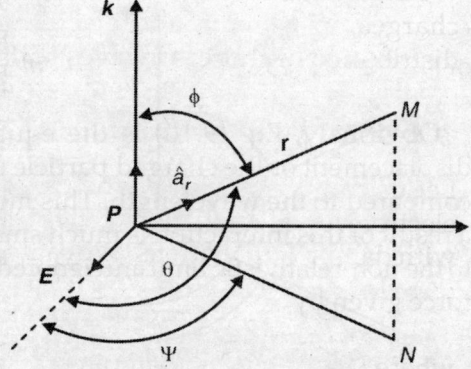

Now, we consider a more general case. Supposing that incident light is unpolarized, let us find the angular distribution of scattered light. For this, we will have to average overall possible azimuthal orientations of field E, i.e. we have to average over the angle ψ (Fig. 9.3). From Fig. 9.3, we have

Fig. 9.3: Scattering of radiation by a free charge

$$\cos\theta = \cos\psi\,\sin\phi$$

or $$\sin^2\theta = 1 - \cos^2\psi\,\sin^2\psi$$

Let $\overline{\sin^2\theta}$ represents the average of $\sin^2\theta$ over all values of ψ, then we have

$$\overline{\sin^2\theta} = 1 - \overline{\cos^2\theta}\,\sin^2\theta = 1 - \frac{1}{2}\sin^2\phi = \frac{1}{2}(1 + \cos^2\phi)$$

∴ For unpolarized electromagnetic radiation, we have

$$\left\langle\frac{d\sigma}{d\Omega}\right\rangle_{\text{unpolarized}} = \left(\frac{q^2}{4\pi\varepsilon_0 mc^2}\right)^2 \overline{\sin^2\theta}$$

$$= \left(\frac{q^2}{4\pi\varepsilon_0 mc^2}\right)^2 \left(\frac{1 + \cos^2\phi}{2}\right) \tag{9.17}$$

ϕ is called the scattering angle, as it is the angle between the directions of the incident radiation and the scattered radiation (Fig. 9.3).

One can obtain the total scattering cross-section by integrating over the entire solid angle. One obtains,

$$\langle \sigma \rangle_{unpol.} = \int \frac{d\sigma}{d\Omega} d\Omega = \left(\frac{q^2}{4\pi\varepsilon_0 mc^2} \right)^2 \int \frac{1 + \cos^2 \phi}{2} \sin \phi \, d\phi \, d\psi$$

$$= \frac{8\pi}{3} \left(\frac{q^2}{4\pi\varepsilon_0 mc^2} \right)^2 \tag{9.18}$$

For electron, $q = e$ and

$$\langle \sigma \rangle = \frac{8\pi}{3} \left(\frac{e^2}{4\pi\varepsilon_0 mc^2} \right)^2 = \frac{8\pi}{3} r_0^2 \tag{9.19}$$

where $r_0 = \left(\frac{e^2}{4\pi\varepsilon_0 mc^2} \right)^2 = \left(\frac{\mu_0 e^2}{4\pi m} \right) = 2.8178 \times 10^{-15}$ m and is known as *classical radius of electron* and the corresponding cross-section is known as *Thomson scattering cross-section*. We must note that the scattering of unpolarized plane electromagnetic wave from free charges, in general, and electrons, in particular, play an important role in the distribution of scattered waves.

9.4 SCATTERING OF RADIATION BY A BOUND CHARGE (RAYLEIGH SCATTERING)

Let us consider the case when an electron system is subjected to a monochromatic electromagnetic radiation. Let us assume that the bound electron is allowed to move with damped oscillations. We can write the equation of motion of such an electron as

$$F = eE = m(\ddot{x} + l\dot{x} + \omega_0^2 x) \tag{9.20}$$

where l measures the damping force, ω_0 is the frequency of bound charge and $E = E_0 \exp(-i\omega t)$ represents the applied field. We can rewrite Eq. (9.20) as

$$\frac{d^2 x}{dt^2} + l\frac{dx}{dt} + \omega_0^2 x = \frac{e}{m} E_0 \exp(-i\omega t) \tag{9.21}$$

One can easily see that the steady state solution of Eq. (9.21) is

$$x(t) = \frac{eE_0}{m} \frac{\exp(-i\omega t)}{(\omega_0^2 - \omega^2) - il\omega} \tag{9.22}$$

For free electron, ω_0 and l both are zero and therefore, we have

$$x(t) = -\frac{eE_0}{m} \frac{\exp(-i\omega t)}{\omega^2} \tag{9.23}$$

The negative sign in Eq. (9.23) is not of importance here as one is not concerned with the relative phases. We note that Eqs. (9.22) and (9.23) are similar. This suggests that one can convert the results for the free charge into the corresponding results for the bound charge by simply multiplying each term in E_0 by $\dfrac{\omega^2}{[(\omega_0^2 - \omega^2) - il\omega]}$.

Now we obtain the time average of the power per unit solid angle radiated by a bound charge,

$$\left\langle \frac{dW}{d\Omega} \right\rangle_{bound} = \frac{e^4 E_0^2}{16\pi^2 m^2 \varepsilon_0 c^2} \frac{\omega^4}{[(\omega_0^2 - \omega^2)^2 + \omega^2 l^2]} \sin^2 \theta \tag{9.24}$$

Differential scattering cross-section for a bound charge is obtained as

$$\left\langle \frac{d\sigma}{d\Omega} \right\rangle_{bound} = \frac{\left\langle \dfrac{dW}{d\Omega} \right\rangle}{\sqrt{\dfrac{\mu_0}{\varepsilon_0}} E_0^2} = \frac{e^4}{16\pi^2 m^2 \varepsilon_0^2 c^4} \frac{\omega^4 \sin^2 \theta}{(\omega_0^2 - \omega^2)^2 + \omega^2 l^2} \tag{9.25}$$

The total scattering cross-section is given by

$$\sigma_{bound} = \frac{e^4}{6\pi \varepsilon_0^2 m^2 c^4} \frac{\omega^4}{[(\omega_0^2 - \omega^2)^2 + \omega^2 l^2]}$$

$$= \frac{\dfrac{8\pi}{3} r_0^2 \omega^4}{(\omega_0^2 - \omega^2)^2 + \omega^2 l^2} \quad \left(\text{where } r_0 = \frac{e^2}{4\pi \varepsilon_0 mc} \right)$$

$$= \sigma \frac{\omega^4}{(\omega_0^2 - \omega^2)^2 + \omega^2 l^2} \tag{9.26}$$

where σ is the scattering cross-section for a free electron. When $\omega = \omega_0$, one obtains the maximum value of scattering cross-section as

$$\sigma_{bound} = \sigma \frac{\omega^2}{l^2} \tag{9.27}$$

When $\omega \ll \omega_0$ and $\omega l \ll \omega_0^2$, one obtains

$$\sigma_{bound} = \sigma \frac{\omega^4}{\omega_0^4} \tag{9.28}$$

This scattering is known as *Rayleigh scattering*. In this case, Eq. (9.28) shows that the scattering varies with the fourth power of the frequency, high frequencies will be more strongly scattered. This is why in the scattering of white light, such as that from sun, blue is more strongly scattered than red. This explains the blue colour of the sky.

We must note that for the result given by Eq. (9.28) to be applicable, the size of the particles causing scattering must be small in comparison to the wavelength of the scattered radiation. When $\omega \approx \omega_0$, then

$$\sigma_{bound} = \frac{8\pi r_0^2}{3} \left(\frac{\omega_0}{l} \right)^2 \tag{9.29}$$

This scattering is known as *resonant scattering*. In this case, the total scattering cross-section is maximum, and the natural frequency of the electron as an oscillator becomes equal to the frequency of incident radiation. The resonance scattering occurs when the sodium vapour is illuminated with the characteristic yellow sodium radiation.

If $\omega \gg \omega_0$ and $l \rightarrow 0$, we have total scattering cross-section as

$$\sigma_{bound} = \frac{8\pi}{3} r_0^2 \qquad\qquad (9.29a)$$

i.e. *Thomson scattering* occurs and the electron behaves as if it was free. Figure 9.4 shows a plot of σ_{bound} as a function of ω.

Rayleigh scattering helps to explain the following:

i. *Sky appears blue:* According to Rayleigh scattering,

$$\sigma_{bound} \propto \omega^4 \propto \nu^4 \propto \frac{1}{\lambda^4}$$

i.e. shorter wavelength is scattered more strongly. The light coming from sun is scattered by air molecules and as we know that shorter visible wavelength corresponds to blue light, hence it is scattered strongly. One observes this scattered light which is blue. This is why the sky appears blue.

Fig. 9.4: σ_{bound} as a function of ω

ii. *Red colour at the time of sunset and sunrise:* At sunset or sunrise, the sun and its neighbouring portion appears red because sunlight travels a greater length of earth's atmosphere and strikes directly to the observer's eye. During this process, the blue light is almost scattered from the direct light reaching to observer's eye, i.e. leaving only the red light to be seen.

iii. *Use of red light for danger signals:* Human eye is most sensitive to yellow green colour even then red light is used for danger signals. The red light has longest wavelength in the visible region and scatters least. Obviously, red light can travel longer distances in atmosphere in comparison to the other colours and can be seen from larger distances. This is why red light is used for danger signals.

9.5 RADIATION DAMPING OR RADIATION REACTION

The motion of charged particles in external force fields necessarily involves the emission of radiation whenever the charged particles are accelerated. The emitted radiation carries off energy, momentum, and angular momentum and so must influence the subsequent motion of the charged particles. Consequently, the motion of the sources of radiation is determined in part by the manner of emission of the radiation. A correct treatment must include the reaction of the radiation on the motion of the sources. The reaction of the radiation on the motion of the particle is called *radiation damping* or *radiation reaction*.

We now examine that how one can include the reactive effects of radiation in the equation of motion of a charged particle. We have the classical equation of motion for a particle of mass m and charge e acted on by a conservative external force F as

$$F_e = m\frac{dv}{dt} \qquad (9.30)$$

While writing Eq. (9.30), we have neglected the emission of radiation. As the particle is accelerated, it will emit radiation. One will have to take account of the reaction of this radiation field on the particle. Obviously, we will have to modify Eq. (9.30) by adding a reaction force F_r, i.e.

$$m\frac{dv}{dt} = F_e + F_r \qquad (9.31)$$

The energy radiated per unit time by a charge particle of charge e moving with a speed $v(\ll c)$, which undergoes acceleration $a\left(=\dfrac{dv}{dt}\right)$ is given by

$$W = \frac{e^2(\dot{v})^2}{6\pi\varepsilon_0 c^3} \qquad (9.32)$$

The energy must be conserved. Obviously, the work done by the force F_r on the particle in a time interval, say t_1 to t_2, be equal to the negative of the energy radiated in this time interval. We have,

$$\int_{t_1}^{t_2} F_r \cdot v\, dt = \frac{e^2}{6\pi\varepsilon_0 c^3} \int_{t_1}^{t_2} \dot{v} \cdot \dot{v}\, dt \qquad (9.33)$$

Integrating the right hand side by parts, one obtains

$$\int_{t_1}^{t_2} F_r \cdot v\, dt = \frac{e^2}{6\pi\varepsilon_0 c^3}[v \cdot \dot{v}]_{t_1}^{t_2} + \frac{e^2}{6\pi\varepsilon_0 c^3} \int_{t_1}^{t_2} v \cdot \ddot{v}\, dt \qquad (9.33a)$$

When the motion of the charge particle is periodic or if the time interval is short, then we can consider that the state of the system will be approximately the same at t_2 as at t_1. This means, one can neglect the integrated term in Eq. (9.33a), which reduces to

$$\int_{t_1}^{t_2}\left(F_r - \frac{e^2}{6\pi\varepsilon_0 c^3}\ddot{v}\right) \cdot v\, dt \approx 0 \qquad (9.34)$$

This leads to

$$F_r = \frac{e^2}{6\pi\varepsilon_0 c^3}\ddot{v} = m\ddot{v}\tau \qquad (9.35)$$

where

$$\tau = \frac{e^2}{6\pi\varepsilon_0 mc^3} \qquad (9.36)$$

Thus, the modified Eq. (9.31) becomes

$$m\frac{dv}{dt} = F_e + m\tau\frac{d^2v}{dt^2}$$

or
$$m(\dot{v} - \tau\ddot{v}) = F_e \tag{9.37}$$

Equation (9.37) is sometimes called the *Abraham–Lorentz equation of motion*. It can be considered as an equation that includes some approximate and time-averaged way the reactive effect of the emission of radiation.

If the external force F_e is a linear restoring force of the type $F_e = -m\gamma r$ (here this γ is different from the one used earlier), one can express Eq. (9.37) as

$$m\ddot{r} = m\tau\dddot{r} + m\gamma r = 0 \tag{9.38}$$

Considering the radiation damping terms as small, the modified equation of motion reads

$$m\ddot{r} + m\gamma r = 0 \tag{9.39}$$

The solution of Eq. (9.39) is

$$r = r_0 \exp[-i\gamma^{\frac{1}{2}}t] = r_0 \exp(-i\omega_0 t) \tag{9.40}$$

where $\omega_0^2 = \gamma$ has been used. We have

$$\dot{r} = -i\omega_0 r(t)$$

and
$$\ddot{r}(t) = -\omega_0^2 r(t) \tag{9.41}$$

Using Eq. (9.41), Eq. (9.38) becomes

$$m\ddot{r} + m\tau\omega_0^2\dot{r} + m\gamma r = 0$$

or
$$m[\ddot{r} + l\dot{r} + \gamma r] = 0 \tag{9.42}$$

where
$$l = \omega_0^2\tau = \frac{e^2\gamma}{6\pi\varepsilon_0 mc^3} \tag{9.43}$$

This shows that the characteristic time is of the order of $\dfrac{1}{(\omega_0\tau)\omega_0}$ and this time is

very long compared to the orbital period $\dfrac{2\pi}{\omega_0}$, provided $\omega_0\tau \ll 1$.

The equations for radiative effects can be used to discuss practical problems such as the moderation time of the *mu* (μ) or *pi* (π) meson is cascading from an orbit of very large quantum number around a nucleus down to the low-lying orbits. Over most of the time interval, the quantum numbers are sufficiently large that the classical description of continuous motion is an adequate approximation.

9.6 DISPERSION

We have read that all media show some dispersion. When the refractive index of a medium varies with frequency, the medium is said to be dispersive and the variation of the refractive index n of the medium with wavelength λ (i.e. frequency) constitutes the phenomenon of dispersion. In the absence of dispersion, an arbitrary wave train travels without distortion. Dispersion is represented by $\dfrac{dn}{d\lambda}$. It is generally observed that n decreases with λ, known as *normal dispersion*. But in the small wavelength ranges, there is often increase of refractive index due to an increased absorption of the passing radiation through the medium called as *anomalous dispersion*.

Dispersion in Gases (Lorentz Theory)

Let us consider the propagation of electromagnetic waves in a dilute gas in which the mutual interactions between the constitutent particles are negligible. As the electromagnetic wave passes through the gas, the electrons in the molecules will be displaced from their equilibrium position and molecules will be polarized, i.e. this induces dipole moment in the gas molecules. As the density of gas is low, one can neglect the difference between the applied electric field and the local field. We assume that the electrons are bound by a linear restoring force and that the damping is proportional to the velocity. Over an atom or a molecule, electric field E is constant in space, i.e.

$$E = E_0 \exp\left[-(\omega t - k \cdot r)\right]$$

or
$$E = E_0 \exp(-i\omega t) \tag{9.44}$$

The equation of motion for an ith electron of charge e and mass m bound by a harmonic force and acted upon an electric field E is

$$m[\ddot{r}_\alpha + l_\alpha r_\alpha + \gamma_\alpha r_\alpha] = eE \tag{9.45}$$

where l_α measures the damping force. We must remember that the γ used in preceding chapters is different from this γ. We can rewrite Eq. (9.45) as

$$\ddot{r}_\alpha + l_\alpha \dot{r}_\alpha + \gamma_\alpha r_\alpha = \frac{e}{m} E_0 \exp(-i\omega t)] \tag{9.46}$$

Magnetic force effects are neglected in Eq. (9.46). We make an additional approximation that the amplitude of oscillation is small enough to permit evaluation of the electric field at the average position of the electron. The steady state solution of Eq. (9.46) is

$$r_\alpha(t) = \frac{\left(\dfrac{e}{m}\right)E_0}{(\omega_\alpha^2 - \omega^2) - il'_\alpha \omega} \exp(-i\omega t) \tag{9.47}$$

where
$$\omega_\alpha^2 = r_{\alpha/m} \quad \text{and} \quad l'_\alpha = l_{\alpha/m} \tag{9.48}$$

If the field varies harmonically in time with frequency ω as $e^{-i\omega t}$, the dipole moment contributed by one electron is

$$p_\alpha = er_\alpha(t) = \frac{\left(\dfrac{e^2}{m}\right)E}{(\omega_\alpha^2 - \omega^2) - il'_\alpha \omega} \tag{9.49}$$

If we suppose that there are N molecules per unit volume with z electrons per molecule in the gas and a fraction f_α of these have the characteristic resonance frequency ω_α, one obtains the total dipole moment per unit volume as

$$P = \sum_\alpha N f_\alpha p_\alpha = E \sum_\alpha \frac{\left(\dfrac{e^2}{m}\right) N f_\alpha}{(\omega_\alpha^2 - \omega^2) - i l'_\alpha \omega} \tag{9.50}$$

where the *oscillator strengths* f_α satisfy the sum rule

$$\sum_\alpha f_\alpha = 1$$

We have $P = \varepsilon_0 \chi E$ and therefore the electric susceptibility χ is given by

$$\chi = \sum_\alpha \frac{\left(\dfrac{e^2}{m\varepsilon_0}\right) N f_\alpha}{(\omega_\alpha^2 - \omega^2) - i l'_\alpha \omega} \tag{9.51}$$

The dielectric constant ε is given by

$$\varepsilon = 1 + \chi = 1 + \sum_\alpha \frac{\left(\dfrac{e^2}{m\varepsilon_0}\right) N f_\alpha}{(\omega_\alpha^2 - \omega^2) - i l'_\alpha \omega} \tag{9.52}$$

The electrical polarizability is given by

$$\frac{P}{E} = \frac{N e^2}{m} \sum_\alpha \frac{f_\alpha}{(\omega_\alpha^2 - \omega^2) - i l'_\alpha \omega} \tag{9.53}$$

The refractive index n is equal to $\sqrt{\varepsilon}$ and therefore

$$n^2 = \varepsilon = 1 + \sum_\alpha \frac{\left(\dfrac{e^2}{m\varepsilon_0}\right) N f_\alpha}{(\omega_\alpha^2 - \omega^2) - i l'_\alpha \omega} \tag{9.54}$$

The relation given by Eq. (9.54) is known as *dispersion relation* for dilute gases. Equation (9.52) shows that the dielectric constant is complex. This means refractive index is also complex. The significance of complex refractive index is that there is an absorption of energy in the medium. The damping constant l'_α is generally small, and therefore one can take (or ε) approximately real for most frequencies for values of $\omega < \omega_\alpha$, one can see that all terms in the sum in Eq. (9.54) are positive and ε is greater than 1. For value of $\omega > \omega_\alpha$, one finds that more and more negative terms occur in the sum and finally is less than 1. However, the behaviour of ω_α is peculiar in the neighbourhood. We find that the real part vanishes and the term is imaginary and large. Figure 9.5 shows the variation of the real and imaginary parts of n^2. These situations are discussed below:

Fig. 9.5: Variation of n^2 with ω

i. Normal dispersion: In the region remote from the natural frequencies of oscillators, i.e. the absorption frequencies of the medium, the term $l'_\alpha \omega$ in the denominator of Eq. (9.54) is so small that one can easily neglect it in comparison to

$$n^2 = 1 + \frac{1}{4\pi\varepsilon_0}\left(\frac{4\pi Ne^2}{m}\right)\sum_\alpha \frac{f_\alpha}{\omega_\alpha^2 - \omega^2} \tag{9.55}$$

This shows that the refractive index is real and increases with frequency of the incident wave. This is called *normal dispersion*. We have $\omega = 2\pi\nu$. Using this value in Eq. (9.55), one obtains

$$n^2 = 1 + \frac{ne^2}{m\varepsilon_0}\sum_\alpha \frac{f_\alpha}{\left(\dfrac{4\pi^2 c^2}{\lambda_\alpha^2} - \dfrac{4\pi^2 c^2}{\lambda^2}\right)}$$

$$= 1 + \frac{Ne^2}{\varepsilon_0 m(4\pi^2 c^2)}\sum_\alpha \frac{f_l \lambda^2 \lambda_\alpha^2}{\left(\lambda^2 - \lambda_\alpha^2\right)}$$

$$= n^2 = 1 + \sum_\alpha \frac{A_\alpha \lambda^2}{\left(\lambda^2 - \lambda_\alpha^2\right)} \tag{9.56}$$

where
$$A_\alpha = \frac{Ne^2 \lambda_\alpha^2 f_\alpha}{\varepsilon_0 m 4\pi^2 c^2} \tag{9.57}$$

Relation given by Eq. (9.57) is known as *Sellmeier's equation*. If we suppose that $\lambda \gg \lambda_\alpha$, then Eq. (9.56) can be expressed as

$$n^2 = 1 + \sum_\alpha A_\alpha \left(1 - \frac{\lambda_\alpha^2}{\lambda^2}\right)^{-1}$$

$$= 1 + \sum_\alpha A_\alpha + \sum_\alpha A_\alpha \frac{\lambda_\alpha^2}{\lambda^2} + \sum_\alpha A_\alpha \frac{\lambda_\alpha^4}{\lambda^4} + \cdots$$

$$= A + \frac{B}{\lambda^2} + \frac{C}{\lambda^4} + \cdots \tag{9.58}$$

where $A = 1 + \sum_\alpha A_\alpha$; $B = \sum_\alpha A_\alpha \lambda_\alpha^2$; $C = \sum_\alpha A_\alpha \lambda_\alpha^4$.

Relation given by Eq. (9.58) is called *Cauchy's dispersion relation*. The coefficient A is the coefficient of refraction while the coefficient B is called coefficient of dispersion. From Eq. (9.58), we have

$$2n\frac{dn}{d\lambda} = -\frac{2B}{\lambda^3} - \frac{4C}{\lambda^5}$$

or

$$\frac{dn}{d\lambda} = -\frac{1}{\lambda^3 n}\left[B + \frac{2C}{\lambda^2}\right]$$

$$= -\frac{1}{\lambda^3}\left[B + \frac{2C}{\lambda^2} + \cdots\right]\left(A + \frac{B}{\lambda^2} + \cdots\right)^{-\frac{1}{2}}$$

$$= -\frac{BA}{\lambda^2} = -\frac{k}{\lambda^3} \tag{9.59}$$

Relation given by Eq. (9.59) gives the *dispersive power*. Obviously, the dispersive power varies inversely as the cube of wavelength The minus sign indicates that the slope of the dispersive power is negative.

ii. Anomalous dispersion: When the frequency of the electromagnetic wave is nearly equal to the natural frequency, one obtains the phenomenon of anomalous dispersion. For simplicity, let us consider that there is only one natural frequency, i.e. $\omega_0 = \omega_{\alpha'}$ then Eq. (9.54) retains complex refractive index say n^* and becomes

$$n^{*2} = 1 + \frac{1}{4\pi\varepsilon_0}\frac{4\pi Ne^2}{m}\frac{1}{(\omega_0^2 - \omega^2) - il'\omega}$$

$$\therefore \quad n^* = \left[1 + \frac{1}{4\pi\varepsilon_0}\frac{4\pi Ne^2}{m}\frac{1}{\omega_0^2 - \omega^2 - il\omega}\right]^{\frac{1}{2}}$$

$$= 1 + \frac{1}{4\pi\varepsilon_0}\frac{2\pi Ne^2}{m}\frac{1}{\omega_0^2 - \omega^2 - il\omega} \tag{9.60}$$

Multiplying the numerator and the denominator of right hand side by $(\omega_0^2 - \omega^2 + il\omega)$, one obtains

$$n^* = 1 + \frac{1}{4\pi\varepsilon_0}\frac{2\pi Ne^2}{m}\left[\frac{\omega_0^2 - \omega^2 + il\omega}{(\omega_0^2 - \omega^2)^2 + l^2\omega^2}\right]$$

or

$$n^* = 1 + \frac{2\pi Ne^2}{4\pi m \varepsilon_0}\left[\frac{\omega_0^2 - \omega^2}{(\omega_0^2 - \omega^2)^2 + l^2\omega^2}\right]$$

$$+ i\frac{2\pi Ne^2}{4\pi m \varepsilon_0}\left[\frac{l\omega}{(\omega_0^2 - \omega^2)^2 + l^2\omega^2}\right] \tag{9.61}$$

Replacing n^* by $n(1 + ik)$ and comparing real and imaginary parts in Eq. (9.61), one obtains

$$n = 1 + \frac{N}{2\varepsilon^2}\left(\frac{e^2}{2\varepsilon^2}\right)\left(\frac{e^2}{m}\right)\frac{\omega_0^2 - \omega^2}{(\omega_0^2 - \omega^2)^2 + l^2\omega^2}$$

and

$$nk = \frac{N}{2\varepsilon^2}\left(\frac{e^2}{m}\right)\frac{l\omega}{(\omega_0^2 - \omega^2)^2 + l^2\omega^2} \tag{9.62}$$

Figure 9.6 represents the variation of real and imaginary parts of n^* (i.e. n and nk) as a function of ω. From the figure it is clear that at low frequencies, n (real) is slightly greater than 1 and increases as ω increases and reaches a maximum and in the neighbourhood of resonance frequency ω_0, n decreases rapidly and at $\omega = \omega_0$ it approaches unity and continues to decrease until it becomes minimum at a certain frequency greater than but in the vicinity of resonance frequency. When frequency further increases, n again increases and approaches a value slightly less than 1, for every large value of frequency. Obviously, in the neighbourhood of absorption region, the refractive index decreases with increasing frequency and this is the anomalous behaviour giving rise to the term *anomalous dispersion*.

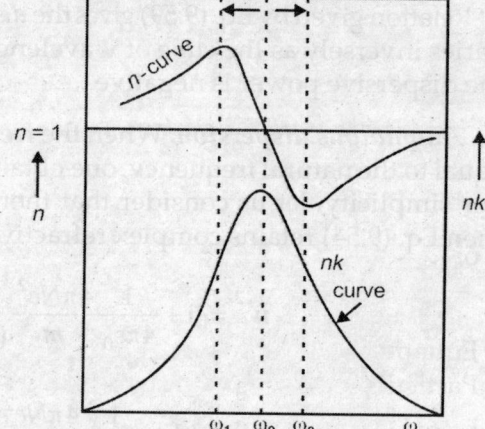

Fig. 9.6: Plot of nk and n as a function of ω

The imaginary part (nk) corresponds to absorption of electromagnetic waves propagating through the gas. From Fig. 9.6, it is evident that k is maximum when $\omega = \omega_0$, where n is 1. It has a width at half maximum approximately equal to 1. This means that in the region where n changes rapidly, the gas is relatively highly absorbing.

We must note that in the case of a real gas, there exist many resonant frequencies and corresponding damping coefficients, the complex refractive index is given by

$$n^* = 1 + \frac{1}{4\pi\varepsilon_0}\frac{2\pi Ne^2}{m}\sum_k \frac{f_k(\omega_k^2 - \omega^2 - il\omega)}{(\omega_k^2 - \omega^2)^2 + l^2\omega^2} \tag{9.63}$$

Similarly, one can consider the behaviour of real and imaginary parts. RW Wood in 1904 provided the experimental demonstration of anamolous dispersion of sodium vapour in the neighbourhood of yellow D-line.

9.7 DISPERSION IN LIQUIDS AND SOLIDS

When the medium is not dilute, i.e. one is dealing with liquids or solids, one finds that the local field is not equal to the external field. Obviously, one cannot neglect the interaction among the molecules within the liquid or gas. One will have to replace the electric field vector E in Eq. (9.50) by

$$E_{eff} = E + \frac{P}{3\varepsilon_0} \tag{9.64}$$

Thus, one obtains the total dipole moment per unit volume as

$$P = \left(E + \frac{P}{3\varepsilon_0}\right) \sum_\alpha \frac{\left(\frac{e}{m}\right)^2 N f_\alpha}{(\omega_0^2 - \omega^2) - i l_\alpha \omega} \tag{9.65}$$

Since $D = \varepsilon_0 E + P = \varepsilon E$, we have

$$\frac{P}{E + \dfrac{P}{3\varepsilon_0}} = \frac{\varepsilon - \varepsilon_0}{\varepsilon + \varepsilon_0} 3\varepsilon_0$$

$$\therefore \quad \frac{\varepsilon - \varepsilon_0}{\varepsilon + 2\varepsilon_0} = \sum_\alpha \frac{\left(\frac{e^2}{m}\right) N f_\alpha}{3\varepsilon_0\{(\omega_\alpha^2 - \omega^2) - i l_\alpha\}} \tag{9.66}$$

or

$$\frac{\dfrac{\varepsilon}{\varepsilon_0} - 1}{\dfrac{\varepsilon}{\varepsilon_0} + 2} = \frac{n^2 - 1}{n^2 + 2} = \sum_\alpha \frac{\left(\dfrac{e^2}{m}\right) N f_\alpha}{3\varepsilon_0\{(\omega_\alpha^2 - \omega^2) - i l\omega\}} \tag{9.67}$$

Equation (9.67) shows that ε_r (i.e. n) is again complex and depends on the frequency. Particular cases of interest are as follows:

i. Clausius–Mossotti Relation: If M is the molecular weight of the substance, ρ its density and N_0 the Avogadro's number, then we have number of molecules per unit volume as

$$N = \frac{N_0 \rho}{M} \tag{9.68}$$

Using Eq. (9.68), Eq. (9.67) becomes

$$\frac{\varepsilon_r - 1}{\varepsilon_r + 2} = \frac{n^2 - 1}{n^2 + 2} = \frac{N_0 e^2}{3m\varepsilon_0} \sum_\alpha \frac{f_\alpha}{(\omega_\alpha^2 - \omega^2 - i l_\alpha \omega)} \tag{9.69}$$

For a given frequency,

$$\left(\frac{\varepsilon_r - 1}{\varepsilon_r + 2}\right)\frac{M}{\rho} = \left(\frac{n^2 - 1}{n^2 + 2}\right)\frac{M}{\rho} = \text{constant} \tag{9.70}$$

This is known as *Clausius–Mossotti relation*.

ii. Lorentz–Lorenz Relation: For transparent materials, one can neglect the damping force, i.e. $l_\alpha \rightarrow 0$. This means that the refractive index is real. Equation (9.69) reduces to

$$\left(\frac{n^2 - 1}{n^2 + 2}\right)\frac{M}{\rho} = \text{constant} \tag{9.71}$$

The relation given by Eq. (9.71) shows that now the refractive index depends on the density of the medium. This is known as *Lorentz–Lorenz relation*. For gases, $n \approx 1$ and therefore one may approximate

$$n^2 - 1 = (n-1)(n+1) \approx 3$$

and $(n^2 + 2) \approx 3$. Therefore, for gases Lorentz–Lorenz relation takes the approximate form

$$\frac{2}{3}(n-1)\frac{M}{\rho} = \text{constant} \tag{9.72}$$

9.8 MEDIUM CONTAINING FREE ELECTRONS

When a medium contains free electrons, there is no current flow due to random motion. When the electric field is applied, the electrons acquire an additional velocity component and a current results. The electrons in the medium collide with the atoms of the medium and are scattered. Obviously, damping results which depends upon the velocity of the electrons. One can write the equation of motion of an electron in the form

$$m\dot{v}_\alpha + ml_\alpha v_\alpha = eE_0 \exp(-i\omega t) \tag{9.73}$$

We must note that here we have assumed that the field varies harmonically with time. One can easily verify that the solution of Eq. (9.73) is

$$v_\alpha = \frac{E_0 \exp(-i\omega t)}{m(l_\alpha - i\omega)} \tag{9.74}$$

Now, the current density J is obtained as

$$J = \sum_\alpha ev_\alpha = Nev = \frac{Ne^2 E_0 \exp(-i\omega t)}{m(l - i\omega)} \tag{9.75}$$

While obtaining Eq. (9.75), we have assumed that there are N electrons per unit volume moving with the common velocity v, we have

$$J = \sigma E$$

$$\therefore \qquad \sigma = \frac{J}{E} = \frac{Ne^2}{m(l - i\omega)} \tag{9.76}$$

ILLUSTRATIVE EXAMPLES

Example 9.1: Calculate the value and dimensions of a Thomson scattering cross-section.

Solution: We have

$$\sigma_T = \frac{8\pi}{3}\left(\frac{1}{4\pi\varepsilon_0}\frac{e^2}{mc^2}\right)^2$$

$$\therefore \qquad \sigma_T = \frac{8}{3} \times 3.14 \times \left[9 \times 10^9 \times \frac{(1.6 \times 10^{-19})^2}{9.1 \times 10^{-31} \times (3 \times 10^8)^2}\right]^2$$

$$= 6.65 \times 10^{-20} \text{ m}^2$$

where $e = 1.6 \times 10^{-19}$ C; $m = 9.1 \times 10^{-31}$ kg; $c = 3 \times 10^8$ m/s. We have $r_0^2 = \left(\dfrac{1}{4\pi\varepsilon_0} \dfrac{e^2}{mc^2} \right)$.

This means the dimensions of σ_T are the same as that of r_0^2. We have

$$r_0 = \frac{1}{4\pi\varepsilon_0} \frac{e^2}{mc^2} \quad \text{and} \quad F = \frac{1}{4\pi\varepsilon_0} \frac{e^2}{r^2} \quad \text{(Coulomb's law)}$$

$$\therefore \qquad [r_0] = \left[\frac{Fr^2}{mc^2} \right]^2 = \frac{[MLT^{-2}]^2[L^2]^2}{[M(LT^{-1})^2]^2}$$

$$= [M^0 L^2 T^0]$$

Thus, the dimension of σ_T is $[M^0 L^2 T^0]$, i.e. that of area.

Example 9.2: Prove that the two frequencies at which the refractive index is maximum and minimum in gases are separated by a distance l.

Solution: We have

$$n^* = 1 + \frac{2\pi N e^2}{4\pi m \varepsilon_0} \left[\frac{\omega_0^2 - \omega^2}{(\omega_0^2 - \omega^2)^2 + l^2 \omega^2} \right] + \frac{i 2\pi N e^2}{4\pi m \varepsilon_0} \left[\frac{l\omega}{(\omega_0^2 - \omega^2)^2 + l^2 \omega^2} \right]$$

$$= n + i\kappa$$

where n is refractive index and κ the coefficient of absorption. The refractive index will be maximum and minimum according to the condition

$$\frac{\partial}{\partial \omega}(n) = 0$$

or $\qquad \dfrac{\partial}{\partial \omega} \left[1 + \dfrac{2\pi N e^2}{4\pi \varepsilon_0 m} \left\{ \dfrac{\omega_0^2 - \omega^2}{(\omega_0^2 - \omega^2)^2 + l^2 \omega^2} \right\} \right] = 0$

or $\qquad \dfrac{\partial}{\partial \omega} \left[\dfrac{\omega_0^2 - \omega^2}{(\omega_0^2 - \omega^2)^2 + l^2 \omega^2} \right] = 0$

We have, near absorption band $\omega \to \omega_0$, i.e.

$$(\omega_0^2 - \omega^2) = (\omega_0 + \omega)(\omega_0 - \omega) \simeq 2\omega_0(\omega_0 - \omega)$$

or $\qquad \dfrac{\partial}{\partial \omega} \left[\dfrac{2\omega_0(\omega_0 - \omega)}{4\omega_0^2(\omega_0 - \omega)^2 + l^2 \omega^2} \right] = 0$

Solving, one obtains

$$\omega = \omega_0 \pm \frac{l}{2}$$

$$\therefore \qquad \omega_1 = \omega_0 - \frac{l}{2} \quad \text{and} \quad \omega_2 = \omega_0 + \frac{l}{2}$$

$$\therefore \qquad \omega_2 - \omega_1 = l$$

Obviously, the frequencies at which refractive index is maximum and minimum are separated by a distance l on the frequency axis.

Example 9.3: Show that in the case of gases the maximum and minimum values of refractive index n occurs at the positions where the absorption coefficient reaches half its maximum value.

Solution: We have to prove that $\omega_0 - \omega_1 = \dfrac{l}{2}$. The absorption coefficient κ is given by

$$\kappa = \frac{2\pi Ne^2}{4\pi m\varepsilon_0}\left[\frac{l\omega}{(\omega_0^2 - \omega^2)^2 + l^2\omega^2}\right]$$

Now, the values of absorption coefficient at ω_1 and ω_2 are expressed as

$$\kappa_{\omega_1} = \frac{2\pi Ne^2}{4\pi m\varepsilon_0}\left[\frac{l\omega_1}{(\omega_0^2 - \omega_1^2)^2 + l^2\omega_2^2}\right]$$

$$\kappa_{\omega_2} = \frac{2\pi Ne^2}{4\pi m\varepsilon_0}\left[\frac{l\omega_2}{(\omega_0^2 - \omega_2^2)^2 + l^2\omega_2^2}\right]$$

Since ω_1 lies near ω_0, i.e. $\dfrac{l}{\omega_0} \ll 1$, one obtains

$$\kappa_{\omega_1} = \frac{2\pi Ne^2}{4\pi\varepsilon_0 m}\left[\frac{l\omega_0}{(\omega_0 - \omega_1)^2\, 4\omega_0^2 + l^2\omega_0^2}\right]$$

Now, using $\omega_0 - \omega_1 = \dfrac{l}{2}$, one obtains

$$\kappa_{\omega_1} = \frac{2\pi Ne^2}{4\pi\varepsilon_0 m}\left[\frac{l\omega_0}{\dfrac{l^2}{4}\, 4\omega_0^2 + l^2\omega_0^2}\right]$$

$$= \frac{2\pi Ne^2}{4\pi\varepsilon_0 m}\left(\frac{1}{2l\omega_0}\right) = \frac{A}{2\omega_0 l} \qquad \left(\text{here } A = \frac{2\pi Ne^2}{4\pi m\varepsilon_0}\right)$$

Similarly, we obtain

$$\kappa_{\omega_2} = \frac{A}{2l\omega_0}$$

$$\therefore \qquad \kappa_{\omega_1} = \kappa_{\omega_2} = \frac{A}{2l\varepsilon_0}$$

Absorption coefficient (κ) is maximum, when $\omega = \omega_0$

$$\therefore \qquad \kappa_{max} = \frac{2\pi Ne^2}{4\pi\varepsilon_0 m}\left[\frac{l\omega_0}{l^2\omega_0^2}\right] = A\left[\frac{1}{\omega_0 l}\right]$$

$$\therefore \qquad \kappa_{\omega_1 = \omega_2} = \frac{1}{2}\kappa_{max}$$

Example 9.4: Determine the value of l using the relation $\sigma = \dfrac{Ne^2}{ml}$, where symbols have their usual meaning. Given $N \simeq 8 \times 10^{28}$ atoms/m^3 at normal temperature and low frequency conductivity $\sigma \simeq 6 \times 10^7$ mho/m.

Solution: We have

$$\sigma = \frac{Ne^2}{ml}$$

$$\therefore \qquad l = \frac{Ne^2}{m\sigma} = \frac{8 \times 10^{28} \times (1.6)^2 \times 10^{-38}}{9 \times 10^{-31} \times 6 \times 10^7} = 0.38 \times 10^{15}$$

Example 9.5: Compute the group velocity for very high frequency electromagnetic radiations such as X-rays.

Solution: If ω is much larger than the characteristic frequency ω_α, we may neglect ω_α s and have

$$n^2 = 1 - \frac{Ne^2}{m\varepsilon_0 \omega^2} \sum_\alpha f_\alpha = 1 - \frac{Ne^2}{m\varepsilon_0 \omega^2}$$

Since $\sum_\alpha f_\alpha = 1$, using the approximation $(1-x)^{\frac{1}{2}} = 1 - \dfrac{1}{2}x + \cdots$ for $x \ll 1$, we have

$$n = 1 - \frac{Ne^2}{2\varepsilon_0 m\omega^2}$$

which is less than 1, giving $v > c$. Now $\dfrac{dn}{d\omega} = \dfrac{Ne^2}{\varepsilon_0 m\omega^2}$. Substituting in the value, we have

$$n^2 = \varepsilon_r = 1 + \frac{Ne^2}{m\varepsilon_0} \left(\sum_\alpha \frac{f_\alpha}{\omega_\alpha^2 - \omega^2} \right)$$

one obtains

$$v_g = \frac{c}{n + \omega\left(\dfrac{dn}{d\omega}\right)}$$

$$= \frac{c}{1 - \left(\dfrac{Ne^2}{2\varepsilon_0 m\omega^2}\right) + \omega\left(\dfrac{Ne^2}{\varepsilon_0 m\omega^3}\right)}$$

$$= \frac{c}{1 + \left(\dfrac{Ne^2}{2\varepsilon_0 m\omega^2}\right)}$$

Therefore, although the phase velocity v is larger than c because n is less than 1, group velocity v_g is less than c. For the phase velocity, we have

$$v = \frac{c}{n} = \frac{c}{1 - \left(\dfrac{Ne^2}{2\varepsilon_0 m\omega^2}\right)} \approx c\left[1 + \left(\frac{Ne^2}{2\varepsilon_0 m\omega^2}\right)\right]$$

Obviously, $v_g v = c^2$

This relation although not of general validity, is satisfied to a very good approximation over a wide range of frequencies.

GLIMPSES

1. Differential scattering cross-section,

$$\frac{d\sigma}{d\Omega} = \left(\frac{q_2}{4\pi\varepsilon_0 mc^2}\right)^2 \sin^2\theta$$

Here θ is the angle between the induced dipole p and the direction of the outgoing radiation, i.e. it is the angle between the electric field E and the unit vector $\hat{a}_r$.

2. The scattering of an unpolarized plane wave by a free charged particle is known as *Thomson scattering*.

$$<\sigma>_{\text{unpolarised}} = \frac{8\pi}{3}\left(\frac{q^2}{4\pi\varepsilon_0 mc^2}\right)^2$$

If the charged particle is an electron $q = e$, and

$$<\sigma> = \frac{8\pi}{3}\left(\frac{e^2}{4\pi\varepsilon_0 mc^2}\right)^2 = \frac{8\pi}{3}r_0^2$$

where $r_0 = \dfrac{e^2}{(4\pi\varepsilon_0 mc^2)}$ is called the *classical radius of the electron*.

3. $\sigma_{\text{bound}} = \sigma\dfrac{\omega^2}{(\omega_0^2 - \omega^2)^2 - l^2\omega^2}$, where σ is the scattering cross-section for a free electron.

$$(\sigma_{\text{bound}})_{\text{max}} = \sigma\frac{\omega^2}{l^2} \qquad (\because \omega = \omega_0)$$

If $\omega \ll \omega_0$ and $\omega_1 \ll \omega_0^2$, then

$$\sigma_{\text{bound}} \overset{\pm}{=} \sigma\frac{\omega^4}{\omega_0^4}$$

This is the result for *Rayleigh scattering*.

4. The reaction of the radiation on the motion of the particle is called *radiation reaction* or *radiation damping*.

5. When electromagnetic waves propagate through a dilute gas in which the mutual interactions between the constituent particles are neglected, the total dipole moment per unit volume is given as

$$P = E\sum_\alpha \frac{(e^2/m)Nf_\alpha}{(\omega_0^2 - \omega^2) - il_d\omega}$$

where f_α is known as oscillator strengths and satisfy the sum rule $\sum_\alpha f_\alpha = 1$.

The dispersion relation for a dilute gas is

$$n^2 = \varepsilon = 1 + \sum_\alpha \frac{(e^2/m\varepsilon_0)Nf_\alpha}{(\omega_\alpha^2 - \omega^2) - il_\alpha\omega}$$

6. *Lorentz–Lorenz formula* for dispersion in liquids and solids

$$\left(\frac{n^2 - 1}{n^2 + 2}\right)\frac{1}{\rho} = A$$

where $A = \left(\dfrac{\varepsilon/\varepsilon_0 - 1}{\varepsilon/\varepsilon_0 + 2}\right)\dfrac{1}{\rho}$ is called the *atomic refractivity*.

$\left(\dfrac{n^2 - 1}{n^2 + 2}\right)\dfrac{W}{e} = A_m$, where $N = \dfrac{\rho N_a}{W}$ and N_a is Avogadro's number. A_m is known as molecular refractivity.

REVIEW QUESTIONS

1. Define scattering cross-section. Discuss different types of scattering by considering the incidence of electromagnetic wave on bound electrons. Why does the sky appears blue?
2. What is the difference between Thomson, Rayleigh and resonance scattering of radiations? Derive the expression for the differential cross-section for Thomson scattering of radiation by electrons.
3. Outline Thomson's theory of scattering of X-rays by free electron and discuss its success and limitations.
4. What is the effect of radiation damping on the monochromatic nature of radiations emitted by a harmonic oscillator?
5. Show that the scattering cross-section of electromagnetic radiation by a bound electron depends upon the inverse fourth power of the incident wavelength.
6. The intensity of X-rays scattered by an electron at a distance r from it and making an angle ϕ with the original direction is given by

$$I_c = \left[\frac{e^2}{4\pi\varepsilon_0 mc^2}\right]\frac{I}{r^2}\frac{1}{2}(1 + \cos^2\phi)$$

Using the above result:
   i. Obtain an expression for the coefficient of scattering and its dimensions.
   ii. Calculate the intensity of scattered radiation I_s in terms of I_e for n-electrons: (a) when they scatter independently and (b) when all of them act as a single scattering centre.

   iii. Plot schematic curves to illustrate the variations in I_s with ϕ in the two cases in part (ii) above for $n = 3$.

7. Obtain the total scattering cross-section for scattering of a plane monochromatic wave by a charge carrying out small vibrations under an elastic force (oscillator type).

8. Explain the following on the basis of Rayleigh scattering:
   i. Red light is used for danger signals while the eye is most sensitive to yellow green
   ii. The clear sky appears blue
   iii. The rising or setting sun appears red.

9. Find the differential scattering cross-section for the scattering of an elliptically polarized wave by a free charged particle.

10. What do you understand by dispersion? Discuss it in the case of gaseous medium having both real and complex refractive indices.

11. Explain normal and anomalous dispersion. Derive Sellmeier's equation for the refractive index of a dielectric medium.

12. What do you understand by anomalous dispersion. Give Lorentz's theory of dispersion and explain the connection between absorption and anomalous dispersion.

13. Give Lorentz's theory of dispersion and show that in the case of non-polar dielectrics

$$\frac{n^2 - 1}{n^2 + 1}\frac{M}{\rho} = \text{constant}$$

where M is the molecular weight of the dielectric, ρ its density and n the refractive index.

14. Explain dispersion. From the theory of dispersion, show that in the case of transparent gaseous substance in the region remote from the natural frequencies the index of refraction is governed approximately by the relation

$$n^2 = A + \frac{B}{\lambda^2} + \frac{C}{\lambda^4}$$

where A, B, and C are constants, and λ is the wavelength of incident light. Obtain dispersive power relation for the medium.

15. Write short notes on the following:
   i. Rayleigh scattering
   ii. Thomson scattering
   iii. Scattering of electromagnetic waves
   iv. Normal and anomalous scattering

16. Show that for gases,

$$n \approx 1 + \frac{Ne^2}{2m\varepsilon_0}\sum_\alpha \left(\frac{f_\alpha}{\omega_\alpha^2 - \omega^2}\right)$$

For only one resonant frequency, the expression becomes

$$n \approx 1 + \frac{Ne^2}{2m\varepsilon_0(\omega_\alpha^2 - \omega^2)}.$$

PROBLEMS

1. The index of refraction of hydrogen gas at STP is $n = 1 + 1.400 \times 10^{-4}$ at $\lambda = 5.46 \times 10^{-7}$m and $n = 1 + 1.547 \times 10^{-4}$ at $\lambda = 2.54 \times 10^{-7}$ m. Assuming a single resonant frequency, compute this frequency and the number of electronic oscillators per unit volume.

   [*Ans.* 2.11×10^{-16}/s and 3.80×10^{25}]

2. Consider a gas whose molecules behave as dipole oscillators, with a restoring constant $k = 3 \times 10^2$ kg/s^2. The oscillating particles are electrons. Compute their characteristic frequency. Write the index of refraction of the gas as a function of the frequency, assuming that the gas is at STP. Obtain the values of the index for $\lambda = 5 \times 10^{-7}$ m and $\lambda = 10^{-2}$ m.

   $$[\textit{Ans. } 1.82 \times 10^{-16}/s, n = \frac{1 + 6.05 \times 10^{28}}{(3.29 \times 10^{-32} - \omega^2)} \text{ and } 1.00019, 1.00018]$$

3. For the *D*-lines of sodium, the following values of constants in the dispersion formula are found

   $$\omega_0 = 3 \times 10^{16}, \quad \alpha = 2 \times 10^{10} \quad \text{and} \quad \frac{Ne^2}{m\varepsilon_0} = 10^{23}$$

   Plot the refractive index (n) and absorption coefficient as a function of frequency of electromagnetic waves. Find the maximum and minimum values of n. Also find the maximum values of absorption coefficient and the half width of the absorption band.

4. Referring to the problem 1 compute the index of refraction of hydrogen for $\lambda = 4 \times 10^{-7}$ m, a pressure of 10 atm and a temperature of 300 K.

SHORT ANSWER QUESTIONS

1. What is absolute index of refraction?

   Ans. The ratio of the velocity of electromagnetic waves in vacuum (c) to the velocity (v) in matter, is called the absolute index of refraction of the substance usually designated by n.

2. What is the relation between the absolute index of refraction (n) and relative permittivity (ε_r) of the medium.

   Ans. $n = \sqrt{\varepsilon_r}$

3. What is classical electron radius?

   Ans. $r_0 = \dfrac{e^2}{4\pi\varepsilon_0 mc^2}$, where symbols have usual meanings.

4. Why sun appears red at sunset?

   Ans. As the light passes through the atmosphere, it looses heavily from the blue end of the spectrum due to scattering.

5. In the absence of atmospheric scattering, how the sky will appear?

   Ans. Dark

6. What is radiation reaction?

   Ans. The reaction of the radiation on the motion of the particle is called radiation reaction or radiation damping.

7. When a dispersion is said to be normal?

Ans. $\dfrac{dn}{d\omega} > 1$

8. What is the condition for anomalous dispersion?

Ans. $\dfrac{dn}{d\omega} < 1$. The ordering of prismatic colours is reversed.

9. What does the condition $\dfrac{dn}{d\omega} > 1$ indicates?

Ans. Dispersion is normal.

10. How will you interpret the regions where $I_m\varepsilon > 0$?

Ans. The regions where $I_m\varepsilon$ is large are called regions of *resonant absorption*.

11. What is the significance of the condition $I_m\varepsilon < 0$?

Ans. This shows that energy is given to the wave by medium resul-ting into the amplification of waves. Examples are masers and lasers.

MULTIPLE CHOICE QUESTIONS

1. Thomson scattering coefficient of an unpolarized plane electromagnetic wave scattered by a free charge particle having charge q and mass m is

    (a) $\langle\sigma\rangle = \dfrac{8\pi}{3}\left(\dfrac{q^2}{4\pi\varepsilon_0 mc^2}\right)$

    (b) $\langle\sigma\rangle = \dfrac{8\pi}{3}\left(\dfrac{q^2}{4\pi\varepsilon_0 mc^2}\right)^2$

    (c) $\langle\sigma\rangle = \dfrac{8\pi}{3}\left(\dfrac{q^2}{4\pi\varepsilon_0 mc^2}\right)^{\frac{3}{2}}$

    (d) $\langle\sigma\rangle = \dfrac{8\pi}{3}\left(\dfrac{q^2}{4\pi\varepsilon_0 mc^2}\right)^{\frac{1}{2}}$ **[b]**

2. The maximum value of scattering coefficient for the scattering of radiation by a bound charge of frequency ω_0 occurs at ($\omega \rightarrow$ frequency of a free electron)

    (a) $\omega > \omega_0$ (b) $\omega < \omega_0$ (c) $\omega = \omega_0$ (c) $\omega = \dfrac{\omega_0}{2}$ **[c]**

3. If ω is the frequency of a free electron, ω_0 is the frequency of bound charge and σ is the scattering cross-section, then Rayleigh scattering cross-section is given by

    (a) $\sigma_{\text{bound}} = \dfrac{\sigma\omega_0^4}{\omega^4}$ (b) $\sigma_{\text{bound}} = \sigma\dfrac{\omega^4}{\omega_0^4}$ (c) $\sigma_{\text{bound}} = \dfrac{\sigma\omega_0^2}{\omega^2}$ (d) $\sigma_{\text{bound}} = \sigma\dfrac{\omega^2}{\omega_0^2}$ **[b]**

4. The dimensions of Thomson scattering coefficient are

    (a) $[M^0 L^2 T^0]$ (b) $[M^1 L^1 T^0]$ (c) $[M^0 L^3 T^0]$ (d) $[M^0 L^{\frac{3}{2}} T^{\frac{1}{2}}]$ **[a]**

5 If r_0 is classical electron radius, ω free electron frequency and ω_0 being the frequency of bound charge, then the resonant scattering cross-section is

    (a) $\sigma = \dfrac{8}{3}\pi r_0^2\left(\dfrac{\omega_0}{\omega}\right)^4$

    (b) $\sigma = \dfrac{8}{3}\pi r_0^3\left(\dfrac{\omega_0}{\omega}\right)^2$

(c) $\sigma = \dfrac{8}{3}\pi r_0^2 \left(\dfrac{\omega_0}{\omega}\right)^2$ \qquad\qquad (d) $\sigma = \dfrac{8}{3}\pi r_0^2 \left(\dfrac{\omega_0}{\omega}\right)$ \hfill **[c]**

6. To a close approximation, dispersion varies inversely as
   (a) wavelength of radiation
   (b) square of the wavelength of radiation
   (c) cube of the wavelength of the radiátion
   (d) none of the above \hfill **[c]**

7. The Lorentz–Lorenz formula for the dispersion in liquids and solids is (symbols have usual meanings)

   (a) $\dfrac{n^2-1}{n^2+2}\dfrac{1}{\rho} = A^2$ \qquad\qquad (b) $\dfrac{n^2-1}{n^2+2}\dfrac{1}{\rho^2} = A$

   (c) $\dfrac{n^2-1}{n^2+2}\dfrac{1}{\sqrt{\rho}} = A$ \qquad\qquad (d) $\dfrac{n^2-1}{n^2+2}\dfrac{1}{\rho} = A$ \hfill **[d]**

8. The dispersion relation for a dilute gas is (symbols have usual meanings)

   (a) $n^2 = 1 + \displaystyle\sum_\alpha \dfrac{\left(\dfrac{e^2}{m\varepsilon_0}\right)N f_\alpha}{(\omega_\alpha^2 - \omega^2) - il_\alpha\omega}$ \qquad (b) $n^2 = \displaystyle\sum_\alpha \dfrac{\left(\dfrac{e^2}{m\varepsilon_0}\right)N f_\alpha}{(\omega_\alpha^2 - \omega^2) - il_\alpha\omega^2}$

   (c) $n^2 = \displaystyle\sum_\alpha \dfrac{\left(\dfrac{e^2}{m\varepsilon_0}\right)^2 N f_\alpha}{(\omega_\alpha^2 - \omega^2)^2 - il_\alpha\omega^2}$ \qquad (d) $n^2 = \displaystyle\sum_\alpha \dfrac{\left(\dfrac{e^2}{m\varepsilon_0}\right)N f_\alpha}{\omega_\alpha^2 - \omega^2}$ \hfill **[a]**

9. Molecular refractivity of a substance with molecular weight W is given by (symbols have usual meanings)

   (a) $A_m = \dfrac{W}{\rho}\left(\dfrac{n^2-1}{n^2+2}\right)^2$ \qquad\qquad (b) $A_m = \dfrac{W}{\rho}\left(\dfrac{n^2-1}{n^2+2}\right)$

   (c) $A_m = \dfrac{W}{\rho}\left(\dfrac{n^2-1}{n^2+2}\right)^{\frac{1}{2}}$ \qquad\qquad (d) $A_m = \sqrt{\dfrac{W}{\rho}}\left(\dfrac{n^2-1}{n^2+2}\right)$ \hfill **[b]**

10. The velocity of light in a particular medium is 10^8 m/s. What is the relative permittivity of the medium?
    (a) 1.732 \qquad\qquad\qquad (b) 3
    (c) 9 \qquad\qquad\qquad (d) 0.333 \hfill **[IEEE] [c]**

[**Hint:** $n \approx \sqrt{\varepsilon_r} = \dfrac{c}{v}$ $\quad \therefore\ \varepsilon_r = \dfrac{c^2}{v^2} = \dfrac{(3\times10^8)^2}{(10^8)^2} = 9$].

Transmission Lines

10.1 INTRODUCTION

Transmission line is the communication link between a transmitter and a receiver. These lines are used to transmit electric energy and signals from one point to another, specifically from a source to a load, e.g. connection between a transmitter and an antenna, connections between computers in a network, etc. Basically, one deals with wave phenomena in transmission lines, in the same manner that one will find with point to point energy propagation in free space or in dielectrics. Depending on the frequency of transmission signal, one may classify transmission lines as follows:

i. *Two-wire transmission line:* This is an open wire line, i.e. two parallel insulationless conductors used for low frequency transmission of signals in telegraphy, and power lines as shown in Fig. 10.1.

ii. *Coaxial type line:* Here the main conductor is inside a hollow conductor, i.e. both the conductors are coaxial (Fig. 10.2). Because of its shielded nature, the coaxial line is preferred for use at microwave frequencies. The section of these coaxial lines are shown in Fig. 10.3.

iii. *Waveguide type line:* Waveguides are single hollow conductors, rectangular or circular in shape and filled with air or any other dielectric, capable of guiding high frequency electromagnetic waves through them.

Fig. 10.1: Two-wire transmission line

Metal waveguides have less loss and can handle greater power than coaxial lines, but they are much larger in size than coaxial lines at microwave frequencies. Figure 10.4 shows a rectangular waveguide.

Fig. 10.2: Coaxial transmission line

Fig. 10.3: Section of line

Fig. 10.4: Rectangular waveguide

10.2 DISTRIBUTED PARAMETERS

The incremental parameters per unit length of line are inductance and capacitance, the resistance of the conductors, and the conductance of the dielectric medium. It was seen that the parameters depend on the geometry of the configuration, the characteristics of the material, and in some cases the frequency. In the following summary list, the dependence on geometry is represented by a geometrical factor GF.

Capacitance:

$$C = \pi \varepsilon_d (\text{GFC}) \text{ (F/m) } [\varepsilon_d = \text{permittivity of dielectric}]$$

Conductance:

$$G = \frac{C}{\varepsilon_d} \sigma_d \text{ (S/m) } [\sigma_d = \text{conductivity of dielectric}]$$

Inductance (external):

$$L_e = \frac{\mu_d}{\pi} (\text{GFL}) \text{ (H/m) } [\mu_d = \text{permeability of dielectric} \approx \mu_0]$$

DC resistance (useful for operation up to 10 kHz):

$$R_d = \frac{1}{\sigma_c \pi} (\text{GER}_d) \text{ (}\Omega\text{/m) } [\sigma_c = \text{conductivity of conductors}]$$

AC resistance (for frequencies above 10 kHz):

$$R_a = \frac{1}{2\pi\sigma_c\delta} (\text{GER}_a) \text{ (}\Omega\text{/m) } \left[\delta = \frac{2}{\sqrt{\pi f \mu_c \sigma_c}} = \text{skin septh} \right]$$

Inductance (internal):

$$L_i = \begin{cases} R_a / 2\pi f \text{ (H/m) for } f > 10 \text{ kHz} \\ \mu_0 / 4\pi \text{ (H/m) } \text{ for } f > 10 \text{ kHz} \end{cases}$$

Inductance (total):

$$L_i = L_e + L_i \approx L_e .$$

For three common line configurations, the geometrical factors are as follows:

Coaxial line (inner radius a, outer radius b, outer thickness t):

$$\text{GFC} = \frac{2}{\ln(b/a)} ; \qquad \text{GFL} = \frac{1}{\text{GFC}}$$

$$\text{GFR}_d = \frac{1}{a^2} + \frac{1}{t(b+t)} \qquad \text{GER}_a = \frac{1}{a} + \frac{1}{b} \quad \text{for } t \gg \delta$$

Parallel wires (radius a, separation d):

$$\text{GFC} = \frac{1}{\text{GFL}} \qquad\qquad \text{GFL} = \cosh^{-1}\frac{d}{2a} \approx \ln\frac{d}{a} \quad \text{for } d \gg a$$

$$\text{GFR}_d = \frac{2}{a^2} \qquad\qquad \text{GFR}_a = \frac{2}{a}$$

Parallel plates (width W, thickness t, separation d):

$$\text{GFC} = \frac{W}{\pi d} \qquad\qquad \text{GFL} = \frac{1}{\text{GFC}}$$

$$\text{GFR}_d = \frac{2\pi}{Wt} \qquad\qquad \text{GFR}_d = \frac{4\pi}{W} \quad \text{for } t \gg \delta$$

10.3 TRANSMISSION LINE EQUATIONS

If a long line consisting of two parallel or coaxial uniform conductors (Figs 10.1 and 10.3) is carrying current, there is a magnetic field around the conductors and a voltage drop along them. The magnetic field, which is proportional to current, indicates that the line has series inductance L and the voltage drop indicates the presence of series resistance R. The voltage applied across the conductors produces an electric field between them and charges on them. This indicates that the line contains shunt capacitance C, and since a capacitor cannot be lossless, thus some shunt conductance G as well. Actually, R, G, C and L are not lumped parameters, but they are distributed along the whole length of the line. The equivalent circuit is shown in Fig. 10.5.

Fig. 10.5: Equivalent circuit of two-wire transmission line

Thus,   total series impedance $Z = R + j\omega L$

and   total shunt admittance $Y = G + j\omega C$

Now, from the equivalent circuit of Fig. 10.3, we may write

$$V(Z + \Delta Z, t) = V(Z,t) + \frac{\partial V(Z,t)}{\partial Z}\Delta Z \tag{10.1}$$

and

$$I(Z + \Delta Z, t) = I(Z,t) + \frac{\partial I(Z,t)}{\partial Z}\Delta Z \tag{10.2}$$

(Using Taylor series and neglecting higher power terms)

Applying Kirchhoff's law and using Eqs. (10.1) and (10.2), we obtain

$$-\frac{\partial V(Z,t)}{\partial Z} = RI(Z,t) + L\frac{\partial I(Z,t)}{\partial t} \tag{10.3}$$

and

$$-\frac{\partial I(Z,t)}{\partial Z} = GV(Z,t) + C\frac{\partial V(Z,t)}{\partial t} \tag{10.4}$$

Differentiating Eq. (10.3) with respect to Z and Eq. (10.4) with respect to t, we obtain

$$-\frac{\partial^2 V(Z,t)}{\partial Z^2} = R\frac{\partial I(Z,t)}{\partial Z} + L\frac{\partial^2 I(Z,t)}{\partial Z \partial t} \qquad (10.5)$$

and

$$-\frac{\partial^2 I(Z,t)}{\partial Z \partial t} = G\frac{\partial V(Z,t)}{\partial t} + C\frac{\partial^2 V(Z,t)}{\partial t^2} \qquad (10.6)$$

Substituting the value of $\dfrac{\partial I}{\partial Z}$ from Eq. (10.4) in Eq. (10.5) and omitting (Z, t), we obtain

$$-\frac{\partial^2 V}{\partial Z^2} = -RGV - RC\frac{\partial V}{\partial t} + L\frac{\partial^2 I}{\partial Z \partial t} \qquad (10.7)$$

Now, substituting the value of $\dfrac{\partial^2 I}{\partial Z \partial t}$ from Eq. (10.6) in Eq. (10.7), we obtain

$$-\frac{\partial^2 V}{\partial Z^2} = -RGV - RC\frac{\partial V}{\partial t} - LG\frac{\partial V}{\partial t} - LC\frac{\partial^2 V}{\partial t^2}$$

or

$$\frac{\partial^2 V}{\partial Z^2} = RGV + (RC + LG)\frac{\partial V}{\partial t} + LC\frac{\partial^2 V}{\partial t^2} \qquad (10.8)$$

Similarly by differentiating Eq. (10.3) with respect to t and Eq. (10.4) with respect to Z, we obtain

$$\frac{\partial^2 I}{\partial Z^2} = RGI + (RC + LG)\frac{\partial I}{\partial t} + LC\frac{\partial^2 I}{\partial t^2} \qquad (10.9)$$

Equations (10.8) and (10.9) are known as **transmission line equations**.

Now, the voltage and current on the line are function of both position Z and time t. Thus, the instantaneous line voltage and current can be expressed as

$$\left. \begin{array}{l} V(Z,t) = \mathrm{Re}\, V(Z)e^{j\omega t} \\ I(Z,t) = \mathrm{Re}\, I(Z)e^{j\omega t} \end{array} \right\} \qquad (10.10)$$

Differentiating Eq. (10.10) with respect to t, we obtain

$$\left. \begin{array}{l} \dfrac{\partial V}{\partial t} = j\omega\, \mathrm{Re}\, V(Z)e^{j\omega t} = j\omega V(Z,t) \\[2mm] \text{then} \quad \dfrac{\partial^2 V}{\partial t^2} = -\omega^2\, \mathrm{Re}\, V(Z)e^{j\omega t} = -\omega^2 V(Z,t) \end{array} \right\} \qquad (10.11)$$

Similarly,

$$\left. \begin{array}{l} \dfrac{\partial I}{\partial t} = j\omega\, \mathrm{Re}\, I(Z)e^{j\omega t} = j\omega I(Z,t) \\[2mm] \text{then} \quad \dfrac{\partial^2 I}{\partial t^2} = -\omega^2\, \mathrm{Re}\, I(Z)e^{j\omega t} = -\omega^2 I(Z,t) \end{array} \right\} \qquad (10.12)$$

Combining Eqs. (10.8) and (10.11), the transmission line equation can be written as

$$\frac{\partial^2 V(Z)}{\partial Z^2} = RGV(Z) + (RC + LG)\, j\omega V(Z) - \omega^2\, LCV(Z)$$

or
$$\frac{\partial^2 V(Z)}{\partial Z^2} = RGV(Z) + j\omega RCV(Z) + j\omega LGV(Z) - \omega^2 LCV(Z)$$

or
$$\frac{\partial^2 V(Z)}{\partial Z^2} = R[G + j\omega C]\, V(Z) + j\omega L\, [G + j\omega C]\, V(Z)$$

or
$$\frac{\partial^2 V(Z)}{\partial Z^2} = [R + j\omega L]\,[G + j\omega C]\,V(Z) \tag{10.13}$$

Similarly, combining Eqs. (10.9) and (10.12), the transmission line equation in terms of current can be written as

$$\frac{\partial^2 I(Z)}{\partial Z^2} = [R + j\omega L]\,[G + j\omega C]\,I(Z) \tag{10.14}$$

Equations (10.13) and (10.14) are second order differential equations and their solutions are given by

$$V(Z) = V^+ e^{-\gamma Z} + V^- e^{\gamma Z} \tag{10.15}$$
and
$$I(Z) = I^+ e^{-\gamma Z} + I^- e^{\gamma Z} \tag{10.16}$$

where V^+, V^-, I^+ and I^- are arbitrary constants and γ is called the propagation constant.

Now, second order derivatives of Eqs. (10.15) and (10.16) give us

$$\frac{\partial^2 V(Z)}{\partial Z^2} = \gamma^2 [V^+ e^{-\gamma Z} + V^- e^{\gamma Z}] = \gamma^2 V(Z) \tag{10.17}$$

and
$$\frac{\partial^2 I(Z)}{\partial Z^2} = \gamma^2 [I^+ e^{-\gamma Z} + I^- e^{\gamma Z}] = \gamma^2 I(Z) \tag{10.18}$$

Combining Eqs. (10.13) and (10.17) or Eqs. (10.14) and (10.18), we obtain

$$\gamma = \sqrt{(R + j\omega L)(G + j\omega C)} \tag{10.19}$$

In general, γ is a complex quantity and may be expressed as $\alpha + j\beta$, where α is called *attenuation constant* and β is called *phase constant*.

$$\gamma = \alpha + j\beta = \sqrt{(R + j\omega L)(G + j\omega C)}$$

This leads to
$$\gamma = \sqrt{(RG - \omega^2 LC) + j\omega(LG + RC)} \tag{10.20}$$

so that

$$\alpha = \frac{1}{\sqrt{2}} \sqrt{\sqrt{(RG - \omega^2 LC)^2 + \omega^2 (RC + LG)^2} + (RG - \omega^2 LC)} \tag{10.21}$$

and $\qquad \beta = \dfrac{1}{\sqrt{2}} \sqrt{\sqrt{(RG - \omega^2 LC)^2 + \omega^2 (RC + LG)^2} + (RG - \omega^2 LC)}$ $\qquad$ (10.22)

Condition for Minimum Attenuation

Minimum α with respect to L

In a transmission line when R, G and C are constants then α depends on L only and the condition for minimum attenuation is expressed as

$$\frac{d\alpha}{dL} = 0 \qquad (10.23)$$

Substituting the value of α from Eq. (10.21) in Eq. (10.23), one obtains

$$\frac{d\alpha}{dL} = \frac{1}{2} \frac{\dfrac{1}{2}\left[\omega^2 L \sqrt{\dfrac{G^2 + \omega^2 C^2}{R^2 + \omega^2 L^2}} - \omega^2 C \right]}{\alpha} = 0$$

This yields

$$C^2 (R^2 + \omega^2 L^2) = L^2 (G^2 + \omega^2 C^2)$$

or $\qquad\qquad RC = LG$

or $\qquad\qquad L = \dfrac{R}{G} C$

Hence $\qquad\qquad \alpha_{min} = \sqrt{RG}$

Minimum α with respect to C

If instead of L, C of the lumped circuit for *transmission* line varies, then condition for minimum attenuation is

$$\frac{d\alpha}{dC} = 0 \qquad (10.24)$$

Substituting the value of α from Eq. (10.21) in Eq. (10.24), one obtains

$$\frac{d\alpha}{dC} = \frac{1}{2} \frac{\dfrac{1}{2}\left[\omega^2 C \sqrt{\dfrac{R^2 + \omega^2 L^2}{G^2 + \omega^2 C^2}} - \omega^2 L \right]}{\alpha} = 0$$

This yields

$$C^2 (R^2 + \omega^2 L^2) = L^2 (G^2 + \omega^2 C^2)$$

or $\qquad\qquad RC = LG$

or $\qquad\qquad C = \dfrac{G}{R} L$

Hence $\qquad\qquad \alpha_{min} = \sqrt{RG}$

Minimum α with respect to R and G

α will be minimum with respect to R and G when

$$\frac{d\alpha}{dR} = \frac{1}{2}\frac{\frac{1}{2}\left[R\sqrt{\frac{G^2 + \omega^2 C^2}{R^2 + \omega^2 L^2}} + G\right]}{\alpha} = 0 \qquad (10.25)$$

and

$$\frac{d\alpha}{dG} = \frac{1}{2}\frac{\frac{1}{2}\left[G\sqrt{\frac{R^2 + \omega^2 L^2}{G^2 + \omega^2 C^2}} + R\right]}{\alpha} = 0 \qquad (10.26)$$

As R and G are non-negative quantities, thus numerators of Eqs. (10.25) and (10.26) can never be equal to zero unless both R and G are zero, then the minimum value of α = α_{min} = 0. But $G = 0$ means no transmission and in practice R cannot be zero. Thus, the minimum value of attenuation constant is given by

$$\alpha_{min} = \sqrt{RG} \qquad (10.27)$$

Condition for No Distortion

The distortion in a transmission line arises due to the facts that α is frequency dependent and β does not vary linearly with frequency. From Eqs. (10.21) and (10.22), it is clear that distortion is absent; if $R = 0$ and $G = 0$, then $\alpha = 0$ and $\beta = \omega\sqrt{LC}$ for all values of ω. In practice, to make $R = 0$ is practically impossible, thus distortion is absent if

$$\sqrt{(RG - \omega^2 LC)^2 + \omega^2(RC - LG)^2} \cong K + \omega^2 LC \qquad (10.28)$$

where K is a constant and does not depend on frequency. Squaring both sides of Eq. (10.28) and equating the coefficients of like powers of ω on the two sides of the resulting equation, we get

$$K^2 = R^2 G^2 \qquad (10.29)$$

and

$$2LCK = R^2 C^2 + L^2 G^2 \qquad (10.30)$$

As R, C, L and G cannot be negative, hence $K = RG$ and substituting it in Eq. (10.30), one obtains

$$(RC - LG)^2 = 0$$

or

$$RC = LG$$

or

$$\frac{R}{G} = \frac{L}{C} \qquad (10.31)$$

Using Eq. (10.31) in Eq. (10.21), we get $\alpha = \sqrt{RG}$. This is the condition for *minimum attenuation* as obtained in Eq. (10.27). Thus, Eq. (10.31) is not only the condition for zero distortion but also the condition for minimum attenuation.

10.4 CHARACTERISTIC IMPEDANCE (Z_0)

Any circuit, which consists of series and shunt impedances, must have input impedance for the transmission line. This input impedance will depend on the type of the line, its length and the termination at the far end, and is termed characteristic impedance Z_0 when it is measured at the input transmission line of infinite length. Under this condition, the termination at the far end of the transmission line has no effect and hence not mentioned in the definition.

From the analysis of the two-port network of circuit theory,

$$Z_0 = \frac{V}{I} = \sqrt{\frac{Z}{Y}} = \sqrt{\frac{R + j\omega L}{G + j\omega C}} \tag{10.32}$$

For good transmission line, the series resistance R and shunt conductance G are negligible. Then Eq. (10.32) reduces to

$$Z_0 = \sqrt{\frac{L}{C}} \tag{10.33}$$

Physically, characteristic impedance is determined by the size and spacing of the conductors, and the dielectric constant (for coaxial line) of the insulator separating them.

For the parallel wire line of Fig. 10.1, we have

$$Z_0 = 276 \log \frac{2S}{d} \, \Omega \tag{10.34}$$

For the coaxial line of Fig. 10.3, one obtains

$$Z_0 = \frac{138}{\sqrt{K}} \log \frac{b}{a} \, \Omega \tag{10.35}$$

where K is the dielectric constant of the insulating material.

The normal range of Z_0 for a parallel wire line is $150\,\Omega$ to $600\,\Omega$, and for a coaxial line is $40\,\Omega$ to $150\,\Omega$ and is limited by the geometry of the line.

10.5 LOSSES IN TRANSMISSION LINE

Copper loss

It is defined as the power dissipated in the form of heat in the metal conductors and expressed in watt. Here the resistance arises due to *skin effect*.

Dielectric loss

It is the heat loss in the insulation.

Radiation loss

It is defined as the power loss during the transfer of energy from the transmission line to free space through an antenna.

All the three losses are frequency dependent and increase with frequency.

10.6 REFLECTION COEFFICIENT

From maximum power transfer theorem, the maximum microwave power can be transferred from the input to the output of a transmission line, when the output is terminated by the characteristic impedance Z_0 of the network and then the network is called properly terminated or matched or flat line or non-resonant line. But if the transmission line is incorrectly terminated or if there is any discontinuity in the line, then the power is not absorbed by the load and is sent back towards the generator so that an obvious inefficiency exists.

The reflection coefficient Γ can be defined as the ratio of the amplitude of the reflected wave to that of the incident wave, i.e.

$$G = \frac{V^-}{V^+}$$

Now, the propagation function is $\gamma = \alpha + j\beta$.

For a lossless transmission line, $\alpha = 0$, then

$$\gamma = j\beta$$

Equations (10.15) and (10.16) reduce to

$$V(Z) = V^+ e^{-j\beta Z} + V^- e^{j\beta Z}$$

$$= V^+ e^{-j\beta Z}[1 + \Gamma e^{2j\beta Z}] \tag{10.36}$$

and

$$I(Z) = I^+ e^{-j\beta Z} + \Gamma e^{j\beta Z}$$

$$= Y_0 V^+ e^{-j\beta Z}[1 - \Gamma e^{2j\beta Z}] \tag{10.37}$$

If the transmission line is terminated by a load impedance of Z_L, then

$$Z_L = \frac{V(Z)}{I(Z)} = \frac{1}{Y_0} \frac{[1 + \Gamma e^{2j\beta Z}]}{[1 - \Gamma e^{2j\beta Z}]}$$

At the load end, $Z = 0$

$$Z_L = Z_0 \frac{1 + \Gamma}{1 - \Gamma} \tag{10.38}$$

or

$$\Gamma = \frac{Z_L - Z_0}{Z_L + Z_0} \tag{10.39}$$

From Eq. (10.39), it is clear that when $Z_L = Z_0$, i.e. when the load impedance is equal to the characteristic impedance, then there is no reflection ($\Gamma = 0$) and hence the whole input power is transferred to output load Z_L without any reflection.

Now, if $Z_L = \infty$, and not equal to Z_0, then

$$\Gamma = \frac{\infty - Z_0}{\infty + Z_0} = \frac{1 - \dfrac{Z_0}{\infty}}{1 + \dfrac{Z_0}{\infty}} = 1$$

And for $Z_L = 0$ (and equal to Z_0),

$$\Gamma = \frac{0 - Z_0}{0 + Z_0} = -1 \text{ or } |\Gamma| = 1$$

Thus, when the output is open-circuited ($Z_L = \infty$) or short-circuited ($Z = 0$), $\Gamma = 1$ means *total reflection*.

In other cases, where Z_L is real and not equal to Z then Γ is less than unity, but greater than zero and the reflection in called *partial reflection*.

10.7 STANDING WAVE RATIO (SWR)

When the output of a transmission line is not correctly terminated or a discontinuity is there, then the part of the entire incident wave is made to travel back towards the input end. Thus, between the input end and the load, there will be two waves travelling in opposite directions. At some points in the line, the two waves will be in phase and will add, while at other points the two will always be out of phase and will cancel. The places where two waves add will be points of maximum voltage, while the points of cancellations will have minimum voltage. Since the positions of maximum and minimum voltage remain motionless, a standing wave is said to exist on the line. Thus, SWR is a measure of the mismatch between the load and the line. Thus,

$$\text{Standing wave ratio } \rho = \frac{\text{Maximum voltage or current}}{\text{Minimum voltage or current}}$$

$$= \frac{|V_{max}|}{|V_{min}|} = \frac{|I_{max}|}{|I_{min}|}$$

Since the standing wave ratios of voltage and current are identical, no distinctions are made between VSWR and ISWR. Depending on the type of mismatch, VSWR may have low, high or moderate value.

Now
$$|V_{max}| = |V^+| + |V^-|$$

and
$$|V_{max}| = |V^+| - |V^-|$$

From definition,

$$\text{VSWR}(\rho) = \frac{|V_{max}|}{|V_{min}|} = \frac{|V^+| + |V^-|}{|V^+| - |V^-|} = \frac{1 + \left|\frac{V^-}{V^+}\right|}{1 - \left|\frac{V^-}{V^+}\right|} = \frac{1 + \Gamma}{1 - \Gamma} \qquad (10.40)$$

or
$$\Gamma = \frac{\rho - 1}{\rho + 1} \qquad (10.41)$$

Again from Eqs. (10.38) and (10.40), one obtains

$$\frac{Z_L}{Z_0} = \frac{1 + \Gamma}{1 - \Gamma} = \rho \qquad (10.42)$$

10.8 POWER

In an AC circuit, the voltage and current are not in phase, therefore the product of the voltage and current will not give us the power. In transmission line, power is determined by measuring voltage and current at a point where they are in phase.

Thus,
$$P = V_{max} \cdot I_{min}$$

Now
$$I_{min} = |I^+| - |I^-| = \frac{|V^+|}{Z_0} - \frac{|V^-|}{Z_0} = \frac{|V_{min}|}{Z_0}$$

∴
$$P = \frac{V_{max} \cdot V_{min}}{Z_0}$$

For a perfectly matching line,
$$V_{max} = V_{min} = |V^+|$$

$$P = \frac{|V^+|^2}{Z_0} \tag{10.43}$$

10.9 IMPEDANCE OF THE LINE

At any point in the transmission line, the ratio of the voltage to the current gives the impedance at that point. Now, from Eqs. (10.15) and (10.16), we obtain

$$V = V^+ e^{-\gamma Z} + V^- e^{\gamma Z}$$

and
$$I = Y_0(V^+ e^{-\gamma Z} - V^- e^{\gamma Z})$$

At the sending end, $Z = 0$, thus the two equations become

$$I_S Z_S = V^+ + V^-$$

and
$$I_S Z_0 = V^+ - V^-$$

Solving for V^+ and V^-, one obtains

$$V^+ = \frac{I_S}{2}(Z_S + Z_0) \tag{10.44}$$

and
$$V^- = \frac{I_S}{2}(Z_S + Z_0) \tag{10.45}$$

Substituting the values of V^+ and V^- in the two starting equations, one obtains

$$V = \frac{I_S}{2}[(Z_S + Z_0)e^{-\gamma Z} + (Z_S - Z_0)e^{\gamma Z}] \tag{10.46}$$

and
$$I = \frac{I_S}{2}[(Z_S + Z_0)e^{-\gamma Z} + (Z_S - Z_0)e^{\gamma Z}] \tag{10.47}$$

Dividing Eq. (10.46) by Eq. (10.47), one obtains

$$Z = \frac{V}{I} = Z_0 \left[\frac{(Z_S + Z_0)e^{-\gamma Z} + (Z_S - Z_0)e^{\gamma Z}}{(Z_S + Z_0)e^{-\gamma Z} - (Z_S - Z_0)e^{-\gamma Z}} \right] \tag{10.48}$$

Now, if the impedance at the load end ($Z = l$) be Z_l, then replacing Z of Eq. (10.48) by l and Z_S by Z_l, we get the impedance at any point w.r.t the load impedance,

i.e.
$$Z = Z_0 \left[\frac{(Z_l + Z_0)e^{-\gamma Z} + (Z_l - Z_0)e^{\gamma Z}}{(Z_l + Z_0)e^{-\gamma Z} - (Z_l - Z_0)e^{-\gamma Z}} \right] \tag{10.49}$$

Now $\qquad e^{-\gamma Z} = \cosh(\gamma Z) \pm \sinh(\gamma Z)$

Thus, Eqs. (10.48) and (10.49) become

$$Z = Z_0 \frac{Z_S \cosh(\gamma Z) - Z_0 \sinh(\gamma Z)}{Z_0 \cosh(\gamma Z) - Z_S \sinh(\gamma Z)}$$

or $\qquad Z = Z_0 \dfrac{Z_S - Z_0 \tanh(\gamma Z)}{Z_0 - Z_S \tanh(\gamma Z)}$ $\qquad\qquad$ (10.50)

and $\qquad Z = Z_0 \dfrac{Z_l \cosh(\gamma l) - Z_0 \sinh(\gamma l)}{Z_0 \cosh(\gamma l) - Z_l \sinh(\gamma l)}$

or $\qquad Z = Z_0 \dfrac{Z_l - Z_0 \tanh(\gamma l)}{Z_0 - Z_l \tanh(\gamma l)}$ $\qquad\qquad$ (10.51)

$\dfrac{Z}{Z_0} = \mathcal{Z}$ is the *nominal impedance* of a transmission line

$$\mathcal{Z} = \frac{Z}{Z_0} = \frac{1+\Gamma}{1-\Gamma} = \frac{Z_L}{Z_0} \qquad\qquad (10.52)$$

10.10 SMITH CHART

Smith chart is a polar impedance diagram which is used frequently to study the impedance or admittance transformation due to transmission line or waveguide. Since Γ is a complex quantity, we may write

$$\Gamma = u + jv \text{ and } \frac{Z}{Z_0} = R + jX$$

From Eq. (10.52),

$$R + jX = \frac{1+u+jv}{1-u-jv} = \frac{(1+u+jv)(1-u+jv)}{\{(1-u)-jv\}\{(1-u)+jv\}}$$

$$= \frac{1+2jv-u^2-v^2}{(1-u)^2+v^2}$$

Equating the imaginary parts on the left and the right hand side, one obtains

$$X = \frac{2v}{(1-u)^2+v^2}$$

or $\qquad\qquad (u-1)^2 + v^2 = \dfrac{2v}{X}$

or $\qquad (u-1)^2 + v^2 - \dfrac{2v}{X} + \dfrac{1}{X^2} = \dfrac{1}{X^2}$

or
$$(u-1)^2 + \left(v - \frac{1}{X}\right)^2 = \frac{1}{X^2} \tag{10.53}$$

Equating the real parts on the left and the right hand side, one obtains

$$R = \frac{1 - u^2 - v^2}{(1-u)^2 + v^2}$$

or $\qquad R(1-u)^2 + Rv^2 = 1 - u^2 - v^2$

or $\quad R(1 - 2u + u^2) + u^2 + v^2 + Rv^2 = 1$

or $\qquad (1+R)u^2 + (1+R)v^2 = 1 - R + 2uR$

or
$$u^2 + v^2 - \frac{2uR}{1+R} = \frac{1-R^2}{(1+R)^2}$$

or
$$u^2 - \frac{2uR}{1+R} + \frac{R^2}{(1+R)^2} + v^2 = \frac{1}{(1+R)^2}$$

or
$$\left(u - \frac{R}{1+R}\right)^2 + v^2 = \frac{1}{(1+R)^2} \tag{10.54}$$

Equations (10.53) and (10.54) represent two families of circles in complex plane

$\Gamma (= u + jv)$. Equation (10.53) represents constant reactance circle with centre at $\left(1, \frac{1}{X}\right)$

and radius $\frac{1}{X}$ (Fig. 10.6). Equation (10.54) represents constant resistance circles with

centres at $\left(\frac{R}{1+R}, 0\right)$ and radius $\frac{1}{1+R}$, and is shown in Fig. 10.7. Superposition of

Figs 10.6 and 10.7 gives us the complete *Smith chart* (Fig. 10.8). The upper half of the constant reactance circle X of the diagram represents $+jX$ and lower half $-jX$. The distance

around the Smith chart for revolution corres-ponds to one-half wavelength $\left(\frac{\lambda}{2}\right)$.

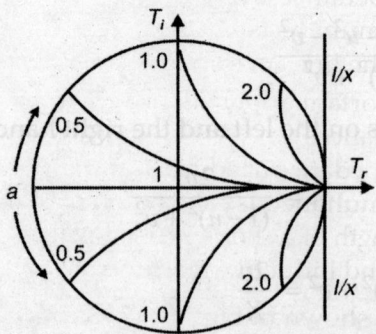

Fig. 10.6: Constant reactance (X) circle

Fig. 10.7: Constant resistance (R) circles

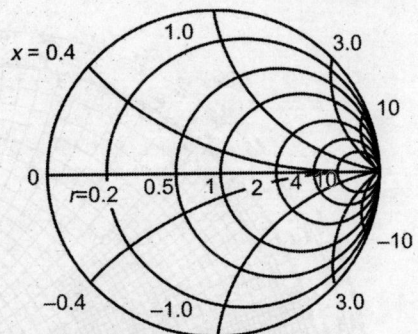

Fig. 10.8: Superposition of R and X circles

It is now evident that if Z_L is given, then on dividing it by Z_0 one can get $Z (= R + jX)$. The appropriate location of R and X circles and Γ are determined by the intersection of two circles. Since the chart does not have concentric circles showing the value of $|\Gamma|$, it is necessary to measure the radial distance from the origin to the intersection with dividers or compass and use an auxiliary scale to find $|\Gamma|$. The angle of $|\Gamma|$ is the counterclockwise from the Γ axis. To avoid the cluttering up of the chart by the angles shown by the radial line, the angles are indicated on the circumference of the circle. A straight line from origin through the intersection may be extended to the perimeter of the chart. A circle is drawn considering origin as centre and the distance between the point of intersection and origin as radius.

A second scale is drawn on the circumference by which distance along the line may be computed. This scale is in wavelength units. This measurement can be done either moving towards generator or towards the source depending on which side of normalized impedance the calculations are to be made.

The line along the x-axis of the Smith chart is SWR scale and the value is determined by the point of intersection between this line and the circle, drawn considering origin as centre and the distance between centre and the point of intersection of R and X circles for the normalized impedance as radius.

The origin is on the zero reactance line, i.e. $X = 0$, but have some value of R. Thus, the coordinate of the origin is $(1, 0)$ and the value of R increases from left to right along the line. The normalized impedance circle cuts the zero reactance line at two points. The point towards the minimum resistance scale corresponds to Z_{min} and V_{min} and that on the opposite side of the line corresponds to Z_{max} and V_{max}.

The following are the important applications of Smith chart:

i. Calculation of admittances

ii. Calculation of the impedance or admittance at any point on any transmission line, with any load and simultaneous calculation of SWR on the line

iii. Calculation of the length of a short-circuited piece of transmission line to give a required capacitive and inductive reactance

A complete Smith chart is shown in Fig. 10.9.

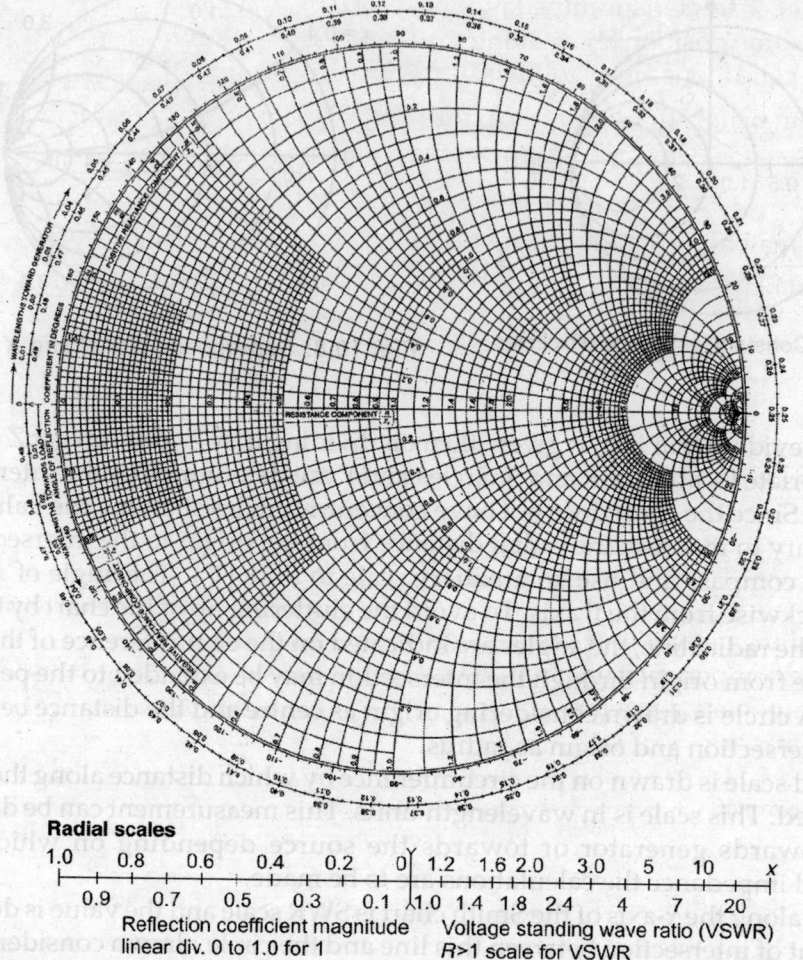

Radial scales

```
1.0    0.8    0.6    0.4    0.2    0   1.2   1.6  2.0    3.0    5     10    x
   0.9    0.7    0.5    0.3    0.1  1.0  1.4  1.8  2.4    4      7     20
   Reflection coefficient magnitude        Voltage standing wave ratio (VSWR)
   linear div. 0 to 1.0 for 1                R>1 scale for VSWR
```

Fig. 10.9: Smith chart: Normalized resistance and reactance coordinates

10.11 IMPEDANCE MATCHING

When the transmission line is not correctly terminated or discontinuity exists or short-circuited, then some portion of the transmitted power to the load (antenna) will be reflected back towards the generator and and results in non-optimization of transmission efficiency. Now to transfer maximum power, matching can be done from load side by adjusting the load impedance or from the generator or input side by adjusting the input impedance.

In a practical transmission line system, the transmitter is ordinarily matched to the coaxial cable for maximum power transfer. Because of the variable loads, however, an impedance matching technique is often required at the load side. Since the matching problem involves parallel connections of the transmission line, it is necessary to work out the problems with admittances rather than impedances.

At high frequencies, it is essential to operate at minimum VSWR (ideally, at VSWR = 1). Several methods are used to match a load Z_R to the line or to match cascaded lines with different characteristic impedances. Matching networks can be placed at the load $(x = 0)$ or at some position $x = x_1$ along the line, as shown in Fig. 10.10. The two sets of normalized conditions are as follows:

(a) Before match: $z(0) = z_R = r_0 + jx_0; y(0) = g_0 + jb_0;$ VSWR > 1

    After match: $\quad z(0) = 1 + j0; y(0) = 1 + j0;$ VSWR $= 1$

(b) At load: $\qquad z(0) = r_0 + jx_0; y(0) = g_0 + jb_0;$ VSWR$(0) > 1$

    Before match: $z(x_1) = r_1 + jx_1; y(x_1) = g_1 + jb_1;$ VSWR $=$ VSWR (0)

    After match: $\quad z(x_1) = 1 + j0; y(x_1) = 1 + j0;$ VSWR $= 1$

(a)   (b)

Fig. 10.10: Matching networks

The matching networks at lower (radio) frequencies can be made with lumped low-loss reactive components; one lumped L-C network is shown in Fig. 10.11. If Z_R has a reactive component, a reactance of opposite sign is added in series so that $Z'_R = R + j0$. Then, for a match,

$$Y_{in} = j\omega C_2 + \frac{1}{R + j\omega L_1} = \frac{1}{R_0}$$

or $\qquad L_1 = \frac{1}{\omega}\sqrt{R(R_0 - R)}$ and $C_2 = \frac{L_1}{RR_0}$

If $R > R_0$, the capacitor should be connected to the other end of the inductor.

To minimise dissipation losses at higher frequencies, a length of open- or short-circuited line is used for matching in either single stub or double stub configuration.

Fig. 10.11: Lumped L–C network

Stubs

A section of short-circuited transmission line can be used in shunt with the main line as impedance matching element by inserting it between the load and source, called stub. The process is known as stub matching and has the following advantages:

    i. Length of the main line remains unaltered

    ii. Characteristic impedance of the line remains constant

    iii. At the higher frequency, the stub may be made adjustable to suit a variety of loads and to operate over a wide range of frequencies. Basically the stub matching is of two types:

    (a) Single stub matching (Fig. 10.12)

    (b) Double stub matching (Fig. 10.13).

Fig. 10.12: Single stub connected across the transmission line

Fig. 10.13: Double stub connected across a transmission line

Single Stub Matching

A short-circuited transmission line is connected in shunt with the line (Fig. 10.12) because a shunt stub can be added to a line without cutting. In practice, an open-circuited line is rarely used at microwaves, since a true open circuit is virtually impossible to obtain. Here, the length l of the stub is adjusted in such a way unless

$$Y_0 = Y_d \pm Y_s = Y_{AA'}$$

In practice, the position AA' of the stub is determined by using a chart, but the short circuit is adjustable. Thus, the length l may be found empirically by moving the shorting plunger until the reflection is minimized. To transfer maximum power, $Y_0 = Y_{AA'}$.

The single stub matching system is useful for a fixed frequency only because as the frequency changes the location of the stub is needed to be altered. Also to adjust Y_s, l may be of very high value, which causes robust mechanical construction and hence single stub is replaced by a double stub.

Double Stub Matching

Here two short-circuited transmission lines are connected in shunt with line as shown in Fig. 10.13. The fixed distance between two stubs is usually one-eighth or five-eighth of a wavelength. The lengths l_1 and l_2 of the two stubs are adjusted alternately by trial and error method until the VSWR is reduced to unity.

Now $$Y_{AA'} = Y_{s1} \pm Y_{d1} \text{ and } Y_{BB'} = Y_{s2} \pm Y_{d2} \tag{10.55}$$

and $$Y_{d1} = Y_{BB'} \tag{10.56}$$

Thus, combining Eqs. (10.55) and (10.56), we get

$$Y_0 = Y_{d2} \pm Y_{s1} \pm Y_{s2} = Y_{AA'}$$

Quarter Wave Transformer

A transmission line having a length equal to one-fourth of the wavelength of the fundamental frequency of the wave propagating through it is called a *quarter wave transformer* and is used to match impedances at high frequencies.

From Eq. (10.50) if the impedance at load end $(Z = l)$ is denoted by Z_l, then

$$Z = Z_0 \left[\frac{Z_l \cosh(\gamma l) - Z_0 \sinh(\gamma l)}{Z_0 \cosh(\gamma l) - Z_l \sinh(\gamma l)} \right]$$

At high frequencies, line has been considered as lossless, thus $\alpha = 0$ and the propagation function $\gamma = j\beta$, then Eq. (10.50) is further modified as

$$Z = Z_0 \left[\frac{Z_l \cosh(j\beta l) - Z_0 \sinh(j\beta l)}{Z_0 \cosh(j\beta l) - Z_l \sinh(j\beta l)} \right] \tag{10.57}$$

For a line which is quarter wavelength,

$$l = \frac{\lambda}{4}, \beta = \frac{2\pi}{\lambda}$$

This gives $$\beta l = \frac{2\pi}{\lambda} \cdot \frac{\lambda}{4} = \frac{\pi}{2}$$

Hence, from Eq. (10.57),

$$Z = Z_0 \left[\frac{Z_l \cosh\frac{\pi}{2} + jZ_0 \sinh\frac{\pi}{2}}{Z_0 \cosh\frac{\pi}{2} + jZ_l \sinh\frac{\pi}{2}} \right]$$

$$= Z_0 \cdot \frac{Z_0}{Z_l} = \frac{Z_0^2}{Z_l}$$

$$Z_0 = \sqrt{Z \times Z_l} \tag{10.58}$$

Equation (10.58) shows that the input impedance depends upon the load impedance Z_l and characteristic impedance Z_0 of the transmission line. If Z_0 varies, then by varying the turns ratio of the transformer, one can obtain the required value of input impedance for any given value of load impedance.

Impedance Measurement

A slotted line is used with high-frequency coaxial lines to measure VSWR and to locate voltage minima on the line. With the aid of the Smith chart, the impedance of an unknown termination can be easily found from the VSWR and the shift of a voltage minimum from a short-circuit reference position.

In Fig. 10.14, the slotted line is inserted at a convenient terminal. With the Z_R in place, a probe is moved along the line to locate and measure maximum and minimum voltages. A suitable amplifier/indicator converts the probe output to a VSWR reading Z_R is replaced by a short circuit, and the reference minima are located for the high-VSWR condition. As would be expected, maxima and minima alternate at intervals of $\lambda/4$.

Fig. 10.14: Insertion of slotted line

To find Z_R with the *Smith chart*, draw the measured VSWR circle as shown in Fig. 10.15 and locate the voltage-minimum line (from 0 to 1 on the $\chi = 0$ line). Convert the measured Δx to wavelengths and mark the points on the VSWR circle that are Δx from the V_{min} line. The correct z_r is capacitive; a rotation through Δx towards the generator takes in into a V_{min} point. (If z_R were inductive, Δx would be greater than a quarter wavelength and a V_{max} point would occur before the V_{min} point.)

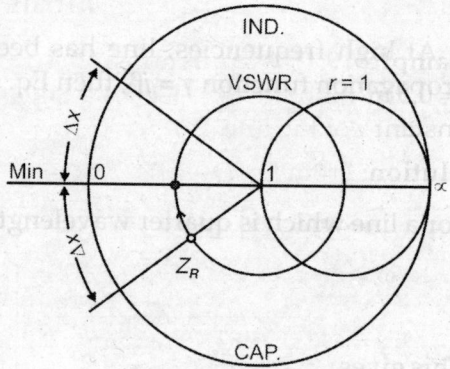

Fig. 10.15: VSWR circle

Transients in Lossless Lines

In switching applications and pulse operation, a change in voltage is suddenly applied to the line. An analysis of this transient condition generally requires recourse to the time PDEs or to their Laplace transforms. However, in the special case of a lossless line $(R = G = 0, R_0 = \sqrt{L/C}, u_p = 1/\overline{LC})$, a simple graphical method is available based on superposition of multiple reflected waves.

Figure 10.16 shows a model for the lossless system, in which the exciting voltage $V_g(t)$ is switched on at $t = 0$ and where R_g is the source resistance. Now an abrupt change at one end of the line has an effect at the other end only after one delay time $t_D = \ell/u_p$ has elapsed. Reflection will occur at the receiving end if the load is not matched to the line $(R_R \neq R_0)$; at the sending end if the source is not matched $(R_g \neq R_0)$.

Fig. 10.16: Model for a lossless system

10.12 INSERTION LOSS

When a network is introduced between a source and for impedance matching, there is a reduction in power level. This is known as *insertion loss*.

Thus, insertion loss = input power – output power

If the receiving end current without the inserted network be I_1 and the receiving end current with the inserted network be I_2, then

$$\text{Insertion loss} = \log_e \left| \frac{I_1}{I_2} \right| = 20 \log_{10} \frac{I_1}{I_2}$$

The insertion loss is governed by the value of the source and load impedances and is not a fixed quantity. In some cases, insertion of a network may cause a phase shift called insertion shift between I_1 and I_2.

ILLUSTRATIVE EXAMPLES

Example 10.1: A transmission line has the following parameters: $R = 2\ \Omega/\text{m}$, $G = 0.5\ \text{m-mho/m}$, $f_g = 1\ \text{GHz}$, $L = 8\ \text{mH/m}$ and $C = 0.23\ \text{pF/m}$. Calculate the propagation constant γ of the line.

Solution: From Eq. (10.19),

$$\gamma = \sqrt{(R + j\omega L)(G + j\omega C)}$$

$$\omega = 2\pi f = 2 \times 3.14 \times 1 \times 10^9 = 6.28 \times 10^9$$

$$\gamma = \sqrt{(50.31 \underline{|87.72°})(15.29 \times 10^{-4} \underline{|70.91°})} = \sqrt{769.24 \times 10^{-4} \underline{|158.63°}}$$

$$= 0.2774 \underline{|79.31°} = 0.051 + j0.273$$

Example 10.2: A coaxial cable has a characteristic impedance of 50Ω with a capacitance of 25 pF/ft. What is the inductance/foot? If the diameter of the inner conductor is 0.025 inch and the dielectric constant of the insulation 2.25, what is the diameter of the outer conductor?

Solution: From Eq. (10.33),

$$Z_0 = \sqrt{\frac{L}{C}} \quad \text{or} \quad L = C Z_0^2$$

$$\therefore \qquad L = 25 \times 10^{-12} \times (50)^2 = 625 \times 10^{-10} = 0.0625\ \mu\text{H/ft}$$

Again

$$Z_0 = \frac{138}{\sqrt{K}} \cdot \log \frac{b}{a} \quad \text{[From Eq. (10.23)]}$$

or

$$\log \frac{b}{a} = \frac{\sqrt{K}}{138} Z_0$$

$$= \frac{\sqrt{2.25}}{138} \times 50 = \frac{1.5 \times 50}{138} = \frac{75}{138} = 0.54$$

$$\therefore \qquad b = a \times \text{antilog}\ (0.54) = 0.025 \times 3.4673 = 0.087\ \text{inch}$$

Example 10.3: A transmission line has a characteristic impedance of $(50 + j0.01)\,\Omega$ and is terminated in a load impedance of $(73 - j42.5)\,\Omega$. Calculate the reflection coefficient.

Solution: From Eq. (10.39), the reflection coefficient is

$$\Gamma = \frac{Z_L - Z_0}{Z_L + Z_0}$$

$$= \frac{73 - j42.5 - (50 + j0.01)}{73 - j42.5 + (50 + j0.01)} = 0.377\underline{|-42.7°}$$

Example 10.4: A low loss coaxial cable of characteristic impedance of $50\,\Omega$ is terminated in a resistive load of $75\,\Omega$. Calculate the VSWR. If the maximum voltage in the SWR is 30 V, calculate the minimum voltage.

Solution: From Eq. (10.42),

$$\text{VSWR} = \frac{Z_L}{Z_0} = \frac{75\,\Omega}{50\,\Omega} = \frac{3}{2} = 1.5$$

Again

$$\rho = \frac{|V_{max}|}{|V_{min}|} = \frac{30}{V_{min}}$$

or

$$1.5 = \frac{30}{V_{min}}$$

or

$$V_{min} = \frac{30}{1.5} = 20\text{ V}$$

Example 10.5: Find the reflection coefficient, standing wave ratio and fraction of incident power delivered to the load when a transmission line of characteristic impedance $100\,\Omega$ is terminated with a load of $(50 + j50)\,\Omega$.

Solution: Normalized load is

$$Z_L = \frac{50 + j50}{100} = 0.5 + j0.5$$

Now

$$\Gamma = \frac{Z_L - 1}{Z_L + 1} = \frac{0.5 + j0.5 - 1}{0.5 + j0.5 + 1} = -\frac{0.5 + j0.5}{1.5 + j0.5} = -0.2 + j0.4$$

$$\therefore \quad |\Gamma| = \sqrt{(0.2)^2 + (0.4)^2} = 0.447$$

The angle of the reflection coefficient is given by

$$\tan\phi = \frac{0.4}{-0.2} = -2$$

$$\therefore \quad \phi = \tan^{-1}(-2)$$

$$\text{VSWR}(\rho) = \frac{1 + |\Gamma|}{1 - |\Gamma|} = \frac{1 + 0.447}{1 - 0.447} \simeq 2.62$$

Fraction of power delivered to load is

$$\frac{P}{P_0} = 1 - |\Gamma|^2 = 1 - 0.2 = 0.8$$

$$\frac{P}{P_0}\% = 80\%$$

Example 10.6: A single stub tuner is to match a lossless line of 400 Ω to a load of $(800 - j300)\,\Omega$. The frequency is 3 GHz.

 i. Find the distance in metres from the load to the turning stub.

 ii. Determine the length in metres of the short-circuited stub.

Solution: Given, $Z = 400\,\Omega$, $Z_L = (800 - j300)\,\Omega$, $f = 3 \times 10^9$ Hz, $\lambda = 10$ cm

Now

$$Z = \frac{Z_L}{Z_0} = \frac{800 - j300}{400} = 2 - j0.75$$

$$\therefore \quad Y_L = \frac{1}{Z} = \frac{1}{2 - j0.75} = 0.46 + j0.17$$

 i. For lossless line shown in Fig. 10.17, $\alpha = 0$, then a SWR circle is drawn through the point Y_L so that the circle intersects $Y_d = 1 + j0.9$.

$$d = 0.16\lambda - 0.33\lambda = 0.127\lambda = 0.0127\,\text{m}$$

Fig. 10.17: A single stub

 ii. $Y_s = j0.9$.
Referring to the Smith chart shown in Fig. 10.18, we get

$$l = 0.116\lambda = 0.0116\,\text{m}$$

Fig. 10.18: Smith chart

Example 10.7: A matched transmission line is shown in Fig. 10.19.

 i. Find l_1 and d which provided a proper match.

Fig. 10.19: Double stub

ii. With the line and load properly matched, determine the VSWR on the section of the line between the stubs.

Solution:

i. Here $Z_l = 1 + j1$, thus $y_l = 0.5 - j0.5$. The spacing circle of $\dfrac{5\lambda}{8}$ is drawn from the unity circle.

$$Y_{s2} = j1.33 \text{ for } l_2 = 0.11$$

Now

$$Z_{s2} = Z_{s2}Z_{02} = \frac{1}{-j1.38} \times 200 = +j14.5$$

At the second stub,

$$Y_{BB'} = Y_{d2} + Y_{s2}$$

$$\frac{Y_{BB'}}{Z_0} = \frac{y_{d2}}{Z_0} + \frac{y_{s2}}{Z_{02}}$$

$$\frac{1}{100} = \frac{y_{d2}}{100} + \frac{-j1.38}{200}$$

$$\therefore \qquad y_{d2} = 1 + j0.69$$

Now, from the Smith chart (Fig. 10.20) we should read

$$Y_{AA'} = 0.52 + j0.125$$

$$Y_{d2} = 0.52 - j0.53$$

Then

$$y_{s1} = y_{11} - y_{d1}$$
$$= 0.52 + j0.125 - 0.52 + j0.53 = j0.655$$
$$l_1 = (0.25 + 0.092)\lambda = 0.342\lambda$$

and

$$d = (0.50 - 0.006)\lambda = 0.494\lambda$$

Fig. 10.20: Smith chart

ii. The SWR on the line between the stubs is about 2.

Example 10.8: An air-filled rectangular waveguide has cross-sectional dimensions of $a = 100$ mm and $b = 50$ mm. Calculate the cut off frequency for the modes TE_{10}, TE_{20}, TE_{01}, TE_{02}, TE_{11}, TE_{12} and TE_{21}, and also the ratio of the guide velocity v_{ps} to the velocity in free space for each of above modes if $f = 2f_c$.

Solution: Since $\quad f_c = \dfrac{v_p}{2ab}\sqrt{(mb)^2 + (na)^2}$

Given $\qquad a = \dfrac{100}{1000}\,\text{m} = \dfrac{1}{10}\,\text{m}, \quad b = \dfrac{50}{1000}\,\text{m} = \dfrac{5}{100}\,\text{m}$

Hence, $\qquad f_c = \dfrac{3\times10^8 \times 10 \times 100}{2\times1\times5}\sqrt{\left(m\times\dfrac{5}{100}\right)^2 + \left(n\times\dfrac{1}{10}\right)^2}$

$$= 3\times10^{10}\sqrt{\dfrac{m^2\times25}{(100)^2} + \dfrac{n^2\times1\times100}{100\times100}}$$

$$= \dfrac{3\times10^{10}}{100}\sqrt{m^2\times25 + n^2\times100}$$

$$= 3\times10^8\sqrt{100(m^2\times0.25 + n^2)}$$

$$= 3\times10^9\sqrt{(m^2\times0.25 + n^2)}$$

Example 10.9: The specifications for rigid air-dielectric coaxial line used in a radar set operating at 3 GHz are: copper material, stub-supported at intervals to maintain the air dielectric; outside diameter, $\dfrac{7}{8}$ inch; wall thickness, 0.032 inch; inner-conductor diameter, 0.375 inch; characteristic impedance, 46.4 Ω; attenuation 0.066 dB/m; maximum peak power, 1.31 kW; operating peak power, 200 kW; lowest safe wavelength, 5.28 cm. Determine the per meter values of L, C, G, and R_a for the line neglecting internal inductance.

Solution: The inner radius a is 4.76 mm and the outer radius b is 10.3 mm. Then $\ln (b/a)$ = 0.771, GFL = 0.386, GFC = 2.59.

$$L = \dfrac{\mu_0}{\pi}\,(\text{GFL}) = 0.154\,\mu\text{H/m}$$
$$C = \pi\varepsilon_0\,(\text{GFC}) = 71.9\,\text{pF/m}$$

For copper and frequency of 3 GHz, $\delta = 1.2\,\mu\text{m}$. Then, we have

$$\text{GFR}_a = \dfrac{1}{a} + \dfrac{1}{b} = 307\,\text{m}^{-1}$$

and $\qquad R_a = \dfrac{1}{2\pi\sigma_c\delta}\,(\text{GFR}_a) = 0.702\,\Omega/\text{m}$

For air dielectric, $G = 0\,\text{S/m}$.

Example 10.10: Show that the voltage $V(x, t) = A\cos(\omega t + \theta)e^{j\beta x}$ satisfies the transmission line equation (3), for a uniform lossless line if $\beta = \omega\sqrt{LC}$.

Solution: For the lossless line, $R = G = 0$, so that the equation reduces to

$$\dfrac{\partial^2 V(x, t)}{\partial x^2} = LC\dfrac{\partial^2 V(x, t)}{\partial t^2}$$

For the given voltage, this requires

$$-\beta^2 V = LC(-\omega^2 V) \quad \text{or} \quad \beta = \omega\sqrt{LC}$$

GLIMPSES

1. Transmission lines are used to transmit electric energy and signals from one point to another, specifically from a source to a load.

2. The basic elements in a circuit, such as resistors, capacitors, inductors, and the connections between them, are considered lumped elements if the time delay in traversing the elements is negligible. On the other hand, if the elements or interconnections are large enough, it may be necessary to consider them as distributive elements. This means that they are resistive, capacitive, and inductive per-unit-distance basis. Transmission lines have this property in general, and thus they become circuit elements in themselves, possessing impedances that contribute to the circuit problem. The basic rule is that one must consider elements as distributed if the propagation delay across the element dimension is of the order of the shortest time interval of interest.

3. If we assume that the existence of voltage and current across and within the transmission-line conductors implies the existence of electric and magnetic fields in the space around the conductors and which are associated with the voltage and current, then we have two possible approaches to the analysis of transmission lines: (i) to solve Maxwell's equations subject to the line configuration to obtain the fields, and with these find general expressions for the wave power, velocity and other parameters of interest, (ii) or one can avoid the fields and solve for the voltage and current using an approximate circuit model.

4. General wave equations for the transmission line are

$$\frac{\partial^2 V}{\partial z^2} = LC\frac{\partial^2 V}{\partial t^2} + (LG + RC)\frac{\partial V}{\partial t} + RGV$$

$$\frac{\partial^2 I}{\partial z^2} = LC\frac{\partial^2 I}{\partial t^2} + (LG + RC)\frac{\partial I}{\partial t} + RGI$$

where L is inductance, C is capacitance, G is shunt conductance and R is resistance. All of these have values that are specified per unit length.

5. Lossless propagation means that power is not dissipated or otherwise deviated as the wave travels down the transmission line; all power at the input end eventually reaches the output end. More realistically, any mechanisms that would causes losses to occur have negligible effect.

The wave velocity (v) for lossless propagation is given by

$$v = 1\sqrt{LC}$$

6. Any signal that is transmitted in practice can be decomposed into a discrete or continuous summation of sinusoids. This is the basis of frequency domain analysis of signals on lines. In such studies, the effect of the transmission line on any signal can be determined by noting the effects on the frequency components. This reveals that one can effectively propagate the spectrum of a given signal, using frequency-

dependent line parameters, and then reassemble the frequency components into the result and signal in time domain.

7. Characteristic impedance in terms of known line parameters

$$Z_0 = \sqrt{\frac{R + j\omega L}{G + j\omega L}}$$

where the phase of the characteristic impedance, $\theta = \phi - \xi$.

8. Conditions of low-loss or distortion-free propagation will usually occur over limited frequency ranges. As a rule, loss increases with increasing frequency, mostly because of the increase in R with frequency. The nature of this latter effect is known as skin effect loss.

9. In transmission line, power is determined by measuring voltage and current at a point where they are in phase.

10. The Smith chart is a graphical aid in solving high-frequency transmission line problems. The chart is essentially a polar plot of the reflection coefficient in terms of the normalized impedance. The basic relationship upon which the chart is constructed is

$$\Gamma = \frac{Z_L + Z_0}{Z_L + Z_0}$$

where $\Gamma = |\Gamma| e^{i\phi}$, $|\Gamma|$ is variable radius. The impedance which is plot on the chart is normalized w.r.t. the characteristic impedance. The normalized load impedance Z is

$$Z = \frac{1 + \Gamma}{1 - \Gamma}$$

11. In a practical transmission line system, the transmitter is ordinarily matched to the coaxial cable for maximum power transmission.

12. A section of short-circuited transmission line can be used in shunt with the main line as impedance matching element by inserting it between the load and the source, called stub.

REVIEW QUESTIONS

1. Derive transmission line equation. Show that propagation constant γ is, in general, a constant quantity and may be expressed as $\gamma = \alpha + j\beta$, where α and β are attenuation constant and phase constant respectively.

2. Derive the condition for minimum attenuation w.r.t. L, C, R and G.

3. What is distortion in a transmission line? Obtain the condition for no distortion.

4. What is characteristic impedance (Z_0)? Show that $Z_0 = \frac{138}{\sqrt{K}} \log \frac{b}{a} \Omega$ for a coaxial line, where K is the dielectric constant of the insulating material.

5. What is reflection coefficient (Γ)? Show that

$$\Gamma = \frac{Z_L - Z_0}{Z_L + Z_0}$$

Obtain conditions for total and partial reflection.

6. Explain standing wave ratio (ρ). Show that $\rho = \dfrac{Z_L}{Z_0} = \dfrac{1+\Gamma}{1-\Gamma}$.

7. Derive an expression for the impedance of a transmission line. Show that the normal impedance ($\mathcal{Z}$) of a transmission line is

$$(\mathcal{Z}) = \frac{Z_L}{Z_0} = \frac{1+\Gamma}{1-\Gamma}$$

8. What do you understand by a Smith chart? Derive equations for two families of circles in complex plane $\Gamma(= u + jv)$. Mention important applications of Smith chart.

9. What is stub? Explain: (i) single stub matching, and (ii) double stub matching.

PROBLEMS

1. The parameters of a certain transmission line operating at 6×10^8 rad/s are $L = 0.4\ \mu\text{H/m}$, $C = 40\ \text{pF/m}$, $G = 80\ \mu\text{S/m}$, and $R = 20\ \Omega/\text{m}$. (a) Find γ, α, β, λ, and Z_0. (b) If a voltage wave travels 20 m down the line, what percentage of the original wave amplitude remains, and by how many degrees is its phase shifted?
   [*Ans.* (a) $\gamma = 0.104 + j2.40\ \text{m}^{-1}$, $\alpha = 0.104\ \text{Np/m}$, $\beta = 2.40\ \text{rad/m}$, $\lambda = 2.62$ m, $Z_0 = 100 - j4.0\ \Omega$ (b) 12.5%, 2.75×10^3 degrees]

2. The characteristic impedance of a lossless transmission line is $72\ \Omega$. If $L = 0.5\ \mu\text{H/m}$, find: (a) C (b) v_p (c) β if $f = 80$ MHz (d) the line is terminated with a load of $60\ \Omega$, find Γ and s.
   [*Ans.* (a) 96 pF/m (b) 1.44×10^8 m/s (c) 3.5 rad/m (d) $\Gamma = -0.09$, $s = 1.2$]

3. Two characteristics of a certain lossless transmission line are $Z_0 = 50$ W and $\gamma = 0 + j0.2\pi\ \text{m}^{-1}$ at $f = 60$ MHz. (a) Find L and C for the line. (b) A load $Z_L = 60 + j80\ \Omega$ is located at $z = 0$. What is the shortest distance from the load to a point at which $Z_{in} = R_{in} + j0$?
   [*Ans.* (a) $83.3\ \mu\text{H/m}$, 33.3 pF/m (b) 65 cm]

4. A transmitter and receiver are connected using a cascaded pair of transmission lines. At the operating frequency, line 1 has a measured loss of 0.1 dB/m and line 2 is rated at 0.2 dB/m. The link is composed of 40 m of line 1 joined to 25 m of line 2. At the joint, a splice loss of 2 dB is measured. If the transmitted power is 100 mW, what is the received power?   [*Ans.* 7.9 mW]

5. For the transmission line represented in Fig. 10.21, find $V_{s,out}$ if f is: (a) 60 Hz and (b) 500 kHz.

   [*Ans.* (a) 104 V (b) $52.6 - j123$ V]

Fig. 10.21: Transmission line

6. Determine the average power absorbed by each resistor in Fig. 10.22.

[*Ans.* $P_{25} = 2.28\,\text{W}, \; P_{100} = 1.16\,\text{W}$]

Fig. 10.22

SHORT ANSWER QUESTIONS

1. How a transmission line may be classified depending on the frequency of transmitted signal?

   Ans. Depending on the frequency of transmitted signal, one can classify transmission line in the following categories: (a) two wire transmission line (b) coaxial type line (c) waveguide type line.

2. Why a coaxial type transmission line is preferred for use at microwave frequencies?

   Ans. Because of shielded nature of this type of line.

3. Write transmission line equations.

   Ans. $\dfrac{\partial^2 V}{\partial Z^2} = RGV + (RC + LG)\dfrac{\partial V}{\partial t} + LC\dfrac{\partial^2 V}{\partial t^2}$

   $\dfrac{\partial^2 I}{\partial Z^2} = RGI + (RC + LG)\dfrac{\partial I}{\partial t} + LC\dfrac{\partial^2 I}{\partial t^2}$

4. Write expressions for attenuation constant (α) and phase constant (β) for a transmission line.

   Ans. $\alpha = \dfrac{1}{\sqrt{2}}\sqrt{\sqrt{(RG - \omega^2 LC)^2 + \omega^2(RC + LG)^2} + (RG - \omega^2 LC)}$

   $\beta = \dfrac{1}{\sqrt{2}}\sqrt{\sqrt{(RG - \omega^2 LC)^2 + \omega^2(RC + LG)^2} + (RG - \omega^2 LC)}$

5. Write the condition for minimum attenuation w.r.t. L, C, R and G.

   Ans. $\alpha_{min} = \sqrt{RG}$, α_{min} w.r.t L, $\dfrac{d\alpha}{dL} = 0$, $L = \dfrac{R}{G}C$, α_{min} w.r.t. C, $\dfrac{d\alpha}{dC} = 0$, $C = \dfrac{G}{R}\alpha$,

   α_{min} w.r.t. R and G $\dfrac{d\alpha}{dR} = 0$, $\dfrac{d\alpha}{dG} = 0$.

   In all cases, minimum value of attenuation constant $\alpha_{min} = \sqrt{RG}$.

6. Write condition for no distortion.

   Ans. For distortion to be absent; if $R = 0$ and $G = 0$, then $\alpha = 0$ and $\beta = \omega\sqrt{LC}$ for all values of ω. Since, in practice it is impossible to make $R = 0$. thus distortion is absent if

   $$\sqrt{(RG - \omega^2 LC)^2 + \omega^2(RC - LG)^2} \equiv K + \omega^2 LC$$

   where K is constant and does not depend upon frequency.

MULTIPLE CHOICE QUESTIONS

1. In the analysis of two conductor transmission line, one has to consider an incremental portion of length Δz on the line because it is a
   (a) distributed line
   (b) lumped line
   (c) lossy line
   (d) dispersive line
   [a]

2. In the equivalent circuit of a transmission line, if we replace the equivalent T-network by a π-network, then
   (a) the telegrapher line equations will change
   (b) the telegrapher line equations will remain same
   (c) the value of propagation constant will change
   (d) the value of circuit impedance will change.
   [b]

3. If the time dependence of voltage is given as $e^{-j\omega x}$, then $V_0 e^{-\gamma z}$ will represent
   (a) forward travelling wave
   (b) backward travelling wave
   (c) standing wave
   (d) does not represent a wave
   [b]

4. The line parameters of a transmission line are expressed as respective units/length. The characteristic impedance Z_0 for a line will have units of
   (a) Ω km
   (b) Ω/km
   (c) km/Ω
   (d) Ω
   [d]

5. A line of length l has characteristic impedance Z_0. The line is cut into half. The value of characteristic impedance becomes
   (a) $Z_0/2$
   (b) $Z_0/4$
   (c) $2Z_0$
   (d) Z_0
   [d]

6. A line is of length l. The electrical length of the line is given as
   (a) l/λ
   (b) l
   (c) $2\pi \dfrac{l}{\lambda}$
   (d) $2\pi \dfrac{\lambda}{l}$
   [c]

7. In a transmission line, the attenuation is given as 0.3 dB/km. After 10 km, the power will be what fraction of input power?
   (a) $\dfrac{1}{4}$
   (b) $\dfrac{1}{10}$
   (c) $\dfrac{1}{3}$
   (d) $\dfrac{1}{2}$
   [d]

8. A lossless line of length 500 m has $L = 10$ mH/m and $C = 0.1$ pF/m at 1 MHz. The electrical length of the line is
   (a) $360°$
   (b) $270°$
   (c) $180°$
   (d) $90°$
   [c]

9. The voltage reflection coefficient at the load is given as $\Gamma(l)$. The current reflection coefficient at the load will be
   (a) $\Gamma(l)$
   (b) $1/\Gamma(l)$
   (c) $-\Gamma(l)$
   (d) $-1/\Gamma(l)$
   [c]

10. For a given line the characteristic impedance is Z_0, load Z_L and VSWR in S. The maximum input impedance is
    (a) Z_L/S (b) Z_0/S
    (c) $Z_0 S$ (d) $Z_L S$ **[c]**

11. For the above line (MCQ 10), the minimum input impedance is
    (a) Z_L/S (b) Z_0/S
    (c) $Z_0 S$ (d) $Z_L S$ **[d]**

12. For a shorted line which one of the following is not true
    (a) $Z_{in} = j Z_0 \tan \beta l$ (b) $\Gamma_l = -1$
    (c) $\Gamma_l = +1$ (d) $S = \infty$ **[c]**

13. For an open-circuited line which of the following is not true
    (a) $Z_{in} = -j Z_0 \cot \beta l$ (b) $\Gamma_l = -1$
    (c) $\Gamma_l = +1$ (d) $S = \infty$ **[b]**

14. For a matched lossless line of characteristic impedance Z_0 which of the following is not true
    (a) input impedance $Z_{in} = Z_0$
    (b) reflection coefficient at load $\Gamma_l = 0$
    (c) VSWR $S = \infty$
    (d) attenuation constant $\alpha = 0$ **[c]**

15. In a lossless line, the adjacent maximum are measured at distance l_1 and l_2. If c is the phase velocity, the operating frequency f is
    (a) $2c/(l_2 - l_1)$ (b) $c/2(l_2 - l_1)$
    (c) $c/(l_2 - l_1)$ (d) $c/\{2\pi(l_2 - l_1)\}$ **[b]**

16. A line is terminated by pure reactance. The reflection coefficient at the load is
    (a) $|\Gamma_l| = 0$ (b) $|\Gamma_l| = 1$
    (c) $|\Gamma_l| = 0.5$ (d) $|\Gamma_l| = \infty$ **[b]**

17. A shorted line has $Z_{in} = +j Z_0$. The length of the line is
    (a) λ (b) $\lambda/2$
    (c) $\lambda/4$ (d) $\lambda/8$ **[d]**

18. An open circuit line has $Z_{in} = -j0$. The length of the line is
    (a) λ (b) $\lambda/2$
    (c) $\lambda/4$ (d) $\lambda/8$ **[c]**

19. A lossy transmission line terminated by load Z_l, has characteristic impedance Z_0 and open circuit input impedance Z_{OC}. The input impedance of the line is
    (a) $Z_0 Z_{OC}/Z_L$ (b) $Z_0^2(Z_L + Z_{OC})/(Z_0^2 + Z_L Z_{OC})$
    (c) $(Z_0 + Z_L)^2/(Z_0 + Z_{OC})$ (d) $(Z_0^2 + Z_L Z_{OC})/Z_{OC} + Z_L)$ **[d]**

20. For a given line attenuation, $\alpha = \sqrt{RG}$ and $\beta = \omega\sqrt{RG}$. Such a line is known as
    (a) lossless line (b) distortionless line
    (c) dispersive line (d) all of the above **[b]**

21. For a low loss line, attenuation constant $\alpha = \frac{1}{2}\left(R\sqrt{\frac{C}{L}} + G\sqrt{\frac{L}{C}} \right)$ and phase constant

$\beta = \omega\sqrt{LC}\left[1 + \frac{1}{8}\left(\frac{G}{\omega C} - \frac{R}{\omega L} \right) \right]$. Such a line will be

(a) lossless and distortionless line
(b) distortionless and dispersionless line
(c) dispersionless and distortion line
(d) dispersive and distortionless line [d]

22. If a lossless line is terminated with load impedance $40 + j0.30\Omega$, then the characteristic of the line for minimum possible standing wave ratio (SWR) will be

(a) $100\,\Omega$ (b) $70\,\Omega$
(c) $50\,\Omega$ (d) $40\,\Omega$ [c]

23. The characteristic impedance of transmission line with inductance $0.29\,\mu H/m$ and capacitance $60\,pF/m$ is

(a) 140 W   (b) 120 W
(c) 60 W   (d) 70 W [d]

24. For a transmission line, the propagation constant for a TEM wave travelling in it, is given by (where the symbols have the usual meaning)

(a) $[(R - j\omega L)(G + j\omega^2 C)]^{1/3}$ (b) $[(R + j\omega L)(G + j\omega C)]$
(c) $[(R + j\omega L)(G + j\omega C)]^{1/2}$ (d) $[(R - j\omega L)(G + j\omega C)]^{1/2}$ [c]

25. A transmission line having a length equal to one-fourth of the wavelength of the fundamental frequency of the wave propagating through it is called

(a) transformer   (b) half wave transformer
(c) quarter wave transformer   (d) none of the above [c]

26. To study the impedance or admittance transformation due to a transmission line or waveguide, polar impedance diagram is used. This is called

(a) spectrograph   (b) lines of force chart
(c) Smith chart   (d) stub [c]

11

Waveguides

11.1 INTRODUCTION

A *waveguide* is basically a structure through which electromagnetic waves can be transmitted from point to point and within which the fields are confined to a certain extent. A transmission line fits this description, but it is a special case that employs two conductors, and it propagates a purely TEM field configuration. Waveguides, in general, depart from these restrictions and may employ any number of conductors and dielectrics, dielectrics alone and no conductors.

A waveguide is constructed of hollow metallic tube of circular or rectangular cross-section and used to guide electromagnetic waves through it. They are basically used for frequency range corresponding to X-band, i.e. from 8 GHz to 12 GHz. In waveguide, both electric and magnetic fields are confined to space within the waveguide. The dielectric loss within the waveguide is negligibly small because the guides are normally air filled. We may note that within a waveguide several modes of electromagnetic waves could be propagated. The mode of propagation of a waveguide is determined from solution to Maxwell's equations. If the frequency of a particular signal will be passed through it or else it gets attenuated. Basically waveguides are of two types: (i) *rectangular waveguide*, and (ii) *circular waveguide*. When a waveguide is closed by end surfaces, it is called a *cavity*.

At UHF and microwave frequencies such a device provides means for transmission of electromagnetic energy. In order that electric and magnetic fields may exist inside such devices there must be the solution of Maxwell's field equations and in addition, should satisfy the boundary conditions at the walls of waveguide. If the walls are mode of perfect conductors there should be no tangential component of electric field at the walls.

The most efficient way of transmitting energy over short distances is by using waveguides. Basically, a waveguide is a completely shielded transmission line, and may be bent and twisted with no radiation loss as long as the cross-section remains uniform. A change in dimensions amounts to a change in characteristic impedance. In transverse electric (TE) modes the electric vector (E) of the field has no component in the propagation. In the transverse magnetic (TM) modes, the magnetic vector has no such component. Physical constraints and the frequency of the wave usually limit the number of modes.

Waveguides were first evolved and used in *practical communication electronics* and now assumed great importance in *optical communication:* photonics and optoelectronics.

11.2 ELECTROMAGNETIC WAVE PROPAGATION BETWEEN TWO PARALLEL CONDUCTING PLATES

Let us consider the propagation of electromagnetic wave between two parallel conducting planes of infinite extent separated by a distance of the order of a wavelength. The free space is assumed between the two conducting parallel plates. The propagation of the electromagnetic wave through the system of two parallel plates obey the Maxwell's field equations for free-space and also satisfy appropriate boundary conditions at all points on the conducting planes.

Figure 11.1 shows two parallel, infinite and conducting plates. One plate is in the plane $y = 0$ and the other in the plane $y = b$. Since the space between the plates is a vacuum, we have $\varepsilon = \varepsilon_0$, $\mu = \mu_0$, $\sigma = 0$ and $\rho = 0$. Let a plane polarized electromagnetic wave is introduced in the region between the plates, the walls will reflect the wave to and fro and as a result the wave will propagate in the direction parallel to the plates. As stated above, the fields must obey Maxwell's equations in

Fig. 11.1: Propagation of electromagnetic wave between two parallel infinite and conduction plates

free space between the plates and must satisfy certain boundary conditions at all points on the wall. The boundary conditions to be satisfied are:

i. the tangential component of E must be zero at all points on the wall, i.e.

$$n \times E = 0 \tag{11.1}$$

ii. the normal component of the magnetic field must be zero at all points on the wall, i.e.

$$n \cdot B = 0 \tag{11.2}$$

where n is the surface normal.

One can obtain derived boundary conditions from the two boundary conditions given by Eqs. (11.1) and (11.2). Substituting for E from Maxwell's equation in Eq. (11.1), one obtains

$$n \times (\nabla \times B) = 0 \tag{11.3}$$

and likewise substituting for B from Maxwell's equation in Eq. (11.2), one obtains

$$n \cdot (\nabla \times E) = 0 \tag{11.4}$$

These conditions imply that the normal component of the electric field vector, E is not necessarily zero, since there can be some charges on the conducting surfaces and likewise the tangential component of the magnetic field vector B is not necessarily zero, since there can be some surface currents in the perfectly conducting plates. Since the electromagnetic waves are transverse, with their electric (E) and magnetic field (H) vectors being transverse to the direction of propagation (k), they are called *TEM waves*.

Let us consider that a linearly plane polarized wave is propagating between the plates in a direction given by the propagation vector k in the plane perpendicular to the x-axis and making an angle θ with y-axis as shown in Fig. 11.1. If E is in the x-direction, B must be in the yz plane (since B is perpendicular to E). Obviously, $B = \hat{a}_y B_y + \hat{a}_z B_z$, has a component in the direction of wave, i.e. z-direction, whereas E does not. We call such a wave as *transverse electric* (TE) wave. If one chooses B to be along the x-axis, then E will have a longitudinal component in the z-direction. Such a wave is called as *transverse magnetic* (TM) wave. We must note that the waves propagating within closed conducting regions are either TE or TM modes. The expression for the plane wave electric field is written as

$$E = \hat{a}_x E_0 \exp[i(k \cdot r - \omega t)] \tag{11.5}$$

where ω and k are the angular frequency and wave vector of the wave in free-space respectively. We can write

$$k \cdot r = ky \cos\theta + kz \sin\theta \tag{11.6}$$

Using Eq. (11.6), one can rewrite Eq. (11.5) as

$$E_I = \hat{a}_x E_{0I} \exp[-i(\omega t - kz \cos\theta - ky \sin\theta)] \tag{11.7}$$

When this wave strikes the conductor plane at $y = 0$, it is totally reflected with a phase change of 180°. For the reflected wave, we have

$$k_2 = \hat{a}_y k \sin\theta + \hat{a}_z k \cos\theta \tag{11.8}$$

The total electric field between the conducting planes

$$E = E_I + E_R = \hat{a}_x E_{oI} \exp[i(ky \cos\theta + kz \sin\theta - \omega t)]$$
$$+ \hat{a}_x E_{oR} \exp[i(-ky \cos\theta + kz \sin\theta - \omega t)] \tag{11.9}$$

$$= \hat{a}_x \exp[i(kz \sin\theta - \omega t)[E_{oI} \exp(iky \cos\theta) + E_{oR} \exp(-iky \cos\theta)] \tag{11.10}$$

one finds that at each point between the two plates there is an interference between the incident and reflected wave so that the resultant field and the associated energy always flows parallel to the z-axis.

At the boundary (i.e. at $Y = 0$) $E = 0$. One can see that this condition will be satisfied if $E_{oI} + E_{oR} = 0$, i.e. $E_{oI} = -E_{oR}$ and E is given by

$$E = \hat{a}_x E_{oI} \exp(ik \sin\theta - t)[e^{iky \cos\theta} - e^{-iky \cos\theta}] \tag{11.11}$$

$$= 2i\hat{a}_x E_{oI} \sin(ky \cos\theta) \exp(i(kz \cos\theta - \omega t] \tag{11.11a}$$

The waves propagating at an arbitrary angle θ do not satisfy simultaneously the boundary conditions at both the planes. For example, the wave reflecting from the upper surface satisfies the boundary condition given by Eq. (11.1). However, at the lower surface, i.e. at $Y = b$, the boundary condition is satisfied by the resultant of the incident and reflected waves. For $E = 0$ at $Y = b$, one obtains

$$d_1 + d_2 = n\lambda \tag{11.12}$$

which leads to

$$2d \sin\theta = n\lambda \tag{11.13}$$

where n takes only integer values. It cannot take value $n = 0$ for which all the components of the wave vanish. There will, thus, be several acceptable fields corresponding to the different values of n. The different waves that correspond to different values of n are called *modes*. For a given wave frequency the maximum number of wave modes satisfying the boundary condition (and since $\cos\theta < 1$) are governed by the following condition,

$$\frac{n\pi}{kb} = \frac{n\lambda}{2b} \leq 1 \tag{11.14}$$

This condition restricts the maximum value of n for a given frequency of radiation. Obviously, Eq. (11.14) restricts the maximum value of n for a given frequency of radiation. If $\frac{n\pi}{kb} \geq 1$, i.e. $\frac{n\pi}{2b} \geq 1$, the boundary conditions are not satisfied and the transverse propagation between the parallel plates is not possible. Obviously, the system of parallel plates behaves like a filter and is characterized by a cut-off frequency below which transverse electric waves do not propagate. Each mode has its own cut-off wavelength λ_c. One can obtain the cut-off frequencies for n order modes by assigning different values to n and taking the equality sign in Eq. (11.14). One can write the general equation for cut-off frequency as

$$f_c = \frac{nc}{2b} \tag{11.15}$$

or

$$\lambda_c = \frac{2b}{n} \tag{11.16}$$

we note that the cut-off frequency of each mode increases as the order of mode increases. f_c also depends upon the separation of plates.

From Eq. (11.13), we get

$$\sin\theta = \frac{n\lambda}{2b} \tag{11.17}$$

$$\cos\theta = \sqrt{1-\sin^2\theta} = \sqrt{1 - \frac{n^2\lambda^2}{4b^2}} \tag{11.18}$$

Using Eqs. (11.17) and (11.18), we obtain from Eq. (11.11) **the form of electric field of** the n-mode waves propagating along the z-axis.

$$E = 2i\hat{a}_x E_{ol} \sin\left(\frac{n\pi y}{b}\right)\exp\left[i\left(kz\sqrt{1-\frac{n^2\pi^2}{k^2b^2}} - \omega t\right)\right]$$

$$= 2i\hat{a}_x E_{ol} \sin\frac{n\pi y}{b}\exp[i(k_g z - \omega t)] \tag{11.19}$$

where

$$k_g = \frac{2\pi}{\lambda_g}\left(k^2 - \frac{n^2\pi^2}{b^2}\right)^{1/2} \tag{11.20}$$

The wave number k_g is known as the *guide wave number* and the corresponding λ_g is known as *guide wave length*. λ_g is given by

$$\lambda_g = \frac{2\pi}{k_g} = \frac{2\pi}{\left(k^2 - \frac{n^2\pi^2}{b^2}\right)^{1/2}} = \frac{\lambda}{\left(1-\frac{f_c}{f}\right)^2} \tag{11.21}$$

The electric field E given by Eq. (11.19) does not have a component in the z-direction and is thus said to be formed by transverse electric waves and is generally denoted by TE. The waves of mode n are usually designated by TE_n, where n is the mode number. One can obtain the magnetic field in TE_n waves using the Maxwell's relation,

$$\nabla \cdot E = -\frac{\partial B}{\partial t} + i\omega B \tag{11.22}$$

The phase velocity of the wave is given by

$$v_p = \frac{\omega}{k_g} = c\frac{k}{k_g} = c\left(1 - \frac{n^2\pi^2}{k^2 b^2}\right)^{-1/2} \tag{11.23}$$

One can see from Eq. (11.23) that the phase velocity of the wave exceeds the velocity of light ($v_p > c$) in free space. Question arises, whether this contradict Einstein's principle of special relativity? We can see that this does not. We must note that the phase velocity v_p is the velocity of nodes of the wave—the velocity of the maxima in the sinusoidal wave, or more generally, v_p is the velocity of the plane of constant phase and not of energy. The latter can be obtained from the group velocity, which is not the same as phase velocity. The *group velocity* associated with the phase velocity is given by

$$v_g = \frac{d\omega}{dk_g} \tag{11.24}$$

where $v_0 = \frac{d\omega}{dk}$, the group velocity of the wave in the medium, filling waveguide. The energy of the group of waves is moving with this velocity. From Eq. (11.20), we have

$$k_g^2 = k^2 - \frac{n^2\pi^2}{b^2} \tag{11.25}$$

Differentiating with respect to k_g, one obtains

$$k_g = k\frac{dk}{dk_g} = \frac{k}{c}\frac{d\omega}{dk_g}$$

and

$$v_g = c\frac{k_g}{k} \tag{11.26}$$

which is less than c since $k_g \le k$. Multiplying Eqs. (11.23) and (11.26) we get

$$v_p v_g = c^2 \tag{11.27}$$

Obviously, the phase velocity may be greater than c, the energy propagation velocity can never exceed the velocity of light. We see that, even if it is empty, an electromagnetic waveguide acts as a dispersive medium with an index of refraction less than one. First three modes of propagation of a wave between two parallel reflecting planes is shown in Fig. 11.2.

One can obtain another set of solutions in the similar way in which the magnetic field acts along the x-axis and it does not have component in the direction of propagation of wave. These are designated as *transverse magnetic waves* or TM_n waves. We must note that these different TM_n modes have cut-off wavelengths at the same value as the TE_n modes given by Eq. (11.16). In practice waveguides have either a rectangular or a circular

The electric field E given by Eq. (11.19) does not have a component in the z-direction and is thus said to be formed by transverse electric waves and is generally denoted by TE. The waves of mode n are usually designated by TE$_n$, where n is the mode number. One can obtain the magnetic field in TE$_n$ waves using the Maxwell's relation,

$$\nabla \cdot A + \frac{\partial B}{\partial t} = \cdots + i\omega B \qquad (11.22)$$

The phase velocity of the wave is given by

$$v_p = \frac{\omega}{k} = c = c \left(1 - \frac{n^2 \pi^2}{k^2}\right) \qquad (11.23)$$

One can see from Eq. (11.23) the phase velocity of the wave exceeds the velocity of light ($v_p > c$) in free space. Question arises, whether this contradict Einstein's principle of special relativity? We can see that this does not. We must note that the phase velocity v_p is the velocity of nodes of the wave—the velocity of the maxima in the sinusoidal wave, or more generally, v_p is the velocity of the plane of constant phase and not of energy. The latter of $\cdots$ is not the same as phase velocity. The group velocity associated with the phase velocity is given by

$$v_g = \frac{d\omega}{dk} \qquad (11.24)$$

where v_g, the group velocity of the wave in the medium filling waveguide. The energy of the group of waves is moving with this velocity. From Eq. (11.20), we have

$$\qquad (11.25)$$

Fig. 11.2: First three modes of propagation of a wave between two parallel reflecting planes

Differentiating with respect to k_x, one obtains

cross-section. The two shapes yield similar results with respect to phase velocity along the axis of the guide and cut off frequency.

A simple example of parallel plane waveguides in the optical region is two parallel mirrors, such as those found in some barber shops.

$$\qquad (11.26)$$

which is less than c since $k_x \le k$. Multiplying Eqs. (11.23) and (11.26) we get

11.3 PROPAGATION OF ELECTROMAGNETIC WAVES IN GUIDES OF ARBITRARY CROSS-SECTION

Obviously, the phase velocity may be greater than c, the energy propagation velocity

We now consider the propagation of electromagnetic waves inside a hollow conductor of any uniform cross-section, the walls of which are assumed to be perfect conductors. Let us obtain the general solution of the wave equation for the electric and magnetic fields within the waveguide, subject to the requirement that they satisfy Maxwell's equations and appropriate boundary conditions.

One can obtain another set of solutions in the similar way in which the magnetic field acts along the x-axis and it does not have x-component in the direction of propagation of wave. These are designated as transverse magnetic or TM$_n$ waves. We must note that these different TM$_n$ modes have cut-off wavelengths this is the same value as the TE$_n$ modes given by Eq. (11.16). In practice waveguides have either a rectangular or a circular

We have the wave equation for the electric field

$$\left(\nabla^2 - \frac{1}{c^2}\frac{\partial^2}{\partial t^2}\right) E = 0 \qquad (11.28)$$

The plane wave solution of Eq. (11.28) is of the form

$$E = E_0(x, y) \exp[i(k_g r - \omega t)] \tag{11.29}$$

One can easily write similar equations for the magnetic field. Substituting Eq. (11.29) in Eq. (11.28), one obtains

$$\left(\frac{\partial^2}{\partial x^2} + \frac{\partial^2}{\partial y^2} - k_g^2 + \frac{\omega^2}{c^2} \right) E_0(x, y) = 0 \tag{11.30}$$

Let us define a transverse Laplacian operator as

$$\nabla_T^2 = \frac{\partial^2}{\partial x^2} + \frac{\partial^2}{\partial y^2} = \nabla^2 - \frac{\partial^2}{\partial z^2} \tag{11.31}$$

Using Eq. (11.31), one can write Eq. (11.30) as

$$(\nabla_T^2 + k_c^2) E_0(x, y) = 0 \tag{11.32}$$

where $\qquad k_c^2 = -k_g^2 + \dfrac{\omega^2}{c^2} \quad \text{or} \quad k_g^2 = k_c^2 - \dfrac{\omega^2}{c^2}$ \hfill (11.33)

One finds that it is more convenient, if we express fields E and B in terms of the components parallel (E_z, B_z) and transverse (E_T, B_T) to the axis of the conductor, i.e. z-axis. We have

$$E = E_z + E_T \quad \text{and} \quad B = B_z + B_T$$

where $\qquad E_z = \hat{a}_z E_{0z}(x, y) \exp(i(k_g z - \omega t)] \hfill (11.34)$

and $\qquad E_T = E_{0T} \exp[i(k_g z - \omega t)]$

$$= \{\hat{a}_x E_{0x}(x, y) + \hat{a}_y E_{oy}(x, y) \exp[i(k_g z - \omega t)]\} \tag{11.35}$$

One can write similar equations for B_z and B_T. The electric and magnetic fields must satisfy Maxwell's equations

$$\nabla \cdot E = 0, \text{ i.e. } \frac{\partial E_{0x}}{\partial x} + \frac{\partial E_{0y}}{\partial y} + ik_g E_{oz} = 0 \tag{11.36}$$

$$\nabla \cdot B = 0, \text{ i.e. } \frac{\partial B_{0x}}{\partial x} + \frac{\partial B_{0y}}{\partial y} + ik_g B_{oz} = 0 \tag{11.37}$$

$$\nabla \times E = -\frac{\partial B}{\partial t} = i\omega B$$

i.e.

$$\frac{\partial E_{0z}}{\partial y} - ik_g E_{0y} = i\omega B_{0x} \tag{11.38}$$

$$ik_g E_{0x} - \frac{\partial E_{0z}}{\partial x} = i\omega B_{0y} \tag{11.39}$$

$$\frac{\partial E_{0y}}{\partial x} - \frac{\partial E_{0x}}{\partial y} = i\omega B_{0z} \tag{11.40}$$

and
$$\nabla \times \boldsymbol{B} = \mu \frac{\partial \boldsymbol{D}}{\partial t} = -\frac{i\omega}{c^2} \boldsymbol{E}$$

$$\frac{\partial B_{0z}}{\partial y} - ik_g B_{0y} = -\frac{i\omega}{c^2} E_{0x} \qquad (11.41)$$

$$ik_g B_{0x} - \frac{\partial B_{0z}}{\partial x} = -\frac{i\omega}{c^2} E_{0y} \qquad (11.42)$$

$$\frac{\partial B_{0y}}{\partial x} - \frac{\partial B_{0x}}{\partial y} = -\frac{i\omega}{c^2} E_{0z} \qquad (11.43)$$

Solving Eqs. (11.39) and (11.41) for E_{0x}, one obtains

$$E_{0x} = \frac{i}{k_c^2} \left(k_g \frac{\partial E_{0z}}{\partial x} + \omega \frac{\partial B_{0z}}{\partial y} \right) \qquad (11.44)$$

Similarly, one obtains

$$E_{0y} = \frac{i}{k_c^2} \left(kg \frac{\partial E_{0z}}{\partial y} - \omega \frac{\partial B_{0z}}{\partial x} \right) \qquad (11.45)$$

$$\therefore \quad E_{0T} = \hat{a}_x E_{0x} + \hat{a}_y E_{0y}$$

$$= \frac{i}{k_c^2} k_g \left(\hat{a}_x \frac{\partial E_{0z}}{\partial x} + \hat{a}_y \frac{\partial E_{0z}}{\partial y} \right) + \frac{i\omega}{k_c^2} \left(\hat{a}_x \frac{\partial B_{0z}}{\partial y} - \hat{a}_y \frac{\partial B_{0z}}{\partial x} \right)$$

$$= \frac{i}{k_c^2} [k_g \nabla_T E_{0z} - \omega \hat{a}_z \times \nabla_T B_{0z}] \qquad (11.46)$$

Here
$$\nabla_T = \hat{a}_x \frac{\partial}{\partial x} + \hat{a}_y \frac{\partial}{\partial y}$$

One can obtain similar structural equations for B_{0x} and B_{0y}. We note that these equations reveal that all the transverse components may be completely specified in terms of longitudinal components.

Now, we examine whether TEM waves can propagate inside a hollow conductor.

We have for TEM waves, $E_{0z} = B_{0z} = 0$. Therefore, from Eqs. (11.36) and (11.37), one obtains

$$\frac{\partial E_{0x}}{\partial x} + \frac{\partial E_{0y}}{\partial y} = 0; \quad \frac{\partial B_{0x}}{\partial x} + \frac{\partial B_{0y}}{\partial y} = 0 \qquad (11.47)$$

and from Eqs. (11.39) and (11.42), one obtains

$$\frac{\partial E_{0y}}{\partial x} + \frac{\partial E_{0x}}{\partial y} = 0; \quad \frac{\partial B_{0y}}{\partial x} - \frac{\partial B_{0x}}{\partial y} = 0 \qquad (11.48)$$

Taking the partial derivatives of Eqs. (11.47) and (11.48) with respect to x and y respectively and combining them suitably, one obtains

$$\nabla_T^2 E_{0x} = 0; \quad \nabla_T^2 B_{0x} = 0 \tag{11.49}$$

The components of E satisfy Laplace's equation. This means that the surface of the guide is an equipotential surface and, hence, E inside the conductor is zero. Obviously, there cannot exist, therefore, a TEM mode within a hollow waveguide with perfectly conducting walls. The result obtained is valid for a singly connected surface. Thus for a TEM mode to exist, it is necessary to have two or more unconnected surfaces. One can see that a TEM mode is a dominant mode in a coaxial cable.

Equations (11.32) and (11.49) reveal that for TEM waves $k_c = 0$, i.e. $k_g = k$. Obviously, the wave number is real for all frequencies. This means that there is no cut off frequency for TEM waves.

$$E_{\text{tangential}} = \hat{a}_n \times E = 0 \quad \text{and} \quad B_{\text{normal}} = \hat{a}_n \cdot B = 0 \tag{11.50}$$

One can see that boundary conditions given by Eq. (11.50) together with the wave equations for the field vectors E and B lead to the eigen value problems. Obviously, for any given frequency ω, only certain values of k will be constant with the boundary conditions and the wave equation.

11.4 RECTANGULAR WAVEGUIDE

The rectangular waveguide is the most commonly used waveguide and can be obtained by imposing two conducting metal plates with a separation a along the x-axis as shown in Fig. 11.3. The configuration thus formed is a hollow metal pipe of rectangular cross-section which allows the wave propagation along the z-axis. This guide is an example of the theory developed in Section 11.3.

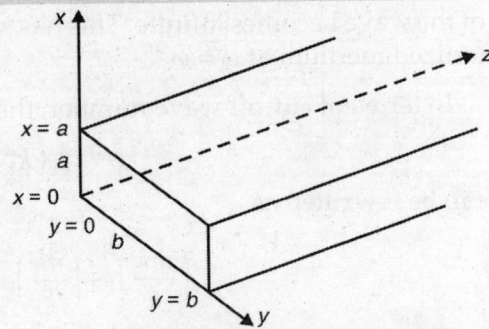

Fig. 11.3: A rectangular waveguide

TE Mode Waves

For such waves $E_{0z} = 0$. Therefore, we will have to determine the field by the solution of the equation

$$(\nabla_T^2 + k_c^2)B_{0z} = 0 \tag{11.51}$$

From Eq. (11.46), one finds that B_{0z} has to satisfy besides Eq. (11.51) the following condition

$$\left.\frac{\partial B_{0z}}{\partial x}\right|_{x=0,a} = 0 \quad \text{and} \quad \left.\frac{\partial B_{0z}}{\partial y}\right|_{y=0,b} = 0 \tag{11.52}$$

This will make $E_{\text{tangential}} = 0$

One can easily verify that the solution

$$B_{0z} = B_o \cos\left(\frac{m\pi x}{a}\right)\cos\left(\frac{n\pi y}{b}\right) \tag{11.53}$$

where m, n are integers, satisfies both Eqs. (11.50) and (11.51) provided

$$k_c^2 = \pi^2 \left(\frac{m^2}{a^2} + \frac{n^2}{b^2} \right) \tag{11.54}$$

i.e.

$$\omega_{mn} = \pi c \left(\frac{m^2}{a^2} + \frac{n^2}{b^2} \right)^{1/2} \tag{11.55}$$

where ω_{mn} is the cut-off frequency corresponding to the mode numbers m, n. The cut-off wavelength is given by

$$(\lambda_c)_{mn} = \frac{2\pi}{k_c} \frac{2}{\left[\left(\frac{m}{a} \right)^2 + \left(\frac{n}{b} \right)^2 \right]^{1/2}} \tag{11.56}$$

We find that the different values of m and n correspond to different modes of wave propagation in the rectangular waveguide. The corresponding mode is designated as R mode. If $m = 0$ and $n = 0$, the case corresponds to TEM mode. Obviously, TE_{00} mode does not exist. In case of TE_{10} mode, $n = 1$ and $m = 0$. If $a < b$, one can obtain the lowest cut off frequency by putting $n = 1$, $m = 0$, i.e.

$$\omega_{01} = \frac{\pi c}{b} \tag{11.57}$$

This is called the *principal* or *dominant mode*. If free space wavelength $\lambda = 2b$, then the guide wavelength becomes infinite and the corresponding $k_c = 0$. The phase velocity of the wave becomes infinite. This has close similarity with wave propagation through ionized medium at $\omega = \omega_p$.

In terms of cut-off wave number, the separation equation

$$k_x^2 + k_y^2 + k_z^2 = k^2 \tag{11.58}$$

can be rewritten as

$$\left(\frac{m\pi}{a} \right)^2 + \left(\frac{n\pi}{b} \right)^2 = k_c^2 - k_g^2 \tag{11.59}$$

This is known as *waveguide equation* and determines the values of waveguide wave number k_g for propagating waves in terms of the integers m and n, the dimensions a and b of the waveguide and the wave number k of the wave in free-space. From Eq. (11.59), one finds that k_g is real if free-space wave number

$$k^2 \geq \left(\frac{m\pi}{a} \right)^2 + \left(\frac{n\pi}{b} \right)^2 = k_{c_s}^2 \tag{11.60}$$

and the wave propagates in the waveguide. On the other hand, k_g is imaginary for free-space wave number

$$k^2 < \left(\frac{m\pi}{a} \right)^2 + \left(\frac{n\pi}{b} \right)^2 \tag{11.61}$$

Under this condition the waves are heavily attenuated and become evanescent. One can restate these conditions in terms of k values as

$$R_e(k_z) = \sqrt{(k_c)_{mn} - k^2} \quad \text{for} \quad k < k_c \tag{11.62}$$

and

$$I_m(k_z) = \sqrt{(k_c)_{mn} - k^2} \quad \text{for} \quad k > k_c \tag{11.63}$$

In terms of wavelength, one obtains

$$R_e(k_z) = k_c\sqrt{1 - \left(\frac{f}{f_c}\right)^2} \quad \text{and} \quad f < f_c \tag{11.64}$$

and

$$I_m(k_z) = k_c\sqrt{1 - \left(\frac{f_c}{f}\right)^2} \quad \text{and} \quad f > f_c \tag{11.65}$$

We must note that for waves propagating through the waveguide, the usual concept of wavelength and phase velocity holds good. The wavelength λ_g is defined as the distance in which the phase of E or H fields increase by 2π. In the case of TE mode waves, one can write the guide wavelength as

$$\lambda_g = \frac{\lambda}{\sqrt{1 - \left(\frac{f_c}{f}\right)^2}} \tag{11.66}$$

The values of f and f_c govern the range of permissible λ_g in practical guides. The f_c is function of waveguide dimensions.

Obviously, the physical size of the waveguide as compared to wavelength determines which wave modes can propagate in a given waveguide. Whenever two or more wave modes have the same cut-off frequency, they are said to be *degenerate modes*. In order to avoid *degeneracy* of wave modes in a rectangular waveguide, allow propagation of only one frequency in waveguide $\frac{b}{a} > 1$. The ideal ratio used in practice is $\frac{b}{a} > 2$. For higher values of $\frac{b}{a}$, the high power operation of the waveguide produces arching. At frequency close to f_c, one finds that the conductor losses become rather large.

One can write the field components of TE_{01}, as

$$E_{0x} = i\omega\frac{\partial B_{0z}}{\partial y} = -\frac{i\omega}{k_c^2}B_0 \sin\left(\frac{\pi y}{b}\right)\frac{\pi}{b}$$

$$= -\frac{i\omega b}{\pi}B_0 \sin\left(\frac{\pi y}{b}\right)$$

$$E_{0y} = 0, \ E_{0z} = 0$$

$$B_{0y} = 0, \quad B_{0y} = \frac{ik_g}{k_c^2} \frac{\partial B_{0z}}{\partial y} = -\frac{ik_g b}{\pi} B_0 \sin\left(\frac{\pi y}{b}\right)$$

$$B_{0z} = B_0 \cos\left(\frac{\pi y}{b}\right) \tag{11.67}$$

Fields given by Eq. (11.67) correspond to the wave propagated in the z-direction. We must note that this is also the direction of the Poynting vector. One obtains the energy flow in the waveguide as

$$<N>_{01} = \frac{1}{2} R_e(E \times H^*)$$

$$= \frac{1}{2\mu} R_e(E \times H^*) = \frac{1}{2\mu} R_e(-\hat{a}_y E_{0x} B_{0z}^* + \hat{a}_z E_{0x} B_{0y}^*)$$

One finds from Eq. (11.67) that $E_{0x} B_{0z}^*$ is purely imaginary.

$$\therefore \qquad <N>_{01} = \frac{\hat{a}_z}{2\mu} \frac{\omega b^2}{\pi^2} B_0^2 k_g \sin^2\left(\frac{\pi y}{b}\right) \tag{11.68}$$

One can obtain the total power by integrating the above expression from $y = 0$ to $y = b$.

TM Mode Waves

One can obtain the TM$_{mn}$ mode waves in an analogous manner by considering the characteristic *mode functions* of the form

$$\psi_{mn}^{TE} = \sin\left(\frac{m\pi x}{a}\right) \sin\left(\frac{m\pi y}{b}\right) e^{ik_z z}$$

where $m, n = 1, 2, 3, \ldots$. These wave functions satisfy the separation parameter of Eq. (11.58). One obtains the TM$_{mn}$ wave mode fields in free space from derivatives of wave mode functions.

$$E_x = -\frac{1}{i\omega\varepsilon_0} \frac{\partial^2 \psi}{\partial x \partial z}, \quad H_x = \frac{\partial \psi}{\partial y} \quad \text{at } y = 0 \text{ and } y = b \tag{11.69}$$

$$E_y = -\frac{1}{i\omega c} \frac{\partial^2 \psi}{\partial y \partial z}, \quad H_y = -\frac{\partial \psi}{\partial x} \quad \text{at } x = 0 \text{ and } x = a \tag{11.70}$$

and

$$E_z = -\frac{1}{i\omega\varepsilon_0}\left(\frac{\partial^2}{\partial z^2} + k^2\right)\psi, H_z = 0 \tag{11.71}$$

Obviously, the TM$_{mn}$ mode waves are characterized by zero H_z component and are known as *transverse magnetic* to z-axis.

TE vs TM: Wave Mode Characteristics

We have seen that in the case of TE mode waves the E_z component is zero. The TE mode field components are designated by integers m and n which characterize their behaviour in x and y directions. For TE mode waves the field components do not vanish for either m or n equal to zero. However, one finds that both m and n cannot be equal to zero at a time. We have seen that the lowest order TE mode wave in the case of rectangular waveguide is the TE_{10} mode. This mode has the lowest cut-off frequency and is said to be the *dominant mode* wave in

Fig. 11.4: Electric and magnetic field configurations for TE_{10} mode in rectangular waveguide

the rectangular waveguide. The subscripts m and n represent the number of half period variations of the field along the x-and y-coordinates respectively. We have considered the x-coordinate to coincide with the larger transverse dimension, so the TE_{10} wave has a lower cut-off frequency than TE_{01}. Figure 11.4 shows the electric and magnetic field configurations for TE_{10} mode in rectangular waveguide.

We have seen that B_z component is zero in the case of TM mode waves. TM mode waves are also designated by integers m and n which characterize their behaviour in x-and y-directions. Further, one finds that if either m or n *is* zero, the field will all be identically zero which is in contrast to the behaviour of TE mode waves in rectangular wave-guides. The TM mode waves obey the waveguide Eq. (11.59) and have identical cut-off frequencies given by Eq. (11.55). There are no magnetic charges and the B field has to vary with the x-coordinate in order to make the lines of the field continuous. As a result of these requirements, one finds that in the case of TM mode waves, the lowest possible value for either m or n is unity. The lowest cut-off frequency for the TM mode corresponds to $m = n = 1$. The wave mode corresponding to these restrictions forms the dominant mode and is designated as TM_{11}. The lowest order magnetic mode in rectangular waveguide is TM_{11} which is

Fig. 11.5: *E* and *H* configurations of dominant magnetic TM_{11} in rectangular mode

significantly different from the dominant electric mode TE_{10} in the rectangular waveguide. Figure 11.5 shows E and H configuration of dominant magnetic TM_{11} in rectangular mode.

11.5 COAXIAL WAVEGUIDE

It consists of an inner and an outer conductor. Let us find the field distribution for such a waveguide. The solution for such a waveguide should be circularly symmetric with a radial electric field and an azimuthal magnetic field (Fig. 11.6). One can write the components of the field in cylindrical coordinates as $(E_R, 0, 0)$ and $(0, H_\phi, 0)$. We can see that these satisfy the boundary conditions. Maxwell's equations are

$$\nabla \times H = J + \frac{\partial D}{\partial t} \quad \text{and} \quad \nabla \times E = -\frac{\partial D}{\partial t}$$

The lead to the equations:

$$-\frac{\partial H_\phi}{\partial z} = \varepsilon \frac{\partial E_R}{\partial t} = -i\omega\varepsilon E_R \qquad (11.72)$$

and

$$\frac{\partial E_R}{\partial z} = -\mu \frac{\partial H_\phi}{\partial t} = -i\mu\omega H_\phi \qquad (11.73)$$

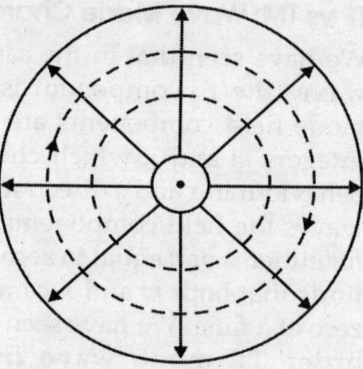

Fig. 11.6: Circularly symmetric with a radial electric field and azimuthal magnetic field

Differentiating Eq. (11.63), one obtains

$$\frac{\partial^2 E_R}{\partial z^2} = i\mu\omega \frac{\partial H_\phi}{\partial z} = i^2\varepsilon\omega^2 E_R$$

or

$$\frac{\partial^2 E_R}{\partial z^2} + \varepsilon\mu\omega^2 E_R = 0$$

and

$$\frac{\partial^2 E_R}{\partial z^2} + k^2 E_R = 0 \qquad (11.74)$$

One can see from Eq. (11.64) that the propagation constant does not depend on the dimensions of the waveguide. This shows that the guide wavelength is equal to the free space wavelength. Obviously, the propagation is similar to that of a plane wave. We must note that there is no cut-off wavelength in a coaxial waveguide.

11.6 CYLINDRICAL WAVEGUIDE

Hollow metallic cylinders are also used as waveguides. Although rectangular waveguides are widely used in practice, cylindrical waveguides have some advantages over the rectangular ones. The propagation of electromagnetic wave in this waveguide also confirms to the scalar Helmholtz equation in cylindrical coordinates given by

$$\frac{1}{\rho}\frac{\partial}{\partial\rho}\left(\rho\frac{\partial\psi}{\partial\rho}\right) + \frac{1}{\rho^2}\frac{\partial^2\psi}{\partial\phi^2} + \frac{\partial^2\psi}{\partial z^2} + k^2\psi = 0 \qquad (11.75)$$

One can show that the solution of this equation confirms with a cylindrical wave function given by

$$\psi = R(\rho)\,\Phi(\phi)\,Z(z) \qquad (11.76)$$

Figure 11.7 shows a cylindrical waveguide and the coordinate system.

Fig. 11.7: Cylindrical waveguide

One can express the ϕ-variation of cylindrical wave function as

$$\psi = \begin{Bmatrix} \sin n\phi \\ \cos n\phi \end{Bmatrix} J_n(k_\rho \rho) \exp(-ik_z z) \qquad (11.77)$$

where $J_n(k_\rho \rho)$ is a Bessel function.

One can choose the wave function either with $\sin n\phi$ or $\cos n\phi$ in order to achieve mode degeneracy except for the case when $n = 0$. In terms of this wave function the components of wave field are given by their derivatives.

The components of TM mode waves in free space can be written as

$$E_r = \frac{1}{i\omega\varepsilon_0} \frac{\partial^2 \psi}{\partial\rho\partial z} \qquad \text{and} \qquad H_\rho = \frac{1}{\rho} \frac{\partial\psi}{\partial\phi} \qquad (11.78)$$

$$E_\phi = \frac{1}{i\omega\varepsilon_0} \frac{\partial^2 \omega}{\partial\phi\partial z} \qquad \text{and} \qquad H_\phi = -\frac{\partial\psi}{\partial\rho} \qquad (11.79)$$

$$E_z = \frac{1}{i\omega\varepsilon_0}\left(\frac{\partial^2}{\partial z^2} + k^2\right)\psi \quad \text{and} \quad H_z = 0 \qquad (11.80)$$

The components of TE mode waves can be written in a similar way as

$$E_\rho = -\frac{1}{\rho} \frac{d\psi}{\partial\phi} \quad \text{and} \quad H_\rho = \frac{1}{i\omega\mu_0} \frac{\partial^2 \psi}{\partial\rho\partial z} \qquad (11.81)$$

$$E_\phi = \frac{\partial\psi}{\partial\phi} \quad \text{and} \quad H_\phi = \frac{1}{i\omega\mu_0} \frac{\partial^2 \psi}{\partial\phi\partial z} \qquad (11.82)$$

$$E_z = 0 \quad \text{and} \quad H_z = \frac{1}{i\omega\mu_0}\left(\frac{\partial^2}{\partial z^2} + k^2\right)\psi \qquad (11.83)$$

The TE mode waves in cylindrical waveguide is characterized by $E_\phi = 0$ at $\rho = a$ which confirms with

$$J_n'(k_\rho a) = 0$$

or

$$\frac{dJ_n(d_\rho a)}{da} = k_\rho\left[\frac{n}{ka} J_n(k_\rho a) - J_{n+1}(k_\rho a)\right] = 0 \qquad (11.84)$$

One finds that this Bessel function is oscillatory and has an infinite number of zeros. One can write the corresponding arguments as

$$k_\rho a = x'_{np} \tag{11.85}$$

where n stands for the order of the Bessel function, p is for order of zero and prime is used to indicate the confirmity with the first derivative of Bessel function. Using this with Eq. (11.78), one obtains for TE mode wave function

$$\psi_{np}^{TM} = J_n\left(\frac{x'_{np}\rho}{a}\right)\{\sin n\phi\}\exp(-ik_z z) \tag{11.86}$$

where $n = 0, 1, 2,...$ and $p = 1, 2, 3,...$. The wave propagation confirming with the wave function is governed by the separation equation

$$\left(\frac{x'_{np}}{a}\right)^2 + k_z^2 = k^2 \tag{11.87}$$

One can define the propagation through waveguide in terms of *guide wave number*.

$$k_g^{TE} = \sqrt{\left(\frac{x'_{np}}{a}\right)^2 - k^2} \tag{11.88}$$

One finds that the wave propagation through cylindrical waveguide stops fo $k_g = 0$. This defines the *critical wave number*, i.e.

$$k_c^{TC} = \frac{x'_{np}}{a} \tag{11.89}$$

Obviously, the cut off frequencies for TE mode wave are proportional to x'_{np}. The zero of the Bessel function for x'_{11}, determines the dominant wave propagation and is designated as TE_{11} mode. This TE_{11} mode is actually a pair of degenerate modes corresponding to $\sin\phi$ and $\cos\phi$ variations. This is why, there is no frequency range for single mode propagation.

We must note that the treatment of TM mode waves is analogous to TE mode. One finds the z-component of electric field in TE mode as

$$E_z = \frac{1}{\omega\varepsilon}(k^2 - k_z^2)\psi \tag{11.90}$$

This must vanish at the conducting walls $\rho = a$. This requires that

$$J_n(k\rho a) = 0 \tag{11.91}$$

One can write the TM mode wave function analogous to Eq. (11.86) as

$$\psi_{np}^{TM} = J_n\left(\frac{x_{np}\rho}{a}\right)\begin{Bmatrix}\sin n\phi\\\cos n\phi\end{Bmatrix}\exp(-ik_z z) \tag{11.92}$$

Proceeding exactly in the same manner as in the case of the TE mode waves, one obtains the guide wave number as

$$k_g^{TM} = \sqrt{\left(\frac{x_{np}}{a}\right)^2 - k^2} \tag{11.93}$$

and the critical wave number as

$$(k_c)_{np}^{TM} = \frac{x_{np}}{a} \tag{11.94}$$

The dominant mode of TM waves is TM_{11}. On comparing the results for TE and TM mode waves, one finds that there exists *degeneracy* between TE_{op} and TE_{1p}. Other TE and TM modes are *nondegenerate* and have different cut-off frequencies and hence different propagation constants. One can define the Z-directed characteristic wave impedances for the TE and TM modes as

$$(Z_0)^{TE} = \left(\frac{E_\rho}{H_\phi}\right) = -\frac{E_\phi}{H_\rho} = \frac{\omega\mu_0}{k_z} \tag{11.95}$$

and

$$(Z_0)^{TM} = \left(\frac{E_\rho}{H_\phi}\right) = -\frac{E_\phi}{H_\rho} = \frac{k_z}{\omega\varepsilon}$$

where Z_0 refers to the cylindrical waveguide with loss free media.

11.7 CAVITY RESONATORS

The purpose of transmission lines and waveguides is to transmit electromagnetic energy efficiently from one point to another. A resonator is an energy storage device and such a device will follow several physical analogies which will make the wave regions to function as 'circuits' particularly at ultra high frequencies. Although an electromagnetic cavity resonator can be of any shape whatsoever, an important class cavities is produced by placing end faces on a length of cylindrical waveguide. The closed chambers of conducting walls have close similarities with the LC circuit and are known as *cavities* or *resonators*. The standing waves are set in the cavities.

Fig. 11.8: A perfect conductor placed into the half infinite space $Z > Z_0$

Let us consider a plane electromagnetic wave with components E_x and H_y, travelling in the z-direction. Let us suppose that a perfect conductor is placed into the half infinite space $Z > Z_0$ as shown in Fig. 11.8.

One can see that the wave incident on the conductor turns back and one find the total electric field as

$$E_x = A_i \exp(i(kz - \omega t)] + A_r \exp[-i(kz + \omega t) \tag{11.96}$$

This must satisfy the boundary condition, $E_x = 0$ at $z = z_0$. We have

$$A_i \exp[i(kx_0 - \omega t)] + A_r \exp[-i(kx_0 + \omega t)] = 0$$

$$\therefore \quad A_r = -A_i \exp(2ikz_0)$$

and

$$\begin{aligned}
E_x &= A_i\{\exp i(kz - \omega t) - \exp(2ikz_0)\exp(-i(kz + \omega t))\} \\
&= A_i \exp(-i\omega t)\{\exp(ikz) - \exp(2ikz_0)\exp(-ikz)\} \\
&= A_i \exp(-i\omega t)\exp(ikz_0)\{\exp ik(z - z_0) - \exp(-ik(z - z_0)\} \\
&= 2iA_i \exp(-i\omega t)\exp(ikz_0)\sin k(z - z_0) \tag{11.97}
\end{aligned}$$

Obviously, Eq. (11.97) represents a standing wave. We must note that the standing wave pattern does not undergo any change if an infinitely conducting plate is placed parallel to the first plate at a position of a node where the electric field is zero. Obviously, the two conducting plates, thus, constitute a resonator in which the electromagnetic energy bounces between the two plates.

Let us consider a closed box formed by placing end faces on a rect-angular waveguide. Let us assume that the end surfaces are plane and perpendicular. Because of the reflection at the end faces, we can see that the waves in the cavity are standing waves, i.e. not progressive. Obviously, this cavity resonator will resonate at a frequency at which the length of the cavity is an integral multiple of half wavelength measured in waveguide.

$$
\left.
\begin{aligned}
E_x &= E_1 \cos(k_1 x)\sin(k_2 y)\sin(k_3 z)\exp(-i\omega t) \\
E_y &= E_2 \sin(k_1 x)\cos(k_2 y)\sin(k_3 z)\exp(-i\omega t) \\
E_z &= E_3 \sin(k_1 x)\sin(k_2 y)\cos(k_3 z)\exp(-i\omega t)
\end{aligned}
\right\}
$$

and $\qquad\qquad$ (11.98)

For the boundary conditions to be satisfied, one finds that k_1, k_2, k_3 must have the values given by

$$
k_1 = \frac{l\pi}{a}, \; k_2 = \frac{n\pi}{b} \text{ and } k_3 = \frac{n\pi}{c} \qquad\qquad (11.99)
$$

where a, b, c are the dimensions of the box and l, m, n are integers.

Substitution of any component in the appropriate wave equation exhibits that the fields given by Eq. (11.98) to be acceptable, the free space wave number has to satisfy the following condition

$$
k^2 = \frac{\omega^2}{c^2} = \pi^2 \left(\frac{l^2}{a^2} + \frac{m^2}{b^2} + \frac{n^2}{c^2} \right) \qquad\qquad (11.100)
$$

Obviously, there is an infinite number of resonant frequencies and, hence, the infinite number of modes of the cavity corresponding to the different values of l, m, n.

One can obtain the magnetic field components from Maxwell's relation $\nabla \times E$ $= -\dfrac{\partial B}{\partial t}$. One can see from Eqs. (11.98) and (11.99) that if $l = 0, m = 1$ and $n = 1$, the electric field is transverse to the direction of propagation. This mode is designated as TE_{011} mode. One can see that besides TE_{lmn} modes there are also other modes possible in the cavity. There are TM_{lmn} modes in which the magnetic field is transverse to the direction of propagation. We must note that for the same cavity, TE_{lmn} waves and TM_{lmn} waves occur at the same frequency. Thus at a particular resonant frequency the standing wave in the cavity is the sum of two resonating waves, i.e. the TE mode and TM mode.

We have mentioned above that for a definite field configuration the cavities have some discrete resonant frequencies. Obviously, if one tries to excite a particular mode of oscillation in a cavity, the right sort of fields will not be built up unless the exciting frequency is equal to the resonant frequency. In practice, one finds that appreciable excitation occurs over a narrow band of frequencies around the resonant frequency. This means that the sharp frequency of oscillation occurs partly, because of the dissipation of energy in the cavity walls. One can express the measure of these losses in terms of the Q factor of the cavity defined as

$$Q = \omega_0 \frac{\text{stored energy in the cavity}}{\text{power loss per cycle to the walls of the cavity}} \qquad (11.101)$$

Here ω_0 is the resonance frequency, assuming no losses. One can readily estimate the power loss in a cavity by computing the time average of the Poynting vector into the wall at the surface

$$<N> = \frac{1}{2} \, \text{Real} \, (E_{11} \times H_{11}) \qquad (11.102)$$

where E_{11} and H_{11} represents the tangential components of the electric and magnetic fields respectively.

An important practical resonant cavity is the *right circular cylinder*, perhaps with a piston to allow tuning by varying the height. The cylinder is shown in Fig. 11.9 with inner radius R and length d. For a TM mode the transverse wave equation for $\psi = E_z$, subject to the boundary condition $E_z = 0$ at $\rho = R$, has the solution

$$\psi(\rho, \phi) = E_0 J_m(\gamma_{mn} \rho) \exp(\pm im\phi) \qquad (11.103)$$

where

$$\gamma_{mn} = \frac{x_{mn}}{R} \qquad (11.104)$$

Fig. 11.9: Cylindrical cavity

x_{nm} is the nth root of the equation, $J_m(x) = 0$. The integers m and n take on the values $m = 0, 1, 2,...$ and $n = 1, 2, 3,.....$. The resonant frequencies are given by

$$\omega_{mn\rho} = \frac{1}{\sqrt{\mu\varepsilon}} \sqrt{\frac{x_{mm}^2}{R^2} + \frac{p^2\pi^2}{d^2}} \qquad (11.105)$$

The lowest TE mode has $m = 0$, $n = 1$, $p = 0$ and so it is designated as TM_{010}. Its resonant frequency is

$$\omega_{010} = \frac{2.405}{\sqrt{\mu\varepsilon}R} \qquad (11.106)$$

One can write the explicit expressions for the field as

$$\left. \begin{array}{l} E_z = E_0 J_0\left(\dfrac{2.405\rho}{R}\right)\exp(-i\omega t) \\[3mm] H_\phi = -\sqrt{\dfrac{\varepsilon}{\mu}} \, E_0 J_1\left(\dfrac{2.405\rho}{R}\right)\exp(-i\omega t) \end{array} \right] \qquad (11.107)$$

Obviously, the resonant frequency for this mode is independent of d and hence simple tuning is impossible.

For TE modes, the basic solution Eq. (11.104) still applies, but the boundary condition on $H_z \left[\left(\dfrac{\partial \psi}{\partial \rho}\right)\Big|_R = 0 \right]$ makes

$$\therefore \quad \gamma_{mn} = \frac{x'_{mn}}{R}$$

where x'_{mn} is the nth root of $J'_n(x) = 0$. The resonant frequencies are given by

$$\omega_{mnp} = \frac{1}{\sqrt{\mu\varepsilon}}\left(\frac{x'^2_{mn}}{R^2} + \frac{p^2\pi^2}{d^2}\right) \tag{11.108}$$

where $m = 0, 1, 2,...$ but $n, p = 1, 2, 3,...$. The lowest TE mode has $m = n = p = 1$ and is designated as TE_{111}. Its resonance frequency is

$$\omega_{111} = \frac{1.841}{\sqrt{\mu\varepsilon}R}\left(1 + 2.912\frac{R^2}{d^2}\right)^{1/2} \tag{11.109}$$

while the fields can be derived from

$$\psi = H_z = H_0 J_1\left(\frac{1.841\rho}{R}\right)\cos\phi\sin\left(\frac{\pi z}{d}\right)\exp(-i\omega t) \tag{11.110}$$

For d large enough ($d > 2.03/R$), the response frequency ω_{111} is smaller than that for the lowest TM mode. Then the TE_{111} mode is the fundamental oscillation of the cavity.

Since the frequency depends on the ratio $\dfrac{d}{R}$, it is possible to provide easy tuning by making the separation of the end faces adjustable.

These cavities are widely used at microwave frequencies to study the nature of filling dielectrics. The resonant frequencies and strength of resonance are diagnostic features of dielectrics filling the cavities.

The quality factor can be obtained from the following expression

$$Q = \frac{1.202\eta}{\text{Real}\,(\eta)\left(1 + \dfrac{R}{d}\right)} \tag{11.111}$$

where η is the intrinsic impedance of cavity metal. If $R = d$, the Q of cavity depends mainly on the property of the conductor used in making the cavity.

The cavity resonators are characterized by their resonant frequencies and quality factors. These parameters depend mainly on: (i) the size of the cavity and (ii) the nature of the dielectric filling the cavity.

Any change in these parameters perturb the normal functioning of the cavity and causes corresponding changes in the performance of the cavity. In practice, the perturbation in the size of a cavity is impractical and is generally not used for diagnostic purposes. However, the perturbation produced by the changes in the dielectric property of the cavity has proved to be of great diagnostic value.

Cavities are widely used as frequency meters. Cavities are also used in experiments where high microwave fields are required, e.g. in electric spin resonance experiments.

There are also open resonators which employ open structures. Such resonators consist of two reflectors facing each other. The reflection may be flat or spherical. These resonators have been found to be very useful at optical wavelengths and are widely used in lasers.

11.8 EARTH AND IONOSPHERE AS A RESONANT CAVITY

A somewhat unusual example of a resonant cavity is provided by the earth itself as one boundary surface and the ionosphere as the other. The region between the earth's surface and the ionosphere, which is approximately 80 km above the earth, forms a waveguide that allows the propagation of radio waves around the curve of the earth, as shown in Fig. 11.10.

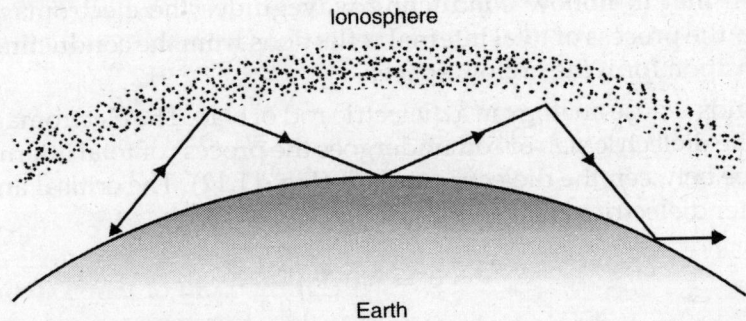

Ionosphere

Earth

Fig. 11.10: The ionosphere and the earth act as a waveguide for radio waves

The lowest resonant modes of such a system are evidently of very low frequency, since the characteristic wavelength must be of the order of magnitude of the earth's surface. In such circumstances the ionosphere and the earth both appear as conductors with real conductivities. Sea water has a conductivity $\sigma \sim 0.1\,\Omega^{-1}\,m^{-1}$, while the ionosphere has $\sigma \sim 10^{-7}$–$10^{-4}\,\Omega^{-1}\,m^{-1}$. The walls of the cavity are thus far from perfectly conducting, especially the outer one.

11.9 OPTICAL FIBER WAVEGUIDE

It consists of transparent fibers with a diameter of a few microns, called optical fibers. These fibers are made of glass or quartz, although other materials, such as nylon, are being tested. A ray entering at one end follows the axis of the fiber as a result of several reflections, emerging at the other end (Fig. 11.11). When the fibers are arranged in bundles, an image can be transmitted from one point to another.

The recent development of fused-silica optical fibers of high purity have proved to be very good waveguides for millimetre waves and light waves, it has a capability of handling a large number of wave modes.

1 micron

Fig. 11.11: Optical fiber waveguide

Acoustical waveguides are also very common. The air ducts in the heating system of a house, for example, act as acoustical waveguides which are capable of transmitting the noises from the furnace or sounds from one room to another. The inner ear is essentially an acoustical waveguide.

11.10 DIELECTRIC WAVEGUIDES

We have seen that in hollow conducting waveguides the electromagnetic energy propagates by the process of total internal reflections from the conducting surface and their superposition form the hollow space.

One also finds a close analogy in a dielectric rod or slab. The electromagnetic energy launched in the dielectric slab or rod undergoes the process of total internal reflections at the interface between the dielectric and air (Fig. 11.12). The critical angle for given inner and outer dielectrics is given by

$$\theta_c = \sin^{-1} \sqrt{\frac{\mu_2 \varepsilon_2}{\mu_1 \varepsilon_1}} \tag{11.112}$$

where μ_2, ε_2 and μ_1, ε_1 are relative permeabilities and dielectric constants of outer and inner media.

Fig 11.12. Slab dielectric waveguide

Let us now consider how a plane slab of a dielectric can serve as a waveguide.

We consider a plane dielectric sheet (A) sandwitched between two layers (B) and (C) as shown in Fig. 11.13. The outer layers B and C forming the cladding are called as substrate (B) and superstrate (C). Between these two layers there is the infinite sheet, a thin film of thickness of the order of the optical wavelength d. Usually these are made from gallium arsenide, zinc oxide, etc. Let us assume that the layers do not absorb the electromagnetic radiation incident on them, or absorb it very weakly, and their refractive indices satisfy the following condition,

Fig. 11.13: Plane dielectric sheet sandwitched between two dielectrics

$$n_1 > n_2 \geq n_3 \tag{11.113}$$

The condition given by Eq. (11.113) fulfils the condition for internal reflection.

TE mode Waves

Let us consider a TE mode wave propagating in the z-direction and field E be a function of transverse coordinates. Let the components of the field be $\{0, E_y \exp i(\beta z - \omega t), 0\}$ where β is the propagation constant in the z-direction. β is given by

$$\beta = k' \cos\theta = n_1 k \cos\theta \qquad (11.114)$$

($\because \ k' = \dfrac{2\pi}{\lambda_d} = \dfrac{2\pi}{\lambda} n_1$, where λ_d is the wavelength in the dielectric).

One finds that the wave equation

$$\left(\nabla^2 - \frac{1}{c^2}\frac{\partial^2}{\partial t^2}E\right) = 0$$

reduces to $\qquad \dfrac{\partial^2 E_y}{\partial x^2} + (n^2 k^2 - \beta^2)E_y = 0 \qquad (11.115)$

The fields must vanish at $x = \pm\infty$. The general solution of Eq. (11.115) for two media in this problem can be written in the form of sines and cosines or in terms of complex exponentials. We have

$$\begin{aligned}
E_y &= ae^{-\delta x}e^{i(\beta z - \omega t)} \ \text{ for } x \geq 0 \\
&= (B\cos kx + c\sin kx)\exp i(\beta z - \omega t) \ \text{ for } 0 \geq x \geq -d \\
&= De^{\gamma x}e^{i(\beta z - \omega t)} \ \text{ for } x \geq -d
\end{aligned} \qquad (11.116)$$

where $\delta^2 = \beta^2 - n_3^2 k^2 ; k^2 = n_1^2 k^2 - \beta^2 ; \gamma^2 = \beta^2 - n_2^2 k^2$; each one of these being a positive quantity. From first of these relations, one obtains

$$\beta^2 > n_3^2 k^2, \ \text{ i.e. } \ n_1^2 k^2 \cos^2\theta > n_3^2 k^2$$

or $\qquad n_1^2 \cos^2\theta > n_3^2$

and from the third relation, one obtains $n_1^2\cos^2\theta > n_2^2$. Obviously, these satisfy conditions for internal reflection at the two boundary conditions.

One obtains the tangential components of the magnetic field H with the help of Maxwell's equations as

$$H_z = -\frac{i\delta}{\omega\mu_0}[Ae^{-\delta x}e^{i(\beta z - \omega t)}] \qquad\qquad \text{for } x \geq 0$$

$$= \frac{ik}{\omega\mu_0}[-A\sin kx + C\cos kx]\exp i(\beta z - \omega t) \qquad \text{for } 0 \geq x \geq -al$$

$$= \frac{i\gamma}{\omega\mu_0}[A\cos kd - C\sin kd]\exp\{\gamma(x+d)\}\exp\{i(\beta z - \omega t) \ \text{ for } x \leq -d$$

The tangential component of H has to satisfy boundary conditions, i.e. H_t must be continuous across the surface at $x = 0$ and $x = -d$. One obtains

$$-\delta A = kc \qquad (11.117)$$

or $\qquad k[A\sin kd + C\cos kd] = \gamma[A\cos kd - C\sin kd] \qquad (11.118)$

i.e.
$$dA + kc = 0$$

and $(kA + \gamma C)\sin kd + (kc - \gamma A)\cos kd = 0$

One can easily see that for these to have a non-trivial solution for A and C, the determinant of the system must vanish. This gives

$$\tan kd = \frac{k(\gamma + \delta)}{k^2 - \gamma\delta} \tag{11.119}$$

When one try to solve Eq. (11.119) graphically or numerically, one obtains the value of β^2 which, in turn, provides the value of the angle θ at which the electromagnetic energy must be coupled into the guide.

We must note that if $\gamma = 0$, there is a loss of total internal reflection at the lower boundary and the mode will no longer be guided.

11.11 TE AND TM MODES: WAVE IMPEDANCES

We have seen that according to $E_z = 0$ or $H_z = 0$, the waves are referred to as transverse electric (TE) or transverse magnetic (TM) waves. When carrying such waves, the guide is said to operate in TE or TM mode.

For any transverse EM wave, the wave impedance (η) in ohms is defined as

$$\eta = \frac{|E_T|}{|H_T|}$$

For a waveguide in a TE mode,

$$|E_1|^2 = |E_x|^2 + |E_y|^2 = \left(\frac{\omega\mu}{k_{TE}}\right)^2 \left(|H_y|^2 + |H_x|^2\right)$$

$$= \left(\frac{\omega\mu}{k_{TE}}\right)^2 |H_T|^2$$

or
$$\eta_{TE} = \frac{\omega\mu}{k_{TE}} \tag{11.120}$$

Equation (11.120) involves only lengths of two-dimensional vectors, η must be independent of the coordinate system. One can easily show that

$$\eta_{TM} = \frac{k_{TM}}{\omega\varepsilon} \tag{11.121}$$

11.12 DETERMINATION OF THE AXIAL FIELDS

For complete description of the TE and TM modes, it is essential to determine the respective axial fields: $F_z = H_z$ for TE; E_z for TM. $F_z e^{-jkz}$ is a cartesian component of F and must satisfy the scalar wave equation

$$\nabla^2(F_z e^{-ikz}) = -\omega^2\mu\varepsilon(F_z e^{-jkz}) \tag{11.122}$$

together with appropriate boundary conditions. We may note that $H_\phi e^{-jkz}$ is not a cartesian component and do not obey a scalar wave equation.

Explicit Solution for TE Modes of a Rectangular Guide

Wave equation can be written as

$$\frac{\partial^2 H_z}{\partial x^2} + \frac{\partial^2 H_z}{\partial y^2} + k_{cTE}^2 H_z = 0 \tag{11.123}$$

where $k_{cTE}^2 = \omega^2 \mu\varepsilon - k_{TE}^2$. Solving Eq. (11.123) by separation of variables, one obtains

$$H_z(x, y) = (A_x \cos k_x x + B_x \sin k_x x)\ (A_y \cos k_y y + B_y \sin k_y y) \tag{11.124}$$

where $k_x^2 + k_y^2 = k_{cTE}^2$. The separation constant k_x and k_y can be determined by the boundary conditions. Considering first the x-conditions $E_y(0, y) = E_y(a, y) = 0$ and $E_z = 0$, one obtains

$$\left.\frac{\partial H_z}{\partial x}\right|_{x=0} = \left.\frac{\partial H_z}{\partial x}\right|_{x=a} = 0$$

Applying these conditions to Eq. (11.124) gives $B_x = 0$ and

$$k_y = \frac{n\pi}{b} \quad (n = 0, 1, 2, ...)$$

Each pair of non-negative integers (m, n) with the exception of $(0, 0)$ which gives a trivial solution, identifies a distinct TE mode, indicated as TE_{mn}. This mode has axial field

$$H_{zmn}(x, y) = H_{mn} \cos\frac{\mu\pi x}{a} \cos\frac{n\pi y}{b} \tag{11.125}$$

from which the transverse field can be obtained. The critical wave number for TE_{mn} is

$$k_{cTE_{mn}} = \sqrt{\left(\frac{m\pi}{a}\right)^2 + \left(\frac{n\pi}{b}\right)^2} \tag{11.126}$$

in terms of which for wave number and wave impedance for TE_{mn} are

$$k_{TE_{mn}} = \sqrt{\omega^2 \mu\varepsilon - k_{cTE_{mn}}^2} \tag{11.127}$$

$$\eta_{TE_{mn}} = \frac{\omega\mu}{\sqrt{\omega^2 \mu\varepsilon - k_{cTE_{mn}}^2}} \tag{11.128}$$

One can show that $k_{cTM_{mn}} = k_{cTE_{mn}}$.

11.13 MODE CUT-OFF FREQUENCIES

In practice one has to deal with frequencies and not wave number; it is then desirable to replace the concept of critical wave number (k_c) by one of cut-off frequency (f_c). We obtain

$$f_c = \frac{u_0}{2\pi} k_c = \frac{1}{2\pi\sqrt{\mu\varepsilon}} k_c \tag{11.129}$$

In terms of cut-off frequency f_c and the operating frequency $f = \omega/2\pi > f_c$, we have

$$f_{cmn} = \frac{u_0}{2}\sqrt{\left(\frac{m}{a}\right)^2 + \left(\frac{n}{b}\right)^2} \quad \text{(rectangular waveguide)} \tag{11.130a}$$

$$k_{mn} = \frac{2\pi}{\mu_0}\sqrt{f^2 - f_{cmn}^2} \quad \text{or} \quad \lambda_{mn} = \frac{\lambda_0}{\sqrt{1 - (f_{cmn}/f)^2}} \tag{11.130b}$$

$$\eta_{\text{TE}mn} = \frac{\eta_0}{\sqrt{1 - (f_{cmn}/f)^2}} \tag{11.130c}$$

where $\lambda_0 = u_0/f$ is the wavelength of an imaginary uniform plane wave at the operating frequency and $\eta_0 = \sqrt{\mu/\varepsilon}$ is the plane wave impedance of the lossless dielectric. Equation (11.130b) exhibits the relation between the operating wavelength λ_0 and the actual guide wavelength λ_{mn}. For TM_{mn} waves, Eq. (11.130c) is replaced by

$$\eta_{\text{TM}mn} = \eta_0\sqrt{1 - \left(\frac{f_{cmn}}{f}\right)^2} \tag{11.131}$$

The phase velocity of a TE_{mn} or TM_{mn} wave is given by

$$u_{mn} = \lambda_{mn} f = \frac{\mu_0}{\sqrt{1 - \left(\dfrac{f_{cmn}}{f}\right)^2}} \tag{11.132}$$

If Eq. (11.130a) is replaced by a similar expression involving a Bessel function, then all relations remain valid for cylindrical guides.

The meaning of cut off is made clear in Eq. (11.132). As the operating frequency drops down to cut-off frequency, the velocity becomes infinite, which is characteristic, not of wave propagation, but of diffusion (instantaneous spread of exponentially small disturbances).

11.14 DOMINANT MODE

The dominant mode of any waveguide is that of lowest cut-off frequency. For a rectangular guide, the coordinate system may always be oriented to make $a \geq b$. Since

$$f_{cmn} = \frac{u_0}{2}\sqrt{\left(\frac{m}{a}\right)^2 + \left(\frac{n}{b}\right)^2}$$

for either TE or TM, but neither m nor n can vanish in TM, the dominant mode of a rectangular guide is invariably TE_{10} with

$$f_{c10} = \frac{u_0}{2a}, \quad \lambda_{10} \frac{\lambda_0}{\sqrt{1 - (\lambda_0/2a)^2}} = \frac{2\pi}{k_{10}}$$

$$u_{10} = \lambda_{10} f, \ \eta_{10} = \frac{\lambda_{10}}{\lambda_0} \eta_0$$

Since $E_{z10} = 0$, and from

$$H_{z10} = H_{10} \cos \frac{\pi x}{a}, \ E_{x10} = 0$$

$$H_{x10} = j\left(\frac{2a}{\lambda_{10}}\right) H_{10} \sin \frac{\pi x}{a}, \ E_{y10} = -\eta_{10} H_{x10}$$

$$= -j\eta_0 H_0 \sin \frac{\pi x}{a} \tag{11.133}$$

For H_{10} real, the three non-zero field components have the time-domain expressions

$$H_{z10} = H_{10} \cos\left(\frac{\pi x}{a}\right) \cos(\omega t - k_{10} z)$$

$$\text{or} \quad = -\left(\frac{2a}{\lambda_{10}}\right) H_{10} \sin\left(\frac{\pi x}{a}\right) \sin(\omega t - k_{10} z)$$

$$E_{y10} = \eta_0 \left(\frac{2a}{\lambda_0}\right) H_{10} \sin\left(\frac{\pi x}{a}\right) \sin(\omega t - k_{01} z) \tag{11.134}$$

Plots of the dominant-mode fields [Eq. (11.134)] at $t = 0$ are shown in Fig. 11.14.

Fig. 11.14: Transverse cross-section $z = \lambda_{10}/4 \ (-k_{10} z = \pi/2)$

Both $|E_y|$ and $|H_x|$ vary as $\sin(\pi x/a)$. This is indicated in Fig. 11.14 by drawing the lines of E close together near $x = a/2$ and for a part near $x = 0$ and $x = a$. The lines of H are shown evenly spaced because there is no variation with y. This same line-density convention is followed to indicate the local value of $|E| = |E_y|$ in Fig. 11.15a and of

$$|H| = \sqrt{H_x^2 + H_z^2}$$

in Fig. 11.15b. We may note that lines of H are closed curves $(\nabla \cdot H = 0)$, the H field may be considered as circulating about the perpendicular displacement current density J_D.

Figure 11.16 illustrates how the TE_{10} mode can be initiated in a rectangular waveguide by inserting a probe halfway across the top wall $(y = b, x = a/2)$ at a distance

Fig. 11.15: Longitudinal cross-sections

Fig. 11.16: TE_{10} mode in a rectangular waveguide

$z = \lambda_{10}/4$ from the end of the guide. Higher order modes are present in the vicinity of the probe, but they will not propagate if the frequency-size condition is selected correctly.

11.15 POWER TRANSMITTED IN A LOSSLESS WAVEGUIDE

The time-averaged power transmitted in the $+Z$ direction is calculated by integration of z-component of the complex Poynting vector over a transverse cross-section of the guide. Thus,

$$\overline{P}_z = \frac{1}{2}\text{Re} \iint_{\text{cross-section}} E_T \times H_T \cdot \hat{a}_z \, ds \tag{11.135}$$

Substituting the field component from Eq. (11.133) and using $A_g = ab$, one obtains the dominant mode of a lossless rectangular waveguide,

$$\overline{P}_{z10} = \frac{\eta_0}{4}|H_{10}|^2\left(\frac{2a}{\lambda_0}\right)\left(\frac{2a}{\lambda_{01}}\right)A_g$$

$$= \frac{\eta_0}{4}|H_{10}|^2 A_g \left(\frac{f}{f_{c10}}\right)^2 \sqrt{1-\left(\frac{f_{c10}}{f}\right)^2} \tag{11.136}$$

For a lossless system, we expect $\overline{P}_{z10}$ independent of z; moreover the power is proportional to the square of the field amplitude and the cross-sectional area of the guide. Since the excitation of a guide is commonly specified through the electric field amplitude, we have

$$|E_{10}| = \eta_0 \left(\frac{2a}{\lambda_0}\right)|H_{10}|$$

One can rewrite Eq. (11.136) as

$$\overline{P}_{z10} = \frac{|E_{10}|^2 A_g}{4\eta_0}\sqrt{1-\left(\frac{f_{c10}}{f}\right)^2} = \frac{|E_{10}|^2 A_g}{4\eta_{TE_{10}}} \qquad (11.137)$$

Relations similar to Eqs. (11.136) and (11.137) exist for higher-order modes.

11.16 POWER DISSIPATION IN A LOSSY WAVEGUIDE

When the conductivity of the waveguide dielectric is nonzero (but small) and/or the conductivity of the guide walls is non finite, the wave in any propagating mode will be attenuated and transmitted power will decrease exponentially with z. An approximate treatment of these dielectric and wall losses is possible based on the assumption that the two types may be analyzed separately and that the fields which interact with the walls are those which would present if the dielectric were lossless. Here, we will restrict to TE_{10} mode of rectangular waveguide.

Dielectric Loss

We have Maxwell's equations

$$\nabla \times H = (\sigma + j\omega\varepsilon)E \qquad (11.138a)$$
$$\nabla \times E = -j\omega\mu H \qquad (11.138b)$$
$$\nabla \cdot E = 0 \qquad (11.138c)$$
$$\nabla \cdot H = 0 \qquad (11.138d)$$

Now replacing $\sigma = \sigma_d$, the dielectric conductivity by zero and $\varepsilon = \varepsilon_d$, the dielectric permittivity by its complex permittivity

$$\varepsilon = \varepsilon_d - \frac{j\sigma_d}{\omega}$$

Therefore, one can obtain the field equations for the lossy dielectric from those for the lossless dielectric by formal substitution of ε for ε_d. In particular, the z-dependence of the field vectors in the lossy TE_{10} mode is exp $(-\gamma_{10} z)$, where

$$\gamma_{10} = \hat{j}\, k_{10}(\varepsilon) = \hat{j}\sqrt{\omega^2\mu\varepsilon - k_{c10}^2}$$

$$= \hat{j}\sqrt{(\omega^2\mu_d\varepsilon_d - k_{c10}^2 - j\omega\mu_d\sigma_d}$$

$$= j\beta_{10}\left(1 - \frac{j\omega\mu_d\sigma_d}{\beta_{10}^2}\right)^2 \simeq \left(\frac{\omega\mu_d\sigma_d}{2\beta_{10}}\right) + j\beta_{10} \qquad (11.139)$$

where
$$\beta_{10} = \sqrt{\omega^2\mu_d\varepsilon_d - k_{10}^2} = k_{10}(\varepsilon_d) = \frac{2\pi}{\lambda_0}\sqrt{(1-(f_{c10}/f)^2} \qquad (11.140)$$

Here we have used binomial approximation presuming that σ_d and ω are small enough to make $\omega \mu_d \sigma_d \ll \beta_{10}^2$. Now the wave number, the imaginary part of γ_{10}, in the lossy dielectric equals the wave number in perfect dielectric, while the attenuation factor, $\alpha_d = \mathrm{Re}\, \gamma_{10}$, which governs the power loss in the dielectric, is as follows:

$$\alpha_d = \frac{\omega \mu_d \sigma_d}{2\beta_{10}} = \frac{(\sqrt{(\mu_d / \varepsilon_d)}\sigma_d)}{2\sqrt{1 - (f_{c10}/f)^2}}$$

$$= \frac{1}{2}\eta_{\mathrm{TE}_{10}}\sigma_d \; (\mathrm{Np/m}) \tag{11.141}$$

Wall Loss

The attenuation factor α_d, governing the wall loss may be determined indirectly. Since power varies as the square of the field strength, the time-average transmitted power in the TE_{10} mode must obey

$$P_{av}(z) = \bar{P}_{z10}\, e^{-2\alpha_0 z}$$

where $\bar{P}_{z10}$, the entrace power is given by Eq. (11.136). Thus, the power dissipated in the walls per unit z-length is

$$P_{\mathrm{loss}}(z) = P'_{av}(z) = 2\alpha_W P_{av}(z)$$

whence
$$\alpha_W = \frac{P_{\mathrm{loss}}(z)}{2P_{av}(z)} = \frac{P_{\mathrm{loss}}(0)}{z\bar{P}_{z10}} \tag{11.142}$$

Now, we have to calculate $P_{\mathrm{loss}}(o)$, the power flowing into the 1m of the wall inner surface. We can easily show that at a wall surface, tangential H which by hypothesis can be obtained from Eq. (11.133) sets up a Poynting vector, of time average magnitude

$$\bar{S}_{\mathrm{loss}} = \frac{1}{2}R_s \,|\,H_{\mathrm{tang}}\,|^2 \tag{11.143}$$

and directed into wall. Here
$$R_s = R_e\, \eta_W = \pi\sqrt{\mu_W / \sigma_W}$$

is the surface resistance (Ω) of the wall material at the given frequency f. Integrating the appropriate expression given by Eq. (11.142) over the first 1m of each wall surface and adding the results yields finally

$$P_{\mathrm{loss}}(o) = R_s\,|\,H_{10}\,|^2 \left[b + \frac{a}{2}(f/f_{c10})^2 \right] (\mathrm{W/m}) \tag{11.144}$$

Now, from Eqs. (11.142), (11.136) and (11.146), one obtains

$$\alpha_W = \frac{R_{sc10}}{\eta_0}\left(\sqrt{\frac{f}{f_{c10}}}\right)\frac{a + 2b(f_{c10}/f)^2}{ab\sqrt{-(f_{c10}/f)^2}} \; (\mathrm{Np/m})$$

where R_{sc10} is the surface impedance at the cut-off frequency of TE_{10} and $\eta_0 = \sqrt{\mu_d/\varepsilon_d}$ is the plane wave impedance of the (lossless) dielectric.

Combined Losses

The total attenuation factor is $\alpha_{tot} = \alpha_W + \alpha_d$. We may note that

$$1\,\text{Np} = 8.686\ \text{dB}.$$

11.17 PLANAR OPTICAL WAVEGUIDE

Planar optical waveguides are important component in integrated optical devices. The simplest optical waveguide may have a planar geometry as shown in Fig. 11.17.

Fig. 11.17: Planar optical waveguide

To fabricate a planar waveguide, usually a film (refractive index n_f) with a cover layer (refractive index n_c) is grown on a substrate (refractive index n_s) such that $n_f > n_c > n_s$. Such waveguides are termed *asymmetric waveguides* and have their refractive index distribution as shown in Fig. 11.18a. For symmetric planar optical waveguides (Fig. 11.18b), the cover and substrate are fabricated with same material and refractive indices are equal, i.e. $n_c = n_s$.

Fig. 11.18: (a) Asymmetric waveguide (b) Symmetric waveguide

11.18 MODES IN PLANAR OPTICAL WAVEGUIDE

We assume that the time and z-dependence of the EM waves are $\sim \exp{(j\omega t)}$ and $\exp{(-j\beta z)}$ respectively. This implies that we can represent the operators $\dfrac{\partial}{\partial t} = -j\omega$ (condition 1) and $\dfrac{\partial}{\partial z} = -j\beta$ (condition 2).

We may note that in most of the books on optical communications, the wave number in the bounded medium is taken b and that in unbounded medium $k_0\,(=2\pi/\lambda)$. We will follow these notations for optical waveguides.

For $b \gg a$, the waveguide in y-direction is assumed to be unbounded and hence the variation along y-axis can be neglected. Mathematically, one can represent this condition as $\sigma/\sigma_y = 0$ (condition 3). Applying 1 to 3 conditions, we obtain the components of EM fields separate out into TE modes (E_y, H_x, H_z and $E_z = 0$) and TM modes (H_y, E_x, and $H_z = 0$) as follows:

TE Mode

$$H_x = -\frac{\beta}{\omega\mu_0} E_y \tag{11.145}$$

$$H_z = \frac{j}{\omega\mu_0} \frac{\partial E_y}{\partial x} \tag{11.146}$$

$$-j\beta H_x - \frac{\partial H_z}{\partial x} = j\omega\varepsilon_0 n^2(x) E_y \tag{11.147}$$

Substituting Eqs. (11.145) and (11.146) in Eq. (11.147), one obtains the wave equation as

$$\frac{d^2\varepsilon_y}{dx^2} + [k_0^2 n^2(x) - \beta^2] E_y = 0 \tag{11.148}$$

Boundary conditions: E_y and H_z should be continuous at $x = 0$ and $x = a$.

TM Mode

$$E_x = -\frac{\beta}{\omega\varepsilon_0 n^2(x)} H_y \tag{11.149}$$

$$E_z = -\frac{j}{\omega\varepsilon_0 n^2(x)} \frac{\partial H_y}{\partial x} \tag{11.150}$$

$$j\beta E_x \frac{\partial E_z}{\partial x} = j\omega\mu_0 H_y \tag{11.151}$$

$$\frac{d^2 H_y}{dx^2} - \left[\frac{1}{n^2(x)} \frac{d_n^2}{dx}\right] \frac{dH_y}{dx} + [k_0^2 n^2(x) - \beta^2] H_y = 0 \tag{11.152}$$

For step index, refractive index distribution is shown in Fig. 11.17, within the three layers the refractive index is constant (i.e. $d_n^2/dx = 0$), Eq. (11.152) reduces to

$$\frac{d^2 H_y}{dx^2} + [k_0^2 n^2(x) - \beta^2] H_y = 0 \tag{11.152a}$$

Boundary conditions: H_y and E_z should be continuous at $x = 0$ and $x = a$.

TE Modes in a Planar Waveguide

Substituting the proper values of refractive indices, the wave equation given by Eq. (11.148) is represented in the three media as:

$$\frac{d^2 E_y}{dx^2} + [k_o^2 n_s^2 - \beta^2] E_y = 0 \quad x \leq a \tag{11.153}$$

$$\frac{d^2 E_y}{dx^2} + [k_o^2 n_f^2 - \beta^2] E_y = 0 \qquad 0 < x < a \tag{11.154}$$

$$\frac{d^2 E_y}{dx^2} + [k_o^2 n_c^2 - \beta^2] E_y = 0 \qquad x \geq a \tag{11.155}$$

One can obtain the electric field E_y by solving Eqs. (11.153) to (11.155) for wave guidance that imposes the following conditions:

i. The field should decay exponentially in the substrate and cover (known as evanescent field or tail).

ii. It should be oscillatory in the film.

One obtains the electric fields in three regions as

$$E_y = A \exp(+\gamma_s x) \qquad\qquad x \leq 0 \tag{11.156}$$
$$= B \cos kx + C \sin kx \qquad 0 \leq x \leq a \tag{11.157}$$
$$= D \exp\{-\gamma_c(x-a)\} \qquad x \leq a \tag{11.158}$$

where $\qquad \gamma_{s,c}^2 = \beta^2 - k_o^2 n_{s,c}^2$ and $k^2 = k_o^2 n_f^2 - \beta^2$

Now, applying the boundary conditions, one obtains a transcendental equation as

$$\tan ka = (\gamma_s/\kappa + \gamma_c/\kappa)/(1 - \gamma_s \gamma_c/\kappa) + m\pi \tag{11.159}$$

or $\qquad ka = \tan^{-1}[(\gamma_s/\kappa + \gamma_c/\kappa)/(1 - \gamma_s \gamma_c/\kappa^2)] + m\pi \tag{11.160}$

Solving Eq. (11.160), one can obtain the propagation characteristics for a particular mode given by the integral value of m ($m = 0$ gives lowest order mode).

In optical domain, we represent the frequency by its equivalent normalized parameter termed *V-parameter* or *normalized frequency* and is given as

$$V = k_o a (n_f^2 - n_s^2)^{1/2} \tag{11.161}$$

Cut-off frequency

One can define the cut-off condition for optical waveguide as

$$\beta = k_o n_s \text{ or } \gamma_s = 0 \tag{11.162}$$

The cut-off frequency $V_c (= \kappa a)$ is given as

$$V_c = \tan^{-1}[(\gamma_c/\kappa)] + n\pi \tag{11.163}$$

TM Modes in a Planar Waveguide

For TM modes in the planar waveguide, the field component H_y can be represented by similar expression as that for E_y [Eqs. (11.156) to (11.158)]. The boundary condition demands the continuity of H_y and E_z and this leads to transcendental equation:

$$\tan \kappa_m a = (\gamma_{sm}/\kappa_m + \gamma_{cm}/\kappa_m)/(1 - \gamma_{sm}\gamma_{cm}/\kappa_m^2) \tag{11.164}$$

or $\qquad \kappa_m a = \tan^{-1}[\gamma_{sm}/\kappa_m + \gamma_{cm}/\kappa_m)/(1 - \gamma_{sm}\gamma_{cm}/\kappa_m^2)] + m\pi \tag{11.164a}$

where $\kappa_m = \kappa/n_f^2$, $\gamma_{cm} = \gamma/n_c^2$ and $\gamma_m = \gamma_s/n_s^2$ $\tag{11.164b}$

The cut-off condition is same as given by Eq. (11.162) and cut-off frequency (V_c) is given by

$$V_c = \tan^{-1}[(\gamma_{cm}/\kappa_m)] + m\pi \qquad (11.165)$$

Symmetric waveguide

The refractive index distribution for a symmetric planar waveguide (Fig. 11.18b) is

$$n(x) = n_f \quad \text{for} \quad 0 < x < a \qquad (11.166a)$$
$$= n_s \quad \text{for} \quad x \leq 0 \text{ and } x \geq a \qquad (11.166b)$$

One can obtain all the field and transcendental equations in the same way as done for asymmetric waveguide.

However, there are differences between the modes of symmetric and asymmetric waveguides. In symmetric waveguide, the TE (or TM) modes divide themselves into two sets of independent modes known as symmetric and asymmetric modes. The symmetric modes are represented in the film or guiding region by cosine function $(B\cos\kappa x)$ and asymmetric modes by sine function $(C\sin\kappa x)$. The modes of an asymmetric waveguide as discussed before can be considered as the superposition of symmetric and asymmetric modes and hence represented by Eq. (11.157). The representations in other two regions are exponentially decaying, $A\exp(+\gamma_x)$ in substrate and $D\exp\{-\gamma(x-a/2)\}$ in cover region with $\gamma_s = \gamma_c = \gamma$.

Transcendental Equations

i. Symmetric: $\qquad\qquad \tan(\kappa a/2) = (\gamma/\kappa)$ $\qquad\qquad\qquad$ (11.167a)
ii. Asymmetric: $\qquad\quad -\cot(\kappa a/2) = (\gamma/\kappa)$ $\qquad\qquad\qquad$ (11.167b)

Cut-off condition: $\gamma = 0$

Cut-off frequency: $V_c (= \kappa a) = m\pi$, with $m = 0, 1, 2, 3, 4$.

The even and odd values of m correspond to the symmetric and asymmetric modes respectively.

TM Modes

i. Symmetric: $\qquad\qquad \tan(\kappa a/2) = (\gamma_m/\kappa_m)$ $\qquad\qquad\qquad$ (11.167c)
ii. Asymmetric: $\qquad\quad -\cot(\kappa a/2) = (\gamma_m/\kappa_m)$ $\qquad\qquad\qquad$ (11.167d)

Cut-off condition: $\gamma = 0$

Cut-off frequency: $V_c (= \kappa a) = m\pi$, with $m = 0, 2, 3, 4,$. The even and odd values of m correspond to symmetric and asymmetric modes respectively.

11.19 OPTICAL FIBER (CYLINDRICAL WAVEGUIDE)

Optical fiber works on the same principle as the dielectric slab wave-guide, except, of course, for the round cross-section (Fig. 11.19). The optical fiber shown in Fig. 11.19a has a high index core of radius a surrounded by a lower-index cladding of radius b. Light is confined to the core through the mechanism of total internal reflection, but again some fraction of power resides in the cladding as well. Figure 11.19b shows the refractive index profile of the fiber.

Fig. 11.19: (a) The optical fiber geometry (b) The refractive index profile

The wave equation together with the boundary conditions for cylindrical waveguides, describes the propagation of EM waves in step-index and graded-index optical fibers. Solving these equations for an ideal step-index fiber, under the weakly guiding approximation, gives a set of solutions:

$$\Psi(r, \phi, z, t) = R(r)e^{i\phi}e^{i(\omega t - \beta z)}$$

where

$$R(r) = \begin{cases} AJ_1\left(\dfrac{ur}{a}\right), r < a \\ BK_1\left(\dfrac{wr}{a}\right), r > a \end{cases}$$

The Bessel functions $J_1(ur/a)$ are oscillatory in nature, and hence there exist m in allowed solutions (corresponding to m roots of J_1) for each value of l. Thus, the propagation phase constant b is characterized by two integers l and m.
The requisite fields in the core ($r \le a$),

$$E_z = AJ_n(ur)\exp(jn\phi)\exp\{(j(\omega t - \beta z)\} \tag{11.168a}$$

$$H_z = BJ_n(ur)\exp(jn\phi)\exp\{(j(\omega t - \beta z)\} \tag{11.168b}$$

and in cladding,

$$E_z = CK_n(Wr)\exp(jn\phi)\exp\{(j(\omega t - \beta z)\} \tag{11.168c}$$

$$H_z = DK_n(Wr)\exp(jn\phi)\exp\{(j(\omega t - \beta z)\} \tag{11.168d}$$

The boundary condition demands the continuity of E_z, E_ϕ, H_z and H_ϕ. Applying these boundary conditions, the transcendental equation is obtained as

$$(J+\kappa)(n^2 J_{core}^2 J + n_{clad}^2 \kappa) = (\beta n/k_o a)^2 \left(\frac{1}{u^2} + \frac{1}{W^2}\right)^2 \tag{11.169}$$

where

$$u^2 = \omega^2 \mu_o n_{core}^2 - \beta^2 = k_o^2 n_{core}^2$$

$$W^2 = \beta^2 - \omega^2 \mu_o \varepsilon_o n_{clad}^2 = \beta^2 - k_o^2 n_{clad}^2 \tag{11.169a}$$

Further, we obtain

$$J = J_n'(ua)/uJ_n(ua) \text{ and}$$

$$\kappa = k_n'(ua)/uk_n(ua) \tag{11.170}$$

For $n = 0$, Eq. (11.169) breaks up in two equations that is satisfied by TM and TE modes as

$$(J + \kappa) = 0 \text{ TM}_{np} \text{ modes}$$

$$(n^2_{\text{core}} J + n^2_{\text{clad}} \kappa) = 0 \quad \text{TE}_{np} \text{ modes}$$

For $n = 1$, the fiber supports hybrid modes designated as EH_{np} or HE_{np} modes. HE_{11} mode is the dominating mode of the fiber and the cut-off V for the first higher mode is 2.405.

For a multimode fiber, the number of supported modes depend on the V parameter as

$$N = V^2/2$$

The cut-off value of the normalized frequency to support a single mode in a graded index fiber is given by

$$V_c = 2.405 \left(1 + \frac{2}{\alpha}\right)^{1/2}$$

The ratio of the power in the cladding to the total power depends on the number of modes supported by the fiber and to a good approximation is as

$$P_{\text{clad}}/P_{\text{total}} = 1/0.75 \, N^{1/2}$$

ILLUSTRATIVE EXAMPLES

Example 11.1: Discuss electromagnetic waves in a plane parallel waveguide.

Solution: We have studied that electromagnetic waves in guides have certain pecularities of their own, which are due to their transverse character and the boundary conditions at the surface of the conductor. These boundary conditions are: (i) the electric field is normal, and (ii) the magnetic field is tangential to the surface of the conductor. One possible solution of Maxwell's equations satisfying these conditions for a plane wave is $E_y = E_0 \sin(\omega t - kx)$ and $B_z = B_0 \sin(\omega t - kx)$ with $B_0 = \dfrac{E_0}{C}$. The lines of force of the electric field are indicated by vertical lines and those of the magnetic field by dots and crosses in Fig. 11.20a. We must note that in this case, waveguide does not change the phase velocity of the wave, which propagates with the same phase velocity $c = \dfrac{\omega}{k}$, corresponding to propagation in free space; the waveguide limits only the wavefront.

However, Maxwell's equations also admit other solutions, which also satisfy our boundary conditions. One possible solution can be expressed as

$$E_x = E_y = 0$$

$$E_z = E_0 \sin k_2 y \cos(\omega t - k_1 x),$$

$$B_x = -\frac{k_2}{\omega} E_0 \cos k_2 y \sin(\omega t - k_1 x),$$

$$B_y = -\frac{k_1}{\omega} E_0 \sin k_2 y \cos(\omega t - k_1 x), \, B_z = 0$$

One can easily verify this by direct substitution in Maxwell's equations. This solution is called the TE solution, because the electric field is *transverse*, but the magnetic field has a component along the effective direction of propagation, or x-axis. The electric and magnetic fields are, however, perpendicular to each other. To satisfy the boundary

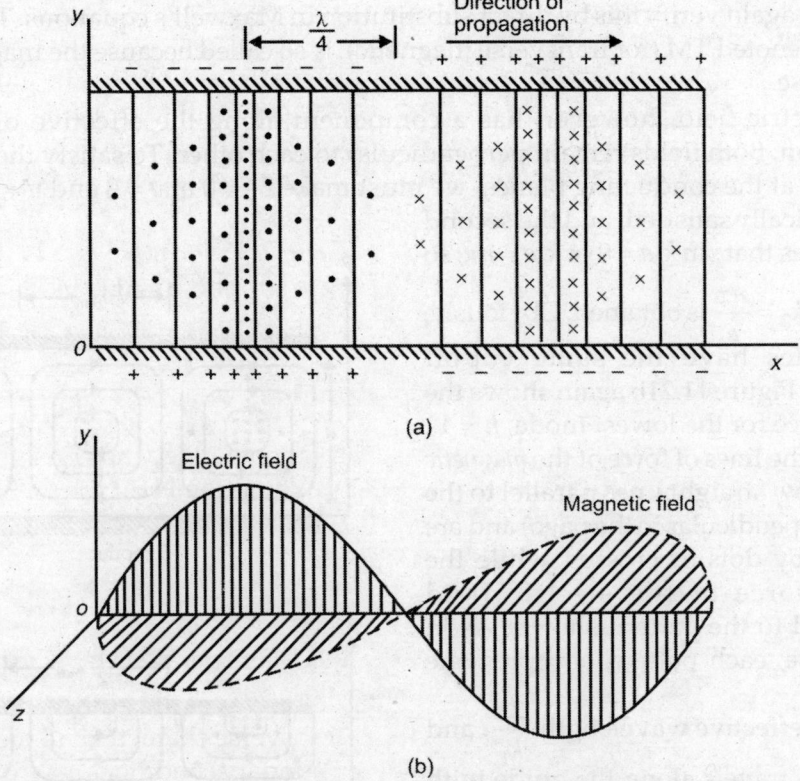

Fig. 11.20: (a) Electric lines of force (vertical lines) and magnetic lines of force (dots and crosses in the *xy* plane for an electromagnetic wave propagating parallel to two reflecting planes parallel to *xz* plane (b) Electric and magnetic fields corresponding to fields depicted in part (a).

conditions at both conducting planes, we must set $E_z = 0$ and $B_y = 0$ for $y = 0$ and $y = a$. The first is automatically satisfied, while the second requires that $\sin k_2 a = 0$ or $k_2 a = n\pi$, so that the condition $k_2 = \dfrac{n\pi}{a}$ is obtained. Figure 11.21a shows the lines of force for the lowest mode, $n = 1$. The lines of force of the electric field are straight lines parallel to the planes (perpendicular to the page) and are thus indicated by dots or crosses, while the lines of force of the magnetic field are the closed curves. Each pattern occupies one-half of the effective wavelength $\dfrac{2\pi}{k}$, and successive patterns have a phase difference of π. The patterns travel along the guide with the phase velocity, $v_p = \dfrac{\omega}{k_1}$.

Another possible solution of Maxwell's equations is

$$E_x = -\frac{k_2}{k_1} E_0 \sin k_2 y \sin(\omega t - k_1 x)$$

$$E_y = E_0 \cos k_2 y \cos(\omega t - k_1 x), \quad E_z = 0$$

$$B_x = B_y = 0, \quad B_z = \frac{\omega}{k_1 c^2} \cos k_2 y \cos(\omega t - k_1 x)$$

One can again verify this by direct substitution in Maxwell's equations. This second solution, denoted TM (for transverse magnetic), is so called because the magnetic field is transverse.

The electric field, however, has a component along the effective direction of propagation. Both fields remain perpendicular to each other. To satisfy the boundary conditions at the conducting planes, we must make $E_x = 0$ at $y = 0$ and $y = a$. The first is automatically satisfied, and the second one requires that $\sin k_2 a = 0$ or $k_2 a = n\pi$, so that again $k_2 = \dfrac{n\pi}{a}$ is obtained. Obviously, both modes have the same cut-off frequency. Figure 11.21b again shows the lines of force for the lowest mode, $n = 1$. However, the lines of force of the *magnetic* field are now straight lines parallel to the planes (perpendicular to the page) and are indicated by dots or crosses, while the lines of force of the electric field correspond to the pattern shown. As in the TE case, each pattern occupies one half of the effective wavelength $\dfrac{2\pi}{k_1}$, and the pattern travels along the guide with the phase velocity $v_p = \dfrac{\omega}{k_1}$.

We must note that the general solution of Maxwell's equation satisfying the boundary conditions of this problem is a linear combination of the TE and TM modes.

TE mode

(a)

TM mode

(b)

Fig. 11.21: Waveguide for electromagnetic waves (a) Electric field perpendicular to the page or TE mode (b) Magnetic field perpendicular to the page, or TM mode

Example 11.2: Consider a parallel plane waveguide with plate separation $20\,cm$ with the TE_{10} mode excited at $1\,GHz$. Determine the propagation constant (γ), the cut-off frequency (f_c) and the wavelength in guide (λ), assuming (i) $\varepsilon = \varepsilon_0$ (ii) $\varepsilon_r = 4$ for the medium of propagation in the guide.

Solution: We have for TE_{mo} mode,

$$\gamma_{mo} = \sqrt{\left(\frac{m\pi}{a}\right)^2 - \omega^2 \mu \varepsilon}$$

and

$$\gamma_1 = \pi \sqrt{\frac{1}{a^2} - \left(\frac{2f}{v_0}\right)^2}$$

For $f = 1$ GHz, width $a = 20$ cm, we have

$$\gamma_{10} = \gamma\sqrt{\left(\frac{1}{0.3}\right)^2 - \left(\frac{2\times10^9}{3\times10^8}\right)^2} = 14 \text{ radians/m}$$

$$f_c = \frac{m}{2a\sqrt{\mu\epsilon}} = \frac{m}{2a}v_0$$

$$\therefore \quad v_0 = \frac{1}{\sqrt{\mu\epsilon}} = \frac{3\times10^8}{\sqrt{\epsilon_r}} \text{ m/s}$$

i. For TE_{10} mode ($m = 1$, $n = 0$) and $\epsilon_r = 1$, we have

$$f_c = \frac{3\times10^8}{2a} = \frac{3\times10^8}{0.4} = 0.75 \text{ GHz}$$

$$\lambda_{10} = \frac{\lambda_0}{\sqrt{1 - \left(\frac{\lambda_0}{\lambda_c}\right)^2}}, \lambda_0 = \frac{v_0}{f} = \frac{3\times10^8}{10^9} = 0.3 \text{ m}$$

$$\therefore \quad \lambda_c = \frac{v_0}{f_c} = \frac{2a}{m} = 0.4 \text{ m}$$

$$\therefore \quad \lambda_{10} = \frac{0.3}{\sqrt{1 - \left(\frac{0.3}{0.4}\right)^2}} = 45.5 \text{ cm} \text{ and } \gamma = 14 \text{ radians/m}$$

ii. For TE_{10} mode and $\epsilon_r = 4$, we have

$$\gamma = \pi\sqrt{\left(\frac{1}{0.2}\right)^2 - \left(\frac{2\times10^9}{3\times10^8}\right)^2 \times 4} = 39.3 \text{ radians/m}$$

$$f_c = \frac{v_0}{2a} = \frac{3\times10^8}{2\times2\times0.2} = 0.375 \text{ GHz}$$

$$\therefore \quad \lambda_{10} = \frac{\lambda_0}{\sqrt{1 - \left(\frac{\lambda_0}{\lambda_c}\right)^2}}, \lambda_0 = \frac{v_0}{f} = \frac{3\times10^8}{2\times10^9} = 0.15 \text{ m}$$

$$\therefore \quad \lambda_c = \frac{v_0}{f_c} 2a = 2\times0.2 = 0.4 \text{ m}$$

$$\therefore \quad \lambda = \frac{0.15}{\sqrt{1 - \left(\frac{0.15}{0.4}\right)^2}} = 0.16 \text{ m}$$

Example 11.3: Find the width of a rectangular guide such that the energy of the electromagnetic waves whose free space wavelength is 3 cm, travels down the guide at 95% of the speed of the light.

Solution: The group velocity is given by

$$v_g = c\sqrt{1 + \left(\frac{\lambda_0}{\lambda_c}\right)^2}$$

where

$$\lambda_c = \frac{2}{\sqrt{\left(\frac{m}{a}\right)^2 + \left(\frac{n}{b}\right)^2}}$$

where the indices m and n specify the mode or

$$(\lambda_c)_{01} = \frac{2}{\sqrt{\frac{1}{a^2} + 0}} = 2a$$

We have $v = 0.95c$ and $\lambda_0 = 3\,\text{cm}$

$$\therefore \qquad 0.95\,c = c\sqrt{1 - \left(\frac{3}{2a}\right)^2}$$

This gives $a = 4.8\,\text{cm}$.

Example 11.4: What guide wavelength does 10 cm radiation (free space wavelength) exhibits in a rectangular guide whose width is 6 cm. Assuming the dominant mode, determine the cut-off wavelength for guide?

Solution: We have

$$\lambda_c = \frac{2}{\left[\left(\frac{m}{a}\right)^2 + \left(\frac{n}{b}\right)^2\right]^{1/2}} = \frac{2}{\left[\left(\frac{0}{a}\right)^2 + \left(\frac{1}{6}\right)^2\right]^{1/2}} = 12\,\text{cm}$$

Further,

$$\frac{1}{\lambda_g^2} = \frac{1}{\lambda_f^2} - \frac{1}{\lambda_c^2} = \frac{1}{(10)^2} - \frac{1}{(12)^2}$$

$$\therefore \qquad \lambda_g = \frac{120}{2\sqrt{11}} \approx 18\,\text{cm}$$

Example 11.5: Radar waves of wavelength 3 cm is propagating in a waveguide of rectangular cross-section with $a = 1\,\text{cm}$, $b = 2\,\text{cm}$. Find the modes and group velocity of the waves.

Solution: We have the cut-off frequency

$$(\omega_c)_{mn} = \pi c\left(\frac{m^2}{a^2} + \frac{n^2}{b^2}\right)^{1/2}$$

and

$$(\omega_c)_{mn} = 2\pi\nu_{mn}$$

We have the frequency of the given radiations $= 10^{10}\,\text{Hz}$. One obtains the cut-off frequencies for different modes as

$$\nu_{01} = \frac{c}{2} \times \frac{1}{2 \times 10^{-2}} = 7.5 \times 10^9\,\text{Hz}$$

$$v_{10} = \frac{c}{2} \times \frac{1}{10^{-2}} = 1.5 \times 10^{10}\ \text{Hz}$$

and

$$\omega_{mn} = \pi c \left[\frac{m^2}{a^2} + \frac{m^2}{b^2} \right]$$

$$2\pi v_{11} = \pi \times 3 \times 10^8 \left[\frac{1}{1 \times 10^{-4}} + \frac{1}{4 \times 10^{-4}} \right]^{1/2}$$

or

$$v_{11} = 1.5 \times 10^{10} \times 1.1 = 1.67 \times 10^{10}\ \text{Hz}$$

Obviously, the lowest cut-off frequency for TM mode, i.e. TM_{11} mode is

$$v_{11} = 1.68 \times 10^{10}\ \text{Hz}$$

Since only TE_{01} mode can have a frequency lower than the frequency of radiation, only TE_{01} mode can be propagated.

Now, the phase velocity of the wave

$$v_p = \frac{\omega}{k_g}$$

$$k_g = \sqrt{k_c^2 - \left(\frac{\omega}{c} \right)^2} = 1.4/\text{cm} \qquad \text{[by Eq. (11.33)]}$$

$$\therefore \qquad v_p = 4.5 \times 10^8\ \text{m/s}$$

Now, we have

$$v_p v_g = c^2$$

$$\therefore \qquad v_g = 2 \times 10^8\ \text{m/s}.$$

Example 11.6: Calculate the maximum thickness of the guide slab of a symmetrical SI planar waveguide so that it supports only the fundamental TE mode. Given $n_1 = 3.6$, $n_2 = 3.56$ and $\lambda = 0.85\ \mu m$.

Solution: For a waveguide to support only the fundamental mode, V should be less than $\pi/2$, i.e. $\dfrac{2\pi a}{\lambda}(n_1^2 - n_2^2)^{1/2} \leq \pi/2$

or

$$2a < \frac{\lambda}{2(n_1^2 - n_2^2)^{1/2}} \leq \frac{0.85\,\mu m}{2[(3.6)^2 - (3.56)^2]^{1/2}} < 0.793\ \mu m$$

Obviously, the thickness of the guide slab should not be more than $0.793\ \mu m$.

Example 11.7: A step-index fiber has a numerical operator (NA) of 0.17 and a core diameter of $100\ \mu m$. Determine the normalized frequency parameter of the fiber when light of wavelength $0.85\ \mu m$ is transmitted through it. Also find the number of guided modes propagating in the fiber.

Solution: We have $V = \dfrac{2\pi a}{\lambda}(n_1^2 - n_2^2)^{1/2}$

$$= \frac{2\pi a(\text{NA})}{\lambda} = \frac{\lambda(100\,\mu m)(0.17)}{(0.85\,\mu m)} = 62.83$$

$$\therefore \qquad M_S = \frac{V^2}{2} = \frac{(62.83)^2}{2} = 1974$$

Example 11.8: A multimode step-index fiber has a relative refractive index difference of 2% and core refractive index of 1.5. The number of modes propagating at a wavelength of 1.3 μm is 1000. Calculate the diameter of the fiber core.

Solution:

$$M_S = \frac{V^2}{2} = \frac{1}{2}\left[\frac{2\pi a}{\lambda} n_1 \sqrt{2\Delta}\right]^2$$

$$= \frac{1}{2}\frac{4\pi^2 a^2 n_1^2 2\Delta}{\lambda^2}$$

$$\therefore \qquad 2a = \frac{\lambda}{\pi n_1}\left(\frac{M_S}{\Delta}\right)^{1/2} = \frac{1.3}{\pi \times 1.5}\left(\frac{1000}{0.02}\right)^{1/2} = 62\,\mu m$$

Example 11.9: Show that E and H are mutually perpendicular in any TE and TM modes (as with ordinary plane waves).

Solution: For either type of wave, we have

$$E_x = \eta H_y \text{ and } E_y = -\eta H_x; \text{ therefore, since } \eta \text{ is real}$$

$$E_T \cdot H_T = \text{Re}(E_x H_x^* + E_y H_y^*)$$

$$= \text{Re}(\eta H_y H_x^* - \eta H_x H_y^*)$$

$$= \eta \text{Re}(H_y H_x^* - H_x H_y^*) = 0$$

As $E_z H_z^*$ also vanishes, $E \cdot H = 0$

Example 11.10: Determine the TM modes of a lossless cylindrical waveguide.

Solution: The wave equation for $E_z(r, \phi)$ is as follows, subject to the boundary conditions (i) $E_z(r, \phi + 2\pi) = E_z(r, \phi)$, (ii) $E_z(0, \phi)$ bounded and (iii) $E_z(a, \phi) = 0$

$$\frac{\partial^2 E_z}{\partial r^2} + \frac{1}{r}\frac{\partial E_z}{\partial r} + \frac{1}{r^2}\frac{\partial^2 E_z}{\partial \phi^2} + k_{c\text{TM}}^2 E_z = 0 \qquad (\because k_{c\text{TM}}^2 = \omega^2 \mu\varepsilon - k_{\text{TM}}^2)$$

Solving by separation of variables to find

$$E_{znp}(r, \phi) = E_{np} J_n(k_{c\text{TM}np} r)\cos n\phi$$

where $n = 0, 1, 2,...$ and $x_{np} = k_{c\text{TM}np} a$ is the pth positive root ($p = 1, 2,...$) of $J_n(x) = 0$.

For $H_z = 0$, one obtains all transverse field components in TM mode. The cut-off frequency of TM_{np} is obtained as

$$f_{c\text{TM}np} = \frac{\mu_0}{2\pi a} x_{np}$$

Similarly, one obtains

$$\eta_{\text{TM}np} = \eta_0 \sqrt{1 - \left(\frac{\lambda_0 x_{np}}{2\pi a}\right)^2}$$

Example 11.11: Determine the TE modes of a lossless cylindrical waveguide.

Solution: In TE mode, the axial field $H_z(r, \phi)$ obeys the wave equation and also obeys the conditions (i) and (ii) of Example 11.10. Condition (iii) must be replaced by

$$\left. \frac{dH_z}{dr} \right|_{r=a} = 0$$

The solution by separation is therefore

$$H_{znp}(r, \phi) = H_{np} J_n(k_{c\,TE\,np}\, r) \cos n\phi$$

where $n = 0, 1, 2...$ and $x'_{np} = k_{c\,TE\,np}$ is the pth positive root $(p = 1, 2, 3,...)$ of $J'_n(x) = 0$. We obtain the cut-off frequency TE$_{np}$ as

$$f_{c\,TE\,np} = \frac{\mu_0}{2\pi a} x'_{np}$$

and

$$\eta_{TM\,np} = \eta_0 \Big/ \sqrt{1 - \left(\frac{\lambda_0 x'_{np}}{2\pi a}\right)^2}$$

Example 11.12: A graded-index fiber with a parabolic profile supports the propagation of 700 guided modes. The fiber has a refractive index difference of 2%, a core refractive index of 1.45 and a core diameter of 75 μm. Calculate the wavelength of light propagating in the fiber. Also, estimate the maximum diameter of the fiber core which can give single-mode operation the same wavelength.

Solution: We have with $\alpha = 2$ for the parabolic profile

$$V = \sqrt{4Mg} = \sqrt{\pi \times 700} = 52.91$$

and

$$\lambda = \frac{2\pi a}{V} n_1 \sqrt{2\Delta} = \frac{\pi \times (75\,\mu m)}{52.91} = (2 \times 0.02)^{1/2} = 1.3\,\mu m$$

The cut-off value of the normalized frequency, V_c for single mode operation in a graded-index fiber is obtained as

$$V_c = 2.405\sqrt{2}$$

The maximum core diameter is obtained as

$$2a = \frac{V_c \lambda}{\pi n_1 \sqrt{(2\Delta)}} = \frac{2.405\sqrt{2} \times (1.3\,\mu m)}{\pi \times 1.45 \times \sqrt{2 \times 0.02}} = 4.85\,\mu m$$

GLIMPSES

1. At high frequencies $(\lambda < 0.1 \text{ m})$, the losses in transmission line become quite large. It is found convenient to use the hollow conducting tube for the propagation of EM energy at such high frequencies. The EM wave is confined to move over in the direction of the tube. Such a wave is called a *guided wave* and the hollow conducting tube as *waveguide*.

2. In the waveguide, the wave is altered from a plane wave, as is required in order that the electric and magnetic boundary conditions at the surface of the conductor be satisfied. There are several different modes (patterns) of EM waves that can satisfy the boundary conditions and propagate EM energy down a waveguide. There are two most commonly used types: *rectangular* and *circular*. The geometry and size of a waveguide allows only one mode of transmission for a limited range of frequencies.

3. The two types of waves according as $E_z = 0$ or $H_z = 0$ are referred to as transverse electric (TE) or transverse magnetic (TM) waves. When carrying such waves, the guide is said to operate in a TM mode.

4. A mode is a stable (electric or magnetic) field pattern in the transverse direction (e.g. the x-direction) with only a periodic z-dependance. In an isotropic and inhomogeneous dielectric medium, the modes of wave propagation are described by the following sets of relations:

$$\left. \begin{array}{l} \beta E_y = -\mu_0 \omega H_x \\[2mm] \dfrac{\partial E_y}{\partial x} = -i\mu_0 \omega H_z \\[2mm] -i\beta H_x = i\varepsilon_0 \omega n^2(x) E_y \end{array} \right\} \text{Transverse electric (TE) modes}$$

$$\left. \begin{array}{l} \beta H_y = -\varepsilon_0 \omega n^2(x) E_x \\[2mm] \dfrac{\partial H_y}{\partial x} = -i\varepsilon_0 \omega n^2(x) E_z \\[2mm] i\beta E_x + \dfrac{\partial E_z}{\partial x} = i\mu_0 \omega H_y \end{array} \right\} \text{Transverse magnetic (TM) modes}$$

where β is the component of the propagation vector ($k = 2\pi/\lambda$) in the z-direction,

i.e. $\beta = \beta_z = \beta_1 \cos\theta = \dfrac{2\pi}{\lambda_m}\cos\theta = \dfrac{2\pi n_1}{\lambda}\cos\theta = k n_1 \cos\theta$.

5. The maximum number of TE modes, M, supported by a symmetrical step-index planar waveguide is an integer close to or greater than $2V/\pi$, where

$$V = \frac{2\pi}{\pi} a \sqrt{n_1^2 - n_2^2}$$

6. The normalized propagation constant (b) is given by

$$b = \frac{\beta^2 - \beta_1^2}{\beta_1^2 - \beta_2^2} = 1 - \left(\frac{ua}{V}\right)^2 = \left(\frac{Wa}{V}\right)^2$$

where $\beta_1 = 2\pi/\lambda_m = \dfrac{2\pi n_1}{\lambda} = k n_1$, $k = 2\pi/\lambda$ is the vacuum propagation constant or the propagation vector. The minimum value of β in z-direction, i.e. β_{min} is obtained as

$$\beta_{min} = \beta_1 \cos\theta_m = \beta_1 \frac{n_2}{n_1} = \beta_2.$$

7. For a given guided mode, the value of propagation constant (b) lies between 0 and 1, and β^2 lies between β_1^2 and β_2^2. The maximum value that β can have is β_1, which corresponds to $\theta = 0$, i.e. the plane TEM waves travelling parallel to waveguide axis, $\beta_{max} = \beta_1$.

8. For any transverse EM wave, the wave impedance (in ohms) is defined as

$$\eta = \frac{|E_T|}{|H_T|}$$

For a waveguide in TE mode,

$$\eta_{TE} = \frac{\omega\mu}{k_{TE}}$$

η is independent of the coordinate system. Further,

$$\eta_{TM} = \frac{k_{TM}}{\omega\varepsilon}$$

9. The critical wave number for TE$_{mn}$, where (m, n) are non-negative integers with the exeption of (0, 0) which gives a trivial solution - identifies a distinct TE mode, indicated as TE$_{mn}$. The critical wave number for TE$_{mn}$ is

$$k_{cTE_{mn}} = \sqrt{\left(\frac{m\pi}{a}\right)^2 + \left(\frac{n\pi}{b}\right)^2} \tag{1}$$

where $\sin k_x a = 0$ or $k_x = \dfrac{m\pi}{a}$ ($m = 0, 1, 2,...$).

In terms of Eq. (1), the wave number and wave impedance for TE$_{mn}$ are

$$k_{TE_{mn}} = \sqrt{\omega^2\mu\varepsilon - k_{cTE_{mn}}^2}$$

and

$$\eta_{TE_{mn}} = \frac{\omega\mu}{\sqrt{\omega^2\mu\varepsilon - k_{cTE_{mn}}^2}}$$

10. Cut-off frequency (f_c)

$$f_c = \frac{u_0}{2\pi}k_c = \frac{1}{2\pi\sqrt{\mu\varepsilon}}k_c$$

The operating frequency (f) is given as

$$f = \frac{\omega}{2\pi} > f_c$$

Now $\quad f_{cmn} = \dfrac{u_0}{2}\sqrt{\left(\dfrac{m}{a}\right)^2 + \left(\dfrac{n}{b}\right)^2}$ (rectangular waveguide)

$$\eta_{TE_{mn}} = \frac{\eta_0}{\sqrt{1-(f_{cmn}/f)^2}}$$

$$\eta_{TM_{mn}} = \eta_0\sqrt{1-(f_{cmn}/f)^2}$$

The phase velocity of a TE_{mn} or TM_{mn} wave is given by

$$u_{mn} = \lambda_{mn}f = \frac{u_0}{\sqrt{1-\left(\dfrac{f_{cmn}}{f}\right)^2}}$$

11. The cut-off frequency of a mode occurs at $b = 0$, $ua = V_c = m\pi/2$, and $Wa = 0$. The fundamental mode has no cut-off frequency.

12. The dominant mode of any waveguide is that of lowest cut-off frequency. The dominant mode of a rectangular guide is invariably TE_{10}, with

$$f_{c10} = \frac{u_0}{2a}, \ \lambda_{10} = \frac{\lambda_0}{\sqrt{1-(\lambda_0/2a)^2}} = \frac{2\pi}{k_{10}}$$

$$u_{10} = \lambda_{10}f, \quad \eta_{10} = \frac{\lambda_{10}}{\lambda_0}\eta_0$$

For H_{10} real, the three non-zero components have the time-domain expressions

$$H_{z10} = H_{10}\cos\left(\frac{\pi x}{a}\right)\cos(\omega t - k_{10}z)$$

$$H_{x10} = -\left(\frac{2a}{\lambda_{10}}\right)H_{10}\sin\left(\frac{\pi x}{a}\right)\sin(\omega t - k_{10}z)$$

$$E_{y10} = \eta_0\left(\frac{2a}{\lambda_0}\right)H_{10}\sin\left(\frac{\pi x}{a}\right)\sin(\omega t - k_{10}z)$$

13. The thickness of guide layer that can support m modes is given by

$$2a = \frac{m\lambda}{2(n_1^2 - n_2^2)^{1/2}}$$

14. The thickness of a guide layer required to support only the fundamental mode is given by

$$2a \leq \frac{\lambda}{2(n_1^2 - n_2^2)^{1/2}}$$

15. The fraction of guided optical power that is confined within a guide layer is called the confinement factor. It depends on the thickness of the guide layer and the mode number.

16. Power transmitted in lossless waveguide is

$$\bar{P}_{z10} = \frac{\eta_0}{4} |H_{10}|^2 \left(\frac{2a}{\lambda_0}\right)\left(\frac{2a}{\lambda_{10}}\right) A_g$$

$$= \frac{\eta_0}{4} |H_{10}|^2 A_g \left(\frac{f}{f_{c10}}\right)^2 \sqrt{1-\left(\frac{f_{c10}}{f}\right)^2}$$

$$= \frac{|E_{10}|^2 A_g}{4\eta_0} \sqrt{1-\left(\frac{f_{c10}}{f}\right)^2} \frac{|E_{10}|^2 A_g}{4\eta_{TE\,10}}$$

where $\qquad A_g = ab$

17. The cut-off value of the normalized frequency V_c to support a single mode in a graded-index fiber is given by

$$V_c = 2.405\left(1+\frac{2}{\alpha}\right)^{1/2}$$

18. Intermodal dispersion in multimode fibers restricts their use in long-haul communications.

REVIEW QUESTIONS

1. What is a waveguide? For transverse electric waves perfectly propagating in a rectangular waveguide with perfectly conducting walls, find: (i) the cut-off wavelength, (ii) the magnetic field induction and (iii) the velocity with which energy is transmitted along the guide.

2. Describe the propagation of electromagnetic waves along a hollow waveguide of uniform cross-section.

3. Explain the principle of operation of waveguides. Explain TE, TM and TEM mode waves. Derive an expression for the guide wavelength of TE mode propagation in hollow rectangular waveguide. Explain how the cut-off mode arises. Show that the phase velocity of waves becomes infinite exactly at cut-off frequency.

4. Discuss the propagation of electromagnetic waves in a rectangular waveguide in TE mode. Show that there is a certain minimum frequency below which no transmission is possible and obtain an expression for that frequency.

5. What is a waveguide? Discuss the theory of waveguides with reference to TM mode of propagation of electromagnetic waves. Show that a waveguide acts as a high pass filter.

6. What are cavity resonators? How do they resonate? How they resemble with electronic resonant circuits? Discuss in detail the theory of a rectangular cavity resonator, showing what lowest modes are possible in TE mode.

7. What is an electromagnetic cavity resonator? Describe the excitation of TEM waves in the class of cavity resonators produced by placing plane end faces on a length of a cylindrical waveguide.

PROBLEMS

1. A waveguide consists of a long tube of rectangular cross-section with sides a and b. Show that the resultant wave is described by $\xi = 4\xi_0 \sin k_2 y \sin k_3 z \cos(\omega t - k_1 x)$ and that the only frequencies transmitted along the waveguide are those satisfying

$$v \geq \frac{1}{2} v \sqrt{\frac{n_1^2}{a^2} + \frac{n_2^2}{b^2}},$$ where n_1 and n_2 are integers. Discuss the nodal planes in the

waveguide for $n_1 = 2$ and $n_2 = 3$.

2. If waves are propagating along the z-axis in a rectangular waveguide, show that the field component for the TE mode is

$$E_x = E_0 \sin\left(\frac{\pi y}{b}\right) \exp[i(\omega t - k_g z)].$$

The width of the waveguide along the y-axis is b. Also show that the E_y component of the wave is zero.

3. Show that in a rectangular waveguide, TE_{00} mode cannot exist.

4. A rectangular waveguide has dimensions 2.5 cm and 5 cm. Determine guide wavelength λ_g, phase velocity and phase constant at a wavelength of 4.5 cm for dominant mode.

   [*Ans.* $\lambda_g = 5$cm, $k_g = 1.256/$cm and $v_g = 3.3 \times 10^8$ m/s]

5. A hollow rectangular waveguide has $a = 6$cm and $b = 4$cm. The impressed signal has the frequency = 3 GHz. Determine TE_{10} mode and (i) cut-off wavelength, (ii) guide wavelength, (iii) phase constant, (iv) intrinsic wave impedence, (v) phase velocity and (vi) group velocity.

   Ans. (i) $(\lambda_c)_{10} = 12$ cm (ii) $\lambda_g = 18.2$ cm (iii) $k_g = 0.345/$cm (iv) intrinsic wave impedance $Z_E = 685 \, \Omega$ (v) $v_g = 5.5 \times 10^8$ m/s and (vi) $v_z = 1.65 \times 10^8$ m/s]

6. In a rectangular cavity with square cross-section a and height b, a thin slab of dielectric with thickness d has been placed. Show that

$$\frac{\omega - \omega_0}{\omega_0} = -\frac{1}{2}(\varepsilon_r - 1)\frac{b}{a}.$$

SHORT ANSWER QUESTIONS

1. What do you understand by TEM modes in circular waveguides?
   [*Ans.* The transverse electric and magnetic modes of a waveguide is basically TEM mode which is otherwise called *transmission-line* modes. For this mode of operation of circular waveguide, the z-components of both E and H vectors are zero, i.e.
   $$E_z = H_z = 0$$

This means that the electric and magnetic fields are completely transverse to the direction of propagation. This mode cannot exist in hollow waveguides, since it requires two conductors such as the coaxial transmission line and two-open wire line.

2. What is a hohlraum resonator?

Ans. If the waveguide is limited by closing the ends of a finite length of the tube in conducting surfaces, then it is known as cavity or hohlraum resonator.

3. What will be the shortest length of a simplest cavity resonator to be made from a rectangular waveguide with $a = 10.16$ mm and $b = 22.86$ mm which resonate at 10 GHz?

Ans. $\lambda_g = \dfrac{\lambda}{\left[1 - \left(\dfrac{\lambda}{2b}\right)^2\right]^{1/2}}$, $\lambda = \dfrac{3 \times 10^8}{10^{10}} = 0.03$ m

$$\lambda_g = \dfrac{0.03}{\left[1 - \left(\dfrac{0.03}{2 \times 22.86 \times 10^{-3}}\right)^2\right]^{1/2}} = 0.0397 \text{ m}$$

$\therefore \lambda_g = 0.0397$ m. Obviously, the length of the shortest cavity ($\lambda_g/2$) is 0.0198 m.

4. In what respects cavity resonators are superior to conventional *L-C* circuits?

Ans. (i) The fraction of the stored energy dissipated per cycle in guide is about 1/20th the fraction dissipated per cycle in *L-C* circuit and (ii) guide have resonant frequencies which range upward from a few hundred megacycles which is almost impossible in case of *L-C* circuit.

5. What is a TE mode?

Ans. In a transverse electric (TE) mode, the electric vector of the field has no component in the direction of propagation.

6. What is TM mode?

Ans. The TM mode is basically called *transverse magnetic mode* which suggests propagation of EM energy in positive *z*-direction such that the magnetic field intensity vector *H* in *z*-direction is tentatively zero, i.e., $H_z = 0$. This implies that *E* vector is non-zero in *z*-direction for TM mode in circular waveguide.

7. What happens when electromagnetic waves propagate through a conductive medium, e.g. metal?

Ans. The conducting medium intensively absorbs electromagnetic waves. The most prominent role in this process is played by the conduction electrons.

8. When the wavelength of the emitter is too large, what will be the relation between the fields of the waveguide walls and the field of the incoming wave?

Ans. Both will be in antiphase. The resultant field is damped at the distance of wavelength, and the wave does not penetrate to any appreciable depth into the waveguide.

9. What are the typical properties of electronic resonant circuits which cavity resonators do not possess?

Ans. (i) Electronic resonant circuits can store energy in oscillating electric and magnetic fields, and (ii) dissipate a fraction of the stored energy in each cycle of oscillation.

10. What constitutes a complete set of fields to describe an arbitrary electromagetic disturbance in a waveguide or cavity?

    Ans. The various TM and TE waves, plus a TEM wave if it can exist, constitute a complete set of fields to describe an arbitrary electromagnetic disturbance in a waveguide or cavity.

11. A filled rectangular waveguide has dimensions $a = 2$ cm and $b = 1$ cm. Show that the operating frequency range over which the guide will be single mode is 7.5 Hz $< f < 15$ GHz.

    Ans. Since the guide is air-filled, $n = 1$. For $m = 1$, we have

    $$f_{c10} \text{ (for TE}_{10}\text{)} = \frac{\omega_c}{2\pi} = \frac{c}{2a} = \frac{3 \times 10^{10}}{2 \times 2} = 7.5 \text{ GHz}$$

    The next higher mode will be TE_{20} or TE_{01}, which will have the same cut-off frequency since $a = 2b$. This frequency will be twice that found for TE_{10}, or 15 GHz. Thus, the operating frequency range over which the guide will be single mode is 7.5 GHz $< f < 15$ GHz.

12. What will be the phase velocity at cut off.

    Ans. Infinite.

13. (a) Explain the notion of cut-off wavelength. (b) Is the cut-off wavelength an upper limit of the guide wavelength, just as the cut-off frequency is a lower limit of the guide frequency?

    Ans. (a) The cut-off wavelength λ_c is the wavelength of an unguided plane wave whose frequency is the cut-off frequency (f_c) (b) No; in fact, the relation

    $$\lambda_{mn} = \frac{\mu_0}{\sqrt{f^2 - f_{cmn}^2}}$$

    reveals that an (m, n) mode can propagate any guide wavelength greater than λ.

14. Write the boundary conditions on E and H at each perfectly conducting wall of the waveguide (Fig. 11.22).

    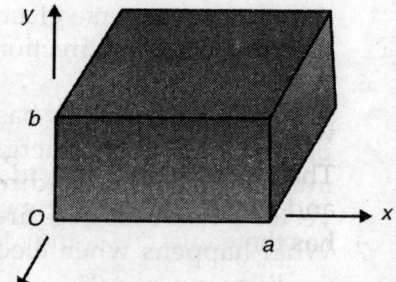

    Ans. For a perfect conductor, tangential E and normal H must vanish. Thus,

    top wall: $E_z(x, b) = E_x(x, b) = 0$ and $H_y(x, y) = 0$

    left wall: $E_z(0, y) = E_y(0, y) = 0$ and $H_x(0, y) = 0$

    right wall: $E_z(a, y) = E_y(a, y) = 0$ and $H_x(0, y) = 0$

    bottom wall: $E_z(x, 0) = E_x(x, 0) = 0$ and $H_y(x, 0) = 0$

    Fig. 11.22

15. What is "critical" about the number k_c?

    Ans. For propagation through a lossless dielectric, the wave number k must be real. But $k = \sqrt{\omega^2 \mu \varepsilon - k_c^2} = \sqrt{k_0^2 - k_c^2}$

    where k_0 is the wave number of a uniform plane wave in the unbounded dielectric at the given ω. Obviously, k_c is a critical wave number in the sense that a guided wave's same frequency "twin" must have a wave number exceeding k_c. This means the frequency f of the guided wave should exceed the quantity $(u_0/2\pi)\,k_c$, where $u_0 = 1/\sqrt{\mu\varepsilon}$ is the wave velocity in the unbounded dielectric.

16. Discuss the relative magnitudes of f_{cTEnp} and f_{cTMnp}.

    [**Ans.** For each fixed n, the zeros x_{np} of $J_p(x)$ and stationary points x'_{np}, where $J_n(x)$ is a maximum or a minimum, alternate along the x-axis (sine wave like behaviour). For $n > 0$, the function starts at 0 and the first stationary point precedes the first positive zero; $x'_{np} < x_{np}$, whence $k_{cTEop} < k_{cTMop}$ and $k_{cTEop} > k_{cTMop}$. For $n = 0$, the function starts at a maximum and the ordering gets reversed as

    $$k_{cTEop} > k_{cTMop} \text{ and } k_{cTEop} > k_{cTMop}.$$

17. Compare the rectangular and cylindrical waveguides as power transmitters when each operates in its dominant mode.

    Ans. We have for the dominant mode of lossless rectangular waveguide

    $$\bar{P}_{z_{10}} = \frac{\eta_0}{4}\,|H_{10}|^2 \left(\frac{2a}{\lambda_0}\right)\left(\frac{2a}{\lambda_{10}}\right) A_g$$

    $$= \frac{\eta_0}{4}\,|H_{10}|^2\, A_g \left(\frac{f}{f_{c10}}\right)^2 \sqrt{1 - \left(\frac{f_{c10}}{f}\right)^2}$$

    Time average power transmission in TE_{11} mode of a lossless cylindrical waveguide is

    $$\bar{P}_{z_{11}} = \frac{\eta_0}{4}\,|H_{11}|^2\, A_g \left(\frac{f}{f_{cTE_{11}}}\right)^2 \sqrt{1 - \left(\frac{f_{cTE_{11}}}{f}\right)^2} \times \left[\frac{(x'_{11})^2 - 1}{(x'_{11})^2} J_1^2(x'_{11})\right]$$

    where $A_g = \pi a^2$ is cross-sectional area.

    These expressions show identical dependence on H-amplitude, cross-sectional area, and normalized frequency. The only difference lies in a geometrical factor, which has the value 1.0 for the rectangular guide and the value

    $$\frac{(1.841)^2 - 1}{(1.841)^2}(0.5814)^2 = 0.239$$

    for the cylindrical waveguide.

18. What do you understand by cut-off wavelength? Is the cut-off wavelength an upper limit of the waveguide wavelength, just as the cut-off frequency is a lower limit of the guide frequency?

    Ans. λ_c is the wavelength of an unguided plane wave whose frequency is the cut-off frequency, $\lambda_c f_c = \mu_0$. We have

    $$\lambda_{min} = \frac{\mu_0}{\sqrt{f^2 - f_{cmn}^2}}$$

This reveals that an (m, n) mode can propagate with any guide wavelength greater than λ.

MULTIPLE CHOICE QUESTIONS

1. The cut-off frequency corresponding to the mode numbers m, n for rectangular cross-section of dimensions a and b, waveguide is given by

   (a) $\omega_{mn} = \pi c \sqrt{\dfrac{m^2}{a^2} + \dfrac{n^2}{b^2}}$

   (b) $\omega_{mn} = \left(\dfrac{m^2}{a^2} + \dfrac{n^2}{b^2}\right)$

   (c) $\omega_{mn} = \pi \sqrt{\dfrac{m}{a} + \dfrac{n}{b}}$

   (d) $\omega_{mn} = \pi c \left(\dfrac{m^2}{a^2} + \dfrac{n^2}{b^2}\right)$ [a]

2. The principal or dominant mode cut-off frequency for rectangular cross-section having dimensions a and b, waveguide is given by

   (a) $\omega_{mn} = \pi c \sqrt{\dfrac{m^2}{a^2} + \dfrac{n^2}{b^2}}$

   (b) $\omega_{01} = \dfrac{\pi c}{b}$

   (c) $\omega_{01} = \dfrac{\pi c}{a}$

   (d) $\omega_{01} = \pi c$ [b]

3. Which one of the following statements is true for a coaxial waveguide:
   (a) guide wavelength is equal to the free space wavelength
   (b) propagation constant depends on the dimension of the waveguide
   (c) there is a cut-off wavelength
   (d) the propagation is not similar to that of a plane wave   [a]

4. Electromagnetic cavity resonators are energy storage devices and used in
   (a) klystron
   (b) wave meters
   (c) band pass filter
   (d) all of the above   [d]

5. The cavity resonators used in lasers are called
   (a) open resonators
   (b) closed resonators
   (c) Raman resonators
   (d) none of the above   [a]

6. In a waveguide, suffix mn of the modes TE/TM denotes
   (a) half wavelength of E field and full wavelength of H field
   (b) half wavelengths of E and H fields in directions other than guide axis
   (c) full wavelength of E field and half wavelength of H field
   (d) half wavelength of E and H fields   [b]

7. Cavity resonators are characterized by their
   (a) resonant frequencies only
   (b) quality factors only
   (c) resonant frequencies and quality factors
   (d) none of the above   [c]

8. The critical angle for given inner and outer dielectrics in a dielectric waveguide with μ_2, ε_2 and μ_1, ε_1 as relative permeabilities and dielectric constants of outer and inner media respectively is given by

(a) $\theta_c = \cos^{-1}\sqrt{\dfrac{\mu_2\varepsilon_2}{\mu_1\varepsilon_1}}$ (b) $\theta_c = \sin^{-1}\sqrt{\dfrac{\mu_2\varepsilon_2}{\mu_1\varepsilon_1}}$

(c) $\theta_c = \tan^{-1}\sqrt{\dfrac{\mu_2\varepsilon_2}{\mu_1\varepsilon_1}}$ (d) $\theta_c = \sin^{-1}\sqrt{\dfrac{\mu_1\varepsilon_1}{\mu_2\varepsilon_2}}$ **[b]**

9. The product of wave impedance of $TE_{11} - TM_{11}$ mode, i.e. $\eta_{TE} \times \eta_{TM}$ is

(a) $\sqrt{\mu/\varepsilon}$ (b) μ/ε

(c) 1.5   (d) $\omega^2\mu\varepsilon - [\pi/a]^2$ **[b]**

10. A rectangular waveguide acts as a
(a) band pass filter   (b) high pass filter
(c) low pass filter   (d) band stop filter   **[b]**

11. For a square waveguide, the dominant TM mode will be
(a) TM_{01} (b) TM_{10}
(c) both TM_{01} and TM_{10} (d) none of the above   **[c]**

12. In a waveguide the operating wavelength λ is less than the cut-off wavelength λ_c,
(a) attenuated
(b) guided
(c) guided in the same attenuation
(d) none of the above   **[b]**

13. For a cylindrical waveguide, the dominant mode is
(a) TE_{11} (b) TM_{11}
(c) TM_{01} (d) TE_{01} **[a]**

14. The minimum frequency for mobile communication to be functional in a cylindrical tunnel of minimum diameter 6 m is
(a) 58.6 MHz   (b) 87.9 MHz
(c) 29.3 MHz   (d) 38.3 MHz   **[c]**

15. A cylindrical waveguide supports TE_{01}, TM_{01}, TE_{11} and TM_{11} modes. The order of the propagation from dominant to higher order mode is
(a) TE_{11}, TM_{11}, TE_{01}, TM_{01} (b) TE_{11}, TM_{01}, TE_{01}, TM_{11}
(c) TE_{01}, TM_{01}, TE_{11}, TM_{11} (d) TM_{11}, TE_{01}, TM_{01}, TE_{11} **[b]**

16. In a step index fiber, the cut-off frequency of the LP_{11} mode is
(a) 6.520   (b) 3.972
(c) 2.505   (d) 1.372   **[b]**

17. In a multimode in fiber, the higher order modes propagate within the fiber with
(a) same group velocity as that of lower order modes
(b) lower group velocity than the lower order modes
(c) higher group velocity than the lower order modes
(d) none of the above   **[b]**

18. Pulse broadening in GI givers is due to
    (a) intramodel dispersion   (b) intermodel dispersion
    (c) both (a) and (b)   (d) none of the above   **[c]**

19. A square waveguide carriers TE_{11} mode whose magnetic field is given by

    $H_z = H_o \cos \dfrac{px}{\sqrt{8}} \cos \dfrac{py}{\sqrt{8}}$ Am^{-1}. The dimensions of waveguide are in cm. The cut-off

    frequency of the mode is
    (a) 3.5 GHz   (b) 7.5 GHz
    (c) 9.5 GHz   (d) 15 GHz   **[b]**

20. A circular waveguide and a rectangular waveguide have the same cut-off frequency
    for dominant mode. The ratio of their area is
    (a) 1 : 1   (b) 2.16 : 1
    (c) 1 : 3   (d) 4 : 1   **[b]**

    [*Hint:* $A_c = \pi a^2$. For TE_{10} mode in rectangular waveguide, $\lambda_c = \dfrac{2a'}{m} = 2a'$ ($A' =$ length
    of waveguide to distinguish from radius a).
    Since cut-off wavelength of circular and rectangular guides are same so
    $$\lambda_c = 2a' = 3.14a$$

    or $\qquad a' = \dfrac{3.41}{2} = 1.705a$

    The area of rectangular waveguide is

    $$A_r = a'b = a' \times \frac{a'}{2} \quad (\because \; b = \frac{a'}{2} \text{ for standard rectangular waveguide})$$

    $$= \frac{(1.705a)^2}{2} = \frac{2.9070a^2}{2} = 1.4535a^2$$

    $$\therefore \quad \frac{A_c}{A_r} = \frac{\pi a^2}{1.4535a^2} = 2.16 : 1].$$

21. The large dimension of cross-section of a rectangular waveguide is 2 cm. The
    cut-off wavelength for the dominant TE mode is
    (a) 0.01 m   (b) 0.02 m
    (c) 0.03 m   (d) 0.04 m   **[d]**

    [*Hint:* $\qquad \lambda_c = \dfrac{2\pi}{\sqrt{\left(\dfrac{m\pi}{a}\right)^2 + \left(\dfrac{n\pi}{b}\right)^2}} = \dfrac{2\pi}{\dfrac{m\pi}{a}}$

    $$= \frac{2a}{m} = \frac{2 \times 0.02}{1} = 0.04 \text{ m.}]$$

Electromagnetic Radiation and Radiating Systems

12.1 INTRODUCTION

We have discussed various features, e.g. properties of electromagnetic waves and their propagation in both bounded and unbounded geometries in previous chapters, but very little was said about generation of such waves. In the present chapter, we shall discuss "radiation", i.e. the unguided transport of waves and energy through empty space and also the emission of radiation by localized systems of oscillating charge and current densities. We know that visible light is also a member of electromagnetic radiations. The question arises, how visible light is produced. It is now clear that the visible light results from the rapid adjustment of the charge and current distribution within the electron clouds of atoms. We must note that the emission of radiation by atoms can only be described by quantum mechanics. We know that the break-down of classical concepts first became evident only in the theory of radiation. However, the classical theory of radiation discussed in this chapter is of immense help in understanding the theory of radiation.

12.2 OSCILLATING CHARGE: A DIPOLE

One can visualize a charge constrained to move back and forth to form an oscillating electric *doublet* or an *electric dipole*. One can easily see that an oscillating charged particle at a given place is equivalent to an oscillating electric dipole. The harmonically oscillating charge gives rise to harmonic current. The charged particle undergoes harmonic oscillations, i.e. the particle is accelerated and hence it radiates electromagnetic energy in space. Obviously, the number of charged particles constituting the magnitude of current thus governs the near and the far electric and magnetic field vectors of the radiated electromagnetic wave. One can easily see that in a complete cycle of oscillating electric dipole, the electric and magnetic field lines are released from the dipole and advance in free space with the velocity of light. The cumulative forces of repulsion between the field lines radiate the electromagnetic energy at the velocity of light in free space.

12.3 RETARDED POTENTIALS

One can easily find the relation between radiation fields and their sources if we express the fields in terms of electromagnetic potentials A and ϕ, i.e.

$$E = -\nabla\phi - \frac{\partial A}{\partial t} \quad \text{and} \quad B = \nabla \times A \qquad (12.1)$$

We have seen that the introduction of Lorentz condition in Maxwell's equations yields the following two inhomogeneous equations for A and ϕ,

$$\nabla^2\phi - \frac{1}{c^2}\frac{\partial^2\phi}{\partial t^2} = -\frac{\rho}{\varepsilon_0} \tag{12.2}$$

and

$$\nabla^2 A - \frac{1}{c^2}\frac{\partial^2 A}{\partial t^2} = -\mu_0 J \tag{12.3}$$

We can see that the operator on the left hand side is identical with that of the homogeneous wave equation, but the source terms J and ρ are appearing now on the right hand side. Now, the problem is to solve Eqs. (12.2) and (12.3) to obtain the expression for A and ϕ in terms of charge and current distribution.

One can obtain the solutions of Eqs. (12.2) and (12.3) by analogy with solutions obtained for steady state problems in previous chapters. We have the solutions of the equations

$$\nabla^2\phi = -\frac{\rho}{\varepsilon_0} \text{ and } \nabla^2 A = -\mu_0 J$$

as

$$\phi(r) = \frac{1}{4\pi\varepsilon_0}\int_V \frac{\rho(r)}{|r-r'|}d\tau$$

$$\left. \right\} \tag{12.4}$$

and

$$A(r) = \frac{\mu_0}{4\pi}\int_V \frac{J(r)}{|r-r'|}d\tau$$

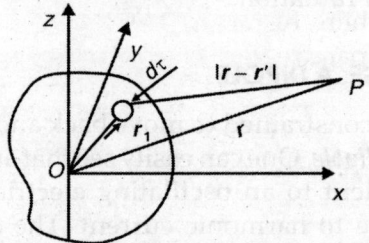

Fig. 12.1: Small volume element ($d\tau$)

where $d\tau$ is the small volume element (Fig. 12.1). One has to compute the potentials at a point P whose position vector is r. One can do this by integrating ρ and J throughout V by considering them as functions of r', the position vector of the volume element $d\tau$. We know that both the expressions of Eq. (12.4) are valid if the charges are at rest and the currents are steady. Let us now consider a situation in which ρ changes with time t. Now, the question arises, what is the potential at P at time t? Let us consider that the charge distribution within the volume element $d\tau$ changes with time. This means the field observed at P at time t must have been launched by the element $d\tau$ at a time t' earlier than t. The electric fields associated with these charges propagate with the finite velocity c. Obviously, the time taken to travel the distance from $d\tau$ to P is $\dfrac{|r-r'|}{c}$. This means that the contribution to the potential at the point P at a time t, due to the charge in $d\tau$ does not depend on the charge at time t but depends upon the charge at the time $t - \dfrac{|r - r'|}{t}$. We can see that this is the time at which the electric fields must have been propagated from the charges in $d\tau$ at r' in order to arrive at time t. Thus, we have

$$\phi(r, t) = \frac{1}{4\pi\varepsilon_0} \int_V \frac{\rho\left(r', t - \dfrac{|r - r'|}{c}\right)}{|r - r'|} d\tau \tag{12.5}$$

and

$$A(r, t) = \frac{\mu_0}{4\pi} \int_V \frac{J\left(r', t - \dfrac{|r - r'|}{c}\right)}{|r - r'|} d\tau \tag{12.6}$$

We have obtained Eqs. (12.5) and (12.6) on physical grounds. Now, the question arises, whether Eqs. (12.5) and (12.6) represent the solutions of non homogeneous equations (12.4). To check this, let us substitute Eqs. (12.5) and (12.6) in Eq. (12.4). Now, one can write

$$\phi(r, t) = \frac{1}{4\pi\varepsilon_0} \int_V \frac{\rho\left(r', t - \dfrac{d}{c}\right)}{d} d\tau \tag{12.7}$$

where $d = |r - r'|$. We can see that it is difficult to evaluate the potential at $d = 0$. To evaluate the integral in Eq. (12.7), one can divide the volume of integration into two regions, viz. (*i*) a small sphere of radius r_0 around the point at which ϕ is to be determined and (*ii*) rest of space region. One can express the potential as

$$\phi = \phi_1 + \phi_2 \tag{12.8}$$

In the case of the region of the small sphere, $\dfrac{d}{c} = \dfrac{|r - r'|}{c}$ is negligible and hence one finds

$$\nabla^2 \phi_1 = \frac{1}{4\pi\varepsilon_0} \nabla^2 \int_V \frac{\rho(r, t)}{d} d\tau = \frac{\rho}{4\pi\varepsilon_0} \int_V \nabla^2 \left(\frac{1}{d}\right) d\tau$$

We have obtained previously that

$$\nabla^2 \phi_1 = \frac{\rho}{4\pi\varepsilon_0} \int_V \nabla^2 \left(\frac{1}{d}\right) d\tau = -\frac{\rho}{\varepsilon_0} \tag{12.9}$$

We find that Eq. (12.9) is independent of r_0, and therefore, we can have the sphere as small as possible shrinking into a point.

We now determine $\nabla^2 \phi_2$. One can express $\nabla^2 \phi_2$ in spherical polar coordinates as

$$\nabla^2 \phi_2 = \frac{1}{d} \frac{\partial^2}{\partial d^2} (d\phi_2) \tag{12.10}$$

$$= \frac{1}{4\pi\varepsilon_0} \int_V \frac{1}{d} \frac{\partial^2}{\partial d^2} \rho\left(r', t - \frac{d}{c}\right) d\tau \tag{12.11}$$

Let us consider any function $f\left(\dfrac{d}{c}\right)$, we have

$$\frac{\partial^2 f}{\partial d^2} = \frac{1}{c^2}\frac{\partial^2 f}{\partial t^2} \qquad (12.12)$$

$$\therefore \quad \nabla^2 \phi_2 = \frac{1}{4\pi\varepsilon_0}\int_V \frac{1}{dc^2}\frac{\partial^2}{\partial t^2}\rho\left(r', t-\frac{d}{c}\right)d\tau$$

$$= \frac{1}{c^2}\frac{\partial^2}{\partial t^2}\left[\frac{1}{4\pi\varepsilon_0}\int_V \frac{\rho\left(r', t-\dfrac{d}{c}\right)}{d}d\tau\right]$$

$$= \frac{1}{c^2}\frac{\partial^2 \phi}{\partial t^2} \qquad (12.13)$$

$$\therefore \quad \nabla^2 \phi = \nabla^2 \phi_1 + \nabla^2 \phi_2 = -\frac{\rho}{\varepsilon_0} + \frac{1}{c}\frac{\partial^2 \phi}{\partial t^2}$$

or $\qquad \nabla^2 \phi - \dfrac{1}{c}\dfrac{\partial^2 \phi}{\partial t^2} = -\dfrac{\rho}{\varepsilon_0}$

This shows that Eq. (12.5) is a solution of Eq. (12.2). Potentials represented by Eq. (12.5) and (12.6) are called *retarded potentials*. We must note that one will have to evaluate the integrands $\rho(r')$ and $J(r')$ at the retarded time $t - \dfrac{|r-r'|}{c}$, where r' is the radius vector at the retarded time. Once the potentials are known, one can find fields E and B. However, the computation of A and ϕ is not as simple as it appears. Difficulties arise because retarded times are different for different parts of the sources. One can determine the fields in some certain simple situations with the help of some sensible approximations. The retarded potentials have shown to be the solution of non-homogeneous wave equation and are very useful in calculating radiated field vectors.

12.4 RADIATION FROM AN ELECTRIC DIPOLE

An electric dipole consists of two equal and opposite charges separated by a finite distance. Now, we calculate the radiation field produced by an electric oscillating dipole. One finds many important practical applications. One may consider many practical radiating systems to be made up by putting together a large number of such dipoles. A short electric dipole fed with a sinusoidal current source radiates electromagnetic energy in space.

Let us consider two small spheres at the two ends of a short wire. Suppose that the charge is transferred periodically from one sphere to the other, and the time variation being harmonic, i.e.

$$q = q_0 \exp(i\omega t) \qquad (12.14)$$

where ω is the angular frequency of the oscillation and q_0 is the amplitude of the oscillating charge. Assuming that the wavelength of the radiation produced is large compared with the length of the wire l, we have

$$\lambda = \frac{2\pi c}{\omega} \gg l \quad \text{or} \quad \frac{2\pi}{\omega} = T \gg \frac{l}{c} \tag{12.15}$$

Obviously, $\dfrac{l}{c}$ is the time taken for a signal to propagate along the wire from one end to the other. We can easily see that this is very much less than that over which the source current changes appreciably. This means, one can easily consider the current I to be the same at all points along its length. Thus, we have

$$I = \frac{dq}{dt} = i\omega q_0 \exp(i\omega t) \tag{12.16}$$

We know that the charge oscillating between two spheres is equivalent to an oscillating electric dipole moment p. We have

$$|p| = ql = q_0 l \exp(i\omega t) = p_0 \exp(i\omega t) = \frac{Il}{i\omega}$$

Here $\qquad\qquad p_0 = q_0 l \tag{12.17}$

We are interested in studying the distribution of the radiation field of this dipole throughout space and total power radiated by it.

Let us consider the dipole shown in Fig. 12.2. We can see from the figure that the wire lies along the z-axis. The origin of the coordinate system coincides with the centre of the wire. Let us find the field values at the point P specified by the position vector r. To determine the field, we need to know the retarded potentials. We have the vector potential,

$$A(r, t') = \frac{\mu_0}{4\pi} \int_V \frac{J\left(r', t - \dfrac{r - r'}{c}\right)}{|r - r'|} d\tau$$

Fig. 12.2: An oscillating dipole

The integration in the above expression is over the volume occupied by the current. We have

$$J\left(r', t - \frac{r - r'}{c}\right) d\tau = I\left(r', t - \frac{r - r'}{c}\right) \hat{a} \, dz$$

The current is always along the z-axis, and hence one may replace r' by z. Thus,

$$A(r, t) = \frac{\mu_0}{4\pi} \int_{-1/2}^{1/2} \frac{I\left(z, t - \frac{r - \hat{a}_z z}{c}\right)}{|r - \hat{a}_z z|} \hat{a}_z dz \tag{12.18}$$

Since $r \gg l$, one may neglect $\hat{a}_z z$ with respect to r. Now, we replace the denominator in Eq. (12.18) as r and $t - \frac{|r - \hat{a}_z z|}{c}$ at which the current density in the wire is to be measured by the time $t - \frac{r}{c}$. One obtains,

$$A(r, t) = \frac{\mu_0}{4\pi} \int_{-1/2}^{1/2} \frac{I\left(z, t - \frac{r}{c}\right)}{r} \hat{a}_z dz = \frac{\mu_0}{4\pi} \hat{a}_z l \frac{I\left(t - \frac{r}{c}\right)}{r} \tag{12.19}$$

as the current is same at all points along the wire. Equation (12.19) reveals that vector potential A is everywhere parallel to l as shown in Fig. 12.2. One obtains the components of the vector potential as

$$A_x(r, t) = 0, \quad A_y(r, t) = 0$$

and

$$A_z(r, t) = \frac{\mu_0 l}{4\pi r} i\omega q_0 \exp\left\{i\omega\left(t - \frac{r}{c}\right)\right\}$$

$$= \frac{\mu_0}{4\pi r} i\omega p_0 \exp\left\{(i\omega t)\exp\left(\frac{-i\omega r}{c}\right)\right\}$$

$$= \frac{\mu_0}{4\pi r} i\omega p(t)\exp(-ikr) = \frac{\mu_0}{4\pi} p(t)\frac{e^{-ikr}}{r} \tag{12.20}$$

Scalar Potential

The Lorentz condition for free space is

$$\nabla \cdot A + \frac{1}{c^2}\frac{\partial \phi}{\partial t} = 0 \tag{12.21}$$

Since only the z-component of A is non-zero, we have

$$\frac{\partial A_z}{\partial z} = -\frac{1}{c^2}\frac{\partial \phi}{\partial t}$$

$$\therefore \quad -\frac{1}{c^2}\frac{\partial \phi}{\partial t} = \frac{\mu_0}{4\pi} p(t)\frac{\partial}{\partial r}\left(\frac{e^{-kr}}{r}\cos\theta\right) \quad \left(\because \frac{\partial r}{\partial z} = \cos\theta\right)$$

$$\frac{\partial \phi}{\partial t} = -\frac{1}{4\pi\varepsilon_0} p(t)\cos\theta\frac{\partial}{\partial r}\left(\frac{e^{-ikr}}{r}\right)$$

This gives

$$\phi = -\frac{1}{4\pi\varepsilon_0} p(t)\cos\theta\left(\frac{e^{-ikr}}{r}\right) \tag{12.22}$$

One can also express Eq. (12.22) as

$$\phi = \frac{1}{4\pi\varepsilon_0}\left[\frac{z}{r^3}q\left(t-\frac{r}{c}\right)+\frac{z}{cr^2}I\left(t-\frac{r}{c}\right)\right]$$ (12.23)

Obviously, we have determined A and ϕ and we can now determine E and B.

Electric Field Strength *(E)*

One can express the electric field strength E as

$$E = -\nabla\phi - \frac{\partial A}{\partial t}$$ (12.24)

Now, we have the components of A in spherical polar coordinates as

$$A_r = \frac{\mu_0}{4\pi r}i\omega p_0\cos\theta\exp\left[i\omega\left(t-\frac{r}{c}\right)\right]$$

$$= \frac{\mu_0}{4\pi r}\cos\theta\, e^{-ikr}$$ (12.25)

$$A_\theta = \frac{\mu_0}{4\pi r}i\omega p_0\sin\theta\exp\left[i\omega\left(t-\frac{r}{c}\right)\right]$$

$$= \frac{\mu_0}{4\pi r}p\sin\theta\, e^{-ikr}$$ (12.26)

and $\qquad A_\phi = 0$ (12.27)

Using Eqs. (12.25) to (12.27), one obtains the components of E as

$$E_r = -\frac{\partial\phi}{\partial r}-\frac{\partial A_r}{\partial t}=\frac{ipk}{2\pi\varepsilon_0 r}\cos\theta\left[1-\frac{i}{kr}\right]\frac{e^{-ikr}}{r}$$ (12.28)

$$E_\theta = -\frac{1}{r}\frac{\partial\phi}{\partial\theta}-\frac{\partial A_\theta}{\partial t}$$

$$= \frac{pk^2}{4\pi\varepsilon_0}\sin\theta\left[1-\frac{i}{kr}\left(1-\frac{i}{kr}\right)\right]\frac{e^{-ikr}}{r}$$ (12.29)

and $\qquad E_\phi = 0$ (12.30)

Magnetic Field Strength *(H)*

Using $B = \nabla\times A$, and Eqs. (12.25) to (12.27), one obtains

$$H_r = 0,\ H_\theta = 0$$

and $\qquad H_\phi = -\frac{cpk^2}{4\pi}\sin\theta\left[1-\frac{i}{kr}\right]\frac{e^{-ikr}}{r}$ (12.31)

Equations (12.28) to (12.31) are known as the *Hertz's relations* for the oscillating dipole. From Eq. (12.31), it is obvious that:

i. the magnetic induction (B) varies as $\sin\theta$. This indicates that the magnetic induction B is minimum along the axis of the dipole and maximum along the equatorial plane.

ii. the magnetic induction (B) contains two terms, one varying as r^{-2} and other varying as r^{-1}. The term varying as r^{-1} dominates for large values of r and obviously, this term is responsible for radiation field. The term varying as r^{-2} dominates for small values of r and hence this term is responsible for induction field.

One can imagine that the space is divided into two regions:

i. $|r| \ll \lambda$. This region is called as the *near zone*. One finds that the field in this case is exactly the same as that for an electrostatic field.

ii. $|r| \gg \lambda$. This region is called the *radiation zone*. One can see that fields in this region varies as r^{-1}.

We must note that the region were $kr \simeq 1$ is quite complicated.

We now examine Eqs. (12.28) to (12.31) in the near zone and the radiation zone.

i. *Near zone:* $|r| \ll \lambda$, i.e. $k_r \ll 1$, we have

$$\left. \begin{array}{l} E_r \simeq \dfrac{p\cos\theta}{2\pi\varepsilon_0 r^3}, E_\theta \simeq \dfrac{p\sin\theta}{4\pi\varepsilon_0 r^3} \text{ and } E_\phi = 0 \\[3mm] H_r = 0, H_\theta = 0 \text{ and } H_\phi = \dfrac{i\omega p\sin\theta}{4\pi r^2} \end{array} \right\} \tag{12.32}$$

One finds that relations expressed by Eq. (12.32) are equivalent to the field of an electrostatic dipole. We obtain

$$\frac{\omega\mu_0}{k}\frac{|H|}{|E|} \simeq \frac{\omega\mu_0}{k}\varepsilon_0\omega r = kr \ll 1$$

Obviously, the electric field dominates in this zone.

ii. *Radiation zone:* $k_r \gg 1$, we have

$$\left. \begin{array}{l} E_r = 0, \quad E_\theta = -\dfrac{pk^2}{4\pi\varepsilon_0}\sin\theta\dfrac{e^{-ikr}}{r}, \quad E_\phi = 0 \\[3mm] H_r = 0, \quad H_\theta = 0, H_\phi = -\dfrac{cpk^2}{4\pi}\sin\theta\dfrac{e^{-ikr}}{r} \end{array} \right\} \tag{12.33}$$

While obtaining these results, we have neglected the terms involving $\dfrac{1}{r^2}$ and its higher orders. One obtains from Eq. (12.33), the two components E_θ and H_ϕ are in time phase in the far field approximation.

$$E_\theta = \sqrt{\frac{\mu_0}{\varepsilon_0}}H_\phi \tag{12.34}$$

The intrinsic impedance of free space can thus be defined as

$$Z = \frac{E_\theta}{H_\phi} = \frac{1}{\varepsilon_0 c} = 376.7\,\Omega \tag{12.35}$$

•

Figure 12.3 shows the directions of the fields. We can see that the fields are mutually perpendicular. For the radiation zone, $E_r = 0$ and we have the field distribution in spherical width the electric field as circles of longitude and the magnetic field as the circles of latitude. One obtains that the field is maximum at the equator and zero at the poles. Since the magnetic field is transverse everywhere, the radiation from an oscillating electric dipole is generally TM. We can see that at large distances, $E_r \rightarrow 0$ and the electric field also is transverse. Obviously, the waves are TEM.

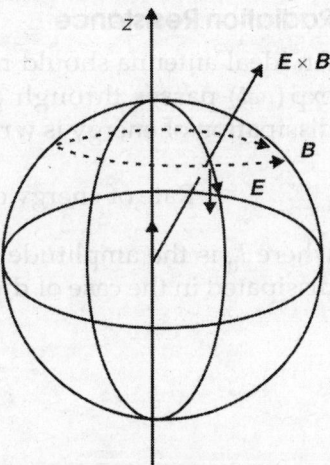

Quasi-Stationary Field

The low frequency radiation pattern has shown special features and is generally referred to as quasi-stationary field.

For very low frequency radiation, $\omega \rightarrow 0$, the field components are non-zero in the vicinity of the dipole and decrease rather fast with increasing distance.

Fig. 12.3: Directions of the electric and magnetic fields

Poynting Vector $\langle N \rangle$ and Radiated Power (*W*)

The energy flow takes place due to radiation fields. We have the time averaged Poynting vector $\langle N \rangle$ as

$$\langle N \rangle = \frac{1}{2} R_e (E \times H^*)$$

Using the values obtained for radiation zone in Eq. (12.33), one obtains

$$\langle N \rangle = \frac{c p_0^2 k^4}{32 \pi^2 \varepsilon_0} \frac{\sin^2 \theta}{r^2} \hat{a}_n \qquad (12.36)$$

Equation (12.36) represents the average energy radiated per second per unit area in a normal direction by an oscillating dipole.

From Eq. (12.36), it is obvious that the time averaged *Poynting vector* varies as $\sin^2 \theta$. This means that the flow of energy is zero along the dipole axis and maximum along the equatorial axis and shown in Fig. 12.4.

The total radiated power is given by

Fig. 12.4: Flow of energy along dipole axis (zero) and along the equatorial axis (maximum)

$$W = \int_0^{2\pi} \int_0^{\pi} \frac{c p_0^2 k^4}{32 \pi^2 \varepsilon_0} \frac{\sin^2 \theta}{r^2} r^2 \sin \theta \, d\theta \, d\phi$$

$$= \frac{c p_0^2 k^4}{32 \pi^2 \varepsilon_0} 2\pi \frac{4}{3} = \frac{c p_0^2 k^4}{12 \pi \varepsilon_0} = \frac{c p_0^2}{12 \pi \varepsilon_0} \left(\frac{2\pi}{\lambda} \right)^4 \qquad (12.36a)$$

From Eq. (12.36a), it is clear that the total power radiated by an oscillating dipole is inversely proportional to the fourth power of the wavelength λ. Examples of electric dipole radiators are the emission of radiation by an excited atom and electromagnetic waves transmitted by an antenna.

Radiation Resistance

An ideal antenna should radiate all the power delivered to it. When a current $I = I_0 \exp(i\omega t)$ passes through a circuit containing a resistance R, the average rate of dissipation of energy is written as

$$\text{Rate of energy dissipated} = \frac{1}{2}I_0^2 R \tag{12.37}$$

where I_0 is the amplitude of the current. We can express the expression for power dissipated in the case of dipole, i.e. Eq. (12.36a) as

$$\frac{cp_0^2}{12\pi\varepsilon_0}\left(\frac{2\pi}{\lambda}\right)^4 = \frac{p_0^2\omega^4}{12\pi\varepsilon_0 c^3} = \frac{1}{12\pi\varepsilon_0}\frac{I_0^2 l^2 \omega^2}{c^3}$$

$$= \frac{1}{12\pi\varepsilon_0}\frac{4\pi l^2}{\lambda^2}\cdot\frac{1}{c}I_0^2$$

$$= \frac{1}{2}\left(\frac{2\pi}{3}\right)\sqrt{\frac{\mu_0}{\varepsilon_0}}\left(\frac{l}{\lambda}\right)^2 I_0^2 \tag{12.38}$$

Comparing Eq. (12.38) with Eq. (12.37), one obtains the radiation resistance as

$$R_r = \frac{2\pi}{3}\sqrt{\frac{\mu_0}{\varepsilon_0}}\left(\frac{l}{\lambda}\right)^2 \tag{12.39}$$

It is obvious from Eq. (12.39) that the radiation resistance of an electric dipole depends on its length as compared to the wavelength of radiation. We must note that Eq. (12.39) is valid provided $l \ll \lambda$.

The radiation resistance governs the radiation efficiency of antennas. If a part of the power delivered to the antenna is lost, the corresponding loss resistance is denoted by R_{loss}. One can define the antenna efficiency as

$$\eta = \frac{\text{power radiated}}{\text{power input}} = \frac{R_r}{R_r + R_{loss}} \tag{12.40}$$

Minimizing the radiation losses enhances the radiation efficiency.

12.5 ELECTRIC QUADRUPOLE RADIATION

A quadrupole is a combination of two electric dipoles. Let us imagine two electric dipoles, each of amplitude p, aligned along a given straight line and oscillating exactly out of phase so that there is zero net dipole moment. This system possess quadrupole moment which varies with time and hence it will emit radiation, called *quadrupole radiation*. One can derive the quadrupole field by considering it to be the superposition of two dipole fields. Figure 12.5 shows a quadrupole.

Fig. 12.5: Quadrupole

The electric field due to a single dipole located at the origin, at a point P in the radiation zone is

$$E_\theta = -\frac{pk^2}{4\pi\varepsilon_0}\sin\theta\frac{e^{-ikr}}{r} = \frac{k^2 p_0}{4\pi\varepsilon_0 r}e^{i\omega t'}\sin\theta \qquad (12.41)$$

The dipoles are at distances $\frac{l}{4}$ and $-\frac{l}{4}$ from the origin (Fig. 12.5). From Fig. 12.5, it is clear that the dipole (1) is close to the point P than is the origin by approximately a distance $\left(\frac{l}{4}\right)\cos\theta$. Obviously, dipole (2) is farther by the same amount. One finds the phases of two dipoles relative to origin and therefore $\frac{kl}{4}\cos\theta$ and $-\frac{kl}{4}\cos\theta - \pi$, where $k = \frac{\omega}{c}$ and $-\pi$ is introduced because two dipoles are oscillating out of phase. Therefore, the field component of the quadrupole is

$$E_\theta = -\frac{k^2 p_0}{4\pi\varepsilon_0 r}\sin\theta\, e^{i\left(\omega t' - \frac{kl}{4}\cos\theta\right)} - \frac{k^2 p_0}{4\pi\varepsilon_0}\sin\theta\, e^{l\left\{\omega t - \left(-\frac{kl}{4}\cos\theta - \pi\right)\right\}}$$

$$= -\frac{k^2 p_0 \sin\theta}{4\pi\varepsilon_0 r}\left[\exp\left(-i\frac{kl}{4}\cos\theta\right) - \exp\left(i\frac{kl}{4}\cos\theta\right)\right]\exp(i\omega t)$$

$$= \frac{k^2 p_0 \sin\theta}{4\pi\varepsilon_0 r}2l\sin\left(\frac{kl}{4}\cos\theta\right)\exp(i\omega t) \qquad (12.42)$$

Using H_ϕ, one can obtain Poynting vector as

$$\langle N \rangle = \frac{1}{2}R_e(E \times H^*) = \frac{1}{2}\sqrt{\frac{\varepsilon_0}{\mu_0}}\,|E_0|^2\,\hat{a}_n$$

$$= \frac{1}{2}\sqrt{\frac{\varepsilon_0}{\mu_0}}\frac{k^4 p_0^2 \sin^2\theta}{16\pi^2\varepsilon_0^2 r^2}\sin^2\left(\frac{kl}{4}\cos\theta\right)\cdot 4 \qquad (12.43)$$

Average Power Radiated Per Unit Solid Angle

$$\left\langle \frac{dW}{d\Omega} \right\rangle = \langle N \rangle \cdot \hat{a}_n r^2 = \frac{\sqrt{\varepsilon_0}k^4 p_0^2 \sin^2\theta}{8\sqrt{\mu_0}\pi^2\varepsilon_0^2}\sin^2\varepsilon_0^2\sin^2\left(\frac{kl}{4}\cos\theta\right)$$

$$= \frac{\omega^4 p_0^2}{8\pi^2 \varepsilon_0 c^3} \sin^2 \theta \sin^2 \left(\frac{kl}{4} \cos \theta \right) \qquad (12.44)$$

Special Cases

i. If $l = \lambda$, i.e. $kl = 2\pi$, then

$$\left\langle \frac{dW}{d\Omega} \right\rangle = \omega^4 p_0^2 \sin^2 \theta \sin^2 \left(\frac{\pi}{2} \cos \theta \right) \qquad (12.45)$$

The quadrupole moment, $Q = 2p_0 l$. Therefore,

$$\left\langle \frac{dW}{d\Omega} \right\rangle = \frac{\omega^4 Q^2}{8\pi^2 \varepsilon_0 c^3 4l^2} \sin^2 \theta \sin^2 \left(\frac{\pi}{2} \cos \theta \right)$$

$$= \frac{\omega^6 Q^2}{128\pi^4 \varepsilon_0 c^5} \sin^2 \theta \sin^2 \left(\frac{\pi}{2} \cos \theta \right) \qquad (12.46)$$

ii. If $l \to 0$ be very small: Obviously, as p_0 increases, Q remains constant. One finds that the dipole pair now reduces to a point quadrupole located at the origin. Since kl is very small, one may write Eq. (12.42) as

$$E_\theta = 2i \frac{k^2 p_0}{4\pi\varepsilon_0 r} \sin \theta \frac{kl}{4} \cos \theta e^{i\omega t'} \qquad (12.47)$$

Hence

$$\left\langle \frac{dW}{d\Omega} \right\rangle = \frac{k^6 p_0^2 l^2 c}{128\pi^2 \varepsilon_0} \sin^2 \theta \cos^2 \theta$$

$$= \frac{\omega^6 p_0^2 l^2}{128\pi^2 \varepsilon_0 c^5} \sin^2 \theta \cos^2 \theta \qquad (12.48)$$

One finds that the maximum power is radiated at $\theta = 45°$.

12.6 RADIATION FROM LINEAR ANTENNA

A practical antenna is a centre-fed thin linear antenna having a length comparable with free space wavelength. The variation of electric field or magnetic vector in space gives an idea of antenna radiation pattern. The antenna used in radio or television transmission, are not short compared to the wavelength of the radiation they transmit. This means that the current in the antenna is not constant and the variation of its amplitude have to be taken into account. The current amplitude of such an antenna decreases uniformly from a maximum at the centre to minimum at the ends. A simple centre driven linear antenna is shown in Fig. 12.6.

Fig. 12.6. Linear, centre-fed antenna

Let the antenna (usually a very thin wire) is to be oriented along the z-axis and has a length l. The current is fed into the antenna via a coaxial cable transmission line. We assume the antenna to have a small gap at its centre so that it can be suitably excited. The current along the antenna vanishes at the end points and is an even function of z. The current density J is a radial current and assumed to vary harmonically in time and space along the antenna. To a good approximation, one can write the current density as

$$J(r, t) = \hat{a}_z I_0 \exp(i\omega t)\sin\left(\frac{kl}{2} - k|z|\right)\delta x \delta y \qquad (12.49)$$

The delta function in Eq. (12.49) ensures that the current flows in z-direction only. We can easily verify that Eq. (12.49) satisfies the required conditions. The exact form of the current at the gap, i.e. the input signal is

$$I(t) = I_0 \exp(i\omega t)\sin\frac{kl}{2} \qquad (12.50)$$

where I_0 is the maximum current. The vector potential at a point P is given by

$$A(r, t) = \frac{\mu_0}{4\pi}\int \frac{J(r',t')}{|r-r'|}dr' \text{ where } t' = t - \frac{|r-r'|}{c}$$

$$= \frac{\mu_0}{4\pi}\hat{a}_z I_0 \int_{-l/2}^{l/2} \frac{\left(t - \frac{|r-r'|}{c}\right)\sin\left(\frac{kl}{2} - k|z|\right)}{|r-r'|}dr'$$

$$= \frac{\mu_0}{4\pi}\hat{a}_z I_0 \int_{-l/2}^{l/2} \frac{\exp\left\{i\omega\left(t - \frac{|r-z|}{c}\right)\right\}\sin\left(\frac{kl}{2} - k|z|\right)}{|r-z|} \qquad (12.51)$$

When the point P is far away, one can always replace the denominator by r and we may substitute $|r - z|$ in the numerator. Here, θ is the angle which the vector r make with z-axis. Thus, we have

$$A(r, t) = \frac{\mu_0}{4\pi}\hat{a}_z \frac{I_0}{r}\exp(i\omega t)\exp(-ikr) \int_{-l/2}^{l/2} \sin\left(\frac{kl}{2} - k|z|\right)\exp(ikz\cos\theta)dz$$

On integrating, one obtains

$$A(r, t) = \frac{\mu_0}{2\pi}\hat{a}_z I_0 \exp(i\omega t)\frac{\exp(-ikr)}{kr}\left[\frac{\cos\left(\frac{kl}{2}\cos\theta\right) - \cos\left(\frac{kl}{2}\right)}{\sin^2\theta}\right] \qquad (12.52)$$

We can easily calculate *electric field strength* and *magnetic field strength* from Eq. (12.52). We have the *average rate of power flow* as

$$\langle N \rangle = \frac{1}{2}R_e(E \times H^*) = \frac{\hat{a}_n}{2c\mu_0}|E_0|^2$$

$$= \frac{1}{2}\sqrt{\frac{\mu_0}{\varepsilon_0}}\frac{I_0^2}{\varepsilon_0}\hat{a}_n\frac{I_0^2}{4\pi^2}\frac{1}{r^2}\left[\frac{\cos\left(\frac{kl}{2}\cos\theta\right)-\cos\left(\frac{kl}{2}\right)}{\sin\theta}\right]^2 \tag{12.53}$$

Now, the *average power radiated* into *unit solid angle* is obtained as

$$\left\langle\frac{dW}{d\Omega}\right\rangle = \frac{4\pi r^2\langle N\rangle}{4\pi}\hat{a}_n = r^2\langle N\rangle\hat{a}_n$$

$$= \sqrt{\frac{\mu_0}{\varepsilon_0}}\frac{I_0^2}{4\pi}\left[\cos\frac{\left(\frac{kl}{2}\cos\theta\right)-\cos\frac{kl}{2}}{\sin\theta}\right]^2 \tag{12.54}$$

Obviously, the angular distribution of radiated power depends upon the value of $\frac{kl}{2}$. Let us consider the case of half wave antenna $\left(l=\frac{2\pi l}{\lambda}\right)$, i.e. $kl=\pi$ (half wave antennas are one of the most widely used form of practical antennas. It is a straight conductor whose length is equal to half of the free space wavelength). When a periodic current of the form $I_0\cos\omega t$ is fed at the centre of an antenna, a standing wave pattern is formed along the length of the conductor. One can express the current pattern set in the antenna as $\left(I=I_0\cos\left(\frac{2\pi l}{\lambda}\right)\cos\omega t\right)$. We have

$$\left\langle\frac{dW}{d\Omega}\right\rangle = \sqrt{\frac{\mu_0}{\varepsilon_0}}\frac{I_0^2}{4\pi^2}\left[\frac{\cos\left(\frac{\pi}{2}\cos\theta\right)}{\sin\theta}\right]^2 \tag{12.55}$$

One can obtain the average total power $\langle W\rangle$ radiated by a half wave antenna from Eq. (12.53) as

$$\langle W\rangle = \frac{1}{2}\sqrt{\frac{\mu_0}{\varepsilon_0}}\frac{I_0^2}{4\pi^2}\frac{1}{r^2}\int_0^{2\pi}\int_0^{\pi}\left[\frac{\cos\left(\frac{\pi}{2}\cos\theta\right)^2}{\sin\theta}\right]^2 r^2\sin\theta\,d\theta\,d\phi$$

$$= \sqrt{\frac{\mu_0}{\varepsilon_0}}\frac{I_0^2}{4\pi}\int_0^{\pi}\frac{\cos^2\left(\frac{\pi}{2}\cos\theta\right)}{\sin\theta}\,d\theta$$

This integral cannot be solved analytically and we find its value from the tables. We have

$$\langle W\rangle \simeq 73.1\frac{I_0^2}{2} \tag{12.56}$$

Thus, we find that the radiation resistance of a half wave antenna is $73.1\,\Omega$, i.e.

$$R_r = 73.1\,\Omega$$

Comparing this value with the corresponding value for an electric dipole ($R_r = 0.08\,\Omega$), we find that the half wave antenna is much more efficient radiator.

When l is equal to the integral number of half wavelengths of the driving oscillations,

$$\left\langle \frac{dW}{d\Omega} \right\rangle = \sqrt{\frac{\mu_0}{\varepsilon_0}} \frac{I_0^2}{8\pi^2} \left[\frac{\cos^2\left(\dfrac{m\pi}{2}\cos\theta\right)}{\sin\theta} \right] \qquad (12.57)$$

for odd values of m

$$= \sqrt{\frac{\mu_0}{\varepsilon_0}} \frac{I_0^2}{8\pi^2} \left[\frac{\sin^2\left(\dfrac{m\pi}{2}\cos\theta\right)}{\sin^2\theta} \right] \qquad (12.58)$$

for even values of m.

12.7 ANTENNA ARRAY

It is a set of suitably spaced, similarly oriented aerials to radiate energy predominently in some given direction. Antenna arrays make use of interference phenomenon that occurs between the radiations from the different elements of the array. Accordingly the resultant field developed by an antenna array at a distant point will be the superposition of field contributions from each constitutents antenna of the array. Two types of antenna arrays are commonly used. These are:

i. *Broadside array—arrays of n sources:* It exhibits the marked directional properties in the horizontal plane and in a direction normal to the line of array, i.e. the beam is radiated at an angle $\theta = 90°$ to the axis of the array.

ii. *End-fire array—arrays of n sources:* It exhibits the marked directional properties in the horizontal plane and in a direction along the line of array, i.e. the beam is radiated at an angle $\theta = 0°$ to the axis of the array.

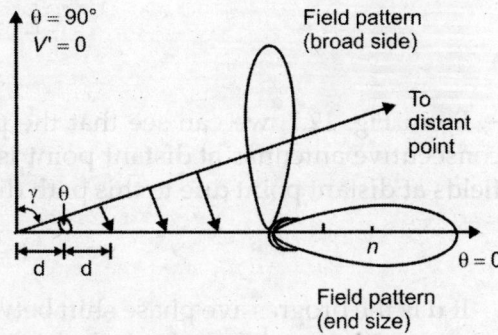

Fig. 12.7: Antenna array of *n* isotropic point radiation with equal spacing, same amplitude and a constant phase difference

Let us consider a linear antenna array (an antenna is said to be linear when the elements of the array are spaced equally along a straight line) consisting of *n* isotropic sources separated by a distance *d* (Fig. 12.7). With respect to the source 1 at the left end, the phase of each of the source vary by an angle $\phi = \dfrac{2\pi}{\lambda}(d\cos\theta + \delta)$, where δ is the progressive phase difference between two consecutive sources. If $E_1, E_2, E_3,...$ are the fields due to antennas 1,2,3,... respectively, then the resultant far-field at an angle θ is given by

$$E = E_0[1 + \exp(i\phi) + \exp(2i\phi) + \exp(3i\phi) + \cdots + \exp\{i(n-1)\phi\}]$$

$$= E_0 \sum_{n=1}^{N} \exp i(n-1)\phi \tag{12.59}$$

Multiplying Eq. (12.59) by exp ($i\phi$) and subtracting the resulting equation from Eq. (12.59), one obtains

$$E = E_0 \frac{[1-\exp(in\phi)]}{[1-\exp(i\phi)]} = E_0 \frac{\sin(n\phi/2)}{\sin(\phi/2)} \bigg/ (n-1)\phi/2 \tag{12.60}$$

We see that the resultant signal acquires a phase difference of $(n-1)\dfrac{\phi}{2}$. If the phase is referred to the central point of the array, then $\phi \to 0$. The electric field at a distant point is characterized by the individual elements of arrays and $\dfrac{\sin(n\phi/2)}{\sin(\phi/2)}$ is called the array factor. Equation (12.60) takes the form

$$E = E_0 \frac{\sin\left(\dfrac{n\phi}{2}\right)}{\sin\left(\dfrac{\phi}{2}\right)} \tag{12.61}$$

For $\phi \to 0$, one finds that the resultant field is maximum and expressed as

$$E = nE_0 \tag{12.62}$$

One can express the normalized field as

$$E_n = \frac{E}{nE_0} = \frac{1}{n} \frac{\sin\left(\dfrac{n\phi}{2}\right)}{\sin\left(\dfrac{\phi}{2}\right)} \tag{12.63}$$

From Fig. 12.7, we can see that the path difference between radiations from the consecutive antennas at distant point is $d\cos\theta$. Thus, the phase difference between fields at distant point due to this path difference is

$$\delta = \frac{2\pi}{\lambda}(d\cos\theta) \tag{12.64}$$

If α is the progressive phase shift between elements, i.e. α is the net angle by which the current in any element leads the current in the preceding element, then the resultant phase difference ψ between the fields from two consecutive antennas is given by

$$\psi = \frac{2\pi}{\lambda}(d\cos\theta) + \alpha \tag{12.65}$$

Broadside array: For broadside array, the maximum power radiated by the array is concentrated perpendicular to the plane of the array (Fig. 12.7). In the direction of the pattern of maximum radiation, the normalized electric field as given by Eq. (12.63) is maximum. If $\sin\dfrac{n\phi}{2} \neq 0$, we must have $\sin\dfrac{\phi}{2} \to 0$ or $\phi = (kd\cos\theta + \delta) = 0$

or $$kd\cos\theta = -\delta \tag{12.66}$$

For broadside array, we have $\theta_{max} = 90°$, and therefore $\delta = 0$. This means the electric fields in all elements of the array are in phase. One obtains that the position of null for broadside array is given by $\sin\dfrac{\phi}{2} \to 0$ and $\sin\dfrac{\phi}{2} \neq 0$. This corresponds to

$$\phi = \frac{2m\pi}{n} = k(d\cos\theta_0 + \delta)$$

Expressing in terms of a new angle measured from the vertical, one obtains

$$\xi_0 = (90 - \theta_0) = \sin^{-1}\left(\frac{m\lambda}{nd}\right) \tag{12.67}$$

For large array, $nd \gg m\lambda$, and we have

$$\xi_0 = \frac{m\lambda}{nd} = \frac{m}{(L/\lambda)} \quad (m = 1, 2, 3) \tag{12.68}$$

where L is the length of the array. The first null corresponds to $m = 1$ and the beam width between first nulls (BWFN) on either sides is obtained as

$$2\xi_0 = \frac{2}{L/\lambda} \tag{12.69}$$

The half power beam width (HPBW) is obtained as

$$\frac{1}{L/\lambda} = \frac{\lambda}{L} \text{ radian} \tag{12.70}$$

Obviously, HPBW is half of the BWFN.

We note that the width of the beam depends inversely on the product and which is virtually the length of the broadside antenna array.

End-fire array: For an end fire array, the maximum radiation is along the line of the array at $\theta = 0$. Proceeding exactly the same way, one obtains,

$$\phi = (kd\cos\theta + \delta) = 0$$

or

$$kd = -\delta \tag{12.71}$$

For maximum power concentration, we have

$$n\phi = \pm 2m\lambda \to 0$$

and

$$\phi = \pm\frac{n2\pi}{n} = (kd\cos\theta_0 + \delta) \to 0 \tag{12.72}$$

Using Eq. (12.71) in Eq. (12:72), one obtains

$$(\cos\theta_0 - 1) = \pm\frac{2m\pi}{n(kd)} = \pm\frac{m}{nd/\lambda}$$

or

$$\frac{\theta_0}{2} = \sin^{-1}\left(\pm\frac{m}{2nd/\lambda}\right) \tag{12.73}$$

We have $nd \gg m\lambda$ for a long array. Thus, we have

$$\theta_0 = \pm\sqrt{\frac{2m}{nd/\lambda}} \simeq \sqrt{\frac{2m}{L/\lambda}} \tag{12.74}$$

Obviously, the beam width between the first two nulls on either side of the maximum corresponds to $m \to 1$. This is given by

$$\text{BWFN} = 2\theta_0 = \pm 2\sqrt{\frac{2}{L/\lambda}} \qquad (12.75)$$

Obviously, the width of the radiated beam depends inversely on the product and which is virtually the length of the antenna array. From Eq. (12.75), it is evident that the beam width is narrower for the high frequency of an antenna array of a given length.

12.8 LIENARD-WIECHERT POTENTIALS

We now consider the application of retarded potentials [Eqs. (12.5) and (12.6)], to compute the radiation from a single charged particle, say, an electron, in arbitrary motion. We have seen that the calculation of the potentials depends upon the position and velocity of the charge at the retarded time $t - \dfrac{|r - r'|}{c}$ and therefore it is essential

Fig. 12.8: Trajectory of the electron

to know the details of the motion of the charge. Figure 12.8 shows a trajectory of the electron described by the radius vector $r_e t'$. We can see that the calculation of the potentials given by Eqs. (12.5) and (12.6) involves a retarded time integration over the entire volume enclosing the charge. We must note that the way the charge is distributed geometrically within the electron is not known to us. We know that the particle has a certain total charge. Let us assume that the electron has a finite radius (the assumption that the electron has zero physical extent can lead to difficulties). However, we shall consider only those properties which are independent of the magnitude of the radius. One can express the retarded potential for an electron in terms of delta function as

$$\phi(r, t) = \frac{e}{4\pi\varepsilon_0} \int_{-\infty}^{\infty} \frac{\delta\left\{t' - \left(t - \dfrac{r - r_e t}{c}\right)\right\}}{|r - r_e|} dt' \qquad (12.76)$$

To make use of delta function property, we will have to put the integral in the form $\int_{-\infty}^{\infty} f(x)\delta(x - x')$. One can see that the integral is equal to $f(x')$.

Let us introduce a new variable t'' such that

$$t'' = t' - t + \frac{|r - r_e(t')|}{c} \qquad (12.77)$$

$\therefore \qquad dt'' = dt' + \dfrac{1}{c}\dfrac{d}{dt'}|r - r_e(t')| dt' \qquad (12.78)$

In obtaining Eq. (12.78), we have taken $dt = 0$, because the observation is made at a fixed time. Let the coordinates of the fixed point P be (x_1, x_2, x_3) and those of the electron be $\left(x_{e_1}(t'), x_{e_2}(t'), x_{e_3}(t')\right)$.

Now, we have

$$|r - r_e(t')| = \sqrt{\sum_i [x_{e_1}(t')^2]} \qquad (12.79)$$

$$\therefore \quad \frac{1}{c}\frac{d}{dt'}|r - r(t')| = \frac{1}{c}\sum_i \frac{\partial}{\partial x_{e_i}}|r - r_e(t')|\frac{dx_{e_i}}{dt'} \qquad (12.80)$$

But $\dfrac{\partial}{\partial x_{e_i}}|r - r_e(t')|$ are the components of the gradient of $|r - r_e(t')|$ and $\dfrac{dx_e}{dt'}$ are the components of $\dfrac{dr_e}{dt}$. One obtains,

$$\frac{1}{c}\frac{d}{dt'}|r - r_e(t') = \frac{1}{c}\mathrm{grad}_{r_e}|r - r_e(t')|\frac{dr_e}{dt'} \qquad (12.81)$$

One can determine the gradient as

$$\mathrm{grad}_{r_e}|r - r_e(t')| = -\frac{r - r_e(t')}{|r - r_e(t')|} = \frac{R}{|R|} \qquad (12.82)$$

Here $R = r - r_e(t)$. Also we have $u = \dfrac{dr_e}{dt}$, i.e. the velocity of the electron. Now, we can write

$$\frac{1}{c}\frac{d}{dt'}|r - r_e(t')| = -\frac{1}{c}\frac{R}{|R|}\cdot u = -\frac{\beta \cdot R}{|R|} \qquad (12.83)$$

where $\beta = \dfrac{u}{c}$. Therefore,

$$dt'' = dt'\left(1 - \frac{\beta \cdot R}{|R|}\right)$$

and

$$dt' = \frac{|R|}{|R| - \beta \cdot R}dt'' \qquad (12.84)$$

Using Eq. (12.84), Eq. (12.76) takes the form as

$$\phi(r, t) = \frac{e}{4\pi\varepsilon_0}\int_{-\infty}^{\infty}\frac{\delta(t'')}{R(t')}\frac{|R(t')|}{|R(t') - \beta(t)\cdot R(t')} \qquad (12.85)$$

$$= \frac{e}{4\varepsilon_0}\left[\frac{1}{|R(t')| - \beta(t')\cdot R(t')}\right]_{t''=0} \qquad (12.86)$$

We have $t'' = 0$ indicates $t' = t - \dfrac{R(t')}{c}$

$$\therefore \qquad \phi(r, t) = \frac{e}{4\pi\varepsilon_0} \frac{1}{|R - \beta \cdot R|_{t' = t - \frac{R(t')}{C}}} \qquad (12.87)$$

Similarly, one finds the vector potential as

$$A(r, t) = \frac{\mu_0 e}{4\pi} \left(\frac{u}{R - \beta \cdot R} \right)_{t' = t - \frac{R(t')}{C}} \qquad (12.88)$$

These potentials are velocity dependent but independent of the extent of the charge, i.e. any detailed electronic model. These were first given by A Lienard and E Wiechert for a moving point charge and after their names they are known as *Lienard–Wiechert potentials*.

12.9 THE ELECTROMAGNETIC FIELDS FROM LIENARD–WIECHERT POTENTIALS OF A MOVING POINT CHARGE

So far we have obtained the electromagnetic potentials A and ϕ for a charge particle undergoing arbitrary motion. Let us now find the electromagnetic fields E and B of a moving charge. These can be found from the relations

$$E = -\nabla\phi - \frac{\partial A}{\partial t}$$

and
$$B = \nabla \times A$$

Let us start with Lienard–Wiechert potentials to determine the fields of a charge moving with uniform velocity in a straight line.

Let the observer be in the rest frame of the charge, i.e. if observer is moving with the charge, the situation as observed by him is electrostatic. The scalar potential according to the observer is given by

$$\phi(r) = \frac{e}{4\pi\varepsilon_0 r}$$

Now, the question arises, what will be the potential if the observer is at rest and the charge is moving? One can obtain the desired expression using relativistic transformations (see Chapter 13). However, the relativity has its foundation in electrodynamics. Here we shall show that Maxwell's equations lead to *Lorentz transformations*.

Let us imagine that a charge e is moving along the x-axis with velocity $u\hat{a}_x$. We are interested in determining the potential at the point $P(x, y, z)$ at time t. At time t the charge will be at the position (2) represented by $x = ut$. We are interested in knowing its position (1) at the retarded time $t' = t - \dfrac{R(t')}{C}$, where $R(t') = BP$, i.e. the distance of the point P from the position of the charge at time t'. From Fig. 12.9, we have

$$R(t') = |r - \hat{a}_x ut'|$$

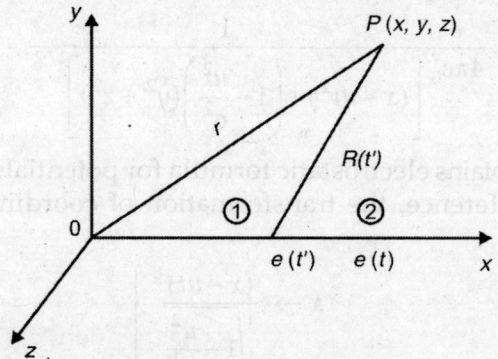

Fig. 12.9: Potential due to moving charge when observer is at rest

where r is the position vector of the point P as shown in Fig. 12.9. We have

$$R(t') = [(x - ut')^2 + y^2 + z^2]^{1/2} \quad \left(\because t' = t - \frac{Rt'}{c} \right)$$

$$\therefore \quad R(t') = c^2(t - t')^2 = (x - ut')^2 + y^2 + z^2$$

$$\text{or} \qquad (c^2 - u^2)t'^2 - 2(c^2t - xu)t' - r^2 + c^2t^2 = 0$$

$$\therefore \qquad t' = \frac{(c^2t - xu) - [(c^2t - xu)^2 + (c^2 - u^2)(r^2 - c^2t^2)]^{1/2}}{(c^2 - u^2)}$$

$$\text{or} \quad \left(1 - \frac{u^2}{c^2}t'\right) = \left(1 - \frac{xu}{c^2}\right) - \frac{1}{c}\left[(x - ut^2) + \left(1 - \frac{u^2}{c^2}\right)(y^2 + z^2)\right]^{1/2}$$

$$\text{or} \quad c(t - t') = \frac{u}{c}(x - ut') + \left[(x - ut^2) + \left(1 - \frac{u^2}{c^2}\right)(y^2 + z^2)\right]^{1/2} \qquad (12.89)$$

We have

$$\beta \cdot R = \frac{u}{c}R\cos\theta = \frac{u}{c}(x - ut')$$

$$\therefore \quad R - \beta \cdot R = c(t - t') - \frac{u}{c}(x - ut')$$

$$= \left[(x - ut)^2 + \left(1 - \frac{u^2}{c^2}\right)(y^2 + z^2)\right]^{1/2} \qquad (12.90)$$

Now, we can express Lienard–Wiechert potentials as

$$A(r, t) = \frac{\mu_0}{4\pi} \frac{u}{\left(1 - \frac{u^2}{c^2}\right)^{1/2}\left[\dfrac{(x - ut)^2}{1 - \dfrac{u^2}{c^2}} + y^2 + z^2\right]^{1/2}} \qquad (12.91)$$

and
$$\phi(\mathbf{r}, t) = \frac{e}{4\pi\varepsilon_0} \frac{1}{\left[(x - ut^2) + \left(1 - \frac{u^2}{c^2} \right)(y^2 + z^2) \right]^{1/2}} \tag{12.92}$$

On putting $u = 0$, one obtains electrostatic formula for potential. We can conclude that in moving frame of reference, the transformation of coordinate should take place according to

$$\left. \begin{aligned} x &\to \frac{(x - ut)}{\sqrt{1 - \frac{u^2}{c^2}}} \\ y &\to y \\ z &\to z \end{aligned} \right\} \tag{12.93}$$

These transformations are known as *Lorentz transformations*.

12.10 ELECTRIC AND MAGNETIC FIELDS OF AN ACCELERATED CHARGE

In the case of an accelerated source charge, it is not possible to express the potential in terms of coordinates of the 'present position' alone. Let the point of observation P be at $x_a = x_1, x_2, x_3$, (Fig. 12.10) and $x'_\alpha(t') = x'_1(t'), x'_2(t'), x'_3(t')$ be the coordinates of the charge (source) at time t' at which a signal propagated with velocity c and arrive at x_α at time t. Thus, the variables of source point and field point are related by

Fig. 12.10: Accelerated charge

$$R = \sum (x_\alpha - x'_\alpha)^{1/2} = c(t - t') \tag{12.94}$$

The electric and magnetic fields can be found using the basic equations

$$E = -\nabla\phi - \frac{\partial A}{\partial t} \tag{12.95}$$

and
$$B = \nabla \times A \tag{12.96}$$

The vector operator ∇ gives partial space derivatives at constant time t (and not t') and the partial time derivatives are with respect to t (and not t') refer to a constant field point x. Since the coordinates and charge descriptions of the source point are known, it is easy to find the electric and magnetic fields at the point of observation. In effect, it is essential to transform $(\partial/\partial t')|_{x_\alpha = \text{constant}}$ and $\nabla|_{x_\alpha = \text{constant}}$ in terms of $(\partial/\partial t')|_{x_\alpha = \text{constant}}$. This transformation is necessary because in the case of an accelerated charge it is not possible, in general, to express the potentials in terms of the 'present position' alone. The potentials are given by

$$A = \frac{\mu_0}{4\pi} e \left[\frac{u}{R - \boldsymbol{\beta} \cdot \mathbf{R}} \right] = \frac{e}{4\pi\varepsilon_0} \frac{u}{c^2 S} \tag{12.97}$$

and
$$\phi = \frac{e}{4\pi\varepsilon_0} \frac{1}{[R - \boldsymbol{\beta} \cdot \boldsymbol{R}]} = \frac{e}{4\pi\varepsilon_0} \frac{1}{S} \tag{12.98}$$

where
$$S = R - \boldsymbol{\beta} \cdot \boldsymbol{R} \tag{12.99}$$

Now, we write

$$R[x_\alpha, x'_\alpha(t')] = [\Sigma(x_\alpha - x'_\alpha)^2]^{1/2} = c(t - t') \tag{12.100}$$

Since x'_α is a function of t', we have R as a function of x_α and t'. We can express Eq. (12.100) as

$$R[x_\alpha, x'_\alpha(t')] = f(x_\alpha t') = c(t - t') \tag{12.101}$$

$$\therefore \qquad \frac{\partial R}{\partial t} = c\left(1 - \frac{\partial t'}{\partial t}\right) \tag{12.102}$$

Also
$$\frac{\partial R}{\partial t} = \frac{\partial R}{\partial t'} \cdot \frac{\partial t'}{\partial t} = -\frac{\boldsymbol{R} \cdot \boldsymbol{u}}{R} \cdot \frac{\partial t'}{\partial t} \tag{12.103}$$

From Eqs. (12.102) and (12.103), we obtain

$$\frac{\partial t'}{\partial t} = \frac{1}{1 - \dfrac{\boldsymbol{R} \cdot \boldsymbol{u}}{Rc}} = \frac{R}{R - \dfrac{\boldsymbol{R} \cdot \boldsymbol{u}}{c}} = \frac{R}{S} \tag{12.104}$$

For brevity, we denote $\left(R - \dfrac{\boldsymbol{R} \cdot \boldsymbol{u}}{c}\right) = S$. This gives the desired transformation for time derivatives

$$\frac{\partial}{\partial t} = \frac{R}{S} \frac{\partial}{\partial t'} \tag{12.105}$$

Let us now proceed to evaluate the transformation of the differential operator (∇) from the coordinates of the field point, to those of the source point. Because R is a function of x_a and t', one can write

$$\nabla R = \nabla_1 R + \frac{\partial R}{\partial t'} \nabla t' = \frac{\boldsymbol{R}}{R} - \frac{\boldsymbol{R} \cdot \boldsymbol{u}}{R} \nabla t' \tag{12.106}$$

where ∇' means differentiation with respect to the coordinates of the field point at constant retarded time t'. From Eq. (12.106), one obtains

$$\nabla R = -c\nabla t'$$

$$\therefore \qquad -c\nabla t' = \frac{\boldsymbol{R}}{R} = \frac{\boldsymbol{R} \cdot \boldsymbol{u}}{R} \nabla t'$$

$$\nabla t' = -\frac{\boldsymbol{R}}{Sc} \tag{12.107}$$

With the help of Eqs. (12.106) and (12.107), one obtains a general relation for transformation of source and field points as

$$\nabla = \nabla_1 - \frac{\boldsymbol{R}}{Sc} \frac{\partial}{\partial t} \tag{12.108}$$

Using these transformation equations, one can easily obtain expressions for electric and magnetic fields. We have

$$E = -\frac{e}{4\pi\varepsilon_0}\nabla\left(\frac{1}{S}\right) - \frac{e}{4\pi\varepsilon_0}\frac{\partial}{\partial t}\left(\frac{u}{Sc^2}\right)$$

$$= \frac{e}{4\pi\varepsilon_0}\left[\frac{1}{S^2}\nabla_1 S - \frac{u}{S^2 c}\frac{\partial S}{\partial t'} - \frac{R}{S^2 c^2}u + \frac{Ru}{c^2 S^3}\frac{\partial}{\partial t'}\right]$$

$$= \frac{e}{4\pi\varepsilon_0}\left[\frac{R}{S^2 R} - \frac{u}{cS^2} + \frac{R}{S^3 c}\left(\frac{R\cdot u}{R}\right) - \frac{R}{S^3 c}\frac{u^2}{c} + \frac{R}{S^3 c}\left(\frac{R\cdot u}{c}\right)\right]$$

$$- \frac{Ru}{S^2 c^2} - \frac{R}{S^3 c^2}u\left(\frac{R\cdot u}{c}\right) + \frac{R}{S^3 c^3}uu^2$$

$$- \frac{R}{c^2 S^3}u\left(\frac{R\cdot u}{c}\right) \qquad \left(\because \nabla_1 S = \frac{R}{R} - \frac{u}{c}\right) \tag{12.109}$$

After rearranging and combining the appropriate terms, one obtains

$$E = \frac{e}{4\pi\varepsilon_0}\left[\frac{1}{S^3}\left(R - \frac{Ru}{c}\right)\left(1 - \frac{u^2}{c^2}\right) + \frac{1}{c^2 S^3}\left\{R\times\left(\left(R - \frac{Ru}{c}\right)\times\dot{u}\right)\right\}\right] \tag{12.110}$$

Similarly, one obtains for magnetic field as

$$B = \frac{e}{4\pi\varepsilon_0 c^2}\left[\frac{u\times R}{S^3}\left(1 - \frac{u^2}{c^2}\right) + \frac{1}{cS^3}\frac{R}{R}\times\left\{R\times\left(\left(R - \frac{Ru}{c}\right)\times\dot{u}\right)\right\}\right] \tag{12.111}$$

Comparison of Eq. (12.110) with Eq. (12.111) leads to the result

$$B = \frac{R\times E}{Rc} \tag{12.112}$$

Obviously, the magnetic field emitted from a moving charge is always perpendicular to E and the retarded radius vector R.

On examination of Eq. (12.110), we note that the electric field consists of two distinct parts. The first component given by the first term is a function of velocity u, whereas the second is a function of acceleration. One can write

$$E = E_v + E_a \tag{12.113}$$

where E_v is the velocity field and E_a the acceleration field. We also see that $E_v \propto \frac{1}{R}$ and $E_a \propto \frac{1}{R}$. Now we compute the Poynting vector for these fields. One finds that the contribution to this vector due to the two components is

$$N_v \propto \frac{1}{R^4}$$

and

$$N_a \propto \frac{1}{R^2} \tag{12.114}$$

For the determination of energy radiated by the charged particle, one will have to integrate the normal component of N over the surface of the sphere of radius R. Because the element of surface area involves R^2, the integral involving N_v varies as $\dfrac{1}{R^2}$ while that containing N_a remains finite. Obviously, for large R, the contribution due to N_v tends to zero and that due to N_a is finite. This means, a particle moving with a uniform velocity cannot radiate energy. This shows that energy can be radiated only by accelerated charges. We must note that deceleration also produce radiation, e.g. when electrons are projected into a block of material, the electrons are stopped and radiation is produced. This radiation is *braking radiation* and called *bremsstrahlung*.

12.11 RADIATION FROM AN ACCELERATED CHARGE AT LOW VELOCITY: LARMOR'S FORMULA

We have seen that accelerated charges radiate electromagnetic energy in space. Now, if the velocity of the charge particle is so small that $\dfrac{u}{c}$ can be neglected, then $S \approx R$ and fields as obtained from Eq. (12.110) and Eq. (12.111) become

$$E_a = \frac{e}{4\pi\varepsilon_0 R^3}[R \times (R \times \dot{u})] \tag{12.115}$$

and

$$B_a = \frac{e}{4\pi\varepsilon_0 c^3 R^3}(\dot{u} \times R) \tag{12.116}$$

One obtains the Poynting vector which contributes to the radiation as

$$N_a = E_a \times E_a = E_a \times \frac{B_a}{\mu_0} = E_a \times \frac{1}{\mu_0 c}\left(\frac{R \times E_a}{R}\right)$$

E_a is perpendicular to R and hence we have

$$N_a = \frac{1}{\mu_0 c}E_a^2 \hat{a}_n = \sqrt{\frac{\varepsilon_0}{\mu_0}}E_a^2 \hat{a}_n \tag{12.117}$$

But

$$E_a = \frac{e}{4\pi\varepsilon_0 c^2 R^3}[R \times (R \times \dot{u})]$$

$$= \frac{e}{4\pi\varepsilon_0 c^2 R^3}[R \times (R \times \dot{u})]$$

$$= \frac{e}{4\pi\varepsilon_0 c^2 R^3}[R\dot{u}\cos\theta R - R^2\dot{u}]$$

Here θ is the angle between R and $\dot{u}$. One obtains finally

$$N_a = \frac{1}{\mu_0 c}\left(\frac{e^2}{16\pi^2\varepsilon_0^2 c^4 R^6}\right)[R\dot{u}\cos\theta R - R^2\dot{u}]^2 \hat{a}_n$$

$$= \left(\frac{e^2}{16\pi^2\varepsilon_0 c^3 R^6}\right)[R^4(\dot{u})^2\cos^2\theta + R^4(\dot{u})^2 - 2R^4(\dot{u})^2\cos^2\theta]\hat{a}_n$$

$$= \frac{(\dot{u})^2 e^2}{16\pi^2 \varepsilon_0 c^3 R^2}(1-\cos\theta)\hat{a}_n = \frac{(\dot{u})^2 e^2}{16\pi^2 \varepsilon_0 c^3 R^2}\sin^2\theta\,\hat{a}_n \tag{12.118}$$

Equation (12.118) gives us the energy flow per unit area per unit time. One can obtain the power radiated per unit solid angle by multiplying by R^2 which is the area per unit solid angle. This can be written as

$$\frac{dW}{d\Omega} = \frac{e^2(\dot{u})^2}{16\pi^2 \varepsilon_0 c^3}\sin^2\theta \tag{12.119}$$

This exhibits the characteristic $\sin^2\theta$ angular dependence, which is well-known result (Fig. 12.11). Integrating over the whole sphere, one obtains the total radiated power as

$$W = \frac{e^2(\dot{u})^2}{16\pi^2 \varepsilon_0 c^3 R^2}\int_0^{2\pi}\int_0^{\pi}(1-\cos^2\theta)R^2\sin\theta\,d\theta\,d\phi$$

$$= 2\pi\frac{e^2(\dot{u})^2}{16\pi^2 \varepsilon_0 c^3}\cdot\frac{4}{3} = \frac{e^2(\dot{u})^2}{6\pi\varepsilon_0 c^3} \tag{12.120}$$

This is the familiar *Larmor result* for a non relativistic accelerated charge.

Fig. 12.11: Angular distribution of energy, i.e. $\sin^2\theta$ distribution

Special Cases

i. *When the velocity and acceleration of the particles are parallel (collinear):*
One can express the radiation fields for this case as

$$E_a = \frac{e}{4\pi\varepsilon_0 c^2 S^2}[R\times(R\times\dot{u})] \tag{i}$$

and

$$B_a = \frac{eR}{4\pi\varepsilon_0 c^2 S^3}(\dot{u}\times R) \tag{ii}$$

$$\therefore \qquad E_a^2 = \frac{e^2 R^4(\dot{u})^2}{16\pi^2 \varepsilon_0^2 c^4 S^6}\sin^2\theta \tag{iii}$$

and

$$N_a = \frac{1}{\mu_0}E_a^2\,\hat{a}_n = \frac{e^2 R^4(\dot{u})^2}{16\pi^2 \varepsilon_0 c^3 S^6}\sin^2\theta\,\hat{a}_n \tag{iv}$$

Now, we obtain the energy radiated into a unit solid angle at θ and measured during the interval dt is

$$-dW(\theta) = \frac{e^2 R^6(\dot{u})^2}{16\pi^2 \varepsilon_0 c^3 S^6}\sin^2\theta\,dt \tag{v}$$

The negative sign on the left hand side shows that this is the energy lost by the charged particle (here electron) in a time interval dt' during the emission of the signal. We find the energy lost per unit time per unit solid angle as

$$-dW(\phi) = -\frac{dW(\theta)}{dt'} = \frac{e^2 R^6(\dot{u}^2)}{16\pi^2 \varepsilon_0 c^3 S^6}\sin^2\theta\frac{dt}{dt'}$$

$$= \frac{e^2 R^6 (\dot{u})^2}{16\pi^2 \varepsilon_0^2 c^4 S^6} \sin^2 \theta \, \frac{S}{R} \qquad \text{(vi)}$$

In obtaining Eq. (vi) we have used Eq. (12.104).

$$\therefore \quad \frac{dW}{d\Omega} = \frac{e^2 R^5 (\dot{u}^2)}{16\pi^2 \varepsilon_0 c^3 S^6} \sin^2 \theta$$

$$= \frac{e^2 R^5 (\dot{u})^2}{16\pi^2 \varepsilon_0 c^3 R^5 \left(1 - \dfrac{\beta \cdot R}{R}\right)^5} \sin^2 \theta$$

$$= \frac{e^2 (\dot{u})^2 \sin^2 \theta}{16\pi^2 \varepsilon_0 c^3 (1 - \beta \cos \theta)^5} \qquad \text{(vii)}$$

Equation (vii) gives the angular distribution of radiated energy.

For $\beta \ll 1$, i.e. $u \ll c$, one obtains Eq. (12.119) for low velocities. However, if $u \to c$, i.e. $\beta \to 1$, the radiated energy increases in the forward direction (Fig. 12.12), but not exactly in the forward direction (i.e. for $\theta = 0$).

ii. *Charged particle moving in a circular orbit:*

Now we consider the case of a charged particle moving in a circle of radius ρ with an angular frequency ω. We have

Fig. 12.12: Radiation patterns of electrons for $u \to c$

$$u = \rho \omega \quad \text{and} \quad \dot{u} = \rho \omega^2$$

This provides another example of angular distribution of radiation for a charge in instantaneously circular motion with its acceleration $\dot{u}$ perpendicular to its velocity u. Figure 12.13 shows the orbit of the particle in the yz plane. We choose a coordinate system such that instantaneously the acceleration $\dot{u}$ is directed towards the centre along the z-direction and velocity u along the y-direction.

Fig. 12.13: Orbit of particle in yz plane

Let θ be the angle between u and R and ϕ be the azimuthal angle. Now, we have

$$u \cdot R = uR \cos \theta \qquad \text{(viii)}$$

or

$$\dot{u} \cdot R = \dot{u} R \sin \theta \cos \phi \qquad \text{(ix)}$$

The radiation field is given by

$$E_a = \frac{e}{4\pi \varepsilon_0 c^2 S^3} \left[R \times \left(R - R \frac{u}{c} \right) \times \dot{u} \right]$$

and Poynting vector is given by

$$N_a = \frac{1}{\mu_0 c} E_a^2 \hat{a}_n$$

Now, we have

$$\left[R \times \left(R - \frac{Ru}{c} \right) \times \dot{u} \right]^2 = \left[(R \cdot u)\left(R - \frac{R\dot{u}}{c} \right) - \left\{ R \cdot \left(R \cdot \frac{Ru}{c} \right) \right\} \dot{u} \right]^2$$

$$= (R \cdot u)^2 \left(R - \frac{Ru}{c} \right)^2 + \left\{ R^2 \dot{u} - \frac{R}{c}(R \cdot u)\dot{u} \right\}^2$$

$$- 2(R \cdot \dot{u})\left(R - \frac{Ru}{c} \right) \cdot \left\{ R^2 \dot{u} - \frac{R}{c}(R \cdot u)\dot{u} \right\}$$

$$= -R^2 \left(1 - \frac{u^2}{c^2} \right)(u \cdot R)^2 + R^4 \left(1 - \frac{R \cdot u}{Rc} \right)(\dot{u})^2 R^4$$

$$= -(\dot{u})^2 R^4 - \frac{2\dot{u}^2 R^4 u \cos\theta}{c} + \frac{(\dot{u})^2 R^4 u^2 \cos^2\theta}{c^2}$$

$$= \left[\left(1 - \frac{u}{c}\cos\theta \right)^2 - \left(1 - \frac{u^2}{c^2} \right)\sin^2\theta\cos^2\phi \right](\dot{u})^2 R^4 \tag{x}$$

In obtaining Eq. (x), we have used Eqs. (viii) and (ix). Now we obtain

$$N_a = \frac{e^2 (\dot{u})^2 R^4}{16\pi^2 \varepsilon_0^2 \mu_0 c^5 S^6}\left[\left(1 - \frac{u}{c}\cos\theta \right)^2 - \left(1 - \frac{u^2}{c^2} \right)\sin^2\theta\cos^2\phi \right]\hat{a}_n \tag{xi}$$

One obtains the angular distribution of the radiated energy as

$$-\frac{dW(\theta)}{dt'}d\Omega = (N_a \cdot \hat{a}_n)R^2 \frac{dt}{dt'}d\Omega = \frac{e^2 (\dot{u})^2}{16\pi^2 \varepsilon_0 c^3}$$

$$\frac{\left[\left(1 - \frac{u}{c}\cos\theta \right)^2 - \left(1 - \frac{u^2}{c^2} \right)\sin^2\theta\cos^2\phi \right]}{\left(1 - \frac{u}{c}\cos\theta \right)^5}d\Omega \tag{xii}$$

In obtaining Eq. (xii), we have used Eq. (12.104) and substituted for S and in Eq. (vii). The radiation pattern in the plane $\phi = 0$ for $\beta = 0.5$ is shown in Fig. 12.14.

Fig. 12.14: Radiation pattern in the plane $\phi = 0$ for $\beta = 0.5$

12.12 RELATIVISTIC GENERALIZATION OF LARMOR'S FORMULA

Larmor's formula [Eq. (12.120)] can be generalized for arbitrary velocities of charge by using the fact that radiated electromagnetic energy behaves under Lorentz transformation like the fourth component of a four vector. We must note that if a pulse of electromagnetic radiation of finite spatial extent exists in charge and current free space, then total electromagnetic momentum and energy transform like a four vector. Since $W = \dfrac{dE_{rad}}{dt}$, $dE_{rad} = W dt$ and time t also behaves under Lorentz transformations like the fourth component of *space-time four vector*. This means that the power W is a Lorentz invariant quantity. Obviously, if one can find a Lorentz invariant reduces to non-relativistic Larmor's relation for $\beta = \dfrac{u}{c} \ll 1$, then one obtains the required relativistic generalization of Larmor's relation. However, from the general form of electromagnetic fields for a moving charge, it is evident that the general result must involve only β and $\dot{\beta}$. With this restriction on the order of derivatives which may appear, one finds unique result.

To obtain the appropriate generalization of Larmor's formula, let us use the familiar fact that $a = \dfrac{F}{m} = \left(\dfrac{dp/dt}{m} \right)$. We can express Larmor's non-relativistic formula [Eq. (12.120)] in the form

$$W = \frac{1}{4\pi\varepsilon_0} \frac{2e^2 (a \cdot a)}{3c^3} = \frac{1}{4\pi\varepsilon_0} \frac{2}{3} \frac{e^2}{m^2 c^2} \left(\frac{dp}{dt} \cdot \frac{dp}{dt} \right) \tag{12.121}$$

where m is the mass of the charge and p its momentum. Equation (12.121) reveals that its Lorentz invariant generalization must be

$$W = \frac{1}{4\pi\varepsilon_0} \frac{2}{3} \frac{e^2}{m^2 c^2} \left(\frac{dp_\mu}{d\tau} \frac{dp_\mu}{d\tau} \right) \tag{12.122}$$

where p_μ is the charged particle's momentum energy for vector given by

$$p_\mu = \left(p, \frac{iE}{c} \right) \tag{12.123}$$

and $d\tau$ is proper time interval given by dt

$$dt = \frac{d\tau}{\sqrt{1 - \dfrac{u^2}{c^2}}} = \frac{d\tau}{\sqrt{1 - \beta^2}} = \gamma d\tau$$

$$\text{with } \gamma = \frac{1}{\sqrt{1 - \beta^2}} \tag{12.124}$$

One can easily check that Eq. (12.122) reduces properly to Eq. (12.121) as $\beta \to 0$. For this, one will have to evaluate the four-vector scalar product. We have

$$\frac{dp_\mu}{d\tau} \frac{dp_\mu}{d\tau} = \left(\frac{dp_1}{d\tau} \right)^2 + \left(\frac{dp^2}{d\tau} \right)^2 + \left(\frac{dp_3}{d\tau} \right)^2 + \left(\frac{dp_4}{d\tau} \right)^2$$

$$= \left(\frac{dp}{d\tau}\right)^2 + \left\{\frac{d}{d\tau}\left(\frac{iE}{c}\right)\right\}^2 = \left(\frac{dp}{d\tau}\right)^2 - \frac{1}{c^2}\left(\frac{dE}{d\tau}\right)^2$$

But $\qquad dt = \dfrac{dt}{\gamma}$

$\therefore \qquad \dfrac{dp_\mu}{d\tau}\dfrac{dp_\mu}{d\tau} = \gamma^2\left(\dfrac{dp}{dt}\right)^2$

Using the above, Eq. (12.122) becomes

$$W = \frac{1}{4\pi\varepsilon_0}\frac{2}{3}\frac{e^2\gamma^2}{m^2c^2}\left[\left(\frac{dp}{dt}\right)^2 - \frac{1}{c^2}\left(\frac{dE}{dt}\right)^2\right] \tag{12.125}$$

To express the above equation in terms of velocity and acceleration, we make use of relativistic expressions

$$E = \frac{mc^2}{\sqrt{1-\beta^2}} \quad \text{and} \quad p = \frac{mu}{\sqrt{1-\beta^2}} \tag{12.126}$$

where m is the rest mass of the charge. From Eq. (12.126), one obtains

$$\frac{dE}{dt} = \frac{d}{dt}\{mc^2(1-\beta^3)^{-1/2}\} = \frac{mc^2\beta\cdot\dot\beta}{(1-\beta^2)^{3/2}} = mc^2\gamma^3\beta\cdot\dot\beta \tag{12.127}$$

and $\qquad \dfrac{dp}{dt} = \dfrac{d}{dt}\left\{\dfrac{mu}{\sqrt{(1-\beta^2)}}\right\} = \dfrac{d}{dt}\left\{\dfrac{mc\beta}{\sqrt{1-\beta^2}}\right\} \dfrac{mc[\dot\beta(1-\beta^2)+\dot\beta(\beta-\dot\beta)]}{(1-\beta^2)^{3/2}}$

or $\qquad \dfrac{dp}{dt} = mc\gamma^3[\beta\cdot(1-\beta^2)+\beta(\beta\cdot\dot\beta)] \tag{12.128}$

Using Eqs. (12.127) and (12.128), Eq. (12.126) becomes

$$W = \frac{1}{4\pi\varepsilon_0}\frac{2}{3}\frac{e^2\gamma^3}{m^3c^3}\left[m^2c^2\gamma^6\left\{\dot\beta(1-\beta^2)+\beta(\beta\cdot\dot\beta)\right\}^2 - m^2c^2\gamma^6(\beta\cdot\beta)^2\right]$$

On simplification, one obtains

$$W = \frac{1}{4\pi\varepsilon_0}\frac{2}{3}\frac{e^2}{c}\gamma^6\left[(\dot\beta)^2 - \beta^2\dot\beta^2 + (\beta\cdot\dot\beta)^2\right]$$

But $\qquad (\beta\times\dot\beta)^2 = (\beta\times\dot\beta)^2 - \beta^2\dot\beta^2$

$\therefore \qquad W = \dfrac{1}{4\pi\varepsilon_0}\dfrac{2}{3}\dfrac{e^2\gamma^6}{c}[\dot\beta^2 - (\beta\times\dot\beta)^2] \tag{12.129}$

Equation (12.129) is the *relativistic generalization of Larmor's formula*.

Special Cases

i. When the motion of the particle is non-relativistic:

We have $\beta \to 0$ and $\gamma = \dfrac{1}{\sqrt{1-\beta^2}} \to 1$. Equation (12.129) takes the form

$$W = \frac{1}{4\pi\varepsilon_0} \frac{2}{3} \frac{e^2}{c} (\dot{\boldsymbol{\beta}})^2$$

$$= \frac{1}{4\pi\varepsilon_0} \frac{2}{3} \frac{e^2}{c^3} \dot{u}^2 \qquad (12.130)$$

This is non-relativistic Larmor's formula.

ii. When the velocity and acceleration of the particles are collinear (parallel):

We have $|\boldsymbol{\beta} \times \dot{\boldsymbol{\beta}}| = \beta\dot{\beta}\sin\theta = 0$. From Eq. (12.129), we have

$$W = \frac{1}{4\pi\varepsilon_0} \frac{2}{3} \frac{e^2\gamma^6\dot{\beta}^2}{c} = \frac{1}{4\pi\varepsilon_0} \frac{2}{3} \frac{e^2\gamma^6\dot{u}^2}{3c^2} \qquad (12.131)$$

This type of radiations are *bremsstrahlung radiations* which are produced from accelerated electrons with the assumption that the direction of motion of electrons does not change.

iii. When the motion of the charged particle is such that its velocity and acceleration are perpendicular:

We have $|\boldsymbol{\beta} \times \dot{\boldsymbol{\beta}}| = \beta\dot{\beta}\sin 90° = \beta\dot{\beta}$

From Eq. (12.129), one obtains

$$W = \frac{1}{4\pi\varepsilon_0} \frac{2}{3} \frac{q^2 r^6}{3c} (\dot{\beta}^2 - \beta^2\dot{\beta}^2)$$

$$= \frac{1}{4\pi\varepsilon_0} \frac{2}{3} \frac{q^2 r^6}{3c} \dot{\beta}^2 (1 - \beta^2)$$

$$= \frac{1}{4\pi\varepsilon_0} \frac{2}{3} \frac{q^2\gamma^4\dot{\beta}^2}{c} \qquad (12.132)$$

Such type of radiations occurs in circular accelerators like cyclotron or synchrotron, where energy loss is drastic.

12.13 CERENKOV RADIATION

We have seen that the uniformly moving charge do not give rise to electromagnetic radiation. The Poynting vector corresponding to the induction field and other magnetic fields fall off rather rapidly with increasing distances and it is observed that it does not contribute to surface integral at large distances. However, this situation changes in the case of material media for which the dielectric constant $\varepsilon_r > 1$. The electromagnetic radiations from charge and current sources, in general, are obtained from the solution of inhomogeneous wave equations. For charges moving in a material medium and constituting a current source, one can express the inhomogeneous wave equation in terms of vector potential as

$$\nabla^2 A - \frac{M^2}{e^2} \frac{\partial^2 A}{\partial t^2} = \mu_0 J \tag{12.133}$$

The solution of Eq. (12.133) confirms to electromagnetic radiation moving with the phase velocity $v_p = \frac{c}{M}$, where $M = \sqrt{\varepsilon_r}$ is the index of refraction. One obtains the retarded

time corresponding to this solution as $t' = \left(t - \frac{r}{v_p} \right) = \left(1 - \frac{r\sqrt{\varepsilon_r}}{c} \right)$. When the intervening medium in a certain frequency range behaves like a medium of high refractive index, the situation may arise where the velocity of the charged particle, u becomes greater than the phase velocity (v_p) of the electromagnetic radiations. We can see that in such

circumstances the retardation denominator may vanish, i.e. $\left(1 - \frac{u\sqrt{\varepsilon_r}}{c} \cos\theta \right) \to 0$. For

$\cos\theta = \frac{c}{\sqrt{\varepsilon_r}} u = \frac{v_p}{u}$, the charged particle velocity is greater than the phase velocity of the

emitted waves in the given medium and the electromagnetic radiation becomes large and confined to cone of angles around the direction of motion of the charged particles, i.e. the electromagnetic waves propagate along conical surfaces, making an angle α with the direction of propagation given by

$$\sin\alpha = \frac{u}{v_p} \tag{12.134}$$

These waves are called *Cerenkov radiations* which were first observed by Cerenkov in 1934.

Cerenkov radiations are similar to the *Mach sonic shock wave* developed when bodies travel at velocities greater than the phase velocities of the elastic waves in the given medium.

From Eq. (12.134), it is obvious that the effective direction of propaga-tion of wavefront is related to the velocity of the charged particle and it may be used to measure it. Devices used for this purpose are called *Cerenkov detectors*, which are widely used in experiments with fundamental particles because they provide direct information about the velocity of the particles.

ILLUSTRATIVE EXAMPLES

Example 12.1: A dipole with moment p vibrates with frequency ω. A molecule with polarizability α is placed at a point with radius vector r (r is perpendicular to p). If $r \ll \lambda$, where λ is the radiation wavelength, show that the power radiated by the system is given by

$$W = \frac{\omega^4 p^2}{12\pi\varepsilon_0 c^3} \left(1 - \frac{\alpha}{4\pi r^3} \right)^2 .$$

Solution: We have the electric field vector of the dipole at a point with the radius vector r (r perpendicular to p) as

$$E_1 = \frac{3(p \cdot r)r}{4\pi\varepsilon_0 r^5} - \frac{p}{4\pi\varepsilon_0 r^3} = -\frac{p}{4\pi\varepsilon_0 r^3}$$

The induced moment p_1 in the molecule due to the field is given by

$$p_1 = \varepsilon_0 \alpha E_1 = -\frac{\varepsilon_0 \alpha p}{4\pi\varepsilon_0 r^3}$$

The resultant dipole moment $p' = p + p_1$. This creates the net field. The power generated in the radiation zone is given by

$$W = \frac{cp^2}{12\pi\varepsilon_0}\left(\frac{2\pi}{\lambda}\right)^4 = \frac{\omega^4 p^2}{12\pi\varepsilon_0 c^3}\left(1 - \frac{\alpha}{4\pi r^3}\right)^2$$

Example 12.2: A Hertzian dipole of length 5 cm is driven at a frequency $\nu = 150$ MHz. The peak current in the antenna is 30 A. Calculate average radiated power and effective resistance of the antenna.

Solution: Total radiated power by a charge q moving with uniform acceleration a is given by $P = \dfrac{q^2 a^2}{6\pi\varepsilon_0 c^3}$, where symbols have usual meanings.

Now, for a dipole of length l and driven at frequency ν, one can substitute $I_0 l \omega$ for the peak value of qa. Thus, the peak power is

$$P_0 = \frac{I^2 l^2 \omega^2}{6\pi\varepsilon_0 c^3}$$

and average power

$$P_{av} = \frac{1}{2}P_0 = \frac{I_0^2 l^2}{12\pi\varepsilon_0 c^3}\left(\frac{2\pi c}{\lambda}\right)^2 = \frac{\pi I_0^2 l^2}{3\varepsilon_0 c \lambda^2} \tag{1}$$

Substituting the following values in Eq. (1),

$$\lambda = c/\nu = \frac{3 \times 10^8}{150 \times 10^6} = 2 \text{ m}$$

$$I_0 = 30 \text{ A}, \quad l = 5 \times 10^{-2}\text{m}$$

One obtains
$$P_{av} = \frac{3.14 \times (30)^2 \times (5 \times 10^{-2})^2}{3 \times (8.85 \times 10^{-12}) \times 3 \times 10^8 \times 2^2} = 221.9 \text{ W}$$

Now, the effective resistance also called radiation resistance, R_{rad} of an antenna is defined as

$$P_{av} = \frac{1}{2}I_0^2 \ R_{rad}$$

or
$$R_{rad} = \frac{2\pi l^2}{3\varepsilon_0 c \lambda^2} = 788\left(\frac{l}{\lambda}\right)^2 = 788\left(\frac{5\times 10^{-2}}{2}\right)^2$$

$$= 0.4925\ \Omega.$$

Example 12.3: (a) Calculate the current required to radiate a power of 100W at 100MHz from a 0.01m Hertzian dipole. (b) Find the magnitudes of E and H at (100 m, 90°, 9°).

Solution:

(a) $\lambda = \dfrac{3\times 10^8}{10^8} = 3\,\text{m}$ $\qquad\qquad R_{rad} = 790\left(\dfrac{d\ell}{\lambda}\right)^2 = 8.78\times 10^{-3}\,\Omega$

$R_{rad} = \dfrac{2P_{rad}}{I^2}$ $\qquad\qquad\qquad I = \sqrt{\dfrac{200}{8.78\times 10^{-3}}} = 151\,\text{A}$

(This extremely high current illustrates that an antenna with length much less than wavelength is not an efficient radiator).

(b) $|E| = \dfrac{\eta\beta Id\ell}{4\pi r}\sin 90° = 0.95\,\text{V/m}$ and $|H| = 2.52\times 10^{-3}\,\text{A/m}$

Example 12.4: A Hertzian dipole of length $L = 2\,\text{m}$ operates at 1 MHz. Find the radiation efficiency if the copper conductor has $\sigma = 57\,\text{MS/m}$, $\mu_r = 1$ and radius $a = 1\,\text{mm}$.

Solution: $\varepsilon_{rad} = \dfrac{P_{rad}}{P_{in}} = \dfrac{P_{rad}}{P_{rad} + P_{loss}} = \dfrac{R_{rad}}{R_{rad} + R_L}$

where R_{rad} is the radiation resistance and R_L is the ohmic resistance. The radius a is much greater than the skin depth,

$$\delta = \frac{1}{\sqrt{\pi f \mu \sigma_c}} \approx \frac{1}{15}\,\text{mm}$$

so that the current may be assumed to be confined to a cylindrical shell of thickness δ.

$$R_L = \frac{1}{\sigma_c}\frac{L}{(2\pi a)\delta} = 0.084\,\Omega$$

$$R_{rad} = (790\,\Omega)\left(\frac{L}{\lambda}\right)^2 = (790\,\Omega)\left(\frac{Lf}{u}\right)^2 = 0.035\,\Omega$$

$$\varepsilon_{rad} = \frac{0.035}{0.119}\times 100 = 29.4\%$$

Example 12.5: Find the radiation resistance of the oscillating magnetic dipole and compare the radiation resistance of the electric dipole.

Solution: The average intensity of the radiations emitted from the magnetic dipole

$$<S> = \frac{\mu_0 m_0^2 \omega^4}{32\pi^2 c^3}\frac{\sin^2\theta}{r^2}$$

∴ Total radiated power

$$P_{mag} = \int_0^\pi \frac{\mu_0 m_0^2 \omega^4}{32\pi^2 c^3} \frac{\sin^2\theta}{r^2} r\sin\theta \, rd\theta \int_0^{2\pi} d\phi$$

$$= \frac{\mu_0 m_0^2 \omega^4}{12\pi c^3} = \frac{1}{4\pi\varepsilon_0} \frac{m_0^2 \omega^4}{3c^5}$$

The magnetic moment $m_0 = \pi a^2 I_0$, $\omega = 2\pi\nu = \frac{2\pi c}{\lambda}$

∴ $$P_{mag} = \frac{1}{4\pi\varepsilon_0} \frac{\pi^2 a^4 I_0^2}{3c^5} \left(\frac{2\pi c}{\lambda}\right)^4 = \frac{1}{4\pi\varepsilon_0} \frac{16\pi^6 I_0^2}{3c} \left(\frac{a}{\lambda}\right)^4$$

The radiation resistance of magnetic dipole can be obtained from the relation

$$P_{mag} = \frac{1}{2} I_0^2 R_{rad}$$

or $$R_{rad} = \frac{1}{4\pi\varepsilon_0} \frac{32\pi^6}{3c} \left(\frac{a}{\lambda}\right)^4 = \frac{9\times10^9 \times 32 \times (3.14)^6}{3\times(3\times10^8)} \left(\frac{a}{\lambda}\right)^4$$

∴ $$\frac{\text{Radiation resistance of the magnetic dipole}}{\text{Radiation resistance of the electric dipole}}$$

$$= \frac{3.07\times10^5 \left(\dfrac{a}{\lambda}\right)^4}{788 \left(l/\lambda\right)^2} = 389 a^4/(2\pi a\lambda)^2 = 9.86 \, (a/\lambda)^2.$$

Example 12.6: A transmitting antenna radiates 100 kW. Determine the field strength at a distance of 3000 km.

Solution: The field strength E can be calculated with the Somerfield's empirical formula

$$E = \frac{300\sqrt{P}}{d} \qquad [\text{Here} \quad P = 100 \text{ kW}, \ d = 3000 \times 10^{-3} \text{ m}]$$

∴ $$E = \frac{300\sqrt{100}}{3000\times10^{-3}} \ 1 \text{ mV/metre}$$

Example 12.7: A dipole of length 3 cm is operated at 1 GHz. The efficiency factor is 0.6. Calculate the radiation resistance, the antenna gain and the effective aperture.

Solution: $$\lambda = \frac{c}{f} \quad | \text{Here} \ f = 1 \text{ GHz} = 1 \times 10^9 \text{ Hz}$$

$$= \frac{3\times10^8}{1\times10^9} \text{ m} = 0.3 \text{ m}$$

Since $$\frac{l}{\lambda} = \frac{3\times10^{-2}}{0.3} = 0.1$$

Considering the antenna to be short dipole, we use the relation

$$R_r = 80\pi^2\left(\frac{l}{\lambda}\right)^2 = 80 \times \pi^2 \times (0.1)^2 = 7.9\ \Omega$$

The directivity of the short dipole is 1.5. Thus, gain $G = kD = 0.6 \times 1.5 = 0.9$

The effective aperture is given by

$$A_e = D\frac{\lambda^2}{4\pi} = 1.5 \times \frac{(0.3)^2}{4\pi} = 0.0107\ \text{m}^2$$

Example 12.8: Calculate the driving impedance of a half wavelength slot antenna radiating into free space. The corresponding dipole has an input impedance of $(73 + \hat{j}43)\ \Omega$.

Solution: Slot impedance is given by

$$Z_s = \frac{Z_i^2}{4Z_d}\quad |\text{Here } Z_i = 377\ \Omega,\ Z_d = (73 + j43)\ \Omega$$

or

$$Z_s = \frac{(377)^2}{4(73+j43)} = \frac{(377)^2(73-j43)}{4(73^2 + 43^2)}$$

$$= (361 - j212.85)\ \Omega$$

Example 12.9: Calculate the approximate gain and beam width of a paraboloidal reflector antenna at operating frequency 4 GHz, diameter 20 m and illumination efficiency 55%.

Solution:

$$G = \frac{4\pi}{\lambda^2}A_e = \frac{4\pi}{\lambda^2}kA$$

$$A = \frac{\pi D^2}{4} = \pi r^2 = \pi\left(\frac{20}{2}\right)^2 = 100\pi$$

$$\lambda = \frac{30}{4\times10^3} = 0.075\ \text{m}$$

Now, gain $G = \dfrac{4\pi \times 0.55 \times 100\pi}{(0.075)^2} = 385619.9$

or

$$(G)_{dB} = 10\log 385619.9$$

$$= 10 \times 5.55862 = 55.862\ \text{dB}$$

$$(BW)_{(-3\,dB)} = \frac{70\lambda}{D}$$

and

$$BWFN = \frac{2\times70\lambda}{D} = \frac{140\lambda}{D}$$

∴

$$BW = \frac{70\times0.075}{20} = 3.5\times0.75$$

$$= 0.2625 \approx 26°.$$

Example 12.10: Find the current required to radiate a power of 100 W at 100 MHz from a 0.01 m Hertzian dipole.

Solution:
$$\lambda = \frac{c}{\nu} = \frac{3 \times 10^8}{10^8} = 3\,\text{m}$$

$$R_{\text{rad}} = \frac{2\pi\eta}{3}\left(\frac{dl}{\lambda}\right)^2 \approx 790\left(\frac{dl}{\lambda}\right)^2 \Omega$$

$$= 790 \times \left(\frac{0.01}{3}\right)^2 = 8.78 \times 10^{-3}\,\Omega$$

Now
$$R_{\text{rad}} = \frac{2P_{\text{rad}}}{I^2}$$

or
$$I = \sqrt{\frac{2P_{\text{rad}}}{R_{\text{rad}}}} = \sqrt{\frac{2 \times 100}{8.78 \times 10^{-3}}} = 151\,\text{A}$$

We may note that this extremely high current reveals that an antenna with length much less than wavelength is not an efficient radiator.

Example 12.11: A Hertzian dipole of length $L = 2$m operates at 1 MHz. Calculate the radiation efficiency if the copper conductor has $\sigma_c = 57$ MS/m, $\mu_r = 1$, and radius $a = 1$ mm.

Solution: The radiation efficiency of an antenna

$$\varepsilon_{\text{rad}} = \frac{P_{\text{rad}}}{P_{\text{in}}} = \frac{P_{\text{rad}}}{P_{\text{rad}} + P_{\text{loss}}} = \frac{R_{\text{rad}}}{R_{\text{rad}} + R_L}$$

where P_{in} is the time averaged power, R_{rad} is the radiation resistance and R_L is ohmic resistance. The radius a is much greater than the skin depth (δ),

$$\delta = \frac{1}{\sqrt{\pi f\,\mu\sigma_c}} \approx \frac{1}{15}\,\text{mm}$$

Obviously, we can assume current to be confined to a cylindrical shell of thickness δ. Now,

$$R_L = \frac{1}{\sigma_c}\frac{L}{(2\pi a)\delta} = \frac{2}{57 \times 10^6 (2\pi \times 10^{-3}) \times \dfrac{1}{15} \times 10^{-3}} = 0.0838\,\Omega$$

$$R_{\text{rad}} = (790\,\Omega)\left(\frac{L}{\lambda}\right)^2 = (790\,\Omega)\left(\frac{Lf}{u}\right)^2 = 0.035\,\Omega$$

$$\therefore \qquad \varepsilon_{\text{rad}} = \frac{0.035}{0.119} \times 100 = 29.4\% .$$

Example 12.12: Calculate the input impedances for two side-by-side half wave dipoles with a separation $d = \lambda/2$. Assume equal-magnitude, opposite-phase feed-point currents. Take for half wave dipoles, $Z_{11} = Z_{22} = (73 + j\,42.5)\Omega$ and $Z_{12} = (-12.5 - j\,28)\Omega$.

Solution: The two feed-point voltages are given by
$$V_1 = I_1 Z_{11} + I_2 Z_{12}, \quad V_2 = I_1 Z_{21} + I_2 Z_{22}$$

where $\qquad Z_{12} = Z_{21}$

Consequently $\qquad Z_1 = \dfrac{V_1}{I_1} = Z_{11} + \left(\dfrac{I_2}{I_1}\right)Z_{12}$

$$Z_2 = \dfrac{V_2}{I_2} = Z_{22} + \left(\dfrac{I_1}{I_2}\right)Z_{12}$$

For half-wave dipoles,

$$Z_{11} = Z_{22} = (73 + j42.5)\,\Omega \quad \text{and } Z_{12} = (-12.5 - j28)\,\Omega$$

Then, we have $\qquad I_1 = -I_2,$

$$Z_1 = Z_2 = [73 + j42.5 - (-12.5 - j28)]\,\Omega$$

$$= (85.5 + j70.5)\,\Omega.$$

Example 12.13: Show that when particles with identical charge and mass collide, there is no dipole radiation.

Solution: In the centre of mass system, the dipole moment of two charged particles having charge q_1 and q_2 and mass m_1 and m_2 respectively and separated by distance r is given by

$$p = q_1 r_1 + q_2 r_2 = \frac{m_1 m_2}{m_1 + m_2}\left(\frac{q_1}{m_1} - \frac{q_2}{m_2}\right)r$$

Here $\qquad q_1 = q_2 = q, \quad m_1 = m_2$

so $\qquad p = 0$

We may note that dipole radiation is proportional to p^2.

Example 12.14: The transmitting and receiving antennas are separated by a distance of 300 λ and have directive gains of 25 and 18 dB, respecti-vely. Calculate the minimum transmitted power when power received is 4 mW.

Solution: $G_{dt} = 10^{2.5} = 316.23 \qquad |$ Here G_{dt} (dB) $= 25$ dB $= 10\log_{10}G_{dt}$

Similarly,

$$G_{dr}(\text{dB}) = 18\,\text{dB}$$

or $\qquad G_{dr} = 10^{1.8} = 63.1$

Now, $\qquad P_r = G_{dr}\,G_{dt}\left[\dfrac{\lambda}{4\pi r}\right]^2 P_t$

or $\qquad P_t = P_r\left[\dfrac{4\pi r}{\lambda}\right]^2 \dfrac{1}{G_{dr}G_{dt}}$

$$= 4\times 10^{-3}\left[\frac{4\pi \times 300\lambda}{\lambda}\right]^2 \times \frac{1}{61.1\times 316.23}$$

$$= 2.849\,\text{W}$$

Example 12.15: Calculate the front to back ratio (FBR) of an antenna in dB, which radiates 3 kW in its most optimum direction and 500 W in opposite direction.

Solution: $G_{dB} = 10 \log_{10} \dfrac{3000}{500} = 10 \log_{10} 6$

$$= 10 \times 0.7782 = 7.782 = \text{FBR}$$

Figure 12.15 shows the front to back ratio of powers of the optimum direction to its opposite direction.

500 W | 3000 W

Fig. 12.15: Antenna element for computation of front to back ratio

GLIMPSES

1. Electromagnetic radiation refers to the entire spectrum of EM waves including gamma rays, X-rays, ultraviolet, infrared radiations, microwaves and radio waves.

2. An antenna is device that radiates electromagnetic radiations into space, where the energy originates from a source that feeds the antenna through a transmission line or waveguide. The antenna, thus, serves as an interface between the confining line and space when used as a transmitter or between space and line when used as a receiver.

3. Dipole antenna is the simplest antenna formed by a charge that moves back and forth in harmonic motion along a line. The radiation emitted is called dipole radiation. Intensity of radiation emitted by a dipole antenna along the direction of θ is proportional to $\sin\theta/r^2$.

4. The time varying dipole generates not only an electric field but also a magnetic field, as the moving charge constitutes a current and current gives rise to magnetic fields. E and B are perpendicular to each other and to the direction of propagation.

5. The retarded potentials describe the potentials of arbitrary charge and current distributions. The fields produced by a moving point charge can be calculated from these potentials.

REVIEW QUESTIONS

1. What are retarded potentials? Explain the concept of retardation and origin of potential functions ϕ and A.

2. What are Lienard–Wiechert potentials? How do you get them from retarded potentials? Discuss the transformations by using δ-function (Dirac).

3. Write down the Lienard–Wiechert potentials produced by a mov-ing point charge. Obtain the corresponding electric and magnetic fields and identify their respective induction and radiation parts. Show that a charge must be accelerated in order to radiate.

4. Starting from the expression for retarded potentials, obtain the electromagnetic field of a point charge in arbitrary motion.

5. Derive an expression for the rate of radiation of energy from an accelerated charge at low velocity.

6. Derive expressions for the field radiated by an accelerated charge at low velocity. Explain bremsstrahlung and Cerenkov radiations.

7. Obtain an expression for the power radiated by an accelerated charged particle in the non-relativistic limit and discuss its angular distribution.

8. Obtain the electric and magnetic fields from an oscillating electric dipole and calculate the time averaged power radiated per unit solid angle and the total power radiated.

9. Obtain expressions for Poynting vector and power radiated by electric quadrupole.

10. Obtain an expression for the radiation resistance of centre-fed half wave antenna. Explain, why it is an efficient radiator in comparison to dipole antenna.

11. What is an antenna array? Explain its directional properties analytically.

12. Discuss the principle of working of an antenna array. Derive an expression for the resultant field in a direction making an angle ϕ with the line of an array consisting of N equally spaced antenna carrying currents of equal magnitude and having successful phase difference.

13. Write short notes on: (i) Lienard–Wiechert potentials, (ii) Brems–Strahlung and (iii) Cerenkov radiations.

PROBLEMS

1. Show that when two particles with identical charge and mass collide, there is no dipole radiation.

   [*Hint:* Let m_1 and m_2 be the masses of two particles, e_1 and e_2 their charges, and r be the separation between them. For identical particles, we have $m_1 = m_2 = m$ and $e_1 = e_2 = e$. In the centre of mass system, the dipole moment of two particles is

   $$p = e_1 r_1 + e_2 r_2 = \frac{m_1 m_2}{m_1 + m_2}\left(\frac{e_1}{m_1} - \frac{e_2}{m_2}\right)r$$

   Here $r_1 = r_2 = r$, $m_1 = m_2 = m$ and $e_1 = e_2 = e$ $\therefore$ $p = 0$.
   Obviously, dipole moment is zero].

2. Prove in the case of uniformly moving charge that the ratio $\dfrac{E_{90°}}{E_{0°}} = \gamma^{-3}$, where $\gamma = \dfrac{1}{\sqrt{1-\beta^2}}$, $E_{90°}$ is the field at right angles to the direction of charge motion, and E_0 is in the direction of charge motion. Interpret numerically the ratio for $\beta \to 0$ and $\beta = 0.3$.

3. Show that sine square distribution law is followed by power radiated by a low velocity accelerated charged particle. Also show that the total power radiated in such a case is given by Larmor formula.

4. Taking the total radiated power by an arbitrary accelerated charge given by Lienard radiation formula

$$W = \frac{1}{4\pi\varepsilon_0} \frac{2e^2}{3c} \gamma^6 [\dot{\beta}^2 - |\beta \times \dot{\beta}|^2]$$

where $\beta = \dfrac{u}{c}$ and $\dot{\beta} = \dfrac{\partial \beta}{\partial t}$ with u as velocity of the charge, discuss: (i) non-relativistic case, (ii) collinear velocity and acceleration, and (iii) perpendicular velocity and acceleration.

5. A radiator approximates to an electric dipole of length 250 m at a frequency of 60 kHz. Assuming that the current is maintained over the length, show that the radiation resistance of the radiator is 1.97 Ω.

6. An antenna of length L carries alternating current of angular frequency ω. Treating it as an oscillating dipole, determine the total power radiated. Show that the equivalent resistance to produce the same powerloss (the radiation resistance) is

$$R = \frac{2\pi}{3} \sqrt{\frac{\mu_0}{\varepsilon_0}} \left(\frac{L\omega}{2\pi c} \right)^2 .$$

7. Show that the resultant field in a direction making an angle ϕ with the line of an array consisting of N aerials equally spaced d apart and carrying equal aerial currents and phase is given by

$$E = E_0 \frac{\sin(N\phi/2)}{N \sin \phi/2}$$

where E_0 is the value of maximum field and $\phi = \dfrac{2\pi}{\lambda}(d \cos \theta)$.

8. A particle short dipole is a centre fed antenna with length much smaller than the wavelength. The current distribution may be assumed to be linear, the current being maximum at the centre and zero at the ends. Show that the radiation resistance is $20\pi^2 \left(\dfrac{L}{\lambda} \right)^2 \Omega$, where L is the length of the antenna.

9. The charges $-e$, $2e$ and $-e$ form a linear quadrupole oscillator. The positive charge $2e$ is stationary at the origin. One of the negative charges is at $z_1 = a \cos \omega t/2$ and the other is at $z_2 = -a \cos \omega t$. Find the fields at large distances.

SHORT ANSWER QUESTIONS

1. What is the basic difference between the fields of magnetic dipole and fields of an oscillating electric dipole?

    Ans. The fields of magnetic dipole are similar to the fields of an oscillating electric dipole, except that the directions of B and E are interchanged.

2. What is the relation between magnetic and electric fields of a uni-formly moving charge moving with velocity u?

    Ans. $B = \dfrac{1}{c^2} (u \times E)$

3. What is non-radiating or induction field?

Ans. While studying the electric and magnetic fields of an accelerated electron or charge, we have seen that electric field can be expressed as $E = E_v + E_a$, where E_v is the velocity field and E_a is a function of acceleration. E_v around the charge is conserved and does not show any loss of electrical energy. This is called *non-radiating* or *induction field*.

4. What is the nature of radiant energy from an accelerated charge?

Ans. The radiation fields show a loss of energy from the accelerated charge and contribute to the energy flow in the surrounding space. The radiant energy has a angular distribution in space. The angular distribution of radiated energy has an angular dependence of $\sin^2 \theta$. The maximum field is radiated at $\theta = 90°$ and minimum field is radiated along $\theta = 0°$, i.e. along the direction of the acce-lerated charge.

5. What are distant sources of electromagnetic radiations in space?

Ans. Radio stars.

6. An antenna is desired to operate at a frequency of 30 MHz whose Q is 40. Show that its bandwidth is 750 kHz.

Ans. $\Delta f = \dfrac{f_r}{Q} = \dfrac{30 \times 10^6}{40} = 0.75 \times 10^6$ Hz = 750 kHz.

MULTIPLE CHOICE QUESTIONS

1. Radiation resistance of a half wave (linear) dipole antenna is
   (a) $377\,\Omega$ (b) $73\,\Omega$
   (c) $37\,\Omega$ (d) $552\,\Omega$ **[b]**

2. Radiation resistance in a given direction is the
   (a) power radiated per square metre
   (b) energy radiated per square metre
   (c) power radiated per unit solid angle in that direction
   (d) none of the above   **[c]**

3. In case of dipole antenna, for large values of r, the ratio of only non zero field components E_θ and H_ϕ is equal to
   (a) $\dfrac{E_\theta}{H_\phi} = \sqrt{\dfrac{\mu_0}{\varepsilon_0}}$ (b) $\dfrac{E_\theta}{H_\phi} = \dfrac{\mu_0}{\varepsilon_0}$

   (c) $\dfrac{E_\theta}{H_\phi} = 1$ (d) $\dfrac{E_\theta}{H_\phi} = \left(\dfrac{\mu_0}{\varepsilon_0}\right)^2$ **[a]**

4. In case of dipole antenna, for $l \ll \lambda$, the radiation resistance is given by
   (a) $R_{rad} = 20\pi^2 \left(\dfrac{l}{\lambda}\right)^2 \Omega$ (b) $R_{rad} = 80\pi^2 \left(\dfrac{l}{\lambda}\right)^2 \Omega$

   (c) $R_{rad} = \pi^2 \dfrac{l}{\lambda^2} \Omega$ (d) $R_{rad} = 80\pi^2 \Omega$ **[b]**

5. The average total power radiated by a half-wave antenna is given by

    (a) $<W>= 36.57 I_0^2$ W

    (b) $<W>= 73 I_0^2$ W

    (c) $<W>= 377 I_0^2$ W

    (d) $<W>= 730 I_0^2$ W **[b]**

6. The radiated power in a travelling wave antenna in comparison to the standing wave antenna for the same length and frequency is

    (a) more

    (b) less

    (c) equal

    (d) infinite **[a]**

7. The radiation resistance for a small planar circular loop antenna of radius a is

    (a) $R_r = \dfrac{(\omega a)^4}{c^4 \pi^2} \Omega$

    (b) $R_r = \dfrac{(\omega a)^4}{20 \pi^2 c^4} \Omega$

    (c) $R_r = 20 \pi^2 \dfrac{(\omega a)^4}{c^4} \Omega$

    (d) $R_r = 20 \pi^2 \left(\dfrac{\omega}{ac}\right)^4 \Omega$ **[b]**

8. If we neglect phase delay term along the line in case of half wave antenna, then the radiation resistance will be

    (a) $73.14 \, \Omega$

    (b) $37 \, \Omega$

    (c) $370 \, \Omega$

    (d) $80 \, \Omega$ **[d]**

9. Power gain of an antenna in a given direction is the ratio of

    (a) radiation intensity in that direction to the total input radiated power

    (b) radiation intensity in that direction to the average radiated power

    (c) average power to maximum power

    (d) none of the above **[b]**

10. Radiation resistance of a current element of length dl is

    (a) $80\left(\dfrac{\lambda dI}{\pi}\right)^2 \Omega$

    (b) $80\left(\dfrac{\lambda dI}{2\pi}\right)^2 \Omega$

    (c) $800\left(\dfrac{\pi dl}{\lambda}\right)^2 \Omega$

    (d) $80\left(\dfrac{\pi dl}{\lambda}\right)^2 \Omega$ **[d]**

11. In a broadside array, the maximum radiation occurs

    (a) at $45°$ to the array

    (b) along the array

    (c) at $60°$ to the array

    (d) perpendicular to the array **[d]**

12. Electrical length of an antenna is

    (a) small than physical length

    (b) greater than physical length

    (c) equal to its physical length

    (d) none of the above **[b]**

13. The vector potential A measured at point z at time t is known as retarded potential because

    (a) it is attenuated while moving in space

    (b) it is generated due to current at the dipole earlier in time

(c) its velocity suffers retardation in space

(d) none of the above [b]

14. The radiation resistance R_{rad} is

(a) resistance offered by free space

(b) resistance offered by antenna

(c) coupling resistance between free space and antenna

(d) coupling resistance offered by objects in free space [c]

15. An antenna of diameter D is operating at wavelength λ. The critical distance r_c beyond which the electrostatic and inductive contribution to the radiation fields can be neglected is

(a) D^3/λ^2

(b) $\dfrac{2D^2}{\lambda}$

(c) $\dfrac{8D^2}{\lambda}$

(d) $\dfrac{D^2}{2\lambda}$ [b]

16. For sinusoidal current distribution, the ratio of effective length and physical length of an antenna is

(a) $\pi/2$

(b) $2/\pi$

(c) $\sqrt{2/\pi}$

(d) $\sqrt{\pi/2}$ [b]

17. The effective length of an antenna depends on

(a) the angle of radiation

(b) the current distribution

(c) the wavelength of radiation

(d) the area of cross-section [b]

18. Antenna A has radiation resistance twice that of antenna B. This implies that

(a) antenna B delivers quarter power to space than antenna A

(b) antenna B delivers half power to space than antenna A

(c) antenna B delivers equal power to space than antenna A

(d) antenna B delivers double power to space than antenna A [b]

19. An antenna has radiation efficiency = 0.9. If input power is increased, the directive gain will

(a) decrease

(b) increase

(c) remain unchanged

(d) none of the above [a]

20. For a lossless antenna, the magnitude of gain is

(a) greater than that of directivity

(b) equal to that of directivity

(c) equal to that of directivity

(d) none of the above [a]

21. Two antennas A and B have directivity 1.5 and 3.8 respectively. The power concentrated into a small solid angle will be

(a) more for antenna B

(b) more for antenna A

(c) same for both A and B

(d) none of the above [a]

22. Two isotropic antennas are separated by a distance of two wavelengths. If both the antennas are fed with currents of equal phase and magnitude, the number of lobes in the radiation pattern in the horizontal plane are

(a) 2   (b) 4

(c) 6   (d) 8   **[b]**

23. In a broadside array of 20 isotopic radiators equally spaced at a distance of $\lambda/2$, the beamwidth between the first nulls is

(a) 51.3°   (b) 11.46°

(c) 22.9°   (d) 102.6°   **[GATE] [b]**

24. Two dissimilar antennas having their maximum directivities equal,

(a) must have their beam widths equal because they are dissimilar antennas

(b) cannot have their beam widths equal because they are dissimilar antennas

(c) may not necessarily have their maximum power gains equal

(d) must have their effective aperture areas also equal   **[GATE] [c]**

25. The beam width between first null of uniform linear array of N equally spaced (element spacing = d), equally excited antennas is determined by

(a) N alone and not by d

(b) d alone and not by N

(c) the ratio N/d

(d) the product (Nd)   **[GATE] [d]**

26. An antenna, when radiating, has a highly directional radiation pattern. When the antenna is receiving, its radiation pattern

(a) is more directive   (b) is less directive

(c) is the same   (d) exhibits no directivity at all   **[GATE] [c]**

27. Consider a lossless antenna with a directive gain of +6 dB. If 1 MW of power is fed to it, the total power radiated by the antenna will be

(a) 4 MW   (b) 1 MW

(c) 7 MW   (d) 1/4 MW   **[GATE] [a]**

28. For a 8 ft (2.4 m) parabolic dish antenna operating at 4 GHz, the minimum distance required for far field is closest to

(a) 7.5 cm   (b) 15 cm

(c) 15 m   (d) 150 m   **[GATE] [d]**

29. A transmitting antenna radiates 251 W isotropically. A receiving antenna located 100 m away from the transmitting antenna has an effective aperture of 500 cm^2. The total power received by the antenna is

(a) 10 μW   (b) 1 μW

(c) 20 μW   (d) 100 μW   **[d]**

30. The far field of an antenna varies with distance r as

(a) $\dfrac{1}{r}$ (b) $\dfrac{1}{r^2}$

(b) $\dfrac{1}{r^3}$ (d) $\dfrac{1}{\sqrt{r}}$ **[GATE] [a]**

31. An antenna in free space receives 2 μW of power when the incident electric field is 20 mV/metre rms. The effective aperture of the antenna is
    (a) 0.005 m²   (b) 0.05 m²
    (c) 1.885 m²   (d) 3.77 m²   **[GATE] [d]**

32. Two identical antennas are placed in θ = π/2 plane as shown in Fig. 12.16. The elements have equal amplitude excitation with 180° polarity difference, operating at wavelength λ. The correct value of the magnitude of the far-zone resultant electric field strength normalized with that of a single element both computed for φ = 0 is

Fig. 12.16

(a) $2\cos\left(\dfrac{2\pi s}{\lambda}\right)$

(b) $2\sin\left(\dfrac{2\pi s}{\lambda}\right)$

(c) $2\cos\left(\dfrac{\pi s}{\lambda}\right)$

(d) $2\sin\left(\dfrac{\pi s}{\lambda}\right)$ **[GATE] [d]**

Antenna

1. INTRODUCTION

All communication and radar system require an essential terminal device known as *antenna*. It has following features:

i. It is a device which receives or transmits EM wave.

ii. It is a device which makes the transition between free space and guided path and *vice versa*.

iii. An antenna gives particular direction to EM wave for propagation.

iv. An antenna is an impedance matching device.

v. Antenna may also be a piece of conducting material in the form of a wire, rod, or any shape of excitation.

Whenever we want to transmit EM wave created by sources, e.g. charges or current, it should pass through the *transmission line* or *waveguide*. This signal is called as *guided wave*.

Transmission line guides the signal inside it. To transmit in space we require another method which makes relation with waveguide and free space called antenna and also receives EM wave through antenna. Obviously, antenna can sense EM wave. According to the characteristics, antennas are of following two types:

1. Transmitting antenna
2. Receiving antenna

2. GENERAL STRUCTURE OF AN ANTENNA

Antenna converts electrical signal into EM wave at the transmitting end and EM wave into electrical signal at the receiving end. This property renames the antenna as *transducer*. Antenna also acts as a *temperature sensor*.

3. DIFFERENT TYPES OF ANTENNA

Wire Antenna

Dipole and loop antenna have identical field patterns, but *E* and *H* are interchanged.

Helical antenna is an antenna which is in the shape of helix in the form of screw thread. This antenna consists of helical loop made of a thick copper conductor. Its polarization and radiation properties depend upon the diameter, pitch, number of turns, wavelength excitation and shaping between the helical loops.

This antenna provides circular polarized waves which are used in extraterrestrial communications such as satellite relaying. It is broad-band VHF and UHF antenna to provide circular polarization characteristics.

Dipole, loop and helix types of antennas are used in automobiles, buildings, aircrafts, spacecrafts and ships.

Aperture Antennas

(a) *Horn antenna:* This may be regarded as flared out or opened out waveguide. A waveguide is capable of radiating radiation into open space provided the same is excited at one end and opened at other end. The radiation is much greater through waveguide than two-wire transmission line. The function of horn is to produce phase front with a large aperture than that of the waveguide and hence, the greater directivity, improved efficiency and reduced VSWR are resulted.

(b) *Rectangular waveguide:* This is also an example of aperture antenna. This is widely used structure that is usually used in the microwave region of EM spectrum. TE_{10} mode is dominant (and most important) mode in the rectangular waveguide because it can propagate alone if the operating frequency is appropriately chosen.

Array Antennas

(a) *Yagi–Uda antenna:* It consists of a driven element and some parasitic elements. The driving element is half-wave dipole or active element where the power from the transmitter is fed. The driving element feeds receiving power to the receiver. The parasitic elements are continuous metallic rod kept parallel to the driving element. It consists of a reflector and one or more directors. Parasitic or passive elements are not connected directly to the transmission line, but electrically coupled.

(b) *Slot antenna:* It is simply an opening cut in a sheet of conductor, which is energized in some appropriate manner. A vertical slot so energized produces horizontal polarization and *vice versa*. The antenna consists of two resonate $\lambda/4$ stobs connected to a two-wire transmission line and forms an inefficient antenna. This antenna makes use of the fact that the energy is radiated when a high frequency field exists across a narrow slot in a conducting plane. Instead of a single slot, large number of slots can be cut to form an array for wide angle coverage for a moving radar system. Single slot antenna is used in aircraft with a part of the aircraft body such as tailfin.

(c) *Aperture array antenna:* It is nothing but the waveguide antenna opening outside. When we use multiple number of such antenna in a metal plate, then it is called aperture array antenna. The naming of aperture antenna depends upon the shape of the waveguide.

(d) *Microstrip array antenna:* In order to increase the directivity of the antenna, multiple microstrip radiators are used in cascade to form an array.

(e) *Log periodic antenna:* The name comes from its geometrical structure when the electrical property repeats periodically with logarithm of frequency. It has number of dipoles of different length spacing and is fed by balanced two-wire line which is transposed between each adjacent pair of dipoles. The dipole length increases along the antenna such that the inclined angle θ is constant.

The behaviour of this antenna implies that the impedance is logarithmically periodic function of the frequency. Not only this, all the electrical properties undergo similar periodic variation particularly radiation pattern, directivity gain, slide lobe level, beam width and direction.

This antenna is used for TV reception and monitoring.

Reflector Antenna

(a) *Parabolic reflector antenna:* A structure of parabola is two-dimensional, but in practice a parabolic reflector is a three dimensional curved surface obtained by rotating a parabola about its axis. The structure is termed as *paraboloid*. Since the mouth of a paraboloid is circular, it generates a parallel beam of circular cross-section.

A suitable directivity can be achieved using suitable shape parabolic reflector behind the main antenna. The main antenna is known as primary antenna, i.e. feed and the reflector as the secondary antenna.

A common feed radiation for paraboloid reflector antenna is a waveguide horn. Moreover, if circular polarization is required, then conical horn and helix antenna can be used as a feed at the focus of the paraboloid.

Another improved type of feed system is case grain feed in which primary feed radiator is positioned around an opening near the vertex of the paraboloid of focus. The feed radiator is aimed at the secondary hyperboloid reflector or sub-reflector. Thus, the radiation emitted from feed radiator is reflected from case grain secondary reflector which illuminates the main paraboloid reflector as if they had originated from the focus. Then the parabolic reflector collimates the rays (render parallel) as usual.

The advantages of case grain, in general, are: (i) it has an ability to get an equivalent focal length much greater than the physical length, and (ii) reduction in spillover and minor lobe radiation.

(b) *Corner reflector antenna:* The flat reflecting sheets meet at an angle or corner form an effective directional antenna. This is a driven antenna. A half wave dipole associated with a reflector consists of two flat conducting sheets which meet at a corner or at an angle to form a corner. This arrangement with corner reflector and driven antenna is known as corner reflector antenna.

(c) *Lens antenna:* It is an antenna which consists of electromagnetic lens with a feed. In other words, a lens antenna is a 3D electromagnetic device whose refractive index is different from unity. It is seen like glass lens in optics. It is usually made of luciti and polystyrene. They are used to control the aperture illumination, collimate the electromagnetic rays, produce directional characteristic and plane spherical waveforms.

Electric (Hertzian) Dipole Antenna

The vector potential set up by the infinitesimal current element of Fig. 1 is

$$A(P) = \frac{\mu e^{-j\beta r}}{4\pi r}(Id\ell)a_z$$

Fig. 1: Hertzian dipole antenna

In spherical coordinates, $a_z = \cos\theta\, a_r - \sin\theta\, a_\theta$; this relation yields

$$H_\phi = \frac{Id\ell}{4\pi}\beta^2 \sin\theta e^{-j\beta r}\left[\frac{j}{\beta r} + \frac{1}{\beta^2 r^2}\right] \tag{1}$$

$$E_r = \eta\frac{2Id\ell}{4\pi}\beta^2 \cos\theta e^{-j\beta r}\left[\frac{1}{\beta^2 r^2} - j\frac{1}{\beta^3 r^3}\right] \tag{2a}$$

$$E_\theta = \eta\frac{Id\ell}{4\pi}\beta^2 \sin\theta e^{-j\beta r}\left[j\frac{1}{\beta r} + \frac{1}{\beta^2 r^2} - j\frac{1}{\beta^3 r^3}\right] \tag{2b}$$

All other components are zero. Attention will be restricted to the field in which terms containing $1/r^2$ or $1/r^3$ are neglected.

$$\text{far field} = \frac{jId\ell\beta}{4\pi r}\sin\theta e^{-j\beta r} \tag{3}$$

$$E_\theta = \eta\frac{jId\ell\beta}{4\pi r}\sin\theta e^{-j\beta r} = \eta H_\phi$$

It is clear that Eq. (3) represents a diverging spherical wave which at any point is travelling in the $+a_r$ direction with an amplitude that falls off as $1/r$.

The power radiated by the Hertzian dipole is obtained by integrating the time-averaged Poynting vector,

$$P_{\text{avg}} = \frac{1}{2}\text{Re}(E \times H^*)$$

of the far field over the surface of a (large) sphere.

$$P_{\text{rad}} = \int_0^{2\pi}\int_0^\pi P_{\text{avg}} \cdot r^2 \sin\theta\, d\theta\, d\phi\, a_r$$

$$= \int_0^{2\pi}\int_0^\pi \left[\frac{1}{2}\text{Re}(E_\theta H_\phi^*)\right]r^2 \sin\theta\, d\theta\, d\phi$$

$$= \frac{\eta(\beta Id\ell)^2}{12\pi} = \frac{\eta\pi I^2}{3}\left(\frac{d\ell}{\lambda}\right)^2 \tag{4}$$

Antenna Parameters

The radiation resistance R_{rad} is defined as the value of a hypothetical resistor that would dissipate a power equal to the power radiated by the antenna when fed by the same current, thus $P_{\text{rad}} = \frac{1}{2}I_0^2 R_{\text{rad}}$ or $R_{\text{rad}} = 2P_{\text{rad}}/I_0^2$, where I_0 is the peak value of the feed point current. For Hertzian dipole, from Eq. (4),

$$R_{\text{rad}} = \frac{2\pi\eta}{3}\left(\frac{d\ell}{\lambda}\right)^2 = 790\left(\frac{d\ell}{\lambda}\right)^2 \Omega$$

The pattern function $F(\theta, \phi)$ gives the variation of the far-zone electric or magnetic field magnitude with direction. For the Hertzian dipole, this reduces to $F(\theta) = \sin\theta$, since $|E|$ and $|H|$ are independent of ϕ.

The radiation intensity $F(\theta, \phi)$ is another measure of antenna performance; it is defined as the time-averaged radiated power per unit solid angle. From Fig. 2,

$$U(\theta, \phi) = \frac{dP_{rad}}{d\Omega} = \frac{|\mathscr{P}_{avg}|\, dS'}{dS'/r^2} = r^2\, |\mathscr{P}_{avg}|$$

Because U is independent of r (by energy conservation), the far field may be used in evaluation. For the Hertzian dipole.

Fig. 2: Time averaged radiated power

$$U(\theta) = \frac{\eta}{8}\left(\frac{Id\ell}{\lambda}\right)^2 \sin^2\theta \tag{5}$$

Polar plots of the pattern function and radiation intensity distribution for the Hertzian dipole are shown in Fig. 3.

In Fig. 3b, the half-power points are at $\theta = 45°$ and $\theta = 135°$ and the half-power beamwidth is $90°$. In general, the smaller the beamwidth (about the direction of U_{max}), the more directive the antenna.

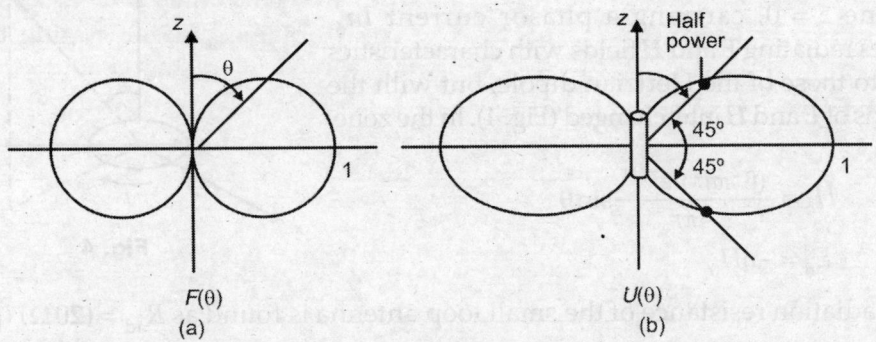

Fig. 3: Polar plots and radiation intensity distribution for the Hertzian dipole

Directive gain $D(\theta, \phi)$ of an antenna is defined as the ratio of the radiation intensity $U(\theta, \phi)$ to that of a hypothetical isotropic radiator that radiates the same total power U_0. For the isotropic radiator,

$$U_0 = \frac{P_{rad}}{4\pi}$$

Then

$$D(\theta, \phi) = \frac{U(\theta, \phi)}{U_0} = \frac{4\pi U(\theta, \phi)}{P_{rad}}$$

The directivity of an antenna is the maximum value of its directive gain,

$$D_{max} = \frac{4\pi U_{max}}{P_{rad}}$$

The Hertzian dipole [Eqs. (4) and (5)] gives

$$D(\theta, \phi) = \frac{(4\pi)\dfrac{\eta}{8}\left(\dfrac{id\ell}{\lambda}\right)^2 \sin\theta}{\left(\dfrac{\eta\pi}{3}\right)\left(\dfrac{id\ell}{\lambda}\right)^2} = 1.5\sin^2\theta$$

and

$$D_{\max} = 1.5 \tag{6}$$

The radiation efficiency of an antenna is $\varepsilon_{rad} = P_{rad}/P_{in}$, where P_{in} is the time-averaged power at the antenna accepts from the feed. The power gain $G(\theta, \phi)$ is defined as the efficiency times the directive gain:

$$G(\theta, \phi) = \varepsilon_{rad}D(\theta,\phi) = \frac{4\pi U(\theta,\phi)}{P_{in}} = \frac{4\pi U(\theta,\phi)}{P_{rad} + P_L}$$

where P_L is the ohmic loss of the antenna. A lossless isotropic radiator has a power gain, $G_0 = 1$. At times the power gain of an antenna is expressed in decibels, where

$$G_{dB} = 10\log_{10}\frac{G}{G_0} = 10\log_{10}G \tag{7}$$

Small Circular Loop Antenna

Also known as the *magnetic dipole*, a small loop in the plane $z = 0$, carrying a phasor current Ia_ϕ, produces radiating E and H fields with characteristics similar to those of the Hertzian dipole, but with the directions of E and H interchanged (Fig. 4). In the zone,

Fig. 4

$$H_\phi = \frac{(\beta^2 \pi a^2)Ie^{-j\beta r}}{4\pi r}\sin\theta$$

$$E_\phi = -\eta H_\theta$$

The radiation resistance of the small loop antenna is found as $R_{ad} = (20\,\Omega)(\beta^2\pi a^2)^2$.

Finite Length Dipole

Equation (4) for the radiated power of the Hertzian dipole contains the term $(d\ell/\lambda)^2$ which suggests that the length should be comparable to the wavelength. The open-circuited two-wire transmission line shown in Fig. 5a has currents in the conductors that are out of phase, so that the far field nearly cancels out. An efficient antenna results when the line is opened is shown in Fig. 5b, producing current phasors

$$I_1(z') = I_m\sin\beta\left(\frac{L}{2} - z'\right) \quad (0 < z' < L/2) \tag{8a}$$

and

$$I_2(z') = I_m\sin\beta\left(\frac{L}{2} + z'\right) \quad (-L/2 < z' < 0) \tag{8b}$$

The two currents are exactly is phase at mirror image points in the y-axis, and they vanish at the end points $z' = \pm L/2$. The two legs form a single dipole antenna of finite

Fig. 5: (a) Open-circuited two-wire transmission line (b) Line is opened

length L. Note that the current at the feed point ($z' = 0$) is related to the maximum current by $I_0 = I_m \sin \dfrac{\beta L}{2}$.

The far field is calculated by means of (2) and (1), under the assumption $r \gg L$ and $r \gg \lambda$.

$$H_\phi = jI_m \frac{e^{-j\beta r}}{2\pi r} F(\theta), \quad E_\theta = \eta H_\phi$$

where the pattern function is given by

$$F(\theta) = \frac{\cos\left(\beta \dfrac{L}{2} \cos\theta\right) - \cos\left(\beta \dfrac{L}{2}\right)}{\sin\theta}$$

The antenna can be assigned an effective length

$$\left(\text{write } I(z') = I_m \sin\beta\left(\frac{L}{2} - |z'|\right) \right)$$

$$h_e(\theta) = \frac{\sin\theta}{I_0} \int_{-L/2}^{L/2} I(z') e^{j\beta z' \cos\theta} dz' = \frac{2I_m}{\beta I_0} F(\theta) \qquad (9)$$

which has the unit of length and contains all the pattern information.

For L up to about 1.2λ, the antenna patterns resemble the figure eight, becoming sharper as L approaches 1.2λ. In the other limit, as $L \ll \lambda$, the pattern is that of the Hertzian dipole as shown in Fig. 3a. As L becomes greater than 1.2λ, the patterns become multilobed (Fig. 6).

Fig. 6: $L > 1.2\lambda$, multilobed pattern

The radiation resistance of a finite dipole of length $(2n - 1)\lambda/2$ $(n = 1, 2, 3)$ can be shown to $R_{rad} = (30\,\Omega)\sin[(4n - 2)\pi)$, where

$$\mathrm{sin}\,(x) = \int_0^x \frac{1 - \cos y}{y}\,dy \tag{10}$$

is a tabulated function. For $n = 1$ (half-wave dipole), $R_{rad} = 30\,(2.438) = 73\,\Omega$ and $D_{max} = 1.64$.

Monopole Antenna

A conductor of length $L/2$ normal to an infinite conducting plane (Fig. 7a) forms a monopole antenna. When fed at the base, the resulting E and H fields are identical to the dipole. This is evident when the image of the monopole is positioned below the conducting plane as shown in Fig. 7b.

As the monopole radiates power only in the region above the conducting plane, the total radiated power is one-half that of the corresponding dipole. From $R_{rad} = 2P_{rad}/I_0^2$, it follows that the radiation resistance is one-half the value for the dipole. Thus, for $L/2 = \lambda/4$ (quarter-wave monopole), $R_{rad} = 36.5\,\Omega$.

(a)   (b)

Fig. 7: Monopole antenna

Self and Mutual Impedances

With respect to its feed, an antenna is equivalent to a load impedance $Z_a = R_a + jX_a$, where $R_a = R_{rad} + R_L$ and R_L is ohmic resistance. The reactance X_a is not easily calculated; it is function of the radius ρ of the conductors for dipoles and monopoles. Figure 8 illustrates the variation of both R_a and X_a for monopoles of length $L/2$; the figure also applies to dipoles of length L if vertical scale values are doubled. Thus, the half-wave dipole has $R_a = 73\,\Omega$ and, roughly independent of ρ, $X_a \approx 40\,\Omega$. (It can be shown that $\rho = 0$, $X_a \rightarrow 42.5\,\Omega$).

When a second antenna is placed adjacent to first antenna, current in one will induce a voltage in the other. Consequently, a mutual impedance $Z_{21} = V/_{21}/I_1 = R_{21} + jX_{21}$ exists in the system for two side-by-side half-wave dipoles with small conductor size, R_{21} and X_{21} vary with separation d (Fig. 9).

Fig. 8: Variation of R_a and X_a for monopoles of length 1/2

Fig. 9: Variation of R_{21} and X_{21} with separation d between two adjacent antenna

Receiving Antenna

An antenna in the far field of a transmitter extracts energy from what is essentially a plane wave and delivers it to a load impedance Z_l. In (Fig. 10a) the dipole antenna lies along the z-axis and the incident wave has a Poynting vector $\mathscr{P}$. The open-circuit voltage

Fig. 10: (a) Dipole antenna along the z-axis (b) Equivalent circuit

is equal to the product of the effective length $h_e(\theta)$ and the magnitude E of the projection of E onto the plane of incidence. [For the coordinate system of Fig. 10a, $E = \sqrt{E_y^2 + E_z^2}$].

$$V_{OC} = h_e(\theta)\,E$$

The pattern for the receiving antenna is identical to that of a similar transmitting antenna. The *available power* P_a is the maximum power which the receiving antenna can deliver to a load, which occurs when $Z_l = Z_a^*$. From the equivalent circuit of Fig. 10b, we have

$$P_a = \frac{h_e(\theta)^2 E^2}{8 R_a} \tag{11}$$

The *effective area* $A_e(\theta)$ for an antenna is hypothetical area such that when multiplied by the power density of the incident wave $E^2/2\eta$, it results in the available power.

$$A_e(\theta)\left(\frac{E^2}{2\eta}\right) = P_a = \frac{h_e(\theta)^2 E^2}{8 R_a}$$

or

$$A_e(\theta) = h_e(\theta)^2\left(\frac{\eta}{4 R_a}\right) \tag{12}$$

One can show that the effective area is related to the directive gain by

$$\frac{A_e(\theta,\phi)}{D(\theta,\phi)} = \frac{\lambda^2}{4\pi} \tag{13}$$

When both a transmitting and a receiving antenna are considered, power $P_{\text{rad}\,1}$ radiated by antenna 1 and the available power P_{a2} at the receiving antenna 2, are related by the *Friss transmission formula*,

$$\frac{P_{a2}}{P_{\text{rad}1}} = \frac{D_1(\theta_1,\phi_1) A_{e2}(\theta_2,\phi_2)}{4\pi r^2} \tag{14}$$

Here, r is the separation of the two antennas. Angles θ_1 and ϕ_1 specify the direction of the receiving antenna as seen from the coordinate system of antenna 1. Similarly, θ_2 and ϕ_2 specify the direction of the transmitting antenna as viewed from the coordinate system of antenna 2.

Linear Arrays

A far-field pattern with a narrow beam width and high gain can be achieved by forming an array of identical antenna elements, each with the same orientation as shown in Fig. 11. The pattern function of the array is equal to the pattern function of an individual element multiplied by an array factor $f(\chi)$. One can show that, for uniformly spaced array of N elements where d is the spacing

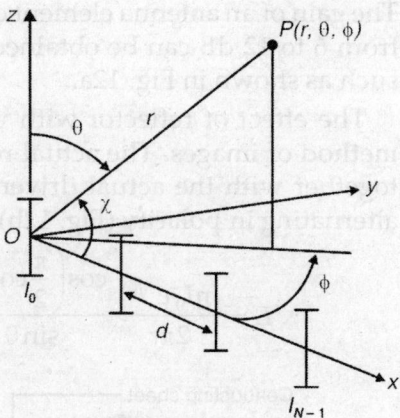

Fig. 11: Linear arrays

$$f(\chi) = \sum_{n=0}^{N-1} I_n e^{j\beta nd \cos \chi} \qquad (15)$$

The angle χ is the angle between the array axis and the line OP; by geometry, $\cos \chi = \sin \theta \cos \phi$. If the elements are progressively phased, then $I_n = a_n e^{jn\alpha}$ ($n = 0, 1,..., N-1$).

$$f(\chi) = \sum_{n=0}^{N-1} a_n e^{jn(\alpha + \beta d \cos \chi)}$$

or, defining $u = \alpha + \beta d \cos \chi$,

$$f_1(u) = \sum_{n=0}^{N-1} a_n e^{jnu} \qquad (16)$$

The overall pattern function will be a maximum when $|f_1(u)|$ is a maximum, which occurs for $u = 0$. If $\alpha = 0$ (the individual antennas are all in phase), then $u = 0$ implies $\chi = \pm 90°$, i.e. peak radiation occurs at right angles to the line of antennas. This is called a *broadside array*. On the other hand if the phasing $\alpha = -\beta t$ is imposed, $u = 0$ implies $\chi = 0°$; this is an end fire array.

A uniform array has all antenna currents equal in magnitude. For $a_0 = a_1 = ... = a_{N-1}$, Eq. (16) becomes

$$f_1(u) = \frac{\sin(Nu/2)}{\sin(u/2)} = e^{j(N-1)u/2} \qquad (17)$$

Thus, the main peak or lobe of the radiation pattern, centered on $u = 0$, has height $|f_1(0)| = N$. The first two nulls of the pattern [zeros of $|f_1(u)|$] occurs at $u = \pm 2\pi/N$. The operation of the two first nulls can be used to define the beamwidth. Concentrating on the plane $\theta = 90°$, one finds

broadside uniform $\qquad \Delta\phi = 2\sin^{-1}\dfrac{2\pi}{\beta Nd} = \dfrac{2\lambda}{Nd} \qquad (18)$

end fire uniform $\qquad \Delta\phi = 4\sin^{-1}\sqrt{\dfrac{\pi}{\beta Nd}} = \sqrt{\dfrac{8\lambda}{Nd}} \qquad (19)$

where the approximations are for the case $Nd \gg \lambda$.

The side lobes occur approximately midway between the nulls. The ratio of the main lobe to the first side lobe is $N\sin(3\pi/2N)$ which approaches the value $3\pi/2$ for large N.

Reflectors

The gain of an antenna element can be enhanced by means of a reflector. Gains ranging from 6 to 12 dB can be obtained by using a half-wave dipole and a corner reflector such as shown in Fig. 12a.

The effect of reflector with $\psi = 180°/N$ ($N = 1, 2, 3,...$) can be calculated by the method of images. The actual reflector is replaced by $(2N - 1)$ image dipoles, which together with the actual driven dipole, constitute an evenly spaced circular array, alternating in polarity (Fig. 12b). Superposition of the far fields yields

$$E = \frac{j\eta I_0\, e^{-j\beta\gamma}}{2\pi r} \frac{\cos\left(\dfrac{\pi}{2}\cos\theta\right)}{\sin\theta} \sum_{n=0}^{2N-1} [(-1)^n\, e^{j\beta s\, \sin\theta\cos(n\psi-\phi)}\hat{a}_\theta] \qquad (20)$$

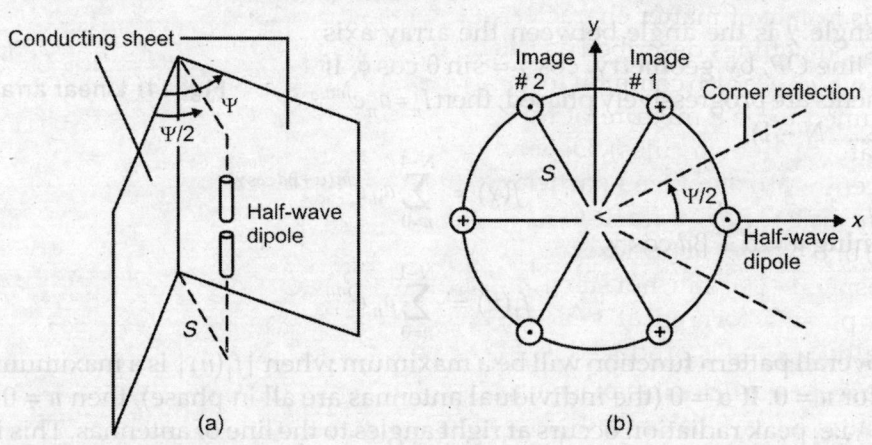

Fig. 12: Reflectors: (a) corner reflector (b) evenly spaced circular array

For high gain applications, the parabolic reflector driven by a source located at its focus, as shown in Fig. 13, is widely used. The directivity of the parabolic reflector is proportional to the aperture radius a and the aperture efficiency E

$$D_{max} = \left(\frac{2\pi a}{\lambda}\right)^2 E$$

The aperture efficiency depends on a variety of design factors, a reasonable value is 55%. The half-power beam width can be estimated from the relation

$$HPBW = 117° \,(\lambda/2a).$$

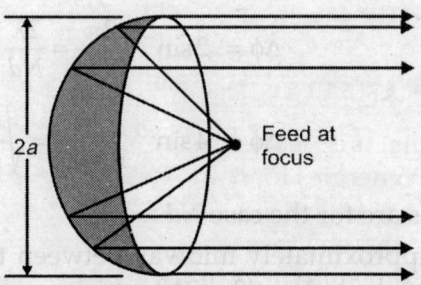

Fig. 13: Parabolic reflector

13

Propagation of Electromagnetic Waves through Ionized Media— Plasma Physics

13.1 INTRODUCTION

Plasma is a state of matter characterized by high, or even complete, ionization of its particles. Sometimes described as *fourth state of matter*, plasma exhibits various basic features which are also common to conductors and semiconductors. Generally, plasma can assumed to be a mixture of three components: free electrons, positive ions and neutral atoms (or molecules). Depending upon the degree of ionization α, the ratio of the concentration of ionized particles to the total concentration, a plasma is classified as *weakly ionized* (α is a fraction of one percent), *moderately ionized* (α equals several percent) or *fully ionized* (α close to 100%). In nature, weakly ionized plasma is found in the ionosphere. The sun, hot stars, and certain interstellar clouds are examples of fully ionized plasma formed at very high temperatures (high temperature plasma). Artificially, plasma is produced in gas discharges and gas discharge tubes.

Because the particles in a plasma are charged, its behaviour differs in some respects from that of a gas. The control of its motion is the basis for the use of plasma as the working medium in various engines and for the direct conversion of its internal energy into electrical energy (magnetohydrodynamics, or MHD, generators, plasma sources of electric energy, etc.).

The concentration of the current carriers in a plasma is very high. Therefore, *plasma is an excellent conductor*. The mobility of the electrons is about three orders of magnitude greater than that of the ions. Hence, the current in a plasma is mainly set up by its electrons.

The propagation of electromagnetic waves in plasma, i.e. ionized medium has gained overwhelming importance because of various wave–wave and wave–particle interactions which have been used as diagnostic techniques for plasma fields and other plasma parameters. The studies in plasma phenomena, therefore, are of great interest and one has to understand the mechanism of various processes occuring in a plasma.

13.2 CONDITIONS FOR PLASMA EXISTENCE

In order to produce plasma, it is necessary to free electrons that are normally bound into atoms. The ionization energies for many elements run from several electron volts (eV) to a few tens of eV, e.g. for hydrogen it is 13.6 eV and for nitrogen it is 15.8 eV. There are various means whereby these energies may be added to the atomic system to produce ionization. A convenient way is to impart the energy by collision events of one sort or another.

On the addition of the ionization energy to a fraction of the atoms of a neutral gas, an *ionized gas* is formed. If sufficient total energy can be added, the gas may be completely ionized with no neutral particles remaining.

One of the most powerful influences on plasma behaviour is the electromagnetic interaction of the charged particles. Even a partially ionized gas can be affected by externally applied electric and magnetic fields and can conduct electric currents. The ions and electrons serve as charge earners in somewhat the same way as they do in an electrolyte. Most importantly, the Coulomb electrostatic fields of the charged particles in the ionized gas will produce the interesting consequences of the collective effects. The presence of collective effects constitutes the primary plasma criterion. Plasma differs from a simple accumulation of charged particles in way that it must have minimum density. This is determined from the condition that $L \gg D$, where L is any length characterizing the plasma and D is the *Debye screening radius* (also called *Debye shielding distance* and *Debye length*). Now, we shall obtain expression for Debye length.

13.3 QUASINEUTRALITY OF A PLASMA: DEBYE SHIELDING

The criterion for the existence of a plasma can be written as

$$n_i \approx n_e \tag{13.1}$$

where n_i is ion density and n_e is electron density in a sample of plasma contained in a sphere of 1cm radius. According to Schottky, one of the definitive properties of plasma is its quasineutrality expressed by Eq. (13.1). In an ensemble of ionized gases having approximately an equal number of electrons and ions, the phenomenon of charge particle shielding occurs. The negative and positive plasma particles are surrounded by particles of opposite charge. The electric forces which bind the oppositely charged particles in plasma provide for its quasineutrality, i.e. its tendency to balance negative and positive space charges in each macroscopic volume element. Since the velocities of electrons and ions are different ($v_e > v_i$, i.e. velocity of electrons, v_e being more than that of positive ions, v_i), it is obvious that more electrons leave the plasma boundary (e.g. discharge tube) than positive ions. Any separation of charges when electrons shift with respect to ions, gives rise to electric fields. These electric fields act in the direction of restoring neutrality. If plasma is not cold and has finite temperature, then some of the charged particles from the edge of the surrounding cloud escape out. This violates quasineutrality in a small volume of plasma. However, the space charge densities adjust themselves so that the major part of the plasma is shielded from the field. For a given temperature and concentration, the Debye shielding distance (δ) is determined by the balance of the potential and kinetic energies of the total number of charged particles taking part in the shielding process. Within a volume with the linear size x, if $x \gg \delta$, then the concentration of the opposite charges in this volume are approximately equal and neutrality is maintained. However, if $x \ll \delta$, one finds that the separation of charges has no significance on the motion of particles and the neutrality is isolated. We must note that the electric field is screened off at a distance equal to Debye length (δ). Now we will estimate δ.

Suppose that the charges within the plasma are completely separated in a volume element with linear size δ. If T be the plasma temperature, then the potential energy of a charged particle in a given volume of plasma will be of the order of thermal motion of the particles, i.e. kT, where k is Boltzmann's constant. If we consider the electric field to be a continuous function of distance, then the electric field in a volume satisfies the Poisson's equation

$$\nabla \cdot E = \frac{\rho}{\varepsilon_0} \qquad (13.2)$$

where ρ is the charge density. We must remember that the condition of quasineutrality, and non-zero value of electric field ($\nabla \cdot E \neq 0$), are known as plasma approximations. If ϕ is the potential, then

$$E = \nabla \phi \qquad (13.3)$$

The charge particle velocity distribution function in a plasma obeys the Maxwell velocity distribution law. If n_i and n_e are the number of ions and electrons per unit volume respectively, then

$$\rho = (n_i e - n_e e) \qquad (13.4)$$

where e is electronic charge. In the light of Eqs. (13.3) and (13.4), Eq. (13.1) becomes

$$\nabla \cdot (-\nabla \phi) = \frac{n_i e - n_e e}{\varepsilon_0}$$

or

$$\nabla^2 \phi = \frac{-(n_i - n_e)e}{\varepsilon_0} \qquad (13.5)$$

If n_0 is the mean density of charged particles in the quasineutral plasma, then one obtains from Boltzmann distribution law,

$$n_i = n_0 e^{-e\phi/kT} \quad \text{and} \quad n_e = n_0 e^{e\phi/kT} \qquad (13.6)$$

where T is the absolute temperature of the plasma. Using Eq. (13.6), Eq. (13.5) takes the form

$$\nabla^2 \phi = \frac{n_0 e}{\varepsilon_0}\left[\exp\left(-\frac{e\phi}{kT}\right) - \exp\left(\frac{e\phi}{kT}\right)\right]$$

$$= \frac{2en_0}{\varepsilon_0} \sinh\left(\frac{e\phi}{kT}\right)$$

In the region $\dfrac{e\phi}{kT} \ll 1$, one can take $\sinh\left(\dfrac{e\phi}{kT}\right) \approx \dfrac{e\phi}{kT}$. Thus, we have

$$\nabla^2 \phi = \frac{2en_0}{\varepsilon_0}\frac{e\phi}{kT} = \frac{2n_0 e^2 \phi}{\varepsilon_0 kT} \qquad (13.7)$$

or

$$\nabla^2 \phi = \frac{\phi}{\delta^2} \qquad (13.8)$$

where

$$\delta = \left(\frac{\varepsilon_0 kT}{2n_0 e^2}\right)^2 \qquad (13.9)$$

If charge distribution is spherically symmetric, it is independent of angular coordinates, i.e. θ and ϕ (this ϕ is different from ϕ we have used earlier). Thus, one can express $\nabla^2\phi$ as

$$\nabla^2\phi = \frac{1}{r^2}\frac{d}{dr}\left(r^2\frac{d\phi}{dr}\right) \tag{13.10}$$

Using Eq. (13.10), Eq. (13.8) takes the form

$$\frac{1}{r^2}\frac{d}{dr}\left(r^2\frac{d\phi}{dr}\right) = \frac{2N_0e^2\phi}{\varepsilon_0 kT} = \frac{\phi}{\delta^2} \tag{13.11}$$

Differential equation (13.11) has finite number of solutions. We are interested in one solution which satisfies the boundary conditions,

$$\phi = \frac{1}{4\pi\varepsilon_0}\frac{Q}{r} \quad [= \phi_0 \text{ (say) at } r \to 0 \text{ and } = \phi \text{ at } r \to \infty]$$

One can express the solution as

$$\phi = \frac{1}{4\pi\varepsilon_0}\frac{Q}{r}\exp\left(-\frac{r}{\delta}\right) = \phi_0\exp\left(-\frac{r}{\delta}\right) \tag{13.12}$$

$\delta = \left(\frac{kT\varepsilon_0}{2n_0e^2}\right)^{1/2}$ is known as *Debye length*. This indicates the shortest distance at which electrons moving at random in the plasma screen, the Coulomb field of the probe particle. At distance δ (also called as *Debye screening radius* or *Debye shielding distance*), the Coulomb field of an arbitrary charge of the plasma is shielded because charge is predominantly surrounded by oppositely charged particles. The Debye shielding distance δ depends directly on electron temperature and inversely on the electron density. Equation (13.12) further reveals that the effect of the plasma is to reduce the potential due to a charge object exponentially. A useful relation often used is

$$\lambda_D\left(=\frac{\delta^2}{2}\right) = 6.9\times10^{-2}\left(\frac{T_e}{n_0}\right)^{1/2} m \tag{13.13}$$

where T_e is in Kelvin, $\lambda_D = 7.4\left(\frac{kT}{n_0}\right)^{1/2} m$ when kT_e is in eV.

Langmuir defined plasma as an ionized gas. For quasineutrality of plasma, $n_e = n_i$ to better than one part in 10^6; it is necessary that the dimension of plasma $L \gg \delta$, i.e. the Debye length is small compared to linear dimension occupied by the gas. The plasma under such condition is *quasineutral* but not exactly neutral so that all the interesting electromagnetic forces also vanish.

13.4 PLASMA PARAMETER

In a plasma of density n_0, the average distance between particles is $n_0^{-1/3}$. The average potential energy $|V|$ is

$$|V| \approx \frac{q^2}{r} \approx n_0^{1/3} e^2 \tag{13.14}$$

The kinetic energy of a particle is given by

$$\frac{1}{2} m_s \langle v \rangle^2 = \frac{3}{2} kT = \frac{3}{2} m_s v_s^2 \tag{13.15}$$

For a plasma to exist, we require that the kinetic energy be much greater than the potential energy, i.e.

$$n_0^{1/3} e^2 \ll kT \tag{13.16}$$

On rearranging terms, we get

$$n_0^{1/3} \frac{e^2}{kT} \ll 1 \tag{13.17}$$

or

$$\frac{kT}{n_0^{1/3} e^2} = \varepsilon_0 n_0^{2/3} \left(\frac{kT}{\varepsilon_0 n_0 e^2} \right) \gg 1 \tag{13.18}$$

We define

$$\wedge_s = 2\sqrt{2n_0} \left(\frac{\varepsilon_0 kT}{2n_0 e^2} \right) = 2\sqrt{2}\, n_0 \delta^3 \gg 1 \tag{13.19}$$

where $\wedge_s$ is the *plasma parameter*, as a necessary condition for a plasma to exist.

13.5 PLASMA OSCILLATIONS

Since a plasma has a tendency to be macroscopically neutral, it will tend to return to its neutral equilibrium state after a local perturbation is introduced in the form of an excess of positive or negative charge. If, for example, in Fig. 13.1 the striped region has an excess of electrons, this excess produces an electric field urging these electrons in the arrows. In this motion the electrons acquire a certain kinetic energy, and at the end of a given time they will pass through the equilibrium position. If too many electrons leave the striped region, there will be a deficiency of electrons here, and the electric field will tend to pull them back. One can thus picture electron oscillations in a plasma. We have assumed in these oscillations that the ions stay fixed.

The frequency of these oscillations can be calculated by means of a simple scheme of Fig. 13.2. In this figure, a plane slice is considered containing a block of immobile

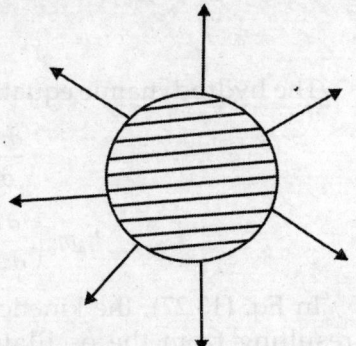

Fig. 13.1: Electrical field produced by a region having an excess of electrons

ions, and the electrons which were associated with them are assumed to move out in masses. An electric field is created, and given by

$$E = \frac{\sigma}{\varepsilon_0} \qquad (13.20)$$

where σ is the surface density in the thin sheath of positive charge on the left hand side of the figure or of negative charge on the right

$$\sigma = n_0 ex \qquad (13.21)$$

and

$$E = \frac{n_0 ex}{\varepsilon} \qquad (13.22)$$

Fig. 13.2: Scheme for calculating the frequency of oscillations

Each electron of the cloud is then subject to a restoring force

$$f = \frac{n_0 e^2 x}{\varepsilon_0} \qquad (13.23)$$

Oscillations will take place, the frequency being given by

$$\omega_p^2 = \frac{n_0 e^2}{m \varepsilon_0} \qquad (13.24)$$

i.e.

$$\omega_p = \left(\frac{n_0 e^2}{m \varepsilon_0} \right)^{1/2} \qquad (13.24a)$$

While obtaining Eq. (13.24) by solving general equations of electron space charge oscillation, we have made following approximations: (i) the ions are assumed fixed and their density uniform, (ii) thermal agitation is negligible, (iii) collisions are negligible, (iv) the oscillations are of small amplitude, and (v) there is no externally imposed mag-netic field.

Rewritting Eq. 13.24(a) in the following form, we have

$$v_p = \frac{\omega}{2\pi} = \frac{1}{2\pi} \left(\frac{n_0 e^2}{m \varepsilon_0} \right)^{1/2} \qquad (13.25)$$

The frequency v_p is called the *plasma frequency*. One obtains

$$v_p = 9 \times 10^3 n_0^{1/2} \text{ cycles/s} \qquad (13.26)$$

The hydrodynamic equations for the electrons can be written as

$$\frac{\partial n_e}{\partial t} + \nabla \cdot n_e v_e = 0 \qquad (13.27)$$

$$n_e m_e \left(\frac{\partial}{\partial t} + v_e \cdot \nabla \right) v_e = n_e q_e E \qquad (13.28)$$

In Eq. (13.27), the kinetic stress term, the collision term, and the magnetic field resulting from the oscillation have been neglected. Using Boltzmann formula, $n_e = n_0 \exp\left(\frac{e\phi}{kT} \right)$, Eqs. (13.26) and (13.27) can be simplified in the form

$$n_e(\boldsymbol{r}, t) = n_0 + a(\boldsymbol{r}, t) \qquad (13.29)$$

where $a(r, t)$ is a small perturbation. Equations (13.26) and (13.27) can therefore be linearized by neglecting the infinitesimals of higher order. If quantities vary at frequency ω, one can finally express Eqs. (13.18) and (13.19) as

$$i\omega a + n_0 \nabla \cdot v_c = 0 \tag{13.30}$$

$$n_e m_e i\omega v_e = n_e q_e E \tag{13.31}$$

But on the other hand, we have

$$\nabla \cdot E = \frac{(n_i q_i + n_e q_e)}{\varepsilon_0} = \frac{q_e a}{\varepsilon_0} \tag{13.32}$$

Eliminating v_e and a between these equations successively, one obtains

$$v_e = q_e (i\omega m_e) E \tag{13.33}$$

$$a = \left(\frac{n_0 q_e}{\omega^2 m_e} \right) \nabla \cdot E \tag{13.34}$$

From Eq. (13.30), one obtains

$$\nabla \cdot E \left(1 - \frac{\omega_p^2}{\omega^2} \right) = 0 \tag{13.35}$$

which requires that $\omega = \omega_p$.

The plasma oscillations represent stationary internal oscillations. If, as supposed, the electrons and ions do not have thermal motion, there is no wave propagation. The study of plasma oscillations in a situation involving thermal agitation has shown that they are able to propagate as a consequence of this agitation with a speed which is of the order of the thermal agitation speed of electrons and that, furthermore, their frequency is slightly modified and is expressed as

$$\omega^2 = \omega_p^2 + \left(\frac{3kT_e}{m} \right) k_1^2 \tag{13.36}$$

with

$$k_1 = \frac{2\pi}{\lambda} \tag{13.37}$$

We have considered electron oscillations in a plasma. Actually, ion oscillations in a plasma could be defined in a similar fashion, but their existence is subject to doubt. In reasoning about the electron oscillations in a plasma, it is validly assumed that the ions are fixed as a consequence of their great inertia. The corresponding assumption made in the case of ion oscillations about a cloud of electrons seems a little unrealistic.

A more general investigation shows that three types of waves can exist in an ionized gas: (i) *electromagnetic waves*, which in the simplest cases are only transverse. These waves are analogus to the usual electromagnetic waves propagating in a vacuum. Their speed is of the order of the speed of light in a vacuum, (ii) *the plasma waves* that are longitudinal and which only propagate due to thermal agitation with a low speed of the order of the speed of thermal agitation, (iii) *magnetohydrodynamic waves*, which are somewhat similar to acoustic waves. They cause motion of electrons and ions with electromagnetic coupling. Their propagation speed is of the order of speeds of thermal agitation.

13.6 OCCURRENCE OF PLASMA

The various systems and situations in which plasma occur fall into three broad categories: various *plasma devices* such as "neon" lights that are common in use, *laboratory plasma* that is used to investigate plasma properties and to develop devices such as thermonuclear fusion reactors, and *cosmic or space plasma* that occur throughout most of the extraterrestrial universe.

For a plasma to be in stationary state, processes are needed that replenish the stocks of ions diminishing as a result of recombination. In high-temperature plasma, this is achieved as a result of thermal ionization, and in gas-discharge plasma, as a result of collision ionization by electrons accelerated by an electric field. The ionosphere (one of the layers of the atmosphere) is a special variety of plasma. The high degree of ionization of molecules (~1%) is maintained in the ionosphere by photoionization due to the sun's short wave radiation.

13.7 PLASMA BEHAVIOUR IN ELECTRIC AND MAGNETIC FIELDS

We have read that charged particles are always present in plasma and the plasma is very frequently exposed to the effect magnetic and electric fields. Obviously, it would be worthwhile to study the behaviour of charge in these fields. When a charged particle is subjected to constant electric field, then it leads to the formation of a thin sheath of space charge, which shields the major part of the plasma. This is not of any particular interest. Now, we consider following cases:

i. *Charged particle in homogeneous magnetic field*

Let us suppose that a particle of mass m, charge q be shot into a magnetic field of induction B with velocity v in a direction perpendicular to the lines of force as shown in Fig. 13.3. The magnetic force F_m on charged particle, i.e. Lorentz force acting at right angles to v as well as to B is

$$F_m = q(v \times B) \tag{13.38}$$

The magnitude of this force is $qvB \sin \theta$, where θ is the angle between v and B and the direction of force is normal to the plane containing v and B. The Lorentz force causes the particle to trace a circular path (as $\theta = 90°$) in the plane of the paper (we

Fig. 13.3: Charged particle in a homogeneous magnetic field

must remember that when the particle enters parallel to the magnetic field, i.e. $\theta = 0$, then $F_m = qvB\sin\theta = 0$, i.e. the path of the particle remains undeflected in the magnetic field).

The magnetic force F_m must be at every instant in equilibrium with the necessary centripetal force for circular path, i.e.

$$qvB = \frac{mv^2}{R} \qquad (13.39)$$

where R is the radius of the orbit. We have

$$R = \frac{mv}{qB} \qquad (13.40)$$

This radius is known as the *Larmor radius* of the orbit. From Eq. (13.38), one obtains the angular frequency

$$\omega_c = \frac{v}{R} = \frac{qB}{m} = 2\pi v_c \qquad (13.41)$$

This is known as *cyclotron frequency*. We can see that ω_c is independent of the speed v of the particle. This provides the physical mechanism for the design of low energy cyclotron. $v_c = \dfrac{\omega_c}{2\pi}$ is known as *linear frequency*.

Let us consider the case when we shoot the charge particle into the magnetic f eld at an angle other than a right angle and $\theta \neq 0$, then one can resolve velocity v in two components, i.e. v_{11} parallel to the magnetic field B and $v_\perp$ in the plane perpendicular to magnetic field B. Since $v_{11} = v\cos\theta$ is unaffected by the field and path of the particle due to this is the straight line, one can write Eq. (13.38) as

$$qv_\perp B = \frac{mv_\perp^2}{R}$$

or

$$R = \frac{mv_\perp}{qB} \qquad (13.42)$$

The resulting path will be the superposition of the straight line and circular paths, i.e. spiral or helix as shown in Fig. 13.4. Obviously, the particle is gyrating about the line of magnetic force. In other words, the particle is "locked" to the field line.

ii. Charged particle in simultaneous electric and magnetic fields

When a charged particle of charge q and mass m enters simultaneously electric and magnetic fields E and B respectively, then the total force on the particle is

$$F = F_e + F_m = qE + q(v \times B) = q[E + v \times B] \qquad (13.43)$$

v being the velocity of the particle.

(a)

(b)

Fig. 13.4: Helix or spiral path of a charged particle in a magnetic field

Let electric and magnetic fields be mutually perpendicular and the velocity v of the particle be perpendicular to both E and B. The force F_e experienced by the particle due to the electric field is along E and that exerted by the magnetic field on the particle, i.e. F_m is perpendicular to B, i.e. parallel or antiparallel to E. If the magnitudes and directions of E and B fields, i.e. F_e and F_m are so adjusted that F_e and F_m are equal and opposite, the path of the charge particle will remain undeflected. Let us consider the case when E is along y-axis and B along z-axis with v along x-axis, then $F_e = qE = qE\hat{j}$ and $F_m = q(v \times B) = q(v\hat{i} + b\hat{k}) = -qvB\hat{j}$. Obviously, $F_m = F_e$. The net result is zero.

If the magnitudes of F_e and F_m are equal, then $qE = qvB$, i.e. $v = \dfrac{E}{B}$. The velocity of the particle perpendicular to both E and B is called *drift velocity* and denoted by

$$v_D = \frac{E}{B}.$$

When the initial velocity of the particle is not equal to v_D or is not perpendicular to B, one obtains more complicated results. However, one can describe all these cases by particle gyration about the direction of B at an angular frequency $\omega_c = \dfrac{qB}{m}$, superimposed upon a linear drift motion of velocity $v_0 = \dfrac{E}{B}$. As an example, let us consider that a positively charged particle of charge q moves with velocity $v = v_D + v'$, then the force

$$F = q(E + v \times B) = q(E + (v_D + v') \times B]$$
$$= q(E + v_D \times B + v' \times B] \tag{13.44}$$

If we choose

$$v_D = \frac{E \times B}{B^2}$$

then the first two terms on right hand side of Eq. (13.44) cancel each other and the force is given by

$$F = q(v' \times B)$$

which is same as studied earlier.

Figure 13.5 shows the trajectory of a charged particle. It is obvious from the figure that at point A, the magnetic and electric forces act in the same sense and so the radius of curvature is smaller; at the point P, the forces partly balance each other and therefore, the radius of curvature is larger.

Fig. 13.5: Trajectory of a charged particle

iii. *Charged particle in non-homogeneous magnetic field*

Let us consider that a charge particle of charge +q enter a non-homogeneous magnetic field expressed by the density of lines of force in Fig. 13.6. One can easily see from the figure that where the field is large, the lines of forces are more densely spaced and the curvature of the trajectory is more marked and *vice versa*. The result of this variation of radius of curvature during particle motion is also to induce a particle drift as can be seen in Fig. 13.6.

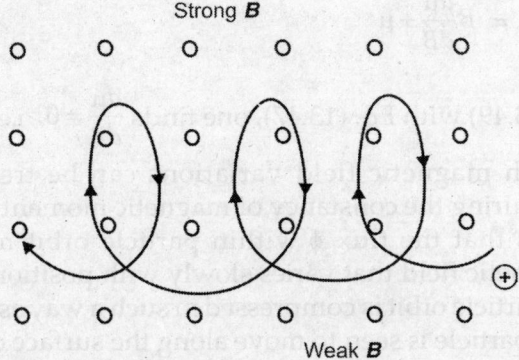

Fig. 13.6: Charged particle in a non-homogeneous magnetic field

A gyrating charged particle gives rise to a magnetic moment μ which is defined as

$$\mu = IA = \text{current due to particle gyration} \times \text{area enclosed by that current}$$

$$= \frac{qv_\perp}{2\pi R}\pi R^2 = \frac{qv_\perp R}{2} = \pi R^2 B\left(\frac{q^2}{2\pi m}\right) = \frac{q^2}{2\pi m}\phi$$

$$\because \qquad B = \frac{mv_\perp}{qR} \text{ and } \phi = \pi R^2 B \qquad\qquad (13.44a)$$

Obviously, the magnetic moment is proportional to the flux ϕ.

One can easily show that for slow variation of B either in space or time, μ is essentially constant.

The induced emf (ε) around a closed path in accordance with Faraday's law is given by

$$\varepsilon = -\frac{d\phi}{dt} = -\pi R^2\frac{dB}{dt} = \pi R^2\frac{dB}{dt} \text{ (numerically)} \qquad\qquad (13.45)$$

When the charge q in the field is small, the work done upon a particle in one orbit is

$$W = q\varepsilon \qquad\qquad (13.46)$$

But this work done must be equal to $\dfrac{dW}{dt}\times t$, where t is the time taken by the particle to go through one orbit, i.e. $t = \dfrac{2\pi R}{v}$. We have

$$q\varepsilon = \frac{2\pi R}{v}\frac{dW}{dt} = \frac{2\pi R}{v}\frac{dW}{dB}\frac{dB}{dt} = q\pi R^2\frac{dB}{dt} \qquad\qquad (13.47)$$

$$\therefore \qquad \frac{dW}{dB} = \frac{qvR}{2} = \mu B \qquad (13.48)$$

We also have

$$W = \frac{1}{2}mv^2 = \mu B \qquad (13.49)$$

$$\therefore \qquad \frac{dW}{dB} = B\frac{d\mu}{dB} + \mu \qquad (13.50)$$

Comparing Eq. (13.49) with Eq. (13.47), one finds $\frac{d\mu}{dB} = 0$, i.e. n is independent of B.

The effect of such magnetic field variations can be treated by knowing the consequences of requiring the constancy of magnetic moment μ. For μ to be constant, Eq. (13.44a) implies that the flux ϕ within particle orbit must remain constant. Obviously, in a magnetic field that varies slowly with position, as the magnetic field lines converge, the particle orbit is compressed in such a way as to maintain ϕ constant. This means that the particle is seen to move along the surface of a flux tube as shown in Fig. 13.7.

Fig. 13.7: Particle moving along the surface of a flux tube

13.8 MAGNETOHYDRODYNAMICS (MHD)

Now, we study the waves developing in a high-conductivity plasma which travels in a magnetic field. The magnetic lines of force in this case are "frozen" into the plasma so that if the plasma current is shifted somewhat so as to displace the magnetic lines of force, the magnetic field acts so as to shift the plasma current back. This gives rise to waves known as magnetohydrodynamic waves. In this section, we shall treat a conducting fluid in particular plasma subject simultaneously to the laws of electromagnetics and hydrodynamics. The subject is usually known as magnetohydrodynamics or simply hydromagnetics.

Let us consider an electrically neutral, conducting non-permeable fluid in electromagnetic fields. One can characterize it by a matter density $\rho(r, t)$, a velocity $v(r, t)$, a pressure $p(r, t)$ taken to be a scalar and real conductivity σ. One can express hydrodynamic equations as follows:

We have the continuity equation

$$\frac{d\rho}{dt} + \nabla \cdot (ev) = 0 \qquad (13.51)$$

The force equation is

$$\rho\frac{dv}{dt} = -\nabla p + (J \times B) + F_v + \rho g \qquad (13.52)$$

where $-\nabla p$ and $J \times B$ are the pressure and magnetic force terms respectively, F_v is the viscous force and ρg is the gravitational force. We have the viscous force for an incompressible fluid as

$$F_v = \eta\nabla^2 v \qquad (13.53)$$

where η is the coefficient of viscosity of the plasma. The time derivative of velocity is the *connective derivative* expressed as

$$\frac{d}{dt} = \frac{\partial}{\partial t} + v \cdot \nabla \qquad (13.54)$$

One obtains from Eq. (13.54) the total rate of change of a quantity moving instantaneously with velocity v.

One can write the electromagnetic fields in the fluid after neglecting the displacement current as

$$\nabla \times E = -\frac{\partial B}{\partial t} \qquad (13.55)$$

$$\nabla \times B = \mu_0 J \qquad (13.56)$$

The condition $\nabla \cdot J = 0$ implies the neglect of displacement current. From Eq. (13.55) and Faraday's law, we have $\frac{\partial}{\partial t}\nabla \cdot B = 0$ and the requirement $\nabla \cdot B = 0$ can be imposed as an initial condition. When displacement current is neglected, one may ignore Coulomb's law. The reason behind this is that the electric field is completely determined by curl equations and Ohm's law. If we retain displacement current in Ampere's law and $\nabla \cdot E = \frac{\rho}{\varepsilon_0}$, where ρ being electric charge density, is take into account, we find that the corrections only of the order of $\frac{v^2}{c^2}$ result. However, these are completely negligible for normal magneto-hydrodynamics problems.

To justify the specification of dynamical equations, it is essential to specify the relation between current density J and E and B. For a simple conducting medium having conductivity σ, Ohm's law is applicable and the current density is expressed as

$$J' = \sigma E' \qquad (13.57)$$

where J' and E' are measured in the rest frame of reference. For a medium moving with respect to laboratory frame, the non-relativistic transformation of field E and J are as follows:

$$E = E' + v \times B' \qquad (13.58)$$

and

$$J = J' + \rho v \qquad (13.59)$$

We have $\rho = 0$ for a one-component conducting fluid. Thus, Ohm's law can be expressed as

$$J = \sigma(E + v \times B) \qquad (13.60)$$

When the conductivity of the fluid is effectively infinite, then one finds that the condition is satisfied

$$E + (v \times B) = 0 \tag{13.61}$$

Equations (13.51), (13.52), (13.53), (13.56) and (13.60) supplemented by an equation of state for the fluid are called *equations of magnetohydrodynamics*.

Behaviour of Fluid in Electromagnetic Fields

The behaviour of a fluid in the electromagnetic fields is mainly governed by electrical conductivity of fluid. One finds that the effects are both electromagnetic and mechanical.

We have from Eq. (13.55), the time dependence of magnetic field as

$$\frac{\partial B}{\partial t} = -\nabla \times E \tag{13.62}$$

From Eq. (13.60), we have

$$E = \frac{J}{\sigma} - v \times B \tag{13.63}$$

Using Eq. (13.63), Eq. (13.62) becomes

$$\frac{\partial B}{\partial t} = -\nabla \times \left(\frac{J}{\sigma} - v \times B \right) = -\nabla \times (v \times B) - \nabla \times \left(\frac{J}{\sigma} \right)$$

Assuming σ to be constant in space and making use of Eq. (13.56), one obtains

$$\frac{\partial B}{\partial t} = \nabla \times (v \times B) - \frac{1}{\mu_0 \sigma} \nabla (\nabla \times B)$$

$$= \nabla \times (v \times B) - \frac{1}{\mu_0 \sigma} [\nabla (\nabla \cdot B) - \nabla^2 B]$$

Since $\nabla \cdot B = 0$, one obtains

$$\frac{\partial B}{\partial t} = \nabla \times (v \times B) + \frac{1}{\mu_0 \sigma} \nabla^2 B \tag{13.64}$$

If fluid is at rest, then $v = 0$. Equation (13.64) reduces to

$$\frac{\partial B}{\partial t} = \frac{1}{\mu_0 \sigma} \nabla^2 B \tag{13.65}$$

This is the *diffusion equation*.

When the conductivity is large enough, one can neglect second term on RHS in Eq. (13.64). In this case, one obtains the temporal behaviour of the magnetic field from

$$\frac{\partial B}{\partial t} = \nabla \times (v \times B) \tag{13.66}$$

Equation (13.66) indicates that magnetic lines of force also move with velocity v with the fluid. This means that the magnetic flux through any circuit moving with local fluid velocity is constant w.r.t time. We have assumed that conductivity is sufficiently large, and therefore Eq. (13.61) is applicable.

With the help of vector analysis, the component of v perpendicular to B and symbolized by w is obtained as

$$w = \frac{(B \times v) \times B}{B^2} \tag{13.67}$$

Using Eq. (13.61), Eq. (13.67) reduces to

$$w = \frac{(E \times B)}{B^2} \tag{13.68}$$

One can understand the so called "$E \times B$" drift of both fluid and lines of force in terms of individual particle orbits of electrons and ions in the crossed electric and magnetic fields.

One can study the mechanical behaviour of the system with the form of Eq. (13.52).

$$\rho \frac{dv}{dt} = F - (J \times B) \tag{13.69}$$

where $F = -\nabla p + F_v + \rho g$ = sum of all the non-electromagnetic forces. Using sum of all the non-electromagnetic forces and using Eq. (13.60), one obtains

$$\rho \frac{dv}{dt} = F - \sigma\{E + (v \times B)\} \times B$$

$$= F - \sigma B^2 (v_\perp - w) \tag{13.70}$$

where $\qquad w = \dfrac{(B \times v) \times B}{B^2}$ and $v_\perp = \dfrac{(B \times v)}{B^2}$

We have for fluids of infinite conductivity,

$$v_\perp = w, \quad \text{i.e.} \quad \rho \frac{dv}{dt} = F \tag{13.71}$$

This shows that flow parallel to B is governed by non-electromagnetic forces only.

In the limit of large conductivity, one can conveniently relate J in force equation to B via Ampere's law and use infinite conductivity condition [Eq. (13.61)] to eliminate E from Faraday's law to yield Eq. (13.66).

Using $\nabla \times B = \mu_0 J$, one can write the magnetic force term in Eq. (13.52) as

$$J \times B = \frac{1}{\mu_0}(\nabla \times B) \times B = -\frac{1}{\mu_0} B \times (\nabla \times B) \tag{13.72}$$

Making use of vector identity,

$$\nabla (B \cdot B) = 2(B \cdot \nabla)B + 2B \times (\nabla \times B)$$

i.e. $\qquad B \times (\nabla \times B) = \dfrac{1}{2}\nabla(B \cdot B) - (B \cdot \nabla)B$

One obtains Eq. (13.72) as

$$J \times B = -\frac{1}{\mu_0}\left[\frac{1}{2}\nabla(B \cdot B) - (B \cdot \nabla)B\right]$$

$$= \left[-\nabla\left(\frac{B^2}{2\mu_0}\right) + \frac{1}{\mu_0}(B \cdot \nabla)B\right] \tag{13.73}$$

From Eq. (13.73), one finds that the *magnetic force* is equivalent to a *magnetic hydrostatic pressure, $p_m = \dfrac{B^2}{2\mu_0}$* plus a term which may be taken as an additional tension along magnetic field direction.

If one neglects the viscous effects and assume that the gravitational force is derivable from a potential, say ψ such that $g = -\nabla\psi$, then one finds that Eq. (13.52) takes the form

$$\rho\frac{dv}{dt} = -\nabla(p + p_m + \rho\psi) + (B \cdot \nabla)\frac{B}{\mu_0} \tag{13.74}$$

When one is dealing with simple geometrical situations, e.g. B having only one component, the last term in Eq. (12.74) representing additional tension, vanishes. One obtains

$$\rho\frac{dv}{dt} = -\nabla(p + p_m + \rho\psi) \tag{13.75}$$

One finds that the static properties of the fluid can be described by

$$p + p_m + \rho\psi = \text{constant} \tag{13.76}$$

This obviously reveals that *apart from gravitational effects, any change in mechanical pressure is balanced by an opposite change in magnetic pressure.* If one wants that the fluid is to be confined within a certain region so that the mechanical pressure p falls rapidly to zero outside the region, then the magnetic pressure will have to rise equally rapidly.

13.9 PLASMA CONFINEMENT

In a high temperature plasma, the nuclei and electrons move in all directions at average speeds of several kilometers per second. Hence, unless confined in some way, all the particles would soon strike the walls of a containing vessel, thereby imparting some of their energy to the walls. The particles would then return to the plasma with less energy, that is, at a lower temperature. Thus, the plasma would be cooled, and the fusion temperature could not be realized or maintained. Two quite different methods are currently being investigated: (i) magnetic confinement, and (ii) inertial confinement. We shall restrict ourself to the study of confinement by self magnetic fields plasma.

Magnetic Confinement: Pinch Effect

The use of various magnetic field arrangements for confining high tem-perature plasma to achieve controlled thermonuclear fusion have been made.

If one neglects the viscous and gravitational effects, the condition for equilibrium is $F = 0$, i.e.

$$\nabla p - J \times B = 0 \quad \text{or} \quad \nabla p = J \times B \tag{13.77}$$

The scalar product with B and J yields

$$B \cdot \nabla p = B \cdot (J \times B) = 0 \tag{13.78}$$

and

$$J \cdot \nabla p = J \cdot (J \times B) = 0 \tag{13.79}$$

Equations (13.78) and (13.79) reveal that both B and J lie on the surfaces of constant pressure. If these surfaces are closed (say), then in accordance with Eqs. (13.78) and (13.79), no B-lines or J-lines can cross them. Obviously, one can view them as made up from a winding of B-lines and of J-lines (Fig. 13.8).

One can see that the pressure increases from outside towards the axis and the force $J \times B$ also points towards the axis in these isobaric surfaces. This shows that the plasma is contained by the force $J \times B$ which is electromagnetic in nature. This is the basic principle of magnetic confinement of plasma.

Fig. 13.8: Pinch effect

The phenomena of confinement of a plasma by self magnetic fields is termed *pinch effect*. To understand the pinch effect, let us consider an infinite cylindrical plasma or conducting fluid with an axial current density $J_z = J(r)$ and a resulting azimuthal magnetic induction $B_\phi = B(r)$. For simplicity, we have assumed the dependence of magnetic field, the current density, pressure, etc. on distance r from the cylindrical axis. Moreover, we will neglect the effects of viscous and gravitation force. Let us first consider whether a steady state condition can exist in which the plasma is mainly confined within a certain radius $r = R$ by the action of self magnetic field. We have $v = 0$ for a steady state and hence one can write the equation of motion for the fluid in the following form:

$$\rho \frac{dv}{ds} = -\nabla(p + p_m + \rho\psi) + \frac{1}{\mu_0}(B \times \nabla)B$$

$$0 = -\frac{dp}{dr} - \frac{d}{dr}\left(\frac{B^2}{2\mu_0}\right) + \frac{B^2}{\mu_0 r} \tag{13.80}$$

We have Ampere's law in integral form relating $B(r)$ to the current enclosed as

$$B(r) = \frac{\mu_0}{r} \int_0^r rJ(r)\,dr \tag{13.81}$$

When the fluid lies almost entirely inside $r = R$, then Eq. (13.81) for the magnetic induction inside the fluid can be expressed as

$$B(r) = \frac{\mu_0 I}{2\pi r} \tag{13.82}$$

where I is the total current flowing in the cylinder and given by

$$I = \int_0^R 2\pi r\, J(r) \tag{13.83}$$

One can write Eq. (13.80) as

$$\frac{dp}{dr} = -\frac{1}{2\pi_0 r^2} \frac{d}{dr}(r^2 B^2) \tag{13.84}$$

Integration of Eq. (13.84) yields

$$p(r) = p_0 - \frac{1}{2\mu_0} \int_0^r \frac{1}{r^2} \frac{d}{dr}(r^2 B^2)\,dr \tag{13.85}$$

where p_0 is the pressure of fluid at $r = 0$. If we take that the matter is confined to $r \leq R$, the pressure drops to zero at $r = R$. One obtains axial pressure as

$$p_0 = \frac{1}{2\mu_0} \int_0^R \frac{1}{r^2} \frac{d}{dr}(r^2 B^2) dr \qquad (13.86)$$

We can see from Eq. (13.82) that Eq. (13.86) vanishes for $r \geq R$. This means that one can replace the upper limit of the integration by infinity. Using Eq. (13.86), Eq. (13.85) reduces to

$$p(r) = \frac{1}{2\mu_0} \int_0^R \frac{1}{r^2} \frac{d}{dr}(r^2 R^2) dr \qquad (13.87)$$

One can express the average pressure inside the cylinder as

$$\langle p \rangle = \frac{2\pi \int_0^R r p(r) dr}{\pi R^2} \qquad (13.88)$$

Integrating and making use of Eqs. (13.82) and (13.87), one obtains

$$\langle p \rangle = \frac{2I^2}{\mu_0 R^2} \qquad (13.89)$$

Obviously, we have obtained a relation between average pressure, total current I and the radius of cylindrical plasma confined by its self magnetic field. We must note that the magnetic pressure, $\dfrac{B^2}{2\mu_0}$, is equal

to the average pressure of matter at the surface of the cylinder.

We must note that the temperature of hot plasma used for the production of controlled thermonuclear fusion is of the order of $10^8 K (\approx 80 \text{keV})$ and density $\approx 10^{21}$ particles/m^3. These parameters correspond to pressure $\approx 1.4 \times 10^6$ N/m^2 (= 14 atmospheres). Calculations reveal that a magnetic induction $B = 1.9$ T at the surface, corresponding to a current $\approx 9 \times 10^2 R$ amperes is required for the confinement of this high temperature plasma. Obviously, very high currents are required to confine plasma at about 10^8 K temperature.

13.10 INSTABILITIES

Studies reveal that there is no stable equilibrium pinch, i.e. there is a slight deviation from the equilibrium state which tends to increase and finally results into the disintegration of the form of plasma. There is a large class of instability phenomena that is important to the dynamics of plasma. Most instabilities involve the collective motions of the plasma particles. Instabilities are important in a wide range of physical situa-tions in which plasma plays a part.

The two important instabilities are: (i) magnetohydrodynamic insta-bilities, which can be explained in terms of magnetic pressure and stresses, and (ii) instabilities associated with particle motions.

Let us consider a *kink instability* which develops when a linear pinch is bent due to some perturbation (Fig. 13.9). We can see that the magnetic lines of force attenuate, i.e. the field is weakened outside of the curve, and they grow denser, i.e. the field is intensified inside.

One finds that the stronger field causes stronger magnetic pressure in the bend which is conductive to further bending. This results ultimately into the breaking up of the linear pinch.

The second type of distortion observerd in plasma is *sausage instability* which occurs if the plasma column is pinched by chance (Fig. 13.10). In the neighbourhood of constriction, the azimuthal induction increases $\left(\text{as } B \propto \dfrac{1}{r}\right)$ causing a greater inward pressure at the neck, i.e. squeezed region than elsewhere. This causes further pinching, ultimately disrupting the column.

Fig. 13.9: Kink instability

In addition to the above mentioned instabilities, there are several other kinetic instabilities which are caused either by chance or artificial deviation of the distribution of particles from Maxwellian distribution.

Attention has been given to inhibiting the sausage and kink instabilities by providing a *frozen in* axial magnetic field along the pinch. This axial field tends to slow the growth of kink instabilities by requiring that they expend energy in stretching this field. Sausage instabilities also tend to be inhibited because of the compression of the axial *frozen in* field required by their development. Nevertheless, it has not been possible in practice to prevent either form of pinch instability by use of such fields.

Various schemes for containing plamsa and overcoming the difficulty of instability are proposed. One such most promising system is 'tokamak' an axially symmetric toroidal system. In this system the plasma is contained by the magnetic field of the current which flows along the

Fig. 13.10: Sausage instability

axis. This makes it possible to achieve values of $N\tau$ less than $10^{19}\,\text{s/m}^3$ and ion temperature less than $10^7\,\text{K}$. A very powerful magnetic field is applied parallel to the current which suppresses the plasma instability.

13.11 PLASMA WAVES

Although plasma, as a consequence of the interaction between particles, is a gas, oscillations and noises play a much greater role than in ordinary gas. First, in a plasma that is located in external fields and is not homogeneous, a wide variety of oscillations can occur because of the long-range interaction between particles. Secondly, these oscillations vary frequently and become amplified to a relatively high energy. In this case, the plasma oscillations determine its parameters and development. We shall analyze briefly the two cases.

i. Plasma-electron oscillations

When electrons are displaced with respect to ions in a system, high energy electron oscillations result. A plasma in which electrons and ions have been slightly but collectively displaced relative to each other is shown in Fig. 13.11. The displacement

Fig. 13.11: Plasma-electron oscillations

distance δ is small compared to plasma thickness L. The displacement of electrons in a neutral plasma disturbs it. The displaced charges produce electric field in the interior of the plasma. This field is given by

$$E = \frac{Ne\delta}{\varepsilon_0} \tag{13.90}$$

where N represents the density of charges in the plasma. In the interior of the plasma, the force per unit area acting on the electrons is given by

$$NeEL = \frac{N^2 e^2 \delta L}{\varepsilon_0} \tag{13.91}$$

If m is the mass of the electron, then the mass per unit area upon which this force acts is equal to LmN. One can write the equation of motion for each electron in the plasma as

$$LmN \frac{d^2\delta}{dt^2} + \frac{N^2 e^2 L \delta}{\varepsilon_0} = 0 \tag{13.92}$$

or

$$\frac{d^2\delta}{dt^2} + \frac{n^2 e^2}{m\varepsilon_0} \delta = 0 \tag{13.93}$$

The solution of Eq. (13.93) is

$$\delta = F \exp[i\omega_p t] \tag{13.94}$$

where F is constant and

$$\omega_p = \left(\frac{Ne^2}{\pi\varepsilon_0}\right)^{1/2} = 2\pi v_p \tag{13.95}$$

$$v_p = \frac{1}{2\pi}\left(\frac{Ne^2}{m\varepsilon_0}\right)^{1/2} \tag{13.96}$$

is the oscillation frequency. v_p is generally known as *plasma frequency*.

We must note that plasma oscillations are longitudinal in contrast to electromagnetic oscillations. Hence, the electric field due to plasma waves is directed along the wave vector.

One can also consider the oscillations that are due to the motion of the ions in a uniform isotropic plasma. The peculiarities of these oscillations are determined by the large mass of the ions. Owing to its small mass, an electron can follow the motion of an ion so that the plasma remains quasineutral in any large volume containing a large number of charged particles, $n_i = n_e$. One obtains the following dispersion relation between the frequency (ω) and the wave number (k)

$$\omega = \left(\frac{T_e}{M}\right)^{1/2} k \tag{13.97}$$

where M is the ion mass.

The oscillations caused by the ion motion are known as *ion sound*. Similar to the plasma oscillations, the ion sound is a longitudinal wave, i.e. the wave vector k is parallel to the oscillation vector of the electric field E. The dispersion relation for the ion sound is similar to that for ordinary sound. This is due to the fact that both types of oscillations are characterized by a short-range interactions. In case of the ion sound, the interaction is short-ranged because the electric field of propagating wave is screened by the plasma. This screening is effective if the wavelength of the ion sound is considerably larger than the *Debye–Huckel radius* (r_D) for the plasma where the sound propagation occurs,

$$kr_D \ll 1 \tag{13.98}$$

The dispersion relation [Eq. (13.97)] is valid for the ion sound if Eq. (13.98) is satisfied.

ii. Alfven waves or hydromagnetic waves

When ions are considered to be massive as compared to electrons, then it is justified to ignore their effect on the high frequency wave propagation. However, for low frequency electromagnetic wave propagation, the gyrofrequency of ions becomes comparable to the propagating wave frequency and the effect of ions play an important role.

Let us study the waves developing in a high conductivity plasma which travels in a magnetic field. The magnetic lines of force in this case are *frozen* into the plasma so that if the plasma current is *shifted* somewhat so as to displace the magnetic lines of force, the magnetic field acts so as to shift the plasma current back. This gives rise to waves known as magnetohydrodynamic waves.

For wave periods that are short compared to the diffusion times, the magnetic fields of the wave are frozen into the fluid. They thereby force the fluid to move along the flux lines. Since the fluid is thus tied to the wave fields, the wave velocity is slowed by the inertial effects of the fluid. This situation is analogous to the propagation of waves along a string where the wave velocity is given in terms of tension T and the linear mass density ρ of the string as

$$V = \left(\frac{T}{\rho}\right)^{1/2} \tag{13.99}$$

Making the analogy to hydrodynamic plasma waves, we replace the string tension T by the magnetic tension along the field lines, $\dfrac{B^2}{\mu_0}$, and ρ becomes the plasma mass density. Thus, for these hydrodynamic waves we might expect the velocity to be

$$v_A = \frac{B}{(\mu_0 \rho)^{1/2}} \tag{13.100}$$

The velocity v_A is known as *the Alfven velocity*, and the hydrodynamic waves are known as the *Alfven waves*, after Hannes Alfven, who first recognized that such hydrodynamic waves are possible.

13.12 IONOSPHERE (REFLECTION FROM A PLASMA)

Aside from lighting and the aurora, perhaps the longest-studied aspect of the plasma environment of the earth is the ionosphere. As early as 1902 the presence of an ionized layer in the upper atmosphere was postulated by Heaviside and Kennelly to account for the anomalously long distances radio waves could be propagated.

Present day techniques for the study of the ionosphere include the reflection and transmission of radio signals by the ionized material. The reflection technique, which has been used since 1925, involves the use of ground-based radio transmitters and receivers that measure the transit time required for signals transmitted from the earth to be reflected from various levels of ionosphere. The reflection occurs from that region at which the local plasma frequency equals that of the electromagnetic signal. Now, we shall obtain the conditions for reflection from an ionized layer.

We have the following relation in the case of gases

$$\chi = \alpha N \tag{13.101}$$

where α is the polarizability of the atom and χ is the electrical susceptibility. Now,

$$\varepsilon_{r-1} = \alpha N \quad \text{or} \quad n^2 = 1 + \alpha N \tag{13.102}$$

The equation of a charged particle having charge q, in the presence of an electric field E is

$$qE = m(\ddot{x} + l\dot{x} + \omega_0^2 x) \tag{13.103}$$

where $m\omega_0^2 x$ is the *restoring force* and $ml\dot{x}$ is the *damping force*. The solution of Eq. (13.103) is

$$x = \frac{qE/m}{-\omega^2 + il\omega + \omega_0^2} \tag{13.104}$$

We obtain the induced dipole moment as

$$p = qx = \frac{q^2 E/m}{-\omega^2 + il\omega + \omega_0^2} \tag{13.105}$$

But

$$p = \varepsilon_0 qE \tag{13.106}$$

$$\therefore \qquad \alpha = \frac{q^2/m\varepsilon_0}{-\omega^2 + il\omega + \omega_0^2} \tag{13.107}$$

We obtain

$$n^2 = 1 + \frac{Nq^2/m\varepsilon_0}{-\omega^2 + il\omega + \omega_0^2} \tag{13.108}$$

One can take $\omega_0 = 0$, as the free charges in the plasma are not tightly bound to ions. Obviously, the restoring force is almost negligible. Equation (13.108) reduces to

$$n^2 = 1 + \frac{Nq^2/m\varepsilon_0}{-\omega^2 - il\omega} = 1 + \frac{(Nq^2/m\varepsilon_0)}{\omega^4 + l^2\omega^2}(-\omega^2 - il\omega)$$

$$= 1 - \frac{Nq^2}{m\varepsilon_0}\left(\frac{\omega^2}{\omega^4 + l^2\omega^2}\right) - i\frac{Nq^2(l\omega)}{m\varepsilon_0(\omega^4 + l^2\omega^2)} \qquad (13.109)$$

One can express the complex refractive index in terms of its real and imaginary parts, i.e.

$$n = n_r - in_i \qquad (13.110)$$

Squaring Eq. (13.110) and comparing with Eq. (13.109), one obtains

$$n_r^2 - n_i^2 = 1 - \frac{Nq^2}{m\varepsilon_0}\left(\frac{1}{\omega^2 + l^2}\right) \qquad (13.111)$$

and

$$2n_r n_i = \frac{Nq^2}{n\varepsilon_0}\left(\frac{1}{\omega^3 + l\omega}\right)$$

or

$$n_i = \frac{Nq^2}{2m\varepsilon_0 n_r}\left(\frac{1}{\omega^2 + l^2\omega}\right) \qquad (13.112)$$

Using Eq. (13.112), Eq. (13.111) becomes

$$n_r^4 - \left(1 - \frac{Nq^2}{m\varepsilon_0(\omega^2 + l^2)}\right)n_r^2 - \left(\frac{Nq^2 l}{2m\varepsilon_0(\omega^3 + l^2\omega)}\right)^2 = 0 \qquad (13.113)$$

Let

$$a_1 = 1 - \frac{Nq^2}{m\varepsilon_0(\omega^2 + l^2)}$$

and

$$a_2 = \frac{Nq^2 l}{2m\varepsilon_0(\omega^3 + l^2\omega)}$$

One obtains

$$n_r^4 - a_1 n_r^4 - a_2^2 = 0$$

$$\therefore \qquad n_r^2 = \frac{a_1 \pm \sqrt{a_1^2 + 4a_2^2}}{2}$$

We can see that for $\omega \ll l$, the term a_2 dominates the calculation of n_r. Obviously, we can take

$$n_r \approx \sqrt{a_2}\left[\frac{Nq^2 l}{2m\omega\varepsilon_0(\omega^2 + l^2)}\right]^{1/2} \approx \left(\frac{Nq^2}{2m\varepsilon_0\omega l}\right)^{1/2} \qquad (13.114)$$

Equation (13.114) is the expression for real part of the refractive index (n_r) at low frequencies. When $\omega \approx l$, i.e. at high frequencies, we can see that the term a_1 dominates the calculation of n_r, and we have

$$n_r \approx \sqrt{a_1} = \left(1 - \frac{Nq^2}{m\varepsilon_0(\omega^2 + l^2)}\right)^{1/2}$$

$$\approx \left(1 - \frac{Nq^2}{m\varepsilon_0\omega^2}\right)^{1/2} \qquad (13.115)$$

This means that at high frequencies the real part of the refractive index will be less than 1 .

We must note that there is no sharp boundary to the ionosphere. Experiments have shown that the degree of ionization and, hence N varies in the ionospheric layer. Obviously, the value of n_r changes with the position. It is found that at each point the local value of n_r is related to the angle of incidence (θ_i) and angle of refraction (θ_T) through the following relation:

$$n_r = \frac{\sin\theta_i}{\sin\theta_T} \qquad (13.116)$$

We can see that as the wave enters the ionospheric layer, n_r decreases because N increases, i.e. θ_T increases. We can further see that for $\theta_T = 90°$, the wave is just reflected. It is interesting to note that the reflection in the ionosphere corresponds to a gradual turning of the incident wave until it is out of the ionospheric layer as shown in Fig. 13.12.

Fig. 13.12: Reflection in ionosphere

At the point of reflection of wave from the layer, we have $\theta_T = 90°$. Therefore, we have

$$n_r = \left(1 - \frac{Nq^2}{m\varepsilon_0\omega^2}\right)^{1/2} = \sin\theta_i$$

$\therefore$

$$\frac{Nq^2}{m\varepsilon_0\omega^2} = \cos^2\theta_i$$

or

$$\omega = \frac{1}{\cos\theta_i}\left(\frac{Nq^2}{m\varepsilon_0}\right)^{1/2} \qquad (13.117)$$

When the wave is making normal incidence, we have

$$\omega = \left(\frac{Nq^2}{m\varepsilon_0}\right)^{1/2} \qquad (13.118)$$

or

$$2\pi v_c = \left(\frac{Nq^2}{m\varepsilon_0}\right)^{1/2}$$

or

$$v_c = \frac{1}{2\pi}\left(\frac{Nq^2}{m\varepsilon_0}\right)^{1/2} \qquad (13.119)$$

where v_c is the *critical or maximum* frequency for which reflection will occur at normal incidence. From Eq. (13.119) it is evident that $v_c \propto \sqrt{N}$, i.e. v_c depends on the degree of ionization. We must note that ionization mainly depends on the radiation from the sun. This means ionization varies from place to place around the earth and from time to time. The time behaviour of the ionization regions is interesting and somewhat complex. Associates with solar flares are sudden ionospheric disturbances. During these disturbances, the D-region ionization is markedly increased. In the polar region the increased ionization may be so extensive as to prevent almost all radio noise from sources beyond the solar system from penetrating the ionosphere.

The day-to-night variation in the E-region ionization is very great, amounting to something like two orders of magnitude in electron density. The F-region is divided in daytime, but the division disappears at night. The variability of this region is much greater than the others.

ILLUSTRATIVE EXAMPLES

Example 13.1: Consider a plasma consisting of two homogeneous streams of electrons with equal densities and equal but opposite velocities of magnitude v'. Show that waves with wave number $k < k_c = \sqrt{2}\dfrac{\omega_p}{v'}$, where ω_p is the plasma frequency, are unstable.

Solution: One can solve this problem easily in a frame of reference where one stream of electrons has velocity zero and the other has veocity $v_0 = 2v'$. We may then linearize about equilibrium with the assumptions

$$n = n_0 + n_1 \tag{1}$$

and
$$v = v_0 + v_1 \tag{2}$$

where n_0 and v_0 do not vary in time or space. The equation of continuity for the stream with velocity $2v'$ is

$$\frac{\partial}{\partial t}(n_{01} + n_{11}) + \nabla \cdot [(n_{01} + n_{11})(v_{01} + v_{11})] = 0 \tag{3}$$

which reduces to

$$\frac{\partial}{\partial t}n_{11} + v_{01}\nabla n_{11} + n_{01}\nabla \cdot v_{11} = 0 \tag{4}$$

Similarly, the equation of motion is

$$m_e(n_{01} + n_{11})\left[\frac{\partial}{\partial t}(v_{01} + v_{11}) + \{(v_{01} + v_{11})\cdot\nabla\}(v_{01} + v_{11})\right]$$
$$= (n_{01} + n_{11})\, qE_1 \tag{5}$$

or
$$m_e n_{01}\frac{\partial}{\partial t}v_{11} + m_1 n_{01}v_{01}\nabla \cdot v_{11} = n_{01}qE_1 \tag{6}$$

where we have assumed $E_0 = 0$. By comparison, the equation of continuity for the stream of electrons at rest is

$$\frac{\partial}{\partial t}n_{12} + n_{02}\nabla \cdot v_{12} = 0 \tag{7}$$

and the equation of motion is

$$m_e n_{02}\frac{\partial}{kdt}v_{12} = n_{02}qE_1 \tag{8}$$

Since we are looking for plane wave solutions, let $\nabla \to ik$, and $\dfrac{\partial}{\partial t} \to i\omega$. Equations (7) and (8) reduces to

$$-i\omega m_e n_{01}v_{11} + ikm_e n_{01}v_{01}v_{11} = n_{01}qE_1 \tag{9}$$

and
$$-i\omega m_e n_{02} v_{12} = n_{02} q E_1 \qquad (10)$$

Using the fact that $v_{01} = 2v'$ and $q = -e$, one obtains

$$v_{12} = \frac{eE_1}{i\omega m_e} \qquad (11)$$

and
$$v_{22} = \frac{eE_1}{-i\omega m_e + ik2m_e v'} \qquad (12)$$

Equations (9) and (10) reduces to

$$-iwn_{11} + ikv_{01}n_{11} + ikn_{01}v_{11} = 0 \qquad (13)$$

and
$$-i\omega n_{12} + ikn_{02}v_{12} = 0 \qquad (14)$$

Using the values for v_{11} and v_{12} given in Eqs. (11) and (12), one finds

$$n_{11} = n_{01}\left(\frac{-ikeE_1}{m_e(\omega - 2v'k)^2}\right) \qquad (15)$$

and
$$n_{12} = n_{02}\left(\frac{-ikeE_1}{m_e\omega^2}\right) \qquad (16)$$

Substituting these expressions in Poisson's equation,

$$ikE_1 = \nabla \cdot E = \frac{\rho}{\varepsilon_0} = -\frac{e}{\varepsilon_0}(n_{11} + n_{12}) \qquad (17)$$

This yields

$$1 = \frac{e^2 n_0}{m_e \varepsilon_0}\left[\frac{1}{\omega^2} + \frac{1}{(\omega - 2v'k)^2}\right]$$

$$= \omega_{p,e}^2\left[\frac{1}{(\omega + v'k)^2} + \frac{1}{(\omega - v'k)^2}\right] \qquad (18)$$

This in turn reduces to

$$(\omega + v'k)^2(\omega - v'k)^2 = \omega_{p,e}^2[(\omega + v'k)^2 + (\omega - v'k)^2] \qquad (19)$$

or
$$\omega^4 - \omega^2(2\omega_{p,e}^2) + (v'^4 k^4 - 2v'^2 k^2 \omega_{p,e}^2) = 0 \qquad (20)$$

Solving for ω, one obtains

$$\omega^2 = \omega_{p,e}^2 \pm \sqrt{\omega_{p,e}^4 + 2\omega_{p,e}^2 v'^2 k^2 - v'^4 k^4} \qquad (21)$$

It is seen that if $k < \sqrt{2}\,\dfrac{\omega_{p,e}}{v'}$, then $\omega^2 < 0$, ω is imaginary, and the wave is unstable.

Example 13.2: Consider a plasma in a uniform static magnetic field B_0. The plasma consists of singly charged ions of mass m_i and electrons of mass m_e with equilibrium plasma density n_0. Assuming that the ions are very cold in comparison to the electrons, $T_i \sim 0$, and that the electrons have a finite temperature T_e, derive the dispersion relation for small amplitude electrostatic ion–cyclotron waves propagating nearly perpendicular to B_0. Assume that the electrons move along the magnetic field and are in Boltzmann equilibrium.

Solution: When we linearize about equilibrium, we have

$$n_e \rightarrow n_{0e} + n_{1e} \tag{1}$$

$$n_i \rightarrow n_{0i} + n_{1i} \tag{2}$$

$$v_e \rightarrow v_{0e} + v_{1e} \tag{3}$$

$$v_i \rightarrow v_{1i} \tag{4}$$

The equations of continuity for the electrons and ions are

$$\frac{\partial}{\partial t} n_{e1} + n_{e0} \nabla \cdot v_{el} = 0 \tag{5}$$

and

$$\frac{\partial}{\partial t} n_{i1} + n_{i0} \nabla \cdot v_{i1} = 0 \tag{6}$$

respectively. The fact that the electrons are in Boltzmann equilibrium implies that

$$n_e = n_{e0} \exp\left(\frac{-eV}{kT_e}\right) \quad \text{or} \quad n_{e1} = n_{e0}\left(\frac{e\phi_1}{kT_e}\right) \tag{7}$$

where $E_1 = -\nabla\phi_1$. Similarly, because we are looking for plane wave solutions, Eq. (6) implies that

$$n_{i1} = n_{i0} \frac{k}{\omega} v_{i1} \tag{8}$$

From the problem definition, we may assume that both n_0 and B_0 are constant and uniform, and that $v_0 = E_0 = 0$. The equations of motion for electrons and ions are

$$m_e n_{0e} \frac{\partial}{\partial t} v_{el} = -e n_{e0}[E_1 + v_{e1} \times B_0] + v \times P_e \tag{9}$$

$$m_i n_{0i} \frac{\partial}{\partial t} v_{i1} = e n_{i0}[E_1 + v_{i1} \times B_0] \tag{10}$$

respectively. Because $m_e \ll m_i$, we assume $m_i = 0$. With the proper choice of geometry, B_0 in the z-direction with k, E in the xy-plane, the vector equation of motion for the electrons reduces to

$$e n_{e0}[E_1 + v_{e1,y} B_0] - ik(k_B)T_e n_{e1} = 0 \tag{11}$$

The vector equation of motion for the ions gives

$$-i\omega m_i n_{0i} v_{i1,x} = -e n_{i0}[-ik\phi_1 + v_{i1,y} B_0] \tag{12}$$

and

$$-i\omega m_i n_{0i} v_{i1,y} = -e n_{i0}[v_{i1,x} B_0] \tag{13}$$

Solving last two equations, i.e. Eqs. (12) and (13) for $v_{i1,x}$ we obtain

$$v_{i1,x} = \frac{ek\phi_1}{\omega m_i}\left[1 - \frac{e^2 B_0^2}{\omega^2 m_i^2}\right]^{-1} \tag{14}$$

By definition, $\Omega_i = \dfrac{eB_0}{m_i}$ is the ion–cyclotron frequency, so

$$v_{il,x} = \frac{ek\phi_1}{\omega m_i}\left(1 - \frac{\Omega_i^2}{\omega^2}\right)^{-1} \tag{15}$$

From Eq. (6),

$$\phi_1 = \frac{n_{e1}}{n_{e0}}\left(\frac{kT_e}{e}\right) \tag{16}$$

$$v_{i1,x} = \left(1 - \frac{\Omega_i^2}{\omega^2}\right) = \frac{ek}{\omega m_i}\frac{n_{e1}}{\omega m_i}\left(\frac{k_B T_e}{e}\right) \tag{17}$$

In the plasma approximation, $n_e \sim n_1$, so

$$\frac{n_{e1}}{n_{e0}} \approx \frac{n_{i1}}{n_{01}} = \frac{k}{\omega}v_{i1,x} \tag{18}$$

Using this in the preceding equation gives

$$v_{i1,x}\left[1 - \frac{\Omega_i^2}{\omega^2}\right] = \frac{ek}{\omega m_i}\frac{k_B T_e}{e}\frac{k}{\omega}v_{i1,x} \tag{19}$$

or

$$\left(1 - \frac{\Omega_i^2}{\omega^2}\right) = \frac{k^2}{\omega^2}\frac{k_B T_e}{m_i} \tag{20}$$

Rearranging terms gives

$$\omega^2 = \Omega_i^2 + k^2\left(\frac{k_B T_e}{m_i}\right) = \Omega_i^2 + k^2 v_s^2 \tag{21}$$

which is the dispersion relation for electrostatic ion–cyclotron waves.

Example 13.3: A cylindrical column of gas of radius R carries a uniform axial current density j. The column is surrounded by vacuum. Show that the pressure distribution necessary for equilibrium is given by

$$p = \frac{\mu_0 j^2}{4}(R^2 - r^2)$$

Solution: The field at a distance r from the axis (Fig. 13.13) by Ampere's law is given by

$$\oint B \cdot dr = 2\pi r B = \mu_0 \pi r^2 j \tag{1}$$

$\therefore$

$$B = \frac{1}{2}\mu_0 jr \tag{2}$$

The force $v \times B$ acts radially inwards. Under the condition as equilibrium, we have

$$-jB - \frac{dp}{dr} = 0$$

or

$$\frac{dp}{dr} = -\frac{1}{2}\mu_0 j^2 r \qquad \text{[using Eq. (2)]}$$

On integration, we obtain

$$p = -\frac{1}{2}\mu_0 j^2 \int_R^r r\,dr$$

$$= \left[-\frac{1}{4}\mu j^2 r^2\right]_R^r$$

$$= \frac{\mu_0 j^2}{4}(R^2 - r^2)$$

Here we have assumed $p = 0$ when $r = R$. This reveals that the pressure is distributed parabolically with radius, reaching maximum at $r = 0$, i.e. on the centre dashed line shown in Fig. 13.13.

Fig. 13.13

GLIMPSES

1. *Plasma* is a quasineutral gas of charged and neutral particles which exhibits collective behaviour. Plasma is found in a natural form in several *cosmic* objects and the upper atmosphere of the earth. Plasma is also called as the *fourth state of matter*.

2. *Natural plasma* exists in some cosmic objects, e.g. interiors and atmospheres of hot stars, planetary nebulae, regions of ionized hydrogen in the interstellar medium, and the upper atmosphere of the earth.

3. The upper atmosphere of the earth, which is about 50km above the earth's surface is known as *ionosphere*. This ionosphere absorbs hazardous radiations (γ-rays, X-rays, UV radiations) coming from the outer space. This ionosphere is used for *communication* purposes, as the radiations of frequency less than the plasma frequency are reflected back by plasma in the ionosphere.

4. Satellites investigations revealed two regions called as the *van Allen radiation belts*, which envelop the earth. One belt is at distance of about 9700 km and second at 22,500 km from the surface of the earth. The thickness of inner and outer belts is respectively about 4800 km and 8000 km. These belts contain charged particles (ions) trapped between the magnetic lines of force.

5. The charged particles have free access to the earth's surface above the earth's magnetic poles, as the lines of magnetic field are concentrated towards the surface. These charged particles interact with the molecules of the upper air, causing a glow from time to time. These glow are the *auroral*. These are also known as *northern lights* and *southern lights*.

6. Sun is a star at the centre of the solar system. The surface of the sun, the photosphere forms the boundary between its opaque interior and its transparent atmosphere. Above the *photosphere* is the *chromosphere* and above this the corona, which extends tenuously into interplanetary space. The temperature of the central core is sufficiently high to maintain the nuclear fusion reactions. Thus, the material in the core of sun or stars is in *plasma* state.

7. *Plasma* is a state of matter characterized by a high, or even complete ionization of its particles. Depending upon the degree of ionization α, i.e. the ratio of the concentration of ionized particles to the total concentration, a plasma is classified as: *weakly ionized* (α is a fraction of 1%), *moderately ionized* (α equals several percent) and *fully ionized* (α close to 100%). In nature, weakly ionized plasma is found in the ionosphere. The sun, hot stars, and certain interstellar clouds are examples of fully ionized plasma formed at very high temperatures (high temperature plasma). Artificially, plasma is produced in gas discharge tubes. The control of its motion is the basis for the use of plasma as the working medium in various engines and for direct conversion of its internal energy into electric energy (magneto-hydrodynamic, or MHD generators plasma sources of electric energy, etc.)

8. The *high electric conductivity* of a plasma brings its properties near to those of conductors.

9. *Electrostatic forces* act between charged particles of a plasma; the forces between charged and neutral particles are of quantum mechanical nature.

10. Plasma differs from a simple accumulation of charge particles in a way that it must have a *minimum density*. This is determined from the condition that $L \gg D$, where L is any length characterizing the plasma, and D is the *Debye screening radius*, (also called as *Debye shielding distance* and *Debye length*).

    The fundamental property of plasma is that it can *shield out* electric potentials that are applied to it. At a distance D, the coulomb field of an arbitrary charge of plasma is shielded because the charge is predominantly surrounded by oppositely charged particles. As a whole, a plasma is a quasineutral system with great number of charged particles arranged in a region of space having various dimensions, $L \gg D$.

11. The total number of charged particles of plasma in a sphere of radius D is called the *Debye number N*. A plasma is said to be gaseous if N is very large, and thermodynamically it is regarded as an ideal gas.

12. *Specific features* of plasma, related to the coulomb long-range interaction of its particles, allow it to be regarded as a *special state of aggregation* of matter. These features are its strong interaction with external magnetic and electric fields, owing to its *electric conductivity*; the singular collective interaction of particles of plasma by means of the self-consistent field and the existence of other properties enabling various vibrations and waves to be excited and propagated in the plasma.

13. The *state of thermodynamic equilibrium* of a highly ionized gas is similar to a plasma at a definite temperature in which the loss of charged particles due to recombination is recompensated by new acts of ionization. The average kinetic energy of the various particles (positive or negative ions and neutral particles in various states of excitation) making up such a plasma is the same. The energy of a blackbody radiation existing in such a plasma corresponds to the same temperature. The

processes of energy exchange between the particles are equilibrium ones. A plasma having the preceding properties is said to be *isothermal*. It exists in the atmospheres of high temperature stars.

14. *A fully ionized plasma* can be produced under the condition that
$$kT \gg e\phi$$
where ϕ is the ionization potential of the atoms of the gas. For hydrogen and deuterium, this corresponds to a temperature $T = 160,000°$ K. Under these conditions, radiation of the plasma, making its adiabatic insulation difficult, becomes of vital importance.

REVIEW QUESTIONS

1. What do you understand by a plasma? Mention and explain the conditions for the existence of a plasma.
2. Derive the magnetohydrodynamic equations and show that the magnetic force is equivalent to magnetic hydrostatic pressure $p_m = B^2/m_0$ plus an additional tension along the magnetic lines of force.
3. What is pinch effect in plasma? Give a brief account of dynamical model of pinch effect.
4. Explain kink, sausage and flute instabilities in a plasma. How can the plasma confinement be overcome due to these instabilities?
5. Explain hydromagnetic waves. Show that the phase velocity for plasma waves approximates to Alfven velocity.
6. Write short notes on:
   i. pinch effect in plasma
   ii. kink, sausage and flute instabilities in plasma
   iii. Alfven waves
   v. reflection from a plasma (ionosphere).

PROBLEMS

1. A cylindrical column of gas of radius R carries a uniform axial current intensity J. The column is surrounded by vacuum. Show that the pressure distribution necessary for equilibrium is distributed parabolically with radius, reaching maximum at $r = 0$, i.e. on the centre-line and can be represented as $P = \dfrac{\mu_0 J^2}{4}(R^2 - r^2)$.

2. The sun's corona consists of ionized hydrogen with approximately 10^{12} particles per cubic meter. If the field in the corona, $B \approx 1000$ gauss, show that the phase velocity of the Alfven waves is 8.9×10^3 m/s.

3. A current flows axially in a cylindrical tube of plasma. It is observed that the plasma pressure vanishes at $r = r_0$ and $P = N(r) kT$.

   If $N = \int_0^{r_0} n(r) 2\pi r dr$, then show that $nkT = \dfrac{\pi}{4\mu_0}(rB)^2_{r=r_0}$.

SHORT ANSWER QUESTIONS

1. What do you understand by a plasma state of matter?

   Ans. It is a state of matter characterized by a high, or even complete, ionization of its particles.

2. Depending upon the degree of ionizations, in how many categories a plasma can be classified?

   Ans. Three categories: (a) weakly ionized, (b) moderately ionized, and (c) fully ionized.

3. What is the basis for the use of plasma as the working medium in various engines and for MHD generators, plasma sources of electric energy, etc.

   Ans. The control of plasma motion.

4. What is the property of plasma which brings it near to conductors?

   Ans. High electric conductivity.

5. How is a plasma regarded thermodynamically?

   Ans. As an ideal gas.

6. What is the condition for the production of a fully ionized plasma?

   Ans. $kT = e\phi_i$, where ϕ_i is the ionization potential of the atoms of the gas. For hydrogen and deuterium, this corresponds to a temperature $T \simeq 160{,}000$ K.

7. What kinds of radiations of plasma are in the optical range and the far ultraviolet?

   Ans. Breaking (bremsstrahlung) radiations of the electrons, occurring when they are retorted by the ions; recombination radiation, accompanying the recombination processes; and ordinary radiation produced by the excited particles of the spectral lines. Moreover, in a magnetic field, betatron (synchrotron) radiation is also possible.

8. Why plasma is considered as fluid?

   Ans. The plasma consists of equal number of negative and positive charges behaving very much like a fluid. A majority of the plasma phenomena in the loboratory and space can be explained in terms of rather crude *fluid model*. The fluid model considers all the plasma particles to be identical.

9. What do you understand by plasma pressure?

   Ans. The isotropic plasma exerts a scalar pressure. The fluid motion in a particular direction experience a pressure gradient force ∇p. In presence of inhomogeneity or plasma anisotropy brought in by the introduction of a magnetic field, the plasma pressure is not a scalar function.

10. What is plasma frequency?

    Ans. The perturbed plasma parameters oscillate with a characterstic frequency known as plasma frequency (ω_p),

    $$\omega_p = \frac{nq^2}{m\varepsilon_0} \text{ (rad } s^{-1})^2$$

    The plasma oscillation frequency depends on: (i) density, and (ii) mass of charged particles.

11. What is pinch effect?

    Ans. In the pinch effect, a pinching pressure is exerted on a plasma in a long circular cylinder, by its own circular magnetic field. A current at the surface of the plasma column produces this circular magnetic field.

12. Explain the origin of plasma frequency.

    Ans. In an equilibrium position, charged particles in a plasma are uniformly distributed in such a manner that neutrality is maintained everywhere. When electrons in the plasma are displaced relative to the uniform background of the ions, an electric field is developed in such a direction that it tries to pull the electrons back to restore the neutrality. While returning back, owing to inertia, the electrons overshoot and now the electric field is developed in the reverse direction which tries to pull back the electrons to their equilibrium positions with a characteristic frequency, known as the plasma frequency.

13. What is natural plasma?

    Ans. Natural plasma exists in some cosmic objects, e.g. interiors and atmospheres of hot stars, planetary nebulae, regions of ionized, and the upper atmosphere of the earth. On the earth, plasma can be produced in laboratories. The natural plasma do not exist on the earth due to very large density and very low temperature on the surface of the earth compared to cosmic objects. Natural plasma is found in the regions, e.g. ionosphere, van Allers belts, aurorae, solar corona, core of the sun, H_{11} regions, etc.

14. Why the properties of the plasma become quite different from those of natural atoms and molecules?

    Ans. When a gas is ionized, its dynamical behaviour is influenced by the external electric and magnetic fields. Further, the separated charged particles within the plasma give rise to new forces between the constituent of the particles. This is why the properties of the plasma become quite different from those of neutral atoms and molecules. This is why plasma is also called as the *fourth state of matter*.

15. Write an expression for the dielectric constant of plasma and explain it.

    Ans. The dielectric constant of a plasma is given by

    $$\varepsilon = 1 - \frac{\omega_p^2}{\omega^2} \tag{1}$$

    where ω_p is electron plasma frequency $\left(\omega_p = \sqrt{\dfrac{ne^2}{m\varepsilon_0}}\right)$ and ω is the frequency of the applied field, $E = E_0 \varepsilon^{i\omega t}$.

    Equation (1) shows that when $\omega_p < \omega$, the dielectric constant is positive and the wave can propagate. But when $\omega_p > \omega$, the dielectric constant is negative and the wave cannot propagate. The cut off takes place when $\omega_p = \omega$.

MULTIPLE CHOICE QUESTIONS

1. The characteristic linear parameter δ (Debye length) for a plasma is given by

   (a) $\delta \simeq \dfrac{\varepsilon_0 kT}{N_0 e^2}$

   (b) $\delta \simeq \left(\dfrac{\varepsilon_0 kT}{N_0 e^2}\right)^{1/2}$

   (c) $\delta \simeq \left(\dfrac{\varepsilon_0 kT}{N_0 e^2}\right)^2$

   (d) $\delta \simeq \left(\dfrac{\varepsilon_0 kT}{N_0 e^2}\right)^{3/2}$

   [b]

2. In magnetic confinement pinch effect, the force which contains the plasma is given by

   (a) $J \times B$

   (b) $E \times B$

   (c) $\dfrac{B^2}{2\mu_0}$

   (d) $\dfrac{E}{B}$

   [a]

3. The condition for stable equilibrium of plasma as a conducting fluid MHD is

   (a) $\dfrac{B^2}{2\mu_0} = \text{constant}$

   (b) $P + \dfrac{B^2}{2\mu_0} = 0$

   (c) $P + \dfrac{B^2}{2\mu_0} = \text{constant}$

   (d) $P = 0$

   [c]

4. Plasma frequency is given by

   (a) $\nu_p = \dfrac{1}{2\pi}\left(\dfrac{Ne^2}{m\varepsilon_0}\right)$

   (b) $\nu_p = \dfrac{1}{2\pi}\left(\dfrac{Ne^2}{m\varepsilon_0}\right)^{3/2}$

   (c) $\nu_p = \dfrac{1}{2\pi}\left(\dfrac{Ne^2}{m\varepsilon_0}\right)^2$

   (d) $\nu_p = \dfrac{1}{2\pi}\left(\dfrac{Ne^2}{m\varepsilon_0}\right)^{1/2}$

   [d]

5. The Alfven velocity is given by

   (a) $v_A = \dfrac{B}{\mu_0 e}$

   (b) $v_A = \dfrac{B}{(\mu_0 e)^{1/2}}$

   (c) $v_A = \sqrt{\dfrac{B}{\mu_0 \rho}}$

   (d) $v_A = \dfrac{B^2}{\mu_0}$

   [b]

6. A particle of mass m and charge q is moving in a uniform magnetic field B and a uniform electric field E with velocity v. The equation of motion of the particle is

   (a) $m\dfrac{dv}{dt} = qE$

   (b) $m\dfrac{dv}{dt} = q(V \times B)$

   (c) $m\dfrac{dv}{dt} = q[E + (v \times B)]$

   (d) $m\dfrac{dv}{dt} = q(E \times B)$

   [c]

7. A particle of mass m and charge q is moving in a time varying magnetic field B with velocity v. The associated electric field E is given by

   (a) $\nabla \times E = -B$

   (b) $E = \nabla \cdot B$

   (c) $\nabla \cdot E = 0$

   (d) $\nabla \times E = 0$

   [a]

Chapter

14

Relativistic Electrodynamics

14.1 INTRODUCTION

We all are aware that the theory of special relativity was developed to prove the invariance of Maxwell's field equations and to unify the electrical and magnetic phenomena. We have seen that Maxwell studied the consistency of existing experimental laws, such as Gauss, Faraday and Ampere and unified these laws by adding a new term in Ampere's law. The special theory of relativity was propounded by Einstein in 1905 with a view to reconcile with certain results of Maxwell's field equations, i.e. Maxwellian electrodynamics. In the present chapter, we will see that the laws of Ampere and Faraday and their generalization to *Maxwellian electrodynamics* are consistent with the implications and predictions of Einstein's special theory of relativity.

14.2 GALILEAN RELATIVITY

According to the *principle of relativity*, all laws of nature are identical in all inertial systems of reference. Obviously, the equations expressing the laws of nature retain their form in different inertial systems, i.e. they are invariant with respect to transformation of coordinates and time from one system to another. Large number of carefully performed experiments has shown that the principle is valid.

Fig. 14.1: Frames of reference in relative motion

Let us consider two inertial frames S and S' with parallel axes as shown in Fig. 14.1 Suppose S' is moving along x-direction with a uniform velocity v relative to frame S. Any event occuring at a point will be specified by an observer in frame S by the coordinates (x, y, z, r). The same event will be described by an observer in S' by (x', y', z', t'). One can write the transformation of coordinates of a point from one system to another as

$$x' = x - vt, \quad y' = y, \quad z' = z \quad \text{and} \quad t' = t \tag{14.1}$$

The last relation in Eq. (14.1) is a consequence of the implicit assumption of classical physics that time is *absolute*, i.e. independent of a frame of reference. These simple transformations are known as *Galilean transformations*. The form of Eq. (14.1) depends, of course, on the relative motion of two frames of reference, but it also depends upon certain assumptions regarding the nature of space and time. As stated earlier, it is assumed that time t is independent of any particular frame of reference, i.e. if t and t' be the time recorded by the observer O and O' of an event occuring at P, then $t' = t$. The other assumption, regarding the nature of space, is that the distance between two points or particles is quite independent of any particular frame of reference, i.e. according to the Galilean transformation the space-interval measurements are absolute, i.e. they are same for all inertial observers. From Eq. (14.1), we have

$$\frac{dx'}{dt'} = \frac{dx'}{dt} = \frac{dx}{dt} - v \quad (\because \ t' = t)$$

$$\frac{dy'}{dt'} = \frac{dy}{dt}$$

$$\frac{dz'}{dt'} = \frac{dz}{dt}$$

and, hence, $\quad \dfrac{d^2x'}{dt'^2} = \dfrac{d^2x}{dt^2}, \quad \dfrac{d^2y'}{dt'^2} = \dfrac{d^2y}{dt^2}, \quad \dfrac{d^2z'}{dt'^2} = \dfrac{d^2z}{dt^2}$

This shows that though velocities vary, but the acceleration of a particle relative to S and S' frames are identical. The Newton's equation of motion in the two systems is

$$F_i = m\ddot{x}_i = m\ddot{x}'_i = F'_i, \quad \text{where} \ i = 1, 2, 3 \tag{14.2}$$

Obviously, the laws of classical mechanics are invariant under the Galilean transformations. The real difficulty was experienced when these transformations were applied to electric and magnetic fields predicted by Maxwell's field equations. Now, we examine the behaviour of laws of electromagnetism under Galilean transformations. We know that the following scalar equation results from Maxwell's theory of electromagnetism

$$\nabla^2 \psi - \frac{1}{c}\frac{d^2\psi}{dt^2} = 0 \tag{14.3}$$

Galilean transformation gives

$$\frac{\partial}{\partial x} = \frac{\partial}{\partial x'} + \frac{1}{v}\frac{\partial}{\partial t'}$$

$$\therefore \qquad \frac{\partial^2}{\partial x^2} = \frac{\partial}{\partial x'^2} + \frac{1}{v^2}\frac{\partial^2}{\partial t'^2} + \frac{2}{v}\frac{\partial^2}{\partial t'\partial x'}$$

and $\qquad \dfrac{\partial^2}{\partial y^2} = \dfrac{\partial^2}{\partial y'^2}, \dfrac{\partial^2}{\partial z^2} = \dfrac{\partial^2}{\partial z'^2}$ and $\dfrac{\partial^2}{\partial t^2} = \dfrac{\partial^2}{\partial t'^2}$ (14.4)

Using Eq. (14.4), Eq. (14.3) gives

$$\nabla^2\psi - \frac{1}{c^2}\frac{\partial^2\psi}{\partial t'^2} + \frac{1}{v^2}\frac{\partial^2\psi}{\partial t'^2} + \frac{2}{v}\frac{\partial^2\psi}{\partial t'\partial x'} \tag{14.5}$$

We see that Eqs. (14.3) and (14.5) are not the same wave equations, i.e. the form of the wave equation is not preserved by the substitution of Eq. (14.1). One can easily see that Maxwell's equations are variant with the Galilean transformations. We obtain the electric field in the moving frame as

$$E' = E + v \times B \tag{14.6}$$

and

$$B' = B - \frac{1}{c^2}(v \times B) \tag{14.7}$$

where

$$v = \hat{x}v_x, \quad v_y = v_z = 0$$

We can easily see that these are inconsistent and reveals that the invariance of physical laws with Galilean transformation [Eq. (14.1)] is not valid. This can also be seen if we recall that the electromagnetic waves, as found from Maxwell's electromagnetic theory propagate with a constant velocity c, while the velocity measured by the observer in frame S' moving with velocity v relative to the frame S will be $c + v$ or $c - v$ depending on the direction of relative motion. Obviously, the velocity of electromagnetic waves is not invariant under the Galilean transformations, i.e. Maxwell's equations will also change their form on transformation from one system to another as evident from Eqs. (14.6) and (14.7). In order to resolve this conflict, the bold new concept of the special theory of relativity was introduced by A Einstein in 1905.

14.3 BASIC POSTULATES OF SPECIAL THEORY OF RELATIVITY

The two postulates are:

1. *All physical phenomena (mechanical, electromagnetic, etc.) can be described by the same physical laws by two observers moving with constant linear velocity with respect to each other* or in other words, it is impossible to ascertain, by means of any experiments, whatsoever, conducted in a closed system of bodies, whether the system is at rest or travelling at uniform velocity in a straight line with respect to some inertial frame of reference.

   The laws of physics may be expressed in equations having the same form in all frames of reference moving at constant velocity with respect to one another.

   This postulate expresses the absence of a universal frame of reference.

2. *The free space speed of light is same for all observers regardless of their motion.*

   or

   The speed of light in vacuum is independent of the speed of the source of this light and is same in all directions.

According to the first postulate, this speed is the same in all inertial frames of reference, being thereby a universal constant. This postulate follows directly from the results of the Michelson–Morley experiment and many others. Recently, it has been decisively verified. For example, by the speed of radiation emitted in the decay of π-meson as measured by Parley *et al* was $(2.9979 \pm 0.0003) \times 10^8$ m/s, although the mesons were moving at a speed greater than $0.99975\,c$.

Of the two postulates, the second represents an experimental fact, whereas the first is a generalization from a wide range of physical experience. The first postulate is in no way self evident; it is a hypothesis to be tested by experiment. Einstein decided that the conflict between them must be due to an imperfection of the classical ideas about *simultaneity*.

14.4 THE THEORY OF SIMULTANEITY

Time in classical relativity (in contrast to the coordinates) is regarded as being independent of the frame of reference. Two events which are simultaneous in one inertial frame are considered to be simultaneous in any other inertial frame of reference. This contradicts the special theory of relativity which leads to relativity of simultaneity (events). According to the principle of special relativity, time elapses different frames of reference, and the specification of an interval of time between two events has a meaning only if the frame of reference is indicated, i.e. *the concept of simultaneity is only a relative concept.*

14.5 LORENTZ TRANSFORMATION EQUATIONS

A concept of inertial frame was developed. Inertial frames are those frames in which laws of inertia hold good. In an inertial frame of reference, all bodies move with a constant velocity unless they are subjected to external force. All inertial frames move with a constant velocity relative to one another. Now, we obtain the equations of transformation between two inertial systems which would replace Galilean transfor-mation and keep the speed of light constant. The new set of trans-formation equations we are going to obtain is known as *Lorentz transformation* equations. These were developed to relate space in different inertial frames of reference. We

Fig. 14.2: S' frame moving along x-axis

consider frames of reference S and S' having a common origin and coincident axes system at $t = 0$, and S' moving along x-axis with a constant relative velocity v as shown in Fig. 14.2.

Now, consider a light signal emitted in free space at the instant $t = 0$ from the coincident origins of the systems S and S'. When the light signal reaches at the point P, let the positions and time measured by the observers O and O' be (x, y, z, t) and (x', y', z', t') respectively. If the velocity of light signal is c, then the time required by the light signal in traversing the distance OP in frame S is

$$t = \frac{OP}{c} = \frac{\sqrt{x^2 + y^2 + z^2}}{c}$$

or
$$x^2 + y^2 + z^2 - c^2 t^2 = 0 \tag{14.8}$$

The time required by the same light signal in travelling the distance $O'P$ in frame S', as seen by the observer at rest at the origin of S' is

$$t' = \frac{O'P}{c} = \frac{\sqrt{x'^2 + y'^2 + z'^2}}{c}$$

or
$$x'^2 + y'^2 + z'^2 - c^2 t'^2 = 0 \tag{14.9}$$

The velocity of light c is same in both the frames of reference (according to second postulate of Einstein's special theory of relativity).

The Galilean transformation connects measurements in the two frames according to the transformation Eq. (14.1). Substituting Eq. (14.1) in Eq. (14.9), one obtains

$$x^2 - 2xvt + v^2 t^2 + y^2 + z^2 = c^2 t^2$$

and this is not in agreement with Eq. (14.8) in any way. This clearly indicates that the Galilean transformation fails. Therefore, we want new transformation equations which transform Eq. (14.9) to Eq. (14.8). The transformation must be linear in x and t, because we want to get a sphere which expand at a uniform rate. Therefore, we try the relations

$$\left.\begin{array}{l} x' = k(x - vt) \\ y' = y \\ z' = z \\ t' = k'(t - bx) \end{array}\right\} \qquad (14.10)$$

where k, k' and b are constants. Making these substitutions in Eq. (14.9), one obtains

$$k^2(x^2 - 2xvt + v^2 t^2) + y^2 + z^2 = c^2 k'^2(t^2 - 2bxt + b^2 x^2)$$

or $(k^2 - k'^2 c^2 b^2)x^2 - 2(2k^2 v + k'^2 c^2 b)xt + y^2 + z^2 = (k'^2 - k^2 v^2 / c^2)c^2 t^2$

This must be identical to Eq. (14.8). Equating these two, one obtains

$$k^2 - k'^2 b^2 c^2 = 1$$
$$k^2 v^2 - k'^2 bc^2 = 0$$
$$k'^2 - k^2 v^2 / c^2 = 1$$

Solving these equations for k, k' and b, one obtains

$$k = k' = \left(1 - \frac{v^2}{c^2}\right)^{-1/2}$$

and

$$b = \frac{v}{c^2}$$

The new transformation equations which are essentially based on the invariance of the velocity of light, are then

$$\left.\begin{array}{l} x' = k(x - vt) = \dfrac{x - vt}{\sqrt{1 - v^2 / c^2}} \\[4mm] y' = y \\ z' = z \\[2mm] t' = k(t - bx) = \dfrac{t - \dfrac{vx}{c^2}}{\sqrt{1 - v^2 / c^2}} \end{array}\right\} \qquad (14.11)$$

This set of transformation equations is known as the Lorentz transformations. The Lorentz transformations are symmetrical and retain their form in the transformation from S' to S except that the sign of v is changed. Thus, the corresponding inverse transformation equations are

$$x = \frac{x' + vt'}{\sqrt{1 - v^2/c^2}}, \quad y = y', z = z'$$

$$\text{and} \quad t = \frac{\left(t' + \dfrac{vx'}{c^2}\right)}{\sqrt{1 - v^2/c^2}} \quad\quad\quad (14.12)$$

The Lorentz transformations are linear, and at low velocities ($v/c \ll 1$) they become the Galilean transformations. In most of the cases, which we encounter on earth, $v \ll c$ so that the results of Lorentz transformations do not differ to any great extent from those of the Galilean transformations. This shows that classical mechanics can be regarded as a special case of relativistic mechanics.

From these equations, we can easily confirm the statement that events which happen at the same place at different times, as viewed from one frame, may be seen from another frame to happen at different place as well. Similarly, a difference in spatial position w.r.t. one frame may correspond to a difference in both space and time w.r.t. another. Thus, a space difference can be converted partly into a time difference, or *vice versa*, merely by changing the frame of reference that is being used. Obviously, Lorentz transformations represent a most profound conceptual change specially with regard to *space and time*.

It is convenient to make use of some standard notations. Abbreviating the quantities v/c and $\left(1 - \dfrac{v^2}{c^2}\right)^{-1/2}$ as β and γ, Lorentz transformations can be expressed as

$$x' = \frac{x - vt}{\sqrt{1 - \beta^2}} = \gamma(x - vt)$$

$$y' = y$$

$$z' = z \quad\quad\quad (14.13)$$

$$\text{and} \quad t' = \frac{\left(t - \dfrac{vx}{c^2}\right)}{\sqrt{1 - \beta^2}} = \gamma\left(t - \frac{vx}{c^2}\right)$$

The transformation equations for the velocity of a moving object can be obtained by taking the derivatives of Eq. (14.13) with respect to t and t'.

14.6 CONSEQUENCES OF LORENTZ TRANSFORMATIONS

Two features of Lorentz transformation equations are particularly important in evaluating the electric and magnetic fields: (i) time dilation, and (ii) length contraction.

i. *Length Contraction*

One consequence of the Lorentz transformation is the contraction of moving rod standard in the direction of its motion (Fig. 14.3).

Let a rod in frame S be carried along the x-axis and be at rest there. If x_1 and x_2 are the abscissae of the ends of the rod, then the length l_0 in the frame S is given by

$$l_0 = x_2 - x_1$$

This is the *proper length* of the rod. Proper length is the linear dimension l_0 of a body in the system of coordinates in which it is at rest. Now, let an observer in frame S' measure the length of the same rod when S' is in motion with a velocity v relative to S along x-axis. If x'_1 and x'_2 are the abscissae of the ends of the rod at the time t of this system (since the rod is in motion and hence it is necessary that the abscissae must be noted simultaneously, i.e. at the same moment by this observer), then the length L of the rod for him is given by

Fig. 14.3: Length contraction

$$l = x'_2 - x'_1$$

Now from Lorentz transformation equations, one obtains

$$x_1 = \frac{(x'_1 + vt')}{\sqrt{1 - v^2/c^2}} \qquad (14\ 14)$$

and

$$x_2 = \frac{(x'_2 + vt')}{\sqrt{1 - v^2/c^2}} \qquad (14.15)$$

Subtracting Eq. (14.14) from Eq. (14.15), one obtains

$$x_2 - x_1 = \frac{x'_2 - x'_1}{\sqrt{1 - v^2/c^2}}$$

or

$$l_0 = \frac{l}{\sqrt{1 - v^2/c^2}}$$

or

$$l = l_0\sqrt{1 - v^2/c^2} \qquad (14.16)$$

Clearly, a moving rod appears shorter or contracted by a factor $\sqrt{1 - \beta^2}$ in its direction of motion, the transverse section remains unaffected as $\Delta y = \Delta y'$ and $\Delta z' = \Delta z$. This is called *Fitzgerald contraction*. It is important to note that the effect is *mutually reciprocal between different observers*.

Let us consider a spherical body of radius a with its centre at O' of the system S' at rest in this system. Its surface is represented by

$$x'^2 + y'^2 + z'^2 = a^2 \qquad (14.16a)$$

Now, if S' is moving with a velocity v in the x-direction w.r.t. S, then the equation of spherical surface in the system S is given by

$$\frac{(x - vt)^2}{1 - \beta^2} + y^2 + z^2 = a^2 \qquad (14.16b)$$

If the two frames of reference S and S' coincide at $t' = t = 0$, then Eq. (14.16b) reduces to

$$\frac{x^2}{1-\beta^2} + y^2 + z^2 = a^2 \tag{14.16c}$$

Equation (14.16c) represents an ellipsoid. Obviously, the sphere appears to be contracted in the x-direction. If $v = c$, and if the sphere travels with the velocity of the light with reference to frame S, an observer would see it like a disc of zero thickness.

The phenomenon of length contraction can hardly be noticed in everyday life because the motions ordinarily encountered are negligibly slow in comparison to that of light ($v \ll c$).

ii. Time Dilation

The time interval measured in two inertial frames in motion are not same. Suppose that a clock at a fixed point x' in the frame measures a time interval $\Delta t' = (t_2' - t_1')$ between two events that occurred at x' at times t_1' and t_2'. To an observer in the frame S, the two events appear to occur at times t_1 and t_2, given by

$$t_1 = \frac{t_1' + \left(\dfrac{vx'}{c^2}\right)}{\sqrt{1-(v^2/c^2)}} \tag{14.17}$$

and

$$t_2 = \frac{t_2'(vx'/c^2)}{\sqrt{1-(v^2/c^2)}} \tag{14.18}$$

or

$$t_2 - t_1 = \Delta t = \frac{t_2' - t_1'}{\sqrt{1-(v^2/c^2)}} = \frac{\Delta t'}{\sqrt{1-\beta^2}} \tag{14.19}$$

Obviously, the time interval in the S frame is larger than the time measured by the clock in the S' frame. Time interval measured on a clock that travels together with a system is called *its proper time*. This means that the proper time noted by a moving observer is always less than the corresponding interval of time in a stationary frame. *A stationary observer finds that a moving clock runs slower than a stationary one.* The effect due to time dilatation has to be considered while considering Doppler shift in radiation.

Time dilatation has been best verified in experiments on a nuclear particle, called meson. These are subatomic particles which can be created in the laboratory by high-energy particle accelerators. Rossi and Hall in 1941 observed that the apparent lifetime of fast moving mesons in the cosmic rays was considerably larger than the lifetime of slow moving mesons in the laboratory. [The mean proper lifetime of a positive μ-meson is 2.20×10^{-6} s. The average path of such a particle, travelling at a velocity $0.98\,c$, should be 660 m in air before decaying. Actually, the average path of μ-meson in air is much longer because its mean lifetime, measured in a stationary frame fixed in the earth (in the air) is substantially larger than 2.20×10^{-6} s]. They attributed this result to the time dilatation effect according to which the meson clock, being in fast motion with respect to the terrestrial clocks, would seem to go relatively slow; consequently, a cosmic ray meson which lives its normal span of life according to its own proper time would appear to have lived longer according to the terrestrial clocks.

Very recently, the dilatation caused by the thermal vibration of the nuclei in certain crystals has also been verified.

We have seen that the lengths and time intervals are changed by Lorentz transformations. However, observers in the two inertial frames S and S' find that the light travels with a constant velocity c. When an electromagnetic signal starts from origin of the reference frame at time $t = 0$ and propagates in space as a spherical wave, then an observer in the S frame describes the locus of spherical wavefront at time by the equation

$$x^2 + y^2 + z^2 - c^2 t^2 = 0 \qquad (14.20)$$

Similarly, as seen by an observer in the S' frame, the locus of the spherical wavefront at time t' is given by

$$x'^2 + y'^2 + z'^2 - c^2 t'^2 = 0 \qquad (14.21)$$

One can easily see that Eqs. (14.20) and (14.21) are consistent with the Lorentz transformation, i.e.

$$x'^2 + y'^2 + z'^2 - c^2 t'^2 = x^2 + y^2 + z^2 - c^2 t^2 \qquad (14.22)$$

A quantity $(x^2 + y^2 + z^2 - c^2 t^2)$ is said to be *Lorentz invariant*, i.e. this has always the same value for observers in S and S' frames.

Equation (14.19) connecting Δt and $\Delta t'$ suggests that any physical process occurring in the moving frame S' appears to have been slowed down to an observer in frame S. The various biological processes such as respiration, digestion, heart beating, ageing, etc. would appear to have slowed down in the moving frame when its speed (v) is comparable with the speed of light (c).

14.7 TRANSFORMATION PROPERTIES OF MOMENTUM, ENERGY, MASS AND FORCE

One of the important consequences of special theory of relativity is that mass increases with velocity. The mass of a particle moving with relativistic speed v in S frame increases and the increased mass is given by

$$m = \frac{m_0}{\sqrt{1 - \beta^2}} \qquad (14.23)$$

where m_0 is the rest mass, $\beta = \dfrac{v}{c}$ and v is the constant relative velocity of the moving particle in the same frame. The total energy of the particle moving with relativistic velocity v consists of two parts: (i) the rest energy $m_0 c^2$ and (ii) the kinetic energy of the particle. In terms of the relativistic mass, one can express the kinetic energy of the particle as

$$\text{KE} = mc^2 - m_0 c^2 \qquad (14.24)$$

Therefore, the total energy of the particle is

$$U = mc^2 = m_0 c^2 + \text{KE} \qquad (14.25)$$

We can define the momentum of a particle in the S frame by $p = mv$, where m is the relativistic mass given by Eq. (14.23). One can write the components of momentum and total energy of the particle moving with a velocity v in the S frame as

$$p_x = \frac{m_0 v_x}{\sqrt{1-\beta^2}}, \qquad p_y = \frac{m_0 v_y}{\sqrt{1-\beta^2}}$$

$$p_z = \frac{m_0 v_z}{\sqrt{1-\beta^2}}, \quad \text{and} \quad U = \frac{m_0 c^2}{\sqrt{1-\beta^2}} \quad (\beta = v/c) \tag{14.26}$$

If v' is the velocity of the particle in S' frame, the components of momentum and total energy of the particle are

$$p'_x = \frac{m_0 v'_x}{\sqrt{1-\beta'^2}}, \qquad p'_y = \frac{m_0 v'_y}{\sqrt{1-\beta'^2}}$$

$$p'_z = \frac{m_0 v'_z}{\sqrt{1-\beta'^2}}, \quad \text{and} \quad U' = \frac{m_0 c^2}{\sqrt{1-\beta'^2}} \quad (\beta' = v'/c) \tag{14.27}$$

We can write the Lorentz transformation for momentum of particles in S frame and S' frame moving with a relative velocity u along the x-axis as

$$
\left.
\begin{aligned}
p_x &= \frac{1}{\sqrt{1-(u^2/c^2)}}\left(p'_x + \frac{U' u_x}{c^2}\right) \\
p_y &= p'_y, \quad p_z = p'_z \\
\text{and} \quad U &= \frac{1}{\sqrt{1-u^2/c^2}}(U' + u_x p'_x)
\end{aligned}
\right\} \tag{14.28}
$$

Changing u to $-u$ and interchanging primed and unprimed quantities, one obtains the inverse transformations as

$$
\left.
\begin{aligned}
p'_x &= \frac{1}{\sqrt{1-(u^2/c^2)}}\left(p_x - \frac{U u_x}{c^2}\right) \\
p'_y &= p_y, \quad p'_z = p_z \\
\text{and} \quad U' &= \frac{1}{\sqrt{1-(u^2/c^2)}}(U - u_x p_x)
\end{aligned}
\right\} \tag{14.29}
$$

Obviously, the momentum has four-dimensional form whose time component is total energy. If total energy and momentum both are conserved in a frame of reference, then the energy and momentum will be conserved in any frame of reference. In case if only momentum is conserved, energy must also be conserved.

One can write the force components experienced by a particle moving with velocity in a reference frame as

$$F_x = \frac{d}{dt}(mv_x), \quad F_y = \frac{d}{dt}(mv_y) \quad \text{and} \quad F_z = \frac{d}{dt}(mv_z) \tag{14.30}$$

The force components in the S' frame are

$$F'_x = \frac{d}{dt}(m'v'_x), \quad F'_y = \frac{d}{dt}(m'v'_y) \quad \text{and} \quad F'_z = \frac{d}{dt'}(m'v'_z) \tag{14.31}$$

The frame S' is moving with a velocity u with respect to frame S. Making use of Lorentz transformations, one can show that the force components in the two frames S and S' are related as

$$\left. \begin{array}{l} F_x = \dfrac{F'_x + (u/c^2)(v \cdot F)}{1 + (v'_x u/c^2)}, \quad F_y = \dfrac{F'_y \sqrt{1 - u^2/c^2}}{1 + (v'_x u/c^2)} \\[4mm] \text{and} \quad F_z = \dfrac{F'_z \sqrt{1 - u^2/c^2}}{1 + (v'_x u/c^2)} \end{array} \right\} \tag{14.32}$$

From Eq. (14.32), it is obvious that the force in frame S is related with the power $v \cdot F$ generated by the force in frame S'. Moreover, one can also see that if a particle is at rest in the frame S' where it is subjected to a force with components F'_x, F'_y and F'_z, and $v' = 0$ in this frame, then the force transformation equations reduce to

$$F_x = F'_x, \ F_y = F'_y \sqrt{1 - (u^2/c^2)} \quad \text{and} \quad F_z = F'_z \sqrt{1 - (u^2/c^2)} \tag{14.33}$$

Equation (14.33) is widely used in studying the motion of a rest particle in electromagnetic fields.

14.8 RELATIVITY OF CHARGES

We have seen that *electric charge is absolutely conserved*. In an atom, the electrons and protons are moving about, yet the atom is electrically neutral, i.e. positive and negative charges exactly cancel, obviously the charge on each elementary particle is $\pm e$ whether or not the particle is moving relative to the observer. This means that the *charge on an elementary particle* and, hence, *the net charge carried by any macroscopic system is Lorentz invariant*. Now, the question arises, whether the charge density is also invariant. To answer this question, let us consider a series of positive charges equally spaced along a line, each with charge $+e$. Let the charges be at rest in frame S and span a distance of unit length. The linear charge density as seen by an observer in frame S is Ne coulomb/m, where N is the number of charges in unit length. In the light of Lorentz transformations, an observer is frame S' moving with a speed u parallel to the charges finds that the charges are squashed up to a distance $\sqrt{1 - (u^2/c^2)}$. Obviously, the linear charge density according to observer in S' frame is $\dfrac{Ne}{\sqrt{1 - (u^2/c^2)}}$ Cm^{-1}.

14.9 RELATIVITY OF MAGNETIC AND ELECTRIC FIELDS

We now consider the forces exerted by electric and magnetic fields, as viewed by observers in frames S and S'. Let us consider a long straight conducting wire in the frame S in which free electrons move with a drift velocity v in the x-direction. Let n be the number of electrons per unit length of the conductor. Obviously, this is also the number of positive ions per unit length. Let us consider that a charge q is moving parallel the conductor at a speed v. The magnetic force experienced by this charge, $F = q(v \times B)$ in the frame S, where B is the magnetic field produced by the current in the conductor.

Now, we consider an observer in frame S', moving with velocity v in the x-direction. For this observer, the charge q is at rest and hence, the magnetic force $F = 0$. However, all inertial frames are equivalent, i.e. observer in frame S' must agree with the observer in frame S. This means that there is a force on charge q though it is not a magnetic force. Now, the question arises, how we should interpret this force. In the frame S, the current is constituted by conduction electrons of linear charge density (λ^-) moving with velocity v, where $\lambda^- = -ne$. The positive charge, i.e. the immobile background charge density due to positive ions is $\lambda^+ = ne$. The net linear charge density is

$$\lambda = \lambda^- + \lambda^+ = 0 \tag{14.34}$$

Now, we obtain the current density in the wire as

$$J_x^- = nev = \lambda^- v, \quad J_x^+ = 0$$

$$\therefore \quad J_x = J_x^+ + J_x^- = \lambda^- v \tag{14.35}$$

The magnetic field B produced by the current is given by

$$B = -\mu_0 \frac{nev}{2\pi r}$$

Obviously, B causes a force F to act on the charge. We have

$$|F| = |q(v \times B)| = -\mu_0 \frac{nev^2 q}{2\pi r} \tag{14.36}$$

Now, we are interested in the observation of the observer in frame S' about this force. The positive charges which were at rest to the observer in frame S are moving at a speed v in the opposite direction to the observer in frame S'. Now, applying Lorentz transformations, the unit length in frame S is equivalent to the length $\sqrt{1-\beta^2}$ in frame S' and the linear charge density due to the positive ions is $\lambda^+ = \dfrac{ne}{\sqrt{1-\beta^2}}$. The electrons are at rest in this frame and they span a distance $\dfrac{1}{\sqrt{1-\beta^2}}$ in frame S'. (We must remember that the distance is of unit length in frame S). Obviously, the linear charge density is $\lambda' = -ne\sqrt{1-\beta^2}$. Therefore, the net charge density in frame S' is given by

$$\lambda' = \lambda'^- + \lambda'^+ = -ne\sqrt{1-\beta^2} + \frac{ne}{\sqrt{1-\beta^2}}$$

$$= ne \frac{\beta^2}{\sqrt{1-\beta^2}} \tag{14.37}$$

We can see that this is positive and not zero as found by the observer in frame S. Thus, the net charge in the moving wire is positive. Obviously, the observer in frame S', therefore, does not think that the wire is electrically neutral. According to the observer in frame S', there is an electrostatic field around the wire given by

$$E' = \frac{\lambda'}{2\pi\varepsilon_0 r} = \frac{ne\beta^2}{2\pi\varepsilon_0 r\sqrt{1-\beta^2}} \tag{14.38}$$

The magnitude of the electric force in frame S' can be easily obtained as

$$|F'| = |qE'| = \frac{qne\beta^2}{2\pi\varepsilon_0 r\sqrt{1-\beta^2}}$$

$$= \frac{\mu_0 nev^2 q}{2\pi r\sqrt{1-\beta^2}} = \frac{|F|}{\sqrt{1-\beta^2}} \quad \left(\text{we have } c = \frac{1}{\sqrt{\varepsilon_0\mu_0}}\right) \tag{14.39}$$

We have obtained a very striking result. The force that was purely magnetic in frame S appears to be purely electrostatic in frame S'. This shows that the *magnetic and electric effects are just two ways of looking at the same thing*. Obviously, *the special theory of relativity provides the unification of electric and magnetic effects.*

The charges moving with velocity v in the frame S' constitute a current. We have

$$J = \lambda^- v = \frac{\lambda_0 v}{\sqrt{1-v^2/c^2}} \tag{14.40}$$

One can write the components of the current density in frame S as

$$J_x = \frac{v_x}{\sqrt{1-(v^2/c^2)}}, \quad J_y = \frac{v_y}{\sqrt{1-(v^2/c^2)}}$$

and

$$J_z = \frac{v_z}{\sqrt{1-(v^2/c^2)}} \tag{14.41}$$

Let us find the relation between current density and the charge density. We can see that this has a close similarity with the relation between momentum and energy. Consistent with these transformations, one can write three invariant equations as

$$x^2 + y^2 + z^2 - c^2 t^2 = c^2 t^2 = \text{constant} \tag{14.42}$$

$$p_x^2 + p_y^2 + p_z^2 - c^2 m^2 = c^2 m_0^2 = \text{constant} \tag{14.43}$$

and

$$J_x^2 + J_y^2 + J_z^2 = c^2 \rho_0^2 = \text{constant} \tag{14.44}$$

$$\lambda = \frac{\lambda_0}{\sqrt{1-(v/c)^2}} \text{ together with Eq. (14.40) yields}$$

$$\lambda = \frac{\lambda_0}{m_0} m \quad \text{and} \quad J = \frac{\lambda_0}{m_0} p \tag{14.45}$$

Obviously, the charge density λ, and current density J transform exactly as m and p respectively. The transformation of current density in inertial frames S and S' moving with a constant velocity u along the x-axis can be expressed as

$$J_x = \frac{J_x + \lambda_x'}{\sqrt{1-(u/c)^2}}, \quad J_y = J_y' \text{ and } J_z = J_z' \tag{14.46}$$

and
$$\lambda = \frac{\lambda' + (uJ'_x / c^2)}{\sqrt{1 - (u/c)^2}}$$
(14.47)

This means that in the frame S, the net charge density is zero and there is no observed electric field. The motion of charges gives rise to a net current density giving rise to a magnetic field around the conducting wire. We have seen that in the frame S there is only the B field and the charge outside the conductor experiences a magnetic force directed towards to the conductor. On the other hand, in the frame S', there is a positive net charge density, as well as, a current density which gives rise to E' and B' fields. However, the net charge in the frame S' is not zero and the charge outside the conducting wire experiences an electrical force directed towards the wire. This leads one to conclude at whether an electromagnetic field is purely magnetic field or electric field or electromagnetic field depends on the inertial frame of reference in which sources are observed. This is an important result of special theory of relativity.

14.10 THE LORENTZ TRANSFORMATION AS AN ORTHOGONAL TRANSFORMATION

Let us consider that a frame of reference is rotated through an angle θ about z-axis. Obviously, the coordinates of point G in the rotated frame will not be the same as those in the original frame of reference.

One can express the coordinates in the rotated frame as (Fig. 14.4)

$$\left. \begin{array}{l} x' = x \cos\theta + y \sin\theta \\ y' = -x \sin\theta + y \cos\theta \\ z' = z \end{array} \right\}$$
(14.48)

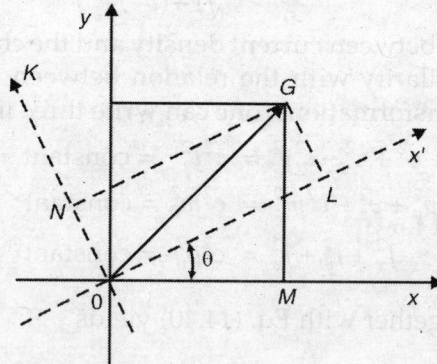

Fig. 14.4: A frame of reference rotating about z-axis

We can see that Eq. (14.48) is a linear combinations of original coordinates. Let the point G be specified by a position vector r, then its length in the original frame is given by

$$|r|^2 = x^2 + y^2 + z^2$$

One can see from Eq. (14.48) that

$$|r|^2 = x^2 + y^2 + z^2 = x'^2 + y'^2 + z'^2$$
(14.48a)

This means that Eq. (14.48a) leaves the length of the vector unchanged. Such transformations are called *orthogonal transformations*.

When the new coordinates can be expressed as a linear combination of the old coordinates, then the coordinate transformation is said to be *linear*. Thus, we have

$$x_1' = a_{11}x_1 + a_{12}x_2 + a_{13}x_3$$
$$x_2' = a_{21}x_1 + a_{22}x_2 + a_{23}x_3$$
$$x_3' = a_{31}x_1 + a_{32}x_2 + a_{33}x_3$$

i.e.
$$x_i' = \sum_j a_{ij}x_j \tag{14.49}$$

is a linear transformation. One can express it in the matrix form as

$$\begin{bmatrix} x_1' \\ x_2' \\ x_3' \end{bmatrix} = \begin{bmatrix} a_{11} & a_{12} & a_{13} \\ a_{21} & a_{22} & a_{23} \\ a_{31} & a_{32} & a_{33} \end{bmatrix} \begin{bmatrix} x_1 \\ x_2 \\ x_3 \end{bmatrix} \tag{14.50}$$

For Eq. (14.50) to be an orthogonal transformation, we have

$$\sum_i (x_i')^2 = \sum_k x_k^2$$

i.e.
$$\sum_{j,k,l} a_{ij}a_{ik}x_j x_k = \sum_k x_k^2 \tag{14.51}$$

Equation (14.51) is possible if

$$\sum_i a_{ij}a_{ik} = \begin{cases} 0 \text{ if } j \neq k \\ 1 \text{ if } j = k \end{cases}$$

In other words, we can express it in terms of delta function as

$$\sum_i a_{ij}a_{ik} = \delta_{jk} \tag{14.52}$$

We can easily see that for the transformation to be orthogonal, the elements of the transformation matrix

$$A = \begin{bmatrix} a_{11} & a_{12} & a_{13} \\ a_{21} & a_{22} & a_{23} \\ a_{31} & a_{32} & a_{33} \end{bmatrix}$$

must satisfy the condition Eq. (14.52).

We have seen that the quantity $x^2 + y^2 + z^2 - c^2 t^2$ is Lorentz invariant, i.e. it has the same value in frames S and S'. Now, we find that the Lorentz transformation, like the rotational transformation, is a linear relation between different coordinate systems, but it mixes up time with the space coordinates. One can specify an event occuring at a particular time and place in a frame S by four coordinates (x, y, z, t).

As compared to three-dimensional scalar product r^2, the S^2 $(=x^2 + y^2 + z^2 - c^2 t^2)$ has the dimension of length. Obviously, one can interpret S^2 as the square of the *length* of the *four vector S*, i.e. one can visualize S^2 as a scalar product of two vectors. The four "coordinates" are $x_\mu = x_1, x_2, x_3,$ and x_4, where

$$x_1 = x, \quad x_2 = y, \quad x_3 = z \text{ and } x_4 = ict \tag{14.53}$$

Obviously, a new time variable namely *icf* in place of time has been choosen. One can write the four-vector scalar product as

$$S^2 = x_\mu x_\mu = x_1^2 + x_2^2 + x_3^2 + x_4^2 = x^2 + y^2 + z^2 - c^2 t^2 \tag{14.54}$$

This non-Euclidean four-dimensional space is known as *Minkowaski space*. The four-vector of an event is generally represented by x_μ. The line element is called a *tensor of rank zero* and is expressed as

$$dS^2 = g_{ij} \, dx^i \, dx^j \tag{14.55}$$

where dx^i and dx^j are contravariant components of basic displacement vector dx. In writing Eq. (14.54), we have used the summation convention. We must remember that whenever a Greek suffix (e.g. μ) appears twice in a single term, the term is assumed to be summed from 1 to 4.

Now, one can express the Lorentz transformation equations in new notations as

$$\left. \begin{aligned} x_1' &= \frac{x_1 + i\beta x_4}{\sqrt{1-\beta^2}} \\ x_2' &= x_2 \\ x_3' &= x_3 \end{aligned} \right\} \tag{14.56}$$

and

$$x_4' = \frac{x_4 - i\beta x_1}{\sqrt{1-\beta^2}}$$

One finds the inverse transformation equations as

$$\left. \begin{aligned} x_1 &= \frac{x_1' - i\beta x_4'}{\sqrt{1-\beta^2}} \\ x_2 &= x_2' \\ x_3 &= x_3' \end{aligned} \right\} \tag{14.57}$$

and

$$x_4 = \frac{x_4' + i\beta x_1'}{\sqrt{1-\beta^2}}$$

One can write the matrix of Lorentz transformation as

$$A = \begin{bmatrix} \dfrac{1}{\sqrt{1-\beta^2}} & 0 & 0 & \dfrac{i\beta}{\sqrt{1-\beta^2}} \\ 0 & 1 & 0 & 0 \\ 0 & 0 & 1 & 0 \\ -\dfrac{i\beta}{\sqrt{1-\beta^2}} & 0 & 0 & \dfrac{1}{\sqrt{1-\beta^2}} \end{bmatrix} \tag{14.58}$$

One can easily see that Eq. (14.58) satisfies Eq. (14.52). This shows that *Lorentz transformation is an orthogonal transformation.*

The momentum and energy of the particles in an inertial frame form a four-vector set and one can express their scalar product as

$$p^2 = p_\mu p_\mu = p_x^2 + p_y^2 + p_z^2 - \frac{E^2}{c^2} = -m_0^2 c^2 \qquad (14.59)$$

One can see that this is also *Lorentz invariant.*

14.11 COVARIANT FORMULATION OF ELECTRODYNAMICS

In order to investigate the relativistic behaviour of Maxwell's equations, it is necessary to write the equations in four-dimensional form and to evaluate the transformation properties of spatial and temporal differential operators. Applying the rules of partial differentiation, one can write

$$\frac{\partial}{\partial x_1'} = \frac{\partial x_1}{\partial x_1'}\frac{\partial}{\partial x_1} + \frac{\partial x_2}{\partial x_1'}\frac{\partial}{\partial x_2} + \frac{\partial x_3}{\partial x_1'}\frac{\partial}{\partial x_3} + \frac{\partial x_4}{\partial x_1'}\frac{\partial}{\partial x_4}$$

$$= \frac{1}{\sqrt{1-\beta^2}}\left(\frac{\partial}{\partial x_1} + i\beta\frac{\partial}{\partial x_4}\right) \qquad (14.60)$$

where we have used Eq. (14.57). This transformation is similar to that shown for x_1' in Eq. (14.56). Similar transformation equations can be obtained for $\frac{\partial}{\partial x_2'}, \frac{\partial}{\partial x_3'}$ and $\frac{\partial}{\partial x_4'}$. This shows that they all transform according to the Lorentz transformation and the differential operators also constitute a set of four-vectors and is represented by $\frac{\partial}{\partial x_\mu}$. One can write the scalar product of this four-vector differential operator as

$$\frac{\partial}{\partial x_\mu}\frac{\partial}{\partial x_\mu} = \frac{\partial^2}{\partial x_1^2} + \frac{\partial^2}{\partial x_2^2} + \frac{\partial^2}{\partial x_3^2} + \frac{\partial^2}{\partial x_4^2}$$

$$= \frac{\partial^2}{\partial x^2} + \frac{\partial^2}{\partial y^2} + \frac{\partial^2}{\partial z^2} - \frac{1}{c^2}\frac{\partial^2}{\partial t^2} = \Box^2 \qquad (14.61)$$

where $\Box^2$ is known as *d'Alembertian operator.* This operator is also a Lorentz invariant and widely used in wave equation, e.g.

$$\nabla^2\psi - \frac{1}{c^2}\frac{\partial^2\psi}{\partial t^2} = \Box^2\psi = 0 \qquad (14.62)$$

Maxwell's Equations in Four-vector Form

We have already seen that the electric and magnetic fields transform into one another in different moving inertial frames. We have achieved this with the help of Lorentz transformation equations and the invariance of total charge in the moving inertial frame. The mathematical statement of charge conservation is contained in the equation of continuity

$$\nabla \cdot J + \frac{\partial\rho}{\partial t} = 0 \qquad (14.63)$$

This equation must, therefore, have the same form in all inertial frames. Let us see how this can be expressed in an invariant form. We must remember that in the theory of relativity, the current density and charge density are not distinct entities since the charge distribution that is static in one reference frame will appear to be a current distribution in a moving reference frame. Let us introduce a four-vector $J_\mu = (J, ic\rho)$, where the current density J is the space-like part of J_μ and the charge density is used to denote the fourth component $J_4 = ic\rho$, the four components of J_μ being J_1, J_2, J_3 and $ic\rho$. In terms of four-vector, the continuity equation can be expressed as

$$\nabla \cdot J + \frac{\partial \rho}{\partial t} = \frac{\partial J_1}{\partial x_1} + \frac{\partial J_2}{\partial x_2} + \frac{\partial J_3}{\partial x_3} + \frac{\partial (ic\rho)}{\partial (ict)} = 0 \qquad (14.64)$$

or

$$\frac{\partial J_\mu}{\partial x_\mu} = 0 \qquad (14.65)$$

Obviously, the current continuity equation expressed in four-vector form is also *Lorentz invariant*.

Now, representing the four dimensional divergance operator by 'div', one may express the equation of continuity in the four-dimension as

$$\text{div } J = 0 \qquad (14.66)$$

We have seen that in free space the potential A and ϕ can be written in the form

$$\nabla^2 A - \frac{1}{c^2} \frac{\partial A}{\partial t^2} = -\mu_0 J \qquad (14.67)$$

$$\nabla^2 \phi - \frac{1}{c^2} \frac{\partial^2 \phi}{\partial t^2} = -\frac{\rho}{\varepsilon_0} \qquad (14.68)$$

We can write these equations in terms of four-dimensional D' Alem-operator as

$$\Box A = \frac{\partial^2 A}{dx_\mu^2} = -\mu_0 J \qquad (14.69)$$

and

$$\Box \phi = \frac{\partial^2 \phi}{dx_\mu^2} = -\frac{\rho}{\varepsilon_0} \qquad (14.70)$$

The operator $\Box$ operates on the scalar and vector potentials. We find that the vector and scalar potential form a set of another four vector $\left(A, \frac{i\phi}{c} \right)$. The four vector potential A_v can be defined as

$$A_v = (A_1, A_2, A_3, A_4) \equiv \left(A_1, A_2, A_3 \frac{i\phi}{c} \right) = \left(A, \frac{i\phi}{c} \right) \qquad (14.71)$$

where v takes values from 1 to 4 to denote four-vectors. The Lorentz condition is also invariant and can be written in the four-dimensional form as

$$\text{div } A_v + \frac{1}{c^2} \frac{\partial \phi}{\partial t} = 0$$

Now

$$\text{div } A_v = \frac{\partial A_1}{\partial x_1} + \frac{\partial A_2}{\partial x_2} + \frac{\partial A_3}{\partial x_3} + \frac{\partial A_4}{\partial x_4}$$

$$= \operatorname{div} A + \frac{\partial(i\phi/c)}{\partial(ict)} = \operatorname{div} A + \frac{1}{c^2}\frac{\partial\phi}{\partial t} = 0 \qquad (14.72)$$

Equations (14.69) and (14.70) can now be expressed in a single equation

$$\frac{\partial^2 A}{\partial x_\mu^2} = -\mu_0 J \qquad (14.73)$$

where the fourth component of Eq. (14.73) is found by making substitution $A_4 \to \dfrac{i\phi}{c}$ and $J_4 \to i\rho c$. From these equations, one finds that the Lorentz condition used for making the potentials unique is also *Lorentz invariant*. One can also write in terms of four vectors

$$\nabla \cdot A = 0, \quad \text{i.e.} \quad \frac{\partial A_\mu}{\partial x_\mu} = 0 \qquad (14.74)$$

We have read that in free space the potential A and ϕ satisfy the equations

$$\nabla^2 A - \frac{1}{c^2}\frac{\partial^2 A}{\partial t^2} = -\mu_0 J \qquad (14.74a)$$

and

$$\nabla^2 \phi - \frac{1}{c^2}\frac{\partial^2 \phi}{\partial t^2} = -\frac{\rho}{\varepsilon_0} \qquad (14.74b)$$

Using J and A defined earlier, the two Eqs. (14.74a) and (14.74b) can be written together in the form of a single four-vector equation

$$\Box^2 \phi = \mu_0 J \qquad (14.74c)$$

The space portion of Eq. (14.74c) gives Eq. (14.74a) and the fourth component gives Eq. (14.74b).

We find that Maxwell's equations are consistent with the basic axioms of special theory of relativity. We have seen that the Maxwell's field equations, the continuity equation and the conditions of uniqueness of potentials and fields are all *Lorentz invariant*. As the electric and magnetic fields are *Lorentz invariant*, the Lorentz force acting on the charged particles is also *Lorentz invariant*. This shows that observers in different reference frames will have a consistent description of the motion of charged particles in electromagnetic fields.

14.12 ELECTROMAGNETIC FIELD TENSOR

Now, we examine whether the Lorentz force is invariant. For this purpose, one has to investigate the transformations of E and B. We know that E and B are not four vectors, but we can first express them in terms of potentials and then we can have transformations. Let us see how the six components E_1, E_2, E_3, B_1, B_2, and B_3 can be used to define a tensor of rank two, called the electromagnetic field tensor. One can easily see that in n-dimensional space, a tensor of the mth rank is a set of n quantities and this transforms as

$$T'_{abcd} = \sum_{i,j,k,l,\ldots} \lambda_{ai}\lambda_{bj}\lambda_{ck}\lambda_{dl}\ldots T_{ijkl}$$

Since we are dealing with a four-dimensional space and hence we need to consider only the tensors of rank two which transform as

$$T'_{ij} = \sum_{k,l} \lambda_{lk}\lambda_{jl}T_{kl} \tag{14.75}$$

We have

$$E_x = -\frac{\partial \phi}{\partial x} - \frac{\partial A_x}{\partial t}, \text{ etc.} \tag{14.76}$$

Since $A_\mu = \left(A_1, A_2, A_3, A_4 = \dfrac{i\phi}{c} \right)$ and $x_4 = ict$, one can write Eq. (14.76) as

$$E_1 = ic\left(\frac{\partial A_4}{\partial x_1} - \frac{\partial A_1}{\partial x_4} \right)$$

$$E_2 = ic\left(\frac{\partial A_4}{\partial x_2} - \frac{\partial A_2}{\partial x_4} \right) \tag{14.77}$$

and

$$E_3 = ic\left(\frac{\partial A_4}{\partial x_3} - \frac{\partial A_3}{\partial x_4} \right)$$

Similarly, one obtains

$$B_1 = \frac{\partial A_3}{\partial x_2} - \frac{\partial A_2}{\partial x_3}$$

$$B_2 = \frac{\partial A_1}{\partial x_3} - = \frac{\partial A_1}{\partial x_3} - \frac{\partial A_3}{\partial x_1} \tag{14.78}$$

and

$$B_3 = \frac{\partial A_2}{\partial x_1} - \frac{\partial A_1}{\partial x_2}$$

Let us define a set of quantities

$$F_{\mu\nu} = \frac{\partial A_\nu}{\partial x_\mu} - \frac{\partial A_\mu}{\partial x_\nu} \tag{14.79}$$

Thus, one obtains

$$E_1 = icF_{14}, \quad E_2 = icF_{24} \quad \text{and } E_3 = icF_{34} \tag{14.80}$$

and

$$B_1 = F_{23}, \quad B_2 = F_{31} \quad \text{and } B_3 = F_{12} \tag{14.81}$$

Since $F_{\mu\nu} = -F_{\nu\mu}$ i.e. tensor is antisymmetric and $F_{\mu\mu} = 0$, one can express the tensor $\{F\}$ in terms of the elements $F_{\mu\nu}$, as

$$\{F\} = \begin{Bmatrix} F_{11} & F_{12} & F_{13} & F_{14} \\ F_{21} & F_{22} & F_{23} & F_{24} \\ F_{31} & F_{32} & F_{33} & F_{34} \\ F_{41} & F_{42} & F_{43} & F_{44} \end{Bmatrix} \tag{14.82}$$

$$= \begin{Bmatrix} 0 & B_3 & -B_2 & -iE_1/c \\ -B_3 & 0 & B_1 & -iE_2/c \\ B_2 & -B_1 & 0 & -iE_3/c \\ iE_1/c & iE_2/c & iE_3/c & 0 \end{Bmatrix}$$

This tensor is usually called as the *electromagnetic* antisymmetric tensor of rank two. Now, we shall see how this tensor can be used to express Maxwell's equations. Let us consider the equation

$$\frac{\partial F_{\lambda\mu}}{\partial x_\nu} + \frac{\partial F_{\mu\nu}}{\partial x_\lambda} + \frac{\partial F_{\nu\lambda}}{\partial x_\mu} = 0 \qquad (14.83)$$

Now assigning λ, μ, ν values as 1, 2, 3 in this or in any other order, one finds that Eq. (14.83) reduces to

$$\frac{\partial F_{12}}{\partial x_3} + \frac{\partial F_{23}}{\partial x_1} + \frac{\partial F_{31}}{\partial x_2} = 0 \qquad (14.83a)$$

which means

$$\frac{\partial B_3}{\partial x_3} + \frac{\partial B_1}{\partial x_1} + \frac{\partial B_2}{\partial x_2} = 0$$

or $\qquad\qquad\qquad\qquad \nabla \cdot \boldsymbol{B} = 0$

Obviously, we have obtained one of the Maxwell's equations.

Let us now assign to one of the indices the value 4, say $\lambda = 2$, $\mu = 3$ and $\nu = 4$, one obtains from Eq. (14.83)

$$\frac{\partial F_{23}}{\partial x_4} + \frac{\partial F_{34}}{\partial x_2} + \frac{\partial F_{42}}{\partial x_3} = 0$$

or $\qquad \dfrac{\partial B_1}{\partial (ict)} + \dfrac{1}{ic}\dfrac{\partial E_3}{\partial x_2} - \dfrac{1}{ic}\dfrac{\partial E_2}{\partial x_3} = 0$

or $\qquad \dfrac{\partial B_1}{\partial t} + \dfrac{\partial E_3}{\partial x_2} - \dfrac{\partial E_2}{\partial x_3} = 0$

or $\qquad (\nabla \times E)_x + \dfrac{\partial B_x}{\partial t} = 0 \quad (x = 1) \qquad (14.84)$

Obviously, Eq. (14.84) is the x-component of $\nabla \times E + \dfrac{\partial \boldsymbol{B}}{\partial t} = 0$. This shows that Eq. (14.83) also represents the other homogeneous Maxwell's equations. One can easily obtain the other two inhomogeneous Maxwell's equations from Eq. (14.83) as follow. We have

$$\frac{\partial F_{\mu\nu}}{\partial x_\nu} = \mu_0 J_\mu \qquad (14.85)$$

Let $\mu = 1$, Eq. (14.85) takes the form

$$\frac{\partial F_{11}}{\partial x_1} + \frac{\partial F_{12}}{\partial x_2} + \frac{\partial F_{13}}{\partial x_3} + \frac{\partial F_{14}}{\partial x_4} = \mu_0 J_1$$

or

$$0 + \frac{\partial B_3}{\partial x_2} - \frac{\partial B_2}{\partial x_3} + \frac{1}{ic}\frac{\partial E_1}{\partial (ict)} = \mu_0 J_1$$

or

$$(\nabla \times B)_x - \frac{1}{c^2}\frac{\partial E_x}{dt} = \mu_0 J_x \quad (x = 1)$$

which is the x-component of

$$\nabla \times B - \frac{1}{c^2}\frac{\partial E}{\partial t} = \mu_0 J \tag{14.86}$$

Now taking $\mu = 4$, one can express Eq. (14.85) as

$$\frac{\partial F_{41}}{\partial x_1} + \frac{\partial F_{42}}{\partial x_2} + \frac{\partial F_{43}}{\partial x_3} + \frac{\partial F_{44}}{\partial x_4} = \mu_0 J_4$$

or

$$-\frac{1}{ic}\frac{\partial E_1}{\partial x_1} - \frac{1}{ic}\frac{\partial E_2}{\partial x_2} - \frac{1}{ic}\frac{\partial E_3}{\partial x_3} + 0 = ic\mu_0 \rho$$

or

$$\frac{\partial E_1}{\partial x_1} + \frac{\partial E_2}{\partial x_2} + \frac{\partial E_3}{\partial x_3} = \nabla \cdot E = \frac{\rho}{\varepsilon_0}$$

Obviously, the four Maxwell's electromagnetic equations are represented by only two tensor equations involving operations on the components of the field tensor.

14.13 LORENTZ TRANSFORMATION OF ELECTRIC AND MAGNETIC FIELDS

Let us now obtain the transformation of electric and magnetic fields. For this, we first evaluate E_x' in the frame S' in terms of E and B in frame S. We have

$$E_x' = -\frac{\partial \phi'}{dx'} - \frac{\partial A_x'}{\partial t'}$$

Since $A_\mu = \left(A_1, A_2, A_3, \dfrac{i\phi}{c}\right)$ and $\dfrac{\partial}{\partial x_\mu} = \left(\dfrac{\partial}{\partial x_1}, \dfrac{\partial}{\partial x_2}, \dfrac{\partial}{\partial x_3}, \dfrac{\partial}{\partial (ict)}\right)$, we have

$$E_x' = icF_{14}' = ic\left\{\frac{\partial A_4'}{\partial x_1'} - \frac{\partial A_1'}{\partial x_4'}\right\}$$

We know that $A_\mu, \dfrac{\partial}{\partial x_\mu}$ are Lorentz covariant, we have

$$A_4' = \frac{1}{\sqrt{1-\beta^2}}(A_4 - i\beta A_1) = \gamma(A_4 - i\beta A_1) \quad \left(\text{Here}\, \gamma = \frac{1}{\sqrt{1-\beta^2}}\right)$$

$$\frac{\partial}{\partial x_4'} = \gamma\left(\frac{\partial}{\partial x_4} - i\beta\frac{\partial}{\partial x_1}\right)$$

$$A_1' = \gamma(A_1 + i\beta A_4); \quad \frac{\partial}{\partial x_1'} = \gamma\left(\frac{\partial}{\partial x_1} + i\beta\frac{\partial}{\partial x_4}\right)$$

Therefore, one obtains

$$E_x' = ic\left[\gamma\left(\frac{\partial}{\partial x_1} + i\beta\frac{\partial}{\partial x_4}\right)\{r(A_4 - i\beta A_1)\} - \gamma\left(\frac{\partial}{\partial x_4} - i\beta\frac{\partial}{\partial x_1}\right)\{\gamma(A_1 + i\beta A_4)\}\right]$$

$$= ic\gamma^2\left[\frac{\partial A_4}{\partial x_1} + i\beta\frac{\partial A_4}{\partial x_4} - i\beta\frac{\partial A_1}{\partial x_1} + \beta^2\frac{\partial A_1}{\partial x_4} - \frac{\partial A_1}{\partial x_4} + i\beta\frac{\partial A_1}{\partial x_1} - i\beta\frac{\partial A_4}{\partial x_4} - \beta^2\frac{\partial A_4}{\partial x_4}\right]$$

$$= ic\gamma^2(1-\beta^2)\left(\frac{\partial A_4}{\partial x_1} - \frac{\partial A_1}{\partial x_4}\right) = ic\left(\frac{\partial A_4}{\partial x_1} - \frac{\partial A_1}{\partial x_4}\right) = E_x$$

Similarly, one can obtain the other components. Finally, one obtains

$$\left.\begin{array}{ll} E_x' = E_x; & B_x' = B_x \\[2mm] E_y' = \gamma[E_y - c\beta B_z]; & B_y' = \gamma\left[B_y + \dfrac{\beta}{c}E_z\right] \\[3mm] E_z' = \gamma[E_z + \beta c B_y]; & B_z' = \gamma\left[B_z - \dfrac{\beta}{c}E_y\right] \end{array}\right\} \qquad (14.87)$$

and

We can see that the components of E and B in the direction of motion are unaffected, but in the transverse direction are modified.

14.14 ELECTROMAGNETIC FIELD DUE TO A POINT CHARGE IN UNIFORM MOTION

Let us consider that a charge is stationary in the frame S. Obviously, $B = 0$ in this frame and E is given by

$$E = \frac{q}{4\pi\varepsilon_0}\frac{r}{|r|^3} \qquad (14.88)$$

Now, we are interested in finding the fields in frame S' which is moving with relative velocity v relative to frame S along positive direc-tion of common x-axis. Let us assume that the charge to be at rest at the origin of frame S and the field is calculated at the instant $t = 0$, i.e. at $t = 0$ two origins of frames S and S' coincide. From Eq. (14.87), we have

$$E_1' = E_1, \quad E_2' = \gamma E_2 \text{ and } E_3' = \gamma E_3 \qquad (14.89)$$
$$(\because \quad B_2 = B_3 = 0 \text{ in the rest frame } S)$$

We have at $t = 0$,

$$r_1 = \gamma x_1'; \quad x_2 = x_2' \text{ and } x_3 = x_3'$$

If r is the distance of the observation point P from the origin, then

$$r = \sqrt{x_\mu x_\mu} = \sqrt{\gamma^2 x_1'^2 + x_2'^2 + x_3'^2} \qquad (14.90)$$

Now, one obtains the components of the electric field in frame S' as

$$E_1' = E_1 = \frac{q}{4\pi\varepsilon_0}\frac{x_1}{r^3} = \frac{q}{4\pi\varepsilon_0}\frac{\gamma x_1'}{(\gamma^2 x_1'^2 + x_2'^2 + x_3'^2)^{3/2}}$$

$$E_2' = \gamma E_2 = \frac{q}{4\pi\varepsilon_0} \frac{\gamma x_2}{r^3} = \frac{q}{4\pi\varepsilon_0} \frac{\gamma x_2}{(\gamma^2 x_1'^2 + x_2'^2 + x_3'^2)^{3/2}}$$

$$E_3' = \gamma E_3 = \frac{q}{4\pi\varepsilon_0} \frac{\gamma x_3}{r^3} = \frac{q}{4\pi\varepsilon_0} \frac{\gamma x_3'}{(\gamma^2 x_1'^2 + x_2'^2 + x_3'^2)^{3/2}}$$

Thus, we have

$$E' = \frac{q\gamma}{4\pi\varepsilon_0} \frac{r'}{(\gamma^2 x_1'^2 + x_2'^2 + x_3'^2)} \tag{14.91}$$

We have $x_1' = r'\cos\theta$, where θ is the angle that r' makes with the x-axis and

$$x_1'^2 + x_2'^2 + x_3'^2 = r'^2$$

This leads to

$$x_2'^2 + x_3'^2 = r'^2 \sin^2\theta$$

Now, we have

$$\gamma^2 x_1'^2 + x_2'^2 + x_3'^2 = \gamma^2 r'^2 \cos^2\theta + r'^2 \sin^2\theta$$

$$= r'^2 \gamma^2 \left(\cos^2\theta + \frac{\sin^2\theta}{\gamma} \right)$$

$$= \gamma^2 r'^2 (1 - \beta^2 \sin^2\theta) \tag{14.92}$$

or

$$E_1 = \frac{q}{4\pi\varepsilon_0} \frac{r'(1-\beta^2)}{r'^3 (1-\beta^2 \sin^2\theta)^{3/2}} \tag{14.93}$$

Equation (14.93) once again illustrates that Maxwell's field equations are relativistically correct.

From Eq. (14.93), one can see that the electric field is no longer spher-ically symmetric. Along the line of motion ($\theta = 0$), one obtains the field as

$$E = \frac{q}{4\pi\varepsilon_0} \frac{r}{r^3} (1 - \beta^2) \tag{14.94}$$

This shows that coulomb field is reduced by a factor $(1 - \beta^2)$. For the field perpendicular to the line of motion, i.e. $\theta = \frac{\pi}{2}$, we have

$$E = \frac{q}{4\pi\varepsilon_0} \frac{r}{r^3} \frac{1}{\sqrt{1-\beta^2}} \tag{14.95}$$

This shows that the electric field perpendicular to the line of motion is enhanced by a factor $\frac{1}{\sqrt{1-\beta^2}}$.

14.15 LAGRANGIAN FORMULATION OF THE MOTION OF A CHARGED PARTICLE IN AN ELECTROMAGNETIC FIELD

One can apply the Lagrangian method of classical mechanics to the motion of a charged particle in an EM field if a suitable Lagrangian function is devised. First we shall study for a non-relativistic case and then develop for a more general relativistic approach.

Lagrangian in a static field can be expressed as the difference between the KE and PE,

$$L = T - U = \frac{1}{2}mv^2 - q\phi$$

When magnetic field is present, we have to suitably modify the Lagrangian. Magnetic field depends upon the velocity of the moving charges and hence L is a scalar function. One will have to add a term in order to modify L as a scalar product of v and vector potential A and this describes the field. Now L can be expressed as

$$L = \frac{1}{2}mv^2 + qv \cdot A - q\phi \tag{14.96}$$

The equation of motion is obtained from Eq. (14.96) as

$$\frac{\partial L}{\partial v_i} = mv_i + qA_i \ (i = 1, 2, 3) \tag{14.97}$$

In vector form,

$$\sum_i \hat{e}_i \frac{\partial L}{\partial v_i} = p + qA \tag{14.98}$$

where $p = mv$ is the linear momentum and $p + qv$ is called the generalized momentum. Further, we also have

$$\frac{\partial L}{\partial x_i} = q\frac{\partial}{\partial x_i}(v \cdot A) - q\frac{\partial \phi}{\partial x_i} \tag{14.99}$$

In vector form,

$$\sum_i \hat{e}_i \frac{\partial L}{\partial x_i} = q\sum_i \hat{e}_i \frac{\partial}{\partial x_i}(v \cdot A) - q\sum_i \hat{e}_i \frac{\partial \phi}{\partial x_i}$$

$$= q\operatorname{grad}(v \cdot A) - q\operatorname{grad}\phi \tag{14.100}$$

Using the following identity

$$\nabla(v \cdot A) = (v \cdot \nabla)A + (A \cdot \nabla)v + A \times \nabla \times v + v \times \nabla \times A$$

one obtains for constant v

$$\sum_i \hat{e}_i \frac{dL}{dx_i} = q(v \cdot \nabla)A + qv \times \nabla \times A - q\nabla\phi \tag{14.101}$$

The Lagrangian equation of motion is

$$\frac{d}{dt}\frac{\partial L}{\partial V_i} = \frac{\partial L}{\partial x_i}$$

One obtains from Eqs. (14.98) and (14.101)

$$\frac{d}{dt}(p + qA) = \frac{dp}{dt} + q\frac{dA}{dt}$$

$$= q(v \cdot \nabla)A + qA \times \nabla \times A - q\nabla\phi \tag{14.102}$$

Now, the time derivative of A is

$$\frac{dA}{dt} = \frac{\partial A}{\partial t} + \sum_i \frac{\partial A}{\partial x_i}\frac{dx_i}{dt} = \frac{\partial A}{\partial t} + \left(\sum_i v_i \cdot \frac{\partial}{\partial x_i}\right)A$$

$$= \frac{\partial A}{\partial t} + (v \cdot \nabla)A$$

Equation (14.102) takes the form

$$\frac{dp}{dt} = -q\frac{dA}{dt} + qv \times \nabla \times A - q\nabla\phi$$

i.e.

$$F = q\left(-\nabla\phi - \frac{\partial A}{\partial t}\right) + qv \times \nabla \times A$$

$$= qE + qv \times \nabla \times A = q(E + v \times B) \tag{14.103}$$

Equation (14.103) is Lorentz force equation.

One can assume the Lagrangian for a relativistic particle in an electromagnetic field as

$$L = -m_0 c^2 \sqrt{1 - \beta^2} + qv \cdot A - q\phi \tag{14.104}$$

where m_0 is the rest mass of the particle in the frame of reference in which particle is at rest.

The position vector of a particle in *Minkowski space* is

$$\varkappa = (x_1, x_2, x_3, ict) \tag{14.105}$$

The differential of $\varkappa$ is also a four-vector, i.e.

$$d\varkappa = (dx_1, dx_2, dx_3, icdt) \tag{14.106}$$

We know that four-dimensional element of length is invariant under Lorentz transformation, i.e.

$$dS = \sqrt{dx_\mu dx_\mu} = \sqrt{dx_1^2 + dx_2^2 + dx_3^2 - c^2 dt^2} \tag{14.107}$$

Let us introduce a quantity $d\tau$ such that

$$d\tau = \sqrt{dt^2 - \frac{1}{c^2}(dx_1^2 + dx_2^2 + dx_3^2)}$$

$$= \sqrt{-\frac{1}{c^2}(dx_1^2 + d_2^2 + dx_3^2 - c^2 dt^2)} = \frac{i}{c}\sqrt{dx_\mu dx_\mu} \tag{14.108}$$

One can also write $d\tau$ as

$$d\tau = dt\sqrt{1 - \frac{1}{c^2}\left\{\left(\frac{dx_1}{dt}\right)^2 + \left(\frac{dx_2}{dt}\right)^2 + \left(\frac{dx_3}{dt}\right)^2\right\}}$$

$$= dt\sqrt{1 - \frac{v^2}{c^2}} = dt\sqrt{1 - \beta^2} = \frac{dt}{\gamma} \tag{14.109}$$

This quantity is called the *element of proper time* in Minkowski space and obviously invariant. The vector

$$u = \frac{d\varkappa}{d\tau} = \left(\frac{dx_1}{d\tau}, \frac{dx_2}{d\tau}, \frac{dx_3}{d\tau}, ic\frac{dt}{d\tau}\right)$$

$$= \left(\gamma \frac{dx_1}{dt}, \gamma \frac{dx_2}{dt}, \gamma \frac{dx_3}{dt}, ic\gamma \right)$$

$$= (\gamma v_1, \gamma v_2, \gamma v_3, ic\gamma) \tag{14.110}$$

is a four-vector velocity.

Now, $\mathfrak{R} = \left(A, \dfrac{i\phi}{c} \right)$, the Lagrangian equation (14.104) may be expressed as

$$L = -\frac{m_0 c^2}{\gamma} + \frac{q u \cdot \mathcal{A}}{\gamma} = \frac{1}{\gamma}(-m_0 c^2 + q u_\mu \mathcal{A}_\mu) \tag{14.111}$$

One can obtain the equation of motion by using the Lagrangian equation [Eq. (14.111)] in the principle of least action.

$$\delta \int_{t_1}^{t_2} L \, dt = 0$$

or

$$\delta \int_{t_1}^{t_2} \gamma L \, d\tau = 0 \tag{14.112}$$

i.e.

$$\delta \int_{t_1}^{t_2} [-m_0 c^2 + q u_\mu \mathfrak{R}_\mu] \delta\tau = 0$$

or

$$\delta \int_{t_1}^{t_2} [-m_0 c^2 d\tau + q \mathfrak{R}_\mu dx_\mu] = 0 \quad \left(\because u_\mu = \frac{dx_\mu}{dt} \right) \tag{14.113}$$

Now, performing the variation, one obtains

$$\int_{t_1}^{t_2} [-m_0 c^2 \delta(d\tau) + q \mathcal{A}_\mu \delta(dx_\mu) + q \delta \mathcal{A}_\nu \, dx_\nu] = 0$$

Now,

$$\delta(d\tau) = \frac{\partial \tau}{\partial x_\mu} \delta(dx_\mu) = \frac{\partial \tau}{\partial x_\mu} d(\delta x_\mu) \tag{14.114}$$

But

$$d\tau = \frac{i}{c} \sqrt{dx_\mu dx_\mu}$$

$\therefore$

$$\frac{d\tau}{dx_\mu} = \frac{i}{c} \frac{dx_\mu}{\sqrt{dx_\mu dx_\mu}} = -\frac{1}{c^2} \frac{dx_\mu}{d\tau}$$

$\therefore$

$$\delta(d\tau) = -\frac{1}{c^2} \frac{dx_\mu}{d\tau} d(\delta x_\mu) = -\frac{1}{c^2} u_\mu d(\delta x_\mu) \tag{14.115}$$

Further,

$$\delta \mathcal{A}_\nu = \frac{d\mathcal{A}_\mu}{dx_\mu} \cdot \delta x_\mu \tag{14.116}$$

Equation (14.114) may be rewritten as

$$\int_{t_1}^{t_2} \left[\{m_0 u_\mu + q \mathcal{A}_\mu\} d(\delta x_\mu) + q \frac{\partial \mathcal{A}_\nu}{\partial x_\mu} \delta x_\mu dx_\nu \right] = 0 \tag{14.117}$$

Integration of first term yields

$$\int_{t_1}^{t_2} \{m_0 u_\mu + q \mathcal{A}_\mu\}(dx_\mu) = \left[\{m_0 u_\mu + q \mathcal{A}_\mu\} \delta x_\mu \right]_{t_1}^{t_2}$$

$$-\int_{t_1}^{t_2} \frac{\partial}{\partial x_v}[m_0 u_\mu + q \mathcal{A}_\mu] \partial x_\mu dx_v$$

$$= -\int_{t_1}^{t_2} \frac{\partial}{\partial x_v}\{m_0 u_\mu + q \mathcal{A}_\mu\} \delta x_\mu dx_v$$

The first term vanishes due to variation of coordinates and should vanish at the end points. Therefore, Eq. (14.117) takes the form

$$\int_{t_1}^{t_2} \left[-\frac{\partial}{\partial x_v}\{m_0 u_\mu + q \mathcal{A}_\mu\} + q \frac{\partial \mathcal{A}_v}{\partial x_\mu} dx_v \right] = 0$$

i.e.

$$\int_{t_1}^{t_2} \left[m_0 \frac{\partial u_\mu}{\partial x_v} + \frac{q \mathcal{A}_\mu}{\partial x_v} - q \frac{\partial \mathcal{A}_v}{dx_\mu} \right] \delta x_\mu dx_v = 0$$

or

$$\int_{t_1}^{t_2} \left[m_0 \frac{\partial u_\mu}{\partial x_v} - q \left(\frac{\partial \mathcal{A}_v}{\partial x_\mu} - \frac{\partial \mathcal{A}_\mu}{\partial x_v} \right) \right] \delta x_\mu dx_v = 0$$

Making use of Eq. (14.79), one obtains

$$\int_{t_1}^{t_2} \left[m_0 \frac{\partial u_\mu}{\partial x_v} - q F_{\mu v} \right] \delta x_\mu dx_v = 0 \tag{14.118}$$

Now,

$$u_v = \frac{dx_v}{d\tau} \quad \text{and} \quad \frac{\partial u_\mu}{\partial x_v} = d u_\mu = \frac{d u_\mu}{d\tau} d\tau$$

Thus, Eq. (14.118) takes the form

$$\int_{t_1}^{t_2} \left[m_0 \frac{d u_\mu}{d\tau} - q F_{\mu v} u_v \right] \delta x_\mu d\tau = 0 \tag{14.119}$$

As δx_μ is an arbitrary variation, one finds that

$$m_0 \frac{d u_\mu}{d\tau} - q F_{\mu v} \mu_v = 0 \tag{14.120}$$

Equation (14.120) is the desired *covariant equation of motion*. One can obtain the space portion of this equation by alloting μ the values 1, 2, 3.

Thus,

$$m_0 \frac{d u_1}{d\tau} = q F_{1v} u_v = q F_{11} u_1 + q F_{12} u_{12} + q F_{13} u_3 + q F_{14} u_4$$

$$= \gamma q v_2 B_3 - \gamma q v_3 B_2 + q \gamma E_1$$

In obtaining the above result, we have made use of Eqs. (14.80), (14.81) and (14.110)

i.e.

$$\frac{d}{dt}(m_0 \gamma u_1) = \gamma q[E_1 + v B_3 - v_3 B_2]$$

Writing in general vectorial form, we have

$$\frac{d\mathcal{P}}{dt} = q(E + V \times B) \tag{14.121}$$

where $\mathcal{P} = m_0 \gamma v$ is the momentum.

We have seen that making use of first three components of Eq. (14.120), equation of motion retains it from in Eq. (14.104). Now, we consider the fourth component

$$m_0 \frac{du_4}{d\tau} = qF_{41}u_1 + qF_{42}u_2 + qF_{43}u_3 + qF_{44}u_4$$

$$= -\frac{\gamma q}{ic}E_1 v_1 - \frac{\gamma q}{ic}E_2 v_2 - \frac{rq}{ic}E_3 v_3$$

$$\therefore \quad \gamma\frac{d}{dt}(m_0 i\gamma_c) = -\frac{\gamma q}{ic}E \cdot v$$

or

$$\frac{d}{dt}(m_0\gamma_c) = qE \cdot v \tag{14.122}$$

The RHS of Eq. (14.122) gives us the rate at which the work is done on the particle by the electric field which, in fact, is equal to the rate of change of KE, T with time. Thus, one finds

$$qE \cdot v = \frac{dT}{dt} = \frac{d}{dt}(m_0\gamma c^2)$$

and

$$\int_{t_1}^{t_2} \frac{dT}{dt} dt = [m_0\gamma c^2]_{t_1}^{t_2}$$

If the particle is at rest at $t = t_1$, the value of $\gamma = 1$ at this time. Thus,

$$T = m_0\gamma c^2 - m_0 c^2 = mc^2 - m_0 c^2 \tag{14.123}$$

where

$$m = m_0\gamma$$

The total energy W is obtained as

$$W = mc^2 = T + m_0 c^2 \tag{14.124}$$

where $m_0 c^2$ is rest energy.

One can also express the total energy in terms of momentum $\mathcal{P}$.

We have

$$W = m_0\gamma c^2$$

$$\therefore \quad \frac{W^2}{c^2} = m_0^2 \gamma c^2$$

Now,

$$\mathcal{P}_2 = m_0 \gamma v^2 \quad \text{[from Eq. (14.121)]}$$

$$\therefore \quad \frac{W^2}{c^2} = \mathcal{P}_2 + m_0^2 \gamma^2 (c^2 - v^2)$$

or

$$W = [m_0^2 c^4 + \mathcal{P}^2 c^2]^{1/2} \tag{14.125}$$

14.16 RADIATION FROM RELATIVISTIC PARTICLES

We have read that only charges undergoing acceleration could produce radiation and the rate of radiation in accordance with Larmor relation is given by

$$-W = \frac{e^2 |\dot{v}|^2}{6\pi\varepsilon_0 c^3} \qquad (14.126)$$

If the velocity of the moving charges is much smaller than c, i.e. $v \ll c$, the formula is exact in the reference frame at rest w.r.t. charges. Now, we investigate the relation for the particles moving with velocities comparable with the velocity of light, i.e. $v \sim c$.

We have
$$u = (\gamma v, i\gamma c)$$

Differentiating w.r.t. τ, one obtains the four-vector acceleration

$$\frac{du}{dt} = \left\{ \gamma^2 \dot{v} + \frac{\gamma^4 v (v \cdot v)}{c^2}, \frac{\gamma^4 i (v \cdot \dot{v})}{c} \right\}$$

where
$$\dot{v} = \frac{dv}{dt} \quad \text{and} \quad \gamma dt = d\tau \qquad (14.127)$$

One can generalize the result [Eq. (14.126)] to relativistic energies by replacing $\dot{v}$ by the four-vector acceleration.

$$\left(\frac{du}{d\tau}\right)^2 = \left[\gamma^4 (\dot{v} \cdot \dot{v}) + 2\gamma^6 \left(\frac{v \cdot \dot{v}}{c^2}\right)^2 + \frac{\gamma^8 (v \cdot v)(v \cdot \dot{v})^2}{c^4} - \frac{\gamma^8 (v \cdot \dot{v})^2}{c^2} \right]$$

$$= \left[\gamma^4 (\dot{v} \cdot \dot{v}) + \frac{2\gamma^6 (v \cdot \dot{v})^2}{c^2} - \frac{\gamma^8 (v \cdot \dot{v})^2}{c^2} \right] \qquad (14.128)$$

$$\therefore \quad -W = \frac{e^2}{6\pi\varepsilon_0 c^3 (1-\beta^2)^2} \left[(\dot{v} \cdot \dot{v}) + \frac{(\dot{v} \cdot v)^2}{c^2 (1-\beta^2)} \right] \qquad (14.129)$$

The particular cases that may be of interest are:

i. When v and $\dot{v}$ are collinear, then

we have
$$-W = \frac{e^2}{6\pi\varepsilon_0 c^2 (1-\beta^2)^2} \left[|\dot{v}|^2 + \frac{v^2 |\dot{v}|^2}{c^2 (1-\beta^2)} \right]$$

$$= \frac{e^2}{6\pi\varepsilon_0 c^2} \frac{|\dot{v}|^2}{(1-\beta^2|^3} = \frac{e^2}{6\pi\varepsilon_0 c} \frac{\dot{\beta}^2}{(1-\beta^2)^3} \qquad (14.130)$$

ii. When $v \perp \dot{v}$, then we have

$$-W = \frac{e^2}{6\pi\varepsilon_0 c^3} \frac{|\dot{v}|^2}{(1-\beta^2)^2} = \frac{e^2}{6\pi\varepsilon_0 c} \frac{\dot{\beta}^2}{(1-\beta^2)^2} \qquad (14.131)$$

ILLUSTRATIVE EXAMPLES

Example 14.1: Show that the three-dimensional volume element $dx\,dy\,dz$ is not invariant, but four-dimensional volume element is invariant under Lorentz transformation.

Solution: According to Lorentz–Fitzgerald length contraction, one finds

$$dx' = dx\sqrt{1-(v/c)^2},$$

$$dy' = dy \quad \text{and } dz' = dz$$

∴ The three-dimensional volume element in frame S' is equal to

$$= dx'\,dy'\,dz' = d\sqrt{1-\beta^2}\,dy\,dz$$

$$= dx\,dy\,dz\sqrt{1-\beta^2}$$

$$\neq dx\,dy\,dz$$

Obviously, three-dimensional volume element is not invariant under Lorentz transformations. We have

$$dt' = \frac{dt}{\sqrt{1-\beta^2}} \quad \text{where } \beta = \frac{v}{c}$$

∴ Four-dimensional volume element in frame S' is

$$= dx'\,dy'\,dz'\,dt' = dx\sqrt{1-\beta^2}\,dy\,dz\,dt / \sqrt{1-\beta^2} = dx\,dy\,dz\,dt$$

Obviously, four-dimensional volume element is invariant under Lorentz transformations.

Example 14.2: Frame S' is moving relative to frame S with velocity $v = (4\hat{i} + 3\hat{j}) \times 10^7$ m/s. What will be the electric field in frame S' when the field due to a stationary charge in frame S is $E = (400\hat{i} + 500\hat{j})$ V/m?

Solution: We have the unit vector along velocity v as

$$\hat{v} = \frac{v}{|v|} = \frac{(4\hat{i} + 3\hat{j}) \times 10^7}{5 \times 10^7} = \frac{4\hat{i} + 3\hat{j}}{5}$$

Now, the component of E parallel to v is given by

$$E_{11} = E \cdot v = (400\hat{i} + 500\hat{j}) \cdot \frac{(4\hat{i} + 3\hat{j})}{5} = 620\,\text{V/m}$$

Representing $E_{||}$ vectorially, we have

$$E_{||} = 620\left(\frac{4\hat{i} + 3\hat{j}}{5}\right) = (496\hat{i} + 372\hat{j})\,\text{V/m}$$

∴

$$E_{\perp} = E - E_{||} = (420\hat{i} + 500\hat{j}) - (496\hat{i} + 372\hat{j})$$

$$= (-96\hat{i} + 128\hat{j})\,\text{V/m}$$

Now, making use of Lorentz transformation equations, one obtains the component of E' (field in the frame S') as

$$E'_\perp = \frac{E_\perp}{\sqrt{1-\dfrac{v^2}{c^2}}} = \frac{(-96\hat{i}+128\hat{j})}{\left[1-\left|\dfrac{(4\hat{i}+3\hat{j})\times10^7}{3\times10^8}\right|^2\right]} = \frac{9\times(-96\hat{i}+128\hat{j})}{8.75}$$

$$= 1.028(-96\hat{i}+128\hat{j})$$

$$= (-98.74\hat{i}+131.6\hat{j})\,V/m.$$

Thus, the electric field in frame S' will be

$$E = E'_\parallel + E'_\perp = (496\hat{i}+372\hat{j})+(-98.74\hat{i}+131.6\hat{j})$$

$$= (397.26\hat{i}+523.6\hat{j})\,V/m$$

Example 14.3. Prove that $E \cdot B$ and $E^2 - c^2 B^2$ are Lorentz scalars.

Solution: In order to prove that $E \cdot B$ is a Lorentz scalar, we have to prove that $E \cdot B$ is invariant under Lorentz transformation, i.e.

$$E \cdot B = E'_1 B'_1 + E'_2 B'_2 + E'_3 B'_3$$

$$= E_1 B_1 + \gamma^2(E_2 - uB_3)(B_2 + uE_3/c^2) + \gamma^2(E_3 + uB_2)(B_3 - uE_2/c^2)$$

or $\quad E \cdot B = E_1 B_1 + E_2 B_2 \gamma^2\left(1 - \dfrac{u^2}{c^2}\right) + E_3 B_3 \gamma^2(1 - u^2/c^2)$

$$= E_1 B_1 + E_2 B_2 + E_3 B_3 = E \cdot B$$

Similarly,

$$E'^2 - c^2 B'^2 = E'^2_1 + E'^2_2 + E'^2_3 - c^2(B'^2_1 + B'^2_2 + B'^2_3)$$

$$= E^2_1 + \gamma^2(E_2 - uB_3)^2 + \gamma^2(E_3 + uB_2)^2$$

$$- c^2[B^2_1 + \gamma^2(B_2 + uE_3/c^2)^2 + \gamma^2\left(B_3 - \frac{uE_2}{c^2}\right)^2$$

$$= E^2_1 + \gamma^2(E^2_2 + u^2 B^2_3 - 2uE_2 B_3 + E^2_3 + u^2 B^2_2$$

$$+ 2uE_2 B_2 - c^2 B^2_2 - u^2 E^2_3/c^2 - 2uB_2 E_3$$

$$- c^2 B^2_3 - u^2 E^2_2/c^2 + 2uB_3 E_2) - c^2 B^2_1$$

$$= E^2_1 + \gamma^2 E^2_2(1 - u^2/c^2) + \gamma^2 E^2_3(1 - u^2/c^2)$$

$$- c^2 B^2_1 - c^2 \gamma^2 B^2_2(1 - u^2/c^2) - c^2 \gamma^2 B^2_3(1 - u^2/c^2)$$

$$= E^2_1 + E^2_2 + E^2_3 - c^2(B^2_1 + B^2_2 + B^2_3) = E^2 - c^2 B^2$$

Example 14.4. Mesons are found to have lifetime 2.2×10^{-6} s, as observed in the laboratory. Fast moving mesons form a part of cosmic rays. Find the apparent lifetime of the mesons moving with speed $0.98c$ relative to the earth. Calculate the distance travelled by them during this time.

Solution: In the laboratory frame (say frame S) mesons have lifetime of 2.2×10^{-6} s, where they are at rest. In cosmic rays, the mesons move with a speed $u = 0.98c$. Assume frame of cosmic ray mesons as S' which has a speed of $0.98c$ w.r.t frame S. Thus, the apparent time T' of the mesons in frame S' as observed by observer in frame S will be

$$T' = \frac{T}{\sqrt{1 - u^2/c^2}} = \frac{2.2 \times 10^{-6}}{\sqrt{1 - (0.98)^2}} = 11.06 \times 10^{-6} \text{s}.$$

Now, the apparent distance travelled by the mesons in this time

$$L' = LT' = 0.98 \times 3 \times 10^8 \times 11.06 \times 10^{-6} = 3252 \, \text{m}.$$

Example 14.5. A spaceship travels along the x-axis in the initial frame of reference. Let the spaceship starts with the initial velocity $v = 0$ when $x = 0$ and has a constant acceleration g. Find the distance x travelled when it has achieved a velocity $v(\sim c)$.

Solution: Let us consider two frames of reference S and S' with the latter moving with constant velocity v along the x-direction, relative to frame S. Let u, $\dfrac{du}{dt}$ and u', $\dfrac{du'}{dt}$ be the velocity and acceleration of the object moving in x-direction in frames S and S' respectively. Lorentz transformation gives

$$x = \gamma (x' + \beta ct'), \quad ct = \gamma (ct' + bx'),$$

where

$$\beta = v/c, \quad \text{and} \quad \gamma = (1 - \beta^2)^{-1/2}$$

The velocity of the object is transformed according to the following relation:

$$u = (u' + v) / \left(1 + \frac{uu'}{c^2}\right) \tag{1}$$

Also, we have

$$dt = \gamma \left(dt' + \frac{v}{c^2} dx'\right) = \gamma \, dt' \left(1 + \frac{vu'}{c^2}\right) \tag{2}$$

On differentiating Eq. (1), one obtains

$$du = \frac{du'}{\gamma^2 \left(1 + \dfrac{vu'}{c^2}\right)^2} \tag{3}$$

The ratio du/dt gives us the relation for transformation of acceleration as

$$a = \frac{a'}{\gamma^3 \left(1 + \dfrac{uu'}{c^2}\right)^3} \tag{4}$$

Now, if frame S (inertial frame) is attached to the fixed stars and S' is the inertial frame in which the spaceship is momentarily at rest, then in frame S', we have

$$u' = 0 \quad \text{and} \quad a' = g \tag{5}$$

$$\therefore \qquad u = v \quad \text{and} \quad a = g/\gamma^3 \tag{6}$$

Now, if the velocity of spaceship is increased from 0 to v in frame S, then the distance travelled is

$$x = \int u \, dt = \int_0^v \frac{u \, dt}{a} = \frac{1}{g} \int_0^v \frac{u \, du}{(1 - \beta^2)^{3/2}}$$

$$= \frac{c^2}{g}(\gamma - 1).$$

Example 14.6. What will be the form of Maxwell's equations in four-vector?

Solution: Maxwell's two equations (out of four) are

$$\nabla \times H = J + \frac{\partial D}{\partial t} \tag{1}$$

and

$$\nabla \cdot D = \rho \tag{2}$$

x_1 components in these equations are

$$\frac{\partial D_1}{\partial t} - \frac{\partial H_3}{\partial x_2} + \frac{\partial H_2}{\partial x_3} = -J_1 \tag{3}$$

$$\frac{\partial D_1}{\partial t} + \frac{\partial D_2}{\partial t} + \frac{\partial D_3}{\partial t} = \rho \tag{4}$$

Components of E and B can be expressed in terms of second rank, anti-symmetric field strength tensor

$$F_{\mu\nu} = \frac{\partial A_\nu}{\partial x_\mu} - \frac{\partial A_\mu}{\partial x_\nu} \tag{5}$$

where

$$(F_{\mu\nu}) = \begin{vmatrix} 0 & B_3 & -B_2 & -iE_1 \\ -B_3 & 0 & B_1 & -iE_2 \\ B_2 & -B_1 & 0 & -iE_3 \\ iE_1 & iE_2 & iE_3 & 0 \end{vmatrix} \tag{6}$$

Now, writing $F_{\mu\nu}$ as $f_{\mu\nu}$, we can write the components of electric and magnetic fields as

$$f_{\mu\nu} = f = \begin{vmatrix} 0 & f_{12} & f_{13} & f_{14} \\ f_{21} & 0 & f_{23} & f_{24} \\ f_{31} & f_{32} & 0 & f_{34} \\ f_{41} & f_{42} & f_{43} & 0 \end{vmatrix} = \begin{vmatrix} 0 & H_3 & -H_z & -icD_1 \\ -H_3 & 0 & H_1 & -icD_2 \\ H_2 & -H_3 & 0 & -icD_3 \\ -icD_1 & -icD_2 & -icD_2 & 0 \end{vmatrix} \tag{7}$$

We see that in all, there are six components of the electric and magnetic fields and all these are the elements of second rank antisymmetric field strength sensor.

$$f_{\mu\nu} = \frac{\partial A_\mu}{\partial x_\nu} - \frac{\partial A_\nu}{\partial x_\mu}$$

Explicitly, the form of $f_{\mu\nu}$ is as given in Eq. (7). Now, Eq. (3) reduces to

$$\frac{\partial f_{12}}{\partial x_2} - \frac{\partial f_{13}}{\partial x_3} - \frac{\partial f_{14}}{\partial x_4} = J_1$$

One can obtain similar equations for x_2 and x_3 components. Hence, in general form, we can write

$$\sum_{v=1}^{4} \frac{\partial f_{\mu v}}{\partial x_v} = J_\mu \quad (\mu = 1, 2, 3, 4) \tag{8}$$

For $\mu = 4$, Eq. (8) gives

$$\frac{\partial f_{41}}{\partial x_1} + \frac{\partial f_{42}}{\partial x_2} + \frac{\partial f_{43}}{\partial x_3} + \frac{\partial f_{44}}{\partial x_4} = J_4$$

$$ic\left[\frac{\partial D_1}{\partial x_1} + \frac{\partial D_2}{\partial x_2} + \frac{\partial D_3}{\partial x_3}\right] = ic\rho$$

or

$$\frac{\partial D_1}{\partial x_1} + \frac{\partial D_2}{\partial x_2} + \frac{\partial D_3}{\partial x_3} = \rho$$

which is Eq. (4).

This shows that the four-vector form of Maxwell's Eqs. (3) and (4) are given by Eq. (8). The other two Maxwell's equations are

$$\left.\begin{array}{ll} \nabla \times \boldsymbol{E} = -\dfrac{\partial \boldsymbol{B}}{\partial t} & \text{(i)} \\[2mm] \nabla \cdot \boldsymbol{B} = 0 & \text{(ii)} \end{array}\right\} \tag{9}$$

This first component of Eq. [(9i)] is

$$\frac{\partial E_3}{\partial x_2} - \frac{\partial E_2}{\partial x_3} - \frac{\partial B_1}{\partial t} = 0 \tag{10}$$

In view of the meaning of field tensor f, which is analogous to Eq. (7), we have

$$f = \begin{vmatrix} 0 & f_{12} & f_{13} & f_{14} \\ f_{21} & 0 & f_{23} & f_{24} \\ f_{31} & f_{32} & 0 & f_{34} \\ f_{41} & f_{42} & f_{43} & 0 \end{vmatrix}$$

$$= \begin{vmatrix} 0 & cB_3 & -cB_2 & -iE_1 \\ -cB_3 & 0 & cB_1 & -iE_2 \\ cB_2 & -cB_3 & 0 & -iE_3 \\ iE_1 & iE_2 & iE_3 & 0 \end{vmatrix} \tag{11}$$

Equation (11) may be expressed in the form

$$i\left(\frac{\partial f_{23}}{\partial x_4} + \frac{\partial f_{34}}{\partial x_2} + \frac{\partial f_{42}}{\partial x_3}\right) = 0 \tag{12}$$

The homogeneous Maxwell's equations (9i) and (9ii) are written as

$$\frac{\partial f_{\mu v}}{\partial x_\lambda} + \frac{\partial f_{v\lambda}}{\partial x_\mu} + \frac{\partial f_{\lambda\mu}}{\partial x_v} = 0 \quad (\mu \neq v \pm \lambda = 1, 2, 3, 4)$$

The lack of symmetry and confusing subscripts of Eq. (12) can be eliminated by considering the 'dual' six-vector of f,

$$f^* = (i\bar{E}, c\bar{B}) \tag{13}$$

which is obtained from f by an exchange of real and imaginary constituents. We may note that the individual components of f^* are fixed by the rule

$$f^*_{mm} = \frac{1}{2}\varepsilon_{klmn} f^{mn} \tag{14}$$

with the prescription that the sequence of subscripts $klmn$ arises from 1 2 3 4, by an even number of interchanges. From this requirement, we have uniquely

$$f_{23} = f^*_{14}, \quad f_{34} = f^*_{12}, \quad f_{42} = f^*_{13} \tag{15}$$

Now, Eq. (11) takes the form

$$\frac{\partial f^*_{12}}{\partial x_2} + \frac{\partial f^*_{13}}{\partial x_3} + \frac{\partial f^*_{14}}{\partial x_4} = 0 \tag{16}$$

which is the first component of

$$\nabla \cdot f^* = 0 \tag{17}$$

One can obtain the similar form of other two components of Eq. (9i). Thus, in terms of the dual field strength tensor, the Maxwell's equations are

$$\frac{df^*_{\mu\nu}}{\partial x_\mu} = 0$$

The fourth component of Eq. (17) is

$$\frac{\partial f^*_{41}}{\partial x_1} + \frac{\partial f^*_{42}}{\partial x_2} + \frac{\partial f^*_{43}}{\partial x_3} = 0$$

which by Eqs. (11) and (13) is transformed to

$$-i\left(\frac{\partial B_1}{\partial x} + \frac{\partial B_2}{\partial x_2} + \frac{\partial B_3}{\partial x_3}\right) = 0$$

which is Eq. (9ii).

We find that it is, thus, identical with the familiar absence of sources of the magnetic field intensity B. Obviously, the four-dimensional relativistic stand point is a formally necessary completion of first Maxwell's triplet.

Example 14.7: A particle with charge q moves in an inertial system Σ with the velocity
$$u = (a, a, a)$$
in a homogeneous magnetic field $B = (B, 0, 0)$ considering Σ' be an inertial system which moves relative to Σ with velocity $V_z = v\hat{e}_z = $ const. Find the force acting on the particle in Σ and Σ', respectively.

Solution: It frame in Σ:

$$F = q\mathbf{u} \times \mathbf{B} = q(0, aB, -aB) = qaB(0, 1, -1)$$

In Σ' one finds:

$$F'_x = \frac{1}{\gamma}\frac{F_x}{1 - \dfrac{vu_z}{c^2}} = 0, \quad \gamma = \left(1 - \frac{v^2}{c^2}\right)^{-1/2}$$

$$F'_y = \frac{1}{\gamma}\frac{F_y}{1 - \dfrac{vu_z}{c^2}} = \frac{1}{\gamma}\frac{qaB}{1 - \dfrac{va}{c^2}}$$

$$\mathbf{F} \cdot \mathbf{u} = qa^2 B(1 - 1) = 0$$

$$F'_x = \frac{F_z - \dfrac{v}{c^2}(\mathbf{F} \cdot \mathbf{u})}{1 - \dfrac{vu}{c^2}} = \frac{-qaB}{1 - \dfrac{va}{c^2}}$$

$$\Rightarrow \mathbf{F'} = \frac{qaB}{1 - \dfrac{va}{c^2}}\left(0, \frac{1}{\gamma}, -1\right)$$

Example 14.8: Show by use of a transformation relation for the electromagnetic field that $\left(\mathbf{B} + \dfrac{i}{c}\mathbf{E}\right)^2$ is a Lorentz invariant.

Solution: Let Σ and Σ' be two arbitrary inertial systems with

$$\Sigma \xrightarrow{\;v\;} \Sigma' \quad (v - v\mathbf{e}_z)$$

We show that

$$\left(\mathbf{B'} + \frac{i}{c}\mathbf{E'}\right)^2 = \left(\mathbf{B} + \frac{i}{c}\mathbf{E}\right)^2$$

The transformation formulae yield:

$$B'_x + \frac{i}{c}E'_x = \gamma\left(B_x + \frac{i}{c}E_x\right) - i\gamma\beta\left(B_y + \frac{i}{c}E_y\right)$$

$$B'_y + \frac{i}{c}E'_y = \gamma\left(B_y + \frac{i}{c}E_y\right) + i\gamma\beta\left(B_x + \frac{i}{c}E_x\right)$$

$$B'_z + \frac{i}{c}E'_z = B_z + \frac{i}{c}E_z$$

From that we get:

$$\left(B' + \frac{i}{c}E'\right)^2 = \left(B_x + \frac{i}{c}E_x\right)^2 \gamma^2(1-\beta^2)$$

$$+ \left(B_y + \frac{i}{c}E_y\right)^2 \gamma^2(1-\beta^2) + \left(B_z + \frac{i}{c}E_z\right)^2$$

$$= \left(B + \frac{i}{c}E\right)^2$$

Example 14.9: Show by explicitly calculation that the components of dual field strength tensor $\bar{F}^{\mu v}$ follow from those of the covariant tensor $F_{\mu v}$ by the substitution.

$$\vec{B} \longleftrightarrow -\frac{1}{c}\vec{E}$$

Solution: Because of $\bar{F}^{\mu v} = -\bar{F}^{v \mu}$ all the diagonal elements are zero.

$$\bar{F}^{\mu \mu} = F^{\mu \mu} = 0, \qquad \mu = 0, 1, 2, 3$$

And we need to calculate only the elements $\bar{F}^{\mu v}$ with $\mu < v$:

$$\bar{F}^{12} = \frac{1}{2}\varepsilon^{12\rho\sigma} F_{\rho\sigma} = \frac{1}{2}\left(\varepsilon^{1230} F_{30} + \varepsilon^{1203}F_{03}\right)$$

$$= \frac{1}{2}(-F_{30} + F_{03}) = F^{30} = \frac{1}{c}E_z$$

$$\bar{F}^{13} = \frac{1}{2}\varepsilon^{13\rho\sigma} F_{\rho\sigma} = \frac{1}{2}\left(\varepsilon^{1230} F_{20} + \varepsilon^{1203}F_{02}\right)$$

$$= \frac{1}{2}(F_{20} + F_{02}) = F^{02} = -\frac{1}{c}E_y$$

$$\bar{F}^{01} = \frac{1}{2}\varepsilon^{01\rho\sigma} F_{\rho\sigma} = \frac{1}{2}\left(\varepsilon^{0123} F_{23} + \varepsilon^{0132}F_{32}\right)$$

$$= \frac{1}{2}(F_{23} - F_{32}) = F^{23} = -B_x$$

$$\bar{F}^{23} = \frac{1}{2}\varepsilon^{23\rho\sigma} F_{\rho\sigma} = \frac{1}{2}\left(\varepsilon^{2310} F_{10} + \varepsilon^{2301}F_{01}\right)$$

$$= \frac{1}{2}(-F_{10} + F_{01}) = F^{10} = -\frac{1}{c}E_x$$

$$\bar{F}^{02} = \frac{1}{2}\varepsilon^{02\rho\sigma} F_{\rho\sigma} = \frac{1}{2}\left(\varepsilon^{0213} F_{13} + \varepsilon^{0231}F_{31}\right)$$

$$= \frac{1}{2}(-F_{13} + F_{31}) = F^{31} = -B_y$$

$$\bar{F}^{03} = \frac{1}{2}\varepsilon^{03\rho\sigma}F_{\rho\sigma} = \frac{1}{2}\left(\varepsilon^{0312}F_{12} + \varepsilon^{0321}F_{21}\right)$$

$$= \frac{1}{2}(F_{12} - F_{21}) = F^{12} = -B_z$$

Example 14.10: Let Σ and Σ' are two inertial systems. The electromagnetic field in Σ is (E, B) and in Σ' is (E', B'). The field E has the same direction in the whole space. Σ' moves relative to Σ with velocity v_0 parallel to E, i.e. $v_0 = \alpha E$. A particle with the charge q lies at the origin of coordinates in Σ. Use the Lorentz force on the particle and show that the component of E' in direction of E is equal to E.

Solution:

• In Σ:

Lorentz force on the point charge q:

$$F = q\left[E + (v \times B)\right]$$

The particle is at rest, i.e. $v = 0$. Thus:

$$F = qE$$

• In Σ':

Velocity of the particle: $v' = -v_0$, thus the Lorentz force reads:

$$F' = q(E' + (-v_0) \times B')$$

• Σ, Σ' are both inertial systems. Therefore, it must hold: $F = F'$. This means here:

$$E' = E + (v_0 \times B') = E + \alpha E \times B'$$

Component in the direction of E:

$$\frac{1}{E}(E' \cdot E) = \frac{1}{E}(E \cdot E) + 0 = E$$

Example 14.11: Let Σ and Σ' be two inertial systems, where Σ' moves relative to Σ with the constant velocity $v = v\hat{e}_z$. In Σ a particle with charge q and velocity $u = (a, b, d)$ propagates in the electromagnetic field:

$$B = (0, B, 0); \quad E = \frac{1}{\sqrt{2}}(E, E, 0)$$

Find the forces acting on the particle in Σ and Σ', respectively?

Solution: Force on charge q in Σ:

$$F = q(E + u \times B)$$

with

$$u = (a, b, d); \quad B = (0, B, 0); \quad E = \frac{1}{\sqrt{2}}(E, E, 0)$$

Therefore:

$$u \times B = (-Bd, 0, Ba)$$

That yields:

$$F = q\left(\frac{1}{\sqrt{2}} E - Bd, \frac{1}{\sqrt{2}} E, Ba\right)$$

Therewith we calculate

$$F \cdot u = q\left(\left(\frac{1}{\sqrt{2}} E - Bd\right)a + \frac{1}{\sqrt{2}} Eb + Bad\right) = \frac{1}{\sqrt{2}} qE(a + b)$$

We now find the forces in Σ':

$$F_x' = \frac{1}{\sqrt{1 - \dfrac{v^2}{c^2}}} \frac{q\left(\dfrac{1}{\sqrt{2}} E - Bd\right)}{1 - \dfrac{vd}{c^2}}$$

$$F_y' = \frac{1}{\sqrt{1 - \dfrac{v^2}{c^2}}} \frac{q\dfrac{1}{\sqrt{2}} E}{1 - \dfrac{vd}{c^2}}$$

$$F_z' = \frac{q\left(Ba - \dfrac{v}{c^2} \dfrac{1}{\sqrt{2}} E(a + b)\right)}{1 - \dfrac{vd}{c^2}}$$

Example 14.12:

1. A magnetic dipole is oriented paralled to the z-axis (magnetic moment $m = m\hat{e}_z$, $m > 0$). What are the Cartesian components of the **B** field?
2. Calculate now the **E**- and **B**-field of a magnetic dipole which moves uniformly in straight-line in z-direction and whose moment is oriented parallel to the z-direction. The dipole at the time $t = 0$ is at thé origin of Σ.
3. Write the fields of moving magnetic dipole by cylindrical coordinates.
4. Which shape do the electric filed lines have in the xy-plane? How does the electric field charge in course of time?
5. The **E**-lines do not start or end at electric charges. Under which conditions can such electric fields occur?

Solution:

1. We have

$$B = \frac{\mu_0}{4\pi}\left(\frac{3(r \cdot m)r}{r^5} - \frac{m}{r^3}\right)$$

$$B_x = \frac{\mu_0}{4\pi} \frac{3mzx}{r^5} ; B_y = \frac{\mu_0}{4\pi} \frac{3mzy}{r^5}$$

$$B_z = \frac{\mu_0}{4\pi} \frac{m}{r^5}\left(2z^2 - x^2 - y^2\right)$$

2. Let Σ' be the rest system of the dipole. As calculated in part 1, we find there the fields:

$$B'_x = \frac{\mu_0}{4\pi} \frac{3mz'x'}{r'^5}; \; B'_y = \frac{\mu_0}{4\pi} \frac{3mz'y'}{r'^5}$$

$$B'_z = \frac{\mu_0}{4\pi} \frac{m}{r'^5}\left(2z'^2 - x'^2 - y'^2\right)$$

$$E' \equiv 0: \; r' = \sqrt{x'^2 + y'^2 + z'^2}$$

In frame Σ we have

$$B_x = \gamma B'_x = \gamma \frac{\mu_0}{4\pi} 3m \frac{z'x'}{r^3}$$

$$x' = x; y' = y; z' = \gamma(z - vt)$$

$$\Rightarrow r' = \sqrt{x^2 + y^2 + \gamma^2(z - vt)^2}$$

$$\Rightarrow B_x = \gamma^2 \frac{\mu_0}{4\pi} 3m \frac{x(z - vt)}{[x^2 + y^2 + \gamma^2(z - vt)^2]^{5/2}}$$

$$B_y = \gamma B'_y = \gamma^2 \frac{\mu_0}{4\pi} 3m \frac{y(z - vt)}{[x^2 + y^2 + \gamma^2(z - vt)^2]^{5/2}}$$

$$B_z = B'_z = \frac{\mu_0}{4\pi} m \frac{2\gamma^2(z - vt)^2 - x^2 - y^2}{[x^2 + y^2 + \gamma^2(z - vt)^2]^{5/2}}$$

$$E_x = +\gamma\beta cB'_y = vB_y$$

$$E_y = -\gamma\beta cB'_x = -vB_x$$

$$E_z = E'_z = 0$$

3. $\quad e_\rho = (\cos\varphi, \sin\varphi, 0); \; e_\varphi = (-\sin\varphi, \cos\varphi, 0); \; e_z = (0, 0, 1)$

Electric field:

$$E_\rho = E \cdot \hat{e}_\rho = E_x \cos\varphi + E_y \sin\varphi = v(B_y \cos\varphi - B_x \sin\varphi)$$

$$= \dots(y\cos\varphi - x\sin\varphi) = \dots(\rho\sin\phi\cos\varphi - \rho\cos\varphi\sin\varphi) = 0$$

$$E_\varphi = E \cdot e_\varphi = -E_x \sin\varphi + E_y \cos\varphi = -v(B_y \sin\varphi + B_x \cos\varphi)$$

$$= \gamma^2 \frac{\mu_0}{4\pi} 3m \frac{(z - vt)}{(x^2 + y^2 + \gamma^2(z - vt)^{5/2}}(-v\,y\sin\varphi - v\,x\cos\varphi)$$

$$= \dots - v\,\rho(\sin^2\varphi + \cos^2\varphi) = \gamma^2 \frac{\mu_0}{4\pi} 3m \frac{-v(z - vt)\rho}{[\rho^2 + \gamma^2(z - vt)^{5/2}}$$

$$E_z = 0$$

$$B_\rho = B_x \cos\varphi + B_y \sin\varphi = \frac{1}{v}(E_x \sin\varphi - E_y \cos\varphi) = -\frac{1}{v}E_\varphi$$

$$B_\varphi = -B_x \sin\varphi + B_y \cos\varphi = \frac{1}{v}(E_y \sin\varphi + E_x \cos\varphi) = \frac{1}{v}E_\rho = 0$$

$$B_z = \frac{\mu_0}{4\pi}m\,\frac{2\gamma^2(z-vt)^2 - \rho^2}{[\rho^2 + \gamma^2(z-vt)^{5/2}}$$

4. The *E*-lines are circles in the *xy*-plane with their centers at the origin of co-ordinates (Fig. 14.5).

Fig. 14.5

$$E = E_\varphi(t)\,e_\varphi$$

$$E_\varphi(t; z = 0) = \gamma^2\,\frac{\mu_0}{4\pi}\,3m\,\frac{v^2 t}{(\rho^2 + y^2 v^2 t^2)^{5/2}}$$

5. Close *E*-lines can appear, because of curl$E = -\dot{B}$, when a time-dependent magnetic field is present.

Example 14.13: Let Σ and Σ' be two inertial systems moving relative to one another with a velocity v = const. Explain why in Σ an electric field appears although in Σ' only a B'-field exists. Also explain in the same manner why in Σ' a magnetic field occurs although in Σ' only a pure electric field is present.

Solution:
1. A charge, which is at first at rest in Σ, experiences in Σ' an acceleration due to the pure B'-field since the charge is moving in this system. In Σ, however, this acceleration must stem from an *E*-field, since the charge is initially at rest.
2. Let the E'-field in Σ' be produced by charges resting in Σ', e.g. by charges on the plates of a capacitor. In Σ these charges are moving thereby generating a B-field.

Example 14.14: Let a charge particle (charge q, mass m) be moving in a homogeneous electric field.
$$B = (0, 0, B)$$
1. Show that its relativistic energy is constant in time.
2. Calculate the time-dependence of the relativistic momentum with the initial condition $v_0 = (v_0, 0, 0)$.
3. Calculate the trajectory $r(t)$ of the particle with $r(t = 0) = (0, y_0, 0)$ where $y_0 = \dfrac{mv_0}{qB}$.

Solution:
1. For the space components of the Minkowski force it holds:

$$\frac{d}{d\tau}p_r = \frac{d}{dt}p_r = \gamma F = \gamma q(v \times B)$$

$$p_r = \gamma m v$$

$$\Rightarrow p_r \cdot \frac{d}{dt} p_r = \gamma m q v \cdot (v \times B) = 0 = \frac{1}{2} \frac{d}{dt} p_r^2$$

$$\Rightarrow p_r^2 = \text{const} \Rightarrow v^2 = \text{const} \Leftrightarrow \gamma(v) = \left(1 - \frac{v^2}{c^2}\right)^{-1/2} = \text{const}$$

The relativistic energy

$$T_r = \sqrt{c^2 p_r^2 + m^2 c^4} = \text{const}\cdot$$

2. Initial condition: $p_r \, (t = 0) = m\gamma \, v_0(1, 0, 0)$

$$\dot{p}_r = q(v \times B) = q(v_y B - v_x B, 0)$$

$$= \frac{I}{\gamma} \frac{qB}{m} (R_{ry} - p_{\hat{r}x}, 0)$$

First partial result: $p_{rz} = \text{const}$

$$\omega \equiv \frac{qB}{m} \frac{1}{2} \equiv \omega_0 \frac{1}{y} \qquad (\gamma = \text{const})$$

$$\frac{d}{dt}(p_x + i p_{ry}) = \omega(p_{ry} - i p_{\hat{r}x}) = -i\omega(p_{\hat{r}c} + i p_{ry})$$

$$\Rightarrow (p_{\hat{r}x} + i p_{ry})(t) = (p_{\hat{r}c} + i p_{ry})(0)e^{-i\omega t}$$

$$\Rightarrow p_{rx}(t) = p_{\hat{r}x}(t)\cos \omega t + p_{ry}(0)\sin \omega t$$

$$p_{ry}(t) = -p_{\hat{r}x}(t)\sin \omega t + p_{ry}(0)\cos \omega t$$

Initial conditions:

$$p_r(0) = m\gamma(v_0, 0, 0)$$

This yields:

$$p_\tau(t) = m\gamma v_0(\cos \omega t, -\sin \omega t, 0)$$

$p_r^2(t) = \text{const}$ is obviously guaranteed.

3. The trajectory we get from

$$r(t) - r(t = 0) = \frac{1}{my}\int_0^t dt' \, p_r(t')$$

$$= \frac{v_0}{\omega}(\sin \omega t', \cos \omega t', 0)$$

$$\Rightarrow \quad r(t) = (0, y_0, 0) + \gamma \frac{mv_0}{qB}\{(\sin \omega t, \cos \omega t, 0) - (0, 1, 0)\}$$

$$\Rightarrow \quad r(t) = \gamma \frac{mv_0}{qB} (\sin wt, \cos \omega t, 0)$$

Example 14.15: A π^+ meson at rest decays within 2.5×10^{-8}s into a μ^+ meson and a neutrino. Let the π^+ meson have a kinetic energy which is equal to its rest energy.

1. What is the velocity of the π^+ meson ?
2. Which distance is covered by the meson before its decay, detected by an observer 'at rest'?

Solution:

$$T_r = T + T_r(0) = 2T_r(0) = \gamma T_r(0) \Rightarrow \gamma = 2$$

This means:

$$\beta = \left(\frac{\gamma^2 - 1}{\gamma^2} \right)^{1/2} = \sqrt{\frac{3}{4}} = 0.866$$

The velocity of the π^+-meson therefore amounts to:

$$v(\pi^+) = 0.866c = 2.598 \times 10^8 \frac{m}{s}$$

3. Decay time in the rest system of the meson:

$$\tau = 2.5 \times 10^{-8} s$$

Decay time in the rest system of the observer:

$$\tau' = \gamma \tau = 5 \times 10^{-8} s$$

'Decay-distance':

$$d = v \tau' = 12.990 \text{ m}$$

Example 14.16: Considering that a photon (rest mass $m = 0$) have the same momentum as a 1 MeV electron. Find the energy of photon.

Solution: Energy of the electron:

$$T_r^2 = c^2 p_r^2 + m^2 c^4 = c^2 p_r^2 + T_r^2(0) \quad \text{... (i)}$$

On the other hand:

$$T_r = T + T_r(0) \quad \text{... (ii)}$$

Combining (i) and (ii), we obtain:

$$T^2 + 2TT_r(0) = c^2 r_r^2$$

$$T \leftrightarrow 1 \text{ MeV}$$

$$T_r(0) \leftrightarrow 0.511 \text{ MeV}$$

$$\Rightarrow c^2 p_r^2 = 2.022(MeV)^2 \Rightarrow cp_\tau = 1.422 \text{ MeV}$$

The photon has no rest mass:

$$T_{\hat{r}}(\gamma) = cp_{\hat{r}} = 1.422\,\text{MeV}$$

Example 14.17: In the case of a non-relativistic collision of two particles with equal masses the trajectories enclose after the collision an angle of $\pi/2$. Show that for the relativistic collision it holds (Fig. 14.6).

$$\tan\phi\tan\theta = \frac{2}{\gamma+1},$$

where

$$\gamma = \left(1 - \frac{v^2}{c^2}\right)^{-1/2}.$$

Fig. 14.6

Show that $(\theta+\phi) \le \dfrac{\pi}{2}$, where the equality sign becomes valid in the non-relativistic limit $(\gamma \to 1)$.

Solution: Law of conservation of energy:

$$T_r(v) + T_r(0) = T_r(v_1) + T_r(v_2)$$

$$\Rightarrow \gamma(v) + 1 = \gamma(v_1) + \gamma(v_2)$$

Law of conservation of momentum:

$$m\gamma v = m\gamma(v_1)v_1 + m\gamma(v_2)v_2$$

$$\Rightarrow \gamma(v)\boldsymbol{\beta} = \gamma(v_1)\boldsymbol{\beta}_1 + \gamma(v_2)\boldsymbol{\beta}_2$$

The last equation may be decomposed into components:

$$\gamma(v)\beta = \gamma(v_1)\beta_1\cos\theta + \gamma(v_2)\beta_2\cos\varphi$$

$$0 = \gamma(v_1)\beta_1\sin\theta - \gamma(v_2)\beta_2\sin\varphi$$

$$\gamma^2 = (v_2)\beta_2^2(\cos^2\varphi + \sin^2\varphi) = \gamma^2(v_1)\beta_1^2\sin^2\theta + (\gamma(v)\beta - \gamma(v_1)\beta_1\cos\theta)^2$$

$$\Rightarrow \gamma^2(v_2)\beta_2^2 = \gamma^2(v_1)\beta_1^2 + \gamma^2(v)\beta^2 - 2\gamma(v)\gamma(v_1)\beta\beta_1\cos\theta$$

We eliminate the β's by the relation:

$$\gamma(v_i)\beta = \sqrt{\gamma^2(v_i) - 1}$$

$$\gamma^2(v_2) - 1 = \gamma^2(v_1) - 1 + \gamma^2(v) - 1 - 2\sqrt{\gamma^2(v) - 1}\ \sqrt{\gamma^2(v_1) - 1}\cos\theta$$

We combine this equation with the law of conservation of energy in order to remove $\gamma(v_2)$:

$$\gamma(v_2) = \gamma(v) + 1 - \gamma(v_1)$$

$$\Rightarrow \gamma^2(v_2) = \gamma^2(v) + 1 + \gamma^2(v_1) + 2\gamma(v) - 2\gamma(v_1) - 2\gamma(v)\gamma(v_1)$$

Now, inserting this into the above equation:

$$2\gamma(v) - 2\gamma(v_1) - 2\gamma(v)\gamma(v_1) = -2 - 2\sqrt{\gamma 2(v) - 1}\ \sqrt{\gamma^2(v_1) - 1}\cos\theta$$

$$\Rightarrow -2(\gamma(v_1) - 1)\,(\gamma(v) + 1) = -2\sqrt{\gamma^2(v) - 1}\,\sqrt{\gamma^2(v_1) - 1}\,\cos\theta$$

$$\Rightarrow (\gamma(v_1) - 1)\,(\gamma(v) + 1) = (\gamma(v) - 1)\,(\gamma(v_1) + 1)\cos^2\theta$$

Solving for $\gamma(v_1)$, we obtain

$$\gamma(v_1) = \frac{(\gamma(v) + 1) + (\gamma(v) - 1)\cos^2\theta}{(\gamma(v) + 1) - (\gamma(v) - 1)\cos^2\theta}$$

With the abbreviation

$$\alpha^2 = \frac{\gamma(v) - 1}{\gamma(v) + 1}$$

It then remains to be solved:

$$\gamma(v_1) = \frac{1 + \alpha^2\cos^2\theta}{1 - \alpha^2\cos^2\theta}$$

Let us now go back to the law of conservation of momentum and build the quotient of the two component equations:

$$\tan\varphi = \frac{\gamma(v_1)\beta_1\sin\theta}{\gamma(v)\beta - \gamma(v_1)\beta_1\cos\theta}$$

It is

$$\gamma(v_1)\beta_1 = \sqrt{\gamma^2(v_1) - 1} = \sqrt{\frac{(1 + \alpha^2\cos^2\theta)^2}{(1 - \alpha^2\cos^2\theta)^2} - 1}$$

$$= \frac{2\alpha\cos\theta}{1 - \alpha^2\cos^2\theta}$$

On the other hand it holds:

$$\gamma(v)\beta = \sqrt{\gamma^2(v) - 1} = \sqrt{(\gamma(v) + 1)(\gamma(v) - 1)} = \alpha(\gamma(v) + 1)$$

On inserting this into the expression for $\tan\varphi$, we obtain

$$\tan\varphi = \frac{\dfrac{2\alpha\cos\theta\sin\theta}{1 - \alpha^2\cos\theta}}{\alpha(\gamma(v) + 1) - \dfrac{2\alpha\cos^2\theta}{1 - \alpha^2\cos^2\theta}}$$

$$= \frac{2\cos\theta\sin\theta}{(\gamma(v) + 1) - (\gamma(v) + 1)\,\alpha^2\cos^2\theta - 2\cos^2\theta}$$

$$= \frac{2\cos\theta\sin\varphi}{\gamma(v)(1 - \cos^2\theta) + (1 - \cos^2\theta)}$$

$$= \frac{2}{\gamma(v) + 1} \frac{\cos\theta \sin\theta}{1 - \cos^2\theta}$$

From this it follows that

$$\tan\varphi \tan\theta = \frac{2}{\gamma(v) + 1}$$

In the non-relativistic limit $\gamma \to 1$, we get:

$$\tan\varphi \tan\theta \to 1$$

Because of

$$\tan(\varphi + \theta) = \frac{\tan\theta + \cos\varphi}{1 - \tan\varphi \tan\theta}$$

This means

$$\tan(\varphi + \theta) \to \infty$$

$$\Rightarrow \quad \varphi + \theta \to \frac{\pi}{2}$$

Example 14.18: The inertial system Σ' moves relative to inertial system Σ with the velocity $v = v\hat{e}_z = $ constant. At the time $t = 0$ the two system coincide. The point charge q is located at the origin of Σ'.

1. Determine the four potentials A'^μ and A_μ in Σ' and Σ respectively.
2. Calculate with A^μ the electromagnetic fields E and B of the point charge in Σ. Which connection exists between E, B and v? Compare the result for (E, B) with the field strength tensor.
3. Does the scalar potential fulfill the wave equation in Σ.

$$\Box\phi = -\frac{\rho}{\varepsilon_0}$$

Solution:

$$\Sigma \xrightarrow{v\hat{e}_z} \Sigma'$$

1. Point charge q 'at rest' at the origin of coordinates in Σ. Hence, the potentials in Σ' are:

$$\phi' = \frac{q}{4\pi\varepsilon_0} \frac{1}{\hat{r}} : A' = 0$$

Here

$$r' = \sqrt{x'^2 + y'^2 + z'^2} = \sqrt{x^2 + y^2 + \gamma^2(z - vt)^2}$$

On the right-hand side $\hat{r}$ is expressed by Σ-coordinates. Therewith the four-potential reads.

$$A'^\mu = \left(\frac{q}{4\pi\varepsilon_0 c} \frac{1}{r'}, 0, 0, 0 \right)$$

Potentials in Σ

$$A^\mu = \left(L'^{-1}\right)_{\mu\alpha} A'^\alpha \quad \left(L'^{-1} \equiv L(v \to -\alpha)\right)$$

Transformation formulae:

$$A^0 = \gamma A'^0 + \gamma\beta A'^3 = \gamma A'^0 = \gamma \frac{q}{4\pi\varepsilon_0 c} \frac{1}{\hat{r}}$$

$$A^{1,2} = A'^{1,2} = 0$$

$$A^3 = \gamma\beta A'^0 + \gamma A'^3 = \gamma\beta A'^0 = \gamma\beta \frac{q}{4\pi\varepsilon_0 c} \frac{1}{\hat{r}}$$

This leads to

$$A^\mu = \frac{q}{4\pi\varepsilon_0} \frac{1}{\hat{r}} \gamma \frac{1}{c} (1, 0, 0, \beta)$$

That means explicitly for the electromagnetic potentials, we have

$$\varphi(x, y, z, t) = \frac{q}{4\pi\varepsilon_0} \frac{\gamma}{\sqrt{x^2 + y^2 + \gamma^2(z - vt)^2}}$$

$$A(x, y, z, t) = \frac{q}{4\pi\varepsilon_0} \frac{v}{c^2} \frac{\gamma}{\sqrt{x^2 + y^2 + \gamma^2(z - vt)^2}} (0, 0, 1)$$

2. Fields in Σ:
   (a) Magnetic induction:

$$B = \text{curl} A = \left(\frac{\partial}{\partial y} A_z, -\frac{\partial}{\partial x} A_z, 0\right)$$

With

$$\frac{\partial}{\partial y} \frac{1}{r'} = -\frac{y}{r'^3}; \quad \frac{\partial}{\partial x} \frac{1}{r'} = \frac{x}{r'^3}$$

it follows

$$B = \frac{q}{4\pi\varepsilon_0} \gamma \frac{\beta}{c} \frac{\beta}{c} (-y, x, 0)$$

(b) Electric field:

$$E = -\nabla\varphi - \dot{A} = -\left(\frac{\partial\varphi}{\partial x}, \frac{\partial\varphi}{\partial y}, \frac{\partial\varphi}{\partial z}, +\dot{A}_z\right)$$

It is

$$\nabla^2 = \frac{q}{4\pi\varepsilon_0} \gamma \left(-\frac{1}{r'^3}\right) (x, y, \gamma^2(z - vt)v)$$

Furthermore:

$$\dot{A} = \frac{q}{4\pi\varepsilon_0} \gamma \frac{v}{c^2} \left(\frac{1}{r'^3}\right) (0, 0, -\gamma^2(z - vt)v)$$

This means eventually:

$$E = \frac{q}{4\pi\varepsilon_0}\, \gamma\, \frac{1}{r'^3}\, (x, y, z - vt)$$

Because of

$$\beta \times E = \beta(-E_y, E_x, 0)$$

$$\Rightarrow \quad B = \frac{1}{c}(\beta \times E)$$

3. One recognizes:

$$\frac{1}{c^2}\frac{\partial^2\varphi}{\partial t^2} = \frac{v^2}{c^2}\frac{\partial^2\varphi}{\partial z^2}$$

and therewith

$$\Box\varphi = \left(\nabla^2 - \frac{1}{c^2}\frac{\partial^2}{\partial t^2}\right)\varphi = \left(\frac{\partial^2}{\partial x^2} + \frac{\partial^2}{\partial y^2} + \left(1 - \frac{v^2}{c^2}\right)\frac{\partial^2}{\partial z^2}\right)\varphi$$

$$= \left(\frac{\partial^2}{\partial x^2} + \frac{\partial^2}{\partial y^2} + \frac{1}{\gamma^2}\frac{\partial^2}{\partial z^2}\right)\varphi$$

We substitute:

$$u = z - vt: \quad v = \frac{1}{\gamma}y \text{ and } w = \frac{1}{\gamma}x$$

This means:

$$\frac{\partial}{dx} = \frac{1}{\gamma}\frac{\partial}{\partial w}; \quad \frac{\partial}{dy} = \frac{1}{\gamma}\frac{\partial}{\partial w}; \quad \frac{\partial}{\partial z} = \frac{\partial}{\partial u}$$

It follows:

$$\varphi = \frac{q}{4\pi\varepsilon_0}\frac{\gamma}{\sqrt{\gamma^2 w^2 + \gamma^2 v^2 + \gamma^2 u^2}} = \frac{q}{4\pi\varepsilon_0}\frac{1}{\sqrt{w^2 + v^2 + u^2}}$$

It remains to be calculated:

$$\left(\frac{\partial^2}{\partial x^2} + \frac{\partial^2}{\partial y^2} + \frac{1}{\gamma^2}\frac{\partial^2}{\partial z^2}\right)\varphi = \frac{1}{\gamma^2}\left(\frac{\partial^2}{\partial w^2} + \frac{\partial^2}{\partial v^2} + \frac{\partial^2}{\partial u^2}\right)$$

On the other hand, one gets with $r \equiv (w, v, u)$

$$\nabla^2_{w,v,u}\frac{1}{r} = -4\pi\delta(r) = -4\pi\delta(w)\delta(v)\delta(u)$$

So one can write:

$$\left(\frac{\partial^2}{\partial x^2} + \frac{\partial^2}{\partial y^2} + \frac{1}{\gamma^2}\frac{\partial^2}{\partial z^2}\right)\varphi = -\frac{q}{4\pi\varepsilon_0}4\pi\frac{1}{\gamma^2}\gamma\delta(x)\gamma\delta(y)\delta(z - vt)$$

$$= -\frac{q}{\varepsilon_0}\delta(x)\delta(y)\delta(z - vt) = -\frac{\rho(\mathbf{r})}{\varepsilon_0}$$

That is valid for the point charge:

$$\rho(x, y, z) = q\delta(x)\delta(y)\delta(z - vt)$$

GLIMPSES

1. *Special theory of relativity* is based on two postulates: (i) the laws of physics are the same in all inertial frames of reference. This is also known as *principle of equivalence*, and (ii) the speed of light is always a constant equal to c, and is independent of relative motion of the inertial system, the source and the observer. This is known as the principle of *constancy of speed of light*.

2. According to the special theory of relativity, time elapses differently in different frames of reference, and the specification of an interval of time between two events has a meaning only if the frame of reference is indicated.

3. Relativistic formulas for the transformation of coordinates that comply with the requirements for the invariance of intervals are called *Lorentz transformations*. They express the transformation from an inertial frame of reference S to a frame S' which travels w.r.t. S at the velocity v in the positive direction along x-axis.
   The transformation equations are of the form $x' = \gamma(x, vt)$, $y' = y$, $z' = z$ and
   $t' = \gamma(t - vx/c^2)$, where $\gamma = 1/\sqrt{1 - v^2/c^2}$.
   The Lorentz transformations are symmetrical and retain their form in the transformation from S' to S except that the sign of v is changed. Thus,

   $$x = \gamma(x' + vt'),\ y = y',\ z = z',\ \text{and } t = \gamma\left(t' + \frac{vx'}{c^2}\right)$$

   Lorentz transformations are linear and reduce to Galilean transformations at $v \to 0$. If $v > c$, the space and time coordinates will be imaginary. This means that no physical object can move with a velocity greater than the velocity of light.

4. The *invariance of any physical theory* w.r.t. the Lorentz transformations (relativistic), or *Lorentz invariance* is a necessary condition for the validity of this theory. The lack of relativistic invariance of any physical law means that it must be differently formulated in different inertial frames of reference. This violates the principle of relativity.

5. Time measured on a clock that travels together with a system is called its *proper time*. In this frame of reference, $dS = cdt'$. It follows from the Lorentz formulas that the following relation exists between the interval of proper time dt' and an interval of time dt in the frame of reference with respect to which motion at the velocity v is observed

$$dt' = \frac{dS}{c} = dt\sqrt{1 - \frac{v^2}{c^2}}$$

Proper time noted by a moving observer is always less than the corresponding interval of time in a stationary frame. A stationary observer finds that a moving clock runs slower than a stationary one.

6. Another consequence of *Lorentz transformation* is the contraction of a moving measuring rule or length standard in the direction of motion (Lorentz contraction). Thus,

$$\Delta x = \Delta x' \bigg/ \sqrt{1 - \frac{v^2}{c^2}}$$

where Δx is the length of the rule at rest in the frame S and $\Delta x'$ is the length of the rule measured in the frame S' which travels at the velocity v w.r.t. frame S.

7. The *proper length* is the linear dimension l_0 of a body in the system of coordinates in which it is at rest ($l_0 = \Delta x$). The length l of the same body, measured in a frame of reference S' which moves relative to the body, is reduced in the ratio $\sqrt{1 - \frac{v^2}{c^2}}$.

Thus, $l = l_0\sqrt{1 - \frac{v^2}{c^2}}$. The transverse dimensions of the body do not change, i.e. $\Delta y = \Delta y'$ and $\Delta z = \Delta z'$.

8. *Transformation of velocities:* The projections of the velocity v of a body on the coordinate axes in a stationary frame of reference S are related to the projections of its velocity v' in the frame S', travelling at the velocity v in the positive direction along x-axis by the equations

$$v_x = \frac{v'_x + v}{1 + \frac{v'_x v}{c^2}}, \quad v_y = \frac{v'_y \sqrt{1 - \frac{v^2}{c^2}}}{1 - \frac{v'_x v}{c^2}}, \quad v_z = \frac{v'_z \sqrt{1 - \frac{v^2}{c^2}}}{1 + \frac{v'_x v}{c^2}}$$

These relations represent the law of the composition of velocities in the theory of relativity. As $c \to \infty$, they are converted into the law of composition of velocities of classical mechanics

$$v_x = v'_x + v; v_y = v'_y \text{ and } v_z = v'_z.$$

When the body travels along the x-axis

$$(v_x = v, v_y = v_z = 0, v'_x = v', v'_y = v'_z = 0), \text{ then}$$

$$v = \frac{v' + v}{1 + \frac{v'v}{c^2}}$$

The relations for the transformations from v to v' differ from those given above only in the sign of v. If, in particular, $v = c$, then $v' = c$. The sum of the two velocities, lower than, or equal to c, is a velocity not higher than c. It follows from Lorentz transformations that $v < c$ in all cases. An exception is velocity of photon which equals c.

9. Velocity 4 vector is a vector with components

$$u_j = \frac{dx_j}{ds} \quad (j = 1, 2, 3, 4)$$

where $x_1 = x$; $x_2 = y$; $x_3 = z$; $x_4 = ict$ and $dS = cdt\sqrt{1 - v^2/c^2}$, and v is the velocity of the body.

The velocity-4 vector is a dimensionless quantity.

10. Acceleration-4 vector is defined as a vector with components

$$a_j = \frac{du_j}{ds} = \frac{d^2 x_j}{ds^2} \quad (j = 1, 2, 3, 4)$$

11. Mechanics based on the special principle of relativity and invariant w.r.t. the Lorentz transformations, is called *relativistic mechanics*.

12. The momentum vector p is

$$p = \frac{m_0 v}{\sqrt{1 - \dfrac{v^2}{c^2}}}$$

13. The relativistic expression for the mass is

$$m = \frac{m_0}{\sqrt{1 - \dfrac{v^2}{c^2}}}$$

where m_0 is the rest mass. As $v \to c$, $m \to \infty$ and $p \to \infty$ provided that $m_0 \neq 0$. In case of photon, $v = c$ and $m_0 \approx 0$. The momentum of a photon is determined on the basis of energy.

14. The Hamiltonian function for a free particle (the total energy expressed in terms of momentum) is

$$E = H = \sqrt{p^2 c^2 + m_0^2 c^4}$$

15. The properties of an electromagnetic field differ in different inertial frames of reference. In particular, one of the fields, either electric or magnetic, may be absent in one coordinate system, and present in another.

16. Electric charge:

$$q : \text{Lorentz invariant}$$

Four-current density:

$$j^\mu = (c\rho, j) = \gamma \rho_0 (c, v) = \rho_0 u^\mu$$

$$\rho_0 \; : \text{rest-charge density}$$

$$u^\mu \; = \; \gamma(c, v) : \text{world-velocity}$$

Continuity equation:

$$\partial_\mu j^\mu = 0; \quad \partial_\mu = \left(\frac{1}{c} \frac{\partial}{\partial t}, \Delta \right)$$

Four-potential:

$$A^\mu = \left(\frac{1}{c} \varphi, A \right)$$

Four-wave equation:

$$\Box A^\mu = -\mu_0 j^\mu \; : \; \Box = \Delta - \frac{1}{c^2} \frac{\partial^2}{\partial t^2}$$

Lorenz gauge:

$$\partial_\mu A^\mu = 0$$

Contravariant field-strength tensor:

$$F^{\mu v} = \partial^\mu A^v - \partial^v A^\mu \; : \left(\frac{1}{c} \frac{\partial}{\partial t}, -\Delta \right)$$

$$F^{\mu v} = \begin{pmatrix} 0 & -\frac{1}{c} E_x & -\frac{1}{c} E_y & -\frac{1}{c} E_z \\ \frac{1}{c} E_x & 0 & -B_z & B_y \\ \frac{1}{c} E_y & B_z & 0 & -B_x \\ \frac{1}{c} E_z & -B_y & B_x & 0 \end{pmatrix}$$

Covariant field-strength tensor:

$$F_{\mu v} = \mu_{\mu \alpha} \mu_{v \beta} F^{\alpha \beta}$$

Dual field-strength tensor:

$$\bar{F}_{\mu v} = \frac{1}{2} \varepsilon^{\mu v \rho \alpha} F_{\rho \sigma}$$

from $F_{\mu v}$ by substitution: $B \leftrightarrow -(1/c)E$.

Maxwell equations:

Homogeneous:

$$\partial_\alpha \bar{F}^{\alpha \beta} = 0 \; : \; \beta = 0, 1, 2, 3$$

or

$$\partial^\alpha F^{\beta \gamma} + \partial^\beta F^{\gamma \alpha} + \partial^\gamma F^{\alpha \beta} = 0$$

α, β, γ arbitrary from $(0, 1, 2, 3)$

Inhomogeneous:

$$\partial_\alpha F^{\alpha\beta} = \mu_0 j^\beta \; ; \beta = 0, 1, 2, 3$$

Transformation of the fields:

$$\beta = \frac{v}{c}$$

$$E' = \gamma[E + c(\beta \times B)] - \frac{\gamma^2}{\gamma + 1}\beta(\beta \cdot E)$$

$$B' = \gamma\left[B - \frac{1}{c}(\beta \times E)\right] - \frac{\gamma^2}{\gamma + 1}\beta(\beta \cdot B)$$

Especially for $v = ve_z$:

$$B'_x = \gamma\left(B_x + \frac{\beta}{c}E_y\right) \quad E'_x = \gamma(E_x - \beta c B_y)$$

$$B'_y = \gamma\left(B_y + \frac{\beta}{c}E_x\right) \quad E'_y = \gamma(E_y - \beta c B_x)$$

$$B'_z = B_z \quad E'_z = E_z$$

Invariants of the electromagnetic field:

$$\left(B^2 - \frac{1}{c^2}E^2\right) \text{ and } E \cdot B$$

Minkowski force:

$$K^\mu \equiv \left(\frac{F \cdot v}{c}, F_x, F_y, F_z\right)$$

$$F = q(E + v \times B) \text{ Lorentz force}$$

$$v : \text{particle velocity}$$

Covariant force equation:

$$K^\mu = m\frac{d}{d\tau}u^\mu = qF^{\mu\alpha}u_\alpha$$

REVIEW QUESTIONS

1. Show that the Maxwell's electromagnetic equations are invariant under the Lorentz transformation.
2. Express Maxwell's equations in covariant form and derive the transformation laws for the electric and magnetic fields.
3. What is an inertial frame? Show that under Lorentz transfor-mations all inertial frames are equivalent for the description of electromagnetic phenomenon.
4. Show that the charge is not altered by the motion relative to an observer.
5. Prove the Lorentz covariance of the first pair of Maxwell's equations of the electromagnetic field: $\nabla \cdot B = 0$ and $\nabla \times E = \frac{\partial B}{\partial t}$.

6. Using the tensor expression for the electromagnetic field components or otherwise, find the transformation equations for the electric and magnetic fields.

7. Write down the equation for field tensor and express Maxwell's equations in 4-tensor form. Using this form show that these are invariant under Lorentz transformations.

8. Show that $c^2B^2 - E^2$ and $E \cdot B$ are invariant under Lorentz transformations.

9. Derive the Lorentz transformations for electromagnetic potential four vector and current density four vector.

10. Write Maxwell's equations in the presence of a point charge in motion. Show that the total charge, energy and momentum of the system are conserved.

11. Write the Maxwell's equations in a covariant form and show that the electric and magnetic induction are elements of a second rank antisymmetric field strength tensors.

PROBLEMS

1. If two reference frames are related to each other by the Lorentz transformations,

$$x' = \frac{(x - vt)}{\sqrt{1 - (v^2/c^2)}}, \, y' = y, \, z' = z \, \text{ and } \, t' = \frac{t - (vx/c^2)}{\sqrt{1 - (v^2/c^2)}}$$

how are the electromagnetic potentials (ϕ, A) and electromagnetic field strengths (E, B) in the two systems related?

2. Let frames S' and S'' move with velocities v_1 and v_2 relative to a frame of reference S along z-axis. Write the Lorentz transformation equations between S' and S and between S'' and S. Find the law of addition for two velocities along z-axis.

$$\left[\textbf{\textit{Ans.}} \, u = \frac{(v_1 - v_2)}{\left[1 - \frac{v_1 v_2}{c^2} \right]} \right]$$

3. If p is momentum of particles and A and ϕ are vector and scalar potentials, what will be the total energy of the charged particles in an electromagnetic field?

$$\left[\textbf{\textit{Ans.}} \, W = \left(\left[\left(p - \frac{eA}{c} \right)^2 c^2 + m_0^2 c^2 \right]^{1/2} + e\phi \right) \right]$$

4. The frequency of vibration of an oscillator is ω_0. When this oscillator is placed in a magnetic field, show that the frequency will be

$$\omega = \sqrt{\omega_0^2 + \left(\frac{eB}{2\mu_0 mc} \right)^2} \pm \frac{eB}{2\mu_0 mc}.$$

Also show that for weak field, $\omega \approx \omega_0 \pm \dfrac{eB}{2\mu_0 mc}$.

5. Show explicitly that two successive Lorentz transformations in the same direction are equivalent to a single Lorentz transformation with a velocity $v = \dfrac{v_1 + v_2}{\left[1 + \dfrac{v_1 v_2}{c^2} \right]}$.

6. Show that if the medium has a velocity $v = c\beta$ with respect to some inertial frame, then the 3-vector current in the frame is

$$J = \lambda\sigma[E + \beta \times B - \beta\,(\beta \cdot E)] + \rho v$$

where ρ is the charge density observed in the frame.

7. Show that for liquid flow at a speed v, parallel or antiparallel to the path of the light, the speed of light as observed in the laboratory, is given to first order in v by

$$u = \frac{c}{n(\omega)} \pm v\left(1 - \frac{1}{n^2} + \frac{\omega}{n}\frac{dn(\omega)}{d\omega}\right)$$

where ω is the frequency of the light in the laboratory (in the liquid and outside it) and $n(\omega)$ is the index of refraction of the liquid.

8. Show that the relativistic form of the 3-dimensional acceleration vector for a charged particle in an electromagnetic field is as follows:

$$a = \frac{du}{dt} = \frac{e}{m\gamma}\left[E + \frac{1}{c}(u \times B) - \frac{1}{c^2}(E \cdot u)u\right]$$

SHORT ANSWER QUESTIONS

1. Is electromagnetic theory invariant under Newtonian mechanics?

Ans. No

2. Are all inertial frames equivalent?

Ans. Yes

3. What do you understand by the postulate of the constancy of the speed of light?

Ans. The speed of light has the same value in all inertial frames.

4. What do you understand by Fitzgerald contraction?

Ans. $L = L_0\sqrt{1-\beta^2}$. The rod appears contracted by a factor $\sqrt{1-\beta^2}$, i.e. every body appears to be longest when at rest relatively to the observer.

5. What is time dilation?

Ans. $\Delta t = \dfrac{\Delta t'}{\sqrt{1-\beta^2}}$, i.e. moving clock appears to run *slow*.

6. A force which is purely magnetic in S frame appears to be purely electrostatic in S' frame moving relative to S frame with velocity v in x-direction. What does this signify?

Ans. The theory of relativity provides the unification of electric and magnetic fields, i.e. the magnetic and electric effects are just two ways of looking at the same thing. This also leads to the conclusion that whether an electromagnetic field is purely magnetic or purely electric or electromagnetic field will depend on the inertial reference frame in which sources are observed.

7. What does the transformation equations of E signify?

Ans. The form of transformation equations of E, in general, asserts that the component of E parallel to the relative velocity of the two frames is unchanged, whereas the components of E perpendicular to the relative velocity transform to a mixed electric and magnetic fields and depend on the magnitude and direction of the relative velocity.

8. **What is Lorentz ionization?**

   Ans. The microscopic object of atomic size can be easily accelerated to a velocity of the order of 5×10^6 m/s. Such an atom moving in a magnetic field of 1 tesla in the S' frame (at rest with respect to the atom) experiences an electric field $\sim 5 \times 10^6$ V/m. The neutral atoms with captured electrons moving at this speed are most likely to be in the excited state. Such atoms subjected to electric fields of 5×10^6 V/m are quickly ionized and form an ion beam. This is termed *Lorentz ionization*. This process seems quite feasible on the laboratory scale and can provide a possible means of injecting ions into fusion reactor plasma.

9. **What is the difference between a four vector and an ordinary vector?**

   Ans. A four vector differs from an ordinary vector in one important respect. The length of a four vector can be zero and yet the four vector can have non-zero components. The length of a photon's four-momentum, i.e. its mass is zero, and yet the photon can have momentum and energy. Thus, a four vector can be called as time like, or null, depending on whether the square of its magnitude is negative, positive or zero respectively.

10. **What is eigen time or proper time?**

    Ans. This is the time measured in inertial frame in which the body is instantaneously at rest.

11. **What do you understand by frame of reference in relativity theory?**

    Ans. A frame of reference in relativity theory is known as the Galilean frame of reference. This corresponds to rigid bodies situated at remote spaces far away from attracting matter and without rotation relative to stars as whole. Thus, a Galilean frame of reference is a rigid body, isotropic w.r.t. mechanical and optical experiments.

12. **In how many parts is theory of relativity divided?**

    Ans. Theory of relativity is divided into two parts:

    (i) *Special theory of relativity:* This deals with phenomena where gravitational attraction plays no part.

    (ii) *General theory of relativity:* In this theory, gravitational attraction is taken into account and generally known as the Einstein's theory of gravitation.

13. **What is the significance of Lorentz transformation?**

    Ans. Lorentz showed that Maxwell's equations are invariant under a transformation known as Lorentz transformation, provided the field strengths were suitably transformed. By supposing that all matters are essentially electromagnetic in origin and so transformed in the same way in Maxwell's equations, Lorentz was able to deduce the contraction law

$$L = L_0 \sqrt{1 - \frac{v^2}{c^2}} = L_0 \sqrt{1 - \beta^2} .$$

14. Why Maxwell's equations of electrodynamics remain valid without any change even for high velocities?

   Ans. Maxwell's equations of electrodynamics remain valid without any change even for high velocities. This is because they are already form-invariant under Lorentz transformations, which is explicitly demonstrated while reformulating them in terms of four-tensors.

15. Show that the continuity equation is Lorentz invariants

   Ans. The experimental observation teaches us that the **electric charge q is a Lorentz invariant.**

   There does not exist the slightest indication that the charge of a particle is dependent on its velocity. However, this does not at all hold for quantities like the **charge density** ρ

   or the *current density* $j = \rho v$.

   The reason is evident and is, in the final analysis, related to the length contraction. Let Σ_0 be a (co-moving) inertial system in which the considered charge is at rest.

   $$dq = \rho_0 \, dV_0$$

   Hence, ρ_0 is the co-moving charge density.

   Let Σ be another inertial system moving relative to Σ_0 with the velocity v parallel to the z-axis. Since the amount of charge cannot have changed in the given volume element, we can set:

   $$dq = \rho dV$$

   For the volume element dV the length contraction leads to:

   $$dV = dxdydz = dx_0 \, dy_0 \, dz_0 \sqrt{1 - v^2/c^2} = \frac{1}{\gamma} dV_0$$

   From $\rho dV = \rho_0 dV_0$ it then follows for the charge density seen from Σ:

   $$\rho = \rho_0 \gamma \tag{1}$$

   According to its meaning as *rest-charge density* ρ_0 must be considered as Lorentz invariant. The charge density ρ produces in Σ a current density j:

   $$j = \gamma \rho_0 v \tag{2}$$

   One recognizes from Eqs. (1) and (2) a contravariant four-vector, the so-called, four-current density

   $$j^\mu = (c\rho, j_x, j_y, j_z) \equiv (c\rho \cdot j) = \gamma(\rho_0(c \cdot v)) = \rho_0 u^\mu \tag{3}$$

   That it is indeed a contravariant four-vector follows from the fact that ρ_0 is a four-scalar. J^μ thus transforms as the world-velocity u^μ, about which we already know that it is such a contravariant four-vector. Nevertheless, we prove it explicitly. In the rest system Σ_0 of the charge we have:

   $$j_0 = \rho_0 v_0 = 0 : j_0^0 = c\rho_0 \rightarrow j_0^\mu = (c\rho_0, 0, 0, 0)$$

The Lorentz transformation then yields for the components of the four-current density in Σ:

$$j^0 = c\rho = \gamma j_0^0 - \beta\gamma j_0^3 = c\gamma\rho_0$$

$$j^1 = j_x = j_{0x} = 0$$

$$j^2 = j_y = j_{0y} = 0$$

$$j^3 = j_z = -\beta\gamma j_0^0 + \gamma j_0^3 = -\gamma\rho_0 v = -\rho v$$

This is obviously the correct result when one bears in mind that $v = (0, 0, -v)$ is the velocity of the charge in Σ.

We now consider the **continuity equation**

$$\frac{\partial\rho}{\partial t} + div\ \dot{J}\ (\text{or } \nabla \cdot \dot{J}\ ) = 0$$

We recognize on the left-hand side the divergence of the four-vector j^μ as

$$\partial_\mu j^\mu = \frac{1}{c}\frac{\partial}{\partial t}j^0 + divj = divj + \frac{\partial}{\partial t}\rho$$

The continuity equation can therefore be written in the very compact form as

$$\partial_\mu j^\mu = 0 \tag{4}$$

Both sides of the equation are four-scalars. The continuity equation is thus Lorentz invariant.

16. Discuss electromagnetic potentials and show that Lorentz condition

$$\partial_\mu A^\mu \equiv 0$$

is a world-scalar Lorentz invariant.

Ans. Let $\phi(r, t)$; and $A(r, t)$ denote the scalar potential and vector potential respectively. We know that the **Maxwell equations** can be grouped into two homogeneous and two inhomogeneous differential equations:

$$\text{homogeneous:} \quad div\mathbf{B} \qquad = 0$$

$$curl\mathbf{E} + \dot{\mathbf{B}} = 0$$

$$\text{inhomogeneous:} \quad div\mathbf{D} \qquad = \rho \tag{1}$$

$$curl\mathbf{H} - \dot{\mathbf{D}} = \mathbf{j}$$

We restrict our considerations to the vacuum for which we have to use:

$$\mathbf{D} = \varepsilon_0\mathbf{E} : \mathbf{B} = \mu_0\mathbf{H} \tag{2}$$

The vector potential is defined by the ansatz

$$\mathbf{B} = curl\mathbf{A} \tag{3}$$

The second homogeneous Maxwell equation then reads

$$curl(\mathbf{E} + \dot{\mathbf{A}}) = 0 \tag{4}$$

which leads to the following ansatz for the electric field E

$$E = -grad\varphi - \dot{A} \tag{5}$$

φ and A are not uniquely determined by the above definition equations. One still has the free choice of a function $\chi(r, t)$ provided it is chosen so that

$$\gamma \to \varphi - \dot{\chi}; \quad A \to A + grad\chi \tag{6}$$

is guaranteed. Because of curlgrad $\chi = 0$ such a gauge transformation does not change the fields E and B. One therefore can fix it under practically aspects.

By the introduction of the electromagnetic potentials j and A the homogeneous Maxwell equations are automatically fulfilled, while the two inhomogeneous equations become now differential equations of second order for the potentials j and A. These equations, however, assume an especially symmetric shape if one chooses the gauge function $\chi(r, t)$ such that the

Lorenz condition

$$divA + \frac{1}{c^2}\dot{\varphi} = divA + \frac{1}{c}\frac{\partial}{\partial t}\left(\frac{1}{c}\varphi\right) = 0 \tag{7}$$

If fulfilled. With $c = (\mu_0\varepsilon_0)^{-1/2}$ the potentials φ and A are solutions of the following *wave equations:*

$$\Box A = \left(\nabla^2 - \frac{1}{c^2}\frac{\partial^2}{\partial t^2}\right)A = -\mu_0 j \tag{8}$$

$$\nabla^2\left(\frac{1}{c}\varphi\right) \equiv \left(\nabla^2 - \frac{1}{c^2}\frac{\partial^2}{\partial t^2}\right)\left(\frac{1}{c}\varphi\right) = -\frac{1}{c}\frac{\rho}{\varepsilon_0} - \mu_0(c\,\rho) \tag{9}$$

On the right-hand sides of these wave equations we recognize the space and time components of the four-current density j^μ. Since the d'Alembert operator $\Box$ is a scalar operator, Eqs. (8) and (9) make it appear reasonable to introduce a further four-vector, the

Four-potential

$$A^\mu \equiv \left(\frac{1}{c}\varphi, A_x, A_y, A_z\right) \equiv \left(\frac{1}{c}\varphi, A\right) \tag{10}$$

The wave equations for φ and A can therewith be combined as
Four-wave equation

$$\Box A^\mu = -\mu_0 j^\mu \tag{11}$$

which is covariant since both sides are four-tensors of the same, i.e. the first rank. The Lorenz condition (7) can finally be written as four-divergence of the potential A^μ. The relation

$$\partial_\mu A^\mu = \frac{1}{c}\frac{\partial}{\partial t}A^0 + divA$$

is obviously identical to the left-hand side of (7).

Lorenz condition

$$\partial_\mu A^\mu = 0 \tag{12}$$

is as *world-scalar Lorentz invariant*.

17. Find an expression for field-strength tensor and explain its importance.

    Ans. The field strengths E and B cannot be written in the relativistic electrodynamics as four-vectors. Instead of that, we have to introduce for the fields, a four-tensor of second rank which incorporates likewise the fields E and B. Starting point is again the connection between fields and potentials.

$$B = \text{curl} A \text{ and } E = -grad\varphi - \dot{A}$$

In terms of four gradients, we have

$$\partial_\mu = \left(\frac{1}{c}\frac{\partial}{\partial t}, \nabla^2\right) = (\partial_0, \partial_1, \partial_2, \partial_3) \tag{1}$$

$$\partial^\mu = \left(\frac{1}{c}\frac{\partial}{\partial t}, -\nabla^2\right) = (\partial^0, \partial^1, \partial^2, \partial^3) \tag{2}$$

It holds obiously:

$$\partial_0 = \partial^0 \ ; \ \partial_{1,2,3} = -\partial^{1,2,3} \tag{3}$$

We use this to reformulate at first the B-field:

$$B_x = \frac{\partial}{\partial y} A_z - \frac{\partial}{\partial z} A_y = \partial_2 A^3 - \partial_3 A^2 = -\left(\partial^2 A^3 - \partial^3 A^2\right)$$

Analogously, it follows for the two other Cartesian components:

$$B_y = -\left(\partial^3 A^1 - \partial^1 A^3\right); \ B_z = \left(\partial^1 A^2 - \partial^2 A^1\right)$$

The E-field can be written quite similarly:

$$E_x = -\frac{\partial}{\partial x}\varphi - \frac{\partial}{\partial t}A_x = -c\left[\frac{\partial}{\partial x}\left(\frac{1}{c}\phi\right) + \frac{1}{c}\frac{\partial}{\partial t}A_x\right]$$

$$= \left(\partial^1 A^0 - \partial^0 A^1\right)$$

Corresponding expressions are valid for E_y and E_z:

$$E_y = c(\partial^2 A^0 - \partial^0 A^2); \ E_z = c(\partial^3 A^0 - \partial^3 A^3);$$

We introduce

$$F^{\mu\upsilon} \equiv \partial^\mu A^\upsilon - \partial^\upsilon A^\mu \tag{4}$$

a new **four-tensor of second rank**. As a tensor of two contravariant four-vectors it is also *contravariant* and is obviously antisymmetric:

$$F^{\mu\upsilon} = -F^{\upsilon\mu} \tag{5}$$

One can regard this tensor as *four-dimensional generalization of the curl of the vector A^μ*:
Field-strength tensor

$$
F^{\mu\nu} \equiv
\begin{pmatrix}
0 & -\dfrac{1}{c}E_x & -\dfrac{1}{c}E_y & \dfrac{1}{c}E_z \\[2mm]
\dfrac{1}{c}E_x & 0 & -B_z & B_y \\[2mm]
\dfrac{1}{c}E_y & B_z & 0 & -B_x \\[2mm]
\dfrac{1}{c}E_z & -B_y & B_x & 0
\end{pmatrix}
\tag{6}
$$

The electromagnetic field is described in the *Minkowski space* not anymore by two fields but rather by one tensor of second rank. Now, we will use the field-strength tensor for the covariant formulation of the Maxwell equations.

The *covariant field-strength tensor* comes out easily by use of the general 'conversion-prescription'

$$
F_{\mu\nu} \equiv \mu_{\mu\alpha}\mu_{\mu\beta}F^{\alpha\beta}
\tag{7}
$$

Since the metric tensor $\mu_{\alpha\beta}$ is diagonal in the special theory of relativity, it simply follows:

$$
F_{0\nu} = -F^{0\nu}; \quad F_{\nu 0} = -F^{\nu 0}; \quad F_{\mu\nu} = F^{\mu\nu}; \quad \mu\nu \in \{1,2,3\}
\tag{8}
$$

So we have to replace in (6) only E by $-E$ in order to come from $F^{\mu\nu}$ to $F_{\mu\nu}$. We have an important **invariant of the electromagnetic field**

$$
F_{\mu\nu}F^{\mu\nu} = 2\left(B^2 - \frac{1}{c^2}E^2 \right)
\tag{9}
$$

Which as a four-scalar remains uninfluenced by Lorentz transformations. It is obviously never possible to transform a **pure B-field** into a **pure E-field**, or vice versa, since the two terms in (9) carry different signs.

18. Derive Maxwell's equations in explicit covariant form with the aid of the field strength tensor (Eq. (6) of Q. 17).

Ans. We start with the inhomogeneous equations which can be written with $c = (\varepsilon_0\mu_0)^{-(1/2)}$ as follows:

$$
\mathrm{div}\left(\frac{1}{c}E \right) = \mu_0 cp = \mu_0 j^0
\tag{1}
$$

$$
\mathrm{curl}B - \frac{1}{c}\frac{\partial}{\partial t}\left(\frac{1}{c}E \right) = \mu_0 j
\tag{2}
$$

On the right-hand side of these equations we identify the components of the four-current j^μ (Eq. (3), Q. 14). The left –hand sides should therefore also be components of a four vector, if, as initially stated, the system of the Maxwell equations is indeed covariant. Let us try to express the left-hand sides by the field-strength tensor.

$\boxed{\mu = 0}$

$$\mu_0 j^0 = div\left(\frac{1}{c}\boldsymbol{E}\right) = \frac{1}{c}\left(\frac{\partial}{\partial x}E_x + \frac{\partial}{\partial y}E_y + \frac{\partial}{\partial z}E_z\right)$$

$$= \partial_1 F^{10} + \partial_2 F^{20} + \partial_3 F^{30} = \partial_\alpha F^{\alpha 0}$$

$\boxed{\mu = 1}$

$$\mu_0 j^1 = \mu_0 j_x = \frac{\partial}{\partial y}B_z - \frac{\partial}{\partial z}B_y + \frac{1}{c}\frac{\partial}{\partial t}\left(-\frac{1}{c}E_x\right)$$

$$= \partial_2 F^{21} + \partial_3 F^{31} + \partial_0 F^{01} = \partial_\alpha F^{\alpha 1}$$

$\boxed{\mu = 2}$

$$\mu_0 j^2 = \mu_0 j_y = \frac{\partial}{\partial z}B_x - \frac{\partial}{\partial x}B_z + \frac{1}{c}\frac{\partial}{\partial t}\left(-\frac{1}{c}E_y\right)$$

$$= \partial_3 F^{32} + \partial_1 F^{12} + \partial_0 F^{02} = \partial_\alpha F^{\alpha 2}$$

$\boxed{\mu = 3}$

$$\mu_0 j^3 = \mu_0 j_z = \frac{\partial}{\partial x}B_y - \frac{\partial}{\partial y}B_x + \frac{1}{c}\frac{\partial}{\partial t}\left(-\frac{1}{c}E_z\right)$$

$$= \partial_1 F^{13} + \partial_2 F^{23} + \partial_0 F^{03} = \partial_\alpha F^{\alpha 3}$$

These relations can be summed up to a compact expression:

Inhomogeneous Maxwell equations

$$\partial_\alpha F^{\alpha\beta} = \mu_0 j^\beta \ ; \quad \beta = 0, 1, 2, 3 \tag{3}$$

On the left we have a *contracted* third-rank tensor, consequently a four-vector like the one on the right-hand side. Covariance is therewith guaranteed. The inhomogeneous Maxwell equations in this form are valid in all inertial systems. We now consider the homogeneous Maxwell equations:

$$div\boldsymbol{B} = 0 \tag{4}$$

$$curl\left(\frac{1}{c}\boldsymbol{E}\right) + \frac{1}{c}\frac{\partial}{\partial t}\boldsymbol{B} = 0 \tag{5}$$

For (4) we can also write with [(3) and (4) Q.17]

$$0 = div\boldsymbol{B} = \frac{\partial}{\partial x}B_x + \frac{\partial}{\partial y}B_y + \frac{\partial}{\partial z}B_z = \partial_1 F^{12} + \partial_2 F^{13} + \partial_3 F^{21}$$

$$= \left(\partial^1 F^{23} + \partial^2 F^{31} + \partial^3 F^{12}\right)$$

The three components of the vector-equation (5) can be reformulated with [(3) Q.17] in the following manner:

$$0 = \mathrm{curl}\left(\frac{1}{c}E\right)_x + \frac{1}{c}\frac{\partial}{\partial t}B_x = \frac{\partial}{\partial y}\left(\frac{1}{c}E_z\right) - \frac{\partial}{\partial z}\left(\frac{1}{c}E_y\right) + \frac{1}{c}\frac{\partial}{\partial t}B_x$$

$$= \partial_2 F^{30} + \partial_3 F^{02} - \partial_0 F^{23} = -\left(\partial^2 F^{30} + \partial^3 F^{02} + \partial^0 F^{23}\right)$$

$$0 = \mathrm{curl}\left(\frac{1}{c}E\right)_y + \frac{1}{c}\frac{\partial}{\partial t}B_y = \frac{\partial}{\partial x}\left(\frac{1}{c}E_x\right) - \frac{\partial}{\partial x}\left(\frac{1}{c}E_z\right) + \frac{1}{c}\frac{\partial}{\partial t}B_y$$

$$= \partial_3 F^{10} + \partial\cdot F^{03} - \partial_0 F^{31} = -\left(\partial^3 F^{10} + \partial^1 F^{03} + \partial^0 F^{31}\right)$$

$$0 = \mathrm{curl}\left(\frac{1}{c}E\right)_z + \frac{1}{c}\frac{\partial}{\partial t}B_z = \frac{\partial}{\partial x}\left(\frac{1}{c}E_y\right) - \frac{\partial}{\partial y}\left(\frac{1}{c}E_x\right) + \frac{1}{c}\frac{\partial}{\partial t}B_z$$

$$= \partial_1 F^{20} + \partial_2 F^{01} - \partial_0 F^{12} = -\left(\partial^1 F^{20} + \partial^2 F^{01} + \partial^0 F^{12}\right)$$

These equations, too, can be brought into a compact form:

Homogeneous Maxwell equations

$$\partial^\alpha F^{\beta\gamma} + \partial^\beta F^{\gamma\alpha} + \partial^\gamma F^{\alpha\beta} = 0, \quad \alpha, \beta, \gamma \text{ arbitrary from } (0, 1, 2, 3). \tag{6}$$

All additive terms in this expression, which differ from one another by a cyclic interchange of the indexes α, β, γ, are four-tensors of the same rank. If two indexes in (6) are equal, then the left-hand side is identical to zero. So it follows, for instance, for $\alpha = \beta$ [(5) Q. 17]

$$\partial^\alpha F^{\alpha\gamma} + \partial^\alpha F^{\gamma\alpha} + \partial^\gamma F^{\alpha\alpha} = \partial^\alpha(F^{\alpha\gamma} - F^{\alpha\gamma}) = 0$$

Hence, only the combinations $(0, 1, 3)$, $(0, 1, 2)$, $(1, 2, 3)$ are of interest. But these are just the above discussed four homogeneous Maxwell equations.

Equations (3) and (6) demonstrate that the system of Maxwell equations can be expressed by four-tensors in a very compact and symmetric form, where the covariance with respect to Lorentz transformations becomes immediately evident. A still more compact representation of the homogeneous Maxwell equations than that in (6) is achieved by introducing the so-called **dual field-strength tensor**.

$$\overline{F}^{\mu\nu} = \frac{1}{2}\varepsilon^{\mu\nu\rho\sigma}F_{\rho\sigma} \tag{7}$$

Here we use the totally antisymmetric four-tensor

$$\varepsilon^{\mu\nu\rho\sigma} = \begin{cases} +1, & \text{if}(\mu, \nu, \rho, \sigma) \text{ is an even permutation of } (0, 1, 2, 3) \\ -1, & \text{if it is an odd permutation} \\ 0, & \text{if two or more indexes are equal} \end{cases} \tag{8}$$

The elements $F_{\mu\nu}$ of the covariant field-strength tensor are related by [(8) Q.17] to those of the contravariant tensor (7). At first, one read off immediately from the definition (7).

$$\overline{F}^{\mu\nu} = \overline{F}^{\nu\mu} \tag{9}$$

Hence the diagonal elements are all zero. Let us calculate as an example under consideration of [(8) Q. 17]

$$\overline{F}^{12} = \frac{1}{2}\varepsilon^{12\rho\sigma}F_{\rho\sigma} = \frac{1}{2}\left(\varepsilon^{1230}F_{30} + \varepsilon^{1203}F_{03}\right)$$

$$= \frac{1}{2}\left(-F_{30} + F_{03}\right) = F^{30} = \frac{1}{c}E_z$$

Fully analogously one finds

$$\overline{F}^{13} = F^{02}; \quad \overline{F}^{01} = F^{23}; \quad \overline{F}^{23} = F^{10}; \quad \overline{F}^{02} = F^{31}; \quad \overline{F}^{03} = F^{12}$$

According to that one obtains the components of the dual field-strength tensor $\overline{F}^{\mu\nu}$ from those of the covariant tensor Fmv by the replacement:

$$B \longleftrightarrow -\frac{1}{c}E \tag{10}$$

This yields with [(6) Q. 17]

$$\overline{F}^{\mu\nu} = \begin{pmatrix} 0 & -B_x & -B_y & -B_z \\ B_x & 0 & \dfrac{1}{c}E_z & -\dfrac{1}{c}E_y \\ B_y & -\dfrac{1}{c}E_z & 0 & \dfrac{1}{c}E_x \\ B_z & \dfrac{1}{c}E_y & -\dfrac{1}{c}E_x & 0 \end{pmatrix} \tag{11}$$

It is now easy to check the following relations:

$$\partial_\alpha \overline{F}^{\alpha 0} = \partial_0\overline{F}^{00} + \partial_1\overline{F}^{10} + \partial_2\overline{F}^{20} + \partial_3\overline{F}^{30}$$

$$= \partial_1 F^{32} + \partial_2 F^{13} + \partial_3 F^{21} = \partial^1 F^{23} + \partial^2 F^{31} + \partial^3 F^{12}$$

$$\partial_\alpha \overline{F}^{\alpha 1} = \partial_0 F^{23} + \partial_2 F^{03} + \partial_3 F^{20} = \partial^0 F^{23} + \partial^2 F^{30} + \partial^3 F^{02}$$

$$\partial_\alpha \overline{F}^{\alpha 2} = \partial^0 F^{31} + \partial^1 F^{03} + \partial^3 F^{10}$$

$$\partial_\alpha \overline{F}^{\alpha 3} = \partial^0 F^{12} + \partial^1 F^{20} + \partial^2 F^{01}$$

We thus can also write in lieu of (6)
Homogeneous Maxwell equations

$$\partial_\alpha \overline{F}^{\alpha\beta} = 0; \quad \beta = 0, 2, 2, 3$$

Eventually, one still recognizes with [(6) q. 17] and (11) a further
Invariant of the electromagnetic field

$$\partial_{\alpha\beta}\overline{F}^{\alpha\beta} = -\frac{4}{c}E \cdot B$$

A *four-tensor of zeroth rank appears* on the left side, i.e., a world-scalar product of the electric and the magnetic field $\mathbf{E} \cdot \mathbf{B}$ therefore does not change under a Lorentz transformation being therewith the same in all inertial systems.

19. Using the result of Q 14 to 18 calculate how now electric and magnetic fields behave under a Lorentz transformation?

Ans. The transformed field strength tensor following the transformation behavior of a covariant second-rank tensor is obtained as

$$(F^{\mu\nu})' = \frac{\partial x'^\mu}{\partial x^\alpha} \frac{\partial x'^\nu}{\partial x^\beta} F^{\alpha\beta} = L_{\mu\alpha} L_{\nu\beta} F^{\alpha\beta}$$

Because of $F^{\alpha\beta} = -F^{\beta\alpha}$ it also follows immediately

$$(F^{\mu\nu})' = -(F^{\nu\mu})'$$

This means in particular that the diagonal elements of the transformed tensor vanish.

Therefore, only six elements have to be explicitly calculated:

$$(F^{01})' = L_{11}(L_{00}F^{01} + L_{03}F^{31}) = \gamma\left(-\frac{1}{c}E_x + \beta B_y\right) \overset{!}{=} -\frac{1}{c}E'_x$$

$$(F^{02})' = L_{22}(L_{00}F^{02} + L_{03}F^{32}) = \gamma\left(-\frac{1}{c}E_y + \beta B_x\right) \overset{!}{=} -\frac{1}{c}E'_y$$

$$(F^{03})' = L_{00}(L_{30}F^{00} + L_{33}F^{03}) + L_{03}\left(L_{30}F^{30} + L_{33}F^{33}\right)$$

$$= \gamma^2(1 - \beta^2)\left(-\frac{1}{c}E_z\right) = -\frac{1}{c}E_z = -\frac{1}{c}E'_z$$

$$(F^{12})' = L_{11}L_{22}F^{12} = -B_z = -B'_z$$

$$(F^{13})' = L_{11}(L_{30}F^{10} + L_{33}F^{13}) = \gamma\left(B_y - \frac{\beta}{c}E_x\right) = B'_y$$

$$(F^{23})' = L_{22}(L_{30}F^{20} + L_{33}F^{13}) = \gamma\left(-\frac{\beta}{c}E_y - B_x\right) = -B'_x$$

The components of the magnetic field transform according to the following formulae:

$$B'_x = \gamma\left(B_x + \frac{\beta}{c}E_y\right) \tag{1}$$

$$B'_y = \gamma\left(B_y - \frac{\beta}{c}E_x\right) \tag{2}$$

$$B'_z = B_z \tag{3}$$

For the components of the electric field we have:

$$E'_x = \gamma \left(E_x - \beta c B_y \right) \tag{4}$$

$$E'_y = \gamma \left(E_y + \beta c B_x \right) \tag{5}$$

$$E'_z = E_z \tag{6}$$

These formulae demonstrate the close entanglement of electric and magnetic fields. That what appears in one system as pure *E*- or *B*-field is in an other inertial system a mixture of both. Under all Lorentz transformations, however, the **Invariants of the electromagnetic field**

$$\left(B^2 - \frac{1}{c^2} E^2 \right) \quad \text{and} \quad E \cdot B$$

Remain unchanged. From that it follows, as we have already stated earlier, that it is impossible to transform a pure *B*-field into a pure *E*-field and vice versa.

For certain purposes it appears reasonable to resolve the fields into components parallel and perpendicular to the relative velocity *v* of the two inertial systems Σ and Σ'. The parallel-component of the electric field is of course identical to the z-component:

$$E'_{\parallel} = E'_x \hat{e}_x = E_z \hat{e}_z = E_{\parallel} \tag{7}$$

The orthogonal component is determined slightly more complicatedly:

$$E'_{\perp} = E'_x e_x + E'_y e_y = \gamma \left[E_x e_x + E_y e_y + \beta c (B_x \hat{e}_y - B_y \hat{e}_x) \right]$$

With

$$\beta = \frac{v}{c} \equiv \left(0, 0, \frac{v}{c} \right) \tag{8}$$

The last summand can be written as vector product:

$$E' = \gamma \left[E_{\perp} + c(\beta \times B) \right] \tag{9}$$

Analogously one finds for the magnetic induction:

$$B' = B_{\parallel}; \quad B'_{\perp} = \gamma \left[B_{\perp} + \frac{1}{c}(\beta \times E) \right] \tag{10}$$

These results for the electromagnetic field can be combined as follows:

$$E' = \gamma \left[E + c(\beta \times B) \right] - \frac{\gamma^2}{\gamma + 1} \beta(\beta \cdot E) \tag{11}$$

$$B' = \gamma \left(B - \frac{1}{e}(\beta \times E) \right) - \frac{\gamma^2}{\gamma + 1} \beta(\beta \cdot B) \tag{12}$$

In this form the transformation formulae are valid for arbitrary velocities $v \cdot v$ need not necessarily be directed parallel to the z-axis. Let us check it:

$$E'_{\parallel} = \frac{\beta}{\beta} \left(E' \cdot \frac{\beta}{\beta} \right) = \frac{\beta}{\beta^2} \left(\gamma \ E \cdot \beta - \frac{\gamma^2}{\gamma + 1} \beta^2 (\beta \cdot E) \right)$$

$$= E_{\parallel}\left(\gamma - \frac{\gamma^2}{\gamma + 1}\beta^2\right) = E_{\parallel}\frac{\gamma^2 + \gamma - \gamma^2\beta^2}{\gamma + 1} = E_{\parallel}$$

$$E'_{\perp} = E' - \frac{\beta}{\beta}\left(E' \cdot \frac{\beta}{\beta}\right) = E' - E_{\parallel}$$

$$= \gamma[E_{\perp} + c(\beta \times B)] + E_{\parallel}\left(\gamma - \frac{\gamma^2}{\gamma + 1}\beta^2 - 1\right) = \gamma[E_{\perp} + c(\beta \times B)]$$

These are for the E-field the correct expressions (7) and (9). The procedure for the B-field succeeds in the same manner.

We discuss as an example of application use the fields of a moving point charge q.

We assume that the charge q is located at the origin of Σ'. Let Σ be the rest system of the charge. On the other hand, let Σ be the rest system of an observer P, where we can assume, without loss of generality, that his position lies on the y-axis in Σ'. We assume that Σ' moves relative to Σ with the constant velocity v in z-direction, where at $t = t' = 0$ both the systems of coordinates coincide (Fig. 14.7). The position of P in Σ is given by

$$r_P = (0, y_P, 0)$$

and in Σ' by

$$r' = (0, y_P, -vt') \tag{13}$$

The distance of the observer from the point charge

$$r' = \sqrt{y_P^2 + v^2 t'^2}$$

is, because of

$$t' = \gamma\left(t - \frac{v}{c^2}z_P\right) = \gamma t$$

in Σ-coordinates:

$$r' = \sqrt{y_P^2 + v^2\gamma^2 t^2} \tag{14}$$

Fig. 14.7: For the calculation of the fields of a moving point charge in two inertial systems moving relative to one another

Which fields are now seen by the observer P when the point charge q moves with $v = $ const along the z-axis?

The fields in Σ' are those of a charge **at rest**, therefore known from non-relativistic electrodynamics

$$B' = 0 \tag{15}$$

$$E'(t') = \frac{q}{4\pi\varepsilon_0}\frac{r'}{r'^3} = \frac{q}{4\pi\varepsilon_0 r'^3}(0, y_P, -vt') \tag{16}$$

The E-field is time-dependent since the position of the observer P in Σ' charges in the course of time, i.e. the field is measured for different times t' at different places r' in Σ' (Fig. 14.8).

For the calculation of the E-field in Σ we use the transformation formulae (4) to (6), where it is to bear in mind that Σ moves relative to Σ' with the velocity $-v$:

$$E_x = \gamma\left(E_x' + \beta c B_y'\right) = 0 \tag{17}$$

$$E_y = \gamma\left(E_y' - \beta c B_x'\right) = \gamma E_y' = \frac{q}{4\pi\varepsilon_0} \frac{\gamma\, y_P}{\left(y_P^2 + \gamma^2 v^2 t^2\right)^{3/2}} \tag{18}$$

$$E_z = E_z' = \frac{q}{4\pi\varepsilon_0} \frac{-\gamma\, vt}{\left(y_P^2 + \gamma^2 v^2 t^2\right)^{3/2}} \tag{19}$$

The **transverse** component E_y is obviously remarkably different from zero only in the time interval

$$-t_{1/2} \le t \le +t_{1/2}$$

where the half width $t_{1/2}$ is defined by

$$\frac{E_y(t = t_{1/2})}{E_y(t = 0)} = \frac{1}{2}$$

It is thus fixed by

$$\frac{1}{2} = \frac{y_P^3}{\left(y_P^2 + \gamma^2 v^2 t_{1/2}^2\right)^{3/2}}$$

as

$$t_{1/2} = \frac{y_P}{\gamma v}\left(2^{2/3} - 1\right)^{1/2}$$

Fig. 14.8: Time-dependence of the transverse component E_y of the electric field of a point charge, which moves relative to the observer in z-direction, seen from the rest system of the observer

The maximum of the E_y-peak appears at $t = 0$, hence at the moment of the minimum spatial distance between charge q and P. In the relativistic region $v \le c$ the peak becomes very sharp because $t_{1/2}$ is then very small.

The **longitudinal** component E_z of the electric field (19) changes its sign (Fig. 14.9) at $t = 0$ and has extrema at

$$\pm t_0 = \frac{y_P}{\gamma v} \frac{1}{\sqrt{2}}$$

This is easily proved by putting the first time-derivative equal to zero. The extreme values,

$$E_z(t = \pm t_0) = \mp \frac{q}{6\sqrt{3}\pi\varepsilon_0 y_P^2}$$

are thereby v-independent.

Fig. 14.9: Time-dependence of the longitudinal component E_z of the electric field of a point charge, which moves relative to the observer in z-direction, seen from the rest system of the observer

How does the spatial distribution of the electric field **E** relative to the point charge look like in the observer-system Σ at a fixed time t? Figure 14.7 illustrates

$$y_p = r \sin \varphi \; ; \; vt = r \cos \varphi$$

where r is the distance between charge an observer in Σ. It follows with

$$\left(y_P^2 + \gamma^2 v^2 t^2 \right) = r^3 (\sin^2 \varphi + \gamma^2 \cos^2 \varphi)^{3/2}$$

$$= r^3 \gamma^3 (1 - \beta^2 \sin^2 \varphi)^{3/2}$$

for the field distribution in Σ around the point charge q:

$$E = \frac{qr}{4\pi\varepsilon_0 r^3 \gamma^2 (1 - \beta^2 \sin^2 \varphi)^{3/2}} \tag{20}$$

r is the space vector form the point charge to the observer. The E-field is radial as for a resting charge, but for $\beta \neq 0$ it is no longer isotropic. In the direction of motion ($\varphi = 0$ or π), we have:

$$\frac{E(\beta)}{E(0)} = \frac{1}{\gamma^2}$$

and, in contrast, **perpendicular** to the direction of motion ($\phi = \tfrac{\pi}{2}$)

$$\frac{E(\beta)}{E(0)} = \gamma$$

Now, we discuss the B-field of the moving point charge. With the transformation formulae (1) to (3) we find for the Cartesian components in Σ:

$$B_x = \gamma \left(B_x' - \frac{\beta}{c} E_y' \right) = -\gamma \frac{\beta}{c} E_y' = -\frac{\beta}{c} E_y'$$

$$B_y = -\gamma \left(B_y' + \frac{\beta}{c} E_x' \right) = 0$$

$$B_z = B_z' = 0$$

For the evaluation, we have used Eqs. (15) and (16). We see that the moving point charge generates a magnetic induction in x-direction. Its time-dependence corresponds to that of Ey, which we have discussed in detail. A bit more generalization leads for the magnetic induction according to (12) with $\beta' \equiv 0$.

$$B = \frac{\gamma}{c} (\beta \times E') = \frac{\mu_0}{4\pi} \frac{q\gamma}{r'^3} (v \times r')$$

For $\gamma = 1$ this is the **Biot and Savart law** for the magnetic field produced by a moving charge. In our special case ($v = v\hat{e}_z$), we have

$$v \times r' = vy_p \hat{e}_z$$

20. Obtain the covariant formulation of Lorentz force.

Ans. We have

$$F = q(E + v \times B) \tag{1}$$

At first we interpret F as the time derivative of the space components of the four-momentum

$$p^\mu = \left(\frac{T_r}{c}, p_r \right) = mu^\mu = m\gamma(c, v) \tag{2}$$

However, $(d/dt)p_r$ are by themselves, not yet the space components of a four-vector since the time t does not transform accordingly. We therefore introduce again the proper time τ ($t = \gamma\tau$) and try to generalize the Lorentz force in such a manner that, as in mechanics, a covariant force law comes out:

$$K^\mu \equiv \frac{d}{d\tau} p^\mu = m \frac{d}{dt} u^\mu \quad , \tag{3}$$

On the left-hand side there should of course appear the **Minkowski force**

$$K^\mu \equiv \gamma \left(\frac{F \cdot v}{c}, F_x, F_y, F_z \right) \equiv (K^0, K) \tag{4}$$

For the space components, we have

$$K \equiv \frac{d}{d\tau} p_r = \gamma F = q(\gamma E + \gamma v \times B)$$

On the right-hand side we introduce the contravariant four-velocity u^μ

$$u^\mu \equiv \gamma(c, v) \equiv (u_0, u)$$

Then we have:

$$K = q \left[u^0 \left(\frac{1}{c} E \right) + u \times B \right] \tag{5}$$

We express the right-hand side by the contravariant field-strength tensor

$$K_x = q \left[u^0 \left(\frac{1}{c} E_x \right) + u_y B_z - u_z B_y \right] = q \left(u^0 F^{10} + u^2 F^{21} + u^3 F^{31} \right) = K^1$$

$$K_y = q \left[u^0 \left(\frac{1}{c} E_y \right) + u_z B_x - u_x B_z \right] = q \left(u^0 F^{20} + u^3 F^{32} + u^1 F^{12} \right) = K^2$$

$$K_z = q \left[u^0 \left(\frac{1}{c} E_z \right) + u_x B_y - u_y B_x \right] = q \left(u^0 F^{30} + u^1 F^{13} + u^2 F^{23} \right) = K^3$$

Now, if we replace in these expressions the terms $F^{\alpha\beta}$ with $\beta \neq 0$ by $-F^{\beta\alpha}$, and the contravariant by the covariant four-velocity ($u^0 \to u_0; u \to -u$), then because of $F^{\mu\mu} = 0$ it obviously holds:

$$K^1 = K_x = qF^{1\alpha} u_\alpha \tag{6}$$

$$K^2 = K_y = qF^{2\alpha} u_\alpha \tag{7}$$

$$K^3 = K_z = qF^{3\alpha} u_\alpha \tag{8}$$

The right-hand sides transform certainly like the space components of a contravariant four-vector. It therefore suggests itself to complement the system by a corresponding time component.

$$K^0 = qF^{0\alpha}u_\alpha$$

$$= q\left[\left(-\frac{1}{c}E_x\right)(-u_x) + \left(-\frac{1}{c}E_y\right)(-u_y)\left(-\frac{1}{c}E_z\right)(-u_z)\right]$$

$$= q\left(\frac{1}{c}E\right)\cdot(\gamma v)$$

This expression is easy to interpret because

$$qv \cdot E = v \cdot F$$

This leads to

$$K^0 = \gamma\left(\frac{F \cdot v}{c}\right) = \gamma\frac{d}{dt}\left(\frac{1}{c}Tr\right) = \frac{d}{d\tau}p^0 \tag{9}$$

This result fits (3) and (4). We have therewith found the covariant representation of the Lorentz force:

$$K^\mu = \frac{d}{d\tau}p^\mu = qF^{\mu\alpha}u_\alpha = \gamma\left(\frac{1}{c}F\cdot v, F\right) \tag{10}$$

K^μ is the Minkowski force for the covariant representation of classical mechanics. Let us investigate the explicit transformation behavior of the Lorentz force F. We consider for the purpose, two inertial systems Σ and Σ', out of which Σ' moves relative to Σ with the velocity v parallel to the z-axis. A particle of the charge q with the velocity u in Σ then experiences the Lorentz force:

$$F = q(E + u \times B) \quad \text{in } \Sigma \tag{11}$$

Due to which force F' does it move as seen by an observer in Σ'? We determine F' from the transformation behavior of the space components of the contravariant Minkowski force (4)

$$K^{1'} = K^1; \quad K^{2'} = K^2$$

$$K^{3'} = \gamma_v\left(-\beta K^0 + K^3\right); \quad \beta = \frac{v}{c}; \quad \gamma_v = (1-\beta^2)^{-1/2}$$

If u' is the particle velocity in F', then it follows from these equations:

$$\gamma_{u'}F'_x = \gamma_u F_x; \quad \gamma_{u'}F'_y = \gamma_u F_y$$

$$\gamma_{u'}F'_z = \gamma_u\left(\gamma_u F_z - \beta\gamma_u\frac{F \cdot u}{c}\right)$$

One finally obtains

$$\gamma_{u'} = \gamma_u\gamma_v\left(1 - \frac{vu_z}{c^2}\right)$$

Therewith, the components of the Lorentz force in F' read

$$F'_x = \frac{1}{\gamma_v} \frac{F_x}{1 - \frac{v u_z}{c^2}} \tag{12}$$

$$F'_y = \frac{1}{\gamma_v} \frac{F_y}{1 - \frac{v u_z}{c^2}} \tag{13}$$

$$F'_z = \frac{F_z - \frac{v}{c^2}(F \cdot u)}{1 - \frac{v u_z}{c^2}} \tag{14}$$

MULTIPLE CHOICE QUESTIONS

1. The net charge density (λ) in frame S is zero. In frame S' moving with a velocity v relative to S along the x-axis, the charge density (λ') will be (n is the number of electrons per unit length of the conducting wire)

   (a) zero (b) $n e \beta^2$

   (c) $\dfrac{n e \beta^2}{\sqrt{1 - \beta^2}}$ (d) $\dfrac{n e \beta^2}{2}$ [c]

2. In MCQ 1, the electrostatic field around the conducting wire according to the observer in S' is

   (a) $E' = 0$ (b) $E' = \dfrac{n e \beta^2}{2 \pi \varepsilon_0 r}$

   (c) $E' = \dfrac{n e \beta^2}{2 \pi \varepsilon_0 r^2 \sqrt{1 - \beta^2}}$ (d) $E' = \dfrac{n e \beta^2}{4 \pi \varepsilon_0 r}$ [c]

3. The square of the length of the four-vector S is

   (a) $x^2 + y^2 + z^2 - c^2 t^2$ (b) $x^2 + y^2 + z^2$

   (c) $(x^2 + y^2 + z^2 - c^2 t^2)^{1/2}$ (d) zero [a]

4. The d'Alembertian operator ($\Box^2$) is

   (a) $\dfrac{\partial^2}{\partial x^2} + \dfrac{\partial^2}{\partial y^2} + \dfrac{\partial^2}{\partial z^2}$ (b) $\dfrac{\partial^2}{\partial x^2} + \dfrac{\partial^2}{\partial y^2} + \dfrac{\partial^2}{\partial z^2} - \dfrac{1}{c^2} \dfrac{\partial^2}{\partial t^2}$

   (c) $\dfrac{\partial}{\partial x} + \dfrac{\partial}{\partial y} + \dfrac{\partial}{\partial z}$ (d) $\left(\dfrac{\partial}{\partial x} + \dfrac{\partial}{\partial y} + \dfrac{\partial}{\partial z} - \dfrac{1}{c} \dfrac{\partial}{\partial t} \right)^2$ [b]

5. Electromagnetic field tensor is a tensor of rank

   (a) four (b) three

   (c) two (d) six [c]

6. When four-vector scalar product is written as
$$S^2 = x_\mu x_\mu = x_1^2 + x_2^2 + x_3^2 + x_4^2$$
then the line element is called a tensor of rank
   (a) four   (b) two
   (c) zero   (d) three   **[c]**

7. In terms of four-vector, the continuity equation is written as

   (a) $\dfrac{\partial J_\mu}{\partial x_\mu} = 0$ (b) $\dfrac{\partial J_\mu}{\partial x_\mu} = \rho$

   (c) $\dfrac{\partial J_\mu}{\partial x_\mu} = E$ (d) $\dfrac{\partial J_\mu}{\partial x_\mu} = \dfrac{1}{2}$ **[a]**

Covariant Lagrange Formulation

The *action functional*

$$S = \int_{t_1}^{t_2} L(q(t), \dot{q}(t), t) dt \tag{1}$$

Over the *Lagrange function* $((q, \dot{q}, t))$ takes among the competitive set M of admitted paths

$$M = \left\{ q(t) : q(t_1) = q_i, q(t_2) = q_e \right\} \tag{2}$$

An extreme value for the real path:

$$\delta S = 0 \tag{3}$$

The paths, which belong to M, have all the same initial and end configurations q_i and q_e, which are adopted in each case at certain given times t_1 and t_2, while the various paths differ from one another by virtual displacements. One of the main consequences of this Hamilton principle are the Lagrange equations of motion:

$$\frac{d}{dt} \frac{\partial L}{\partial \dot{q}_j} - \frac{\partial L}{\partial q_j} = 0; \quad j = 1 \dots s \tag{4}$$

The *equivalence principle* can now be interpreted such that the action functional S, out of which the fundamental equations of motion arise, has to be **Lorentz invariant**. Furthermore, the Lagrange function L should, in particular, be a four-scalar and should depend on the contravariant four-vectors x^μ and u^μ as well as the four-scalar τ, instead of $q, \dot{q}$ and t. We therefore try to formulate with the substitution

$$L \rightarrow \bar{L}(x^\mu, u^\mu, \tau) \tag{5}$$

(a) Covariant Hamilton principle

$$\delta S = \delta \int_{\tau_1}^{\tau_2} d\tau \bar{L}(x^\mu, u^\mu, \tau) = 0 \tag{6}$$

Out of which the covariant Lagrange equation arise:

$$\frac{d}{d\tau} \frac{\partial \bar{L}}{\partial u^\mu} - \frac{\partial \bar{L}}{\partial u^\mu} = 0 \tag{7}$$

Now, we find the correct relativistic Lagrange function $\bar{L}$. As a rule one has to apply analogy-considerations to non-relativistic mechanics. In particular, the *familiar* relations should be reproduced for $v \ll c$. The main problem lies in the fact that because $L = T - V$, statements must be available about the potential V and therewith about forces in covariant four-representation. Electromagnetic forces are, as we have seen, rather unproblematic, while, for instance, hardly any noteworthy theory exists about nuclear forces. We content ourself to a single particle.

1. FORCE-FREE PARTICLE

We look for a Lagrange function $\bar{L}$ from which we get the equation of motion:

$$m \frac{d}{d\tau} u^\mu = 0 \tag{8}$$

Non-relativistically one would investigate

$$S = \int_{t_1}^{t_2} L_0(x(t), v(t), t)dt$$

where the particle-momentum must hold

$$p_i = \frac{\partial L_0}{\partial v_i}; \quad i = x, y, z$$

It we understand in this relation by p_i directly the relativistically correct expression,

$$\frac{\partial L_0}{\partial v_i} = p_{ri} = \frac{mv_i}{\sqrt{1 - v^2/c^2}}$$

then it follows by integration:

$$L_0 = -mc^2 \sqrt{1 - v^2/c^2} \tag{9}$$

This is the relativistically correct form. But one has to bear in mind that L_0 is no longer identical to the kinetic energy.

$$H_0 = \sum_i p_{ri} v_i - L_0 = \frac{mv^2}{\sqrt{1 - \dfrac{v^2}{c^2}}} + mc^2 \sqrt{1 - v^2 / c^2} = \gamma mc^2$$

$$\Rightarrow \quad H_0 = \frac{mc^2}{\sqrt{1 - v^2/c^2}} = T_r \tag{10}$$

We substitute, in the action functional S the time t by the proper time τ:

$$S = \int_{\tau_1}^{\tau_2} d v \gamma L_0$$

The comparison with (6) yields:

$$\bar{L}_0 = \gamma L_0 = -mc^2 \tag{11}$$

We have

$$c^2 = u^\mu u_\mu$$

Thus, the correct functional dependence of the Lagrange function $\bar{L}_0$ on u^μ could be given by

$$\bar{L}_0 = L_0(u^\mu) = -mu^\mu u_\mu \tag{12}$$

or somewhat more generally:

$$\bar{L}_0 = -mc^\eta (u^\mu u_\mu)^{\frac{1}{2}(2-\eta)} \tag{13}$$

We can determine η by the requirement for '*correct*' equations of motion. Since u^μ is contravariant,

$$u'^\mu = \frac{\partial x'^\mu}{\partial x^\alpha} u^\alpha = L_{\mu\alpha} u^\alpha$$

The *velocity-gradient*

$$\frac{\partial}{\partial u'^\mu} = \frac{\partial x^\alpha}{\partial u'^\mu} \frac{\partial}{\partial u^\alpha} = (L^{-1})_{\alpha\mu} \frac{\partial}{\partial u^\alpha} \tag{14}$$

Transforms like covariant four-vector. We define the **covariant canonical momentum**

$$p_\mu = \frac{\partial \bar{L}}{\partial u^\mu} = (p^0, -p) \tag{15}$$

with a minus sign in order to guarantee that the space components match the non-relativistic definition for $v \ll c$. We then have for the **contravariant canonical momentum**.

$$p^\mu = \mu^{\mu\alpha} p_\alpha = (p^0 \cdot p) \tag{16}$$

Finally it should be valid for the **free** particle discussed here:

$$p_\mu^{(0)} = -\frac{\partial \bar{L}_0}{\partial u^\mu} = mu_\mu \Leftrightarrow p_{(0)}^\mu = mu^\mu \tag{17}$$

This means with respect to the anastz (13), we have

$$p_\mu^{(0)} = -\frac{\partial}{\partial u^\mu}\left(-mc^\eta (u^\beta u_\beta)^{\frac{1}{2}(2-\eta)}\right)$$

$$= mc^\eta \frac{\partial}{\partial u^\mu}\left(\mu_{\beta\gamma} u^\beta u^\gamma\right)^{\frac{1}{2}(2-\eta)}$$

$$= (2-\eta)mc^\eta \mu_{\mu\gamma} u^\gamma \left(\mu_{\beta\gamma} u^\beta u^\gamma\right)^{-\frac{1}{2}\eta}$$

$$= (2-\eta)mc^\eta \mu_\mu \left(u^\beta u_\beta\right)^{-\frac{1}{2}\eta}$$

$$= (2-\eta)mu_\mu$$

The comparison with (17) leads to $n = 1$. That gives us the required functional dependency of the Lagrange function $\bar{L}_0$ on the four-velocity u^μ.

$$\bar{L}_0 = -mc(u^\mu u_\mu)^{\frac{1}{2}} \tag{18}$$

The covariant Hamilton principle was the starting point for our considerations which led to the equations of motion (7). That means here for the force-free particle:

$$\frac{\partial \bar{L}_0}{dx^\mu} = 0 \rightarrow \frac{d}{d\tau} p_\mu^{(0)} = m\frac{d}{d\tau} u_\mu = -\frac{d}{d\tau}\frac{\partial \bar{L}_0}{\partial u^\mu} = 0 \tag{19}$$

$$p_{(0)}^\mu = m\frac{d}{d\tau} u^\mu = 0 \tag{20}$$

This clearly reveals that the ansatz (18) for $\bar{L}_0$ appears to be correct.

2. CHARGED PARTICLE IN THE ELECTROMAGNETIC FIELD

We now obtain the covariant Lagrange function $\bar{L}$, from which the Lorentz-force equation,

$$F = q(E + v \times B)$$

can be derived. Now-relativistically we have

$$L(x, v, t) = \frac{m}{2} v^2 + q(v \cdot A) - q\varphi \tag{21}$$

$\varphi(x, t)$ is the scalar potential and $A(x, t)$ the vector potential. The kinetic part should be identical to L_0, while, supposedly, the two electromagnetic parts behave already realtivistically correctly. With (9) for L_0 we use in the action functional

$$L = L_0 + q(v \cdot A) - q\varphi$$

and replace, the time t by the proper time τ. The comparison with (6) then lead to the following anasatz:

$$\bar{L} = \gamma L = \gamma L_0 + q(\gamma v \cdot A) - q\gamma\varphi$$

$$= \bar{L}_0 + q\left[\gamma v \cdot A - (\gamma c)\left(\frac{1}{c}\varphi\right)\right]$$

One can now recognize on the right-hand side the scalar product of four-velocity u^μ and four-otential A^μ.

$$\bar{L}(x^\mu, u^\mu, \tau) = \bar{L}_0(u^\mu) - q(u^\mu A_\mu) \tag{22}$$

Let us test whether the known equation of motion results from this more or less *guessed* ansatz for $\bar{L}$. For this purpose we insert $\bar{L}$ into the covariant Lagrange equation (7)

$$\frac{d}{d\tau}\frac{\partial \bar{L}}{\partial u^\mu} = -m\frac{d}{d\tau} u_\mu - q\frac{d}{d\tau} A_\mu = \frac{\partial \bar{L}}{\partial x^\mu} = -q\frac{\partial}{\partial x^\mu}(u^\alpha A_\alpha)$$

Hence, we obtain

$$m \frac{d}{d\tau} u_\mu = q \left[\partial_\mu (u^\alpha A_\alpha) - \frac{d}{d\tau} A_\mu \right] = K_\mu \tag{23}$$

If (22) indeed proves to be correct then we have found with (23) a way as a by-product, a novel representation for the Minkowski force on a charge particle in the electromagnetic field.

Let us check at first the **space components,** for which we assume that they lead to the familiar Lorentz force. As usual we want the **contravariant** version of the four-vector (23).

$$K^\mu = q \left[\partial^\mu (u^\alpha A_\alpha) - \frac{d}{d\tau} A^\mu \right] \tag{24}$$

We obtain

$$K_i = q \left[-\frac{\partial}{\partial x_i} (-\gamma v \cdot A + \gamma \varphi) - \gamma \frac{d}{dt} A_i \right], \qquad i \in (x, y, z)$$

The Cartesian components of the Lorentz force are:

$$F_i = \frac{1}{\gamma} K_i = q \left[\frac{\partial}{\partial x_i} (v \cdot A) - \frac{\partial \varphi}{\partial x_i} - \frac{d}{dt} A_i \right]$$

Using

$$\frac{d}{dt} A_i = v \cdot \nabla A_i + \frac{\partial A_i}{\partial t}$$

one can perform the following rearrangement

$$(v \times \mathrm{curl} A)_x = v_y (\mathrm{curl} A)_z - v_z (\mathrm{curl} A)_y$$

$$= v_y \left(\frac{\partial}{\partial x} A_y - \frac{\partial}{\partial y} A_x \right) - v_z \left(\frac{\partial}{\partial z} A_x - \frac{\partial}{\partial x} A_z \right)$$

$$= \frac{\partial}{\partial x} (v \cdot A) - v_x \frac{\partial}{\partial x} A_x - v_y \frac{\partial}{\partial y} A_x - v_z \frac{\partial}{\partial z} A_x$$

$$= \frac{\partial}{\partial x} (v \cdot A) - v \cdot \nabla A_x$$

$$= \frac{\partial}{\partial x} (v \cdot A) - \frac{d}{dt} A_x + \frac{\partial}{\partial t} A_x$$

Analogously, one can calculate the other two components:

$$\frac{\partial}{\partial x_i} (v \cdot A) - \frac{d}{dt} A_i = (v \times \mathrm{curl} A)_i - \frac{d}{dt} A_i$$

This yields for the force components F_i:

$$F_i = q\left[-\frac{\partial \varphi}{\partial x_i} - \frac{\partial A_i}{\partial t} + (v \times \mathrm{curl}A)_i\right]$$

The first two summands just represent the i-th component of the electric field E, so that the space components of the four-force (24) lead indeed to the correct Lorentz force.

$$F = q\left(-\nabla\varphi - \dot{A} + v \times \mathrm{curl}A\right) = q(E + v \times B) \tag{25}$$

Now, we check the **time component**:

$$K^0 = q\left[\frac{1}{c}\frac{\partial}{\partial t}(-\gamma A \cdot v + \gamma\varphi) - \gamma \frac{d}{dt}\left(\frac{1}{c}\varphi\right)\right]$$

We insert

$$\frac{d}{dt}\varphi = \nabla\varphi \cdot v + \frac{\partial}{\partial t}\varphi$$

and obtain:

$$K^0 = \frac{1}{c}q\gamma\left(-v \cdot \frac{\partial A}{dt} - \nabla\varphi \cdot v\right) = \frac{1}{c}q\,\gamma(E \cdot v)$$

With (25) this can be written as

$$K^0 = \gamma \frac{F \cdot v}{c} \tag{26}$$

We have therefore deduced from our ansatz (22) for $\bar{L}$. Now, the correct equation of motion reads as

$$K^\mu = m\frac{d}{d\tau}u^\mu = \gamma\left(\frac{F \cdot v}{c}, F_x, F_y, F_z\right) \tag{27}$$

Now, we can conclude further on:

Canonical momentum

$$p_\mu = \frac{\partial \bar{L}}{\partial u^\mu} = mu_\mu + qA_\mu \tag{28}$$

This is the *covariant version*; the contravariant vector arises simple by transposing the indexes:

$$p^\mu = mu^\mu + qA^\mu \tag{29}$$

The **space components**

$$p_i = m\gamma v_i + qA_i \tag{30}$$

contain, besides the relativistic mechanical momentum $p_\tau^{(i)} = m\gamma v_i$, still an additional term qA_i, which, however, does **not** represent a relativistic effect. One has therefore to distinguish here also canonical and mechanical moments. We consider finally the **time component** of the canonical momentum:

$$p^0 = m\gamma c + q\frac{1}{c}\varphi = \frac{1}{c}(m\gamma c^2 + q\varphi)$$

The first summand on the right-hand side is just the relativistic kinetic energy T_r, which we introduced in (2.54). The bracket therefore represents the **total energy of the particle**.

$$E = T_r + q\varphi = m\gamma c^2 + q\varphi \tag{31}$$

For the **canonical momentum** it therewith holds:

$$p^\mu = \left(\frac{1}{c} E, m\gamma v + qA \right) \tag{32}$$

$$= \left(\frac{1}{c} E, p \right) \tag{33}$$

From this one has to distinguish the **mechanical momentum**

$$p^\mu_m = \left(\frac{T_r}{c}, m\gamma v \right) = \left(\frac{T_r}{c}, p_r \right) = \left[\frac{1}{c}(E - q\varphi), p - qA \right] \tag{34}$$

Finally, we obtain

$$p^\mu_m p_{m\mu} = m^2 c^2$$

i.e.

$$m^2 c^2 = -(p - qA)^2 + \frac{1}{c^2}(E - q\varphi)^2$$

Thus, the relativistic energy of a charged particle in the electromagnetic field:

$$E = \sqrt{(p - qA)^2 c^2 + m^2 c^4} + q\varphi \tag{35}$$

Chapter

15

Interaction of Charged Particles with Matter

15.1 INTRODUCTION

Whenever a charged particle passes through matter, it collides with the constitutents of the medium, atoms, molecules, the electrons and atomic nuclei. In case if the incident particle is more massive than an electron, then its collision with them produces different effects. The incident particle loses most of the energy to the lighter electrons, which do not cause any significant deflection of the incident particle. On the other hand, the nuclei, due to their greater charge, cause deflection of the incident particle and gain very little energy due to collision. This reveals that the incident particle keeps up more or less the straight path after collision and lose energy till it nears the end of its range. On the other hand, for incident electrons, the loss of energy as well as scattering from their path, is caused by the electrons of the medium and as a consequence the path of incident electron is much less rectilinear. Obviously, after traversing a small range in the material, the incident electrons tend to diffuse into the material.

We may note that the problem of the deflection, identification and energy determination of the radiations emitted in any *nuclear reaction* and decay, is mainly based on the ionization caused by the emitted charged particles due to collision. Further, the detection of neutral particles such as neutrons, neutrinos, and also of high energy photons is achieved through an intermediate reaction releasing a charged particle, which is then rendered detectable. In this chapter, we propose to study the energy transfer to a free electron in a coulomb collision and then to an electron which is bound harmonically to positive centre of force. We will also discuss the *classical energy-loss formula* and its modification due to quantum mechanical considerations.

15.2 ENERGY LOSS TO A FREE ELECTRON IN A COULOMB COLLISION

Let us consider an ion of charge Ze and mass M, moving with velocity v along a path and passes within a distance b (called *impact parameter*) of an electron at O, which for the present, we assume to be free and essentially also at rest (Fig. 15.1).

We may note that when ion moves rapidly as compared to the characteristic velocity of the electron in the atom, the electron can be assumed to be essentially free and also at

Fig. 15.1: An ion (charge Ze and mass M) is moving with velocity v passes an impact paramter b from a free electron at O

rest during collision. Further, we also assume that the momentum impulse is quite small so that the incident ion moves undeflected from its rectilinear path. Moreover, the electron does not move significantly due to the said collision. Now, to calculate the energy transfer during collision, we find the momentum impulse due to the electric field of the incident ion at the position of the electron. Since the electron is at rest during the said collision, the magnetic field of the incident particle will not exert any force on the electron.

Now, to calculate the electric and magnetic fields caused by the incident ion at the position of the electron, we conside that the ion having charge Ze is moving in the $+x_3$ direction and its closest distance of approach to the electron at rest at O in the reference system K is b (impact parameter). In reference system K' ion is at rest (Fig. 15.2). At $t = t' = 0$, the origins O and O' of K and K' reference frames respectively coincide and the incident ion is at a distance b from the electron.

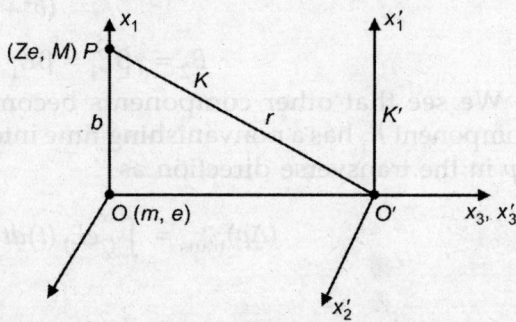

Fig. 15.2: An ion of charge Ze and mass M moving at constant velocity v passes an electron (m, e) at an impact parameter b at observation point B

The coordinates of the electron in the frame K' are

$$x_1' = b, \ x_2' = 0, \ x_3' = -vt' \tag{15.1}$$

and its distance from the ion is

$$r' = \sqrt{b^2 + (vt')^2} \tag{15.2}$$

Let us express r' in terms of unprimed coordinates of system K and the only coordinate which is to be transformed is the time

$$t' = \gamma \left(t - vx_3/c^2 \right) = \gamma t,$$

as $x_3 = 0$ for P in frame of reference K.

Now, the electric and magnetic fields in the rest frame of reference K' of the incident ion are:

$$E_1' = \frac{Zeb}{r'^3} \qquad\qquad B_1' = 0$$

$$E_2' = 0 \qquad\qquad B_2' = 0 \tag{15.3}$$

$$E_3' = \frac{Zevt'}{r'^3} \qquad\qquad B_3' = 0$$

The nonzero field components expressed in terms of unprimed coordinates of frame K, are

$$E_1' = \frac{Zeb}{(b^2 + \gamma^2 v^2 t^2)^{3/2}} \tag{15.4}$$

$$E_3' = -\frac{Ze\gamma vt}{(b^2 + \gamma^2 v^2 t^2)^{3/2}} \tag{15.5}$$

Now, making use of the field transformation formulae, these fields in the frame K reads as

$$E_1 = \gamma E_1' = \frac{\gamma Zeb}{(b^2 + \gamma^2 v^2 t^2)^{3/2}} \tag{15.6}$$

$$E_3 = E_3' = -\frac{Ze\gamma vt}{(b^2 + \gamma^2 v^2 t^2)^{3/2}} \tag{15.7}$$

$$B_2 = \gamma \beta E_1' = \beta E_1 \tag{15.8}$$

We see that other components become zero. We also see that only the field component E_1 has a nonvanishing time integral and gives rise to momentum impulse Δp in the transverse direction as

$$(\Delta p)_{\text{trans}} = \int_{-\infty}^{+\infty} eE_1(t)\,dt$$

$$= \frac{Ze^2 \gamma b}{b^3} \int_{-\infty}^{+\infty} \frac{dt}{\left(1 + \dfrac{\gamma^2 v^2 t^2}{b^2}\right)^{3/2}}$$

Putting $\gamma vt/b = \cot \theta$, one obtains

$$(\Delta p)_{\text{trans}} = \frac{Ze^2}{bv} \int_0^\pi \sin \theta\,d\theta = \frac{2Ze^2}{bv} \tag{15.9}$$

The longitudinal impulse vanishes as

$$(\Delta p)_{\text{long}} = \int_0^\pi eE_3(t)\,dt$$

$$= \frac{Ze^2 v\gamma}{b^3} \int_{-\infty}^{+\infty} \frac{-t\,dt}{\left(1 + \dfrac{\gamma^2 v^2 t^2}{b^2}\right)}$$

$$= \frac{Ze^2}{\gamma bv} \int_{-\pi}^0 \cos \theta\,d\theta = 0 \tag{15.10}$$

The energy transferred to the electron is given by

$$\Delta \mathcal{E} = \frac{(\Delta p)^2}{2m}$$

$$= \frac{2Z^2 e^4}{mv^2}\left(\frac{1}{b^2}\right) \tag{15.11}$$

The angular deflection of the incident ion is

$$\theta \approx \frac{\Delta p}{p}$$

$$= \frac{2Ze^2}{pbv} \tag{15.12}$$

provided $\Delta p << p$. We see from Eq. (15.11) that the energy transfer depends on the charge of the incident particle and also on its velocity, whereas it is independent of its mass. Moreover, it also depends inversely on the square of the impact parameter b, obviously, the energy transfer being greater for close collisions. We may also note that the transfer of energy is maximum for head-on collision and therefore there must be an upper limit on the energy transferred.

One can obtain an estimate of the lower limit on b from the calculation of the distance moved by the electron during the collision. We may note that while deducing Eq. (15.11), we have assumed that the electron did not move significantly during the collision, so the Eq. (15.11) will remain valid so far as the distance d moved by the electron is small compared to b, i.e. $d < b$. One can obtain the time of collision by the measure of the time interval over which the field E_1 is appreciable and is given by

$$\Delta t \sim \frac{b}{\gamma v} \tag{15.13}$$

Thus, the distance moved by the electron during the collision is obtained as

$$d \sim \frac{\Delta p}{2m} \times \Delta t = \frac{Ze^2}{\gamma m v^2} = b_{min} \tag{15.14}$$

Clearly, as long as $b << d$, Eq. (15.11) holds.

Although we have derived the energy transfer result by treating the electrons as free, but actually these are bound to the atoms. So far as the collision time is short as compared to the orbital period of motion, one can safely assume that the collision is so sudden that the electron may be treated as free. However, if the collision time is very long compared to the orbital period of the electron, one finds that the latter will undergo many cycles of orbital motion as the incident ion passes at the distance of approach. No doubt, the electron will be influenced adiabatically by the field given by Eqs. (15.6) to (15.8) and also the net transfer of energy will be zero. However, the demarcation will occur at a point where the collision time becomes comparable to the orbital time. This means that if ω is the characteristic frequency of motion, then this situation occurs when

$$\Delta t(b_{max}) \sim \frac{1}{\omega}$$

$$\Rightarrow \qquad b_{max} = \frac{\gamma v}{\omega} \tag{15.15}$$

We see that for impact parameters greater than b_{max}, the energy transfer falls rapidly to zero and the collisions become adiabatic.

However, in an actual medium, the electrons will be lying at various distances from the path of the incident particle. Now, assuming that there are N atoms per unit volume and z electrons per atom, the number of electrons in a thickness dx of the absorber lying at impact parameters b and $b + db$, is

$$dn = Nz2\pi b \, db dx \tag{15.16}$$

The energy loss is given by

$$\frac{dE}{dx} \approx 4\pi Nz \frac{Z^2 e^4}{mv^2} \int_{b_{min}}^{b_{max}} \frac{1}{b} \, db$$

which yields

$$\approx 4\pi Nz \frac{Z^2 e^4}{mv^2} \ln B \tag{15.17}$$

with

$$B = \frac{b_{max}}{b_{min}} \approx \frac{\gamma^2 m v^3}{Ze^2 \omega} \tag{15.18}$$

We see that the expression for energy loss has all the characteristics of the exact calculation by Bohr in 1915. However, there is some uncertainty in B since the estimates of both the limits b_{min} and b_{max} are approximate. We may also note that B occurs in logarithm (Eq. 15.17) and as such this factor is not that important numerically.

15.3 TRANSFER OF ENERGY TO A HARMONICALLY BOUND ELECTRON

We now consider a simple model of harmonically bound charge to understand the loss of energy of incident particle passing through matter. We will see that this will land credence and justify the value of $b_{max} = \gamma v/\omega$ of impact parameter assumed earlier as dividing the coulomb collisions with free energy transfer for $b < b_{max}$ (Eq. 15.11) and adiabatic collisions for $b > b_{max}$, with no net transfer of energy.

Let us consider that an incident particle of charge Ze, mass M and velocity v pass a harmonically bound electron of mass m and charge e at an impact parameter b, from the origin O of the binding force of the electron as shown in Fig. 15.3.

Let us assume that the energy transfer is small which is possible only in case of large impact parameter b. Interestingly, it is only under these

Fig. 15.3: A charged particle of mass M and charge Ze passing an electron at O at an impact parameter b with velocity v. The electron is bound harmonically to the center of force O

conditions that the binding of the electron is important. Further, we assume that the motion of the electron remains non-relativistic during collision. The amplitude of oscillation of bound electron is considered is considered small as compared to impact parameter b, and this assumption enables one to neglect the spatial variation of the electric field over the position of the electron. Moreover, as the motion of the electron is nonrelativistic, the effect of the magnetic field of the incident particle has negligible effect on the motion of electron. Now, the equation of motion for the electron reads as

$$x + \Gamma x + \omega_0^2 = \frac{e}{m} \mathbf{E}(t) \tag{15.19}$$

Here $\mathbf{E}(t)$ is the electric field at the origin of the binding force and its components are given by Eqs. (15.6) – (15.8). ω_0 is the characteristic frequency of binding of the electron and Γ is a *damping constant* which is characteristic of any resistive or dissipative force. We may note that the inclusion of the damping effect not only generalizes the treatment but also eliminates the possibility of the amplitude of motion becoming infinite.

Now, we try to seek the solution of Eq. (15.19) through Fourier transforms. The Fourier transforms of $\mathbf{E}(t)$ and $x(t)$ are as follows:

$$x(t) = \frac{1}{\sqrt{2\pi}} \int_{-\infty}^{+\infty} x(\omega)\, e^{-i\omega t}\, d\omega \tag{15.20}$$

$$\mathbf{E}(t) = \frac{1}{\sqrt{2\pi}} \int_{-\infty}^{+\infty} \mathbf{E}(\omega)\, e^{-i\omega t}\, d\omega \tag{15.21}$$

Now, the reality condition on $x(t)$ and $\mathbf{E}(t)$ gives the following relations on the positive and negative parts of the transforms :

$$x(-\omega) = x^*(\omega) \tag{15.22}$$

$$\mathbf{E}(-\omega) = \mathbf{E}^*(\omega) \tag{15.23}$$

Now, substituting the Eqs. (15.20) and (15.21) into the force Eq. (15.19), one obtains

$$x(\omega) = \frac{e}{m} \frac{\mathbf{E}(\omega)}{\omega_0^2 - i\omega\Gamma - \omega^2} \tag{15.24}$$

From the given $\mathbf{E}(t)$, one can find $\mathbf{E}(\omega)$ using Eq. (15.23) and $\mathbf{E}(\omega)$ in turn when substituted in Eq. (15.24) will yield $\mathbf{E}(t)$. However, rather than knowing the detailed motion, the quantity of interest is the energy transfer in a collision. Now, we calculate the energy transfer.

The rate of work done by the incident particle on the electron, is as follows:

$$\frac{d\mathcal{E}}{dt} = \int \mathbf{E} \cdot \mathbf{J} d^3 x \tag{15.25}$$

and the energy transfer during the collision is

$$\Delta \mathcal{E} = \int_{-\infty}^{+\infty} dt \int \mathbf{E} \cdot \mathbf{J} d^3 x' \tag{15.26}$$

Putting the expression for the current density **J** for the bound charge

$$\mathbf{J} = e\mathbf{v}\delta(x' - x(t)) \tag{15.27}$$

in Eq. (15.26), one obtains

$$\Delta\mathcal{E} = e\int_{-\infty}^{+\infty} \mathbf{v} \cdot \mathbf{E}\,dte \tag{15.28}$$

Now, making use of Eqs. (15.20) and (15.21) along with the Fourier representation of delta function given by

$$\delta(k - k') = \frac{1}{2\pi}\int_{-\infty}^{+\infty} e^{i(k-k')x}\,dx \tag{15.29}$$

and the reality conditions, Eqs. (15.22) and (15.23), one may rewrite the expression for energy transfer, Eq. (15.28), as follows:

$$\Delta\mathcal{E} = e\int_{-\infty}^{+\infty}\frac{1}{\sqrt{2\pi}}(-i\omega)\,x(\omega)\,e^{-i\omega t}\,d\omega\frac{1}{\sqrt{2\pi}}\int_{-\infty}^{+\infty}\mathbf{E}^*(\omega')\,e^{i\omega' t}\,d\omega'\,dt$$

$$= 2e\,\mathrm{Re}\,\frac{1}{2\pi}\int_{0}^{+\infty}\int_{-\infty}^{+\infty} e^{-i(\omega-\omega')t}\,dt\,x(\omega)\,(-i\omega)\,\mathbf{E}^*(\omega')\,d\omega'$$

$$= 2e\,\mathrm{Re}\int_{0}^{\infty} -i\omega\,x(\omega)\cdot\mathbf{E}^*(\omega)\,d\omega \tag{15.30}$$

Recasting the expression for energy transfer $\Delta\mathcal{E}$ in view of Eq. (15.24), one obtains

$$\Delta\mathcal{E} = \frac{e^2}{m}\int_{0}^{\infty}|\mathbf{E}(\omega)|^2\,\frac{2\omega^2\Gamma}{(\omega_0^2 - \omega^2)^2 + \omega^2\Gamma^2}\,d\omega \tag{15.31}$$

Now, we consider the case of light damping, i.e., Γ being small, we see that the integrand in Eq. (15.31) peaks around $\omega = \omega_0$ and for the factor involving electric field, the maximum of its contribution will come from the region $\omega = \omega_0$. Thus the above equation takes the form

$$\Delta\mathcal{E} = \frac{2e^2}{m}|\mathbf{E}(\omega_0)|^2\int_{0}^{\infty}\frac{\omega^2\Gamma}{(\omega_0^2 - \omega^2)^2 + \omega^2\Gamma^2}\,d\omega \tag{15.32}$$

Putting $\omega/\Gamma = x$, one can transform it into the equation

$$\Delta\mathcal{E} = \frac{2e^2}{m}|\mathbf{E}(\omega_0)|^2\int_{0}^{\infty}\frac{\omega^2\Gamma}{\Gamma^4\left[\left(\dfrac{\omega_0^2}{\Gamma^2} - \dfrac{\omega^2}{\Gamma^2}\right) + \left(\dfrac{\omega}{\Gamma}\right)^2\right]}\,d\omega$$

$$= \frac{2e^2}{m}|\mathbf{E}(\omega_0)|^2\int_{0}^{\infty}\frac{x^2\,dx}{\left(\dfrac{\omega_0^2}{\Gamma^2} - x^2\right) + x^2}$$

But

$$\int_{0}^{\infty}\frac{x^2\,dx}{\left(\dfrac{\omega_0^2}{\Gamma^2} - x^2\right) + x^2} = \frac{\pi}{2}$$

so the energy transfer is obtained as $\Delta E = \dfrac{\pi e^2}{m} |E(\omega_0)|^2$ (15.33)

Equation (15.33) gives the energy transfer to a non-relativistic oscillator by the external electromagnetic field.

The fields produced by a charged particle of mass M, charge Ze and velocity v at the origin O which is at an impact parameter b, are given by Eqs. (15.6)–(15.8). Now, we take the case of $E_1(t)$, its Fourier transform is

$$E_1(\omega) = \frac{Zeb\gamma}{\sqrt{2\pi}} \int_{-\infty}^{+\infty} \frac{e^{i\omega t} dt}{(b^2 + \gamma^2 v^2 t^2)^{3/2}}$$ (15.34)

Putting $x = \gamma v t / b$, the expression for $E_1(\omega)$ reads as

$$E_1(\omega) = \frac{Zebv}{\sqrt{2\pi}} \int_{-\infty}^{+\infty} \frac{e^{i\omega bx/\gamma v}}{(1+x^2)^{3/2}} dx$$ (15.35)

Using the table of Fourier transforms, one can show that the integral is proportional to the modified Bessel function. Then, one finds

$$E_1(\omega) = \frac{Ze}{bv} \left(\frac{2}{\pi}\right)^{1/2} \left[\frac{\omega b}{\gamma v} K_1\left(\frac{\omega b}{\gamma v}\right)\right]$$ (15.36)

In an analogous manner, the Fourier transform of $E_3(t)$ is obtained as

$$E_3(\omega) = -i \frac{Ze}{\gamma bv} \left(\frac{2}{\pi}\right)^{1/2} \left[\frac{\omega b}{\gamma v} K_0\left(\frac{\omega b}{\gamma v}\right)\right]$$ (15.37)

Now, rewriting Eq. (15.33) for the energy transfer in view of Eqs. (15.36) and (15.37), one obtains

$$\Delta E(b) = \frac{2Z^2 e^4}{mv^2} \left(\frac{1}{b^2}\right) \left[\zeta^2 K_1^2(\zeta) + \frac{1}{\gamma^2} \zeta^2 K_0^2(\zeta)\right]$$ (15.38)

where $\zeta = \dfrac{\omega_0 b}{\gamma b} = \dfrac{b}{b_{max}}$.

The factor outside the square brackets in Eq. (15.38) is the approximate result. Through the use of the limiting forms of the modified Bessel functions for small and large values of ζ, one can show that Eq. (15.38) reduces to

$$\Delta E(b) = \frac{2Z^2 e^4}{mv^2} \frac{1}{b^2} \quad \text{for} \quad \zeta \ll 1$$

$$= \frac{2Z^2 e^4}{mv^2} \left(\frac{1}{b^2}\right) \left(1 + \frac{1}{\gamma^2}\right) \frac{\pi}{2} \zeta e^{-2\zeta} \quad \text{for } \zeta \gg 1$$ (15.39)

Obviously, for $b \ll b_{max}$, the energy transfer is given by Eq. (15.11) and for $b \gg b_{max}$, the energy transfer falls exponentially to zero.

15.4 CLASSICAL THEORY OF ENERGY LOSS: BOHR'S FORMULA, QUANTUM MECHANICAL CONSIDERATIONS AND BETHE'S GENERALIZATION

One can calculate the classical energy loss per unit length for a massive charged particle on passing through a material medium by making use of the expression for the energy transfer to a harmonically bound charge i.e. Eq. (15.38). Now, we assume that there are N atoms per unit volume and z electrons per atom, f_j electrons bound by the harmonic binding frequency ω_j, the oscillator strength f_j will now satisfy the

sum rule $\sum_j f_j = z$.

The loss of energy per unit length is given by

$$\frac{dE}{dx} = 2\pi N \sum_j f_j \int_{b_{min}}^{\infty} \Delta E(b)\, b\, db \tag{15.40}$$

where $\Delta E(b)$ is given by Eq. (15.38) with $\zeta = \omega_j b/\gamma b$. The lower limit is specified with the prescription that the energy transfer reduces to the following expression as $b \to 0$,

$$\Delta E(b) \approx \frac{2Z^2 e^4}{mv^2}\left(\frac{1}{b_{min}^2 + b^2}\right) \tag{15.41}$$

Here, the upper limit has been taken as since Eq. (15.38) falls to zero for large b.

With the help of mathematical tables, one finds that the integral over the modified Bessel functions, leads to the following result in a closed form

$$\frac{dE}{dx} = \frac{4\pi N Z^2 e^4}{mv^2} \sum_j f_j \left[\zeta_{min} K_1(\zeta_{min}) K_0(\zeta_{min}) - \frac{v^2}{2c^2}\zeta^2_{min} K_1^2(\zeta_{min}) - K_0(\zeta_{min}) \right] \tag{15.42}$$

where $\zeta_{min} = \dfrac{\omega_j b_{min}}{\gamma v}$. Since $\zeta_{min} \ll 1$, one may use the limiting forms of the modified

forms Bessel functions and as a consequence, one finds the classical energy loss formula as

$$\frac{dE}{dx} = 4\pi N z \frac{Z^2 e^4}{mv^2}\left[\ln B_c - \frac{v^2}{2c^2}\right] \tag{15.43}$$

where $$B_c = \frac{1.123}{Ze^2}\frac{\gamma^2 mv^3}{<\omega>} \tag{15.44}$$

The average frequency $<\omega>$ is defined as a geometric mean as follows:

$$z \ln <\omega> = \sum_j f_j \ln \omega_j \tag{15.45}$$

Equation (15.43) is the *classical energy loss formula* deduced by Bohr in 1915. We may note that (15.43) is adequate in describing the energy loss from slowly moving alpha particles and heavier nuclei. However, it is reported that for other lighter particles like electrons, mesons, protons and even relativistic alpha, Eq. (15.43) overestimates the energy loss. This is primarily due to the fact that for lighter particles, quantum-mechanical considerations cause a failure. We know that the quantum effects include the fact that the energy transfer has to be discrete and the requirement that the wave nature of the particle has to be taken into account.

Now, the smallest energy transfer will result form, $b = b_{max}$ as

$$\Delta E\,(b_{max}) \approx \frac{2}{\gamma^2} Z^2 \left(\frac{v_0}{v}\right)^4 \hbar \omega_0 \tag{15.46}$$

where ω_0 is the only one binding frequency and $v_0 = c/137$ is the *velocity of electron in the first Bohr orbit*.

Since $\hbar \omega_0$ is of the order of the ionization potential of the atom, one finds that for a fast particle ($v << v_0$), the energy transfer is very small as compared to the ionization potential of the atom or the smallest excitation energy of the atom. Obviously, the classical result fails in the realm of small excitation energies. However, one can reconcile the classical result with the quantum consideration, provided we interpret the energy loss in a statistical sense. In most of the collisions, an excitation occurs, which gives a small average value over a large number of collisions. However, for establishing an exact agreement, one has to resort to the quantum-mechanical definitions of f_j and ω_j.

There is also the other quantum effect that refers to the consideration of *uncertainty principle*, according to which the trajectory of a particle can be known within an uncertainty $\Delta x \geq \hbar/p$. Clearly, the classical approach have to be abandoned for an impact parameter less than Δx. Since the incorporation of the wave nature of the particle, causes a smearing of the path over distances of the order of Δx, the quantum-mechanical energy loss will be much less than the classical result (Eq. (15.17)). Obviously, $\Delta x \geq \hbar/p$ is the analogue of b_{min} (Eq. (15.41)).

For a collision of two particles, the wave nature of both has to be accounted for. However, for a given relative velocity, the limiting uncertainty will come from the lighter particle. For a massive incident particle colliding with an electron, the momentum of the electron in the rest frame of the incident particle is $p' = \gamma m v$, where m is the mass of the electron. This leads to the following quantum-mechanical impact parameter

$$b_{min}^{(q)} = \frac{\hbar}{\gamma m v} \tag{15.47}$$

This clearly indicates that one has to use the larger of the two minimum impact parameters, Eqs. (15.14) and (15.47) for defining the argument B of the logarithm in dE/dx (Eq. (15.18)).

Now, defining the ratio of the classical to the quantum value of b_{min} as

$$\eta = \frac{Ze^2}{\hbar v} \tag{15.48}$$

One finds that if $\eta > 1$, the classical Bohr formula Eq. (15.43) will be valid. However, this refers to the case for slow and highly charged incident particles as observed experimentally. We may note that for $\eta < 1$, the quantum impact parameter (Eq. 15.47) have to be employed and Eq. (15.18) reads as

$$B_q = \frac{b_{max}}{b_{min}} = \eta B = \frac{\gamma^2 m v^2}{\hbar < \omega >} \tag{15.49}$$

Now, we see that Eq. (15.17) with B_q substituted in it, gives reasonable approximation to the quantum theoretical result of Bethe (1930). Bethe's formula, including the close collisions, is as follows:

$$\frac{d\mathcal{E}_q}{dx} = 4\pi N z \frac{Z^2 e^4}{m^2 v^4} \left[\ln \frac{2\gamma^2 m v^2}{\hbar < \omega >} - \frac{v^2}{c^2} \right] \tag{15.50}$$

We see that Bethe's result differs from the approximate result in having a factor 2 in the argument of the log and the presence of an additional term.

REVIEW QUESTIONS AND PROBLEMS

1. Show that the energy loss to a free electron of charge e and mass m by a simple particle of charge Ze and mass M moving with velocity in a coulomb collision is

$$\frac{d\mathcal{E}}{dx} = 4\pi N z \frac{Z^2 e^4}{m v^2} \ln B$$

where symbols have usual meanings and

$$B = \frac{b_{max}}{b_{min}} = \frac{\gamma^2 m v^3}{Z e^2 \omega}$$

Here b_{min} determines the transfer of maximum energy and b_{max} divides the collisions into adiabatic and those with free electron transfer. z is the number of electrons per atom and N is the number of atoms per unit volume.

2. Find the expression for the energy transfer to a non-relativistic harmonically bound electron by a incident particle of charge Ze, mass M and moving with velocity v at an impact parameter b.

3. Using the expression for the energy transfer to a harmonically bound charge, calculate the classical energy loss per unit length for a massive charged particle. How, the classical result can be reconciled with the quantum consideration?

Multipole Radiation

16.1 INTRODUCTION

We have read that the study of the interaction of electromagnetic radiation with atoms through spectroscopic techniques has played a key role in understanding the atomic structure. However, when we consider nuclei, we find that the said interaction could not be as important a tool, since the wavelengths of nuclear gamma rays (γ) being so short, i.e. they cannot be measured by usual optical systems. Moreover, the indirect methods which are used for energy are relatively inaccurate as compared to the spectral standards and also the resolution of energy available is lower. This is why, the study of gamma rays from nuclei, has been on an elementary level. Theoretical analysis of the available data by various researchers has made available only a general classification of possible gamma-ray transitions within nuclei, rough estimates of their lifetimes and their comparison with experimental observations.

In this chapter, we propose to show how to express an arbitrary electromagnetic field in terms of an infinite vector spherical harmonics multiplied by appropriate spherical Hankel functions. We will see that these are related to radiation quanta with definite angular momentum. We will also show that for small sources, each component of the field is related to a definite multipole moment in the source, i.e. the multipole field with quantum numbers (l, m) results from a specific 2^l pole of charge (or of magnetization) density. We will also see that these radiation-oscillating sources of definite l result from γ-ray transitions within the nucleus in which the total angular momentum of the nucleus changes exactly by $\pm m$, i.e., the multipolarity l of the γ-ray transion between two states of specified angular momentum J_i and J_f of the nucleus satisfies $\mid J_i - J_f \mid\, \leq l \leq\, \mid J_i + J_f \mid$.

Multipole Expansion of the Electromagnetic Fields

Let us consider the electromagnetic field in a uniform isotropic and non-dissipative medium suchas free space, i.e. in a region which is free of source of radiation. Maxwell's equations for the said situation are

$$\nabla \times E = -\frac{1}{c}\frac{\partial B}{\partial t} \tag{16.1}$$

$$\nabla \times B = \frac{1}{c}\frac{\partial E}{\partial t} \tag{16.2}$$

$$\nabla \cdot E = 0 \tag{16.3}$$

$$\nabla \cdot B = 0 \tag{16.4}$$

Assuming the time depence, $e^{-j\omega t}$, Maxwell's equations (16.1) to (16.4) reads as

$$\nabla \times E = ikB \tag{16.5}$$

$$\nabla \times B = -ikE \tag{16.6}$$

$$\nabla \cdot E = 0 \tag{16.7}$$

$$\nabla \cdot B = 0 \tag{16.8}$$

Now, eliminating E between the equations (16.1) and (16.2), one obtains the following relation for B

$$(\nabla^2 + k^2)B = 0 \tag{16.9}$$

$$\nabla \cdot B = 0 \tag{16.10}$$

and the defining relation for

$$E = \frac{i}{k}(\nabla \times B) \tag{16.11}$$

Alternatively, we see that one could also eliminate B between the equations (16.1) and (16.2) and get the following equations for E

$$(\nabla^2 + k^2)E = 0 \tag{16.12}$$

$$\nabla \cdot E = 0 \tag{16.13}$$

with the corresponding defining relation

$$B = -\frac{i}{k}\nabla \times E \tag{16.14}$$

This clearly reveals that either of the sets, Eqs. (16.9)–(16.11) or Eqs. (16.12)–(16.14) as equivalent to Maxwell's equations.

Now, we are interested to determine the multipole solutions for E and B. From Eq. (16.9), it is obvious that each component of B satisfies the scalar *Helmholtz wave equation*. Thus the general solution can be written in terms of spherical Hankel function $h_l^{(1,\,2)}(kr)$ multiplied by the spherical harmonics $Y_{lm}(\theta, \varphi)$. We see that any magnetic field will have the following general vector form

$$B = \sum_{l,m}\left[A_{lm}^{(1)} h_l^{(1)}(kr) + A_{lm}^{(2)} h_l^{(2)}(kr)\right] Y_{lm}(\theta, \varphi) \tag{16.15}$$

where A_{lm} are arbitrary vectors.

Since the divergence condition $\nabla \cdot B = 0$ has to be satisfied, and the radial functions $h_l^{(1,\,2)}(kr)$ are linearly independent, each of the two terms have to satisfy the requirement $\nabla \cdot B = 0$ separately. This means that the coefficients A_{lm} are to be chosen so that the following relation is obeyed

$$\nabla \cdot \sum_{l\,m} h_l(kr) A_{lm} Y_{lm}(\theta, \varphi) = 0 \tag{16.16}$$

One can express the gradient operator in terms of orbital angular momentum operator as

$$\hat{L} = \frac{1}{h}(r \times p) = -i(r \times \nabla)$$

$$\nabla = \frac{r(r \cdot \nabla)}{r^2} - \frac{r \times (r \times \nabla)}{r^2}$$

$$= r \frac{1}{r} \frac{\partial}{\partial r} - \frac{i}{r^2} r \times \hat{\mathbf{L}} \tag{16.17}$$

since $\hat{\mathbf{L}} F(r) = 0$ for any function of the radius r only. Now, rewriting Eq. (16.16) keeping in view of Eq. (16.17), one obtains the condition

$$r \cdot \sum_l \left[\frac{\partial h_l}{\partial r} \sum_m A_{lm} Y_{lm} - \frac{i h_l}{r} \hat{\mathbf{L}} \times \sum_m A_{lm} Y_{lm} \right] = 0 \tag{16.18}$$

From recursion formulae for spherical Bessel functions, we see that the coefficients A_{lm} for a given l will be coupled with those for $l' = \pm l$. This will be the case unless the $(2l + 1)$ vector coefficients for each l satisfy the following condition

$$r \cdot \sum_m A_{lm} Y_{lm} = 0 \tag{16.19}$$

In view of this special situation, the condition on the coefficients from Eq. (16.18) reads as

$$r \cdot \left(\hat{\mathbf{L}} \times \sum_m A_{lm} Y_{lm} \right) = 0 \tag{16.20}$$

Now, we see that the condition of transversality of the field B to the radius vector, (Eq. (16.19)) alongwith Eq. (16.20) will suffice to determine a unique set of vector angular functions of order l, one for each m value. However, all these conditions can be easily fulfilled if one assume the appropriate angular solution in terms of vector spherical harmonics as follows:

$$\sum_{m'} A_{lm'} Y_{lm'} = \sum_m a_{lm} \hat{\mathbf{L}} Y_{lm} \tag{16.21}$$

We see that the solution of Eq. (16.21) satisfies the transversality condition, Eq. (16.19), since $r \cdot \hat{\mathbf{L}} = 0$. Further, the condition [Eq. (16.20)] is also fulfilled since the commutation relation

$$\hat{\mathbf{L}} \times \hat{\mathbf{L}} = i \hat{\mathbf{L}} \tag{16.22}$$

makes it possible. Moreover, the commutation relation

$$L_j \nabla^2 = \nabla^2 L_j \tag{16.23}$$

holds for the operators $\hat{\mathbf{L}}$ and ∇^2, it is obvious that the function $f_l(r) \hat{\mathbf{L}} Y_{lm}$ satisfy the wave equation (Eq. 16.9). Finally the set of electromagnetic multipole fields are obtained as

$$B_{lm} = f_l(kr) \hat{\mathbf{L}} Y_{lm}(\theta, \varphi) \tag{16.24}$$

$$E_{lm} = \frac{i}{k} \nabla \times B_{lm} \tag{16.25}$$

where

$$f_1(kr) = A_l^{(1)} h_l^{(1)}(kr) + A_l^{(2)} h_l^{(2)}(kr) \tag{16.26}$$

Obviously, any linear combination of fields summed over l and m can be a solution of Eq. (16.9).

However, we see that this solution can be characterized by the fact that magnetic induction will be perpendicular to the radius vector and so will not constitute the general solution of Eq. (16.9). This is the spherical analogue of the transverse magnetic (TM), or electric cylindrical fields.

If we start with Eq. (16.12), we will obtain the equivalent of the transverse electric (TE), or magnetic cylindrical fields as follows:

$$E_{lm} = f_l(kr)\,\hat{L}Y_{lm}(\theta, \varphi) \tag{16.27}$$

$$B_{lm} = -\frac{i}{k}\nabla \times E_{lm} \tag{16.28}$$

We see that the two sets of multipole fields (16.24)–(16.25) and (16.27)–(16.28) form a complete set of vector solutions to Maxwell's Eqs. (16.1)–(16.4). The general solutions are found as

$$B = \sum_{l,m}\left[a_E(l, m)\,f_l(kr)\,X_{lm} - \frac{i}{k}a_M(l, m)\nabla \times g_l(kr)\,X_{lm}\right] \tag{16.29}$$

$$E = \sum_{l,m}\left[\frac{i}{k}a_E(l, m)\nabla \times f_l(kr)\,X_{lm} + a_M(l, m)g_l(kr)\,X_{lm}\right] \tag{16.30}$$

where $a_E(l, m)$ and $a_M(l, m)$ are the amounts of electric (l, m) multipole and magnetic (l, m) multipole fields.

One may determine these by the sources and boundary conditions. X_{lm} are the normalized vector spherical harmonics defined by

$$X_{lm}(\theta, \varphi) = \frac{1}{\sqrt{l(l+1)}}\hat{L}Y_{lm}(\theta, \varphi) \tag{16.31}$$

subject to the orthogonality condition

$$\int[X_{lm}(\theta, \varphi)]^* \cdot [X_{l'm'}(\theta, \varphi)]\,d\Omega = \delta_{ll'}\,\delta_{mm'} \tag{16.32}$$

16.2 ENERGY AND ANGULAR MOMENTUM OF MULTIPOLE RADIATIONS

We know propose to discuss the properties of the individual multipole fields, Eqs. (16.24)–(16.25) and Eqs. (16.27)–(16.28).

$$B_{lm} = f_l(kr)\,\hat{L}Y_{lm}(\theta, \varphi) \tag{16.33}$$

$$E_{lm} = \frac{i}{k}\nabla \times B_{lm} \tag{16.34}$$

and

$$E_{lm} = f_l(kr)\,\hat{L}Y_{lm}(\theta, \varphi) \tag{16.35}$$

$$B_{lm} = -\frac{i}{k} \nabla \times E_{lm} \tag{16.36}$$

where
$$f_l(kr) = A_l^{(1)} h_l^{(1)}(kr) + A_l^{(2)} h_l^{(2)}(kr) \tag{16.37}$$

For the case of near zone, we have $kr \ll 1$, the radial function $f_l(kr)$ is proportional to n_l given by the following relation

$$n_l \to -\frac{(2l-1)!!}{x^{l+1}} \tag{16.38}$$

and, in view of it, the limiting behavior of the magnetic induction for an electric (l, m) multipole is obtained as

$$B_{lm} \to -\frac{k}{l} \hat{L} \frac{Y_{lm}}{r^{l+1}} \tag{16.39}$$

where the proportionality coefficient (k/l) is chosen for later convenience. In order to find the electric field E_{lm} from Eq. (16.39), one may employ an operator identity

$$i\nabla \times \hat{L} = r\nabla^2 - \nabla\left(1 + r\frac{\partial}{\partial r}\right) \tag{16.40}$$

and get from Eq. (16.34), the electric field for an electric (l, m) multipole for $kr \ll 1$, as follows:

$$E_{lm} \to -\frac{i}{l} \nabla \times \hat{L}\left(\frac{Y_{lm}}{r^{l+1}}\right)$$

$$\to \frac{1}{l}\left[-r\nabla^2\left(\frac{Y_{lm}}{r^{l+1}}\right) + \left(\frac{Y_{lm}}{r^{l+1}} + r\frac{\partial}{\partial r}\right)\left(\frac{Y_{lm}}{r^{l+1}}\right)\right]$$

$$\to \frac{1}{l}\left[\nabla\left(\frac{Y_{lm}}{r^{l+1}} - \frac{(l+1)Y_{lm}}{r^{l+1}}\right)\right]$$

$$\to \frac{1}{l} - \nabla\left(\frac{Y_{lm}}{r^{l+1}}\right) \tag{16.41}$$

Since Y_{lm}/r^{l+1} is a solution of Laplace's equation, one may assume the expression $\nabla^2(Y_{lm}/r^{l+1})$ to be zero.

We see that the magnetic induction B_{lm}, Eq. (16.39) is smaller than E_{lm}, Eq. (16.41) by a factor kr. Thus in the near zone, the magnetic induction of an electric multipole is much smaller than the electric field. On the other hand for the case of magnetic multipole fields, Eq. (16.28), the roles of E and B are interchanged according to the transformation given by

$$E_E \to -B_M \tag{16.42a}$$
$$B_E \to E_M \tag{16.42b}$$

We now consider the behavior of the multipole fields in the far (or radiation) zone.

We now consider the case of an outgoing wave, radiated by a localized source, we see that the radial function $f_l(kr)$ is proportional to the spherical Hankel function $h_l^{(1)}(kr)$. The asymptotic form of the spherical Hankel function is given by the following expression

$$k_l^{(1)}(x) \to (-i)^{l+1} \frac{e^{ix}}{x} \tag{16.43}$$

and in view of it, B_{lm} for an electric (l, m) multipole reads as

$$B_{lm} \to (-i)^{l+1} \frac{e^{ikr}}{kr} \hat{L} Y_{lm} \tag{16.44}$$

The electric field, E_{lm} (Eq. (16.25)) now reads as

$$E_{lm} = \frac{i}{k} \nabla \times B_{lm}$$

$$= \frac{(-i)^l}{k^2} \left[\nabla \left(\frac{e^{ikr}}{r} \right) \times \hat{L} Y_{lm} + \frac{e^{ikr}}{r} \nabla \times \hat{L} Y_{lm} \right] \tag{16.45}$$

Since we have already used the asymptotic form of Hankel function (Eq. (16.45)), we may ignore all those terms which are of a degree higher than first in $(1/r)$. Making use of the identity (16.40), in the second term, one obtains

$$E_{lm} \to -\frac{(-i)^l}{k^2} \left[\frac{i r e^{ikr}(ik) - e^{ikr}}{r^2} \times \hat{L} Y_{lm} + \frac{e^{ikr}}{ir} \left\{ r \nabla^2 - \nabla \left(1 + r \frac{\partial}{\partial r} \right) Y_{lm} \right\} \right]$$

$$= -(-i)^{l+1} \frac{e^{ikr}}{r} \left[\hat{n} \times \hat{L} Y_{lm} - \frac{1}{k} (r \nabla^2 - \nabla) Y_{lm} \right] \tag{16.46}$$

where $\hat{n} = r/r$ is a unit vector in the radial direction. We see that the second term in the bracket is $1/kr$ times some dimensionless function of angles and can be easily ignored in the limit $kr \gg 1$. Therefore the electric field in the radiation zone now reads as

$$E_{lm} = B_{lm} \times \hat{n} \tag{16.47}$$

As expected these are radiation fields transverse to the radius vector and falling off as $1/r$.

We now show how to calculate the energy and angular momentum carried off by the radiation from an electric (l, m) multipole. Rewriting the fields as

$$B_{lm} = a_E(l, m) h_l^{(1)}(kr) X_{lm} e^{-i\omega t} \tag{16.48}$$

$$E_{lm} = \frac{i}{k} \nabla \times B_{lm} \tag{16.49}$$

For harmonically varying fields, the time-averaged energy density is obtained as

$$u = \frac{1}{16\pi} (E \cdot E^* + B \cdot B^*) \tag{16.50}$$

and in the radiation zone, both the terms are equal. The energy in a spherical shell lying between r and $(r + dr)$ for $kr \geq 1$ is obtained as

$$du = \frac{|a_E(l, m)|^2}{8\pi} |h_l^{(1)}(kr)|^2 r^2 dr \int \mathbf{X}^*_{lm} \cdot \mathbf{X}_{lm} \, d\Omega \tag{16.51}$$

Now, using the asymptotic form of the spherical hankel function, [Eq. (16.43)], and the orthogonality condition for the vector spherical harmonics, [Eq. (16.32)], one obtains

$$\frac{du}{dr} = \frac{|a_E(l, m)|^2}{8\pi k^2} \tag{16.52}$$

For the magnetic multipole, the corresponding result is obtained by replacing $a_E(l, m)$ by $a_M(l, m)$ as follows:

$$\frac{du}{dr} = \frac{|a_M(l, m)|^2}{8\pi k^2} \tag{16.53}$$

The density of linear momentum in the field is $1/c^2$ of the Poynting vector, i.e. $(4\pi c)^{-1} (\mathbf{E} \times \mathbf{B})$. The density of angular momentum is the cross product of the radius vector $\mathbf{r}$ and the linear-momentum density. Thus the time-averaged angular momentum density is obtained as

$$\mathbf{m} = \frac{1}{8\pi c} \text{Re} \, [\mathbf{r} \times (\mathbf{E} \times \mathbf{B}^*)] \tag{16.54}$$

Substituting for $\mathbf{E}$ from Eq. (16.49) and expanding the triple vector product, one obtains the angular momentum density for electric multipoles as follows:

$$\mathbf{m} = \frac{1}{8\pi\omega} \text{Re} \, [\mathbf{B}^*(\hat{\mathbf{L}} \cdot \mathbf{B})] \tag{16.55}$$

The angular momentum in a spherical shell lying between r and $r + dr$ is given by

$$d\mathbf{M} = \frac{|a_E(l, m)|^2}{8\pi\omega} |h_l^{(1)}(kr)|^2 r^2 dr \int \text{Re} \, [\mathbf{X}^*_{lm} \hat{\mathbf{L}} \mathbf{X}_{lm}] \, d\Omega \tag{16.56}$$

Using the form of $\mathbf{X}_{lm}$ (Eq. (16.31)), one obtains from Eq. (16.56) in the radiation zone

$$\frac{d\mathbf{M}}{dr} = \frac{|a_E(l, m)|^2}{8\pi\omega k^2} \int \text{Re} \, [Y^*_{lm} \hat{\mathbf{L}} Y_{lm}] \, d\Omega \tag{16.57}$$

It is only the z-component of $d\mathbf{M}$ that exists, so we have

$$\frac{dM_z}{dr} = \frac{m}{\omega} \frac{|a_E(l, m)|^2}{8\pi k^2} \tag{16.58}$$

The radio of the z-component of angular momentum to the energy radiated is as follows:

$$\frac{M_z}{u} = \frac{m}{\omega} = \frac{m\hbar}{\hbar\omega} \tag{16.59}$$

Obviously, the radiation from a multipole of order (l, m) carries of $m\hbar$ units of the z-component of angular momentum per photon of energy $\hbar\omega$.

16.3 PARITY OF MULTIPLE RADIATIONS AND SELECTION RULES

Generally, the parity of the multipole radiation is specified by the parity of the magnetic field B_{lm} and the argument goes as follows. The parity of the initial state is given by the product of the parities of the final state and the multipole field. The interaction of the charged particle and the electromagnetic field is given by $v \cdot A$ where A is the vector potential. Since A is proportional to the electric field E in a light wave of definite frequency, it has the same parity as E. v has odd parity, hence the parity of $(v \cdot A)$ is the opposite of the parity of E since E and B have opposite parities being connected by curl operation $c\nabla \times B = -i\omega E$ which changes the parity. Therefore, one may conclude that $(v \cdot A)$ has the same parity as the magnetic field B_{lm} in the emitted wave.

We have for electric multipoles, the magnetic induction B_{lm} given by Eq. (16.25),

$$B_{lm} = f_l (kr) \hat{L} Y_{lm} (\theta, \varphi) \tag{16.60}$$

$$E_{lm} = \frac{i}{k} \nabla \times B_{lm} \tag{16.61}$$

We see that the parity transformation $(r \to -r)$ is equivalent to $(r \to r, \theta \to \pi - \theta, \varphi \to \varphi + \pi)$ in spherical coordinates.

We may note that the operator $\hat{L}$ is invariant under parity operation. Therefore the parity of B_{lm} of electric multipole is specified by the parity of $Y_{lm} (\theta, \varphi)$. The parity of $Y_{lm} (\theta, \varphi)$ is $(-1)^l$ and so the magnetic induction B_{lm} has parity $(-1)^l$ while the electric field E_{lm} has the parity $(-1)^{l + 1}$, since $E_{lm} \sim \nabla \times B_{lm}$. For a magnetic multipole of order (l, m) the parity is $(-1)^{l + 1}$, since the electric field E_{lm} is of the same form as B_{lm} for electric multipoles.

Now, if the nucleus undergoes a transition from the initial state Ψ_i to the final state Ψ_f, from the conservation of parities, π_i and π_f of the initial and the final nuclear states, we see that they are related as follows:

$$\pi_i = (-1)^l \pi_f \quad \text{for} \quad E\text{-radiation} \tag{16.62}$$

$$\pi_i = (-1)^{l+1} \pi_f \quad \text{for} \quad M\text{-radiation} \tag{16.63}$$

However, except when the initial and final total angular momenta J_i and J_f are $J_i = J_f = 0$ in which case there is no emission of a multipole radiation since the transition between the two states $J_i = J_f = 0$ is totally forbidden.

Now, according to the law of conservation of total angular momentum, one obtains a vectorial relation and a relation between their z-components M_i, M_f, m. These are as follows:

$$J_i = J_f + 1 \tag{16.64}$$

$$M_i = M_f + m \tag{16.65}$$

Eq. (16.64) clearly implies that

$$\left| J_i - J_f \right| \le l \le \left| J_i + J_f \right| \tag{16.66}$$

16.4 ANGULAR DISTRIBUTION OF THE MULTIPOLE RADIATION

We know that the instantaneous energy flux is given by the Poynting vector

$$S = \frac{c}{4\pi}(E \times B)$$

$$= \frac{c}{4\pi}|\mathbf{E}_a|^2\,\hat{\mathbf{n}} \tag{16.67}$$

and the angular distribution of the power radiated per unit solid angle is

$$\frac{dP}{d\Omega} = \frac{c}{4\pi}|RE_a|^2 \tag{16.68}$$

For a localized source, the fields in the radiation zone are given by the superposition as follows:

$$B \to \frac{e^{i(kr-\omega t)}}{kr}\sum_{l,m}(-1)^{l+1}[a_E(l,m)X_{lm} + a_M(l,m)\,\hat{\mathbf{n}} \times X_{lm}] \tag{16.69}$$

$$E \to B \times \hat{\mathbf{n}} \tag{16.70}$$

The time averaged power radiated per unit solid angle is as follows:

$$\frac{dP}{d\Omega} = \frac{c}{8\pi k^2}\left|\sum_{l,m}(-1)^{l+1}[a_E(l,m)X_{lm} \times \hat{\mathbf{n}} + a_M(l,m)X_{lm}]\right|^2 \tag{16.71}$$

However, the polarization of radiation is specified by the directions of vectors. Further, from Eq. (16.71), we see that electric and magnetic multipoles of a given (l, m) have the same angular dependence, but their polarizations are orthogonal to each other. The order of the multipole (l, m) can be determined from the angular distribution of the radiated power but the character of the radiation whether electric or magnetic, can only be established on the basis of the polarization measurement.

For a pure electric multipole of order (l, m), the angular distribution reduces to the form as follows:

$$\frac{dP(l,m)}{d\Omega} = \frac{c}{8\pi k^2}|a_E(l,m)|^2\,|X_{lm}|^2 \tag{16.72}$$

Using the following expression for $|X_{lm}|^2$

$$|X_{lm}|^2 = X_{lm} \cdot X_{lm}^*$$

$$= \frac{1}{2}\left[1 - \frac{m(m+1)}{l(l+1)}\right]|Y_{l,m+1}|^2 + \frac{1}{2}\left[1 - \frac{m(m-1)}{l(l+1)}\right]|Y_{l,m-1}|^2 + \frac{m^2}{l(l+1)}|Y_{lm}|^2$$

in Eq. (16.72), one obtains

$$\frac{dP(l,m)}{d\Omega} = \frac{c\,|a(l,m)|^2}{8\pi k^2\,l(l+1)}\left[\frac{1}{2}(l-m)(l+m+1)|Y_{l,m+1}|^2\right.$$

$$\left. + \frac{1}{2}(l+m)(l-m+1)|Y_{l,m-1}|^2 + m^2|Y_{lm}|^2\right] \tag{16.73}$$

For reference, one may give below the values of $\left|X_{lm}(\theta)\right|^2$ for a dipole ($l = 1$) and a quadrupole ($l = 2$).

$$\left|X_{1,0}\right|^2 = \frac{3}{8\pi}\sin^2\theta$$

$$\left|X_{1,\pm 1}\right|^2 = \frac{3}{16\pi}(1 + \cos^2\theta)$$

$$\left|X_{2,0}\right|^2 = \frac{15}{8\pi}\sin^2\theta\cos^2\theta \tag{16.74}$$

The total power P radiated by a pure multipole of order (l, m) can be expressed as

$$P(l, m) = \frac{c}{8\pi k^2}\left|a(l, m)\right|^2 \tag{16.75}$$

where X_{lm} are normalized to unity.

However, for a general source, one finds that the angular distribution is given by the coherent sum. One may easily show that on integration over angles, the interference terms do not contribute. Thus the total power radiated from a general source is just an incoherent sum of contributions from the different multipoles as

$$P = \frac{c}{8\pi k^2}\sum_{l, m}\left[\left|a_E(l, m)\right|^2 + \left|a_M(l, m)\right|^2\right] \tag{16.76}$$

At this stage, one can easily show that the radiation distribution from an atomic or nuclear source which consists of a set of multipoles of order l with coefficients a (l, m) independent of m, added incoherently, will be isotropic. This is because the absolute squares of the vector spherical harmonics obey the following sum rule

$$\sum_{m=-l}^{l}\left|X_{lm}(\theta, \varphi)\right|^2 = \frac{2l + 1}{4\pi} \tag{16.77}$$

16.5 SOURCES OF MULTIPOLE RADIATION

We now try to establish the relationship of the fields with the sources, charge density $\rho(x, t)$, current density $J(x, t)$ and magnetization density $M(x, t)$. Further, we assume that these sources are varying harmonically in time as

$$\rho(x)e^{-i\omega t}, \quad J(x)e^{-i\omega t} \quad M(x)e^{-i\omega t} \tag{16.78}$$

It is implied that one should take the real part of any of these complex quantities. Further, any arbitrary time dependence can be generated by the linear superposition. However, the presence of the magnetization density requires that the distinction between B and H be retained. Now, Maxwell's equations for harmonically varying sources reads as

$$\nabla \times E = ikB \tag{16.79}$$

$$\nabla \times H + ikE = \frac{4\pi}{c}J \tag{16.80}$$

$$\nabla \cdot \mathbf{B} = 0 \tag{16.81}$$

$$\nabla \cdot \mathbf{E} = 4\pi\rho \tag{16.82}$$

Now, the continuity equation takes the form

$$i\omega\rho = \nabla \cdot \mathbf{J} \tag{16.83}$$

For being able to use the general solution of source free Maxwell's equation, one may consider the field variables

$$\mathbf{B} \text{ and } \mathbf{E}' = \mathbf{E} + \frac{4\pi i}{\omega} \mathbf{J} \tag{16.84}$$

Let us eliminate $\mathbf{E}$ from the Eqs. (16.70)-(16.83), one obtains

$$\mathbf{E} = \frac{4\pi}{ikc} \mathbf{J} - \frac{1}{ik} \nabla \times \mathbf{H}$$

$$= \frac{4\pi}{ikc} \mathbf{J} - \frac{1}{ik} (\nabla \times \mathbf{B}) + \frac{4\pi}{ik} \nabla \times \mathbf{M}$$

and inserting it in Eq. (16.79), one obtains

$$\nabla \times \frac{4\pi}{ikc} \mathbf{J} - \frac{1}{ik} \nabla \times (\nabla \times \mathbf{B}) + \frac{4\pi}{ik} \nabla \times (\nabla \times \mathbf{M}) = ik\mathbf{B}$$

$$\Rightarrow \quad (\nabla^2 + k^2)\mathbf{B} = -\frac{4\pi}{c} [\nabla \times \mathbf{J} + c\nabla \times (\nabla \times \mathbf{M})] \tag{16.85}$$

$$\nabla \cdot \mathbf{B} = 0 \tag{16.86}$$

$$\mathbf{E}' = \frac{i}{k} [(\nabla \times \mathbf{B}) - 4\pi (\nabla \times \mathbf{M})] \tag{16.87}$$

In an analogous manner, one can eliminate $\mathbf{B}$ from the Eqs. (16.79)–(16.83) and obtain the set of following equations

$$(\nabla^2 + k^2)\mathbf{E}' = -\frac{4\pi ik}{c} \left[c\nabla \times \mathbf{M} + \frac{1}{k^2} \nabla \times (\nabla \times \mathbf{J}) \right] \tag{16.88}$$

$$\nabla \cdot \mathbf{E}' = 0 \tag{16.89}$$

$$\mathbf{B} = -\frac{i}{k} \left[(\nabla \times \mathbf{E}') - \frac{4\pi i}{\omega} (\nabla \times \mathbf{J}) \right] \tag{16.90}$$

In the region outside the source, the set of Eqs. (16.85)–(16.87) and (16.88)–(16.90) are found to reduce to the Eqs. (16.9)–(16.10) and (16.11)–(16.12). The general solution for $\mathbf{B}$ and $\mathbf{E}'$ in the outside region is given by Eqs. (16.29) and (16.30). However, even inside the source, one finds that both the fields $\mathbf{B}$ and $\mathbf{E}'$ have zero divergence. Therefore the same general solution is expected to hold with the only difference that the radial functions $f_l(r)$ and $g_l(r)$ are modified.

Let us now consider the magnetic induction

$$\mathbf{B} = \sum_{l,m} \left[f_{lm}(r) \mathbf{X}_{lm} - \frac{i}{k} \nabla \times g_{lm}(r) \mathbf{X}_{lm} \right] \tag{16.91}$$

which may agree with the multipole fields from an electric (l, m) multipole (Eq. (16.49)) provided

$$f_{lm}(r) \rightarrow a_E(l, m) h_l^{(1)}(kr) \tag{16.92}$$

$$g_{lm}(r) \rightarrow a_M(l, m) h_l^{(1)}(kr) \tag{16.93}$$

Substituting Eq. (16.91) into Eq. (16.85) and taking the scalar product of both sides with a typical X^*_{lm} and integrating over all angles, one obtains

$$\left[\frac{d^2}{dr^2} + \frac{2}{r}\frac{d}{dr} + k^2 - \frac{l(l+1)}{r^2}\right] f_{lm}(r)$$

$$= -\frac{4\pi}{c} \int d\Omega X^*_{lm} \cdot [\nabla \times J + c\nabla \times (\nabla \times M)] \tag{16.94}$$

We see that the terms involving $g_{lm}(r)$ will vanish on account of orthogonality. Analogously, one can obtain the equation for $g_{lm}(r)$ by substituting the expression for E' into the Eq. (16.88) and through similar manipulations, one obtains

$$\left[\frac{d^2}{dr^2} + \frac{2}{r}\frac{d}{dr} + k^2 - \frac{l(l+1)}{r^2}\right] g_{lm}(r)$$

$$= -4\pi ik \int d\Omega X^*_{lm} \cdot \left[\nabla \times M + \frac{1}{ck^2}\nabla \times (\nabla \times J)\right] d\Omega \tag{16.95}$$

Now, one can write the solutions for $f_{lm}(r)$ and $g_{lm}(r)$ of the inhomogeneous Eqs. (16.94) and (16.95) using the Green's functions.

Denoting the R.H.S. of Eq. (16.94) by $-K_E(r)$, one obtains the solution for $f_{lm}(r)$ as

$$f_{lm}(r) = ik \int_0^\infty r'^2 j_l(kr_<) h_l^{(1)}(kr_>) K_E(r') dr' \tag{16.96}$$

For a field point outside the source, $r_< = r'$ and $r_> = r$ and so, one obtains

$$f_{lm}(r) \rightarrow ik h_l^{(1)}(kr) \int_0^\infty r'^2 j_l(kr') K_E(r') dr' \tag{16.97}$$

On comparing with the Eq. (16.92), one can identify the expression for electric multipole coefficient.

With the form of $K_E(r)$ from Eq. (16.94), one obtains

$$a_E(l, m) = \frac{4\pi ik}{c} \int j_l(kr) X^*_{lm} \cdot [\nabla \times J + c\nabla \times (\nabla \times M)] d^3x \tag{16.98}$$

Analogously, one obtains the expression for magnetic multipole coefficients as

$$a_M(l, m) = -4\pi k^2 \int j_l(kr) X^*_{lm} \cdot \left[\nabla \times M + \frac{1}{ck^2}\nabla \times (\nabla \times J)\right] d^3x \tag{16.99}$$

The Eqs. (16.98) and (16.99) can be put into a useful form through the use of an identity.

$$\int j_l(kr) X^*_{lm} \cdot (\nabla \times A) d^3x = \frac{i}{\sqrt{l(l+1)}} \int Y^*_{lm} \left\{ (\nabla \cdot A) \frac{\partial}{\partial r} [rj_l(kr)] - k^2 r \cdot A j_l(kr) \right\} d^3x$$

$$(16.100)$$

where A is a vector field which is well behaved throughout the range and vanishes faster than r^{-2}. Applying the identity repeatedly to $J, M, \nabla \times J, \nabla \times H$, one obtains the following expressions for the electric and magnetic multipole coefficients

$$a_E(l, m) = \frac{4\pi k^2}{i\sqrt{l(l+1)}} \int Y^*_{lm} \left\{ \rho \frac{\partial}{\partial r} [rj_l(kr)] + \frac{ik}{c}(r \cdot J) j_l(kr) - ik\nabla \cdot (r \times M) j_l(kr) \right\} d^3x$$

$$(16.101)$$

$$a_M(l, m) = \frac{4\pi k^2}{i\sqrt{l(l+1)}} \int Y^*_{lm} \left\{ \nabla \cdot \left(\frac{r \times J}{c} \right) j_l(kr) + \nabla \cdot M \frac{\partial}{\partial r} [rj_l(kr)] - k^2 (r \cdot M) j_l(kr) \right\} d^3x$$

$$(16.102)$$

We see that so far these expressions for multipole coefficients are exact for any source size. However, in problems involving atomic and nuclear systems, the source dimensions are small as compared to a wavelength ($kr_{max} \ll 1$), thus one have to seek the approximate expressions for the multipole coefficients in the limit $kr \ll 1$. Using the values of spherical Bessel functions for small argument limit

$$j_l(kr) \to \frac{(kr)^l}{(2l+1)!!}$$

$$(16.103)$$

and keeping the lowest powers in kr for terms involving ρ, J and M, one obtains the approximate expressions for $a_E(l, m)$ as

$$a_E(l, m) = \frac{4\pi k^2}{i\sqrt{l(l+1)}} \int Y^*_{lm} \left[\rho \frac{\partial}{\partial r} \left(\frac{k^l r^{l+1}}{(2l+1)!!} \right) - ik \nabla \cdot (r \times M) \frac{(kr)^l}{(2l+1)!!} \right] d^3x$$

$$= \frac{4\pi k^{l+2}}{i(2l+1)!!} \left(\frac{l+1}{l} \right)^{1/2} \left[\int r^l Y^*_{lm} \rho d^3x - \frac{ik}{l+1} \int r^l Y^*_{lm} \nabla \cdot (r \times M) \right] d^3x$$

$$= \frac{4\pi k^{l+2}}{i(2l+1)!!} \left(\frac{l+1}{l} \right)^{1/2} (Q_{lm} + Q'_{lm})$$

$$(16.104)$$

where

$$Q_{lm} = \int r^l Y^*_{lm} \rho d^3x$$

$$(16.105)$$

$$Q'_{lm} = -\frac{ik}{l+1} \int r^l Y^*_{lm} \nabla \cdot (r \times M) d^3x$$

$$(16.106)$$

The moment Q_{lm} (Eq. 16.105) is the *real electrostatic multipole moment* whereas Q'_{lm} (Eq. 16.106) is *induced electric multipole moment* due to magnetization and Q'_{lm} smaller than Q_{lm} by a factor kr.

Identically, one can see that the magnetic multipole coefficients $a_M\,(l,m)$ in the long wavelength approximation is given by

$$a_M(l,m) \cong \frac{4\pi i k^{l+2}}{(2l+1)!!}\left(\frac{l+1}{l}\right)^{1/2}\;(M_{lm} + M'_{lm}) \tag{16.107}$$

where

$$M_{lm} = -\frac{1}{l+1}\int r^l\,Y^*_{lm}\,\nabla\cdot\left(\frac{r\times J}{c}\right)d^3x \tag{16.108}$$

and

$$M'_{lm} = \int r^l\,Y^*_{lm}\,\nabla\cdot M\,d^3x \tag{16.109}$$

We see that the magnetic moments M_{lm} and M'_{lm} are of the same order of magnitude and are determined by the magnetic moment densities $(1/2c)\,(r\times J)$ and M. The electric moments are found to be related to the charge density r as expected.

16.6 MULTIPOLE TRANSITIONS IN ATOMIC AND NUCLEAR SYSTEMS

Now, we discuss the essential features of the multipole radiations from atomic and nuclear systems based on classical treatment. However, one can get deeper insight only from quantum mechanical treatment.

We see that for a pure multipole of order (l, m), the angular distribution of multipole radiation reduces to a single term given by (Eq. (16.72))

$$\frac{dP(l,m)}{d\Omega} = \frac{c}{8\pi k^2}\left|a_E(l,m)\right|^2\left|X_{lm}\right|^2 \tag{16.110}$$

One can obtain the total power radiated by a pure multipole of order (l, m) by integrating the expression Eq. (16.110) over all angles. Since X_{lm} are normalized, the total power radiated is obtained as

$$P(l,m) = \frac{c}{8\pi k^2}\left|a(l,m)\right|^2 \tag{16.111}$$

Substituting the value of $a_E\,(l, m)$ for the pure electric multipole (l, m), into Eq. (16.104), one obtains the expression for the total power by an electric multipole as

$$P_E(l,m) = \frac{2\pi c}{\left[(2l+1)!\right]^2}\left(\frac{l+1}{l}\right)k^{2l+2}\left|Q_{lm} + Q'_{lm}\right|^2 \tag{16.112}$$

In a similar manner, for a pure magnetic multipole, the total power radiated is obtained from Eq. (16.107) as

$$P_M(l,m) = \frac{2\pi c}{\left[(2l+1)!\right]^2}\left(\frac{l+1}{l}\right)k^{2l+2}\left|M_{lm} + M'_{lm}\right|^2 \tag{16.113}$$

Now, we define the transition probability (the reciprocal mean life) as the power divided by the energy of the radiated photon

$$\frac{1}{\tau} = \frac{P}{\hbar\omega} \tag{16.114}$$

However, to calculate the expression for total power $P_E(l, m)$ from a pure electric multipole and $P_M(l, m)$ from a pure magnetic multipole, one has to obtain the estimates of the moments Q_{lm}, Q'_{lm} and M_{lm}, M'_{lm} respectively. One can estimate these only from the charge density $\rho(x)$ and the divergence of the magnetization. Since these are not known exactly for atomic or nuclear systems, one will have to content by making the following schematic model of the source.

The oscillating charge density can be assumed as

$$\rho(x) = \frac{3e}{a^3} Y_{lm}(\theta, \varphi) \qquad r < a$$

$$= 0 \qquad r > a \qquad (16.115)$$

and the divergence of magnetization is assumed to be

$$\nabla \cdot M + \frac{1}{l+1} \nabla \cdot \left(\frac{r \times J}{c} \right) = \frac{2g}{a^3} Y_{lm}(\theta, \varphi) \left(\frac{e\hbar}{mcr} \right) \qquad r < a$$

$$= 0 \qquad r > a \qquad (16.116)$$

where g is the effective g factor for the magnetic moment of the particles in the atomic or nuclear system and $e\hbar/mc$ is twice the *Bohr magneton* for these particles.

From Eq. (16.115), we obtain the electric multipole moment Q_{lm} to be

$$Q_{lm} = \int r^l Y_{lm}^* \rho d^3 x$$

$$= \int_0^\infty r^l Y_{lm}^* \frac{3e}{a^3} Y_{lm}(\theta, \varphi) d\Omega$$

$$= \frac{3}{l+3} e a^l \qquad (16.117)$$

which clearly depends only on l and not m.

Now, we obtain the magnetic multipole moments sum from Eq. (16.107) as

$$M_{lm} + M'_{lm} = -\frac{2}{l+1} e a^l \left(\frac{g\hbar}{mca} \right) \qquad (16.118)$$

One can easily show that the induced electric multipole moment Q'_{lm} is ineligible compared to Q_{lm}. From the definition of Q'_{lm}, Eq. (16.106), one obtains

$$Q'_{lm} \sim g \left(\frac{\hbar\omega}{mc^2} \right) Q_{lm} \qquad (16.119)$$

We know that the energies radiated from atoms and nuclei are always smaller than the rest energies of the particles involved; thus Q'_{lm} is much less than Q_{lm}. The transition probability for an electric multipole of order l is obtained from Eqs. (16.111), (16.113) and (16.116) as follows:

$$\frac{1}{\tau_E(l)} \cong \frac{e^2}{\hbar c} \frac{2\pi}{[(2l+1)!!]^2} \left(\frac{l+1}{l} \right) \left(\frac{3}{l+3} \right) (ka)^{2l} \omega \qquad (16.120)$$

The transition probability for magnetic multipoles of order l, apart from factors of the order of unity is obtained as follows:

$$\frac{1}{\tau_M(l)} \cong \left(\frac{g\hbar}{mca}\right)^2 \frac{1}{\tau_E(l)} \tag{16.121}$$

From Eq. (16.120), one finds that in the long wavelength approximation $ka \ll 1$, the transition rate falls off rapidly with the multipole order for a fixed radiated energy. Thus, we see that in such systems, only the lowest nonvanishing multipole will be the only one of significance. The ratio of transition probabilities of successive orders of multipoles for either electric or magnetic radiation is obtained as

$$\frac{[\tau(l+1)]^{-1}}{[\tau(l)]^{-1}} \sim \frac{(ka)^2}{4l^2} \tag{16.122}$$

We see that this is an approximate result since other factors of the order of $(1/l)$ have been omitted.

Let us now estimate the ratio of transition probabilities of successive orders for the case of atomic systems. Since it is the electrons which are involved in the radiation process, the dimensions of the source can be easily taken as

$$a \sim \frac{a_0}{Z_{\text{eff}}} \tag{16.123}$$

where a_0 is the Bohr radius and Z_{eff}, the effective nuclear charge. $Z_{\text{eff}} \sim 1$ for optical transitions and $Z_{\text{eff}} \leq Z$ for x-ray transitions. The atomic transition energy is of the order of $\hbar\omega \leq Z_{\text{eff}}^2 e^2/a_0$. With these data, one can easily estimate ka as follows:

$$ka = \frac{\omega}{c} a$$

$$\Rightarrow \qquad ka \leq \frac{Z_{\text{eff}}}{\hbar c/e^2}$$

$$\Rightarrow \qquad ka \leq \frac{Z_{\text{eff}}}{137} \tag{16.124}$$

Now, from Eq. (16.122), it is evident that for atomic radiation the successive multipoles will be in the ratio $(Z_{\text{eff}}/137)^2$. Further, let us estimate the ratio of magnetic to electric-multipole rates for atomic radiation. For electrons, the g factor is of the order of unity. Since $a \approx a_0/Z_{\text{eff}} = 137\hbar/(mcZ_{\text{eff}})$, we find

$$\frac{1}{\tau_M(l)} \approx \left(\frac{Z_{\text{eff}}}{137}\right)^2 \frac{1}{\tau_E(l)} \tag{16.125}$$

The magnetic lth multipole rate is smaller than the lth electric multipole rate by a factor $(Z_{\text{eff}}/137)^2$. This clearly reveals that in atomic systems, the electric dipole radiation will be most intense to be followed by electric quadrupole and magnetic dipole transitions, a factor $(Z_{\text{eff}}/137)^2$ weaker. Obviously, only in the case of x-ray transitions in heavy elements, there is some likelihood of competition from other than the lowest order electric multipole.

Let us consider the case of nuclear transitions from atomic nuclei. In this case, the transition energies span a wide range from, say, 10 keV to many MeV and so the ka values will also vary to a wide range. This reveals that for a given multipole order, the transition probability will vary over many order of magnitude depending upon the transition energy. However, rough estimates of the transition probabilities are useful for cataloguing nuclear multipole transitions, because for a definite energy the transition probabilities (Eqs. (16.120) and (16.121)) vary greatly. Figure 16.1 shows a log-log plot of the lifetime versus photon energy for an electric multipole of order l (Eq. (16.120), for nuclear radius $a \approx 5.6 \times 10^{-15}$ m corresponding to mass number $A \sim 100$. Obviously, the curves for different l values converge at high energies but the lifetimes for different l at the same energy differ by factors of the order of 10^5. This clearly reveals that our estimates serve as guides for assigning multipole orders even if the multipole moments involved in the given multipole transitions vary greatly from our estimates. In an actual experiment, the lifetime-energy plot shows broad bands lying near a straight line in the plot of Fig. 16.1. Normally, one finds that the electric multipole moment (Eq. (16.117)), serves as an upper limit on the multipole moment and the lifetime estimate (Eq. (16.120)) in turn, serving as a lower bound on the measured value. However, in so-called enhanced electric quadrupole transitions, the lifetimes have been observed to be two orders of magnitude smaller than shown in Fig. 16.1.

Fig. 16.1: Schematic of a log-log plot exhibiting lifetime versus photon energy for an electric multipole of order l for nuclear radius $a \approx 5.6 \times 10^{-15}$ m corresponding to mass number $A \sim 100$

At this stage one can compare the magnetic and electric multipoles of the same order with the help of Eq. (16.121). However, due to the anomalous magnetic moment of nucleus, the effective g factor is of the order of 3 and assuming the source size $a \sim R = 1.2A^{1/3} \times 10^{-15}$ m, one obtains from Eqs. (16.120) and (16.121)

$$\frac{1}{\tau_M(l)} \sim \frac{0.3}{A^{2/3}} \frac{1}{\tau_E(l)} \tag{16.126}$$

Further, we see that the numerical factor for $20 < A < 250$ ranges from 4×10^{-2} to $0.3\, 5\, 10^{-2}$ and thus for a given multipole order l, one finds that electric transitions will be 25–120 times more intense as compared to magnetic transitions. However, this statement holds for all multipoles except $l = 1$ since strong, attractive charge-independent nuclear forces inhibit electric-dipole transitions at low energies. Now, one can see that for that case, magnetic-dipole transitions will be as intense as electric-dipole transitions for other multipoles.

As mentioned earlier, in a given multipole transition a mixture of multipoles such as magnetic l, $(l + 2)$, ... pole and electric $(l + 1)$, $(l + 3)$, ... pole could occur. In the long wavelength approximation $ka \ll 1$, and hence considering the lowest order multipoles, one obtains from Eqs. (16.120) and (16.121) as

$$\frac{[\tau_E(l+1)]^{-1}}{[\tau_M(l)]^{-1}} \sim \left(\frac{A^{1/3}\mathcal{E}}{200l} \right) \tag{16.127}$$

where $\mathcal{E}$ denotes photon energy in MeV. In heavy elements, the electric-quadrupole amplitude is approximately one-twentieth of the magnetic-dipole amplitude. However, one finds that as in rare earth and transuranic elements, the quadrupole moment is larger by a factor of ten or so, the electric-quadrupole transitions become as intense as the magnetic-dipole transitions.

For a mixture of magnetic $(l+1)$ pole and electric l-pole, one finds that the transition rates can be related to each other as follows:

$$\frac{[\tau_M(l+1)]^{-1}}{[\tau_E(l)]^{-1}} \sim \left(\frac{\mathcal{E}}{600l} \right)^2 \tag{16.128}$$

Obviously, for magnetic transitions, a magnetic $(l + 1)$ pole is far less intense as compared to electric l-pole transitions.

REVIEW QUESTIONS AND PROBLEMS

1. Obtain the electric and magnetic coefficients $a_E(l, m)$ and $a_M(l, m)$ for localized distribution of charge density $\rho(x, t)$, current density $J(x, t)$ and intrinsic magnetization $M(x, t)$. How these expressions are modified in the limit $kr_{max} \ll 1$?

2. Obtain the time-averaged power radiated per unit solid angle, the angular distribution for a pure multipole order (l, m) and the total power radiated by considering the fields in the radiation zone for a generalized source distribution as

$$B \to \frac{e^{i(kr - \omega t)}}{kr} \sum_{l, m} (-1)^{l+1} \left[a_E(l, m) X_{lm} + a_M(l, m) \hat{n} \times X_{lm} \right]$$

$$E \to B \times \hat{n}$$

3. Show that the radiation from a multipole of order (l, m) carries off m units of the z-component of angular momentum per photon of energy $\hbar\omega$. Give its quantum-mechanical interpretation and also deduce the selection rules for multipole transitions between quantum states.

4. Write the general solutions to $r \cdot B$ and $r \cdot E$ in the presence of sources and comment on these results.

5. Show that $L_f(r) = 0$ for any function of r only.

6. Show that L_x and L_y operating on Y_{lm} give linear combination of $Y_{l,\,m+1}$ and $Y_{l,\,m-1}$.

17

Hamiltonian Formulation of Field Equation

17.1 INTRODUCTION

It is well established that in much the same way as the quantum theory of particle is based on the canonical form of classical dynamics, the transition from *classical to quantum electrodynamics*, is faciliated through the canonical formulation of the equation of electromagnetic field. One can accomplish this by expressing the field equations as Hamiltonian equations with a single Hamiltonian function depending on the coordinates of the particles alongwith some other variables. We know that the number of degrees of freedom of an electromagnetic field is non-denumerably infinite, in the mechanical sense and hence we first study the transition from a *discrete* to a *continuous* system.

17.2 LAGRANGE AND HAMILTON EQUATIONS

We have read that one of the main consequences of the *Hamilton principle* are the *Lagrange equations of motion*. One can describe a system with n degrees of freedom by Lagrange equations of motion,

$$\frac{d}{dt}\frac{\partial L}{\partial \dot{q}_i} - \frac{\partial L}{\partial q_i} = 0 \quad (i = 1, 2, ..., n) \tag{17.1}$$

where the Lagrangian $L(q_i, \dot{q}_i, t)$ is given by

$$L = T - V \tag{17.2}$$

$T(q_i, \dot{q}_i, t)$ and $V(q_i)$ respectively in Eq. (17.2), being the kinetic and potential energies of the system. The variables q_i, $\dot{q}_i$ and t represent, respectively, the generalized coordinates of the system, the generalized velocities and time.

Now, defining a generalized particle momentum p_i by

$$p_i = \frac{\partial L}{\partial \dot{q}} \tag{17.3}$$

the Lagrange equations (17.1), reads as

$$\dot{p}_i = \frac{\partial L}{\partial q_i} \quad (i = 1, 2, ..., n) \tag{17.4}$$

Now, introducing the Hamiltonian as

$$\mathcal{H}(p_i, q_i, t) = \sum_i p_i q_i - L(q_i, \dot{q}_i, t) \tag{17.5}$$

one obtains the total differential as follows:

$$dH = \sum_i (p_i dq_i + \dot{q}_i \, dp_i) - \frac{\partial L}{\partial q_i} dq_i - \frac{\partial L}{\partial \dot{q}_i} dq_i - \frac{\partial L}{\partial t} dt$$

$$= \sum_i (\dot{q}_i \, dp_i - p_i d\dot{q}_i) - \frac{\partial L}{\partial t} dt \qquad (17.6)$$

Let us consider the implicit dependence of H as a function of the variables p_i, q_i and t. One obtains

$$dH = \sum_i \left(\frac{\partial H}{\partial p_i} dp_i + \frac{\partial H}{\partial q_i} dq_i \right) + \frac{\partial H}{\partial t} dt \qquad (17.7)$$

Now, comparing the Eqs. (17.6) and (17.7), one obtains the following set of $(2n + 1)$ equations

$$\dot{q}_i = \frac{\partial H}{\partial p_i} \qquad (17.8)$$

$$\dot{p}_i = -\frac{\partial H}{\partial q_i} \qquad (i = 1, 2, ..., n) \qquad (17.9)$$

and
$$\frac{\partial L}{\partial t} = \frac{\partial H}{\partial t} \qquad (17.10)$$

The set of $2n$ equations represented by Eqs. (17.8) and (17.9) are called *Hamilton's canonical equations of motion*. We may note that the coordinates q_i and the generalized momenta p_i are implied to be canonically conjugate to each other.

We see that incidently, if the Lagrangian L is independent of time, then one finds in Eq. (17.10) that Hamiltonian $\mathcal{H}$ also does not depend on time. Obviously, the equations, which define the generalized coordinates in terms of the cartesian coordinates, do not explicitly depend on time, the Hamiltonian $\mathcal{H}$ is not only a constant of motion but also the total energy $(T + V)$ of the system.

17.3 TRANSITION FROM A DISCRETE TO A CONTINUOUS SYSTEM

(i) *One-dimensional continuum.* One method to accomplish a transition from a discrete to a continuous system is (a) to approximate the continuous system by one containing discrete units, (b) to analyze the discrete system, and (c) to examine the changes taking place which result as the continuous limit is approached. As an example, let us study the free vibrations of a string of finite length and constant mass per unit length ρ, under the action of tension T.

Fig. 17.1: A set of mass points connected by springs of length a and force constant k

Let us consider a set of n point particles, each of mass m, connected by springs of length a and force constant k (Fig. 17.1).

If η_i is the displacement from equilibrium of the ith mass, then one can write the Lagrangian L as

$$L = T - V$$

$$= \frac{1}{2} \sum_{i=1}^{n} \left[m\dot{\eta}_i^2 - k(\beta\eta_{i+1} - \eta_i)^2 \right] = \sum_{i=1}^{n} a\mathcal{L}_i \tag{17.11}$$

where $\mathcal{L}_i$, the Lagrangian density, is given by

$$\mathcal{L}_i = \frac{1}{2} \left[\frac{m}{a} \eta_i^2 - ka \left(\frac{\dot{\eta}_{i+1} - \eta}{a} \right)^2 \right] \tag{17.12}$$

The transition to a continuous system is affected by the transformations as follows:

$$a \to dx$$

$$\frac{m}{a} \to \rho \tag{17.13}$$

$$ka \to \mathcal{T}$$

$$\frac{\eta_{i+1} - \eta_i}{a} \to \frac{\partial \eta}{\partial x}$$

where ρ is the linear mass density and $\mathcal{T}$ is the tension. Therefore, the Lagrangian now reduces to

$$L = \frac{1}{2} \int \left[\rho\dot{\eta}^2 - \mathcal{T} \left(\frac{\partial \eta}{\partial x} \right)^2 \right] dx = \int \mathcal{L} dx \tag{17.14}$$

with

$$\mathcal{L}\left(\eta, \dot{\eta}, \frac{\partial \eta}{\partial x}, t \right) = \frac{1}{2} \left[\rho\eta^2 - \mathcal{T} \left(\frac{\partial \eta}{\partial x} \right)^2 \right] \tag{17.15}$$

The *variational principle* states that

$$\delta \int L \left(\eta, \dot{\eta}, \frac{\partial \eta}{\partial x}, t \right) dt = \delta \iint \mathcal{L} dx \, dt \tag{17.16}$$

Now, confining the treatment to the case that $\mathcal{L}$ does not explicitly depend on t, one obtain by substituting Eq. (17.15) into Eq. (17.16) and integrating by parts

$$\delta \int L dt = \iint \left[\frac{\partial \mathcal{L}}{\partial \eta} \delta\eta + \frac{\partial \mathcal{L}}{\partial \left(\frac{\partial \eta}{\partial x} \right)} \delta \left(\frac{\partial \eta}{\partial x} \right) + \frac{\partial \mathcal{L}}{\delta \left(\frac{\partial \eta}{\partial x} \right)} \delta \left(\frac{\partial \eta}{\partial x} \right) \right] dx \, dt \tag{17.16a}$$

Now, employing the relation

$$\delta \left(\frac{\partial \eta}{\partial x} \right) = \frac{\partial}{\partial x} (\partial\eta)$$

and integrating the second and third terms of the Equation (17.16a), one obtains

$$\int \frac{\partial L}{\partial \left(\frac{\partial \eta}{\partial x}\right)} \frac{\partial}{\partial x} (\delta \eta) dx = \frac{\partial L}{\partial \left(\frac{\partial \eta}{\partial x}\right)} \delta \eta \Big|_1^2 - \int \frac{\partial}{\partial x} \left[\frac{\partial L}{\partial \left(\frac{\partial \eta}{\partial x}\right)} \right] \delta \eta \, dx$$

$$\Rightarrow \int \frac{\partial L}{\partial \left(\frac{\partial \eta}{\partial x}\right)} \frac{\partial}{\partial x} (\delta \delta) dt = \frac{\partial L}{\partial \left(\frac{\partial \eta}{\partial x}\right)} \delta \eta \Big|_1^2 - \int \frac{\partial}{\partial t} \left[\frac{\partial L}{\partial \left(\frac{\partial \eta}{\partial t}\right)} \right] \delta \eta \, dt$$

Thus, we have $\delta \int L dt = \iint \left\{ \frac{\partial L}{\partial \eta} - \frac{\partial}{\partial x} \left[\frac{\partial L}{\partial \left(\frac{\partial \eta}{\partial x}\right)} \right] - \frac{\partial}{\partial t} \left[\frac{\partial L}{\partial \left(\frac{\partial \eta}{\partial x}\right)} \right] \right\} \delta \eta \, dx \, dt$ (17.17)

We see that the integrated part becomes zero by the requirement that $\delta \eta = 0$ at the end points of integration over t. $\delta \eta$ being an arbitrary function of x, one obtains from Eq. (17.17)

$$\frac{\partial L}{\partial \eta} - \frac{\partial}{\partial x} \left[\frac{\partial L}{\partial \left(\frac{\partial \eta}{\partial x}\right)} \right] - \frac{\partial}{\partial t} \left[\frac{\partial L}{\partial \left(\frac{\partial \eta}{\partial t}\right)} \right] = 0 \tag{17.18}$$

One can write, Eq. (17.18) in the following form

$$\frac{\partial L}{\partial \eta} - \frac{\partial}{\partial t} \left[\frac{\partial L}{\partial \left(\frac{\partial \eta}{\partial t}\right)} \right] = 0 \tag{17.19}$$

where

$$\frac{\partial L}{\partial \eta} = \frac{\partial L}{\partial \eta} - \frac{\partial}{\partial x} \left[\frac{\partial L}{\partial \left(\frac{\partial \eta}{\partial t}\right)} \right] \tag{17.20}$$

is the 'variational or functional derivative'. We may note that in the transition to a continuous system, the set of n Eqs. (17.11), has been reduced to a partial differential equation, Eq. (17.19).

We see that the Eq. (17.19) gives the correct equation of motion on substituting the expression for Lagrangian density, [Eq. (17.15)]. The various derivatives are obtained as

$$\frac{\partial L}{\partial \eta} = 0; \qquad \frac{\partial L}{\partial \left(\frac{\partial \eta}{\partial t}\right)} = \frac{\partial L}{\partial \dot{\eta}} = \rho \dot{\eta}$$

$$\Rightarrow \quad \frac{\partial \mathcal{L}}{\partial \left(\frac{\partial \eta}{\partial x} \right)} = -T \frac{\partial \mathcal{L}}{\partial \eta}$$

Thus, the Eq. (17.19) reduces to $\rho \dfrac{\partial^2 \eta}{\partial t^2} - T \dfrac{\partial^2 \eta}{\partial x^2} = 0$

$$\Rightarrow \quad \frac{\partial^2 \eta}{\partial x^2} = \frac{1}{v^2} \frac{\partial^2 \eta}{\partial t^2} \tag{17.21}$$

where $v = \sqrt{T/\rho}$ is the velocity of the *compressional wave*.

Now we derive the expression for Hamiltonian H and Hamiltonian density H for a continuous system. For a discrete system, the Hamiltonian H is written as

$$H = \sum_i p_i \eta_i - L$$

$$\Rightarrow \quad = \frac{1}{2} \sum_i \left[\frac{p_i^2}{ma} + T \left(\frac{\eta_{i+1} - \eta_i}{a} \right)^2 \right] a$$

$$= \sum_i H_i a \tag{17.22}$$

where Eq. (17.11) has been used and also the definition

$$p_i = \frac{\partial \mathcal{L}}{\partial \dot{\eta}_i} = m \dot{\eta}_i \tag{17.23}$$

Defining the variable p in the limit of continuous system as follows:

$$p = \lim_{a \to 0} \frac{p_i}{a} \tag{17.24}$$

The Hamiltonian for a uniform string, in the light of Eq. (17.14), reads as

$$H = \frac{1}{2} \int_0^a \left[\frac{p^2}{\rho} + T \left(\frac{\partial \eta}{\partial x} \right)^2 \right] dx$$

$$= \int_0^a H \, dx \tag{17.25}$$

which reveals that $\quad H = \dfrac{1}{2} \left[\dfrac{p^2}{\rho} + T \left(\dfrac{\partial \eta}{\partial x} \right)^2 \right]$

$$= \frac{\rho}{2} \left[\frac{p^2}{\rho^2} + v^2 \left(\frac{\partial \eta}{\partial x} \right)^2 \right] \tag{17.26}$$

Interestingly, one can also write down the Hamiltonian density $\mathcal{H}$ from the Lagrangian density as follows:

$$\mathcal{H}\left(p, \frac{\partial \eta}{\partial x}\right) = p\,\dot{\eta} - \mathcal{L}\left(\dot{\eta}, \frac{\partial \eta}{\partial x}\right) \tag{17.27}$$

where the generalized momentum density p is defined as

$$p = \frac{\partial \mathcal{L}}{\partial \dot{\eta}} \tag{17.28}$$

Obviously, *Hamilton's canonical equations* in the case of continuous string takes the form

$$\dot{\eta} = \frac{\partial \mathcal{H}}{\partial \rho} = \frac{p}{\rho} \tag{17.29}$$

$$\dot{p} = \frac{\partial p}{\partial t}$$

$$= \frac{\partial}{\partial t}\left(\frac{\partial \mathcal{L}}{\partial \dot{\eta}}\right)$$

$$= \frac{\partial}{\partial t}\frac{\partial \mathcal{L}}{\partial\left(\dfrac{\partial \eta}{\partial t}\right)}$$

Now, rewriting Eq. (17.18) as follows

$$\frac{\partial \mathcal{L}}{\partial \eta} - \frac{\partial}{\partial x}\left[\frac{\partial \mathcal{L}}{\partial\left(\dfrac{\partial \eta}{\partial t}\right)}\right] - \frac{\partial}{\partial t}\left[\frac{\partial \mathcal{L}}{\partial\left(\dfrac{\partial \eta}{\partial t}\right)}\right] = 0$$

But, we have $\qquad \dfrac{\partial \mathcal{L}}{\partial \eta} = 0$

therefore $\qquad \dot{p} = \dfrac{\partial}{\partial t}\dfrac{\partial \mathcal{L}}{\partial\left(\dfrac{\partial \eta}{\partial t}\right)}$

$$= -\frac{\partial}{\partial x}\frac{\partial \mathcal{L}}{\partial\left(\dfrac{\partial \eta}{\partial x}\right)}$$

$$= \frac{\partial}{\partial x}\frac{\partial \mathcal{H}}{\partial\left(\dfrac{\partial \eta}{\partial x}\right)}$$

$$= \mathcal{T} \frac{\partial^2 \eta}{\partial x^2} \tag{17.30}$$

In obtaining (17.30) we have used Eqs. (17.27) and (17.21).

(ii) *Three-diemnsional Continuum*: We now summarize the results for the one-dimensional continuum prior to attempting the transition to the three-dimensional case. The Lagrangian density in a one-dimensional continuum reads as

$$\mathcal{L}\left(\eta, \dot{\eta}, \frac{\partial \eta}{\partial x}, t\right) = \frac{1}{2}\left[\rho \dot{\eta}^2 - \mathcal{T}\left(\frac{\partial \eta}{\partial x}\right)^2\right]$$

$$= \frac{\rho}{2}\left[\dot{\eta}^2 - v^2\left(\frac{\partial \eta}{\partial x}\right)^2\right] \tag{17.31}$$

The Lagrange equation of motion is as follows:

$$\frac{\partial \mathcal{L}}{\partial \eta} - \frac{\partial}{\partial x}\left[\frac{\partial \mathcal{L}}{\partial\left(\frac{\partial \eta}{\partial x}\right)}\right] - \frac{\partial}{\partial t}\left[\frac{\partial \mathcal{L}}{\partial\left(\frac{\partial \eta}{\partial t}\right)}\right] = 0 \tag{17.32}$$

and the substitution of Eq. (17.31) into Eq. (17.32) leads to the equation for wave motion as follows:

$$\frac{\partial^2 \eta}{\partial x^2} - \frac{1}{v^2}\frac{\partial^2 \eta}{\partial t^2} = 0 \tag{17.33}$$

Defining the generalized momentum density as

$$p = \frac{\partial \mathcal{L}}{\partial \dot{\eta}} \tag{17.34}$$

One can write the equation of motion (17.32) as

$$\dot{p} = -\frac{\partial}{\partial x}\frac{\partial \mathcal{L}}{\partial\left(\frac{\partial \eta}{\partial x}\right)} \tag{17.35}$$

Further defining the Hamiltonian density by the following relation

$$\mathcal{H}\left(p, \frac{\partial \eta}{\partial x}\right) = p\dot{\eta} - \mathcal{L}\left(\eta, \frac{\partial \eta}{dx}\right) \tag{17.36}$$

and using Eqs. (17.34) and (17.29) in Eq. (17.36), Eq. (17.36) reads as

$$\mathcal{H}\left(p, \frac{\partial \eta}{\partial x}\right) = \frac{\rho}{2}\left[\frac{p^2}{\rho^2} + v^2\left(\frac{\partial \eta}{dx}\right)^2\right] \tag{17.37}$$

The Hamiltonian equations of motion are as follows:

$$\dot{\eta} = \frac{\partial \mathcal{H}}{\partial p} = \frac{p}{\rho} \tag{17.38}$$

$$\dot{p} = \frac{\partial}{\partial x} \frac{\partial \mathcal{H}}{\partial \left(\dfrac{\partial \eta}{\partial x} \right)} = \rho v^2 \frac{\partial^2 \eta}{\partial x^2} \tag{17.39}$$

One can accomplish the extension to the three-dimensional continuum by the following prescription

$$x \to x_i \, ; \; \eta \to \eta_i \, ; \; p \to p_i \, ; \quad (i = 1, 2, 3)$$

then the Lagrange equation of motion (17.32) reads as

$$\frac{\partial \mathcal{L}}{\partial \eta_i} - \sum_{j=1}^{3} \frac{\partial}{\partial x_j} \left[\frac{\partial \mathcal{L}}{\partial \left(\dfrac{\partial \eta_i}{\partial x_j} \right)} - \frac{\partial}{\partial t} \left(\frac{\partial \mathcal{L}}{\partial \dot{\eta}_i} \right) \right] = 0 \quad (i = 1, 2, 3) \tag{17.40}$$

The generalized momentum densities p_i can then be defined as follows:

$$p_i = \left(\frac{\partial \mathcal{L}}{\partial \dot{\eta}_i} \right) \quad (i = 1, 2, 3) \tag{17.41}$$

and the equation of motion now reads as

$$\dot{p}_i = \sum_{j=1}^{3} \frac{\partial}{\partial x_j} \left[\frac{\partial \mathcal{L}}{\partial \left(\dfrac{\partial \eta_i}{\partial x_j} \right)} \right] \quad (i = 1, 2, 3) \tag{17.42}$$

Now, from the requirement that the equation of motion must be linear, we see that the Lagrange density cannot contain higher powers than the second of either $\dot{\eta}_i$ or $\dfrac{\partial \eta_i}{\partial x_j}$. One possible trial function which satisfies these requirements is as follows:

$$\mathcal{L} = \frac{\rho}{2} \left[\dot{\eta} \cdot \dot{\eta} - v^2 \, (\nabla \times \eta) \cdot (\nabla \times \eta) \right] \tag{17.43}$$

where $\eta = \eta_1 \hat{\mathbf{i}}_1 + \eta_2 \hat{\mathbf{i}}_2 + \eta_3 \hat{\mathbf{i}}_3$.

The Hamiltonian density is as follows:

$$\mathcal{H} = p \cdot \dot{\eta} - \mathcal{L} \tag{17.44}$$

$$= \frac{\rho}{2} \left[\frac{p \cdot p}{\rho^2} + v^2 v^2 \, (\nabla \times \eta) \cdot (\nabla \times \eta) \right]$$

where $p = p_1 \hat{i}_1 + p_2 \hat{i}_2 + p_3 \hat{i}_3$, and we have made use of Eq. (17.27)

The Hamiltonian equations now reads as

$$\dot{\eta}_i = \frac{\partial \mathcal{H}}{\partial p_i} = \frac{p_i}{\rho} \tag{17.45}$$

$$\dot{p}_i = \sum_{j=1}^{3} \frac{\partial}{\partial x_j} \left[\frac{\partial \mathcal{H}}{\partial \left(\dfrac{\partial \eta_i}{\partial x_j} \right)} \right]$$

$$= \rho v^2 (\nabla \times \nabla \times \boldsymbol{\eta})_i \quad (i = 1, 2, 3) \tag{17.46}$$

17.4 CLASSICAL HAMILTONIAN FOR THE MAXWELL-LORENTZ FIELD

One can express the field strengths E and B resulting from a classical field of e.m. waves and charged particles in terms of potential A' and ϕ' as

$$E = -\frac{1}{c} \frac{\partial A'}{\partial t} - \nabla \phi' \tag{17.47}$$

$$B = \nabla \times A' \tag{17.48}$$

where

$$\nabla \cdot A' + \frac{1}{c} \frac{\partial \phi'}{\partial t} = 0 \tag{17.49}$$

Now, if we make the following substitutions in Eqs. (17.47) to (17.49)

$$A' = A + \nabla \chi \tag{17.50}$$

$$\phi' = \phi - \frac{1}{c} \frac{\partial \chi}{\partial t} \tag{17.51}$$

then one obtains

$$E = -\frac{1}{c} \frac{\partial A}{\partial t} - \nabla \phi \tag{17.52}$$

$$B = \nabla \times A \tag{17.53}$$

and

$$\nabla \cdot A = \nabla \cdot A' - \nabla^2 \chi \tag{17.54}$$

We may note that the choice of χ is still arbitrary, one may choose it so that

$$\nabla^2 \chi = \nabla \cdot A \tag{17.55}$$

which, however, leads to the following condition for the vector potential

$$\nabla \cdot A = 0 \tag{17.56}$$

In addition, we have

$$4\pi \rho = \nabla \cdot E = -\frac{1}{c} \nabla \cdot \frac{\partial A}{\partial t} - \nabla^2 \phi$$

which in the light of Eq. (17.56) reduce as

$$\nabla^2\phi = -4\pi\rho \tag{17.57}$$

If v denotes the velocity of a charged particle, then, we have

$$\frac{4\pi\rho v}{c} = \nabla \times B - \frac{1}{c}\frac{\partial E}{\partial t}$$

$$= \nabla(\nabla \cdot A) - \nabla^2 A + \frac{1}{c^2}\frac{\partial^2 A}{\partial t^2} + \frac{1}{c}\nabla\frac{\partial\phi}{\partial t} \tag{17.58}$$

$$\Rightarrow \quad \nabla^2 A - \frac{1}{c^2}\frac{\partial^2 A}{\partial t^2} = -\frac{4\pi\rho v}{c} + \frac{1}{c}\nabla\frac{\partial\phi}{\partial t} \tag{17.59}$$

Equation (17.57) clearly reveals that the scalar potential ϕ is the *electrostatic potential* for the instantaneous configuration of the chrages, whose mutual interaction energy reads as $\frac{1}{2}\sum_{i \neq j}\frac{e_i e_j}{r_{ij}}$.

Let us first consider a pure radiation field. Now, one can assume the coulomb gauge (17.56) to hold and the vector potential A will satisfy the equation

$$\nabla^2 A - \frac{1}{c^2}\frac{\partial^2 A}{\partial t^2} = 0 \tag{17.60}$$

A is a function defined at all points of space and time and in order to express A in terms of canonical variables, one ideally requires an infinite set: however, it is possible to choose an enumerable set, provided we assume that the entire radiation field is enclosed in a volume, say, a cube of volume L^3 and is subject to a certain set of boundary conditions. Let us also assume that A and its derivatives have the same value on two opposite planes of the volume, viz., A is periodic on the surface and this ensures that both running as well as standing waves will be solutions of the Eq. (17.60). However, it is necessary to assume that L is large compared with the dimensions of the material system, i.e., the radius of the atomic or nuclear system placed in the cube or the wavelength of radiation, so that the physical behavior of the system does not depend on L.

We know that electromagnetic field consists of light waves and no charges are present, the scalar potential and the electrostatic energy vanish and the field E and B are as follows:

$$E = -\frac{1}{c}\frac{\partial A}{\partial t} \tag{17.61}$$

$$B = \nabla \times A \tag{17.62}$$

and the vector potential satisfies the equations:

$$\nabla^2 A - \frac{1}{c}\frac{\partial^2 A}{\partial t^2} = 0 \tag{17.63}$$

$$\nabla \cdot A = 0 \tag{17.64}$$

Since A is divergenceless, it is obvious that it describes transverse waves and the electric intensity E, Eq. (17.61), is in the transverse direction.

Now, we solve Eq. (17.60) and assume that

$$A = A(x)\, \chi(t) \tag{17.65}$$

On substituting Eq. (17.65) in Eqs. (17.63) and (17.64), one obtains

$$\chi(t)\nabla^2 A(x) - \frac{1}{c^2}\ddot{\chi}(t) A(x) = 0$$

$$\nabla \cdot A(x) = 0$$

so that

$$\nabla^2 A(x) + k^2 A(x) = 0 \tag{17.66}$$

Its solution is as follows:

$$A = a e^{i(k \cdot x)} \tag{17.67}$$

where a is a real constant. Since $\nabla \cdot A(x) = 0$, $A(x)$ is a transverse vector and if a and a' describe two independent linear polarizations, then, we have

$$(a \cdot a') = 0 \tag{17.68}$$

In order to make periodic on the surfaces of the cube, one will have to see that it satisfy the following boundary conditions;

$$k_x = \frac{2\pi}{L}n_x \,;\; k_y = \frac{2\pi}{L}n_y \,;\; k_z = \frac{2\pi}{L}n_z \tag{17.69}$$

where n_x, n_y, n_z may be either zero or integer, positive or negative. Obviously, the value of k constitute a denumberable set. Now, for such a combination, the *propagation vector* is denoted by k_l, where

$$ck_l = 2\pi\nu_l \tag{17.70}$$

The propagation vector $-k_l$ also corresponds to the value ν_l. The time-dependent part of the differential equation is given by

$$\ddot{\chi}(t) + 4\pi^2 \nu_l^2 \chi(t) = 0 \tag{17.71}$$

and if its solution is represented by q_l whose time dependence is given by the factor $e^{-2\pi i\nu_l t}$, then solution of Eq. (17.61) reads as

$$A = \sum_l q_l\, a_l\, e^{i(k_l \cdot x)} \tag{17.72}$$

However, we see that the general solution is obtained by superimposing the partial solutions and is given by

$$A = \frac{1}{\sqrt{V}}\sum_i a_l (q_l e^{i(k_l \cdot x)} + q_l'^{*} e^{-i(k_l \cdot x)}) \tag{17.73}$$

Since A is real, each partial wave is real and the reality condition requires that

$$q_l = q_l^{*}$$

Thus, we have

$$A = \frac{1}{\sqrt{V}}\sum_i a_l (q_l e^{i(k_l \cdot x)} + q_l'^{*} e^{-i(k_l \cdot x)}) \tag{17.74}$$

One may normally express Eq. (17.74) as follows:

$$A = \frac{1}{\sqrt{V}}\sum_i a_l \left(Q_l \cos(k_l \cdot x) - \frac{1}{2\pi\nu_l} P_l \sin(k_l \cdot x) \right) \tag{17.75}$$

where

$$Q_l = q_l + q_l^* \tag{17.76}$$

$$P_l = \dot{Q}_l = -2\pi i v_l (q_l - q_l^*) \tag{17.77}$$

Now, we see that P_l and Q_l are both real,. If the quantities referring to the propagation vector $-k_l$ are desginated by subscript $-l$, then Q_{-l} is independent of Q_l and taking $a_l = a_{-l}$, one obtains

$$A = \frac{1}{\sqrt{V}} \sum_i{}' a_l \left[(Q_l + Q_{-l}) \cos(k_l \cdot x) - \frac{1}{2\pi v_l} (P_l - P_{-l}) \sin(k_l \cdot x) \right] \tag{17.78}$$

We may note that the summation is over a hemisphere.

The equation of motion for Q_l, viz.,

$$\ddot{Q}_l + 4\pi^2 v_l^2 Q_l = 0 \tag{17.79}$$

is given by the Hamiltonian

$$H_l = \frac{1}{2} P_l^2 + 2\pi^2 v_l^2 Q_l^2 \tag{17.80}$$

Since

$$\dot{Q}_l = \frac{\partial H_l}{\partial P_l} = P_l$$

$$\Rightarrow \quad \ddot{Q}_l = \dot{P}_l = -\frac{\partial H_l}{\partial Q_l} = -4\pi^2 v_l^2 Q_l \tag{17.81}$$

Now, making use of the expansion of A, Eq. (17.78), one can easily calculate the expressions for the electric field E and magnetic field B. The field strengths, according to Eqs. (17.61) and (17.62), reads as

$$E = -\frac{1}{c\sqrt{V}} \sum{}' a_l \left[(P_l + P_{-l}) \cos(k_l \cdot x) + 2\pi v_l (Q_l - Q_{-l}) \sin(k_l \cdot x) \right] \tag{17.82}$$

$$B = -\frac{1}{c\sqrt{V}} \sum{}' (k_l \times a_l) \left[(Q_l + Q_{-l}) \sin(k_l \cdot x) - \frac{1}{2\pi v_l} (P_l - P_{-l}) \cos(k_l \cdot x) \right]$$

$$= -\frac{1}{c\sqrt{V}} \sum{}' a_l \left[2\pi v_l (Q_l + Q_{-l}) \sin(k_l \cdot x) + (P_l - P_{-l}) \cos(k_l \cdot x) \right] \tag{17.83}$$

(by putting $|a_l'| = |a_l|$)

The electromagnetic energy density contained within the cube is expressed by the following expression (keeping in view the condition expressed by [Eq. (17.68)],

$$= \frac{1}{8\pi c^2} \sum_i a_l^2 (P_l^2 + 4\pi^2 v_l^2 Q_l^2) \tag{17.84}$$

One finds that this is equal to $\sum_l H_l$, if we impose the normalization condition

$$a_l^2 = 4\pi c^2 \tag{17.85}$$

Introducing $a_1^2 = \sqrt{4\pi c^2}\hat{e}_l$ where $\hat{e}_l$ is a unit polarization vector Eq. (17.73) reads as

$$A = \frac{1}{\sqrt{V}} \sum_i \left[q_l A_l(x) + q_l^* A_l^*(x) \right] \tag{17.86}$$

with

$$A_l(x) = \sqrt{4\pi c^2}\,\hat{e}_l\, e^{i(\mathbf{k}_l \cdot \mathbf{x})} \tag{17.87}$$

Denoting by H_f, the Hamiltonian of the entire field of e.m. radiation, reads as

$$H_f = \sum_l \left(\frac{1}{2} P_l^2\, 2\pi^2 v_l^2 Q_l^2 \right) \tag{17.88}$$

which evidently gives the correct equation of motion for Q_l. Now, rewriting Eq. (17.88) in view of Eq. (17.77), one obtains

$$H_f = \sum_l \left[-2\pi^2 v_l^2 (q_l - q_l^*)^2 + 2\pi^2 v_l^2 (q_l + q_l^*)^2 \right]$$

$$= \sum_l 8\pi^2 v_l^2\, q_l^* q_l \tag{17.89}$$

Now, we are interested to see that what changes one may brought about when charged particles are introduced in the radiation field. The Hamiltonian of a particle of rest mass m_s and charge e_s in the presence of an electromagnetic field, which is characterized by A', ϕ' is expressed by Eq. (17.90). The electromagnetic field may be due to other charged particles and to e.m. waves coming from outside sources. We have

$$H_s = c\sqrt{m_s^2 c^2 + \left| p_s - (e_s/c) A^s \right|^2} + e_s \phi^s \tag{17.90}$$

where ρ_s is the momentum of the particle.

The total Hamiltonian for the radiation field and the changes present in it, now reads as

$$H = H_f + \sum_S c\sqrt{m_s^2 c^2 + \left| p_s - e/c A^s \right|^2} + e_s \phi^s$$

$$= H_f + \sum_S c_1 \sqrt{m_s^2 c^2 + \left| p_s - e/c A^s \right|^2} + e_s \phi^s + \frac{1}{2} \sum_{S \neq m} \sum \frac{e_s e_m}{r_{sm}} \tag{17.91}$$

However, the contribution to energy due to the scalar potential is as follows,

$$\sum_S e_s \phi^s = \frac{1}{2} \sum_{S \neq m} \sum \frac{e_s e_m}{r_{sm}} \tag{17.92}$$

We may note that the above treatment is based on the assumption that coulomb gauge holds, viz., $\nabla \cdot A = 0$ and as such the theory appears to be non-relativistic. However, one actually finds that it is not so, since the Lorentz condition on the potentials A and ϕ is invariant under Lorentz transformation. This clearly implies, that if $\nabla \cdot A$ and ϕ vanish in one frame of reference, these will not vanish when transformed to another arbitrary frame. However, one finds that the Lorentz condition could be made to vanish by choosing a suitable χ and redefining A and ϕ. We may note that the theory is a relativistic one since it is based on relativistically invariant Maxwell's equations of electromagnetic field.

Example 17.1

Verify that the Lagrangian density given by

$$\mathcal{L} = \frac{E^2 - B^2}{8\pi} - \rho\phi + \frac{J \cdot A}{c}$$

satisfies the following Maxwell's equations

$$\nabla \cdot B = 0 \qquad\qquad \nabla \times E + \frac{1}{c}\frac{\partial B}{\partial t} = 0$$

$$\nabla \cdot E = 4\pi\rho \qquad\qquad \nabla \times B - \frac{1}{c}\frac{\partial E}{\partial t} = \frac{4\pi}{c}J$$

where E and B are the electric and magnetic field intensities. [PU, DU]

Solution

We see that the homogeneous equations serve to define E and B in terms of ϕ and A as

$$E = -\nabla\phi - \frac{1}{c}\frac{\partial A}{\partial t}$$

$$B = \nabla \times A$$

The inhomogeneous equations describe the generation of the fields by external charges and currents and as such they are the desired field equations. The six components are not suitable as generalized coordinates since these are not independents, but one can express in terms of the four potential components. Here A and ϕ are treated as generalized coordinates of the field.

Let us now write the Lagrange's equation for ϕ. The various derivatives involved are as follows:

$$\frac{\partial \mathcal{L}}{\partial \phi} = -\rho$$

$$\frac{\partial \mathcal{L}}{\partial\left(\dfrac{\partial \phi}{\partial x_k}\right)} = \frac{E_k}{4\pi}\frac{\partial E_k}{\partial\left(\dfrac{\partial \phi}{\partial x_k}\right)} = -\frac{E_k}{4\pi}$$

since

$$\frac{\partial E_k}{\partial\left(\dfrac{\partial \phi}{\partial x_k}\right)} = -1$$

The Lagrange equation for ϕ reads as

$$\frac{1}{4\pi}\sum\frac{\partial E_k}{\partial x_k}-\rho=0$$

$$\Rightarrow \qquad\qquad \nabla\cdot E=4\pi\rho$$

Furthermore, let us now consider a typical component of A say A_1. The expression for Lagrangian density contains both the spatial and time derivatives of A_1, with the component itself occuring in the term $J\cdot A/c$. The various derivatives of $\mathcal{L}$ are as follows:

$$\frac{\partial\mathcal{L}}{\partial A_1}=\frac{J_1}{c},\qquad \frac{\partial\mathcal{L}}{\partial A_1}=\frac{E_1}{4\pi},\qquad \frac{\partial E_1}{\partial A_1}=-\frac{E_1}{4\pi c}$$

$$\frac{\partial\mathcal{L}}{\partial\left(\dfrac{\partial A_1}{\partial x_2}\right)}=-\frac{1}{4\pi},\qquad B_3\frac{\partial B_3}{\partial\left(\dfrac{\partial A_1}{\partial x_2}\right)}=\frac{B_3}{4\pi}$$

and

$$\frac{\partial\mathcal{L}}{\partial\left(\dfrac{\partial A_1}{\partial x_3}\right)}=-\frac{B_2}{4\pi}$$

Using these expressions, Lagrange's equation for A_1 reads as

$$\frac{1}{4\pi}\left(\frac{\partial B_3}{\partial x_2}-\frac{\partial B_2}{\partial x_3}\right)-\frac{1}{4\pi c}\frac{\partial E_1}{\partial t}-\frac{J_1}{c}=0$$

which is the 1-component of the equation

$$\nabla\times B-\frac{1}{c}\frac{\partial E}{\partial t}\frac{4\pi}{c}J$$

Analogously, one can also obtain the other two components, i.e. A_2, and A_3.

Example 17.2

Using the following expression for Lagrangian density

$$\mathcal{L}=\frac{E^2-B^2}{8\pi}-\rho+\frac{J\cdot A}{c}$$

write the total Lagrangian for the field when the charges and currents arise from a system of particles. [Cohin Univ.]

Solution

The volume integral of the Lagrangian density for the electromagnetic field reads as

$$L=\int\left[\frac{E^2-B^2}{8\pi}-\rho\left(\phi-\frac{v\cdot A}{c}\right)\right]d^3x \qquad (1)$$

where we have made use of the relation

$$J=\rho v \qquad\qquad (2)$$

v is the velocity of the charges and is a function of position in space. The charge density function for a system of n particles is represented as follows:

$$\rho(r) = \sum_{i=1}^{n} q_i \delta(x - x_i(t)) \tag{3}$$

where q_i is the charge on the ith particle, r_i its position vector. Obviously, the integral for the charge density ρ can be expressed as

$$\int \rho(r) \left(\phi(x) - \frac{v \cdot A(x)}{c} \right) d^3 x = \sum_i q_i \left[\phi_i(x_i) - \frac{v_i \cdot A_i(x_i)}{c} \right] \tag{4}$$

We can see that it leads to the following expression for the total Lagrangian for the field

$$L = \int \frac{E^2 - B^2}{8\pi} d^3 x - \sum_i q_i \left[\phi_i(x_i) - \frac{v_i \cdot A_i}{c} \right] \tag{5}$$

where ϕ_i and A_i have to be evaluated at the position of the particles. We may note that the summation pertains to that part of the Lagrangian for the n particles, which produces the electromagnetic forces acting on the particles. Now, adding the kinetic energy of the particles to the above expression, one obtains the final form of the total Lagrangian L which induces the mechanical motion, as

$$L = \int \frac{E^2 - B^2}{8\pi} d^3 x - \sum_i q_i \left[\phi_i(r_i) - \frac{v_i \cdot A_i}{c} \right] + \frac{1}{2} \sum_i m_i v_i^2 \tag{6}$$

Alternatively, one may also express Eq. (vi) as

$$L = \int \frac{E^2 - B^2}{8\pi} d^3 x - \sum_i q_i \delta(x - x_i) \left[\phi_i - \frac{v_i \cdot A_i}{c} \right] + \frac{1}{2} \sum_i m_i v_i^2 \tag{7}$$

We see that the first term in (5) represents the Lagrangian for the pure radiation field, the last term represents the Lagrangian for the particles alone and the middle term describes the mutual interaction of the particles and field.

Example 17.3

Show that the Hamiltonian

$$\mathcal{H} = \sum_l 8\pi^2 v_l^2 q_l^* q_l + \sum_S c_1 \sqrt{m_s^2 c^2 + \left| p_s - \frac{e}{c} A^s \right|^2} + e_s \phi^s + \frac{1}{2} \sum_{S \neq m} \sum \frac{e_s e_m}{r_{sm}}$$

where symbols have usual meanings, is the correct Hamiltonian for a system of charged particles in the electromagnetic field. [MU, MDSU]

Solution

We know that when charged particles are present in the electromagnetic field, the wave equation for vector potential A reads as

$$\nabla^2 A - \frac{1}{c^2} \frac{\partial^2 A}{\partial t^2} = -\frac{4\pi}{c} \rho v + \frac{1}{c} \nabla \phi \tag{1}$$

Now, expanding A into a Fourier series, we have

$$A = \sqrt{\frac{8\pi c^2}{V}} \sum{}' \hat{\mathbf{e}}_l \left[b_l \cos(k_l \cdot x) + d_l \sin(k_l \cdot x) \right] \tag{2}$$

In the absence of charges, it will give

$$\sqrt{2} b_l = Q_l + Q_{-l} \tag{3}$$

$$\sqrt{2} d_l = -\frac{1}{2\pi v_l}(P_l + P_{-l}) \tag{4}$$

and the corresponding conjugate momenta B_l, D_l defined by

$$B_l = b_l \; ; \; D_l = d_l \tag{5}$$

satisfies the following relations

$$\sqrt{2} B_l = P_l + P_{-l} \tag{6}$$

$$\sqrt{2} D_l = 2\pi v_l + (Q_l - Q_{-l}) \tag{7}$$

The Hamiltonian H_f in terms of these can be written as

$$H_f = \sum{}' \left[\frac{1}{2}(B_l^2 + D_l^2) + 2\pi^2 v_l^2(b_l^2 + d_l^2) \right] \tag{8}$$

and the equations of motion

$$\dot{b}_l = \frac{\partial H_f}{\partial B_l} \; ; \; -B_l = \frac{\partial H_f}{\partial b_l}$$

$$\dot{d} = \frac{\partial H_f}{\partial D_l} \; ; \; -D_l = \frac{\partial H_f}{\partial d_l}$$

reads as

$$\ddot{b}_l + 4\pi^2 v_l^2 b_l = 0 \tag{9}$$

$$\ddot{d}_l + 4\pi^2 v_l^2 d_l = 0 \tag{10}$$

However, in the presence of charges, the expansion of A reads as

$$A = -\sqrt{\frac{8\pi c^2}{V}} \sum{}' \hat{\mathbf{e}}_l \left[\left(k_l^2 b_l + \frac{1}{c^2}\ddot{b}_l \right) \cos(k_l \cdot x) + \left(k_l^2 d_l + \frac{1}{c^2}\ddot{d}_l \right) \sin(k_l \cdot x) \right]$$

$$= -\frac{4\pi}{c}\rho v + \frac{1}{c}\nabla\dot{\phi} \tag{11}$$

On taking the scalar product successively with $\hat{\mathbf{e}}_m \cos(k_l \cdot x)$ and $\hat{\mathbf{e}}_m \sin(k_l \cdot x)$ and integrating over the cube, one obtains

$$\ddot{b}_m + 4\pi^2 v_l^2 b_m = \sqrt{\frac{8\pi}{V}} \int \rho(v \cdot \hat{\mathbf{e}}_m) \cos(k_m \cdot x) d\tau$$

$$-\frac{1}{\sqrt{2V}}\frac{\partial}{\partial t} \int (\nabla\phi \cdot \hat{\mathbf{e}}_m) \cos(k_m \cdot x) d\tau \tag{12}$$

Analogously, one finds

$$\ddot{d}_m + 4\pi^2 v_l^2 d_m = \sqrt{\frac{8\pi}{V}} \int \rho(v \cdot \hat{\mathbf{e}}_m) \sin(k_m \cdot x) d\tau$$

$$-\frac{1}{\sqrt{2\pi V}} \frac{\partial}{\partial t} \int (\nabla\phi \cdot \hat{\mathbf{e}}_m) \sin(k_m \cdot x) d\tau \qquad (13)$$

We now consider the first term of Eq. (12).

$$\int (\nabla\phi \cdot \hat{\mathbf{e}}_m) \cos(k_m \cdot x) d\tau = \int \nabla \cdot [\phi\hat{\mathbf{e}}_m \cos(k_m \cdot x)] - \phi\nabla \cdot [\hat{\mathbf{e}}_m \cos(k_m \cdot x)] d\tau$$

$$= \int \nabla \cdot [\phi\hat{\mathbf{e}}_m \cos(k_m \cdot x)] d\tau \qquad (14)$$

$$= \int \nabla \cdot [\phi\hat{\mathbf{e}}_m \cos(k_m \cdot x)]_N \, da$$

where we have used the fact that $(k_m \cdot \hat{\mathbf{e}}_m)$ vanishes. However, the surface integral does not contribute anything since $\hat{\mathbf{e}}_m \cos(k_m \cdot x)$ has the same value at corresponding points on opposite faces of the cube but the normals are in opposite directions. Therefore, one obtains from Eq. (12)

$$\ddot{b}_l + 4\pi^2 v_l^2 d_l = \sqrt{\frac{8\pi}{V}} \sum_S e_s(v_s \cdot \hat{\mathbf{e}}_l) \cos(k_l \cdot x_s) \qquad (15)$$

$$\ddot{d}_l + 4\pi^2 v_l^2 d_l = \sqrt{\frac{8\pi}{V}} \sum_S e_s(v_s \cdot \hat{\mathbf{e}}_l) \cos(k_l \cdot x_s) \qquad (16)$$

where x_s and v_v are the position and velocity vectors of the sth charged particle carrying charge e_s. Further, the suffix m has been changed to l. We may note that one should be able to deduce these equations of motion from the Hamiltonian given by Eq. (17.91). From Eq. (3) and the canonical equations, one obtains

$$\ddot{b}_l = \dot{B}_l = -\frac{\partial H}{\partial b_l}; \quad \ddot{d}_l = -\frac{\partial H}{\partial_l} \qquad (17)$$

The additional term in $-\dfrac{\partial H}{\partial b_l}$ reads as

$$\sum_S \left(\frac{\partial_i H}{\partial A^s} \frac{\partial A^s}{\partial b_l} \right) = \sum_S \frac{e_s}{c} \left(\frac{\partial H}{\partial P_s} \frac{\partial A^s}{\partial b_s} \right)$$

$$= \sqrt{\frac{8\pi c^2}{V}} \sum_S \frac{e_s}{c} (v_s \cdot \hat{\mathbf{e}}_l) \cos(k_l \cdot x_s)$$

which is the desired result in the Eq. (14). $\qquad (18)$

REVIEW QUESTIONS AND PROBLEMS

1. Show that the Hamiltonian for a plane wave representation of pure radiation fields is

$$H_f = \sum_l \left(\frac{1}{2} P_l^2 + 2\pi^2 \nu_l^2 Q_l^2 \right)$$

where P_l and Q_l are real canonical variables. [MU, PU]

2. Write the equations of motions of motion of a system of charged particles and its field in the Hamiltonian form and then show that the term which depends on the longitudional waves, corresponds to the coulomb interaction between the particles. [DU, MDSU]

3. Show that the Hamiltonian

$$H = \sum_l 8\pi^2 \nu_l^2 q_l^* q_l + \sum_s \left(c\sqrt{m_s^2 c^2 + \left| p_s - e/cA^s \right|} + \frac{1}{2} \sum_{s \neq m} \sum \frac{e_s e_m}{r_{sm}} \right)$$

describes both the mechanical motion of charged particles of charge with rest mass e_s and momentum p_s in the presence of electromagnetic field.

4. Show that the Lagrangian density for the electromagnetic field given by the relativistic covariant form

$$\mathcal{L} = -\frac{1}{16\pi} \sum_{\mu,\nu} \left(\frac{\partial A_\nu}{\partial x^\mu} - \frac{\partial A_\mu}{\partial x^\nu} \right) \left(\frac{\partial A^\nu}{\partial x_\mu} - \frac{\partial A^\mu}{\partial x_\nu} \right)$$

$$-\frac{1}{8\pi} \sum_\mu \left(\frac{\partial A^\mu}{\partial x^\mu} \right) \left(\frac{\partial A_\mu}{\partial x_\mu} \right) - \sum_\mu \frac{J^\mu A_\mu}{c}$$

leads directly to the wave equation for the vector potential

$$\Box^2 A^\mu = -\frac{4\pi}{c} J^\mu$$

where A_μ is the four potential and J_μ is the current density. [MLSU]

5. Show that the Lagrangian density in Q.No. 4 is identical to the following expression for the Lagrangian density

$$\mathcal{L} = \frac{E^2 - B^2}{8\pi} - \phi + \frac{J \cdot A}{c}$$

except for the middle term, which has the zero value as a result of gauge condition.

Basics of Quantum Electrodynamics

18.1 INTRODUCTION

Quantum electrodynamics (QED) is the study of the properties of electromagnetic radiation and the way in which it interacts with charged matter in terms of quantum mechanics. Basically, QED is the quantum field theory of electrons and positrons (the electron-positron field) and photons (the electromagnetic or radiations field). The theory also applies to the known heavy leptons (μ and τ), and in general, can be used to describe the electromagnetic interaction of other charged elementary particles. However, these particles are also subject to non-electromagnetic forces, i.e. the strong and weak interactions. Strongly interacting particles (hadrons) are found to be composed of other particles, the *quarks*, so the new degrees of freedom become important (colour, flavour). It is believed that on this level the strong and weak interactions can be described by *"non-Abelian"* gauge theories modelled on QED, which is the prototype of an "Abelian" gauge theory. These are theories of *"quantum chromodynamics"* (QCD)[*] for the strong interaction and *quantum flavour dynamics* for the weak interaction. QED is not only the archetype for all modern field theories, but it also is of great importance in its own right since it provides the theoretical foundation for atomic physics.

There are two approaches to QED. The more formal one relies on a general appratus for the quantization of wave fields; the other more illustrative, way originates from *Stuckelberg* and *Feynman*, and uses the *propagator formalism*. Nowadays, we must know both, but it is better, both in terms of physics and teaching, if it is obvious at an early stage why a formalism was developed and to what it can be applied. We are keen to know how different processes are actually calculated. Although begun by Paul Dirac, it was finally brought to consistent form in 1948 by Richard Feynman, Sin-Itiro Tomonaga, and Julian Schwinger. The fundamental event

---

[*] *QCD is a gauge theory that describes the strong interactions in terms of quarks and antiquarks and the exchange of massless* gluons *between them. QCD is similar to QED, with colour being analogous to electric charge and the gluon being the analogue of the photon. The gauge group of QCD is non-Abelian and the theory is much more complicated than QED; the gauge symmetry in QCD is not a* broken symmetry.

QCD has the important property of *asymptotic freedom*-that at very high, energies (and hence, for short distances the interactions between quarks tend to zero as the distance between them tends to zero. Because of asymptotic freedom, perturbation theory may be used to calculate the high energy aspects of strong interactions, such as those described by *parton model. Parton* is a point like, almost free, particle postulated as a component of nucleons. The parton model enabled the results of very high-energy experiments on nucleons to be understood.

in QED is the interaction of a photon with an electrically charged particle. Two such events joined by a common virtual photon give the quantum electrodynamical description of electromagnetic force (Fig. 18.1). Feyman's propagator formalism is the best way to achieve this.

Fig. 18.1: A Feynman diagram, showing the interaction of two electrons through the exchange of a virtual photon.

We may note that the predictions of QED have been confirmed experimentally to a remarkably high precision, and today QED is best verified theory of physical reality. QED is a relativistic quantum field theory and explain the *Lambshift*, the anomalous magnetic moment of electron, and so on using the Feynman diagram technique of the covariant perturbation approach. QED is the quantum field theory of electrons and positrons (the electron-positron field) and photons (the electromagnetic or radiation field).

As stated above, there are two approaches to QED: (i) the more formal one relies on a general apparatus for the quantization of wave fields; (ii) the other, more illustrative, way originates from Stuckelberg and Feynman, and uses the propagator formalism. Former is less intutive but more systematic treatment of QED based on the quantum field theory. However, Feynman's propagation formalism is generally preferred for the study of QEDs.

In Feynman's propagator method, scattering processes, e.g. coulomb scattering of electrons; scattering of an electron off a free proton; the effect of recoil; scattering of identical fermions; electron-positron scattering; scattering of polarized Dirac particles; Bremsstrahlung; Compton scattering-The Klein-Nishina formula; Annihilation of particle and antiparticle, etc. are described by means of *integral equations*. The guiding idea is, that positrons are to be interpreted as electrons with negative energy which move in the reverse direction. This chapter deals with the basics of quantum electrodynamics. More details are beyond the scope of this book.

18.2 BASICS OF QUANTUM ELECTRODYNAMICS

18.2.1 Quantization of the Free Electromagnetic Field (Coulomb Gauge)

We have read that the electromagnetic field constitutes a continuous system with an infinite number of degrees of freedom. This puts it apart from discrete systems with a finite number of degrees of freedom. The quantization of discrete systems, which is carried out in quantum mechanics, consists in imposing certain commutation relations on the operators that represent physical quantities. These operators must be defined in an appropriate way before quantization is enforced. The common approach to quantization is to take the following commutation relation between the momentum p_j and coordinate q_i operators:

$$[q_i, p_j] = i\hbar \delta_{ij} \tag{18.1}$$

where i and j label the degrees of freedom corresponding to the coordinate and the momentum.

However, Eq. (18.1) cannot be used to describe fields, since the variables in the field functions $u = u(x)$ depend on the "number" of the particular degree of freedom, that is, coordinate x, in a continuous manner. Two approaches exist for overcoming this difficulty. One consists in partitioning the three-dimensional space in which the field is "moving" into discrete cells, imposing a periodicity condition on the field functions, with the period encompassing a sufficiently large (but finite) number of cells, and quantizing in a standard manner the resulting discrete system with a finite number of degrees of freedom over a single period.

The other approach consists in going over from the *coordinate representation* to the *momentum representation*, in which picture the periodicity condition results in the momentum values becoming discrete. The standard quantization procedure can then be applied to such a system. Here we use the second approach since it is more pictorial than the first and reduces the problem of *field quantization* to the well-known problem of a system of *harmonic oscillators*.

$$A^0 = 0, \quad \text{div } A = 0, \tag{18.2}$$

the vector potential of the free electromagnetic field, $A(r, t)$, obeys d'Alembert's equation.*

$$\Box A(r, t) = 0 \tag{18.3}$$

The real-valued general solution of this wave equation can be represented by the following Fourier series:

$$A(r, t) = \frac{1}{L^{3/2}} \sum_{k} \sqrt{\frac{2\pi}{\omega}} \left(a_k(t) e^{ik \cdot r} + a_k^*(t) e^{-ik \cdot r} \right). \tag{18.4}$$

Here L is the length of the principal periodicity cube (the three-dimensional space has been partitioned into such cubes), the wave vector k assumes discrete values with components

$$k_x = 2\pi n_x/L, \quad k_y = 2\pi n_y/L, \quad k_z = 2\pi n_z/L, \tag{18.5}$$

$n_x, n_y, n_z = 0, \pm 1, \pm 2, ...,$ and the meaning of the factor $\sqrt{2\pi/\omega}$ will become clearer from what follows.

---

* $\Box$ is d'Alembert operator

$$\Box = \partial_\mu \partial^\mu \equiv \partial^\mu \partial_\mu = \frac{1}{c^2} \frac{\partial^2}{\partial t^2} - \nabla^2$$

Note: $\Delta = \nabla^2$ is the Laplace operator. As a scalar product the d'Alembert operator, too is Lorentz invariant.

To satisfy d'Alembert's equation [Eq. (18.3)], the Fourier coefficients must obey the following equations:

$$\frac{d^2 a_k}{dt^2} + k^2 \cdot a_k = 0, \qquad \frac{d^2 a_k^*}{dt^2} + k^2 \cdot a_k^* = 0. \tag{18.6}$$

The fact that potential [Eq. (18.4)] is real implies that the solutions to Eqs. (18.6) are complex-valued, namely,

$$a_k(t) = a_k(0)\, e^{-i\omega t}, \quad \omega = |k|, \tag{18.7}$$

and

$$a_k^*(t) = a_k^*(0)\, e^{i\omega t} = a_{-k} \,|0|^{i\omega t}. \tag{18.8}$$

The *Coulomb gauge*[*] [Eq. (18.2)] forces us to conclude that monochromatic components of the field are transverse, that is,

$$k \cdot ak = 0. \tag{18.9}$$

Therefore, in the plane orthogonal to vector k we can select two base vectors e_1 and e_2,

$$e_1 \perp e_2, \; e_1 \perp k, \; e^2 \perp k, \tag{18.10}$$

and represent ak in the form

$$ak = a_{k_1} e_1 + a_{k_2} e_2 = e_{k_s} e_s. \tag{18.11}$$

The energy of the electromagnetic field,

$$H = \int d^3 x \mathcal{H}, \quad \mathcal{H} = \frac{E^2 + H^2}{8\pi} \tag{18.12}$$

can be expressed in terms of the Fourier coefficients if we recall that $E = -A_0$ and $H = \text{curl } A$ and use the expansion [Eq. (18.4)]. The integral over the volume yields Kronecker deltas,

$$\frac{1}{L^3} \int d^3 x\, e^{\pm i(k - k') \cdot r} = \delta_{kk'} \tag{18.13}$$

$$\left(\delta_{kk'} \equiv \delta_{n_x n_x'}\, \delta_{n_y n_y'}\, \delta_{n_z n_z'} \right)$$

---

[*] *Coulomb gauge*: One chooses the gauge function χ such that $\Delta \cdot A = 0 \; (\Delta = \nabla^2)$
The scalar potential then fulfills a differential equation,

$$\Delta \phi = \frac{-\rho}{\epsilon_0}$$

which, provided that there are no boundary conditions in the finiteness, is formally identical to the Poisson equation of the electrostatics. Its solution is

$$\phi(r, t) = \frac{1}{4\pi \epsilon_0} \int d^3 r' \frac{\rho(r, t)}{|r - r'|}$$

$\phi(r, t)$ is the *instantaneous* Coulomb *potential* of the charge density $\rho(r, t)$. One therefore speaks of *Coulomb-gauge.*

while the transversality condition (18.9) leads to

$$(k \times a_k) \cdot (k \times a_k^*) = \omega^2 a_k \cdot a_k^*, \tag{18.14}$$

The result is

$$H = \sum_{k, s = 1, 2} \frac{\omega}{2} (a_{ks} a_{ks}^* + a_{ks}^* a_{ks}). \tag{18.15}$$

As one can see, the dynamical description of the electromagnetic field has reduced itself to the problem of describing an infinite discrete set of dynamical systems, with each system described by the oscillation equations [Eq. (18.6)]. Since only complex-valued solutions [Eq. (18.7)] and [Eq. (18.8)] are allowed, the quantization procedure requires imposing on the operators a_{ks} and a_{ks}^+, which correspond to the Fourier components a_{ks} and a_{ks}^*, the following conditions:

$$[a_{ks}, a_{k's'}^+] = \delta_{kk'} \delta_{ss'}. \tag{18.16}$$

The real combinations

$$q_{ks} = \frac{1}{\sqrt{2}} (a_{ks} + a_{ks}^*),$$

$$p_{ks} = -\frac{i}{\sqrt{2}} (a_{ks} + a_{ks}^*). \tag{18.17}$$

have corresponding to them the operators q_{ks} and p_{ks} satisfying the common commutation relations [Eq. (18.1)] for the coordinate and momentum:

$$[q_{ks}, p_{k's'}] = i\delta_{kk'} \delta_{ss'}. \tag{18.18}$$

As is known, the quantization of an oscillator with the Hamiltonian

$$H_{ks} = \frac{\omega}{2} (a_{ks} a_{ks}^+ + a_{ks}^+ a_{ks}) \tag{18.19}$$

under condition [Eq. (18.6)] leads to the following eigenvalues of the energy

$$E(N_{ks}) = \omega(N_{ks} + 1/2), \tag{18.20}$$

where $N_{ks} = 0, 1, 2, \ldots$. The operators a_{ks} and a_{ks}^+ are represented by the following matrices with an infinite number of rows and columns:

$$a_{ks} = \begin{pmatrix} 0 & \sqrt{1} & 0 & 0 & \cdot \\ 0 & 0 & \sqrt{2} & 0 & \cdot \\ 0 & 0 & 0 & \sqrt{3} & \cdot \\ \cdot & \cdot & \cdot & \cdot & \end{pmatrix}, \quad a_{ks}^+ = \begin{pmatrix} 0 & 0 & 0 & \cdot \\ \sqrt{1} & 0 & 0 & \cdot \\ 0 & \sqrt{2} & 0 & \cdot \\ 0 & 0 & \sqrt{3} & \cdot \\ \cdot & \cdot & \cdot & \end{pmatrix}, \tag{18.21}$$

that is, the only nonzero matrix elements are

$$\langle N_{ks} - 1 | a_{ks} | N_{ks} \rangle = \langle N_{ks} | a_{ks}^+ | N_{ks} - 1 \rangle = \sqrt{N_{ks}}. \tag{18.22}$$

For a system of oscillators with the Hamiltonian

$$H = \sum_{k,s=1,2} \frac{\omega}{2} (a_{ks} a_{ks}^+ + a_{ks}^+ a_{ks})$$ (18.23)

we obtain the following *eigen values* of the energy:

$$E = \sum_{k,s=1,2} \omega (N_{ks} + 1/2),$$ (18.24)

corresponding to various sets of nonnegative integers N_{ks}. Equation (18.24), which is simply the sum of energies of quantum oscillators, or the so-called modes characterized by k and s, can be interpreted differently. The energy quanta corresponding to the oscillators ω can be interpreted as the energies of particles, or photons, while the numbers $N_{ks} = 0, 1, 2, \ldots$ can be interpreted as the numbers of such photons with definite values of momentum k and polarization s, that is, the occupation numbers of the states k, s. Then Eq. (18.24) stands for the total energy of the quantized electromagnetic field in a state with a definite number of photons, N_{ks}. The infinitely large positive (constant) quantity corresponding to the ground state of the system. $N_{ks} = 0$, can be removed by redefining the Hamiltonian H.

The ground, or vacuum, state of the electromagnetic field $|0\rangle$ satisfies the equation

$$a_{ks} |0\rangle = 0$$ (18.25)

for all k and s. The derivation of this condition is known from the theory of a quantum harmonic oscillator. The excited states with a definite number of photons N_{ks} are obtained, via a well-known procedure, from the vacuum state. The result is

$$|N_{ks}\rangle = \frac{(a_{ks}^+)^{N_{ks}}}{\sqrt{N_{ks}!}} |0\rangle.$$ (18.26)

These states are the eigenstates of the occupation number operator

$$N_{ks} = a_{ks}^+ a_{ks}.$$ (18.27)

Of course, vector [Eq. (18.26) describes only a single mode k, s. The total state vector of the field is obtained by acting with operators a_{ks}^+ of each mode raised to the corresponding powers (corresponding to the occupation number N_{ks}) on the vacuum state $|0\rangle$. The expression for the momentum of the electromagnetic field is obtained following the general definition. One obtains,

$$P = \int d^3x \frac{\partial L}{\partial A_{\mu,t}} \nabla A_\mu = \frac{1}{4\pi} \int E \times H d^3x.$$ (18.28)

In the *Fourier representation*, (18.28) yields

$$P = \sum_{k,s=1,2} k(a_{ks}^* a_{ks} + a_{ks} a_{ks}^*).$$ (18.29)

Going over to operators, one obtains

$$P = \sum_{k,s=1,2} k(a_{ks}^+ a_{ks} + a_{ks} a_{ks}^+),$$ (18.30)

which is the operator of the electromagnetic-field momentum. Its eigenvalues are

$$P = \sum_{k,s=1,2} k(N_{ks} + 1/2).$$ (18.31)

This result corroborates the above interpretation of k as the photon momentum. Recalling the relationship $\omega = |k|$, which links the photon energy to the photon momentum, we conclude that the photon is massless, which is in complete agreement with the fact that the electromagnetic field, being a vector gauge field, is massless.

Now let us investigate the question of defining the energy of the electromagnetic field. First we note that the energy of the classical field, Eq. (18.15), is positive definite and obviously, does not depend on the order in which the factors a_{ks} and a_{ks}^* follow in Eq. (18.15). The quantum definition of the *Hamiltonian operator* [Eq. (18.23)] copies the classical expression Eq. (18.15), with the classical Fourier coefficients replaced with the operators a_{ks} and a_{ks}^+. The commutation relation [Eq. (18.16)] makes it possible to write Eq. (18.23) in the form

$$H = \sum_{k,s=1,2} \omega \left(a_{ks}^+ a_{ks} + \frac{1}{2} \right),$$ (18.32)

where $a_{ks}^+ a_{ks} = N_{ks}$ is the occupation number operator, which has nonnegative integral eigenvalues. Thus, the energy of the quantized electromagnetic field assumes nonnegative values, just as the energy of the classical electromagnetic field does. However, if we define operator H in such a manner, the energy acquires an infinitely large positive term

$$E_0 = \sum_{k,s=1,2} \frac{1}{2} \omega,$$ (18.33)

which corresponds to the ground state of the field, or the vacuum state. This term, which has no physical meaning, can be eliminated and thus the energy of the vacuum state can be made equal to zero, $E_0 = 0$, if we define H differently. Indeed, the definition of the Hamiltonian operator [(Eq. 18.23)], which copies Eq. (18.15), can be replaced with another definition in which the order in which a_{ks} and a_{ks}^+ operate is changed. If we start with the classical definition [Eq. (18.15)], the Fourier coefficients can be freely interchanged, say, to obtain

$$H = \sum_{k,s=1,2} \omega a_{ks}^* a_{ks}.$$ (18.34)

In accordance with Eq. (18.34), instead of Eq. (18.23), we can introduce an alternative definition:

$$H = \sum_{k,s=1,2} \omega a_{ks}^+ a_{ks}.$$ (18.35)

Then the infinitely large term in the field's energy does not appear, the energy of the vacuum state, E_0, is zero, and the energy of an excited state of the field is

$$E = \sum_{k,s=1,2} N_{ks} \omega.$$ (18.36)

Definition (18.35) corresponds to the so-called normal order in which the operators act (the normal product, see below), that is, an order in which the *creation operators* a^+_{ks} stand to the left of the *annihilation operators* a_{ks} in a product.

Thus, quantization of the electromagnetic field results in a replacement of the classical expansion of the vector potential [Eq. (18.4)] with the quantum operator

$$A(r) = \frac{1}{L^{3/2}} \sum_{k,s=1,2} \sqrt{\frac{2\pi}{\omega}} (e_s a_{ks} e^{ik\cdot r} + e^*_s a^+_{ks} e^{-ik\cdot r}), \tag{18.37}$$

where the e_s are the vectors of transverse polarization ($s = 1, 2$), and a^+_{ks} and a_{ks} are the photon creation and annihilation operators (these operators obey the commutation relations (18.16)). The operators a_{ks} and a^+_{ks} and the vector potential operator $A(r)$ do not depend on time explicitly (in contrast to the classical case where, say, $A = A(r, t)$), since they are given in the *Schrödinger representation*. Time dependence of the operators is achieved if we go over to the *Heisenberg representation*.

18.2.2 Photon Spin

In accordance with the general definition of the spin of a vector field, the electromagnetic-field spin tensor is

$$S_{ij} = \int d^3x \, (u_i \pi_j - u_j \pi_i), \tag{18.38}$$

where $u_i = A_i$ and $\pi_j = -E_j/4\pi$. Then the electromagnetic-field spin vector proves to be

$$S = \frac{1}{4\pi} \int d^3x \, E \times A \tag{18.39}$$

Substituting the expansion (18.37), we get the *spin operator*

$$S = \sum_k \frac{i}{2} (a_k \times a^+_k - a^+_k \times a_k), \tag{18.40}$$

where $a_k = \sum_s a_{ks} \hat{e}_s$, or in explicit form,

$$S = -i \sum_k (a^+_{k1} a_{k2} - a^+_{k2} a_{k1}) n \tag{18.41}$$

We introduce complex combinations of the creation and annihilation operators,

$$c_k(1) = \frac{a_{k1} - i a_{k2}}{\sqrt{2}}, \quad c_k(-1) = \frac{a_{k1} + i a_{k2}}{\sqrt{2}}, \tag{18.42}$$

in terms of which the spin operator is expressed in a diagonal manner:

$$S = \sum_k (c^+_k(1) c_k(1) - c^+_k(-1) c_k(-1)) n, \tag{18.43}$$

where $n = k/\omega$ is a unit vector, $|n| = 1$.

The new operator $c_k(i)$ and $c^+_k(i)$ defined in [Eq. (18.42)] obey the same commutation relations [Eq. (18.16)] as the operators a_{ks} and a^+_{ks}, and the Hamiltonian operator H has the same form [Eq. (18.35)]:

$$H = \sum_{k, i = +1, -1} c^+_k(i) c_k(i) \, \omega. \tag{18.44}$$

Thus, the new operators have the same matrix elements [Eq. (18.22)] and the operator $c_k^+ (i) c_k (i)$ possesses nonnegative integral eigenvalues and can be identified with the photon occupation number operator. The diagonal spin operator, [Eq. (18.43)], possesses the following eigenvalues:

$$S = \sum_k (N_k (1) - N_k (-1)) n. \tag{18.45}$$

It is now clear that $N_k (1)$ is the number of photons with spin 1 directed along n, while $N_k (-1)$ is the number of photons with spin 1 directed along $- n$.

In accordance with the fact that the electromagnetic field is transverse, vector A has only two linear polarizations $s = 1, 2$ (from which two circular polarizations are constructed), one parallel to the direction of wave propagation and the other opposite to that direction. Therefore, the operators $c_k^+ (1)$ and $c_k^+ (-1)$ correspond to the creation of photons with right-handed circular polarization (spin directed along n) and left-handed circular polarization (spin directed along $- n$), respectively. The expression (18.45) shows also that the spin of the electromagnetic field is unity, just as of any vector field. Moreover, quantization combined with the commutation relations (18.16) leads to arbitrary values of the occupation numbers $N_{ks} = 0, 1, 2, 3, ...,$ a situation characteristic of the *Bose-Einstein statistics*. This fact demonstrates that there is a connection between spin and statistics, with the result that vector fields are quantized with due regard for the Bose statistics, or that vector particles are bosons.

One can also show that for *integral-spin particles (bosons)* the quantization rules are fixed by the commutation relations of Eq. (18.16) type; hence, states are possible in which there is an arbitrary number of particles with the same quantum numbers (the Bose-Einstein statistics). For particles with *half-integral spin (fermions)* the quantization rules obey anticommutation relations; hence, in this case the occupation numbers cannot exceed unity, in accord with *Pauli's exclusion principle*. This fundamental conclusion concerning the relationship between spin and statistics is known as *Pauli's theorem*.

18.3 QUANTIZATION OF THE ELECTROMAGNETIC FIELD (LORENTZ GAUGE)

Quantization of the three-dimensional transverse electromagnetic field in Coulomb gauge is convenient in the study of nonrelativistic processes and in problems involving electromagnetic radiation. More convenient, from the standpoint of building a general quantum theory of electromagnetic interaction, is the approach that uses *Lorentz gauge*[*], since this gauge is Lorentz-invariant.

---

[*] *Lorentz-Gauge*: This gauge leads to a complete decoupling of both the differential equations for scalar and vector potential ϕ and A respectively, which then in addition assume very symmetric forms
Lorentz Condition

$$\nabla \cdot A + \frac{1}{c^2} \dot{\phi} = 0$$

which then leads to,

$$\Box A(r, t) = - \mu_0 J$$

$$\Box \phi(r, t) = \frac{-\rho}{\epsilon_0}$$

one can show that this gauging is independent of reference system (inertial system) therefore being Lorentz-invariant.

18.3.1 The Energy of the Electromagnetic Field

In the general case, neither the temporal component A^0 nor the longitudinal component A^3 of the 4-potential A^μ is zero. For the free field we can write four d'Alembert's equations

$$\Box\, A_\mu\, (x) = 0,$$

(18.46)

which are obtained from the Lagrange-Euler equations with the Lagrangian density

$$L = -\frac{1}{8\pi} A_{\mu,\nu}\, A^{\mu,\nu}.$$

(18.47)

However, Eq. (18.46) are not equivalent to the Maxwell equation. To obtain the Maxwell equations, it is required in the classical theory that the Lorentz condition be satisfied, or

$$\partial A^\mu / \partial x^\mu = 0.$$

(18.48)

The general solution to *d'Alembert's equation* can be represented by a Fourier series for the 4-potential A_μ, similar to the three-dimensional case:

$$A_\mu\,(x) = \frac{1}{L^{3/2}} \sum_{k\lambda} \sqrt{\frac{2\pi}{\omega}}\, e_\mu^{(\lambda)}\,(c_{k\lambda} e^{-ikx} + c_{k\lambda}^{*} e^{ikx}),$$

(18.49)

where $kx = \omega t - k \cdot r$, $\omega = |\,k\,|$, while the polarization unit vectors $e^{(\lambda)}$, or explicitly,

$$e^{(0)} = (1, 0, 0, 0), \quad e^{(1)} = (0, 1, 0, 0),$$
$$e^{(2)} = (0, 0, 1, 0), \quad e^{(3)} = (0, 0, 0, 1),$$

(18.50)

possess the following properties:

$$e_\mu^{(\lambda)}\, e^{(\lambda')\mu} = e_0^{(\lambda)} e_0^{(\lambda')} - e^{(\lambda)} \cdot e^{(\lambda')} = \begin{cases} 0 & \text{if } \lambda \neq \lambda', \\ 0 & \text{if } \lambda \neq \lambda' = 0, \\ -0 & \text{if } \lambda \neq \lambda' = 1, 2, 3; \end{cases}$$

(18.51)

$$e_\mu^{(\lambda)}\, k^\mu = \begin{cases} k^0 & \text{if } \lambda \neq 0, \\ -k^0 & \text{if } \lambda = 3, \\ 0 & \text{if } \lambda = 1, 2. \end{cases}$$

The 4-potential [Eq. (18.49)] can be represented as the sum of two terms,

$$A_\mu\,(x) = A_\mu^{(+)}\,(x) + A_\mu^{(+)}\,(x),$$

(18.52)

where $A_\mu^{(+)}$ is the positive-frequency part, which consists of terms proportional to exp $(-ikx)$, and $A_\mu^{(-)}$ is the negative-frequency part, which consists of terms proportional to exp (ikx). These two parts differ in the sign in front of frequency ω in the exponents:

$$A_\mu^{(+)} = \sum_k A_{\mu k}^{(+)} e^{-i\omega t + i k \cdot r}, \quad A_\mu^{(-)} = \sum_k A_{\mu k}^{(-)} e^{+i\omega t + i k \cdot r},$$

where in the sum for $A_\mu^{(-)}$ we have replaced k with $-k$. As is known, the decomposition according to the sign of the temporal component of an isotropic 4-vector, with $k^2 = \omega^2 - k^2 = 0$, is Lorentz-invariant.

Now let us find the energy of the field,

$$H = \int d^3x \mathcal{H}, \quad \mathcal{H} = A_{\mu,t} \frac{\partial L}{\partial A_{\mu,t}} - L = -\frac{1}{8\pi} [A_{\mu,0} A^{\mu,0} + (\nabla A_\mu)^2]. \tag{18.53}$$

Allowing for (18.49), one obtains

$$H = -\sum_k \frac{1}{2} \omega (c^*_{k0} c_{k0} + c_{k0} c^*_{k0}) + \sum_{k, s = 1, 2, 3} \frac{1}{2} \omega (c^*_{ks} c_{ks} + c_{ks} c^*_{ks}). \tag{18.54}$$

We see that the energy of the field is not positive definite. To make it positive definite, we must impose the Lorentz condition on the 4-potentials.

18.3.2 Canonical Quantization

Let us now turn to the problem of quantizing the electromagnetic field. In contrast to the case of Coulomb gauge, we will not use the analogy with quantum oscillators. Instead we employ the method of canonical quantization, consisting in the replacement of the classical Poisson brackets $[H, u]_{c_l}$ with the quantum Poisson brackets $i [H, u]$ in the expression for the time derivative of a dynamical variable u,

$$\frac{du}{dt} = [H, u]_{c_l}, \tag{18.55}$$

where $[H, u]$ is the commutator involving the Hamiltonian operator H and the operator u representing the dynamical variable, or

$$\frac{du}{dt} = i [H, u]. \tag{18.56}$$

Here the operator u is given in the Heisenberg representation, that is, is time dependent, while the state vectors do not depend on time. Let us write the quantum Heisenberg operator of the electromagnetic field 4-potential in the form of a Fourier series:

$$A_\mu (r, t) = \frac{1}{L^{3/2}} \sum_{k\lambda} \sqrt{\frac{2\pi}{\omega}} e_\mu^{(\lambda)} (c_{k\lambda} (t) e^{i k \cdot r} + c^+_{k\lambda} (t) e^{-i k \cdot r}), \tag{18.57}$$

where $c_{k\lambda} (t)$ is the Heisenberg (i.e. time dependent) operator representing a Fourier coefficient.

For the operator representing the energy of the field we employ the definition that corresponds to the classical expression Eq. (18.54)

$$H = -\frac{1}{2} \sum_k \omega (c^+_{k0} c_{k0} + c_{k0} c^+_{k0}) + \frac{1}{2} \sum_{k, s = 1, 2, 3} \omega (c^+_{ks} c_{ks} + c_{ks} c^+_{ks}). \tag{18.58}$$

Then the dynamical variable $u = c_{k\lambda} (t) e^i_k \cdot r$ will satisfy Eq. (18.56). Substituting expansion of Eq. (18.58) into this equation, we get the following expression for $u = c_{ks} (t) e^i_k \cdot r$ ($s = 1, 2, 3$):

$$\frac{du}{dt} = \frac{i\omega}{2} [u^+ u + u u^+, u]. \tag{18.59}$$

We may note that out of the entire sum of Eq. (18.58) only the term with quantum numbers k, s that coincide with those of operator u is not commutative with u.

We now use an identity that is valid for any three operators A, B and C:
$$[AB, C] = A[B, C] + [A, C]B.$$

Then instead of Eq. (18.59), one obtains

$$\frac{du}{dt} = -\frac{i\omega}{2}([u, u^+]u + u[u, u^+]). \tag{18.60}$$

To linearize this equation we require that the following quantization rules hold true:

$$[u, u^+] = 1. \tag{18.61}$$

Having in mind the remark made after formula of Eq. (18.59), we can write in general form

$$[c_{ks}(t), c_{k's'}^+(t)] = \delta_{kk'}\delta_{ss'}. \tag{18.62}$$

Then instead of Eq. (18.60), we obtain the equation

$$\frac{du}{dt} = -i\omega u, \tag{18.63}$$

whose general solution is

$$u = c_{ks}(t)e^{ik \cdot r} = c_{ks}e^{-i\omega t + ik \cdot r}, \tag{18.64}$$

where c_{ks} is a constant operator coefficient.

In this case the field operator $A_\mu(r, t)$ proves to satisfy homogeneous d'Alembert's equation Eq. (18.46), just as the classical 4-potential does, with Eq. (18.64) corresponding to the positive-frequency part of this operator. As for temporal polarization, that is, $u = c_{k0}(t)\,e^i{}_k \cdot r$, we arrive at equation for it that differs from Eq. (18.60) in that the equation contains a "plus" sign in front of the right-hand side. This means that to obtain a positive-frequency solution of the Eq. (18.64) type we must fix quantization rules that have a sign opposite to that in Eq. (18.61), that is,

$$[u, u^+] = -1, \tag{18.65}$$

or, in general form,

$$[c_{k0}(t), c_{k0}^+(t)] = -\delta_{kk'}. \tag{18.66}$$

Let us define the vacuum state of the electromagnetic field, $|0\rangle$, by the equation

$$c_{k\lambda}|0\rangle = 0, \tag{18.67}$$

where $\lambda = 0, 1, 2, 3$, and k assumes any value. By analogy with the Coulomb case, we preserve the meaning of $c_{k\lambda}$ and $c_{k\lambda}^+$ as the operators of annihilation and creation of particles (photons):

$$c_{k\lambda}^+|0\rangle = |1_{k\lambda}\rangle, \tag{18.68}$$

where $|1_{k\lambda}\rangle$ is the state of a single photon with momentum k and polarization λ. Then, if we assume that the norm of the vacuum state vector is positive definite, that is,

$$\langle 0 | 0 \rangle = 1, \tag{18.69}$$

we are forced to conclude that the single-particle state vector of the temporal polarization has a negative norm,

$$\langle 1_{k0} | 1_{k0} \rangle = \langle 0 | c_{k0} c_{k0}^+ | 0 \rangle = \langle 0 | -1 + c_{k0}^+ c_{k0} | 0 \rangle = -1, \qquad (18.70)$$

as a result of the commutation relations Eq. (18.66). Thus, generally, the metric in the space of state vectors $| N_{k\lambda} \rangle$ proves to be indefinite, if we define the state with $N_{k\lambda}$ photons as

$$| N_{k\lambda} \rangle = \frac{(c_{k\lambda}^+)^{N_{k\lambda}}}{\sqrt{N_{k\lambda}!}} | 0 \rangle, \qquad (18.71)$$

then because of the commutation relations Eq. (18.66), we obtain

$$\langle N_{ks} | N_{ks} \rangle = 1,$$

$$\langle N_{k0} | N_{k0} \rangle = (-1)^{N_{k0}}. \qquad (18.72)$$

We may note that the states $| N_{k0} \rangle$ are eigenvectors of the operator $c_{k0}^+ c_{k0}$:

$$c_{k0}^+ c_{k0} | N_{k0} \rangle = -N_{k0} | N_{k0} \rangle. \qquad (18.73)$$

Indeed, if we denote, for the sake of brevity, c_{k0} by c, then for $N_{k0} = 1$ we have

$$c^+ c | 1 \rangle = c^+ c c^+ | 0 \rangle = c^+ (c^+ c - 1) | 0 \rangle = -c^+ | 0 \rangle = -| 1 \rangle.$$

If we assume that Eq. (18.73) holds true for $| N_{k0} - 1 \rangle$, for $| N_{k0} \rangle$, we get

$$c^+ c | N \rangle = c^+ c \frac{c^+}{\sqrt{N}} | N-1 \rangle = \frac{1}{\sqrt{N}} c^+ (c^+ c - 1) | N-1 \rangle$$

$$= (-1 + 1 - N) | N \rangle = (-N) | N \rangle.$$

We have thus proved that (18.73) is valid in the general case.

In the same manner we can show that the operators c and c^+ act on the state vectors as follows:

$$c_{ks} | N_{ks} \rangle = \sqrt{N_{ks}} | N_{ks} - 1 \rangle,$$

$$c_{ks}^+ | N_{ks} \rangle = \sqrt{N_{ks} + 1} | N_{ks} + 1 \rangle,$$

$$c_{k0} | N_{k0} \rangle = -\sqrt{N_{k0}} | N_{k0} - 1 \rangle, \qquad (18.74)$$

$$c_{k0}^+ | N_{k0} \rangle = \sqrt{N_{k0} + 1} | N_{k0} + 1 \rangle,$$

from which, incidentally, Eq. (18.73) follows directly. Thus, the operators $c_{k\lambda}^+$ and $c_{k\lambda}$ are indeed the photon creation and annihilation operators. Note that up till now we have imposed no additional conditions on the field operators A_μ and the state vectors. For this reason the states with temporal photons, $| \dots N_{k0} \dots \rangle$, where the dots stand for the occupation numbers of states with other possible quantum numbers, can provide, due to the indefinite metrics [Eq. (18.72)], a negative contribution to the total energy of the electromagnetic field

$$E = \langle H \rangle.$$

For this reason the energy of the field is not positive definite. For E to be nonnegative, we must ensure that the admissible states $|\Phi\rangle$ of the field obey the condition

$$\left(\frac{\partial A_\mu}{\partial x_\mu}\right)^{(+)}|\Phi\rangle=0, \tag{18.75}$$

where by $(\partial A^\mu/\partial x^\mu)^{(+)}$ we denote the positive-frequency part of the divergence $\partial A^\mu/\partial x^\mu$. For the negative-frequency part we obtain, as a corollary to (18.75),

$$\langle\Phi|\left(\frac{\partial A_\mu}{\partial x_\mu}\right)^{(-)}=0. \tag{18.76}$$

Using expansion (18.57) and solution (18.64), for each value of k we arrive at the following gauge condition:

$$\sum_\lambda e_\mu^{(\lambda)}k^\mu c_{k\lambda}|\Phi\rangle=0, \tag{18.77}$$

or, if we allow for (18.51),

$$\omega(c_{k0}-c_{k3})|\Phi\rangle=0. \tag{18.78}$$

For $\omega\neq0$ the last condition, when combined with Eq. (18.76), leads to

$$(c_{k0}-c_{k3})|\Phi\rangle=0, \quad \langle\Phi|(c_{k0}^+-c_{k3}^+)=0, \tag{18.79}$$

which conditions the state vectors of the quantized electromagnetic field must satisfy. Then, as a corollary to these conditions, we obtain

$$\langle\Phi|c_{k0}^+c_{k0}-c_{k3}^+c_{k3}|\Phi\rangle\equiv\frac{1}{2}\langle\Phi|(c_{k0}^++c_{k3}^+)(c_{k0}-c_{k3})$$

$$+(c_{k0}^+-c_{k3}^+)(c_{k0}+c_{k3})|\Phi\rangle=0,$$

that is,

$$\langle\Phi|c_{k0}^+c_{k0}|\Phi\rangle=\langle\Phi|c_{k3}^+c_{k3}|\Phi\rangle. \tag{18.80}$$

Thanks to this property, the contributions provided by the temporal and longitudinal photons to the value of the field energy averaged over the state $|\Phi\rangle$ of the field cancel out, and we find that

$$E=\langle H\rangle=\frac{1}{2}\sum_k\omega\langle c_{k1}^+c_{k1}+c_{k1}c_{k1}^+\rangle$$

$$+\frac{1}{2}\sum_k\omega\langle c_{k2}^+c_{k2}+c_{k2}c_{k2}^+\rangle. \tag{18.81}$$

This value is, obviously, positive definite for those states that contain neither longitudinal nor temporal photons. Such physical states, which possess a positive norm (see Eq. (18.72)), we will use in what follows. Let us eliminate the indefinitely large part, proportional to $\Sigma\omega$, by requiring that the creation and annihilation operators follow in normal order. As a result we arrive at the following expression for the eigenvalues of the field energy:

$$E = \sum_{k, s = 1, 2} N_{ks}\omega, \tag{18.82}$$

We may note that conditions Eqs. (18.75) and (18.76) ensure that the Lorentz condition is satisfied in the mean:

$$\left\langle \frac{\partial A^\mu}{\partial x^\mu} \right\rangle = \left\langle \left(\frac{\partial A^\mu}{\partial x^\mu}\right)^{(+)} + \left(\frac{\partial A^\mu}{\partial x^\mu}\right)^{(-)} \right\rangle = 0. \tag{18.83}$$

The same conditions, obviously, lead to the Maxwell equations in the mean. Indeed, since the field operators A_μ satisfy d'Alembert's equation, we have

$$\Box A^\mu = \frac{\partial^2 A^\mu}{\partial x^\nu \partial x_\nu} \equiv \frac{\partial}{\partial x^\nu}\left(\frac{\partial A^\mu}{\partial x_\nu} - \frac{\partial A^\nu}{\partial x_\mu}\right) + \frac{\partial}{\partial x^\nu}\frac{\partial A^\nu}{\partial x_\mu}$$

$$= \frac{\partial F^{\nu\mu}}{\partial x^\nu} + \frac{\partial}{\partial x_\mu}\frac{\partial A^\nu}{\partial x^\nu} = 0.$$

Combining this with (18.83) yields

$$\left\langle \frac{\partial F^{\nu\mu}}{\partial x^\nu} \right\rangle = 0, \tag{18.84}$$

which are the *Maxwell equations for the free electromagnetic field.*

Thus, longitudinal and temporal photons cannot be observed and contribute nothing to the energy and momentum of the field,

$$E = \sum_{k, s = 1, 2} \omega N_{ks} \geq 0, \quad P = \sum_{k, s = 1, 2} k N_{ks}. \tag{18.85}$$

Now, the question arises, why cannot we restrict along the same lines as we do in the classical theory, that is, impose the Lorentz condition directly on the field operators A_μ. The answer is as follows: such a condition would have led to a dependence between the operator coefficients c_{k0} and c_{k3}, say, $c_{k0} = c_{k3}$, and this contradicts the commutation relations (18.62) and (18.66).

A final remark is in order. Since longitudinal and temporal photons cannot be observed, we could redefine the initial vacuum in such a manner that it would contain an arbitrary number of temporal photons and also satisfy the conditions

$$c_{k1}|\text{vac}\rangle = c_{k2}|\text{vac}\rangle = 0, \quad (c_{k0} - c_{k3})|\text{vac}\rangle = 0. \tag{18.86}$$

Such a new vacuum $|\text{vac}\rangle$ can be obtained from the initial vacuum $|0\rangle$, which contains no photons, via a gauge transformation U:

$$|\text{vac}\rangle = U|0\rangle. \tag{18.87}$$

The explicit form of operator U will not be given here, but it is important to note that

$$U = U (c_{k0} - c_{k3}, \; c_{k0}^+ - c_{k3}^+)$$

and $[U, c_{k0} - c_{k3}] = 0$, so that Eqs. (18.86) are satisfied.

Let us assume that there exists a state of the field with a single photon specified by a polarization fixed by the 4-vector e_μ. Then the gauge condition [Eq. (18.77)] yields

$$e^\mu k_\mu = 0. \tag{18.88}$$

This means, for one, that there can be no pure states with a single temporal photon $e^\mu = (1, 0, 0, 0)$ or a single longitudinal photon $e^\mu = (0, 0, 0, 1)$. A photon whose polarization satisfies condition [Eq. (18.88)] is known as a *four-dimensional transversal photon*. This condition is not violated the 4-vector of polarization is subjected to a gradient (or gauge) transformation

$$e^\mu \rightarrow e'_\mu = e_\mu + fk_\mu, \tag{18.89}$$

where f is an arbitrary scalar function, and $k^2 = k_\mu k^\mu = 0$. In particular, we can select f in such a way that in a given reference frame we obtain the condition of three-dimensional transversality

$$e_0 = 0, e \cdot {}_k = 0, \; e^2 = 1 \tag{18.90}$$

instead of Eq. (18.88). We can easily see, the normalization condition of the unit 4-vector e^μ,

$$e_\mu e^\mu = -1 \tag{18.91}$$

does not change under Eq. (18.89).

18.4 THE COMMUTATION AND CAUSAL FUNCTIONS OF THE ELECTROMAGNETIC FIELD

Now we study of quadratic combinations of electromagnetic-field operators $A_\mu (x)$ and their averages, which are expressed in terms of various singular solutions of the wave equation for the electromagnetic potentials.

18.4.1 The Commutation Relation for Electromagnetic-Field Operators

Let us write the electromagnetic-field operator, following Eqs. (18.57) and (18.64), in the form

$$A_\mu (x) = \frac{1}{L^{3/2}} \sum_{k\lambda} \sqrt{\frac{2\pi}{\omega}} \, e_\mu^{(\lambda)} \, (c_{k\lambda}^+ e^{ikx} + c_{k\lambda} e^{-ikx}). \tag{18.92}$$

The photon creation and annihilation operators, c^+ and c, in which the operator A_μ is expanded, satisfy the commutation relations (18.62) and (18.66). We now wish to calculate the commutator of the field operators A_μ at point x and x', or $[A_\mu (x), A_\nu (x')]$. To this end, we employ expansion (18.92) and the commutators involving c and c^+,

$$[c_{k\lambda}, c_{k'\lambda'}^+] = -\delta_{kk'} \, g_{\lambda\lambda'}. \tag{18.93}$$

Then

$$[A_\mu (x), A_\nu (x')] = \frac{2\pi}{L^3} \sum_k \sum_{\lambda\lambda'} \frac{1}{\omega} e_\mu^{(\lambda)} e_\nu^{(\lambda')} g_{\lambda\lambda'} (e^{ik(x - x')} - e^{-ik(x - x')})$$

Here, as it follows from Eq. (18.50), the sum over the polarizations yields

$$\sum_{\lambda\lambda'} e_\mu^{(\lambda)} e_\nu^{(\lambda')} g_{\lambda\lambda'} = g_{\mu\nu},$$

so that

$$[A_\mu(x), A_\nu(x')] = \frac{2\pi}{L^3} g_{\mu\nu} \sum_k \frac{1}{\omega}(e^{ik(x-x')} - e^{-ik(x-x')}). \tag{18.94}$$

Going over from summation to integration via

$$\frac{1}{L^3} \sum_k = \frac{1}{(2\pi)^3} \int d^3k. \tag{18.95}$$

and replacing k with $-k$ in the first integral, we find that

$$[A_\mu(x), A_\nu(x')] = 4\pi i g_{\mu\nu} D_0(x - x') \tag{18.96}$$

where

$$D_0(x - x') = \frac{1}{(2\pi)^3} \int \frac{d^3k}{\omega} e^{ik \cdot (r - r')} \sin \omega(t - t'). \tag{18.97}$$

The invariant function $D_0(x - x')$ is known as the commutation function for the electromagnetic field. Since $A_\mu(x)$ and $A_\nu(x')$ satisfy homogeneous d'Alembert's equation, so does $D_0(x - x')$:

$$\Box_x D_0(x - x') = \Box_{x'} D_0(x - x') = 0. \tag{18.98}$$

As definition (18.97) implies, the function $D_0(x) = D_0(r, t)$, being the solution to Eq. (18.98), satisfies the following initial conditions:

$$D_0(r, 0) = 0, \quad \frac{\partial}{\partial t} D_0(r, t)\big|_{t=0} = \delta(r). \tag{18.99}$$

Hence, at $t = t'$ Eq. (18.97) yields the *equal-time commutator*

$$\left[A_\mu(x), \frac{\partial}{\partial t'} A_\nu(x')\right]_{t = t'} = -4\pi i g_{\mu\nu} \delta(r - r') \tag{18.100}$$

The meaning of this relationship can easily be clarified if we go over to the canonical field-momentum operator

$$\pi_\nu = -\frac{1}{4\pi} \frac{\partial A_\nu}{\partial t}.$$

Then instead of Eq. (18.100), we get

$$[A_\mu(x), \pi_\nu(x')]_{t = t'} = i g_{\mu\nu} \delta(r - r'). \tag{18.101}$$

The operators A_μ and π_ν are canonically conjugate quantities and correspond to the generalized coordinate and momentum of the continuous system, the electromagnetic field. Hence, the commutator Eq. (18.101) constitutes a direct generalization of the commutation relation (18.1) for a quantum mechanical discrete system. Note that we could also put Eq. (18.101) at the base of the quantum theory of the electromagnetic field, but the substantiation of this step would require partitioning the space into cells and thus going over to a discrete system, which procedure is in itself rather awakward.

Let us now show that the commutation function $D_0(x)$ is relativistically invariant. To this end, we evaluate in Eq. (18.97) the integral with respect to k. The result is

$$D_0(x) = \frac{2\pi}{(2\pi)^3} \int_0^\infty \omega \, d\omega \int_0^\pi \sin\theta \, d\theta \, e^{i\omega r \cos\theta} \sin\omega t$$

$$= \frac{1}{2ir(2\pi)^2} \int_\infty^\infty d\omega \, (e^{i\omega r} - e^{-i\omega r}) \sin\omega t.$$

Integration with respect to ω yields

$$D_0(x) = \frac{1}{4\pi r}[\delta(r-t) - \delta(r+t)]. \tag{18.102}$$

Since r is always nonnegative, we have

$$\delta(r^2 - t^2) = \frac{\delta(r - |t|)}{2r}.$$

The final result is

$$D_0(x) = \frac{1}{2\pi} \delta(x^2) \operatorname{sgn} x^0, \tag{18.103}$$

where $\operatorname{sgn} x^0$ is the sign function

As is known, the sign of $t = x^0$, that is, $\operatorname{sgn} x^0$, is invariant for the isotropic vector $x^2 = 0$, which proves the invariance of $D_0(x)$. We may note the singular solutions of the inhomogeneous wave equation, $D_{ret}(x)$ and $D_{adv}(x)$, are the retarded and advanced Green functions, D_{rad} is their half-difference.

Comparing these expressions with (18.103), one obtains
$$D_0(x) = D_{ret}(x) - D_{adv}(x) = 2D_{rad}(x). \tag{18.104}$$

Let us now find the vacuum average of quadratic combinations of potentials $A_\mu(x) A_\nu(x')$. Using expansion of Eq. (18.92) and the property of the vacuum, one obtains

$$\langle 0|A_\mu(x) A_\nu(x')|0\rangle = -4\pi g_{\mu\nu} D_+(x - x'), \tag{18.105}$$

where

$$D_+(x) = \frac{1}{(2\pi)^3} \int \frac{d^3k}{2\omega} e^{-ik(x - x')}, \tag{18.106}$$

and, similarly,

$$\langle 0|A_\nu(x') A_\mu(x)|0\rangle = -4\pi g_{\mu\nu} D_-(x - x'), \tag{18.107}$$

where

$$D_-(x) = D_+(-x) = D_+^*(x). \tag{18.108}$$

Average of Eq. (18.96) over the vacuum, are arrive at a relationship that links the three functions:

$$D_0(x) = i(D_+(x) - D_-(x)). \tag{18.109}$$

The functions D_+, D_-, and D_0 are singular solutions of the homogeneous wave equation. It can be shown that these invariant functions have the following form:

$$D_\pm(x) = -\frac{1}{(2\pi)^2 x^2} \mp \frac{i}{4\pi} \delta(x^2) \operatorname{sgn} x^0. \tag{18.110}$$

18.4.2 The Normal and Chronologically Ordered Products of Electromagnetic-Field Operators

We have already introduced the concept of normal product of photon creation and annihilation operators. The definition we gave is also valid for the field operators $A_\mu(x)$. Precisely, we define the normal product of two operators, $A_\mu(x)$ and $A_\nu(x')$, in such a manner that all creation operators are to the left of all the annihilation operators. Since we know that $A_\mu(x)$ can be decomposed in an invariant manner,

$$A_\mu(x) = A_\mu^{(-)}(x) + A_\mu^{(+)}(x), \tag{18.111}$$

where, as Eq. (18.92) implies, the negative-frequency part $A_\mu^{(-)}(x)$ contains only creation operators and the positive-frequency part $A_\mu^{(+)}(x)$, only annihilation operators, the normal product (denoted by the letter N) is

$$N A_\mu(x) A_\nu(x') = A_\mu^{(+)}(x) A_\nu^{(+)}(x') + A_\mu^{(-)}(x) A_\nu^{(-)}(x')$$

$$+ A_\mu^{(-)}(x) A_\nu^{(+)}(x') + A_\nu^{(-)}(x') A_\nu^{(+)}(x). \tag{18.112}$$

We now introduce another ordering scheme for field operators, namely, the chronologically ordered product. We define the chronologically ordered product, or T-product, as follows:

$$T A_\mu(x) A_\nu(x') = \begin{cases} A_\mu(x) A_\nu(x'), & t > t', \\ A_\nu(x') A_\mu(x), & t' > t, \end{cases} \tag{18.113}$$

where the operators are placed in the order of increasing times from right to left.

Just as in the case of the normal product, the chronologically ordered product is defined in an invariant manner. Indeed, for $(x - x')^2 \geq 0$ the sign of $t - t'$ is invariant, while for $(x - x')^2 < 0$ the operators $A(x)$ and $A(x')$ are commutative.

The definitions of the T- and the N-product of any number of operators can be given in a similar manner. Note that the reason why the N-product was introduced is that the average of the N-product of any number of operators over the vacuum is zero:

$$\langle 0 | N A_\mu(x) A_\nu(x') A_\lambda(x'') \ldots | 0 \rangle = 0. \tag{18.114}$$

Let us define the contraction, or chronological pairing, of two operators as follows:

$$\overline{A_\mu(x) A_\nu(x')} = T A_\mu(x) A_\nu(x') - N A_\mu(x) A_\nu(x'). \tag{18.115}$$

We can easily see that this is a c-number, that is, not an operator. Suppose that $x^0 > x'^0$. Then

$$T A_\mu(x) A_\nu(x') - N A_\mu(x) A_\nu(x') = A_\mu^{(+)}(x) A_\nu^{(-)}(x') - A_\nu^{(-)}(x') A_\mu^{(+)}(x)$$

$$= -4\pi g_{\mu\nu} D_+(x - x'), \tag{18.116}$$

and similarly for $x^0 < x'^0$. Averaging Eq. (18.115) over the vacuum and employing Eq. (18.114), we find that the pairing is the average of the T-product over the vacuum state:

$$\overline{A_\mu(x)\,A_\nu(x')} = \langle 0\,|\,T\,A_\mu(x)\,A_\nu(x')\,|\,0\rangle \equiv \langle T\,A_\mu(x)\,A_\nu(x')\rangle_0, \tag{18.117}$$

whence, employing Eq. (18.105) and (18.106), we get

$$\overline{A_\mu(x)\,A_\nu(x')} = \begin{cases} \langle A_\mu(x)\,A_\nu(x')\rangle_0 = -4\pi g_{\mu\nu}\,D_+(x-x') & t > t', \\ \langle A_\nu(x')\,A_\mu(x)\rangle_0 = -4\pi g_{\mu\nu}\,D_-(x-x') & t' > t. \end{cases} \tag{18.118}$$

We introduce the function

$$D_c(x) = \theta(x)\,D_+(x) + \theta(-x)\,D_-(x), \tag{18.119}$$

which is known as the causal function, or propagator. The pairing is then expressed in terms of the propagator as follows:

$$\overline{A_\mu(x)\,A_\nu(x')} = \langle T\,A_\mu(x)\,A_\nu(x')\rangle_0 = -4\pi g_{\mu\nu}\,D_c(x-x'). \tag{18.120}$$

From Eqs. (18.118) and (18.119) it follows that a propagator describes the following chain of events: for $t > t'$ a photon is created at point x' and is then destroyed at point x, while for $t < t'$ the order is reversed, that is, a photon is created at point x and destroyed at point x'.

The integral representation for $D_c(x)$ follows from the definitions (18.106) and (18.108) for $D_+(x)$ and $D_-(x)$, respectively, where in the integral for $D_-(x)$ we must replace k with $-k$:

$$D_c(x) = \frac{1}{(2\pi)^3}\int \frac{d^3k}{2\omega}\,e^{i\mathbf{k}\cdot\mathbf{r} - i\omega|t|}. \tag{18.121}$$

The propagator can be written in the form of a quadruple integral,

$$D_c(x) = \frac{i}{(2\pi)^4}\int \frac{d^4k}{k^2 + i\varepsilon}\,e^{-ikx} \tag{18.122}$$

where $\varepsilon \to +0$, that is, a particular solution of the inhomogeneous wave equation

$$\square\, D_c(x) = -i\delta(x). \tag{18.123}$$

Some particular solutions of such an equation (without the factor $-i$ in front of the delta function) exist. As was established that the particular solutions differ in the choice of the sense of traversal of the poles of the integrand of the integral representation. Introduction of a small imaginary term $i\varepsilon$ in Eq. (18.122) means that the integral with respect to k^0 contains two poles in the integrand at $k^0 = \pm(\omega - i\varepsilon)$, $\omega = |k|$, one of which (the left one) lies above the real axis, and the other (the right one) lies below the real axis. If $t > 0$, the integration contour is closed in the lower half-plane (Figure 18.2 (a)), while if $t < 0$, the integration contour is closed in the upper half-plane (Figure 18.2(b)). As a result we find that

$$D_c(x) = D_+(x),\ t > 0,\ D_c(x) = D_-(x),\ t < 0. \tag{18.124}$$

(a)   (b)

Fig. 18.2

We have thus proved that Eq. (18.122) indeed coincides with definition Eq. (18.119). Note that, as follows from Eq. (18.104) for D_{rect} and D_{adv} in terms of contour integrals, the commutation function $D_0(x)$ can also be written as a contour integral:

$$D_0(x) = -\frac{1}{(2\pi)^4} \int_{C_0} \frac{d^4 k}{k^2} e^{-ikx}, \tag{18.125}$$

where contour C_0 is depicted in Fig. 18.3.

Fig. 18.3

The expression for the commutation function [Eq. (18.96)] was obtained through expanding the electromagnetic-potential operators, Eq. (18.92), in a gauge specified by Eq. (18.50). In accordance with Eq. (18.89), we could have also considered other gauges for which

$$e_\mu^{(\lambda)} \to e_\mu'^{(\lambda)} = e_\mu^{(\lambda)} + f^{(\lambda)} k_\mu, \tag{18.126}$$

where $f^{(\lambda)}$ are arbitrary scalar functions, with $k^2 \neq 0$, generally speaking. In this case the commutation function [Eq. (18.96)] is replaced with functions of the following type:

$$D_{\mu\nu}(x) = g_{\mu\nu} D_0(x) \to D'_{\mu\nu}(x) = g_{\mu\nu} D_0(x)$$

$$+ \frac{1}{(2\pi)^4} \int_{C_0} \frac{d^4 k}{k^2} e^{-ikx}(f_\nu k_\mu + f_\mu k_\nu + f k_\mu k_\nu), \tag{18.127}$$

where

$$f_\nu = \sum_{\lambda\lambda'} g_{\lambda\lambda'} f^{(\lambda)} e_\nu^{(\lambda')}, \quad f_\mu = \sum_{\lambda\lambda'} g_{\lambda\lambda'} f^{(\lambda')} e_\mu^{(\lambda)},$$

$$f = \sum_{\lambda\lambda'} g_{\lambda\lambda'} f^{(\lambda)} f^{(\lambda')}. \tag{18.128}$$

In the same manner, one finds that the *gauge transformation* results in a change in propagator:

$$D_{\mu\nu}^c(x) = g_{\mu\nu} D_c(x) \to g_{\mu\nu} D_c(x)$$

$$+ \frac{i}{(2\pi)^4} \int \frac{d^4 k}{k^2} e^{-ikx}(f_\nu k_\mu + f_\mu k_\nu + f k_\mu k_\nu), \tag{18.129}$$

Say, if we put $f^{(\lambda)} = -k^{\lambda}/k^2$, we arrive at the following relation for the propagator:

$$D^c_{\mu\nu}(x) = \frac{i}{(2\pi)^4} \int \frac{d^4k}{k^2 + i\varepsilon} e^{-ikx} \left(g_{\mu\nu} - \frac{k_\mu k_\nu}{k^2} \right). \tag{18.130}$$

We may note that the concrete choice of gauge is determined by the features of the problem being solved. At the same time, in view of gauge invariance, the final results do not depend on this choice.

18.5 QUANTIZATION OF THE DIRAC FIELD

The Dirac equation describes massive charged particles with half-integral spin, electrons and positrons. Quantization of the electron-positron field differs from quantization of the electromagnetic field primarily in that particles described by the Dirac equation are fermions, that is, obey the *Fermi-Dirac statistics**. This fact is also expressed by *Pauli's exclusion principle****: there can be no more than one fermi-particle in a single quantum state.

18.5.1 Charge Conjugation

The free electron is described by the Dirac equation

$$(i\gamma_\mu \partial^\mu - m)\psi = 0$$

which after being multiplied by the matrix γ^0 can be expressed in the following Hamilton form:

$$i\frac{\partial}{\partial t}\psi(\mathbf{r}, t) = H_D \psi(\mathbf{r}, t), \tag{18.131}$$

where

$$H_D = -i\alpha \cdot \nabla + \beta m \tag{18.132}$$

is the Hamiltonian of the Dirac equation, and α and $\beta = \gamma^0$ are the Diract matrices. We represent the Dirac wave function of a stationary state by

$$\psi(\mathbf{r}, t) = Ae^{-iEt + i\mathbf{r}\cdot\mathbf{p}}u, \tag{18.133}$$

where E is an eigenvalue of H_D, $\mathbf{p}$ is an eigenvalue of the momentum operator $-i\nabla = \mathbf{p}$, u is a constant bispinor, and A is the normalization constant.

Let us write the bispinor u in terms of the two-component spinors φ and χ,

$$u = \begin{pmatrix} \varphi \\ \chi \end{pmatrix} \tag{18.134}$$

---

* *Fermi Dirac (FD) Statistics* gives the probability that in an ideal electron gas in thermal equilibrium at a temperature T an electron state with an energy E will be occupied

$$f(E) = \frac{1}{\exp[(E-\mu)/k_B T] + 1},$$

where μ is the chemical potential. In contrast to Bose-Einstein (BE) statistics the FD distribution also takes into account the Pauli exclusion principle.

** *Pauli exclusion principle* states that there can never exist two or more equivalent electrons in the atom for which the values of all quantum numbers coincide.

and use the standard representation for the Dirac matrices. Then for φ and χ we have the system of equations

$$(E - m)\, \varphi = (\boldsymbol{\sigma} \cdot \boldsymbol{p})\, \chi,$$
$$(E + m)\, \chi = (\boldsymbol{\sigma} \cdot \boldsymbol{p})\, \varphi, \tag{18.135}$$

whose compatibility is ensure by the condition

$$E^2 = \boldsymbol{p}^2 + m^2.$$

This leads to a relationship that links the eigenvalues of the Hamiltonian operator and the momentum:

$$E = \pm\, \varepsilon p, \quad \varepsilon p = \sqrt{\boldsymbol{p}^2 + m^2}. \tag{18.136}$$

Thus, the possible solution split into two branches, the positive-frequency branch with $E = \varepsilon p$ and the negative-frequency branch with $E = -\,\varepsilon p$, and

$$\psi^{(+)} \propto e^{-i\varepsilon_{\boldsymbol{p}} t}, \quad \psi^{(-)} \propto e^{i\varepsilon_{\boldsymbol{p}} t}. \tag{18137}$$

Since the two branches correspond to different eigenvalues of the Hamiltonian, the corresponding wave functions must be orthogonal to each other:

$$\int (\psi^{(+)})^+\, \psi^{(-)} d^3 x = 0. \tag{18.138}$$

Note that the separation (18.136) is Lorentz-invariant, since the 4-vector $p^\mu = (E, \boldsymbol{p})$ is timelike: $p^2 = E^2 - \boldsymbol{p}^2 = m^2 > 0$.

If the spinor φ is given, the other spinor χ can be determined from the second equation of the system [Eq. (18.134)]:

$$\chi = \frac{(\boldsymbol{\sigma} \cdot \boldsymbol{p})}{E + m}\, \varphi. \tag{18.139}$$

Specifying spinor is accomplishes by imposing an additional condition that fixes the spin state of the particle. For instance, we can require that the wave function be an eigenfunction of the helicity operator* $(1/2)\,(\boldsymbol{\Sigma} \cdot \boldsymbol{p})$:

$$\frac{1}{2}\,(\boldsymbol{\Sigma} \cdot \boldsymbol{p})\, \psi = \sigma |\boldsymbol{p}|\, \psi, \quad \sigma = \pm \frac{1}{2}. \tag{18.140}$$

This operator, obviously, is commutative with H_D, and therefore the Dirac equation and Eq. (18.140) have common solutions.

The solutions to the Diract equation for a free particle, Eq. (18.131), corresponding to two different branches, $E = \varepsilon p$ and $E = -\,\varepsilon p$, can be written finally as follows:

$$\psi_{\mathbf{p}}^{(-)}(\mathbf{r}, t) = \frac{1}{\sqrt{2\varepsilon_{\mathbf{p}} V}}\, u\,(\sigma, -p)\, e^{-ipx}, \tag{18.141}$$

$$\psi_{\mathbf{p}}^{(-)}(\mathbf{r}, t) = \frac{1}{\sqrt{2\varepsilon_{\mathbf{p}} V}}\, u\,(\sigma, -p)\, e^{ipx}. \tag{18.142}$$

---

* This operator is proportional to the temporal component of the 4-vector spin operator

$$S^\mu = (S^0, \mathbf{S}), \quad S^0 = \frac{1}{m}\boldsymbol{\Sigma} \cdot \mathbf{p}, \quad \mathbf{S} = \frac{1}{m}(m\gamma^0 \boldsymbol{\Sigma} \cdot p\gamma^5) \tag{18.143}$$

where $\gamma^5 = -i\gamma^0\gamma^1\gamma^2\gamma^3 \equiv -\rho_1$ is the pseudoscalar Dirac matrix.

Here V is the normalization volume, and $p^\mu = (p^0, \boldsymbol{p})$ is the 4-momentum, where $p^0 \equiv \varepsilon p$, with the negative-frequency solution $\psi^{(-)}$ chosen in such a way that it corresponds to the value of momentum $-\boldsymbol{p}$ taken with a sign opposite to that of $\psi^{(+)}$. The bispinors $u\,(\sigma, \pm p)$ satisfy the equations

$$(\pm \gamma p - m)\, u\,(\sigma, \pm p) = 0. \tag{18.144}$$

The normalization coefficient $A = (2\varepsilon p V)^{-1/2}$ is selected such that the bispinors are normalized by the conditions

$$\bar{u}\,(p)\, \gamma^\mu\, u\,(p) = 2p^\mu, \quad \bar{u}\,(-p)\, \gamma^\mu\, u\,(-p) = 2p^\mu. \tag{18.145}$$

Then at $\mu = 0$, we get

$$u^+\,(p)\, u\,(p) = 2\varepsilon p, \quad u^+\,(-p)\, u\,(-p) = 2\varepsilon p \tag{18.146}$$

whence

$$\int_V (\psi_{\mathbf{p}}^{(+)})^+\, \psi_{\mathbf{p}}^{(+)}\, d^3x = 1, \quad \int_V (\psi_{\mathbf{p}}^{(-)})^+\, \psi_{\mathbf{p}}^{(-)}\, d^3x = 1.$$

The positive-frequency solutions of the Dirac equation Eq. (18.141), can be interpreted as being the wave functions of an electron whose energy is $E = \varepsilon p > 0$. To give meaning to the negative-frequency branch of solutions, Eq. (18.143), with $E = -\varepsilon p$, we consider the so-called *charge conjugation operation*, or C-conjugation, to which the Dirac functions can be subjected. Let us assume that the Dirac particles are not free but interact with the electromagnetic field $A_\mu\,(x)$. In this case the Direct equation is obtained by "extending" the derivative $\partial_\mu \to \partial_\mu + ieA_\mu$ in the same manner as was done earlier:

$$(i\gamma\partial - e\gamma A - m)\, \psi(x) = 0, \quad \bar{\psi}\,(x)\,(i\gamma\partial + e\gamma A + m) = 0. \tag{18.147}$$

We introduce the charge conjugation operation as follows

$$\psi\,(x) \to \psi^C\,(x) = C\bar{\psi}(x),$$

$$\bar{\psi}\,(x) \to \bar{\psi}^C\,(x) = C^{-1}\,\bar{\psi}(x), \tag{18.148}$$

$$A_\mu\,(x) \to A_\mu^C\,(x) = -A_\mu\,(x).$$

Suppose that the operator C possesses the following property:

$$C^{-1}\, \gamma^\mu C = -\gamma_\mu^T. \tag{18.149}$$

Then, obviously, Eqs. (18.147) will remain invariant under C-conjugation (18.148):

$$(i\gamma\partial - e\gamma A^C - m)\, \psi^C\,(x) = 0, \quad \bar{\psi}^C\,(x)\,(i\gamma\partial + e\gamma A^C + m) = 0. \tag{18.150}$$

Condition [Eq. (18.149)] can be met if we select the C-conjugation operator in the form $C = \gamma^2\gamma^0$. Matrix C possesses the following properties:

$$C^+ = C^{-1}, C^T = -C, \tag{18.151}$$

that is, it is *unitary* and *skew-symmetric*.

Using the charge conjugation operation, we can construct a positive-frequency part out of the negative-frequency part of the solution:

$$\psi^{C(+)}\,(x) = \overline{C\psi^{(-)}\,(x)}. \tag{18.152}$$

As Eq. (18.150) implies, the function $\psi^{C(+)}$ satisfies the same equation as $\psi^{(+)}$, but with the opposite sign of the charge (since $A^C = -A$) and has a positive frequency, that is, the normal sign of the energy $E = \varepsilon p > 0$. This function, $\psi^{C(+)}$, describes a new particle, the positron, that is, the antiparticle of the electron, which carries a charge of opposite sign. If the electron charge is negative, $e = -e_0 < 0$, the charge of the positron is positive, $e = e_0 > 0$. Thus, the wave function of the electron, ψ_{el}, and that of the positron, ψ_{pos}, are constructed from the positive- and negative-frequency parts of the solution to the *Dirac equation*,

$$\psi_{el} = \psi^{(+)}, \quad \psi_{pos} = \overline{C\psi^{(-)}}. \tag{18.153}$$

18.5.2 Commutation Relations, Anticommutators

Now let us consider the Dirac equation Eq. (9.1) not as a quantum mechanical equation for single-particle states but as an equation for a spinor Dirac field. The general solution for the free case, $A_\mu(x) = 0$, can be split in an invariant manner into the positive- and negative-frequency parts $\psi^{(+)}$ and $\psi^{(-)}$ as follows:

$$\psi(x) = \psi^{(+)}(x) + \psi^{(-)}(x) = \sum_s (b_s \psi_s^{(+)}(x) + b_s^* \psi_s^{(-)}(x)), \tag{18.154}$$

where s stands for the set of quantum numbers specifying the stationary states

$$\psi_s^{(+)} \propto e^{-i\varepsilon_s^{(+)}t}, \quad \psi_s^{(-)} \propto e^{i\varepsilon_s^{(-)}t}, \tag{18.155}$$

$\pm \varepsilon_s^{(\pm)}$ are the eigenvalues of the Dirac Hamiltonian, and a_s and b_s^* are expansion coefficients.

Let us find the total energy and charge of the field

$$H = \int d^3x\, \mathcal{H}, \quad \mathcal{H} = \frac{i}{2}(\overline{\psi}\gamma^0\psi_{,t} - \overline{\psi}_{,t}\gamma^0\psi) = \frac{i}{2}(\psi^+\psi_{,t} - \psi_{,t}^+\psi),$$

$$Q = e\int d^3x\, \overline{\psi}\gamma^0\psi = e\int d^3x\, \psi^+\psi. \tag{18.156}$$

Substituting expansion (18.154) and employing the fact that the solution $\psi_s^{(\pm)}$ are orthonormalized, we obtain

$$H = \sum_s (\varepsilon_s^{(+)} a_s^* a_s - \varepsilon_s^{(-)} b_s b_s^*), \quad Q = e\sum_s (a_s^* a_s + b_s b_s^*). \tag{18.157}$$

As we see, the energy of the field does not possess a definite sign, while the charge has the sign of e. We may recall that the expansion coefficients a_{ks} were subjected, when the electromagnetic field was quantized, to commutation relations (18.16) so that they would become operators. We will now define the operators of energy and charge for the Dirac field using the classical expressions (18.157) as a model, replacing a_s and b_s in these expressions with operators. The result is

$$H = \sum_s (\varepsilon_s^{(+)} a_s^+ a_s - \varepsilon_s^{(-)} b_s b_s^+), \quad Q = e\sum_s (a_s^+ a_s - b_s b_s^+), \tag{18.158}$$

where a_s^+ and b_s^+ are the Hermitian-conjugate operators. However, in contrast to the electromagnetic field (a boson field!), we cannot require the validity of the commutation relations of the Eq. (18.16) type for a_s and b_s, since if these relations were valid, the energy would not be positive definite and the electrons and positrons would contribute terms of the same sign to the charge of the field. Consequently, for the quantized expansion coefficients of the operator of the Dirac field

$$\psi(x) = \sum_s (a_s \psi_s^{(+)}(x) + b_s^+ \psi_s^{(-)}(x)),$$

$$\overline{\psi}(x) = \sum_s (a_s^+ \overline{\psi_s^{(+)}}(x) + b_s \overline{\psi_s^{(-)}}(x)), \tag{18.159}$$

that is, operators a_s and $b_{s'}$ we postulate the following commutation relations given in the form of anticommutators:

$$\{a_s, a_{s'}^+\} \equiv a_s a_{s'}^+ + a_{s'}^+ a_s = \delta_{ss'}, \quad \{a_s, a_{s'}\} = \{a_s^+, a_{s'}^+\} = 0 \tag{18.160}$$

and similarly for operators b and b⁺:

$$\{b_s, b_{s'}^+\} = \delta_{ss'}, \text{etc.,} \tag{18.161}$$

with

$$\{a_s, b_{s'}\} = \{a_s^+, b_{s'}\} = \{a_s, b_{s'}^+\} = 0. \tag{18.162}$$

Now let us define the operators

$$N_s^{(+)} = a_s^+ a_s, \quad N_s^{(-)} = b_s^+ b_s. \tag{18.163}$$

By virtue of the commutation relations (18.160),

$$(N_s^{(+)})^2 = a_s^+ a_s a_s^+ a_s = -(a_s^+)^2 (a_s)^2 + a_s^+ a_s = N_s^{(+)}, \tag{18.164}$$

that is, the operator $N_s^{(+)}$ has only two eigenvalues, $N_s^{(+)} = 0, 1$, and similarly, $N_s^{(-)} = 0, 1$.

When describing the states of the Dirac field in the occupation number space, that is, using the language of particles, it is natural to assume that $N_s^{(+)}$ and $N_s^{(-)}$ are the electron and positron occupation number operators, under the assumption that $\psi_s^{(+)}$ and $\psi_s^{(-)}$ describe the electron and the positron, respectively. Then the conditions $N_s^{(\pm)} = 0, 1$ express the well-known *Pauli exclusion principle*, which limits the occupation numbers for fermion states to values not exceeding unity. If we use the commutation relations (18.160) and (18.161), we can reduce the energy and charge operators Eq. (18.158) to

$$H = \sum_s (\varepsilon_s^{(+)} N_s^{(+)} + \varepsilon_s^{(-)} N_s^{(-)}) - \sum_s \varepsilon_s^{(-)},$$

$$Q = e \sum_s (N_s^{(+)} - N_s^{(-)}) - e \sum_{s,(-)} 1, \tag{18.165}$$

where (–) under the summation sign means that we must sum only over the negative-frequency states. If we drop the *c*-number terms in Eq. (18.165), we arrive at the following formulas for the energy and charge values of the electron-positron field:

$$E = \sum_s (\varepsilon_s^{(+)} N_s^{(+)} + \varepsilon_s^{(-)} N_s^{(-)}), \quad Q = e \sum_s (N_s^{(+)} - N_s^{(-)}), \tag{18.166}$$

which have a definite physical meaning, with the energy positive definite and the charge consisting of electron and positron terms with opposite signs. The exclusion of c-number divergent terms is achieved by an appropriate redefinition of the operators H and Q.

Using the commutation relations for a_s and b_s, we can obtain the following matrix elements between states with a single electron $|1_s^{(+)}\rangle$ or positron $|1_s^{(-)}\rangle$ and states in which there is no electron $|0_s^{(+)}\rangle$ or positron $|0_s^{(-)}\rangle$:

$$\langle 1_s^{(+)} | a_s^+ | 0_s^{(+)} \rangle = 1, \quad \langle 0_s^{(+)} | a_s | 1_s^{(+)} \rangle = 1,$$

$$\langle 0_s^{(+)} | a_s^+ | 1_s^{(+)} \rangle = 0, \quad \langle 1_s^{(+)} | a_s | 0_s^{(+)} \rangle = 0, \tag{18.167}$$

and similarly for b_s and b_s^+. Indeed, for vacuum states $|0\rangle$ we have $a|0\rangle = 0$, whereby $\langle 1|a|0\rangle = \langle 0|a^+|1\rangle = 0$ (for the sake of brevity the subscripts are dropped). Moreover, $a_s^+ a_s^+ = 0$, whereby

$$0 = \langle 1|a^+ a^+|1\rangle = \langle 1|a^+|0\rangle \langle 0|a^+|1\rangle$$

$$+ \langle 1|a^+|1\rangle \langle 1|a^+|1\rangle = (\langle 1|a^+|1\rangle)^2,$$

whence $\langle 1|a^+|1\rangle = \langle 1|a|1\rangle = 0$, and similarly

$$\langle 0|a^+|0\rangle = \langle 0|a|0\rangle = 0.$$

Finally,

$$\langle 1|a^+ a|1\rangle = \langle 1|a^+|0\rangle \langle 0|a|1\rangle + \langle 1|a^+|1\rangle \langle 1|a|1\rangle$$

$$= \langle 0|a|1\rangle^2 = 1,$$

whence $\langle 0|a|1\rangle = 1$, etc.

Thus, a_s^+ is the electron creation operator and b_s^+ is the positron creation operator at state s, while a_s and b_s are the electron and positron annihilation operators:

$$a_s^+|0\rangle = |1_s^{(+)}\rangle, \quad b_s^+|0\rangle = |1_s^{(-)}\rangle. \tag{18.168}$$

In general, the *many-particle state* with $N^{(+)}$ electrons and $N^{(-)}$ positrons is obtained through acting $N^{(+)}$ times with operators a_s^+ and $N^{(-)}$ times with operators b_s^+ on the vacuum state. Note, however, that in contrast to multiphoton states, multifermion states are constructed with due regard for the order in which the operators a, b, a^+, and b^+ act (because a and b are anticommutative), which means changing the sign of the state every time the operator cofactors are permuted. For instance, if

$$|1_{s_1}^{(+)}, 0_{s_2}^{(-)}\rangle = a_{s_1}^+|0\rangle,$$

then

$$b_{s_2}^+|1_{s_1}^{(+)}, 0_{s_2}^{(-)}\rangle = -a_{s_1}^+ b_{s_2}^+|0\rangle = -|1_{s_1}^{(+)}, 1_{s_2}^{(-)}\rangle.$$

The field operators $\bar{\psi}(x)$ (18.159) unite the electron creation operator and the positron annihilation operator:

$$\langle 1_s^{(+)} | \bar{\psi}(x) | 0_s^{(+)} \rangle = \overline{\psi_s^{(+)}}(x), \quad \langle 0_s^{(-)} | \bar{\psi}(x) | 1_s^{(-)} \rangle = \overline{\psi_s^{(-)}}(x). \tag{18.169}$$

On the other hand, the field operators $\bar{\psi}(x)$ unite the positron creation operator and the electron annihilation operator:

$$\langle 0_s^{(+)}|\psi(x)|1_s^{(+)}\rangle = \psi_s^{(+)}(x), \quad \langle 1_s^{(-)}|\psi(x)|0_s^{(-)}\rangle = \psi_s^{(-)}(x). \tag{18.170}$$

Finally, let us write out explicitly the expansion of the field Dirac operators in the operators of creation and annihilation of electrons and positrons in free states:

$$\psi(x) = \sum_{\mathbf{p},\sigma} \frac{1}{\sqrt{2\varepsilon_\mathbf{p}V}} (a_{\mathbf{p}\sigma}\, u(p,\sigma)\, e^{-ipx} + b_{\mathbf{p}\sigma}^+ u(-p,\sigma)\, e^{ipx}),$$

$$\bar{\psi}(x) = \sum_{\mathbf{p},\sigma} \frac{1}{\sqrt{2\varepsilon_\mathbf{p}V}} (a_{\mathbf{p}\sigma}^+ \bar{u}(p,\sigma)\, e^{ipx} + b_{\mathbf{p}\sigma}\, \bar{u}(-p,\sigma)\, e^{-ipx}). \tag{18.171}$$

But why for the charged Dirac field must we introduce two pairs of operators, a, a$^+$ and b, b$^+$, while the electromagnetic field requires only one pair, say, c$^+$ c? The reason for this is that the bispinors $u(p)$ and $u(-p)$ satisfy different equations, while the Fourier components of the electromagnetic field $A_k^{(+)}$ and $A_k^{(-)}$ satisfy one equation. In other words, two different particles correspond to the Dirac field, the electron and the positron, while only one particle corresponds to the electromagnetic field, the photon. We note, finally, that quantization of the Dirac field via anticommutators and of the electromagnetic field via commutators ensures the positive definiteness of the field energy in both cases. At the same time, the difference in quantization of the electron-positron (fermion) field and the electromagnetic (boson) field leads to different types of statistics: in the first case this is the *Fermi-Dirac statistics* and its basis, Pauli's exclusion principle, while in the second case this is the *Bose-Einstein statistics*[1], with arbitrary occupation numbers. Pauli's theorem, which relates the spin (integral spin for bosons and half-integral for fermions) and statistics, manifests itself through the above-noted facts.

18.6 THE COMMUTATION AND CAUSAL FUNCTIONS OF THE DIRAC FIELD

The goal of this section is, first, to show that canonical quantization leads to the same rules of quantization of the Dirac field as the ones arrived at in the previous section and, second, to introduce the normal and chronologically ordered products. We then discuss the vacuum averages of quadratic combinations of the Dirac field operators.

18.6.1 Canonical Quantization

Let us write the Dirac field operators as

$$\psi(\mathbf{r}, t) = \sum_s (a_s(t)\, \psi_s^{(+)}(\mathbf{r}) + b_s^+(t)\, \psi_s^{(-)}(\mathbf{r})), \tag{18.172}$$

---

1. *Bose-Einstein (BE) Statistics* gives the number of photons, with an energy E at a temperature T:

$$n(E) = \frac{1}{\exp(E/k_B T) - 1}.$$

In general, it is relevant to particles, which are indistinguishable and occupy definite quantum states, i.e., *bosons*.

where $a_s(t)$ and $b_s^+(t)$ are the Heisenberg (time dependent) operators representing the coefficients in the expansion in positive-frequency and negative-frequency Dirac wave functions. For the energy operator we take the definition [Eq. (18.158)]:

$$H = \sum_s (\varepsilon_s^{(+)} a_s^+ a_s - \varepsilon_s^{(-)} b_s b_s^+). \tag{18.173}$$

Let us write the equation of motion for the Heisenberg operators $a_s(t)$ and $b_s(t)$ in form Eq. (18.57):

$$\frac{da_s(t)}{dt} = i [H, a_s(t)], \quad \frac{db_s(t)}{dt} = i [H, b_s(t)]. \tag{18.174}$$

We will consider the first equation in greater detail:

$$\frac{da_s(t)}{dt} = \sum_{s'} (i\varepsilon_{s'}^{(+)} [a_{s'}^+ a_{s'}, a_s] - i\varepsilon_{s'}^{(-)} [b_{s'} b_{s'}^+, a_s]). \tag{18.175}$$

To transform the commutators on the right-hand side of this equation we employ the following identity, which holds true for arbitrary operators A, B and C:

$$[AB, C] = A \{B, C\} - \{A, C\} B. \tag{18.176}$$

As a result, one obtains

$$\frac{da_s(t)}{dt} = i \sum_{s'} \varepsilon_{s'}^{(+)} (a_{s'}^+ \{a_{s'}, a_s\} - \{a_{s'}^+, a_s\} a_{s'})$$

$$- i \sum_{s'} \varepsilon_{s'}^{(-)} (b_{s'} \{b_{s'}^+, a_s\} - \{b_{s'}, a_s\} b_{s'}^+), \tag{18.177}$$

where the anticommutators are designated by braces.

The last operator equation can be linearized if we subject a and b to commutation relations (18.160) and (18.162). The summation signs are eliminated because of the Kronecker deltas* $\delta_{ss'}$, and we find that

$$\frac{da_s}{dt} = -i\varepsilon_s^{(+)} a_s. \tag{18.178}$$

This equation has positive-frequency solutions,

$$a_s(t) = a_s(0) e^{-i\varepsilon_s^{(+)} t}. \tag{18.179}$$

In the same manner, for b_s^+ we can write the equation

$$\frac{db_s^+}{dt} = i \sum_{s'} \varepsilon_{s'}^{(+)} (a_{s'}^+ \{a_{s'}, b_s^+\} - \{a_{s'}^+, b_s^+\} a_{s'})$$

$$- i \sum_{s'} \varepsilon_{s'}^{(-)} (b_{s'} \{b_{s'}^+, b_s^+\} - \{b_{s'}, b_s^+\} b_{s'}^+), \tag{18.180}$$

---

* Kronecker delta function $\delta_{ss'}$ has the value 1 when $s = s'$ and 0 otherwise. It is also known as Kronecker symbol.

which is linearized via the quantization conditions Eqs. (18.161) and (18.162):

$$\frac{db_s^+}{dt} = -i\varepsilon_s^{(-)} b_s^+.$$

(18.181)

As expected, the solutions to this equation are negative-frequency:

$$b_s^+(t) = b_s^+(0)\, e^{i\varepsilon_s^{(-)}t}.$$

(18.182)

Substituting Eq. (18.179) and Eq. (18.182) into Eq. (18.172), we find the earlier expansions, (18.159) and (18.171).

Thus the quantization rules fixed by the anticommutation relations [Eqs. (18.161)-(18.163)] do, indeed, ensure that the field operators are decomposed into positive-frequency and negative-frequency parts, Eqs. (18.172), (18.179), and (18.182). These rules, as the previous section has shown, also ensure that the *Dirac field energy* is positive definite. If instead of Eq. (18.176) we were to use Eq. (18.60) and fix the quantization rules via commutators, then for b_s^+ we would obtain a positive frequency solution instead of a negative-frequency one. Another consequence of this would be the indefiniteness of the sign of the field energy.

18.6.2 The Commutation Function

Let us consider the commutation relations for the field operators $\psi(x)$ and $\bar\psi(x)$ Bearing in mind the expansion [Eq. (18.160)] and the commutation relations [Eqs. (18.161–18.163)] for a and b, we find that

$$\{\psi(x), \psi(x')\} = 0, \quad \{\bar\psi(x), \bar\psi(x')\} = 0.$$

(18.183)

At the same time,

$$\{\psi(x), \bar\psi(x')\} = \sum_s \psi_s^{(+)}(x)\, \overline{\psi_s^{(+)}}(x') + \sum_s \psi_s^{(-)}(x)\, \overline{\psi_s^{(-)}}(x')$$

$$= S^{(+)}(x, x') + S^{(-)}(x, x').$$

(18.184)

For free electrons, using expansion [Eq. (18.171)], one obtains

$$S^{(+)}(x - x') = \sum_{\mathbf{p},\sigma} \psi_{\mathbf{p}\sigma}^{(+)}(x)\, \overline{\psi_{\mathbf{p}\sigma}^{(+)}}(x')$$

$$= \sum_{\mathbf{p},\sigma} \frac{1}{2\varepsilon_{\mathbf{p}}V}\, u(\sigma, p)\, \bar u(\sigma, p)\, e^{-ip(x - x')}$$

(18.185)

$$S^{(-)}(x - x') = \sum_{\mathbf{p},\sigma} \psi_{\mathbf{p},\sigma}^{(-)}(x)\, \overline{\psi_{\mathbf{p},\sigma}^{(-)}}(x')$$

$$= \sum_{\mathbf{p},\sigma} \frac{1}{2\varepsilon_{\mathbf{p}}V}\, u(\sigma, -p)\, \bar u(\sigma, -p)\, e^{ip(x - x')}$$

(18.186)

In these equations the products of bispinors form a matrix. For instance, writing out explicitly the spinor indices, we obtain

$$S_{\alpha\beta}^{(+)}(x - x') = \sum_{\mathbf{p},\sigma} \frac{1}{2\varepsilon_{\mathbf{p}}V}\, u_\alpha(\sigma, p)\, \bar u_\beta(\sigma, p)\, e^{-ip(x - x')}$$

By direct calculation we can establish the following equations involving the bispinors $u(\sigma, p)$ and $u(\sigma, -p)$ normalized by Eq. (18.145):

$$\sum_\sigma u(\sigma, p)\,\bar{u}(\sigma, p) = \gamma p + m \tag{18.187}$$

$$\sum_\sigma u(\sigma, -p)\,\bar{u}(\sigma, -p) = \gamma p - m \tag{18.188}$$

These can be verified in the following manner. The right-hand side of the first equation contains only the unit matrix and the γ-matrices and, being a scalar quantity, must have the form $A\gamma p + Bm$. The left-hand side vanishes if the operator $\gamma p - m$ acts on it, so in the right-hand side A must be equal to B. Finally, in the rest frame of reference, $p = 0$ and $p^0 = m$, we have (see Eq. 18.139) and Eq. (18.146):

$$u\left(\frac{1}{2}\right) = \sqrt{2m}\begin{pmatrix} 1 \\ 0 \\ 0 \\ 0 \end{pmatrix}, \quad u\left(-\frac{1}{2}\right) = \sqrt{2m}\begin{pmatrix} 0 \\ 1 \\ 0 \\ 0 \end{pmatrix},$$

whence

$$\sum_{\sigma = 1/2, -1/2} u(\sigma, p)\,\bar{u}(\sigma, p)\big|_{\mathbf{p}=0} = 2m\begin{pmatrix} I & 0 \\ 0 & 0 \end{pmatrix},$$

with

$$\gamma p + m = 2m\begin{pmatrix} I & 0 \\ 0 & 0 \end{pmatrix}.$$

Consequently, $A = 1$ and the validity of Eq. (18.187) has been proved. In a similar manner we can prove Eq. (18.188).

Thus, for the positive- and negative-frequency parts of the commutation function

$$S(x - x') = \{\psi(x), \bar{\psi}(x')\}, \tag{18.189}$$

we get

$$S^{(+)}(x - x') = \sum_{\mathbf{p}} \frac{\gamma p + m}{2\varepsilon_{\mathbf{p}} V} e^{-ip(x - x')},$$

$$S^{(-)}(x - x') = \sum_{\mathbf{p}} \frac{\gamma p - m}{2\varepsilon_{\mathbf{p}} V} e^{ip(x - x')}. \tag{18.190}$$

Let us recall the definitions [Eqs. (18.106) and (18.108)] of the functions $D_+(x)$ and $D_-(x)$ for the electromagnetic field and define similar functions for the spinor field:

$$\Delta_\pm(x) = \frac{1}{(2\pi)^3}\int \frac{d^3p}{2\varepsilon_{\mathbf{p}}} e^{\mp ipx}. \tag{18.191}$$

Then, going over from sums to integrals in Eq. (18.190), we find that

$$S^{(+)}(x - x') = (i\gamma\partial + m)\Delta_+(x - x'),$$

$$S^{(-)}(x - x') = -(i\gamma\partial + m)\Delta_-(x - x'). \tag{18.192}$$

The total commutation function

$$S(x) = S^{(+)}(x) + S^{(-)}(x) \tag{18.193}$$

is expressed in terms of the invariant function

$$\Delta_0 = i(\Delta_+ - \Delta_-) \tag{18.194}$$

in the following manner:

$$S(x) = -i(i\gamma\partial + m)\Delta_0(x). \tag{18.195}$$

The function $\Delta_0(x)$ differs from the electromagnetic commutation function $D_0(x)$ in that k is replaced with p in the exponential in Eq. (18.127) and k^2 with $p^2 - m^2$ in the denominator. Just as in the case with D_0, we have

$$\Delta_0(r, t)\big|_{t=0} = 0, \quad \frac{\partial}{\partial t}\Delta(r, t)\big|_{t=0} = \delta(r), \tag{18.196}$$

whence

$$S(r, t)\big|_{t=0} = \gamma^0 \delta(r), \tag{18.197}$$

and therefore

$$\{\psi(r, t), \overline{\psi}(r'\ t')\}_{t=t'} = \gamma^0 \delta(r - r'), \tag{18.198}$$

Function $S(x)$ constitutes a (singular) solution of the Dirac equation

$$(i\gamma\partial - m)\ S(x) = 0, \tag{18.199}$$

since

$$(\square + m^2)\Delta_0(x) = 0. \tag{18.200}$$

18.6.3 The Causal Green Function

Let us introduce the normal and chronologically ordered products of Dirac operators. However, a permutation of operators must result in a change in sign of the product.

The normal product of two operators is defined in such a manner that creation operators stand to the left of annihilation operators. For instance,

$$N\psi(x)\overline{\psi}(x') = N(\psi^{(+)}(x) + \psi^{(-)}(x))(\overline{\psi^{(+)}}(x') + \overline{\psi^{(-)}}(x'))$$

$$= -(\overline{\psi^{(+)}}(x')\psi^{(+)}(x) + \psi^{(+)}(x)\overline{\psi^{(-)}}(x')$$

$$+ \psi^{(-)}(x)\overline{\psi^{(+)}}(x') + \psi^{(-)}(x)\overline{\psi^{(-)}}(x')), \tag{18.201}$$

where we have allowed for the fact that $\psi^{(+)}$ and $\overline{\psi^{(-)}}$ are annihilation operators and $\psi^{(-)}$ and $\overline{\psi^{(+)}}$ are creation operators.

The chronologically ordered product of two Dirac operators $\psi_1(x)$ and $\psi_2(x')$ is defined exactly in the same manner as for electromagnetic operators, with a change in sign under permutation:

$$T\psi_1(x)\psi_2(x') = \begin{cases} \psi_1(x)\psi_2(x'), & t > t', \\ -\psi_2(x')\psi_1(x), & t' > t. \end{cases} \tag{18.202}$$

Let us now define the contraction, or chronological pairing, of two Dirac operators:

$$\overline{\psi_1(x)\psi_2(x')} \equiv T\psi_1(x)\,\psi_2(x') - N\psi_1(x)\,\psi_2(x'). \tag{18.203}$$

One can easily verify that

$$\overline{\psi(x)\psi(x')} = \overline{\overline{\psi}(x)\overline{\psi}(x')} \tag{18.204}$$

(*Hint*: The operators $\psi^{(-)}(x)$ and $\psi^{(+)}(x')$ are anticommutative for all x and x'.)

For $\psi(x)$ and $\overline{\psi}(x')$, the pairing constitutes a c-number. Indeed, let t be greater than t'. Then

$$\overline{\psi(x)\overline{\psi}(x')} = \psi^{(+)}(x)\overline{\psi^{(+)}(x')} + \overline{\psi^{(+)}(x')\psi^{(+)}(x)},$$

or an anticommutator, which is a c-number. A similar result is obtained when t is smaller than t'.

Since the vacuum average of an N-product is equal to zero, the vacuum average of a T-product is equal to the pairing:

$$\langle T\psi(x)\overline{\psi}(x')\rangle_0 = \overline{\psi(x)\overline{\psi}(x')}, \tag{18.205}$$

which implies

$$\overline{\psi_\alpha(x)\overline{\psi}_\beta(x')} = \begin{cases} \langle \psi_\alpha(x)\overline{\psi}_\beta(x')\rangle_0, & t > t', \\ -\langle \overline{\psi}_\beta(x')\psi_\alpha(x)\rangle_0, & t < t, \end{cases} \tag{18.206}$$

where α and β are *spinor** indices.

We have thus arrived at the causal function, which is commonly known as the propagator:

$$S_c(x-x') = \langle T\psi(x)\overline{\psi}(x')\rangle_0 = \overline{\psi(x)\,\overline{\psi}(x')} \tag{18.207}$$

This function describes, for $t > t'$, the creation of an electron at point x' and the annihilation of this electron at point x, while for $t < t'$ it describes the creation of a positron at point x and the annihilation of this positron at point x'.

We can now easily express $S_c(x)$ in terms of the functions $S^{(+)}(x)$ and $S^{(-)}(x)$ introduced in (22.192). Indeed,

$$\langle \psi_\alpha(x)\overline{\psi}_\beta(x')\rangle_0 = S_{\alpha\beta}^{(+)}(x-x')$$

$$\langle \overline{\psi}_\beta(x')\psi_\alpha(x)\rangle_0 = S_{\alpha\beta}^{(-)}(x-x') \tag{18.208}$$

whereby

$$S_c(x-x') = \theta(x-x')\,S^{(+)}(x-x') - \theta(x'-x)\,S^{(-)}(x-x'). \tag{18.209}$$

The derivatives ∂_μ can be taken outside the theta functions, since ∇ is commutative with them, and $\partial_0\theta(x) = \delta(x^0)$ and $\partial_0\theta(-x) = -\delta(x^0)$, with

$$\Delta_+(x-x')\big|_{x_0=x_0} = \Delta_-(x-x')\big|_{x_0=x_0}$$

---

* *Spinor*—a vector with two complex components, which undergoes a unitary unimodular transformation when the three-dimensional coordinate system is rotated. It can represent the spin state of a particle of spin 1/2. More generally, a spinor of order (or rank) n is an object with 2^n components, which transform as products of components of n spinors of rank one.

Thus,

$$S_c(x - x') = (i\gamma\partial_x + m)\, \Delta_c(x - x'), \tag{18.210}$$

where

$$\Delta_c(x - x') = \theta(x - x')\, \Delta_+(x - x') + \theta(x' - x)\, \Delta_-(x - x') \tag{18.211}$$

is the propagator of a massive scalar field; the integral representation of this function is

$$\Delta_c(x) = \frac{1}{(2\pi)^3} \int \frac{d^3 p}{2\varepsilon_p}\, e^{i\mathbf{p}\cdot\mathbf{r} - i\varepsilon_p|t|} = \frac{i}{(2\pi)^4} \int \frac{e^{-ipx}\, d^4 p}{p^2 - m^2 + i\varepsilon},$$

$$\varepsilon p = \sqrt{\mathbf{p}^2 + m^2} \tag{18.212}$$

The validity of this representation can be established in the same manner as we did for the massless electromagnetic field earlier.

The above leads us to the following integral representation of $S_c(x)$:

$$S_c(x) = \frac{i}{(2\pi)^4} \int d^4 p\, e^{-ipx}\, \frac{\gamma p + m}{p^2 - m^2 + i\varepsilon} \tag{18.213}$$

Applying the Dirac operator to $S_c(x)$, one obtains

$$(i\gamma\partial - m)\, S_c(x) = i\delta(x) \tag{18.214}$$

This follows directly from the integral representation (18.213), as well as from (18.210) if we allow for the fact that

$$(\Box + m^2)\, \Delta_c(x) = -i\delta(x) \tag{18.215}$$

Thus, the propagator $S_c(x)$ is the *causal Green function* of the Dirac equation.

18.6.4 Redefining Energy and Current for the Dirac Equation

The definitions of the energy and charge of the Dirac field given in Section 5, to infinitely large negative constant terms in the eigenvalues of the Hamiltonian and the charge operator Eq. (18.165). These terms can be eliminated if both operators are redefined in an appropriate manner. For the Hamiltonian operator the definition is usually given in the form of a normal product, that is, all creation operators stand to the left of the annihilation operators, and $b_s b_s^+ \to - b_s^+ b_s$. Then instead of Eq. (18.158), we have the definition

$$H = \sum_s (\varepsilon_s^{(+)} a_s^+ a_s + \varepsilon_s^{(-)} b_s^+ b_s) \tag{18.216}$$

The current density operator is specified by a commutator instead of the usual definition:

$$j_\mu(x) = \frac{e}{2} [\bar{\psi}(x), \gamma_\mu \psi(x)] = \frac{e}{2} (\bar{\psi}_\alpha (\gamma_\mu)_{\alpha\beta} \psi_\beta - (\gamma_\mu)_{\alpha\beta} \psi_\beta \bar{\psi}_\alpha)$$

$$= \frac{e}{2} (\bar{\psi} \gamma_\mu \psi - \psi \gamma_\mu^T \bar{\psi}) \tag{18.217}$$

Then the charge, too, will rid itself of the infinite term:

$$Q = \frac{1}{2} e \int (\overline{\psi} \gamma^0 \psi - \gamma^0 \psi \overline{\psi}) \, d^3x = e \sum_s (N_s^{(+)} - N_s^{(-)}). \qquad (18.218)$$

Obviously, if the current density operator is defined in terms of a normal product,

$$j_\mu (x) = e N \overline{\psi} (x) \gamma_\mu \psi (x), \qquad (18.219)$$

we arrive at the same expression (18.218) for the charge operator. It can also be shown that not only the charge but the operators $j_\mu (x)$ given by Eqs. (18.217) and (18.219) coincide. We may note, finally, that j_μ can be written in charge-symmetric form. To this end we recall the definition of C-conjugation,

$$\psi^C (x) = C \overline{\psi} (x), \quad \overline{\psi}^C (x) = C^{-1} \psi (x), \qquad (18.220)$$

and the definition of the C-operator: $C = -C^T$ and $C^{-1} \gamma_\mu C = -\gamma T_\mu$. Then from definition Eq. (18.217) follows the charge-symmetric relationship

$$j_\mu (x) = \frac{e}{2} (\overline{\psi} \gamma_\mu \psi - \overline{\psi}^C \gamma_\mu \psi^C). \qquad (18.221)$$

We see that $j_\mu (x)$ is invariant under a C-transformation and a simultaneous change in sign $e \rightarrow - e$.

18.7 ELECTROMAGNETIC INTERACTION

We have so far discussed two main quantized free fields, the electromagnetic and the charged Dirac. The study of the interaction of these fields constitutes the subject of quantum electrodynamics. However, there is no way in which an exact solution to the equations of quantum electrodynamics can be obtained in general form. Nevertheless, a perturbation method has been developed, which, thanks to the smallness of the coupling constant of the electromagnetic interaction, or the fine structure constant $\alpha = e^2 / \hbar c = 1/137$, makes it possible to find an asymptotic solution for practically any interaction between electrons and photons. But before we discuss this method, let us formulate the exact equations of quantum electrodynamics.

18.7.1 The Heisenberg Field Equations

Let us start with two interacting fields, the electron-positron field and the electromagnetic field. To each field we assign an operator in the Heisenberg representation,

$$\psi (x), \quad \overline{\psi} (x), \quad A_\mu (x), \qquad (18.222)$$

which depends explicitly not only on r but on time t as well. For these operators we can write the same equations as for classical interacting fields:

$$(i\gamma \partial - e\gamma A (x) - m) \, \psi (x) = 0, \quad \overline{\psi} (x) (i\gamma \partial + e\gamma A (x) + m) = 0, \qquad (18.223a)$$

$$\Box \, A_\mu (x) = 4\pi j_\mu (x), \qquad (18.223b)$$

where the 4-current density can be written in the charge-symmetric form,

$$j_\mu (x) = \frac{e}{2} [\overline{\psi} (x), \gamma_\mu \psi(x)] = \frac{e}{2} (\overline{\psi} (x) \gamma_\mu \psi (x) - \overline{\psi}^C (x) \gamma_\mu \psi^C (x)). \qquad (18.224)$$

These equations can be obtained by employing the variational principle for the action integral of the *Lagrangian density*. It can be proved, in general form, that Eqs. (18.223) are gauge invariant and that the total 4-momentum of the fields and charges is conserved, which follows from the Lorentz invariance and gauge invariance of the theory. We will not do this here for the general case, but in what follows we will often use these results and one can easily demonstrate their validity using concrete examples.

Equations (18.223a) and (18.223b) for the interacting fields are nonlinear and inhomogeneous. Therefore, we cannot solve them in general form, that is, write out the explicit dependence of the operator functions $\psi, \bar{\psi}$, and A on x, in contrast to the case of free fields.

For the same reason it is impossible to write out the commutation relations for these fields at arbitrary moments of time, since we do not know the time dependence of the fields. However, for coincident times such commutation relations can be written; these are known as equal-time commutation relations. It is expedient to fix these relations in such a manner that they coincide with the commutation relations for free fields:

$$\left[A_\mu(x), \frac{\partial A_\nu}{\partial t}(x') \right]_{t=t'} = -4\pi i g_{\mu\nu} \delta(\mathbf{r} - \mathbf{r}'),$$

$$[A_\mu(x), A_\nu(x')]_{t=t'} = 0, \tag{18.225}$$

$$\{\psi_\alpha(x), \psi_\beta^+(x')\}_{t=t'} = \delta_{\alpha\beta} \delta(\mathbf{r} - \mathbf{r}').$$

All the other commutation relations are equal to zero at $t = t'$:

$$\{\psi, \psi\}_{t=t'} = \{\psi^+, \psi^+\}_{t=t'} = 0,$$

$$[\psi, A]_{t=t'} = \{\psi, \partial A/\partial t'\}_{t=t'} = 0,$$

etc. As for the state vectors $|\Phi\rangle$ for a system of fields, in the Heisenberg representation they are time independent:

$$\frac{\partial}{\partial t} |\Phi\rangle = 0. \tag{18.226}$$

As noted earlier, the *electromagnetic current* is conserved:

$$\partial j_\mu / \partial x^\mu = 0. \tag{18.227}$$

Therefore, calculating the 4-divergence of the right- and left-hand sides of Eq. (18.223b) for the vector potential A_μ, one obtains

$$\frac{\partial}{\partial x_\mu} \Box A_\mu = \Box \frac{\partial A_\mu}{\partial x_\mu} = 4\pi \frac{\partial j_\mu}{\partial x_\mu} = 0,$$

that is, the divergence of A_μ satisfies homogeneous d'Alembert's equation

$$\Box A^\mu_{,\mu} = 0. \tag{18.228}$$

This means that $A^{\mu}_{,\mu}$ can be decomposed into negative- and positive-frequency parts, as it was done for free fields:

$$A^{\mu}_{,\mu} = A^{\mu(+)}_{,\mu} + A^{\mu(-)}_{,\mu}. \tag{18.229}$$

Hence, we can impose the Lorentz condition on the state vectors:

$$A^{\mu(+)}_{,\mu}|\Phi\rangle = 0, \quad \langle\Phi|A^{\mu(-)}_{,\mu} = 0. \tag{18.230}$$

Thus, the wave equation [Eq. (18.223b)] leads to the Maxwell equations in the mean:

$$\left\langle\frac{\partial F_{\mu\nu}}{\partial x_{\nu}}\right\rangle = -4\pi\langle j_{\mu}\rangle. \tag{18.231}$$

Let us write the Hamiltonian operator for the Dirac field:

$$H_D = \int d^3x \frac{i}{2}(\overline{\psi}\gamma^0\psi_{,t} - \overline{\psi}_{,t}\gamma^0\psi) = \int d^3x \frac{i}{2}(\overline{\psi}\gamma^0\psi_{,t} + \overline{\psi}^C\gamma^0\psi^C_{,t}). \tag{18.232}$$

This expression corresponds to the classical definition of H and the charge-symmetric form is obtained in the same manner as we did for the 4-current density. Excluding the time derivatives, that is, going over to canonical variables via Dirac equations,

$$i\gamma^0\psi_{,t} = (e\gamma A + m)\psi - i\gamma\nabla\psi,$$

$$i\gamma^0\psi^C_{,t} = (-e\gamma A + m)\psi^C - i\gamma\nabla\psi^C,$$

(We may note that the charge enters the equation for ψ^C with the opposite sign), we obtain

$$H_D = \int\left[\frac{1}{2}\overline{\psi}(-i\gamma\nabla + m)\psi + \frac{1}{2}\overline{\psi}^C(-i\gamma\nabla + m)\psi^C\right.$$

$$\left. + \frac{e}{2}(\overline{\psi}\gamma^{\mu}\psi - \overline{\psi}^C\gamma^{\mu}\psi^C)A_{\mu}\right]d^3x = H^0_D + H_{int}, \tag{18.233}$$

where

$$H_{int} = \int d^3x\, j_{\mu}A^{\mu} = \frac{e}{2}\int d^3x\,(\overline{\psi}\gamma^{\mu}\psi - \overline{\psi}^C\gamma^{\mu}\psi^C)A_{\mu} \tag{18.234}$$

is the so-called *interaction Hamiltonian*, and H^0_D is the *free Dirac Hamiltonian*.

If we also take into account the free Hamiltonian of the electromagnetic field, H^0_f, we will arrive at the total Hamiltonian for a system of *interacting electron-positron and electromagnetic field*:

$$H = H^0_f + H^0_D + H_{int} = H^0 + H_{int}, \tag{18.235}$$

where H^0 does not depend on the products of components of the various fields, that is, is the free Hamiltonian. Since H_{int} defined by Eq. (18.234) depends on the product of three cofactors corresponding to the charged Dirac and electromagnetic fields, it is called the interaction Hamiltonian. It can easily be verified, this here, that the field equations [Eq. (18.223a) and (18.223b)] are *Heisenberg equations* of motion for the operators $\psi, \bar{\psi}$ and A and may be obtained from the general equation for Heisenberg operators,

$$\frac{d\mathbf{u}(x)}{dt} = i\,[\mathrm{H}, \mathbf{u}(x)], \tag{18.236}$$

in which we must put $\mathbf{u}(x) = \psi(x), \bar{\psi}(x)$, and $\mathbf{A}(x)$, and for H we must take Eq. (18.235). We must also use equal-time commutation relations (18.225).

18.7.2 The interaction Representation (Dirac Picture)

For an approximate solution of the system of equations for interacting fields it often proves convenient to transfer from the Heisenberg representation to the interaction representation, in which the operators obey the free equations. This representation is also known as the *Dirac picture*. In the Schrödinger representation the state vectors obey the Schrödinger equation

$$i\frac{\partial}{\partial t}|\Phi(t)\rangle = \mathrm{H}|\Phi(t)\rangle, \tag{18.237}$$

where $\mathrm{H} = \mathrm{H}_0 + \mathrm{H}_{\mathrm{int}}$, and the operators $\mathbf{u}(r)$ are time independent.

Let us transfer to the interaction representation, assuming that the operators $\mathbf{u}_\mathrm{I}(x)$ and the state vectors $|\Phi_\mathrm{I}(t)\rangle$ in this representation are

$$\mathbf{u}_\mathrm{I}(x) = e^{i\mathrm{H}_0 t}\mathbf{u}(\mathbf{r})\,e^{-i\mathrm{H}_0 t},$$

$$|\Phi_\mathrm{I}(t)\rangle = e^{i\mathrm{H}_0 t}|\Phi(t)\rangle. \tag{18.238}$$

In accord with this definition, the operators $\mathbf{u}_\mathrm{I}(x)$ satisfy the equation

$$\frac{d\mathbf{u}_\mathrm{I}}{dt} = i\,[\mathrm{H}_0, \mathbf{u}_\mathrm{I}], \tag{18.239}$$

while state vectors, by virtue of (18.237), obey the equation

$$i\frac{\partial}{\partial t}|\Phi_\mathrm{I}(t)\rangle = \mathrm{H}_\mathrm{I}(t)|\Phi_\mathrm{I}(t)\rangle, \tag{18.240}$$

where

$$\mathrm{H}_\mathrm{I} = e^{i\mathrm{H}_0 t}\,\mathrm{H}_{\mathrm{int}}\,e^{-i\mathrm{H}_0 t} \tag{18.241}$$

is the interaction operator, that is, the interaction Hamiltonian in the Dirac picture.

We may recall that the Heisenberg and Schrödinger operators and the state vectors are related through the following formulas $(\partial\mathrm{H}/\partial t = 0)$:

$$\mathbf{u}(x) = e^{i\mathrm{H}t}\mathbf{u}(\mathbf{r})\,e^{-i\mathrm{H}t}, \tag{18.242a}$$

$$|\Phi(t)\rangle = e^{-i\mathrm{H}t}|\Phi\rangle, \tag{18.242b}$$

with the Heisenberg operators obeying the equation

$$\frac{d\mathbf{u}}{dt} = i\,[\mathrm{H}, \mathbf{u}], \tag{18.243}$$

and the Heisenberg state vectors being time independent,

$$\frac{\partial}{\partial t}|\Phi\rangle = 0. \tag{18.244}$$

We see that the interaction representation is an intermediate, so to say, between the Schrödinger and Heisenberg representations: the operators satisfy the Heisenberg equation of motion with the free Hamiltonian, while the state vectors obey the Schrödinger equation with the interaction operator as the Hamiltonian.

Equation (18.242b) constitutes a formal solution to the Schrödinger equation for the state vector $|\Phi(t)\rangle$ and can be rewritten thus:

$$|\Phi(t)\rangle = U(t)|\Phi\rangle, \tag{18.245}$$

where

$$U(t) = e^{-iHt} \tag{18.246}$$

is the evolution operator, which connects the state vectors $|\Phi\rangle$ at $t = 0$ and $|\Phi(t)\rangle$ at $t \neq 0$. Obviously, for arbitrary times t' and t'' ($t'' > t'$), we obtain

$$|\Phi(t'')\rangle = U(t'' - t')|\Phi(t')\rangle, \tag{18.247}$$

where

$$U(t'' - t') = e^{-iH(t'' - t')}. \tag{18.248}$$

Formula (18.247) is no more than a formal solution to the Schrödinger equation since we know neither how operator H acts on the state vectors nor the commutation relations involving the operators that constitute H. We will therefore go over in Eq. (18.247) to the interaction representation via Eq. (18.238):

$$|\Phi_I(t'')\rangle = e^{iH_0 t''} U(t'' - t') e^{-iH_0 t'}|\Phi_I(t')\rangle = U_I(t'', t')|\Phi_I(t')\rangle, \tag{18.249}$$

where

$$U(t'' - t') = e^{iH_0 t''} U(t'' - t') e^{-iH_0 t'}. \tag{18.250}$$

is the evolution operator in the interaction representation. This equation is again formal, however. It does not enable us to determine the explicit form of the evolution operator, since it expresses this operator in terms of unknown operators u (r) in the Schrödinger representation, which operators are incorporated in H and H_0. Note, however, that the evolution operator [Eq. (18.250)] satisfies the same equation [Eq. (18.240)] as the state vector in the Dirac picture does,

$$i\frac{\partial}{\partial t} U_I(t, t') = H_I(t) U_I(t, t'). \tag{18.251}$$

Whence, to find the evolution operator in explicit form, we must solve this equation: Note that the operator $H_I(t)$ here depends on time explicitly, and therefore we are unable to solve the nonlinear operator equation exactly, as we did with the Schrödinger equation [Eq. (18.237)] when we had to find operator [Eq. (18.249)]. At the same time, Eq. (18.251) are given in the interaction representation. The explicit form of operator $H_I(t)$, obviously, is the same as H_{int}, only it is expressed in terms of field operators in the interaction representation instead of in terms of Heisenberg fields. The same is true of the free operator H_0 expressed in terms of fields in the interaction representation. Thus, from Eq. (18.239) we obtain the ordinary free equations for the field operators:

$$(i\gamma\partial - m)\,\psi(x) = 0, \quad \overline{\psi}(x)(i\gamma\partial + m) = 0, \quad \Box A_\mu(x) = 0. \tag{18.252}$$

Moreover, the equal-time commutation relations for $\psi, \bar{\psi}, A_\mu$ and $\pi_\mu = -(1/4\pi)$ $\partial A_\mu/\partial t$, obviously, are the same as in the Heisenberg representation, Eq. (18.225). On the basis of the equal-time commutation relations Eq. (18.225) and the solutions to Eqs. (18.252) one can obtain non equal-time $(t \neq t')$ commutation relations that coincide with those for free fields.

Thus, the theory of free fields developed in the preceding sections can be applied to describe the evolution of operators in the Dirac picture. At the same time, the equation of motion for the state vector, Eq. (18.240), or for the evolution operator, Eq. (18.248), contains only the interaction operator and in cases when the interaction is weak can be solved approximately.

Instead of Eq. (18.248), let us take the equivalent integral equation

$$U_I(t, t') = 1 - i \int_{t'}^{t} H_I(t_1) U_I(t_1, t') \, dt_1. \tag{18.253}$$

Successive interactions lead to the following series:

$$U_I(t, t') = 1 - i \int_{t'}^{t} H_I(t_1) \, dt_1 + (-i)^2 \int_{t'}^{t} dt_2 \, H_I(t_2) \int_{t'}^{t_2} dt_1 \, H_I(t_1) +$$
$$... + (-i)^n \int_{t'}^{t} dt_n \int_{t'}^{t_n} dt_{n-1} ... \int_{t'}^{t_2} dt_1 \, H_I(t_n) H_I(t_{n-1}) ... H_I(t_1) + ..., \tag{18.254}$$

where the operators $H_I(t_1)$, $H_I(t_2)$, etc. are not commutative.

In expansions (18.254),

$$t' < t_1 < t_2 < ... < t_{n-1} < t_n < t \tag{18.255}$$

for any $n = 1, 2, 3, ...$. In each term of the series the operators follow in chronological order. We can write the series in compact form thus:

$$U_I(t'', t') = T \exp\left(-i \int_{t'}^{t''} H_I(t) \, dt\right), \tag{18.256}$$

and this series is known as the *T-exponential*. If the interaction is weak, we can take a finite number of terms in series Eq. (18.254) (or "truncate" the series). The matrix elements of the evolution operators in this approximation calculated between free field states are the transition amplitudes in the respective perturbation-theory order. Electrodynamics proved to be the only theory for which such an expansion is justified and provides excellent results. This is due to the relatively simple Abelian structure of electrodynamics and the smallness of the coupling constant, the fine structure constant $\alpha = 1/137$.

18.8 THE S-MATRIX FOR ELECTROMAGNETIC INTERACTION

The solution of the equation of motion for the evolution operator in the form of a *T*-exponential obtained in the previous section makes it possible, at least in principle, to describe electromagnetic processes in any perturbation-theory order in the fine structure constant. This requires, however, a refinement of the statement of the problem, which statement differs considerably from the well-known nonrelativistic perturbation theory used, for example, in scattering theory.

18.8.1 Dyson's Formula

In electrodynamics the interaction operator has the form

$$H_I\,(t) = \int d^3x j_\mu\,(x)\,A^\mu\,(x), \qquad (18.257)$$

where the current and electromagnetic-potential operators are given in the interaction representation, with charge e incorporated in the definition of current Eq. (18.224).

Let us consider a statement of the problem in which the particles are observed only in the asymptotic states, that is, long before and after the interaction. In other words, we observe the states of particles at $t' \to -\infty$ and $t'' \to +\infty$. This means we must calculate the appropriate limit of the evolution operator,

$$S = \lim_{\substack{t' \to -\infty \\ t'' \to +\infty}} e^{iH_0 t''}\, e^{-iH(t''-t')}\, e^{-iH_0 t'} = \lim_{\substack{t' \to -\infty \\ t'' \to +\infty}} T \exp\left(-i \int_{t'}^{t''} H_I(t)\,dt\right), \qquad (18.258)$$

which limit is known as the *S-matrix*. We must also assume that the interaction is switched off in some manner in these asymptotic states, that is,

$$\lim_{t \to \pm\infty} H_I\,(t) = 0. \qquad (18.259)$$

Let us assume, at the same time, that the sets of states of the system of particles as $t \to +\infty$ and $t \to -\infty$, that is, the vectors of asymptotic states in the interaction representation,

$$|\Phi_s\,(+\infty)\rangle = |\Phi_s^{(+)}\rangle \text{ and } |\Phi_s\,(-\infty)\rangle = |\Phi_s^{(-)}\rangle,$$

coincide with the eigenvectors of the free Hamiltonian.

The described situation seems, at least at first glance, to be completely equivalent to the problem of potential scattering in nonrelativistic quantum mechanics. This is not the case, however. The fact is that even for asymptotic states with $t \to \pm\infty$ we cannot include the interaction completely. Although real particles prove to be far apart in such states, so far that their interaction with each other can be ignored, there still remains the interaction with the so-called vacuum, that is, with virtual, or to-be-created, particles. This last interaction leads, for instance, to a change in the electron mass and to renormalization of the electron charge, just like an electric charge of a particle placed in a medium is screened thanks to the polarization of the medium.

To allow for these effects it is common particles to assume that at the beginning the interaction is switched on very slowly, so that the noninteracting particles, which possess bare masses and charges, are transformed into the "dressed", real, particles that are in the same states $|\Phi_s\rangle$ with the same quantum numbers as the initial "bare" particles but possess physically observable masses and charges. This assumption is known as the *adiabatic hypothesis*. The dressed particles then scatter because of the interaction or change their state in some other manner. For the time being we will not touch on the subject of how particles acquire physical charges and masses. In accord with what we have said above, let us assume that the system of interacting fields as $t \to -\infty$ is in a certain initial stationary state described by an eigenvector

$|\Phi^{(-)}\rangle = |\Phi_i\rangle$ of the free Hamiltonian H_0 with physically meaningful masses and charges. The final state

$$|\Phi^{(+)}\rangle = \sum_f c_f |\Phi_f\rangle = S|\Phi^{(-)}\rangle \qquad (18.260)$$

is a series expansion in the initial set of eigenvectors of H_0 with physically meaningful masses and charges. The expansion coefficients are the probability amplitudes of the transition of the system from state $|\Phi_i\rangle$ to one of the possible states $|\Phi_f\rangle$:

$$c_f = \langle f|S|i\rangle = S_{fi}. \qquad (18.261)$$

We see that these amplitudes coincide with the matrix elements of the S-matrix: $S_{fi} = \langle f|S|i\rangle$. According to the solution of Eq. (18.254) for the evolution operator, we arrive at the following expansion for the S-matrix:

$$S = \sum_{n=0}^{\infty} S_n, S_0 = 1 \qquad (n = 0)$$

$$S_n = (-i)^n \int_{-\infty}^{\infty} dt_1 \int_{-\infty}^{t_1} dt_2 \dots \int_{-\infty}^{t_{n-1}} dt_n \, H_I(t_1) H_I(t_2) \dots H_I(t_n) \quad (n \neq 0). \quad (18.262)$$

In n-dimensional space, the domain of integration for the nth term of the series is restricted by the conditions

$$-\infty < t_n \leq t_{n-1} \leq \dots \leq t_2 \leq t_1 < \infty, \qquad (18.263)$$

where in the integrands the interaction operators are positioned from right to left in the order of increasing time, that is, in chronological order.

Let us take the term in the series with $n = 2$. It is clear that the result will be the same if we redesignate the integration variables, $t_2 \to t_1, t_1 \to t_2$:

$$\int_{-\infty}^{\infty} dt_1 \int_{-\infty}^{t_1} dt_2 \, H_I(t_1) H_I(t_2) = \int_{-\infty}^{\infty} dt_2 \int_{-\infty}^{t_2} dt_1 \, H_I(t_2) H_I(t_1). \qquad (18.264)$$

Adding the right term to the left, we get

$$\int_{-\infty}^{\infty} dt_1 \int_{-\infty}^{t_1} dt_2 \, H_I(t_1) H_I(t_2) + \int_{-\infty}^{\infty} dt_2 \int_{-\infty}^{t_2} dt_1 \, H_I(t_2) H_I(t_1).$$

$$= \int_{-\infty}^{\infty} dt_1 \int_{-\infty}^{t_2} dt_2 \, T H_I(t_1) H_I(t_2), \qquad (18.265)$$

or, from Eq. (18.264),

$$\int_{-\infty}^{\infty} dt_1 \int_{-\infty}^{t_1} dt_2 \, H_I(t_1) H_I(t_2) = \frac{1}{2} \int_{-\infty}^{\infty} dt_1 \int_{-\infty}^{\infty} dt_2 \, T H_I(t_1) H_I(t_2). \qquad (18.266)$$

We can rewrite any other term in Eq. (18.262) in exactly the same manner by carrying out a permutation of variables $t_1, \dots, t_n \to t_{i_1}, t_{i_2}, \dots, t_{i_n}$. Adding all such integrals, which are equal to each other, we get

$$(-i)^n \int_{-\infty}^{\infty} dt_1 \int_{-\infty}^{\infty} dt_2 \dots \int_{-\infty}^{\infty} dt_n \, T H_I(t_1) H_I(t_2) \dots H_I(t_n). \qquad (18.267)$$

Since the number of such permutations is $n!$, we can find the value of S_n by dividing (18.267) by $n!$:

$$S_n = \frac{(-i)^n}{n!} \int_{-\infty}^{\infty} dt_1 \int_{-\infty}^{\infty} dt_2 \dots \int_{-\infty}^{\infty} dt_n \, T H_I(t_1) H_I(t_2) \dots H_I(t_n). \qquad (18.268)$$

The S-matrix expansion (18.262) in which the separate terms are defined via (18.268) can be symbolically represented in the form of a T-exponential (Dyson's formula):

$$S = T \exp\left(-i \int_{-\infty}^{\infty} dt \, H_I(t)\right), \qquad (18.269)$$

where the electromagnetic-interaction operator is

$$H_I(t) = \int d^3x j_\mu(x) A^\mu(x) = e \int d^3x N\overline{\psi}(x) \gamma_\mu A^\mu(x) \psi(x), \qquad (18.270)$$

the field operators are taken in the *interaction representation*, and the electromagnetic current is defined via the normal product.

18.8.2 Wick's Theorem

The probabilities of transitions induced by an interaction are expressed in terms of elements of the S-matrix between free physical states, [Eq. (18.261)]. The representation Eq. (18.268) of the terms of the S-matrix is inconvenient for calculating these matrix elements.

Indeed, for real calculations we arrive at quantities of the form

$$\langle f | T u_1 u_2 \dots u_n | i \rangle, \qquad (18.271)$$

where u_k $(k = 1, 2, \dots, n)$ are field operators, and the initial and final states,

$$|i\rangle = A_I^{(-)}(-\infty)|0\rangle, \quad \langle f| = \langle 0| B_I^{(+)}(\infty) \qquad (18.272)$$

are expressed in terms of creation and annihilation operators in the Dirac picture at $t = -\infty$ and $t = +\infty$, respectively, that is, $A_I^{(-)}(-\infty)$ and $B_I^{(+)}(\infty)$, as acting on the physical vacuum $|0\rangle$. For this reason, it would be more convenient if we could represent S_n in such a form that instead of the T-product the operators would be under the sign of an N-product, that is, all creation operators would be placed to the left of all annihilation operators. Then, if the number of annihilation operators for particles of a definite type in the N-product is equal to the number of the corresponding creation operators in $A_I(-\infty)$ we finally obtain, by interchanging them a necessary number of times and applying them to the vacuum state, a c-number as the factor of the vacuum state $|0\rangle$. A similar procedure with the $\langle f|$ state will lead to a c-number as the factor of $\langle 0|$. These factors, obviously, consist of various commutation functions of the field operators. As a result we get rid of all the operators and, in the end, arrive at the sought transition amplitude.

The procedure described above requires first of all the reduction of the S-matrix to normal form. This is carried out via Wick's theorem, which states that any chronologically ordered product is equal to the sum of all possible normal products that can be formed with all possible pairings.

Symbolically, the essence of the theorem can be formulated by the following equation:

$$T u_1 u_2 \ldots u_n = N u_1 u_2 \ldots u_n + \sum N \overline{u_1 u_2} \ldots u_n + \sum \overline{u_1} \overline{u_2} \overline{u_3} \ldots u_n, \qquad (18.273)$$

where the summation signs stand for summation over all possible pairings of a definite number of pairs of operators (only two operators are paired in the second term, while in the last term the maximum number of operators are paired), with the operators, taken in the interaction representation, being the free field operators.

For $n = 2$, Wick's theorem has actually been proved, since according to the definition of chronological pairing, or contraction, of two operators one obtains

$$T u_1 u_2 = N u_1 u_2 + \overline{u_1 u_2}. \qquad (18.274)$$

Suppose that Wick's theorem is valid for an arbitrary n. Let us prove that it is valid for an $(n + 1)$st operator under the sign of the T-product. Since all the u_k can be decomposed into creation and annihilation operators, we assume that the u_k are either creation or annihilation operators. Let us now rearrange the operators on the right and left of Eq. (18.273) so that they stand in chronological order. The sign factors may change if we are dealing with femion operators, but this change will be the same on both sides and Eq. (18.273) will not be violated. Thus, let us assume that

$$t_1 > t_2 > \ldots > t_{n'} \qquad (18.275)$$

whereby

$$T u_1 u_2 \ldots u_n = u_1 u_2 \ldots u_n. \qquad (18.276)$$

Suppose u_{n+1} is a creation operator, that is, $u_{n+1} = u_{n+1}^{(-)}$, with $t_n > t_{n+1}$. Then

$$T u_1 \ldots u_n u_{n+1}^{(-)} = T (u_1 \ldots u_n) u_{n+1}^{(-)}. \qquad (18.277)$$

For the T-product on the right-hand side we have expansion Eq. (18.273). We will interchange $u_{n+1}^{(-)}$ with all the $u_k^{(+)}$ moving to the left until we meet $u_k^{(-)}$ operators. As a result of such permutations the following pairings occur:

$$u_k^{(+)} u_{n+1}^{(-)} = u_{n+1}^{(-)} u_k^{(+)} + \overline{u_k^{(+)} u_{n+1}^{(-)}}, \qquad (18.278)$$

since the operators were chronologically ordered, that is, $t_k > t_{n+1}$. The operator $u_{n+1}^{(-)}$ is commutative (anticommutative) with all the $u_k^{(-)}$. As a result we obtain

$$\pm u_{n+1}^{(-)} T u_1 u_2 \ldots u_n, \qquad (18.279)$$

and also the sum of terms with various pairings of $u_{n+1}^{(-)}$ and $u_k^{(+)}$ of the form

$$\pm \overline{u_k^{(+)} u_{n+1}^{(-)}} \, T u_1 u_2 \ldots u_{k-1} u_{k+1} \ldots u_n \qquad (18.280)$$

The sign in front of the terms in Eqs. (18.279) and (18.280) corresponds to the parity of the number of permutations of fermion operators. Interchanging the order of these fermion operators under the sign of the N-product so as to restore chronological order, we obtain a "plus" sign.

Wick's theorem is valid for the T-product of the product of $n - 1$ operators in Eq. (18.280) because it is valid for n operator cofactors. The remaining pairings, that is, those that are not present in Eq. (18.280), are all zero,

$$\overline{u_k^{(-)} u_{n+1}^{(-)}} = 0,$$

since operators that are either both positive-frequency or both negative-frequency are commutative, or, in the case of fermions, are anticommutative. Thus, we can assume that the expansion contains all possible pairings, which means that Wick's theorem has been proved for the case at hand.

If time t_{n+1} is not the earliest possible, as it was assumed, but lies somewhere between t_1 and t_n, say, $t_k > t_{n+1} > t_{k+1}$, all that we have just proved concerning operators $u_1, u_2, ..., u_k$ remains valid, while the pairings of $u_{n+1}^{(-)}$ with the "earlier" operators, say, u_{k+l} are zero, since

$$\overline{u_{n+1}^{(-)} u_{k+l}^{(-)}} \equiv 0, \quad \overline{u_{n+1}^{(-)} u_{k+l}^{(+)}} = 0 \qquad (t_{n+1} > t_{k+l}).$$

Similar reasoning can be carried out for the case where $u_{n+1} = u_{n+1}^{(+)}$. Thus, Wick's theorem has been proved for the general case as well.

Note that if under the T-product sign there are N-products (a mixed T-product), the operators under the N-product sign would not be rearranged and there would be no pairings between them in the expansion. If we bear in mind this remark, Wick's theorem is valid for mixed T-products, too.

18.9 RADIATION EMITTED BY A CLASSICAL CURRENT

There are only a few problems in quantum electrodynamics for which exact solutions can be found. These are model problems in which certain conditions or restrictions that considerably simplify the problem are made. One of such problems deals with finding the quantized electromagnetic field generated by an external classical source, the 4-current density $j_\mu(x)$, which constitutes a known function of position and time.

9.1 The Electromagnetic Field of a Classical Source

We will obtain the operator that represents the interaction of the classical source with the quantum field from the general formula [Eq. (18.257)] in which the 4-current density operator is replaced with the classical function

$$H_I = \int d^3 x \, j_\mu(x) \, A^\mu(x). \tag{18.281}$$

In essence, the term "interaction" does not reflect the true picture, since here we are speaking most likely of how a given source acts on a dynamical system, the quantized electromagnetic field, without reaction from that system, that is, the system splits and ceases to be self-consistent. The S-matrix in this case will be simpler than Eq. (18.268), since all classical currents j_μ can be taken outside the T-product sign:

$$S = \sum_{n=0}^{\infty} \frac{(-i)^n}{n!} \int d^4 x_1 ... d^4 x_n j_\mu(x_1) ... j_\nu(x_n) \, T A^\mu(x_1) ... A^\nu(x_n). \tag{18.282}$$

For chronological pairing, we have

$$\overline{A_\mu(x_1)\,A_\nu(x_2)} = -4\pi g_{\mu\nu}\,D_c(x_1 - x_2) = -i D_{\mu\nu}(x_1 - x_2), \qquad (18.283)$$

where instead of the causal propagator $D_c(x_1 - x_2)$ we have introduced the function

$$D_{\mu\nu}(x) = g_{\mu\nu}\,D(x), \quad D(x) = -4\pi i D_c(x). \qquad (18.284)$$

Now, using Wick's theorem, we write out the T-product explicitly:

$$T A(1)\,...\,A(n) = \sum (-i)^k\,D(12)\,...\,D(2k-1,2k)\,A^{(-)}(2k+1)$$

$$...\,A^{(-)}(2k+r)\,A^{(+)}(2k+r+1)\,...\,A^{(+)}(n). \qquad (18.285)$$

For the sake of simplicity we did not write out the vector indices and instead of the arguments x_k we simply wrote their numbers, say, $D(12) \equiv D(x_1 - x_2)$. Summation is carried out over all values of k and r that satisfy the condition $2k + r \le n$. For example, for $n = 2$ with $t_1 > t_2$, we obtain

$$T A(1)\,A(2) = A^{(+)}(1)\,A^{(+)}(2) + A^{(-)}(1)\,A^{(+)}(2) + A^{(+)}(1)\,A^{(-)}(2)$$

$$+ A^{(-)}(1)\,A^{(-)}(2) = A^{(+)}(1)\,A^{(+)}(2) + A^{(-)}(1)\,A^{(+)}(2)$$

$$+ A^{(-)}(1)\,A^{(-)}(2) + A^{(-)}(2)\,A^{(+)}(1) + \overline{A^{(+)}(1)\,A^{(-)}(2)}. \qquad (18.286)$$

For the first four terms we have $k = 0$, $r = 0, 1, 1, 2$, while for the last term we have $k = 1, r = 0$.

Upon integrating the S-matrix expansion (18.281) at fixed k and r, we arrive at identical terms that differ only in the permutation of the arguments. In all there are $n!$ permutations some of which are not admissible, namely, 2^k permutations of the arguments of the D-functions, $k!$ permutations of the D-functions, $r!$ permutations of the $A^{(-)}$ operators, and $(n - 2k - r)!$ permutations of the $A^{(+)}$ operators prove to be excluded. The example of Eq. (18.287) illustrates this combinatorics fairly well. Thus, in all there are

$$\frac{n!}{2^k\,k!\,r!\,(n-2k-r)!}$$

identical terms, and the S-matrix expansion assumes the form

$$S = \sum_{n=0}^{\infty} \sum_{k=0}^{[n/2]} \sum_{r=0}^{n-2k} \frac{(-i)^{n+k}}{2^k\,k!\,r!\,(n-2k-r)!} \int d^4x_1\,...\,d^4x_n \prod_{s=1}^{n} j(s)$$

$$\times \prod_{t=1}^{k} D(2t-1,2t) \prod_{u=2k+1}^{2k+r} A^{(-)}(u) \prod_{v=2k+r+1}^{n} A^{(+)}(v). \qquad (12.287)$$

18.9.2 Soft-Photon Emission

Let us calculate the transition amplitude from the vacuum state $|0\rangle$ to the state with N photons, $|N\rangle$, which amplitude corresponds to N-photon emission by a classical current. Then a nonzero contribution will be provided only by those terms in the

S-matrix expansion that contain no photon annihilation operator $A^{(+)}$ but contain N creation operators $A^{(-)}$ (i.e. $r = N = n - 2k$), with the result that

$$\langle N | S | 0 \rangle = \frac{(-i)^N}{N!} \int dx_1 \dots dx_N \langle N | \prod_{i=1}^{N} j(i) A^{(-)}(i) | 0 \rangle$$

$$\times \sum_{k=0}^{\infty} \frac{i^k}{2k!} \int dx_1 \dots dx_{2k}\, j(1) \dots j(2k)\, D(12) \dots D(2k-1, 2k). \tag{18.288}$$

Although we do not write out the vector indices explicitly, it is understood that, for example (see Eq. (18.283)),

$$j(1)\, D(12)\, j(2) = j_\mu(1)\, D^{\mu\nu}(12)\, j_\nu(2) = j_\mu(1)\, D(12)\, j_\mu(2). \tag{18.289}$$

In the expansion in k in (18.288) we can easily see a series expansion for the exponential:

$$\sum_{k=0}^{\infty} \frac{1}{k!} \left[\frac{i}{2} \int d^4x_1\, d^4x_2\, j(1)\, D(12)\, j(2) \right]^k = e^{-\nu/2}, \tag{18.290}$$

where

$$\nu = -i \int d^4x_1\, d^4x_2\, j(1)\, D(12)\, j(2) = \nu_1 + i\nu_2 \tag{18.291}$$

is a complex-valued quantity, with ν_1 and ν_2 the real and imaginary parts of the quantity.

Then instead of Eq. (18.288) we obtain

$$\langle N | S | 0 \rangle = \frac{(-i)^N}{N!} e^{-\nu/2} \int d^4x_1 \dots d^4x_N \langle N | \prod_{i=1}^{N} j(i) A^{(-)}(i) | 0 \rangle \tag{18.292}$$

The probability of N transversal photons with all possible momenta and with polarizations $\lambda = 1, 2$ being created is determined by the square of the modulus of amplitude [Eq. (18.292)]

$$P(N) = \sum_{N_{k\lambda}\,(\lambda=1,2)} |\langle N_{k\lambda} | S | 0 \rangle|^2, \tag{18.293}$$

where summation is carried out over all sets of *occupation numbers* $N_{k\lambda}$ for all values of momenta and transverse polarizations $\lambda = 1, 2$ restricted by the condition

$$\sum_{k} \sum_{\lambda=1,2} N_{k\lambda} = N. \tag{18.294}$$

Substituting the emission amplitude of N real photons, Eq. (18.291), into Eq. (18.292), we arrive at the sum

$$\sum_{N_{k\lambda}\,(\lambda=1,2)} \langle 0 | A^{(+)}(1) \dots A^{(+)}(N) | N_{k\lambda} \rangle \langle N_{k\lambda} | A^{(-)}(N+1) \dots A^{(-)}(2N) | 0 \rangle, \tag{18.295}$$

with condition Eq. (18.294) implied.

Moreover, we note that formula Eq. (18.292) is gauge invariant, that is, permits transformations of polarization vectors (see Eq. (18.88)) of the form

$$e_\mu \to e'_\mu = e_\mu + f k_\mu, \tag{18.296}$$

under which the amplitude does not change. This type of transformation results in the photons acquiring longitudinal and temporal components in the polarization vectors, which are linked through the Lorentz condition [Eq. (18.88)]

$$e_\mu k^\mu = e'_\mu k^\mu = 0. \tag{18.297}$$

Thus, among the states $|N_{k\lambda}\rangle$ we may find states with longitudinal and temporal photons, states that, as we know, cannot be observed and whose number, therefore, is not restricted by condition [Eq.(18.294)].

Thus, summation in Eq. (18.293) can be extended to all kinds of states with arbitrary occupation numbers $N_{k\lambda}$:

$$0 \le \sum_{k\lambda = 1,2,3,0} N_{k\lambda} < \infty. \tag{18.298}$$

The system of such states is complete, that is,

$$\sum_{N_{k\lambda}} |N_{k\lambda}\rangle \langle N_{k\lambda}| = I, \tag{18.299}$$

where summation is extended to all values of $N_{k\lambda}$, (18.298), for all k and $\lambda = 1, 2, 3, 0$.

Thus, using the property of gauge invariance, extending the summation in (18.295) to all $N_{k\lambda}$ ($\lambda = 1, 2, 3, 0$), and employing the completeness condition [Eq. (18.299)], we obtains

$$\sum_{N_{k\lambda}} \langle 0 | A^{(+)}(1) \dots A^{(+)}(N) | N_{k\lambda} | A^{(-)}(N+1) \dots A^{(-)}(2N) | 0 \rangle$$

$$= \langle 0 | \prod_{i=1}^{N} A^{(+)}(i) \prod_{i'=N+1}^{2N} A^{(-)}(i') | 0 \rangle. \tag{18.300}$$

Then probability [Eq. (18.293)] assumes the form

$$P(N) = \frac{e^{-\nu_1}}{(N!)^2} \int dx_1 \dots dx_N \, dx_{N+1} \dots dx_{2N}$$

$$\times \langle 0 | \prod_{i=1}^{N} j^*(i) A^{(+)}(i) \prod_{i'=N+1}^{2N} j(i') A^{(-)}(i') | 0 \rangle. \tag{18.301}$$

Shifting all the operators $A^{(-)}$ to the left and all the operators $A^{(+)}$ to the right and using the property of the vacuum state, namely,

$$A^{(+)} | 0 \rangle = 0, \quad \langle 0 | A^{(-)} = 0,$$

we arrive at the following expression for the commutation function:

$$[A_\mu^{(+)}(x), A_\nu^{(-)}(x')] = -4\pi g_{\mu\nu} D_+(x-x').$$

There are $N!$ such terms. As a result we find that

$$P(N) = \frac{e^{-\nu_1}}{(N!)^2} N! \left(-4\pi \int dx_1 \int dx_2 \, j^*(1) D_+(12) j(2) \right)^N. \tag{18.302}$$

Recalling formula [Eq. (18.106)] for the commutation function D_+, we can evaluate the integrals in Eq. (18.301):

$$-4\pi \int dx_1 \int dx_2 j^*(x_1) D_+ (x_1 - x_2) j(x_2) = \frac{1}{(2\pi)^2} \int \frac{d^3x}{\omega} |j^\mu(k)|^2, \qquad (18.303)$$

where

$$|j^\mu(k)|^2 = |j^1|^2 + |j^2|^2 + |j^3|^2 - |j^0|^2, \qquad (18.304)$$

and

$$j_\mu(k) = \int d^4x\, e^{ikx} j_\mu(x) \qquad (18.305)$$

is the Fourier transform of the *4-current density*.

Let us find the real part of Eq. (18.291),

$$v_1 = \operatorname{Re} v = \operatorname{Re}\left(-i \int dx_1\, dx_2\, j(x_1) D(x_1 - x_2) j(x_2)\right). \qquad (18.306)$$

Using the integral representation Eq. (18.284) of the D-function and the fact that

$$\frac{1}{k^2 + i\varepsilon} = P\,\frac{1}{k^2} - i\pi\delta(k^2)$$

one obtains

$$v_1 = \frac{1}{(2\pi)^2} \int \delta^4 k\, \delta(k^2) |j(k)|^2. \qquad (18.307)$$

Integration with respect to k_0 in the last integral is carried out over the two halves of the light cone $k^2 = 0$, that is, $k^0 > 0$ and $k^0 < 0$, which yields

$$\int \delta^4 k\, \delta(k^2) = \int \frac{d^3k}{\omega} \qquad (\omega = |\mathbf{k}|), \qquad (18.308)$$

whereby

$$v_1 = \frac{1}{(2\pi)^2} \int \frac{d^3k}{\omega} |j(k)|^2, \qquad (18.309)$$

and this coincides with Eq. (18.303).

Thus, for the probability, we have

$$P(N) = \frac{e^{-v_1}}{N!}\, v_1^N. \qquad (18.310)$$

We see that the probability of a classical current emitting N photons is determined by the well-known Poisson distribution for random processes. The Poisson distribution is valid for statistically independent discrete processes, which means that the emission of each separate photon does not depend on whether other photons were emitted or not. This is possible, obviously, only when the photons have no effect on the source (no recoil), which is not surprising since the source is considered classical. In a real problem with a quantum source, an electron-positron current, such a model may be used to describe the emission of soft photons, whose effect on the source can be ignored.

The *Poisson distribution* [Eq. (18.310)] is normalized to unity,

$$\sum_{N} P(N) = 1,$$ (18.311)

while the average number of photons prove to be

$$\sum_{N=1}^{\infty} N P(N) \sum_{N=1}^{\infty} N e^{-v_1} \frac{v_1^N}{N!} = v_1,$$ (18.312)

that is,

$$\overline{N} = \frac{1}{(2\pi)^2} \int \frac{d^3k}{\omega} |j(k)|^2.$$ (18.313)

Using Eq. (18.95), one can rewrite this formula as follows:

$$\overline{N} = \frac{2\pi}{L^3} \sum_{k} \frac{1}{\omega} |j(k)|^2,$$ (18.314)

from which we can find the average number of photons with momentum k per unit volume,

$$\overline{N}_k = \frac{1}{L^3} \cdot \frac{2\pi}{\omega} |j(k)|^2.$$ (18.315)

The average number of photons with k in the interval d^3k is

$$d\overline{N} = \frac{1}{(2\pi)^2} \frac{d^3k}{\omega} |j(k)|^2.$$ (18.316)

The angular spectral distribution of the radiation energy can be found by multiplying the number of photons by the photon energy ω:

$$d\varepsilon(\mathbf{n}, \omega) = \frac{d^3k}{(2\pi)^2} |j(k)|^2,$$ (18.317)

with $n = k/\omega$.

The solution of the problem of emission of photons by a classical current has been obtained without resorting to perturbation methods, that is, the solution is exact in the interaction. This manifests itself in the fact that Eq. (18.310) is written in the form of an exponential, exp $(-v_1)$, which can be expanded in a series for all $v_1 = \overline{N} < \infty$, since an exponential function has an infinite convergence radius. Perturbation theory holds only for $\overline{N} < 1$, that is, only when the average number of emitted photons is less than unity and processes in which one, two, or several photons are emitted can be separated, with one-photon emission being the most probable.

Let us consider an example in which the above results will be used. Suppose that a charged particle moves according to the following law specified in invariant form:

$$x_\mu(\tau) = \tau u_\mu \theta(-\tau) + \tau u'_\mu \theta(\tau),$$ (18.318)

where τ is the proper time, and u and u^* are values of the 4-velocity at $\tau < 0$ and $\tau > 0$, respectively. The velocity changes abruptly at $\tau = 0$. In such an approximate manner we can describe, say, the scattering of electrons on a nucleus, since the time it takes for soft photons to form is much longer than the characteristic scattering time.

The Fourier transform of the current density, $j_\mu(k)$, that corresponds to the law of motion [Eq. (18.318)] is calculated as,

$$j_\mu(k) = eu \int_{-\infty}^{0} d\tau e^{iku\tau} + eu' \int_{0}^{\infty} d\tau e^{iku'\tau}. \tag{18.319}$$

Using formulas

$$\int_{0}^{\infty} e^{i\tau(x+i\varepsilon)} d\tau = i\mathcal{P}\frac{1}{x} + \pi\delta(x),$$

$$\int_{-\infty}^{0} e^{i\tau(x-i\varepsilon)} d\tau = -i\mathcal{P}\frac{1}{x} + \pi\delta(x), \tag{18.320}$$

and bearing in mind that

$$\delta(uk) = \delta(u'k) = 0$$

at $k^2 = 0$ and $u^2 = u'^2 = 1$, we obtain

$$j_\mu(k) = ie\left(\frac{p'_\mu}{p'k} - \frac{p_\mu}{pk}\right), \tag{18.321}$$

where $p^\mu = mu^\mu$ and $p'^\mu = mu'^\mu$ are the 4-momenta before and after the scattering process. Note that $j_\mu(k)$ given by (18.321) obeys the continuity condition

$$kj(k) = 0. \tag{18.322}$$

Substituting Eq. (18.321) into Eq. (18.313), we get the average number of photons

$$\overline{N} = -\frac{e^2}{(2\pi)^2} \int_{\omega_{min}}^{\omega_{max}} \frac{d^3k}{\omega}\left(\frac{p}{pk} - \frac{p'}{p'k}\right)^2, \tag{18.323}$$

where $\overline{N}$ is positive since $j(k)$ is a spacelike vector.

What are the limits of integration in (18.323)? The upper limit, obviously, is restricted by the initial energy of the particle, $\varepsilon = (p^2 + m^2)^{1/2}$, precisely, ω_{max} must be less than ε. As for the lower limit ω_{min}, by sending it to zero we arrive at a divergent integral for $\overline{N}$:

$$\overline{N} \propto \int_{\omega_{min}} \frac{d\omega}{\omega} \to \infty. \tag{18.324}$$

At the same time, as $\omega_{min} \to 0$, the probability of a finite number of photons, $N < \infty$, being emitted tends to zero:

$$P(N) = \frac{e^{-\overline{N}}(\overline{N})^N}{N!} \to 0.$$

Therefore, there is a nonzero probability that an infinite number of photons are emitted, which agrees with the normalization condition for the probability,

Eq. (18.311). It can be demonstrated that ω_{min} can always be restricted by a small, but nonzero, quantity. Let us take the nonrelativistic case of $v \ll 1$ and transfer to the reference frame in which the scattered particle is at rest, $v' = 0$. Then

$$\left(\frac{p}{pk} - \frac{p'}{p'k} \right)^2 \simeq \frac{(\mathbf{n} \cdot \mathbf{v})^2 - v^2}{\omega^2} = -\frac{v^2}{\omega^2} \sin^2 \theta,$$

where θ is the angle between the direction of propagation of radiation, n, and velocity v. Substituting into [Eq. (18.323)], one obtains

$$\overline{N} = \frac{\alpha}{(2\pi)^2} v^2 \int_0^\pi d\theta \sin^3 \theta \int_0^{2\pi} d\varphi \int_{\omega_{min}}^{\omega_{max}} \frac{d\omega}{\omega}$$

$$= \frac{2}{3} \frac{\alpha v^2}{\pi} \ln \frac{\omega_{max}}{\omega_{min}} = \frac{2}{3} \frac{\alpha v^2}{\pi} \ln \frac{m}{\omega_{min}}, \qquad (18.325)$$

where we have put $\omega_{max} \sim m$ and $\alpha = e^2 = 1/137$. Equation (18.325) enables estimating the conditions of applicability of perturbation-theory techniques in the given problem. Putting $\overline{N} < 1$, we find that

$$\overline{N} \propto \alpha \ln \frac{m}{\omega_{min}} < 1. \qquad (18.326)$$

Setting ω_{min} to zero, we find that $\overline{N} \to \infty$. We can now estimate the values of ω_{min} at which the magnitude of $\overline{N}$ is roughly unity:

$$\omega_{min} \lesssim m e^{-137}$$

(e here is the base of natural logarithms), or for the wavelength,

$$\lambda_{max} = \frac{2\pi}{\omega_{min}} \geq e^{137} \,(1/m). \qquad (18.327)$$

Since $1/m$ is the *Compton wavelength* (i.e. $\hbar/mc$), the last inequality yields the estimate $\lambda_{max} > 10^{44}$ cm, which is much greater than the radius of the Universe. Thus, actually we can always assume ω_{min} to be a nonzero quantity and $\overline{N}$ not to be too large. Emission of many soft photons accompanies every process that involves charged particles. Since emission of a soft photon is not followed by recoil, it does not influence the main process. In this case the main process and the emission process are statistically independent, and the differential probability of the total process taking place (per unit time) with emission of N photons is determined by the product

$$dw_N = dw_0 \, P\,(N), \qquad (18.328)$$

where dw_0 is the probability of the main process without photon emission. When the frequency ω is low, the resolving power of the instrumentation proves to be insufficient for recording the soft photons, with the result that the probability must be summed over such photons:

$$dw = dw_0 \sum_N P\,(N) \qquad (18.329)$$

Summing over a large number of photons, we obtain

$$\sum_N P(N) = 1, \quad d\omega = d\omega_0, \tag{18.330}$$

that is, the probability of a process that allows for the emission of all the soft photons coincides with the probability calculated without allowing for photon emission.

18.10 FEYNMAN DIAGRAMS

In the general case, in contrast to the model problem discussed in the preceding section, the series for the S-matrix cannot be summed exactly. We are therefore forced to employ perturbation-theory techniques, considering this series to be asymptotic and ignoring all terms in the series except a few, which correspond to the chosen order of the perturbation theory. The structure of the S-matrix expansion is such that to analyze it, it has proved convenient to use the graphical method developed by Richard Feynman, who established a number of rules for assigning definite graphs, known as *Feynman diagrams**, to the different terms in the S-matrix expansion.

18.10.1 The Structure of the S-matrix Expansion

Let us discuss the properties of the transition matrix element in the *n*th-order perturbation theory,

$$\langle f | S_n | i \rangle = \frac{(-ie)^n}{n!} \langle f | \int dx_1 \dots dx_n$$

$$\times T [N \bar{\psi}_1 \gamma A_1 \psi_1)(N \bar{\psi}_2 \gamma A_2 \psi_2) \dots (N \bar{\psi}_n \gamma A_n \psi_n)] | i \rangle. \tag{18.331}$$

Here the initial and final states, $|i\rangle$ and $|f\rangle$, are obtained through applying a certain number of *fermion*** operators a_s and b_s and *boson**** (electromagnetic) operators c_s to the vacuum state:

$$|i\rangle = \prod_i c_i^+ a_i^+ b_i^+ |0\rangle, \quad \langle f| = \langle 0| \prod_f b_f a_f c_f \tag{18.332}$$

where the number of creation and annihilation operators corresponds to the number of particles with given values of the quantum numbers, the particles being in the initial and final states.

---

* *Feynman diagram:* an intuitive picture of a term in the perturbation expansion of a *scattering matrix* element or other physical quantity associated with interactions of particles; in the diagrams each line represents a particle, each vortex an interaction.

S-matrix = scattering matrix-This expresses what comes out from a system as a function of what goes into it in a matrix form:

$$\begin{pmatrix} C \\ D \\ \dots \end{pmatrix} = S \begin{pmatrix} A \\ B \\ \dots \end{pmatrix}$$

where $A, B, \dots$ represent the incoming influence on the system and $C, D, \dots$ represent the system reaction.

** *Fermions:* quantum particles whose energy distribution obey the *Pauli exclusion principle* and *Fermi-Dirac statistics.*

*** *Bosons:* quantum particles whose energy distribution obeys *Bose-Einstein statistics*. They have an integer or zero spin.

To transform the matrix elements still further, we must employ Wick's theorem. The S-matrix expansion is conveniently represented by Feynman diagrams. In such diagrams, each vertex corresponds to a variable of integration x; a ray ending at a vertex corresponds to operator $\psi(x)$; a ray going out of a vertex corresponds to operator $\bar{\psi}(x)$; and, finally, a dashed ray (without direction being specified) starting at a vertex corresponds to operator $A_{\mu}(x)$ (Fig. 18.4). Thus, electron-positron lines are solid while photon lines are dashed. An nth-order diagram has a vertices.

Fig. 18.4

Expanding the S-matrix via Wick's theorem, we arrive at separate terms which contain several normally ordered creation and annihilation operators and several pairs of coupled operators, chronological pairings, which are expressed in terms of propagators. On a diagram the rays going out of and ending at a vertex correspond to creation and annihilation operators, while propagators correspond to the lines linking vertices. The lines corresponding to electron-positron propagators are directional, and their direction continues that of the fermion lines entering or leaving the vertices.

Let us take for an example the second-order perturbation theory, $n = 2$. Then, by Wick's theorem, for the second term in the S-matrix expansion, S_2, we get a sum of eight addends. One of these addends has an integrand without pairings, or

$$N(\bar{\psi}_1\gamma A_1\psi_1)(\bar{\psi}_2\gamma A_2\psi_2), \qquad (18.333)$$

the second contains one pairing, $\overline{A_1A_2}$, the third contains $\overline{\bar{\psi}_1\psi_2}$, the fourth contains $\overline{\psi_1\bar{\psi}_2}$, the fifth contains two pairings, $\overline{\bar{\psi}_1\psi_2}$ and $\overline{A_1A_2}$, the sixth contains $\overline{A_1A_2}$ and $\overline{\psi_1\bar{\psi}_2}$, the seventh contains $\overline{\bar{\psi}_1\psi_2}$ and $\overline{\psi_1\bar{\psi}_2}$, and the eighth contains three pairings, $\overline{\bar{\psi}_1\psi_2}$, $\overline{\psi_1\bar{\psi}_2}$, and $\overline{A_1A_2}$. There also remains an appropriate number of free, uncoupled, operators. The last addend does not contain free operators. The Feynman diagrams show the structure of S_2 most vividly (Figure 18.5, where the diagrams correspond to the addends in the same order). We can see that diagrams 3 and 4 differ only in the permutation of their vertices. The corresponding addends are not distinct, since one is obtained from the other by interchanging the variables of integration: $x_1 \to x_2$ and $x_2 \to x_1$. The same is true for diagrams 5 and 6. For the reason we can always consider only one diagram in each pair and drop the common factor $1/2! = 1/2$ in S_2.

The higher-order terms in the S-matrix, $n > 2$, have a more complicated structure, but can always be made simple by using the Feynman diagrammatic technique.

18.10.2 Feynman's Rules

To calculate the matrix elements characterizing the transition of a system of particles from one state to another, we must apply plane-wave expansions to the *Dirac and photon operators.*

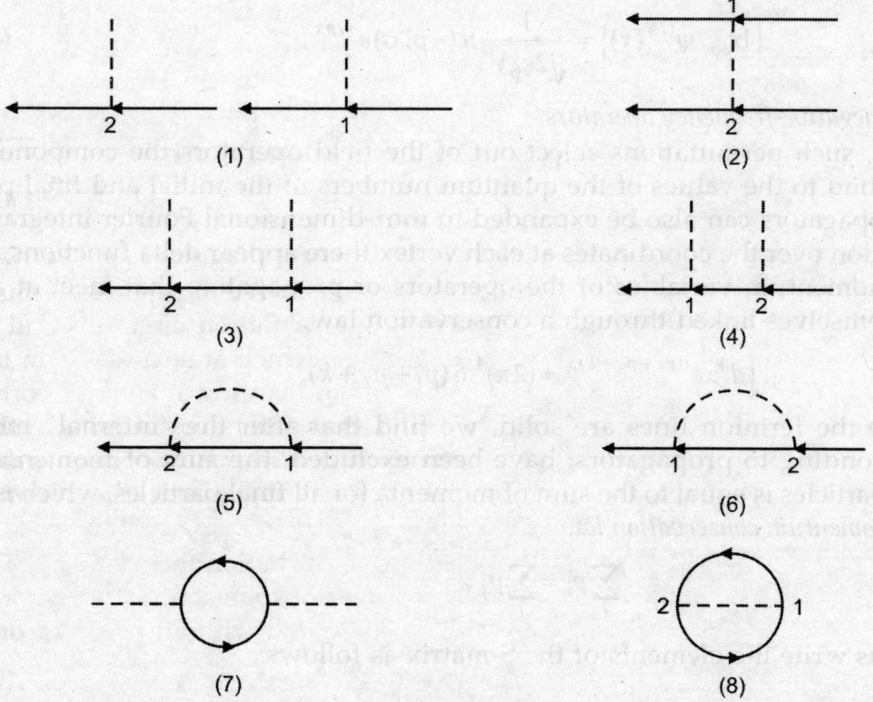

Fig. 18.5

The operators A, ψ, and $\bar{\psi}$, normally ordered in accord with Wick's theorem, must be interchanged with the operators a, a^+, b, b^+, c, and c^+ present in the definition (18.332) of the initial and final states. After this permutation the annihilation operators $A^{(+)}$, $\psi^{(+)}$, and $\overline{\psi^{(-)}}$ will act directly on the vacuum state on the right and the corresponding negative-frequency operators, on the vacuum state on the left, giving zero, while the remaining commutators and anti-commutators will be

$$\left[A_\mu^{(+)}(x), c_{k\lambda}^+\right] = \sqrt{\frac{2\pi}{\omega V}}\, e_\mu^{(\lambda)}\, e^{-ikx}\ (\lambda = 1, 2),\qquad (18.334a)$$

$$\left\{\psi^{(+)}(x), a_{p\sigma}^+\right\} = \frac{1}{\sqrt{2\varepsilon_p V}}\, u(p, \sigma) e^{-ipx},\qquad (18.334b)$$

$$\left\{\overline{\psi^{(+)}}(x), b_{p\sigma}^+\right\} = \frac{1}{\sqrt{2\varepsilon_p V}}\, \bar{u}(-p, \sigma) e^{-ipx},\qquad (18.334c)$$

for positive-frequency operators and

$$\left[c_{k\lambda}, A_\mu^{(-)}(x)\right] = \sqrt{\frac{2\pi}{\omega V}}\, e_\mu^{*(\lambda)}\, e^{ikx}\ (\lambda = 1, 2),\qquad (18.335a)$$

$$\left\{a_{p\sigma}, \overline{\psi^{(+)}}(x)\right\} = \frac{1}{\sqrt{2\varepsilon_p V}}\, \bar{u}(p, \sigma) e^{-ipx},\qquad (18.335b)$$

$$\left\{ b_{p\sigma}, \psi^{(-)}(x) \right\} = \frac{1}{\sqrt{2\varepsilon_p V}} \, u(-p,\sigma) e^{-ipx}, \tag{18.335c}$$

for the *negative-frequency operators.*

Thus, such permutations select out of the field operators the components that correspond to the values of the quantum numbers of the initial and final particles. The propagators can also be expanded in four-dimensional Fourier integrals. After integration over the coordinates at each vertex there appear delta functions, and the three momentum variables of the operators or propagators that meet at a vertex find themselves linked through a conservation law,

$$\int d^4 x \, e^{i(p_1 + p_2 + k)x} = (2\pi)^4 \, \delta\,(p_1 + p_2 + k).$$

Since the fermion lines are solid, we find that after the "internal" momenta, corresponding to propagators, have been excluded, the sum of momenta for all initial particles is equal to the sum of momenta for all final particles, which is simply the 4-*momentum conservation law*

$$\sum_i p_i = \sum_f p_f. \tag{18.336}$$

Let us write the elements of the S-matrix as follows:

$$\langle f | S | i \rangle = \delta_{fi} + i(2\pi)^4 \, \delta\left(\sum_i p_i - \sum_f p_f \right) T_{fi}. \tag{18.337}$$

The first term, δ_{fi}, corresponds to the absence of an interaction within the system of particles, that is, to case where the initial and final states coincide. In the second term we have isolated the delta function responsible for 4-momentum conservation in the course of interaction. In what follows T_{fi} will always be called the amplitude of the process.

Thus, the matrix elements, or amplitudes, we arrive at have a structure that follows directly from the structure of the S-matrix. For this reason it is convenient to express the amplitudes in the form of Feynman diagrams, just as we did for the separate terms of the S-matrix expansion. But these new Feynman diagrams differ from the old in that now each external line (ray) is characterized by a definite set of quantum numbers of the particle—momentum and polarization, while the internal lines may have a definite momentum whose value follows from momentum conservation at a vertex or, if more than one internal line meet at a vertex, may not have such a value. Then we must evaluate the integral over the 4-momentum, which integral was left over from the integral representation of the propagator.

The correspondence between diagrams and amplitudes is established by rules that were formulated by Feynman.

i. Each diagram corresponding to an amplitude has so many external incoming and outgoing lines (rays) as there are initial and final particles. The number of vertices in a diagram corresponds to the order of the perturbation theory (the number of the term in the S-matrix expansion).

ii. To each vertex there corresponds a matrix γ^μ.

iii. In accord with Eqs. (18.334b) and (18.335c), to each incoming solid line there corresponds a positive-frequency bispinor

$$\frac{1}{\sqrt{2\varepsilon_p V}}\, u\,(p,\sigma),$$

which describes an electron in the initial state, or negative-frequency bispinor

$$\frac{1}{\sqrt{2\varepsilon_p V}}\, u\,(-p,\sigma),$$

which describes a positron in the initial state. To each outgoing solid line, in accord with Eqs. (18.334c) and (18.335b), there corresponds either a conjugate negative-frequency bispinor

$$\frac{1}{\sqrt{2\varepsilon_p V}}\, \bar{u}\,(-p,\sigma)$$

for a final positron or a conjugate positive-frequency bispinor

$$\frac{1}{\sqrt{2\varepsilon_p V}}\, \bar{u}\,(p,\sigma)$$

for a final electron.

iv. To each incoming dashed line three corresponds, in accord with (18.334a), a polarization 4-vector

$$\frac{\sqrt{4\pi}}{\sqrt{2\omega V}}\, e_\mu^{(\lambda)},$$

while to each outgoing dashed line there corresponds, in accord with (18.335a), a polarization 4-vector

$$\frac{\sqrt{4\pi}}{\sqrt{2\omega V}}\, e_\mu^{*(\lambda)}$$

with these vectors contracting with the appropriate g-matrices at the vertices: $e_\mu \gamma^\mu$ and $e^*_\mu \gamma^\mu$. Note that in this case the dashed lines acquire direction.

v. To each internal solid line there corresponds the Fourier transform of the spinor-field propagator,

$$\frac{i\,(\gamma p + m)}{p^2 - m^2 + i\varepsilon} = S_c\,(p), \tag{18.338}$$

while to each internal dashed line there corresponds the Fourier transform of the electromagnetic-field propagator taken with the opposite sign,

$$-\frac{4\pi i}{k^2 + i\varepsilon}\, g_{\mu\nu} = -D^c_{\mu\nu}(k). \tag{18.339}$$

The photon propagator is contracted with the γ-matrices at the vertices that it connects: $\gamma^\mu D^c_{\mu\nu} \gamma^\nu$. The electron propagator is contracted in the bispinor indices with the γ-matrices at the vertices. Finally, a closed electron loop corresponds to the trace of the product of the bispinor matrices along this loop.

vi. The value of the 4-*momentum* of each internal line is determined from the momentum conservation law at the vertex. If more than one internal line gather at a vertex, the momentum may have no definite value, and then we must evaluate the integral with respect to this momentum with a factor $1/(2\pi)^4$. We may note that the internal lines correspond to the so-called virtual particles, which are described by propagators instead of wave functions., Generally speaking, the momenta of such particles do not lie on the mass surface, that is, $p^2 \neq m^2$ for electron-positron lines and $k^2 \neq 0$ for photon lines. On the contrary, the momenta of the incoming and outgoing lines are fixed in such a manner that they satisfy the conservation law (18.336). In accord with Feynman's rule (3), the positron momentum is taken equal to the argument of the *bispinor* with the opposite sign: $p_{pos} = -p$.

vii. To an nth-order amplitude we assign a factor $(-ie)^n$. An additional factor equal to -1 appears if the diagram has a closed fermion loop. The example in diagram 7 in Figure 18.5 clearly demonstrates the origin of this factor. As we said earlier, one of the pairings for this diagram is taken with the order in which the operators must follow reserved, $\overline{\psi}_1\psi_2$. Returning to the correct order, with which the electron propagator was defined, $\overline{\psi}_1\psi_2 = -\psi_2\overline{\psi}_1$ we arrive at the factor -1. The same factor appears in diagrams with two external positron ends connected by a single solid line. It originates in the rearrangement in normal order of the positron parts of operators $\overline{\psi}(x)$ and $\psi(x)$ in the S-matrix.

One amplitude may have corresponding to it several different diagrams of the same order and with the same number of external lines but with interchanged vertices of lines between them. All these diagrams must be added to obtain the total amplitude of the physical process. And only the total amplitude is gauge invariant. It can easily be verified that a correctly constructed amplitude satisfies the condition of gauge invariance. For this we must make sure that the substitution of $e_\mu + fk_\mu$ for e_μ, where k_μ is the 4-momentum of the photon, does not change the value of the amplitude.

If there are several identical fermions participating in the process, the diagrams with interchanged fermion lines enter the amplitude with opposite signs.

18.11 THE INVARIANT REACTION CROSS SECTION

In this section we will show how, using the reaction amplitudes, one can determine the physically measurable quantities, probabilities and cross sections.

18.11.1 Probability and Cross Section

The probability that a system that initially was in state $|i\rangle$ will be found in state $|f\rangle$ is equal, as we know, to the square of the modulus of the transition matrix element, $|\langle f|S|i\rangle|^2$. If we ignore the transitions with no change in state, then the first term, the unit matrix, in the expression (18.337) for the S-matrix can be dropped. The square of the delta function can be written, if we employ the representation of the delta function

$$\delta(p) = \frac{1}{(2\pi)^4} \int d^4 x e^{ipx},$$ (18.340)

in the form

$$[\delta(p)]^2 = \delta(p)\,\delta(0) = \delta(p)\frac{VT}{(2\pi)^4},$$ (18.341)

where V is the three-dimensional volume and T is the time interval over which integration in (18.340) is carried out; both V and T are taken sufficiently large but finite. Note that the observable probabilities and cross sections must be independent of V and T.

The probability divided by the time interval T is the transition rate (probability of transition per unit time)

$$w_{fi} = \frac{|\langle f|S|i\rangle|^2}{T}.$$ (18.342)

According to what has been said, for the transition rate we have the following formula:

$$w_{fi} = (2\pi)^4 \delta\left(\sum_i p_i - \sum_f p_f\right)|T_{fi}|^2 V,$$ (18.343)

where T_{fi} is the amplitude of the process.

We may recall that, according to Feynman's rules, each initial or final particle contributes a factor $(2\varepsilon V)^{-1/2}$ to the amplitude, where is the particle energy ($\varepsilon = \omega$ for photons). Since these factors contain the normalization volume V, which has no direct physical meaning, it is advisable to isolate these factors in the amplitude and introduce a new amplitude A_{fi} as follows:

$$T_{fi} = \frac{A_{fi}}{\prod_i \sqrt{2\varepsilon_i V} \prod_i \sqrt{2\varepsilon_f V}},$$ (18.344)

where the products in the denominator are over all initial and final quantum numbers.

In accord with this definition, there is no need to include the factors $(2\varepsilon V)^{1/2}$ in the A_{fi} when we construct these new amplitudes via Feynman's rules, whereas for T_{fi} we must do this.

Summation over the final stages in Eq. (18.342) is carried out as follows. As is known, the states of free particles are quasi-classical. The number of quasi-classical states of a particle per momentum interval d^3p in volume V is $Vd^3p/(2\pi)^3$. For this reason, the probability of a system of particles transforming into a finite state with momenta within the interval $\prod_f d^3p_f$ in volume V is

$$dw = (2\pi)^{4\delta}\left(\sum_i p_i - \sum_f p_f\right)|T_{fi}|^{2V}\prod_f \frac{Vd^3p_f}{(2\pi)^3},$$ (18.345)

or, in terms of the new amplitude A_{fi},

$$dw = (2\pi)^4 \delta \left(\sum_i p_i - \sum_f p_f \right) \frac{|A_{fi}|^2 V}{\prod_i 2\varepsilon_i V} \prod_f \frac{d^3 p_f}{2\varepsilon_f (2\pi)^3}. \tag{18.346}$$

Here are two characteristics reactions. The first is the decay of a particle into several other particles. In this reaction there is one particle in the initial state, and the probability (18.346) assumes the form

$$dw = (2\pi)^4 \delta \left(p_i - \sum_f p_f \right) \frac{|A_{fi}|^2 V}{2\varepsilon_i} \prod_f \frac{d^3 p_f}{2\varepsilon_f (2\pi)^3}. \tag{18.347}$$

We can see that the decay probability does not depend on the (nonphysical) normalization volume.

The second reaction has to do with the collision of two particles with energies ε_1 and ε_2 and momenta p_1 and p_2. In this case it has proved expedient to introduce the notion of cross section. The differential cross section of a collision is defined as

$$d\sigma = j^{-1} dw, \tag{18.348}$$

where j is the flux of particles. The definition (18.348) refers to a special reference frame, namely the reference frame connected with the center of inertia (or center of mass) of the system of the colliding particles ($p_1 = -p_2$). The flux is then

$$j = \frac{v_1 + v_2}{V}, \tag{18.349}$$

where $v_1 = v_2$ are the velocities of the particles in the given frame.

A covariant generalization of the definition of flux can be given in the form

$$j = \frac{I}{V \varepsilon_1 \varepsilon_2}, \tag{18.350}$$

where

$$I = \sqrt{(p_1 p_2)^2 - m_1^2 m_2^2} \tag{18.351}$$

is a scalar quantity. It can easily be verified that the definitions (18.350) and (18.351) coincide in the *center-of-mass frame*. The definition (18.348) then yields the following formula for the cross section:

$$d\sigma = (2\pi)^4 \delta \left(p_1 + p_2 - \sum_f p_f \right) \frac{|A_{fi}|^2}{4I} \prod_f \frac{d^3 p_f}{(2\pi)^3 2\varepsilon_f}. \tag{18.352}$$

The expression obtained defines the invariant cross section of a reaction and does not depend on the normalization volume. The invariance of this expression is obvious, since the amplitude A_{fi} and the scalar quantity I are invariant. So are the volume elements in the momentum space and the *four-dimensional delta function*

$$\delta \left(p_1 + p_2 - \sum_f p_f \right) \prod_f \frac{d^3 p_f}{2\varepsilon_f} = \text{in } v,$$

since the three-dimensional integral $\int (2\varepsilon)^{-1} d^3 p$ can be transformed into the four-dimensional integral $\int d^4 p \delta (p^2 - m^2)$ taken over the upper surface of the light cone, $p^0 > 0$. The four-dimensional delta function is also Lorentz-invariant: $\delta(0) = VT = \text{inv.}$

Up till now we have dealt only with summation over spatial degrees of freedom of particles, the momenta, but we must also carry out summation over the spin states of the particles, the polarizations.

If the initial state is nonpolarized, we must also average the result over the initial spin states.

To carry out these operations, let us write the amplitude of the process as

$$A_{fi} = \bar{u}_f A u_i, \tag{18.353}$$

where u_i is the bispinor of the initial state and u_f is the bispinor of the final state. The square of the modulus is

$$|A_{fi}|^2 = u_i^+ A^+ \bar{u}_f^+ \bar{u}_f A u_i = \bar{u}_i \bar{A} u_f \bar{u}_f A u_i, \tag{18.354}$$

where $\bar{A} = \gamma^0 A^+ \gamma^0$. If we write out all the bispinor indices in (18.354), we get

$$|A_{fi}|^2 = (\bar{u}_i)_{\alpha\beta} (\bar{A})_{\beta\gamma} (u_f)_{\gamma\mu} (\bar{u}_f)_{\mu\nu} A_{\nu\rho} (u_i)_{\rho\alpha},$$

which is the trace of a matrix product:

$$|A_{fi}|^2 = \text{Tr}(\bar{u}_i A u_f \bar{u}_f A u_i). \tag{18.355}$$

By virtue of a property of a trace, the cofactors in Eq. (18.355) allow for cyclic permutations:

$$|A_{fi}|^2 = \text{Tr}(u_i \bar{u}_i \bar{A} u_f \bar{u}_f A). \tag{18.356}$$

Let us now recall the following identity for positive-frequency *bispinors*

$$\sum_\sigma u(p, \sigma) \bar{u}(p, \sigma) = \gamma p + m. \tag{18.357}$$

Thus, if the initial and final states have one electron each, then summing over the final polarizations and averaging over the initial polarizations yield

$$\frac{1}{2} \sum_{\sigma_i, \sigma_f} |A_{fi}|^2 = \frac{1}{2} \text{Tr}\{(\gamma p_i + m) \bar{A} (\gamma p_f + m) A\}. \tag{18.358}$$

For negative-frequency bispinors the identity Eq. (18.357) has a "minus" sign in front of m. For this reason, if among the initial or final particles there is a positron, we must change the sign in front of m in the appropriate factor in Eq. (18.3579). When there is more than one spin or particle in the initial and/or final states, summation over polarizations leads to expressions that are more complicated than Eq. (18.358), but the method of calculation remains the same.

As for polarization of photons, it is described by the 4-vectors e_μ. Each photon contributes a factor $e_\mu e_\nu^*$ to the square of the amplitude. We may recall that, in the case of three-dimensional transverse gauge, summation over polarizations is carried out via,

$$\sum_{\lambda = 1, 2} e_i^{(\lambda)} e_k^{*(\lambda)} = \delta_{ik} - n_i n_k, \tag{18.359}$$

where $i, k = 1, 2, 3$, $n_i = k_i/\omega$, and, obviously, $e_0^{(\lambda)} = 0$ (complex conjugation refers to the case of circular polarization, which is described by complex-valued vectors).

According to Feynman's rules, the polarization vectors are encountered in the amplitude in combination with the transition currents $j_{21}(k) = u_2(p_2) \gamma u_1(p_1)$, that is, in the form

$$ej_{21}(k) = \bar{u}_2(p_2) e_\mu \gamma^\mu u_1(p_1), \tag{18.360}$$

where $k = p_1 - p_2$. For this reason, the sum (18.359) yields

$$\sum_\lambda (\mathbf{j}_{21} \cdot \mathbf{e}^{(\lambda)})(\mathbf{j}_{21}^* \cdot \mathbf{e}^{*(\lambda)}) = |\mathbf{j}_{21}|^2 - |\mathbf{j}_{21} \cdot \mathbf{n}|^2. \tag{18.361}$$

Expression (18.361) is gauge invariant, that is, does not change under transformations $e \to e' = e + fk$. This means that we must have

$$(j_{21} k) = 0, \tag{18.362}$$

which is simply the continuity condition for the transition current, $\partial_\mu j^\mu = 0$, written in the momentum representation. From (18.362) we arrive at

$$j_{21}^0 k^0 = \mathbf{j}_{21} \cdot \mathbf{k}, \tag{18.363}$$

whereby (18.361) assumes the form

$$\sum_{\lambda=1,2} (\mathbf{j}_{21} \cdot \mathbf{e}^{(\lambda)})(\mathbf{j}_{21}^* \cdot \mathbf{e}^{*(\lambda)}) = -(j_{21})^\mu (j_{21}^*)_\mu = |j_{21}|^2. \tag{18.364}$$

In this expression we can go over to any other gauge of the polarization vectors in which the temporal coordinates may be nonzero, too. Then, in view of the *gauge invariance*, one obtains

$$\sum_\lambda (j_{21} e^{(\lambda)})(j_{21}^* e^{*(\lambda)}) = |j_{21}|^2, \tag{18.365}$$

that is,

$$\sum_\lambda e_\mu^{(\lambda)} e_\nu^{*(\lambda)} = -g_{\mu\nu}. \tag{18.366}$$

Averaging over two transverse polarizations differs from Eq. (18.359) in that we have a factor of $1/2$, that is,

$$\overline{e_i e_k^*} = \frac{1}{2} \sum_{\lambda=1,2} e_i^{(\lambda)} e_k^{*(\lambda)} = \frac{1}{2}(\delta_{ik} - n_i n_k). \tag{18.367}$$

Transition to other gauges does not, obviously, change this factor, since temporal and longitudinal photons cannot be observed. Thus, for the general case we have

$$\overline{e_\mu e_\nu^*} = -\frac{1}{2} g_{\mu\nu}. \tag{18.368}$$

18.11.2 Optical Theorem

The transition matrix $\langle f|S|i \rangle$ must satisfy the obvious normalization condition for the probability

$$\sum_f |\langle f|S|i \rangle|^2 = 1, \tag{18.369}$$

or

$$\sum_f \langle i|S|f\rangle\langle f|S|i\rangle = 1. \tag{18.370}$$

This condition follows from the general condition of unitarity of the S-matrix operator

$$S^+S = I, \tag{18.371}$$

which ensure the conservation of probability and the orthogonality of the states.

Let us write Eq. (18.371) in the form

$$\sum_f \langle i|S^+|f\rangle\langle f|S|i'\rangle = \delta_{ii'}. \tag{18.372}$$

Substituting the matrix elements Eq. (18.337), we obtain

$$T_{ii'} - T^*_{i'i} = i(2\pi)^4 \sum_f \delta\left(\sum_i p_i - \sum_f p_f\right) T^*_{fi} T_{fi'}. \tag{18.373}$$

This condition, obviously, relates the matrix elements in different orders of the perturbation theory. It can be employed to determine the probabilities or cross sections expressed in terms of quadratic combinations of amplitudes through linear combinations of higher-order amplitudes. Take, for example, the case where $i = i'$. This means that amplitude T_{ii} describes an elastic process in which the state of the system does not change. Then Eq. (18.373) yields

$$2\,\mathrm{Im}\,T_{ii} = (2\pi)^4 \sum_f |T_{fi}|^2 \,\delta\left(\sum_i p_i - \sum_f p_f\right). \tag{18.374}$$

As we can see, on the right we have, to within a factor, the total probability of a system transferring from state $|i\rangle$ to all states $|f\rangle$, while on the left we have the imaginary part of the amplitude of the elastic process with no change in state, $|i\rangle \rightarrow |i\rangle$.

According to Eq. (18.343), we have

$$2\,\mathrm{Im}\,T_{ii} = \frac{1}{V}\sum_f w_{fi}, \tag{18.375}$$

that is, $\mathrm{Im}\,T_{ii}$ is expressed in terms of the total transition probability per unit time (total transition rate). Going over to the cross section via Eq. (18.348), one obtains

$$\frac{2V}{j}\,\mathrm{Im}\,T_{ii} = \sigma_{\mathrm{tot}}, \tag{18.376}$$

where σ_{tot} is the total cross section. On the left we have the imaginary part of the elastic scattering amplitude without change in state, or scattering by a zero angle. This result, which express the total probability or cross section in terms of the imaginary part of the diagonal matrix element, is known as the *optical theorem*.

SHORT ANSWER QUESTIONS

1. How positrons are viewed within the propagator theory?

Ans. Positrons with positive energy moving forward in space-time are viewed within the propagator theory as electrons with negative energy propagating forward in space-time.

2. What is Bremsstrahlung?

Ans. When electrons scatter at protons or in the field of nucleus, they can emit real photons. The process is called bremsstrahlung because it involves an acceleration or deceleration of the projectile. The emitted real photons fulfill the Einstein condition

$$q^2 = 0.$$

3. In the language of quantum electrodynamics explain the term Compton scattering.

Ans. The name Compton scattering refers to the scattering of photons by free electrons. In the language of QEDs an incoming photon with four momentum k and polarization vector ε is absorbed by an electron (or another charged particle) and a second photon with four momentum k' and polarization vector ε' is emitted.

4. Write the Feynman rules of QED in momentum space.

Ans. The Feynman Rules of QED in momentum space are as follows:

1. In the nth order of perturbation theory one has to draw all possible topologically distinct Feynman diagrams with n vertices that have the prescribed number of particles in the initial and final states (external lines).

2. With each external line one has to associate the following factors:

   (a) incoming electron: $u(p, s)$

   (b) incoming positron: $\bar{v}(p, s)$
       (outgoing electron with negative energy)

   (c) incoming photon: $\varepsilon_\mu (k, \lambda)$

   (d) outgoing electron: $\bar{u}(p, s)$

   (e) outgoing positron: $v(p, s)$
       (incoming electron of negative energy)

   (f) outgoing photon: $\varepsilon_\mu^* (k, \lambda)$

3. Each internal line connecting two vertices has to be associated with a propagator

(a) electron: $i\,S_F(p) = \dfrac{i(\not{p} + m_0)}{p^2 - m_0^2 + i\varepsilon}$

(b) photon: $i\,D_F^{\mu\nu}(k) = \dfrac{-i\,4\pi\,g^{\mu\nu}}{k^2 + i\varepsilon}$

4. Each vertex is associated with a factor

$$-ie\gamma_\mu$$

The index μ has to be multiplied with that of the photon line and summed over.

5. The conservation of four-momentum holds at each vertex. One has to integrate over all momentum variables p that cannot be fixed (internal loops): $\int \dfrac{d^4 p}{(2\pi)^4}$

6. The amplitudes of all graphs have to be added coherently, taking into account the following phase factors:

   (a) a factor of -1 for each incoming positron (outgoing electron with negative energy);

   (b) a factor of -1 in the case that two graphs differ only by the exchange of two fermion lines – this also holds for the exchange of an incoming (outgoing) particle line with an outgoing (incoming) antiparticle line, since the latter is an incoming (outgoing) particle line with negative energy;

   (c) a factor of -1 for each closed fermion loop.

Here we add the following remarks:

To 1: For the construction of Feynman graphs, only the topological structure is important. Since the theory was formulated in a relativistically covariant way, all possible time orderings are automatically taken into account. As long as the ordering of the vertices along the fermion lines is kept, the graphs can be arbitrarily deformed without changing their meaning.

To 2f: If ε_μ is complex the plane wave has to be written as

$$A^\mu(x, k) = N_k \left(\varepsilon^\mu e^{-ik\cdot x} + \varepsilon^{*\mu} e^{ik\cdot x} \right)$$

in order to make A^μ real. The second term, which describes photon emission, contains the complex conjugate polarization vector. This becomes important if photons with *circular polarization* are considered. Using the radiation gauge, the basis vectors for left/right-handed circular polarization are

$$\varepsilon(k, \pm) = \frac{1}{2}(e_1 \pm i e_2),$$

where e_1 and e_2 are unit vectors orthogonal to the direction of propagation, $e_1 \cdot k = e_2 \cdot k = e_1 \cdot e_2 = 0$.

To 5: First one associates a four-momentum δ-function to each vertex and then integrates over the momenta of all internal lines. Since one δ-function is needed for the conservation of the total momentum, for a graph of order n with I internal lines there remain $I - (n - 1)$ integrations over internal lines. We illustrate this for a graph of eighth order (Fig. 1).

Fig. 1. A graph of eighth order with 10 internal lines. One must integrate over $10 - (8 - 1) = 3$ momenta of internal loops

To 6: The sign factor in a) has been discussed for the S matrix. There it is denoted by ε_f. The minus sign in b) can be explained in the examples of electron-electron and electron-positron scatterings in first order. It originates from the antisymmetry of the wave function required by Fermi-Dirac statistics. Rule 6c is new. Nevertheless, it can be derived directly from 6b. Let us consider, e.g. the graph of Fig. 2a which contains an electron loop and may be part of a bigger graph. Exchanging the two electron lines one gets the diagram 2b, which can also be drawn in the way shown in 2c. Consequently the graph with the closed loop gets a phase factor of –1 compared to Fig. 2c.

Fig. 2. In graph (b) the lines (1) and (2) of graph (a) are exchanged explaining the minus sign which accompanies closed fermion loops

5. Explain the general Lorentz transformations.

Ans. The special Lorentz transformation, which corresponds to a transition from one inertial reference frame to another, is only a special case of a more general Lorentz transformation, that is, general linear transformations that preserve the square of the 4-vector in *Minkowski's space-time*. Let us consider such transformations in greater detail. Suppose the "old" basis e_v is linearly transformed into a "new" basis e'_μ. Then the new coordinates x'^μ of a vector are connected with the old coordinates x^v via the linear relation

$$x'^\mu = A_v^\mu x^v, \tag{1}$$

where $A = \| A_v^\mu \|$ is the matrix of the general Lorentz transformation:

$$A = \begin{pmatrix} A_0^0 & A_1^0 & A_2^0 & A_3^0 \\ A_0^1 & A_1^1 & A_2^1 & A_3^1 \\ A_0^2 & A_1^2 & A_2^2 & A_3^2 \\ A_0^3 & A_1^3 & A_2^3 & A_3^3 \end{pmatrix}.$$

Since vector x can be written both in the old basis and in the new one,

$$x = x^\nu e_\nu = x'^\mu e'_\mu = A^\mu_\nu x^\nu e'_\mu, \tag{2}$$

we can write

$$e_\nu = A^\mu_\nu e'_\mu,$$

which yields the law of transformation of basis:

$$e'_\mu = (A^{-1})^\nu_\mu e_\nu. \tag{3}$$

Thus, if the components of a vector x are transformed by means of a matrix $A = \| A^\mu_\nu \|$, the base vectors are transformed via the transpose of the inverse matrix $(A^{-1})^T = \| (A^{-1})^\nu_\mu \|$.

A covariant vector may be constructed out of a contravariant vector as follows:

$$x_\mu = g_{\mu\nu} x^\nu = e_\mu e_\nu x^\nu = e_\mu \cdot x. \tag{4}$$

Indeed, then $x'_\mu = e'_\mu \cdot x$, and the components x'_μ are related to x_μ by the same law as e'_μ is related to e_μ. Taking into account the explicit form of the metric tensor $g_{\mu\nu}$, we get

$$x_0 = x^0, \ x_1 = -x^1, \ x_2 = -x^2, \ x_3 = -x^3. \tag{5}$$

In addition to the metric tensor $g_{\mu\nu}$ we can introduce the contravariant tensor $g^{\mu\nu}$, which is the inversel of $g_{\mu\nu}$, or $g_{\mu\sigma} g^{\sigma\nu} = \delta^\nu_\mu$. The components of the two tensors coincide: $g_{\mu\nu} = g^{\mu\nu}$.

Employing the tensor $g^{\mu\nu}$, we can raise the indices of a covariant vector, thus, transforming it into a contravariant vector:

$$x^\mu = g^{\mu\sigma} x_c. \tag{6}$$

We will now establish what condition must the matrix A^μ_ν of the linear transformation satisfy so that the invariance of the square of a vector, x^2, is ensured. To this end we write the condition $x^2 = \text{inv}$ in explicit form:

$$x'^2 = x^2, \tag{7}$$

that is,

$$A^\mu_\nu A^\lambda_\rho g_{\mu\lambda} x^\nu x^\rho = x^\alpha x^\beta g_{\alpha\beta} \tag{8}$$

From this follows the basic result:

$$A^\mu_\nu g_{\mu\lambda} A^\lambda_\rho = g_{\nu\rho}, \tag{9}$$

which means that $g_{\mu\nu}$ is not transformed when we go over to new co-ordinates—this is simply another way of expressing the fact that x^2 is an invariant. Multiplying (9) into $g^{\nu\sigma}$ yields

$$A^\mu_\nu g_{\mu\lambda} g^{\nu\sigma} A^\lambda_\rho = \delta^\sigma_\rho, \tag{10}$$

that is, the matrix $A'' = \| A^\mu_\nu g_{\mu\lambda} g^{\nu\sigma} \|$ is the inverse of $\| A^\mu_\nu \|$. Let us find the explicit form of A''. First we note that the transition from matrix A to matrix

$A' = \| A_v^\mu g_{\mu\lambda} \|$ is reduced to lowering one index by applying $g_{\mu\lambda}$ and, according to (5), reversing the sign of all matrix rows except the first. The transition from A' to A'' is reduced to transposing A' and reversing the sign of all matrix columns except the first. The result is a matrix A'' which has the form

$$A'' = \begin{pmatrix} A_0^0 & -A_0^1 & -A_0^2 & -A_0^3 \\ -A_1^0 & A_1^1 & A_1^2 & A_1^3 \\ -A_2^0 & A_2^1 & A_2^2 & A_2^3 \\ -A_3^0 & A_3^1 & A_3^2 & A_3^3 \end{pmatrix}. \tag{11}$$

We arrive at the following property of the matrix of the general Lorentz transformation: transposing the initial matrix and reversing the signs of matrix elements in the upper row and left column transforms A into the inverse matrix: $A'' = A^{-1}$.

Such matrices are said to be pseudo-orthogonal (in contrast to orthogonal matrices in which the transition to the inverse matrix means only transposing the initial matrix; such is the matrix of spatial rotations).

Formula (10) implies that

$$\det A' \cdot \det A = 1, \tag{12}$$

but det $A'' =$ det A (transposition of a matrix and subsequent reversal of sign of a column and a row do not change the value of the determinant), whereby $\det^2 A = 1$. This equation has two solutions:

$$\det A = +1 \quad \text{and} \quad \det A = -1. \tag{13}$$

In both cases det $A \neq 0$, and this is proof that each transformation A has an inverse whose matrix is A''.

Note that the condition $| \det A | = 1$ implies the invariance of an element of volume of 4-space:

$$d^4x' = \left| \frac{\partial x'}{\partial x} \right| d^4x = | \det A | \, d^4x = d^4x.$$

The special Lorentz transformation needed for transferring to a reference frame that moves along the x^1 axis with velocity v, corresponds to a symmetric matrix A with the following matrix elements being nonzero:

$$A_0^0 = A_1^1 = \gamma, \quad A_2^2 = A_3^3 = 1, \quad A_1^0 = A_0^1 = -\beta\gamma.$$

For the matrix A'' we have

$$(A'')_0^0 = (A'')_1^1 = \gamma, \quad (A'')_2^2 = (A'')_3^3 = 1, \quad (A'')_1^0 = (A'')_0^1 = \beta\gamma.$$

It is clear that A'' is the matrix of the inverse transformation since it differs from A in the sign of the velocity, $\beta \rightarrow -\beta$, and, besides, direct multiplication of the two matrice yields $A''A = I$. The determinant of the matrix of this transformation is unity, or det $A = +1$.

Let us now consider the main relationship (9) for $v = \rho = 0$;

$$A_0^\mu \, g_{\mu\lambda} \, A_0^\lambda = 1,$$

or in explicit form,

$$(A_0^0)^2 - (A_0^1)^2 - (A_0^2)^2 - (A_0^3)^2 = 1. \tag{14}$$

From this it follows that

$$(A_0^0)^2 = 1 + (A_0^1)^2 + (A_0^2)^2 + (A_0^3)^2 \geq 1,$$

that is,

$$A_0^0 \leq -1 \quad \text{or} \quad A_0^0 \geq 1. \tag{15}$$

Thus, the general Lorentz transformation, denoted here by L, constitutes four sets of transformations, and the choice of a particular transformation depends on the sign of det A and the inequality in (15):

	$L_+^\uparrow$	$L_+^\downarrow$	$L_-^\uparrow$	$L_-^\downarrow$
A_0^0	≥ 1	≤ -1	≥ 1	≤ -1
det A	$+1$	$+1$	-1	-1

Clearly, by continuously varying the elements of any one of the four sets we cannot go over to another set, that is, these are not connected by a continuous transformation. Only $L_+^\uparrow$ contains the unit, or identity, transformation $A = I$, for which, obviously, $A_0^0 = 1$ and det $A = 1$. By continuously changing the unit transformation $A = I$ we can obtain all the transformations in set $L_+^\uparrow$. These are known as *proper Lorentz transformations* and constitute *four-dimensional pseudorotations*. The invariance under such transformations is the reflection of the fact that *space-time is isotropic* (if we ignore gravitational interaction).

All the other transformations contains discrete transformations (space-time reflection) and cannot be reduced to continuous transformations.

Let us consider the timelike vector

$$x^2 = (x^0)^2 - x^2 > 0 \tag{16}$$

(four a spacelike vector we have $x^2 = (x^0)^2 - x^2 < 0$). Suppose that the sign of the temporal component is fixed, say, $x^0 > 0$. Then the Lorentz transformations with $A_0^0 \geq 1$ do not change the sign of the temporal component, or $x'^0 > 0$. Indeed,

$$x'^0 = A_0^0 x^0 + A_1^0 x^1 + A_2^0 x^2 + A_3^0 x^3. \tag{17}$$

If we combine the *Cauchy-Buniakowski inequality* with (14) and (16), we get

$$(A_1^0 x^1 + A_2^0 x^2 + A_3^0 x^3)^2 < [(A_0^0)^2 - 1] \, (x^0)^2 < (A_0^0)^2 \, (x^0)^2].$$

This implies that if $A_0^0 \geq 1$, then x'^0 given by (17) retains the sign of x^0, that is, $x'^0 > 0$ in the case at hand. Lorentz transformations with $A_0^0 \geq 1$, which do not change the sign of the temporal component of a timelike vector, are known as orthochronous. Lorentz transformations with $A_0^0 \leq -1$ do not retain the sign of

x^0 and are called nonorthochronous. Thus, the set $L_+^\uparrow$ consists of proper ortho-chronous transformations, also known as restricted Lorentz transformations. The special Lorentz transformations, which fix a transformation from one inertial reference frame to another, obviously belong to the set $L_+^\uparrow$, or proper orthochronous transformations.

Here are some examples of improper (discrete) Lorentz transformations.

1. Reflection of one of the spatial axes (the P_1 operation). For matrix A we have: $A_0^0 = 1$, det A = – 1. Therefore, $P_1 \subset L_-^\uparrow$.

2. Space inversion (the P operation): $x \to x' = - x$. For this transformation we also have $A_0^0 = 1$, det A = –1, $P \subset L_-^\uparrow$. Note that the reflection of any two spatial axes is equivalent to a finite and continuous rotation transformation, which is a particular case of $L_+^\uparrow$. Therefore, P_1 can be obtained continuously from P, and vice versa.

3. Time inversion (the T operation): $t \to t' = -t$. Here $A_0^0 = -1$, det A = –1, that is, $T \subset L_-^\downarrow$.

4. Space and time inversion (the PT operation). Now $A_0^0 = -1$, det A = +1, and hence $PT \subset L_+^\uparrow$.

We may note that proper Lorentz transformations are used in classical mechanics and in the classical and quantum field theories. The improper, discrete, operations P, T, and PT find use primarily in quantum mechanics, quantum field theory, and the theory of elementary particles. Any improper trans-formation, $L_-^\uparrow$ or $L_-^\downarrow$ or $L_+^\downarrow$, can be obtained from the P, T, or PT operations by applying the proper continuous transformation $L_+^\uparrow$, as can easily be seen.

6. Explain Lorentz group and Poincare group.

Ans. Lorentz transformations constitute an example of transformations that from a group. We have to often consider various groups of transformations of field functions; we therefore outline of the theory of continuous, or topological, groups.

Lie Groups

Mathematics requires a following definition of an abstract group. A group G is a set of elements $A, B, C, ... \in G$ for which the composition law (the product) is defined for any two elements; $AB = C$. The product must obey the following conditions: (1) if $A \in G$ and $B \in G$, then $C = AB \in G$; (2) if $A, B, C, \in G$, then $(AB)C = A(BC) \equiv ABC$; (3) there is an element $I \in G$ such that $AI = IA = A$ for any $A \in G$; and (4) for each $A \in G$ there is a $B \in G$ such that $AB = BA = I$. The element $B = A^{-1}$ is said to be the inverse (or inverse element) of A.

The *general Lorentz transformations constitute a group.* Indeed, a subsequent application of two transformations that do not change the quadratic form x^2 also changes nothing in the quadratic form, there is an identity, or unit, element, and each element has an inverse. Thus, the linear transformations that preserve the invariance of the quadratic form $x^2 = (x^0)^2 - x^2$ constitute a group known as the general Lorentz group. or the homogeneous *Lorentz group*.

If to this group we add translations, that is, translation of the origin of the coordinate system by a constant vector a^μ, we arrive at nonhomogeneous linear transformations $x^\mu \to x'^\mu = A^\mu_\nu x^\nu + a^\mu$. These transformations form the nonhomogeneous Lorentz group, or the Poincare group. Obviously, the length of the vector connecting two points x and y in Minkowski's space-time remains unchanged under all translation transformations: $x \to x + a$, $y \to y + a$, $(x - y)^2$ = inv. This property reflects the *homogeneity of space-time*.

Out of the four sets of transformations into which the general Lorentz group splits, only $L^\uparrow_+$ forms a group, since it contains a unit element. This is the proper Lorentz group.

A more simple example is the $O(3)$ group, which is the group of rotations in three-dimensional space. The matrices of such transformations, A_{ik}, which preserve the length of vector $\mathbf{x}$, are orthogonal: $A_{ik} A_{kj} = \delta_{ij}$. This group constitutes a subgroup of the proper Lorentz group.

When we speak of the $O(3)$ group or the Lorentz group, we have in mind certain objects that undergo the transformations, that is, the coordinates of vectors in three-dimensional Euclidean space and in four-dimensional Minkowski's space-time, as well as matrices of a certain type: three-dimensional orthogonal and four-dimensional pseudo-orthogonal matrices, respectively. All this fixes the concrete (vector) representations of these groups. At the same time, the elements of these groups may have other realizations.

In general, if to each element A of group G we assign a linear operator T_A, that is $A \in G$, $A \to T_A$, which acts in a linear n-dimensional space R, then it is said that an n-dimensional linear representation of group G is fixed. Space R is called the representation space of group G.

The above-noted correspondence means that (1) if $A \in G$ and $B \in G$, then $T_A T_B = T_{AB}$, (2) the operator of the unit element is the unit operator, $T_I = I$, and (3) the operator of the inverse element is the inverse operator, $T_{A^{-1}} = T_A^{-1}$.

In a finite-dimensional space, operators can be represented by matrices, and then we talk of a matrix representation. If e_α is a basis in R, that is, $x = e_\alpha x^\alpha$ and T is the operator of the representation, then

$$e'_\alpha = Te_\alpha = e_\beta D^\beta_\alpha, \quad x'^\beta = D^\beta_\alpha x^\alpha, \tag{1}$$

where $\| D^\beta_\alpha \|$ is the matrix of operator T. If we go over to a new basis in R, or

$$e'_\alpha = e_\beta V^\beta_\alpha, \tag{2}$$

we arrive at a new matrix for operator T, a matrix connected with $\| D^\beta_\alpha \|$ via the similarity transformation

$$D'^\beta_\alpha = (V^{-1} D V)^\beta_\alpha. \tag{3}$$

Indeed, from (2) we obtain

$$e_\alpha = (V^{-1})^\beta_\alpha e'_\beta,$$

and then

$$Te'_\alpha = Te_\beta V^\beta_\alpha = e_\nu D^\nu_\beta V^\beta_\alpha = e'_\delta (V^{-1})^\delta_\gamma D^\nu_\beta V^\beta_\alpha = e'_\delta (V^{-1} D V)^\delta_\alpha.$$

Two representations, D and $V^{-1}DV$, related through a similarity transformation are said to be *equivalent*.

Thus, a representation of the Lorentz group can be considered as being a set of linear operators acting in the finite-dimensional space of the components of field functions $u^A(x)$ and not only in the space of 4-vectors x^μ. An important idea in the theory of the representations of groups is that of reducibility. A representation of a group G in a space R is said to be irreducible if in R there are no subspaces invariant with respect to this representation except the null space and R itself, otherwise it is called reducible. For instance, an example of a reducible representation is the $(n + m)$-dimensional representation whose matrices are of the diagonal-block type:

$$D = \left(\begin{array}{c|c} \overset{n}{D_1} & \overset{n}{0} \\ \hline 0 & D_2 \end{array} \right) \begin{array}{l} \}n \\ \}m \end{array}, \tag{4}$$

where D_1 and D_2 are n-by-n and m-by-m representation matrices, respectively. Separating the representation space into invariant subspaces, we obtain irreducible representations of the group, whereby studying a reducible representation results in studying all the irreducible representations of this group. By studying the finite-dimensional representations of the Poincare group we can obtain the possible types of field functions $u(x)$ and the laws that govern their transformation

$$u^A(x) \to u'^A(x') = D^A_B u^B(x), \tag{5}$$

when the coordinates are transformed thus:

$$x^\mu \to x'^\mu = A^\mu_\nu x^\nu + a^\mu. \tag{6}$$

As an example of the above-said, let us consider the tensor representation

$$u'^{\mu\nu\cdots\rho}(x') = A^\mu_{\alpha_1} A^\nu_{\alpha_2} \cdots A^\rho_{\alpha_N} u^{\alpha_1\alpha_2\cdots\alpha_N}(x),$$

where N Lorentz transformation matrices A are multiplied by $u^{\alpha_1\alpha_2\cdots\alpha_N}(x)$, the components of a contravariant tensor of rank N that are even with respect to coordinate inversion. The corresponding pseudotensors change sign under reflection of an odd number of spatial coordinate axes.

Particular cases of a tensor representation are

(1) the scalar (pseudoscalar) representation

$$u'(x') = u(x), \tag{7}$$

(2) the contravariant vector (pseudovector) representation

$$u'^\nu(x') = A^\nu_\mu u^\mu(x), \tag{8}$$

(3) the covariant vector (pseudovector) representation

$$u'_\nu(x') = (A^{-1})^\mu_\nu u_\mu(x). \tag{9}$$

A special case is the *spinor representation* of the Lorentz group. This representation, together with the spinor representation of the rotation group, can be obtained.

The Poincare and $O(3)$ groups discussed above belong to what is known as *Lie group*. A Lie group is characterized by the fact that all the transformations result in the most general transformation that depends analytically on the r independent (or essential) parameters a^i, for instance,

$$x'^\nu = f^\nu(x^1, ..., x^n, a^1, ..., a^r).$$

A more general definition of a Lie group consists in the following.,

Let us consider a continuous, or topological, group, that is, a group such that (1) in the set of the group elements we can introduce the concepts of proximity (an infinitely small neighborhood), convergence, etc., and (2) from the smallness in the variation in one multiplicand of the group follows the smallness of the variation of the product.

Let us assume that to each element A of a given topological group we can assign, in a one-to-one manner, a set of essential parameters $a = (a^1, ..., a^r)$; these sets form the so-called group, or parametric, space. Such parametrization is fixed to within an arbitrary N-times differentiable transformation $a' = f(a)$.

The one-to-one correspondence

$$A \leftrightarrow (a^1, ..., a^r), B \leftrightarrow (b^1, ..., b^r), AB = C, C \leftrightarrow (c^1, ..., c^r) \qquad (10)$$

implies the group composition law

$$c^i = \varphi^i(a, b), \qquad (11)$$

where φ^i is an analytic function of a and b.

Such a topological group constitutes an r-parameter Lie group.

Suppose that in a certain representation of a Lie group there are operators T_a depending on parameter a. Then the one-to-one correspondence (10) leads to the following restrictions on φ^i:

(1) the *associativity*

$$T_a(T_b T_c) = (T_a T_b) T_c,$$

whence

$$\varphi[a, \varphi(b, c)] = \varphi[\varphi(a, b), c]; \qquad (12)$$

(2) the existence of a *unit element*:

$$T_{a_0} T_a = T_a T_{a_0} = T_a,$$

where $T_{a0} = 1$ is the unit operator, whence

$$a = \varphi(a, a_0) = \varphi(a_0, a), \qquad (13)$$

and in what follows we put $a_0 = 0$;

(3) the existence of an *inverse element*:

$$T_{\bar{a}} T_a = T_a T_{\bar{a}} = 1,$$

where $T_{\bar{a}} = T_a^{-1}$, i.e.

$$a_0 = \varphi(a, \bar{a}) = \varphi(\bar{a}, a) = 0, \qquad (14)$$

which implies that we can find $\bar{a}$ if $(\partial \varphi^i / \partial a^j) \neq 0$.

Infinitely small (or infinitesimal) transformations occupy a special place in the theory of *Lie groups*. Suppose we have a representation of a *Lie group* in which the elements are the operators T_a that are analytic in parameters a, that is, $T_a = A\,(a)$. Let us introduce the operators

$$X_i = \left(\frac{\partial A}{\partial a^i}\right)_{a=0} \quad (i = 1, 2, ..., r), \tag{15}$$

which are known as the *infinitesimal operators* of the group, or generators of the group. Then in the neighborhood of the unit element for small values of a we obtain the following expression for a group element in the linear approximation:

$$A = I + a^i X_i. \tag{16}$$

As we recall from the definition of a Lie group, parametrization is defined to within the transformation

$$a^i \to a'^i = f^i\,(a), \ \det\,(\partial f^i / \partial a^j) \neq 0.$$

Then for the new parameters a'^i we obtain the generators

$$X_i' = \left(\frac{\partial A}{\partial a'^i}\right)_0 = \left(\frac{\partial A}{\partial a_j}\right)_0 \left(\frac{\partial a^j}{\partial a'^i}\right)_0 = X_j \left(\frac{\partial a^j}{\partial a'^i}\right)_0, \tag{17}$$

where the indices 0 mean that the derivatives are taken at $a = 0$ and $a' = 0$, respectively.

We will now derive a remarkable relationship connecting the generators of a Lie group. Suppose that $A = A\,(a)$ and $B = B\,(b)$ are operators belonging to a certain representation of a Lie group. We set up the product $C = BAB^{-1}$. If the group is noncommutative, that is, $BA \neq AB$, then $C \neq A$. For a commutative, or Abelian, group, for which the order in which the factors occur in a product is unimportant, that is, $BA = AB$, we get, obviously, $C = A$. The product C has corresponding to it a parameter

$$c = \varphi\,(b, \varphi\,(a, \bar{b})]. \tag{18}$$

For infinitely small transformations,

$$A = I + a^i X_i, \quad B = I + b^i X_i. \tag{19}$$

We write the product as follows:

$$C = BAB^{-1} = I + a^i X_i - a^i b^j\,(X_i X_j - X_j X_i) = I + a^i X_i', \tag{20}$$

where

$$X_i' = X_i - b^j\,(X_i X_j - X_j X_i). \tag{21}$$

On the other hand,

$$C = I + c^i X_i, \tag{22}$$

and according to (17),

$$X_i' = \frac{\partial c^k}{\partial a^i} X_k. \tag{23}$$

Let us now find parameter c^k. Allowing for (18), we have, to within second-order terms,

$$\varphi^k\,(a, \bar{b}) = a^k + \bar{b}^k + f_{ji}^k\, a^j \bar{b}^i + f_{ji}'^k\, a^j a^i + f_{ji}''^k\, \bar{b}^j \bar{b}^i.$$

Since $\varphi(a, 0) = a$ and $\varphi(0, \bar{b}) = \bar{b}$, the coefficients f' and f'' vanish, and therefore

$$c^k = a^k + f^k_{ji}(b^j a^i - b^j b^i - a^j b^i), \tag{24}$$

where we have allowed for the fact that $\bar{b} = -b$ near the origin o coordinates.
If we differentiate (24) with respect to a^i, we get

$$\frac{\partial c^k}{\partial a^i} = \delta^k_i + (f^k_{ji} - f^k_{ij}) b^j = \delta^k_i + c^k_{ij} b^j, \tag{25}$$

with

$$c^k_{ij} = f^k_{ji} - f^k_{ij}$$

being the antisymmetric combinations $(c^k_{ij} = -c^k_{ji})$ known as structure constants.

If we now substitute (25) into (23), we get

$$X'_i = X_i - X_k c^k_{ij} b^j \tag{26}$$

Comparing (26) with (21), we arrive at an important relation:

$$X_i X_j - X_j X_i = c^k_{ij} X_k, \tag{27}$$

or, in brief form,

$$[X_i, X_j] = c^k_{ij} X_k. \tag{28}$$

The expression

$$[X_i, X_j] \equiv X_i X_j - X_j X_i \tag{29}$$

is called a commutator and it satisfies the obvious antisymmetry condition

$$[X_i, X_j] = - [X_j, X_i], \tag{30}$$

and the *Jacobi identity*

$$[[X_i, X_j], X_k] + [[X_k, X_i], X_j] + [[X_j, X_k], X_i] = 0. \tag{31}$$

The Jacobi identity can easily be verified on the basis of the definition of a commutator. It provides an obvious restriction on the structure constants.

Formula (28) does not depend on the representation. This is clear if only from the fact that the explicit form of a representation was not used when we derived (28). This implies that the structure constants, too, are independent of the representation. These constants are the individual characteristics of a Lie group, whose generators satisfy the commutation relations (28). We can see that knowing the operators of infinitesimal transformations makes it possible to reconstruct the entire group of finite transformations.

Note, however, that the explicit form of the commutation relations and the structure constants depends on the choice of the parameters, which, as we know, are not determined uniquely. The structure constants c^k_{ij} are *tensors* with respect to linear transformations a^i, as follows from the definition of these constants.

The linear space formed by the generators of a Lie group is known as the Lie algebra of that group. In this algebra the product of elements is given in the form of commutators (29) and the *Jacobi identity* (31) is the associativity condition.

Knowing the Lie algebra of a certain Lie group, that is, knowing the generators and the structure constants, is sufficient for building the Lie group, that is, for finding all the finite elements that form the group. To verify this statement, let us assume that in G there is a one-parameter *subgroup**, that is, a subgroup whose elements T_t are additively dependent on a single parameter t:

$$T_{t_1} T_{t_2} = T_{t_1 + t_2}. \tag{32}$$

This means that in the neighbourhood of the origin in the parameter space we have specified a curve

$$a^i = a^i(t), \quad a^i(0) = 0 \tag{33}$$

and the law of multiplication has the form

$$a^i(t_1 + t_2) = \varphi^i(a(t_1), \, a(t_2)). \tag{34}$$

The last relationship can be considered as the functional equation for determining (33).

Let us consider a representation of the one-parameter subgroup in the parameter space

$$T_\tau a^i(t) = a^i(t + \tau). \tag{35}$$

It is clear that in this representation the generator has the form $X_t = d/dt$. We take the derivatives of both sides of (35) with respect to parameter τ and use the fact that $d/dt = d/d\tau$, which follows from the additivity of the group. As a result we get

$$\frac{dT_\tau}{d\tau} a^i(t) = \frac{d}{dt}[T_\tau a^i(t)], \tag{36}$$

$$\Rightarrow \quad \frac{dT_\tau}{d\tau} a^i(t) = X_t T_\tau a^i(t). \tag{37}$$

This leads to the following operator equation

$$\frac{dT_\tau}{d\tau} = X_t T_\tau \tag{38}$$

The solution of this equation with the initial condition $T_0 = I$ can be written in the form of a series:

$$T_\tau = I + \tau X_t + \frac{\tau^2}{2} X_t^2 + \ldots = \sum_{n=0}^{\infty} \frac{\tau^n}{n!} (X_\tau)^n, \tag{39}$$

which has an infinite radius of convergence in τ. In brief form we can write this series as an operator exponential function:

$$T_\tau = \exp(\tau X_\tau), \tag{40}$$

which clearly satisfies the additivity condition (32). This solution, obtained in representation (35), must hold in an arbitrary representation as well, as follows from the aforesaid. Both (39) and (40) enable finding the operator of an

---

* A subgroup is a group whose elements constitute a part of the elements of group G, that is, elements plus a multiplication law and a unit element.

arbitrary finite transformation in a given one-parameter subgroup knowing the appropriate generator.

Using the fact that the choice of the parameters of a group is not unique, we now specify the parametrization of a group in such a manner that the law of composition becomes additive

$$\varphi^i (a, b) = a^i + b^i$$

for each $i = 1, 2, ..., r$. Such a system of parameters is said to be canonical.

Assigning to one parameter, say a^i, arbitrary admissible values $a^i = \tau$ and setting all the other parameters zero, $a^k = 0$ $(k \neq i)$, we arrive at a one-parameter subgroup. Its operators can be represented, in accord with (40), as

$$T_a i = \exp (a^i X_i). \tag{41}$$

This formula does not contain summation over i, and the generator is $X_i = \partial / \partial a^i$.

In the same way we can construct the one-parameter subgroups corresponding to the other parameters a^k. The total number of such subgroups is equal to the number of parameters r.

In group theory it is proved in general form that any element of a Lie group is either an element of a one-parameter subgroup or can be represented in the form of a product of such elements (Lie's theorem). Thus, knowing the operators of infinitely small transformations (generators), one can reconstruct the entire group. We will not prove this theorem in its most general form. Instead we will illustrate the theorem by constructing the group from one-parameter subgroups using the concrete examples of groups of coordinate transformations: the groups of translations and three-dimensional rotations and the Lorentz group.

Let us first write out the generators explicitly, in the coordinate representation. Take the coordinate transformation

$$x^\mu \to x'^\mu = f^\mu (x, a) = T_a x^\mu,$$

depending on r parameters a^i. Near the identity transformation $x^\mu = f^\mu (x, 0)$ we have

$$x'^\mu = x^\mu + dx^\mu = x^\mu + \left[\frac{\partial f^\mu (x, a)}{\partial a^i} \right]_{a = 0} a^i. \tag{42}$$

Let us consider an arbitrary scalar function of coordinates, $\psi (x)$.

Under a coordinate transformation $x \to Tx$ this function does not change:

$$\psi (x) \to \psi' (x') = \psi' (Tx) = \psi (x).$$

From this follows the law of change of the form of the function that compensates for coordinate transformation:

$$\psi' (x) = \psi (T^{-1} x).$$

For infinitesimal transformations we have

$$\psi' (x) = \psi (x) + dx^\mu \frac{\partial \psi (x)}{\partial x^\mu},$$

where dx^μ corresponds to a transformation that is the inverse to (42)

$$dx^\mu = -\left[\frac{\partial f^\mu}{\partial a^i}\right]_{a=0} a^i,$$

so that

$$\psi'(x) = \psi(x) - a^i \left[\frac{\partial f^\mu}{\partial a^i}\right]_{a=0} \frac{\partial}{\partial x^\mu} \psi(x).$$

Thus, we can write the generators of the transformation as follows:

$$X_i = -\left[\frac{\partial f^\mu(x, a)}{\partial a^i}\right]_{a=0} \frac{\partial}{\partial x^\mu}. \tag{43}$$

Using these concepts, one can construct the basic groups, e.g. (i) Translation group, (ii) Rotation-in-a plane group.

7. State and Prove Noether's theorem.

Ans. *Noether's theorem.* The theorem states that under certain conditions there exists a set of *integrals of motion* (dynamical invariants) that characterize a *field* (or a system of fields). This theorem provides a method for constructing the complete set of such integrals for any systems of fields with a given Lagrangian invariant with respect to a certain group of continuous transformation.

Let us consider a continuous (topological) group of coordinate transformations: $x \to x' = f(x, a)$. Suppose that we have a system of field functions $u^A = u^A(x)$ that realize the linear representation of this group:

$$x \to x', \quad u^A(x) \to u'^A(x'), \quad A = 1, 2, ..., N. \tag{1}$$

A function (or functional) I for which

$$I(x, u(x), \partial u/\partial x, ...) = I(x', u'(x'), \partial u'/\partial x', ...)$$

is said to be an invariant of the group. In the action integral we can write

$$S = \int d^4x L(x, u, \partial u/\partial x), \tag{2}$$

where integration is carried out over any real range x and we assume that L depends only on derivatives whose order is no higher than the first, $\partial u/\partial x$. Thus, S is action with variable limits of integration, and L is the density of the Lagrangian. The invariance of functional S means that

$$S = \int d^4x L(x, u, \partial u/\partial x) = \int d^4x' L(x', u', \partial u'/\partial x') \tag{3}$$

for any transformations of the (1) type.

Let us consider the variation of S:

$$\delta S = \int d^4x' L(x', u'(x'), \partial u'/\partial x') - \int d^4x L(x, u(x), \partial u/\partial x). \tag{4}$$

First we separate the portion of the variation that is linked with the variation in coordinates, $x \to x' = x + \delta x$, and the variation caused by a change in the form of $u^A(x)$,

$$u^A(x) \to u'^A(x) = u^A(x) + \overline{\delta}u^A.$$

The variation of $u^A(x)$ that is not due to the variation in coordinates is called local. The total variation

$$\delta u^A = u'A(x') - u^A(x) = \overline{\delta}u^A(x) + \frac{\partial u^A}{\partial x^\nu}\delta x^\nu \qquad (5)$$

is said to be the essential variation of u^A. The variation in the Lagrangian density can be written in the form

$$\delta L = \overline{\delta}L + \frac{\partial L}{\partial x^\nu}\delta x^\nu, \qquad (6)$$

where $\overline{\delta}L$ is the variation in L due to variations in the form of function u, and the procedure for finding the derivatives of L presupposes both the explicit and implicit dependences of L on x. The elements of a 4-volume in the new and old coordinates are related through the formula

$$d^4x' = \det\left\|\frac{\partial x'^\mu}{\partial x^\nu}\right\|d^4x.$$

Let us estimate the magnitude of the Jacobian to within linear terms. We use the identity

$$\det A = e^{\mathrm{Tr}\ln A}, \qquad (7)$$

which is valid for every matrix A. In the linear approximation for matrix $A \doteq \|\partial x'^\mu/\partial x^\nu\|$ we have

$$A = \|\delta_\nu^\mu + \partial\delta x^\mu/\partial x^\nu\|,$$
$$(\ln A)_\nu^\mu = \ln(\delta_\nu^\mu + \partial\delta x^\mu/\partial x^\nu) = \partial\delta x^\mu/\partial x^\nu.$$

Then identity (7) yields (in the same approximation)

$$\det A = \exp(\partial\delta x^\mu/\partial x^\mu) = 1 + \partial\delta x^\mu/\partial x^\mu.$$

This gives us the desired relation:

$$d^4x' = \left(1 + \frac{\partial\delta x^\nu}{\partial x^\nu}\right)d^4x.$$

On the whole, for the action integral one obtains

$$\delta S = \int d^4x\left\{\overline{\delta}L + \frac{\partial}{\partial x^\nu}(L\,\delta x^\nu)\right\}. \qquad (8)$$

The variation of the form of L is

$$\overline{\delta}L = \frac{\partial L}{\partial u^A}\overline{\delta}u^A + \frac{\partial L}{\partial u^A_{,\nu}}\overline{\delta}u^A_{,\nu}. \qquad (9)$$

Since the variation $\bar{\delta} u^A$ is local, one can write

$$\bar{\delta} u^A_{,\nu} = \frac{\partial}{\partial x^\nu} \bar{\delta} u^A. \tag{10}$$

Substituting (9) into (8), allowing for (10), and performing simple identity transformations, we can rewrite δS thus:

$$\delta S = \int d^4 x \left\{ \frac{\delta S}{\delta u^A} \bar{\delta} u^A + \frac{\partial}{\partial x^\nu} \left(L \, \delta x^\nu + \frac{\partial L}{\partial u^A_{,\nu}} \bar{\delta} u^A \right) \right\}, \tag{11}$$

where the functional derivative is

$$\frac{\delta S}{\delta u^A} \equiv \frac{\partial L}{\partial u^A} - \frac{\partial}{\partial x^\nu} \frac{\partial L}{\partial u^A_{,\nu}}.$$

Suppose action is invariant, that is, condition (3) is satisfied. Then $\delta S = 0$ and, in view of the fact that the domain of integration is arbitrary, (11) yields

$$\frac{\delta S}{\delta u^A} \bar{\delta} u^A + \frac{\partial}{\partial x^\nu} \left(L \, \delta x^\nu + \frac{\partial L}{\partial u^A_{,\nu}} \bar{\delta} u^A \right) = 0. \tag{12}$$

We may recall that the transformations (1) constitute a group. There-fore,

$$\delta x^\mu = (X_i)^\mu_\nu \, x^\nu \, \delta a^i, \tag{13}$$

$$\delta u^A (x) = (X_i)^A_B \, u^B (x) \, \delta a^i, \quad i = 1, 2, ..., r.$$

Here $(X_i)^\mu_\nu$ and $(X_i)^A_B$ are the generators of the group in the x- and the u-representation. Since all the a^i are independent, the coefficients of each variation δa^i in (12) must vanish independently. We have therefore arrived at the theorem formulated by Noether as follows.

If the functional S is invariant with respect to an r-parameter Lie group, then r linearly independent combinations of the functional derivatives $\delta S / \delta u^A$ turn into divergences. Actually, most often we have to do with a variational problem, that is, the field functions u^A satisfy the Lagrange equations

$$\delta S / \delta u^A = 0 \quad (A = 1, 2, ..., N).$$

Therefore, (12) implies that

$$\frac{\partial}{\partial x^\nu} \left[L \, \delta x^\nu + \frac{\partial L}{\partial u^A_{,\nu}} \bar{\delta} u^A \right] = 0.$$

Taking into account the expressions for variations, (5) and (13), we find that

$$\frac{\partial}{\partial x^\nu} \left\{ L \, (X_i)^\nu_\mu \, x^\mu + \frac{\partial L}{\partial u^A_{,\nu}} [(X_i)^A_B \, u^B - u^A_{,\lambda} \, (X_i)^\lambda_\mu \, x^\mu] \right\} \delta a^i = 0.$$

In view of the fact that the variations of the parameters, δa^i, are independent, from the last equation we get r continuity equations

$$\frac{\partial}{\partial x^\nu} J_i^\nu = 0 \qquad (i = 1, 2, ..., r) \tag{14}$$

for the Noether currents defined thus:

$$J_i^\nu = -L(X_i)_\mu^\nu x^\mu + \frac{\partial L}{\partial u_{,\nu}^A}[u_{,\lambda}^A (X_i)_\mu^\lambda x^\mu - (X_i)_B^A u^B]. \tag{15}$$

Equation (14) can also be written in integral form:

$$\int d^4x \frac{\partial}{\partial x^\nu} J_i^\nu = 0, \tag{16}$$

where integration is carried out over an arbitrary volume of Minkowski's space-time. Next we employ the divergence theorem and rewrite (16) in the form of a surface integral over the boundary of the volume:

$$\int d^4x \frac{\partial}{\partial x^\nu} J_i^\nu = \int_\sigma d\sigma_\nu J_i^\nu = 0. \tag{17}$$

Here $d\sigma_\nu$ is the projection of the element of hypersurface σ encompassing the 4-volume onto the 3-plane that is perpendicular to the x^ν axis, namely, $d\sigma_0 = dx^1 dx^2 dx^3$, $d\sigma_1 = dx^0 dx^2 dx^3$, etc. Let us assume that the fields and the currents J_i^ν ($\nu = 1, 2, 3$) vanish at spatial infinity, that is,

$$J_i^\nu|_{|\mathbf{r}| \to \infty} \to 0 \qquad (\nu = 1, 2, 3).$$

Then, selecting the volume of integration in such a way that in spacelike directions it extends to infinity, we get, from (17),

$$\int_{t=t_1} J_i^0 d^3x = \int_{t=t_2} J_i^0 d^3x. \tag{18}$$

In these integrals, $d^3x = dx^1 dx^2 dx^3 = d\sigma_0$ are elements of 3-volumes, that is, in essence the area elements of the hyperplanes corresponding to the constant values $t = t_1$ and $t = t_2$, respectively. If we introduce the so-called Noether charges,

$$Q_i = \int J_i^0 d^3x,$$

we will find, as a consequence of (18), that they are time independent: $dQ_i/dt = 0$ ($i = 1, 2, ..., r$).

We therefore arrive at another formulation of Noether's theorem: if the action integral is an invariant of an r-parameter Lie group, then, in view of the field equations, we have r integrals of motion that do not vary with time (dynamical invariants, Noether charges):

$$Q_i = \int J_i^0 d^3x = \text{constant} \qquad (i = 1, 2, ..., r). \tag{19}$$

Equation (19) is not written in covariant form since time is explicitly isolated. We can write covariant equations if while integrating in (17) we restrict the volume of integration not by hyperplanes $t = $ constant but by spacelike surfaces* in the timelike directions, assuming that in spacelike directions the volume extends to infinity, with the current J_i^ν falling off to zero in these directions. Then instead of (19), one obtains

$$Q_i \equiv \int_\sigma J_i^\nu \, d\sigma_\nu = \text{constant.} \tag{20}$$

Here the charges are defined as integrals over *spacelike surfaces**, and we can easily see that they do not depend on the spacelike surface σ.

Note that Noether currents are not uniquely defined since we can always add to them divergences of any antisymmetric tensors $f_i^{\mu\nu} = - f_i^{\mu\nu}$. Indeed, in this case the continuity equation (14) are not violated, since for an antisymmetric tensor we have

$$\frac{\partial^2}{\partial x^\nu \partial x^\mu} f_i^{\mu\nu} \equiv 0.$$

More than that, the charge Q_i (which are conserved) do not change their form since

$$\int d^3x \, \frac{\partial f_i^{0\mu}}{\partial x^\mu} = \int d^3x \, \frac{\partial}{\partial x^k} f_i^{0k} = \int ds_k f_i^{0k} = 0 \quad (k = 1, 2, 3),$$

where the fields and f^{0k} are assumed to vanish on the surface encompassing the three-dimensional volume (the last integral is taken over this surface).

8. Explain the concept of energy-momentum tensor.

Ans. Let us consider the integrals of motion that emerge as a result of the requirement that action be invariant under transformations belonging to the translation group and the proper Lorentz group.

Under translation transformations, $x'^\nu = x^\nu + \delta a^\nu$, the field functions do not transform: $u'^A(x') = u^A(x)$. Since

$$\delta x^\nu = (X_\alpha)_\mu^{\ \nu} x^\mu \delta a^\alpha = \delta a^\nu,$$

we can write

$$(X_\alpha)_\mu^{\ \nu} x^\mu = \delta_\alpha^{\ \nu}. \tag{21}$$

At the same time $\delta u^A = 0$, and therefore

$$(X_\alpha)_B^{\ A} = 0. \tag{22}$$

Substituting (21) and (22) into the definition (15) of the *Noether current* (Q. No. 7) we arrive at a new quantity,

$$T_\alpha^{\ \nu} = \frac{\partial L}{\partial u_{,\nu}^A} u_{,\alpha}^A - L \delta_\alpha^{\ \nu}, \tag{23}$$

---

* Spacelike surface: A spacelike surface possesses the following property: a normal to any point on this surface is a timelike vector, with the result that any two points on the surface are separated by a spacelike interval.

known as the canonical energy-momentum tensor of the system (the field). Note that the upper index is placed somewhat to the right of the lower index so as to leave room for raising the lower index when we go over from the mixed tensor (23) to the contravariant tensor

$$T^{\alpha v} = \frac{\partial L}{\partial u^A_{,v}} u^{A,\alpha} - L g^{\alpha v}. \tag{24}$$

By Noether's theorem, the invariance of action under translations, $\delta S = 0$, implies the continuity equation

$$\frac{\partial T^{\alpha v}}{\partial x^v} = 0 \qquad (\alpha = 0, 1, 2, 3). \tag{25}$$

The integral conservation law has the form

$$\int T^{\alpha 0} d^3 x = \text{constant} = P^\alpha \qquad (\alpha = 0, 1, 2, 3), \tag{26}$$

where integration is carried out over the entire volume occupied by the field. The P^α are the components of the 4-momentum of the field. Indeed, let us write

$$T^{00} = \frac{\partial L}{\partial u^A_{,t}} u^A_{,t} - L = H. \tag{27}$$

As we can see, the T^{00} component of the energy-momentum tensor coincides with the Hamiltonian density of the field, $\mathcal{H}$. Therefore, the volume integral

$$P^0 = \int d^3 x \, \mathcal{H} = \int d^3 x T^{00} \tag{28}$$

determines the total energy of the field. In view of the fact that the P^α are covariant quantities, the other components are those of the total momentum of the field:

$$P^k = \int d^3 x T^{k0} = \int d^3 x \frac{\partial L}{\partial u^A_{,t}} u^{A,k} \qquad (k = 1, 2, 3). \tag{29}$$

Thus, in view of the homogeneity of space-time, which property is expressed by the condition $\delta S = 0$ as $x \to x + \delta a$, we arrive at the conservation law for the total 4-momentum of the field: $P^\alpha = \text{constant}$. It must be noted at this point that when there are several interacting fields of varying nature, the 4-momentum of the entire system is conserved. In this case, the indices A, in the definition (23) of the energy-momentum tensor, over which indices summation is performed, stand not only for the field components but for the type of the field as well.

Let us now consider the proper Lorentz transformations. Using the notation by which the parameters of the generators of a group are included in two indices, we can write the variations of the coordinates and field functions as

$$\delta x^v = (X_{\alpha\beta})^v_\mu x^\mu \delta a^{\alpha\beta},$$

$$\delta u^A = (X_{\alpha\beta})^A_B u^B \delta a^{\alpha\beta}, \tag{30}$$

where $a^{12} = \varphi^3$, etc., and $a^{10} = \tilde{\varphi}^1$, etc. For the spatial part of the Noether current, which is not involved in the transformations of the field functions, the definition (15) yields

$$L_{\alpha\beta}^{\nu} = u_{,\sigma}^A (X_{\alpha\beta})_{\mu}^{\sigma} x^{\mu} \frac{\partial L}{\partial u_{,\nu}^A} - (X_{\alpha\beta})_{\mu}^{\nu} x^{\mu} L = (X_{\alpha\beta})_{\mu}^{\sigma} x^{\mu} T_{\sigma}^{\nu},$$

where T_{σ}^{ν} is the energy-momentum tensor defined in (23). Recalling the explicit form of the generator of the Lorentz group,

$$(X_{\alpha\beta})_{\mu}^{\sigma} = \delta_a^{\sigma} g_{\beta\mu} - \delta_{\beta}^{\alpha} g_{\alpha\mu}, \tag{31}$$

we find that

$$L_{\alpha\beta}^{\nu} = T_{\alpha}^{\nu} x_{\beta} - T_{\beta}^{\nu} x_{\alpha}. \tag{32}$$

The quantities $L_{\alpha\beta}^{\nu}$ (32) form a tensor of rank three, the tensor of the angular momentum of the field, which is antisymmetric in the lower indices:

$$L_{\alpha\beta}^{\nu} = - L_{\beta\alpha}^{\nu}.$$

Let us now consider the field part of the Noether current, the part that is involved in the transformations of the field functions:

$$S_{\alpha\beta}^{\nu} = - (X_{\alpha\beta})_B^A u^B \frac{\partial L}{\partial u_{,\nu}^A}. \tag{33}$$

These quantities form a tensor of rank three, the spin, or intrinsic, angular momentum tensor of the field, which characterizes the transformation properties of the field with respect to spatial-temporal rotations (polarization properties).

To calculate $S_{\alpha\beta}^{\nu}$ we must know the representation of the generators $X_{\alpha\beta}$ in the space of the field functions u^A. If the field is scalar (one-component), that is $u'(x') = u(x)$, then $(X_{\alpha\beta})_B^A \equiv 0$ and, hence,

$$S_{\alpha\beta}^{\nu} \equiv 0.$$

If the field is a vector field, that is, $u^{\nu}(x)$ ($\nu = 0, 1, 2, 3$), the generators $(X_{\alpha\beta})_{\mu}^{\nu}$ are defined by (31). For a spinor field we must know the spinor representation of the Lorentz group (see below).

The continuity equation

$$\frac{\partial}{\partial x^{\nu}} (L_{\alpha\beta}^{\nu} + S_{\alpha\beta}^{\nu}) = 0$$

describes the conservation law for the total angular momentum of the field, or the sum

$$L_{\alpha\beta}^{\nu} + S_{\alpha\beta}^{\nu}.$$

If the field is a scalar field, $S_{\alpha\beta}^{\nu} = 0$ and, hence,

$$\frac{\partial}{\partial x^{\nu}} L_{\alpha\beta}^{\nu} = 0,$$

whence

$$\frac{\partial}{\partial x^\nu} L^{\nu[\alpha\beta]} = \frac{\partial}{\partial x^\nu}(T^{\alpha\nu}x^\beta - T^{\beta\nu}x^\alpha) = T^{\alpha\beta} - T^{\beta\alpha} = 0,$$

that is,

$$T^{\alpha\beta} = T^{\beta\alpha}.$$

Thus, in the case of a scalar field the canonical energy-momentum tensor is symmetric. In the general case, however, this is not so, but if we add the divergence of an antisymmetric tensor $f^{[\nu\beta]\alpha}$,

$$T^{\nu\alpha} \to T^{\nu\alpha} + \frac{\partial}{\partial x^\beta} f^{[\nu\beta]\alpha},$$

the resulting tensor will be symmetric. The new tensor is known as the metric energy-momentum tensor.

Integrating the components $L^0_{\alpha\beta}$ and $S^0_{\alpha\beta}$ over the 3-space, we get

$$L_{\alpha\beta} = \int d^3x\, L^0_{\beta\alpha}, \quad S_{\alpha\beta} = \int d^3x\, S^0_{\beta\alpha}, \tag{34}$$

which are, respectively, the orbital angular momentum tensor and the spin, or intrinsic, angular momentum tensor of the field, both being antisymmetric:

$$L_{\alpha\beta} = -L_{\beta\alpha}, \quad S_{\alpha\beta} = -S_{\beta\alpha}.$$

In the general case of an arbitrary field their sum is conserved:

$$\frac{d}{dt}(L_{\alpha\beta} + S_{\alpha\beta}) = 0.$$

The spatial components, L_{ik},

$$L_{23} = L_1, \quad L_{31} = L_2, \quad L_{12} = L_3, \tag{35}$$

form an axial vector of the field angular momentum:

$$L_i = \frac{1}{2}\varepsilon_{ikl}L_{kl} \quad (i, k, l = 1, 2, 3). \tag{36}$$

According, for the intrinsic angular momentum of the field we get

$$S_i = \frac{1}{2}\varepsilon_{ikl}S_{kl}, \tag{37}$$

which is the spin vector of the field. Thus, the isotropy of the field, that is, the fact that $\delta S = 0$ under spatial rotations, leads to the conservation of the total angular momentum of all the interacting fields in a system:

$$L_i + S_i = \text{constant} \quad (i = 1, 2, 3). \tag{38}$$

9. What is a spinor? Give complete description of $SU(2)$ group.

Ans. *Spinor* is a mathematical entity similar to a vector but having the property that it changes sign on each rotation through 360°. The wave function of a spin 1/2 particle such as an electron, in relativistic quantum mechanics is an example of spinor.

The $SU(2)$ group

Let us consider the two-dimensional linear complex space of vectors (spinors)

$$c = c^1 e_1 + c^2 e_2 \tag{1}$$

with complex-valued base vectors e_1 and $\dot{e}_2$. We denote the components of vector c, or spinor, by a column and the components of a base vector by a row:

$$\|c^i\| = \begin{pmatrix} c^1 \\ c^2 \end{pmatrix}, \quad \|e_i\| = (e_1, e_2). \tag{2}$$

Let us define the scalar product of two vectors, c and d, in the following manner:

$$(c, d) = c^{*1}d^1 + c^{*2}d^2 = c^+d, \tag{3}$$

where

$$c^+ = (c^{*1}, c^{*2}). \tag{4}$$

Let us consider a linear transformation in this space,

$$c' = Uc. \tag{5}$$

We assume that this transformation is unitary, that is, preserves the invariance of the scalar product:

$$c' = Uc, \quad d' = Ud, \quad (c, d) = (c', d') = \text{inv}. \tag{6}$$

Let us write condition (6) explicitly:

$$(c', d') = (Uc)^+Ud = c^+U^+Ud = c^+d.$$

From this immediately follows

$$U^+ = U^{-1}. \tag{7}$$

Such matrices are known as unitary. Next we assume that the transformation matrices are unimodular, or

$$\det U = 1. \tag{8}$$

The matrix of a general linear transformation has the form

$$U = \begin{pmatrix} \alpha & \beta \\ \gamma & \delta \end{pmatrix}, \tag{9}$$

where α, β, γ, and δ are arbitrary complex numbers. Then the linear transformation of vector components can be written as follows:

$$c'^1 = \alpha c^1 + \beta c^2,$$
$$c'^2 = \gamma c^1 + \delta c^2. \tag{10}$$

Accordingly, for the base vectors we can write out the transformation $e' = eU$ explicitly:

$$e_1' = \alpha e_1 + \gamma e_2,$$
$$e_2' = \beta e_1 + \delta e_2. \tag{11}$$

The requirement (8) that U be unimodular leads to the following condition:

$$\alpha\delta - \beta\gamma = 1. \tag{12}$$

It can easily be verified that the inverse of U, provided (12) is satisfied, has the form

$$U^{-1} = \begin{pmatrix} \delta & -\beta \\ -\gamma & \alpha \end{pmatrix}.$$

Since

$$U^+ = \begin{pmatrix} \alpha^* & \gamma^* \\ \beta^* & \delta^* \end{pmatrix},$$

the unitarity condition (7) yields

$$\alpha^* = \delta, \quad \gamma^* = -\beta.$$

Thus, the unitary unimodular transformation has the form

$$c'^1 = \alpha c^1 + \beta c^2,$$
$$c'^2 = -\beta^* c^1 + \alpha^* c^2, \tag{13}$$

and its matrix

$$U = \begin{pmatrix} \alpha & \beta \\ -\beta^* & \alpha^* \end{pmatrix} \tag{14}$$

depends on three real parameters, since the four real numbers α_1, α_2, β_1, β_2, where $\alpha = \alpha_1 + i\alpha_2$ and $\beta = \beta_1 + i\beta_2$, obey the single condition

$$\alpha\alpha^* + \beta\beta^* = 1. \tag{15}$$

We can easily see that the set of matrices of the (14) type augmented by condition (15) forms a group. Indeed, the product of any two unitary unimodular matrices is a matrix of the same type, each matrix U has an inverse, U^{-1}, since $\det U = 1 \neq 0$, and the unit matrix I belongs to the same set. This group of special (unimodular) unitary matrices of the transformation of the linear complex two-dimensional space is denoted by $SU(2)$. The group space of $SU(2)$ is the surface of the unit sphere in four-dimensional real space:

$$|\alpha|^2 + |\beta|^2 = \alpha_1^2 + \alpha_2^2 + \beta_2^1 + \beta_2^2 = 1.$$

Let us investigate the corollaries that follow from the unimodularity and unitarity of the transformations that constitute this group. For two vectors c and d we get

$$c' = Uc, \quad d' = Ud.$$

Let $\det U = 1$. We construct the quadratic combination

$$c'^1 d'^2 - c'^2 d'^1 = (\alpha c^1 + \beta c^2)(\gamma d^1 + \delta d^2) - (\gamma c^1 + \delta c^2)(\alpha d^1 + \beta d^2)$$
$$= (\alpha\delta - \beta\gamma)(c^1 d^2 - c^2 d^1),$$

and in view of (12) we have

$$c'^1 d'^2 - c'^2 d'^1 = c^1 d^2 - c^2 d^1 = \text{inv}. \tag{16}$$

From this it follows that the components c^2 and $-c^1$ are transformed covariantly with respect to the components of the basis, that is,

$$(c^2, -c^1) \sim (e_1, e_2), \tag{17}$$

and form a covariant vector.

Next, in view of the unitary condition we have

$$c^{*1} c^1 + c^{*2} c^2 = \text{inv},$$

and therefore c^{*1} and c^{*2} are transformed covariantly with respect to e_1 and e_2, that is, also form a covariant vector

$$(c^{*1}, c^{*2}) \sim (e_1, e_2). \tag{18}$$

Finally, the components c^{*2} and $-c^{*1}$ are transformed under unitary transformations contravariantly with respect to c^2 and $-c^1$ and, hence, covariantly with respect to c^1 and c^2, that is, form a contravariant vector

$$\begin{pmatrix} c^{*2} \\ -c^{*1} \end{pmatrix} \sim \begin{pmatrix} c^1 \\ c^2 \end{pmatrix}. \tag{19}$$

Linear Representations of SU (2)

Let us consider the linear space whose base vectors are homogeneous products of vectors e_i of order N = 1, 2, 3, 4, ...:

$$e_1^N, e_1^{N-1} e_2, ..., e_2^N. \tag{20}$$

When $N = 1$, we have the initial two-dimensional space. In general, the linear space of vectors

$$c^0 e_1^N c^1 e_1^{N-1} e_2 + ... + c^N e_2^N \tag{21}$$

has a dimension of $N + 1$. Under linear transformations of the basis of the initial space $e \to eU$, (11), the base vector (20) are also transformed linearly, since the result,

$$e_1^{N-s} e_2^s \to (\alpha e_1 + \gamma e_2)^{N-s} (\beta e_1 + \delta e_2)^s, \tag{22}$$

forms a linear combination of homogeneous terms of the (20) type. Thus, we have arrived at linear representations of the SU (2) group. Commonly, N is replaced by the quantity $j = N/2 = 1/2, 1, 3/2, ...$, with the result that the dimension of the representation space is $N + 1 = 2j + 1$. For $N = 1$ we have $j = 1/2$; this is the representation of SU (2) in itself, or the spinor representation. For $N = 2$ ($j = 1$) the dimension of the representation space is $2j + 1 = 3$; this is the vector representation of the SU (2) group.

Let us consider the three-dimensional vector representation ($j = 1$) in greater detail. The representation space consists of vectors

$$c = c^0 e_1^2 + c^1 e_1 e_2 + c^2 e_2^2 = \xi^0 \varepsilon_0 + \xi^1 \varepsilon_1 + \xi^2 \varepsilon_2, \tag{23}$$

where we have introduced a new basis $\varepsilon_0 = e_1^2/\sqrt{2}$, $\varepsilon_1 = e_1 e_2$, $\varepsilon_2 = e_2^2/\sqrt{2}$ and new components $\xi^0 = \sqrt{2} c^0$, $\xi^1 = c^1, \xi^2 = \sqrt{2} c^2$.

Vector (23) can be represented in the form of a product of two vectors $d = d^1 e_1 + d^2 e_2$ and $g = g^1 e_1 + g^2 e_2$. Therefore, the quadratic combination

$$(c^1)^2 - 4c^0 c^2 \sim (d^1 g^2 + d^2 g^1)^2 - 4d^1 g^1 d^2 g^2 = (d^1 g^2 - d^2 g^1)^2 = \text{inv}, \tag{24}$$

in accord with (16), is an invariant of the vector representation of the SU (2) group. The following quantity is also an invariant, in view of the unitarity of the spinor representation of the SU (2) group (6):

$$| \xi^0 |^2 + | \xi^1 |^2 + | \xi^2 |^2 \sim 2 (| d^1 |^2 + | d^2 |^2) = \text{inv}. \tag{25}$$

Thus, the vector representation (23) of the SU (2) group is also unitary. Let us consider the transformation $\xi^0, \xi^1, \xi^2 \to x, y, z$:

$$x = \frac{1}{\sqrt{2}}(\xi^2 - \xi^0), \quad y = -\frac{i}{\sqrt{2}}(\xi^2 + \xi^0), \quad z = \xi^1. \tag{26}$$

This is simply a unitary similarity transformation in the representation space:

$$\begin{pmatrix} x \\ y \\ z \end{pmatrix} = V \begin{pmatrix} \xi^0 \\ \xi^1 \\ \xi^2 \end{pmatrix}, \quad V = \begin{pmatrix} -\dfrac{1}{\sqrt{2}} & 0 & \dfrac{1}{\sqrt{2}} \\ -\dfrac{i}{\sqrt{2}} & 0 & -\dfrac{i}{\sqrt{2}} \\ 0 & 1 & 0 \end{pmatrix}. \tag{27}$$

The inverse matrix is $V^{-1} = V^+$. The transformation (27) preserves the scalar product

$$\mid \xi^0 \mid^2 + \mid \xi^1 \mid^2 + \mid \xi^2 \mid^2 = x2 + y^2 + z^2. \tag{28}$$

If the matrix of transformation (22) in the space ξ^0, ξ^1, ξ^2 is denoted by D_1, then an equivalent transformation in the space x, y, z is carried out by a matrix related to D_1 in the following manner

$$D_1' = VD_1V^+. \tag{29}$$

In view of (24) and (25), the sum of the squares of the new components,

$$x^2 + y^2 + z^2 = (c^1)^2 - 4c^0c^2 = \mid \xi^0 \mid^2 + \mid \xi^1 \mid^2 + \mid \xi^2 \mid^2, \tag{30}$$

is also an invariant of the group. Thus, the transformation D_1', the vector representation of SU (2), is a rotation. What are the conditions that ensure that this rotation is real? Allowing for the representation (23) of c in the form of the product of d and g and going over to the components (26), we get

$$x \sim d^2g^2 - d^1g^1, \quad y \sim -i(d^1g^1 + d^2g^2), \quad z \sim d^1g^2 + d^2g^1. \tag{31}$$

In view of (19), for the transformations belonging to SU (2) group we have $g^1 \sim d^{*2}$ and $g^2 \sim -d^{*1}$, whereby (31) is replaced by the following:

$$x \sim -(d^2d^{*1} + d^1d^{*2}), \quad y \sim i(d^2d^{*1} - d^1d^{*2}), z \sim -d^1d^{*1} + d^2d^{*2}. \tag{32}$$

We see that, x, y, z are transformed as real combinations of the components of spinor d. Hence, the transformation coefficients must also be real, that is, x, y, z undergo real rotations belonging to the O (3) group under transformations D_1'.

Let us find the explicit form of the matrix of real rotation D_1'. Applying directly the law of transformations of base vectors e_1 and e_2,

$$e_1' = \alpha e_1 - \beta^* e_2,$$
$$e_2' = \beta e_2 + \alpha^* e_2,$$

we get

$$(\varepsilon_0', \varepsilon_1', \varepsilon_2') = (\varepsilon_0, \varepsilon_1, \varepsilon_2) D_1,$$

where

$$D_1 = \begin{pmatrix} \alpha^2 & \sqrt{2}\alpha\beta & \beta^2 \\ -\sqrt{2}\alpha\beta^* & \alpha\alpha^* - \beta\beta^* & \sqrt{2}\alpha^*\beta \\ \beta^{*2} & -\sqrt{2}\alpha^*\beta^* & \alpha^{*2} \end{pmatrix}. \tag{33}$$

If we then carry out the similarity transformation (29) using the matrix (27), we finally obtain

$$
D_1' = \begin{pmatrix} \dfrac{\alpha^2 - \beta^{*2} + \alpha^{*2} - \beta^2}{2} & \dfrac{\alpha^2 - \beta^{*2} - \alpha^{*2} + \beta^2}{2i} & -(\alpha\beta + \alpha^*\beta^*) \\[3mm] \dfrac{\alpha^{*2} + \beta^2 - \alpha^2 - \beta^{*2}}{2i} & \dfrac{\alpha^2 + \beta^2 + \alpha^{*2} + \beta^{*2}}{2} & \dfrac{\alpha\beta - \alpha^*\beta^*}{i} \\[3mm] \alpha\beta^* + \alpha^*\beta & \dfrac{\alpha\beta^* - \alpha^*\beta}{i} & \alpha\alpha^* - \beta\beta^* \end{pmatrix}. \tag{34}
$$

As we see, all the coefficients of (34) are real. Let us now show that each real rotation is contained in the representation D_1', that is, we wish to build the operator of an arbitrary rotation, the general element of the $O(3)$ group. Let us consider two matrices belonging to the $SU(2)$ group of the type

$$
A(\alpha_1) = \begin{pmatrix} \cos\alpha_1 & -i\sin\alpha_1 \\ -i\sin\alpha_1 & \cos\alpha_1 \end{pmatrix}, \quad C(\alpha_3) = \begin{pmatrix} e^{-i\alpha_3} & 0 \\ 0 & e^{i\alpha_3} \end{pmatrix}. \tag{35}
$$

In representation D_1', from the general expression we obtain

$$
A(\alpha_1) \to T_{2\alpha_1} = \begin{pmatrix} 1 & 0 & 0 \\ 0 & \cos 2\alpha_1 & -\sin 2\alpha_1 \\ 0 & \sin 2\alpha_1 & \cos 2\alpha_1 \end{pmatrix}, \tag{36}
$$

$$
C(\alpha_3) \to T_{2\alpha_3} = \begin{pmatrix} \cos 2\alpha_3 & -\sin 2\alpha_3 & 0 \\ \sin 2\alpha_3 & \cos 2\alpha_3 & 0 \\ 0 & 0 & 1 \end{pmatrix}. \tag{37}
$$

The matrix (36) coincides with the matrix (57) representing rotations about the first axis through the angle $\varphi^1 = 2\alpha_1$, while the matrix (37) coincides with the matrix (63) representing rotations about the third axis through the angle $\varphi^3 = 2\alpha_3$. We have already demonstrated that an arbitrary rotation belonging to the $O(3)$ group can be constructed out of the above-mentioned rotations if we introduce the Euler angles φ^1, φ^3, φ'^3:

$$
T_{\varphi^3 \varphi^2 \varphi'^3} = T_{\varphi^3} T_{\varphi^1} T_{\varphi'^3}.
$$

Therefore, the general element of the $SU(2)$ group parametrized by the Euler angles and having representation in $O(3)$ can be found in the following manner:

$$
U = C\left(\frac{1}{2}\varphi^3\right) A\left(\frac{1}{2}\varphi^1\right) C\left(\frac{1}{2}\varphi'^3\right).
$$

As a result one obtains

$$
U_{\varphi^3\varphi^1\varphi'^3} =
\begin{pmatrix}
\exp\left(\dfrac{\varphi^3 + \varphi'^3}{2i}\right)\cos\dfrac{\varphi^1} {2} & -i\exp\left(\dfrac{\varphi^3 - \varphi'^3}{2i}\right)\sin\dfrac{\varphi^1}{2} \\[4mm]
-i\exp\left(\dfrac{\varphi'^3 - \varphi^3}{2i}\right)\sin\dfrac{\varphi^1}{2} & \exp\left(-\dfrac{\varphi^3 - \varphi'^3}{2i}\right)\cos\dfrac{\varphi^1}{2}
\end{pmatrix}.
\tag{38}
$$

However, the correspondence between SU (2) and O (3) is not one-to-one. Indeed, according to the general formula (34), the substitution of $-U$ for U, that is, $-\alpha$ and $-\beta$ for α and β, does not change the sign of D_1', in view of the quadratic dependence of D_1' on α and β. Therefore, the representation of D_1' via matrices U, or the spinor representation of O (3) $\rightarrow SU$ (2), is two-valued. Obviously, there are two different matrices in SU (2) that correspond to the unit element in O (3):

$$
I = \begin{pmatrix} 1 & 0 \\ 0 & 1 \end{pmatrix} \text{ and } -I = \begin{pmatrix} -1 & 0 \\ 0 & -1 \end{pmatrix}.
$$

In a small neighborhood of I or $-I$ a one-to-one correspondence with the rotations from O (3) can be established, that is, to each rotation D_1' through a small angle there will correspond a unique unitary transformation U close to I or $-I$. Under a continuous variation in the angle, the matrix U continuously changes with the rotation D_1'. However, when the rotation completes a full cycle $D_1' \rightarrow D_1'$, that is, $\varphi^1 \rightarrow \varphi^1 + 2\pi$ or $\varphi^3 \rightarrow \varphi^3 + 2\pi$ or $\varphi'^3 \rightarrow \varphi'^3 + 2\pi$, matrix U is replaced with $-U$. Thus, U is a continuous two-valued function of rotation D_1'.

Let us now find the generators of SU (2) that realize the spinor representation of the corresponding generators of O (3). According to the definition of generators, we get, from (35),

$$
X_1 = \left(\frac{\partial A(\varphi^1/2)}{\partial \varphi^1}\right)_{\varphi^1 = 0} = -\frac{i}{2}\begin{pmatrix} 0 & 1 \\ 1 & 0 \end{pmatrix},
$$

$$
X_3 = \left(\frac{\partial C(\varphi^3/2)}{\partial \varphi^3}\right)_{\varphi^3 = 0} = -\frac{i}{2}\begin{pmatrix} 1 & 0 \\ 0 & -1 \end{pmatrix}.
\tag{39}
$$

Similarly to (36), we obtain the matrix of rotations about the second axis:

$$
B\left(\frac{1}{2}\varphi^2\right) = \begin{pmatrix} \cos(\varphi^2/2) & -\sin(\varphi^2/2) \\ \sin(\varphi^2/2) & \cos(\varphi^2/2) \end{pmatrix}.
\tag{40}
$$

From this we obtain the generator

$$
X_2 = -\frac{i}{2}\begin{pmatrix} 0 & -i \\ i & 0 \end{pmatrix}.
\tag{41}
$$

It is obvious that in the spinor representation the generators X_i satisfy the group commutation relation

$$[X_i, X_k] = \varepsilon_{ikj} Xj.$$

Let us now construct the hermitian generators by following the general rule:

$$S_k = iX_k = \frac{1}{2} \sigma_k. \tag{42}$$

Here we have introduce the hermitian matrices

$$\sigma_1 = \begin{pmatrix} 0 & 1 \\ 1 & 0 \end{pmatrix}, \quad \sigma_2 = \begin{pmatrix} 0 & -i \\ i & 0 \end{pmatrix}, \quad \sigma_3 = \begin{pmatrix} 1 & 0 \\ 0 & -1 \end{pmatrix}, \tag{43}$$

which are known as the Pauli matrices. They possess the following properties:

$$\sigma_1\sigma_2 = i\sigma_3, \quad \sigma_2\sigma_3 = i\sigma_1, \quad \sigma_3\sigma_1 = i\sigma_2,$$

$$\sigma_i \sigma_j + \sigma_j \sigma_i = 0, \quad i \neq j, \tag{44}$$

$$(\sigma_i)^2 = I,$$

where I is the unit 2-by-2 matrix. In compact form these relations can be written as follows:

$$\sigma_i \sigma_k = I\delta_{ik} + i\varepsilon_{ikj} \sigma_j. \tag{45}$$

We may note that the Pauli matrices are *traceless*, i.e.

$$\text{Tr } \sigma_i = 0, \tag{46}$$

which follows from the unimodularity of the transformation, det U = 1. The operators of finite rotations about the kth axis can be written in the form

$$U_k = \exp(\varphi^k X_k) = \exp(-i\varphi^k S_k) = \exp\left(-\frac{i}{2} \varphi^k \sigma_k\right). \tag{47}$$

In quantum mechanics, the hermitian operators (42) correspond to the components of the spin operators of particles with spin 1/2, while the spinors are vectors in the space in which these operators act. The eigenvectors of the operators S_k describe the quantum mechanical states with definite values of the projection of the spin on the kth axis, +1/2 or –1/2.

10. What are intrinsic symmetries of fields?

Ans. Besides the transformations that involved coordinates (spatial coordinates) and time in the case of the Lorentz group, the field functions realize the representation of the transformation group and transform according to a certain rule together with the coordinates. But coordinate transformations do not exhaust the possible transformations that the field function can undergo. There exists *intrinsic symmetries* of physical objects, or elementary particles, that reflect the internal properties of the objects, properties that are in no way connected with possible displacements in space. These symmetries manifest themselves in transformations that do not involve coordinates; such transformations are known as *gauge transformations*.

(i) Isotopic Rotations

Suppose that a field is described by complex-valued functions $u(x) = u_1 + iu_2$. Then, since the observables, such as, say, the Noether charges, which are integrals of the Lagrange equations, must be real-valued, they must depend only on the real quadratic combinations of field functions. This implies that the Lagrangian and, obviously, the action integral of this field must be invariant with respect to transformations of the type

$$u(x) \to u'(x) = e^{i\alpha} u(x),$$
$$u^*(x) \to u^{*\prime}(x) = e^{-i\alpha} u^*(x), \tag{1}$$

where α is a continuous real parameter that is independent of coordinates x. The gauge transformations (1) do not involve the coordinates; they only change the form of the field functions, which are multiplied by a phase factor equal to unity in absolute value. Such transformations are unitary, that is, they do not change the norm of the field functions, and constitute a one-parameter unitary group, $U(1)$.

The infinitesimal transformations have the form

$$u'(x) = (1 + i\alpha) u(x), \quad u^{*\prime}(x) = (1 - i\alpha) u^*(x). \tag{2}$$

If we consider u' and $u^{*\prime}$ to be independent components of the field, we can write (2) in the form

$$\begin{pmatrix} u'(x) \\ u^{*\prime}(x) \end{pmatrix} = \left(1 + i \begin{pmatrix} 1 & 0 \\ 0 & -1 \end{pmatrix} \alpha \right) \begin{pmatrix} u(x) \\ u^*(x) \end{pmatrix}, \tag{3}$$

from which follows the value of the generator:

$$X = i \begin{pmatrix} 1 & 0 \\ 0 & -1 \end{pmatrix}. \tag{4}$$

According to *Noether's theorem*, the Noether current that is connected with the transformations that do not involve the coordinates and is conserved is

$$J^\nu = -\frac{\partial L}{\partial u_{,\nu}} Xu \tag{5}$$

Here we have a matrix product of three factors, the first being a row and the last a column. Then, substituting (4) into (5), we get

$$j^\nu = -i \left(\frac{\partial L}{\partial u_{,\nu}} u - \frac{\partial L}{\partial u^*_{,\nu}} u^* \right). \tag{6}$$

This Noether current, which we have denoted by j^ν, is assumed to be the electromagnetic current associated with the complex-valued (and therefore charged) field u. By virtue of Noether's theorem, this current satisfies the continuity equation

$$\frac{\partial}{\partial x^\nu} j^\nu = 0, \tag{7}$$

and the associated charge, the electric charge, is conserved:

$$Q = \int d^3 x j^0 = -i \int d^3 x \left(\frac{\partial L}{\partial u,t} u - \frac{\partial L}{\partial u^*,t} u^* \right) = \text{constant}. \tag{8}$$

Thus, the gauge symmetry associated with $U(1)$ group leads to electric charge conservation, and the electric charge is an additive quantity, in view of the fact that the one-parameter group is Abelian.

If instead of u and u^* we take the real part u_1 and the imaginary part u_2 of the field functions as the independent components of the field, then we have the following transformation law (instead of (1)):

$$u'_1 = u_1 \cos \alpha - u_2 \sin \alpha,$$
$$u'_2 = u_1 \sin \alpha + u_2 \cos \alpha. \tag{9}$$

This transformation is nothing other than the rotation of the basis about the third axis through an angle α, the rotation belonging to the $O(1)$ group. Obviously, this rotation occurs not in the common space, the space where the position of material objects is specified, which is the *configuration space*, but in a special, *intrinsic space* of the field.

In similar manner one can specify the gauge symmetries connected with other additive charges, such as the *hypercharge*, the *baryon number*, and the *lepton number*.

Transformation (9), as is known, is a subgroup of the group of three-dimensional orthogonal rotations, $O(3)$, whereby instead of (1) or (9) we can consider more general transformations in the intrinsic space of the fields, rotations $O(3)$ or (for complex-valued fields) the transformations of the $SU(2)$ group.

In the physics of elementary particles at present several intrinsic spaces of varying nature are known. One of these, which was introduced earlier than all other such spaces and which is related to the isotopic invariance of nuclear forces, is known as the *isotopic space*, with the corresponding orthogonal transformations called *isotopic rotations*. Since at present we need not know the particular nature of an intrinsic space, we will focus on isotopic rotations.

As an example, let us take the triplet of pions: π^-, π^0, and π^+.

The charged components of the triplet are described by the complex-valued functions $\pi^+ = \xi^0$ and $\pi^- = \xi^2$, while the neutral component is described by the real-valued function $\pi^0 = \xi^1$.

On the whole they form a three-dimensional representation of the $SU(2)$ group. Going over from the complex components ξ^0, ξ^1, and ξ^2 to the real components π^1, π^2, and π^3 via the well-known similarity transformation

$$\pi = V\xi, \quad V = \frac{1}{\sqrt{2}} \begin{pmatrix} -1 & 0 & 1 \\ -i & 0 & -i \\ 0 & \sqrt{2} & 0 \end{pmatrix}, \tag{10}$$

one obtains

$$\pi^1 = \frac{1}{\sqrt{2}}(\xi^2 - \xi^0), \quad \pi^2 = \frac{1}{i\sqrt{2}}(\xi^0 + \xi^2), \quad \pi^3 = \xi^1, \tag{11}$$

and vice versa,

$$\xi^0 = -2^{-1/2}(\pi^1 - i\pi^2), \quad \xi^1 = \pi^3, \quad \xi^2 = 2^{-1/2}(\pi^1 + i\pi^2). \tag{12}$$

Suppose that

$$\xi^0 \to \xi'^0 = e^{-i\alpha}\xi^0, \quad \xi^1 \to \xi'^1 = \xi^1, \quad \xi^2 \to \xi'^2 = e^{i\alpha}\xi^2. \tag{13}$$

Then

$$\begin{aligned} \pi^1 \to \pi'^1 &= \pi^1 \cos\alpha - \pi^2 \sin\alpha, \\ \pi^2 \to \pi'^2 &= \pi^1 \sin\alpha + \beta^2 \cos\alpha, \\ \pi^3 \to \pi'^3 &= \pi^3. \end{aligned} \tag{14}$$

This transformation constitutes an isotopic rotation about the third axis and is the equivalent representation of complex-valued vectors, (13). The corresponding generator, as we found there, is

$$X_3 = \begin{pmatrix} 0 & -1 & 0 \\ 1 & 0 & 0 \\ 0 & 0 & 0 \end{pmatrix}.$$

Returning to the complex representation (13), we obtain the matrix of the Hermitian generator

$$I_3 = iX_3' = iV^+ X_3 V, \tag{15}$$

where V is the unitary matrix of the similarity transformations (10). In explicit form we have

$$I_3 = \begin{pmatrix} 1 & 0 & 0 \\ 0 & 0 & 0 \\ 0 & 0 & -1 \end{pmatrix}. \tag{16}$$

The other two generators can be found if we employ X_1 and X_2

$$I_1 = \frac{1}{\sqrt{2}} \begin{pmatrix} 0 & 1 & 0 \\ 1 & 0 & 1 \\ 0 & 1 & 0 \end{pmatrix}, \quad I_2 = \frac{1}{\sqrt{2}} \begin{pmatrix} 0 & -i & 0 \\ i & 0 & -i \\ 0 & i & 0 \end{pmatrix}. \tag{17}$$

The Hermitian matrices $I_i = I_i^+ (i = 1, 2, 3)$ form the isospin operator in the vector representation of the isotopic rotation group $SU(2)$. They obey the common commutation relations for angular momentum operators:

$$[I_i, I_j] = i\varepsilon_{ijk} I_k. \tag{18}$$

The eigenvalues of the I_3 matrix are

$$I_3 = -1, 0, 1. \tag{19}$$

The charge Q of the particles, the components of the triplet π^-, π^0, π^+, can be associated with these eigenvalues:

$$Q = I_3. \tag{20}$$

Noether's charge of the isovector field (12) in this case is not additive (but it is conserved), since it consists of three components of the isospin vector of the field ($k = 1, 2, 3$):

$$I_k = i \int \frac{\partial L}{\partial \xi^r_{,t}} (I_k)^r_s \, \xi^s \, d^3x + \text{c.c.}, \tag{21}$$

where c.c. stands for a term that is the complex conjugate of the first term. In particular, for $k = 3$, we obtain

$$I_3 = i \int \left(\frac{\partial L}{\partial \xi^0_{,t}} \xi^0 - \frac{\partial L}{\partial \xi^2_{,t}} \xi^2 \right) d^3x = \text{c.c.} \tag{22}$$

According to (12), $\xi^2 = -\xi^*_0$, and therefore the first term in (22) is real and equal to the second term, so that instead of (22) we can write

$$I_3 = 2i \int \left(\frac{\partial L}{\partial \xi^0_{,t}} \xi^0 - \frac{\partial L}{\partial \xi^2_{,t}} \xi^2 \right) d^3x = 2i \int \frac{\partial L}{\partial \xi_{,t}} I_3 \xi d^3x. \tag{23}$$

Similar equalities exist for the other isospin components of the field, so that in general form, we have

$$I_k = 2i \int \frac{\partial L}{\partial \xi_{,t}} I_k \xi d^3x. \tag{24}$$

Both in (23) and in (24) the integrand is the matrix product of the row $\partial L / \partial \xi_{,t}$ of matrix I_k and the column ξ. We may note that the total projection of the isospin vector of the field, (24), does not coincide with the eigenvalues (19), since the field is a continuous system, while (19) corresponds to single-particle quantum mechanical states. Only second quantization makes it possible to determine the states of a field involving a definite number, $N = 0, 1, 2, ...,$ of particles; in such states the total isospin of the field is the sum of single-particle values (19).

Let us now consider the spinor representation of the group of isotopic rotations, $SU(2)$, in the space

$$c = c^1 e_1 + c^2 e_2. \tag{25}$$

An example of such a representation is the nucleon doublet: $c_1 = p$ (proton) and $c^2 = n$ (neutron). The corresponding generators,

$$X_k = -i\sigma_k/2 \quad (k = 1, 2, 3), \tag{26}$$

are connected with the Pauli matrices σ_k, which operate in the isotopic space (25). The Hermitian generators

$$I_k = iX_k = \sigma_k/2 \tag{27}$$

define the isospin operator in the spinor representation of the isotopic group with the same commutation relations (18) as in the vector representation.

Let us find the isotopic vector of the isospin field (25):

$$I_k = -\int \frac{\partial L}{\partial c^r_{,t}} (X_k)^r_s c^s\, d^3x + \text{c.c.} = \frac{i}{2}\int \frac{\partial L}{\partial c^r_{,t}} (\sigma_k)^r_s c^s\, d^3x + \text{c.c.} \tag{28}$$

(this quantity is conserved). The fact that L is real and isotopically invariant implies, obviously, that

$$\frac{\partial L}{\partial c^1_{,t}} c^1 = \frac{\partial L}{\partial c^{*2}_{,t}} c^{*2},$$

whereby for the third component, we find

$$I_3 = \frac{i}{2}\int \left(\frac{\partial L}{\partial c^1_{,t}} c^1 - \frac{\partial L}{\partial c^2_{,t}} c^2 \right) d^3x + \text{c.c.} = i\int \left(\frac{\partial L}{\partial c^1_{,t}} c^1 - \frac{\partial L}{\partial c^2_{,t}} c^2 \right) d^3x, \tag{29}$$

and similarly for the other components. The general result is

$$I_k = 2i \int \frac{\partial L}{\partial c,t} I_k c\, d^3x, \tag{30}$$

where the isospin operator is

$$I_k = \frac{1}{2}\sigma_k \tag{31}$$

Thus, for the isospin field we arrive at an expression for the isospin of the field, (30), that coincides with that for the isovector field (24).

The eigenvalues of the matrix $I_3 = (1/2)\,\sigma_3$ are equal to $+1/2$ and $-1/2$, as is well known. They determine the single-particle quantum mechanical states of the isospin field, $c^1 = p$ and $c^2 = n$, respectively. The total isospin of the field can be meaningfully interpreted only after second quantization of the field is carried out.

The neutron and the proton have an electric charge of 0 and +1, respectively. We see that these values are shifted in relation to the isospin projections $I_3 = -1/2$ and $+1/2$ by $1/2$:

$$Q = I_3 + 1/2. \tag{32}$$

(ii) Dynamical Symmetries

Today elementary particle theory employs broader symmetries than SU (2). For instance, one generalization of SU (2) is the group of special (unimodular) unitary matrices in three-dimensional complex space, SU (3). The group space for this group is eight-dimensional. Indeed, eighteen real numbers, which parametrize the 3-by-3 general matrix of the SU (3) group, are connected by one unimodularity condition and nine unitarity conditions (nine equations for real numbers). The representation of the lowest dimension (not including the scalar representation) is realized by a triplet of complex components of the vectors of the space in which the SU (3) matrices operate.

Such groups have been related to dynamical symmetries, which manifest themselves in the interaction of fields, precisely, the weak, the electromagnetic, and the strong interaction to demonstrate how group-theoretic methods can be applied to the classification of strongly interacting elementary particles, or hadrons. The known *hadrons* can be grouped into different multiplets corresponding to the various representations of the $SU(3)$ group. The isotopic group $SU(2)$ is a more simple subgroup of $SU(3)$. Earlier we had the opportunity to see the connection between the charge of the components with the projection I_3 of the isospin, formulas (20) and (32), when we studied the multiplets of the $SU(2)$ triplet of pions (π^-, π^0, π^+) and the doublet of nucleons (n, p). There are also other $SU(2)$ isomultiplets of hardons: the doublets of xi particles (Ξ^0, Ξ^-), K mesons (K^+, K^0) and their antiparticles, that is, $\overline{K}$ mesons ($\overline{K^0}$, K^-), the triplet of Σ baryons (Σ^+, Σ^0, Σ^-), and the singlets of the Λ^0 baryon and the η^0 meson.

In the $SU(3)$ group these particles are grouped into octets, the vector representations. The baryon octet consists of the following particles: p, n, Λ^0, Σ^+, Σ^0, Σ^-, Ξ^0, and Ξ^-, while the meson octet consists of the following particles: K^+, K^0, K^-, $\overline{K^0}$, π^+, π^-, π^0, and η^0. There are also representations of higher dimensions, notably the decuplet of unstable particles, or resonances.

In the $SU(3)$ group one is forced to introduce another additive quantum number, the hypercharge Y, which together with I_3 determines the electric charge of the components in a multiplet:

$$Q = I_3 + Y/2. \tag{33}$$

For pions the hypercharge Y is zero, and for nucleons it is 1, in accordance with (20) and (32). The hypercharge is the sum of two other quantum numbers, the baryon number B and the strangeness S, or $Y = B + S$, Strangeness was introduced to account for the behaviour of particles in weak interactions, slow decay of "strange" hadrons into "nonstrange" particles. Heavy hadrons, or baryons, are assigned a baryon number of unity, while the relatively light hadrons, or mesons, are assigned a baryon number of zero.

Today the theory of elementary particles is based on the assumption (which indirectly has been verified by experiments) that all matter consists of elementary building blocks known as quarks: the light quarks u and d (the "up" quark and the "down" quark), the heavy quarks s and c (the "strange" quark and the "charm" quark), and superheavy b and t quarks (the "bottom", or "beauty", quark and the "top", or "truth", quark; the last quark has not manifested itself even indirectly). The u, d, and s quarks have fractional electric charges: 2/3, –1/3, and –1/3, respectively; fractional hypercharges: 1/3, 1/3, and –2/3, respectively; and fractional baryon numbers equal to 1/3 (–1/3 for the antiquarks). From these quarks we can build the common hadrons π, K, p, n, etc., with baryons consisting of three quarks and mesons of one quark and one antiquark.

In *strong interactions* the $SU(2)$ and $SU(3)$ symmetries prove to be broken, that is, inside a given multiplet the masses of particles prove to be the same only approximately. This is due to the difference in masses of the quarks: the greater the difference, the stronger the corresponding symmetry is broken. The masses of quarks are due to spontaneous symmetry breaking in weak interactions.

The c quarks possess a new quantum number C called "charm". They form hadrons with hidden charm, such as the J/ψ meson (it consists of c and $\bar{c}$), and states where the charm property is clearly evident, such as the D meson and the F meson. The b quarks form the massive upsilon particle $\gamma \sim (b\bar{b})$. The introduction of this quantum number, charm, changes the basic formula (33) for the charges of quarks to

$$Q = I_3 + \frac{1}{2}Y + \frac{2}{3}C. \tag{34}$$

Each quark comes in one of the three varieties identical in all measurable qualities but different in an additional quantum number known as *color*. The introduction of this quantum number is necessary for Pauli's exclusion principle to be satisfied when fermions are constructed out of quarks. The color group is an $SU(3)$ and is denoted by $SU(3)_c$. A principle that has yet to be proved theoretically but which lies at the base of the theory of hadrons is that all observable hadron states must be colorless. Thus, quarks cannot be observed in a free state; this constitutes what is known as the quark confinement hypothesis. An important property of the color group $SU(3)_c$ is that it is a local group. This enables building a theory of strong interactions along the same lines as the electromagnetic interaction theory by introducing an octet of vector fields, gluons, which are assumed to be the mediators in the strong interaction of elementary particles, just as photons are mediators in the electromagnetic interactions. The corresponding theory has become known as *quantum chromodynamics* (QCD).

REVIEW QUESTIONS AND PROBLEMS

1. Derive the general conditions that the Lagrangian density L must satisfy in any field theory.

2. Mention important properties of Green's function.

3. Explain the quantization of the free electromagnetic field (Coulomb gauge).

4. Explain canonical quantization and show that the normalization condition of unit 4-vector is
$$e_\mu e^\mu = -1$$

5. Explain Dirac field and discuss quantization of the Dirac field.

6. What are fermions and bosons? How they differ from each other.

7. Obtain the S-matrix for electromagnetic field.

8. Prove Wick's theorem for the general case.

9. What are Feynman diagrams? Write Feynman's rules.

10. Explain, how using the reaction amplitudes one can determine the physically measurable quantities, i.e. probabilities and cross sections.

Basic Interactions

1. FUNDAMENTAL INTERACTIONS IN NATURE

The number of elementary particles so far discovered is very large, exceeding 200. The interactions amongst these particles may be considered as the direct source of the natural phenomena. Only four fundamental interactions known in the nature, e.g. strong, electromagnetic, weak and gravitational are apparently enough to account for all the physical processes and structures in the universe on all scale of size from hadrons to galaxies of stars. Table 1 summarizes salient freatures of these basic interactions.

(i) **Gravitational Inteaction:** Gravitation was the first interaction to be known and is usually discussed in terms of the forces between macroscopic objects or in terms of the general relativity. It operates quite certainly on elementary particle level, but it is so weak that there if any other interactions are present, it is over-whelmed. The strength of gravitational interaction is commonly stated in terms of gravitational constant $G = 6.7 \times 10^{-11}$ N m^2 kg^2. The dimension less quantity $\frac{G m_e M_p}{\hbar c} \sim 3 \times 10^{-42}$ is usually taken as a constant, characterizing this interaction. The gravitational interaction between two bodies is believed to be mediated through the quantum of the this interaction called the *graviton*. Graviton has not been observed experimentally as a unique particle. The extremely small value of the gravitational interaction has so far kept it out of the detailed theory of the elementary particles, although any complete theory obtained in future must include it.

(ii) **Electromagnetic Interaction:** This is much stronger than gravitational interaction. This is the classical interaction of Maxwell-Lorentz field between electrically charged particles. The quantum of the field is photon. It is a long range square type interaction, whose strength is determined by Sommerfeld's fine structure constant $\alpha = \frac{e^2}{4\pi \varepsilon_0 \hbar} \sim \frac{1}{137}$. This electromagnetic interaction between a proton and an electron is about 10^{37} times stronger than gravitational force between them at the same distance.

This interaction is charge dependent. In terms of isobaric spin, this interaction depends on T and is governed by the isospin rule $\Delta T = 0, \pm 1$. All other quantities, e.g. charge, baryon number, lepton number, hypercharge, parity, strangeness numbers are conserved.

Table 1 Fundamental interactions in nature

Interaction	Gravitational	Electromagnetic	Strong	Weak
Relative magnitude	10^{-39}	10^{-3}	1	$10^{-13} - 10^{-11}$
Range	infinite	infinite	Short $\sim 10^{-15}$ m	Short ($\sim 10^{-16}$ m)
Characteristic Time	10^{-16} s	10^{-20} s	10^{-23} s	10^{-10} s
Associated particles	Graviton	Photon	Pion, Kaon; Gluons	Intermediate boson, $W^{\pm}$, Z^0
Particles acted upon	Every particle having mass	Charged particles	Hadrons	Hadrons, Leptons
Examples	Astronomical forces	Atomic forces	Nuclear forces	Forces involved in β-decay
Strength (Natural units)	G (Newton) $= 6 \times 10^{-39}$	$g_e = \dfrac{1}{137}$	$g^2 \approx 1$	G Fermi $= 5 \times 10^{-7}$
Role in universe	Assembles matter into planets, stars and galaxies	Determines structure of atoms, molecules, solids and liquids. An important factor in astronomical universe	i. Gluons holds quarks together to form nucleons ii. Pions holds nucleons together to form atomic nuclei	Mediates transformation of quarks and leptons; helps determine composition of atomic nuclei
Conservation	—	Isospin violation	all	Flavour, CPT violation
Spin	2	1	1	1
Mass	0	0	0	90 Ge V

The electromagnetic interaction is menifested in the chemical behaviours of atoms and molecules, Rutherford scattering and so on.

Photon capture can affect the production of mesons or hyperons by an electromagnetic interaction

$$\gamma + p \begin{cases} \nearrow \pi^0 + p \text{ (photo- pion production)} \\ \searrow \pi^0 + K^+ \text{ (low strange particles)} \end{cases} \tag{1}$$

These decay processes illustrates particularly a important point which at first sight may appear paradoxical namely that uncharged particles can also be subjected to electromagnetic interaction, for example,

$$\left. \begin{array}{l} \pi^0 \rightarrow \gamma + \gamma \\ \eta^0 \rightarrow \pi^+ + \pi^- + \pi^+ , \eta^0 \rightarrow \gamma + \gamma \\ \sum{}^0 \rightarrow \wedge^0 + \gamma \end{array} \right\} \tag{2}$$

In these processes no change of strangeness (S) is involved. The decay process such as $\Sigma^+ \to P^+ + \gamma$ are forbidden because the change $\Delta S = 1$ is required. The paradox that the decay of natural particle is by electromagnetic interaction is resolved by introducing an intermediate step in the overall nuclear reaction, i.e. virtual production of nucleon-antinucleon. We have

$$\pi^0 \xrightarrow{\text{strong}} \left(N + \bar{N}\right) \xrightarrow{\text{Virtual electromagnetic}} \gamma + \gamma \tag{3}$$

An example of the electromagnetic interaction is the process of mutual annihilation of particles and antiparticles.

Unlike the gravitational force the electromagnetic interaction may be either attractive or replusive.

(iii) **Weak Interaction:** This interaction is responsible for the nuclear β-decay and the weak decay of certain elementary particles, e.g. muons, pions, K-mesons and some hyperons. This interaction is about 10^{10} times weaker than the electromagnetic interaction and ulike gravitaional or e.m. interaction, this is a very short range force. The range of weak interaction is given by

$\dfrac{\hbar}{m_w\, c} \sim 2.4 \times 10^{-18}$ m, where m_w is the mass of the W boson, mediating the weak

force (half life $\sim 10^{-17}$ s). The strength of the weak interaction may also be expressed in terms of the universal weak interaction coupling constant g_w, although this constant is not dimensionless as are others. The value of this constant is 1.4×10^{-62} J-m^3 (from Fermi theory of β-decay). When expressed in dimensionless form, g_w is found to be small compared to the fine structural

constant $\alpha \sim \dfrac{1}{137}$. Here $g_w = \left(\dfrac{g}{\hbar c}\right)\left(\lambda^2\right) \simeq 3 \times 10^{-12}$; $\lambda_c = \dfrac{\lambda_c}{2\pi}$, λ_c being the Compton

wavelength for the electron.

The weak interaction is mediated through the heavy vector bosons, known as $W^\pm$ and Z^0 bosons. They have the masses $m_w\, c^2 = 80.9$ GeV and $m_z\, c^2 = 93$ GeV. When one compares the characteristic times for the operation of the weak ($\sim 10^{-10}$ s) and strong ($\sim 10^{-23}$ s) interactions, one finds the ratio of the former to the latter $\sim 10^{-13}$.

Few examples of weak interaction are

$$\begin{aligned}
\bar{\mu} + p &\to n + \nu_\mu \\
\mu &\to \bar{e} + \nu_\mu + \bar{\nu}_e \quad \text{(Muon decay)} \\
\bar{\nu}_e + p &\to n + e^+ \qquad \text{(Proton capture of anti neutrino)}
\end{aligned} \tag{4}$$

Other types of weak interactions which require coupling between other pairs of fermions,

$$\wedge^0 (S = -1) \to \pi (S = 0) + p (S = 0) \tag{5}$$

We must note that ν and $\bar{\nu}$ are associated exclusively with weak interactions just as photons are associated only with e.m. interaction.

An important characteristics of weak interaction which distinguishes it from other fundamental interactions, is that *parity* is violated (i.e. parity is not conserved) in weak interaction.

Recently, weak and electromagnetic interactions have been unified and the unified interaction is known as *electro-weak* interaction.

(iv) Strong Interaction: This is the dominant nuclear interaction responsible for most nuclear phenomena, nuclear energy levels, and so on. It is the interaction between protons and neutrons, having short range like the weak nuclear interaction, the range being $\dfrac{\hbar}{m_\pi c} \sim 10^{-15}$ m. Within this range it predominates over all the other fundamental interactions between the neutron and the proton- with a characteristics strength parameter $\alpha \sim 0.3$. Obviously, the value of α of strong interactions is much larger than the strength parameter $\alpha \sim \dfrac{1}{137}$ of e.m. interaction. The time for the operation of the strong interaction is about $\sim 10^{-23}$ s, which is also the order of the characteristic nuclear time.

The quantum of the strong interaction is the *pi* meson or pion ($\pi^\pm$, π^0).

The nuclear strong interaction is responsible for the formation of the so called meson and baryon resonances, or excited states of the mesons and baryon particles. It is also the coupling that drives their decay, and for this reasion their life times are short. A term usually used for the strongly interacting particles is hadron. The quantitative connection between the strong interactions and the structure of the nucleus is not yet fully clear.

Strong interaction is a short range force ($\sim 10^{-15}$ m) conserves baryon number (B), charge (Q), hypercharge (Y), Parity (π), isopin (T) and its component T_x. The following is an example of a strong interaction.

$$\pi^- + p \rightarrow \pi^0 + n \tag{6}$$

One can investigate the nature of the strong interaction by studing the properties of the hypernuclei in which a nucleon in a nucleus is replaced by a hyperon. The first hypernucleus discovered was a boron nucleus in which a neutron was replaced by a $\wedge^0$-hyperon. So far several hypernuclei have been produced and their properties investigated. An example is $\wedge^0$-nucleus of tritium. This decays according to $\wedge^3 H \rightarrow {}^3He + \pi^-$. The Q value for this is 41.5 MeV.

One can also obtain information about these systems through the study of the exotic atoms in which Σ^- hyperon is captured by the electric field of the nucleus. Sigma has the large mass and hence its orbit round the nucleus has a small radius. One obtain important information about the strong interaction by the measurement of the energies of the X-rays emitted during their transition from one orbit to another. This interaction between elementary particles is responsible for the total cross-section is a function of energy.

Recent studies reveal that protons and neutrons are made up of more fundamental entities known as quarks. It is believed that strong interaction is mediated through the exchange of massless quanta of spin $\hbar$ known as *gluons* between the quarks. Quarks and gluons have not been observed in the free state so far.

Examples of particles and their methods of interaction are shown in Table 2.

Table 2 Elementary particles and corresponding interaction modes

Weak	Electromagnetic	Strong
e	e	π
μ	μ	K
ν	π	P
π	K	n
K	p	$\wedge$
p	n	
n	$\wedge$	
$\wedge$		

(v) Fifth Inteaction: There is an indication of the existence of fifth interaction besides four fundamental interactions discussed so far. It has been reported that the force between two masses be better represented by

$$F = G_\infty \frac{m_1\, m_2}{r^2} \left[1 + \left(1 + \frac{r}{\lambda} \right) \alpha e^{-\frac{r}{\lambda}} \right] \qquad (7)$$

With $\alpha = -\,0.007$ and $\lambda = 200$ m. As α is negative, the second term in the square bracket represents a repulsive force. For $r \gg 200$ m,

$$F = G \frac{m_1\, m_2}{r^2} \qquad (8)$$

Which is the force operative between two masses. For $r \ll 200$ m,

$$F = G_\infty \frac{m_1\, m_2\, (1 + \alpha)}{r^2} = G' \frac{m_1\, m_2}{r^2} \qquad (9)$$

where $\qquad G' = G_\infty\, (1 + \alpha)$

This is the force one measures in a Cavendish experiment. The value of G for small distances is about 1% less than the value of G for large distances.

Unified Approach: There have been efforts to unify the various interactions and deduce them as different manifestations of the same underlying theory. Maxwell in 1864 first achieved such a unification for electric and magnetic forces. Maxwell's equation for the fields in free space, are

$$\nabla \cdot \mathbf{E} = \frac{\rho}{\varepsilon_0}$$

$$\nabla \cdot \mathbf{B} = 0$$

$$\nabla \times \mathbf{E} = -\frac{\partial \mathbf{B}}{\partial t}$$

$$\nabla \times \mathbf{B} = \mu_0 \left(\mathbf{J} + \frac{\partial \mathbf{D}}{\partial t} \right) \tag{10}$$

These four equations describe classical electromagnetic phenomena. Every thing electric or magnetic-from polar molecules to electric current; resistors, capacitors, inductors, and transistors; solar flares and cell membranes; electric generators and thunderstorms; computers and TVs; iPods and the northern lights can be described using Maxwell's equations. Clearly, Maxwell's equations describe completely the behaviors of electric and magnetic fields in *Classical Physics*. Maxwell's equations show that electromagnetic waves are possible and that their speed in vacuum, the speed of light c is related to electric and magnetic constants ε_0 and μ_0 respectively;

$$c = \frac{1}{\sqrt{\varepsilon_0 \mu_0}}$$

The value of c is very nearly 3.00×10^8 m/s.

This shows that electric and magnetic forces fall under the single umbrella of electromagnetism. The unification of electricity and magnetism was a major steps forward in our understanding of physical reality. Salam and Weinberg have now succeeded in unifying electromagnetic and weak forces and further efforts are being made to bring in strong and gravitational interactions. We now present the ideas which led to the unification of electromagnetic and weak interactions.

The electron-antineutron pair in the β-decay is produced in the same angular momentum state as the photon. This similarity between weak and electro-magnetic processes can be further extended by postulating that the neutron changes into a proton by emitting a massive W^- (which carries spin) and this then decays into an electron-anti neutrino pair. The basic weak process here is a vector boson W^- interacting with a fermion pair similar to a photon interacting with a fermion pair. The important differences are (i) the W^- is charged and is accompanied by its antiparticle the W^+ whereas the photon is neutral and is its own antiparticle. (ii) the $W^\pm$ is massive. Since weak interaction is a short range force. From the analysis of the weak processes. It is estimated that

$$m_w > 7.5 \times 10^4 \text{ MeV}$$

(iii) The interaction of the $W^\pm$ violates parity invariance whereas the interaction of the photon conserves parity.

Inspite of the above differences, the interaction of the W bosons and the photon with matter can be combined. This is done by allowing a triplet and a singlet of vector bosons to interact with the lepton and quark doublets. Demanding that the photon does not interact directly with the neutrinos and that its interaction conserves parity identifies the photon and relates the strengths of the weak and the electromagnetic interactions. Around 1967–68, Salam, Weinberg and Glashow showed that electromagnetism and weak interactions are two faces of a unified electroweak force. At this step the number of basic forces or interactions in nature came down to three. The problem though is that this unit can be seen only in processes at energies of about one hundred GeV or higher, i.e., at and above one hundred proton masses. The model developed by Weinberg and Salam has the following interesting features:

(i) There are three massive bosons W^+, W^-, Z (which is neutral), with masses

$$m_w \geq 3.7 \times 10^4 \, \text{MeV}$$

$$m_z \geq 7.3 \times 10^4 \, \text{MeV}$$

and have parity violating interaction. It is large mass of these which suppresses the usual weak interaction. The photon has the usual electromagnetic interaction. In high energy processes (energies comparable with $m_w c^2$) the two interactions would become comparable. These boson with $m_w \approx 8.1 \times 10^4$ MeV and $m_z \approx 9.5 \times 10^4$ MeV been observed recently at CERN.

(ii) The neutrinos and leptons have another interaction with matter, in addition to the electromagnetic and β-decay type of interactions. This interaction has been observed in 1973 in processes of the type

$$\nu_\mu + N \rightarrow \nu_\mu + \text{hadrons}$$

$$\bar{\nu}_\mu + N \rightarrow \bar{\nu}_\mu + \text{hadrons}$$

where N is a nucleus. It has also been observed in 1978 in the scattering of polarized electrons by a deutron target. It was reported that there was difference in the scattering of left-handed and right-handed electrons, indicating violation of the right-left symmetry or the violation is in agreement with the prediction of Weinberg-Salam model.

With the successful unification of weak and electromagnetic interactions, efforts have been directed towards unifying strong weak and electromagnetic interactions. Soon after came a theory of the strong interaction among quarks *quantum chromodynamics (QCD)*, and the two together make up the so-called *Standard model* of particle physics. In most of the theories, leptons and quarks are put in the same multiplet. Therefore, there is possibility of quarks transforming into leptons which means that the protons would be unstable against decay into leptons. Search for such decays (the life time of protons expected in these theories is greater than 10^{30} years) is being pursued vigorously by different groups of experimentalists. Everything except gravity seems to be satisfactorily handled by standard model, but yet it cannot by the last word since there are some twenty independent parameters in it be taken from experiment. So there are constant efforts to test and see what may lie beyond the standard model.

One such attempt is to unify the *QCD* theory with the electroweak one into a *Grand unified Theory* (GUT). If this happens, the theoretical indication is that this unity will be seen only at energies of 10^{15} GeV and higher. To get a feeling for this, recall that the proton mass is about 1 GeV. Present accelerators and those expected in the next decade or so are in the range of a few thousand GeV. But grand unification is only expected at 10^{15} GeV – so no hope of reaching such scales in laboratories, and in nature too such events must be extremely rare. How one can lest such ideas at all?

This brings one to cosmology. The problem of unifying quantum theory and gravity is still open. The question arises, under what conditions might such a combination be relevant? Quantum theory involves Planck's constant h, while general relativity, being a relativistic theory of gravitation, involves both the Newtonian constant G and the speed of light c.

Planck time $\left(\dfrac{hG}{c^5} \right)^{\frac{1}{2}} \approx 5 \times 10^{-14}\,\text{s}$

Planck mass $\left(\dfrac{hC}{G} \right)^{\frac{1}{2}} \approx 1.2 \times 10^{19}\,\text{GeV}/c^2$

Planck temperature $\dfrac{1}{k} \left(\dfrac{hC}{G} \right)^{\frac{1}{2}} \approx 1.4 \times 10^{32}\,\text{K}$

In a process, characterized by these scales, quantum gravity would be relevant.

Suggested Readings

1. Kakani SL and Hemrajani C, *Mathematical Physics*, 3 edn, CBS Publishers and Distributors, New Delhi (2019).
2. Griffiths DJ, *Introduction to Electrodynamics*, Englewood Cliffs, NJ: Prentice Hall (1981).
3. Jackson JD, *Classical Electrodynamics*, 3 edn, John Wiley and Sons (1999).
4. Argence E and Kahan T, *Theory of Waveguides and Cavity Resonators*, Blackie (1967).
5. Konopinski EJ, *Electromagnetic Fields and Relativistic Particles*, Mc-Graw Hill (1981).
6. Boyd TJM and Sandreson JJ, *Plasma Dynamics*, Nelson and Barnes (1969).
7. Lorrain P and Corson DR, *Electromagnetic Fields and Waves*, 2 edn, WH Freeman and Co. (1970).
8. Collin RE, *Field Theory of Guided Waves*, Mc-Graw Hill (1966).
9. Rossiter V, *Electromagnetism*, Heydon and Sons (1979).
10. Landau LD and Lifshitz EM, *The Classical Theory of Fields*, 4 edn, Pergamon Press (1975).
11. Hallen E, *Electromagnetic Theory*, Chapman-Hall (1962).
12. Grant IS and Phillips WR, *Electromagnetism*, John Wiley and Sons (1978).
13. deGroot SR, *The Maxwell Equations*, North Holland (1969).
14. Becker R, *Electromagnetic Fields and Interactions*, Dover Publications, New York (1992).
15. Ramo S, Whinnery JR, and van Duzer T, *Fields and Waves in Communication Electronics*, 3 edn, John Wiley (1990).
16. Brown RG *et al*, *Lines, Waves and Antenna*, 2 edn, John Wiley (1970).
17. Magnusson PC *et al*, *Transmission Lines and Wave Propagation*, 4 edn, CRC Press (2001).
18. Kakani SL and Kakani Shubhra, *Photonics: Optoelectronics*, CBS Publishers and Distributors, New Delhi (2016).
19. W Notling, *Theoretical Physics 4: Special Theory of Relativity*, Springer (2017).
20. W Notling, *Theoretical Physics 3: Electrodynamics*, Springer (2016).
21. W Greiner and J. Reinhardt, *Quantum Electrodynamics*, 4th ed. Springer (2009).
22. A Tachibana, *New Aspects of Quantum Electrodynamics*, Springer (2017).
23. Barry R Clarke, *The Quantum Puzzle*, World Scientific (2017).
24. A Das and T Ferbel, *Introduction to Nuclear and Particle Physics*, World Scientific (2003).
25. G Kellen, *Quantum Electrodynamics*, Springer (1972).
26. J Schwinger, *Quantum Electrodynamics*, Dover (1958).
27. J Schwinger, et al. *Classical Electrodynamics*, Persevs Books, Reading, Mass (1998).
28. YK Lim, *Introduction to classical Electrodynamics*, World Scientific (1985).
29. H Goldstein, *Calssical Mechanis*, 2nd ed. Addison-Wesley, Reading, Mass (1980.
30. AZ Caprt and PV Panat, *Introduction to Electrodynamics*, Narosa Publishing House, New Delhi (2002).
31. RP Feynman, et al. *The Feynman Lectures on Physics*, 3 Vols., Addison-Wesley (1963).

32. SL Kakani and HM Chandalia. *Qunatum Mechanics,* Sultan Chand and Sons, 4th ed. New Delhi (2004).

33. AL Fetter and JD Salecka, *Quantum Theory of Many-Particle Systems,* McGraw Hill (1971).

34. JJ Sakurai, *Advanced Quantum Mechanics,* Pearson Education (1967).

35. A Seiden, *Particle Physics,* Pearson Education (2005).

36. JJ Sakurai, *Modern Quantum Mechanics,* Pearson Education (1994).

37. AL Kuzemsky, *Statistical Mechanics and the Physics of Many-Particle Model Systems,* World Scientific (2017).

38. SL Kakani and C Hemrajani, *Electrostatics,* CBS Publishers, New Delhi (2016).

39. SL Kakani and C Hemrajani, *Statistical Mechanics,* Viva Books, New Delhi (2017).

40. SL Kakani and Shubhra Kakani, *Nuclear and Particle Physics,* 2nd ed. Viva Books, New Delhi (2013).

41. SL Kakani and Shubhra Kakani, *Optical Fiber Communications,* CBS Publishers, New Delhi (2019).

APPENDICES

GATE Physics: Electromagnetic Theory (Examination Questions)

1. Two point charges $Q_1 = 1\,nC$ and $Q_2 = 2\,nC$ are kept in free space such that the distance between them is 0.1 m.
   (a) The force on Q_2 is along the direction from Q_2 to Q_1
   (b) The force on Q_2 is the same in magnitude as that on Q_1
   (c) The force on Q_1 is attractive
   (d) A point charge $Q_2 = -3\,nC$, placed at the midpoint between Q_1 and Q_2, experiences no net force. **[b]**

2. A current I flows in the anticlockwise direction through a square loop of side a lying in the xy plane with its center at the origin. The magnetic induction at the center of the square loop is
   (a) $\dfrac{2\sqrt{2}\,\mu_0 I}{\pi a}\hat{e}_x$
   (b) $\dfrac{2\sqrt{2}\,\mu_0 I}{\pi a}\hat{e}_z$
   (c) $\dfrac{2\sqrt{2}\,\mu_0 I}{\pi a^2}\hat{e}_z$
   (d) $\dfrac{2\sqrt{2}\,\mu_0 I}{\pi a^2}\hat{e}_x$ **[a]**

3. A thin conducting wire is bent into a circular loop of radius r and placed in a time dependent magnetic field of magnetic induction
   $$B(t) = B_0 e^{-\alpha t}\hat{e}_z, \ (B_0 > 0 \text{ and } \alpha > 0)$$
   such that the plane of the loop is perpendicular to $B(t)$. Then the induced emf in the loop is
   (a) $\pi r^2 \alpha B_0 e^{-\alpha t}$
   (b) $\pi r^2 B_0 e^{-\alpha t}$
   (c) $-\pi r^2 \alpha B_0 e^{-\alpha t}$
   (d) $-\pi r^2 B_0 e^{-\alpha t}$ **[a]**

4. Consider an electric field E existing in the interface between a conductor and free space Then the electric field E is
   (a) external to the conductor and normal to the conductor's surface
   (b) internal to the conductor and normal to the conductor's surface
   (c) external to the conductor and tangential to the conductor's surface
   (d) both external and internal to the conductor and normal to the conductor's surface **[a]**

5. A coaxial cable of uniform cross-section contains an insulating material of dielectric constant 3.5. The radius of the central wire is 0.01 m and that of the sheath is 0.02 m. The capacitance per kilometer of a cable is
   (a) 280.5 nF
   (b) 28.05 nF
   (c) 56.10 nF
   (d) 2.805 nF **[a]**

6. The xoy-plane carries a uniform surface current of density $K = 50\hat{e}_z$ A/m. The magnetic field at the point $Z = 0.5$ m is
   (a) 10×10^6 Wb/m^2 (b) 1×10^{-6} Wb/m^2
   (c) $\pi \times 10^6$ Wb/m^2 (d) $\pi \times 10^{-5}$ Wb/m^2 [d]

7. Four point charges are placed at the corners of a square whose center is at the origin of a cartesian coordinate system. A point dipole p is placed at the center of the square as shown in Fig. A.1. Then
   (a) there is no force acting on the dipole
   (b) there is no torque about the center at O on the dipole
   (c) the dipole has minimum energy if it is in $\hat{e}_x$ direction
   (d) the force on the dipole is increased if the medium is replaced by another medium with larger dielectric constant. [b]

Fig. A.1: Four point charges at the corners of a square

8. The electric field $E(r, t)$ at a point r at time t in a metal due to the passage of electrons can be described by the equation

$$\nabla^2 E(r,t) = \frac{1}{c^2}\left[\frac{\partial^2 E(r,t)}{\partial t^2} + \omega'^2 E(r,t)\right]$$

where ω' is a characteristic frequency associated with the metal and c is the speed of light in vacuum. The dispersion relation corresponding to the plane wave solutions of the form $\exp[j(k \cdot r - \omega t)]$ is given by
   (a) $\omega^2 = c^2 k^2 - \omega'^2$ (b) $\omega^2 = c^2 k^2 + \omega'^2$
   (c) $\omega^2 = ck - \omega'$ (d) $\omega = ck + \omega'$ [b]

9. A copper wire of uniform cross-sectional area 1.0×10^{-6} m^2 carries a current of 1 A. Assuming that each copper atom contributes one electron gas, the drift velocity of the free electrons (density of copper is 8.94×10^3 kg/m^3 and its atomic mass is 1.05×10^{-25} kg) is
   (a) 7.4×10^{-4} m/s (b) 74×10^{-4} m/s
   (c) 74×10^{-3} m/s (d) 7.4×10^{-5} m/s [d]

10. Consider an infinitely long straight cylindrical conductor of radius R with its axis along the z-direction which carries a current of 1 A uniformly distributed over its cross-section. Which of the following statements is correct?

(a) $\nabla \times B = 0$ everywhere (b) $\nabla \times B = \dfrac{\mu_0}{\pi R^2}\hat{z}$ everywhere

(c) $\nabla \times B = 0$ for $r > R$ (d) $\nabla \times B = \dfrac{\mu_0}{\pi R^2}\hat{z}$ for $r > R$ [c]

11. A large circular coil of N turns and radius R carries a time varying current $i = i_0 \sin(\omega t)$. A small circular coil of n turns and radius ($r \ll R$) is placed at the center of the large coil such that the coils are concentric and coplanar. The induced emf in the small coil
    (a) leads the current in the large coil by $\pi/2$
    (b) lags the current in the large coil by π
    (c) is in phase with the current in the large coil
    (d) lags the current in the large coil by $\pi/2$ [d]

12. Four charges are placed at the four corners of a square of side a as shown in the Fig. A.2. The electric dipole moment of this configuration is
    (a) $p = qa\hat{i} + qa\hat{j}$ (b) $p = -qa\hat{i} + qa\hat{j}$
    (c) $p = -qa\hat{i} - qa\hat{j}$ (d) $p = +qa\hat{i} - qa\hat{j}$ [c]

Fig. A.2: Four charges at the corners of a square

13. A particle with an initial velocity $v_0\hat{i}$ enters a region with an electric field $E_0\hat{j}$ and a magnetic field $B_0\hat{j}$. The trajectory of the particle will be
    (a) an ellipse
    (b) a cycloid
    (c) a helix with constant pitch
    (d) a helix with variable pitch [d]

14. An electric charge, $+Q$ is placed on the surface of a solid, conducting sphere of radius a. The distance measured from the centre of the sphere is denoted as r. Then
    (a) the charge gets distributed uniformly throughout the volume of the sphere
    (b) the electrostatic potential has the same value for $r < a$
    (c) an equal and opposite charge gets induced in the bottom half of the sphere
    (d) the electric field is given by $1/4\pi\varepsilon_0 r^2$ for $r < a$. [b]

15. An electric field applied along the length of a long cylinder produces a polarization P. The depolarization field produced in this configuration is
    (a) $4\pi P/3$ (b) $-4\pi P/3$
    (c) $2\pi P$ (d) 0   **[b]**

16. Which one of the following Maxwell's equations implies the absence of magnetic monopoles?
    (a) $\nabla \cdot E = \pi / \varepsilon_0$ (b) $\nabla \cdot B = 0$
    (c) $\nabla \times E = -\partial B / \partial t$ (d) $\nabla \times B = (1/c^2)\partial B/\partial t + \mu_0 J$ **[b]**

17. An electromagnetic wave is propagating in free space in the z-direction. If the electric field is given by $E = \cos(\omega t - kz)\,i$, where $\omega t = ck$, then the magnetic field is given by
    (a) $B = (1/c)\cos(\omega t - kz)\hat{j}$
    (b) $B = (1/c)\sin(\omega t - kz)\hat{j}$
    (c) $B = (1/c)\cos(\omega t + kz)\hat{j}$
    (d) $B = (1/c)\cos(\omega t - kz)\hat{i}$ **[b]**

18. The electric field strength at a point P due to a point charge of q C located at the origin is $100\ \mu\mathrm{Vm}^{-1}$. If the point charge is now enclosed by a perfectly conducting metal sheet sphere whose centre is at origin, then the electric field strength at point P outside the sphere becomes
    (a) zero   (b) $100\ \mu\mathrm{V\,m}^{-1}$
    (c) $-100\ \mu\mathrm{V\,m}^{-1}$ (d) $50\ \mu\mathrm{V\,m}^{-1}$ **[b]**

19. Electromagnetic waves are propagating along a hollow, metallic waveguide whose cross-section is a square of side ω. The minimum frequency of the electromagnetic waves is
    (a) c/ω (b) $2c/\omega$
    (c) $\pi c/\omega$ (d) $\sqrt{2}\pi c/\omega$ **[c]**

20. Consider the given statements about $E(r, t)$ and $B(r, t)$, the electric and magnetic vectors respectively in a region of free space.
    (P) Both E and B are conservative vector fields.
    (Q) Both E and B are central force fields.
    (R) E and B are mutually perpendicular in the region.
    (S) Work done by B on a moving charge in the region is zero.
    Which of the above statements are correct?
    (a) P and R   (b) R and S
    (c) S only   (d) P and Q   **[b]**

21. An infinite conducting sheet in the xy-plane carries a surface current density K along the y-axis. The magnetic field B for $Z > 0$ is
    (a) $B = 0$ (b) $B = \dfrac{\mu_0 k}{2}\hat{k}$
    (c) $B = \dfrac{\mu_0 k}{2}\hat{i}$ (d) $B = \dfrac{\mu_0 k}{(x^2 + z^2)^{1/2}}\hat{j}$ **[c]**

22. In the infinite plane, $y = 6$, there exists a uniform surface charge density of 1600π $\mu C\, m^{-2}$. The associated electric field strength is

(a) $30\hat{i}\ Vm^{-1}$ (b) $30\hat{j}\ Vm^{-1}$

(c) $30\hat{k}\ Vm^{-1}$ (d) $60\hat{i}\ Vm^{-1}$ **[b]**

23. An infinitely long, closely wound solenoid carries a sinusoidally varying current. The induced electric field is

(a) zero everywhere

(b) non-zero inside and zero outside the solenoid

(c) non-zero inside as well as outside the solenoid

(d) zero inside and non-zero outside the solenoid **[b]**

24. A charge $+q$ is kept at a distance $2R$ from the centre of a grounded conducting sphere of radius R. The image charge and its distance from the centre are, respectively,

(a) $\dfrac{-q}{2}$ and $\dfrac{R}{2}$ (b) $\dfrac{-q}{2}$ and $\dfrac{-R}{4}$

(c) $-q$ and $\dfrac{R}{2}$ (d) $+\dfrac{q}{2}$ and $\dfrac{R}{2}$ **[c]**

25. A conducting sphere of radius R has charge $+Q$ on its surface. If the charge on the sphere is doubled and its radius is halved, the energy associated with the electric field will

(a) increase four times

(b) increase eight times

(c) remain the same

(d) decrease four times **[b]**

26. A conducting sphere of radius R is placed in a uniform electric field E_0 directed along $+z$-axis. The electric potential for outside points is given as

$$V_{out} = -E_0\left[1 - \frac{R^3}{r^3}\right] r\cos\theta,$$ where r is the distance from the center and θ is the polar

angle. The charge density on the surface of the sphere is

(a) $3\varepsilon_0 E_0 \cos\theta$ (b) $\varepsilon_0 E_0 \cos\theta$

(c) $3\varepsilon_0 E_0 \cos\theta$ (d) $\dfrac{\varepsilon_0}{3} E_0 \cos\theta$ **[c]**

27. A circular arc QTS is kept in an external magnetic field $B_0\hat{j}$ as shown in Fig. A.3. The arc carries a current I. The magnetic field is directed normal and into the page. The force acting on the arc is

(a) $2IB_0 R\hat{k}$ (b) $IB_0 R\hat{k}$

(c) $-2IB_0 R\hat{k}$ (d) $-IB_0 R\hat{k}$ **[b]**

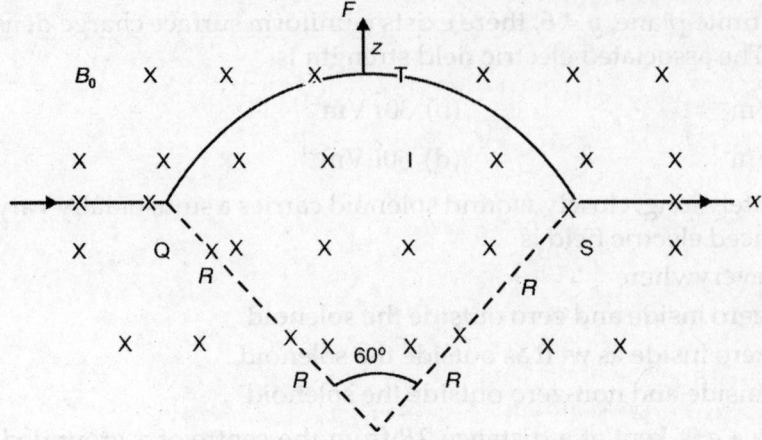

Fig. A.3: A circular arc in external field

28. A plane electromagnetic wave of frequency ω is incident normally on an air–dielectric interface, the dielectric is homogeneous, isotropic and non-magnetic and its refractive index is n. The reflectance (R) and transmittance (T) from the interface are

    (a) $R = \left[\dfrac{n-1}{n+1}\right]^2,\ T = \dfrac{4n}{(n+1)^2}$ (b) $R = -\left[\dfrac{n-1}{n+1}\right],\ T = \dfrac{2}{(n+1)^2}$

    (c) $R = \left[\dfrac{n-1}{n+1}\right]^3,\ T = \dfrac{4n^3}{(n+1)^3}$ (d) $R = \left[\dfrac{(n-1)^2}{n+1}\right],\ T = \dfrac{4n^2}{(n+1)^2}$ [a]

29. The electric field of a plane EM wave is

    $$E = E_0 \exp[i(\hat{x}k\cos\alpha + \hat{y}k\sin\alpha - \omega t)]$$

    If $\hat{x}, \hat{y}$ and $\hat{z}$ are cartesian unit vectors, the wave vector k of the EM wave is

    (a) $\hat{z}k$ (b) $\hat{x}k\sin\alpha + \hat{y}k\cos\alpha$

    (c) $\hat{x}k$ (d) $-\hat{z}k$ [b]

30. The dispersion relation for a low density plasma is $\omega^2 = \omega_0^2 + c^2 k^2$, where ω_0 is the plasma frequency and c is the speed of light in free space. The relationship between the group velocity (V_g) and phase velocity (V_p) is

    (a) $V_p = V_g$ (b) $V_p = V_g^{1/2}$

    (c) $V_p V_g = c^2$ (d) $V_g = V_p^{1/2}$ [c]

31. Which one of the following statements is TRUE?
    (a) The magnitude of the electric field is attenuated as the wave propagates.
    (b) The energy of the EM wave flows along the x-direction.
    (c) The magnitude of the electric field of the wave is a constant.
    (d) The speed of the wave is the same as c (speed of light in free space).   [d]

32. The magnetic field **B** of the wave is

    (a) $\hat{y}\dfrac{k}{\omega}E_0 \exp(-zk\sin\phi)\exp[j(zk\cos\phi - \omega t)]$

    (b) $\hat{y}\dfrac{k}{\omega}E_0 \exp(-zk\sin\phi)\exp[i(zk\cos\phi - \omega t + \phi)]$

    (c) $\hat{y}\dfrac{k}{\omega}E_0 \exp[j(-zk\cos\phi - \omega t + \phi)]$

    (d) $\hat{y}\dfrac{k}{\omega}E_0 \exp(-zk\cos\phi)\exp[i(zk\sin\phi - \omega t)]$ **[b]**

33. Three infinitely long wires are placed equally apart on the circumference of a circle of radius a, perpendicular to its plane. Two of the wires carry current I each, in the same direction, while the third carries current $2I$ along the direction opposite to the other two. The magnitude of the magnetic induction **B** at a distance r from the centre of the circle, for $r > a$, is

    (a) 0

    (b) $\dfrac{2\mu_0}{\pi}\dfrac{I}{r}$

    (c) $-\dfrac{2\mu_0}{\pi}\dfrac{I}{r}$

    (d) $\dfrac{2\mu_0}{\pi}\dfrac{Ia}{r^2}$ **[a]**

34. A solid sphere of radius R carries a uniform volume charge density ρ. The magnitude of electric field inside the sphere at a distance r from the centre is

    (a) $\dfrac{r\rho}{3\varepsilon_0}$

    (b) $\dfrac{R\rho}{3\varepsilon_0}$

    (c) $\dfrac{R^2\rho}{r\varepsilon_0}$

    (d) $\dfrac{R^3\rho}{r^2\varepsilon_0}$ **[a]**

35. The electric field $E(r, t)$ for a circularly polarized electromagnetic wave propagating along the positive z-direction is

    (a) $E_0(\hat{x} + \hat{y})\exp[i(kz - \omega t)]$

    (b) $E_0(\hat{x} + i\hat{y})\exp[i(kz - \omega t)]$

    (c) $E_0(\hat{x} + i\hat{y})\exp[i(kz + \omega t)]$

    (d) $E_0(\hat{x} + \hat{y})\exp[i(kz + \omega t)]$ **[b]**

36. The electric (E) and magnetic (B) field amplitudes associated with an electromagnetic radiation from a point source behave at a distance r from the source as

    (a) $E = \text{constant}, B = \text{constant}$

    (b) $E \propto \dfrac{1}{r}, B \propto \dfrac{1}{r}$

    (c) $E \propto \dfrac{1}{r^2}, B \propto \dfrac{1}{r^2}$

    (d) $E \propto \dfrac{1}{r^3}, B \propto \dfrac{1}{r^3}$ **[c]**

37. Consider the following three independent cases:

    i. Particle A of charge $+q$ moves in free space with constant velocity v ($v \ll$ speed of light)

ii. Particle B of charge $+q$ moves in free space in a circle of radius R with same speed v as in case (i)

iii. Particle having charge $-q$ moves as in case (ii). If the powers radiated by A, B and C are P_A, P_B and P_C respectively, then

(a) $P_A = 0$, $P_B > P_C$ (b) $P_A = 0$, $P_B = P_C$

(c) $P_A > P_B > P_C$ (d) $P_A = P_B = P_C$ **[b]**

38. If the electrostatic potential were given by $\phi = \phi_0(x^2 + y^2 + z^2)$, where ϕ_0 is constant, then the charge density giving rise to the above potential would be

(a) 0 (b) $-6\phi_0\varepsilon_0$

(c) $-2\phi_0\varepsilon_0$ (d) $-\dfrac{6\phi_0}{\varepsilon_0}$ **[b]**

39. The work done in bringing a charge $+q$ from infinity in free space, to a distance d in front of a semi-infinite grounded metal surface is

(a) $\dfrac{q^2}{4\pi\varepsilon_0(d)}$ (b) $-\dfrac{q^2}{4\pi\varepsilon_0(2d)}$

(c) $-\dfrac{q^2}{4\pi\varepsilon_0(4d)}$ (d) $-\dfrac{q^2}{4\pi\varepsilon_0(6d)}$ **[c]**

40. A plane electromagnetic wave travelling in vacuum is incident normally on a non-magnetic, non-absorbing medium of refractive index n. The incident (E_i), reflected (E_r) and transmitted (E_t) electric fields are given as

$$E_i = E\exp[\hat{j}(kz - \omega t)], \quad E_r = E_{or}\exp(\hat{i}z - \omega t)]$$

$$E_t = E_{ot}\exp[\hat{i}(k_t z - \omega t)]$$

If $E = 2\,\text{V/m}$ and $n = 1.5$, then the application of appropriate boundary conditions leads to

(a) $E_{or} = -\dfrac{3}{5}\,\text{V/m}$, $E_{ot} = \dfrac{7}{5}\,\text{V/m}$ (b) $E_{or} = -\dfrac{1}{5}\,\text{V/m}$, $E_{ot} = \dfrac{8}{5}\,\text{V/m}$

(c) $E_{or} = \dfrac{2}{5}\,\text{V/m}$, $E_{ot} = \dfrac{8}{5}\,\text{V/m}$ (d) $E_{or} = -\dfrac{4}{5}\,\text{V/m}$, $E_{ot} = \dfrac{6}{5}\,\text{V/m}$ **[c]**

41. The electric field E (in Vm^{-1}) at the point $(1, 1, 0)$ due to a point charge of $+1\,\mu\text{C}$ located at $(-1, 1, 1)$ (coordinates in metres) is

(a) $\dfrac{10^{-6}}{20\sqrt{5}\pi\varepsilon_0}(2\hat{i} - \hat{k})$ (b) $\dfrac{10^{-6}}{20\pi\varepsilon_0(2\hat{i} - \hat{k})}$

(c) $\dfrac{-10^{-6}}{20\sqrt{5}\pi\varepsilon_0}(2\hat{i} - \hat{k})$ (d) $\dfrac{-10^{-6}}{20\pi\varepsilon_0(2\hat{i} - \hat{k})}$ **[d]**

42. An infinitely long wire carrying a current $I(t) = I_0\cos(\omega t)$ is placed at a distance a from a square loop of side a as shown in Fig. A.4. If the resistance of the loop is R_D, then the amplitude of the induced current in the loop is

(a) $\dfrac{\mu_0}{2\pi}\dfrac{\alpha I_0\omega}{R}\ln 2$ (b) $\dfrac{\mu_0}{\pi}\dfrac{\alpha I_0\omega}{R}\ln 2$

(c) $\dfrac{2\mu_0}{\pi}\dfrac{\alpha I_0\omega}{R}\ln 2$ (d) $\dfrac{\mu_0}{2\pi}\dfrac{\alpha I_0\omega}{R}$ **[a]**

Fig. A.4: Infinitely long current carrying wire

43. A particle with charge q, mass m and linear momentum p enters an electromagnetic field of vector potential A and scalar potential ϕ. The Hamiltonian of the particle is

(a) $\dfrac{p^2}{2m} + q\phi + \dfrac{A^2}{2m}$

(b) $\dfrac{1}{2m}\left[p - \dfrac{q}{c}A\right]^2 + q\phi$

(c) $\dfrac{1}{2m}\left[p - \dfrac{q}{c}A\right]^2 + p \cdot A$

(d) $\dfrac{p^2}{2m} q\phi - p \cdot A$ **[b]**

44. In an electromagnetic field, which one of the following remains invariant under Lorentz transformation?

(a) $E \times B$

(b) $E^2 - c^2 B^2$

(c) B^2

(d) E^2 **[b]**

45. A sphere of radius R has uniform volume charge density. The electric potential at a point $r\,(r < R)$ is

(a) due to the charge inside a sphere of radius r only

(b) due to the entire charge of the sphere

(c) due to charge in the spherical shell of inner and outer radius r and R, only

(d) independent of r **[b]**

46. A medium 1 has the electrical permittivity $\varepsilon_1 = 1.5\varepsilon_0$ farad/m and occupies the region to the left of $x = 0$ plane. A medium 2 has the electrical permittivity $\varepsilon_2 = 2.5\varepsilon_0$ farad/m and occupies the region to the right of $x = 0$ plane. If E_1 in medium 1 is $E_1 = (2\hat{u}_x - 3\hat{u}_y + 1\hat{u}_z)$ then E_2 is

(a) $2.0\hat{u}_x - 7.5\hat{u}_y + 2.5\hat{u}_z$

(b) $2.0\hat{u}_x - 2.0\hat{u}_y + 0.6\hat{u}_z$

(c) $1.2\hat{u}_x - 3.0\hat{u}_y + 1.0\hat{u}_z$

(d) $1.2\hat{u}_x - 2.0\hat{u}_y + 0.6\hat{u}_z$ **[c]**

47. Three point charges q, q and $-2q$ are located at $(0, -a, a)$ $(0, a, a)$ and $(0, 0, -a)$ respectively. The net dipole moment of this charge distribution is

(a) $4qa\hat{j}$

(b) $2qa\hat{k}$

(c) $-4qa\hat{i}$

(d) $-2qa\hat{j}$ **[a]**

48. A material has conductivity of 10^{-2} mho/m and a relative permittivity of 4. The frequency at which conduction current in the medium is equal to the displacement current is

(a) 45 MHz   (b) 90 MHz   (c) 450 MHz   (d) 900 MHz   **[a]**

49. The vector potential in a region is $A(x,y,z) = -y\hat{i} + 2x\hat{j}$. The associa-ted magnetic induction $\boldsymbol{B}$ is

(a) $\hat{j} + \hat{k}$

(b) $3\hat{k}$

(c) $-\hat{j} + 2\hat{j}$

(d) $-\hat{i} + \hat{j} + \hat{k}$ **[b]**

50. At the interface between two linear dielectrics (with dielectric constants ε_1 and ε_2), the electric field lines bend, as shown in Fig. A.5. Assume that there are no free charges at the interface. The ratio $\varepsilon_1/\varepsilon_2$ is

(a) $\dfrac{\tan\theta_1}{\tan\theta_2}$

(b) $\dfrac{\cos\theta_1}{\cos\theta_2}$

(c) $\dfrac{\sin\theta_1}{\sin\theta_2}$

(d) $\dfrac{\cot\theta_1}{\cot\theta_2}$ **[a]**

Fig. A.5: Electric field line at the interface of two linear dielectrics

51. The intrinsic impedance of a lossy dielectric medium is given by

(a) $\dfrac{j\omega\mu}{\sigma}$

(b) $\dfrac{j\omega\varepsilon}{\mu}$

(c) $\sqrt{\dfrac{j\omega\mu}{\sigma + j\omega\mu}}$

(d) $\sqrt{\dfrac{\mu}{\varepsilon}}$ **[c]**

52. A transverse electromagnetic wave with circular polarization is received by a dipole antenna. Due to polarization mismatch the power transfer efficiency from the wave to the antenna is reduced to about

(a) 50%   (b) 35.3%

(c) 25%   (d) 0%   **[d]**

53. If the plane electromagnetic wave satisfies the equation $\dfrac{\partial^2 E_x}{\partial z^2} = c^2 \dfrac{\partial^2 E_x}{\partial t^2}$, the wave propagates in the

    (a) x-direction
    (b) y-direction
    (c) z-direction
    (d) xy plane at an angle of 45° between x- and z-direction   **[c]**

54. The depth of penetration of a wave in a lossy dielectric increases with increasing
    (a) conductivity
    (b) permeability
    (c) wavelength
    (d) permittivity   **[a]**

55. Four point charges are placed in a plane at the following positions: $+Q$ at $(1, 0)$, $-Q$ at $(-1, 0)$, $+Q$ at $(0, 1)$ and $-Q(0, -1)$. At large distances the electrostatic potential due to this charge distribution will be dominated by the
    (a) monopole moment
    (b) dipole moment
    (c) quadrupole moment
    (d) octopole moment   **[b]**

56. A charged capacitor (C) is connected in series with an inductor (L). When the displacement current reduces to zero, the energy of the LC circuit is
    (a) stored entirely in its magnetic field
    (b) stored entirely in its electric field
    (c) distributed equally among its electric and magnetic fields
    (d) radiated out of the circuit   **[b]**

57. Match the following:

    P. Franck–Hertz experiment   1. Electronic excitation of molecules
    Q. Hartree–Fock method   2. Wave function of atoms
    R. Stern–Gerlach experiment   3. Spin angular momentum of atoms
    S. Franck–Condon principle   4. Energy levels in atoms

    (a) P-4   (b) P-1   (c) P-3   (d) P-4
        Q-2   Q-4   Q-2   Q-1
        R-3   R-3   R-4   R-3
        S-1   S-2   S-1   S-2   **[a]**

58. The electromagnetic field due to a point charge must be described by Lienard–Wiechert potentials when
    (a) the point charge is highly accelerated
    (b) the electric and magnetic fields are not perpendicular
    (c) the point charge is moving with velocity close to that of light
    (d) the calculation is done for the radiation zone, i.e. far away from the charge.   **[c]**

59. A toroidal coil has N closely wound turns. Assume the current through the coil to be I and the toroid is filled with a magnetic material of relative permittivity μ_r. The magnitude of magnetic induction B inside the toroid, at a radial distance r from the axis is given by

(a) $\mu_r \mu_0 NI$

(b) $\dfrac{\mu_r \mu_0 NI}{r}$

(c) $\dfrac{\mu_r \mu_b NI}{2\pi r}$

(d) $2\pi \mu_r \mu_0 NI$

$\qquad$ **[c]**

60. An electromagnetic wave with $E(z,t) = E_0 \cos(\omega t - kz)\hat{i}$ is travelling in free space and crosses a disc of radius $2\,\mathrm{m}$ placed perpendicular to the z-axis. If $E_0 = 60\,\mathrm{Vm^{-1}}$, the average power, in watt, crossing the disc along the z-direction is

(a) 30

(b) 60

(c) 120

(d) 270

$\qquad$ **[a]**

61. Can the following scalar and vector potentials describe an electromagnetic field?

$$\phi = (x,t) = 3xyz - 4t$$

$$A(x,t) = (2x - \omega t)\hat{i} + (y - 2z)\hat{j} + (z - 2xe^{-kx})\hat{k}$$

where ω is a constant

(a) yes, in the Coulomb gauge (b) yes, in the Lorentz gauge

(c) yes, provided $\omega = 0$ (d) no

$\qquad$ **[b]**

62. A piece of paraffin is placed in a uniform magnetic field H_0. The sample contains hydrogen nuclei of mass m_p which interact only with external magnetic field. An additional oscillating magnetic field is applied to observe resonance absorption. If g_1 is the g-factor of the hydrogen nucleus, the frequency, at which resonance absorption takes place, is given by

(a) $\dfrac{3g_1 e H_0}{2\pi m_p}$

(b) $\dfrac{3g_1 e H_0}{4\pi m_p}$

(c) $\dfrac{g_1 e H_0}{2\pi m_p}$

(d) $\dfrac{g_1 e H_0}{4\pi m_p}$

$\qquad$ **[d]**

63. A sphere of radius R carries a polarization $P = kr$, where k is a constant and r is measured from the centre of the sphere. The bound surface and volume charge densities are given, respectively, by

(a) $-k|r|$ and $3k$

(b) $k|r|$ and $-3k$

(c) $k|r|$ and $-4\pi kR$

(d) $-k|r|$ and $4\pi kR$

$\qquad$ **[b]**

64. The electric field E at a point r outside the sphere is given by

(a) $E = 0$

(b) $E = \dfrac{kR(R^2 - r^2)}{\varepsilon_0 r^3}\hat{r}$

(c) $E = \dfrac{kR(R^2 - r^2)}{\varepsilon_0 r^5}\hat{r}$

(d) $E = \dfrac{3k(r - R)}{4\pi \varepsilon_0 r^4}\hat{r}$

$\qquad$ **[a]**

SOLUTIONS

1. Option (a) is incorrect as the repelling force acts between them. Newton's third law suggests that they exert equal and opposite force on each other. Hence, option (b) is correct.

2. Magnetic field at point O due to line segment AB

$$= \frac{\mu_0 i}{4\pi(a/2)}(\sin\alpha + \sin\beta)$$

Here $\qquad \alpha = 45° = \beta$

Fig. A.6: Magnetic field at O due to line segment AB

$\therefore$ Magnetic field at O due to $AB = \frac{\mu_0 i}{2\pi a} \cdot \sqrt{2} = \frac{\mu_0 i}{\sqrt{2}\pi a}$

Every segment contributes equally, then

total magnetic induction at the centre $= \dfrac{2\sqrt{2}\,\mu_0 i}{\pi a}$

Hence, option (a) is correct.

3. $\phi = B \cdot \pi r^2$

$\varepsilon = -\dfrac{d\phi}{dt} = \alpha B_0 \pi r^2 \exp(-\alpha t)$. Hence, option (a) is correct.

4. As conductor cannot have charge except on its surface and hence field E is external and normal to the surface. Option (a) is correct.

5. Capacitance per km $= \dfrac{2\pi\varepsilon_0\varepsilon_r}{\ln\left(\dfrac{b}{a}\right)} \times 10^3 = \dfrac{1\times3.5\times10^3}{18\times10^9 \times \ln 2} = 280.5\,\text{nF}$

Option (a) is correct.

6. $B = -\dfrac{\mu_0 k}{2}\hat{y} = \dfrac{4\pi\times10^{-7}\times50}{2} = 10\pi \times 10^{-6}\,\text{Wb/m}^2$.

Option (d) is correct.

7. Electric field is zero at centre. Thus, torque about O is zero. Option (b) is correct.

8. $(ik)^2 = \dfrac{1}{c^2}[(-i\omega)^2 + \omega'^2]$

$-k^2 = \dfrac{1}{c^2}[-\omega^2 + \omega'^2)$

This leads to

$$\omega'^2 = \omega^2 - c^2 k^2$$

or

$$\omega^2 = \omega'^2 + c^2 k^2.$$

Option (b) is correct.

9. Drift speed $= \dfrac{J_d}{ne} = \dfrac{I}{Ane}$

$$= \frac{I}{Ae} \frac{m}{d} = \frac{1 \times 1.05 \times 10^{-25}}{10^{-6} \times 1.6 \times 10^{-19} \times 8.94 \times 10^3}$$

$$= 7.4 \times 10^{-5} \text{ m s}^{-1}$$

Option (d) is correct.

10. Maxwell's electromagnetic equation suggests

$$\nabla \times \boldsymbol{B} = 0 \text{ if } \boldsymbol{J} = 0$$

For outside conductor, $\boldsymbol{J} = 0$

$$\nabla \times \boldsymbol{B} = 0 \text{ for } r > R. \text{ Option (c) correct.}$$

11.

$$\phi = \frac{\mu_0 j}{2R} \pi r^2 = \frac{\mu_0 \pi r^2}{2R} i$$

and

$$\varepsilon = \frac{-d\phi}{dt} = -\frac{\mu_0 \pi r^2}{2R} i_0 \omega \cos(\omega t)$$

or

$$\varepsilon = \frac{\mu_0 \pi r^2}{2R} \cdot i_0 \omega \sin\left(\omega t - \frac{\pi}{2}\right).$$

Hence, induced emf lags current by $\pi/2$. Option (d) is correct.

12. Dipole moment $\boldsymbol{p} = \sum_i q_i \boldsymbol{r}_i$

$$\boldsymbol{p} = q\left[0 + a\hat{i} + a\hat{j} - 2a(\hat{i} + \hat{j})\right] = -qa(\hat{i} + \hat{j}).$$

Option (c) is correct.

13. Trajectory will be a helix with varying pitch. So it will not be confined to any plane.

Option (d) is correct.

14. Charge gets distributed uniformly over the sphere. Because there is no electric field inside the sphere, potential will be same for $r < a$. Hence, option (b) is correct.

15. The depolarization field should be applied in opposite direction and its magnitude is $\dfrac{4\pi}{3} P$.

$\therefore$ Depolarization field $= -\dfrac{4\pi P}{3}$. Option (b) is correct.

16. $$\nabla \cdot B = 0$$

Thus $\iiint_V \nabla \cdot B\, dV = 0$

or $\oiint_S B \cdot dS = 0$

There cannot be any magnetic monopole. Option (b) is correct.

17. Option (b) is correct.

18. Option (b) is correct.

19. $\dfrac{\pi c}{\omega}$. Option (c) is correct.

20. Electric and magnetic fields are perpendicular to each other.

As $F_{mag} = q(v \times B)$ is perpendicular to velocity.

$F \cdot v = 0 \Rightarrow$ Power 0

No work is done by B. Thus, option (b) is correct.

21. Option (c) is correct.

22. Option (b) is correct.

23. Inside solenoid, magnetic field exists so $\dfrac{\partial B}{\partial t}$ gives induced electric field while

outside solenoid, $\dfrac{\partial B}{\partial t} = 0$ as $B = 0$. Option (b) is correct.

24. Image charge lies at $\dfrac{(\text{radius})^2}{(\text{distance})}$, i.e. $\dfrac{R^2}{2R} = \dfrac{R}{2}$

Charge of image $= (-1)$ original charge $\times \dfrac{\text{radius}}{\text{distance}}$

$$= -q \times \dfrac{R}{2R} = -\dfrac{q}{2}.$$

Hence, option (c) is correct.

25. $U \propto \dfrac{q}{R}$

$\therefore$ Energy $\propto \dfrac{q^2}{R}$. Now energy $\propto \dfrac{(2q)^2}{R/2} \propto \dfrac{8q^2}{R}$. Obviously, energy associated with

electric field is 8 times. Option (b) is correct.

26. $E = -\nabla V$

$$\dfrac{\sigma}{\varepsilon_0} = E$$

or
$$\sigma = -\varepsilon_0\left(-E_0\cos\theta - \frac{2R^3}{r^3}E_0\cos\theta\right)\Bigg|_{r=R}$$

$$= 3\varepsilon_0 E_0\cos\theta.\text{ Option (c) is correct.}$$

27. Option (b) is correct.

28. We have

$$R = \left(\frac{n-1}{n+1}\right)^2, T = 1-R$$

∴
$$T = \frac{4n}{(n+1)^2}.\text{ Option (a) is correct.}$$

29. Option (b) is correct.

30 Phase velocity $V_p = \dfrac{\omega}{k}$

Group velocity, $V_g = \dfrac{d\omega}{dk}$

We have $\omega^2 = \omega_0^2 + c^2 k^2$

Differentiating both sides, one obtains

$$2\omega\,d\omega = 2c^2 k\,dk$$

or
$$\frac{\omega}{k}\cdot\frac{d\omega}{dk} = c^2$$

$$V_p, V_g = c^2.\text{ Option (c) is correct.}$$

31. Option (d) is correct.

32.
$$\mathbf{B} = \frac{\hat{r}\times\mathbf{E}}{v}$$

$$= \frac{\hat{z}\times\hat{x}}{\omega/k}\cdot E_0\exp\left[i(kz\cos\phi-\omega t]\cdot\exp(-kz\sin\phi)\right.$$

$$= \hat{y}\frac{k}{\omega}E_0\exp(-kz\sin\phi)\cdot\exp\left[i(kz\cos\phi-\omega t)+\phi\right]$$

Option (b) is correct.

33. Ampere's law $\oint \mathbf{B}\cdot d\mathbf{l} = \mu_0 j_{en}$

or $\quad B\cdot 2\pi r = 0$

or $\qquad B = 0.$ Thus, option (a) is correct.

34. Gauss's theorem gives

$$\oint \mathbf{E}\cdot d\mathbf{s} = \frac{q_{in}}{\varepsilon_0}$$

or $\quad E4\pi r^2 = \frac{\rho}{\varepsilon_0}\cdot\frac{4\pi}{3}r^3$

Fig. A.7

or $\qquad E = \dfrac{\rho r}{3\varepsilon_0}$. Option (a) is correct.

35. $E = E_0(\hat{x} + i\hat{y})\exp[i(kz - \omega t)]$. Option (b) is correct.

36. $E \propto \dfrac{1}{r^2}$ and $B = \dfrac{E}{C}$

$\therefore \qquad\qquad B \propto \dfrac{1}{r^2}$. Option (c) is correct.

37. Particle moving with no acceleration, i.e. with constant velocity does not radiate EM radiation

$\qquad P_A = 0, \quad P_B = P_C$. Option (b) is correct.

38. $\qquad\qquad E = \nabla\phi$

$\qquad\qquad = -2\phi_0(\hat{i}x + \hat{j}y + \hat{k}z) = -2\phi_0 r$

$\qquad \nabla \cdot E = \dfrac{\rho}{\varepsilon_0}$

or $\qquad \rho = \varepsilon_0 \nabla \cdot E$

$\qquad\qquad = -2\phi_0\varepsilon_0 \times 3 = -6\phi_0\varepsilon_0$. Option (b) is correct.

39. Work done is half of the work done when the plate is at infinity.

$\therefore \dfrac{1}{2} \times \dfrac{q^2}{4\pi\varepsilon_0(2d)} = -\dfrac{q^2}{4\pi\varepsilon_0(4d)}$. Option (c) is corect.

40. $\qquad E_{0r} = \dfrac{V_1 - V_2}{V_1 + V_2}E_{0i}$

$\qquad\qquad = \dfrac{\dfrac{V_1}{V_2} - 1}{\dfrac{V_1}{V_2} + 2}E_{oi} = \dfrac{1.5 - 1}{1.5 + 1} \times 2 = \dfrac{2}{5}$ V/m

or $\qquad E_{0t} = \dfrac{2V_2}{V_1 + V_2}E_{0j} \dfrac{2}{\dfrac{V_1}{V_2} + 2}E_{0i} = \dfrac{8}{5}$ V/m.

Option (c) is correct.

41. Option (d) is correct.

42. As magnetic field induction is not constant throughout the area of the loop, so we can have differential element as

$\qquad\qquad d\phi = B \cdot adx = \dfrac{\mu_0 i}{2\pi x}adx$

or $\qquad \int d\phi = \dfrac{\mu_0 ia}{2\pi}\ln\left(\dfrac{a + a}{a}\right)$

Induced emf, $\varepsilon = \dfrac{-d\phi}{dt} = \dfrac{\mu_0}{2\pi} a \ln 2 \left(\dfrac{-di}{dt} \right) = \dfrac{\mu_0 a}{2\pi} \ln 2 \cdot I_0 \omega \sin \omega t$

Induced current $= \dfrac{\varepsilon}{R} = \dfrac{\omega \mu_0 a I_0 \ln 2}{2\pi R} \sin \omega t$. Option (a) is correct.

Fig. A.8

43. Hamiltonian, $H = KE + PE$

In presence of magnetic field momentum p changes to $p - \dfrac{q}{c} A$.

$\therefore$ Kinetic energy $\quad KE = \dfrac{1}{2m} \left(P - \dfrac{q}{c} A \right)^2$

Potential energy, $\quad PE = q\phi$

$$H = \dfrac{1}{2m} \left(P - \dfrac{q}{c} A \right)^2 + q\phi$$

Option (b) is correct.

44. In electromagnetic field,

$\quad\quad E^2 - c^2 B^2$ remains invariant

$\quad\quad E'^2 - c^2 B'^2 = E^2 - c^2 B^2$

Option (b) is correct.

45. $E = -\dfrac{kqr}{a^3}$ and $V = \dfrac{kq}{2a} \left(\dfrac{3r^2}{a^2} \right)$

Option (c) is correct.

46. Option (c) is correct.

47. Net dipole moment $= qa(\hat{i} + \hat{j}) + qa(-\hat{i} + \hat{j}) - 2q(-a\hat{j})$

$\quad\quad = 2qa\hat{j} + 2qa\hat{j} = 4qa\hat{j}$. Option (a) is correct.

48. Option (a) is correct.

49. $$B = \nabla \times A = \begin{vmatrix} \hat{i} & \hat{j} & \hat{k} \\ \dfrac{\partial}{\partial x} & \dfrac{\partial}{\partial y} & \dfrac{\partial}{\partial z} \\ -y & 2x & 0 \end{vmatrix}$$

$$= \hat{i}(0) - \hat{j}(0) + \hat{k}\{2 - (-1)\} = 3\hat{k}.$$ Option (b) is correct.

50. Normal of B and D is same on the boundary. Tangential component of E and H is same on the boundary. Thus,

$$E_1 \sin \theta_1 = E_2 \sin \theta_2$$
$$D_1 \cos \theta_1 = D_2 \cos \theta_2$$

$$\therefore \qquad \frac{E_1}{D_1} \tan \theta_1 = \frac{E_2}{D_2} \tan \theta_2$$

$$\frac{\tan \theta_1}{\varepsilon_1} = \frac{\tan \theta_2}{\varepsilon_2}$$

or $$\frac{\varepsilon_1}{\varepsilon_2} = \frac{\tan \theta_1}{\tan \theta_2}.$$ Option (a) is correct.

51. Option (c) is correct.
52. Option (a) is correct.
53. Option (c) is correct.
54. Option (a) is correct.
55. Sum of charges is zero (i.e. $\Sigma q_i = 0$). Dipole moment is non-zero ($\Sigma q_i r_i \neq 0$). So dipole moment contribution *dominates*. Option (b) is correct.
56. Displacement current tends to zero when rate of charge accumulation is almost zero. It happens only when capacitor is almost fully charged, energy of the LC circuit is stored in its electric field. Option (b) is correct.
57. Franck–Hertz experiment $\rightarrow$ energy level in atoms

    Hartree–Fock Method $\rightarrow$ wave function of atoms

    Stern–Gerlach experiment $\rightarrow$ spin angular momentum of atoms

    Franck–Condon principle $\rightarrow$ electronic excitation of molecules

    Option (a) is correct.
58. Lienard–Weichert potentials are used for a moving point charge. Option (c) is correct.

59. $$\oint B \cdot dl = \mu_r \mu_0 N i_{enclosed}$$

    or $$B = \frac{\mu_r \mu_0 NI}{2\pi r}.$$ Option (c) is correct.

60. Intensity, $I = \dfrac{1}{2}\varepsilon_0 E^2 c$

$$\text{Power} = I \cdot A = \frac{1}{2}\varepsilon_0 E^2 c \times 4\pi = \frac{4\pi\varepsilon_0 E^2 c}{2}$$

or

$$P = \frac{60 \times 60 \times 3 \times 10^8}{9 \times 10^9 \times 2} = 60\,\text{W}$$

$$\text{Average power} = \frac{60}{2}\,\text{W} = 30\,\text{W}$$

Option (a) is correct.

61. Lorentz gauge is satisfied when

$$\nabla \cdot A + \frac{1}{c^2}\frac{\partial \phi}{\partial t} = 0$$

Option (b) is correct.

62. Option (d) is correct.

63. Bound surface density $= \sigma_b = p \cdot \hat{n} = k\,|r|$

Bound volume density $= \rho_b = -\nabla \cdot \overline{P} = 3k$

Option (b) is correct.

64. $-\nabla \cdot P = -3\hat{k} = \rho_b$: Volume density

$p \cdot \hat{r} = k_r = \sigma_b$: Surface bound density

$$\therefore \qquad \text{Total charge} = \frac{4\pi}{3}R^3 \times (-3k) + 4\pi R^2 (kR) = 0$$

Now $\qquad \oiint E \cdot ds = 0$

$\therefore \qquad E = 0$. Option (a) is correct.

Dirac's δ-Function

The center of mass theorem states that the center of mass of a system of mass points moves as if the total mass of the system were concentrated in this point and all external forces would act exclusively on it. The mass M of a body can be expressed in terms of the mass density $\rho(r)$ as

$$M = \int_V d^3r\rho(r) \tag{1a}$$

But how does the mass density of a mass point look like? It can be unequal zero only in a single point:

$$\rho(r) = 0 \quad \forall r \neq r_0$$

The volume integral, however,

$$\int_V d^3r\rho(r)$$

Shall be nevertheless finite provided r_0 lies within the volume V. We thus 'symbolize' $\rho(r)$ as follows:

$$\rho(r) = M\delta(r - r_0) \tag{1b}$$

and require:

$$\int_V d^3r\delta(r - r_0) = \begin{cases} 1, & \text{if} \quad r_0 \in V \\ 0 & \text{otherwise} \end{cases} \tag{2}$$

$$\delta(r - r_0) = 0 \quad \forall r \neq r_0 \tag{3}$$

Equations (2) and (3) are the defining equations of the Dirac's δ-function (short: δ-function). Obviously, one must not understand the integral of Eq. (2) as a usual Riemannian integral. Since, because of Eq. (3), the effective integration interval is of zero-width the integral should actually vanish. Sometimes one helps oneself with the idea that for $r = r_0$ the δ-function takes the value ∞ so that something finite results out of $0 \cdot \infty$. This is of course only an auxiliary view. The δ-function is not a function in the ordinary mathematical sense which ascribes to each value of its definition range uniquely a certain value of the function. Instead it is defined by Eqs. (2) and (3). It is therefore denoted as an improper function or as a distribution. The related exact mathematical theory is called the *distribution theory*. It goes beyond the framework of our introductory presentation here and therefore must be settled by some plausibility considerations. For that we restrict ourselves at first to the one-dimensional case.

Let us consider a sequence of Lorentz curve (Fig. B.1)

$$L_\eta(x - a) = \frac{1}{\pi}\frac{\eta}{\eta^2 + (x - a)^2}, \quad (\eta > 0) \tag{4}$$

Fig. 1: Illustration of the δ-function as a limiting function of a sequence of Lorentzians

For the height of the maximum at $x = 0$ we have

$$\frac{1}{\pi\eta} \overrightarrow{\eta \to 0^+} \infty$$

And for the width of the peak (*'full width at half maximum'*)

$$2\eta \overrightarrow{\eta \to 0^+} \infty$$

The area under the Lorentz curve amounts to

$$\int_\alpha^\beta dx \left[\frac{1}{\pi} \frac{\eta}{\eta^2 + (x-a)^2} \right] = \frac{1}{\pi} \left[\arctan\left(\frac{\beta - a}{\eta} \right) - \arctan\left(\frac{\alpha - a}{\eta} \right) \right]$$

$$\underset{q \to 0^+}{\to} \begin{cases} 1, & \text{if } \alpha < a < \beta \\ \dfrac{1}{2}, & \text{if } \alpha \text{ or } a = \beta \\ 0 & \text{otherwise } (a \neq \alpha, \beta) \end{cases}$$

For $\eta \to 0^+$ L_η becomes arbitrarily narrow. Thus it is:

$$\lim_{\eta \to 0^+} L_\eta(x-a) = 0 \quad \forall \, x \neq a \tag{5}$$

$$\lim_{\eta \to 0^+} \int_\alpha^\beta L_\eta(x-a)dx = \begin{cases} 1, & \text{if } \alpha < a < \beta \\ \dfrac{1}{2}, & \text{if } \alpha \text{ or } a = \beta \\ 0 & \text{otherwise } (a \neq \alpha, \beta) \end{cases} \tag{6}$$

The sequence of integration and limiting process in Eq. (6) must not be interchanged. If we strictly obey that, then we can write *'abbreviatorily'*:

$$\delta(x - a) = \lim_{\eta \to 0^+} \frac{1}{\pi} \frac{\eta}{\eta^2 + (x-a)^2} \tag{7}$$

with

$$\delta(x - a) = 0 \quad \forall \, x \neq a$$

$$\int_\alpha^\beta dx \, \delta(x - a) = \begin{cases} 1, & \text{if } \alpha < a < \beta \\ \dfrac{1}{2}, & \text{if } \alpha \text{ or } a = \beta \\ 0 & \text{otherwise } (a \neq \alpha, \beta) \end{cases} \tag{8}$$

The δ-function can be represented also by other limiting processes (see exercises!) where these processes have to fulfill only (5) and (6).

On the basis of Eqs. (5) to (7) one verifies the following properties of the δ-function:
1. Let $f(x)$ be a function which is continuous in the neighborhood of $x = a$. Then it holds:

$$\int_\alpha^\beta f(x)\,\delta(x-a)dx = \begin{cases} f(a) & \text{if } \alpha < a < \beta \\ \dfrac{1}{2}f(a), & \text{if } \alpha \text{ or } a = \beta \\ 0 & \text{otherwise } (a \neq \alpha, \beta) \end{cases} \tag{9}$$

Proof from the *mean value theorem of integral calculus* we first have:

$$F_\eta(a) = \int_\alpha^\beta L_\eta(x-a)f(x)dx = f(\xi)\int_\alpha^\beta L_\eta(x-a)dx$$

For $\eta \to 0^+$ $L_\eta(x-a)$ becomes an arbitrarily sharp peak around a and $F_\eta(a)$ does not change if the interval of integration is restricted to the region for which L_η is unequal to zero. ξ must lie within this effective region of integration which for $\eta \to 0^+$ contracts to the point a:

$$\lim_{\eta \to 0^+} F_\eta(a) = f(a)\lim_{\eta \to 0^+}\int_\alpha^\beta L_\eta(x-a)dx$$

With (6) we then get the assertion (9).

2.

$$\delta[f(x)] = \sum_i \frac{1}{|f'(x_i)|}\delta(x-x_i) \tag{10}$$

x_i: **simple zero** of $f(x)$; $f(x_i) = 0$; $f'(x_i) \neq 0$

One easily recognizes the following special cases:
(a)

$$\delta(ax) = \frac{1}{|a|}\delta(x) \tag{11}$$

(b)

$$\delta(x^2 - a^2) = \frac{1}{2|a|}[\delta(x-a) + \delta(x+a)] \tag{12}$$

3.

$$g(x)\delta(x-a) = g(a)\delta(x-a) \tag{13}$$

$$x\delta(x) = 0 \tag{14}$$

4.

$$\int_{-\infty}^x d\bar{x}\delta(\bar{x}) = \Theta(x) = \begin{cases} 1 & \text{for } x > 0 \\ 0 & \text{for } x < 0 \end{cases} \tag{15}$$

'step function'

5. **Derivative** of the δ-function $[a \in (\alpha; \beta)]$:

$$\int_\alpha^\beta \delta'(x-a)f(x)dx = f(x)\delta(x-a)\Big|_\alpha^\beta - \int_\alpha^\beta \delta(x-a)f'(x)dx = -f'(a)$$

Since $f(x)$ must only be differentiable, being otherwise arbitrary, this *formal* integration by parts lead to the following identity:

$$f(x)\delta'(x - a) = -f'(a)\delta(x - a) \tag{16}$$

6. One can consider the δ-function also as the derivative of the step function:

$$\delta(x - a) = \frac{d}{dx} \Theta (x - a) \tag{17}$$

because:

$$\int_\alpha^\beta \frac{d}{dx} \Theta (x - a)\, dx = \Theta(x - a)\Big|_\alpha^\beta = \begin{cases} 1, & \text{if } \alpha < a < \beta \\ 0 & \text{otherwise} \end{cases}$$

$$\frac{d}{dx} \Theta (x - a) = 0 \quad \forall x \neq 0$$

But these are just the two defining equations of the δ-function.

7. **Multi-dimensional δ-function**

The three-dimensional δ-function is defined by (2) and (3).

(a) **Cartesian coordinates:**

$$\boldsymbol{r} = (x, y, z); \quad \boldsymbol{r}_0 = (x_0, y_0, z_0)$$

$$\int_v d^3r \dots \to \iiint_v dx\, dy\, dz \dots$$

ansatz:

$$\delta(\boldsymbol{r} - \boldsymbol{r}_0) = \gamma(x, y, z)\, \delta(x - x_0)\delta(y - y_0)\, \delta(z - z_0)$$

$\gamma(x, y, z)$ must be chosen such that (2) is fulfilled:

$$\int_v d^2r\, \delta(\boldsymbol{r} - \boldsymbol{r}_0) = \iiint_v dx\, dy\, dz\, \delta(x, y, z)\, \delta(x - x_0)\, \delta(y - y_0)\, \delta(z - z_0)$$

$$= \gamma(x_0, y_0, z_0) \iiint_v \delta(x - x_0)\, \delta(y - y_0)\, \delta(z - z_0)\, dx\, dy\, dz$$

$$= \begin{cases} \gamma(x_0, y_0, z_0), & \text{if } \boldsymbol{r}_0 \in V \\ 0 & \text{otherwise} \end{cases}$$

That means:

$$\gamma = 1$$

and therewith:

$$\delta(\boldsymbol{r} - \boldsymbol{r}_0) = \delta(x - x_0)\, \delta(y - y_0)\delta(z - z_0) \tag{18}$$

(b) **Curvilinear coordinates** (u, v, w): The volume element reads:

$$d^3r = dx\, dy\, dz = \underbrace{\frac{\partial(x, y, z)}{\partial(u, v, w)}}_{\text{Jacobian determinant}} du\, dv\, dw$$

We choose a similar ansatz as in (a):

$$\delta(\boldsymbol{r} - \boldsymbol{r}_0) = \gamma(u, v, w)\, \delta(u - u_0)\delta(v - v_0)\, \delta(w - w_0) \tag{19}$$

Because of (2) we then have to fulfill:

$$\int_v d^3r \, \delta(\boldsymbol{r} - \boldsymbol{r}_0) = \iiint_v du \, dv \, dw \, \frac{\partial(x, y, z)}{\partial(u, v, w)} \, \gamma(u, v, w) \, \delta(u - u_0)$$

$$\delta(v - v_0) \, \delta(w - w_0) = 1 \quad \text{if} \quad \boldsymbol{r}_0 \in V$$

That leads to:

$$\gamma = \left(\left. \frac{\partial(x, y, z)}{\partial(u, v, w)} \right|_{r_0} \right)^{-1} \tag{20}$$

Examples:

Spherical coordinates (r, θ, φ)

$$\delta\left(\boldsymbol{r} - \boldsymbol{r}_0 \right) = \frac{1}{r_0^2 \sin \theta_0} \delta(r - r_0) \, \delta(\theta - \theta_0) \, \delta(\varphi - \varphi_0) \tag{21}$$

Cylindrical coordinates (ρ, φ, z)

$$\delta\left(\boldsymbol{r} - \boldsymbol{r}_0 \right) = \frac{1}{\rho_0} \delta(\rho - \rho_0) \, \delta(\varphi - \varphi_0) \, \delta(z - z_0) \tag{22}$$

System of Units

Gauss and Weber gave the concept of the absolute system of units to describe all quantities in terms of millimeters, milligrams and seconds. With the passage of time Gaussian, Heavyside-Lorentz, Electrostatic, Electromagnetic and MKS systems came into existence. However, the most commonly used one, are the Gaussian and MKS systems. Here, we briefly describe these systems and discuss inter-conversion between them.

1. GAUSSIAN SYSTEM OR SYMMETRICAL SYSTEM OF UNITS

This system uses the electrical quantities from the e.s.u. system and magnetic quantities from the e.m.u system and, as a consequence, we see that the velocity of light (c) in vacuum appears explicitly in some of the equations which interrelate the electric and magnetic quantities.

In Gaussian system, the field quantities **E,P,D** and **B,M,H** have the same units and as such this system is more convenient to the requirements of relativistic invariance. Furthermore, the system is more adequate to the treatment of microscopic problem involving the electrodynamics of individual charged particles.

In Gaussian system, Maxwell's equations are as follows:

$$\nabla \times E = -\frac{1}{c}\frac{\partial B}{\partial t}$$

$$\nabla \times H = \frac{4\pi}{c}J + \frac{1}{c}\frac{\partial D}{\partial t} \tag{1}$$

$$\nabla \cdot B = 0$$

$$\nabla \cdot D = 4\pi\rho$$

Other relations are

$$D = E + 4\pi P \tag{2}$$
$$B = H + 4\pi M \tag{3}$$

The Lorentz force equation is

$$F = q\left(E + \frac{v}{c} \times B\right) \tag{4}$$

All the field vectors are measure and expressed in terms of a fundamental unit called gauss. From Eq. (4), we see that the dimensions of guass are force/unit charge and gauss is defined as

$$1 \text{ guass} = 1 \text{ dyne}/(1 \text{ e.s.u. of charge})$$

1 e.s.u. of charge is defined as charge the repels and equal charge at a distance of one cm with a force of one dyne. In the Gaussian system, the electrostatic force between two charges q_1 and q_2 is given by Coulomb's law as

$$F = \frac{q_1 q_2}{|x_1 - x_2|^2} \tag{5}$$

and the magnetostatic force/unit length of parallel currents I_1 and I_2 separated by $|x_1 - x_2|$ given by

$$\frac{dF}{dl} = \frac{2}{c^2} \frac{I_1 I_2}{|x_1 - x_2|} \tag{6}$$

In Gaussian system, the fundamental dimensions are chosen to be *mass*, *length* and *time* and all derived quantities are expressed in terms of these three. The dimensions of charge are

$$q = LF^{1/2}$$
$$= M^{1/2} L^{3/2} T^{-1} \tag{7}$$

One can deduce the dimensions of E. We have

$$E = \frac{F}{q}$$
$$= (ML/T^2)/M^{1/2} L^{3/2} T^{-1}$$
$$= \left[M^{1/2} L^{-1/2} T^{-1} \right] \tag{8}$$

The basic units in the Gaussian system are the gram, the centimeter and the second.

2. HEAVISIDE-LORENTZ UNIT

This system of units differs from the Gaussian system only in that a factor of 4π is associated with the definition of charge. The new electromagnetic quantities are defined in terms of Gaussian quantities through the relations as follows:

$$\bar{\rho} = \sqrt{4\pi}\, \rho \qquad\qquad \bar{E} = \frac{1}{\sqrt{4\pi}} E$$

$$\bar{q} = \sqrt{4\pi}\, q \qquad\qquad \bar{B} = \frac{1}{\sqrt{4\pi}} B$$

$$\bar{J} = \sqrt{4\pi}\, J \qquad\qquad \bar{D} = \frac{1}{\sqrt{4\pi}} D \tag{9}$$

$$\bar{I} = \sqrt{4\pi}\, I \qquad\qquad \bar{H} = \frac{1}{\sqrt{4\pi}} H$$

$$\bar{P} = \sqrt{4\pi}\, P \qquad\qquad \bar{M} = \sqrt{4\pi}\, M$$

In view of Eq. (9), Maxwell's equations in the Heaviside-Lorentz system reads as

$$\nabla \times \bar{E} = -\frac{1}{c} \frac{\partial \bar{B}}{\partial t}$$

$$(\nabla \times \bar{H}) = \frac{1}{c}\bar{J} + \frac{1}{c}\frac{\partial \bar{D}}{dt} \tag{10}$$

$$\nabla \cdot \bar{B} = 0$$

$$\nabla \cdot \bar{D} = \bar{\rho}$$

In addition, Eqs. (2) and (3) takes the form

$$\bar{D} = \bar{E} + \bar{P} \tag{11}$$

$$\bar{B} = \bar{H} + \bar{M}$$

$$F = \bar{q}\left(E + \frac{v \times B}{c}\right)$$

The corresponding force equations are as follow

$$F = \frac{\bar{q}_1 \bar{q}_2}{4\pi |x_1 - x_2|^2} \tag{12}$$

$$\frac{dF}{dl} = \frac{2}{4\pi c^2}\frac{\bar{I}_1 \bar{I}_2}{|x_1 - x_2|^2} \tag{13}$$

We see that the ultimate effect of the Heaviside-Lorentz system turns out to be the elimination of the factor 4π for Maxwell's equations and its appearance in the force equations (I.12) and (I.13). They system of units is said to be rationalized system of units, due to the removal of the factor of 4π form Maxwell's equations.

3. ELECTROSTATIC UNIT (e.s.u.)

In e.s.u., unit of charge is defined through *Coulomb's law* of force between static charges, with the consequence that the quantities $\bar{\rho}, \bar{J}, \bar{E}$ and $\bar{P}$ are exactly the same as in Gaussian units. The magnetic quantities are defined in a manner that lead to the elimination of the factor c from Maxwell's equations.. One may achieve this through the following definitions

$$\bar{B} = \frac{1}{c}B$$

$$\bar{H} = cH$$

$$\bar{M} = cM \tag{14}$$

Where the quantities with bars refer to the *e.s.u. system* in terms of the Gaussian vectors.

Maxwell's equations reads as

$$\nabla \times E = -\frac{\partial \bar{B}}{\partial t}$$

$$\nabla \times \bar{H} = 4\pi J + \frac{\partial D}{\partial t} \tag{15}$$

$$\nabla \cdot \bar{B} = 0$$

$$\nabla \cdot \bar{D} = 4\pi\rho$$

Along with the relations

$$D = E + 4\pi P$$

$$c^2\bar{B} = \bar{H} + 4\pi\bar{M}$$

$$F = q(E + v \times \bar{B})$$ (16)

We may note that the above quantities have the same names as the Guasssian ones; magnetic quantities are not assigned any special names. The magnitude of B, for instance, would be referred to as some number of e.s.u. of magnetic induction. The force laws between charges and currents also retain the same form as in the Gaussian system.

4. ELECTROMAGNETIC UNIT (e.m.u)

Electromagnetic units also constitute another systematic extension of c.g.s. units to electromagnetic theory. However, the magnetic quantities **B,H,M** are the same as in Gaussian system, but the electric quantities are defined through the following definitions in terms of the Gaussian quantities:

$$\bar{E} = cE$$

$$\bar{D} = \frac{1}{c}D$$ (17)

$$\bar{P} = \frac{1}{c}P$$

$$\bar{J} = \frac{1}{c}J$$

$$\bar{\rho} = \frac{1}{c}\rho$$

In view of these definitions. Maxwell's equations in electromagnetic units reads as

$$\nabla \times \bar{E} = -\frac{\partial B}{\partial t} \qquad \nabla \times \bar{H} = 4\pi\bar{J} + \frac{\partial \bar{D}}{\partial t}$$

$$\nabla \cdot \bar{D} = 4\pi\bar{\rho} \qquad \nabla \cdot B = 0$$ (18)

Along with relations

$$\bar{D} = \frac{1}{c^2}\bar{E} + 4\pi\bar{P}$$ (19)

$$B = H + +4\pi M$$

And Lorentz force equation being

$$F = \bar{q}(\bar{E} + v \times B)$$ (20)

As is evident from Eq. (18), there is no factor of c appearing in Maxwell's equations. According to e.m.u. units, the force law now reads as

$$F = c^2 \frac{\bar{q}_1\bar{q}_2}{4\pi|x_1 - x_2|^2}$$

$$\frac{dF}{dl} = \frac{2\bar{I}_1 \bar{I}_2}{|x_1 - x_2|} \tag{21}$$

The e.m.u. unit of charge is called an *abcoulomb* and is defined as one abampere second. Now, we deduce the dimensions of abampere. Two straight parallel wires carrying a current of one abampere each and one cm apart, exert a force of two dysnes/cm upon each other. Thus

$$2 \text{ dyne cm}^{-1} = 2 \text{ (abampere)}^2 \text{ cm}^{-1}$$
$$1 \text{ abampere} = 1(\text{dyne})^{1/2} \tag{22}$$
$$= 1 \text{ gm } \frac{1}{2} \text{ cm } \frac{1}{2} \text{ s}^{-1}$$

Obviously, the dimensional of abcoulomb is therefore 1 gm ½ cm ½

5. METRE-KILOGRAM-SECOND (MKS) UNITS

The conversion of the basic equations of electromagnetic theory from the Gaussian to MKS units is a two-step process. The first step involves setting up a rationalized electromagnetic system and the second step involves the inclusion of charge as independent dimension into the fundamental dimensions M, L and T. This is achieved through the following definitions of electric and magnetic quantities in terms of Gaussian quantities.

$$\bar{\rho} = \frac{\sqrt{4\pi}}{c} \rho \qquad\qquad \bar{E} = \frac{c}{\sqrt{4\pi}} E$$

$$\bar{q} = \frac{\sqrt{4\pi}}{c} q \qquad\qquad \bar{B} = \frac{1}{\sqrt{4\pi}} B$$

$$\bar{J} = \frac{\sqrt{4\pi}}{c} J \qquad\qquad \bar{D} = \frac{1}{c\sqrt{4\pi}} D$$

$$\bar{I} = \frac{\sqrt{4\pi}}{c} I \qquad\qquad \bar{H} = \frac{1}{\sqrt{4\pi}} H \tag{23}$$

$$\bar{P} = \frac{\sqrt{4\pi}}{c} P \qquad\qquad \bar{M} = \sqrt{4\pi}\, M$$

In the light of these equations, we see that Maxwell's and associated equations reads as

$$\nabla \times \bar{E} = -\frac{\partial \bar{B}}{\partial t}$$

$$\nabla \times \bar{H} = \bar{J} + \frac{\partial \bar{D}}{\partial t}$$

$$\nabla \cdot \bar{B} = 0 \tag{24}$$

$$\nabla \cdot \bar{D} = \bar{\rho}$$

$$\bar{D} = \frac{1}{c^2} \bar{E} + \bar{P}$$

$$F = \bar{q}(\bar{E} + v \times \bar{B})$$

Now, the force laws between charges and currents takes the form

$$F = \frac{c}{4\pi} \frac{\bar{q}_1 \bar{q}_2}{|x_1 - x_2|^2}$$

$$\frac{dF}{dl} = \frac{2}{4\pi} \frac{\bar{I}_1 \bar{I}_2}{|x_1 - x_2|^2} \tag{25}$$

So far the fundamental dimensions have been mass, length and time. However, with the introduction of charge as another independent dimension leads to the introduction of another constant, which may be chose with a view to making the basic units of convenient magnitude. Keeping this in view, Maxwell's equations and the Lorentz force-rule are preserved and alterations are made in the other equations so that the expressions for force assume the following forms

$$F = \frac{1}{4\pi\varepsilon_0} \frac{\bar{q}_1 \bar{q}_2}{|x_1 - x_2|^2}$$

$$\frac{dF}{dl} = \frac{2\mu_0}{4\pi} \frac{\bar{I}_1 \bar{I}_2}{|x_1 - x_2|^2} \tag{26}$$

We see that there are two constants ε_0 and μ_0 instead of one constant c in these equations. The dimensions of ε_0 and μ_0 are $q^2 T^2 / ML^3$ and ML/q^2 respectively. Since, $\nabla \cdot \bar{D} = \bar{\rho}$ one find $\bar{D}(r) = q/4\mu r^2$. Since the Lorentz force-equation is retained, the electric field is necessarily given by $E(r) = F/\bar{q}_2 = \bar{q}_1/4\pi\varepsilon_0 r$. In the absence of polarizable material, one must have $\bar{D} = \varepsilon_0 E$, where ε_0 is the permittivity of free space. $\bar{P}$ has the dimensions of those of dipole moment per unit volume, or q/L, the same as for $\bar{D}$. Therefore in the MKS system, one writes

$$\bar{D} = \varepsilon_0 \bar{E} + \bar{P} \tag{27}$$

Analogously, one finds that magnetic field $\bar{H}$ at a distance from straight wire carrying current I_1 curls about the wire according to the right hand screw (R.H.S.) rule and its magnitude is $I_1 / 2\pi |x_1 - x_2|$.

The force per unit length exerted on a test current $\bar{I}_2$ parallel to $\bar{I}_1$ at $r = |x_1 - x_2|$ from the Lorentz force is $dF = \rho v \times \bar{B}$ and the force per unit length is towards $\bar{I}_1$, and is expressed as

$$\frac{dF}{dl} = I_2 \bar{B} \tag{28}$$

Obviously $\bar{B}$ is given by

$$\bar{B} = \frac{dF/dl}{\bar{I}_2}$$

Comparing it with Eq. (26), one obtains

$$\bar{B} = \frac{2\mu_0}{4\pi} \frac{\bar{I}}{|x_1 - x_2|}$$

$$= \mu_0 \bar{H} \tag{29}$$

The units of magnetization vector $\bar{M}$ are taken the same as those of $\bar{H}$, out of analogy and parallelism that exist between $\bar{E}$ and $\bar{B}$, $\bar{P}$ and $\bar{M}$, $\bar{D}$ and $\bar{H}$. Thus, one obtains the following relation.

$$\frac{1}{\mu_0}\bar{B} = \bar{H} + \bar{M} \tag{30}$$

Where μ_0 is the permeability of free space.

Finally, **Maxwell's and allied equations** in the MKS system are written after dropping the bars as follows:

$$\nabla \times E = -\frac{\partial B}{\partial t}$$

$$\nabla \times H = J + \frac{\partial D}{\partial t}$$

$$\nabla \cdot B = 0$$

$$\nabla \cdot D = \rho$$

$$D = \varepsilon_0 E + P$$

$$\frac{1}{\mu_0}B = H + M$$

$$F = q(E + v \times B) \tag{31}$$

Values of ε_0 and μ_0 are:

Permittivity of free space $\varepsilon_0 = 8.8542 \times 10^{-12}\ C^2\,N^{-1}m^2$ (or $F\,m^{-1}$)

Permeability of free space $\mu_0 = 4\pi \times 10^{-7}\ WbA^{-1}m^{-1}$ (or $H\,m^{-1}$)

Table A.1 provides Maxwell's equations and some commonly used expressions in SI and Gaussian systems.

Table A.1: Commonly used equations in SI (MKS) and Gaussian (CGS) systems of Units

Equation/Quantity	SI System (MKS)	Gaussian System (cgs)
• *Maxwell's equations*	$\nabla \cdot E = \dfrac{1}{\varepsilon_0 \rho}$	$\nabla \cdot E = 4\pi\rho$
In vacuum	$\nabla \times E = -\dfrac{\partial B}{\partial t}$	$\nabla \times E = -\dfrac{1}{c}\dfrac{\partial B}{\partial t}$
	$\nabla \cdot B = 0$	$\nabla \cdot B = 0$
	$\nabla \times B = \mu_0 J + \mu_0 \varepsilon_0 \dfrac{\partial E}{\partial t}$	$\nabla \times B = \dfrac{4\pi}{c}J + \dfrac{1}{c}\dfrac{\partial E}{\partial t}$
	$\nabla \cdot D = \rho$	$\nabla \cdot D = 4\pi\rho$
	$\nabla \times E = -\dfrac{\partial B}{\partial t}$	$\nabla \times E = -\dfrac{1}{c}\dfrac{\partial B}{\partial t}$
In material media	$\nabla \cdot B = 0$	$\nabla \cdot B = 0$
	$\nabla \times H = J + \dfrac{\partial D}{\partial t}$	$\nabla \times H = \dfrac{4\pi}{c}J + \dfrac{1}{c}\dfrac{\partial D}{\partial t}$

Contd...

- **D and H**

  $$D = \varepsilon_0 E + P \qquad\qquad D = E + 4\pi P$$

  Definitions

  $$H = \frac{1}{\mu_0} B - M \qquad\qquad H = B - 4\pi M$$

  In linear media

  $$P = \varepsilon_0 \chi_e E,\, D = \varepsilon E \qquad\qquad P = \chi_e E,\, D = \varepsilon E$$

  $$M = \chi_m H,\, H = \frac{1}{\mu} B \qquad\qquad M = \chi_m H,\, H = \frac{1}{\mu} B$$

  $$\frac{\varepsilon}{\varepsilon_0} = 1 + \chi_e \qquad\qquad \varepsilon = 1 + 4\pi \chi_e$$

- **Electrostatics**

  Coulomb's law

  $$F = \frac{q_1 q_2}{4\pi\varepsilon_0 r^2}\,\hat{r} \qquad\qquad F = \frac{q_1 q_2}{r^2}\,\hat{r}$$

  Electric field of a stationary
     point charge

  $$E = \frac{1}{4\pi\varepsilon_0}\frac{q}{r^2}\,\hat{r} \qquad\qquad E = \frac{q}{r^2}\,\hat{r}$$

  Gauss's law

  $$\nabla \cdot E = \frac{\rho}{\varepsilon_0} \qquad\qquad \nabla \cdot E = 4\pi\rho$$

  Electric field (static)

  $$E = -\nabla\phi \qquad\qquad E = -\nabla\phi$$

  Electric field (general)

  $$E = -\nabla\phi - \frac{\partial A}{\partial t} \qquad\qquad E = -\nabla\phi - \frac{1}{c}\frac{\partial A}{\partial t}$$

  Electromagnetic energy
     in vacuum

  $$W = \frac{1}{2}\int d^3 r \rho(r)\phi(r) \qquad\qquad W = \frac{1}{2}\int d^3 r \rho(r)\phi(r)$$

- **Magnetostatics**

  The Biot-savart law describing the magnetic field created by a moving point charge is

  $$B = \frac{\mu_0 q v \times r}{4\pi r^2} \qquad\qquad B = \frac{q v \times \hat{r}}{r^2}$$

  $$B = \frac{\mu_0}{4\pi}\oint\frac{I dl \times \hat{r}}{r^2} \qquad\qquad B = \oint\frac{I dl \times \hat{r}}{r^2}$$

  Magnetic field

  $$B = \nabla \times A \qquad\qquad B = \nabla \times A$$

- **Electromagnetic waves**

  Poyting vector

  $$S = \frac{1}{\mu_0} E \times B \qquad\qquad S = \frac{c}{4\pi} E \times B$$

  Impedance of free space

  $$Z_0 = \sqrt{\frac{\mu_0}{\varepsilon_0}} \approx 377\,\Omega \qquad\qquad Z_0 = \frac{4\pi}{c}$$

  Skin depth (δ)

  $$\delta = \sqrt{\frac{2}{\mu\sigma\omega}} \qquad\qquad \delta = \sqrt{\frac{c^2}{2\mu\sigma\omega}}$$

- **Energy and Power**

  Energy

  $$E = \frac{1}{2}\int\left(\varepsilon_0 E^2 + \frac{1}{\mu_0}B^2\right)d^3 x \qquad\qquad E = \frac{1}{8\pi}\int\left(E^2 + B^2\right)d^3 x$$

  Larmor formula

  $$P = \frac{1}{4\pi\varepsilon_0}\frac{2}{3}\frac{q^2 a^2}{c^3} \qquad\qquad P = \frac{2}{3}\frac{q^2 a^2}{c^3}$$

Contd...

Total power radiated

From an oscillating electric dipole

$$P = \frac{c^2 Z_0 k^4}{12\pi} |p|^2 \qquad\qquad P = \frac{ck^4}{3} |p|^2$$

- *Lorentz force*

$$F = q(E + v \times B) \qquad\qquad F = q\left[E + \frac{v \times B}{c} \right]$$

- *Relativity*

The E, M, Tensor

$$F_{\mu v} = \begin{pmatrix} 0 & \dfrac{E_x}{c} & \dfrac{E_y}{c} & \dfrac{E_z}{c} \\ -\dfrac{E_x}{c} & 0 & B_z & -B_y \\ -\dfrac{E_y}{c} & -B_z & 0 & B_x \\ -\dfrac{E_z}{c} & B_y & -B_x & 0 \end{pmatrix} \qquad F_{\mu v} = \begin{pmatrix} 0 & E_x & E_y & E_z \\ -E_x & 0 & B_z & -B_y \\ -E_y & -B_z & 0 & B \\ -E_z & B_y & -B_x & 0 \end{pmatrix}$$

Field invariants

$$\sum_{\mu, v = 0}^{3} F_{\mu v} F^{\mu v} = 2(B^2 - E^2/c^2) \qquad \sum_{\mu, v = 0}^{3} F_{\mu v} F^{\mu v} = 2(B^2 - E^2)$$

Field transformations

$$E'_{\parallel} = E_{\parallel} \qquad\qquad\qquad E'_{\parallel} = E_{\parallel}$$

$$B'_{\parallel} = B_{\parallel} \qquad\qquad\qquad B'_{\parallel} = B_{\parallel}$$

$$E'_{\perp} = \gamma(E_{\perp} + v \times B_{\perp}) \qquad E'_{\perp} = \gamma(E_{\perp} + \beta \times B_{\perp})$$

$$B'_{\perp} = \gamma\left(B_{\perp} - \frac{v}{c^2} \times E_{\perp} \right) \qquad B'_{\perp} = \gamma\left(B_{\perp} - \beta \times E_{\perp} \right)$$

e.m . four potential

$$A^{\mu} = \left(\frac{\phi}{c}, A \right) \qquad\qquad A^{\mu} = (\phi, A)$$

- *Plasma*

Plasma frequency

$$\omega_P^2 = \frac{NZe^2}{\varepsilon_0 m} \qquad\qquad \omega_P^2 = \frac{4\pi NZe^2}{m}$$

Conversion Table for Electric and Magnetic Units

Quantity	Symbol	SI	Gaussian
Charge	q, e	1 C	3×10^9 stat Coul
Current	$I\left(=\dfrac{dq}{dt}\right)$	$1\ A = 1\ Cs^{-1}$	3×10^9 stat amp
Charge density	ρ	$1\ C/m^3$	3×10^3 stat Coul/cm^3
Current density	J	$1\ A/m^3$	3×10^5 stat amp/cm^2
Electric field	$E\ (=-\nabla\phi)$	$1\ V/m$	$(1/3) \times 10^{-4}$ stat volt/cm
Electric potential	$\phi(V)$	1 V	$1/300$ stat volt
Electric displacement	D	$1\ C/m^2$	$12\pi \times 10^5$ stat volt/cm
Polarization	P	$1\ C/m^2$	3×10^5 stat volt/cm
Magnetization	M	$1\ A/m$	$(1/4\pi) \times 10^4$ gauss
Magnetic field	H	$1\ A/m$	$4\pi \times 10^{-3}$ oersted
Vector potential	A	$1\ Wb/m$	$(1/3) \times 10^{-10}$ gauss cm
Magnetic flux density	B	$1\ Wb/m^2$ (tesla)	10^4 gauss
Magnetic permittivity	μ	$1\ H/m$	$(1/4\pi) \times 10^7$ gauss/oersted
Dielectric permittivity	ε	$1\ farad/m$	$36\pi \times 10^9$ stat farad/cm
Capacity	C	1 farad	9×10^{11} stat farad
Electrical conductivity	σ	$1\ mho/m$	$9 \times 10^9/s$
Inductance	L	1 henry (H)	$\dfrac{1}{9} \times 10^{-11}$
Resistance	R	1 ohm (Ω)	$\dfrac{1}{9} \times 10^{-11}$ stat ohms
Power	P	1 watt (W)	10^7 erg/s
Frequency	ν	1 hertz (Hz)	1 Hz
Force	F	1 newton (N)	10^5 ergs

Tensor Analysis

The laws of nature are *covariant*, which means that they have the same forms in all coordinate system. A systematic method of investigating the behavior of quantities that undergo a *coordinate transformation* is the subject matter of tensor analysis.

A *tensor* consists of set of quantities called *components* whose properties are independent of the coordinate system used to describe them. The components of a tensor in two different coordinate systems are related by the characteristic transformation laws.

A collection of indices (subscripts and/or superscripts) is used to make the mathematical development of tensor analysis compact. The superscripts, *contravariant* indices, are used to denote the contravariant components of a tensor $T^{ij}...$. The subscripts covariant indices, are used to represent the covariant components of a tensor $T_{ij}...$ The components of a mixed tensor are specified by indicating both subscripts and superscripts $T^{ij...}_{lmn...}$.

The *rank* (*order*) of a tensor is the number (without counting an index which appears once as a subscript and once as a superscript) of indices in the symbol representing a tensor, ignoring the dummy indices. A tensor of zero rank is a scalar and a tensor of rank one is a vector. A tensor of rank 2 has two indices. In *n*-dimensional space, a vector has n components and a tensor of rank 2 has n^2 components. In electrodynamics, we deal with only tensors of the second rank in three- or four-dimensional space.

In 3-space, a second rank tensor has the matrix of its components as

$$T_{ij} = \begin{pmatrix} T_{11} & T_{12} & T_{13} \\ T_{21} & T_{22} & T_{23} \\ T_{31} & T_{32} & T_{33} \end{pmatrix} \tag{1}$$

Now T_{ij} is symmetric if $T_{ij} = T_{ji}$ and is antisymmetric if $T_{ij} = -T_{ji}$. In 3-space, symmetric tensor T_{ij} has six independent elements, since the terms on the principal diagonal of the matrix T_{ij} are not zero. In 4-space, the electromagnetic field tensor $F_{\mu v}$ has only six independent components, since it is an antisymmetris tensor. On the other hand, the electromagnetic energy-momentum tensor $T_{\mu v}$ being symmetric, has 10 independent elements.

δ^α_β *kronecker delta* in three-space, has the following matrix representation

$$\delta^\alpha_\beta = \begin{pmatrix} 1 & 0 & 0 \\ 0 & 1 & 0 \\ 0 & 0 & 1 \end{pmatrix} \tag{2}$$

The symbols $\delta_{\alpha\beta}$ and $\delta^{\alpha\beta}$ have the same meaning as δ^α_β.

Since x^α and x^β are independent variables,

$$\frac{\partial x^\alpha}{\partial x^\beta} = \delta^\alpha_\beta \tag{3}$$

The line element dS is given by

$$dS^2 = \sum_{\alpha,\beta} g_{\alpha\beta} dx^\alpha dx^\beta \qquad (\alpha, \beta = 1, 2, \ldots n) \tag{4}$$

Where $g_{\alpha\beta}$ is the fundamental tensor. The components $g_{\alpha\beta}$ are determined by the geometrical properties of space. The sum (or difference) of two tensors is determined by the sum (or difference) of their corresponding components such that

$$R_{ij} = S_{ij} + T_{ij} \tag{5}$$

Multiplication of tensors is exemplified by the following relations

$$A^\sigma B_\beta = C^\alpha_\beta \tag{6}$$

$$A_{\mu\nu} B^\alpha_\beta = C^\alpha_{\mu\nu\beta} \tag{7}$$

Now, we show that in rectangular coordinates, $g_{\alpha\beta} = \delta_{\alpha\beta}$. For the sake of simplicity, we consider 2 space. Thus rewriting Eq. (4), one obtains

$$dS^2 = g_{11} dx^1 dx^1 + g_{21} dx^2 dx^1 + g_{12} dx^1 dx^2 + g_{22} dx^2 dx^2 \tag{8}$$

The fundamental tensor $g_{\alpha\beta}$ is symmetric, i.e. $g_{12} = g_{21}$, so Eq. (8) reads as

$$dS^2 = g_{11}(dx^1)^2 + 2g_{12} dx^1 dx^2 + g_{22}(dx_2)^2 \tag{9}$$

However, we know that in Cartesian space,

$$dS^2 = (dx^1)^2 + (dx^2)^2 \tag{10}$$

Now, comparing Eqs. (9) and (10) one obtains

$$g_{12} = g_{22} = 1 \tag{11}$$
$$g_{12} = 0$$

and the matrix representation of (11) is

$$(g_{\alpha\beta}) = \begin{pmatrix} 1 & 0 \\ 0 & 1 \end{pmatrix} \tag{12}$$

In a similar manner, in 3-space, one finds

$$(g_{\alpha\beta}) = \begin{pmatrix} 1 & 0 & 0 \\ 0 & 1 & 0 \\ 0 & 0 & 1 \end{pmatrix} \tag{13}$$

This proves the statement that $g_{\alpha\beta} = \delta_{\alpha\beta}$.

dS^2 in Eq. (4) is an *invariant*. A quantity is said to be invariant if it preserves its form on transformation to a different coordinate system. We may note that it is the purpose of tensor analysis to find the invariants associated with various transformations.

The components of a contravariant vector transform in a change of coordinates as the coordinate differentials and the law of transformation is given by

$$A'^j = A^i \frac{\partial x'^j}{\partial x^\nu} \tag{14}$$

Where the primed quantity pertains to the new coordinates. Now, rewriting Eq. (14) as

$$A'^i \frac{\partial x^k}{\partial x'^j} = A^j \frac{\partial x'^i}{\partial x^j} \frac{\partial x^k}{\partial x'^i} \tag{15}$$

$$= A^j \frac{\partial x^k}{\partial x^i}$$

$$= A^j \delta_j^k$$

$$= A^k$$

Interchanging k by i and i by j, one can express Eq. (15) as

$$A^i = A'^j \frac{\partial x^i}{\partial x'^j} \tag{16}$$

We now find the *law of transformation of the components of a covariant vector*. The law is chosen in such a manner so that the product of a covariant vector and a contravariant vector is a scalar invariant, so that

$$A_\beta A^\beta = \phi \tag{17}$$

Where ϕ is a scalar. This requirement lead to the following law of transformation for a covariant vector

$$A'_j = A_i \frac{\partial x^i}{\partial x'^j} \tag{18}$$

An *inverse transformation* can be written as

$$A_j = A'_i \frac{\partial x'^i}{\partial x^j} \tag{19}$$

We may note that geometrically F^α the contravariant vectors are the component of F in the directions of the tangents to the coordinate curves through P and F_α are the components of F in the directions of the normals to the coordinate curves through P. It is obvious that in rectangular coordinates $F^\alpha = F_\alpha$

Corresponding to the dot product in vector analysis, the scalar product of the vectors A^α and B_α is $A^\alpha B_\alpha$ defined as (Using Eqs. (14) and (18), one obtains

$$A'^\alpha B'_\alpha = A^i B_i \tag{20}$$

Obviously $A^i B_i$ is a *scalar invariant*

The law of transformation of mixed tensor A_β^α is

$$A'^\mu_\nu = A'^\mu A'_\nu$$

$$= \left(A^\alpha \frac{\partial x'^\mu}{\partial x^\mu} \right) \left(A_\beta \frac{\partial x^\beta}{\partial x'^\nu} \right)$$

$$= A_\beta^\alpha \frac{\partial x'^\mu}{\partial x^\alpha} \frac{\partial x^\beta}{\partial x'^{\hat{\nu}}} \tag{21}$$

Similarly one can find the laws of transformation for the contravariant and covariant tensors as

$$A'^{\mu\nu} = A_\beta^\alpha \frac{\partial x'^\mu}{\partial x^\alpha} \frac{\partial x'^\nu}{\partial x^\beta} \tag{22}$$

$$A'_{\mu\nu} = A_{\alpha\beta} \frac{\partial x^\alpha}{\partial x'^\mu} \frac{\partial x^\beta}{\partial x'^\nu} \tag{23}$$

From Eqs. (21) and (22) one finds that if a tensor vanishes in one coordinate system, it will also vanish in any other coordinate system.

Few examples of finding the invariants of tensors are as follows. Putting $\nu = \mu$ in Eq. (21), one obtains

$$A'^\mu_\mu = A_\beta^\alpha \delta_\alpha^\beta = A_\alpha^\alpha \tag{24}$$

Thus, A_α^α is an invariant of tensor A_β^α. Further, one finds that $S_{mn} T^{mn}$ is an invariant. Using eqs. (22) and (23), one obtains

$$S'_{jk} T'^{jk} = S_{\mu\nu} \frac{\partial x^\mu}{\partial x'^j} \frac{\partial x^\nu}{\partial x'^k} T^{mn} \frac{\partial x'^j}{\partial x^m} \frac{\partial x'^k}{\partial x^n}$$

$$= S_{\mu\nu} T^{mn} \delta_m^\mu \delta_n^\nu$$

$$= S_{mn} T^{mn} \tag{25}$$

Spherical Harmonics

The Laplacian in spherical polar coordinate is

$$\nabla^2 = \nabla_r^2 + \frac{1}{r^2}\nabla_{\theta,\phi}^2 \tag{1}$$

where $\nabla_r^2 = \dfrac{1}{r^2}\dfrac{\partial}{\partial r}\left(r^2\dfrac{\partial}{\partial r}\right)$ and

$$\nabla_{\theta,\phi}^2 = \frac{1}{\sin\theta}\frac{\partial}{\partial\theta}\left(\sin\theta\frac{\partial}{\partial\theta}\right) + \frac{1}{\sin^2\theta}\frac{\partial^2}{\partial\phi^2} \tag{2}$$

Helmhotz's equation in spherical polar coordinates reads as

$$\left[\nabla_r^2 + \frac{1}{r^2}\nabla_{\theta,\phi}^2\right]\Psi\,(r,\theta,\phi) + k^2(r)\,\psi\,(r,\theta,\phi) = 0 \tag{3}$$

Using the method of separation of variables, we assume that the desired solution can be represented as a product of the radial and angular parts, i.e.

$$\psi\,(r,\theta,\phi) = R(r)Y(\theta,\phi) \tag{4}$$

Using (4) in (3) and multiplying by r^2/RY, one obtains

$$\frac{r^2}{R}\nabla_r^2 R + r^2k^2(r) = -\frac{\nabla_{\theta,\phi}^2 Y}{Y} \tag{5}$$

LHS of Eq. (5) is a function of r only and RHS is a function of θ and ϕ only, both sides of Eq. (5) can be satisfied only if both sides are separately equal to a constant (say λ). Thus,

$$\nabla_r^2\,R(r) + \left[k^2(r) - \frac{\lambda}{r^2}\right]R(r) = 0$$

and

$$\nabla_{\theta,\phi}^2\,Y(\theta,\phi) + \lambda Y(\theta,\phi) = 0 \tag{6}$$

The solutions of Eq. (6) are called *spherical harmonics*

Properties

The spherical harmonics satisfy the eigenvalue equations

$$L^2 Y_{lm} = l(l+1)Y_{lm} \tag{7}$$

$$L_z Y_{lm} = mY_{lm} \tag{8}$$

where
$$\hat{L} = \frac{1}{i} r \times \nabla \qquad (9)$$

$$L^2 = L_x^2 + L_y^2 + L_z^2 \qquad (10)$$

The operators expressed in spherical coordinates are as follows:

$$L_x = -\sin\varphi \frac{1}{i}\frac{\partial}{\partial\theta} - \cot\theta \cos\varphi \frac{1}{i}\frac{\partial}{\partial\varphi} \qquad (11)$$

$$L_y \doteq \cos\varphi \frac{1}{i}\frac{\partial}{\partial\theta} - \cot\theta \sin\varphi \frac{1}{i}\frac{\partial}{\partial\varphi} \qquad (12)$$

$$L_z = \frac{1}{i}\frac{\partial}{\partial\varphi} \qquad (13)$$

$$L^2 = -\frac{1}{\sin\theta}\frac{\partial}{\partial\theta}\sin\theta\frac{\partial}{\partial\theta} - \frac{1}{\sin^2\theta}\frac{\partial^2}{\partial\varphi^2} \qquad (14)$$

Spherical harmonics are orthogonal with respect integration over the solid angle, i.e.

$$\int_0^{2\pi} d\varphi \int_0^{\pi} \sin\theta d\theta Y_{l'm'}^*(\theta,\varphi) Y_{lm}(\theta,\varphi) = \delta_{ll'}\delta_{mm'} \qquad (15)$$

and are related to their complex conjugates as follows:

$$Y_{lm}^*(\theta,\varphi) = (-1)^m Y_{l,-m}(\theta,\varphi) \qquad (16)$$

Spherical harmonics obey and *addition theorem* which expresses the sum over a product of the harmonics in two different set of angles in terms of the Legendre polynomials. If θ, φ and θ', φ' are the polar angles of the unit vectors $\hat{x}$ and $\hat{x}'$, then, we have

$$P_l(\hat{x}\cdot\hat{x}') = \frac{4gp}{2l+1}\sum_{m=-l}^{+l} Y_{lm}^*(\theta',\varphi') Y_{lm}(\theta,\varphi) \qquad (17)$$

One can obtain expression for $1/|x-x'|^3$ with the help of Eq. (17) and the generating function for Legendre polynomials. We know $P_l(x)$, i.e.,

$$\frac{1}{\sqrt{1-2tx+t^2}} = \sum_{l=0}^{\infty} P_l(x)t^l \quad |t| < 1 \qquad (18)$$

Now, writing the result as follows:

$$\frac{1}{|x-x'|} = \frac{1}{\sqrt{r^2 + r'^2 - 2rr'x\cdot x'}}$$

$$= \frac{1}{r\sqrt{1-(2r'/r)x\cdot x' + (r/r')^2}}$$

$$= \frac{1}{r'\sqrt{1 - (2r/r')\mathbf{x}\cdot\mathbf{x}' + (r/r')^2}} \tag{19}$$

Denoting the larger and smaller $|\mathbf{x}|$ and $|\mathbf{x}'|$ by $r_>$ and $r_<$, one obtains

$$\frac{1}{|\mathbf{x} - \mathbf{x}'|} = \frac{1}{r_>\sqrt{1 - (2r_</r_>)\hat{\mathbf{x}}\cdot\hat{\mathbf{x}}' + (r/r')^2}}$$

$$= \frac{1}{r_>}\sum_{l=0}^{\infty} P_l(\hat{\mathbf{x}}\cdot\hat{\mathbf{x}'})\left(\frac{r_<}{r_>}\right)^l \tag{20}$$

Now, putting the values of $P_l(\hat{\mathbf{x}}\cdot\hat{\mathbf{x}'})$ from Eq. (17) one obtains

$$\frac{1}{|\mathbf{x} - \mathbf{x}'|} = \sum_{l=0}^{\infty}\sum_{m=-l}^{\infty} \frac{4\pi}{2l+1} \frac{r_<^l}{r_>^{l+1}} Y_{lm}^*(\theta', \varphi') Y_{lm}(\theta, \varphi) \tag{21}$$

Fourier Transforms

When the method of separation of variables is applied to certain partial differential equations of mathematical physics, integrals of the form

$$F(\alpha) = \int_a^b f(x)k(\alpha, x)dx \qquad (1)$$

often occur, the function $F(\alpha)$ is said to be the *integral transform* of $f(x)$ by the kernel $k(\alpha, x)$. The kernels associated with Fourier transforms are

$$F(\alpha) = \frac{1}{\sqrt{2\pi}} \int_{-\infty}^{\infty} f(x)e^{i\alpha x}dx \qquad (2)$$

The *periodic function f(x)*, in the limit when the period increases indefinitely, may be expanded in a Fourier integral as

$$f(x) = \frac{1}{2\pi} \int_{-\infty}^{\infty} e^{-ikx}dk \int_{-\infty}^{\infty} f(u)e^{iku}du$$

Which may be written as

$$f(x) = \frac{1}{\sqrt{2\pi}} \int_{-\infty}^{\infty} F(k)e^{-ikx}dk \qquad (3)$$

$$F(k) = \frac{1}{\sqrt{2\pi}} \int_{-\infty}^{\infty} f(x)e^{ikx}dx \qquad (4)$$

The Eqs (3) and (4) for $f(x)$ and $F(k)$ respectively are reciprocally related and are called Fourier transforms. There are several ways of defining *Fourier transforms*, but the differences among the various forms are not significant. Fourier transforms for frequency (ω) and time (t) are

$$f(t) = \frac{1}{\sqrt{2\pi}} \int_{-\infty}^{\infty} F(\omega)e^{-i\omega t}d\omega \qquad (5)$$

$$F(\omega) = \frac{1}{\sqrt{2\pi}} \int_{-\infty}^{\infty} F(t)e^{i\omega t}dt \qquad (6)$$

were $F(\omega)$ denotes the frequency spectrum of the wave.

Fourier transforms $f(t)$ and $F(\omega)$ has the following property

$$\int_{-\infty}^{\infty} |f(t)|^2 dt = \int_{-\infty}^{\infty} |F(\omega)|^2 d\omega \qquad (7)$$

This is *Parseval's theorem.*

Physical Constants

Quantity	Symbol	Value	Unit
Speed of light in vacuum	c	2.998×10^8	m/s
Elementary charge	e	1.602×10^{-19}	coulomb (C)
Electron mass	m_e	9.109×10^{-31}	kg
Proton mass	m_p	1.673×10^{-27}	kg
Planck's constant	h	6.626×10^{-34}	joule second (Js)
Vacuum permittivity	ε_0	$\dfrac{1}{\mu_0 c^2} = 8.854 \times 10^{-12}$	farad/m
Vacuum permeability	μ_0	$4\pi \times 10^{-7} = 1.257 \times 10^{-6}$	henry/m (H m^{-1})
Classical electron radius	a_0	2.818×10^{-15}	m
Electron volt	eV	1.602×10^{-19}	joule (J)
Bohr's radius	r	5.292×10^{-11}	m
Boltzmann constant	k_B	1.381×10^{-23}	J/°K
Avogadro's number	N_A	6.023×10^{23}	per mole
Newton's constant of gravitation	G	6.67259×10^{-11}	m^3 kg^{-1} s^{-2}
Rydberg constant $(m_e c \alpha^2 / 2h)$	R_∞	10973731.534	m^{-1}

Some Common Practical Units

1 fermi = 1 fm = 10^{-15} m

1 angstrom = 1 Å = 10^{-10} m = 10^{-8} cm

1 nanometre = 1 nm = 10^{-9} m

1 micron = 1 μm = 10^{-6} m

1 light year = 1 ly = 9.467×10^{15} m

1 astronomical unit = 1 AU = 1.496×10^{11} m

1 parsec (parallactic second) = 3.8×10^{16} m = 3.26 ly

Index